Many-Particle Physics

SECOND EDITION

PHYSICS OF SOLIDS AND LIQUIDS

AMORPHOUS SOLIDS AND THE LIQUID STATE
Edited by Norman H. March, Robert A. Street, and Mario P. Tosi

CHEMICAL BONDS OUTSIDE METAL SURFACES
Norman H. March

CRYSTALLINE SEMICONDUCTING MATERIALS AND DEVICES
Edited by Paul N. Butcher, Norman H. March, and Mario P. Tosi

ELECTRON SPECTROSCOPY OF CRYSTALS
V. V. Nemoshkalenko and V. G. Aleshin

FRACTALS
Jens Feder

INTERACTION OF ATOMS AND MOLECULES WITH SOLID SURFACES
Edited by V. Bortolani, N. H. March, and M. P. Tosi

MANY-PARTICLE PHYSICS, Second Edition
Gerald D. Mahan

ORDER AND CHAOS IN NONLINEAR PHYSICAL SYSTEMS
Edited by Stig Lundqvist, Norman H. March, and Mario P. Tosi

THE PHYSICS OF ACTINIDE COMPOUNDS
Paul Erdös and John M. Robinson

POLYMERS, LIQUID CRYSTALS, AND LOW-DIMENSIONAL SOLIDS
Edited by Norman H. March and Mario P. Tosi

THEORY OF THE INHOMOGENEOUS ELECTRON GAS
Edited by Stig Lundqvist and Norman H. March

Many-Particle Physics

SECOND EDITION

Gerald D. Mahan

University of Tennessee
and Oak Ridge National Laboratory

PLENUM PRESS • NEW YORK AND LONDON

Library of Congress Cataloging-in-Publication Data

Mahan, Gerald D.
Many-particle physics / Gerald D. Mahan. -- 2nd ed.
p. cm.
Includes bibliographical references.
ISBN 0-306-43423-7
1. Solid state physics. 2. Many-body problem. 3. Green's functions. I. Title.
QC176.M24 1990
530.4'1--dc20 89-48850
CIP

A Division of Plenum Publishing Corporation
233 Spring Street, New York, N.Y. 10013

Printed in the United States of America

Preface

This textbook is for a course in advanced solid-state theory. It is aimed at graduate students in their third or fourth year of study who wish to learn the advanced techniques of solid-state theoretical physics. The method of Green's functions is introduced at the beginning and used throughout. Indeed, it could be considered a book on practical applications of Green's functions, although I prefer to call it a book on physics. The method of Green's functions has been used by many theorists to derive equations which, when solved, provide an accurate numerical description of many processes in solids and quantum fluids. In this book I attempt to summarize many of these theories in order to show how Green's functions are used to solve real problems. My goal, in writing each section, is to describe calculations which can be compared with experiments and to provide these comparisons whenever available.

The student is expected to have a background in quantum mechanics at the level acquired from a graduate course using the textbook by either L. I. Schiff, A. S. Davydov, or I. Landau and E. M. Lifshiftz. Similarly, a prior course in solid-state physics is expected, since the reader is assumed to know concepts such as Brillouin zones and energy band theory. Each chapter has problems which are an important part of the lesson; the problems often provide physical insights which are not in the text. Sometimes the answers to the problems are provided, but usually not. It is hoped that the student can learn the subject by using the book as a study guide, since small enrollments often restrict the availability of courses at the advanced level.

I am often asked why I wrote this book. The questioner usually has an understanding of the work involved and so regards my effort as reflecting poorly upon my sanity. On the whole I agree, and if I knew at the beginning the actual investment of time, I probably would not have started. The actual time is about four to five hours per page, which is roughly divided equally among writing, editing, and library searching. My reason for undertaking this project was the great need for someone to do it and the disinclination of anyone else to write a comprehensive advanced textbook on solid-state theory. My own graduate students kept asking me for the standard reference on a number of topics, and I became tired of replying that none exist. My objective was to take standard subjects, such as the electron gas or polaron theory, and to summarize what is generally known. All the steps are retained in the derivation, so that the answers are obtained by starting from the beginning and working through to the end.

The volume is restricted to a description of the many-particle theory of solids. There is also some discussion of quantum fluids, which have historically been part of solid-state theory. The important subject of classical fluids is omitted entirely. In solids and liquids, the forces between pairs of particles are well understood, and the starting Hamiltonian for the problem is accurate. Here we are better off than our brethren in nuclear or stellar physics, since they are often groping for the Hamiltonian. In solids, the only problem is that there are usually 10^{23} particles in the system. Thus we have a well-defined many-body problem which is easy to state: We have simple forces between particles, and the only complication is the large number of particles. In this regard, the atomic theorists have an easier task since they usually have fewer than one hundred electrons in their theoretical system. Consequently, they have been more successful in achieving a quantitative description of atoms. However, solid-state theory has a richer variety of phenomena: magnetism, superconductivity, superfluidity, phase transitions, etc. We have been successful in describing most of these phenomena with great accuracy—magnetism is probably the greatest exception. There is no doubt that solid-state theory has been the center of developments in many-body theory, and our successes are followed by exporting these ideas to other disciplines; e.g., the plasma theory of metals becomes the giant dipole resonance of nuclei.

The topics chosen for discussion were selected on the basis of what every well-rounded theorist should know. Thus basic subjects such as the electron gas, electron–phonon interactions, transport theory, linear response, superconductivity, and superfluidity are covered. Other subjects were deemed equally important, but I ran out of energy, and the book became too long anyway; the important subjects which were omitted are the Kondo effect, Hubbard models, Anderson models, and magnetic systems. Also omitted were two important subjects which deserve large textbooks of their own: renormalization group and the Hohenberg–Kohn–Sham theory of the inhomogeneous electron gas.

Anyone writing an advanced book is going to receive some criticism. No one is expert in all advanced subjects. There are many topics on which I am not an expert and not even well informed. In this situation any choice of action will be criticized. If the topic is omitted, I am criticized for regarding it as unimportant. If it is included, then I am criticized for not providing the expert viewpoint. The only solution is to do one's best and to challenge critics to write their own book.

Historians of science have described numerous and competing models for how science advances. One model is the Ortega hypothesis which suggests that science advances by a large number of mediocre scientists each making a small incremental contribution. On the opposite end of the spectrum is a quite different view that science advances by great leaps forward by the intellectual giants such as Newton and Einstein. The rest of us merely fill in the details they overlook. In preparing this manuscript, I became aware that solid-state theory has advanced by a process intermediate between these extremes. In each chapter, usually six or eight theorists seem to dominate the subject and to provide most of the major concepts and understanding. But in the next chapter, on another subject, it is usually another entirely different group of six to eight theorists who provide the progress. Thus the advance of science appears to occur by a large number of talented workers, although in each topic only a few are important.

Quite often in the text it is necessary to evaluate standard integrals from tables. I use *Table of Integrals, Series and Products* by I. S. Gradshteyn and I. M. Ryzhik (Academic Press, New York, 1965). It seems to be available worldwide. All special integrals are referred to in this present book as "G & R" followed by the integral number.

It is a pleasure to thank many associates for the substantial assistance I have received in the preparation of the book. About half of the writing was done while I was on leave, from Indiana University, as visiting Professor at Chalmers University of Technology in Gothenburg, Sweden. I wish to thank Professor A. Sjölander and S. Lundqvist for this stimulating and very pleasant year. My financial support was provided by the Nordic Institute for Theoretical Astrophysics (NORDITA) in Copenhagen, and I wish to thank Professors A. Bohr, A. Luther, and J. W. Wilkins for arranging my visiting professorship. The entire draft was read by S. M. Girvin, who deserves special thanks for catching many lapses in the first draft. Additional proofreading was also provided by M. Jonson and P. Tua.

Finally, I wish to thank my family—Sally, Chris, Susie, and Roy—for their understanding and quietness during the eighteen months of preparation.

Indiana University Gerald D. Mahan

Contents

Chapter 1

Introductory Material

1.1. HARMONIC OSCILLATORS AND PHONONS

First quantization in physics refers to the property of particles that certain operators do not commute:

$$[x, p_x] = i\hbar$$
$$E \rightarrow \hbar i \frac{\partial}{\partial t} \tag{1.1.1}$$

Later it was realized that forces between particles were caused by other particles: Photons caused electromagnetic forces, pions caused some nuclear forces, etc. These particles are also quantized, and this leads to second quantization. The basic idea is that forces are caused by particles and that the number of particles is quantized: one, two, three, etc. This imparts a quantum nature to the classical force fields.

In solids the vibrational modes of the atoms are quantized because of first quantization (1.1.1). These quantized vibrational modes are called *phonons*. An electron can interact with a phonon, and this phonon can travel to another electron, interact, and thereby cause an indirect interaction between electrons. Indeed, the phonon does not need to move but can vibrate until the next electron comes by. The induced interaction between electrons is an example of second quantization. The phonons play a role in solids similar to the classical fields of particle physics. They cause quantized interactions between electrons.

Phonons in solids can usually be described as harmonic oscillators. Later we shall have a fuller description of the effects of anharmonicity. But, for the moment, this should be sufficient motivation to study the harmonic oscillator. The one-dimensional harmonic oscillator has the Hamiltonian

$$H = \frac{p^2}{2m} + \frac{K}{2} x^2$$

To solve it we introduce some dimensionless coordinates ξ:

$$\omega^2 = \frac{K}{m}$$

$$\xi = x\left(\frac{m\omega}{\hbar}\right)^{1/2}$$

$$\frac{1}{i}\frac{\partial}{\partial\xi} = p(\hbar m\omega)^{-1/2}$$

and

$$H = \frac{\hbar\omega}{2}\left(-\frac{\partial^2}{\partial\xi^2} + \xi^2\right) \tag{1.1.2}$$

The harmonic oscillator Hamiltonian has a solution in terms of Hermite polynomials. The states are quantized such that

$$H\psi_n = \hbar\omega(n + \tfrac{1}{2})\psi_n \tag{1.1.3}$$

where n is an integer. One can also learn by direct calculation that the following matrix elements exist for the operators x and p:

$$\begin{aligned}\langle n' \mid x \mid n\rangle &= \left(\frac{\hbar}{2m\omega}\right)^{1/2}[(n')^{1/2}\delta_{n'=n+1} + (n)^{1/2}\delta_{n'=n-1}] \\ \langle n' \mid p \mid n\rangle &= -\frac{1}{i}\left(\frac{m\hbar\omega}{2}\right)^{1/2}[(n')^{1/2}\delta_{n'=n+1} - (n)^{1/2}\delta_{n'=n-1}]\end{aligned} \tag{1.1.4}$$

It is customary to define two dimensionless operators as follows:

$$\begin{aligned}a &= \frac{1}{2^{1/2}}\left(\xi + \frac{\partial}{\partial\xi}\right) = \left(\frac{m\omega}{2\hbar}\right)^{1/2}\left(x + \frac{ip}{m\omega}\right) \\ a^\dagger &= \frac{1}{2^{1/2}}\left(\xi - \frac{\partial}{\partial\xi}\right) = \left(\frac{m\omega}{2\hbar}\right)^{1/2}\left(x - \frac{ip}{m\omega}\right)\end{aligned} \tag{1.1.5}$$

They are Hermitian conjugates of each other. They are sometimes called *raising* and *lowering operators*, but we shall call them *creation* ($a^\dagger$) and *destruction operators* (a). The Hamiltonian (1.1.2) may be written with them as

$$
\begin{aligned}
H &= \frac{\hbar\omega}{2}[aa^\dagger + a^\dagger a] \\
&= \frac{\hbar\omega}{2}\left[\frac{1}{2}\left(\xi + \frac{\partial}{\partial\xi}\right)\left(\xi - \frac{\partial}{\partial\xi}\right) + \frac{1}{2}\left(\xi - \frac{\partial}{\partial\xi}\right)\left(\xi + \frac{\partial}{\partial\xi}\right)\right] \\
&= \frac{\hbar\omega}{2}\left(-\frac{\partial^2}{\partial\xi^2} + \xi^2\right)
\end{aligned}
$$

A very important property of these operators is called *commutation relations*. These are derived by considering how they act, sequentially, on any function $f(\xi)$. The two operations a and $a^\dagger$ in turn give

$$aa^\dagger f(\xi) = \frac{1}{2}\left(\xi + \frac{\partial}{\partial\xi}\right)\left(\xi - \frac{\partial}{\partial\xi}\right)f(\xi) = \frac{1}{2}(\xi^2 f + f - f'')$$

while the reverse order gives

$$a^\dagger a f(\xi) = \frac{1}{2}\left(\xi - \frac{\partial}{\partial\xi}\right)\left(\xi + \frac{\partial}{\partial\xi}\right)f(\xi) = \frac{1}{2}(\xi^2 f - f - f'')$$

These two results are subtracted,

$$[aa^\dagger - a^\dagger a]f(\xi) = f(\xi) \tag{1.1.6}$$

and yield the original function. The operator in brackets is replaced by a bracket with a comma,

$$[aa^\dagger - a^\dagger a] \equiv [a, a^\dagger]$$

which means the same thing. The relationship (1.1.6) is usually expressed by omitting the function $f(\xi)$:

$$[a, a^\dagger] = 1 \tag{1.1.7}$$

In a similar way, one can prove that

$$
\begin{aligned}
[a, a] &= 0 \\
[a^\dagger, a^\dagger] &= 0
\end{aligned} \tag{1.1.8}
$$

These three commutators, plus the Hamiltonian

$$H = \frac{\hbar\omega}{2}[aa^\dagger + a^\dagger a] = \frac{\hbar\omega}{2}[aa^\dagger - a^\dagger a + 2a^\dagger a] = \hbar\omega[a^\dagger a + \tfrac{1}{2}] \quad (1.1.9)$$

completely specify the harmonic oscillator problem in terms of operators. With these four relationships, one can show that the eigenvalue spectrum is indeed (1.1.3), where n is an integer. The eigenstates are

$$|n\rangle = \frac{(a^\dagger)^n}{(n!)^{1/2}}|0\rangle$$

where $|0\rangle$ is the state which obeys

$$a|0\rangle = 0$$

and where the $n!$ is for normalization. If one operates on this state by a creation operator, one gets

$$a^\dagger|n\rangle = \frac{1}{(n!)^{1/2}}(a^\dagger)^{n+1}|0\rangle = \frac{(n+1)^{1/2}}{[(n+1)!]^{1/2}}(a^\dagger)^{n+1}|0\rangle$$
$$= (n+1)^{1/2}|n+1\rangle$$

the state with the next highest integer. Thus the only matrix element between states is

$$\langle n'|a^\dagger|n\rangle = (n+1)^{1/2}\delta_{n'=n+1}$$

If we take the Hermitian conjugate of this matrix element,

$$\langle n|a|n'\rangle = (n+1)^{1/2}\delta_{n'=n+1}$$

and exchange dummy variables n and n', we obtain

$$\langle n'|a|n\rangle = (n)^{1/2}\delta_{n'=n-1}$$

or

$$a|n\rangle = (n)^{1/2}|n-1\rangle$$

So the destruction operator a lowers the quantum number. Thus operating by the sequence

$$a^\dagger a|n\rangle = a^\dagger(n)^{1/2}|n-1\rangle) = (n)^{1/2}a^\dagger|n-1\rangle = n|n\rangle$$

gives an eigenvalue n, which verifies the eigenvalue (1.1.3). Furthermore, using the original definitions (1.1.5) permits us to express x and p in terms of these operators,

$$x = \left(\frac{\hbar}{2m\omega}\right)^{1/2} (a + a^{\dagger})$$

$$p = i\left(\frac{\hbar m\omega}{2}\right)^{1/2} (a^{\dagger} - a)$$

and the matrix elements (1.1.4) follow immediately:

$$\langle n' \mid x \mid n\rangle = \left(\frac{\hbar}{2m\omega}\right)^{1/2} (\langle n' \mid a \mid n\rangle + \langle n' \mid a^{\dagger} \mid n\rangle)$$

$$= \left(\frac{\hbar}{2m\omega}\right)^{1/2} [(n)^{1/2}\delta_{n'=n-1} + (n+1)^{1/2}\delta_{n'=n+1}]$$

$$\langle n' \mid p \mid n\rangle = \left(\frac{\hbar m\omega}{2}\right)^{1/2} (\langle n' \mid a^{\dagger} \mid n\rangle - \langle n' \mid a \mid n\rangle)$$

$$= i\left(\frac{\hbar m\omega}{2}\right)^{1/2} [(n+1)^{1/2}\delta_{n'=n+1} - (n)^{1/2}\delta_{n'=n-1}]$$

The description of the harmonic oscillator in terms of operators is equivalent to the conventional method of using wave functions $\psi_n(\xi)$ of position.

We shall frequently need to know the time dependence of these operators. In the Heisenberg representation of quantum mechanics, the time development of operators is given by ($\hbar = 1$)

$$O(t) = e^{iHt} O e^{-iHt}$$

so that the operator obeys the equation

$$\frac{\partial O(t)}{\partial t} = i[H, O(t)]$$

For the destruction operator, this becomes

$$\frac{\partial}{\partial t} a = i[H, a] = i\omega[a^{\dagger}aa - aa^{\dagger}a] = i\omega[a^{\dagger}, a]a = -i\omega a$$

which has the simple solution

$$a(t) = e^{-i\omega t} a$$

The reference point of time may be selected arbitrarily, so that the operators have an arbitrary phase factor associated with them. This phase is unimportant, since it cancels out of all final results. The Hermitian conjugate of this expression is

$$a^\dagger(t) = e^{i\omega t}a^\dagger$$

Thus we can represent the time development of the position operator as

$$x(t) = \left(\frac{\hbar}{2m\omega}\right)^{1/2}(ae^{-i\omega t} + a^\dagger e^{i\omega t}) \tag{1.1.10}$$

This result for $x(t)$ will be used often in discussing phonon problems.

Another familiar problem which can be solved with operators is a charged harmonic oscillator in a constant electric field F:

$$H = \frac{p^2}{2m} + \frac{K}{2}x^2 + eFx = \omega(a^\dagger a + \tfrac{1}{2}) + \lambda(a + a^\dagger)$$

$$\lambda = eF\left(\frac{\hbar}{2m\omega}\right)^{1/2} \tag{1.1.11}$$

This Hamiltonian may be solved exactly. First consider the equation of motion for the time development of the destruction operator:

$$\frac{\partial a}{\partial t} = i[H, a] = -i(\omega a + \lambda)$$

The right-hand side is no longer just proportional to a, since there is the constant term. However, let us define a new set of operators by the relationships

$$A = a + \frac{\lambda}{\omega}$$

$$A^\dagger = a^\dagger + \frac{\lambda}{\omega}$$

They obey the equation

$$\frac{\partial A}{\partial t} = -i\omega A$$

so they have the simple time development

$$A(t) = e^{-i\omega t}A$$
$$A^{\dagger}(t) = e^{i\omega t}A^{\dagger}$$

Indeed, one can show that they have the following properties:

$$[A, A^{\dagger}] = \left(a + \frac{\lambda}{\omega}, a^{\dagger} + \frac{\lambda}{\omega}\right) = 1$$
$$[A, A] = 0$$
$$[A^{\dagger}, A^{\dagger}] = 0$$

$$\begin{aligned} H &= \omega\left[\left(A^{\dagger} - \frac{\lambda}{\omega}\right)\left(A - \frac{\lambda}{\omega}\right) + \frac{1}{2}\right] + \lambda\left(A + A^{\dagger} - \frac{2\lambda}{\omega}\right) \\ &= \omega\left(A^{\dagger}A + \frac{1}{2}\right) - \frac{\lambda^2}{\omega} \end{aligned} \tag{1.1.12}$$

Now we remarked above that any set of operators with these properties had a solution in terms of harmonic oscillator states:

$$H \mid n\rangle = \left[\omega\left(n + \frac{1}{2}\right) - \frac{\lambda^2}{\omega}\right] \mid n\rangle$$
$$\mid n\rangle = \frac{1}{(n!)^{1/2}} (A^{\dagger})^n \mid 0\rangle$$

The operator for the position is

$$x(t) = \left(\frac{\hbar}{2m\omega}\right)^{1/2}\left(Ae^{-i\omega t} + A^{\dagger}e^{i\omega t} - \frac{2\lambda}{\omega}\right)$$

The physics of the Hamiltonian (1.1.11) is very simple. The spring stretches to a new equilibrium point which is displaced a distance

$$x_0 = -\left(\frac{\hbar}{2m\omega}\right)^{1/2} \frac{2\lambda}{\omega} = -\frac{eF}{K}$$

from the original one. It oscillates about this new equilibrium with the same frequency ω as before. These oscillations are still quantized, in units of ω. The energy $-\lambda^2/\omega$ is that gained by the spring from the displacement along the electric field. One can get the same result directly in coordinate space. The Hamiltonian is written as

$$H = \frac{p^2}{2m} + \frac{K}{2}\left(x + \frac{eF}{K}\right)^2 - \frac{e^2F^2}{2K}$$

and a new coordinate

$$x' = x + \frac{eF}{K} = x - x_0$$

is defined which still obeys

$$[x', p] = i\hbar$$

In a solid there are many atoms, which mutually interact. The vibrational modes are collective motions involving many atoms. A simple introduction to this problem is obtained by studying the normal modes of a one-dimensional harmonic chain:

$$H = \sum_i \frac{p_i^2}{2m} + \frac{K}{2}\sum_i (x_i - x_{i+1})^2 \tag{1.1.13}$$

The classical solution is obtained by solving the equation of motion:

$$-m\ddot{x}_l = m\omega^2 x_l = K(2x_l - x_{l-1} - x_{l+1})$$

A solution is assumed of the form

$$x_l = x_0 \cos(kal)$$

and the force term becomes

$$\begin{aligned} 2x_l - x_{l-1} - x_{l+1} &= x_0[2\cos(kal) - \cos(kal + ka) - \cos(kal - ka)] \\ &= x_0 2\cos(kal)(1 - \cos ka) \end{aligned}$$

Thus the normal modes have the solution

$$\omega_k^2 = \frac{K}{m}2(1 - \cos ka) = \frac{4K}{m}\sin^2\!\left(\frac{ka}{2}\right)$$

The quantum mechanical solution begins by defining some normal coordinates, assuming periodic boundary conditions:

$$x_l = \frac{1}{N^{1/2}}\sum_k e^{ikal}x_k; \qquad x_k = \frac{1}{N^{1/2}}\sum_l e^{-ikal}x_l$$

$$p_l = \frac{1}{N^{1/2}}\sum_k e^{-ikal}p_k; \qquad p_k = \frac{1}{N^{1/2}}\sum_m e^{ikam}p_m$$

This choice maintains the desired commutation relations in either real space or wave vector space:

$$[x_l, p_m] = i\delta_{lm}$$

$$[x_k, p_{k'}] = \frac{1}{N}\sum_{l,m} e^{-ikal}e^{ik'am}[x_l, p_m]$$

$$= \frac{i}{N}\sum_l e^{ial(k'-k)} = \delta_{k,k'}i$$

From the general result

$$\sum_l x_l x_{l+m} = \frac{1}{N}\sum_{kk'} x_k x_{k'} \sum_l e^{ila(k+k')}e^{imak'} = \sum_k x_k x_{-k} e^{iamk}$$

$$\sum_l p_l^2 = \sum_k p_k p_{-k}$$

it is straightforward to show that the potential energy term of (1.1.13) is

$$\frac{K}{2}\sum_l (x_l - x_{l+1})^2 = \frac{K}{2}\sum_k x_k x_{-k}(2 - e^{ika} - e^{-ika}) = \frac{m}{2}\sum_k \omega_k^2 x_k x_{-k}$$

Thus the Hamiltonian may be written in wave vector space as

$$H = \frac{1}{2m}\sum_k p_k p_{-k} + \frac{m}{2}\sum_k \omega_k^2 x_k x_{-k}$$

The Hamiltonian has the form of a simple harmonic oscillator for each wave vector. If we define the creation and destruction operators as

$$a_k = \left(\frac{m\omega_k}{2\hbar}\right)^{1/2}\left(x_k + \frac{i}{m\omega_k} p_{-k}\right)$$

$$a_k^\dagger = \left(\frac{m\omega_k}{2\hbar}\right)^{1/2}\left(x_{-k} - \frac{i}{m\omega_k} p_k\right)$$

they obey the commutation relations

$$[a_k, a_{k'}^\dagger] = \frac{-i}{2\hbar}\{[x_k, p_{k'}] - [p_{-k}, x_{-k'}]\} = \delta_{kk'}$$

$$[a_k, a_{k'}] = 0$$

$$[a_k^\dagger, a_{k'}^\dagger] = 0$$

and the Hamiltonian may be written as

$$H = \sum_k \omega_k(a_k^{\dagger}a_k + \tfrac{1}{2})$$

These collective modes of vibration are called *phonons*. They are the quantized version of the classical vibrational modes in the solid. These are the same commutator relations, and Hamiltonian, as in the simple harmonic oscillator. Each wave vector state behaves independently, as a harmonic oscillator, with a possible set of quantum numbers $n_k = 0, 1, 2, \ldots$. The state of the system at any time is

$$\psi = |n_{k1}, n_{k2}, \ldots, n_{kn}) = \prod_k |n_k\rangle = \prod_k \frac{[a_k^{\dagger}]^{n_k}}{(n_k!)^{1/2}} |0\rangle$$

so that the expectation value of the Hamiltonian is

$$H = \sum_k \omega_k(n_k + \tfrac{1}{2})$$

In thermal equilibrium the states have an average value of n_k which is given in terms of the temperature $\beta = 1/k_BT$.

$$\langle n_k\rangle \equiv N_k = \frac{1}{e^{\beta\omega_k} - 1} \equiv n_B(\omega_k)$$

The system fluctuates around this average value.

The position operator in wave vector space, and real space, is

$$\begin{aligned} x_k(t) &= \left(\frac{\hbar}{2m\omega_k}\right)^{1/2}(a_k e^{-i\omega_k t} + a_{-k}^{\dagger}e^{i\omega_k t}) \\ x_l(t) &= \sum_k \left(\frac{\hbar}{2mN\omega_k}\right)^{1/2} e^{ikal}(a_k e^{-i\omega_k t} + a_{-k}^{\dagger}e^{i\omega_k t}) \end{aligned} \tag{1.1.14}$$

Often we shall replace mN by the equivalent quantity

$$mN = \varrho\nu$$

where ϱ is the mass density and ν is the volume. At some point we shall have a summation over the discrete set of eigenstates for the system of finite volume ν. Then it is convenient to change the summation to an integration:

$$\lim_{\nu\to\infty} \frac{1}{\nu}\sum_{\mathbf{k}} f(\mathbf{k}) = \int \frac{d^3k}{(2\pi)^3} f(\mathbf{k}) \tag{1.1.15}$$

During this change, any delta functions must change from discrete delta functions (called *Kronecker* deltas) to continuous ones. Since

$$\frac{1}{v} f(\mathbf{k}') = \frac{1}{v} \sum_{\mathbf{k}} f(\mathbf{k})\delta_{\mathbf{k},\mathbf{k}'} = \int \frac{d^3k}{(2\pi)^3} f(\mathbf{k})\delta_{\mathbf{k},\mathbf{k}'}$$
$$= \frac{(2\pi)^3}{v} \int \frac{d^3k}{(2\pi)^3} f(\mathbf{k})\delta(\mathbf{k} - \mathbf{k}')$$

we conclude that

$$\lim_{v\to\infty} \delta_{\mathbf{k},\mathbf{k}'} = \frac{(2\pi)^3}{v} \delta^3(\mathbf{k} - \mathbf{k}')$$

In general, our preference is to write wave vector summations as discrete summations until it is time to do the integrals and only then make the changes (1.1.15).

The quantum mechanical solution has the same frequencies as found in the classical solution. Quantum mechanics only enters in a quantization of the amplitude of the oscillation. The phonons occur in discrete numbers with zero, one, two, etc., phonons in each state k. When the average number of phonons is large, $n_k \gg 1$, the quantization is irrelevant, since the system behaves classically. The quantum nature of the field is more important when the average number of phonons in each state k is small.

In three-dimensional solids, the theory is nearly identical except there are more indices. Suppose there is a potential function between atoms or ions of the form

$$\sum_{ij} V(\mathbf{R}_i - \mathbf{R}_j)$$

where $\mathbf{R}_i$ is the position of an atom. If it is vibrating, then denote $\mathbf{R}_i^{(0)}$ as the equilibrium position and $\mathbf{Q}_i$ as the displacement from equilibrium:

$$\mathbf{R}_i = \mathbf{R}_i^{(0)} + \mathbf{Q}_i$$

The potential function is expanded in a Taylor series about the equilibrium position:

$$\sum_{ij} V(\mathbf{R}_i - \mathbf{R}_j) = \sum_{ij} V(\mathbf{R}_i^{(0)} - \mathbf{R}_j^{(0)}) + \sum_{ij} (\mathbf{Q}_i - \mathbf{Q}_j) \cdot \nabla V(\mathbf{R}_i^{(0)} - \mathbf{R}_j^{(0)})$$
$$+ \tfrac{1}{2} \sum_{ij} (\mathbf{Q}_i - \mathbf{Q}_j)_\mu (\mathbf{Q}_i - \mathbf{Q}_j)_\nu \frac{\partial^2}{\partial R_\mu \, \partial R_\nu} V(\mathbf{R}_i^{(0)} - \mathbf{R}_j^{(0)})$$
$$+ O(Q^3) \tag{1.1.16}$$

The term linear in displacement vanishes,

$$\sum_{ij} (\mathbf{Q}_i - \mathbf{Q}_j) \cdot \nabla V(\mathbf{R}_i^{(0)} - \mathbf{R}_j^{(0)}) = 0$$

because one defines the equilibrium position $R_j^{(0)}$ as the place where the sum of the forces on an ion j is zero:

$$\mathbf{F}_j = \sum_i \nabla \cdot V(\mathbf{R}_i^{(0)} - \mathbf{R}_j^{(0)}) = 0$$

The first important term is the one which is quadratic in the displacement:

$$\sum_{ij} (\mathbf{Q}_i - \mathbf{Q}_j)_\mu (\mathbf{Q}_i - \mathbf{Q}_j)_\nu \Phi_{\mu\nu}(\mathbf{R}_i^{(0)} - \mathbf{R}_j^{(0)})$$

$$\Phi_{\mu\nu}(\mathbf{R}) = \frac{\partial^2}{\partial R_\mu \, \partial R_\nu} V(R)$$

By now we know enough to solve this in wave vector space by trying an expansion of the form

$$\mathbf{Q}_i(t) = i \sum_{\mathbf{k},\lambda} \left(\frac{\hbar}{2NM\omega_{\mathbf{k}\lambda}} \right)^{1/2} \boldsymbol{\xi}_{\mathbf{k},\lambda}(a_{\mathbf{k},\lambda} e^{-i\omega_{\mathbf{k}\lambda}t} + a^\dagger_{-\mathbf{k}\lambda} e^{i\omega_{\mathbf{k}\lambda}t}) e^{i\mathbf{k}\cdot\mathbf{R}_i^{(0)}} \tag{1.1.17}$$

where M is the ion mass. The factor of i on the right-hand side of the equation is required to make $\mathbf{Q}_i$ Hermitian, $\mathbf{Q}_i^\dagger = \mathbf{Q}_i$. Take the Hermitian conjugate of (1.1.17) and change $\mathbf{k} \to -\mathbf{k}$. Nearly the same result is obtained as found in $\mathbf{Q}_i$. To make $\mathbf{Q}_i$ Hermitian, we need to have

$$-i\boldsymbol{\xi}^*_{\mathbf{k},\lambda} = i\boldsymbol{\xi}_{-\mathbf{k},\lambda}$$

We specify that the polarization vectors $\xi_{\mathbf{k},\lambda}$ are real but change sign with $\mathbf{k}$ direction, $\boldsymbol{\xi}_{-k} = -\boldsymbol{\xi}_k$, and the above identity is satisfied. Since the displacement is in three dimensions, there are $3L$ normal modes for each value of wave vector. Here L is the number of atoms per unit cell of the crystal. The index λ runs over these $3L$ values of normal mode. Each mode will have its own eigenfrequency $\omega_{\mathbf{k},\lambda}$. It will also have a polarization vector $\xi_{\mathbf{k},\lambda}$ which specifies the vibrational direction of the ion for each wave vector and λ. If there are more than one atom per unit cell, one should add further subscripts to M and $\xi_{\mathbf{k},\lambda}$ to specify the values for each atom per unit cell.

The right-hand side of (1.1.16) may be written as

$$\sum_{ij} V(\mathbf{R}_i^{(0)} - \mathbf{R}_j^{(0)}) + \sum_{\mathbf{k},\lambda} Q_{\mathbf{k}\lambda} Q_{-\mathbf{k}\lambda} \omega^2_{\mathbf{k}\,\lambda} \frac{M}{2}$$

The first term is a constant which will be neglected in our discussion of vibrational modes. The eigenstates $\omega_{\mathbf{k},\lambda}$ are those solved in the *harmonic approximation*. In this approximation, one retains the quadratic term only in the displacements in the Hamiltonian. To be more careful, one writes

(for one atom per unit cell)

$$H = H_0 + V$$

where

$$H_0 = \frac{1}{2M} \sum_{\mathbf{k},\lambda} (P_{\mathbf{k},\lambda}^{*} P_{-\mathbf{k},\lambda} + M^2 \omega_{\mathbf{k},\lambda}^2 Q_{\mathbf{k},\lambda} Q_{-\mathbf{k},\lambda}^{*}) = \sum_{\mathbf{k},\lambda} \omega_{\mathbf{k},\lambda} (a_{\mathbf{k},\lambda}^{\dagger} a_{\mathbf{k},\lambda} + \tfrac{1}{2})$$

$$V = \sum_{ij} (\mathbf{Q}_i - \mathbf{Q}_j)_\mu (\mathbf{Q}_i - \mathbf{Q}_j)_\nu (\mathbf{Q}_i - \mathbf{Q}_j)_\sigma \frac{\partial^3 V}{\partial R_\mu \, \partial R_\nu \, \partial R_\sigma} + O(Q^4) \tag{1.1.18}$$

and one solves H_0 for the modes. These are harmonic oscillator states for each wave vector $\mathbf{k}$ and mode λ. The harmonic approximation applies to any theory of phonons which retains only the terms which are quadratic in the displacements $\mathbf{Q}_j$. Actual solids are described by potential functions which are more complicated than the central force field $V(\mathbf{R}_i - \mathbf{R}_j)$ which we have assumed. For example, in semiconductors there are usually bond bending forces between nearest neighbors. Nevertheless, the Hamiltonian is still written as (1.1.18) in the harmonic approximation. A complete description of these calculations is given by Born and Huang (1954) or Maradudin *et al.* (1963).

The terms in the Taylor series higher than quadratic are treated as perturbations. These are called anharmonic effects and are very important in solids with light atomic masses: hydrogen, helium, lithium, etc. They are often important in some other solids. The first term is usually cubic in the displacements and has the form

$$V = \sum_{\substack{\mathbf{k},\mathbf{q} \\ \lambda_1 \lambda_2 \lambda_3}} Q_{\mathbf{k},\lambda_1} Q_{\mathbf{q},\lambda_2} Q_{-\mathbf{k}-\mathbf{q},\lambda_3} M_{\mathbf{k},\mathbf{q},\lambda_1,\lambda_2,\lambda_3}$$

The matrix element in the cubic term is quite complicated, and we shall not write it out. It is difficult to determine from first principles anyway. If the first two displacements Q have wave vectors $\mathbf{k}$ and $\mathbf{q}$, the third has $-(\mathbf{k} + \mathbf{q})$ to ensure wave vector conservation. This interaction may be written in terms of creation and destruction operators by using (1.1.17). In this representation, it is apparent that these cubic terms permit one phonon to decay into two and vice versa.

For solids in which the anharmonic terms are important, one must try to include the effects of the cubic perturbation V and perhaps also higher terms such as quadratic. This is a many-body problem. The effects are quite temperature dependent, so it is necessary to use Green's functions at finite temperatures.

A word about notation. We dislike subscripts and superscripts. Thus in discussing phonons, we shall usually omit the subscript λ, although it should be carried in every expression. The summation over phonon modes is really meant to imply summation over wave vectors and modes λ.

1.2. SECOND QUANTIZATION FOR PARTICLES

There are two ways to introduce the subject of creation and destruction operators for particles. The first is to describe their properties and then to omit any proofs. One could just remark that they work, which is why we use them. The second way is to go through elaborate justification arguments. These tend to leave the reader more confused than convinced. Here we shall try an intermediate approach. A short justification will be attempted. Our discussion follows Schiff (1968).

The first treatment is for boson particles and those which cannot be destroyed. It is hard to think of a fundamental particle with this property. The method is usually applied to composite particles such as ^{4}He which contain even numbers of fermions, so that it has bosonlike properties. In any case, we shall assume that we are discussing a point particle. If it is in a potential $U(r)$, the one-particle Schrödinger equation is

$$i\hbar\dot{\psi}(\mathbf{r}) = \left[-\frac{\hbar^2}{2m}\nabla^2 + U(r)\right]\psi(\mathbf{r}) = H\psi(\mathbf{r})$$

This equation may be derived from the Lagrangian density

$$L = i\hbar\psi^\dagger\dot{\psi} - \frac{\hbar^2}{2m}\nabla\psi^\dagger \cdot \nabla\psi - U(r, t)\psi^\dagger\psi$$

The wave function is complex, with real and imaginary parts. These can be treated as independent variables in the Lagrangian. An alternate procedure is to treat $\psi(\mathbf{r})$ and $\psi^\dagger(\mathbf{r})$ as independent variables. Then the usual variations give

$$\frac{\partial L}{\partial \psi} = -\ U\psi^\dagger$$

$$\frac{\partial L}{\partial \psi_x} = -\ \frac{\hbar^2}{2m}\psi_x{}^\dagger$$

$$\frac{\partial L}{\partial \dot{\psi}} = i\hbar\psi^\dagger$$

When these are put into Lagrange's equation,

$$0 = \frac{\partial L}{\partial \psi} - \sum_{\mu} \frac{\partial}{\partial x_\mu} \left(\frac{\partial L}{\partial(\partial \psi / \partial x_\mu)} \right) - \frac{\partial}{\partial t} \frac{\partial L}{\partial \dot{\psi}}$$

$$0 = -U\psi^\dagger + \frac{\hbar^2}{2m} \nabla^2 \psi^\dagger - i\hbar \frac{\partial}{\partial t} \psi^\dagger$$

we recover the Hermitian conjugate of Schrödinger's equation. If the same manipulations are tried with $\psi^\dagger$ as the variable, then Schrödinger's equation itself is derived. In the Lagrangian formulation, the momentum which is conjugate to the variable ψ is

$$\pi = \frac{\partial L}{\partial \dot{\psi}} = i\hbar \psi^\dagger(\mathbf{r}) \tag{1.2.1}$$

The Hamiltonian density is given by

$$\mathscr{H} = \pi \dot{\psi} - L = \frac{\hbar^2}{2m} \nabla \psi^\dagger \cdot \nabla \psi + U \psi^\dagger \psi$$

where one integrates over all volume to obtain the Hamiltonian

$$H = \int d^3 r \mathscr{H} = \int d^3 r \psi^\dagger \left(-\frac{\hbar^2}{2m} \nabla^2 + U \right) \psi \tag{1.2.2}$$

where we integrated by parts on the kinetic energy term. Since π and ψ are conjugate variables, they obey commutation relations of the form

$$[\psi(\mathbf{r}, t), \pi(\mathbf{r}', t)] = i\hbar \delta(\mathbf{r} - \mathbf{r}')$$

or using (1.2.1), we obtain

$$[\psi(\mathbf{r}, t), \psi^\dagger(\mathbf{r}', t)] = \delta(\mathbf{r} - \mathbf{r}') \tag{1.2.3}$$

A commutation relation of this type is the fundamental basis of second quantization. Although we have made it plausible by the derivation from a Lagrangian, it really is a basic premise. These commutation relations may be satisfied by introducing creation and destruction operators. Let H have eigenstates and eigenvalues of the form

$$H\phi_\lambda = \varepsilon_\lambda \phi_\lambda$$

$$H = -\frac{\hbar^2}{2m} \nabla^2 + U(r)$$

We expand the wave function $\psi(\mathbf{r})$ and its conjugate $\psi^{\dagger}(\mathbf{r})$ in terms of this basis set:

$$\psi(\mathbf{r}, t) = \sum_{\lambda} a_{\lambda}(t)\phi_{\lambda}(\mathbf{r})$$

$$\psi^{\dagger}(\mathbf{r}, t) = \sum_{\lambda} a_{\lambda}^{\dagger}(t)\phi_{\lambda}^{\dagger}(\mathbf{r})$$

The original field commutators (1.2.3) are satisfied if we call a and $a^{\dagger}$ operators with their own commutation relations:

$$\begin{aligned} [a_{\lambda}(t), a_{\lambda'}^{\dagger}(t)] &= \delta_{\lambda\lambda'} \\ [a_{\lambda}(t), a_{\lambda'}(t)] &= 0 \\ [a_{\lambda}^{\dagger}(t), a_{\lambda'}^{\dagger}(t)] &= 0 \end{aligned} \tag{1.2.4}$$

Thus we obtain the commutation relations for the field variables:

$$\begin{aligned} [\psi(\mathbf{r}, t), \psi(\mathbf{r}', t)] &= 0 \\ [\psi^{\dagger}(\mathbf{r}, t), \psi^{\dagger}(\mathbf{r}', t)] &= 0 \\ [\psi(\mathbf{r}, t), \psi^{\dagger}(\mathbf{r}', t)] &= \sum_{\lambda\lambda'} [a_{\lambda}(t), a_{\lambda'}^{\dagger}(t)]\phi_{\lambda}(\mathbf{r})\phi_{\lambda'}^{\dagger}(\mathbf{r}') \\ &= \sum_{\lambda} \phi_{\lambda}(\mathbf{r})\phi_{\lambda}^{\dagger}(\mathbf{r}') = \delta(\mathbf{r} - \mathbf{r}') \end{aligned}$$

One might also ask about the commutation relations at different times. What happens if we try to evaluate

$$[a_{\lambda}(t), a_{\lambda'}^{\dagger}(t')] = ?$$

The answer is that this is a difficult many-body problem. In fact, that is one of the goals of Green's function theory. The commutator at different times is related to the retarded Green's function, which is defined and discussed in Sec. 3.3. The commutator at different times is one property of the time development of the many-body system. Simple commutation relations such as (1.2.4) are valid only if the operators are at the same time.

The Hamiltonian is

$$\begin{aligned} H &= \int d^3r\psi^{\dagger}(\mathbf{r})\mathscr{H}\psi(\mathbf{r}) = \sum_{\lambda\lambda'} a_{\lambda}^{\dagger}a_{\lambda'} \int d^3r\phi_{\lambda}^{\dagger}(\mathbf{r})\, H\phi_{\lambda'}(\mathbf{r}) \\ &= \sum_{\lambda\lambda'} a_{\lambda}^{\dagger}a_{\lambda'}\varepsilon_{\lambda'}\delta_{\lambda\lambda'} = \sum_{\lambda} \varepsilon_{\lambda}a_{\lambda}^{\dagger}a_{\lambda} \end{aligned} \tag{1.2.5}$$

In Sec. 1.1 we noted that any system which had these commutation relations, and a Hamiltonian (1.2.5), behaved as harmonic oscillators for each state λ. The eigenstates for each value of λ have a discrete set of occupation numbers $n_\lambda = 0, 1, 2, 3, \ldots$ All bosons have harmonic oscillator eigenstates. For phonons, we interpreted the number n_λ as the number of phonons in state λ. For particles, the interpretation is the same. The number n_λ tells how many particles in the system are in the same state λ. However, for particles, unlike phonons, the total number of particles is conserved.

The many-particle wave function has the form

$$\prod_\lambda \frac{(a_\lambda^\dagger)^{n_\lambda}}{(n_\lambda!)^{1/2}} \,|\, 0\rangle$$

so that the Hamiltonian (1.2.5) has the eigenvalue of

$$E = \sum_\lambda \varepsilon_\lambda n_\lambda$$

In the thermal equilibrium, the average number of particles in a state λ is given by the usual boson occupation factor:

$$\langle n_\lambda \rangle = \frac{1}{e^{\beta(\varepsilon_\lambda - \mu)} - 1} \equiv N_\lambda = n_B(\varepsilon_\lambda - \mu)$$

Now there is a chemical potential μ which can vary with temperature and concentration. It is absent in the phonon, and photon, cases because these excitations do not conserve particle number. One may make as many phonons or photons as one wishes. Another operator of interest is the density operator:

$$\varrho(\mathbf{r}) = \psi^\dagger(\mathbf{r})\psi(\mathbf{r}) = \sum_{\lambda\lambda'} a_\lambda^\dagger a_{\lambda'} \phi_\lambda^*(\mathbf{r})\phi_{\lambda'}(\mathbf{r})$$

The integral of this is just the number operator:

$$N = \int d^3r \varrho(\mathbf{r}) = \sum_\lambda a_\lambda^\dagger a_\lambda$$

Its thermal average is obtained simply by taking the thermal average of n_λ:

$$\langle N \rangle = \sum_\lambda \langle n_\lambda \rangle = \sum_\lambda N_\lambda$$

This serves as a definition of the chemical potential and determines its variations with temperature and particle number N.

The Hamiltonian (1.2.5) and number operator are *bilinear* in creation and destruction operators. They contain only two operators, one of each kind. Hamiltonians of these kinds may always be solved, at least in principle. The problem may always be reduced to the diagonalization of a matrix. For example, consider the solution of our Hamiltonian:

$$H = -\frac{\hbar^2}{2m}\nabla^2 + U(r)$$

Suppose that we were unable to solve it exactly—this is actually an improbable assumption, since Schrödinger's equation for one-particle potentials may be solved in milliseconds on the computer. Anyway, suppose there were another complete set of states ϕ_n which are the solution to some other Hamiltonian. Expand the wave function in terms of these states,

$$\psi(\mathbf{r}) = \sum_n b_n \phi_n(\mathbf{r})$$

where the creation and destruction operators b_n have the usual commutation relations:

$$[b_n, b_m^\dagger] = \delta_{n,m}$$

$$[b_n, b_m] = 0$$

$$[b_n^\dagger, b_m^\dagger] = 0$$

For the Hamiltonian and number operators we obtain

$$H = \sum_{n,m} b_n^\dagger b_m H_{nm}$$

$$N = \sum_n b_n^\dagger b_n$$

$$H_{nm} = \int d^3r \varphi_n^\dagger H \varphi_m$$

This Hamiltonian may be solved in the following fashion. We examine the equation of motion for the destruction operator:

$$-i\frac{\partial}{\partial t} b_n = [H, b_n] = -\sum_m H_{nm} b_m$$

We assume that an operator has the time development

$$b_n(t) = b_n e^{-iEt}$$

so that we wish solutions of the form

$$0 = \sum_m (H_{nm} - E\delta_{nm})b_m$$

The eigenvalues E are the solution to

$$\det(H_{nm} - E\delta_{nm}) = 0$$

Thus we only need to find the eigenvalues of the Hamiltonian matrix. Usually the matrix is of infinite dimensionality, since there are an infinite number of states in the set φ_n. But one may often diagonalize it exactly for many problems. Computers allow very accurate solutions for any case of interest. If all Hamiltonians had only bilinear operators, then many-body theory would only be an exercise in matrix diagonalization. Fortunately, it is more fun than that.

Many-body theory is used to study Hamiltonians which have additional terms. These may be interactions with phonons, spin effects, or particle–particle interactions. The effects of particle–particle interactions may be understood by examining a many-particle Hamiltonian of the form

$$H = \sum_i H_i + \tfrac{1}{2}\sum_{ij} V(\mathbf{r}_i - \mathbf{r}_j)$$

$$H_i = -\frac{\hbar^2}{2m}\nabla_i^2 + U(\mathbf{r}_i)$$

The first term contains a summation of one-particle Hamiltonians H_i. This term by itself is just as simple to solve as H_i alone. A collection of particles which do not mutually interact makes for a trivial problem. One solves the dynamics of one particle, and the total properties are the summation of the individual ones. The term which makes this hard is the particle–particle interactions, which is written in terms of creation and destruction operators:

$$H = \sum_i H_i + \tfrac{1}{2}\sum_{ij} V(\mathbf{r}_i - \mathbf{r}_j) = \sum_{lm} b_l^\dagger b_m H_{lm} + \tfrac{1}{2}\sum_{\substack{lm\\ij}} V_{lmij} b_j^\dagger b_m^\dagger b_l b_i$$

$$V_{lmij} = \int d^3r_1 \int d^3r_2 \phi_m^*(\mathbf{r}_1)\phi_l(\mathbf{r}_1) V(\mathbf{r}_1 - \mathbf{r}_2)\phi_j^*(\mathbf{r}_2)\phi_i(\mathbf{r}_2) \tag{1.2.6}$$

The interaction term contains two creation and two destruction particles. This term is interpreted as describing two-particle scattering events.

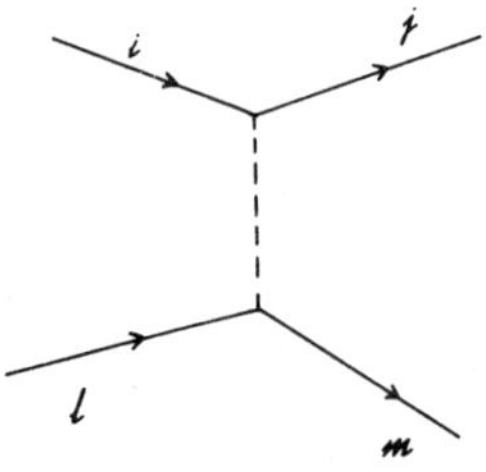

FIGURE 1.1

One particle in state i scatters to state j, while another in state l scatters to state m. The process is illustrated in Fig. 1.1. Each index i, j, l, m runs over all possible sets of values. Thus one has processes where, say, $i = j$, which describes a process where one particle scatters from l to m, while the other does not change its state. We were also careful to write this term so that both destruction operators are to the right and the creation operators to the left. The reason for this is that we usually wish to eliminate processes whereby a particle interacts with itself. For example, if we were to write this as

$$b_j^{\dagger} b_i b_m^{\dagger} b_l$$

then we could have $m = i$, and this would describe how one particle interacts with itself. For example, if $|0\rangle$ is the particle vacuum where all $n_m = 0$, then $b_j |0\rangle = 0$. The state $b_\alpha^{\dagger} |0\rangle$ contains one particle in state α. The operator

$$\tfrac{1}{2} \sum_{ijlm} V_{lmij} b_j^{\dagger} b_m^{\dagger} b_l b_i b_\alpha^{\dagger} |0\rangle = 0$$

on this state gives zero, since two particles cannot interact if there is only one particle in the system. However, the operator

$$\tfrac{1}{2} \sum_{ijlm} V_{lmij} b_j^{\dagger} b_i b_m^{\dagger} b_l b_\alpha^{\dagger} |0\rangle = \tfrac{1}{2} \sum_{mj} V_{\alpha mmj} b_j^{\dagger} |0\rangle$$

is finite, and this term ($l = \alpha$, $i = m$) must just be the particle interacting with itself. These terms are avoided by writing the pairwise interaction in the form (1.2.6).

For a gas or liquid of ^{4}He particles, a common basis set for the starting point of a many-body calculation is just free-particle wave functions,

$$\psi(\mathbf{r}) = \frac{1}{\nu^{1/2}} \sum_{\mathbf{k}} e^{i\mathbf{k}\cdot\mathbf{r}} a_{\mathbf{k}}$$

in which case the Hamiltonian has the form

$$H = \sum_{\mathbf{k}} \varepsilon_{\mathbf{k}} a_{\mathbf{k}}{}^{\dagger} a_{\mathbf{k}} + \frac{1}{2v} \sum_{\mathbf{k}\mathbf{k}'\mathbf{q}} V(q) a^{\dagger}_{\mathbf{k}+\mathbf{q}} a^{\dagger}_{\mathbf{k}'-\mathbf{q}} a_{\mathbf{k}'} a_{\mathbf{k}}$$

$$\varepsilon_k = \frac{k^2}{2m} \tag{1.2.7}$$

$$V(\mathbf{q}) = \int d^3 r e^{i\mathbf{q}\cdot\mathbf{r}} V(\mathbf{r})$$

The operator $\varrho_{\mathbf{q}}$ is the particle density operator in the plane-wave representation:

$$\varrho_{\mathbf{q}} = \sum_{\mathbf{k}} a^{\dagger}_{\mathbf{k}+\mathbf{q}} a_{\mathbf{k}}$$

The simplest way to write the interaction term in (1.2.7) is in the form

$$\frac{1}{2v} \sum_{\mathbf{q}} V(\mathbf{q}) \varrho_{\mathbf{q}} \varrho_{-\mathbf{q}}$$

but this is defective because both destruction operators are not to the right of the creation operators so the term has a particle interacting with itself.

A possible difficulty with (1.2.7) is that $V(\mathbf{q})$ may not exist. The potential $V(r)$ may not possess a Fourier transform if the particle–particle potential is too divergent at small values of r. This happens, for example, with the Lennard-Jones potential, which is often used to represent the helium–helium potential. Later we shall see that this difficulty may be avoided by summing subsets of diagrams to get a T-matrix interaction, which is always well behaved.

So far our discussion has concerned boson operators and boson Hamiltonians. Now we shall start discussing fermions. These are usually electrons, although occasionally one studies holes, positrons, or ^{3}He particles. Fermions have the property that any state may only contain zero or one particle, which is the famous exclusion principle. Jordan and Wigner (1928) discovered that this could be accomplished by making the fields anticommute, which is represented by curly brackets:

$$\psi(\mathbf{r})\psi^{\dagger}(\mathbf{r}') + \psi^{\dagger}(\mathbf{r}')\psi(\mathbf{r}) \equiv \{\psi(\mathbf{r}), \psi^{\dagger}(\mathbf{r}')\} = \delta(\mathbf{r} - \mathbf{r}')$$

$$\{\psi(\mathbf{r}), \psi(\mathbf{r}')\} = 0$$

$$\{\psi^{\dagger}(\mathbf{r}), \psi^{\dagger}(\mathbf{r}')\} = 0$$

Thus, if these wave functions are expanded in a basis set $\phi_\lambda(\mathbf{r})$,

$$\psi(\mathbf{r}) = \sum_\lambda c_\lambda \phi_\lambda$$

$$\psi^\dagger(\mathbf{r}) = \sum_\lambda c_\lambda^\dagger \phi_\lambda^*(\mathbf{r})$$

the coefficients $c_\lambda^\dagger$ and c_λ become creation and destruction operators which obey anticommutation relations:

$$\{c_\lambda, c_{\lambda'}^\dagger\} = \delta_{\lambda\lambda'}$$

$$\{c_\lambda, c_{\lambda'}\} = 0$$

$$\{c_\lambda^\dagger, c_{\lambda'}^\dagger\} = 0$$

For example, consider $\{c_\lambda, c_\lambda\} = 2c_\lambda c_\lambda = 0$. The operator $c_\lambda c_\lambda$ acting upon anything gives zero. We interpret c_λ as a destruction operator, which destroys a particle from a state λ. A state may only contain zero or one particle, which we call $|\,0\rangle_\lambda$ and $|\,1\rangle_\lambda$, respectively. Thus

$$c_\lambda\,|\,1\rangle_\lambda = |\,0\rangle_\lambda, \qquad c^\dagger\,|\,1\rangle_\lambda = 0$$

$$c_\lambda\,|\,0\rangle_\lambda = 0, \qquad c_\lambda^\dagger\,|\,0\rangle_\lambda = |\,1\rangle_\lambda$$

Thus $c_\lambda c_\lambda$ acting upon either $|\,1\rangle_\lambda$ or $|\,0\rangle_\lambda$ gives zero. Similarly, we have the combination $c_\lambda^\dagger c_\lambda^\dagger = 0$. It is zero because two particles cannot be created in the same state. Another way to see this is to consider the number operator for a state,

$$N_\lambda = c_\lambda^\dagger c_\lambda$$

and its square,

$$N_\lambda^2 = c_\lambda^\dagger c_\lambda c_\lambda^\dagger c_\lambda$$

Using the anticommutation relations gives

$$N_\lambda^2 = c_\lambda^\dagger(1 - c_\lambda^\dagger c_\lambda)c_\lambda = c_\lambda^\dagger c_\lambda - c_\lambda^\dagger c_\lambda^\dagger c_\lambda c_\lambda = N_\lambda$$

Thus we get $N_\lambda^2 = N_\lambda$. The only numbers which are equal to its square are 0 and 1. The numbers N_λ^2 may only be 0 or 1. The anticommutation relations have built into them the fermion property that no two particles may be in the same state λ.

Quite often the Hamiltonians for fermion calculations are of the form

$$H = \sum_i \left[\frac{p_i^2}{2m} + U(r_i)\right] + \tfrac{1}{2}\sum_{ij} V(\mathbf{r}_i - \mathbf{r}_j) \tag{1.2.8}$$

with the particles interacting with a potential $U(r)$ and with each other through particle–particle interactions $V(\mathbf{r}_i - \mathbf{r}_j)$. The Hamiltonian is written in terms of creation and destruction operators exactly as in the boson case:

$$H = \sum_{ij} H_{ij} c_i^{\dagger} c_j + \tfrac{1}{2} \sum_{ijlm} c_j^{\dagger} c_m^{\dagger} c_l c_i V_{ijlm}$$

$$H_{ij} = \int d^3 r \phi_i(\mathbf{r})^* \left[-\frac{\hbar^2}{2m} \nabla^2 + U(r) \right] \phi_j(\mathbf{r}) \tag{1.2.9}$$

$$V_{ijlm} = \int d^3 r_1 \int d^3 r_2 \phi_j^*(\mathbf{r}_1) \phi_i(\mathbf{r}_1) V(\mathbf{r}_1 - \mathbf{r}_2) \phi_m^*(\mathbf{r}_2) \phi_l(\mathbf{r}_2)$$

The difference in behavior between fermions and bosons is often due to the difference in the commutation relations of the operators. The starting Hamiltonians are of similar form, e.g., for liquid ^{3}He and ^{4}He.

The particle–particle interaction term is still interpreted in terms of Fig. 1.1. Great care must be used in writing this term, or else one makes a sign mistake. For example, the term

$$\tfrac{1}{2} \sum_{ijlm} V_{ijlm} c_j^{\dagger} c_m^{\dagger} c_i c_l \tag{1.2.10}$$

is wrong. We just exchanged the order of the two destruction operators. But because of the anticommutation relation

$$\{c_i, c_l\} = 0 = c_i c_l + c_l c_i$$

or

$$c_i c_l = -c_l c_i$$

this interchange causes a sign change. Thus (1.2.10) would be correct if we were to multiply the entire term by -1.

For the study of electrons in solids, a popular basis set is plane waves. Thus eigenstates are described by $(\mathbf{p}, \sigma)$, where σ is the spin index which is ± 1 for spin up or down. The Hamiltonian then has the form

$$H = \sum_{\mathbf{p}\sigma} \varepsilon_{\mathbf{p}} c^{\dagger}_{\mathbf{p}\sigma} c_{\mathbf{p}\sigma} + \sum_{\mathbf{q}} U(\mathbf{q}) \varrho_{\mathbf{q}} + \frac{1}{2\nu} \sum_{\substack{\mathbf{q}\mathbf{k}\mathbf{k}' \\ \sigma\sigma'}} v_{\mathbf{q}} c^{\dagger}_{\mathbf{k}+\mathbf{q}\sigma} c^{\dagger}_{\mathbf{k}'-\mathbf{q},\sigma'} c_{\mathbf{k}'\sigma'} c_{\mathbf{k}\sigma} \tag{1.2.11}$$

The second term represents the interaction between the electrons and the atoms or ions of the solid. This interaction $U(\mathbf{q})$ is often represented by a pseudopotential (Harrison, 1966; Heine, 1970).

The electron density operator

$$\varrho(\mathbf{q}) = \sum_{\mathbf{k}\sigma} c^{\dagger}_{\mathbf{k}+\mathbf{q}\sigma} c_{\mathbf{k}\sigma}$$

is the same as for bosons except for the additional summation over spin index. The last term in (1.2.11) contains the electron–electron interaction, which is just a Coulomb potential. The Fourier transform v_q of a Coulomb potential e^2/r will occur often, so that we have given it a special symbol with lowercase "v_q." The Fourier transform is

$$\begin{aligned} v_q &= e^2 \int \frac{d^3r}{r} e^{i\mathbf{q}\cdot\mathbf{r}} = 2\pi e^2 \int_0^\infty dr r \int_{-1}^{1} d(\cos\theta) e^{iqr\cos\theta} \\ &= \frac{2\pi e^2}{iq} \int_0^\infty dr (e^{iqr} - e^{-iqr}) = \frac{4\pi e^2}{q} \int_0^\infty dr \sin qr \\ &= \frac{4\pi e^2}{q^2} (1 - \lim_{r\to\infty} \cos qr) \end{aligned}$$

The integral is not well defined, since it oscillates at infinity. One assumes these oscillations damp out, so that the result is

$$v_q = \frac{4\pi e^2}{q^2} \tag{1.2.12}$$

These electron–electron interactions are a significant part of many-body theory. Most of what we do involves worrying about electron–electron interactions in one form or another.

The full electron gas Hamiltonian (1.2.11) is often too complicated to use for the more elaborate many-body theory. Quite often this is approximated by a model Hamiltonian which has a simpler form. Usually these model Hamiltonians look very simple but still are impossible to solve exactly. Often they are even difficult to solve approximately! We shall now discuss some of these popular models.

The *homogeneous electron gas* is a model which is frequently studied to learn about correlation effects. It has the Hamiltonian

$$H = \sum_{\mathbf{p},\sigma} \varepsilon_{\mathbf{p}} c^{\dagger}_{\mathbf{p}\sigma} c_{\mathbf{p}\sigma} + \frac{1}{2\nu} \sum_{\substack{q\neq 0\\ \mathbf{k}\mathbf{k}'}} v_q c^{\dagger}_{\mathbf{k}+\mathbf{q},\sigma} c^{\dagger}_{\mathbf{k}'-\mathbf{q},\sigma'} c_{\mathbf{k}'\sigma'} c_{\mathbf{k}\sigma} \tag{1.2.13}$$

The basic premise is to get rid of the atoms and to replace them with a uniform positive background charge of density n_0. The homogeneous electron gas is also called the *jellium* model. One can think of taking the positive charge of the ions and spreading it uniformly about the unit cell of the crystal. Of course, the homogeneous electron gas then has no crystal structure. To preserve charge neutrality, the average particle density of the electron gas must also be n_0. The average density of the electrons is just the $\mathbf{q} = 0$ value of the density operator,

$$n_0 = \frac{1}{\nu} \langle \varrho(0) \rangle = \frac{1}{\nu} \sum_{\mathbf{k},\sigma} \langle c^\dagger_{\mathbf{k}\sigma} c_{\mathbf{k}\sigma} \rangle = \frac{1}{\nu} \sum_{\mathbf{k}\sigma} N_{\mathbf{k}\sigma} = \frac{N}{\nu}$$

since the number operator may be summed to give the number of particles N. In writing the Hamiltonian (1.2.13), the $\mathbf{q} = 0$ term in the interaction term was omitted from the summation. This omitted term has the form

$$\lim_{\mathbf{q}\to 0} \frac{N(N-1)}{2\nu} (v_{\mathbf{q}})$$

This term was canceled by two other terms. One of these is the Coulomb interaction of the uniform positive background with itself:

$$\tfrac{1}{2} e^2 \int \frac{n_0^2 d^3 r}{|\mathbf{r} - \mathbf{r}'|} d^3 r' = \lim_{q \to 0} \frac{2\pi e^2}{q^2 \nu} N^2$$

The other is the Coulomb interaction of the uniform positive background with the electrons:

$$-e^2 \int \frac{n_0 \varrho(r)}{|\mathbf{r} - \mathbf{r}'|} d^3 r \, d^3 r' = -\lim_{q \to 0} \frac{4\pi e^2}{\nu q^2} N^2$$

The sum of these three terms cancels the N^2 term, and the other term may be neglected. Thus the Hamiltonian (1.2.13) describes a system which has charge neutrality. The interaction terms with $\mathbf{q} \neq 0$ describe the fluctuations which occur because of electrons interacting with themselves.

The plane-wave model is often a poor approximation of electron behavior in ionic solids. In many solids the electrons are localized on atomic sites and only occasionally hop to neighboring sites. This is called the *tight binding model.* One simple form of this model is bilinear in the operators:

$$H = \sum_{i\delta\sigma} W_\delta c^\dagger_{i\sigma} c_{i+\delta,\sigma}$$

The index i denotes a site at point $\mathbf{R}_i$, while $i + \delta$ represents the nearest neighbor atoms. One can think of the term

$$W_\delta = \int d^3r\phi^*(\mathbf{r} - \mathbf{R}_i)\left[-\frac{\hbar^2}{2m}\nabla^2 + U(r)\right]\phi(\mathbf{r} - \mathbf{R}_{i+\delta})$$

as arising from the matrix elements between orbitals $\phi(r)$ which are localized on site $\mathbf{R}_i$ and $\mathbf{R}_{i+\delta}$. The term W_δ for $\delta \neq 0$ represents processes where the electron jumps from site $i + \delta$ to i, and W_0 is the site energy. Simple versions of the model usually have only a single orbital state for each atomic site. More realistic versions of the tight binding model allow for the multiple orbitals characteristic of p or d electrons. Our discussion will assume a single orbital state per atomic site.

The bilinear form of the Hamiltonian is trivial to solve exactly. One defines the operator in wave vector space in the usual fashion,

$$c_{j\sigma} = \frac{1}{N^{1/2}} \sum_{\mathbf{k}} e^{i\mathbf{k}\cdot\mathbf{R}_j} c_{\mathbf{k},\sigma}$$

$$c_{\mathbf{k},\sigma} = \frac{1}{N^{1/2}} \sum_{j} e^{-i\mathbf{k}\cdot\mathbf{R}_j} c_{j\sigma}$$

and the Hamiltonian may be rewritten as

$$H = \sum_{\mathbf{k}\sigma} W(\mathbf{k}) c^\dagger_{\mathbf{k}\sigma} c_{\mathbf{k}\sigma}$$

If the $c_{j\sigma}$ are fermion operators, then so are the $c_{\mathbf{k}\sigma}$. They obey anticommutation relations $\{c_{\mathbf{k}}, c^\dagger_{\mathbf{k}'}\} = \delta_{\mathbf{kk}'}$. Thus each mode $(\mathbf{k}, \sigma)$ becomes an independent Fermi system, which may be treated separately in thermodynamic averages. The particle energy is

$$W(\mathbf{k}) = \sum_{\delta} \exp(i\mathbf{k}\cdot\boldsymbol{\delta}) W_{\boldsymbol{\delta}}$$

A very common model is the *nearest neighbor model.* Here the hopping term is limited to just the nearest neighbor ions, which are presumed to be all alike. Usually the site energy is W_0 set equal to zero, which is just an arbitrary energy renormalization, and $W_\delta = w$. Then the energy is

$$W(\mathbf{k}) = Zw\gamma_{\mathbf{k}}$$

where the factor

$$\gamma_{\mathbf{k}} = \frac{1}{Z}\sum_{\delta} e^{i\mathbf{k}\cdot\boldsymbol{\delta}}$$

is summed over the Z nearest neighbors. This solution is exact for the tight binding model. For example, we may write the partition function Z_p for fermions as

$$Z_p = e^{-\beta\Omega} = \mathrm{Tr}\, e^{-\beta(H-\mu N)} = \prod_{\mathbf{k},\sigma} \{1 + e^{-\beta[W(\mathbf{k})-\mu]}\} \tag{1.2.14}$$

The many-particle partition function (1.2.14) is obtained by treating each state $\mathbf{k}$ as independent and averaging its thermodynamic properties separately. Thus it appears as if each electron in each state $\mathbf{k}$ is behaving independently. This picture is deceptive. The electrons are not just whizzing around independently.

For example, if one were to calculate the probability that any two electrons of the same spin are on the same atomic site, a zero answer is obtained. Two electrons of the same spin are never on the same site in the single orbital models since they would be in the same state. Of course, it is a basic feature of fermion many-particle wave functions that two electrons of the same spin can never be in the same location. Thus the motion of the electrons is correlated, which arises from the antisymmetry of the wave function under exchange of particle position. Thus the motion of the electrons is not really independent. This correlation is built into the eigenstates but does not affect the energy in the simple bilinear model, which ignores interactions between particles. Thus the partition function (1.2.14) has the appearance of independent particle form, even with correlation in the wave function. The antisymmetrization of the many-particle wave function does not affect the expectation value of one-particle operators. It does affect the expectation value of two(or more)-particle operators.

The tight binding model may also contain the Coulomb interaction between electrons. In its most general form (1.2.9), the interaction term is

$$\frac{1}{2} \sum_{\substack{ijml \\ \sigma\sigma'}} V_{ij,ml} c_j^{\dagger} c_m^{\dagger} c_l c_i$$

$$V_{ij,ml} = \int d^3r_1\, d^3r_2 \phi^*(\mathbf{r}_1 - \mathbf{R}_j)\phi(\mathbf{r}_1 - \mathbf{R}_i) \frac{e^2}{|\mathbf{r}_1 - \mathbf{r}_2|} \times \phi^*(\mathbf{r}_2 - R_m)\phi(\mathbf{r}_2 - \mathbf{R}_l)$$

The four orbitals could be centered on four different sites. These are called *four-center integrals*. They are usually small and nearly always neglected in many-body calculations. The terms which are sometimes included are just the largest Coulomb terms. One possible term is the direct interaction between two particles on different atomic sites. For example, setting $m = l$

and $i = j$ gives

$$\tfrac{1}{2} \sum_{j \neq m} V_{jm} n_j n_m$$

$$V_{jm} = \int d^3r_1\, d^3r_2 \,|\, \phi(\mathbf{r} - \mathbf{R}_j) \,|^2 \frac{e^2}{|\, \mathbf{r}_1 - \mathbf{r}_2 \,|} \,|\, \phi(\mathbf{r} - \mathbf{R}_m) \,|^2$$

Note that we can rearrange the operators into $n_j n_m$ as long as we specify that $m \neq j$, so that one does not have a particle interacting with itself. The interaction terms V_{jm} become just a Coulomb potential,

$$\lim_{|R_j - R_m| \to \infty} V_{jm} = \frac{e^2}{|\, \mathbf{R}_j - \mathbf{R}_m \,|}$$

at separation beyond where the orbitals overlap and if the orbitals have s symmetry. A Hamiltonian with these nearest neighbor Coulomb interactions is similar to the lattice gas Hamiltonians we shall introduce in Sec. 1.4.

The Hubbard model (1963) retains only the Coulomb integral which is the very largest. All four orbitals $\phi(\mathbf{r})$ are centered on the same site m. This describes the interaction between two electrons which are on the same atom. Since two electrons cannot be in the same state, the two on the same atom must be in different atomic states. In the simplest model, which considers only a single orbital state on each atom, the two electrons must have different spin configurations. One has spin up, while the other has spin down. The Hubbard model considers the following Hamiltonian:

$$H = w \sum_{\delta i \sigma} c_{i\sigma}^{\dagger} c_{i+\delta\sigma} + W_0 \sum_{i\sigma} n_{i\sigma} + U \sum_{j} n_{j\uparrow} n_{j\downarrow} \tag{1.2.15}$$

The hopping term is usually limited to nearest neighbors. This Hamiltonian was also introduced by Gutzwiller (1963), who studied the properties of electrons in d bands in ferromagnets. It was then extensively studied by Hubbard (1963–1966). It is thought to be a good model for electron conduction in narrow band materials, for example, in transition metal oxides (Adler, 1967). In spite of much effort, the Hubbard model is still not well understood. This prevents detailed comparisons of the model to experimental systems.

The parameter U is the Coulomb interaction between two electrons on the same atom. Usually the model is applied to tightly bound orbitals such as d states. Then U is quite large, perhaps 10 eV. The bandwidth w is sometimes taken to be smaller. However, some of the most interesting phenomena seem to occur for $U \sim w$.

The Hubbard model can be solved exactly only in one dimension, as shown by Lieb and Wu (1968). But there are two limiting cases where exact solutions can be obtained in other dimensions. One is where $U = 0$, which is the nearest neighbor tight binding model. The other case is where the hopping bandwidth $w = 0$. This is the *atomic limit*, since here each atom is considered individually. The energy in this atomic limit is

$$E = W_0 N + \tilde{n} U \tag{1.2.16}$$

where N is the number of electrons and $\tilde{n}$ is the number of sites with two electrons.

Creation and destruction operators are used to describe other kinds of operators besides Hamiltonians. We shall frequently use density and current operators. The density operator is summed over the position of all particles:

$$\varrho(\mathbf{r}) = \sum_i \delta(\mathbf{r} - \mathbf{r}_i)$$

$\varrho(\mathbf{r})$ may be expressed in terms of creation and destruction operators as

$$\varrho(\mathbf{r}) = \psi(\mathbf{r})^\dagger \psi(\mathbf{r}) = \sum_{\lambda,\eta} c_\lambda^\dagger c_\eta \phi_\lambda^*(\mathbf{r}) \phi_\eta(\mathbf{r})$$

The Fourier transform of the density operator is also needed:

$$\varrho(\mathbf{q}) = \int d^3r e^{-i\mathbf{q}\cdot\mathbf{r}} \varrho(\mathbf{r}) = \sum_{\lambda,\eta} c_\lambda^\dagger c_\eta \int d^3r e^{-i\mathbf{q}\cdot\mathbf{r}} \phi_\lambda^*(\mathbf{r}) \phi_\eta(\mathbf{r})$$

The two most popular representations are the free-particle model,

$$\varrho(\mathbf{q}) = \sum_{\mathbf{k},\sigma} c^\dagger_{\mathbf{k}+\mathbf{q},\sigma} c_{\mathbf{k}\sigma}$$

and the tight binding model when omitting overlap between neighbors,

$$\varrho(\mathbf{q}) = \sum_{i\sigma} n_{i\sigma} e^{i\mathbf{q}\cdot\mathbf{r}_i}$$

Another important operator is the *electrical current*. It is the summation over all particles and their velocities:

$$\mathscr{j}_e(\mathbf{r}) = \tfrac{1}{2} \sum_i e_i [\mathbf{v}_i \delta(\mathbf{r} - \mathbf{r}_i) + \delta(\mathbf{r} - \mathbf{r}_i) \mathbf{v}_i]$$

The summation above is over different groups of particles. Each group of particles contains identical particles with the same charge e_i. Let j_i be the *particle current* for each kind of particle species. Then the electrical current operator is

$$j_e(\mathbf{r}) = \sum_l e_l j_l(\mathbf{r})$$

where one sums over each particle species l and the particle current j_l for each species. The standard quantum mechanical representation for the particle current is

$$j_l(\mathbf{r}) = \frac{1}{2mi} \{\psi^\dagger(\mathbf{r})\nabla\psi(\mathbf{r}) - [\nabla\psi^\dagger(\mathbf{r})]\psi(\mathbf{r})\}$$

The Fourier transform of the current operator has the form

$$j_l(\mathbf{q}) = \frac{1}{2mi} \sum_{\lambda\eta} c_\lambda{}^\dagger c_\eta \int d^3r e^{-i\mathbf{q}\cdot\mathbf{r}}[\phi_\lambda{}^*(\mathbf{r})\nabla\phi_\eta(\mathbf{r}) - \phi_\eta(\mathbf{r})\nabla\phi_\lambda{}^*(\mathbf{r})] \quad (1.2.17)$$

For free particles, this has the form

$$j_l(\mathbf{q}) = \frac{1}{m} \sum_{k\sigma} (\mathbf{k} + \tfrac{1}{2}\mathbf{q}) c^\dagger_{\mathbf{k}+\mathbf{q},\sigma} c_{\mathbf{k}\sigma} \quad (1.2.18)$$

Another case of interest is the current operator for the tight binding model. In this case, it is easier to consider an alternate, but equivalent, formula for the current operator. The derivation starts from the definition of the *polarization* operator,

$$\mathbf{P} = \int d^3r \mathbf{r} \varrho(\mathbf{r})$$

which is a summation over all the particles and their positions. One then recalls that the time derivative of the polarization is just the particle current:

$$\frac{\partial}{\partial t} \mathbf{P} = \int d^3r \mathbf{r} \frac{\partial}{\partial t} \varrho(\mathbf{r}, t)$$

This relationship can be proved easily by using the equation of continuity,

$$\frac{\partial}{\partial t} \varrho(\mathbf{r}, t) = -\nabla \cdot j(\mathbf{r}, t)$$

and an integration by parts,

$$\frac{\partial \mathbf{P}}{\partial t} = -\int d^3r \mathbf{r} \nabla \cdot j(\mathbf{r}, t) = \int d^3r j(\mathbf{r}) \cdot \nabla(\mathbf{r}) = \int d^3r j(\mathbf{r})$$

In the tight binding model, the polarization operator has the form

$$\mathbf{P} = \sum_i \mathbf{R}_i n_i$$

and the time derivative is

$$\mathbf{j} = \frac{\partial \mathbf{P}}{\partial t} = i[H, \mathbf{P}]$$

For example, if the Hamiltonian has the form

$$H = w \sum_{i\sigma\delta} c_{i+\delta,\sigma}^{\dagger} c_{i\sigma} + \sum_{\substack{ij \\ \sigma\sigma'}} n_{i\sigma} n_{j\sigma'} V_{ij}$$

then only the first term contributes to the current operator. The other term contains only the position operator $n_{i\sigma}$, and this commutes with itself. Thus in this case the current operator is

$$\mathbf{j} = -iw \sum_{i\delta\sigma} \boldsymbol{\delta} c_{i+\delta,\sigma}^{\dagger} c_{i\sigma} \tag{1.2.19}$$

This current operator is used in calculations on localized electrons. It applies to organic solids and narrow band ionic solids.

The *energy current* operator is needed to calculate energy transport in solids, which occurs, for example, in discussions of thermal conductivity or thermoelectrical effects. Energy currents flow whenever heat is generated or dissipated nonuniformly in the solid. The energy current $\mathbf{j}_E$ is defined as the energy flow through a surface. It obeys an equation of energy conservation,

$$\frac{\partial}{\partial t} H + \boldsymbol{\nabla} \cdot \mathbf{j}_E = 0 \tag{1.2.20}$$

where the energy change $\partial H/\partial t$ equals the variation in the energy flux. An equation for the energy current may be derived by formally introducing an operator which is the integral over the position and Hamiltonian density:

$$\mathbf{R}_E = \tfrac{1}{2} \int d^3r [\mathbf{r} \mathscr{H}(r) + \mathscr{H}(r) \mathbf{r}]$$

The equation of heat continuity (1.2.20) may be used to show that the time derivative of this quantity is just the energy current:

$$\frac{\partial \mathbf{R}_E}{\partial t} = \frac{1}{2} \int d^3r \left[\mathbf{r} \frac{\partial}{\partial t} \mathscr{H} + \left(\frac{\partial}{\partial t} \mathscr{H} \right) \mathbf{r} \right] = \mathbf{j}_E$$

For a free-particle system, the energy current is

$$\mathbf{j}_E = \sum_{\mathbf{p}\sigma} \mathbf{v}_{\mathbf{p}} \varepsilon_{\mathbf{p}} c^\dagger_{\mathbf{p}\sigma} c_{\mathbf{p}\sigma}$$

This result is sensible. It is just the energy $\varepsilon_p = p^2/2m$ of each particle multiplied times its velocity $v_{\mathbf{p}} = \mathbf{p}/m$. In the nearest neighbor tight binding model, one writes $\mathbf{R}_E$ in terms of the site Hamiltonian h_i and position $\mathbf{R}_i$:

$$\mathbf{R}_E = \sum_i \mathbf{R}_i h_i$$

$$H = \sum_i h_i$$

$$\mathbf{j}_E = i \sum_{l,m} \mathbf{R}_l [h_m, h_l]$$

$$h_i = \frac{w}{2} \sum_{\delta\sigma'} (c^\dagger_{i+\delta,\sigma} c_{i\sigma} + c^\dagger_{i\sigma} c_{i+\delta,\sigma}) + \frac{1}{2} \sum_{\substack{j \\ \sigma\sigma'}} n_{i\sigma} n_{j\sigma'} V_{ij}$$

For example, the current operator from just the hopping term for electrons is in one dimension:

$$\mathbf{j}_E = -i \frac{w^2}{2} \sum_{i\sigma\delta\delta'} (\boldsymbol{\delta} + \boldsymbol{\delta}') c^\dagger_{i+\delta+\delta',\sigma} c_{i,\sigma} \tag{1.2.21}$$

The other terms can become quite complicated, and will be introduced only as they are needed.

There are several other comments. First, the operators **P** and $\mathbf{R}_E$ are not defined in infinite systems, since the integral over position will diverge. One can devise an alternate definition based on a limiting process (Beni, 1974):

$$\mathbf{P} = -i \lim_{\mathbf{q}\to 0} \nabla_{\mathbf{q}} \int d^3r \varrho(r) e^{i\mathbf{q}\cdot\mathbf{r}}$$

$$\mathbf{R}_E = -\frac{i}{2} \lim_{\mathbf{q}\to 0} \nabla_q \int d^3r \{e^{i\mathbf{q}\cdot\mathbf{r}}, \mathscr{H}(\mathbf{r})\}$$

The second comment concerns the energy current. As we shall discuss extensively in Sec. 3.8, the energy current is often *not* the current which describes thermal conductivity or thermoelectric power. In metals, it is customary to use the *heat current*, which is defined as

$$\mathbf{j}_Q = \mathbf{j}_E - \mu \mathbf{j}$$

$\mathbf{j}_Q$ is given in terms of operators $\mathbf{j}_E$ and $\mathbf{j}$, which we have already defined.

Another observation is that these operator definitions are the same for all particles, regardless of whether they are bosons or fermions. Of course, spinless bosons do not have the summation over spin index σ. Otherwise, everything is the same. However, calculations using these operators depend significantly on whether the particles are bosons or fermions.

1.3. ELECTRON–PHONON INTERACTIONS

In the first two sections we described the Hamiltonians for phonons and electrons, respectively. In this section we shall discuss their mutual interaction. This topic is important in many-body theory. The electron–phonon interaction causes superconductivity in many metals and influences the transport properties of every metal. In pure semiconducting and ionic solids, the electron–phonon interaction usually dominates the transport properties. The word *polaron* is used to describe a single electron which is coupled to phonons. The modern formulation of the polaron problem was due to Fröhlich *et al.* (1950), and its study is an important part of the history of many-body theory.

A. Interaction Hamiltonian

The basic Hamiltonian is assumed to have the form

$$
\begin{aligned}
H &= H_p + H_{\mathrm{e}} + H_{\mathrm{ei}} \\
H_p &= \sum_{\mathbf{q}\lambda} \omega_{\mathbf{q}\lambda}(a_{\mathbf{q}\lambda}^\dagger a_{\mathbf{q}\lambda} + \tfrac{1}{2}) \\
H_{\mathrm{e}} &= \sum_i \frac{p_i^2}{2m} + \frac{1}{2} e^2 \sum_{ij} \frac{1}{r_{ij}} \\
H_{\mathrm{ei}} &= \sum_i \tilde{V}(\mathbf{r}_i)
\end{aligned}
$$

The atom part H_p describes the normal modes of vibration of the solid and is the phonon Hamiltonian of Sec. 1.1. The second term is the electron–electron part H_{e}, which was discussed in Sec. 1.2. The third part is the electron–ion interaction. It is assumed that H_{ei} is the summation of the interaction from the individual atoms or ions:

$$
\tilde{V}(\mathbf{r}_i) = \sum_j V_{\mathrm{ei}}(\mathbf{r}_i - \mathbf{R}_j)
$$

The word *ion* is not meant to imply a particular charge state. In metals, the atoms are ions, while in covalently bonded semiconductors they are something else. We use *ion* to encompass all these possibilities. Each ion is at a position $\mathbf{R}_j = \mathbf{R}_j^{(0)} + \mathbf{Q}_j$, which is the sum of the equilibrium position $\mathbf{R}_j^{(0)}$ and the displacement $\mathbf{Q}_j$. The displacements are usually small, so that one can expand in powers of them:

$$V_{\text{ei}}(\mathbf{r}_i - \mathbf{R}_j^{(0)} - \mathbf{Q}_j) = V_{\text{ei}}(\mathbf{r}_i - \mathbf{R}_j^{(0)}) + \mathbf{Q}_j \cdot \boldsymbol{\nabla} V_{\text{ei}}(\mathbf{r}_j - \mathbf{R}_j^{(0)}) + O(Q^2)$$

The linear electron–phonon interaction term is obtained from the first term in $\mathbf{Q}_j$. We shall neglect the terms in $O(Q^2)$, although they are retained in some circumstances. The constant term

$$\sum_j V_{\text{ei}}(\mathbf{r}_i - \mathbf{R}_j^{(0)})$$

is the potential function for the electrons when the atoms are in their equilibrium positions, which forms a periodic potential in a crystal. The solution of the Hamiltonian for electron motion in this periodic potential gives the Bloch states of the solids. They are usually assumed to be known. Thus in many problems we begin by writing the Hamiltonian as $H = H_0 + V$, where H_0 is a Hamiltonian we have solved and V is the perturbation. Quite often the eigenstates of H_0 are just the Bloch states of the solid, calculated by assuming that the atoms are in their equilibrium positions. The sequence of approximations we are making is called, collectively, the *Born–Oppenheimer approximation.* As always, there are circumstances where these approximations are inadequate, and other approaches are necessary, e.g., in the dynamical Jahn–Teller effect.

Here our interest is in the electron–phonon interaction:

$$\tilde{V}(\mathbf{r}) = \sum_j \mathbf{Q}_j \cdot \boldsymbol{\nabla} V_{\text{ei}}(\mathbf{r} - \mathbf{R}_j^{(0)})$$

This interaction is to be written in terms of operators. We assume that the electron–atom potential possesses a Fourier transform:

$$V_{\text{ei}}(\mathbf{r}) = \frac{1}{N} \sum_q V_{\text{ei}}(\mathbf{q}) e^{i\mathbf{q}\cdot\mathbf{r}}$$

$$\boldsymbol{\nabla} V_{\text{ei}}(\mathbf{r}) = i \frac{1}{N} \sum_q \mathbf{q} V_{\text{ei}}(\mathbf{q}) e^{i\mathbf{q}\cdot\mathbf{r}}$$

We need to evaluate the combination

$$\tilde{V}(\mathbf{r}) = \frac{i}{N} \sum_q e^{i\mathbf{q}\cdot\mathbf{r}} V_{\text{ei}}(\mathbf{q}) \mathbf{q} \cdot \left(\sum_j \mathbf{Q}_j e^{-i\mathbf{q}\cdot\mathbf{R}_j^{(0)}} \right)$$

We can use our arlier definition (1.1.17) of Q_j to show that

$$\frac{i}{N}\sum_j \mathbf{Q}_j e^{-i\mathbf{q}\cdot\mathbf{R}_j^{(0)}} = \frac{i}{N^{1/2}}\sum_{\mathbf{G}} \mathbf{Q}_{\mathbf{q}+\mathbf{G}}$$
$$= -\sum_{\mathbf{G}}\left(\frac{\hbar}{2MN\omega_{\mathbf{q}+\mathbf{G}}}\right)^{1/2} \xi_{\mathbf{q}+\mathbf{G}}(a_{\mathbf{q}+\mathbf{G}} + a^{\dagger}_{-\mathbf{q}-\mathbf{G}})$$

where the summation **G** is over the reciprocal lattice vectors of the solid. Of course, the phonon states $\mathbf{q}+\mathbf{G}$ are defined only within the first Brillouin zone of the solid. But the Fourier transform $V_{\mathrm{ei}}(\mathbf{q})$ is defined over all q values, not just the first Brillouin zone. Thus we write the interaction Hamiltonian in the form ($MN = \varrho v$, $\varrho =$ density of solid in grams per cubic centimeter)

$$\tilde{V}(\mathbf{r}) = -\sum_{\mathbf{q},\mathbf{G}} e^{i\mathbf{r}\cdot(\mathbf{q}+\mathbf{G})} V_{\mathrm{ei}}(\mathbf{q}+\mathbf{G})(\mathbf{q}+\mathbf{G})\cdot\hat{\xi}_{\mathbf{q}}\left(\frac{\hbar}{2\varrho\omega_{\mathbf{q}}v}\right)^{1/2}(a_{\mathbf{q}} + a^{\dagger}_{-\mathbf{q}})$$

Here we have restricted the summation over **q** to be within the first Brillouin zone of the crystal. The phonons are defined only in this space, so $\omega_{\mathbf{q}}$, $\mathbf{a}_q$, $\hat{\boldsymbol{\xi}}_q$ have only q labels. But the summation over reciprocal lattice vectors **G** permits the potential $V_{\mathrm{ei}}(\mathbf{q}+\mathbf{G})$ to interact with higher Fourier components.

The potential $V_{\mathrm{ei}}(\mathbf{r})$ is defined as the unscreened electron–atom potential. Later we shall learn that electron–electron interactions in metals cause a significant reduction of this potential, which is called screening. The potential V_{ei} is sometimes calculated from first principles, but more often it is obtained from a pseudopotential (Heine, 1970; Harrison, 1966). Modern calculations usually solve force constant models to obtain accurate phonon energies $\omega_{\mathbf{q}}$, and polarizations $\xi_{\mathbf{q}}$ throughout the Brillouin zone, for each mode of polarization and use these in calculating electron–phonon properties (Tomlinson and Swihart, 1975).

The potential $\tilde{V}(r)$ acts upon the electrons and also upon other particles such as positrons. The electron–phonon interaction is obtained by integrating this over the charge density of the solid $\varrho(r)$:

$$H_{\mathrm{ep}} = \int d^3r\varrho(\mathbf{r})\tilde{V}(\mathbf{r})$$
$$= -\sum_{\mathbf{q},\mathbf{G}} \varrho(\mathbf{q}+\mathbf{G})V_{\mathrm{ei}}(\mathbf{q}+\mathbf{G})(\mathbf{q}+\mathbf{G})\cdot\xi_{\mathbf{q}}\left(\frac{\hbar}{2\varrho v\omega_{\mathbf{q}}}\right)^{1/2}(a_{\mathbf{q}} + a^{\dagger}_{-\mathbf{q}})$$

The particle density operator $\varrho(\mathbf{q})$ was defined in the prior section. Several examples of it were given there. Quite often we shall abbreviate this matrix element by the symbol

$$\begin{aligned} M_{\mathbf{q}+\mathbf{G}} &= -V_{\mathrm{ei}}(\mathbf{q}+\mathbf{G})(\mathbf{q}+\mathbf{G})\cdot\hat{\xi}_{\mathbf{q}}\left(\frac{\hbar}{2\varrho\omega_{\mathbf{q}}}\right)^{1/2} \\ H_{\mathrm{ep}} &= \frac{1}{\nu^{1/2}}\sum_{\mathbf{q},\mathbf{G}} M_{\mathbf{q}+\mathbf{G}}\varrho(\mathbf{q}+\mathbf{G})(a_{\mathbf{q}}+a^{\dagger}_{-\mathbf{q}}) \end{aligned} \tag{1.3.1}$$

B. Localized Electron

There is one problem which can be solved immediately: the electron-phonon Hamiltonian when the electrons are fixed in space at positions $\mathbf{r}_i$. We assume that they cannot recoil, which neglects the electron kinetic energy term. This model is often applied for localized electrons in solids. A localized electron occurs in deep core states and in some impurity levels. Then the Hamiltonian has the form

$$\begin{aligned} H &= H_p + H_{\mathrm{ep}} \\ &= \sum_{0}\left[\omega_q(a_{\mathbf{q}}{}^{\dagger}a_{\mathbf{q}}+\tfrac{1}{2})+\sum_i (a_{\mathbf{q}}+a^{\dagger}_{-\mathbf{q}})\frac{e^{i\mathbf{q}\cdot\mathbf{r}_i}}{\nu^{1/2}}\sum_{\mathbf{G}} M_{\mathbf{q}+\mathbf{G}}\varrho_0(\mathbf{q}+\mathbf{G})e^{i\mathbf{r}_i\cdot\mathbf{G}}\right] \end{aligned} \tag{1.3.2}$$

The electron density operator $\varrho_0(\mathbf{q}+\mathbf{G})$ is just the Fourier transform of the localized charge density:

$$\begin{aligned} \varrho(\mathbf{q}+\mathbf{G}) &= \int d^3r e^{i\mathbf{r}\cdot(\mathbf{q}+\mathbf{G})}\sum_i |\phi_0(\mathbf{r}-\mathbf{r}_1)|^2 \\ &= \sum_i e^{i\mathbf{r}_i\cdot(\mathbf{q}+\mathbf{G})}\varrho_0(\mathbf{q}+\mathbf{G}) \\ \varrho_0(\mathbf{q}+\mathbf{G}) &= \int d^3r e^{i\mathbf{r}\cdot(\mathbf{q}+\mathbf{G})}\,|\phi_0(r)|^2 \end{aligned}$$

The various interaction terms are collected into an effective matrix element:

$$F_{\mathbf{q}}(\mathbf{r}) = \sum_G \varrho_0(\mathbf{q}+\mathbf{G})e^{i\mathbf{G}\cdot\mathbf{r}}M_{\mathbf{q}+\mathbf{G}}$$

$$H = \sum_{\mathbf{q}}\left[\omega_{\mathbf{q}}(a_{\mathbf{q}}{}^{\dagger}a_{\mathbf{q}}+\tfrac{1}{2})+\frac{1}{\nu^{1/2}}(a_{\mathbf{q}}+a_{\mathbf{q}}{}^{\dagger})\sum_i e^{i\mathbf{q}\cdot\mathbf{r}_i}F_{\mathbf{q}}(\mathbf{r}_i)\right]$$

The function $F_{\mathbf{q}}(\mathbf{r})$ is periodic in the lattice, since increasing it by a lattice vector $\mathbf{a}$ does not change its value. We assume that all localized electrons in different unit cells are in the same position within the cell. That is, we assume that $F_{\mathbf{q}}(\mathbf{r})$ is the same for all localized electrons.

We have solved this problem before. It is just the harmonic oscillator in an electric field [Eq. (1.1.11)]. Now each wave vector and polarization state $\mathbf{q}$ is a separate harmonic oscillator, which finds its own equilibrium configuration. It we follow exactly the steps used to solve (1.1.11), then the creation and destruction operators are transformed to the new set

$$A_{\mathbf{q}} = a_{\mathbf{q}} + \frac{1}{\nu^{1/2}} \frac{F_{\mathbf{q}}}{\omega_{\mathbf{q}}} \sum_i e^{i\mathbf{q}\cdot\mathbf{r}_i}$$

$$A_{\mathbf{q}}^{\dagger} = a_{\mathbf{q}}^{\dagger} + \frac{1}{\nu^{1/2}} \frac{F_{\mathbf{q}}^*}{\omega_{\mathbf{q}}} \sum_i e^{-i\mathbf{q}\cdot\mathbf{r}_i}$$

and the Hamiltonian with these operators is

$$H = \sum_{\mathbf{q}} \omega_{\mathbf{q}}(A_{\mathbf{q}}^{\dagger} A_{\mathbf{q}} + \tfrac{1}{2}) - \frac{1}{\nu} \sum_{\mathbf{q}} \frac{|F_{\mathbf{q}}|^2}{\omega_{\mathbf{q}}} \left| \sum_i e^{i\mathbf{q}\cdot\mathbf{r}_i} \right|^2$$

Furthermore, the new operators still obey the harmonic oscillator commutation relations:

$$[A_{\mathbf{q}}, A_{\mathbf{q}'}^{\dagger}] = \delta_{\mathbf{q}\mathbf{q}'}$$

$$[A_{\mathbf{q}}, A_{\mathbf{q}'}] = 0$$

$$[A_{\mathbf{q}}^{\dagger}, A_{\mathbf{q}'}^{\dagger}] = 0$$

We know immediately, as in Sec. 1.1, that the eigenstates and eigenvalues for this Hamiltonian are

$$\frac{(A_{\mathbf{q}}^{\dagger})^{n_{\mathbf{q}}}}{(n_{\mathbf{q}}!)^{1/2}} |0\rangle$$

$$E = \sum_{\mathbf{q}} \omega_{\mathbf{q}}(n_{\mathbf{q}} + \tfrac{1}{2}) - \frac{1}{\nu} \sum_{\mathbf{q}} \frac{|F_{\mathbf{q}}|^2}{\omega_{\mathbf{q}}} \left| \sum_i e^{i\mathbf{q}\cdot\mathbf{r}_i} \right|^2 \qquad (1.3.3)$$

The eigenstates are interpreted in the same way we used for the simple spring. Each normal mode $\mathbf{q}$ has stretched to a new equilibrium configuration,

$$Q_{\mathbf{q}}^{(0)} = -2\left(\frac{\hbar}{2\varrho\omega_{\mathbf{q}}\nu}\right)^{1/2} \frac{F_{\mathbf{q}}}{\omega_{\mathbf{q}}} \sum_i e^{i\mathbf{q}\cdot\mathbf{r}_i}$$

and now oscillates about this new equilibrium point. The oscillation frequencies do not change. The last term in (1.3.3) is the *relaxation energy*. It is the potential energy gained by stretching the springs—the phonon normal modes—to the new equilibrium positions. This energy term may be expanded in terms of the electron coordinates:

$$\text{Relaxation energy} = \frac{1}{\nu}\sum_q \frac{F_{\mathbf{q}}^2}{\omega_{\mathbf{q}}}\sum_{ij} e^{i\mathbf{q}\cdot(\mathbf{r}_i-\mathbf{r}_j)} = +\sum_{i,j}\tfrac{1}{2}V_R(\mathbf{r}_i-\mathbf{r}_j)$$

$$= \tfrac{1}{2}\sum_i V_R(0) + \sum_{i>j} V_R(\mathbf{r}_i-\mathbf{r}_j) \tag{1.3.4}$$

$$V_R(\mathbf{r}) = -\frac{2}{\nu}\sum_{\mathbf{q}} \frac{|F_{\mathbf{q}}|^2}{\omega_{\mathbf{q}}} e^{i\mathbf{q}\cdot\mathbf{r}} = -2\sum_\lambda \int \frac{d^3q}{(2\pi)^3}\frac{|F_{\lambda\mathbf{q}}|^2}{\omega_{\mathbf{q}\lambda}} e^{i\mathbf{q}\cdot\mathbf{r}}$$

The relaxation energy consists of two types of terms. The first is $V_R(\mathbf{r}_i-\mathbf{r}_i) = V_R(0)$, which is the relaxation energy of a single particle by itself. This energy is caused by the electron inducting a static polarization in the phonon field which acts back upon the electron. It is a self-energy effect. We shall often call it a polaron self-energy or electron self-energy. The energy is not just with the electron. As with the stretched spring, it is in the combined particle–oscillator system.

The other type of term is the interaction between pairs of fixed particles $V_R(\mathbf{r}_i-\mathbf{r}_j)$. Here the physical picture is that one particle polarizes the medium, and this polarization field changes the energy of other particles which are nearby. This potential $V_R(\mathbf{r})$ has different r dependence for different types of phonons in solids. In some cases, it is very short-ranged, so that two particles interact only when they are in the same unit cell of the crystal. In other cases, the potential falls as slowly as r^{-1}, as if it were a Coulomb potential. Several of these cases will be presented later.

We shall discuss the same many-body problem again in Chapter 4. The harmonic oscillator in a linear potential is an important model, if only because there are so few models we can solve exactly. We shall see that it is also the solution to the independent boson model.

C. Deformation Potential

In semiconductors and ionic solids, the excited electron states are usually confined to a small location in wave vector space. In thermal equilibrium, the excited states are at an energy band minimum, which is often at the zone center or edge. Usually the polaron effects for these electrons involve only phonons of long wavelength. In this case the tradition has been to parameterize the interaction rather than compute it from first principles. Most electron–phonon interactions in semiconductors use only three types of interactions—deformation potential coupling to acoustical phonons, piezoelectric coupling to acoustical phonons, and polar coupling to optical phonons. Another possible coupling is the deformation coupling to optical phonons, but the strength of this is poorly understood (Ehrenreich and Overhauser, 1958). These interactions are valid only at long wavelength. When the electron–phonon matrix element is needed at short-wavelength phonons, the usual method is to calculate them from pseudopotentials (Shuey, 1965).

The deformation potential coupling to acoustical phonons is just the long-wavelength limit of (1.3.1). Only the $\mathbf{G} = 0$ is retained, since the terms $\mathbf{G} \neq 0$ are of wavelengths that are too short. The potential $V_A(\mathbf{q}) \to D$ at $\mathbf{q} \to 0$, where D is the *deformation constant*. At long wavelength, $\hat{\xi}_{\mathbf{q}} \to \hat{q}$, and only longitudinal phonons are important if the band is nondegenerate. Thus the interaction has the form

$$H_{\mathrm{ep}} = D \sum_{\mathbf{q}} \left(\frac{\hbar}{2\varrho\nu\omega_{\mathbf{q}}} \right)^{1/2} | \mathbf{q} | \, \varrho(\mathbf{q})(a_{\mathbf{q}} + a^{\dagger}_{-\mathbf{q}}) \tag{1.3.5}$$

However, valence bands in semiconductors are often degenerate at the band maximum. Then electron or hole excitations have a deformation coupling to transverse phonons, and this is a very large polaron correction (Mahan, 1965).

The deformation constants are obtained by measuring how energy bands shift with increasing pressure on the solid. The value of D_n for a band n is simply the rate of change of band energy with pressure (Thomas, 1961).

D. Piezoelectric Interaction

Many semiconductors are piezoelectric. The macroscopic effect is that an electric field is generated when a crystal is squeezed and vice versa. Acoustical phonons, which are periodic density modulations, make periodic electric fields. The crystal must lack an inversion center to be piezoelectric. The group IV semiconductors Ge and Si are not piezoelectric. The III–V semiconductors such as GaAs are very weakly piezoelectric, while the II–VI materials such as CdS and ZnO are extremely piezoelectric. A very detailed derivation of the electron–phonon interaction has been given elsewhere (Mahan, 1972). Here we shall provide only a quick sketch. If S_{ij} is the stress on the crystal, then the electric field is proportional to the stress,

$$E_k = \sum_{ij} M_{ijk} S_{ij}$$

where the matrix M_{ijk} is a constant which gives the proportionality. The stress is defined as the symmetric derivative of the displacement field:

$$S_{ij} = \frac{1}{2}\left(\frac{\partial Q_i}{\partial x_j} + \frac{\partial Q_j}{\partial x_i}\right) = \frac{1}{2}\sum_q \left(\frac{\hbar}{2\varrho\nu\omega_{\mathbf{q}}}\right)^{1/2} (\xi_i q_j + \xi_j q_i)(a_{\mathbf{q}} + a_{-\mathbf{q}}^{\dagger})e^{i\mathbf{q}\cdot\mathbf{r}}$$

The electric field may be shown to be longitudinal and to point in the direction $\mathbf{q}$ of the phonon. Thus it may be written as the gradient of a potential $\phi(r)$:

$$E_k = -\frac{\partial}{\partial x_k}\phi(\mathbf{r}) = -\frac{1}{\nu^{1/2}}\sum_{\mathbf{q}} iq_k\phi_{\mathbf{q}}e^{i\mathbf{q}\cdot\mathbf{r}}$$

Thus we end up with the observation that the potential is proportional to the displacement:

$$\phi(\mathbf{r}) \propto Q(\mathbf{r})$$

$$\phi(\mathbf{r}) = i\sum_{\mathbf{q}\lambda}\left(\frac{\hbar}{2\varrho\nu\omega_{\mathbf{q}\lambda}}\right)^{1/2} M_\lambda(\hat{q})e^{i\mathbf{q}\cdot\mathbf{r}}(a_{\mathbf{q}\lambda} + a_{-\mathbf{q}\lambda}^{\dagger}) \tag{1.3.6}$$

Thus for the piezoelectric interaction we can write an electron–phonon interaction:

$$H_{\text{ep}} = i\sum_{\mathbf{q}\lambda}\left(\frac{\hbar}{2\varrho\nu\omega_{\mathbf{q}\lambda}}\right)^{1/2} M_\lambda(\hat{q})\varrho(\mathbf{q})(a_{\mathbf{q}\lambda} + a_{-\mathbf{q}\lambda}^{\dagger}) \tag{1.3.7}$$

The matrix element $M_\lambda(\hat{q})$ does not depend on the magnitude of $\mathbf{q}$, but it very much depends on the direction of $\mathbf{q}$. It also has the property that $M(-\hat{q}) = -M_\lambda(\hat{q})$ so that (1.3.5) is Hermitian. In fact, the piezoelectric interaction is quite anisotropic. The matrix element is also very dependent on the polarization λ of the acoustical phonon—whether it is LA (longitudinal acoustic) or TA (transverse acoustic). Most many-body calculations have tended to take a constant value for the matrix element, where this constant is obtained by averaging over the various angular directions in the crystal. We shall also adopt this approximation.

An interesting result is obtained when we use (1.3.7) to calculate the effective potential energy between two fixed electrons:

$$V_R(r) = -2\int\frac{d^3q}{(2\pi)^3}\sum_\lambda M_\lambda{}^2\frac{1}{2\varrho\omega_{\mathbf{q}\lambda}^2}e^{i\mathbf{q}\cdot\mathbf{r}}$$

In a Debye model, the phonon energy is proportional to wave vector $\omega_{\mathbf{q}\lambda} = c_\lambda q$. Since M_λ and ϱ are constants, the potential $V_R(r)$ is the Fourier transform of q^{-2}. The transform was worked out earlier, in (1.2.12), and just gives a potential varying as r^{-1}:

$$
\begin{aligned}
V_R(r) &= -\frac{1}{r}\frac{1}{4\pi\varrho}\sum_\lambda \frac{M_\lambda^2}{c_\lambda^2} = -\frac{e^2}{r}\gamma \\
\gamma &= \frac{1}{e^2 4\pi\varrho}\sum_\lambda \frac{M_\lambda^2}{c_\lambda^2}
\end{aligned}
\tag{1.3.8}
$$

We seem to have derived a form of Coulomb's law. Actually, we have derived a law of dielectric screening. We know in dielectric materials that the potential between two fixed charges is

$$
V_{\text{total}}(r) = \frac{e^2}{r\varepsilon_{\text{total}}} \tag{1.3.9}
$$

where $\varepsilon_{\text{total}}$ is the total, static, dielectric constant. In our piezoelectric calculation, we have derived the piezoelectric contribution to this dielectric screening. The total dielectric function is the summation of many contributions:

$$
\begin{aligned}
\varepsilon_{\text{total}} &= \varepsilon_\infty + \varepsilon_{\text{piezo}} + \varepsilon_{\text{polar}} + \varepsilon_{\text{e-e}} \\
\varepsilon_{\text{total}} &= \varepsilon_0 + \varepsilon_{\text{piezo}}
\end{aligned}
\tag{1.3.10}
$$

The first term ε_∞ is from interband electronic transitions. The piezoelectric contribution is the term we are now trying to understand. The polar and electron–electron contributions will be explained later. We follow convention and define all other contributions as ε_0. All these contributions are functions of q and ω. In the present problem, with a static, fixed charge, we want these quantities in the limit that $\mathbf{q} \to 0$ and $\omega \to 0$. We hasten to add that the electron–electron term is not finite in this limit, so probably we should not have included it in the list (1.3.10). It may be omitted if there are no mobile charges present, so we shall assume that this is the case.

If there were no piezoelectric contribution, presumably the interaction potential between two fixed charges would be

$$
\frac{e^2}{\varepsilon_0 r}
$$

The piezoelectric contribution (1.3.8) represents the difference between the above result and total screening (1.3.9):

$$
\frac{e^2}{\varepsilon_{\text{total}} r} = \frac{e^2}{r}\left(\frac{1}{\varepsilon_0} - \gamma\right)
$$

so that we get

$$\frac{1}{\varepsilon_0 + \varepsilon_{\text{piezo}}} = \frac{1}{\varepsilon_0} - \gamma$$

We can solve these equations to obtain the piezoelectric contribution to the static dielectric function:

$$\varepsilon_{\text{piezo}} = \frac{\varepsilon_0^{\,2}\gamma}{1 - \varepsilon_0\gamma}$$

The interaction term (1.3.8) is negative because the screening lowers the potential energy. The unscreened potential is just e^2/r, and each bit of screening lowers this by an amount proportional to e^2/r.

The same acoustical phonon may interact with an electron by both the deformation and piezoelectric interactions. These two interactions do not interfere, to second order, because they are out of phase. The sum of the two interactions (1.3.5) and (1.3.7) gives

$$H_{\text{ep}} = \sum_{\mathbf{q}} \left(\frac{\hbar}{2\varrho\nu\omega_{\mathbf{q}}}\right)^{1/2} \bar{M}(\mathbf{q})\varrho(\mathbf{q})(a_{\mathbf{q}} + a^{\dagger}_{-\mathbf{q}})$$

$$\bar{M}(\mathbf{q}) = D\,|\,\mathbf{q}\,| + iM_{\lambda}(\hat{q})$$

The deformation potential is real, while the piezoelectric is imaginary. Thus, to second order, they do not interfere:

$$|\,\bar{M}(\mathbf{q})\,|^2 = D^2q^2 + M_{\lambda}^{\,2}$$

E. Polar Coupling

The polar coupling between electrons and optical phonons can be very large in ionic crystals. The form of the Hamiltonian has been derived often, e.g., Fröhlich (1954) or Evrard (1972). In ionic crystals some of the atoms are positively charged, while others are negatively charged. An optical phonon has the different ions in the crystal vibrating out of phase. When the plus ions and minus ions oscillate in the opposite directions, they set up a polarization field. The polarization causes an electric field which scatters the electrons. The electric field is the source of the polar coupling.

The polar coupling is only to LO (longitudinal optical) phonons and not to TO (transverse optical) phonons (Mahan, 1972), because only the LO phonons set up strong electric fields when they vibrate. These electric fields are in the direction of vibration, which at long wavelength is in the direction of the phonon wave vector $\mathbf{q}$. For a system of no free charges,

$$\nabla \cdot \mathbf{D} = 0 = \sum_{\mathbf{q}} \mathbf{q} \cdot (\mathbf{E}_{\mathbf{q}} + 4\pi \mathbf{P}_{\mathbf{q}}) e^{i\mathbf{q}\cdot\mathbf{r}}$$

For an LO phonon mode of wave vector $\mathbf{q}$, we have the electric field $\mathbf{E}_{\mathbf{q}}$ and polarization $\mathbf{P}_{\mathbf{q}}$ both parallel to $\mathbf{q}$. Thus we get the relation for the electric field produced by the polarization wave:

$$\mathbf{E}_{\mathbf{q}} = -4\pi \mathbf{P}_q$$

The next assumption we make is that the polarization is proportional to the displacement:

$$\mathbf{P}_{\mathbf{q}} = Ue\mathbf{Q}_{\mathbf{q}}$$

$$\mathbf{E}_{\mathbf{q}} = -4\pi Ue\mathbf{Q}_{\mathbf{q}} = -4\pi Ue\left(\frac{\hbar}{2\varrho\nu\omega_{\mathrm{LO}}}\right)^{1/2} \hat{q} i (a_{\mathbf{q}} + a^{\dagger}_{-\mathbf{q}})$$

where the coefficient U is to be determined. The phonon energy ω_{LO} is assumed to be constant. Since the electric field points in the direction of $\mathbf{q}$, it may be expressed by a potential:

$$\mathbf{E} = -\nabla\phi = -i \sum_{\mathbf{q}} e^{i\mathbf{q}\cdot\mathbf{r}} \mathbf{q} \phi_{\mathbf{q}}$$

$$\phi(\mathbf{r}) = \sum_{\mathbf{q}} e^{i\mathbf{q}\cdot\mathbf{r}} \frac{4\pi e U}{q} \left(\frac{\hbar}{2\varrho\nu\omega_{\mathrm{LO}}}\right)^{1/2} (a_{\mathbf{q}} + a^{\dagger}_{-\mathbf{q}})$$

which gives the potential produced by the LO phonons. The interaction constant U still needs to be determined. Its value is obtained by considering the potential between two fixed electrons that was calculated from our relaxation energy (1.3.4):

$$V_R(r) = -\frac{2}{\hbar\omega_{\mathrm{LO}}} (4\pi Ue)^2 \left(\frac{\hbar}{2\varrho\omega_{\mathrm{LO}}}\right) \int \frac{d^3q}{(2\pi)^3} \frac{e^{i\mathbf{q}\cdot\mathbf{r}}}{q^2}$$

The Fourier transform integral again produces a coulomb potential:

$$V_R(r) = -\Gamma \frac{e^2}{r}$$

$$\Gamma = \frac{4\pi U^2}{\varrho\omega_{\mathrm{LO}}^2}$$

This interaction energy is again interpreted as a contribution to the dielectric screening of the solid. This term represents the contribution from the optical phonons. It represents the difference between screening with just the electronic interband part ε_∞ and the interband plus optical phonons ε_0. We deduce that

$$\frac{e^2}{r\varepsilon_0} = \frac{e^2}{r}\left(\frac{1}{\varepsilon_\infty} - \Gamma\right)$$

or

$$\Gamma = \frac{1}{\varepsilon_\infty} - \frac{1}{\varepsilon_0}$$

The unknown factor U is

$$U^2 = \frac{\varrho\omega_{\text{LO}}^2}{4\pi}\left(\frac{1}{\varepsilon_\infty} - \frac{1}{\varepsilon_0}\right)$$

It is possible to write the electron-phonon Hamiltonian in the form

$$H_{\text{ep}} = \sum_q \frac{M}{q(\nu^{1/2})}\,\varrho(\mathbf{q})(a_{\mathbf{q}} + a_{-\mathbf{q}}^\dagger)$$

where the matrix element is

$$M^2 = 2\pi e^2\hbar\omega_{\text{LO}}\left(\frac{1}{\varepsilon_\infty} - \frac{1}{\varepsilon_0}\right)$$

This is a very sensible form in which to express the matrix element. The dielectric constants ε_0 and ε_∞ are both measurable: ε_0 is the low-frequency dielectric function measured by putting the solid between the parallel plates of a capacitor at low frequency, and ε_∞ is the square of the refractive index. Besides these quantities, the matrix element depends only on the charge e and the LO phonon energy $\hbar\omega_{\text{LO}}$. In spite of the elegance of this simple form, the matrix element is not usually expressed this way. Instead, it is customary to introduce the dimensionless polaron constant α, defined as

$$\alpha = \frac{e^2}{\hbar}\left(\frac{m}{2\hbar\omega_{\text{LO}}}\right)^{1/2}\left(\frac{1}{\varepsilon_\infty} - \frac{1}{\varepsilon_0}\right)$$

Correspondingly, the interaction matrix element is

$$M^2 = \frac{4\pi\alpha\hbar(\hbar\omega_{\text{LO}})^{3/2}}{(2m)^{1/2}}$$

The reason for introducing α is simple. We really want the self-energy expressions to have a simple and elegant form. They do with this choice of α.

Electrons in crystalline energy bands have their motion determined by an effective band mass m_b. The band mass enters into the definition of α, which has the obvious disadvantage that one does not know the value of α until one knows the value of m_b. These values have been tabulated by Kartheuser (1972).

1.4. SPIN HAMILTONIANS

The study of spin systems forms a very large part of many-body theory. There are many solids which display magnetic ordering among the electrons. There are many impurity problems where spin plays an important role. In attempting to explain these phenomena, many different types of spin models have been introduced. Some of these involve localized spins interacting among themselves, while others have localized spins interacting with free electrons. It is a subject in which there are few exactly solvable models which are nontrivial. Only a few of the models have solutions which are well understood, in spite of the fact that many of them have been intensely studied. The transference of model results to real solids which are strongly interacting systems has still not been very successful.

The commutation relations for spin one-half operators are ($\hbar = 1$)

$$\begin{aligned}
[S_l^{(x)}, S_j^{(y)}] &= iS_l^{(z)}\delta_{lj} \\
[S_l^{(y)}, S_j^{(z)}] &= iS_l^{(x)}\delta_{lj} \qquad (1.4.1)\\
[S_l^{(z)}, S_j^{(x)}] &= iS_l^{(y)}\delta_{lj}
\end{aligned}$$

The subscript label l or j denotes spin site. Spins on different sites, or with different electrons, commute. The superscripts (x), (y), (z) refer to space coordinates. These spin operators are often represented by Pauli spin matrices,

$$S^{(x)} = \frac{1}{2}\begin{pmatrix} 0 & 1 \\ 1 & 0 \end{pmatrix}$$

$$S^{(y)} = \frac{1}{2}\begin{pmatrix} 0 & -i \\ i & 0 \end{pmatrix}$$

$$S^{(z)} = \frac{1}{2}\begin{pmatrix} 1 & 0 \\ 0 & -1 \end{pmatrix}$$

but these representations are unnecessary for many-body calculations—the commutation relations are sufficient. It is customary to introduce spin raising and lowering operators:

$$S_l^{(+)} = S_l^{(x)} + iS_l^{(y)}$$
$$S_l^{(-)} = S_l^{(x)} - iS_l^{(y)}$$

The names *raising* and *lowering* are applied to these operators because they raise or lower the magnetic quantum number m of the spin state. For spin one-half, we only have spin up $|+\rangle$ and spin down $|-\rangle$, and the operators go from one to the other:

$$\begin{aligned} S^{(+)}|-\rangle &= |+\rangle, \qquad & S^{(-)}|+\rangle &= |-\rangle \\ S^{(+)}|+\rangle &= 0, \qquad & S^{(-)}|-\rangle &= 0 \end{aligned} \tag{1.4.2}$$

By direct multiplication, we find the following for two operators on the same site:

$$\begin{aligned} S^{(+)}S^{(-)} &= S^{(x)^2} + S^{(y)^2} - i(S^{(x)}S^{(y)} - S^{(y)}S^{(x)}) = S^{(x)^2} + S^{(y)^2} + S^{(z)} \\ S^{(-)}S^{(+)} &= S^{(x)^2} + S^{(y)^2} + i[S^{(x)}, S^{(y)}] = S^{(x)^2} + S^{(y)^2} - S^{(z)} \end{aligned} \tag{1.4.3}$$

By subtracting these two results, we find the commutation relation among the raising and lowering operators:

$$\begin{aligned} [S_l^{(+)}, S_j^{(-)}] &= 2S_l^{(z)}\delta_{lj} \\ [S_l^{(z)}, S_j^{(+)}] &= S_l^{(+)}\delta_{lj} \\ [S_l^{(z)}, S_j^{(-)}] &= -S_l^{(-)}\delta_{lj} \end{aligned} \tag{1.4.4}$$

One can easily obtain the last two commutators from the direct definition of the operators; for example,

$$[S_l^{(z)}, S_l^{(+)}] = [S_l^{(z)}, S_l^{(x)}] + i[S_l^{(z)}, S_l^{(y)}] = iS_l^{(y)} - i^2S_l^{(x)} = (S_l^{(x)} + iS_l^{(y)})$$

Spins are neither bosons nor fermions. These commutation relations are unlike any which we have encountered previously. It is precisely these commutation relations which make spin problems so difficult.

A. Homogeneous Spin Systems

The *Heisenberg Hamiltonian* puts one spin on each site of a lattice and has the spins interact with a vector interaction. If the interaction is only between nearest neighbor spins, the Hamiltonian has the form

$$H = -J \sum_{j\delta} \mathbf{S}_j \cdot \mathbf{S}_{j+\delta} = -J \sum_{j\delta} (S_j^{(x)} S_{j+\delta}^{(x)} + S_j^{(y)} S_{j+\delta}^{(y)} + S_j^{(z)} S_{j+\delta}^{(z)}) \qquad (1.4.5)$$

The Heisenberg Hamiltonian is often solved for spin greater than one-half or for coupling between spins which may be further neighbors. But the above form is the most common. The word *solve* means "approximately solve," since it cannot be solved exactly, except in one dimension. Often the coupling constant in one direction, say z, is taken to be different from those in the other directions. This case is the *anisotropic Heisenberg model*:

$$\begin{aligned} H &= -J_{\parallel} \sum_{j\delta} S_j^{(z)} S_{j+\delta}^{(z)} - J_{\perp} \sum_{j\delta} (S_j^{(x)} S_{j+\delta}^{(x)} + S_j^{(y)} S_{j+\delta}^{(y)}) \\ H &= -J_{\parallel} \sum_{j\delta} S_j^{(z)} S_{j+\delta}^{(z)} - J_{\perp} \sum_{j\delta} S_j^{(+)} S_{j+\delta}^{(-)} \end{aligned} \qquad (1.4.6)$$

In this case the operators $S_l^{(+)}$ and $S_{l+\delta}^{(-)}$ can be arranged in any order—they commute since they refer to different sites. There are two limiting cases of this Hamiltonian which have their own names. The *Ising model* has $J_{\perp} = 0$:

$$H_I = -J_{\parallel} \sum_{j\sigma} S_j^{(z)} S_{j+\sigma}^{(z)}$$

It may be solved exactly in one dimension, even if one adds a magnetic field to the Hamiltonian:

$$H_I = -J_{\parallel} \sum_{\delta} S_j^{(z)} S_{j+\delta}^{(z)} - H_0 \sum_j S_j^{(z)}$$

where H_0 is the magnetic field in units of ergs. In two dimensions, it may be solved exactly without the magnetic field, as shown by Onsager (1944). Very accurate three-dimensional results have been obtained by a variety of techniques, including Green's functions (Callen, 1966) and critical point and renormalization group techniques (Domb and Green, 1972–1977). The *XY model* has only $J_{\perp}$:

$$H_{XY} = -J_{\perp} \sum_{j\delta} S_j^{(+)} S_{j+\delta}^{(-)} \qquad (1.4.7)$$

It may be solved exactly only in one dimension.

It is conventional to write the Hamiltonian, as we have in (1.4.5), with the negative sign in front of the interaction term. Then for $J > 0$, the spins tend to line up all parallel, so this is the ferromagnetic arrangement.

For $J < 0$, the ordering has alternate spins up and down, if the lattice permits, and this is called antiferromagnetic.

The difficulty with solving spin problems is well illustrated by defining collective operators. We transform the operators into wave vector space:

$$S_{\mathbf{k}}^{(+)} = \frac{1}{N^{1/2}} \sum_j e^{i\mathbf{k}\cdot\mathbf{r}_j} S_j^{(+)}, \qquad S_j^{(+)} = \frac{1}{N^{1/2}} \sum_k e^{-i\mathbf{k}\cdot\mathbf{r}_j} S_{\mathbf{k}}^{(+)}$$

$$S_{\mathbf{k}}^{(-)} = \frac{1}{N^{1/2}} \sum_j e^{-i\mathbf{k}\cdot\mathbf{r}_j} S_j^{(-)}, \qquad S_j^{(-)} = \frac{1}{N^{1/2}} \sum_k e^{+i\mathbf{k}\cdot\mathbf{r}_j} S_{\mathbf{k}}^{(-)}$$

This transformation appears to be a reasonable approach. We used it successfully in solving fermion and boson problems. But when we examine the commutation relations for these operators,

$$\begin{aligned}
[S_{\mathbf{k}}^{(+)}, S_{\mathbf{k}'}^{(-)}] &= \frac{1}{N} \sum_{lj} e^{i\mathbf{k}\cdot\mathbf{r}_l} e^{-i\mathbf{k}'\cdot\mathbf{r}_j} [S_l^{(+)}, S_j^{(-)}] \\
&= \frac{1}{N} \sum_{lj} e^{i\mathbf{k}\cdot\mathbf{r}_l} e^{-i\mathbf{k}'\cdot\mathbf{r}_l} \delta_{lj} 2S_l^{(z)} \\
[S_{\mathbf{k}}^{(+)}, S_{\mathbf{k}'}^{(-)}] &= \frac{2}{N} \sum_l e^{i\mathbf{r}_l\cdot(\mathbf{k}-\mathbf{k}')} S_l^{(z)}
\end{aligned} \tag{1.4.8}$$

The operators $S_l^{(+)}$ and $S_j^{(-)}$ commute except on the same site, and then we have used (1.4.4). The right-hand side of (1.4.8) is not simple. We would have preferred to find something like

$$[S_k^{(+)}, S_{k'}^{(-)}] = C\delta_{kk'}$$

which would indicate that $S_k^{(+)}$ and $S_k^{(-)}$ were independent operators except for $k = k'$. They would behave just like bosons, and we could use boson statistics. Unfortunately, (1.4.8) does not have this property. One common approximation (Callen, 1966) is to replace (1.4.8) by the approximate expression

$$[S_{\mathbf{k}}^{(+)}, S_{\mathbf{k}'}^{(-)}] = 2\delta_{\mathbf{k}\mathbf{k}'} \frac{1}{N} \sum_l S_l^{(z)} = 2\delta_{\mathbf{k}\mathbf{k}'} \langle S^{(z)} \rangle$$

One then tries to find a self-consistent equation for the average magnetization $\langle S^{(z)} \rangle$, but this approach is approximate. Except in special circumstances, the operators $S_{\mathbf{k}}^{(+)}$ and $S_{\mathbf{k}'}^{(-)}$ do not describe independent eigenstates of the system. This lack of collective eigenstates is the difficulty with solving spin systems and why innocent-looking Hamiltonians like (1.4.7) are difficult to solve.

If we add together the two equations (1.4.3), we obtain

$$S_l^{(+)}S_l^{(-)} + S_l^{(-)}S_l^{(+)} = 2(S_l^{(x)^2} + S_l^{(y)^2}) = 2(\mathbf{S} \cdot \mathbf{S} - S_l^{(z)^2})$$
$$= 2[S(S+1) - S_l^{(z)^2}]$$

For spin one-half, $S^{(z)^2} = \frac{1}{4}$, so that we obtain the result

$$\{S_l^{(+)}, S_l^{(-)}\} = 2\left[\frac{1}{2}\left(\frac{3}{2}\right) - \frac{1}{4}\right] = 1$$

Thus the $S^{(+)}$ and $S^{(-)}$ operators, on the same site, obey an anticommutation relation. We associate anticommutation relations with fermion operators. Thus the spin one-half operators behave as a fermion on the same site. For fermions, each site may have zero or one particle for each orbital and spin state. These two possibilities correspond to the two possible spin states of up and down. For example, spin down is equivalent to zero particles and spin up to one particle on the site. Thus (1.4.2) is analogous to the fermion relations

$$\begin{aligned} C_l \,|\, 1\rangle &= |\, 0\rangle, \qquad & C_l^\dagger \,|\, 0\rangle &= |\, 1\rangle \\ C_l \,|\, 0\rangle &= 0, \qquad & C_l^\dagger \,|\, 1\rangle &= 0 \end{aligned} \tag{1.4.9}$$

Unfortunately, one cannot carry this analogy too far. On different sites, the spin operators commute, while the fermion operators anticommute. Thus the spin operators are not fermions either.

However, in one dimension, the spin one-half operators can be made into exact fermion operators. This transformation was discovered by Jordan and Wigner (1928). In one dimension, the spins are aligned along a chain. A new set of operators is defined: the old raising and lowering operators are multiplied by a phase factor which is dependent on spin site:

$$\begin{aligned} d_l &= e^{i\phi_l} S_l^{(-)} \\ d_l^\dagger &= e^{-i\phi_l} S_l^{(+)} \\ d_l^\dagger d_l &= S_l^{(+)}S_l^{(-)} = S(S+1) - S_l^{(z)^2} + S_l^{(z)} = \tfrac{1}{2} + S_l^{(z)} \end{aligned} \tag{1.4.10}$$

The phase factor ϕ_l is chosen to be π times an operator which measures the number of spin-up operators to the right of that position:

$$l > 1: \qquad \phi_l = \pi \sum_{k=1}^{l-1} (\tfrac{1}{2} + S_k^{(z)}) = \pi \sum_{k=1}^{l-1} d_k^\dagger d_k$$

The chain is numbered $\phi_1 = 0$ from one end, say the right, with site index $l = 1, 2, 3, \ldots$. We shall interpret the $d^\dagger$ operators as creating fermion particles, and

$$d_l^\dagger d_l = n_l$$

is the number operator for each site. Thus $\phi_l = \pi \sum_{k=1}^{l-1} n_k$ is the number of such fermions to the right of the site we are labeling. The phase factor commutes with $S_l^{(-)}$,

$$d_l = e^{i\phi_l} S_l^{(-)} = S_l^{(-)} e^{i\phi_l}$$

since the operator ϕ_l is the number to the right and does not involve the number operator on the same site. On the same site, these operators anti-commute,

$$\{d_l, d_l\} = \{S_l^{(-)}, S_l^{(+)}\} = 1$$

since that is the property of the spin operators themselves. On different sites, they also anticommute, with the help of this new phase factor. By taking the anticommutator,

$$\{d_l, d_m^\dagger\} = e^{i\phi_l} S_l^{(-)} e^{-i\phi_m} S_m^{(+)} + e^{-i\phi_m} S_m^{(+)} e^{i\phi_l} S_l^{(-)}$$

Since $l \neq m$, assume that $l > m$ so that we can write the phase factor ϕ_l as

$$\phi_l = \pi(\tfrac{1}{2} + S_m^{(z)}) + \pi \sum_{\substack{k=1 \\ k\neq m}}^{l-1} (\tfrac{1}{2} + S_k^{(z)})$$

Thus for the anticommutator we obtain

$$\{d_l, d_m^\dagger\} = e^{i(\phi_l - \phi_m)}(S_l^{(-)} S_m^{(+)} + e^{-i\pi n_m} S_m^{(+)} e^{i\pi n_m} S_l^{(-)})$$

The right-hand term contains the operator combination

$$e^{-i\pi[(1/2)+S^{(z)}]} S^{(+)} e^{+i\pi[(1/2)+S^{(z)}]} = e^{i\pi} S^{(+)} = -S^{(+)}$$

The $S^{(+)}$ operator must always raise the magnetic quantum number m by unity, so that $S^{(z)}$ on the left always measures one integer higher value than the $S^{(z)}$ on the right of $S^{(+)}$. Thus one gets an extra factor of $i\pi$, which changes the sign of the term. This phase factor ϕ_l was chosen to produce this sign change. Now we have the anticommutator of the d operators:

$$\{d_l, d_m^\dagger\} = e^{i(\phi_l - \phi_m)}[S_l^{(-)}, S_m^{(+)}] = 0$$

This is equal to the commutator of the $S^{(+)}$, $S^{(-)}$ operators, for different sites, and this is zero. The d operators anticommute for both the same site and different sites. They are pure fermion operators and obey fermion statistics. Thus in one dimension the XY model (1.4.7) may be transformed into a Hamiltonian in terms of the d operators,

$$H_{XY} = -J\sum_j (S_j^{(+)}S_{j+1}^{(-)} + S_j^{(+)}S_{j-1}^{(-)})$$

$$= -J\sum_j (e^{i\phi_j}\, d_j^{\dagger} e^{-i\phi_{j+1}}\, d_{j+1} + e^{i\phi_j}\, d_j^{\dagger} e^{-i\phi_{j-1}}\, d_{j-1})$$

$$H_{XY} = -J\sum_j (d_j^{\dagger} e^{-i\pi n_j}\, d_{j+1} + d_j^{\dagger} e^{i\pi n_{j-1}}\, d_{j-1}) = -J\sum_j (d_j^{\dagger}\, d_{j+1} + d_j^{\dagger}\, d_{j-1})$$

which is just the tight binding model for fermions. The phase factors vanish because in the first term n_j is zero if it precedes a raising operator, and in the second term it is zero if it follows a lowering operator. Now the Hamiltonian is changed to collective coordinates:

$$d_j = \frac{1}{N^{1/2}}\sum_k e^{-ik\cdot r_j}\, d_k, \qquad d_k = \frac{1}{N^{1/2}}\sum_j e^{ik\cdot r_j}\, d_j$$

$$d_j^{\dagger} = \frac{1}{N^{1/2}}\sum_k e^{ik\cdot r_j}\, d_k^{\dagger}, \qquad d_k^{\dagger} = \frac{1}{N^{1/2}}\sum_j e^{-ik\cdot r_j}\, d_j^{\dagger}$$

$$H = -2J\sum_k \gamma_k\, d_k^{\dagger}\, d_k$$

and the collective operators now obey anticommutation relations as well:

$$\{d_k, d_{k'}^{\dagger}\} = \frac{1}{N}\sum_{jl} e^{ik\cdot r_j} e^{-ik'\cdot r_l}\{d_j, d_l^{\dagger}\} = \frac{1}{N}\sum_j e^{ir_j\cdot(k-k')} = \delta_{k,k'}$$

Thus the Hamiltonian may be written as a simple fermion problem. The exact partition function is

$$Z = \mathrm{Tr}\, e^{-\beta H} = \prod_k (1 + e^{2J\gamma_k\beta})$$

Note that there is no chemical potential for spin systems. One can also work out other properties of this model. To prove that the transformation (1.4.10) is valid, we should show that all the commutation relations (1.4.4) are preserved. They are.

The Jordan–Wigner transformation shows that in one dimension the spin one-half operators may be represented exactly as fermions. This result is not valid for higher dimensions. No one has been able to find an equivalent transformation for two or three dimensions. Indeed, most approximate analyses assume that in two or three dimensions the spin excitations behave approximately as bosons rather than as fermions.

In two and three dimensions, one may still transform spin one-half operators into particle operators. In this case, the particle operators have funny commutation relations. They are, like spins, neither fermions nor bosons. This transformation associates the creation operator with $S^{(+)}$ and a destruction operator with $S^{(-)}$:

$$\begin{aligned} C_l^\dagger &= S_l^{(+)} \\ C_l &= S_l^{(-)} \\ n_l = C_l^\dagger C_l &= S_l^{(+)} S_l^{(-)} = \tfrac{1}{2} + S_l^{(z)} \end{aligned} \tag{1.4.11}$$

This transformation preserves all the commutation relations (1.4.4). For example, one has that

$$\begin{aligned} \{C_l, C_l^\dagger\} &= 1 \\ [C_l, C_m^\dagger] &= -2(\tfrac{1}{2} - n_l)\delta_{lm} \\ [C_l, n_l] &= [S_l^{(-)}, S_l^{(z)}] = C_l \end{aligned}$$

The particle operators C_l and $C_m^\dagger$ anticommute if they are on the same site and commute if they are on different sites. Thus they are neither fermions nor bosons. Collective operators such as

$$C_{\mathbf{k}} = \frac{1}{N^{1/2}} \sum_j e^{i\mathbf{k}\cdot\mathbf{r}_j} C_j$$

have funny commutation relations, similar to the spin case in (1.4.8). Nevertheless, this is a popular many-body model for certain systems. These are *lattice gas* models for atoms on lattices. The atoms may be considered as classical particles, which commute on different sites. However, there may not be more than one atom, the atoms being large, substantial objects, on each site. "No more than one atom on each site" is an exclusion principle, which is represented by the anticommutation relations on the same site. The same physics is contained in a model which has the particle obey purely boson statistics but with the provision that there is a strong repulsive interaction U if two particles were on the same atomic site. The lattice gas results would be obtained in the limit $U \to \infty$. The statistics of anticommutation relations on the same site merely represent the strong repulsive interaction between atoms at close separation.

The *lattice gas* (LG) *model* has pairwise interactions U between particles in nearest neighbor positions. A chemical potential μ is also introduced for the particles, since they may have variable concentration:

$$K_{\mathrm{LG}} = H_{\mathrm{LG}} - \mu N = +\tfrac{1}{2}U \sum_{j\delta} n_j n_{j+\delta} - \mu \sum_j n_j$$

The lattice gas Hamiltonian may be transformed into an equivalent magnetic problem by inverting the transformation of (1.4.11):

$$K_{\mathrm{LG}} = \tfrac{1}{2}U \sum_{j\delta} (S_j^{(z)} + \tfrac{1}{2})(S_{j+\delta}^{(z)} + \tfrac{1}{2}) - \mu \sum_j (S_j^{(z)} + \tfrac{1}{2})$$

which may be collected and rearranged into

$$K_{\mathrm{LG}} = \frac{1}{2} U \sum_{j\delta} S_j^{(z)} S_{j+\delta}^{(z)} + \left(\frac{UZ}{2} - \mu\right) \sum_j S_j^{(z)} - \frac{1}{2} N_0 \left(\frac{UZ}{4} - \mu\right)$$

where Z is the coordination number, i.e., the number of nearest neighbors. The spin version of K_{LG} is identical to the Ising model with magnetic field. The magnetic field is

$$H_0 = \frac{UZ}{2} - \mu$$

If exactly one-half of the sites of the lattice gas are occupied, then one has $\mu = \tfrac{1}{2}UZ$ so that the effective magnetic field is zero (Hill, 1956). In this case, for a half-filled band, the chemical potential is temperature independent. For concentrations other than one-half, the chemical potential varies with temperature.

The *quantum lattice gas* (QLG) model adds a nearest neighbor hopping term to the lattice gas model:

$$K_{\mathrm{QLG}} = H_{\mathrm{QLG}} - \mu N = \tfrac{1}{2}U \sum_{j\delta} n_j n_{j+\delta} + w \sum_{l\delta} C_l^{+} C_{l+\delta} - \mu \sum_j n_j \qquad (1.4.12)$$

One can show that the equivalent magnetic problem is the anisotropic Heisenberg model with magnetic field. The quantum lattice gas was suggested by Matsubara and Matsuda (1956) as a model for quantum fluids such as ^{4}He. The superfluid transformation for this system occurs in the liquid state. Nevertheless, a quantum lattice gas model appears to be a good description of its critical properties. The parameter U may be taken to be either positive or negative, depending on whether the nearest neighbors repulse or attract each other. The equivalent magnetic problems are then antiferromagnetic and ferromagnetic, respectively.

B. Impurity Spin Models

The models we have mentioned so far are for homogeneous magnetic systems. The same spin was on each site, and we tried to deduce the magnetic properties of the entire system. Other kinds of popular models are for impurity spin problems. Here the spin is an isolated impurity in an otherwise homogeneous electron gas. One can study, for example, the conditions for the formation of a local moment on the impurity or the scattering properties of the free electrons from the localized spin.

Our derivation follows Kondo (1969). An impurity atom is located at the position $\mathbf{R}_n$ and has a localized electron orbital $\phi_L(\mathbf{r} - \mathbf{R}_n)$ when the electron is on that site. Otherwise the electron is in a continuum state $\mathbf{k}$ with wave function $\phi_{\mathbf{k}}(\mathbf{r})$ and energy $\varepsilon_{\mathbf{k}}$. The wave functions may be considered as plane waves or alternately as Bloch functions of the crystal. A generalized state function is the summation over all possible states,

$$\psi(\mathbf{r}) = \sum_{\mathbf{k}\sigma} \phi_{\mathbf{k}}(\mathbf{r}) X_\sigma C_{\mathbf{k}\sigma} + \sum_{\sigma} \phi_L(\mathbf{r} - \mathbf{R}_n) X_\sigma C_n$$

where the X_σ are the spin wave functions, which denote spin up $X_\uparrow$ or down $X_\downarrow$. We do not assume that the wave functions $\phi_{\mathbf{k}}(\mathbf{r})$ and $\phi_L(\mathbf{r} - \mathbf{R}_n)$ are orthogonal. This step will be taken later, since their orthogonality is a many-body problem of sorts. The Hamiltonian is taken to have terms such as

$$H = \sum_i \left[\frac{\mathbf{p}_i^2}{2m} + U(r_i) \right] + \frac{1}{2} \sum_{ij} \frac{e^2}{r_{ij}}$$

where $U(r)$ is the potential for each electron, which may include the impurity potential as well as the usual potential of the host lattice. The last term is the electron–electron interactions. The first terms of the Hamiltonian are evaluated for the state function, and one gets

$$\begin{aligned} \int \psi^\dagger(\mathbf{r}) \left[\frac{p^2}{2m} + U(r) \right] \psi(\mathbf{r})\, d^3r &= \sum_{\mathbf{k}\sigma} \varepsilon_{\mathbf{k}} C^\dagger_{\mathbf{k}\sigma} C_{\mathbf{k}\sigma} + \sum_{\sigma} \varepsilon_L C^\dagger_{L\sigma} C_{L\sigma} + \sum_{\mathbf{k}\mathbf{k}'\sigma} U_{\mathbf{k}\mathbf{k}'} C^\dagger_{\mathbf{k}\sigma} C_{\mathbf{k}'\sigma} \\ &\quad + \sum_{\mathbf{k}\sigma} M_{\mathbf{k}} (C^\dagger_{\mathbf{k}\sigma} C_{L\sigma} + C^\dagger_{L\sigma} C_{\mathbf{k}\sigma}) \end{aligned} \tag{1.4.13}$$

First we have the unperturbed parts of the Hamiltonian for the impurity and for the continuum states. Second, there are the terms

$$\sum_{\mathbf{k}\mathbf{k}'\sigma} U_{\mathbf{k}\mathbf{k}'} C^\dagger_{\mathbf{k}\sigma} C_{\mathbf{k}'\sigma}$$

which involve the scattering of the continuum functions from the impurity potential. This problem is simple to solve by ordinary scattering theory. The solution will be given in Sec. 4.1. The last term is a mixing term called H_M:

$$H_M = \sum_{\mathbf{k}\sigma} M_\mathbf{k}(C^\dagger_{\mathbf{k}\sigma} C_{L\sigma} + C^\dagger_{L\sigma} C_{\mathbf{k}\sigma})$$

The interaction H_M is sometimes called H_{sd}, but we shall use that name for another contribution which will be discussed shortly. The mixing term describes processes whereby the electron hops off of the impurity and becomes a continuum state or vice versa. This term essentially arises from the nonorthogonality of the continuum and local wave functions. A Hamiltonian of the type

$$H = \sum_{\mathbf{k}\sigma} \varepsilon_\mathbf{k} C^\dagger_{\mathbf{k}\sigma} C_{\mathbf{k}\sigma} + \sum_{\sigma} \varepsilon_{L\sigma} C^\dagger_{L\sigma} C_{L\sigma} + \sum_{\mathbf{k}\sigma} M_\mathbf{k}(C^\dagger_{\mathbf{k}\sigma} C_{L\sigma} + C^\dagger_{L\sigma} C_{\mathbf{k}\sigma}) \qquad (1.4.14)$$

will be called a *Fano–Anderson model*, since it was introduced simultaneously by Fano (1961) and Anderson (1961). It should not be confused with the famous Anderson model, which will be described below. The Fano–Anderson model may be solved exactly, and this solution will be presented in Sec. 4.2. There is no real conceptual difference between (1.4.13) and (1.4.14). If the wave functions $\phi_\mathbf{k}(\mathbf{r})$ are chosen to be eigenstates of the Hamiltonian which includes the impurity scattering $U_{\mathbf{kk}'}$, then the two Hamiltonians become identical. One may always take this to be the case.

The interesting magnetic phenomenon comes from the terms involving electron–electron interactions. Thus we consider

$$\int d^3r_1\, d^3r_2 \frac{e^2}{|\mathbf{r}_1 - \mathbf{r}_2|} \psi^\dagger(\mathbf{r}_1)\psi(\mathbf{r}_1)\psi^\dagger(\mathbf{r}_2)\psi(\mathbf{r}_2) \qquad (1.4.15)$$

Many types of terms are generated by this expression. We shall discuss two of them. The first contains operators from two continuum functions and two localized wave functions:

$$C^\dagger_{L\sigma} C_{L\sigma'} C^\dagger_{\mathbf{k}\sigma''} C_{\mathbf{k}\sigma'''}$$

The four spin operators must occur in pairs. One possible pairing is

$$\sum_{\substack{\mathbf{kk}'\\ \sigma\sigma'}} V_{\mathbf{kk}'} C^\dagger_{L\sigma} C_{L\sigma} C^\dagger_{\mathbf{k}\sigma'} C_{\mathbf{k}'\sigma'} \qquad (1.4.16)$$

$$V_{\mathbf{kk}'} = e^2 \int d^3r_1 \phi_\mathbf{k}{}^*(\mathbf{r}_1)\phi_{\mathbf{k}'}(\mathbf{r}_1) \int \frac{d^3r_2}{|\mathbf{r}_1 - \mathbf{r}_2|} |\phi_L(\mathbf{r}_2 - \mathbf{R}_n)|^2$$

This term is usually ignored, since it does not cause magnetic phenomena. It states that the conduction electrons interact with the impurity in a different way when the localized orbital is occupied. It is quite a reasonable term, since the electron is charged and one expects that the presence of this charge will influence the other electrons. We shall follow custom and ignore this term. Another possible contribution is the exchange term for the above process, which has the form

$$-\sum_{\substack{\mathbf{kk}'\\ \sigma\sigma'}} J_{\mathbf{kk}'} C^{\dagger}_{\mathbf{k}\sigma} C_{\mathbf{k}'\sigma'} C^{\dagger}_{L\sigma'} C_{L\sigma} \tag{1.4.17}$$

$$J_{\mathbf{kk}'} = e^2 \int d^3r_1 \phi_{\mathbf{k}}^*(\mathbf{r}_1)\phi_L(\mathbf{r}_1 - \mathbf{R}_n) \int \frac{d^3r_2}{|\mathbf{r}_1 - \mathbf{r}_2|} \phi_L^*(\mathbf{r}_2 - \mathbf{R}_n)\phi_{\mathbf{k}}(\mathbf{r}_2)$$

In this term, the electron which is scattering may change its spin state during the scattering process. This spin change is always accompanied by an opposite spin change of the impurity spins, so that the total spin angular momentum is conserved during the process. The spin conservation is illustrated by writing out in detail the terms which can occur:

$$C^{\dagger}_{\mathbf{k}\uparrow} C_{\mathbf{k}'\uparrow} C^{\dagger}_{L\uparrow} C_{L\uparrow} + C^{\dagger}_{\mathbf{k}\uparrow} C_{\mathbf{k}'\downarrow} C^{\dagger}_{L\downarrow} C_{L\uparrow} + C^{\dagger}_{\mathbf{k}\downarrow} C_{\mathbf{k}'\uparrow} C^{\dagger}_{L\uparrow} C_{L\downarrow} + C^{\dagger}_{\mathbf{k}\downarrow} C_{\mathbf{k}'\downarrow} C^{\dagger}_{L\downarrow} C_{L\downarrow}$$

The first and last terms are regrouped as

$$\begin{aligned} &\tfrac{1}{2}(C^{\dagger}_{\mathbf{k}\uparrow} C_{\mathbf{k}'\uparrow} + C^{\dagger}_{\mathbf{k}\downarrow} C_{\mathbf{k}'\downarrow})(C^{\dagger}_{L\uparrow} C_{L\uparrow} + C^{\dagger}_{L\downarrow} C_{L\downarrow}) + \tfrac{1}{2}(C^{\dagger}_{\mathbf{k}\uparrow} C_{\mathbf{k}'\uparrow} - C^{\dagger}_{\mathbf{k}\downarrow} C_{\mathbf{k}'\downarrow}) \\ &\quad \times (C^{\dagger}_{L\uparrow} C_{L\uparrow} - C^{\dagger}_{L\downarrow} C_{L\downarrow}) \end{aligned} \tag{1.4.18}$$

The first term in (1.4.18) contains the factor

$$C^{\dagger}_{L\uparrow} C_{L\uparrow} + C^{\dagger}_{L\downarrow} C_{L\downarrow} = \sum C^{\dagger}_{L\sigma} C_{L\sigma}$$

which is always unity if the localized state is occupied. In fact this entire term has the form

$$\tfrac{1}{2} \sum_{\mathbf{kk}'\sigma\sigma'} J_{\mathbf{kk}'} C^{\dagger}_{\mathbf{k}\sigma} C_{\mathbf{k}'\sigma} C^{\dagger}_{L\sigma'} C_{L\sigma'}$$

which is exactly the same as (1.4.16). We shall follow convention and combine these terms, so that it is easy to ignore both of them. The other term in (1.4.18) has the combination

$$\tfrac{1}{2}(C^{\dagger}_{L\uparrow} C_{L\uparrow} - C^{\dagger}_{L\downarrow} C_{L\downarrow}) = S_L^{(z)}$$

which is just the z component of the localized spin. Similarly, we can identify some of the other combinations as the raising and lowering operators for the localized spin:

$$S_L^{(+)} = C_{L\uparrow}^\dagger C_{L\downarrow}$$

$$S_L^{(-)} = C_{L\uparrow}^\dagger C_{L\downarrow}$$

The spin-dependent terms in (1.4.17) may be collected as

$$H_{sd} = -\sum_{\mathbf{kk}'} J_{\mathbf{kk}'}[S_L^{(z)}(C_{\mathbf{k}\uparrow}^\dagger C_{\mathbf{k}'\uparrow} - C_{\mathbf{k}\downarrow}^\dagger C_{\mathbf{k}'\downarrow}) + S_L^{(+)} C_{\mathbf{k}\downarrow}^\dagger C_{\mathbf{k}'\uparrow} + S_L^{(-)} C_{\mathbf{k}\uparrow}^\dagger C_{\mathbf{k}'\downarrow}]$$

This interaction is called H_{sd}. It is a model of the localized d electrons interacting with the s-like continuum wave functions. This form of the interaction is valid even if the localized spin has $S > \frac{1}{2}$. The last two terms flip the spin of the continuum electron while flipping the localized spin of the impurity in the opposite direction. The first term does not flip spin, but the interaction does depend on the z component of the spin of both the impurity and continuum wave function. A Hamiltonian of the type

$$H = \sum_{\mathbf{k}\sigma} \varepsilon_{\mathbf{k}} C_{\mathbf{k}\sigma}^\dagger C_{\mathbf{k}\sigma} + \varepsilon_L(n_{L\uparrow} + n_{L\downarrow}) + H_{sd}$$

is called the *Kondo problem*. It was not formulated by Kondo but rather much earlier by Zener (1951). Kondo's contribution was recognizing that the spin-flip scattering processes could cause unusual low-temperature behavior in the scattering properties. These low-temperature anomalies had been long observed in resistivities (Gerritsen and Linde, 1951, 1954).

Another type of term which may arise from the electron–electron interaction (1.4.15) has four local operators:

$$U C_{L\sigma}^\dagger C_{L\sigma'}^\dagger C_{L\sigma''} C_{L\sigma'''}$$

$$U = \int d^3r_1 \, | \phi_L(\mathbf{r}_1 - \mathbf{R}_n) |^2 \int d^3r_2 \frac{e^2}{| \mathbf{r}_1 - \mathbf{r}_2 |} | \phi_L(\mathbf{r}_2 - \mathbf{R}_n) |^2$$

If the orbital is nondegenerate, then the two electrons can be on the same site only in opposite spin states. Thus one gets a term similar to the one we found earlier in the Hubbard model,

$$U n_{L\uparrow} n_{L\downarrow} \tag{1.4.19}$$

although it was historically introduced first in the *Anderson model*. This famous model considers the model Hamiltonian

$$H = \sum_{\mathbf{k}\sigma} \varepsilon_{\mathbf{k}} C^{\dagger}_{\mathbf{k}\sigma} C_{\mathbf{k}\sigma} + \varepsilon_L \sum_{\sigma} C^{\dagger}_{L\sigma} C_{L\sigma} + \sum_{\mathbf{k}\sigma} M_{\mathbf{k}} (C^{\dagger}_{\mathbf{k}\sigma} C_{L\sigma} + C^{\dagger}_{L\sigma} C_{\mathbf{k}\sigma}) + U n_{L\uparrow} n_{L\downarrow}$$

The Hamiltonian is exactly solvable without the last term. The last term is very important, since it causes magnetic instabilities in some circumstances.

The Anderson model and the Kondo model both describe the interaction of a continuum electron with a localized one. The two models are not totally different. There is a canonical transformation which when applied to the Anderson model will transform it into a form similar to the Kondo model. This transformation on the Anderson model produces quite a few terms, of which the Kondo model is a subset. Thus the transformation does not produce exactly the Kondo model, and the two models are not identical. If we write the Anderson model as ($C_{Ls} \equiv C_s$, $n_s = C_s^{\dagger} C_s$)

$$H = H_0 + H_M, \qquad H_M = \sum_{\mathbf{k}s} M_{\mathbf{k}} (C^{\dagger}_{\mathbf{k}s} C_s + C_s^{\dagger} C_{\mathbf{k}s})$$

$$H_0 = \sum_{\mathbf{k}\sigma} \varepsilon_{\mathbf{k}} n_{\mathbf{k}\sigma} + \varepsilon_L \sum_{s} n_s + U n_{\uparrow} n_{\downarrow}$$

then we seek a canonical transformation (Schrieffer and Wolff, 1966)

$$\begin{aligned}\tilde{H} = e^S H e^{-S} = H_0 + H_M + [S, H_0] + [S, H_M] + \tfrac{1}{2}[S, [S, H_0]] \\ + \tfrac{1}{2}[S, [S, H_M]] + \cdots \end{aligned} \tag{1.4.20}$$

which will eliminate all terms which are linear in $M_{\mathbf{k}}$. This is accomplished by choosing S such that it obeys

$$0 = H_M + [S, H_0] \tag{1.4.21}$$

and then the canonical transformation produces the series

$$\tilde{H} = H_0 + \tfrac{1}{2}[S, H_M] + \tfrac{1}{3}[S, [S, H_M]] + \cdots \tag{1.4.22}$$

There are an infinite number of terms. We shall only evaluate those which come out to be proportional to M_k^2, which is the second term in the series (1.4.22). The transformation factor S turns out to be

$$S = \sum_{\mathbf{k}s} M_{\mathbf{k}} \left(\frac{1 - n_{-s}}{\varepsilon_L - \varepsilon_{\mathbf{k}}} + \frac{n_{-s}}{\varepsilon_L + U - \varepsilon_{\mathbf{k}}} \right) (C_s^{\dagger} C_{\mathbf{k}s} - C^{\dagger}_{\mathbf{k}s} C_s)$$

so that one gets the commutators

$$\left[S, \sum_{\mathbf{k}\sigma} \varepsilon_{\mathbf{k}} n_{\mathbf{k}\sigma}\right] = \sum_{\mathbf{k}s} M_{\mathbf{k}} \left(\frac{1 - n_{-s}}{\varepsilon_L - \varepsilon_{\mathbf{k}}} + \frac{n_{-s}}{\varepsilon_L + U - \varepsilon_{\mathbf{k}}} \right) (C_s^{\dagger} C_{\mathbf{k}s} + C_{\mathbf{k}s}^{\dagger} C_s)\, \varepsilon_{\mathbf{k}}$$

$$\left[S, \varepsilon_L \sum_s n_s\right] = -\varepsilon_L \sum_{\mathbf{k}s} M_{\mathbf{k}} \left(\frac{1 - n_{-s}}{\varepsilon_L - \varepsilon_{\mathbf{k}}} + \frac{n_{-s}}{\varepsilon_L + U - \varepsilon_{\mathbf{k}}} \right) (C_s^{\dagger} C_{\mathbf{k}s} + C_{\mathbf{k}s}^{\dagger} C_s)$$

$$[S, U n_{\uparrow} n_{\downarrow}] = -U \sum_{\mathbf{k}s} M_{\mathbf{k}} \left(\frac{1 - n_{-s}}{\varepsilon_L - \varepsilon_{\mathbf{k}}} + \frac{n_{-s}}{\varepsilon_L + U - \varepsilon_{\mathbf{k}}} \right) (C_{\mathbf{k}s}^{\dagger} C_s + C_s^{\dagger} C_{\mathbf{k}s}) n_{-s}$$

When these are added, one exactly satisfies (1.4.21). Although the form of S appears unwieldly, it gets the job done. The next step is to take the commutator $[S, H_M]$ to generate the terms which are proportional to M_k^2. As a preliminary step, we define the following effective exchange constant:

$$\bar{J}_{\mathbf{k}\mathbf{k}'} = M_{\mathbf{k}} M_{\mathbf{k}'} \left(\frac{1}{\varepsilon_L - \varepsilon_{\mathbf{k}}} + \frac{1}{\varepsilon_L - \varepsilon_{\mathbf{k}'}} - \frac{1}{\varepsilon_L + U - \varepsilon_{\mathbf{k}}} - \frac{1}{\varepsilon_L + U - \varepsilon_{\mathbf{k}'}} \right)$$

In terms of this constant, one finds the following for the commutator:

$$\begin{aligned} \tfrac{1}{2}[S, H_M] = \sum_{\mathbf{k}\mathbf{k}'s} \bar{J}_{\mathbf{k}\mathbf{k}'} [& n_{-s} C_{\mathbf{k}s}^{\dagger} C_{\mathbf{k}'s} + C_{\mathbf{k}'s}^{\dagger} C_{\mathbf{k}-s} C_{-s}^{\dagger} C_s - \delta_{\mathbf{k}\mathbf{k}'} n_s n_{-s} \\ & + \tfrac{1}{4}(C_{\mathbf{k},-s}^{\dagger} C_{\mathbf{k}'s}^{\dagger} C_{-s} C_s + \text{h.c.})] + 2 \sum_{\mathbf{k}} \frac{M_k^2}{\varepsilon_L - \varepsilon_{\mathbf{k}}} n_s \\ & - \sum_{\mathbf{k}\mathbf{k}'s} C_{\mathbf{k}s}^{\dagger} C_{\mathbf{k}'s} M_{\mathbf{k}} M_{\mathbf{k}'} \left(\frac{1}{\varepsilon_L - \varepsilon_{\mathbf{k}}} + \frac{1}{\varepsilon_L - \varepsilon_{\mathbf{k}'}} \right) \end{aligned}$$

The second term flips the spin of the electron and impurity electron while scattering—it has exactly the Kondo form. There are many other terms. The last term which appears is simple potential scattering of the continuum electron, which must be added to our earlier result (1.4.13). There is a term which is the non-spin-flip interaction between a continuum electron and an impurity electron—and this must be added to (1.4.16). There is also a term which contains the interaction between two impurity electrons, one with spin up and the other with spin down. This renormalizes U in (1.4.19). Finally, there is a new term which has not been previously encountered. It describes the process

$$C_{\mathbf{k},-s}^{\dagger} C_{\mathbf{k}'s}^{\dagger} C_{-s} C_s$$

whereby two impurity electrons hop off of the impurity site to become two continuum electrons and vice versa. The Anderson model certainly describes a rich set of phenomena. Of course, many more terms are generated by the additional commutators in the series (1.4.22).

1.5. PHOTONS

Throughout this book we shall always use the following type of Hamiltonian for discussion of the interaction of charges e_j with each other and with a radiation field:

$$H = \sum_i \frac{1}{2m}\left[\mathbf{p}_i - \frac{e_i}{c}\mathbf{A}(r_i)\right]^2 + \frac{1}{2}\sum_{ij}\frac{e_i e_j}{|\mathbf{r}_i - \mathbf{r}_j|} + \sum_{k\lambda}\omega_{\mathbf{k}} a^\dagger_{\mathbf{k}\lambda} a_{\mathbf{k}\lambda} \quad (1.5.1)$$

The radiation field is represented by the vector potential

$$\frac{1}{c}A_\mu(\mathbf{r}, t) = \sum_{\mathbf{k}\lambda}\left(\frac{2\pi}{\nu\omega_{\mathbf{k}}}\right)^{1/2}\xi_\mu(\mathbf{k}, \lambda)(a_{\mathbf{k}\lambda}e^{i(\mathbf{k}\cdot\mathbf{r}-\omega_{\mathbf{k}}t)} + \text{h.c.})$$

$$= \frac{1}{\nu^{1/2}}\sum_{\mathbf{k}} e^{i\mathbf{k}\cdot\mathbf{r}}A_\mu(\mathbf{k}, t)$$

The term $\sum \omega_{\mathbf{k}} a^\dagger a$ represents its unperturbed photon Hamiltonian in the absence of charges. The unit polarization vector is ξ_μ. One feature of this Hamiltonian is the term

$$\frac{1}{2}\sum_{ij}\frac{e_i e_j}{|\mathbf{r}_i - \mathbf{r}_j|}$$

which is the Coulomb interaction between charges. The Coulomb interaction is instantaneous in time, since the potential has no retardation, or speed of light, built into it. The lack of retardation is not an approximation but is rigorously correct in the Coulomb gauge. Of course, there is retardation in the total interaction, which arises through the vector potential fields.

A. Gauges

Although the topic of gauges is treated correctly in a number of texts, it still seems to be poorly understood by students. Thus it seems appropriate to start at the beginning and reproduce some standard material. Maxwell equations are

$$\nabla \cdot \mathbf{B} = 0 \tag{1.5.2}$$

$$\nabla \times \mathbf{E} = -\frac{1}{c}\frac{\partial \mathbf{B}}{\partial t} \tag{1.5.3}$$

$$\nabla \cdot \mathbf{E} = 4\pi\varrho \tag{1.5.4}$$

$$\nabla \times \mathbf{B} = \frac{1}{c}\frac{\partial \mathbf{E}}{\partial t} + \frac{4\pi \mathbf{j}}{c} \tag{1.5.5}$$

We also use the important theorem that any vector function of position can be written as the sum of two terms: One is the gradient of a potential, and the other is the curl of a vector:

$$\mathbf{S}(r) = \nabla g + \nabla \times \mathbf{m}(r) = \mathbf{S}_l + \mathbf{S}_t$$

$$\mathbf{S}_l = \nabla g$$

$$\mathbf{S}_t = \nabla \times \mathbf{m}$$

The term $\mathbf{S}_l$ is called the *longitudinal* part of S, and $\mathbf{S}_t$ is called the *transverse*. If we assume that $\mathbf{B}(\mathbf{r})$ has this form, then Eq. (1.5.2) becomes

$$\nabla \cdot (\nabla g + \nabla \times \mathbf{A}) = \nabla^2 g = 0$$

We usually assume that $g = 0$, so that the vector potential is defined as

$$\mathbf{B} = \nabla \times \mathbf{A}$$

However, this does not uniquely define $\mathbf{A}(\mathbf{r})$. The definition $\mathbf{B} = \nabla \times \mathbf{A}$ is put into (1.5.3):

$$\nabla \times \left\{\mathbf{E} + \frac{1}{c}\frac{\partial}{\partial t}\mathbf{A}\right\} = 0$$

Now the factor in curly braces is also the sum of a longitudinal and transverse part:

$$\nabla \times \{\nabla\psi + \nabla \times \mathbf{M}\} = 0 = \nabla \times (\nabla \times \mathbf{M})$$

The equation is satisfied if $\mathbf{M} = 0$, so that for the electric field we obtain

$$\mathbf{E} = -\frac{1}{c}\frac{\partial \mathbf{A}}{\partial t} - \nabla\psi$$

where ψ is the scalar potential. When these two forms for $\mathbf{B}(\mathbf{r})$ and $\mathbf{E}(\mathbf{r})$ are put into (1.5.4) and (1.5.5), we obtain the equations for the scalar and vector potentials:

$$\nabla^2\psi + \frac{1}{c}\frac{\partial}{\partial t}\boldsymbol{\nabla}\cdot\mathbf{A} = -4\pi\varrho \tag{1.5.6}$$

$$\boldsymbol{\nabla}\times(\boldsymbol{\nabla}\times\mathbf{A}) + \frac{1}{c^2}\frac{\partial^2}{\partial t^2}\mathbf{A} + \frac{1}{c}\boldsymbol{\nabla}\frac{\partial\psi}{\partial t} = \frac{4\pi}{c}\boldsymbol{j} \tag{1.5.7}$$

At first it appears that we have four equations for the four unknowns (A_y, A_y, A_z, ψ): Eq. (1.5.6) and the three vector components of (1.5.7). If this were true, we would then uniquely determine the four unknown functions in terms of the sources $(\boldsymbol{j}, \varrho)$. However, these four equations are not linearly independent; only three of them are independent. To show this, we operate on (1.5.6) by $(1/c)(\partial/\partial t)$ and on (1.5.7) by $\boldsymbol{\nabla}$ and then subtract the two equations. We have that

$$\boldsymbol{\nabla}\cdot[\boldsymbol{\nabla}\times(\boldsymbol{\nabla}\times\mathbf{A})] = \frac{4\pi}{c}\left(\frac{\partial\varrho}{\partial t} + \boldsymbol{\nabla}\cdot\boldsymbol{j}\right)$$

The left-hand side is zero because it is the gradient of a curl. The right-hand side vanishes since it is just the equation of continuity. Thus the four equations are not independent.

Of course, this means that the four unknown functions (A_y, A_y, A_z, ψ) are not uniquely determined. It is necessary to stipulate one additional condition, or constraint, on their values. This is called the *gauge condition.* The condition we shall impose is that the Coulomb field $\psi(r, t)$ act instantaneously, which is accomplished by imposing the condition that

$$\boldsymbol{\nabla}\cdot\mathbf{A} = 0 \tag{1.5.8}$$

Equation (1.5.8) defines the *Coulomb gauge*, sometimes called the *transverse gauge*. The latter name arises because (1.5.8) implies that $\boldsymbol{\nabla}\cdot\mathbf{A}_l = 0$, so that $\mathbf{A}$ is purely transverse. One should realize that any arbitrary constraint may be imposed as long as one can satisfy (1.5.6) and (1.5.7). As long as these two equations are satisfied, one always obtains the same value for $\mathbf{E}(\mathbf{r})$ and $\mathbf{B}(\mathbf{r})$. The arbitrary choice of gauge does not alter the final value of observable quantities.

In the Coulomb gauge, Eq. (1.5.6) simplifies to

$$\nabla^2\psi = -4\pi\varrho$$

which is easily solved to give

$$\psi(\mathbf{r}, t) = \int d^3r' \frac{\varrho(\mathbf{r}', t)}{|\mathbf{r} - \mathbf{r}'|} \tag{1.5.9}$$

Thus the potential $\psi(\mathbf{r}, t)$ is instantaneous and is not retarded. We repeat that this is not an approximation but an exact result for our choice of gauge. Later we shall show that a different choice of gauge leads to a retarded scalar potential.

Next we evaluate the other equation, (1.5.7). We use the identity

$$\nabla \times (\nabla \times \mathbf{A}) = -\nabla^2\mathbf{A} + \nabla(\nabla \cdot \mathbf{A}) \tag{1.5.10}$$

The second term vanishes in the Coulomb gauge. Thus we have

$$\nabla^2\mathbf{A} - \frac{1}{c^2}\frac{\partial^2}{\partial t^2}\mathbf{A} = -\frac{4\pi}{c}\mathbf{j} + \frac{1}{c}\nabla\frac{\partial\psi}{\partial t} \tag{1.5.11}$$

We need to operate a bit on the second term on the right-hand side. Using (1.5.9), for this term we obtain

$$\begin{aligned}\frac{1}{c}\nabla\frac{\partial\psi}{\partial t} &= \frac{1}{c}\nabla\int d^3r' \frac{(\partial/\partial t)\varrho(\mathbf{r}, t)}{|\mathbf{r} - \mathbf{r}'|} \\ &= -\frac{1}{c}\nabla\int d^3r' \frac{1}{|\mathbf{r} - \mathbf{r}'|}\nabla' \cdot \mathbf{j}(r')\end{aligned}$$

where the last identity uses the equation of continuity. We integrate by parts in the last term,

$$\begin{aligned}\frac{1}{c}\nabla\frac{\partial\psi}{\partial t} &= \frac{1}{c}\nabla\int d^3r'\,\mathbf{j}(r') \cdot \nabla' \frac{1}{|\mathbf{r} - \mathbf{r}'|} \\ &= -\frac{1}{c}\nabla\left[\nabla \cdot \int d^3r' \frac{\mathbf{j}(r')}{|\mathbf{r} - \mathbf{r}'|}\right]\end{aligned}$$

and then pull the gradient out by letting it operate on $\mathbf{r}$ instead of $\mathbf{r}'$; the latter step requires a sign change. We also need to operate on the current term itself by using the identity

$$4\pi\mathbf{j}(r, t) = -\nabla^2\int dr' \frac{\mathbf{j}(r', t)}{|\mathbf{r} - \mathbf{r}'|}$$

If we combine these results, we obtain

$$\nabla^2\mathbf{A} - \frac{1}{c^2}\frac{\partial^2}{\partial t^2}\mathbf{A} = (\nabla^2 - \nabla\nabla)\frac{1}{c}\int d^3r' \frac{\mathbf{j}(r')}{|\mathbf{r} - \mathbf{r}'|}$$

Finally, using the identity (1.5.10), we obtain

$$\nabla^2\mathbf{A} - \frac{1}{c^2}\frac{\partial^2}{\partial t^2}\mathbf{A} = -\frac{1}{c}\boldsymbol{\nabla}\times\left[\boldsymbol{\nabla}\times\int d^3r'\,\frac{\mathscr{j}(r')}{|\mathbf{r}-\mathbf{r}'|}\right] \tag{1.5.12}$$

The point of this exercise is that the right-hand side of (1.5.12) is now a transverse vector, since it is the curl of something. Thus if we write the current as a longitudinal plus a transverse part,

$$\mathscr{j}(\mathbf{r}) = \mathscr{j}_l(\mathbf{r}) + \mathscr{j}_t(\mathbf{r})$$

Then our vector potential obeys the equation

$$\nabla^2\mathbf{A} - \frac{1}{c^2}\frac{\partial^2}{\partial t^2}\mathbf{A} = -\frac{4\pi}{c}\mathscr{j}_t(r)$$

where $\mathscr{j}_t(r)$ is defined as the right-hand side of (1.5.12). The final equation for **A** is very reasonable. Since the vector potential **A** is purely transverse, it should respond only to the transverse part of the current. If it were to respond to the longitudinal part of the current, it would develop a longitudinal part. The longitudinal component of **A** does not occur in the coulomb gauge.

As a simple example, consider a current of the form

$$\mathscr{j}(\mathbf{r}) = \mathbf{J}_0 e^{i\mathbf{k}\cdot\mathbf{r}}$$

whose transverse part is

$$\mathscr{j}_t = \frac{-\hat{\mathbf{k}}\times(\mathbf{k}\times\mathbf{J}_0)e^{i\mathbf{k}\cdot\mathbf{r}}}{k^2}$$

$\mathscr{j}_t$ is just the component of **J** which is perpendicular to **k**; see Fig. 1.2. In homogeneous materials, the transverse and longitudinal parts are just those components which are perpendicular and parallel to **k**, the direction of the motion. Unfortunately, solids are not homogeneous but periodic. Along major symmetry directions in the crystal, it is often true that transverse components are perpendicular to the wave vector. However, it is

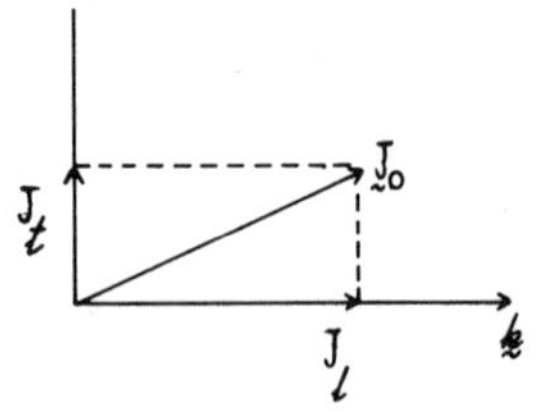

FIGURE 1.2

generally not true for arbitrary points in the Brillouin zone, even in cubic crystals. Thus the words *transverse* and *longitudinal* do not necessarily mean perpendicular and parallel to **k**.

Two charges interact by the sum of the two interactions: Coulomb plus photon. The net interaction may not have a component which is instantaneous. In fact, for a frequency-dependent charge density, at distances large compared to c/ω, one finds that the photon part of the interaction produces a term $-e^2/r$ which exactly cancels the instantaneous Coulomb interaction. The remaining parts of the photon contribution are the net retarded interaction. In solids we are usually concerned with interaction over short distances. Then retardation is unimportant for most problems. In the study of the homogeneous electron gas, for example, the photon part is small and may be neglected. In real solids, the photon part causes some crystal field effects, which is an unexciting many-body effect. The main effect of retardation is the polariton effects at long wavelength (Hopfield, 1958). In general, we have chosen the Coulomb gauge because the instantaneous Coulomb interaction is usually a large term which forms a central part of the analysis, while the photon parts are usually secondary. Like most generalizations, this one has its exceptions.

The Coulomb gauge $\nabla \cdot \mathbf{A} = 0$ is not the only gauge condition popular in physics. Many physicists use the *Lorentz gauge*:

$$\nabla \cdot \mathbf{A} + \frac{1}{c}\frac{\partial \psi}{\partial t} = 0$$

This causes (1.5.6) and (1.5.7) to have the forms

$$\nabla^2 \psi - \frac{1}{c^2}\frac{\partial^2}{\partial t^2}\psi = -4\pi\varrho$$

$$\nabla^2 \mathbf{A} - \frac{1}{c^2}\frac{\partial^2}{\partial t^2}\mathbf{A} = -\frac{4\pi\mathbf{j}}{c}$$

Now both the vector and scalar potentials obey the retarded wave equation. One can also show that they combine to produce a four vector which is invariant under a Lorentz transformation. The Lorentz invariance is very useful in many branches of physics, and the Lorentz gauge is used frequently. We shall not give the Green's functions in this case, but they are provided in books on field theory. Obviously both the scalar and vector parts are retarded.

Another gauge which is often used, but which does not have a formal name, is the condition that $\psi = 0$. The scalar potential is set equal to zero.

In this case we find that the longitudinal vector potential is not zero but now plays an important role. In fact, the longitudinal part of the vector potential leads to an interaction between charges which is just the instantaneous Coulomb interaction. Thus when we set $\psi = 0$, the longitudinal part of the vector potential plays a role that is identical to that of the scalar potential in the Coulomb gauge.

The Hamiltonian (1.5.1) is written in a gauge which has the scalar potential acting instantaneously, so that the Coulomb interaction is unretarded. This form of the Hamiltonian is consistent with either gauge $\nabla \cdot \mathbf{A} = 0$ or $\psi = 0$. One gets a different Hamiltonian for other choices of gauge.

B. Lagrangian

So far we have shown that the Coulomb gauge makes the vector potential transverse and that the scalar potential acts instantaneously. Now we would like to show that the Hamiltonian has the form indicated in Eq. (1.5.1). This is done by starting from the following Lagrangian (Schiff, 1968):

$$L = \frac{1}{2} \sum_i m\mathbf{v}_i^2 + \int \frac{d^3r}{8\pi} [\mathbf{E}(\mathbf{r})^2 - \mathbf{B}(\mathbf{r})^2] - \sum_i e_i \psi(\mathbf{r}_i) + \sum_i \frac{e_i}{c} \mathbf{v}_i \cdot \mathbf{A}(\mathbf{r}_i)$$

This Lagrangian was chosen to produce Maxwell's equations as well as the classical equations of motion for a particle at $\mathbf{r}_i$, with charge e_i and velocity $\mathbf{v}_i$, in a magnetic and electric field. The three variables are $\eta = \psi$, $\mathbf{A}$, and $\mathbf{r}_i$. Each of these is used to generate an equation from

$$\frac{d}{dt} \frac{\delta L}{\delta \dot{\eta}} + \sum_{\alpha=1}^{3} \frac{d}{dr_\alpha} \frac{\partial L}{\partial(\partial \eta / \partial r_\alpha)} - \frac{\delta L}{\delta \eta} = 0 \tag{1.5.13}$$

where the conjugate momentum is

$$P_\eta = \frac{\partial L}{\partial \dot{\eta}}$$

When the scalar potential ψ is chosen as the variable, we obtain

$$\frac{\delta L}{\delta \psi} = -\varrho(\mathbf{r}) = -\sum_i e_i \delta(\mathbf{r} - \mathbf{r}_i)$$

$$\frac{\delta L}{\delta(\partial \psi / \partial x)} = -\frac{E_x}{4\pi}$$

$$P_\psi = \frac{\delta L}{\delta \dot{\psi}} = 0$$

This produces the Maxwell equation (1.5.4) from (1.5.13):

$$\nabla \cdot \mathbf{E} = -4\pi\varrho$$

When one component of the scalar potential A_x is chosen as the variable in the Lagrangian, we obtain

$$\frac{\delta L}{\delta A_x} = \mathscr{j}_x = \frac{1}{c} \sum_i e_i v_{ix} \delta(\mathbf{r} - \mathbf{r}_i)$$

$$\frac{\delta L}{\delta \dot{A}_x} = -\frac{E_x}{4\pi c}$$

$$\frac{\delta L}{\delta(\partial A_x/\partial y)} = \frac{B_z}{4\pi}$$

When this is used in Lagrange's equation (1.5.13), we obtain the x component of the Maxwell equation (1.5.5):

$$\nabla \times \mathbf{B} = \frac{1}{c} \frac{\partial \mathbf{E}}{\partial t} + \frac{4\pi}{c} \mathscr{j}$$

Choosing either A_y or A_z as the active variable will generate the y and z components of this equation. Thus the Lagrangian does generate the two Maxwell equations which depend on particle properties. Another important feature is that the momentum conjugate to the scalar potential is zero:

$$P_\psi = 0$$

The momentum variable conjugate to the vector potential is

$$\mathbf{P}_A = -\frac{\mathbf{E}}{4\pi c}$$

which is just proportional to the electric field. This relationship is important, since the quantization of the fields will require that the vector potential no longer commute with the electric field, since they are conjugate variables.

The other equations generated by this Lagrangian are the equations for particle motion in electric and magnetic fields. Here we use as the variable in the Lagrangian the particle coordinate r_α:

$$\frac{\delta L}{\delta r_{i\alpha}} = -e_i \left[\nabla_\alpha \psi(\mathbf{r}_i) - \frac{v_{i\delta}}{c} \nabla_\alpha \cdot A_\delta(\mathbf{r}_i) \right]$$

$$\frac{\delta L}{\delta v_{i\alpha}} = p_{i\alpha} = m v_{i\alpha} + \frac{e_i}{c} A_\alpha(\mathbf{r}_i)$$

And the equation deduced from Lagrange's equation is

$$0 = \frac{d}{dt}\left[m\mathbf{v}_i + \frac{e_i}{c}\mathbf{A}(\mathbf{r}_i)\right] + e_i\nabla\phi - \frac{e_i}{c}\nabla(\mathbf{v}\cdot\mathbf{A})$$

The total time derivative on the position dependence of the vector potential is interpreted as a hydrodynamic derivative:

$$\frac{d}{dt}A(\mathbf{r}_i) = \frac{\partial}{\partial t}A(\mathbf{r}_i) + \mathbf{v}_i\cdot\nabla A(\mathbf{r}_i)$$

Then the equation may be rearranged into

$$m\frac{d\mathbf{v}_i}{dt} = e_i\left\{E(\mathbf{r}_i) + \frac{1}{c}[\nabla(\mathbf{v}\cdot\mathbf{A}) - \mathbf{v}\cdot\nabla\mathbf{A}]\right\} = e_i\left[E(\mathbf{r}_i) + \frac{1}{c}\mathbf{v}_i\times\mathbf{B}\right]$$

The last term is just the Lorentz force on a particle in a magnetic field since

$$\mathbf{v}\times\mathbf{B} = \mathbf{v}\times(\nabla\times\mathbf{A}) = \nabla(\mathbf{v}\cdot\mathbf{A}) - (\mathbf{v}\cdot\nabla)\mathbf{A}$$

The equation for $m\dot{v}$ is the proper one for a spinless particle in an electric and magnetic field. The Lagrangian is a suitable starting point for the quantization of the interacting system of particles and fields.

C. Hamiltonian

The Hamiltonian is derived from

$$H = \sum_i \dot{\eta}_i p_\eta - L$$

where the first summation is over all the variables and their conjugate moments. In our case this includes the vector potential and the particle momentum; the scalar potential has no momentum. The Hamiltonian has the form

$$\begin{aligned} H = {} & \frac{1}{m}\sum_i \mathbf{p}_i\left[\mathbf{p}_i - \frac{e_i}{c}\mathbf{A}(\mathbf{r}_i)\right] + \int\frac{d^3r}{4\pi c}\left[\dot{\mathbf{A}}(\mathbf{r})\cdot\left(\frac{1}{c}\dot{\mathbf{A}} + \nabla\psi\right) - \frac{E^2 - B^2}{8\pi}\right] \\ & + \sum_i e_i\psi(\mathbf{r}_i) - \frac{1}{2m}\sum_i\left[\mathbf{p}_i - \frac{e_i}{c}\mathbf{A}(\mathbf{r}_i)\right]^2 \\ & - \sum_i \frac{e_i}{mc}\mathbf{A}(\mathbf{r}_i)\cdot\left[\mathbf{p}_i - \frac{e_i}{c}\mathbf{A}(\mathbf{r}_i)\right] \end{aligned}$$

It may be collected into the form

$$H = \frac{1}{2m} \sum_i \left[\mathbf{p}_i - \frac{e_i}{c} \mathbf{A}(\mathbf{r}_i) \right]^2 + \sum_i e_i \psi(\mathbf{r}_i)$$
$$+ \int \frac{d^3r}{8\pi} \left[E^2 + B^2 - 2\nabla\psi \cdot \left(\frac{1}{c} \dot{\mathbf{A}} + \nabla\psi \right) \right]$$

Terms which are the cross product between the scalar and vector potential parts of the electric field always vanish after an integration by parts,

$$\int d^3r (\nabla\psi \cdot \dot{\mathbf{A}}) = -\int d^3r \psi (\nabla \cdot \dot{\mathbf{A}}) = 0$$

since we are using the Coulomb gauge wherein $\nabla \cdot \mathbf{A} = 0$. The terms involving the square of the scalar potential may also be reduced to an instantaneous interaction between charges using (1.5.9):

$$\int \frac{d^3r}{4\pi} \nabla\psi \cdot \nabla\psi = -\int \frac{d^3r}{4\pi} \psi \nabla^2 \psi = \int d^3r \psi(\mathbf{r}) \varrho(\mathbf{r}) = \sum_i e_i \psi(\mathbf{r}_i)$$
$$= \sum_{ij} \frac{e_i e_j}{r_{ij}}$$

With these simplifications, we may write the Hamiltonian as

$$H = \frac{1}{2m} \sum_i \left[\mathbf{p}_i - \frac{e_i}{c} \mathbf{A}(\mathbf{r}_i) \right]^2 + \int \frac{d^3r}{8\pi} (E^2 + B^2)$$

The last terms are just the energy density of the electromagnetic fields. The first terms arise from the charged particles. This form of H is simple and instructive. However, our interest is in obtaining a quantized version. To do this, we write the electric field energy density as the separate parts of vector and scalar potential,

$$\int \frac{d^3r}{8\pi} E^2 = \int \frac{d^3r}{8\pi} \left[\frac{1}{c^2} \dot{\mathbf{A}}^2 + (\nabla\psi)^2 + \frac{2}{c} \nabla\psi \cdot \dot{\mathbf{A}} \right]$$
$$= \frac{1}{2} \sum_{ij} \frac{e_i e_j}{r_{ij}} + \int \frac{d^3r}{8\pi c^2} (\dot{\mathbf{A}})^2$$

and the scalar potential just gives the interaction between charges.

Now we use the fact that the electric field is the conjugate momentum density of the vector potential. These two field variables must obey the following equal time commutation relation:

$$\left[A_\alpha(\mathbf{r}, t), -\frac{E_\beta}{4\pi}(\mathbf{r}', t)\right] = i\delta_{\alpha\beta}\delta(\mathbf{r} - \mathbf{r}') \tag{1.5.14}$$

This commutation relation is satisfied by defining the vector potential in terms of creation and destruction operators in the form

$$A_\alpha(\mathbf{r}, t) = \sum_{\mathbf{k}\lambda} \left(\frac{2\pi\hbar c^2}{v\omega_{\mathbf{k}}}\right)^{1/2} \xi_\alpha(\mathbf{k}, \lambda) e^{i\mathbf{k}\cdot\mathbf{r}}(a_{\mathbf{k}\lambda}e^{-i\omega_{\mathbf{k}}t} + a^\dagger_{-\mathbf{k}\lambda}e^{i\omega_{\mathbf{k}}t}) \tag{1.5.15}$$

There are two transverse modes, one for each transverse direction, and λ is the summation over these two modes. The unit polarization vector ξ_α gives the direction for each mode. The operators obey the commutation relations

$$[a_{\mathbf{k}\lambda}, a^\dagger_{\mathbf{k}'\lambda'}] = \delta_{\mathbf{k}\mathbf{k}'}\delta_{\lambda\lambda'}$$

$$[a_{\mathbf{k}\lambda}, a_{\mathbf{k}'\lambda'}] = 0$$

The time derivative of the vector potential is

$$\dot{A}_\alpha(\mathbf{r}, t) = -i\sum_{\mathbf{k}\lambda} \left(\frac{2\pi\hbar c^2\omega_k}{v}\right)^{1/2} \xi_\alpha(\mathbf{k}, \lambda) e^{i\mathbf{k}\cdot\mathbf{r}}(a_{\mathbf{k}\lambda}e^{-i\omega_{\mathbf{k}}t} - a^\dagger_{-\mathbf{k}\lambda}e^{i\omega_{\mathbf{k}}t})$$

The electric field is

$$E_\alpha(\mathbf{r}, t) = -\frac{1}{c}\dot{A}_\alpha - \nabla_\alpha\psi$$

The scalar potential is not expressed in terms of operators, so it does not influence the commutator. The basic commutation relation (1.5.14) must be only the commutator of the vector potential and its time derivative:

$$[A_\alpha(\mathbf{r}, t), \dot{A}_\beta(\mathbf{r}', t)] = 4\pi ci\delta_{\alpha\beta}\delta(\mathbf{r} - \mathbf{r}')$$

Our vector potential in (1.5.15) has been chosen to give just this result:

$$[A_\alpha(\mathbf{r}, t), \dot{A}_\beta(\mathbf{r}', t)] = i\sum_{\substack{\mathbf{k}\lambda \\ \mathbf{k}'\lambda'}} \left(\frac{2\pi\hbar c^2\omega_{\mathbf{k}}}{v}\right)^{1/2}\left(\frac{2\pi\hbar c^2}{v\omega_{\mathbf{k}'}}\right)^{1/2} e^{i(\mathbf{k}\cdot\mathbf{r}+\mathbf{k}'\cdot\mathbf{r}')}$$

$$\xi_\alpha(\mathbf{k}, \lambda)\xi_\beta(\mathbf{k}', \lambda') \times 2\delta_{\mathbf{k},-\mathbf{k}'} = 4\pi ic\delta(\mathbf{r} - \mathbf{r}')\delta_{\alpha\beta}$$

The various factors of 2π and $\omega_{\mathbf{k}}$ which enter (1.5.15) are selected so that the commutation relation (1.5.14) is satisfied when we choose the usual commutation relations for the operators $a_{\mathbf{k}\lambda}$ and $a^\dagger_{\mathbf{k}\lambda}$. The expressions for the energy density of electric and magnetic fields have the form

$$\int \frac{d^3r}{8\pi}\,\frac{1}{c^2}\,\dot{\mathbf{A}}(\mathbf{r},t)^2 = -\sum_{\mathbf{k}\lambda}\frac{\hbar\omega_k}{4}\,(a_{\mathbf{k}\lambda}a_{-\mathbf{k}\lambda}e^{-2it\omega_{\mathbf{k}}} + a^\dagger_{\mathbf{k}\lambda}a_{-\mathbf{k}\lambda}e^{2it\omega_{\mathbf{k}}}$$
$$-a_{\mathbf{k}\lambda}a^\dagger_{\mathbf{k}\lambda} - a^\dagger_{\mathbf{k}\lambda}a_{\mathbf{k}\lambda})$$

$$\int \frac{d^3r}{8\pi}\,(\nabla\times\mathbf{A})^2 = \sum_{\mathbf{k}\lambda}\frac{\hbar c^2}{4\omega_{\mathbf{k}}}\,(\mathbf{k}\times\boldsymbol{\xi})^2(a_{\mathbf{k}\lambda}a_{-\mathbf{k}\lambda}e^{-2it\omega_{\mathbf{k}}} + a^\dagger_{\mathbf{k}\lambda}a^\dagger_{-\mathbf{k}\lambda}e^{2it\omega_{\mathbf{k}}}$$
$$+ a_{\mathbf{k}\lambda}a^\dagger_{\mathbf{k}\lambda} + a^\dagger_{\mathbf{k}\lambda}a_{\mathbf{k}\lambda})$$

In the second term, the photon energy is $\omega_k = ck$. This term may be added to the first, and the aa and $a^\dagger a^\dagger$ terms both cancel. Thus we get

$$H = \sum_i \frac{1}{2m}\left[\mathbf{p}_i - \frac{e_i}{c}\,\mathbf{A}(r_i)\right]^2 + \frac{1}{2}\sum_{ij}\frac{e_ie_j}{r_{ij}} + \sum_{\mathbf{k}\lambda}\frac{\hbar\omega_{\mathbf{k}}}{2}\,(a^\dagger_{\mathbf{k}\lambda}a_{\mathbf{k}\lambda} + a_{\mathbf{k}\lambda}a^\dagger_{\mathbf{k}\lambda}) \tag{1.5.1}$$

This form of the Hamiltonian is just the result which was asserted in the beginning in (1.5.1). The vector potential in (1.5.15) is expressed in terms of the creation and destruction operators of the photon field. The photon states behave as bosons—as harmonic oscillators. Each photon state of wave vector $\mathbf{k}$ and polarization λ has eigenstates of the form

$$|\,n_{\mathbf{k}\lambda}\rangle = \frac{(a^\dagger_{\mathbf{k}\lambda})^{n_{k\lambda}}}{(n_{\mathbf{k}\lambda}!)^{1/2}}\,|\,0\rangle$$

where $n_{\mathbf{k}\lambda}$ is an integer which is the number of photons in that state. The state $|\,0\rangle$ is the photon vacuum. The total energy in the free-photon part of the Hamiltonian is

$$H_0\,|\,n_{k,\lambda_1}\cdots n_{k_n\lambda_n}\rangle = \sum_{\mathbf{k}\lambda}\hbar\omega_{\mathbf{k}}(n_{\mathbf{k}\lambda}+\tfrac{1}{2})\,|\,n_{k,\lambda_1}\cdots n_{k_n\lambda_n}\rangle$$

Our derivation of the Hamiltonian (1.5.1) has been for spinless, nonrelativistic particles. Certainly the most important relativistic term is the spin–orbit interaction. The effects of spin also enter through the direct interaction of the magnetic moment with an external magnetic field.

1.6. PAIR DISTRIBUTION FUNCTION

In crystalline solids, the atoms are arranged in a regular array. If $\mathbf{R}_j$ denotes the position of the atoms, the summation of $\exp(i\mathbf{q}\cdot\mathbf{R}_j)$ over all the atoms yields zero unless $\mathbf{q}$ has particular values. These values are

$\mathbf{q} = 0$ or else one of the reciprocal lattice vectors $\mathbf{G}$ of the solid. In either case, the factor $\exp(i\mathbf{q} \cdot \mathbf{R}_j) = 1$ so that the summation yields the number of atoms in solid, which is N. This result is written as

$$\sum_j e^{i\mathbf{q}\cdot\mathbf{r}_j} = N \sum_{\mathbf{G}}{}' \delta_{\mathbf{q}+\mathbf{G}} + N\delta_{\mathbf{q}=0}$$

Many materials are not crystalline, e.g., liquids, gases, and disordered solids. In discussing these quantities we often encounter similar summations over particle locations. These summations still equal N for $\mathbf{q} = 0$, but we must learn to evaluate them for $\mathbf{q} \neq 0$. This evaluation is done in terms of a function $S(\mathbf{q})$, which is called the *static structure factor* or the *static form factor.*

Let us first define the density operator for the atomic locations:

$$\varrho(\mathbf{q}) = \sum_j \exp(i\mathbf{q} \cdot \mathbf{R}_j)$$

$\varrho(\mathbf{q})$ is an operator because the $\mathbf{R}_j$ describe the instantaneous location of each particle. We assume the particles are classical objects, so that we do not need to deal with the quantum statistics associated with fermions or bosons. The average of this operator is defined as

$$\langle \varrho(\mathbf{q}) \rangle = \left\langle \sum_j e^{i\mathbf{q}\cdot\mathbf{R}_j} \right\rangle$$

where the average is taken over the various configurations of the atoms. Here we shall just define some of these averages, without explaining how to find them from first principles. The latter is done in standard references (Hill, 1956; Percus, 1964).

The first average is over a single density operator:

$$\langle \varrho(\mathbf{q}) \rangle = N\delta_{\mathbf{q}=0} \tag{1.6.1}$$

$\langle \varrho \rangle$ is zero for finite $\mathbf{q}$ because there is no restriction on the location of each atom. That is, in the averaging, each $\mathbf{R}_j$ can be anywhere in the material with equal likelihood—since we ignore edge effects. Thus the averaging is equivalent to the integral over a continuous distribution of values:

$$\langle \varrho(\mathbf{q}) \rangle = \frac{N}{\nu} \int d^3r e^{i\mathbf{q}\cdot\mathbf{r}} = N\delta_{\mathbf{q}=0}$$

The integral on the right just gives a delta function at $\mathbf{q} = 0$, which is the same result as (1.6.1).

The second interesting average is over a product of two density operators:

$$\langle \varrho(\mathbf{q})\varrho(\mathbf{q}')\rangle = \sum_{ij} \langle \exp(i\mathbf{q} \cdot \mathbf{R}_i) \exp(i\mathbf{q}' \cdot \mathbf{R}_j)\rangle$$

$$\langle \varrho(\mathbf{q})\varrho(\mathbf{q}')\rangle = N^2\delta_{\mathbf{q}=0}\delta_{\mathbf{q}'=0} + N\delta_{\mathbf{q}+\mathbf{q}'=0}S(\mathbf{q})$$

Of course $\langle \varrho(\mathbf{q})\varrho(\mathbf{q})\rangle = N^2$ when both $\mathbf{q}$ and $\mathbf{q}'$ are zero. If both $\mathbf{q}$ and $\mathbf{q}'$ are finite, the average vanishes unless $\mathbf{q} + \mathbf{q}' = 0$. To see this, go to a center-of-mass coordinate system,

$$\mathbf{Q} = \mathbf{q} + \mathbf{q}', \qquad \mathbf{q} = \mathbf{k} + \frac{\mathbf{Q}}{2}$$

$$\mathbf{k} = \tfrac{1}{2}(\mathbf{q} - \mathbf{q}'), \qquad \mathbf{q}' = -\mathbf{k} + \frac{\mathbf{Q}}{2}$$

so that we need to average:

$$\langle \varrho(\mathbf{q})\varrho(\mathbf{q}')\rangle = \sum_{ij} \left\langle \exp\left\{ i\left[\mathbf{Q} \cdot \frac{\mathbf{R}_i + \mathbf{R}_j}{2} + \mathbf{k} \cdot (\mathbf{R}_i - \mathbf{R}_j) \right] \right\} \right\rangle$$

There is a strong correlation between the difference $\mathbf{R}_i - \mathbf{R}_j$ of two particle locations, but there is no correlation for the center of mass $\mathbf{R}_i + \mathbf{R}_j$. Thus when we average, the term $\exp[\frac{1}{2}i\mathbf{Q} \cdot (\mathbf{R}_i + \mathbf{R}_j)]$ averages to zero unless $\mathbf{Q} = 0$ or $\mathbf{q} = -\mathbf{q}' = \mathbf{k}$. Thus we are left with the average of $\mathbf{q} \neq 0$:

$$\langle \varrho(\mathbf{q})\varrho(\mathbf{q}')\rangle = N\delta_{\mathbf{q}+\mathbf{q}'=0} \sum_{ij} \langle \exp[i\mathbf{q} \cdot (\mathbf{R}_i - \mathbf{R}_j)]\rangle$$

There is strong correlation between the relative positions of two atoms at short distances. If one is at a spot, the other will not be at the same spot because atoms repulse each other at short separations. The static structure factor $S(\mathbf{q})$ is defined as

$$N\delta_{\mathbf{q}=0} + S(\mathbf{q}) = \frac{1}{N} \sum_{ij} \langle \exp[i\mathbf{q} \cdot (\mathbf{R}_i - \mathbf{R}_j)]\rangle = \frac{1}{N} \left(\sum_{i=j} 1 + \sum_{i \neq j} e^{i\mathbf{q} \cdot \Delta\mathbf{R}_{ij}} \right)$$

One subtracts the term $N\delta_{\mathbf{q}=0}$. The right-hand side contains a summation over all particles at $\mathbf{R}_i$ and then a summation over all its neighbors $\mathbf{R}_{ij}$. After averaging, this should give the same result for each particle $\mathbf{R}_i$. Thus the summation over $\mathbf{R}_i$ just gives N:

$$N\delta_{\mathbf{q}=0} + S(\mathbf{q}) = 1 + \left\langle \sum_{\Delta\mathbf{R}}{}' e^{i\mathbf{q} \cdot \Delta\mathbf{R}} \right\rangle$$

The prime on the summation means that $\Delta\mathbf{R} = 0$ is excluded from the summation. The crucial average is the second term on the right. It averages over the relative positions $\Delta\mathbf{R}$ of two atoms. This average can also be expressed as a function of r. Define $g(\mathbf{r})$ as the *pair distribution function.* It is defined as the probability that another particle is at $\mathbf{r}$ if there is already one at $\mathbf{r} = 0$. The normalization is chosen to give $g(\mathbf{r}) \to 1$ at large distances. In terms of the pair distribution function, this may be written as

$$N\delta_{\mathbf{q}=0} + S(\mathbf{q}) = 1 + n \int d^3r g(\mathbf{r}) e^{i\mathbf{q}\cdot\mathbf{r}}$$

where $n = N/v$ is the average density of the system. We may also write

$$N\delta_{\mathbf{q}=0} = n \int d^3r e^{i\mathbf{q}\cdot\mathbf{r}}$$

and transpose this to the right to give the equivalent expressions

$$\begin{aligned} S(\mathbf{q}) - 1 &= n \int d^3r e^{i\mathbf{q}\cdot\mathbf{r}}[g(\mathbf{r}) - 1] \\ g(\mathbf{r}) - 1 &= \frac{1}{n} \int \frac{d^3q}{(2\pi)^3} e^{-i\mathbf{q}\cdot\mathbf{r}}[S(\mathbf{q}) - 1] \end{aligned} \tag{1.6.2}$$

The inverse Fourier transform may now be taken to give $g(\mathbf{r})$ as a function of $S(\mathbf{q})$. Both $g(\mathbf{r})$ and $S(\mathbf{q})$ go to unity at large values of their arguments. Thus $[g(\mathbf{r}) - 1]$ and $[S(\mathbf{q}) - 1]$ are both quantities which are short-ranged in their arguments. Thus their mutual Fourier transforms are well defined.

The function $S(\mathbf{q})$ is determined experimentally by scattering from targets. Afterwards, $g(\mathbf{r})$ is found by the numerical Fourier transform. The well-known results for liquid ^{4}He are shown in Fig. 1.3, which is from Mozer *et al.* (1974). The pair distribution function $g(\mathbf{r})$ should vanish according to the dashed line at short distances because the atoms do not penetrate each other at the low thermal energies of the liquid. The numerical transform shown by the solid line has a spurious peak for $R < 2$ Å because of inaccuracies in $S(\mathbf{k})$ for $k \to \infty$.

The higher moments such as $\langle \varrho(\mathbf{q})\varrho(\mathbf{q}')\varrho(\mathbf{q}'')\rangle$ would also be interesting were they known. But they are not easily obtainable from experiments or from theory.

In the model of the homogeneous electron gas, the electrons are not ordered. In this case one can also describe relative behavior of pairs of particles by a pair distribution function $g(\mathbf{r})$. This quantity is always

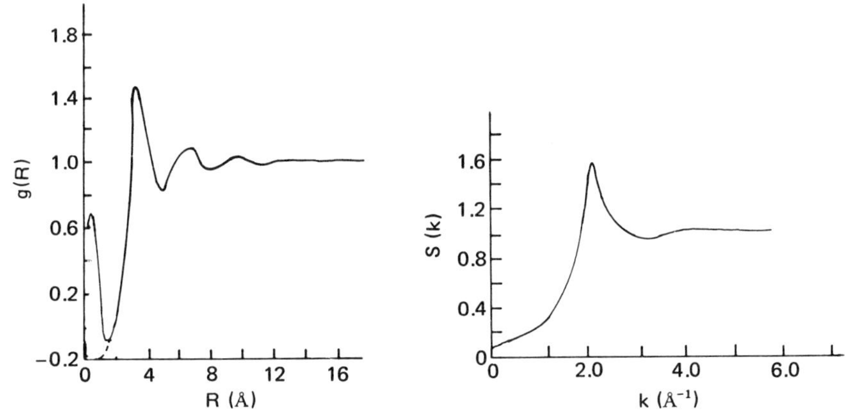

FIGURE 1.3. $g(R)$ (left) and $S(k)$ (right) for liquid ^{4}He. $S(k)$ is obtained by X-ray scattering measurements, and $g(R)$ is found by the numerical transform. Uncertainties in $S(k)$ at large k produce the extraneous structure in $g(R)$ at low R, which is to be ignored. The dashed line shows the proper extrapolation. $\varrho = 0.1628$ g/cm^3, $T = 2.86$ K. *Source*: Mozer *et al.* (1974) (used with permission).

defined as the average over the motion of the particles. In a quantum system the diagonal density matrix is defined as the square of the N-particle wave function:

$$\varrho_N(\mathbf{r}_1 \cdots \mathbf{r}_N) = |\Psi(\mathbf{r}_1 \cdots \mathbf{r}_N)|^2$$

It is normalized so that unity is obtained when integrating over all the coordinates:

$$1 = \int dr_1 \cdots dr_N \varrho_N(\mathbf{r}_1 \cdots \mathbf{r}_N)$$

Since the many-particle wave function Ψ is either even or odd under the interchange of two identical particles, the density matrix is unchanged by this operator. Thus, any two particles can be called $\mathbf{r}_1$ and $\mathbf{r}_2$. Their relative behavior, after averaging over all the other particles, is obtained by integrating over the coordinates of all the other particles:

$$\varrho_2(\mathbf{r}_1, \mathbf{r}_2) = \int dr_3\, dr_4 \cdots dr_N \varrho_N(\mathbf{r}_1 \cdots \mathbf{r}_N)$$

$$1 = \int dr_1\, dr_2 \varrho_2(\mathbf{r}_1, \mathbf{r}_2)$$

$\varrho_2(\mathbf{r}_1, \mathbf{r}_2)$ is the average two-particle density matrix. The "integration" over other particles included spatial integration, spin averages, etc. Similarly, the one-particle density matrix is that obtained by integrating over all but one of the particles:

$$\varrho_1(\mathbf{r}_1) = \int dr_2\, dr_3 \cdots dr_N \varrho_N(\mathbf{r}_1 \cdots \mathbf{r}_N)$$

$$\varrho_1(\mathbf{r}_1) = \int dr_2 \varrho_2(\mathbf{r}_1, \mathbf{r}_2)$$

$$1 = \int \varrho_1(\mathbf{r}_1)\, dr_1$$

For homogeneous systems, the two-particle density matrix $\varrho_2(\mathbf{r}_1, \mathbf{r}_2)$ can only depend on the relative positions $(\mathbf{r}_1 - \mathbf{r}_2)$ of the two coordinates. Thus this quantity must be related to the pair distribution function. They are in fact proportional. The constant of proportionality is determined from the condition that $g(\mathbf{r}) \to 1$, so

$$N(N-1)\varrho_2(\mathbf{r}_1 - \mathbf{r}_2) = n^2 g(\mathbf{r}_1 - \mathbf{r}_2) \tag{1.6.3}$$

where ν is the volume of the system.

As an example of particular behavior, we evaluate $g(\mathbf{r})$ in the Hartree–Fock approximation for electrons. The many-particle wave function is just an N-dimensional Slater determinant, so the N-particle density matrix is just the square of this determinant:

$$\varrho_N(\mathbf{r}_1 \cdots \mathbf{r}_N) = \frac{1}{N!} \begin{vmatrix} \psi_{\lambda_1}(\mathbf{r}_1) & \psi_{\lambda_2}(\mathbf{r}_1) & \cdots & \psi_{\lambda_N}(\mathbf{r}_1) \\ \psi_{\lambda_1}(\mathbf{r}_2) & \psi_{\lambda_2}(\mathbf{r}_2) & \cdots & \psi_{\lambda_N}(\mathbf{r}_2) \\ \psi_{\lambda_1}(\mathbf{r}_N) & & \cdots & \psi_{\lambda_N}(\mathbf{r}_N) \end{vmatrix}^2$$

The single-particle orbitals $\psi_\lambda(\mathbf{r})$ are assumed to be orthogonal for different states λ. A theorem states (see Parr, 1964) that if one integrates over all but two coordinates, the two-particle density matrix, from this Slater determinant, is the sum of all pairs of two-particle wave functions:

$$\varrho_2(\mathbf{r}_1, \mathbf{r}_2) = \frac{1}{N(N-1)} \sum_{\substack{\lambda_1, \lambda_2 \\ \lambda_1 \neq \lambda_2}} \frac{1}{2!} \begin{vmatrix} \psi_{\lambda_1}(\mathbf{r}_1) & \psi_{\lambda_2}(\mathbf{r}_1) \\ \psi_{\lambda_1}(\mathbf{r}_2) & \psi_{\lambda_2}(\mathbf{r}_2) \end{vmatrix}^2$$

The summation is taken over all pairs of occupied states. The two-dimensional Slater determinant can be expanded to give the following equivalent result:

$$\varrho_2(\mathbf{r}_1, \mathbf{r}_2) = \frac{1}{2N(N-1)} \sum_{\substack{\lambda_1, \lambda_2 \\ \lambda_1 \neq \lambda_2}} [\psi_{\lambda_1}(\mathbf{r}_1)\psi_{\lambda_2}(\mathbf{r}_2) - \psi_{\lambda_1}(\mathbf{r}_2)\psi_{\lambda_2}(\mathbf{r}_1)]^2$$

A further integration gives the one-particle density matrix as the summation over all occupied states of the square of the wave function:

$$\varrho_1(\mathbf{r}_1) = \frac{1}{2N(N-1)} \sum_{\substack{\lambda_1, \lambda_2 \\ \lambda_1 \neq \lambda_2}} [\psi_{\lambda_1}(\mathbf{r}_1)^2 + \psi_{\lambda_2}(\mathbf{r}_1)^2]$$

$$\varrho_1(\mathbf{r}_1) = \frac{1}{N} \sum_{\lambda_1} | \psi_{\lambda_1}(\mathbf{r}_1) |^2$$

These theorems apply to any Hartree–Fock system with any orbitals. For homogeneous systems, we assume the wave functions are translationally invariant. If they are plane waves with spin up α, the two-particle density matrix is

$$\psi = \frac{1}{\nu^{1/2}} \exp(i\mathbf{k} \cdot \mathbf{r})\alpha$$

$$\varrho_2(\mathbf{r}, \mathbf{r}') = \frac{1}{\nu^2 N(N-1)} \sum_{\mathbf{k},\mathbf{k}'} (1 - e^{i(\mathbf{k}-\mathbf{k}')\cdot(\mathbf{r}-\mathbf{r}')})\alpha_1\alpha_2$$

Using (1.6.3) gives a pair distribution function of the form

$$g(\mathbf{r} - \mathbf{r}') = \frac{1}{N^2} \sum_{kk'} (1 - e^{i(\mathbf{k}-\mathbf{k}')\cdot(\mathbf{r}-\mathbf{r}')})\alpha_1\alpha_2$$

In most systems, the electrons do not all have spin up, and one also has to average over all spin coordinates to get $g(\mathbf{r})$. This is discussed in Chapter 5.

PROBLEMS

1. Solve the classical vibrational modes of a one-dimensional chain of atoms of type A and B. They alternate on the chain with masses m_A and m_B. The harmonic spring between atoms has spring constant K.

2. Write down the Hamiltonian of Problem 1. Solve it, and show that it may be reduced to the form

$$H = \sum_{k\lambda} \omega_{k\lambda}(a_{k\lambda}^\dagger a_{k\lambda} + \tfrac{1}{2})$$

where the $\omega_{k,\lambda}$ are the classical normal modes.

3. Find the exact solution to

$$H = E_0 a^\dagger a + E_1(a^\dagger a^\dagger + aa)$$

where E_0 and E_1 are constants and a and $a^\dagger$ are boson operators.

4. Solve the Hamiltonian below with a canonical transformation:

$$H = E_0 a^\dagger a + F(a + a^\dagger)$$

$$\bar{H} = e^s H e^{-s}, \qquad S = \lambda(a - a^\dagger)$$

a. Show that $\lambda = \lambda^*$ since $\bar{H}$ is Hermitian.

b. Use the expansion (1.4.22), and show that only a few terms in the series are finite—the remainder vanish.

c. Find the choice of λ which reduces $\bar{H}$ to (1.1.12).

5. Consider a fermion system which has three energy states with eigenvalues E_1, E_2, and E_3. There also exist matrix elements which connect these states and permit transitions between them: M_{12}, M_{23}, and M_{13}.

a. Write down the Hamiltonian for this system in terms of creation and destruction operators.

b. Determine the eigenvalue equation for this system.

6. Consider a tight binding solid which has alternate atoms of type A and B. The electron Hamiltonian in the nearest neighbor model has the form

$$H = W \sum_{i\delta} (a_i^\dagger b_{i+\delta} + b_i^\dagger a_{i+\delta}) + B \sum_i b_i^\dagger b_i + A \sum_j a_j^\dagger a_j$$

where a_i and b_j are electron operators for atoms of type A and B. Find the exact eigenvalues of this Hamiltonian.

7. Calculate the exact partition function

$$e^{-\beta\Omega} = Z = \mathrm{Tr}\, e^{-\beta(H-\mu N)}$$

for the Hubbard model (1.2.15) in the atomic limit. Then give the expression for the average number of electrons $\bar{N} = -(\partial\Omega/\partial\mu)$.

8. Take a three-atom chain with periodic boundary conditions. Each atom has one orbital state with two spin configurations.

a. Solve the nearest neighbor tight binding model for zero, one, two, or three electrons with spin up, and give the partition function. Show that it is the same as (1.2.14).

b. Give the partition function for the XY model, and show that it is different.

9. Give the partition function for the Hubbard model for a two-atom chain with periodic boundary conditions.

10. What is the energy current operator j_E for the Hubbard model (1.2.15)?

11. Show that the energy current in the harmonic chain (1.1.12) is

$$\mathcal{J}_E = \frac{1}{2} \sum_n \frac{d}{dk} (\omega_k{}^2) a_k{}^\dagger a_k$$

12. Assume that the Hamiltonian in the tight binding model has additional terms

$$\omega_0 \sum_i b_i{}^\dagger b_i + M \sum_i n_i (b_i + b_i{}^\dagger)$$

describing how phonons on a site i interact with electrons n_i when they are on the site. How do these terms affect the heat current?

13. Explicitly verify the equation of continuity,

$$\frac{\partial}{\partial t} \varrho(\mathbf{r}, t) + \nabla \cdot \mathcal{J}(\mathbf{r}, t) = 0$$

for a gas of free fermion particles which have

$$H = \sum_{\mathbf{k}\sigma} \varepsilon_{\mathbf{k}} C^\dagger_{\mathbf{k}\sigma} C_{\mathbf{k}\sigma}$$

$$\varrho(\mathbf{r}, t) = \frac{1}{\nu} \sum_{\mathbf{kq}\sigma} e^{-i\mathbf{q}\cdot\mathbf{r}} C^\dagger_{\mathbf{k}+\mathbf{q},\sigma} C_{\mathbf{k}\sigma}$$

$$\mathcal{J}(\mathbf{r}, t) = \frac{1}{m\nu} \sum_{\mathbf{kq}\sigma} e^{-i\mathbf{q}\cdot\mathbf{r}} (\mathbf{k} + \tfrac{1}{2}\mathbf{q}) C^\dagger_{\mathbf{k}+\mathbf{q},\sigma} C_{\mathbf{k}\sigma}$$

14. Find the effective interaction $V_R(r)$ between two fixed electrons (1.3.4) using the deformation potential interaction (1.3.5) and a Debye spectrum.

15. Find the exact solution in one dimension and the partition function for the XY model with a magnetic field in the z direction.

16. The Holstein–Primakoff transformation between spin one-half operators and bosons is

$$S_l{}^{(z)} = \tfrac{1}{2} - a_l{}^\dagger a_l$$

$$S_l{}^{(+)} = (1 - a_l{}^\dagger a_l)^{1/2} a_l$$

$$S_l{}^{(-)} = a_l{}^\dagger (1 - a_l{}^\dagger a_l)^{1/2}$$

where the a_m are boson operators.

a. Use these definitions to evaluate the following commutators:

$$[S_l{}^{(+)}, S_m{}^{(-)}] : [S_l{}^{(z)}, S_m{}^{(-)}]$$

b. Write out the Heisenberg Hamiltonian in terms of the as.

c. At low temperature, where the number of excitations $\bar{n}_l \leq \langle a_l{}^\dagger a_l \rangle$ is small, one can simplify the Hamiltonian by neglecting all terms of the form $a^\dagger a^\dagger aa$ which describe the scattering of two excitations. Then one can solve the Hamiltonian exactly. What are the eigenvalues?

Chapter 2

Green's Functions at Zero Temperature

Many-body calculations are often done for model systems at zero temperature. Of course, real experimental systems are never at zero temperature, although they are often at low temperature. Many quantities are not very sensitive to temperature, particularly at low temperature. Thus, zero temperature calculations are useful even for describing real systems. Furthermore, the zero temperature property of a system is an important conceptual quantity—the ground state of an interacting system. We often describe a system as its ground state plus its excitations, and the ground state may be deduced from a zero temperature calculation. Many zero temperature calculations have been done to deduce, for example, the ground state of the homogeneous electron gas or the ground state of superfluid ^{4}He.

Some workers believe that zero temperature calculations are easier to perform. There are usually more terms at finite temperature, which is a nuisance. We do not completely share this viewpoint. The Matsubara methods, which are described in Chapter 3, for finite temperature are very easy to use. We usually do zero temperature calculations by first finding the finite temperature formulas and then taking the zero temperature limit. However, sometimes it is easier to do zero temperature calculations from the beginning. The zero temperature formalism is a necessary part of one's calculational machinery.

It is presumed that one is trying to solve a Hamiltonian which can not be solved exactly. One does not need Green's functions if the problem may be solved exactly. Very few exact results were obtained by Green's functions which were not first obtained by conventional theoretical techniques. Thus we assume that one is trying to deduce the properties of some system described by a Hamiltonian H which may not be solved exactly. The usual approach is to set

$$H = H_0 + V$$

where H_0 is a Hamiltonian which may be solved exactly. The term V represents all the remaining parts of H. One tries to chose H_0 so that the effects of V are small. The basic procedure is to start with a system completely described by H_0. The effects of V are introduced, and we then try to find how it changes the system we understand. This is the basic procedure in many-body theory.

2.1. INTERACTION REPRESENTATION

Most readers of this book should be familiar with the interaction representation. A short discussion of this topic will be presented anyway, as a refresher. There are three representations which will be discussed: Schrödinger, Heisenberg, and interaction.

A. Schrödinger

Elementary quantum mechanics is taught in the *Schrödinger representation*, which is based on the formula ($\hbar = 1$)

$$i \frac{\partial}{\partial t} \psi(t) = H\psi(t)$$

which has the operator formal solution

$$\psi(t) = e^{-iHt}\psi(0)$$

The use of this formula requires some assumptions:

1. The wave functions are time dependent, even if this dependence is only a simple factor of $\exp(-iEt)$.
2. Operators such as the Hamiltonian H in the above equation are taken to be independent of time.

B. Heisenberg

It is possible to solve quantum mechanical problems another way which gives exactly the same answers yet uses methods that look quite different. The *Heisenberg representation* has the following properties:

1. The wave functions are independent of time.
2. The operators are time dependent, and this dependence is given by

$$O(t) = e^{iHt}O(0)e^{-iHt}$$

or, equivalently, one is trying to solve the equation which is derived from this:

$$i\frac{\partial}{\partial t}O(t) = [O(t), H]$$

Note that the time dependences of these two representations appear to be contrary. Yet we asserted that they give the same answer. To prove this in detail takes more space than we have here, so we shall just make the equivalency plausible.

In physics one is usually trying to evaluate matrix elements in order to determine transition rates. In the Schrödinger representation, the matrix element of the operator $O(0)$ between two states is

$$\langle\psi_1^{\dagger}(t)O(0)\psi_2(t)\rangle = \langle\psi_1^{\dagger}(0)e^{iHt}O(0)e^{-iHt}\psi_2(0)\rangle$$

In the Heisenberg representation one obtains the result

$$\langle\psi_1^{\dagger}(0)O(t)\psi_2(0)\rangle = \langle\psi_1^{\dagger}(0)e^{iHt}O(0)e^{-iHt}\psi_2(0)\rangle$$

The two representations produce the same result. To see this, first recall that we want to include in the Hamiltonian H some interaction terms. In the Heisenberg representation, these interactions act upon the operators, thereby changing them. When taking matrix elements, we are projecting these changed operators back upon the unchanged states in order to see how much the operators are changed by the interactions. On the other hand, the Schrödinger picture leaves the operators fixed and instead has the interactions affect the wave functions. Then the new wave functions are used to evaluate the matrix elements of the unchanged operators. The result is the same either way.

C. Interaction

The interaction representation is another way of doing things. Here both the wave functions and the operators are time dependent. This is done by separating the Hamiltonian into two parts,

$$H = H_0 + V \tag{2.1.1}$$

where H_0 is the unperturbed part, while the V are the interactions. At the moment we leave the exact form of V unspecified and merely note that it may be either, some, or all of the interactions discussed in Chapter 1. Usually H_0 is selected as a Hamiltonian which is exactly solvable.

Operators and wave functions in the interaction representation will be denoted by a caret. This notation will distinguish their time dependence from that of the other representations:

1. Operators have a time dependence

$$\hat{O}(t) = e^{iH_0t}Oe^{-iH_0t} \tag{2.1.2}$$

2. Wave functions have a time dependence

$$\hat{\psi}(t) = e^{iH_0t}e^{-iHt}\psi(0) \tag{2.1.3}$$

We are assuming that $[H_0, V] \neq 0$. If these operators do commute, the problem is usually too trivial to require many-body theory. When these operators do not commute, the exponentials cannot be combined, because

$$e^Ae^B = e^{A+B}$$

only if $[A, B] = 0$.

Before going any further, we check to see that this choice of time dependence does produce the same matrix elements as before:

$$\begin{aligned}\langle\hat{\psi}_1(t)^\dagger\hat{O}(t)\hat{\psi}_2(t)\rangle &= \langle\psi_1{}^\dagger(0)e^{iHt}e^{-iH_0t}(e^{iH_0t}Oe^{-iH_0t})e^{iH_0t}e^{-iHt}\psi_2(0)\rangle\\ &= \langle\psi_1{}^\dagger(0)e^{iHt}Oe^{-iHt}\psi_2(0)\rangle\end{aligned}$$

It does. The time dependence of the operators is governed by the unperturbed Hamiltonian. Now we wish to show that the time dependence of the wave functions is governed by the interactions

$$\begin{aligned}\frac{\partial}{\partial t}\hat{\psi}(t) &= ie^{iH_0t}(H_0 - H)e^{-iHt}\psi(0)\\ &= -ie^{iH_0t}Ve^{-iH_0t}[e^{iH_0t}e^{-iHt}\psi(0)]\end{aligned}$$

so that

$$\frac{\partial}{\partial t}\hat{\psi}(t) = -i\hat{V}(t)\hat{\psi}(t)$$

This proves the assertion that the time dependence of $\hat{\psi}(t)$ is determined by $\hat{V}(t)$.

We have introduced into Eq. (2.1.3) an operator which we shall define as $U(t)$:

$$U(t) = e^{iH_0t}e^{-iHt} \tag{2.1.4}$$

This function has the value of unity at $t = 0$:

$$U(0) = 1$$

Furthermore, it obeys a differential equation which can be written in the interaction representation:

$$\begin{aligned}\frac{\partial}{\partial t} U(t) &= ie^{iH_0t}(H_0 - H)e^{-iHt} \\ &= -ie^{iH_0t}V(e^{-iH_0t}e^{iH_0t})e^{-iHt} \\ &= -i\hat{V}(t)U(t)\end{aligned}$$

We wish to solve this equation. One way of proceeding is by integrating both sides of the equation with respect to time:

$$U(t) - U(0) = -i\int_0^t dt_1 \hat{V}(t_1)U(t_1)$$

Rearranging gives

$$U(t) = 1 - i\int_0^t dt_1 \hat{V}(t_1)U(t_1)$$

If this equation is repeatedly iterated, we get

$$\begin{aligned}U(t) &= 1 - i\int_0^t dt_1 \hat{V}(t_1) + (-i)^2 \int_0^t dt_1 \int_0^{t_1} dt_2 \hat{V}(t_1)\hat{V}(t_2) + \cdots \\ U(t) &= \sum_{n=0}^{\infty} (-i)^n \int_0^t dt_1 \int_0^{t_1} dt_2 \cdots \int_0^{t_{n-1}} dt_n \hat{V}(t_1)\hat{V}(t_2) \cdots \hat{V}(t_n)\end{aligned} \tag{2.1.5}$$

At this point it is convenient to introduce the time-ordering operator T, which should not be confused with the temperature. The T operator acts upon a group of time-dependent operators,

$$T[\hat{V}(t_1)\hat{V}(t_2)\hat{V}(t_3)]$$

and is just an instruction to arrange the operators with the earliest times to the right. For example,

$$T[\hat{V}(t_1)\hat{V}(t_2)\hat{V}(t_3)] = \hat{V}(t_3)\hat{V}(t_1)\hat{V}(t_2) \qquad \text{if } t_3 > t_1 > t_2$$

It helps to introduce the following step function:

$$\theta(x) = 1 \quad \text{if } x > 0$$
$$= 0 \quad \text{if } x < 0$$
$$= \tfrac{1}{2} \quad \text{if } x = 0$$

Thus for two operators, the explicit definition of T ordering gives

$$T[\hat{V}(t_1)\hat{V}(t_2)] = \theta(t_1 - t_2)\hat{V}(t_1)\hat{V}(t_2) + \theta(t_2 - t_1)\hat{V}(t_2)\hat{V}(t_1)$$

Of course, if $\hat{V}(t_1)$ and $\hat{V}(t_2)$ commuted with each other, then the order of the operators is unimportant. Thus the T ordering needs to be applied only to operators which do not commute at different times. Now we rearrange the integral by using the above identity:

$$\frac{1}{2!}\int_0^t dt_1 \int_0^t dt_2 T[\hat{V}(t_1)\hat{V}(t_2)]$$
$$= \frac{1}{2!}\int_0^t dt_1 \int_0^{t_1} dt_2 \hat{V}(t_1)\hat{V}(t_2) + \frac{1}{2!}\int_0^t dt_2 \int_0^{t_2} dt_1 \hat{V}(t_2)\hat{V}(t_1)$$

The second term on the right-hand side is equal to the first, which is easy to see by just redefining the integration variables $t_1 \to t_2$, $t_2 \to t_1$. Thus we get

$$\frac{1}{2!}\int_0^t dt_1 \int_0^t dt_2 T[\hat{V}(t_1)\hat{V}(t_2)] = \int_0^t dt_1 \int_0^{t_1} dt_2 \hat{V}(t_1)\hat{V}(t_2)$$

Similarly, we can show that

$$\frac{1}{3!}\int_0^t dt_1 \int_0^t dt_2 \int_0^t dt_3 T[\hat{V}(t_1)\hat{V}(t_2)\hat{V}(t_3)]$$
$$= \int_0^t dt_1 \int_0^{t_1} dt_2 \int_0^{t_2} dt_3 \hat{V}(t_1)\hat{V}(t_2)\hat{V}(t_3)$$

Thus if we return to our expansion of $U(t)$, we obtain

$$U(t) = 1 + \sum_{n=1}^{\infty} \frac{(-i)^n}{n!} \int_0^t dt_1 \int_0^t dt_2 \cdots \int_0^t dt_n T[\hat{V}(t_1)\hat{V}(t_2) \cdots \hat{V}(t_n)] \tag{2.1.6}$$

This expansion may be abbreviated by writing it as

$$U(t) = T \exp\left[-i \int_0^t dt_1 \hat{V}(t_1)\right] \tag{2.1.7}$$

However, it should always be kept in mind that the exponential form is really just a shorthand for the series definition (2.1.6). And the T-ordered series definition is really equivalent to our original expansion (2.1.5).

2.2. S MATRIX

We showed in the previous section that the wave function in the interaction representation had a time dependence given by

$$\hat{\psi}(t) = U(t)\hat{\psi}(0)$$

We now define the S matrix as the operator $S(t, t')$ which changes the wave function $\hat{\psi}(t')$ into $\hat{\psi}(t)$:

$$\hat{\psi}(t) = S(t, t')\hat{\psi}(t')$$

From our original definition we have

$$\hat{\psi}(t) = U(t)\hat{\psi}(0) = S(t, t')U(t')\hat{\psi}(0)$$

which produces the result

$$S(t, t') = U(t)U^{\dagger}(t')$$

Let us now examine some properties of this operator. The first two of the following identities may be proved in a trivial way:

1. $S(t, t) = 1 = U(t)U^{\dagger}(t) = e^{iH_0t}e^{-iHt}(e^{+iHt}e^{-iH_0t})$
2. $S^{\dagger}(t, t') = U(t')U^{\dagger}(t) = S(t', t)$
3. $S(t, t')S(t', t'') = S(t, t'')$

The third identity can be shown by appealing to the original definition:

$$\hat{\psi}(t) = S(t, t')\hat{\psi}(t') = S(t, t')S(t', t'')\hat{\psi}(t'') = S(t, t'')\hat{\psi}(t'')$$

Finally, we wish to show that $S(t, t')$ can also be expressed as a time-ordered operator,

$$\frac{\partial}{\partial t} S(t, t') = \frac{\partial}{\partial t} U(t)U^{\dagger}(t') = -i\hat{V}(t)S(t, t')$$

which has the solution

$$S(t, t') = T\left[\exp\left(-i \int_{t'}^{t} dt_1 \hat{V}(t_1)\right)\right] \tag{2.2.1}$$

The function $\hat{\psi}(0) \equiv \psi(0)$ was introduced in the discussion of the previous section. It is a wave function in the Heisenberg representation, so that it is independent of time. If we define a Schrödinger wave function as

$$\psi_s(t) = e^{-iHt}\psi(0)$$

then $\psi(0)$ is also the Schrödinger wave function at $t = 0$. If we define

$$\hat{\psi}(t) = U(t)\psi(0)$$

then $\psi(0)$ is also the $t = 0$ wave function in the interaction representation. At zero temperature the only wave function of special interest is the ground state wave function. For our Green's functions we shall need to define $\psi(0)$ as the exact ground state wave function. Since the total Hamiltonian is H, the exact ground state must have the lowest eigenvalue of this Hamiltonian. This presents an immediate problem because we do not know initially any of the eigenvalues or eigenstates of the Hamiltonian H. In fact, that is exactly the kind of information we are trying to obtain by using Green's functions.

Thus we have a problem in that all our formalism is based on the wave function $\psi(0)$ which we do not yet know. Now in the interaction representation we set $H = H_0 + V$, where we have chosen H_0 to be sufficiently simple that we know its eigenvalues and eigenstates. Let the lowest eigenstate of H_0—its ground state—be denoted ϕ_0. Somehow we have to determine the unknown wave function $\psi(0)$ in terms of the known wave function ϕ_0.

The relationship between the two ground states $\psi(0)$ and ϕ_0 at zero temperature was established by Gell-Mann and Low (1951):

$$\psi(0) = S(0, -\infty)\phi_0 \tag{2.2.2}$$

We shall not try to prove this, but instead we shall try to make it plausible. One result we do know is that

$$\hat{\psi}(t) = S(t, 0)\psi(0)$$

We operate by $S(0, t)$ and get

$$\psi(0) = S(0, t)\hat{\psi}(t)$$

since $S(0, t)S(t, 0) = 1$ by the use of our previous theorems. If we let $t \to -\infty$, we get

$$\psi(0) = S(0, -\infty)\hat{\psi}(-\infty) \tag{2.2.3}$$

Thus we are really asserting that $\hat{\psi}(-\infty)$ is equal to ϕ_0. The traditional argument is that one starts in the dim past ($t \to -\infty$) with a wave function ϕ_0 which does not contain the effects of the interaction V. The operator $S(0, -\infty)$ brings this wave function adiabatically up to the present $t = 0$. Thus now we have a wave function which does contain the effects of the interaction V, so that it is an eigenstate of H.

There is an additional property of these states which we shall need for our discussion of Green's functions. As $t \to +\infty$, we get

$$\hat{\psi}(\infty) = S(\infty, 0)\psi(0)$$

One possible assumption is that $\hat{\psi}(\infty)$ must be related to ϕ_0. If they are equal except for a phase factor:

$$\phi_0 e^{iL} = \hat{\psi}(\infty) = S(\infty, 0)\psi(0) = S(\infty, -\infty)\phi_0 \tag{2.2.4}$$

Alternatives to this assumption are discussed in Sec. 2.9.

$$e^{iL} = \langle \phi_0 \mid S(\infty, -\infty) \mid \phi_0 \rangle \tag{2.2.5}$$

2.3. GREEN'S FUNCTIONS

For most of this book we shall be concerned with only three types of Green's functions: electron, phonon, and photon. The electron and phonon cases will be discussed here. The photon case is badly muddled by the several available choices of electromagnetic gauge. Section 2.10 is devoted to discussing these complications, and the photon Green's functions will be treated then. But one should keep in mind that the basic proofs for the photon case are essentially identical to the phonon case.

At zero temperature the electron Green's function is defined as

$$G(\lambda, t - t') = -i\langle | TC_\lambda(t)C_\lambda^\dagger(t') | \rangle \tag{2.3.1}$$

The quantum number λ can be anything depending on the problem of interest. Quite often we shall take it to be the quantum numbers of the free-electron gas $\lambda = (\mathbf{p}, \sigma)$. At zero temperature the state $|\rangle$ must be the ground state. If we have chosen the Hamiltonian of the problem to be H, then $|\rangle$ is the ground state of H, and therefore it is an eigenstate of H.

Of course we do not initially know the ground state or any other eigenstate of H since that is what we are trying to determine by using Green's functions. Thus we write $H = H_0 + V$, where H_0 is the unperturbed part, while V is the interaction. We choose H_0 so that we know its eigenstates. In Sec. 1.3 we defined the C_λ in terms of a complete set of states ψ_λ. We now select this complete set of states as the eigenstates of the unperturbed Hamiltonian H_0. Thus in the definition of the Green's function the C_λ represent states of H_0, while the ground state $|\rangle$ is an eigenstate of H. Furthermore, (2.3.1) is defined in the Heisenberg representation, so that $|\rangle$ is independent of time, while $C_\lambda(t)$ is given by the usual result

$$C_\lambda(t) = e^{iHt}C_\lambda e^{-iHt}.$$

One way of understanding the Green's function is to observe that it describes a certain Gedanken experiment. For $t > t'$ we have

$$G(\lambda, t > t') = -i\langle| C_\lambda(t)C_\lambda^\dagger(t') |\rangle$$

Here one takes the real ground state, and at a time t' one creates an excitation λ. At a later time t one destroys the same excitation. Now if λ were an eigenstate of H, with $HC_\lambda^\dagger |\rangle = \varepsilon_\lambda C_\lambda^\dagger |\rangle$ and $H |\rangle = \varepsilon_0 |\rangle$, then this state would propagate with a simple exponential time dependence:

$$G(\lambda, t > t') = -ie^{-i(t-t')(\varepsilon_\lambda - \varepsilon_0)}$$

Because λ is not usually an eigenstate of H, the particle in the state λ gets scattered, shifted in energy, etc., during the time interval $t - t'$. Thus when one measures at a later time t to see how much amplitude is left in the state λ, the measurement provides information about the system.

For the other time arrangement $t' > t$ we have

$$G(\lambda, t' > t) = +i\langle| C_\lambda^\dagger(t')C_\lambda(t) |\rangle$$

where we have to change the sign whenever we interchange the position of two fermion operators. Now an electron is destroyed from the ground state at time t and recreated at the later time t', which is possible only if there are electrons in the ground state at zero temperature. One case where this is true is in the Fermi sea of a metal. The initial destruction of an electron at time t must, roughly speaking, remove an electron from the filled Fermi sea. This destruction creates a vacancy, often called a *hole*, and the hole can interact and scatter in the interval $t - t'$. Then $C_\lambda^\dagger$ acting at t' destroys the hole state λ, and this measurement provides information about the hole excitation.

We now wish to take the Green's function, defined in the Heisenberg representation, and convert it to the interaction representation. We shall call $|\rangle_0 \equiv \phi_0$ the ground state of H_0, so that

$$|\rangle = S(0, -\infty)\,|\rangle_0$$

Next we change the operators to this representation:

$$\begin{aligned} C_\lambda(t) &= e^{iHt}e^{-iH_0t}\hat{C}_\lambda(t)e^{iH_0t}e^{-iHt} = U^\dagger(t)\hat{C}_\lambda(t)U(t) \\ &= S(0, t)\hat{C}_\lambda(t)S(t, 0) \end{aligned}$$

$$\begin{aligned} G(\lambda, t - t') = {} & -i\theta(t - t')\,{}_0\langle|\, S(-\infty, 0)S(0, t)\hat{C}_\lambda(t)S(t, 0)S(0, t')\hat{C}_\lambda^\dagger(t') \\ & \times S(t', 0)S(0, -\infty)\,|\rangle_0 + i\theta(t' - t)\,{}_0\langle|\, S(-\infty, 0)S(0, t') \\ & \times \hat{C}_\lambda^\dagger(t')S(t', 0)S(0, t)\hat{C}_\lambda(t)S(t, 0)S(0, -\infty)\,|\rangle_0 \end{aligned}$$

This expression may be regrouped by using the properties of the S matrix developed in the previous section. We also replace the left-hand bracket by

$${}_0\langle|\, S(-\infty, 0) = e^{-iL}\,{}_0\langle|\, S(\infty, -\infty)S(-\infty, 0) = \frac{{}_0\langle|\, S(\infty, 0)}{{}_0\langle|\, S(\infty, -\infty)\,|\rangle_0}$$

so that the Green's function becomes

$$\begin{aligned} G(\lambda, t - t') = {} & \frac{-i}{{}_0\langle|\, S(\infty, -\infty)\,|\rangle_0}\,[\theta(t - t')\,{}_0\langle|\, S(\infty, t)\hat{C}_\lambda(t)S(t, t')\hat{C}_\lambda^\dagger(t') \\ & \times S(t', -\infty)\,|\rangle_0 - \theta(t' - t)\,{}_0\langle|\, S(\infty, t')\hat{C}_\lambda^\dagger(t')S(t', t) \\ & \times \hat{C}_\lambda(t)S(t, -\infty)\,|\rangle_0] \end{aligned}$$

The first term can be simplified by writing

$$\begin{aligned} & \theta(t - t')_0\langle|\, S(\infty, t)\hat{C}_\lambda(t)S(t, t')\hat{C}_\lambda^\dagger(t')S(t', -\infty)\,|\rangle_0 \\ & \quad = \theta(t - t')_0\langle|\, T\hat{C}_\lambda(t)\hat{C}_\lambda^\dagger(t')S(\infty, -\infty)\,|\rangle_0 \end{aligned}$$

The operator $S(\infty, -\infty)$ contains operators which act in the three time intervals (∞, t), (t, t'), and $(t', -\infty)$. The T operator automatically sorts these so that they act in their proper sequences, which are, respectively, to the left of $\hat{C}_\lambda(t)$, between $\hat{C}_\lambda(t)$ and $\hat{C}_\lambda^\dagger(t')$, and to the right of $\hat{C}_\lambda^\dagger(t')$. We can express the total Green's function as

$$G(\lambda, t - t') = \frac{-i\,{}_0\langle|\, T\hat{C}_\lambda(t)\hat{C}_\lambda^\dagger(t')S(\infty, -\infty)\,|\rangle_0}{{}_0\langle|\, S(\infty, -\infty)\,|\rangle_0} \tag{2.3.2}$$

which is the desired result. It does not matter where we write $S(\infty, -\infty)$

in the numerator, since the time ordering operator puts the pieces in the right place.

A Green's function can also be defined for the special case where the interactions $V = 0$ and hence the S matrix is unity. This Green's function plays a special role in the formalism, and we designate it by $G^{(0)}$:

$$G^{(0)}(\lambda, t - t') = -i_0\langle| T\hat{C}_\lambda(t)\hat{C}_\lambda^\dagger(t') |\rangle_0$$

$G^{(0)}$ is often called the *unperturbed Green's function*, or sometimes the name free propagator is used.

There are two quite different types of electronic systems in which we want to employ the Green's function analysis. These two have quite different ground states, $|\rangle_0$, and also quite different real ground states, $|\rangle$. These two systems are the following.

1. *An Empty Band*. Here we wish to study the properties of an electron in an energy band in which it is the only electron. An example is when we put an electron in the conduction band of a semiconductor or an insulator. In this case the ground state is the particle vacuum, which we denote as $| 0\rangle$. This state has the property that

$$C_{\mathbf{p}} \,|\, 0\rangle = 0$$
$$a_{\mathbf{q}} \,|\, 0\rangle = 0$$

Therefore both H_0 and V give zero when operating upon the vacuum. It follows that the S matrix also gives unity when operating upon the vacuum:

$$S(t, -\infty) \,|\, 0\rangle = |\, 0\rangle$$

This means that both of the ground states, $|\rangle_0$ and $|\rangle$, are the vacuum. The Green's function can exist only for the time ordering

$$G(\lambda, t - t') = -i\theta(t - t')\langle C_\lambda(t)C_\lambda^\dagger(t')\rangle$$

The unperturbed Green's function $G^{(0)}$ is particularly easy to evaluate:

$$\begin{aligned} G^{(0)}(\lambda, t - t') &= -i\theta(t - t')e^{-i\varepsilon_\lambda(t-t')}\langle 0 \,|\, C_\lambda C_\lambda^\dagger \,|\, 0\rangle \\ &= -i\theta(t - t')e^{-i\varepsilon_\lambda(t-t')} \end{aligned}$$

The Fourier transform of $G(\lambda, t)$ is defined as

$$G(\lambda, E) = \int_{-\infty}^{\infty} dt e^{iEt} G(\lambda, t)$$

To make the integrals converge, we need to add the infinitesimal quantity $i\delta$ to the exponents:

$$G^{(0)}(\lambda, E) = -i \int_0^\infty dt e^{it(E-\varepsilon_\lambda+i\delta)}$$
$$G^{(0)}(\lambda, E) = \frac{1}{E - \varepsilon_\lambda + i\delta} \tag{2.3.3}$$

2. *A Degenerate Electron Gas.* Our second case is where the electrons are in a Fermi sea at zero temperature. The standard example is a simple metal. The system has a chemical potential μ, and all electron states with $E < \mu$ are occupied. If the unperturbed electrons (eigenstates of H_0) are characterized by an energy $\varepsilon_{\mathbf{k}}$, the ground state $|\rangle_0$ has all states $\varepsilon_{\mathbf{k}} < \mu$ filled and states $\varepsilon_{\mathbf{k}} > \mu$ empty. It is convenient and conventional to measure the electron's energy relative to the chemical potential, to define $\xi_{\mathbf{k}} = \varepsilon_{\mathbf{k}} - \mu$. For a spherical Fermi surface with Fermi wave vector p_F, we obtain

$${}_0\langle| C_{\mathbf{k}}^\dagger C_{\mathbf{k}} |\rangle_0 = \theta(p_F - k)$$
$${}_0\langle| C_{\mathbf{k}} C_{\mathbf{k}}^\dagger |\rangle_0 = \theta(k - p_F)$$

A more general way to write this is

$${}_0\langle| C_{\mathbf{k}}^\dagger C_{\mathbf{k}} |\rangle_0 = \theta(-\xi_{\mathbf{k}}) = \lim_{\beta\to\infty} \frac{1}{e^{\beta\xi_{\mathbf{k}}} + 1} \equiv n_F(\xi_{\mathbf{k}})$$

The unperturbed Green's function is now

$$\begin{aligned} G^{(0)}(\mathbf{k}, t - t') &= -i {}_0\langle| T C_{\mathbf{k}}(t) C_{\mathbf{k}}^\dagger(t') |\rangle_0 \\ &= -i[\theta(t - t')\theta(\xi_{\mathbf{k}}) - \theta(t' - t)\theta(-\xi_{\mathbf{k}})]e^{-i\xi_{\mathbf{k}}(t-t')} \end{aligned}$$

The Fourier transform is

$$G^{(0)}(\mathbf{k}, E) = -i\left[\theta(\xi_{\mathbf{k}}) \int_0^\infty dt e^{it(E-\xi_{\mathbf{k}}+i\delta)} - \theta(-\xi_{\mathbf{k}}) \int_{-\infty}^0 dt e^{it(E-\xi_{\mathbf{k}}-i\delta)}\right]$$
$$G^{(0)}(\mathbf{k}, E) = \frac{\theta(\xi_{\mathbf{k}})}{E - \xi_{\mathbf{k}} + i\delta} + \frac{\theta(-\xi_{\mathbf{k}})}{E - \xi_{\mathbf{k}} - i\delta} \tag{2.3.4}$$

The energy $\xi_{\mathbf{k}} = \varepsilon_{\mathbf{k}} - \mu$ is measured with respect to the Fermi surface $\xi_{\mathbf{k}_F} = 0$, $\varepsilon_{\mathbf{k}_F} = \mu$. Another way to write $G^{(0)}$ is

$$G^{(0)}(\mathbf{k}, E) = \frac{1}{E - \xi_{\mathbf{k}} + i\delta_{\mathbf{k}}}$$
$$\delta_{\mathbf{k}} = \delta \operatorname{sgn} \xi_{\mathbf{k}} \tag{2.3.5}$$

where $\delta_{\mathbf{k}}$ is a small infinitesimal part which changes sign at the chemical potential.

3. *Phonons.* The Green's function for phonons is defined as

$$D(\mathbf{q}, \lambda; t - t') = -i\langle| TA_{\mathbf{q},\lambda}(t)A_{-\mathbf{q},\lambda}(t') |\rangle$$
$$A_{\mathbf{q},\lambda} = a_{\mathbf{q}\lambda} + a^{\dagger}_{-\mathbf{q},\lambda}$$

The subscripts λ refer to the polarization of the phonons. Usually we are interested in just one kind of phonon with Hamiltonians which do not mix polarizations, so we shall omit these subscripts entirely. In the interaction representation one obtains the result

$$D(\mathbf{q}; t - t') = \frac{-i\, {}_0\langle| T\hat{A}_{\mathbf{q}}(t)\hat{A}_{-\mathbf{q}}(t')S(\infty, -\infty) |\rangle_0}{{}_0\langle| S(\infty, -\infty) |\rangle_0}$$

At zero temperature there are no phonons. Thus the ground states $|\rangle$ and $|\rangle_0$ are again the particle vacuum $|0\rangle$. Note that in an electron–phonon system the notation $|\rangle_0$ means the combination of ground states for electrons, phonons, etc. Although the phonon system has the vacuum as its ground state, either of the two electron ground states can be used.

The unperturbed phonon Green's function is defined as

$$D^{(0)}(\mathbf{q}, t - t') = -i\, {}_0\langle| T\hat{A}_{\mathbf{q}}(t)\hat{A}_{-\mathbf{q}}(t') |\rangle_0$$
$$= -i\, {}_0\langle| T(a_{\mathbf{q}}e^{-i\omega_{\mathbf{q}}t} + a^{\dagger}_{-\mathbf{q}}e^{i\omega_{\mathbf{q}}t})(a_{-\mathbf{q}}e^{-i\omega_{\mathbf{q}}t'} + a_{\mathbf{q}}^{\dagger}e^{i\omega_{\mathbf{q}}t'}) |\rangle_0$$

At zero temperature

$${}_0\langle| a_{\mathbf{q}}a_{\mathbf{q}}^{\dagger} |\rangle_0 = 1$$
$${}_0\langle| a_{\mathbf{q}}^{\dagger}a_{\mathbf{q}} |\rangle_0 = 0$$

and we have

$$D^{(0)}(\mathbf{q}, t - t') = -i[\theta(t - t')e^{-i\omega_{\mathbf{q}}(t-t')} + \theta(t' - t)e^{i\omega_{\mathbf{q}}(t-t')}]$$

The Fourier transform gives

$$D^{(0)}(\mathbf{q}, \omega) = \int_{-\infty}^{\infty} dt e^{i\omega t}D^{(0)}(\mathbf{q}, t)$$

$$D^{(0)}(\mathbf{q}, \omega) = \frac{1}{\omega - \omega_{\mathbf{q}} + i\delta} - \frac{1}{\omega + \omega_{\mathbf{q}} - i\delta} \tag{2.3.6}$$

$$D^{(0)}(\mathbf{q}, \omega) = \frac{2\omega_{\mathbf{q}}}{\omega^2 - \omega_{\mathbf{q}}^2 + i\delta}$$

Sometimes it is useful to have the phonon Green's function at finite temperature. In this case, the thermal average is taken of the phonon occupation numbers,

$$\langle| a_{\mathbf{q}} a_{\mathbf{q}}^{\dagger} |\rangle = N_{\mathbf{q}} + 1$$

$$\langle| a_{\mathbf{q}}^{\dagger} a_{\mathbf{q}} |\rangle = N_{\mathbf{q}} = \frac{1}{e^{\beta\omega_{\mathbf{q}}} - 1}$$

and the Green's function of time is

$$D^{(0)}(\mathbf{q}, t - t') = -i[(N_{\mathbf{q}} + 1)e^{-i\omega_{\mathbf{q}}|t-t'|} + N_{\mathbf{q}} e^{i\omega_{\mathbf{q}}|t-t'|}]$$

In some many-body systems the interactions cause changes in the phonon energies. For these cases, the finite temperature thermal average should be over the frequencies which result from the interactions. The use of the unperturbed frequencies is valid only if they are not altered; this happens when the perturbation is either localized or confined to a small number of particles.

2.4. WICK'S THEOREM

The Green's function is evaluated by expanding the S matrix $S(\infty, -\infty)$ in (2.2.1) in a series such as (2.1.6):

$$G(\mathbf{p}, t - t') = \sum_{n=0} \frac{(-i)^{n+1}}{n!} \int_{-\infty}^{\infty} dt_1 \cdots \int_{-\infty}^{\infty} dt_n \times \frac{{}_0\langle| T\hat{C}_{\mathbf{p}}(t)\hat{V}(t_1)\hat{V}(t_2) \cdots \hat{V}(t_n)\hat{C}_{\mathbf{p}}^{\dagger}(t') |\rangle_0}{{}_0\langle| S(\infty, -\infty) |\rangle_0} \tag{2.4.1}$$

Let us, for the moment, ignore the factor ${}_0\langle| S(\infty, -\infty) |\rangle_0$. We shall take care of it in Sec. 2.6. Our immediate aim is to learn how to evaluate time-ordered brackets like

$${}_0\langle| T\hat{C}_{\mathbf{p}}(t)\hat{V}(t_1)\hat{V}(t_2)\hat{V}(t_3)\hat{C}_{\mathbf{p}}^{\dagger}(t') |\rangle_0 \tag{2.4.2}$$

Suppose that $\hat{V}(t_1)$ is the electron–electron interaction (1.2.11):

$$\hat{V}(t_1) = \frac{1}{2} \sum_{\mathbf{k}'\mathbf{k},\mathbf{q}} \frac{4\pi e^2}{q^2} C_{\mathbf{k}+\mathbf{q}}^{\dagger} C_{\mathbf{k}'-\mathbf{q}}^{\dagger} C_{\mathbf{k}'} C_{\mathbf{k}} e^{it_1(\xi_{\mathbf{k}+\mathbf{q}} + \xi_{\mathbf{k}'-\mathbf{q}} - \xi_{\mathbf{k}} - \xi_{\mathbf{k}'})}$$

In this case the time-ordered bracket (2.4.2) contains seven creation operators and seven destruction operators. It is a very arduous task to evaluate this bracket: There are many possible time orderings and many possible pairings between creation and destruction operators. What is meant by *pairings* between operators? First note that these brackets always contain the same number of creation and destruction operators. Thus one is always trying to evaluate the product of n creation operators and n destruction operators between the ground state $|\rangle_0$:

$$_0\langle| T\hat{C}_1(t_1)\hat{C}_{1'}^\dagger(t_2) \cdots \hat{C}_n(t_n)\hat{C}_{n'}^\dagger(t') |\rangle_0$$

Now the effect of a creation operator $\hat{C}_{n'}^\dagger(t')$ is to put an electron into the state n'. So that the system will be back in the ground state before the final operator by $_0\langle|$, one of the destruction operators $\hat{C}_m(t)$ must destroy the state n' so that $m = n'$ for some m. For example,

$$_0\langle| T\hat{C}_\alpha(t)\hat{C}_\beta^\dagger(t') |\rangle_0$$

equals zero unless $\alpha = \beta$, while

$$_0\langle| T\hat{C}_\alpha(t)\hat{C}_\beta^\dagger(t_1)\hat{C}_\gamma(t_2)\hat{C}_\delta^\dagger(t') |\rangle_0$$

equals zero unless $\alpha = \beta$ and $\gamma = \delta$ *or* unless $\alpha = \delta$ and $\beta = \lambda$. In spite of the many possible time orderings and pairings in a bracket like (2.4.2), only a limited number of these combinations are physically interesting. Our aim is to sort these in a simple way, which is achieved with the help of some theorems which simplify the procedures. The first of these is Wick's theorem.

This theorem is really just an observation that the time ordering can be taken care of in a simple way. It states that in making all the possible pairings between creation and destruction operators each pairing should be time-ordered. The time ordering of each pair gives the proper time ordering to the entire result. For example, we get

$$\begin{aligned}
&_0\langle| T\hat{C}_\alpha(t)\hat{C}_\beta^\dagger(t_1)\hat{C}_\gamma(t_2)\hat{C}_\delta^\dagger(t') |\rangle_0 \\
&\quad = {}_0\langle| T\hat{C}_\alpha(t)\hat{C}_\beta^\dagger(t_1) |\rangle_0\, {}_0\langle| T\hat{C}_\gamma(t_2)\hat{C}_\delta^\dagger(t') |\rangle_0 \\
&\quad\quad - {}_0\langle| T\hat{C}_\alpha(t)\hat{C}_\delta^\dagger(t') |\rangle_0\, {}_0\langle| T\hat{C}_\gamma(t_2)\hat{C}_\beta^\dagger(t_1) |\rangle_0 \\
&\quad = +\, \delta_{\alpha\beta}\delta_{\gamma\delta}\, {}_0\langle| T\hat{C}_\alpha(t)\hat{C}_\alpha^\dagger(t_1) |\rangle_0\, {}_0\langle| T\hat{C}_\gamma(t_2)\hat{C}_\gamma^\dagger(t') |\rangle_0 \\
&\quad\quad - \delta_{\alpha\delta}\delta_{\gamma\beta}\, {}_0\langle| T\hat{C}_\alpha(t)\hat{C}_\alpha^\dagger(t') |\rangle_0\, {}_0\langle| T\hat{C}_\gamma(t_2)\hat{C}_\gamma^\dagger(t_1) |\rangle_0
\end{aligned}$$

Note that there is a time-ordering operator T in each pairing bracket. For $n = 3$ creation and destruction operators there are six possible pairings; for n operators of each kind there are obviously $n!$ possible pairings.

A few simple rules should be kept in mind when making these pairings. The first is that a sign change occurs each time the positions of two neighboring Fermi operators are interchanged. Thus one keeps count of the number of interchanges needed to achieve the desired pairing. An odd number of interchanges is the origin of the minus sign in the second term of the example above.

The second rule concerns the time ordering of combinations of operators representing different excitations. For example, consider the following mixture of phonon and electron operators:

$$_0\langle| T\hat{C}_{\mathbf{p}}(t)\hat{C}^\dagger_{\mathbf{p}_1}(t_1)\hat{A}_{\mathbf{q}_1}(t_1)\hat{C}_{\mathbf{p}_2}(t_2)\hat{C}^\dagger_{\mathbf{p}_3}(t_3)\hat{A}_{\mathbf{q}_2}(t_2)\ |\rangle_0$$

Because electron operators commute with phonon operators, we do not care how they are ordered with respect to each other. Thus we can immediately factor the bracket into separate electron and phonon parts:

$$_0\langle 0\ |\ T\hat{C}_{\mathbf{p}}(t)\hat{C}^\dagger_{\mathbf{p}_1}(t_1)\hat{C}_{\mathbf{p}_2}(t_2)\hat{C}^\dagger_{\mathbf{p}_3}(t_3)\ |\ 0\rangle_0\ {}_0\langle|\ T\hat{A}_{\mathbf{q}_1}(t_1)\hat{A}_{\mathbf{q}_2}(t_2)\ |\rangle_0$$

This separation is always possible with different kinds of operators, i.e., whenever operators commute. Wick's theorem also applies to brackets of phonon operators; for example,

$$\begin{aligned}
&_0\langle|\ T\hat{A}_{\mathbf{q}_1}(t_1)\hat{A}_{\mathbf{q}_2}(t_2)\hat{A}_{\mathbf{q}_3}(t_3)\hat{A}_{\mathbf{q}_4}(t_4)|\ \rangle_0 \\
&\quad = {}_0\langle|\ T\hat{A}_{\mathbf{q}_1}(t_1)\hat{A}_{\mathbf{q}_2}(t_2)\ |\rangle_0\ {}_0\langle|\ T\hat{A}_{\mathbf{q}_3}(t_3)\hat{A}_{\mathbf{q}_4}(t_4)\ |\rangle_0 \\
&\qquad + {}_0\langle|\ T\hat{A}_{\mathbf{q}_1}(t_1)\hat{A}_{\mathbf{q}_3}(t_3)\ |\rangle_0\ {}_0\langle|\ T\hat{A}_{\mathbf{q}_2}(t_2)\hat{A}_{\mathbf{q}_4}(t_4)\ |\rangle_0 \\
&\qquad + {}_0\langle|\ T\hat{A}_{\mathbf{q}_1}(t_1)\hat{A}_{\mathbf{q}_4}(t_4)\ |\rangle_0\ {}_0\langle|\ T\hat{A}_{\mathbf{q}_2}(t_2)\hat{A}_{\mathbf{q}_3}(t_3)\ |\rangle_0
\end{aligned}$$

and each bracket vanishes unless the two wave vectors are equal and opposite:

$$\begin{aligned}
&= \delta_{\mathbf{q}_1+\mathbf{q}_2=0}\delta_{\mathbf{q}_2+\mathbf{q}_4=0}\ {}_0\langle|\ T\hat{A}_{\mathbf{q}_1}(t_1)\hat{A}_{-\mathbf{q}_1}(t_2)\ |\rangle_0\ {}_0\langle|\ T\hat{A}_{\mathbf{q}_3}(t_3)\hat{A}_{-\mathbf{q}_3}(t_4)\ |\rangle_0 \\
&\quad + \delta_{\mathbf{q}_1+\mathbf{q}_3=0}\delta_{\mathbf{q}_2+\mathbf{q}_4=0}\ {}_0\langle|\ T\hat{A}_{\mathbf{q}_1}(t_1)\hat{A}_{-\mathbf{q}_1}(t_3)\ |\rangle_0\ {}_0\langle|\ T\hat{A}_{\mathbf{q}_2}(t_2)\hat{A}_{-\mathbf{q}_2}(t_4)\ |\rangle_0 \\
&\quad + \delta_{\mathbf{q}_1+\mathbf{q}_4=0}\delta_{\mathbf{q}_2+\mathbf{q}_3=0}\ {}_0\langle|\ T\hat{A}_{\mathbf{q}_1}(t_1)\hat{A}_{-\mathbf{q}_1}(t_4)\ |\rangle_0\ {}_0\langle|\ T\hat{A}_{\mathbf{q}_2}(t_2)\hat{A}_{-\mathbf{q}_2}(t_3)\ |\rangle_0
\end{aligned}$$

The third rule we need is a method of treating the "time ordering" of two operators which occur at the same time, such as

$$_0\langle|\ T\hat{C}^\dagger_{\mathbf{k}_1}(t_1)\hat{C}_{\mathbf{k}_2}(t_1)\ |\rangle_0$$

In these cases the destruction operator always goes to the right,

$$= \delta_{\mathbf{k}_1=\mathbf{k}_2}\ {}_0\langle|\ \hat{C}^\dagger_{\mathbf{k}_1}(t_1)\hat{C}_{\mathbf{k}_1}(t_1)\ |\rangle_0 = \delta_{\mathbf{k}_1=\mathbf{k}_2} n_F(\xi_{\mathbf{k}_1})$$

and the term is just the number operator which is independent of time. This convention is dependent on the way in which we wrote down the Hamiltonian. In constructing H we were careful to put the destruction operators to the right of the creation operators in all terms in the Hamiltonian.

When two electron operators have *different* time arguments in a pairing, we conventionally put the creation operator to the right:

$$_0\langle| T\hat{C}_{\mathbf{k}_1}(t_1)\hat{C}^\dagger_{\mathbf{k}_2}(t_2) |\rangle_0 = \delta_{\mathbf{k}_1=\mathbf{k}_2}\, {}_0\langle| T\hat{C}_{\mathbf{k}_1}(t_1)\hat{C}^\dagger_{\mathbf{k}_1}(t_2) |\rangle_0$$

This term can be immediately identified as the unperturbed Green's function $iG^{(0)}(\mathbf{k}_1, t_1 - t_2)$. Our previous examples can also be written in terms of Green's functions:

$$\begin{aligned}
&_0\langle| T\hat{C}_\alpha(t)\hat{C}_\beta{}^\dagger(t_1)\hat{C}_\gamma(t_2)\hat{C}_\delta{}^\dagger(t') |\rangle_0 \\
&\quad = (i)^2\delta_{\alpha,\beta}\delta_{\gamma,\delta}G^{(0)}(\alpha, t - t_1)G^{(0)}(\gamma, t_2 - t') \\
&\quad\quad - (i)^2\delta_{\alpha,\delta}\delta_{\gamma,\beta}G^{(0)}(\alpha, t - t')G^{(0)}(\gamma, t_2 - t_1) \\
&(-i)^3\, {}_0\langle| T\hat{A}_{\mathbf{q}_1}(t_1)\hat{A}_{\mathbf{q}_2}(t_2)\hat{A}_{\mathbf{q}_3}(t_3)\hat{A}_{\mathbf{q}_4}(t_4) |\rangle_0 \\
&\quad = \delta_{\mathbf{q}_1=-\mathbf{q}_2}\delta_{\mathbf{q}_3=-\mathbf{q}_4}D^{(0)}(\mathbf{q}_1, t_1 - t_2)D^{(0)}(\mathbf{q}_3, t_3 - t_4) \\
&\quad\quad + \delta_{\mathbf{q}_1=-\mathbf{q}_3}\delta_{\mathbf{q}_2=-\mathbf{q}_4}D^{(0)}(\mathbf{q}_1, t_1 - t_3)D^{(0)}(\mathbf{q}_2, t_2 - t_4) \\
&\quad\quad + \delta_{\mathbf{q}_1=-\mathbf{q}_4}\delta_{\mathbf{q}_2=-\mathbf{q}_3}D^{(0)}(\mathbf{q}_1, t_1 - t_4)D^{(0)}(\mathbf{q}_2, t_2 - t_3)
\end{aligned}$$

In summary, Wick's theorem tells us that a time-ordered bracket may be evaluated by expanding it into all possible pairings and that each of these pairings will be a Green's function or a number operator n_F or n_B.

We shall now do a comprehensive example. We shall consider the $n = 2$ term in the S matrix expansion in (2.4.1). The $n = 0$ term is always $G^{(0)}$:

$$\begin{aligned}
G(\mathbf{p}, t - t') &= G^{(0)}(\mathbf{p}, t-t') + (-i)^2 \int_{-\infty}^{\infty} dt_1\, {}_0\langle T\hat{C}_\mathbf{p}(t)\hat{V}(t_1)\hat{C}_\mathbf{p}{}^\dagger(t')\rangle_0 \\
&\quad + \frac{(-i)^3}{2!}\int_{-\infty}^{\infty} dt_1 \int_{-\infty}^{\infty} dt_2\, {}_0\langle TC_\mathbf{p}(t)\hat{V}(t_1)\hat{V}(t_2)C_\mathbf{p}{}^\dagger(t')\rangle_0
\end{aligned}$$

The interaction will be taken as the electron–phonon interaction:

$$V = \sum_{\mathbf{q},\mathbf{k}} M_\mathbf{q} A_\mathbf{q} C^\dagger_{\mathbf{k}+\mathbf{q}} C_\mathbf{k} \tag{2.4.3}$$

First note that the $n = 1$ term must vanish because it contains the factor

$$_0\langle| T\hat{A}_\mathbf{q} |\rangle_0$$

which is zero since the factors $\langle 0 \mid a_q \mid 0\rangle$ and $\langle 0 \mid a_q^\dagger \mid 0\rangle$ are zero. Similarly, all the terms where n is odd vanish because their time-ordered bracket for phonons contains an odd number of $A_{\mathbf{q}}$ factors. Only the terms even in n contribute to the S-matrix expansion for the electron–phonon interaction. We obtain

$$\begin{aligned} G(\mathbf{p}, t - t') &= G^{(0)}(\mathbf{p}, t - t') + \frac{(-i)^3}{2!} \int_{-\infty}^{\infty} dt_1 \int_{-\infty}^{\infty} dt_2 \\ &\times \sum_{\mathbf{q}_1\mathbf{q}_2} M_{\mathbf{q}_1} M_{\mathbf{q}_2} \langle T\hat{A}_{\mathbf{q}_1}(t_1)\hat{A}_{\mathbf{q}_2}(t_2)\rangle_0 \\ &\times \sum_{\mathbf{k}_1\mathbf{k}_2} {}_0\langle T\hat{C}_{\mathbf{p}}(t)\hat{C}^\dagger_{\mathbf{k}_1+\mathbf{q}_1}(t_1)\hat{C}_{\mathbf{k}_1}(t_1)\hat{C}^\dagger_{\mathbf{k}_2+\mathbf{q}_2}(t_2)\hat{C}_{\mathbf{k}_2}(t_2)\hat{C}^\dagger_{\mathbf{p}}(t')\rangle_0 \\ &+ \cdots \end{aligned}$$

The phonon bracket gives a single-phonon Green's function:

$$-i\, {}_0\langle T \mid \hat{A}_{\mathbf{q}_1}(t_1)\hat{A}_{\mathbf{q}_2}(t_2)\rangle_0 = \delta_{-\mathbf{q}_1=\mathbf{q}_2} D^{(0)}(\mathbf{q}_1, t_1 - t_2)$$

The electron bracket, unfortunately, has six possible combinations of pairings. We shall give these six terms and use the fact that $\mathbf{q}_1 = -\mathbf{q}_2$. Wick's theorem gives the result

$$\begin{aligned} &{}_0\langle T\hat{C}_{\mathbf{p}}(t)\hat{C}^\dagger_{\mathbf{k}_1+\mathbf{q}_1}(t_1)\hat{C}_{\mathbf{k}_1}(t_1)\hat{C}^\dagger_{\mathbf{k}_2+\mathbf{q}_2}(t_2)\hat{C}_{\mathbf{k}_2}(t_2)\hat{C}_{\mathbf{p}}^\dagger(t')\rangle_0 \\ &\quad = {}_0\langle T\hat{C}_{\mathbf{p}}(t)\hat{C}^\dagger_{\mathbf{k}_1+\mathbf{q}_1}(t_1)\rangle_0\, {}_0\langle T\hat{C}_{\mathbf{k}_1}(t_1)\hat{C}^\dagger_{\mathbf{k}_2+\mathbf{q}_2}(t_2)\rangle_0\, {}_0\langle T\hat{C}_{\mathbf{k}_2}(t_2)\hat{C}_p^\dagger(t')\rangle_0 \\ &\qquad + {}_0\langle T\hat{C}_{\mathbf{p}}(t)\hat{C}^\dagger_{\mathbf{k}_2+\mathbf{q}_2}(t_2)\rangle_0\, {}_0\langle T\hat{C}_{\mathbf{k}_2}(t_2)\hat{C}^\dagger_{\mathbf{k}_1+\mathbf{q}_1}(t_1)\rangle_0\, {}_0\langle T\hat{C}_{\mathbf{k}_1}(t_1)\hat{C}_{\mathbf{p}}^\dagger(t')\rangle_0 \\ &\qquad + {}_0\langle T\hat{C}_{\mathbf{p}}(t)\hat{C}^\dagger_{\mathbf{k}_1+\mathbf{q}_1}(t_1)\rangle_0\, {}_0\langle T\hat{C}_{\mathbf{k}_1}(t_1)\hat{C}_{\mathbf{p}}^\dagger(t')\rangle_0\, {}_0\langle \hat{C}^\dagger_{\mathbf{k}_2+\mathbf{q}_2}(t_2)\hat{C}_{\mathbf{k}_2}(t_2)\rangle_0 \\ &\qquad + {}_0\langle T\hat{C}_{\mathbf{p}}(t)\hat{C}_{\mathbf{p}}^\dagger(t')\rangle_0\, {}_0\langle \hat{C}^\dagger_{\mathbf{k}_1+\mathbf{q}_1}(t_1)\hat{C}_{\mathbf{k}_1}(t_1)\rangle_0\, {}_0\langle \hat{C}^\dagger_{\mathbf{k}_2+\mathbf{q}_2}(t_2)\hat{C}_{\mathbf{k}_2}(t_2)\rangle_0 \\ &\qquad + {}_0\langle T\hat{C}_{\mathbf{p}}(t)\hat{C}^\dagger_{\mathbf{k}_2+\mathbf{q}_2}(t_2)\rangle_0\, {}_0\langle \hat{C}^\dagger_{\mathbf{k}_1+\mathbf{q}_1}(t_1)\hat{C}_{\mathbf{k}_1}(t_1)\rangle_0\, {}_0\langle T\hat{C}_{\mathbf{k}_2}(t_2)\hat{C}_{\mathbf{p}}^\dagger(t')\rangle_0 \\ &\qquad - {}_0\langle T\hat{C}_{\mathbf{p}}(t)\hat{C}_{\mathbf{p}}^\dagger(t')\rangle_0\, {}_0\langle T\hat{C}_{\mathbf{k}_1}(t_1)\hat{C}^\dagger_{\mathbf{k}_2+\mathbf{q}_2}(t_2)\rangle_0\, {}_0\langle T\hat{C}_{\mathbf{k}_2}(t_2)\hat{C}^\dagger_{\mathbf{k}_1+\mathbf{q}_1}(t_1)\rangle_0 \end{aligned}$$

Next we turn each bracket into either $G^{(0)}$ functions or number operators n_F. The six terms, taken in the same order, give

$$\begin{aligned} &i^3\delta_{\mathbf{p}=\mathbf{k}_1+\mathbf{q}_1=\mathbf{k}_2} G^{(0)}(\mathbf{p}, t - t_1)G^{(0)}(\mathbf{p} - \mathbf{q}_1, t_1 - t_2)G^{(0)}(\mathbf{p}, t_2 - t') \\ &\quad + i^3\delta_{\mathbf{p}=\mathbf{k}_1=\mathbf{k}_2-\mathbf{q}_1} G^{(0)}(\mathbf{p}, t - t_2)G^{(0)}(\mathbf{p} + \mathbf{q}, t_2 - t_1)G^{(0)}(\mathbf{p}, t_1 - t') \\ &\quad + i^2\delta_{\mathbf{q}_1=0}\delta_{\mathbf{p}=\mathbf{k}_1} n_F(\xi_{\mathbf{k}_2})G^{(0)}(\mathbf{p}, t - t_1)G^{(0)}(\mathbf{p}, t_1 - t') \\ &\quad + i\delta_{\mathbf{q}_1=0} n_F(\xi_{\mathbf{k}_1})n_F(\xi_{\mathbf{k}_2})G^{(0)}(\mathbf{p}, t - t') \\ &\quad + i^2\delta_{\mathbf{q}_1=0}\delta_{\mathbf{p}=\mathbf{k}_2} n_F(\xi_{\mathbf{k}_1})G^{(0)}(\mathbf{p}, t - t_2)G^{(0)}(\mathbf{p}, t_2 - t') \\ &\quad - i^3\delta_{\mathbf{k}_1=\mathbf{k}_2-\mathbf{q}_1} G^{(0)}(\mathbf{p}, t - t')G^{(0)}(\mathbf{k}_1, t_1 - t_2)G^{(0)}(\mathbf{k}_1 + \mathbf{q}_1, t_2 - t_1) \end{aligned} \tag{2.4.4}$$

2.5. FEYNMAN DIAGRAMS

Feynman introduced the idea of representing the kind of terms in (2.4.4) by drawings. These drawings, called diagrams, are extremely useful for providing an insight into the physical process which these terms represent. These diagrams can be drawn both for the Green's function depending on time $G(\mathbf{p}, t)$ as well as for functions which are Fourier-transformed and depend on energy $G(\mathbf{p}, E)$.

The diagrams in time space are drawn by representing the electron Green's function $G^{(0)}(\mathbf{p}, t - t')$ by a solid line which goes from t' to t, as shown in Fig. 2.1. An arrow is often included to represent the direction. The arrow is mostly for convenience, and it does not imply or require that $t > t'$. The phonon Green's function is represented by a dashed line. This does not have a directional arrow because

$$D^{(0)}(\mathbf{q}, t - t') = D^{(0)}(-\mathbf{q}, t' - t)$$

and the sign of $\mathbf{q}$ is irrelevant. Thus phonons can be viewed as going either direction in time. Next we must decide how to treat the factor

$$\langle C_{\mathbf{p}}^{\dagger}(t) C_{\mathbf{p}}(t) \rangle = n_F(\xi_p)$$

The occupation number n_F is drawn as a solid line which loops and represents an electron line which starts and ends at the same point in time. By using these rules, we can construct the diagrams which represent each of the six terms in the S-matrix expansion (2.4.4). These six terms are shown in Fig. 2.2, and the diagrams (a) to (f) are the six terms in (2.4.4) in the same order. Each term has the phonon line connecting the times t_1 and t_2.

In Fig. 2.2 the terms (c), (d), and (e) are zero. They vanish because they exist only if the phonon wave vector $\mathbf{q}$ is zero, but there are no phonons with $\mathbf{q} = 0$. A $\mathbf{q} = 0$ phonon is either a translation of the crystal or a

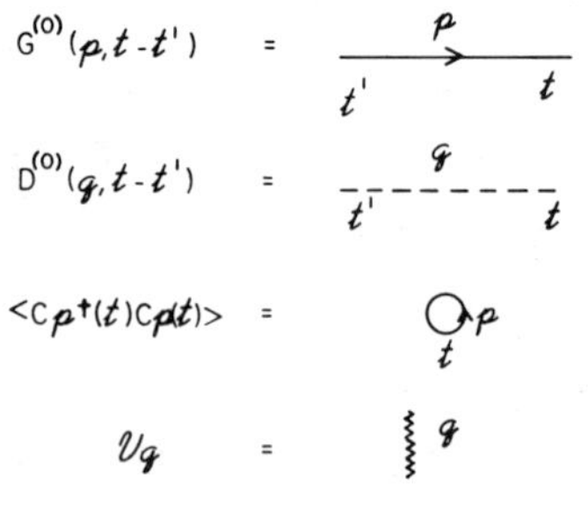

FIGURE 2.1

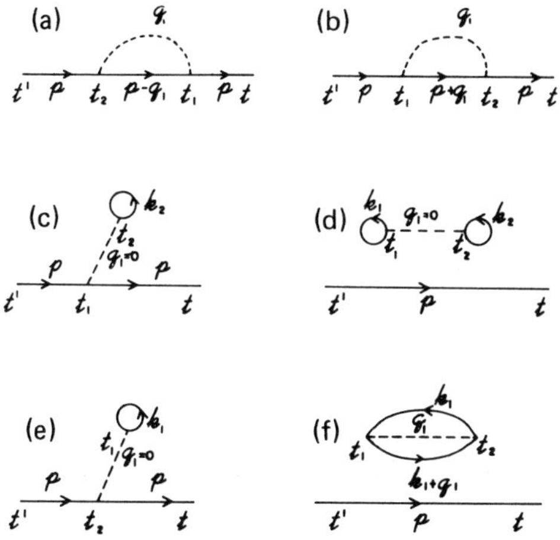

FIGURE 2.2

permanent strain, and neither of these is meant to be in the Hamiltonian. The sum over $\mathbf{q}$ in (2.4.3) should exclude the $\mathbf{q} = 0$ term.

The two terms (a) and (b) in Fig. 2.2 are not zero. They are the contributions of primary interest. By collecting the results of (2.4.4), for their contribution we obtain

$$\frac{i}{2!}\int_{-\infty}^{\infty} dt_1 \int_{-\infty}^{\infty} dt_2 \sum_{\mathbf{q}} |M_{\mathbf{q}}|^2 D^{(0)}(\mathbf{q}, t_1-t_2)[G^{(0)}(\mathbf{p}, t-t_1)G^{(0)}(\mathbf{p}-\mathbf{q}, t_1-t_2)$$
$$\times G^{(0)}(\mathbf{p}, t_2 - t') + G^{(0)}(\mathbf{p}, t - t_2)G^{(0)}(\mathbf{p}+\mathbf{q}, t_2 - t_1)G^{(0)}(\mathbf{p}, t_1 - t')] \tag{2.5.1}$$

The two drawings (a) and (b) look alike. They differ only in the labeling of the variables t_1, t_2, $\pm\mathbf{q}$. But these are variables of integration in (2.5.1) and may be relabeled. By doing this, it is easy to show that the two terms are equal.

The term (f) in Fig. 2.2 is

$$G^{(0)}(\mathbf{p}, t - t')F_1 \tag{2.5.2}$$

where

$$F_1 = \frac{-i}{2!}\int_{-\infty}^{\infty} dt_1 \int_{-\infty}^{\infty} dt_2 \sum_{\mathbf{k},\mathbf{q}} |M_{\mathbf{q}}|^2 D^{(0)}(\mathbf{q}, t_1 - t_2)G^{(0)}(\mathbf{k}, t_1 - t_2)$$
$$\times G^{(0)}(\mathbf{k}+\mathbf{q}, t_2 - t_1) \tag{2.5.3}$$

This drawing has the property that part of it is not topologically connected to the Green's function line $G^{(0)}(\mathbf{p}, t - t')$. Diagrams in which *all* parts are

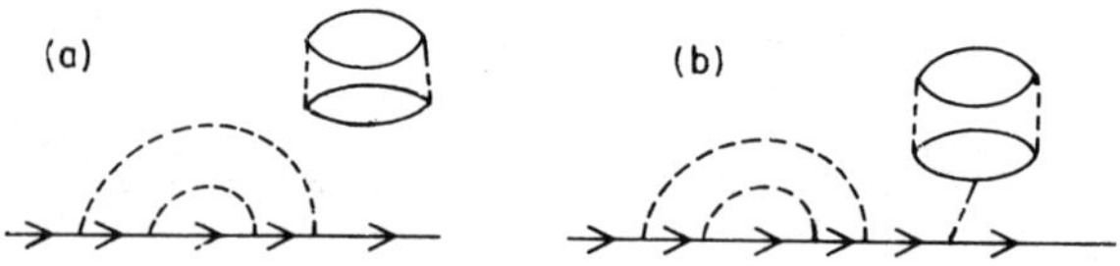

FIGURE 2.3

not connected are called *disconnected* diagrams. For example, in Fig. 2.3, part (a) is disconnected, while part (b) is connected. Note that the disconnected parts, as in (2.5.2), provide just a multiplicative constant like F_1 which multiplies the contribution from the connected parts.

2.6. VACUUM POLARIZATION GRAPHS

We now turn our attention to the factor which to this point we have been ignoring:

$$_0\langle| S(\infty, -\infty) |\rangle_0 = \sum_{n=0}^{\infty} \frac{(-i)^n}{n!} \int_{-\infty}^{\infty} dt_1 \cdots \int_{-\infty}^{\infty} dt_n \, _0\langle| T\hat{V}(t_1) \cdots \hat{V}(t_n) |\rangle_0$$

Let us again consider the electron–phonon interaction and evaluate the term $n = 2$. The $n = 1$ term vanishes as it did for the Green's function expansion,

$$_0\langle S\rangle_0 = 1 + \frac{(-i)^2}{2!} \int_{\infty}^{\infty} dt_1 \int_{\infty}^{\infty} dt_2 \, _0\langle| T\hat{V}(t_1)\hat{V}(t_2) |\rangle_0 + \cdots \tag{2.6.1}$$

where

$$_0\langle| T\hat{V}(t_1)\hat{V}(t_2) |\rangle_0 = \sum_{\substack{\mathbf{q}_1\mathbf{q}_2\\ \mathbf{k}_1\mathbf{k}_2}} M_{\mathbf{q}_1} M_{\mathbf{q}_2} \, _0\langle| T\hat{A}_{\mathbf{q}_1}(t_1)\hat{A}_{\mathbf{q}_2}(t_2) |\rangle_0$$

$$\times \, _0\langle| T\hat{C}^{\dagger}_{\mathbf{k}_1+\mathbf{q}_1}(t_1)\hat{C}_{\mathbf{k}_1}(t_1)\hat{C}^{\dagger}_{\mathbf{k}_2+\mathbf{q}_2}(t_2)\hat{C}_{\mathbf{k}_2}(t_2) |\rangle_0$$

By using Wick's theorem, we get

$$_0\langle| T\hat{A}_{\mathbf{q}_1}(t_1)\hat{A}_{\mathbf{q}_2}(t_2) |\rangle_0 \; i\delta_{\mathbf{q}_1+\mathbf{q}_2} D^{(0)}(\mathbf{q}_1, t_1 - t_2)$$

$$\begin{aligned} &_0\langle| T\hat{C}^{\dagger}_{\mathbf{k}_1+\mathbf{q}_1}(t_1)\hat{C}_{\mathbf{k}_1}(t_1)\hat{C}^{\dagger}_{\mathbf{k}_2-\mathbf{q}_1}(t_2)\hat{C}_{\mathbf{k}_2}(t_2) |\rangle_0 \\ &\quad = \delta_{\mathbf{q}_1=0} n_{\mathbf{k}_1} n_{\mathbf{k}_2} + \delta_{\mathbf{k}_1=\mathbf{k}_2-\mathbf{q}} G^{(0)}(\mathbf{k}_1, t_1 - t_2) G^{(0)}(\mathbf{k}_1 + \mathbf{q}_1, t_2 - t_1) \end{aligned} \tag{2.6.2}$$

The Feynman diagrams for the two terms in Eq. (2.6.2) are shown in Figs. 2.4(a) and 2.4(b). The (a) term is zero because there are no $\mathbf{q} = 0$ phonons.

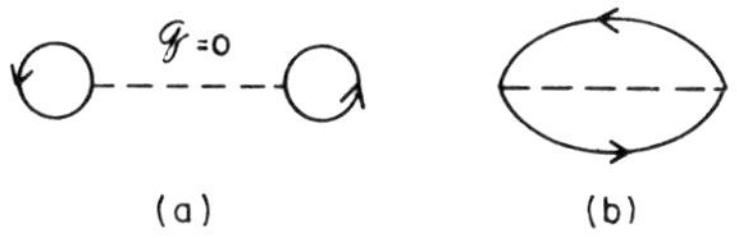

FIGURE 2.4

The (b) term is finite and gives a contribution

$$_0\langle|\, S(\infty, -\infty)\,|\rangle_0 = 1 + F_1 + \cdots$$

where F_1 is defined in (2.5.3). The constant F_1 appears whenever the closed bubble of Fig. 2.4(b) occurs, regardless of whether the term arises in the disconnected diagrams of $G(\mathbf{p}, t - t')$ or in the expansion of $_0\langle|\, S(\infty, -\infty)\,|\rangle_0$.

The terms in the series for $_0\langle|\, S(\infty, -\infty)\,|\rangle_0$ are called *vacuum polarization terms*. Some terms for $n = 4$, where there are two phonon lines, are shown in Fig. 2.5. Each of these diagrams or terms represents a constant F_j which one can evaluate by doing the required time and wave vector integrals. Thus the constant $_0\langle|\, S(\infty, -\infty)\,|\rangle_0$ could be evaluated by computing all the F_j and then summing them:

$$_0\langle|\, S(\infty, -\infty)\,|\rangle_0 = \sum_{j=0}^{\infty} F_j$$

This procedure is unnecessary because of a cancellation theorem.

The next theorem which we shall state also simplifies the calculation of the Green's function expansion (2.4.1). This theorem is that the vacuum polarization diagrams exactly cancel the disconnected diagrams in the expansion for $G(\mathbf{p}, t - t')$. The net result is that in calculating $G(\mathbf{p}, t - t')$ one need only evaluate the connected diagrams. The other contributions, from the disconnected diagrams and from $_0\langle|\, S(\infty, -\infty)\,|\rangle_0$, exactly cancel one another.

We shall not prove this theorem but only try to explain it. If we call $G_c(\mathbf{p}, t - t')$ the summation of all connected diagrams, then the basic theorem is that

$$_0\langle|\, T\hat{C}_{\mathbf{p}}(t)\hat{C}_{\mathbf{p}}^{\dagger}(t')S(\infty, -\infty)\,|\rangle_0 = i\, G_c(\mathbf{p}, t - t')\, {}_0\langle|\, S(\infty, -\infty)\,|\rangle_0$$

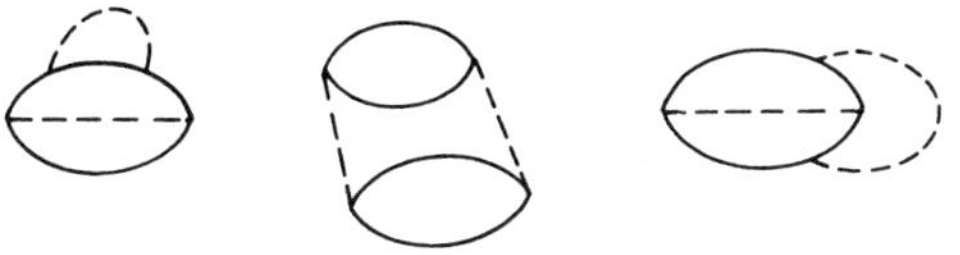

FIGURE 2.5

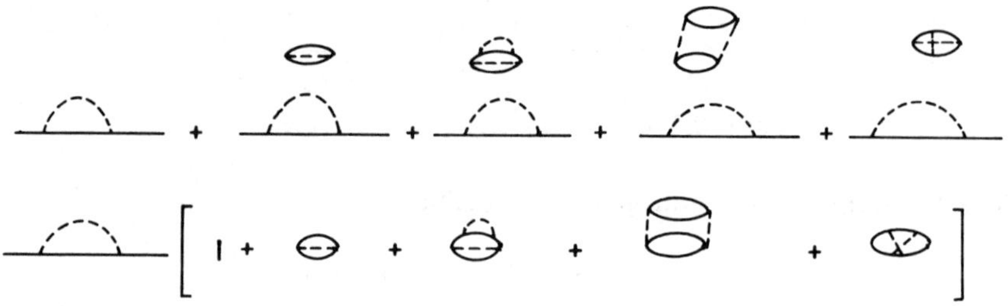

FIGURE 2.6

Thus the Green's function (2.3.1) is just the summation of all the connected diagrams:

$$G(\mathbf{p}, t - t') = G_c(\mathbf{p}, t - t')$$

The proof of this theorem is just a counting problem. One must convince oneself that each connected diagram has, in higher-order terms in the S-matrix expansion, all disconnected parts which exactly add up to ${}_0\langle| S(\infty, -\infty) |\rangle_0$. For example, the self-energy diagram in Fig. 2.2(a) has in higher order the vacuum polarization terms shown in Fig. 2.6. The summation of all these terms, to all order, is just the factor ${}_0\langle| S(\infty, -\infty) |\rangle_0$. The important point is that each disconnected part is just a constant factor F_j.

This theorem is very convenient, since it states that one can just ignore the disconnected diagrams. They do not need to be calculated. It is just as well, since when they are evaluated they often turn out to be infinity. In fact, it is easy to show that they are infinity. The disconnected term F_1 is defined in (2.5.3). The integrand is only a function $f(t_1 - t_2)$ of $t_1 - t_2$. Thus if we change integration variables to

$$\tau = t_1 - t_2$$

$$s = \tfrac{1}{2}(t_1 + t_2)$$

then Eq. (2.5.3) becomes

$$F_1 = \int_{-\infty}^{\infty} ds \int_{-\infty}^{\infty} d\tau f(\tau)$$

The important point is that there is no dependence on s, so

$$\int_{-\infty}^{\infty} ds = \infty$$

One can show that each disconnected part has an "extra" time integral and has the same infinity. Thus we have shown that the one-particle Green's functions consist of just connected diagrams:

$$G(\mathbf{p}, t - t') = -i \sum_{n=0}^{\infty} \frac{(-i)^n}{n!} \int_{-\infty}^{\infty} dt_1 \cdots \int_{-\infty}^{\infty} dt_n$$
$$\times\ {}_0\langle| T\hat{C}_{\mathbf{p}}(t)\hat{C}_{\mathbf{p}}^{\dagger}(t')\hat{V}(t_1) \cdots \hat{V}(t_n) |\rangle_0 \qquad \text{(connected)}$$

Next we get rid of the $1/n!$ factor. It is eliminated because there are just $n!$ terms exactly alike in each bracket of the nth term in the expansion. Thus if we consider only different terms, we obtain the result

$$G(\mathbf{p}, t - t') = -i \sum_{n=0}^{\infty} (-i)^n \int_{-\infty}^{\infty} dt_1 \cdots \int_{-\infty}^{\infty} dt_n\ {}_0\langle| T\hat{C}_{\mathbf{p}}(t)\hat{C}_{\mathbf{p}}^{\dagger}(t')$$
$$\times \hat{V}(t_1) \cdots \hat{V}(t_n) |\rangle_0 \qquad \text{(different connected)} \tag{2.6.3}$$

The obvious question is then, How do we tell when terms are different? Usually one can tell by inspection, although sometimes terms must be examined carefully. In Fig. 2.2, terms (a) and (b) are the same and provide the $2 = 2!$ necessary for this $n = 2$ term. Similary, (c) and (e) are identical. For example, (a) and (b) differ only in the variables t_1, t_2, and $\mathbf{q}_1$. But these are dummy variables of integration and so may be relabeled, respectively, to t_2, t_1, and $-\mathbf{q}_1$, so that the (a) term is obviously the same as (b).

The next terms in the electron–phonon expansion have $n = 4$, which are diagrams with two phonons. Here each different connected diagram is found $4! = 24$ times.

2.7. DYSON'S EQUATION

The Green's function of energy is defined by taking the usual Fourier transform with respect to the time variable:

$$G(\mathbf{p}, E) = \int_{-\infty}^{\infty} dt e^{iE(t-t')} G(\mathbf{p}, t - t') \tag{2.7.1}$$

This time integral has already been evaluated for the unperturbed Green's function with the following results: For a single particle in a band,

$$G^{(0)}(\mathbf{p}, E) = \frac{1}{E - \varepsilon_{\mathbf{p}} + i\delta} \tag{2.3.3}$$

and for a fermion in a degenerate electron gas,

$$G^{(0)}(\mathbf{p}, E) = \frac{\theta(p - k_F)}{E - \xi_{\mathbf{p}} + i\delta} + \frac{\theta(k_F - p)}{E - \xi_{\mathbf{p}} - i\delta} \equiv \frac{1}{E - \xi_{\mathbf{p}} + i\delta_{\mathbf{p}}} \tag{2.3.5}$$

The Fourier transform in time is applied to each term in the S-matrix summation:

$$G(\mathbf{p}, E) = -i \sum_{n=0}^{\infty} (-i)^n \int_{-\infty}^{\infty} dt e^{iE(t-t')} \int_{-\infty}^{\infty} dt_1 \cdots \int_{-\infty}^{\infty} dt_n$$
$$\times {}_0\langle| T\hat{C}_{\mathbf{p}}(t)\hat{C}_{\mathbf{p}}^{\dagger}(t')\hat{V}(t_1) \cdots \hat{V}(t_n) |\rangle_0 \qquad \text{(different connected)}$$

To see what sort of terms develop, consider the example of the electron–phonon interaction which we have been using above. The first two terms [Figs. 2.2(c) and 2.2(e) are zero] are $G^{(0)}$ plus the *self-energy term* in Fig. 2.2(a):

$$G(\mathbf{p}, E) = G^{(0)}(\mathbf{p}, E) + (-i)^3 \sum_{\mathbf{q}} M_{\mathbf{q}}^2 \int_{-\infty}^{\infty} dt e^{iE(t-t')} \int_{-\infty}^{\infty} dt_1 \int_{-\infty}^{\infty} dt_2$$
$$\times G^{(0)}(\mathbf{p}, t - t_1)G^{(0)}(\mathbf{p} - \mathbf{q}, t_1 - t_2)G^{(0)}(\mathbf{p}, t_2 - t')D^{(0)}(\mathbf{q}, t_1 - t_2)$$

The phonon Green's function of energy is defined the same way:

$$D(\mathbf{q}, \omega) = \int_{-\infty}^{\infty} dt e^{it\omega} D(\mathbf{q}, t)$$
$$D(\mathbf{q}, t) = \int_{-\infty}^{\infty} \frac{d\omega}{2\pi} e^{-it\omega} D(\mathbf{q}, \omega) \tag{2.7.2}$$

If we use the unperturbed phonon Green's function,

$$D^{(0)}(\mathbf{q}, t_1 - t_2) = \int_{-\infty}^{\infty} \frac{d\omega}{2\pi} e^{-i\omega(t_1 - t_2)} D^{(0)}(\mathbf{q}, \omega)$$

then the remaining time integrals are easy:

$$\int_{-\infty}^{\infty} dt G^{(0)}(\mathbf{p}, t - t_1) e^{i(t-t_1)E} \int_{-\infty}^{\infty} dt_1 e^{i(t_1 - t_2)(E-\omega)} G^{(0)}(\mathbf{p} - \mathbf{q}, t_1 - t_2)$$
$$\times \int_{-\infty}^{\infty} dt_2 e^{i(t_2 - t')E} G^{(0)}(\mathbf{p}, t_2 - t') = G^{(0)}(\mathbf{p}, E)^2 G^{(0)}(\mathbf{p} - \mathbf{q}, E - \omega)$$

Thus the first two terms are

$$G(\mathbf{p}, E) = G^{(0)}(\mathbf{p}, E) + G^{(0)}(\mathbf{p}, E)^2 \Sigma^{(1)}(\mathbf{p}, E) \tag{2.7.3}$$

where the *self-energy* of the electron due to one-phonon processes is

$$\Sigma^{(1)}(\mathbf{p}, E) = i\int_{-\infty}^{\infty} \frac{d\omega}{2\pi} \sum_{\mathbf{q}} M_{\mathbf{q}}{}^2 D^{(0)}(\mathbf{q}, \omega) G^{(0)}(\mathbf{p} - \mathbf{q}, E - \omega) \tag{2.7.4}$$

This quantity will be evaluated in Chapter 6.

The electron–phonon interaction has four connected diagrams in $n = 4$; they are self-energy diagrams with two phonons and are shown in Fig. 2.7. These four terms give, respectively, the contribution to the Green's function series (2.6.3):

$$\begin{aligned}
&\int_{-\infty}^{\infty} dt_1 \int_{-\infty}^{\infty} dt_2 \int_{-\infty}^{\infty} dt_3 \int_{-\infty}^{\infty} dt_4 \Big\{ \sum_{\mathbf{q}\mathbf{q}'} M_{\mathbf{q}}{}^2 M_{\mathbf{q}'}^2 D^{(0)}(\mathbf{q}, t_1 - t_2) D^{(0)}(\mathbf{q}', t_3 - t_4) \\
&\quad \times [G^{(0)}(\mathbf{p}, t - t_1) G^{(0)}(\mathbf{p} + \mathbf{q}, t_1 - t_2) G^{(0)}(\mathbf{p}, t_2 - t_3) G^{(0)}(\mathbf{p} + \mathbf{q}', t_3 - t_4) \\
&\quad \times G^{(0)}(\mathbf{p}, t_4 - t') + G^{(0)}(\mathbf{p}, t - t_1) G^{(0)}(\mathbf{p} + \mathbf{q}, t_1 - t_3) \\
&\quad \times G^{(0)}(\mathbf{p} + \mathbf{q} + \mathbf{q}', t_3 - t_2) G^{(0)}(\mathbf{p} + \mathbf{q}', t_2 - t_4) G^{(0)}(\mathbf{p}, t_4 - t') \\
&\quad + G^{(0)}(\mathbf{p}, t - t_1) G^{(0)}(\mathbf{p} + \mathbf{q}, t_1 - t_3) G^{(0)}(\mathbf{p} + \mathbf{q} + \mathbf{q}', t_3 - t_4) \\
&\quad \times G^{(0)}(\mathbf{p} + \mathbf{q}, t_4 - t_2) G^{(0)}(\mathbf{p}, t_2 - t')] + \sum_{\mathbf{q}} M_q{}^4 D^{(0)}(\mathbf{q}, t_1 - t_2) \\
&\quad \times D^{(0)}(\mathbf{q}, t_3 - t_4) G^{(0)}(\mathbf{p}, t - t_1) G^{(0)}(\mathbf{p} + \mathbf{q}, t_1 - t_4) G^{(0)}(\mathbf{p}, t_4 - t') \\
&\quad \times \sum_{\sigma\mathbf{k}} G^{(0)}(\mathbf{k}, t_2 - t_3) G^{(0)}(\mathbf{k} + \mathbf{q}, t_3 - t_2) \Big\}
\end{aligned}$$

Figure 2.7 shows the labeling of electron and phonon Green's functions and the time label of each vertex. The fourier transform of these terms is taken to give their contribution to the Green's function of energy,

$$G^{(0)}(\mathbf{p}, E)^3 \Sigma^{(1)}(\mathbf{p}, E)^2 + G^{(0)}(\mathbf{p}, E)^2 [\Sigma^{(2a)}(\mathbf{p}, E) + \Sigma^{(2b)}(\mathbf{p}, E) + \Sigma^{(2c)}(\mathbf{p}, E)]$$

so that combining this with (2.7.3) gives

$$\begin{aligned}
G(\mathbf{p}, E) = G^{(0)}(\mathbf{p}, E)[1 &+ G^{(0)}\Sigma^{(1)} + (G^{(0)}\Sigma^{(1)})^2 + G^{(0)}\Sigma^{(2a)} \\
&+ G^{(0)}\Sigma^{(2b)} + G^{(0)}\Sigma^{(2c)} + \cdots]
\end{aligned} \tag{2.7.5}$$

The first term, which comes from Fig. 2.7(a), contains three unperturbed

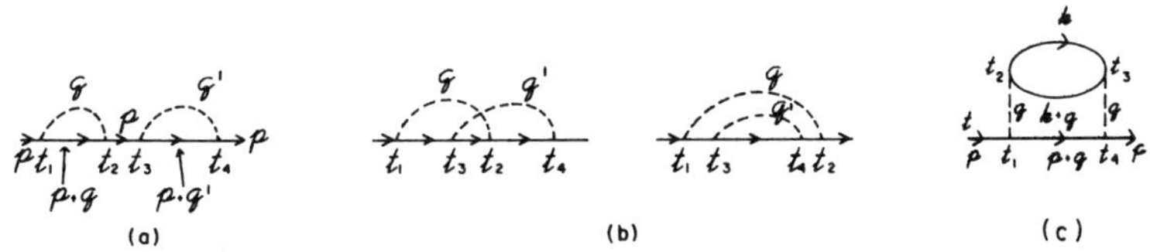

FIGURE 2.7

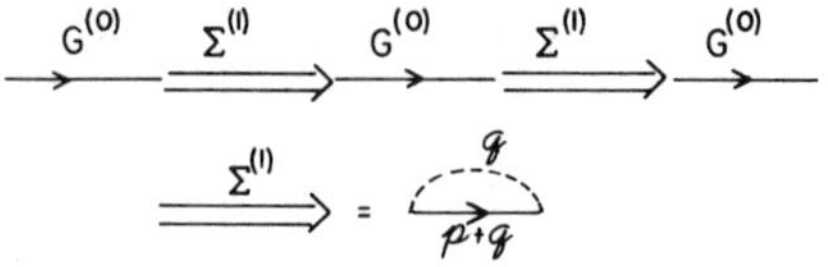

FIGURE 2.8

Green's functions $G^{(0)}$ and two one-phonon self-energy terms $\Sigma^{(1)}$. Each of these factors can be associated with a piece of Fig. 2.7(a), as illustrated in Fig. 2.8. Similarly, the last three terms in Fig. 2.7 contain the three diagrams which represent the two-phonon self-energy terms:

$$\Sigma^{(2a)}(p, E) + \Sigma^{(2b)}(p, E)$$
$$= \int_{-\infty}^{\infty} \frac{d\omega}{2\pi} \int_{-\infty}^{\infty} \frac{d\omega'}{2\pi} \sum_{qq'} M_q^2 M_{q'}^2 D^{(0)}(q, \omega) D^{(0)}(q', \omega')$$
$$\times [G^{(0)}(\mathbf{p}+\mathbf{q}, E+\omega) G^{(0)}(\mathbf{p}+\mathbf{q}+\mathbf{q}', E+\omega+\omega') G^{(0)}(\mathbf{p}+\mathbf{q}', E+\omega')$$
$$+ G^{(0)}(\mathbf{p}+\mathbf{q}, E+\omega) G^{(0)}(\mathbf{p}+\mathbf{q}+\mathbf{q}', E+\omega+\omega') G^{(0)}(\mathbf{p}+\mathbf{q}, E+\omega)] \tag{2.7.6}$$

$$\Sigma^{(2c)}(p, E) = \int_{-\infty}^{\infty} \frac{d\omega}{2\pi} \sum_{\mathbf{q}} M_q^4 D^{(0)}(\mathbf{q}, \omega)^2 G^{(0)}(\mathbf{p}+\mathbf{q}, E+\omega)$$
$$\times \int_{-\infty}^{\infty} \frac{d\omega'}{2\pi} \sum_{\mathbf{k},\sigma} G^{(0)}(\mathbf{k}, \omega') G^{(0)}(\mathbf{k}+\mathbf{q}, \omega+\omega')$$

Dyson's equation is obtained by formally summing the series in (2.7.5),

$$G(\mathbf{p}, E) = \frac{G^{(0)}(\mathbf{p}, E)}{1 - G^{(0)}(\mathbf{p}, E)\Sigma(\mathbf{p}, E)} \tag{2.7.7}$$

where the total self-energy $\Sigma(p, E)$ is the summation of all *different* self-energy contributions:

$$\Sigma(\mathbf{p}, E) = \sum_j \Sigma^{(j)}(\mathbf{p}, E)$$

So far our example of the electron–phonon interaction has yielded the following four terms for the self-energy:

$$\Sigma(\mathbf{p}, E) = \Sigma^{(1)}(\mathbf{p}, E) + \Sigma^{(2a)}(\mathbf{p}, E) + \Sigma^{(2b)}(\mathbf{p}, E) + \Sigma^{(2c)}(\mathbf{p}, E) + \cdots$$

Contributions such as Fig. 2.7(a) which contain $(\Sigma^{(1)})^2$ do not mean that Σ contains $(\Sigma^{(1)})^2$. The Green's function expansion contains terms in $(\Sigma^{(1)})^2$ because when we expand G in the power series

$$G = \frac{G^{(0)}}{1 - G^{(0)}\Sigma} = G^{(0)} + G^{(0)^2}\Sigma + G^{(0)^3}\Sigma^2 + G^{(0)^4}\Sigma^3 + \cdots$$

the successive terms contain higher powers of each $\Sigma^{(i)}$ plus cross terms. Dyson's equation is really a theorem which states that one may sum the series of self-energy terms which develops in higher order, and the form (2.7.7) is obtained with each distinct contribution to Σ occurring just once.

The derivation of Dyson's equation has been rather complicated and consisted of a large number of steps and theorems. Each step has to be understood before the final result is understood. Yet the final result—Dyson's equation—has achieved a great simplification. It states that the exact Green's function is obtained from (2.7.7) by just calculating the self-energy $\Sigma(\mathbf{p}, E)$. The self-energy is a summation of an infinite number of distinct diagrams. This method is only useful if we can approximate $\Sigma(\mathbf{p}, E)$ by the lowest few terms in the series. Alternately, sometimes one can sum subsets of diagrams in the series. However, except in a few rare cases, it is impossible to get $\Sigma(\mathbf{p}, E)$ exactly, and one must be content with an approximate result. If the approximate result is not a very good approximation, one should not try to solve the problem in this fashion. Some of the alternate methods will be developed in subsequent chapters. But basically one should realize that Dyson's equation is usually useful only in weak coupling theory—where the perturbation is sufficiently weak that an adequate approximation is obtained with a few terms in $\Sigma(\mathbf{p}, E)$. Later we shall encounter some *strong coupling theories* which use Dyson's equation.

Thus a great simplification has been achieved. We need only to evaluate a few self-energy diagrams in Σ. If they are sufficient, we are finished. If they are not, then we discard the results and try something else. Thus the formidable-looking series (2.4.1), which served as our starting point, has been reduced to the evaluation of a few terms.

Dyson's equation is often written in a slightly different but equivalent form. It is obtained by using the algebraic form for $G^{(0)}$. Thus for one electron in a band,

$$\begin{aligned} G^{(0)}(\mathbf{p}, E) &= \frac{1}{E - \varepsilon_{\mathbf{p}} + i\delta} \\ G(\mathbf{p}, E) &= \frac{1}{E - \varepsilon_{\mathbf{p}} + i\delta - \Sigma(\mathbf{p}, E)} \end{aligned} \tag{2.7.7a}$$

When the electron is in a Fermi sea at zero temperature,

$$\begin{aligned} G^{(0)}(\mathbf{p}, E) &= \frac{1}{E - \varepsilon_{\mathbf{p}} + i\delta_{\mathbf{p}}} \\ G(\mathbf{p}, E) &= \frac{1}{E - \varepsilon_{\mathbf{p}} + i\delta_{\mathbf{p}} - \Sigma(\mathbf{p}, E)} \end{aligned} \tag{2.7.8}$$

One aspect of this result deserves special mention. The infinitesimal part $\delta_{\mathbf{p}}$ switches sign depending on whether $\xi_{\mathbf{p}} > 0$ or $\xi_{\mathbf{p}} < 0$, which makes $\operatorname{Im} G^{(0)}(\mathbf{p}, E)$ change sign depending on whether $E > 0$ or $E < 0$. The self-energy has real and imaginary parts which we write as

$$\Sigma(\mathbf{p}, E) = \Sigma_R(\mathbf{p}, E) + i\Sigma_I(\mathbf{p}, E)$$

We shall show later that the imaginary part of Σ, or Σ_I, also switches sign at $E = 0$ in the same manner. Thus

$$\begin{aligned} \Sigma_I(\mathbf{p}, E) &< 0, \qquad E > 0 \\ \Sigma_I(\mathbf{p}, E) &> 0, \qquad E < 0 \end{aligned}$$

The switching of signs of δ_p was caused by the distinction between electron excitations with $\xi_{\mathbf{p}} > 0$ and hole excitations with $\xi_{\mathbf{p}} < 0$. This distinction is maintained even when $\Sigma(\mathbf{p}, E)$ is included, i.e., in the presence of interactions. The one-electron Green's function (2.7.7a) has the feature that $\Sigma_I < 0$, so the imaginary part of the denominator always has the same sign. The electron self-energy is sometimes called a *mass operator*.

The phonon Green's function has the same type of Dyson equation:

$$D(\mathbf{q}, \omega) = \frac{D^{(0)}(\mathbf{q}, \omega)}{1 - D^{(0)}(\mathbf{q}, \omega)\pi(\mathbf{q}, \omega)} \tag{2.7.9}$$

This equation may also be written in an alternate form by utilizing the following result for $D^{(0)}$:

$$D^{(0)}(\mathbf{q}, \omega) = \frac{2\omega_{\mathbf{q}}}{\omega^2 - \omega_{\mathbf{q}}^2 + i\delta} \tag{2.7.10}$$

$$D(\mathbf{q}, \omega) = \frac{2\omega_{\mathbf{q}}}{\omega^2 - \omega_{\mathbf{q}}^2 + i\delta - 2\omega_{\mathbf{q}}\pi(\mathbf{q}, \omega)} \tag{2.7.11}$$

The phonon self-energy term $\pi(\mathbf{q}, \omega)$ is sometimes called a *polarization operator*. This name is quite descriptive, since the self-energy effects arise from the phonons causing polarization in the medium.

The real and imaginary parts of the self-energies Σ and π each have interpretations. The imaginary part Σ_I or π_I is interpreted as causing the damping of the particle motion. They are related to the finite mean free path of the excitation or its energy and momentum uncertainty. The real parts are actual energy shifts of the excitation, which may also change its dynamical motion. The excitation may alter its effective mass or group velocity because of the self-energy contributions.

2.8. RULES FOR CONSTRUCTING DIAGRAMS

It is certainly worthwhile for the student to evaluate some self-energy diagrams by the method we have outlined above. One should expand the S matrix, decide which terms are connected and which are zero, and finally obtain the self-energies by a Fourier transform. But this laborious procedure can be avoided because the self-energy diagrams can be written down directly by following a few simple rules. That is, self-energy expressions such as (2.7.4) and (2.7.5) are easily written down in the form shown in these equations. The evaluation of the wave vector integrals in these expressions remains a formidable task. These rules are as follows:

1. Draw the Feynman diagram for the self-energy term, with all phonon, Coulomb, and electron lines.

2. For each electron line, introduce the following Green's function:

$$G_{\alpha\beta}^{(0)}(\mathbf{p}, E) = \frac{\delta_{\alpha\beta}}{E - \xi_{\mathbf{p}} + i\delta_{\mathbf{p}}}$$

The $\delta_{\alpha\beta}$ is a spin index and indicates that the electron line must have the same spin at both ends of the propagator line. This feature is important in spin problems. The factor δ_p is δ for one electron in a band and $\delta \operatorname{sgn} \xi_{\mathbf{p}}$ for degenerate Fermi systems, i.e., those with a Fermi sea at zero temperature.

3. For each phonon line, introduce the following phonon propagator:

$$D^{(0)}(\mathbf{q}, \omega) = \frac{2\omega_{\mathbf{q}}}{\omega^2 - \omega_{\mathbf{q}}{}^2 + i\delta}$$

Also add a factor of $M_q{}^2/\nu$ for each phonon Green's function, where $M_{\mathbf{q}}$ is the matrix element for electron and phonon interaction and ν is the volume.

4. Add a Coulomb potential $v_q = 4\pi e^2/q^2\nu$ for each Coulomb interaction. Note that we always draw the Coulomb line as a wiggly vertical line. The Coulomb interaction is regarded as happening instantaneously in time, and time flows horizontally, from left to right, in our diagrams. One could, of course, have time flow upwards and draw the Coulomb interactions as horizontal wiggly lines.

5. Conserve energy and momentum at each vertex. Thus each electron line, phonon line, and Coulomb line have their variables labeled to conform with this rule.

6. Sum over internal degrees of freedom: momentum, energy, and spin. Thus if one is calculating a self-energy term $\Sigma(\mathbf{p}, E)$, then all momentum and energies except p and E are *internal* and must be summed over.

7. Finally, we multiply the result by the factor

$$\frac{(i)^m}{(2\pi)^{4m}} (-1)^F (2S + 1)^F$$

where F is the number of closed fermion loops. The index m is chosen as follows:

a. For electron self-energies, m is the number of internal phonon and Coulomb lines. Thus $m = 1$ for (2.7.4), and $m = 2$ for (2.7.6).

b. For phonon self-energies, m is one-half the number of vertices. The spin of the particle is S, and the $2S + 1$ factor is from the summation over spin quantum number m_s. Usually we have electrons with $2S + 1 = 2$. The factor $(2\pi)^{-4}$ assumes that we have taken the limit $\nu \to \infty$, so that the wave vector summations are integrals. For box normalization in a finite volume ν, the factor is

$$\frac{i^m}{(2\pi\nu)^m} (-1)^F (2S + 1)^F$$

and then wave vector summations are discrete summations.

The photon Green's functions will not be discussed until the last section. But it seems tidy to present the rules for constructing diagrams with photons at this point, so that all the rules are together. One draws photon lines between lines which represent charged particles. The photon lines are usually dotted and usually are represented just like phonon lines. Charged particles interact with the photons through two terms in the interaction Hamiltonian. The first is the $\mathbf{j} \cdot \mathbf{A}$ term. For free particles, this has the form

$$\begin{aligned} \frac{1}{c} e \sum_i \mathbf{j}(\mathbf{r}_i) \cdot \mathbf{A}(\mathbf{r}_i) &= e \sum_{\mathbf{q},\mu} j_\mu(\mathbf{q}) A_\mu(\mathbf{q}) \\ &= \frac{e}{m} \sum_{\mathbf{q}\mu} A_\mu(\mathbf{q}) \sum_{\mathbf{k},\sigma} (\mathbf{k} + \tfrac{1}{2}\mathbf{q})_\mu C^\dagger_{\mathbf{k}+\mathbf{q},\sigma} C_{\mathbf{k},\sigma} \end{aligned}$$

This brings us to the next rule for constructing diagrams.

8. For each photon line which interacts with particles through the $\mathbf{j} \cdot \mathbf{A}$ interaction, insert a factor

$$\frac{e}{m} \sum_{\mu\nu} (\mathbf{k} + \tfrac{1}{2}\mathbf{q})_\mu D_{\mu\nu}(\mathbf{q}, \omega)(\mathbf{k}' + \tfrac{1}{2}\mathbf{q})_\nu$$

where $D_{\mu\nu}(\mathbf{q}, \omega)$ is the photon Green's function and $\mathbf{k}$ and $\mathbf{k}'$ are the wave vectors of particles scattered at the two vertices. The other possible interaction of a charged particle with photons occurs through the term

$$\frac{e^2}{2mc^2} \sum_i A(r_i)^2 = \frac{e^2}{2m} \sum_{\mathbf{q}\mathbf{k}_\mu}{}' \varrho(\mathbf{q}) A_\mu(\mathbf{k}) A_\mu(\mathbf{q} - \mathbf{k})$$

In Sec. 4.5 it is shown that this interaction contributes a self-energy term of $e^2 n_0/m$ to the self-energy of the photon, where n_0 is the density of charged particles.

Other texts often define the phonon Green's function differently from the form we have selected here. However, all these cases have the final product of the vertex and the Green's function as

$$\frac{1}{\nu} \frac{2\omega_{\mathbf{q}} M_{\mathbf{q}}{}^2}{\omega^2 - \omega_{\mathbf{q}}{}^2 + i\delta}$$

For example, one choice has a Green's function of the form

$$D^{(0)}(\mathbf{q}, \omega) = \frac{\omega_{\mathbf{q}}{}^2}{\omega^2 - \omega_{\mathbf{q}}{}^2 - i\delta}$$

Then, in the rules for constructing diagrams, one multiplies by the factor $2M_{\mathbf{q}}{}^2/\omega_{\mathbf{q}}\nu$ for each vertex pair. Of course, this gives the same result.

Now we wish to give some examples. Electron–phonon examples are given previously in (2.7.4) and (2.7.6). Some phonon self-energies are shown in Fig. 2.9(a). The first example is from the electron–phonon interaction, and the self-energy contribution is a closed fermion loop, so that $F = 1$ in rule 7. This contribution is

$$\begin{aligned} \pi(\mathbf{q}, \omega) &= M_q{}^2 P^{(1)}(\mathbf{q}, \omega) \\ P^{(1)}(\mathbf{q}, \omega) &= -i2 \int \frac{dE}{2\pi} \int \frac{d^3p}{(2\pi)^3} G^{(0)}(\mathbf{p}, E) G^{(0)}(\mathbf{p} + \mathbf{q}, E + \omega) \end{aligned} \tag{2.8.1}$$

The other phonon self-energy contribution arises from the lattice anharmonicity, which leads to interaction terms in the ion Hamiltonian proportional to the third power of the phonon displacement:

$$V = \sum_i Q_i{}^3 V_i = \sum_{\mathbf{q},\mathbf{q}'} M_{\mathbf{q},\mathbf{q}'} A_{\mathbf{q}} A_{\mathbf{q}'} A_{-\mathbf{q}-\mathbf{q}'}$$

Here the self-energy term is

$$\pi^{(1)}(\mathbf{q}, \omega) = i \int \frac{d\omega'}{2\pi} \int \frac{d^3q'}{(2\pi)^3} |M_{\mathbf{q},\mathbf{q}'}|^2 D^{(0)}(\mathbf{q}', \omega') D^{(0)}(\mathbf{q} + \mathbf{q}', \omega + \omega') \tag{2.8.2}$$

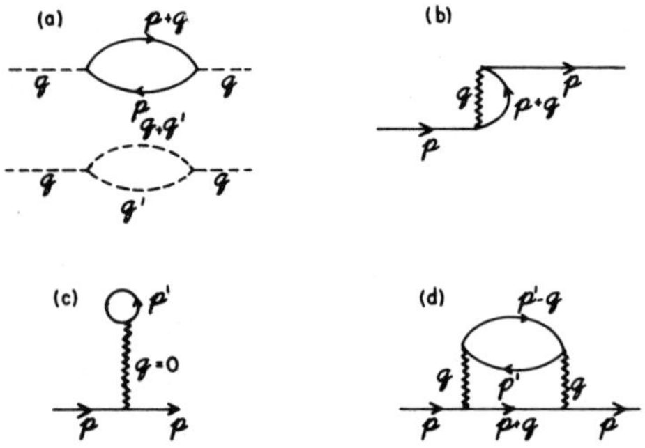

FIGURE 2.9

Next we consider terms in the electron self-energy arising from electron–electron interactions. The self-energy term in Fig. 2.9(b) is called the *unscreened exchange energy* and is a very important contribution to the electron's energy. Its self-energy is

$$\Sigma(\mathbf{p}, E) = \frac{i}{(2\pi)^4} \int d\omega \int d^3q v_{\mathbf{q}} G^{(0)}(\mathbf{p} + \mathbf{q}, \omega + E) \tag{2.8.3}$$

This result can be simplified immediately using an important identity:

$$i \int_{-\infty}^{\infty} \frac{d\omega}{2\pi} G^{(0)}(\mathbf{p} + \mathbf{q}, E + \omega) = -n_F(\xi_{\mathbf{p}+\mathbf{q}}) \tag{2.8.4}$$

This identity is proved by replacing the Green's function by its Fourier definition,

$$i \int_{-\infty}^{\infty} \frac{d\omega}{2\pi} \int_{-\infty}^{\infty} dt e^{it(\omega+E)} G^{(0)}(\mathbf{p} + \mathbf{q}, t)$$

and then inverting the order of integrations. The frequency integral gives a delta function in time,

$$\int_{-\infty}^{\infty} \frac{d\omega}{2\pi} e^{it\omega} = \delta(t)$$

and thus we obtain

$$i \int_{-\infty}^{\infty} \frac{d\omega}{2\pi} G^{(0)}(\mathbf{p} + \mathbf{q}, \omega + E) = iG^{(0)}(\mathbf{p} + \mathbf{q}, t = 0)$$

The right-hand side is ambiguous. Since the Green's function is a time-ordered product, then as $t = 0$ is approached from plus time and negative time, we obtain the different results

$$G^{(0)}(\mathbf{p}+\mathbf{q}, t \to 0^+) = -i\langle C_{\mathbf{p}+\mathbf{q}} C^\dagger_{\mathbf{p}+\mathbf{q}}\rangle = -i[1 - n_F(\xi_{\mathbf{p}+\mathbf{q}})]$$

$$G^{(0)}(\mathbf{p}+\mathbf{q}, t \to 0^-) = i\langle C^\dagger_{\mathbf{p}+\mathbf{q}} C_{\mathbf{p}+\mathbf{q}}\rangle = i n_F(\xi_{\mathbf{p}+\mathbf{q}})$$

In our identity (2.8.4) we chose the $t = 0^-$ result. We agreed in Sec. 2.5 that equal time operators were to be taken in this order. Thus we obtain the identity (2.8.4). The self-energy (2.8.3) may be written in the simple form

$$\Sigma_x(\mathbf{p}) = -\frac{1}{\nu}\sum_{\mathbf{q}} v_{\mathbf{q}} n_F(\xi_{\mathbf{p}+\mathbf{q}}) \tag{2.8.5}$$

The self-energy is no longer a function of the energy E of the particle, since all E dependence has vanished from the right-hand side. This exchange energy is easily evaluated at zero temperature (see Problem 6 at the end of the chapter).

The next electron self-energy from electron–electron interactions is shown in Fig. 2.9(c):

$$\Sigma_H(\mathbf{p}, E) = \frac{-2i}{(2\pi)^4}\int dE' \int d^3p' v_{\mathbf{q}=0} G^{(0)}(\mathbf{p}', E')$$

Our identity (2.8.4) may be used again to produce the result

$$\Sigma_H = 2v_{\mathbf{q}=0}\sum_{\mathbf{p}'} n_F(\xi_{\mathbf{p}'}) = v_{\mathbf{q}=0}N$$

where N is the number of electrons. Of course, when $v_q = 4\pi e/q^2\nu$, then the limit $\mathbf{q} \to 0$ gives infinity. This term is the unscreened coulomb energy from one electron interacting with all the other electrons in our system. This potential energy is truly a large number, which becomes infinity in the limit of an infinite system. But there must be an equal amount of positive charge in the system, and the electron interaction with the positive charge yields another large number which cancels the present divergence. The Hartree energy is defined as the net interaction energy of an electron from both of the negative and positive charge sources.

The third electron self-energy diagram is shown in Fig. 2.9(d). Its evaluation yields

$$\Sigma(\mathbf{p}, E) = i\int \frac{dE'}{2\pi}\int \frac{d^3q}{(2\pi)^3} v_q^2 P^{(1)}(\mathbf{q}, E') G^{(0)}(\mathbf{p}+\mathbf{q}, E+E')$$

The factor $P^{(1)}(\mathbf{q}, E')$ is given in (2.8.1). It corresponds to the closed fermion

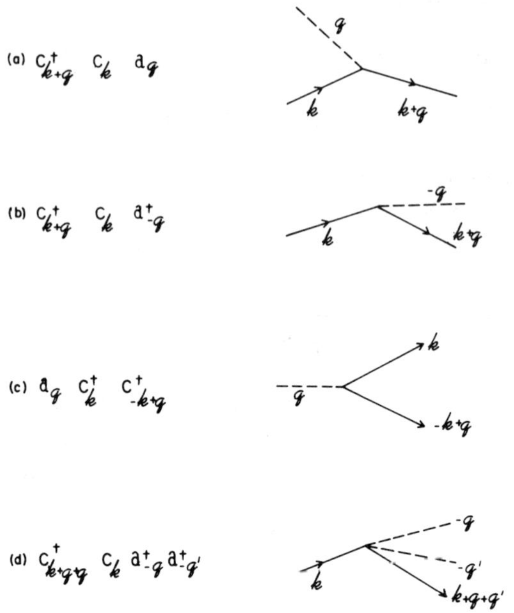

FIGURE 2.10

loop. This polarization diagram occurs frequently, is very important, and will be evaluated in Chapter 5.

Finally, we end this section with a short lecture on the way to draw diagrams. The types of diagrams we draw are very intimately connected with the types of terms in the Hamiltonian we are trying to solve. Figures 2.10(a) and 2.10(b) show the two types of electron–phonon vertices and the terms they correspond with in the Hamiltonian. In each case, an electron is destroyed in state $\mathbf{k}$ (incoming arrow) and created in $\mathbf{k} + \mathbf{q}$ (outgoing arrow). All our electron–phonon diagrams involve only diagrams which can be constructed from these basic building blocks. We do not draw diagrams in which one phonon makes two electrons, as shown in Fig. 2.9(c), since such terms do not occur in our Hamiltonian. Later we shall encounter such terms for other problems. Also, we do not include terms in which two phonons are emitted while scattering one electron. These are anharmonic terms in the electron–phonon interaction. However, it has recently been suggested that they are important in some materials, so that in fact people do draw diagrams using them. The important point is that one must start with a Hamiltonian, decide which kind of vertex terms are permitted by each term, and draw Feynman diagrams using only these basic building blocks.

2.9. TIME-LOOP S MATRIX

The S matrix is defined in Sec. 2.2. The time in (2.2.4) is taken over the interval $(-\infty, +\infty)$. The state at $t = -\infty$ is well defined as the ground state of the noninteracting system. The interactions are turned on slowly. At $t \sim 0$ the fully interacting ground state is $\psi(0) = S(0, -\infty)\phi_0$. In condensed matter physics the state at $t \to \infty$ must be defined carefully. If the interactions remain on, then this state is not well described by the noninteracting ground state ϕ_0. Alternately, one could require that the interactions turn off at large times, which returns the system to the ground state ϕ_0.

Schwinger (1961) suggested another method of handling the asymptotic limit $t \to \infty$. He proposed that the time integral in the S matrix has two pieces: one goes from $(-\infty, \tau)$ while the second goes from $(\tau, -\infty)$. Eventually $\tau \to \infty$. The integration path is a time loop, which starts and ends at $t = -\infty$. The advantage of this method is that one starts and ends the S matrix expansion with a known state $\psi(-\infty) = \phi_0$. Usually it is the only ground state one knows exactly.

For equilibrium phenomena the time-loop method of evaluating the S matrix gives results that are identical to the other methods such as those described earlier in this chapter. A small advantage of the time-loop is that the formalism has a sounder philosophical basis since the state $\psi(\infty)$ is avoided. However, the main advantage of the time-loop method is in describing nonequilibrium phenomena using Green's functions. Nonequilibrium theory is entirely based upon this formalism, or equivalent methods. The equation of motion for the Green's function can be cast into the form of a quantum Boltzmann equation for transport theory. This application will be developed in Chapter 7.

A disadvantage of the time-loop method is that it employs six different Green's functions. They are discussed next, as a preliminary to the expansion of the S matrix. Two of these Green's functions, called retarded and advanced, are also needed for the later discussion of equilibrium theory at finite temperature.

The need for six Green's functions is explained by considering the time-loop expression for the S matrix:

$$S(-\infty, -\infty) = T_s \exp\left[-i \int_{\text{loop}} ds_1 \hat{V}(s_1) \right] \tag{2.9.1}$$

The integration path is the time-loop shown in Fig. 2.11. The variable s_1 goes $(-\infty, \tau)$ and then $(\tau, -\infty)$. The operator T_s orders along the entire loop, with earliest values of s_1 occurring first. In expanding the S matrix, we will encounter Green's functions of the form

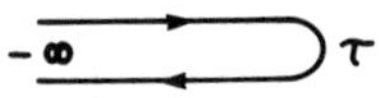

FIGURE 2.11. The time-loop integration path in the S matrix. Eventually set $\tau \to \infty$.

$$G(\lambda, s_1 - s_2) = -i_0\langle T_s C_\lambda(s_1)\, C_\lambda^\dagger(s_2)\rangle_0$$

If both s_1 and s_2 are on the top, or outward leg, of the loop then s ordering is identical to time ordering. However, if both s_1 and s_2 are on the lower, or backward leg, of the loop then s ordering is the opposite of time ordering. This case is called anti-time-ordering. Another case is when s_1 and s_2 are on different legs. Then they are automatically ordered, and independently of the values of the s arguments.

Besides the time-ordered Green's function, it is convenient to also define one that is anti-time-ordered, and others that have no time ordering. These different cases give rise to the four Green's functions. The other two are called retarded and advanced. They are linear combinations of these four.

A. Six Green's Functions

For the time-loop expansion, it is necessary to define six different Green's functions. It is possible to employ fewer than six since they are not independent, but using six simplifies the notation. They are all correlation functions which relate the field operator $\psi(x)$ of the particle at one point $x_1 = (\mathbf{r}_1, t_1)$ in space-time to the conjugate field operator $\psi^\dagger(x_2)$ at another point $x_2 = (\mathbf{r}_2, t_2)$. The six functions are the advanced G_{adv}, retarded G_{ret}, time-ordered G_t, anti-time-ordered $G_{\bar{t}}$, and $G^<$, $G^>$ which have no name:

$$\begin{aligned}
G^>(x_1, x_2) &= -i\langle\psi(x_1)\psi^\dagger(x_2)\rangle \\
G^<(x_1, x_2) &= i\langle\psi^\dagger(x_2)\psi(x_1)\rangle \\
G_t(x_1, x_2) &= \theta(t_1 - t_2)\, G^>(x_1, x_2) + \theta(t_2 - t_1)\, G^<(x_1, x_2) \\
G_{\bar{t}}(x_1, x_2) &= \theta(t_2 - t_1)\, G^>(x_1, x_2) + \theta(t_1 - t_2)\, G^<(x_1, x_2) \\
G_{\text{ret}}(x_1, x_2) &= G_t - G^< = G^> - G_{\bar{t}} \\
G_{\text{adv}}(x_1, x_2) &= G_t - G^> = G^< - G_{\bar{t}}
\end{aligned} \tag{2.9.2}$$

The brackets $|\rangle$ and $\langle|$ have the same meaning of Sec. 2.3 as the ground state of the interacting system. The time-ordered Green's function is the same one in (2.3.1). Here it has been written in the field operator representation.

For homogeneous systems in equilibrium, the Green's functions depend only upon the difference of their arguments $(x_1, x_2) = (x_1 - x_2)$. Then the most simple, and useful, quantities are the Fourier transforms of these quantities:

$$G(\mathbf{k}, \omega) = \int d^3r\, e^{-i\mathbf{k}\cdot\mathbf{r}} \int dt\, e^{i\omega t} G(\mathbf{r}, t) \tag{2.9.3}$$

where the symbol G represents any of the six functions. Explicit expressions for these quantities are given below.

Often the Hamiltonian H can be solved exactly in terms of eigenfunctions $\psi_\lambda(r_1)$ and eigenvalues ε_λ. Two examples are the electron in a magnetic field or a free particle. Then it is useful to have the expressions for the Green's functions in terms of these eigenfunctions. They are derived by expanding the field operators in terms of the eigenfunctions and creation $C_\lambda{}^\dagger$ and destruction C_λ operators:

$$\begin{aligned} \psi(x_1) &= \sum_\lambda C_\lambda \psi_\lambda(\mathbf{r}_1) e^{-i\varepsilon_\lambda t_1} \\ \psi^+(x_2) &= \sum_\lambda C_\lambda{}^+ \psi_\lambda{}^*(\mathbf{r}_2) e^{i\varepsilon_\lambda t_2} \end{aligned} \tag{2.9.4}$$

The Green's functions in (2.9.2) are evaluated with the occupation factor $n_\lambda = \langle C_\lambda{}^\dagger C_\lambda \rangle$ and $t = t_1 - t_2$. At zero temperature $n_\lambda = \theta(-\xi_\lambda)$ is a step function that is zero or one depending upon whether $\xi_\lambda = \varepsilon_\lambda - \mu$ is positive or negative:

$$\begin{aligned} G^>(x_1, x_2) &= -i \sum_\lambda (1 - n_\lambda) \psi_\lambda(\mathbf{r}_1) \psi_\lambda{}^*(\mathbf{r}_2) e^{-i\varepsilon_\lambda t} \\ G^<(x_1, x_2) &= i \sum_\lambda n_\lambda \psi_\lambda(\mathbf{r}_1) \psi_\lambda{}^*(\mathbf{r}_2) e^{-i\varepsilon_\lambda t} \\ G_t(x_1, x_2) &= -i \sum_\lambda [\theta(t) - n_\lambda] \psi_\lambda(\mathbf{r}_1) \psi_\lambda{}^*(\mathbf{r}_2) e^{-i\varepsilon_\lambda t} \\ G_{\bar{t}}(x_1, x_2) &= -i \sum_\lambda [\theta(-t) - n_\lambda] \psi_\lambda(\mathbf{r}_1) \psi_\lambda{}^*(\mathbf{r}_2) e^{-i\varepsilon_\lambda t} \\ G_{\text{ret}}(x_1, x_2) &= -i\theta(t) \sum_\lambda \psi_\lambda(\mathbf{r}_1) \psi_\lambda{}^*(\mathbf{r}_2) e^{-i\varepsilon_\lambda t} \\ G_{\text{adv}}(x_1, x_2) &= i\theta(-t) \sum_\lambda \psi_\lambda(\mathbf{r}_1) \psi_\lambda{}^*(\mathbf{r}_2) e^{-i\varepsilon_\lambda t} \end{aligned} \tag{2.9.5}$$

The above formulas are also valid in equilibrium at finite temperatures if $n_\lambda = 1/[\exp(\beta\xi_\lambda) + 1]$ is the thermodynamic average ($\beta = 1/k_BT$) of the occupation number.

The starting point for any calculation, at least conceptually, is the behavior of the Green's functions for systems without interactions. Then the wave functions are those for plane wave or noninteracting Bloch states, if such can be defined. The quantum number λ becomes the wave vector $\mathbf{k}$, and a spin index σ, which is usually not written. The eigenvalue combination is $\psi_k(\mathbf{r}_1)\psi_k{}^*(\mathbf{r}_2) = \exp[i\mathbf{k}\cdot(\mathbf{r}_1 - \mathbf{r}_2)]/\nu$. The superscript "(0)" or the subscript "0" on the Green's functions means to use those for a noninteracting system in equilibrium. Fourier-transforming the $\mathbf{r}$-variable to $\mathbf{k}$ as in (2.9.3) gives the free particle Green's functions $G^{(0)}(\mathbf{k}, t)$. For Fermions of band energy ε_k and occupation number $n_k = n_F(\varepsilon_k)$ they are

$$\begin{aligned}
G_t{}^{(0)}(\mathbf{k}, t) &= -i[\theta(t) - n_\mathbf{k}]\exp(-i\varepsilon_\mathbf{k} t) \\
G_{\bar{t}}{}^{(0)}(\mathbf{k}, t) &= -i[\theta(-t) - n_\mathbf{k}]\exp(-i\varepsilon_\mathbf{k} t) \\
G_0{}^{<}(\mathbf{k}, t) &= i n_\mathbf{k}\exp(-i\varepsilon_\mathbf{k} t) \\
G_0{}^{>}(\mathbf{k}, t) &= -i(1 - n_\mathbf{k})\exp(-i\varepsilon_\mathbf{k} t) \\
G_{\text{ret}}^{(0)}(\mathbf{k}, t) &= -i\theta(t)\exp(-i\varepsilon_\mathbf{k} t) \\
G_{\text{adv}}^{(0)}(\mathbf{k}, t) &= i\theta(-t)\exp(-i\varepsilon_\mathbf{k} t)
\end{aligned} \tag{2.9.6}$$

The t variable can be Fourier transformed, which gives the noninteracting Green's function of frequency; the quantity δ is infinitesimal:

$$\begin{aligned}
G_{\text{ret}}^{(0)}(\mathbf{k}, \omega) &= \frac{1}{\omega - \varepsilon_\mathbf{k} + i\delta} \\
G_{\text{adv}}^{(0)}(\mathbf{k}, \omega) &= \frac{1}{\omega - \varepsilon_\mathbf{k} - i\delta} \\
G_0{}^{<}(\mathbf{k}, \omega) &= 2\pi i n_\mathbf{k}\delta(\omega - \varepsilon_\mathbf{k}) \\
G_0{}^{>}(\mathbf{k}, \omega) &= -2\pi i(1 - n_\mathbf{k})\delta(\omega - \varepsilon_\mathbf{k}) \\
G_t{}^{(0)} &= G_{\text{ret}}^{(0)} + G_0{}^{<} = \frac{1}{\omega - \varepsilon_\mathbf{k} + i\delta_\mathbf{k}} \\
G_{\bar{t}}{}^{(0)} &= -G_{\text{adv}}^{(0)} + G_0{}^{<} = \frac{-1}{\omega - \varepsilon_\mathbf{k} - i\delta_\mathbf{k}}
\end{aligned} \tag{2.9.7}$$

The time-ordered function $G_t{}^{(0)}$ is exactly the same one in (2.3.5). Note the two kinds of infinitesimal deltas: δ is always positive, while δ_k is positive for $k > k_F$ and negative for $k < k_F$ as defined earlier in (2.3.5). The retarded functions always have a positive δ, even for electrons in a partially filled band. The noninteracting advanced Green's function resembles that of an empty band in (2.3.3). These two Green's functions could differ as soon as interactions are introduced, since they have different self-energy functions in degenerate Fermi systems. Also note that expressions such as $G_{\rm ret} = G_t - G^<$ are obeyed for interacting and noninteracting functions. They are obeyed for both cases of arguments $(\mathbf{k}, t)$ and $(\mathbf{k}, \omega)$.

The above Green's functions are suitable for particles such as electrons, or holes in semiconductors. Another type of Green's functions is needed for Boson fields such as phonons or photons. For phonons let $Q(x)$ be the displacement from equilibrium of the ions in the solid at position $x = (\mathbf{r}, t)$ in space-time. The phonon Green's functions are defined as follows:

$$\begin{aligned}
D^>(x_1, x_2) &= -i\langle| Q(x_1)Q(x_2) |\rangle \\
D^<(x_1, x_2) &= -i\langle| Q(x_2)Q(x_1) |\rangle \\
D_t(x_1, x_2) &= \theta(t_1 - t_2)D^>(x_1, x_2) + \theta(t_2 - t_1)D^<(x_1, x_2) \\
D_{\bar{t}}(x_1, x_2) &= \theta(t_2 - t_1)D^>(x_1, x_2) + \theta(t_1 - t_2)D^<(x_1, x_2) \\
D_{\rm ret} &= D_t - D^< = \theta(t_1 - t_2)[D^> - D^<] \\
D_{\rm adv} &= D_t - D^> = -\theta(t_2 - t_t)[D^> - D^<]
\end{aligned} \tag{2.9.8}$$

These expressions are rather similar to those in (2.9.2) for particles. The main difference is that $D^<$ and $D^>$ have the same sign, since no sign change is made when interchanging the positions of Boson operators. Also the displacement operator is Hermitian $[Q^+ = Q]$, which introduces some redundancy such as $D^<(x_1, x_2) = D^>(x_2, x_1)$.

The displacement operators Q are usually represented in terms of phonon raising $(a^\dagger)$ and lowering (a) operators. The usual case is to use $A_q = a_q{}^\dagger + a_{-q}$ instead of $Q(x)$ in the phonon operator. In this representation, the phonon Green's functions in equilibrium are expressed in terms of the phonon occupation number $N_q = \langle a_q{}^\dagger a_q\rangle$, which equals $1/[\exp(\beta\omega_q) - 1]$ in thermal equilibrium at finite temperature, and equals zero at $T = 0$:

$$\begin{aligned}
D^>(\mathbf{q}, t) &= -i[(N_q + 1)e^{-i\omega_q t} + N_q e^{i\omega_q t}] \\
D^<(\mathbf{q}, t) &= -i[(N_q + 1)e^{i\omega_q t} + N_q e^{-i\omega_q t}] \\
D_{\rm ret}(\mathbf{q}, t) &= -2\,\theta(t)\sin(\omega_q t) \\
D_{\rm adv}(\mathbf{q}, t) &= -2\,\theta(-t)\sin(\omega_q t) \\
D_t(\mathbf{q}, t) &= -i\{[N_q + \theta(-t)]e^{i\omega_q t} + [N_q + \theta(t)]e^{-i\omega_q t}\} \\
D_{\bar{t}}(\mathbf{q}, t) &= -i\{[N_q + \theta(t)]e^{i\omega_q t} + [N_q + \theta(-t)]e^{-i\omega_q t}\}
\end{aligned} \tag{2.9.9}$$

This completes the description of the various Green's functions. The next step is to learn how to use them to expand the S matrix.

B. Dyson's Equation

Each of the six Green's functions can be evaluated for an interacting system. They can be expressed in the interaction representation in terms of the time-loop S matrix. For example, one of them is

$$G^<(x_1, x_2) = i\langle| S\hat{\psi}^\dagger(x_2)\hat{\psi}(x_1) |\rangle$$

where the S matrix is given in (2.9.1). This argument has no time-ordering operator since the order of the two state operators is fixed. However, the S matrix is time ordered. The time t_1 is on the upper loop, while t_2 is on the return loop. The Green's function in the above equation has an S matrix expansion of the form

$$G^<(x_1, x_2) = i\sum_n \frac{(-i)^n}{n!}\int ds_1 \int ds_2 \cdots \int ds_{n0}\langle| T_s\hat{V}(s_1)\cdots\hat{V}(s_n)\hat{\psi}^\dagger(x_2)\hat{\psi}(x_1) |\rangle_0 \tag{2.9.10}$$

where all s integrals are over the time loop. We must learn to evaluate expressions of this type.

The potential V is composed of electron, phonon, or photon operators. The operators are paired using Wick's theorem. Each pair will have a time argument such as $G(s_i, s_j)$. If both s_i and s_j are in the top loop, the expression is just the time-ordered Green's function. If they are both in the return loop, the expression is the anti-time-ordered Green's function. If one s variable is in the top loop and the other is in the bottom loop, then the T_s operator makes this expression be either $G^<$ or $G^>$. These relationships are shown in Fig. 2.12. Thus the n term in the Green's function expansion is a product of $n+1$ factors, where each factor is one of the four Green's functions in Fig. 2.12.

It helps to have a simple example. Below is given a potential term V of the type found for electrons scattering from impurities. The first term in the S matrix expansion for $G^<$ with this interaction is

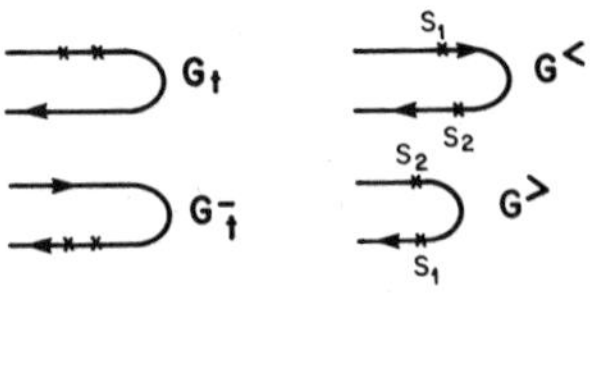

FIGURE 2.12. The four Green's functions $G(s_1, s_2)$ depend upon whether the time variables (s_1, s_2) are on the outgoing or return parts of the time-loop.

$$V = \sum_{\alpha\beta} M_{\alpha\beta} C_\alpha^{\dagger} C_\beta$$

$$G^<(\lambda, t_1 - t_2) = G_0^<(\lambda, t_1 - t_2)$$
$$+ \sum_{\alpha\beta} M_{\alpha\beta} \int ds_1 \langle TC_\lambda^{\dagger}(t_2) C_\beta(s_1)\rangle \langle TC_\alpha^{\dagger}(s_1) C_\lambda(t_1)\rangle$$

The s_1 integral runs over the time loop. For $G^<$ remember that t_1 is on the top loop while t_2 is on the return loop. There are two possibilities. If s_1 is on the top loop, the first Green's function in the s integral is $G^<$ and the second is time-ordered. If s_1 is on the return loop, the first is anti-time-ordered while the second is $G^<$. The two terms are

$$G^<(\lambda, t_1 - t_2) = G_0^<(\lambda, t_1 - t_2) + M_{\lambda\lambda} \int_{-\infty}^{\infty} dt' [G_t^{(0)}(\lambda, t_1 - t') G_0^<(\lambda, t' - t_2\rangle$$
$$- G_0^<(\lambda, t_1 - t') G_{\bar{t}}^{(0)}(\lambda, t' - t_2)] \qquad (2.9.11)$$

A sign change occurred in the last term when the direction of the time integration was changed from $(\infty, -\infty)$ to $(-\infty, \infty)$. The above expression contains only the first two terms in the S matrix expansion, which has an infinite number of terms.

In the expansion of the S matrix, each time integral produces one set of terms for the outward s leg, and another for the return leg. The n term in the S matrix expansion produces 2^n arrangements. All of these terms can be managed by using a matrix formulation.

Keldysh (1963) was the first to develop the time-loop theory for solid state applications. Here the theory is presented using a matrix notation suggested by Craig (1968). He expressed four of these Green's functions as the elements of a 2×2 matrix. The self-energy terms are also a matrix:

$$\tilde{G} = \begin{bmatrix} G_t & -G^< \\ G^> & -G_{\bar{t}} \end{bmatrix}$$
$$\tilde{\Sigma} = \begin{bmatrix} \Sigma_t & -\Sigma^< \\ \Sigma^> & -\Sigma_{\bar{t}} \end{bmatrix} \qquad (2.9.12)$$

The actual form for the self-energy functions is discussed in the following chapters.

For systems either in equilibrium or nonequilibrium, *Dyson's equation* is most easily expressed by using the matrix notation:

$$\tilde{G}(x_1, x_2) = \tilde{G}_0(x_1 - x_2) + \int_{-\infty}^{\infty} dx_3 \int_{-\infty}^{\infty} dx_4 \tilde{G}_0(x_1 - x_3)\, \tilde{\Sigma}\,(x_3, x_4) \tilde{G}(x_4, x_2) \qquad (2.9.13)$$

The matrix formulation comes directly from the time-loop. Each s-integral in the S matrix has an outward and return leg. Each of these legs gives a different Green's function. So each time integral generates two Green's functions. Hence the usefulness of the 2×2 matrix formalism.

Some simple expressions can be obtained for the Green's functions. First write the above equation in a notation where the product of two functions implies an integration over the four-variable dx, which condenses the same equation to

$$\tilde{G} = \tilde{G}_0 + \tilde{G}_0 \tilde{\Sigma} \tilde{G}$$

Then the equations are iterated. The following exact expressions are derived for the equations obeyed by the various Green's functions, using the same product notation:

$$\begin{aligned}
G_{\text{ret}} &= G_{\text{ret}}^{(0)}[1 + \Sigma_{\text{ret}} G_{\text{ret}}] \\
G_{\text{adv}} &= G_{\text{adv}}^{(0)}[1 + \Sigma_{\text{adv}} G_{\text{adv}}] \\
G^{\lessgtr} &= [1 + G_{\text{ret}} \Sigma_{\text{ret}}] G_0^{\lessgtr} [1 + \Sigma_{\text{adv}} G_{\text{adv}}] + G_{\text{ret}} \Sigma^{\lessgtr} G_{\text{adv}} \\
G_t &= [1 + G_{\text{ret}} \Sigma_{\text{ret}}] G_t^{(0)} [1 + \Sigma_{\text{adv}} G_{\text{adv}}] + G_{\text{ret}} \Sigma_{\bar{t}} G_{\text{adv}} \\
G_{\bar{t}} &= [1 + G_{\text{ret}} \Sigma_{\text{ret}}] G_{\bar{t}}^{(0)} [1 + \Sigma_{\text{adv}} G_{\text{adv}}] + G_{\text{ret}} \Sigma_t G_{\text{adv}}
\end{aligned} \tag{2.9.14}$$

These equations represent multiple integrals in d^3r and dt.

Equations (2.9.14) simplify considerably for homogeneous systems in steady state where the arguments of the Green's functions and self-energies depend only upon $(x_1 - x_2)$. If the equations are Fourier transformed, then all quantities depend only upon $(\mathbf{k}, \omega)$. The above equations are, after Fourier transforming, just algebraic quantities, which are easily solved. The time-ordered Green's function has the same equation as given in (2.7.8). Some of the other Green's functions are given below. The expressions are presented for finite temperature. The zero-temperature cases are found by setting the Fermion occupation factor $n_F(\omega) = \theta(-\omega)$:

$$\begin{aligned}
G_{\text{ret}}(\mathbf{k}, \omega) &= \frac{1}{\omega - \varepsilon_k - \Sigma_{\text{ret}}}, \qquad & \sigma &= \omega - \varepsilon_k - \operatorname{Re} \Sigma_{\text{ret}} \\
G_{\text{adv}}(\mathbf{k}, \omega) &= \frac{1}{\omega - \varepsilon_k - \Sigma_{\text{adv}}}, \qquad & \Sigma_{\text{adv}} &= \Sigma_{\text{ret}}^* \\
A(\mathbf{k}, \omega) &= -2 \operatorname{Im} G_{\text{ret}} = \frac{2\Gamma}{\sigma^2 + \Gamma^2}, \qquad & \Gamma &= -\operatorname{Im} \Sigma_{\text{ret}} > 0 \\
\Sigma^< &= 2i n_F(\omega) \Gamma(\mathbf{k}, \omega), \qquad & n_F(\omega) &= \frac{1}{e^{\beta\omega} + 1}
\end{aligned} \tag{2.9.15}$$

$$G^< = in_F(\omega)A(\mathbf{k}, \omega)$$

$$\Sigma^> = -2i(1 - n_F)\Gamma$$

$$G^> = -i(1 - n_F)A$$

The notation $\Gamma = -\operatorname{Im}\Sigma_{\text{ret}}$ will be used throughout, and rather often. The quantity $A(\mathbf{k}, \omega)$ is called the *spectral function.* It is an important quantity, which is discussed further in the next chapter. These formulas are incomplete without a prescription for calculating the different self-energy functions $\Sigma^<$, Σ_{ret}, etc. They will be presented in the next chapter.

2.10. PHOTON GREEN'S FUNCTIONS

In Sec. 1.5 we discussed the interaction of charges with themselves and with the photon field. For spinless particles, this interaction has the Hamiltonian in the nonrelativistic limit:

$$H = \sum_i \frac{1}{2m}\left[\mathbf{p}_i - \frac{e_i}{c}\mathbf{A}(\mathbf{r}_i)\right]^2 + \frac{1}{2}\sum_{ij}\frac{e_i e_j}{r_{ij}} + \sum_{\mathbf{k}\lambda}\omega_{\mathbf{k}} a^\dagger_{\mathbf{k}\lambda} a_{\mathbf{k}\lambda} \tag{2.10.1}$$

The vector potential is given by the expansion

$$\frac{1}{c}A_\mu(\mathbf{r}) = \frac{1}{\nu^{1/2}}\sum_{\mathbf{k},\lambda} e^{i\mathbf{k}\cdot\mathbf{r}} A_\mu(\mathbf{k}, t)$$

$$A_\mu(\mathbf{k}, t) = \left(\frac{2\pi}{\omega_k}\right)^{1/2} \xi_\mu(\mathbf{k}, \lambda)(a_{\mathbf{k}\lambda}e^{-it\omega_{\mathbf{k}}} + a^\dagger_{-\mathbf{k}\lambda}e^{it\omega_{\mathbf{k}}}) \tag{2.10.2}$$

The creation and destruction operators $a^\dagger_{\mathbf{k}\lambda}$, $a_{\mathbf{k}\lambda}$ obey boson statistics. Each state with wave vector $\mathbf{k}$ and polarization λ has its own harmonic oscillator statistics. The vector potential represents the photon field. Two charges may interact via their common photon field or more directly through the instantaneous Coulomb interaction $e_i e_j/r_{ij}$. The division of the interaction between photons and Coulomb field is arbitrary—both interactions come from the same basic processes. The Hamiltonian we have in (2.10.1) is in the Coulomb gauge where $\nabla \cdot \mathbf{A} = 0$. Another choice of gauge will result in a different division between photon and Coulomb. The basic forces between the particles are the same regardless of how the gauge is selected.

Now it is time to talk about Green's functions. The scalar potential $\psi(\mathbf{r}, t)$ has a Green's function. The potential from a point charge is

$$\psi(r) = \frac{e^2}{r}$$

The factor

$$v_q = \frac{4\pi e^2}{q^2 \nu}$$

which we have already been using for the Coulomb interaction, is in fact just the Green's function of the longitudinal potential. It has no frequency dependence because it is instantaneous. After all, in our rules for constructing diagrams, in Sec. 2.8, we treated it as a Green's function. If two electrons interacted by phonons, we said to put in the phonon Green's function and vertex,

$$\frac{M_q^{\ 2}}{\nu} D^{(0)}(\mathbf{q}, \omega)$$

while if two electrons interact by electron–electron interactions, we said to put in the factor

$$v_q = \frac{4\pi e^2}{q^2 \nu}$$

where we might regard $4\pi/q^2$ as the Green's function and e^2 as the vertex. Thus we were treating the Coulomb interaction as a Green's function, on equal footing with the phonon Green's function.

Since v_q is a Green's function, it has a Dyson equation of the usual form:

$$v_{\mathbf{q}}(\omega) = \frac{v_{\mathbf{q}}}{1 - v_{\mathbf{q}} P(\mathbf{q}, \omega)\nu} \tag{2.10.3}$$

The factor $P(\mathbf{q}, \omega)$ is the self-energy or polarization operator. We shall discuss its properties extensively in Chapter 5. There is one result which can be obtained with very little effort. We consider the form of Maxwell's equations in a homogeneous material with an isotropic dielectric constant ε:

$$\nabla \cdot \mathbf{B} = 0$$

$$\nabla \times \mathbf{E} = -\frac{1}{c}\frac{\partial \mathbf{B}}{\partial t}$$

$$\varepsilon \nabla \cdot \mathbf{E} = 4\pi\varrho$$

$$\nabla \times \mathbf{B} = \frac{\varepsilon}{c}\frac{\partial \mathbf{E}}{\partial t} + \frac{4\pi \mathbf{j}}{c}$$

If we solve this set in the usual way, we obtain equations for the scalar and vector potentials:

$$\begin{aligned}\psi(\mathbf{r}) &= \frac{1}{\varepsilon}\int \frac{d^3r'\varrho(\mathbf{r}')}{|\mathbf{r}-\mathbf{r}'|} \\ \nabla^2\mathbf{A} - \frac{\varepsilon}{c^2}\frac{\partial^2}{\partial t^2}\mathbf{A} &= -\frac{4\pi}{c}\mathbf{j}_t\end{aligned} \tag{2.10.4}$$

Here the Coulomb Green's function is

$$\bar{\bar{v}}_q = \frac{v_q}{\varepsilon}$$

If this result is regarded as equivalent to (2.10.3), this gives a formula for the dielectric function:

$$\varepsilon(\mathbf{q}, \omega) = 1 - \frac{4\pi e^2}{q^2}P(\mathbf{q}, \omega) \tag{2.10.5}$$

On the left-hand side we have generalized this result to include the case where the dielectric function depends on both $\mathbf{q}$ and ω. Equation (2.10.5) will serve as our definition of the *longitudinal dielectric function.* It arises from the self-energy parts of the Coulomb potential.

The Green's function for the vector potential is

$$\begin{aligned}D_{\mu\nu}(\mathbf{k}, t-t') &= -i\sum_\lambda \langle TA_\mu(\mathbf{k}, \lambda, t)A_\nu(-\mathbf{k}, \lambda, t')\rangle \\ A_\mu(\mathbf{k}, \lambda, t) &= \xi_\mu(k, \lambda)\left(\frac{2\pi}{\omega_k}\right)^{1/2}[a_{\mathbf{k}\lambda}(t) + a^\dagger_{-\mathbf{k}\lambda}(t)]\end{aligned} \tag{2.10.6}$$

where μ, ν are the x, y, z components. The sum over λ is the sum over the two transverse polarizations of the light, while the ξ_μ are the polarization vectors for each component. The free propagator at zero temperature is evaluated with the states $|\rangle$ and $\langle|$ as the photon vacuum:

$$\begin{aligned}D^{(0)}_{\mu\nu}(\mathbf{k}, t-t') &= -\frac{2\pi i}{\omega_k}\sum_\lambda \xi_\mu(\mathbf{k}, \lambda)\xi_\nu(-\mathbf{k}, \lambda)\langle T\,|\,(a_{\mathbf{k}\lambda}e^{-i\omega_{\mathbf{k}}t} + a^\dagger_{-\mathbf{k}\lambda}e^{i\omega_{\mathbf{k}}t}) \\ &\quad \times (a_{-\mathbf{k}\lambda}e^{-i\omega_{\mathbf{k}}t'} + a^\dagger_{\mathbf{k}\lambda}e^{i\omega_{\mathbf{k}}t'})\rangle \\ &= -\frac{2\pi i}{\omega_k}\sum_\lambda \xi_\mu\xi_\nu[\theta(t'-t)e^{i\omega_{\mathbf{k}}(t-t')}\langle a_{-\mathbf{k}\lambda}a^\dagger_{-\mathbf{k}\lambda}\rangle \\ &\quad + \theta(t-t')e^{i\omega_{\mathbf{k}}(t'-t)}\langle a_{\mathbf{k}\lambda}a^\dagger_{\mathbf{k}\lambda}\rangle] \\ &= -\frac{2\pi i}{\omega_k}e^{-i\omega_k|t-t'|}\sum_\lambda \xi_\mu\xi_\nu\end{aligned}$$

and its Fourier transform is

$$D_{\mu\nu}^{(0)}(\mathbf{k}, \omega) = \int_{-\infty}^{\infty} dt e^{i\omega(t-t')} D_{\mu\nu}^{(0)}(\mathbf{k}, t - t')$$

$$= \frac{4\pi}{\omega^2 - \omega_{\mathbf{k}}^2 + i\delta} \sum_{\lambda} \xi_\mu \xi_\nu$$

Next, consider the factor

$$\sum_{\lambda} \xi_\mu \xi_\nu \tag{2.10.7}$$

The unit tensor is

$$\delta_{\mu\nu} = \hat{x}\hat{x} + \hat{y}\hat{y} + \hat{z}\hat{z}$$

Or, if we go to a coordinate system in the direction of **k**,

$$\sum_{\lambda} \xi_\mu \xi_\nu + \hat{k}_\mu \hat{k}_\nu = \delta_{\mu\nu}$$

since (2.10.7) is the unit dyadic for directions perpendicular to **k**. The Green's function is

$$D_{\mu\nu}^{(0)} = \frac{4\pi[\delta_{\mu\nu} - (k_\mu k_\nu / k^2)]}{\omega^2 - \omega_{\mathbf{k}}^2 - i\delta} \tag{2.10.8}$$

We shall refer to this as the *photon Green's function*. One should keep in mind that the interaction between two charges occurs via both the scalar and vector potentials. How we divide the interaction between scalar and vector potentials is somewhat arbitrary and is determined by the gauge condition. After making this choice, we assigned the word *photon* to the vector potential part. This division between photon and Coulomb is arbitrary, and both parts should really be viewed as arising from photons.

The photon Green's function also obeys a Dyson equation. Since it is a matrix quantity, we must be careful about the treatment of indices. Previously we have only treated scalar quantities. Since the Green's function is defined as a series expansion, we actually derive an equation, say, for electrons, of the form

$$G(\mathbf{p}, E) = G^{(0)}(\mathbf{p}, E) + G^{(0)}(\mathbf{p}, E)\Sigma(\mathbf{p}, E)G(\mathbf{p}, E)$$

Since all the quantities are scalar functions, we could immediately solve this equation for $G(\mathbf{p}, E)$ and obtain (2.7.7). But for the photon Green's function, we find the equation

$$D_{\mu\nu} = D_{\mu\nu}^{(0)} + \sum_{\lambda\delta} D_{\mu\lambda}^{(0)} \pi_{\lambda\delta} D_{\delta\nu} \tag{2.10.9}$$

where $\pi_{\lambda\delta}$ is the self-energy function, which is now a 3×3 matrix. Each term in the equation is a function of $(\mathbf{k}, \omega)$. However, in homogeneous materials all matrices are of the form

$$
\begin{aligned}
D^{(0)}_{\mu\nu} &= \left(\delta_{\mu\nu} - \frac{k_\mu k_\nu}{k^2}\right) D^{(0)} \\
D_{\mu\nu} &= \left(\delta_{\mu\nu} - \frac{k_\mu k_\nu}{k^2}\right) D \\
\pi_{\mu\nu} &= \delta_{\mu\nu}\pi^{(1)} + \frac{k_\mu k_\nu}{k^2}\pi^{(2)}
\end{aligned}
\tag{2.10.10}
$$

where the factors $D^{(0)}$, D, $\pi^{(1)}$, and $\pi^{(2)}$ are scalars. The self-energy function $\pi_{\mu\nu}(\mathbf{k}, \omega)$ has the form shown, which is the most general dependence on $\mathbf{k}$ of a matrix function. Now it is simple to do the summations over the matrix components:

$$
\sum_{\lambda\delta} (\delta_{\mu\lambda} - \hat{k}_\mu \hat{k}_\lambda)(\delta_{\lambda\delta}\pi^{(1)} + \hat{k}_\lambda \hat{k}_\delta \pi^{(2)})(\delta_{\delta\nu} - \hat{k}_\delta \hat{k}_\nu) = (\delta_{\mu\nu} - \hat{k}_\mu \hat{k}_\nu)\pi^{(1)}
$$

and we really have the scalar equation

$$
D = \frac{D^{(0)}}{1 - D^{(0)}\pi^{(1)}}
$$

The transverse photon has no dependence in its self-energy on the longitudinal part of the self-energy $\pi^{(2)}$. Actual solids are periodic, rather than homogeneous, and the matrix form of $\pi_{\mu\nu}$ may be more complicated than (2.10.10).

The Dyson equation for the photon Green's function is

$$
D_{\mu\nu}(\mathbf{k}, \omega) = \frac{4\pi[\delta_{\mu\nu} - k_\mu k_\nu/k^2)]}{\omega^2 - \omega_{\mathbf{k}}{}^2 - 4\pi\pi^{(1)}(\mathbf{k}, \omega)} \tag{2.10.11}
$$

Again we emphasize that these results are appropriate for a homogeneous medium. In a real crystal, the self-energy function $\pi_{\mu\nu}$ will be transverse but not necessarily perpendicular to $\mathbf{k}$. Then one must start from (2.10.9) and actually solve for the various components of $D_{\mu\nu}$.

In a homogeneous medium with dielectric constant ε, (2.10.4) is the equation obeyed by the vector potential. The double time derivative is multiplied by ε. Since this yields the ω^2 term in the Green's function, the appropriate Green's function in a medium with dielectric function ε is

$$
D_{\mu\nu} = \frac{4\pi[\delta_{\mu\nu} - k_\mu k_\nu/k^2)]}{\varepsilon\omega^2 - \omega_{\mathbf{k}}{}^2 + i\delta} \tag{2.10.12}
$$

If we equate this $D_{\mu\nu}$ with Dyson's equation (2.10.11), for the dielectric function this gives

$$\varepsilon = 1 - \frac{4\pi}{\omega^2}\,\pi^{(1)}(\mathbf{k}, \omega)$$

This result is not very useful for crystals, since the general $\varepsilon_{\mu\nu}(\mathbf{k}, \omega)$ is a function of $(\mathbf{k}, \omega)$ and is not just the scalar function which we have given above. The scalar result is correct in crystals in the limit of $\mathbf{k} \to 0$:

$$\lim_{\mathbf{k}\to 0} \varepsilon_{\mu\nu}(\mathbf{k}, \omega) = \delta_{\mu\nu}\left[1 - \frac{4\pi}{\omega^2}\,\pi^{(1)}(\mathbf{k}, \omega)\right] \tag{2.10.13}$$

It is interesting to note that in the limit of $k \to 0$, this transverse dielectric function becomes exactly equal to the longitudinal one (2.10.5) at $\mathbf{k} \to 0$. This is not obvious yet and will be proved in Sec. 3.7.

PROBLEMS

1. Show explicitly that

$$\frac{1}{3!}\int_0^t dt_1 \int_0^t dt_2 \int_0^t dt_3 T[V(t_1)V(t_2)V(t_3)] = \int_0^t dt_1 \int_0^{t_1} dt_2 \int_0^{t_2} dt_3 V(t_1)V(t_2)V(t_3)$$

2. For the phonon Green's function $D(\mathbf{q}, t - t')$, let $V(t)$ be the electron–phonon interaction and evaluate all the $n = 2$ diagrams. Which are connected, and which are disconnected? Draw the Feynman graphs for each term.

3. Let $V(t)$ be the electron–phonon interaction in the expansion for the electron Green's function $G(\mathbf{p}, t - t')$. What are the contributions from the different connected diagrams for $n = 4$ (two phonons). Just draw the graphs. Also draw all the graphs for the disconnected diagrams.

4. Let $V(t)$ be the electron–electron interaction in the expansion for the electron Green's function. Evaluate the term for $n = 1$, including all equations and Feynman graphs. Also draw the connected diagrams for $n = 2$.

5. Prove the Feynman result $e^{A+B} = e^A e^B e^{-(1/2)[A,B]}$, which is true only if $[A, B]$ commutes with both A and B. *Hint*: Recall that

$$e^{-it(H_0+V)} = e^{-itH_0} T \exp\left[-i\int_0^t dt_1 (e^{it_1 H_0} V e^{-it_1 H_0})\right]$$

Use the same method to prove that

$$e^{s(A+B)} = e^{sA} T_s \exp\left(\int_0^s ds_1 e^{-s_1 A} B e^{s_1 A}\right)$$

and evaluate this for $s = 1$.

6. Evaluate the wave vector integrals at zero temperature for the exchange energy in (2.8.5).

7. Express $U_1(t) = e^{iH_0t}e^{-iHt}$ and $U_2(t) = e^{iHt}e^{-iH_0t}$ as time-ordered exponential integrals for both cases $t > 0$ and $t < 0$. You will need to use the operator T^{-1}, which arranges operators in their inverse time ordering.

8. Evaluate the first interaction term in Dyson's equation (2.9.13) and show that it does give (2.9.11) when $\Sigma(x_3, x_4)_{\mu\nu} = \delta^4(x_3 - x_4)V(x_3)\delta_{\mu\nu}$.

9. Show that Eq. (2.9.14) for G_t is identical to (2.7.8) using the result in (2.9.15). Self-energies obey the same relations as the Green's functions: i.e., $\Sigma_t = \Sigma^r + \Sigma^<$, etc.

Chapter 3

Green's Functions at Finite Temperatures

3.1. INTRODUCTION

Experiments are done at finite temperatures. Since one goal of many-body theory is to explain experiments (another is to predict them), we should do our theories at finite temperatures. This is often unnecessary if the temperature is small compared to other energies in the problem. But often temperature is important, and here we shall learn how to incorporate it into Green's functions. The finite temperature formalism was originated by Matsubara. It will actually be easier to use than the zero temperature theory of Chapter 2, so that we shall use the Matsubara method throughout the remainder of the book. The zero temperature result is always easily obtained from the finite temperature result by just setting $T = 0$.

At finite temperatures, we assume that there is something with a finite temperature. That is, our particle, whether electron, phonon, or spin, is interacting with a bath of other particles which have an average energy. We do not know, or need to know, the exact state of all these other particles, since they are fluctuating between different configurations. All we know is the temperature, which is related to the mean energy.

Thus when we define the Green's function, we must average over all possible configurations of the system. A possible Green's function for the electron is

$$\frac{\mathrm{Tr}[e^{-\beta H} C_p(t) C_p^{\dagger}(t')]}{\mathrm{Tr}(e^{-\beta H})}$$
$$C_{\mathbf{p}}(t) = e^{itH} C_{\mathbf{p}} e^{-itH} \tag{3.1.1}$$

where the symbol "Tr" denotes trace and is the summation over some complete set of states:

$$\mathrm{Tr} \equiv \sum_n \langle n \mid \cdots \mid n \rangle$$

The definition (3.1.1) would be suitable for a Green's function and is $iG^>(\mathbf{p}; t, t')$. However, it has one drawback which makes its use unwieldy. Usually we are going to write the Hamiltonian

$$H = H_0 + V$$

as a part H_0 that we can solve exactly and a part V which remains and becomes the perturbation. However, V now appears in two different places. First it is in $\exp(\pm iHt)$, which can be expanded in the usual S matrix. But it also occurs in the factor $\exp(-\beta H)$. Thus we must also do a perturbation expansion on this thermodynamic weighting factor. Of course, it is a nuisance to be doing two different expansions at once.

The Hamiltonian enters both terms as an exponential factor. Thus we can loosely think of $\beta = 1/k_BT$ as just a complex time. The Matsubara (1955) method does just the converse; it treats time as a complex temperature. The object is to treat t and β as the real and imaginary parts of a complex variable, which will require only one S-matrix expansion.

Another motivation for the Matsubara method is provided by examining the thermal occupation numbers for bosons $(e^{\beta\omega_\mathbf{q}} - 1)^{-1}$ and fermions $(e^{\beta\xi_\mathbf{p}} + 1)^{-1}$. Each of these can be expanded in a series ($\xi_\mathbf{p} = \varepsilon_\mathbf{p} - \mu$):

$$n_F(\xi_\mathbf{p}) = \frac{1}{e^{\beta\xi_\mathbf{p}} + 1} = \frac{1}{2} + \frac{1}{\beta} \sum_{n=-\infty}^{\infty} \frac{1}{(2n+1)i\pi/\beta - \xi_\mathbf{p}} \tag{3.1.2}$$

$$n_B(\omega_\mathbf{q}) = \frac{1}{e^{\beta\omega_\mathbf{q}} - 1} = -\frac{1}{2} + \frac{1}{\beta} \sum_{n=-\infty}^{\infty} \frac{1}{2n\pi i/\beta - \omega_\mathbf{q}} \tag{3.1.3}$$

These series can be derived from a theorem which states that any meromorphic function may be expanded as a summation over its poles and residues at those poles. Thus the boson occupation factor $(e^{\beta\omega_\mathbf{q}} - 1)^{-1}$ has poles at $\omega_q = 2ni\pi/\beta$ and the fermion factor $(e^{\beta\xi_\mathbf{p}} + 1)^{-1}$ has poles at $\xi_p = (2n+1)i\pi/\beta$. It is convenient to define the frequencies at the pole

$$\omega_n = \frac{(2n+1)\pi}{\beta}, \qquad \text{fermions}$$

$$\omega_n = \frac{2n\pi}{\beta}, \qquad \text{bosons}$$

where the fermions have poles at odd multiples of π/β, while bosons have poles at even multiples, including zero. Thus both summations above can be written as

$$\sum_n \frac{1}{i\omega_n - \omega_{\mathbf{q}}} \qquad \text{or} \qquad \sum_n \frac{1}{i\omega_n - \xi_{\mathbf{p}}}$$

where for fermions we sum over only odd integers and for bosons over even integers. The factor

$$\frac{1}{i\omega_n - \xi_{\mathbf{p}}}$$

has the nature of a Green's function. Indeed, it is the unperturbed Green's function in the Matsubara method.

In the Matsubara method, time becomes a complex quantity which is usually called τ, where $\tau = it$. Green's functions are functions of τ with domain

$$-\beta \leq \tau \leq \beta$$

Fourier transform theory states that if a function $f(\tau)$ is defined over the range $-\beta \leq \tau \leq \beta$, then its Fourier expansion is

$$f(\tau) = \frac{1}{2} a_0 + \sum_{n=1}^{\infty} \left[a_n \cos\left(\frac{n\pi\tau}{\beta}\right) + b_n \sin\left(\frac{n\pi\tau}{\beta}\right) \right]$$

where

$$a_n = \frac{1}{\beta} \int_{-\beta}^{\beta} d\tau f(\tau) \cos\left(\frac{n\pi\tau}{\beta}\right)$$

$$b_n = \frac{1}{\beta} \int_{-\beta}^{\beta} d\tau f(\tau) \sin\left(\frac{n\pi\tau}{\beta}\right)$$

Another way to write this is to define

$$f(i\omega_n) = \frac{\beta(a_n + ib_n)}{2}$$

and hence

$$f(\tau) = \frac{1}{\beta} \sum_{n=-\infty}^{\infty} e^{-in\pi\tau/\beta} f(i\omega_n) \tag{3.1.4}$$

$$f(i\omega_n) = \frac{1}{2} \int_{-\beta}^{\beta} d\tau f(\tau) e^{in\pi\tau/\beta} \tag{3.1.5}$$

There is still a further simplification which can be achieved. Later we shall find that boson Green's functions have the additional property that

$$\text{Boson:}\quad f(\tau)=f(\tau+\beta)\quad\text{when } -\beta<\tau<0 \text{ (and } 0<\tau+\beta<\beta) \tag{3.1.6}$$

If we divide the integral (3.1.5) into its negative and positive regions,

$$f(i\omega_n)=\frac{1}{2}\left[\int_0^\beta d\tau f(\tau)e^{in\pi\tau/\beta}+\int_{-\beta}^0 d\tau f(\tau)e^{in\pi\tau/\beta}\right]$$

and change variables in the second term from τ to $\tau+\beta$, this gives

$$f(i\omega_n)=\tfrac{1}{2}(1+e^{-in\pi})\int_0^\beta d\tau f(\tau)e^{in\pi\tau/\beta}$$

Thus we get that $f(i\omega_n)=0$ whenever n is odd. Thus, for bosons,

$$\left.\begin{aligned} f(i\omega_n)&=\int_0^\beta d\tau f(\tau)e^{i\omega_n\tau}\\ f(\tau)&=\frac{1}{\beta}\sum_n e^{-i\omega_n\tau}f(i\omega_n)\\ \omega_n&=\frac{2n\pi}{\beta}\end{aligned}\right\}\quad\text{bosons} \tag{3.1.7}$$

This result agrees with the previous observation (3.1.3) that boson frequencies contain only even integers.

Similarly, the fermion Green's functions will have the property that

$$\text{Fermions:}\quad f(\tau)=-f(\tau+\beta)\quad\text{when } -\beta<\tau<0 \tag{3.1.8}$$

The same manipulations on the integral in (3.1.5) give

$$f(i\omega_n)=\tfrac{1}{2}(1-e^{in\pi})\int_0^\beta d\tau e^{in\pi\tau/\beta}f(\tau)$$

In this case $f(i\omega_n)=0$ if n is even, while for n an odd integer,

$$\left.\begin{aligned} f(i\omega_n)&=\int_0^\beta d\tau f(\tau)e^{i\omega_n\tau}\\ f(\tau)&=\frac{1}{\beta}\sum_n e^{-i\omega_n\tau}f(i\omega_n)\\ \omega_n&=\frac{(2n+1)\pi}{\beta}\end{aligned}\right\}\quad\text{fermions} \tag{3.1.9}$$

These equations are identical in form to those in (3.1.7). The only difference is whether the frequency has even or odd integers. This pair of equations will be used often to define the Fourier transforms of Green's functions.

At this point the reader is probably overwhelmed with complex frequencies and complex times. One might argue that this suffering is necessary when including finite temperatures. Actually, the forecast is not as gloomy as that. We shall find that the Matsubara method is very easy to use. It is particularly good for evaluating high-order diagrams with many internal lines. Thus it is a remarkably easy method to use in practice. Except for the notion of complex times and frequencies, it is not that much different from the zero temperature Green's functions which we defined in Chapter 2.

A second great merit of the Matsubara method is that it leads us directly to physical results. In Secs. 3.7 and 3.8 we shall derive some Kubo formulas for the exact definitions of physical quantities such as the electrical conductivity, thermal conductivity, magnetic susceptibility, etc. In Sec. 3.3 we shall show that these are just the retarded Green's functions. Finally, we shall show that the Matsubara Green's functions lead directly to the retarded functions. Our Matsubara functions will be functions of the complex frequencies $i\omega_n$ such as $f(i\omega_n)$. We shall show that the equivalent retarded function is obtained by replacing $i\omega_n$ by $\omega + i\delta$, where δ is infinitesimal. This is called analytical continuation. In practice, one just takes the formula one has derived for $f(i\omega_n)$, erases $i\omega_n$ everywhere, and replaces it by $\omega + i\delta$. This simple procedure yields the retarded function, which is what we want for the physically measurable quantities. Thus the Matsubara technique is a direct method of calculating the quantities which can be compared with experiment.

3.2. MATSUBARA GREEN'S FUNCTIONS

The electron Green's function is defined as

$$\mathscr{G}(\mathbf{p}, \tau - \tau') = -\langle T_\tau C_\mathbf{p}(\tau) C_\mathbf{p}^\dagger(\tau')\rangle \tag{3.2.1}$$

$$\mathscr{G}(\mathbf{p}, \tau - \tau') = -\,\mathrm{Tr}(e^{-\beta(H-\mu N-\Omega)} T_\tau e^{\tau(H-\mu N)} C_\mathbf{p} e^{-(H-\mu N)(\tau-\tau')} \times C_\mathbf{p}^\dagger e^{-\tau'(H-\mu N)}) \tag{3.2.2}$$

$$e^{-\beta\Omega} = \mathrm{Tr}(e^{-\beta(H-\mu N)}) \tag{3.2.3}$$

These definitions have several features and conventions which need to be explained. First, the bracket $\langle \cdots \rangle$ in (3.2.1) has the definition implied by the equivalent equation (3.2.2). The bracket $\langle O \rangle$ on an operator O means that the thermodynamic average is made. Second, the Hamiltonian is now replaced by $H - \mu N$, where μ is the chemical potential and N is the particle number operator. We have introduced a grand canonical ensemble, where the number of particles is variable. This definition of the Green's function applies to a many-particle system. We shall see that it can also be used very successfully for one particle in an empty band. In the latter case, we take the last analytical continuation as $i\omega_n \to E + \mu + i\delta$, and the chemical potential will vanish from all expressions. One is not bothered by the fact that $\beta\mu \ll 0$ in one-particle systems at finite temperatures. In the many-electron system we shall take $i\omega_n \to E + i\delta$ and thereby measure energy from the chemical potential (Fermi energy). The factor

$$T_\tau$$

is a τ ordering operator, which arranges operators with earliest τ (closest to $-\beta$) to the right. It serves the same function as the time ordering operator in the zero temperature Green's functions. The subscript τ is affixed to T to distinguish this operator from the temperature. The thermodynamic potential Ω in $\exp -\Omega\beta$ is the usual normalization factor for a thermodynamic average. The script symbol $\mathscr{G}$ has been used for these Matsubara functions. This will always be done to alert the reader that these are Green's functions of complex time and complex frequency.

In (3.2.1) the Green's function on the left has been written as a function of the difference $\tau - \tau'$, although the right-hand side is not obviously a function of only the difference. We shall prove this to be the case. First we explicitly write the Green's function for the separate cases for $\tau > \tau'$ and $\tau < \tau'$:

$$\begin{aligned} K &= H - \mu N \\ \mathscr{G}(\mathbf{p}, \tau - \tau') &= -\theta(\tau - \tau') \operatorname{Tr}(e^{-\beta(K-\Omega)} e^{+\tau K} C_\mathbf{p} e^{-K(\tau-\tau')} C_\mathbf{p}^\dagger e^{-\tau' K}) \\ &\quad + \theta(\tau' - \tau) \operatorname{Tr}(e^{-\beta(K-\Omega)} e^{+\tau' K} C_\mathbf{p}^\dagger e^{-K(\tau'-\tau)} C_\mathbf{p} e^{-\tau K}) \end{aligned} \tag{3.2.4}$$

The sign change in the second term appears whenever we interchange two fermion operators. Next we use the theorem that the trace is unchanged by a cyclic variation of the operators

$$\operatorname{Tr}(ABC \cdots YZ) = \operatorname{Tr}(BC \cdots XYZA) \tag{3.2.5}$$

so that (3.2.4) can be rewritten as

$$\begin{aligned} \mathscr{G}(p, \tau - \tau') &= -\theta(\tau - \tau') \operatorname{Tr}(e^{-\tau' K} e^{-\beta(K-\Omega)} e^{\tau K} C_\mathbf{p} e^{-K(\tau-\tau')} C_\mathbf{p}^\dagger) \\ &\quad + \theta(\tau' - \tau) \operatorname{Tr}(e^{-\tau K} e^{-\beta(K-\Omega)} e^{\tau' K} C_\mathbf{p}^\dagger e^{-K(\tau'-\tau)} C_\mathbf{p}) \end{aligned}$$

Finally, we can commute the exponential operators,

$$e^{-\tau' K} e^{-\beta(K-\Omega)} = e^{-\beta(K-\Omega)} e^{-\tau' K}$$

since they both contain the same operators [the thermodynamic potential Ω is not an operator but is a scalar function of β and μ, as defined in (3.2.3)]:

$$\begin{aligned}\mathscr{G}(p, \tau - \tau') = &-\theta(\tau - \tau')\,\mathrm{Tr}(e^{-\beta(K-\Omega)} e^{K(\tau-\tau')} C_p e^{-K(\tau-\tau')} C_p{}^\dagger) \\ &+\theta(\tau' - \tau)\,\mathrm{Tr}(e^{-\beta(K-\Omega)} e^{K(\tau'-\tau)} C_{\mathbf{p}}{}^\dagger e^{-K(\tau'-\tau)} C_{\mathbf{p}})\end{aligned}$$

The right-hand side of this equation is now a function only of the combination $\tau - \tau'$. Thus we can write the Green's function as a function of this difference. It also enables us to drop one of the time variables entirely since it is unnecessary. Thus an equivalent definition of the Green's function is

$$\begin{aligned}\mathscr{G}(\mathbf{p}, \tau) &= -\langle T_\tau C_{\mathbf{p}}(\tau) C_{\mathbf{p}}{}^\dagger(0)\rangle \\ &= -\mathrm{Tr}[e^{-\beta(K-\Omega)} T_\tau(e^{\tau K} C_{\mathbf{p}} e^{-\tau K} C_{\mathbf{p}}{}^\dagger)]\end{aligned} \tag{3.2.6}$$

Next we wish to examine the behavior of the Green's function for $\tau < 0$ to verify that it does have the property asserted in (3.1.8):

$$\tau < 0: \qquad \mathscr{G}(\mathbf{p}, \tau) = +\mathrm{Tr}(e^{-\beta(K-\Omega)} C_{\mathbf{p}}{}^\dagger e^{\tau K} C_{\mathbf{p}} e^{-\tau K})$$

By using the cyclic property of the trace several times, this equation can be rearranged into

$$\tau < 0: \qquad \mathscr{G}(\mathbf{p}, \tau) = \mathrm{Tr}(e^{\beta\Omega} e^{\tau K} C_{\mathbf{p}} e^{-\tau K} e^{-\beta K} C_{\mathbf{p}}{}^\dagger)$$

We do not need to cycle exp $\beta\Omega$ since it is not an operator. These terms can be regrouped by adding exp $\pm\beta K$ to the first terms to give

$$\tau < 0: \qquad \mathscr{G}(\mathbf{p}, \tau) = \mathrm{Tr}(e^{-\beta(K-\Omega)} e^{(\tau+\beta)K} C_{\mathbf{p}} e^{-(\tau+\beta)K} C_{\mathbf{p}}{}^\dagger)$$

The term on the right is $-\mathscr{G}(p, \tau + \beta)$ when $0 < \tau + \beta < \beta$. Thus we have shown that

$$-\beta < \tau < 0: \qquad \mathscr{G}(\mathbf{p}, \tau) = -\mathscr{G}(\mathbf{p}, \tau + \beta)$$

as asserted earlier in (3.1.8). We know immediately that the Green's function can be expanded in a Fourier series of the type in (3.1.9):

$$\begin{aligned}\mathscr{G}(\mathbf{p}, i\omega_n) &= \int_0^\beta d\tau\, \mathscr{G}(\mathbf{p}, \tau) e^{i\tau\omega_n} \\ \mathscr{G}(\mathbf{p}, \tau) &= \frac{1}{\beta} \sum_n e^{-i\omega_n \tau} \mathscr{G}(\mathbf{p}, i\omega_n)\end{aligned} \tag{3.2.7}$$

Equation (3.2.7) serves as the definition of $\mathscr{G}(\mathbf{p}, i\omega_n)$, where ω_n is always an odd multiple of π/β for fermions.

The unperturbed Green's function, or free-particle Green's function, is obtained from (3.2.6) by using for the Hamiltonian

$$H = H_0 = \sum_{\mathbf{p}} \varepsilon_{\mathbf{p}} C_{\mathbf{p}}^{\dagger} C_{\mathbf{p}}$$
$$K_0 = H_0 - \mu N = \sum_{\mathbf{p}} (\varepsilon_{\mathbf{p}} - \mu) C_{\mathbf{p}}^{\dagger} C_{\mathbf{p}} = \sum_{\mathbf{p}} \xi_{\mathbf{p}} C_{\mathbf{p}}^{\dagger} C_{\mathbf{p}}$$

The τ evolution of the operators is just

$$\begin{aligned} C_{\mathbf{p}}(\tau) &= e^{\tau K_0} C_{\mathbf{p}} e^{-\tau K_0} = e^{-\tau \xi_{\mathbf{p}}} C_{\mathbf{p}} \\ C_{\mathbf{p}}^{\dagger}(\tau) &= e^{\tau K_0} C_{\mathbf{p}}^{\dagger} e^{-\tau K_0} = e^{\tau \xi_{\mathbf{p}}} C_{\mathbf{p}}^{\dagger} \end{aligned} \tag{3.2.8}$$

which is easily derived from the Baker–Hausdorff theorem:

$$e^A C e^{-A} = C + [A, C] + \frac{1}{2!}[A, [A, C]] + \frac{1}{3!}[A, [A, [A, C]]] + \cdots$$

Thus the τ form of the Green's function is

$$\mathscr{G}^{(0)}(\mathbf{p}, \tau) = -\theta(\tau) e^{-\tau \xi_{\mathbf{p}}} \langle C_{\mathbf{p}} C_{\mathbf{p}}^{\dagger} \rangle + \theta(-\tau) e^{-\tau \xi_{\mathbf{p}}} \langle C_{\mathbf{p}}^{\dagger} C_{\mathbf{p}} \rangle$$

or

$$\mathscr{G}^{(0)}(\mathbf{p}, \tau) = -e^{-\tau \xi_{\mathbf{p}}} \{\theta(\tau)[1 - n_F(\xi_{\mathbf{p}})] - \theta(-\tau) n_F(\xi_{\mathbf{p}})\}$$

or

$$\mathscr{G}^{(0)}(\mathbf{p}, \tau) = -e^{-\tau \xi_{\mathbf{p}}} [\theta(\tau) - n_F(\xi_{\mathbf{p}})] \tag{3.2.9}$$

where $n_F(\xi_{\mathbf{p}})$ is the expectation of the number operator:

$$n_F(\xi_{\mathbf{p}}) = \langle C_{\mathbf{p}}^{\dagger} C_{\mathbf{p}} \rangle$$

which from elementary statistical mechanics has the form

$$n_F(\xi_{\mathbf{p}}) = \frac{1}{e^{\beta \xi_{\mathbf{p}}} + 1} \tag{3.2.10}$$

It is also easy to obtain the Green's function of frequency:

$$\mathscr{G}^{(0)}(\mathbf{p}, i\omega_n) = \int_0^\beta d\tau e^{+i\omega_n \tau} \mathscr{G}^{(0)}(\mathbf{p}, \tau) = -(1 - n_F) \int_0^\beta d\tau e^{+\tau(i\omega_n - \xi_{\mathbf{p}})}$$

The exponential integral is simple and yields

$$\mathscr{G}^{(0)}(\mathbf{p}, i\omega_n) = \frac{-(1 - n_F)(e^{\beta(i\omega_n - \xi_{\mathbf{p}})} - 1)}{i\omega_n - \xi_{\mathbf{p}}}$$

The second term in the numerator may be simplified by remembering that

$$\beta i\omega_n = i(2n+1)\pi$$
$$e^{\beta i\omega_n} = -1$$

and the exponential of this is -1. Thus we have

$$\mathscr{G}^{(0)}(\mathbf{p}, i\omega_n) = \frac{(1-n_F)(e^{-\beta\xi_{\mathbf{p}}}+1)}{i\omega_n - \xi_{\mathbf{p}}}$$

From (3.2.10) it is simple to show that

$$1 - n_F = \frac{1}{e^{-\beta\xi_{\mathbf{p}}}+1}$$

and hence we obtain the following simple form for the Green's function:

$$\mathscr{G}^{(0)}(\mathbf{p}, i\omega_n) = \frac{1}{i\omega_n - \xi_{\mathbf{p}}} \tag{3.2.11}$$

This result for $\mathscr{G}^{(0)}$ does have the form suggested in Sec. 3.1. Temperature information is still contained in this expression but now only in the frequency $(2n+1)\pi/\beta$. We shall see later how the occupation factors reenter the expressions when we actually evaluate diagrams and correlation functions.

The phonon and photon Green's functions are defined in the same fashion. They are obviously similar to each other, so we shall present only the derivation of the phonon Green's function. The photon results will be stated at the end. For phonons for $-\beta \le \tau \le \beta$, we have the definition

$$\mathscr{D}(\mathbf{q}, \tau - \tau') = -\langle T_\tau A(\mathbf{q}, \tau) A(-\mathbf{q}, \tau')\rangle \tag{3.2.12}$$

where

$$A(\mathbf{q}, \tau) = e^{\tau H}(a_{\mathbf{q}} + a_{-\mathbf{q}}^\dagger)e^{-\tau H}$$
$$\langle \cdots \rangle = \mathrm{Tr}(e^{-\beta(H-\Omega)} \ldots)$$

Phonons have no chemical potential, since one can make an arbitrary number of them, and the tau dependence is just governed by the Hamiltonian. Again one can show that the right-hand side of (3.2.12) is only a function of $\tau - \tau'$. Thus it is not necessary to keep two τ variables, and one can instead define

$$\mathscr{D}(\mathbf{q}, \tau) = -\langle T_\tau A(\mathbf{q}, \tau) A(-\mathbf{q}, 0)\rangle \tag{3.2.12a}$$

Next we examine the behavior for negative tau:

$$\tau < 0: \quad \mathscr{D}(\mathbf{q}, \tau) = -\langle A(-\mathbf{q}, 0)A(\mathbf{q}, \tau)\rangle$$
$$= -\mathrm{Tr}[e^{-\beta(H-\Omega)}A(-\mathbf{q}, 0)e^{\tau H}A(\mathbf{q}, 0)e^{-\tau H}]$$

Using cyclic permutations of the trace turns this into

$$\tau < 0: \quad \mathscr{D}(\mathbf{q}, \tau) = -\mathrm{Tr}[e^{\beta\Omega}e^{\tau H}A(\mathbf{q}, 0)e^{-\tau H}e^{-\beta H}A(-\mathbf{q}, 0)]$$

or

$$\tau < 0: \quad \mathscr{D}(\mathbf{q}, \tau) = -\mathrm{Tr}[e^{-\beta(H-\Omega)}e^{H(\tau+\beta)}A(\mathbf{q}, 0)e^{-H(\tau+\beta)}A(-\mathbf{q}, 0)]$$

so that we have proved

$$-\beta \leq \tau \leq 0: \quad \mathscr{D}(\mathbf{q}, \tau) = \mathscr{D}(\mathbf{q}, \tau + \beta)$$

where the right-hand side of the equation is the Green's function when $0 < \tau + \beta < \beta$. This satisfies the general conditions we asserted in (3.1.6) for boson correlation functions. Thus the Fourier transform has the following form in (3.1.7):

$$\mathscr{D}(\mathbf{q}, i\omega_n) = \int_0^\beta d\tau e^{i\omega_n\tau}\mathscr{D}(\mathbf{q}, \tau)$$
$$\mathscr{D}(\mathbf{q}, \tau) = \frac{1}{\beta}\sum_n e^{-i\omega_n\tau}\mathscr{D}(\mathbf{q}, i\omega_n) \tag{3.2.13}$$
$$\omega_n = \frac{2\pi n}{\beta}$$

Equation (3.2.13) provides the definition of the frequency-dependent Green's function.

The difference between (3.1.6) and (3.1.8) is just a sign change. The fermion functions have a sign change because their operators obey anticommutation relations, while the bosons have no sign change because their operators obey commutation relations. Of course, this sign change is the fundamental difference between bosons and fermions. This sign change is also responsible for the sign change between ± 1 in their two forms of thermal distributions: $(e^{\beta\xi_\mathbf{p}} + 1)^{-1}$ vs. $(e^{+\beta\omega_\mathbf{q}} - 1)^{-1}$. This sign change is very important, and one has to keep track of it carefully for fermions in problems with many operators.

For phonons, the unperturbed or free-phonon Green's function is obtained by taking $H = H_0 = \sum_\mathbf{q} \omega_\mathbf{q} a_\mathbf{q}^\dagger a_\mathbf{q}$, which for the τ variation of the operators yields

$$e^{\tau H_0} a_{\mathbf{q}} e^{-\tau H_0} = e^{-\tau\omega_{\mathbf{q}}} a_{\mathbf{q}}$$

$$e^{\tau H_0} a_{\mathbf{q}}^{\dagger} e^{-\tau H_0} = e^{\tau\omega_{\mathbf{q}}} a_{\mathbf{q}}^{\dagger}$$

Always remember that $[a_{\mathbf{q}}(\tau)]^{\dagger} \neq a_{\mathbf{q}}^{\dagger}(\tau)$! The unperturbed Green's function is

$$\begin{aligned}\mathscr{D}^{(0)}(\mathbf{q}, \tau) = &-\theta(\tau)\langle(a_{\mathbf{q}} e^{-\tau\omega_{\mathbf{q}}} + e^{\tau\omega_{\mathbf{q}}} a_{-\mathbf{q}}^{\dagger})(a_{-\mathbf{q}} + a_{\mathbf{q}}^{\dagger})\rangle \\ &-\theta(-\tau)\langle(a_{\mathbf{q}} + a_{\mathbf{q}}^{\dagger})(e^{-\tau\omega_q} a_{\mathbf{q}} + e^{\tau\omega_q} a_{-\mathbf{q}}^{\dagger})\rangle\end{aligned}$$

We use capital letters to signify the thermal expectation value of boson number operators:

$$\begin{aligned} N_{\mathbf{q}} &= \langle a_{\mathbf{q}}^{\dagger} a_{\mathbf{q}}\rangle \\ N_{\mathbf{q}} + 1 &= \langle a_{\mathbf{q}} a_{\mathbf{q}}^{\dagger}\rangle \\ N_{\mathbf{q}} &= \frac{1}{e^{\beta\omega_{\mathbf{q}}} - 1} = N_{-\mathbf{q}} \equiv n_B(\omega_{\mathbf{q}}) \end{aligned} \tag{3.2.14}$$

Averages such as $\langle a_{\mathbf{q}} a_{-\mathbf{q}}\rangle$ and $\langle a_{\mathbf{q}}^{\dagger} a_{-\mathbf{q}}^{\dagger}\rangle$ yield zero since they vanish for each term in the trace.

The Green's function of τ can be written as

$$\begin{aligned}\mathscr{D}^{(0)}(\mathbf{q}, \tau) = &-\theta(\tau)[(N_{\mathbf{q}} + 1)e^{-\tau\omega_q} + N_{\mathbf{q}} e^{\tau\omega_{\mathbf{q}}}] \\ &-\theta(-\tau)[N_{\mathbf{q}} e^{-\tau\omega_{\mathbf{q}}} + (N_{\mathbf{q}} + 1)e^{+\tau\omega_{\mathbf{q}}}]\end{aligned}$$

which may be expressed in an equivalent but shorter formula as

$$\mathscr{D}^{(0)}(\mathbf{q}, \tau) = -[e^{-|\tau|\omega_{\mathbf{q}}} + 2N_{\mathbf{q}} \cosh(\omega_{\mathbf{q}}\tau)] \tag{3.2.15}$$

The Green's function of frequency is

$$\begin{aligned}\mathscr{D}^{(0)}(\mathbf{q}, i\omega_n) &= \int_0^{\beta} d\tau e^{i\omega_n\tau} \mathscr{D}^{(0)}(\mathbf{q}, \tau) \\ &= -\left[\frac{(N_q + 1)(e^{\beta(i\omega_n - \omega_{\mathbf{q}})} - 1)}{i\omega_n - \omega_{\mathbf{q}}} + \frac{N_q(e^{\beta(i\omega_n + \omega_{\mathbf{q}})} - 1)}{i\omega_n + \omega_{\mathbf{q}}}\right]\end{aligned}$$

The second terms in the numerators may be simplified by noting that for bosons

$$e^{i\omega_n\beta} = 1$$

so that the Green's function is

$$\mathscr{D}^{(0)}(\mathbf{q}, i\omega_n) = -\left[\frac{(N_\mathbf{q}+1)(e^{-\beta\omega_\mathbf{q}}-1)}{i\omega_n - \omega_\mathbf{q}} + \frac{N_\mathbf{q}(e^{\beta\omega_\mathbf{q}}-1)}{i\omega_n + \omega_\mathbf{q}}\right]$$

The first numerator equals -1, and the second is $+1$ by simply using the expression (3.2.14):

$$\mathscr{D}^{(0)}(\mathbf{q}, i\omega_n) = -\left(\frac{-1}{i\omega_n - \omega_\mathbf{q}} + \frac{1}{i\omega_n + \omega_\mathbf{q}}\right)$$
$$\mathscr{D}^{(0)}(\mathbf{q}, i\omega_n) = +\frac{2\omega_\mathbf{q}}{(i\omega_n)^2 - \omega_\mathbf{q}^2} = -\frac{2\omega_\mathbf{q}}{\omega_n^2 + \omega_\mathbf{q}^2} \tag{3.2.16}$$

This Green's function also has a simple form. It is almost identical to the zero temperature case (2.3.6), and the only difference is the use of complex frequencies instead of real ones. The photon Green's function is also identical to its zero temperature result, except for complex frequencies. The fundamental definition is

$$\mathscr{D}_{\mu\nu}(\mathbf{k}, \tau) = -\sum_\lambda \langle T_\tau A_\mu(\mathbf{k}, \tau) A_\nu(-\mathbf{k}, 0)\rangle$$
$$A_\mu(\mathbf{k}, \tau) = e^{\tau H}\xi_{\mathbf{k}\mu}\left(\frac{2\pi}{\omega_\mathbf{k}}\right)^{1/2}(a_{\mathbf{k}\lambda} + a^\dagger_{-\mathbf{k}\lambda})e^{-\tau H} \tag{3.2.17}$$

where the operator A_μ is the usual vector potential operator in (2.9.2). The free-photon Green's function is

$$\mathscr{D}^{(0)}_{\mu\nu}(\mathbf{k}, i\omega_n) = -\frac{4\pi(\delta_{\mu\nu} - k_\mu k_\nu/k^2)}{\omega_n^2 + \omega_\mathbf{k}^2}$$

which should be compared with (2.9.17).

We shall end this section with a comment on notation. The following three forms for the Green's function are equivalent and will be used interchangeably:

$$\mathscr{G}(\mathbf{p}, ip_n) \equiv \mathscr{G}(\mathbf{p}, ip) \equiv \mathscr{G}(p)$$
$$\mathscr{D}(\mathbf{q}, i\omega_n) \equiv \mathscr{D}(\mathbf{q}, i\omega) \equiv \mathscr{D}(q)$$

The form on the left has been used so far. In the second form, we have used ip instead of ip_n. They mean the same thing, since the i in ip is enough information to alert the reader that we are using complex frequencies, which are always discrete. Hence the n subscript is redundant. In the last form, we use a four-vector notation $p \equiv (\mathbf{p}, ip)$, and the script form of $\mathscr{G}$ is sufficient to alert the reader that we are using Matsubara Green's functions.

3.3. RETARDED AND ADVANCED GREEN'S FUNCTIONS

The retarded and advanced Green's functions were introduced in Sec. 2.9. They play an important role in the nonzero temperature theory. Their properties are discussed in this section. Their importance comes from the fact that all measurable quantities, such as conductivities or susceptibilities, are actually retarded correlation functions. The goal of many calculations is to calculate a retarded function. There are several different ways to obtain them. One is to use a real time theory even at nonzero temperatures. This method is obvious to the beginner, but is the hardest way. The second method, which we use most often, is to first calculate the equivalent Matsubara function of imaginary frequency. It is shown below that the retarded function is obtained from the Matsubara function by simply changing $i\omega_n$ to $\omega + i\delta$ where δ is infinitesimal. The Matsubara function is the easiest one to calculate because its S matrix expansion is simple. The retarded function is most easily found from the Matsubara function.

The retarded Green's functions may be defined for both zero and finite temperatures. The retarded Green's function for an electron in state p is

$$\begin{aligned}
G_{\text{ret}}(\mathbf{p}, t - t') &= -i\theta(t - t')\langle[C_{\mathbf{p}}(t)C_{\mathbf{p}}^{\dagger}(t') + C_{\mathbf{p}}^{\dagger}(t')C_{\mathbf{p}}(t)]\rangle \\
&= -i\theta(t - t')\,\mathrm{Tr}\{e^{-\beta(K-\Omega)}[C_{\mathbf{p}}(t)C_{\mathbf{p}}^{\dagger}(t') + C_{\mathbf{p}}^{\dagger}(t')C_{\mathbf{p}}(t)]\} \\
K &= H - \mu N \qquad\qquad (3.3.1) \\
C_{\mathbf{p}}(t) &= e^{itK}C_{\mathbf{p}}e^{-itK}
\end{aligned}$$

The brackets $\langle\cdots\rangle$ indicate thermodynamic average, as is explicitly shown on the second line. The square brackets mean nothing in particular; they are used to group symbols together. The retarded Green's function depends on real time, not tau. The tip-off for this is the $-i$ factor in front which belongs with all real-time Green's functions. The Green's function operates only for $t > t'$, and this makes it causal. One starts a signal at one time t' and measures it later at t. Of course, actual systems are causal, which is why these Green's functions are the ones of physical interest. The argument of the Green's function is an anticommutator at different times. In the limit that the times become equal, the anticommutator becomes unity,

$$\lim_{t\to t'} C_{\mathbf{p}}(t)C_{\mathbf{p}}^{\dagger}(t') + C_{\mathbf{p}}^{\dagger}(t')C_{\mathbf{p}}(t) = 1$$

since it just becomes the usual fermion anticommutator. Thus the plus sign in the middle of the two terms is an important feature for retarded Green's functions of fermion operators.

For phonons, the retarded Green's function is

$$D_{\text{ret}}(\mathbf{q}, t-t') = -i\theta(t-t')\langle[A(\mathbf{q}, t)A(-\mathbf{q}, t') - A(-\mathbf{q}, t')A(\mathbf{q}, t)]\rangle \quad (3.3.2)$$

It is very similar to (3.3.1) in that it is for real time, is also thermodynamically averaged, and is defined only for $t > t'$. However, the sign in the middle is now minus, which corresponds to the fact that bosons obey commutation relations. For both electron and phonon retarded functions, the right-hand side can be shown to be a function only of $t - t'$, as is indicated in the argument of the Green's function on the left-hand side of the definition.

We shall need to define retarded Green's functions for many types of operators. These operators will usually be products of electron or boson operators. For example, let us define the operators

$$U = \sum_{ij} M_{ij} C_i^{\dagger} C_j$$

$$V = \sum_{ijn} M_{ijn} C_i^{\dagger} C_j C_n$$

The operator U is bilinear in the operators C_i, where M_{ij} is just a matrix element. The operator U is regarded as having boson properties, regardless of whether both Cs are fermion or both boson operators; we exclude the case where one C is boson and one is fermion. U is boson because it acts like a composite particle. This bilinear form will be encountered quite often, since it is characteristic of some important operators such as the current and density operators. The retarded Green's function for the operator U is denoted $\bar{U}$ and is defined as

$$\bar{U}_{\text{ret}}(t - t') = -i\theta(t - t')\langle[U(t)U^{\dagger}(t') - U^{\dagger}(t')U(t)]\rangle \quad (3.3.3)$$

This definition is similar to (3.3.2), with the important feature that it has the minus sign in the center of the bracket, which is the case for all boson operators, that is, for any operator which is a product of bosons or an even number of fermions.

However, an operator such as V above is considered fermion if it is a product of an odd number of fermions. Its retarded function is

$$\bar{V}_{\text{ret}}(t - t') = -i\theta(t - t')\langle[V(t)V^{\dagger}(t') + V^{\dagger}(t')V(t)]\rangle$$

which now has the plus sign in the bracket.

All these retarded functions have Fourier transforms defined by the usual convention:

$$G_{\text{ret}}(p, E) = \int_{-\infty}^{\infty} dt e^{iE(t-t')} G_{\text{ret}}(p, t-t')$$

$$D_{\text{ret}}(q, \omega) = \int_{-\infty}^{\infty} dt e^{i\omega(t-t')} D_{\text{ret}}(q, t-t') \tag{3.3.4}$$

$$\bar{U}_{\text{ret}}(\omega) = \int_{-\infty}^{\infty} dt e^{i\omega t} \bar{U}(t)$$

The advanced Green's function for each of these is defined by

$$G_{\text{adv}}(\mathbf{p}, t-t') = i\theta(t'-t)\langle[C_{\mathbf{p}}(t)C_{\mathbf{p}}^{\dagger}(t') + C_{\mathbf{p}}^{\dagger}(t')C_{\mathbf{p}}(t)]\rangle$$

$$D_{\text{adv}}(\mathbf{q}, t-t') = i\theta(t'-t)\langle[A(\mathbf{q}, t)A(-\mathbf{q}, t') - A(-\mathbf{q}, t')A(\mathbf{q}, t)]\rangle \tag{3.3.5}$$

$$\bar{U}_{\text{adv}}(\mathbf{q}, t-t') = i\theta(t'-t)\langle[U(t)U^{\dagger}(t') - U^{\dagger}(t')U(t)]\rangle$$

The only two differences are the sign change in front and the fact that the time domain is now $t' > t$, which is just opposite of that for retarded functions. Their Fourier transforms with respect to frequency are defined in the usual way as in (3.3.4).

The advanced functions of energy turn out to be just the complex conjugate of the corresponding retarded function. To prove this, first note that if we start with the advanced function, take its Hermitian conjugate, and then invert its time variables, we get the retarded function

$$\bar{U}_{\text{adv}}(t'-t) = i\theta(t-t')\langle[U(t')U^{\dagger}(t) - U^{\dagger}(t)U(t')]\rangle$$

$$\bar{U}_{\text{adv}}(t'-t)^{\dagger} = \bar{U}_{\text{ret}}(t-t')$$

Now take the Fourier transform of both sides,

$$\bar{U}_{\text{ret}}(\omega) = \int_{-\infty}^{\infty} dt e^{i\omega(t-t')} \bar{U}_{\text{adv}}(t'-t)^{\dagger} = \int_{-\infty}^{\infty} dt_1 e^{-i\omega t_1} \bar{U}_{\text{adv}}(t_1)^{\dagger}$$

where the last step is obtained by changing variables $t_1 = t' - t$. Hence we have

$$\bar{U}_{\text{ret}}(\omega) = \bar{U}_{\text{adv}}(\omega)^* \tag{3.3.6}$$

This result can be generalized to any of the retarded and advanced Green's functions. Thus it is sufficient to find the retarded function, since a simple complex conjugation then derives the advanced one.

Now we wish to introduce a particular representation for these Green's functions. This representation is a formal one, which is not generally useful for calculating physical quantities and determining numbers. However, it is useful for proving theorems and in particular for relating one Green's function to another. This representation uses the complete set of states $|\, m\rangle$ which are the exact eigenstates of $K = H - \mu N$. Usually we do not know how to solve K exactly and do not know these states. However, in principle they exist, which is sufficient for proving theorems. The eigenvalues of K are E_m:

$$K\,|\, n\rangle = E_n\,|\, n\rangle$$

This complete set of states will be used in the thermodynamic average: The symbol Tr denotes trace, and we shall use the set $|\, n\rangle$ for this summation:

$$\begin{aligned}\bar{U}_{\text{ret}}&(t-t')\\ &= -i\theta(t-t')e^{\beta\Omega}\sum_n \langle n\,|\, e^{-\beta K}[U(t)\wedge U^\dagger(t') - U^\dagger(t')\wedge U(t)]\,|\, n\rangle\end{aligned}$$

The above equation has two points marked by an insertion sign "$\wedge$." In both places, we insert another complete set of states which is unity,

$$1 = \sum_m |\, m\rangle\langle m\,|$$

which gives

$$\begin{aligned}\bar{U}_{\text{ret}}(t-t') = -i\theta(t-t')e^{\beta\Omega}\sum_{n,m} e^{-\beta E_n}[&\langle n\,|\, U(t)\,|\, m\rangle\langle m\,|\, U^\dagger(t')\,|\, n\rangle\\ &-\langle n\,|\, U^\dagger(t')\,|\, m\rangle\langle m\,|\, U(t)\,|\, n\rangle]\end{aligned}$$

Now we can evaluate an expression such as

$$\langle n\,|\, U(t)\,|\, m\rangle = \langle n\,|\, e^{iKt}Ue^{-iKt}\,|\, m\rangle = e^{it(E_n-E_m)}\langle n\,|\, U\,|\, m\rangle$$

and so we obtain the double summation

$$\begin{aligned}\bar{U}_{\text{ret}}(t-t') = -i\theta(t-t')e^{\beta\Omega}\sum_{n,m} e^{-\beta E_n}(&e^{i(t-t')(E_n-E_m)}\,|\,\langle n\,|\, U\,|\, m\rangle\,|^2\\ &-e^{-i(t-t')(E_n-E_m)}\,|\,\langle m\,|\, U\,|\, n\rangle\,|^2)\end{aligned}$$

In the second term we exchange the dummy summation variables n and m so that the matrix elements $\langle n\,|\, U\,|\, m\rangle$ are the same in each term:

$$\begin{aligned}\bar{U}_{\text{ret}}&(t-t')\\ &= -i\theta(t-t')e^{\beta\Omega}\sum_{n,m} |\,\langle n\,|\, U\,|\, m\rangle\,|^2 e^{i(E_n-E_m)(t-t')}(e^{-\beta E_n} - e^{-\beta E_m})\end{aligned}$$

This formula is the result for the retarded Green's function of time. The Fourier transform is taken to give the frequency function,

$$\begin{aligned}\bar{U}_{\text{ret}}(\omega) &= -i\int_0^\infty dt e^{i(\omega+i\delta)t} e^{\beta\Omega} \sum_{n,m} |\langle n | U | m\rangle|^2 e^{it(E_n-E_m)}(e^{-\beta E_n} - e^{-\beta E_m}) \\ &= e^{\beta\Omega}\sum_{n,m} |\langle n | U | m\rangle|^2 (e^{-\beta E_n} - e^{-\beta E_m}) \frac{1}{\omega + E_n - E_m + i\delta}\end{aligned} \tag{3.3.7}$$

where we have added the $i\delta$ to the frequency to ensure convergence at large times.

The equivalent Matsubara function for the operator U is defined by a script symbol:

$$\begin{aligned}\mathscr{U}(\tau) &= -\langle T_\tau U(\tau)U^\dagger(0)\rangle \\ \mathscr{U}(i\omega_n) &= \int_0^\beta d\tau e^{i\omega_n\tau}\mathscr{U}(\tau)\end{aligned} \tag{3.3.8}$$

We also apply our representation technique to this expression for $\tau > 0$,

$$\tau > 0: \qquad \mathscr{U}(\tau) = -e^{\beta\Omega}\sum_{n,m}\langle n | e^{-\beta K}U(\tau) | m\rangle\langle m | U^\dagger(0) | n\rangle$$

which gives the result

$$\tau > 0: \qquad \mathscr{U}(\tau) = -e^{\beta\Omega}\sum_{n,m} |\langle n | U | m\rangle|^2 e^{\tau(E_n-E_m)} e^{-\beta E_n}$$

The frequency transform is

$$\begin{aligned}\mathscr{U}(i\omega_n) &= -e^{\beta\Omega}\sum_{n,m} |\langle n | U | m\rangle|^2 e^{-\beta E_n}\int_0^\beta d\tau e^{i\omega_n\tau}e^{\tau(E_n-E_m)} \\ &= +e^{\beta\Omega}\sum_{n,m}\frac{|\langle n | U | m\rangle|^2(e^{-\beta E_n} - e^{-\beta E_m})}{i\omega_n + E_n - E_m}\end{aligned} \tag{3.3.9}$$

and again we have used the fact that $\exp(\beta i\omega_n) = 1$ for bosons. This result should be compared with the retarded function in (3.3.7). They differ only in the frequencies in the energy denominator since the Matsubara result has $i\omega_n$ where the retarded function has $\omega + i\delta$. Thus we can change the Matsubara function to a retarded one with just this alteration:

$$\underset{i\omega_n\to\omega+i\delta}{\text{change}} \ \mathscr{U}(i\omega_n) = \bar{U}_{\text{ret}}(\omega) \tag{3.3.10}$$

This step is called an analytic continuation. The same method can be used to show the same identity for the other Green's functions:

$$\begin{aligned} \underset{i\omega_n \to \omega + i\delta}{\text{change}} \mathscr{G}(\mathbf{p}, i\omega_n) &= G_{\text{ret}}(\mathbf{p}, \omega) \\ \underset{i\omega_n \to \omega + i\delta}{\text{change}} \mathscr{D}(\mathbf{q}, i\omega_n) &= D_{\text{ret}}(\mathbf{q}, \omega) \end{aligned} \tag{3.3.11}$$

This relationship to the retarded function is one of the primary reasons that the Matsubara functions are so useful. After they are evaluated, this simple analytical continuation then yields the retarded function, which is the function of physical interest. The advanced functions are obtained by the analytic continuation $i\omega_n \to \omega - i\delta$, which is obviously true since the advanced function is the complex conjugate of the retarded function.

Another quantity of great importance is the *spectral function*, which is also called the *spectral density function*. It is the imaginary part of any retarded function multiplied by -2; for example,

$$\begin{aligned} R(\omega) &= -2 \operatorname{Im} \bar{U}_{\text{ret}}(\omega) \\ B(\mathbf{q}, \omega) &= -2 \operatorname{Im} D_{\text{ret}}(\mathbf{q}, \omega) \\ A(\mathbf{p}, \omega) &= -2 \operatorname{Im} G_{\text{ret}}(\mathbf{p}, \omega) \end{aligned} \tag{3.3.12}$$

There is not a formal symbol for this quantity, so we shall use a variety. From the representation of the retarded function (3.3.7), the only complex part is

$$\frac{1}{\omega + E_n - E_m + i\delta} = P \frac{1}{\omega + E_n - E_m} - i\pi\delta(\omega + E_n - E_m)$$

Thus we obtain

$$R(\omega) = e^{\beta\Omega} \sum_{n,m} | \langle n \mid U \mid m \rangle |^2 (e^{-\beta E_n} - e^{-\beta E_m}) 2\pi\delta(\omega + E_n - E_m)$$

The temperature factors can be regrouped to give

$$\exp(-\beta E_n)\{[1 - \exp[-\beta(E_m - E_n)]\}$$

or

$$R(\omega) = 2\pi(1 - e^{-\beta\omega}) e^{\beta\Omega} \sum_{n,m} | \langle n \mid U \mid m \rangle |^2 e^{-\beta E_n} \delta(\omega + E_n - E_m) \tag{3.3.13}$$

It is now possible to write the retarded or Matsubara functions as integrals over these expressions:

$$\bar{U}_{\text{ret}}(\omega) = \frac{1}{2\pi} \int_{-\infty}^{\infty} \frac{d\omega' R(\omega')}{\omega - \omega' + i\delta}$$

$$\mathscr{U}(i\omega_n) = \frac{1}{2\pi} \int_{-\infty}^{\infty} \frac{d\omega' R(\omega')}{i\omega_n - \omega'} \tag{3.3.14}$$

These identities follow directly from the prior results (3.3.7), (3.3.9), and (3.3.13). An expression of this form is called a Lehmann representation, and it was first used in quantum electrodynamics.

Much of this book is devoted to studying electrons, and the spectral function for electrons will be calculated often in a variety of problems. This is a good place to discuss some of its general features. First we give the representation for the retarded function:

$$G_{\text{ret}}(\mathbf{p}, \omega) = e^{\beta\Omega} \sum_{n,m} | \langle n \mid C_{\mathbf{p}} \mid m \rangle |^2 \frac{e^{-\beta E_n} + e^{-\beta E_m}}{\omega + E_n - E_m + i\delta}$$

This equation is similar to the boson result (3.3.7), the only difference is the plus sign in $e^{-\beta E_n} + e^{-\beta E_m}$. The plus sign follows directly from the plus sign between the two terms in the definition (3.3.1). The spectral function for the electron is then

$$A(\mathbf{p}, \omega) = 2\pi e^{\beta\Omega} \sum_{n,m} | \langle n \mid C_{\mathbf{p}} \mid m \rangle |^2 (e^{-\beta E_n} + e^{-\beta E_m}) \delta(\omega + E_n - E_m) \tag{3.3.15}$$

This quantity is absolutely positive for all values of the variables $\mathbf{p}, \omega$ since the right-hand side of (3.3.15) contains only positive factors:

$$A(\mathbf{p}, \omega) \geq 0$$

This positiveness is an important feature, since we shall interpret $A(\mathbf{p}, \omega)$ as a probability function. The spectral functions for bosons do not have this property, since they are sometimes plus and sometimes minus. One can show, however, that they are always plus for $\omega > 0$ and always minus for $\omega < 0$, which follows from Eq. (3.3.13).

Another important feature of the electron spectral function is obtained by integrating over all frequencies:

$$1 = \int_{-\infty}^{\infty} \frac{d\omega}{2\pi} A(\mathbf{p}, \omega) \tag{3.3.16}$$

This important theorem, actually a sum rule, is proved by integrating the representation (3.3.15):

$$\int_{-\infty}^{\infty} \frac{d\omega}{2\pi} A(\mathbf{p}, \omega) = e^{\beta\Omega} \sum_{n,m} | \langle n \mid C_{\mathbf{p}} \mid m \rangle |^2 (e^{-\beta E_n} + e^{-\beta E_m})$$

This expression can be simplified by eliminating the summations over n and m. This elimination is achieved by reversing the steps by which we derived (3.3.7). First one relabels n and m in the second term:

$$e^{\beta\Omega} \sum_{n,m} e^{-\beta E_n} (\langle n \mid C_{\mathbf{p}} \mid m \rangle \langle m \mid C_{\mathbf{p}}^{\dagger} \mid n \rangle + \langle n \mid C_{\mathbf{p}}^{\dagger} \mid m \rangle \langle m \mid C_{\mathbf{p}} \mid n \rangle)$$

Then one can eliminate the summation over m to give

$$\int_{-\infty}^{\infty} \frac{d\omega}{2\pi} A(\mathbf{p}, \omega) = e^{\beta\Omega} \sum_{n} e^{-\beta E_n} \langle n \mid (C_{\mathbf{p}} C_{\mathbf{p}}^{\dagger} + C_{\mathbf{p}}^{\dagger} C_{\mathbf{p}}) \mid n \rangle$$

But the anticommutator in brackets yields unity, so that we obtain the proof:

$$\int_{-\infty}^{\infty} \frac{d\omega}{2\pi} A(\mathbf{p}, \omega) = e^{\beta\Omega} \operatorname{Tr}(e^{-\beta K}) = 1$$

The spectral function may be obtained for a free electron or an unperturbed one. In the definition (3.3.1) we use $K_0 = H_0 - \mu N$ and obtain

$$C_{\mathbf{p}}(t) = e^{-i\xi_{\mathbf{p}} t} C_{\mathbf{p}}$$

so that

$$\begin{aligned} G_{\text{ret}}^{(0)}(\mathbf{p}, t - t') &= -i\theta(t - t') e^{-i\xi_{\mathbf{p}}(t-t')} \langle (C_{\mathbf{p}} C_{\mathbf{p}}^{\dagger} + C_{\mathbf{p}}^{\dagger} C_{\mathbf{p}}) \rangle \\ &= -i\theta(t - t') e^{-i\xi_{\mathbf{p}}(t-t')} \end{aligned}$$

The Fourier transform is

$$G_{\text{ret}}^{(0)}(\mathbf{p}, \omega) = \frac{1}{\omega - \xi_{\mathbf{p}} + i\delta}$$

The $i\delta$ factor was inserted for convergence. It has one sign $\delta > 0$, even in many-electron systems with a Fermi surface. The retarded functions do not have δ changing sign at the Fermi surface, which makes them easier to use then than the zero temperature Green's functions introduced in Chapter 2.

The spectral function for the noninteracting Green's function is

$$A^{(0)}(\mathbf{p}, \omega) = 2\pi\delta(\omega - \xi_{\mathbf{p}}) = -2 \operatorname{Im} G_{\text{ret}}^{(0)}(\mathbf{p}, \omega) \tag{3.3.17}$$

It is just a delta function. The spectral function $A(\mathbf{p}, \omega)$ is interpreted as a probability function. It is the probability that an electron has momentum $\mathbf{p}$ and energy ω. For a free, or noninteracting, particle we have that $\omega = \xi_\mathbf{p}$ so the probability distribution is a delta function: Here there is only one value ω for each $\xi_\mathbf{p}$ and vice versa. Equation (3.3.17) is plotted graphically in Fig. 3.1. For a fixed value of $\mathbf{p}$, a plot of the spectral function vs. ω is a sharp delta function (here given a small width to aid the eye) at $\omega = \xi_p$. When we compute $A(\mathbf{p}, \omega)$ for interacting systems, we typically find the broad distribution labeled A in Fig. 3.1. There is a band of ω values for each p, which is not surprising. When the electron scatters, it has a finite mean free path, and there is some uncertainty in its momentum or energy or both. Thus we must treat $\mathbf{p}$ and ω as separate variables and sum over them both when evaluating physical quantities. The spectral function $A(\mathbf{p}, \omega)$ appears in these summations and gives the proper probability weighting between these variables.

Another quantity to evaluate, for an interacting electron system, is the number of electrons in a momentum state p, which is

$$n_\mathbf{p} = \langle C_\mathbf{p}^\dagger C_\mathbf{p} \rangle$$

For a noninteracting electron system, the number is trivially given by (3.2.10). In the interacting system, we again introduce our representation

$$\begin{aligned} n_\mathbf{p} &= e^{\beta\Omega} \sum_{n,m} \langle m \mid e^{-\beta K} C_\mathbf{p}^\dagger \mid n \rangle \langle n \mid C_\mathbf{p} \mid m \rangle \\ &= e^{\beta\Omega} \sum_{n,m} \mid \langle n \mid C_\mathbf{p} \mid m \rangle \mid^2 e^{-\beta E_m} \end{aligned} \tag{3.3.18}$$

This result $n_\mathbf{p}$ should be compared with $A(\mathbf{p}, \omega)$ in (3.3.15). The two expressions differ in the factors

$$\begin{aligned} (e^{-\beta E_n} + e^{-\beta E_m})\delta(\omega + E_n - E_m) &= e^{-\beta E_m}(e^{-\beta(E_n - E_m)} + 1)\delta(\omega + E_n - E_m) \\ &= e^{-\beta E_m}(e^{\beta\omega} + 1)\delta(\omega + E_n - E_m) \end{aligned}$$

Thus if we evaluate the integral

$$\int_{-\infty}^{\infty} \frac{d\omega}{2\pi} \frac{1}{e^{\beta\omega} + 1} A(\mathbf{p}, \omega) = e^{\beta\Omega} \sum_{n,m} \mid \langle n \mid C_\mathbf{p} \mid m \rangle \mid^2 e^{-\beta E_m}$$

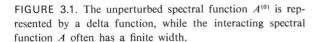

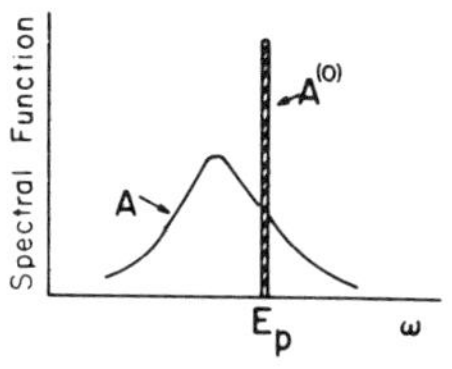

FIGURE 3.1. The unperturbed spectral function $A^{(0)}$ is represented by a delta function, while the interacting spectral function A often has a finite width.

the right-hand side is (3.3.18) for $n_\mathbf{p}$:

$$n_\mathbf{p} = \int_{-\infty}^{\infty} \frac{d\omega}{2\pi} n_F(\omega) A(\mathbf{p}, \omega)$$
$$n_F(\omega) = \frac{1}{e^{\beta\omega} + 1} \tag{3.3.19}$$

The factor $n_F(\omega)$ is the fermion occupation factor at finite temperatures. Remember that the energy ω is measured with respect to the chemical potential. The thermal occupation probability is determined by the energy ω of the particles. Thus the number of particles in a state $\mathbf{p}$ is obtained by summing over all energies ω, weighted by the spectral function which gives the probability that a particle in state $\mathbf{p}$ has ω, and also by multiplying by the thermal occupation factor $n_F(\omega)$. This reasonable expression provides further examples of the use of the spectral function. In the limit of no interactions, then $A(\mathbf{p}, \omega)$ becomes the noninteracting spectral function (3.3.17), and we again recover (3.2.10).

For phonons, the average number of phonons in a state $\mathbf{q}$ is $N_\mathbf{q} = \langle a_\mathbf{q}^\dagger a_\mathbf{q} \rangle$, and one can similarly show that

$$2N_\mathbf{q} + 1 = \langle A_\mathbf{q}^\dagger A_\mathbf{q} \rangle = \int_{-\infty}^{\infty} \frac{d\omega}{2\pi} n_B(\omega) B(\mathbf{q}, \omega)$$
$$n_B(\omega) = \frac{1}{e^{\beta\omega} - 1} \tag{3.3.20}$$

The phonon spectral function $B(\mathbf{q}, \omega)$ was defined earlier in (3.3.12). The factor $n_B(\omega)B(\mathbf{q}, \omega)$ is always positive for phonons and could serve as the temperature-dependent probability of having phonons with q and ω. The unperturbed phonon spectral function is

$$B^{(0)}(\mathbf{q}, \omega) = 2\pi[\delta(\omega - \omega_\mathbf{q}) - \delta(\omega + \omega_\mathbf{q})]$$

The retarded Green's functions will be calculated from the Matsubara functions. We shall show in the next section that the Matsubara functions have a Dyson equation of the form

$$\mathscr{G}(\mathbf{p}, i\omega_n) = \frac{1}{i\omega_n - \xi_\mathbf{p} - \Sigma(\mathbf{p}, i\omega_n)}$$
$$\mathscr{D}(\mathbf{q}, i\omega_n) = \frac{-2\omega_\mathbf{q}}{\omega_n^2 + \omega_\mathbf{q}^2 + 2\omega_\mathbf{q} P(\mathbf{q}, i\omega_n)}$$

The self-energy functions will be calculated according to rules described later. We can define the retarded self-energies according to (3.3.11):

$$\underset{i\omega_n \to \omega + i\delta}{\text{change}} \Sigma(\mathbf{p}, i\omega_n) = \Sigma_{\text{ret}}(\mathbf{p}, \omega) = \text{Re}\, \Sigma_{\text{ret}}(\mathbf{p}, \omega) + i\, \text{Im}\, \Sigma_{\text{ret}}(\mathbf{p}, \omega)$$

$$\underset{i\omega_n \to \omega + i\delta}{\text{change}} P(\mathbf{q}, i\omega_n) = P_{\text{ret}}(\mathbf{q}, \omega)$$

and the retarded Green's function will have a Dyson equation also,

$$G_{\text{ret}}(\mathbf{p}, \omega) = \frac{1}{\omega + i\delta - \xi_{\mathbf{p}} - \Sigma_{\text{ret}}(\mathbf{p}, \omega)}$$

$$D_{\text{ret}}(\mathbf{p}, \omega) = \frac{2\omega_q}{\omega^2 - \omega_{\mathbf{q}}{}^2 - i\delta - 2\omega_{\mathbf{q}} P_{\text{ret}}(\mathbf{q}, \omega)}$$

where we have used (3.3.11). Thus we can write the spectral function for the electron in terms of the retarded self-energies,

$$A(\mathbf{p}, \omega) = \frac{-2\, \text{Im}\, \Sigma_{\text{ret}}(\mathbf{p}, \omega)}{[\omega - \xi_{\mathbf{p}} - \text{Re}\, \Sigma_{\text{ret}}(\mathbf{p}, \omega)]^2 + [\text{Im}\, \Sigma_{\text{ret}}(\mathbf{p}, \omega)]^2} \tag{3.3.21}$$

where $\text{Im}\, \Sigma < 0$ so that $A > 0$. One way to obtain the spectral function is to evaluate the self-energies. The method for doing this is described in the next section.

The formal distinctions between the Matsubara, retarded, and advanced Green's functions are best understood by some simple examples. We shall give some simple functions which have the correct analytical properties. For example, consider a self-energy operator which has the following functional form,

$$\Sigma(\mathbf{p}, Z) = C \ln[f(p) - Z]$$

where Z is a complex variable representing the frequency. We take C as a constant and $f(p)$ as some function of momentum. The Matsubara self-energy is evaluated at the points ip_n:

$$\Sigma(\mathbf{p}, ip_n) = C \ln[f(p) - ip_n]$$

The analytic continuation $ip_n \to \omega \pm i\delta$ to the real axis has the following values. For the retarded function we get, $ip_n \to \omega + i\delta$,

$$\Sigma_{\text{ret}}(\mathbf{p}, \omega) = C \ln|f(p) - \omega| - i\pi C\theta[\omega - f(p)]$$

while for the advanced function we have, $ip_n \to \omega - i\delta$,

$$\Sigma_{\text{adv}}(\mathbf{p}, \omega) = C \ln|f(p) - \omega| + i\pi C\theta[\omega - f(p)]$$

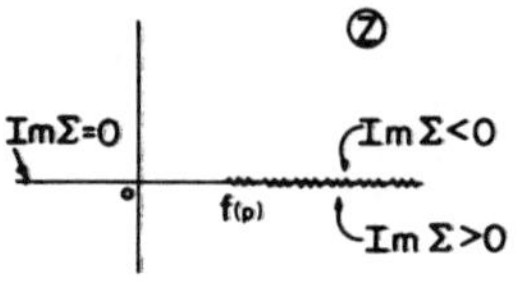

FIGURE 3.2

These two differ in the region $\omega > f(p)$, because their imaginary parts have the opposite sign. This agrees with our general theorem that

$$G_{\text{adv}}(\mathbf{p}, \omega)^* = G_{\text{ret}}(\mathbf{p}, \omega)$$

which also implies that

$$\Sigma_{\text{adv}}(\mathbf{p}, \omega)^* = \Sigma_{\text{ret}}(\mathbf{p}, \omega)$$

This functional behavior is shown in Fig. 3.2. There is a *branch cut* on the real axis for $\omega > f(p)$. This branch cut just expresses the fact that $\ln(f - Z)$ is not a continuous function of Z across the real axis for $\omega > f$, since the imaginary part changes sign.

Another example which has similar analytical properties is

$$\Sigma(\mathbf{p}, Z) = C[f(p) - Z]^{1/2}$$

This also has a branch cut for $\omega > f(p)$, with $\text{Im}\,\Sigma < 0$ above the branch cut and $\text{Im}\,\Sigma > 0$ below. In fact, a branch cut is a necessary feature whenever $\text{Im}\,\Sigma \neq 0$, for whenever $\text{Im}\,\Sigma \neq 0$, we have that

$$\Sigma(\mathbf{p}, \omega + i\delta) \neq \Sigma(\mathbf{p}, \omega - i\delta)$$

which requires a branch cut in the analytic function. When we actually evaluate self-energy functions, we shall find that they are often given by logarithmic or square root functions.

When a branch cut occurs and $\text{Im}\,\Sigma \neq 0$, then the spectral function is given by (3.3.21). In frequency regions where $\text{Im}\,\Sigma = 0$ and there is no branch cut, then we take the limit of $\text{Im}\,\Sigma \to 0$ and obtain

$$\text{Im}\,\Sigma = 0: \qquad A(\mathbf{p}, \omega) = 2\pi\delta\{\omega - \xi_{\mathbf{p}} - \text{Re}[\Sigma(\mathbf{p}, \omega)]\}$$

Here the spectral function is again a delta function, but now the real part of the self-energy may be finite, and usually is, so that it affects the spectral function. Let us denote by E_p the solution to the equation

$$E_p - \mu = \xi_p + \text{Re}[\Sigma(\mathbf{p}, E_p - \mu)] \tag{3.3.22}$$

Let us assume that we have a problem in which (3.3.22) is satisfied when $\text{Im}\,\Sigma = 0$. Recall that if $g(x) = 0$ at $x = x_0$, then delta functions have the property

$$\delta[g(x)] = \frac{\delta(x - x_0)}{|g'(x_0)|}$$

Thus we can write the spectral function as

$$\begin{aligned} A(\mathbf{p}, \omega) &= 2\pi Z(\mathbf{p})\delta(\omega - E_{\mathbf{p}} + \mu) \\ Z(\mathbf{p}) &= \left|1 - \frac{\partial}{\partial \omega} \text{Re}[\Sigma(\mathbf{p}, \omega)]\right|^{-1}_{\omega = E_{\mathbf{p}} - \mu} \end{aligned} \tag{3.3.23}$$

The factor $Z(\mathbf{p})$ is called a *renormalization factor*. If we recall the facts that (1) $A(\mathbf{p}, \omega) > 0$ and (2) $\int_{-\infty}^{\infty} (d\omega/2\pi)A(\mathbf{p}, \omega) = 1$, then it is easy to show that $Z(\mathbf{p}) \leq 1$. The strength of the delta function peak is always less than or equal to unity. An example is illustrated in Fig. 3.3. The spectral function is shown with a sharp delta function peak at $\omega = E_p - \mu$ and a continuous spectra in the region $\omega_1 < \omega < \omega_2$. The latter region is where $\text{Im}\,\Sigma \neq 0$. Since the total integrated area under the entire spectra is unity, the existence of any areas where $\text{Im}\,\Sigma \neq 0$ implies that the renormalization factor $Z(p)$ for the delta function is less than unity.

Equation (3.3.22) may be used to define the *effective mass*. We assume that the unperturbed states are free particles, so that

$$\xi_{\mathbf{p}} = \frac{p^2}{2m} - \mu$$

Furthermore, we assume that at low momentum E_p varies quadratically with momentum,

$$E_p = E_0 + \frac{p^2}{2m^*} + O(p^4)$$

and this proportionality constant is the inverse effective mass m^*. Thus we define

$$\frac{m}{m^*} = \frac{\partial E_p}{\partial \varepsilon_p} \tag{3.3.24}$$

FIGURE 3.3. The spectral function $A(p, \omega)$ will have a delta function peak whenever $\text{Im}\,\Sigma = 0$, which is shown here at $\omega = E_p - \mu$. The spectral function also has values in regions where $\text{Im}\,\Sigma \neq 0$, as shown by the crosshatched distribution.

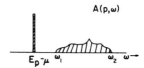

and the derivative of (3.3.22) is

$$\frac{\partial E_p}{\partial \varepsilon_p} = \lim_{\varepsilon_p \to 0} \left(1 + \left\{\frac{\partial}{\partial \varepsilon_p} \operatorname{Re}[\Sigma(\mathbf{p}, \omega)]\right\} + \left\{\frac{\partial}{\partial \omega} \operatorname{Re}[\Sigma(\mathbf{p}, \omega)]\right\} \frac{\partial E_p}{\partial \varepsilon_p}\right)_{\omega = E_p - \mu}$$

The last term on the right contains the factor (3.3.24), so we bring this to the left again to finally obtain

$$\frac{m}{m^*} = \lim_{\varepsilon_p \to 0} \left\{\frac{1 + (\partial/\partial \varepsilon_p) \operatorname{Re}[\Sigma(\mathbf{p}, E_0 - \mu)]}{1 - (\partial/\partial E_0) \operatorname{Re}[\Sigma(\mathbf{p}, E_0 - \mu)]}\right\} \tag{3.3.25}$$

This formula will be used frequently to obtain the effective mass from self-energy calculations.

3.4. DYSON'S EQUATION

The Matsubara Green's functions are evaluated by the same method of Feynman diagram techniques that were introduced in Chapter 2 for the case of zero temperature. These methods are slightly modified to account for complex times and frequencies, but otherwise the S-matrix expansion appears to be very similar. Here we shall sketch the ideas behind expanding the S matrix and rederive Dyson's equation for the Matsubara Green's functions.

We consider as an example the case of the electron Green's function:

$$\begin{aligned} \mathscr{G}(\mathbf{p}, \tau) &= -e^{\Omega\beta} \operatorname{Tr}[e^{-\beta K} T_\tau(e^{\tau K} C_\mathbf{p} e^{-\tau K}) C_\mathbf{p}^\dagger] \\ e^{-\beta\Omega} &= \operatorname{Tr}(e^{-\beta K}) \end{aligned} \tag{3.4.1}$$

Again we consider the general case where

$$\begin{aligned} K &= K_0 + V = H_0 - \mu N + V \\ H &= H_0 + V \end{aligned}$$

where K_0 is a problem we can solve, so that we know its complete set of states. The Hamiltonians we solve usually have the property that they commute with the number operator:

$$\begin{aligned} [H_0, N] &= 0 \\ [H, N] &= 0 \end{aligned}$$

Thus we can define simultaneous eigenstates of H_0 and N and also of H and N. We call H_0 or K_0 the unperturbed problem, and V is the perturbations whose effects we are trying to evaluate. Consider the operators

$$\begin{aligned} U(\tau) &= e^{\tau K_0} e^{-\tau K} \\ \bar{U}^{-1}(\tau) &= e^{+\tau K} e^{-\tau K_0} \end{aligned} \tag{3.4.2}$$

where we are again in the interaction representation. We put a caret above operators to denote their τ development with respect to the unperturbed operators:

$$\hat{C}_p(\tau) = e^{\tau K_0} C_p e^{-\tau K_0}$$

Thus we can write our Green's function (3.4.1) as

$$\begin{aligned} \tau > 0: \quad \mathscr{G}(\mathbf{p}, \tau) &= -e^{+\Omega\beta} \operatorname{Tr}[e^{-\beta K_0}(e^{\beta K_0} e^{-\beta K})(e^{\tau K} e^{-\tau K_0})(e^{+\tau K_0} C_{\mathbf{p}} e^{-\tau K_0}) \\ &\quad \times (e^{\tau K_0} e^{-\tau K}) C_{\mathbf{p}}^{\dagger}] \\ &= \frac{-\operatorname{Tr}[e^{-\beta K_0} U(\beta) U^{-1}(\tau) \hat{C}_{\mathbf{p}}(\tau) U(\tau) \hat{C}_{\mathbf{p}}^{\dagger}(0)]}{\operatorname{Tr}[e^{-\beta K_0} U(\beta)]} \end{aligned} \tag{3.4.3}$$

We have replaced the thermodynamic potential in the denominator by the equivalent factor

$$e^{-\beta\Omega} = \operatorname{Tr}(e^{-\beta K}) = \operatorname{Tr}[e^{-\beta K_0}(e^{\beta K_0} e^{-\beta K})] = \operatorname{Tr}[e^{-\beta K_0} U(\beta)]$$

A similar substitution has been made on the $\exp(-\beta K)$ factor in the numerator.

The operator $U(\tau)$ can be solved in terms of τ-ordered products. Consider the derivative

$$\frac{\partial}{\partial\tau} U(\tau) = e^{\tau K_0}(K_0 - K) e^{-\tau K} = -e^{\tau K_0} V e^{-\tau K}$$

which can be expressed in the interaction representation as

$$\frac{\partial}{\partial\tau} U(\tau) = -e^{\tau K_0} V e^{-\tau K_0}(e^{\tau K_0} e^{-\tau K}) = -\hat{V}(\tau) U(\tau)$$

This equation for $U(\tau)$ may be solved, at least formally, by repeated integrations and using $U(0) = 1$:

$$\begin{aligned} U(\tau) &= 1 - \int_0^\tau d\tau_1 \hat{V}(\tau_1) U(\tau_1) = 1 - \int_0^\tau d\tau_1 \hat{V}(\tau_1) + (-1)^2 \int_0^\tau d\tau_1 \int_0^{\tau_1} d\tau_2 \\ &\quad \times \hat{V}(\tau_1) \hat{V}(\tau_2) U(\tau_2) \\ &= \sum_{n=0}^{\infty} (-1)^n \int_0^\tau d\tau_1 \cdots \int_0^{\tau_{n-1}} d\tau_n \hat{V}(\tau_1) \cdots \hat{V}(\tau_n) \end{aligned}$$

This summation can be expressed as an ordered product,

$$U(\tau) = \sum_{n=0}^{\infty} \frac{(-1)^n}{n!} \int_0^\tau d\tau \cdots \int_0^\tau d\tau_n [T_\tau \hat{V}(\tau_1)\hat{V}(\tau_2)\hat{V}(\tau_3)\hat{V}(\tau_4) \cdots \hat{V}(\tau_n)] \tag{3.4.4}$$

$$= T_\tau \exp\left[-\int_0^\tau d\tau_1 \hat{V}(\tau_1)\right] \tag{3.4.5}$$

and finally as a τ-ordered exponential integral, although the strict definition of (3.4.5) is just (3.4.4). Next consider the definition

$$S(\tau_1, \tau_2) = T_\tau \exp\left[-\int_{\tau_1}^{\tau_2} d\tau \hat{V}(\tau)\right]$$

It is easy to prove the following operator identities:

$$S(\tau_1, \tau_2) = S(\tau_2 - \tau_1)$$

$$S(\tau_1, \tau_2)S(\tau_2, \tau_3) = S(\tau_1, \tau_3)$$

$$S(\tau_1, \tau_2) = U(\tau_1)U^{-1}(\tau_2)$$

In this notation, the Green's function (3.4.3) may be rewritten as

$$\tau > 0: \qquad \mathscr{G}(\mathbf{p}, \tau) = \frac{-\mathrm{Tr}[e^{-\beta K_0} S(\beta, \tau)\hat{C}_\mathbf{p}(\tau)S(\tau)\hat{C}_\mathbf{p}^\dagger(0)]}{\mathrm{Tr}[e^{-\beta K_0} S(\beta)]} \tag{3.4.6}$$

Using the above properties of $S(\beta, \tau)$ and taking advantage of the freedom to rearrange terms within the τ ordering operator, we may also express the numerator as

$$\mathscr{G}(\mathbf{p}, \tau) = \frac{-\mathrm{Tr}[e^{-\beta K_0} T_\tau S(\beta)\hat{C}_\mathbf{p}(\tau)\hat{C}_\mathbf{p}^\dagger(0)]}{\mathrm{Tr}[e^{-\beta K_0} S(\beta)]} \tag{3.4.7}$$

From now on we shall write the trace over $\exp(-\beta K_0)$ with a subscript 0,

$$\mathrm{Tr}[e^{-\beta K_0}\theta] = {}_0\langle\theta\rangle \tag{3.4.8}$$

where θ is any operator. Thus our Green's function is

$$\mathscr{G}(\mathbf{p}, \tau) = \frac{-{}_0\langle T_\tau[S(\beta)\hat{C}_\mathbf{p}(\tau)\hat{C}_\mathbf{p}^\dagger(0)]\rangle}{{}_0\langle S(\beta)\rangle}$$

We continue to write a caret over $C_\mathbf{p}(0)$. Although $C_p(0)$ has no τ dependence when $\tau = 0$, the $\tau = 0$ label is important for ordering the operator. We have really only proved (3.4.7) for the case $\tau > 0$, but this is the only case of interest. The T_τ operator in (3.4.7) really means (3.4.6). The S

matrix is divided into parts, which are ordered with respect to the other operators $\hat{C}_{\mathbf{p}}(\tau)$ and $\hat{C}_{\mathbf{p}}^{\dagger}(0)$. This form of the Green's function is similar to the zero temperature result (2.3.2).

The Green's function is evaluated, at least formally, by expanding the S matrix in the numerator:

$$
\begin{aligned}
&{}_0\langle T_\tau S(\beta)\hat{C}_{\mathbf{p}}(\tau)\hat{C}_{\mathbf{p}}^{\dagger}(0)\rangle \\
&= \sum_{n=0}^{\infty} \frac{(-1)^n}{n!} \int_0^\beta d\tau_1 \cdots \int_0^\beta d\tau_n \, {}_0\langle T_\tau \hat{C}_{\mathbf{p}}(\tau)\hat{V}(\tau_1)\cdots\hat{V}(\tau_n)\hat{C}_{\mathbf{p}}^{\dagger}(0)\rangle
\end{aligned} \tag{3.4.9}
$$

Each of the nth terms are evaluated by applying Wick's theorem to the brackets and thereby expressing the brackets as combinations of the unperturbed Green's functions $\mathscr{G}^{(0)}$ and $\mathscr{D}^{(0)}$. Thus, as before, we apply Wick's theorem to give, for example,

$$
\begin{aligned}
&{}_0\langle T_\tau \hat{C}_{\mathbf{p}}(\tau)\hat{C}_{\mathbf{k}}^{\dagger}(\tau_1)\hat{C}_{\mathbf{k}'}(\tau_1)\hat{C}_{\mathbf{p}}^{\dagger}(0)\rangle \\
&\qquad = \delta_{\mathbf{p},\mathbf{k}}\delta_{\mathbf{p},\mathbf{k}'}\mathscr{G}^{(0)}(\mathbf{p}, \tau - \tau_1)\mathscr{G}^{(0)}(\mathbf{p}, \tau_1) - \delta_{\mathbf{k}\mathbf{k}'}n_{\mathbf{k}}\mathscr{G}^{(0)}(\mathbf{p}, \tau)
\end{aligned}
$$

Wick's theorem has a double meaning for finite temperature Green's functions. The first meaning is just the pairing feature: If we start with a state $|n\rangle$ (in our trace) and operate on this state by some C_λ, then to get back to $\langle n|$ at the other side of the trace, we must have another operator of the form $C_\lambda^{\dagger}$ appearing in the product of operators. This pairing was the use made of Wick's theorem for zero temperatures, and this use is made here. Whenever we have a bracket with M creation and M destruction operators, each of the creation operators must correspond to the same state as one of the destruction operators and vice versa.

The other feature of Wick's theorem concerns the thermodynamic average. The brackets ${}_0\langle\cdots\rangle$ mean the average in (3.4.8), and usually

$$
{}_0\langle AB\rangle \neq {}_0\langle A\rangle_0\langle B\rangle
$$

where A and B are two arbitrary operators. Usually, the product of averages is *not* equal to the average of products. However, in Wick's theorem, we do just that—we take a product of operators, pair them up, and thermodynamically average each pair separately. This procedure does give the right answer as long as H_0 is only bilinear in the operators. Any errors that are made vanish in the limit of infinite volume. One exception is for macroscopic quantum states as occur in superfluids. As an example, consider the evaluation of

$$
W = \frac{1}{\nu^2}\left\langle T_\tau\left[\sum_{\mathbf{p}} C_{\mathbf{p}}^{\dagger}(\tau)C_{\mathbf{p}}(\tau)\sum_{\mathbf{k}} C_{\mathbf{k}}^{\dagger}(\tau')C_{\mathbf{k}}(\tau')\right]\right\rangle
$$

where the operators refer to fermions. First, do it exactly. The combination

$$\sum_p C_{\mathbf{p}}^{\dagger} C_{\mathbf{p}} = \sum_{\mathbf{p}} n_{\mathbf{p}}$$

is the number operator which we assume commutes with the Hamiltonian and thus has no τ dependence. We must separately average when $\mathbf{p} \neq \mathbf{k}$ and $\mathbf{p} = \mathbf{k}$:

$$W = \frac{1}{\nu^2} \sum_{\mathbf{p} \neq \mathbf{k}} {}_0\langle C_{\mathbf{p}}^{\dagger} C_{\mathbf{p}} C_{\mathbf{k}}^{\dagger} C_{\mathbf{k}} \rangle + \frac{1}{\nu^2} \sum_{\mathbf{p}} {}_0\langle C_{\mathbf{p}}^{\dagger} C_{\mathbf{p}} C_{\mathbf{p}}^{\dagger} C_{\mathbf{p}} \rangle$$

If we define

$$n_{\mathbf{p}} = {}_0\langle C_{\mathbf{p}}^{\dagger} C_{\mathbf{p}} \rangle$$

then we get exactly that

$$W = \frac{1}{\nu^2} \sum_{\mathbf{p} \neq \mathbf{k}} n_{\mathbf{p}} n_{\mathbf{k}} + \frac{1}{\nu^2} \sum_{\mathbf{p}} n_{\mathbf{p}} = \left(\frac{1}{\nu} \sum_{\mathbf{p}} n_{\mathbf{p}} \right)^2 + \frac{1}{\nu^2} \sum_{\mathbf{p}} n_{\mathbf{p}} (1 - n_{\mathbf{p}}) \qquad (3.4.10)$$

Solving the same problem by the application of Wick's theorem gives

$$W = \frac{1}{\nu^2} \left(\sum_{\mathbf{p}} n_{\mathbf{p}} \right)^2 - \frac{1}{\nu^2} \sum_{\mathbf{p},\mathbf{k}} \delta_{\mathbf{pk}} \mathscr{G}^{(0)}(\mathbf{p}, \tau - \tau') \mathscr{G}^{(0)}(\mathbf{p}, \tau' - \tau)$$

If we recall that

$$\mathscr{G}^{(0)}(\mathbf{p}, \tau) = -e^{-\xi_{\mathbf{p}} \tau} [\theta(\tau) - n_{\mathbf{p}}]$$

then for the combination we get

$$\mathscr{G}^{(0)}(\mathbf{p}, \tau) \mathscr{G}^{(0)}(\mathbf{p}, -\tau) = [\theta(\tau) - n_{\mathbf{p}}][\theta(-\tau) - n_{\mathbf{p}}] = -n_{\mathbf{p}}(1 - n_{\mathbf{p}})$$

and this method yields the same answer (3.4.10) as the exact method. Thus Wick's theorem gives the right answer, even, in this example, for ordering the inverse volume $1/\nu$; since the first term

$$\left(\frac{1}{\nu} \sum_{\mathbf{p}} n_{\mathbf{p}} \right)^2 = \left[\int \frac{d^3\mathbf{p}}{(2\pi)^3} n_{\mathbf{p}} \right]^2 = n_0^2$$

is the square of the particle density, while the second is $O(\nu^{-1})$,

$$\frac{1}{\nu^2} \sum_{\mathbf{p}} n_{\mathbf{p}}(1 - n_{\mathbf{p}}) = \frac{1}{\nu} \int \frac{d^3p}{(2\pi)^3} n_{\mathbf{p}}(1 - n_{\mathbf{p}})$$

since the integral is finite.

Diagrams in the S-matrix expansion are classified as connected or disconnected according to the conventions discussed in Sec. 2.6. Only connected diagrams are retained, since the disconnected diagrams are canceled by the vacuum polarization diagrams. For Matsubara Green's functions, the vacuum polarization terms come from the denominator:

$$e^{-\beta\Omega} = \text{Tr}[e^{-\beta K_0}S(\beta)] = \sum_{n=0}^{\infty} \frac{(-1)^n}{n!} \int_0^\beta d\tau_1 \cdots \int_0^\beta d\tau_n \, {}_0\langle T_\tau \hat{V}(\tau_1) \cdots \hat{V}(\tau_n)\rangle \tag{3.4.11}$$

The expansion of this quantity produces a series of diagrams which just cancel the disconnected parts of the expansion in (3.4.9). In the zero temperature case of Chapter 2, we found that each vacuum polarization graph gave infinity, which was perhaps not a worry since they canceled out of the answer. In the Matsubara formalism, the vacuum polarization terms are all finite. The "time" integrals, which are now τ integrals, are only over the finite range $0 \leq \tau \leq \beta$, so that the extra integration does not diverge. The Matsubara formalism eliminates this one divergence, which is another small advantage of the method.

The summation in (3.4.11) actually evaluates a useful quantity—the thermodynamic potential Ω. One might actually wish to calculate it. The method of doing this, which is presented in Sec. 3.6, is called a linked cluster expansion.

The Matsubara Green's function can be reduced to an evaluation of all connected, different diagrams:

$$\mathscr{G}(\mathbf{p}, \tau) = -\sum_{n=0}^{\infty} (-1)^n \int_0^\beta d\tau_1 \cdots \int_0^\beta d\tau_n \times {}_0\langle \text{Tr}[\hat{C}_\mathbf{p}(\tau)\hat{V}(\tau_1)\hat{V}(\tau_2) \cdots \hat{V}(\tau_n)\hat{C}_\mathbf{p}^\dagger(0)]\rangle_{\substack{\text{different}\\ \text{connected}}} \tag{3.4.12}$$

Each diagram is evaluated as a function of τ, and then one takes the Fourier transform:

$$\mathscr{G}(\mathbf{p}, ip_n) = \int_0^\beta d\tau e^{ip_n\tau}\mathscr{G}(\mathbf{p}, \tau)$$

$$p_n = \frac{(2n+1)\pi}{\beta}$$

The terms in the series yield self-energy diagrams, which may be collected into Dyson equations:

$$\begin{aligned} \mathscr{G}(\mathbf{p}, ip_n) &= \frac{\mathscr{G}^{(0)}(\mathbf{p}, ip_n)}{1 - \mathscr{G}^{(0)}(\mathbf{p}, ip_n)\Sigma(\mathbf{p}, ip_n)} \\ \mathscr{D}(\mathbf{q}, i\omega_n) &= \frac{\mathscr{D}^{(0)}(\mathbf{q}, i\omega_n)}{1 - \mathscr{D}^{(0)}(\mathbf{q}, i\omega_n)\pi(\mathbf{q}, i\omega_n)} \end{aligned} \tag{3.4.13}$$

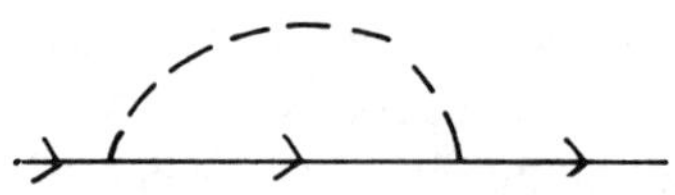

FIGURE 3.4

As an example, consider the basic electron self-energy contribution from the electron–phonon interaction shown in Fig. 3.4. This self-energy was evaluated in (2.7.4) using zero temperature methods. We shall do it again using Matsubara methods. The interaction is

$$V = \frac{1}{\nu^{1/2}} \sum_{\mathbf{q}} M_{\mathbf{q}} A_{\mathbf{q}} \sum_{\mathbf{k}} C^{\dagger}_{\mathbf{k}+\mathbf{q}} C_{\mathbf{k}}$$

so that the $n = 2$ term in the S-matrix expansion is

$$\begin{aligned}\mathscr{G}_2 = -\frac{1}{\nu} \sum_{\mathbf{q}\mathbf{q}'} M_{\mathbf{q}} M_{\mathbf{q}'} \int_0^\beta d\tau_1 \int_0^\beta d\tau_2 \, {}_0\langle T_\tau \hat{C}_{\mathbf{p}}(\tau) \sum_{\mathbf{k}} \hat{C}^{\dagger}_{\mathbf{k}+\mathbf{q}}(\tau_1) \hat{C}_{\mathbf{k}}(\tau_1) \\ \times \sum_{\mathbf{k}'} \hat{C}^{\dagger}_{\mathbf{k}'+\mathbf{q}'}(\tau_2) \hat{C}_{\mathbf{k}'}(\tau_2) \hat{C}_{\mathbf{p}}^{\dagger}(0)\rangle \, {}_0\langle T_\tau \hat{A}_{\mathbf{q}}(\tau_1) \hat{A}_{\mathbf{q}'}(\tau_2)\rangle\end{aligned}$$

The factor

$${}_0\langle T_\tau \hat{A}_{\mathbf{q}}(\tau_1) \hat{A}_{\mathbf{q}'}(\tau_2)\rangle = -\,\delta_{\mathbf{q}+\mathbf{q}'} \mathscr{D}^{(0)}(\mathbf{q}, \tau_1 - \tau_2)$$

gives the phonon Green's function. Wick's theorem yields six terms when applied to the electron bracket, but there is only one different disconnected term, and it gives the contribution:

$$\begin{aligned}\mathscr{G}_2(p, \tau) = -\frac{1}{\nu} \sum_{q} M_q^2 \int_0^\beta d\tau_1 \int_0^\beta d\tau_2 \mathscr{D}^{(0)}(\mathbf{q}, \tau_1 - \tau_2) \mathscr{G}^{(0)}(\mathbf{p}, \tau - \tau_1) \\ \times \mathscr{G}^{(0)}(\mathbf{p} + \mathbf{q}, \tau_1 - \tau_2) \mathscr{G}^{(0)}(\mathbf{p}, \tau_2)\end{aligned}$$

The τ integrals may be performed easily, since we have previously derived the τ dependence of the unperturbed Green's functions in (3.2.9) and (3.2.15). However, this step is rarely performed. Instead, the Fourier transform is taken:

$$\mathscr{G}(\mathbf{p}, ip_n) = \int_0^\beta d\tau \mathscr{G}(\mathbf{p}, \tau) e^{ip_n\tau}$$

The τ integrals are done by using the Fourier expansion, given in (3.2.7) and (3.2.13), such as

$$\mathscr{D}^{(0)}(\mathbf{q}, \tau_1 - \tau_2) = \frac{1}{\beta} \sum_{i\omega_n} e^{-i\omega_n(\tau_1 - \tau_2)} \mathscr{D}^{(0)}(\mathbf{q}, i\omega_n)$$

$$\mathscr{G}^{(0)}(\mathbf{p}, \tau - \tau_1) = \frac{1}{\beta} \sum_{ip_{n'}} e^{-ip_{n'}(\tau - \tau_1)} \mathscr{G}^{(0)}(\mathbf{p}, ip_{n'})$$

Thus we can write

$$\mathscr{G}(p, ip_n) = -\frac{1}{\nu}\sum_{\mathbf{q}} M_{\mathbf{q}}^2 \frac{1}{\beta^4} \sum_{n',n'',n''',m} \mathscr{D}^{(0)}(\mathbf{q}, i\omega_m)\mathscr{G}^{(0)}(\mathbf{p}, ip_{n'})$$
$$\times\, \mathscr{G}^{(0)}(\mathbf{p}-\mathbf{q}, ip_{n''})\mathscr{G}^{(0)}(\mathbf{p}, ip_{n'''}) \int_0^\beta d\tau \int_0^\beta d\tau_1 \int_0^\beta d\tau_2$$
$$\times \exp[ip_n\tau - i\omega_m(\tau_1-\tau_2) - ip_{n'}(\tau-\tau_1) - ip_{n''}(\tau_1-\tau_2) - ip_{n'''}\tau_2]$$

All three τ integrals are of the form

$$\frac{1}{\beta}\int_0^\beta d\tau e^{i\tau(p_n-p_{n'})} = \frac{1}{\beta i(p_n - p_{n'})}(e^{i\beta(p_n-p_{n'})} - 1)$$
$$= \frac{1}{2i\pi(n-n')}(e^{i2\pi(n-n')} - 1) = 0 \qquad \text{if } n \neq n'$$

which we write in the shorthand notation

$$\frac{1}{\beta}\int_0^\beta d\tau e^{i\tau(p_n-p_{n'})} = \delta_{p_n=p_{n'}}$$
$$\frac{1}{\beta}\int_0^\beta d\tau_1 e^{i\tau_1(ip_{n'}-p_{n''}-\omega_m)} = \delta_{p_{n'}=p_{n''}+\omega_m}$$
$$\frac{1}{\beta}\int_0^\beta d\tau_2 e^{i\tau_2(\omega_m+p_{n''}-p_{n'''})} = \delta_{p_{n'''}=p_{n''}+\omega_m}$$

Combining these results yields the final result

$$\mathscr{G}_2(p, ip_n) = \mathscr{G}^{(0)}(\mathbf{p}, ip_n)^2\Sigma^{(1)}(\mathbf{p}, ip_n)$$

where the one-phonon self-energy term is now

$$\Sigma^{(1)}(\mathbf{p}, ip_n) = -\frac{1}{\beta\nu}\sum_{\mathbf{q}}\sum_{i\omega_n} M_{\mathbf{q}}^2\mathscr{D}^{(0)}(\mathbf{q}, i\omega_n)\mathscr{G}^{(0)}(\mathbf{p}-\mathbf{q}, ip_n - i\omega_n) \quad (3.4.14)$$

The self-energy (3.4.14) in the Matsubara notation is similar to (2.7.4). The only difference is the use of complex frequencies rather than real ones. Also note that there is frequency conservation in this expression. An electron starts with frequency ip_n, and it emits or absorbs a phonon with frequency $i\omega_n$, so the electron goes to an intermediate state with frequency $ip_n - i\omega_n$. Again the summation is taken over internal variables, in this case the momentum $\mathbf{q}$ and the frequency $i\omega_n$.

In general, the complex frequency is conserved at each vertex in the Feynman diagram. This conservation does maintain the oddness of fermion frequencies and the evenness for bosons. For example, the electron in (3.4.13) has the frequency $ip_n - i\omega_n$. Now ip_n is an odd integer, since it was associated with the original electron line, while $i\omega_n$ is an even integer, since it was associated with the phonon. But an odd plus an even is odd, which preserves the oddness of electron frequencies.

The exact Matsubara Green's function may be obtained by finding the exact self-energies. The exact solution is rarely possible, and usually we are content to write down a few diagrams and evaluate the first few terms in the expansion for $\Sigma(\mathbf{p}, ip_n)$. To obtain these terms, we draw the Feynman diagrams, conserve momentum and frequency at each vertex, and then integrate or sum over internal variables. All one needs now are some *rules for constructing diagrams*:

1. With each internal electron line, associate a quantity $\mathscr{G}^{(0)}(\mathbf{p}, ip_n)$.
2. With each internal phonon line, associate a quantity $M_{\mathbf{q}}{}^2\mathscr{D}^{(0)}(\mathbf{q}, i\omega_n)$.
3. With each internal coulomb line, associate a quantity $v_{\mathbf{q}} = 4\pi e^2/\mathbf{q}^2$.
4. Conserve momentum and complex frequency at each vertex. Keep in mind that fermion frequencies are odd integers $(2n + 1)\pi/\beta$ and boson frequencies are even integers $2n\pi/\beta$. Their oddness and evenness will be maintained in the energy conservation.
5. Sum over internal degrees of freedom: momentum and frequency. Internal variables are all those except the $(\mathbf{p}, ip_n)$ of the self-energy.
6. Multiply the expression by

$$\frac{(-1)^{m+F}(2S + 1)^F}{(\nu\beta)^m}$$

where F is the number of closed fermion loops. The $2S + 1$ factor in $(2S + 1)^F$ is a summation over spin degrees of freedom, and $2S + 1 = 2$ for electrons. The integer m is the order of the diagram, as defined earlier in Sec. 2.8.

We shall give several examples. The basic fermion loop in Fig. 3.5(a) occurs in the phonon self-energy and also the coulomb self-energy. It is

$$P^{(1)}(\mathbf{q}, i\omega_n) = \frac{2}{\beta\nu} \sum_{ip_n} \sum_{\mathbf{p}} \mathscr{G}^{(0)}(\mathbf{p}, ip_n)\mathscr{G}^{(0)}(\mathbf{p} + \mathbf{q}, ip_n + i\omega_n) \quad (3.4.15)$$

In Fig. 3.5(b) we have added a phonon line between the two electron lines,

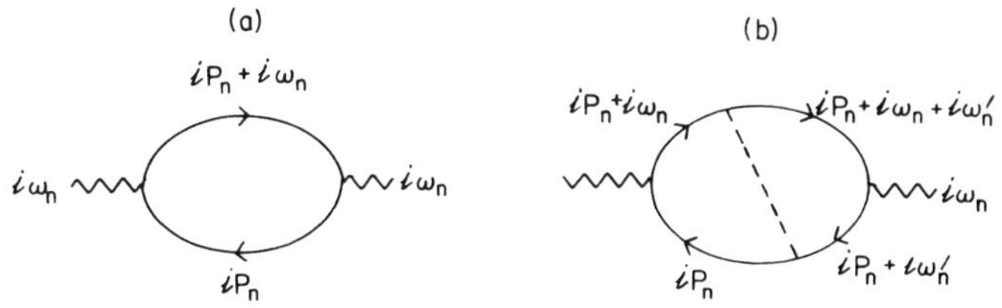

FIGURE 3.5

which is called a vertex correction. This diagram gives, according to the rules,

$$P^{(2)}(\mathbf{q}, i\omega_n) = -\frac{2}{\beta^2\nu^2}\sum_{\mathbf{k},\mathbf{p}} M_{\mathbf{k}}^2 \sum_{i\omega_{n'}, ip_n} \mathscr{G}^{(0)}(\mathbf{p}, ip_n)\mathscr{G}^{(0)}(\mathbf{p}+\mathbf{q}, ip_n + i\omega_n)$$
$$\times \mathscr{D}^{(0)}(\mathbf{k}, i\omega_{n'})\mathscr{G}^{(0)}(\mathbf{p}+\mathbf{k}+\mathbf{q}, ip_n + i\omega_n + i\omega_{n'})$$
$$\times \mathscr{G}^{(0)}(\mathbf{p}+\mathbf{k}, ip_n + i\omega_{n'}) \qquad (3.4.16)$$

3.5. FREQUENCY SUMMATIONS

When using the Matsubara Green's functions, one must often evaluate frequency summations over combinations of unperturbed Green's functions. We shall discuss the technique for evaluating these summations for both the case of unperturbed functions and also Green's functions with self-energies. First we shall present a table of results for combinations which often occur and then explain how to derive them:

$$-\frac{1}{\beta}\sum_{i\omega_n} \mathscr{D}^{(0)}(\mathbf{q}, i\omega_n)\mathscr{G}^{(0)}(\mathbf{p}, ip_n + i\omega_n) = \frac{N_{\mathbf{q}} + n_F(\xi_{\mathbf{p}})}{ip_n + \omega_{\mathbf{q}} - \xi_{\mathbf{p}}} + \frac{N_{\mathbf{q}} + 1 - n_F(\xi_{\mathbf{p}})}{ip_n - \omega_{\mathbf{q}} - \xi_{\mathbf{p}}} \qquad (3.5.1)$$

$$\frac{1}{\beta}\sum_{ip_n} \mathscr{G}^{(0)}(\mathbf{p}, ip_n)\mathscr{G}^{(0)}(\mathbf{k}, ip_n + i\omega_n) = \frac{n_F(\xi_{\mathbf{p}}) - n_F(\xi_{\mathbf{k}})}{i\omega_n + \xi_{\mathbf{p}} - \xi_{\mathbf{k}}} \qquad (3.5.2)$$

$$-\frac{1}{\beta}\sum_{ip_n} \mathscr{G}^{(0)}(\mathbf{p}, ip_n)\mathscr{G}^{(0)}(\mathbf{k}, i\omega_n - ip_n) = \frac{1 - n_F(\xi_{\mathbf{p}}) - n_F(\xi_{\mathbf{k}})}{i\omega_n - \xi_{\mathbf{p}} - \xi_{\mathbf{k}}} \qquad (3.5.3)$$

$$\frac{1}{\beta}\sum_{ip_n} \mathscr{G}^{(0)}(\mathbf{p}, ip_n) = n_F(\xi_{\mathbf{p}}) \qquad (3.5.4)$$

where

$$N_{\mathbf{q}} = \frac{1}{e^{\beta\omega_q} - 1}, \qquad n_F(\xi) = \frac{1}{e^{\beta\xi} + 1}$$

The combination (3.5.1) occurs in the electron self-energy (3.4.14) from the electron–phonon interaction, while the combination (3.5.2) is found in the basic polarization diagram (3.4.15). The other combinations will be encountered later.

First consider the summation over a boson series, where the summation is over even integer combinations $\omega_n = 2n\pi/\beta$. For example, consider (3.5.1)

$$S = +\frac{1}{\beta}\sum_{n=-\infty}^{\infty}\frac{2\omega_{\mathbf{q}}}{\omega_n^{\,2}+\omega_{\mathbf{q}}^{\,2}}\,\frac{1}{ip_n + i\omega_n - \xi_{\mathbf{p}}} \tag{3.5.5}$$

First, write this as

$$S = -\frac{1}{\beta}\sum_n f(i\omega_n)$$

where $f(i\omega_n)$ is just the product of Green's functions in (3.5.5). This summation is evaluated by a contour integration. We shall do an integral of the form

$$I = \lim_{R\to\infty}\int\frac{dz}{2\pi i}\,f(z)n_B(z) \tag{3.5.6}$$

where the contour is a large circle of radius R in the limit as $R \to \infty$. The function $n_B(z)$ is chosen to generate poles at the points $i\omega_n$ for all even integer n. The function which does this is

$$n_B = \frac{1}{e^{\beta z} - 1} \tag{3.5.7}$$

The poles of $n_B(z)$ were discussed earlier at the beginning of the chapter, in (3.1.3). The function $n_B(z)$ has poles at the points $i2n\pi/\beta$ for all positive and negative integer n and $n = 0$. The residue at these poles is $1/\beta$. In Fig. 3.6, these poles are shown as **X** marks which are evenly spaced on the

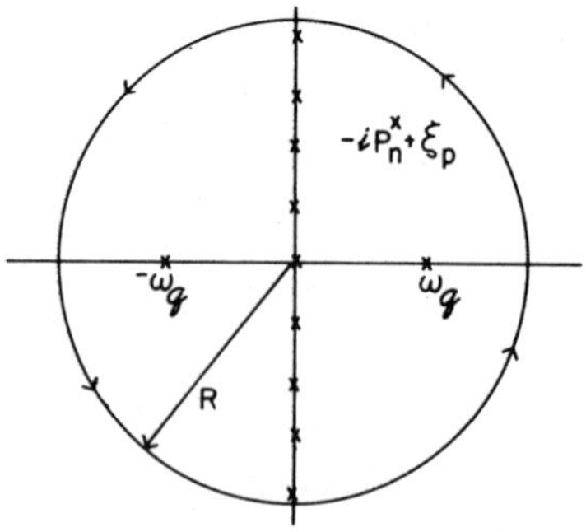

FIGURE 3.6

vertical axis. The large circle is the contour of integration, which is a circle of radius R. The function $f(z)$ is

$$f(z) = \frac{2\omega_q}{z^2 - \omega_{\mathbf{q}}^2} \frac{1}{ip_n + z - \xi_{\mathbf{p}}}$$

It has poles at the points $+\omega_{\mathbf{q}}$ and $-\omega_{\mathbf{q}}$, which originate from the phonon Green's function. Another pole is at the point $\xi_p - ip_n$, and this pole comes from the electron Green's function. Thus the poles, and their residues, of the integral I in (3.5.6) are

$$\begin{aligned}
z_i &= \frac{i2n\pi}{\beta}, & R_i &= \frac{1}{\beta} f(i\omega_n) \\
z_1 &= \omega_{\mathbf{q}}, & R_1 &= \frac{N_{\mathbf{q}}}{ip_n + \omega_{\mathbf{q}} - \xi_{\mathbf{p}}} \\
z_2 &= -\omega_{\mathbf{q}}, & R_2 &= \frac{N_{\mathbf{q}} + 1}{ip_n - \omega_{\mathbf{q}} - \xi_{\mathbf{p}}} \\
z_3 &= \xi_{\mathbf{p}} - ip_n, & R_3 &= -\frac{2\omega_{\mathbf{q}} n_F(\xi_p)}{(ip_n - \xi_{\mathbf{p}})^2 - \omega_{\mathbf{q}}^2}
\end{aligned}$$

where our notation is that $N_{\mathbf{q}} \equiv n_B(\omega_{\mathbf{q}})$. In the last residue, we have

$$n_B(\xi_{\mathbf{p}} - ip_n) = \frac{1}{e^{\beta(\xi_{\mathbf{p}} - ip_n)} - 1} = -\frac{1}{e^{\beta\xi_{\mathbf{p}}} + 1} = -n_F(\xi_{\mathbf{p}})$$

since $\exp(i\beta p_n) = -1$ because p_n is an odd integer (fermion). The last residue may be rewritten as

$$R_3 = \frac{n_F(\xi_{\mathbf{p}})}{ip + \omega_{\mathbf{q}} - \xi_{\mathbf{p}}} - \frac{n_F(\xi_{\mathbf{p}})}{ip - \omega_{\mathbf{q}} - \xi_{\mathbf{p}}}$$

so that the integral is evaluated by adding all these residues:

$$I = \frac{1}{\beta} \sum_{\omega_n} f(i\omega_n) + \frac{N_q + n_F(\xi_{\mathbf{p}})}{ip_n - \xi_{\mathbf{p}} + \omega_{\mathbf{q}}} + \frac{N_q + 1 - n_F(\xi_{\mathbf{p}})}{ip_n - \xi_{\mathbf{p}} - \omega_{\mathbf{q}}}$$

In the limit as $R \to \infty$, the integral vanishes, $I = 0$, which gives the final result,

$$S = \frac{N_q + n_F(\xi_{\mathbf{p}})}{ip_n - \xi_{\mathbf{p}} + \omega_{\mathbf{q}}} + \frac{N_q + 1 - n_F(\xi_{\mathbf{p}})}{ip_n - \xi_{\mathbf{p}} - \omega_{\mathbf{q}}}$$

which is (3.5.1). Thus the method of evaluating these boson series is quite simple. To evaluate a series such as

$$S = -\frac{1}{\beta}\sum_n f(i\omega_n)$$

one just finds all the simple poles of $f(z)$ and at these poles z_j finds the residues r_j of $f(z)$, and

$$S = \sum_j R_j$$

$$R_j = r_j n_B(z_j)$$

The same procedure is used to evaluate fermion series. Then the summations are of the form

$$S = -\frac{1}{\beta}\sum_{ip_n} f(ip_n)$$

where $p_n = (2n+1)\pi/\beta$ contains odd integers. One constructs the same contour integral as in (3.5.6), except now the function $n_F(z)$ is chosen to have poles at the points ip_n, and this function is

$$n_F(z) = \frac{1}{e^{\beta z}+1}$$

where the residue is $-(1/\beta)$ at these points. Again we find that the integral $I = 0$, so that the summation is just the residue r_i at the poles z_i of $f(z)$:

$$S = -\sum_i R_i$$
$$R_i = r_i n_F(z_i) \tag{3.5.8}$$

The minus sign in front occurs because of the feature that the residue of the fermion $n_F(z)$ is $-(1/\beta)$, whereas it is $+(1/\beta)$ for the boson $n_B(z)$.

As an example of summing a fermion series, we do the same summation as before, (3.5.1). The summation variable is changed to $p_{n'} = p_n + \omega_n$ so that the summation is now

$$S = -\frac{1}{\beta}\sum_{p_{n'}} \mathscr{D}^{(0)}(\mathbf{q}, ip_{n'} - ip_n)\mathscr{G}^{(0)}(\mathbf{p}, ip_{n'})$$

$$= \frac{1}{\beta}\sum_{n'} \frac{2\omega_q}{(p_{n'}-p_n)^2+\omega_{\mathbf{q}}^2}\,\frac{1}{ip_{n'}-\xi_{\mathbf{p}}}$$

$p_{n'}$ contains odd integers, since it is a fermion frequency. Thus we have that

$$f(z) = \frac{2\omega_q}{(z - ip_n)^2 - \omega_{\mathbf{q}}^2} \frac{1}{z - \xi_{\mathbf{p}}}$$

which has the following poles and residues:

$$z_1 = \xi_p, \qquad R_1 = \frac{n_F(\xi_{\mathbf{p}})2\omega_{\mathbf{q}}}{(\xi_{\mathbf{p}} - ip_n)^2 - \omega_{\mathbf{q}}^2}$$

$$= n_F(\xi_{\mathbf{p}})\left(\frac{1}{ip_n - \xi_{\mathbf{p}} - \omega_q} - \frac{1}{ip_n - \xi_{\mathbf{p}} + \omega_{\mathbf{q}}}\right)$$

$$z_2 = ip - \omega_{\mathbf{q}}, \qquad R_2 = -\frac{n_F(ip_n - \omega_{\mathbf{q}})}{ip_n - \xi_{\mathbf{p}} - \omega_{\mathbf{q}}}$$

$$z_3 = ip_n + \omega_{\mathbf{q}}, \qquad R_3 = +\frac{n_F(ip_n + \omega_{\mathbf{q}})}{ip_n + \omega_{\mathbf{q}} - \xi_{\mathbf{p}}}$$

The last two thermal factors may be simplified

$$n_F(ip_n - \omega_{\mathbf{q}}) = \frac{1}{e^{\beta(ip_n - \omega_{\mathbf{q}})} + 1} = \frac{1}{1 - e^{-\beta\omega_{\mathbf{q}}}} = N_{\mathbf{q}} + 1$$

$$n_F(ip_n + \omega_{\mathbf{q}}) = -N_{\mathbf{q}}$$

The final result is obtained, as in (3.5.8), by adding these three residues,

$$S = \frac{N_{\mathbf{q}} + n_F(\xi_{\mathbf{p}})}{ip - \xi_{\mathbf{p}} + \omega_{\mathbf{q}}} + \frac{N_{\mathbf{q}} + 1 - n_F(\xi_{\mathbf{p}})}{ip - \xi_{\mathbf{p}} - \omega_{\mathbf{q}}}$$

which gives the same answer again.

At the beginning of the section we listed our results. The first we have derived twice. The next two, (3.5.2) and (3.5.3), we leave as an exercise for the student. We shall now explain the fourth result,

$$\frac{1}{\beta} \sum_{ip_n} \mathscr{G}^{(0)}(\mathbf{p}, ip_n) = n_F(\xi_{\mathbf{p}}) \tag{3.5.4}$$

The left-hand side of this equation is most familiar as the Fourier transform of the Matsubara Green's function:

$$\mathscr{G}^{(0)}(\mathbf{p}, \tau) = \frac{1}{\beta} \sum_{ip_n} e^{-ip_n\tau} \mathscr{G}^{(0)}(\mathbf{p}, ip_n) = -\langle T_\tau C_{\mathbf{p}}(\tau) C_{\mathbf{p}}^\dagger(0)\rangle$$

Our result (3.5.4) is just the limit as $\tau \to 0$. This limit is ambiguous, since a different result is obtained if $\tau = 0$ is approached from the positive or negative direction:

$$\mathscr{G}^{(0)}(\mathbf{p}, \tau = 0^+) = -\langle C_\mathbf{p} C_\mathbf{p}^\dagger \rangle = -[1 - n_F(\xi_\mathbf{p})]$$

$$\mathscr{G}^{(0)}(\mathbf{p}, \tau = 0^-) = \langle C_\mathbf{p}^\dagger C_\mathbf{p} \rangle = n_F(\xi_\mathbf{p})$$

Our result (3.5.4) is merely the convention that we adopt the limit $\tau = 0^-$. This choice has been discussed several times before. The same result at zero temperature was expressed in (2.8.4). It is our convention that two operators at equal time are taken in the order which gives the number operator:

$$-\langle T_\tau C_\mathbf{p}(\tau) C_\mathbf{p}^\dagger(\tau) \rangle = \langle C_\mathbf{p}^\dagger C_\mathbf{p} \rangle = n_F(\xi_\mathbf{p})$$

The Matsubara sum in the preceding example was very easy. Summations which contain only $\mathscr{G}^{(0)}$ and $\mathscr{D}^{(0)}$ always cause the contour integral I to have just poles. If the summation contained $\mathscr{G}$, instead of $\mathscr{G}^{(0)}$, then the I integral has branch cuts. We shall show how to treat this more difficult case.

Consider the example of evaluating the polaron self-energy $\Sigma^{(1)}$ for an electron in a crystal containing impurities. Assume that the self-energy $\Sigma_i(p, ip_n)$ from impurity scattering has already been evaluated, giving the Green's function

$$\mathscr{G}(\mathbf{p}, ip_n) = \frac{1}{ip - \xi_\mathbf{p} - \Sigma_i(\mathbf{p}, ip_n)}$$

In the presence of this impurity scattering, the lowest-order polaron self-energy is

$$\Sigma^{(1)}(\mathbf{p}, ip_n) = -\frac{1}{\beta} \int \frac{d^3q}{(2\pi)^3} M_\mathbf{q}^2 \sum_{i\omega_n} \mathscr{D}^{(0)}(\mathbf{q}, i\omega_n) \mathscr{G}(\mathbf{p} + \mathbf{q}, ip_n + i\omega_n)$$

Thus we wish to evaluate the summation

$$S = +\frac{1}{\beta} \sum_{i\omega_n} \frac{2\omega_\mathbf{q}}{\omega_n^2 + \omega_\mathbf{q}^2} \frac{1}{ip_n + i\omega_n - \xi_{\mathbf{p}+\mathbf{q}} - \Sigma_i(\mathbf{p} + \mathbf{q}, ip_n + i\omega_n)}$$

$$= -\frac{1}{\beta} \sum_{i\omega_n} f(i\omega_n)$$

where the function $f(z)$ is

$$f(z) = \frac{2\omega_\mathbf{q}}{z^2 - \omega_\mathbf{q}^2} \frac{1}{ip_n + z - \xi_{\mathbf{p}+\mathbf{q}} - \Sigma_i(\mathbf{p} + \mathbf{q}, ip_n + z)}$$

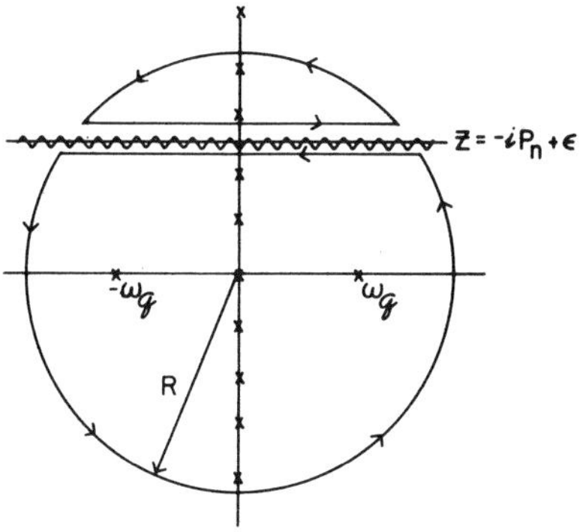

FIGURE 3.7

Since we are doing a boson series, we again construct the contour integral:

$$I = \int_{\Gamma} \frac{dz}{2\pi i} n_B(z) f(z)$$

The contour Γ is shown in Fig. 3.7. The self-energy function $\Sigma_i(\mathbf{p} + \mathbf{q}, ip_n + z)$ has a possible branch cut at $ip_n + z = \varepsilon =$ real. Thus the contour has two separate pieces. Each piece has a part which runs along the branch cut, either above or below, and then connects to the large circle of radius R, which is taken in the limit of $R \to \infty$. The branch cut just extends over those regions where $\text{Im}\,\Sigma \neq 0$, but for now put it clear across the axis, for if $\text{Im}\,\Sigma = 0$ over part of our contour, then the $\rightleftarrows$ parts just cancel. The I integral is now nonzero as $R \to \infty$. It just equals the contribution along the branch cuts, since the circular arcs of radius R give a vanishing contribution as $R \to \infty$. The integration along the branch cut gives

$$\begin{aligned} I = & \int_{-\infty}^{\infty} \frac{d\varepsilon}{2\pi i} n_B(\varepsilon - ip_n) \mathscr{D}^{(0)}(\mathbf{q}, \varepsilon - ip_n) \mathscr{G}(\mathbf{p} + \mathbf{q}, \varepsilon + i\delta) \\ & + \int_{\infty}^{-\infty} \frac{d\varepsilon}{2\pi i} n_B(\varepsilon - ip_n) \mathscr{D}^{(0)}(\mathbf{q}, \varepsilon - ip_n) \mathscr{G}(\mathbf{p} + \mathbf{q}, \varepsilon - i\delta) \end{aligned}$$

where the first term is the integral above the branch cut at $ip_n + i\omega_n = \varepsilon + i\delta$, while the second is the integral below. Now we recognize the following simplifications,

$$\begin{aligned} n_B(\varepsilon - ip_n) &= -n_F(\varepsilon) \\ \mathscr{G}(\mathbf{p} + \mathbf{q}, \varepsilon + i\delta) &= G_{\text{ret}}(\mathbf{p} + \mathbf{q}, \varepsilon) \\ \mathscr{G}(\mathbf{p} + \mathbf{q}, \varepsilon - i\delta) &= G_{\text{adv}}(\mathbf{p} + \mathbf{q}, \varepsilon) \end{aligned}$$

so that we can rewrite the integral as

$$I = -\int_{-\infty}^{\infty} \frac{d\varepsilon}{2\pi i} n_F(\varepsilon)\mathscr{D}^{(0)}(\mathbf{q}, \varepsilon - ip_n)[G_{\text{ret}}(\mathbf{p}+\mathbf{q}, \varepsilon) - G_{\text{adv}}(\mathbf{p}+\mathbf{q}, \varepsilon)]$$

The minus sign on the last term comes from changing the direction of integration. Since the advanced function is the complex conjugate of the retarded function, the part in brackets is just the spectral function

$$G_{\text{ret}}(\mathbf{p}+\mathbf{q}, \varepsilon) - G_{\text{adv}}(\mathbf{p}+\mathbf{q}, \varepsilon) = 2i \operatorname{Im} G_{\text{ret}}(\mathbf{p}+\mathbf{q}, \varepsilon) = -iA(\mathbf{p}+\mathbf{q}, \varepsilon)$$

Thus for the integral we finally obtain

$$I = \int_{-\infty}^{\infty} \frac{d\varepsilon}{2\pi} n_F(\varepsilon) A(\mathbf{p}+\mathbf{q}, \varepsilon) \frac{2\omega_{\mathbf{q}}}{(ip_n - \varepsilon)^2 - \omega_{\mathbf{q}}^2}$$

Of course, I also equals the summation of the residues of the poles contained within the contour. These include the phonon poles at $z = \pm\omega_{\mathbf{q}}$ and the thermal poles at $z = i2n\pi/\beta$. None of these poles lies on the branch cut. The branch cut was on the line $z = -ip_n + \varepsilon$, so that its complex values were odd multiples of $\pi i/\beta$, while the thermal poles are at even multiples of $\pi i/\beta$. As shown in Fig. 3.7, the contour does enclose all these other poles. One could have the situation where the branch cut did fall on one of the thermal poles $i\omega_n$. This would happen if, for example, $G(\mathbf{p}+\mathbf{q}, ip + i\omega)$ described a boson particle such as ^{4}He, so that the ip_n were also a boson frequency. This situation is treated by deforming the contours along the branch cut into small semicircles around the pole.

The integral I equals the sum of the residues of the poles,

$$I = -S + \frac{N_q}{ip_n + \omega_{\mathbf{q}} - \xi_{\mathbf{p}+\mathbf{q}} - \Sigma_i(\mathbf{p}+\mathbf{q}, ip_n + \omega_q)} + \frac{N_q + 1}{ip_n - \omega_{\mathbf{q}} - \xi_{\mathbf{p}+\mathbf{q}} - \Sigma_i(\mathbf{p}+\mathbf{q}, ip_n - \omega_{\mathbf{q}})}$$

where S comes from the thermal poles and the other two terms from the poles of the phonon Green's function. If we equate our two results for I, we obtain the result for the summation:

$$S = N_q\mathscr{G}(\mathbf{p}+\mathbf{q}, ip_n + \omega_{\mathbf{q}}) + (N_{\mathbf{q}} + 1)\mathscr{G}(\mathbf{p}+\mathbf{q}, ip_n - \omega_{\mathbf{q}}) - \int_{-\infty}^{\infty} \frac{d\varepsilon}{2\pi} n_F(\varepsilon) A(\mathbf{p}+\mathbf{q}, \varepsilon)\mathscr{D}^{(0)}(\mathbf{q}, \varepsilon - ip_n) \tag{3.5.9}$$

This equation is the final answer, which has been reduced as much as possible. The next step would be to know the form for Σ_i, so that one could proceed with the integration over ε. Also note that in the limiting case that $\Sigma_i \to 0$, the spectral function becomes

$$\lim_{\Sigma_i \to 0} A(\mathbf{p} + \mathbf{q}, \varepsilon) = A^{(0)}(\mathbf{p} + \mathbf{q}, \varepsilon) = 2\pi\delta(\varepsilon - \xi_{\mathbf{p}+\mathbf{q}})$$

and the last term in (3.5.9) becomes

$$-n_F(\xi_{\mathbf{p}+\mathbf{q}})\left(\frac{1}{ip_n - \omega_{\mathbf{q}} - \xi_{\mathbf{p}+\mathbf{q}}} - \frac{1}{ip_n + \omega_{\mathbf{q}} - \xi_{\mathbf{p}+\mathbf{q}}}\right)$$

This result then agrees with the previous derivations using $\mathscr{G}^{(0)}$. The effect of the impurity interactions, which cause Σ_i, is to "smear out" the Green's function poles: The pole is replaced by a distribution given by the spectral density. If the phonons had a scattering mechanism which one wished to include in the calculations, then one would also use $\mathscr{D}$ instead of $\mathscr{D}^{(0)}$. This would cause the phonon poles to be replaced by a branch cut along the $Z =$ real axis; the phonon spectral density would enter the result.

At zero temperature, the summation procedure is particularly simple. The discrete summation is replaced by a continuous integral,

$$\lim_{T \to 0} S = \lim_{T \to 0} \frac{1}{\beta} \sum_{i\omega_n} f(i\omega_n) = \int_{-\infty}^{\infty} \frac{d\omega}{2\pi} f(i\omega)$$

which may be evaluated by contour integration. This change into a continuous integral is valid for both fermion and boson series.

The most general possibility when evaluating a summation over Matsubara frequencies is to have all the Green's functions fully dressed. Then it is easiest to proceed by first expressing all the Green's functions in the Lehmann representation. For example, if we wish to evaluate

$$S = -\frac{1}{\beta} \sum_{i\omega_n} \mathscr{D}(\mathbf{q}, i\omega_n)\mathscr{G}(\mathbf{p} + \mathbf{q}, ip + i\omega_n)$$

we then express each Green's function as a frequency integral over its respective spectral functions:

$$\mathscr{D}(\mathbf{q}, i\omega_n) = \int_{-\infty}^{\infty} \frac{d\omega'}{2\pi} \frac{B(\mathbf{q}, \omega')}{i\omega - \omega'}$$

$$\mathscr{G}(\mathbf{p} + \mathbf{q}, ip + i\omega) = \int_{-\infty}^{\infty} \frac{d\varepsilon'}{2\pi} \frac{A(\mathbf{p} + \mathbf{q}, \varepsilon')}{ip + i\omega - \varepsilon'}$$

$$S = \int_{-\infty}^{\infty} \frac{d\omega'}{2\pi} B(\mathbf{q}, \omega') \int_{-\infty}^{\infty} \frac{d\varepsilon'}{2\pi} A(\mathbf{p} + \mathbf{q}, \varepsilon') S_0(\omega', \varepsilon')$$

$$S_0(\omega', \varepsilon') = -\frac{1}{\beta} \sum_{i\omega} \frac{1}{i\omega - \omega'} \frac{1}{ip + i\omega - \varepsilon'}$$

The summation $S_0(\omega', \varepsilon')$ is now the easy type we have already evaluated over noninteracting Green's functions:

$$S_0 = \frac{n_B(\omega') + n_F(\varepsilon')}{ip + \omega' - \varepsilon'}$$

$$S = \int_{-\infty}^{\infty} \frac{d\omega'}{2\pi} B(\mathbf{q}, \omega') \int_{-\infty}^{\infty} \frac{d\varepsilon'}{2\pi} A(\mathbf{p} + \mathbf{q}, \varepsilon') \frac{n_B(\omega') + n_F(\varepsilon')}{ip + \omega' - \varepsilon'}$$

This provides the result for the most general case of fully interacting Green's functions. In general, any frequency summation can be done by expressing all the Green's functions in their Lehmann representation, which results in a summation of the easy type listed in (3.5.1) to (3.5.4).

The result (3.5.1) has now been derived, in this section, by several different methods. Now we should like to explain the physics behind this result. It is a bit subtle. The virtue of the Matsubara method is that these subtleties are automatically handled correctly. It is a machinery which cranks out the correct thermal occupation factors n_F and $N_\mathbf{q}$ even in complicated cases.

The summation (3.5.1) arises from the electron–phonon interaction and is a phonon contribution to the self-energy of the electron. From our rules for constructing diagrams, the one-phonon self-energy of the electron is

$$\Sigma^{(1)}(\mathbf{p}, ip_n) = \frac{1}{\nu} \sum_\mathbf{q} M_\mathbf{q}^2 \left[\frac{N_\mathbf{q} + n_F(\xi_{\mathbf{p}+\mathbf{q}})}{ip_n + \omega_\mathbf{q} - \xi_{\mathbf{p}+\mathbf{q}}} + \frac{N_\mathbf{q} + 1 - n_F(\xi_{\mathbf{p}+\mathbf{q}})}{ip_n - \omega_\mathbf{q} - \xi_{\mathbf{p}+\mathbf{q}}} \right] \tag{3.5.10}$$

This result is similar to what one would get from second-order perturbation theory. The standard quantum mechanical expression for this quantity is

$$\Delta E^{(2)} = \sum_I \frac{|\langle I | H_{\text{ep}} | i \rangle|^2}{E_i - E_I}$$

where $|i\rangle$ is the initial state of the system, and the summation $|I\rangle$ is over possible intermediate states. We assume that the initial state is an electron in state $\mathbf{p}$ with energy $\varepsilon_\mathbf{p}$. The electron–phonon interaction has two terms. One of these is $M_q C^\dagger_{\mathbf{p}+\mathbf{q}} C_\mathbf{p} a^\dagger_{-\mathbf{q}}$, which describes a process where an electron scatters from $\mathbf{p}$ to $\mathbf{p} + \mathbf{q}$ while creating a phonon of momentum $-\mathbf{q}$. This contribution from second-order perturbation theory is

$$\frac{1}{\nu} \sum_q \frac{M_q^2(1 + N_q)[1 - n_F(\xi_{\mathbf{p}+\mathbf{q}})]}{\varepsilon_\mathbf{p} - \varepsilon_{\mathbf{p}+\mathbf{q}} - \omega_\mathbf{q}}$$

The factor $1 + N_q$ comes from the phonon creation operator, $a_{-\mathbf{q}}^{\dagger} \mid n_{-\mathbf{q}}\rangle = (1 + n_{-\mathbf{q}})^{1/2} \mid n_{\mathbf{q}} + 1\rangle$, and we use the thermal average of $n_{-\mathbf{q}}$, which is N_q. Similarly, the factor $1 - n_F$ is the probability that the electron state $\mathbf{p} + \mathbf{q}$ is empty, so that the operator $C_{\mathbf{p}+\mathbf{q}}^{\dagger}$ can create an electron in that state. There is also a factor $n_F(\xi_{\mathbf{p}})$, which is the probability that $\mathbf{p}$ is occupied with an electron. We assume this is unity if we are trying to calculate the properties of an electron in that state. The energy denominator gives the difference between the initial energy ε_p and the energy in the intermediate state $\varepsilon_{\mathbf{p}+\mathbf{q}} + \omega_{\mathbf{q}}$ which has an additional phonon. Actually, there is an average phonon energy $\sum_q (N_{\mathbf{q}} + \frac{1}{2})\omega_{\mathbf{q}}$ which is common to both initial and final states and cancels in the difference.

Similarly, the interaction term which destroys a phonon is $M_q C_{\mathbf{p}+\mathbf{q}}^{\dagger} C_{\mathbf{p}} a_{\mathbf{q}}$. In second order it contributes an energy term

$$\frac{1}{\nu} \sum_q \frac{M_{\mathbf{q}}^2 N_{\mathbf{q}}[1 - n_F(\xi_{\mathbf{p}+\mathbf{q}})]}{\varepsilon_{\mathbf{p}} - \varepsilon_{\mathbf{p}+\mathbf{q}} + \omega_{\mathbf{q}}}$$

Here the factor N_q is the matrix element of the phonon destruction operator $a_{\mathbf{q}} \mid n_{\mathbf{q}}\rangle = (n_{\mathbf{q}})^{1/2} \mid n_{\mathbf{q}} - 1\rangle$. The electron occupation factor $1 - n_F$ is the same as before. The phonon energy in the denominator has changed sign, because there is one less phonon in the intermediate state.

These are the only two terms in the Hamiltonian. The summation of these two contributions, however, does not equal (3.5.10). In fact, we have omitted some important processes. There are two more terms. They arise from the other electrons in the system with the same spin state σ. These other electrons are also trying to emit and absorb phonons, to alter their energy. The other electrons may not, in this process, use the state $\mathbf{p}$ as their intermediate state. For example, if another electron starts in $\mathbf{p}'$ and scatters to $\mathbf{p}' + \mathbf{q}$, we may not have $\mathbf{p} = \mathbf{p}' + \mathbf{q}$. They may not use $\mathbf{p}$ because our electron, whose energy we are trying to calculate, is occupying it already. Thus the other electrons of the system have a reduced second-order self-energy because our electron is in the state $\mathbf{p}$. This energy reduction is associated with $\mathbf{p}$. We assign it to the self-energy of the electron in $\mathbf{p}$, because the system would not have it if there were no electron in $\mathbf{p}$. Thus we must calculate how much energy is deprived the other electrons. We assume that they start in $\mathbf{p} + \mathbf{q}$ initially and get scattered into $\mathbf{p}$. In analogy with the two terms above, the self-energy of a particle in the state $\mathbf{p} + \mathbf{q}$, for emitting or absorbing a phonon and going into $\mathbf{p}$, is

$$\frac{1}{\nu} \sum_q M_{\mathbf{q}}^2 n_F(\xi_{\mathbf{p}+\mathbf{q}})[1 - n_F(\xi_{\mathbf{p}})]\left(\frac{N_{\mathbf{q}} + 1}{\varepsilon_{\mathbf{p}+\mathbf{q}} - \varepsilon_{\mathbf{p}} - \omega_{\mathbf{q}}} + \frac{N_{\mathbf{q}}}{\varepsilon_{\mathbf{p}+\mathbf{q}} - \varepsilon_{\mathbf{p}} + \omega_{\mathbf{q}}}\right)$$

The factor $1 - n_F$ is the probability that $\mathbf{p}$ is empty. This factor is dropped, since p is occupied. The factor $n_F(\xi_{\mathbf{p}+\mathbf{q}})$ is the probability that there is an electron in $\mathbf{p} + \mathbf{q}$; if there is none, then the electron in $\mathbf{p}$ is not depriving any electron of energy. Our notation assumes that the energy does not depend on spin, so the occupation factors n_F should not depend on spin either. In a magnetic system, one would need to worry about the spin dependence of energy and occupation numbers. These last two terms are then subtracted from the first two to give the result, from second-order perturbation theory,

$$\Delta E^{(2)} = \frac{1}{\nu} \sum_{\mathbf{q}} M_{\mathbf{q}}^2 \left[\frac{1 + N_q - n_F(\xi_{\mathbf{p}+\mathbf{q}})}{\varepsilon_{\mathbf{p}} - \varepsilon_{\mathbf{p}+\mathbf{q}} - \omega_{\mathbf{q}}} + \frac{N_q + n_F(\xi_{\mathbf{p}+\mathbf{q}})}{\varepsilon_{\mathbf{p}} - \varepsilon_{\mathbf{p}+\mathbf{q}} + \omega_{\mathbf{q}}} \right]$$

The Green's function theory (3.5.10) changes this result only a little bit. It allows that the original electron may have had wave vector $\mathbf{p}$, but its "energy" is $ip_n + \mu$ and not $\varepsilon_{\mathbf{p}}$. It keeps the energy and momentum as separate variables, at least at this stage in the calculation. An obvious improvement in the calculation is to make the intermediate state energy not $\varepsilon_{\mathbf{p}+\mathbf{q}}$ but some other energy variable which reflects the possible effects of interaction. The difficulty with such a procedure is that changes in $\varepsilon_{\mathbf{p}+\mathbf{q}}$ also require changes in the occupation probability $n_F(\xi_{\mathbf{p}+\mathbf{q}})$. Later we shall show that one subset of diagrams, when summed to infinity, yield exactly this approximation; the intermediate state energies are treated exactly also.

In many-particle systems, the interactions which alter energy states of particles usually lead to subtle counting problems about where energy has been added or subtracted from the system. The Matsubara method has the virtue that it cranks out the correct answers automatically.

3.6. LINKED CLUSTER EXPANSIONS

Another method for evaluating correlation functions is called the *linked cluster* method or the *cumulant expansion* (Abrikosov *et al.*, 1963; Brout and Carruthers, 1963). It is the method which is used to evaluate the thermodynamic potential Ω. However, the method has also been applied to evaluating Green's functions of time $G(\mathbf{p}, t)$ for a few problems, which represents an alternative to the usual Dyson equation approach. Both methods, Dyson's equation and linked cluster, would give the same result if all terms were summed exactly. However, in the usual case, when only a few terms in a perturbation expansion are evaluated and the rest neglected, Dyson's equation and the linked cluster method are both approximate solutions. In some special cases the linked cluster expansion may be the better approximation. The art of choosing which method to apply to a given problem is a matter of judgment, which comes from experience.

A. Thermodynamic Potential

First we shall discuss the thermodynamic potential:

$$e^{-\Omega\beta} = \mathrm{Tr}(e^{-\beta K}) = \mathrm{Tr}[e^{-\beta K_0}S(\beta)] \tag{3.6.1}$$

This quantity may be useful to evaluate by itself. After it is found, perhaps only approximately, one may determine some thermodynamic results by taking derivatives of it; for example,

$$-\frac{\partial\Omega}{\partial\mu} = \langle N\rangle = \bar{N} \tag{3.6.2}$$

$$\frac{\partial(\Omega\beta)}{\partial\beta} = +\langle H-\mu N\rangle = U-\mu\bar{N} = \Omega+TS \tag{3.6.3}$$

$$\frac{\partial\Omega}{\partial m} = +\frac{1}{m}\left\langle\sum_i \frac{p_i^2}{2m}\right\rangle = \frac{1}{m}\sum_{\mathbf{p}} n_{\mathbf{p}}\frac{p^2}{2m} \tag{3.6.4}$$

In the last identity (3.6.4), the function $n_{\mathbf{p}}$ is meant to be the number of electrons with momentum $\mathbf{p}$. This quantity was defined earlier, in (3.3.19), for an interacting system.

Usually we are working with a Hamiltonian which we write as

$$H = H_0 + V$$
$$K = K_0 + V = (H_0 - \mu N) + V$$

where H_0 is something we can solve. In the interaction representation, the thermodynamic potential is given by the right-hand side of (3.6.1). If there are no interactions and $V = 0$, then we have the definition of Ω_0:

$$e^{-\beta\Omega_0} = \mathrm{Tr}(e^{-\beta K_0})$$

For a system of electrons and phonons,

$$K_0 = \sum_{\mathbf{p},\sigma} \xi_{\mathbf{p}} C^\dagger_{\mathbf{p},\sigma}C_{\mathbf{p},\sigma} + \sum_q \omega_{\mathbf{q}}(a_{\mathbf{q}}^\dagger a_{\mathbf{q}} + \tfrac{1}{2})$$

the trace is now simple since one can expand in the unperturbed eigenstates:

$$e^{-\beta\Omega_0} = \prod_{\mathbf{p}} (1 + e^{-\beta\xi_{\mathbf{p}}})^2 \prod_{\mathbf{q}} \left(\frac{e^{-(1/2)\beta\omega_{\mathbf{q}}}}{1 - e^{-\beta\omega_{\mathbf{q}}}} \right)$$

If we take the logarithm of both sides of this equation,

$$\beta\Omega_0 = -2 \sum_{\mathbf{p}} \ln(1 + e^{-\beta\xi_{\mathbf{p}}}) + \sum_{\mathbf{q}} [\ln(1 - e^{-\beta\omega_{\mathbf{q}}}) + \tfrac{1}{2}\beta\omega_{\mathbf{q}}]$$

The factor of 2 in front of the electron term is for the spin degeneracy. The last term on the right is the zero-point energy of the phonons. If we change the summations to integrals,

$$\Omega_0 = -\frac{\nu}{\beta} 2 \int \frac{d^3p}{(2\pi)^3} \ln(1 + e^{-\beta\xi_{\mathbf{p}}}) + \nu \int \frac{d^3q}{(2\pi)^3} \left[\frac{1}{2} \omega_{\mathbf{q}} + \frac{1}{\beta} \ln(1 - e^{-\beta\omega_{\mathbf{q}}}) \right] \tag{3.6.5}$$

The thermodynamic potential is proportional to the volume ν of the system. This was also apparent from Eqs. (3.6.2) to (3.6.4), since the right-hand side in each case is proportional to the volume ν or to the number of particles.

Our task is to evaluate (3.6.1) with $S(\beta)$ included. The interaction V will add some correction terms to Ω_0 which change it into Ω. So far we have only learned one method of evaluating this correlation function, and that is by the S-matrix expansion given earlier:

$$e^{-\beta\Omega} = e^{-\beta\Omega_0} \sum_{n=0}^{\infty} \frac{(-1)^n}{n!} \int_0^\beta d\tau_1 \cdots \int_0^\beta d\tau_n \, {}_0\langle T_\tau \hat{V}(\tau_1) \cdots \hat{V}(\tau_n) \rangle \tag{3.4.11}$$

In evaluating the right-hand side, we must now include all diagrams, whether connected or disconnected, since there is now no other series at hand to cancel the disconnected diagrams. We shall find that the disconnected diagrams are necessary for the resummation procedure. Basically, we wish to examine the terms which occur in the S-matrix series and to find some convenient method of resumming them. This method must be such that we need to evaluate each different type of diagram only once. The way of doing this is the linked cluster, or cumulant, expansion.

We shall begin by stating the answer. Then the following discussion will try to make the answer appear plausible. First we introduce the parameter λ. It multiplies the potential V, so that everywhere we formerly wrote

V, or $\hat{V}(\tau)$, we now write λV or $\lambda\hat{V}(\tau)$. The parameter λ is actually just unity, but it is used to keep track of the number of times the potential occurs in each term in the S matrix. Thus our S-matrix expansion in (3.4.11) above may be written as

$$e^{-\beta\Omega} = e^{-\beta\Omega_0} \sum_{n=0}^{\infty} \lambda^n W_n$$
$$W_n = \frac{(-1)^n}{n!} \int_0^\beta d\tau_1 \cdots \int_0^\beta d\tau_n \langle T_\tau \hat{V}(\tau_1) \cdots \hat{V}(\tau_n)\rangle \tag{3.6.6}$$

The basic linked cluster theorem is that this series can be resummed into

$$e^{-\beta\Omega} = \exp\left(-\beta\Omega_0 + \sum_{l=1}^{\infty} \lambda^l U_l\right) \tag{3.6.7}$$

where U_l contains just different, connected diagrams:

$$U_l = \frac{(-1)^l}{l} \int_0^\beta d\tau_1 \cdots \int_0^\beta d\tau_l \langle T_\tau \hat{V}(\tau_1) \cdots \hat{V}(\tau_l)\rangle_{\substack{\text{different}\\ \text{connected}}} \tag{3.6.8}$$

Of course, the thermodynamic potential is obtained by setting $\lambda = 1$ in (3.6.7):

$$\Omega = \Omega_0 - \frac{1}{\beta} \sum_{l=1}^{\infty} U_l \tag{3.6.9}$$

Thus the basic theorem is that the thermodynamic potential is the summation of the different, connected diagrams. The factor $1/l$ in (3.6.8) turns out to be quite a nuisance, as we shall discuss later. It makes further summation difficult, except by a trick.

The theorem will be proved below. First it is helpful to give a simple example. Consider the perturbation V to be the electron–phonon interaction:

$$V = \sum_{q\mathbf{k}} \frac{M_q}{\nu^{1/2}} A_{\mathbf{q}} C^\dagger_{\mathbf{k}+\mathbf{q}} C_{\mathbf{k}}$$
$$A_{\mathbf{q}} = a_{\mathbf{q}} + a^\dagger_{-\mathbf{q}}$$

For this example, only terms with even n occur in the series, since the expectation value of $A_{\mathbf{q}}{}^n$ is zero if n is odd. The first term in the series, after $n = 0$, is $n = 2$:

$$W_2 = U_2 = \tfrac{1}{2} \int_0^\beta d\tau_1 \int_0^\beta d\tau_2 \langle T_\tau \hat{V}(\tau_1)\hat{V}(\tau_2)\rangle$$

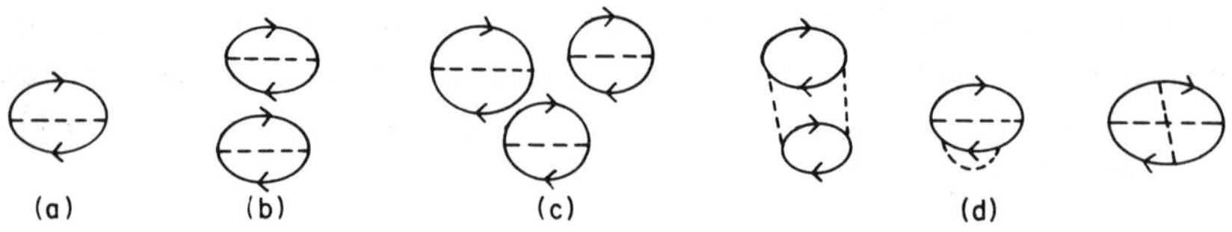

FIGURE 3.8

We shall not evaluate this disconnected diagram here. Its Feynman diagram is shown in Fig. 3.8(a). There are two electron lines, and the two vertices are connected by a phonon line. Next consider the term with $n = 4$:

$$W_4 = \frac{1}{4!} \int_0^\beta d\tau_1 \int_0^\beta d\tau_2 \int_0^\beta d\tau_3 \int_0^\beta d\tau_4 \langle T_\tau \hat{V}(\tau_1)\hat{V}(\tau_2)\hat{V}(\tau_3)\hat{V}(\tau_4)\rangle$$

All contributions for $n = 4$ consist of diagrams with four electron lines and two phonon lines. Some are shown in Fig. 3.8(d). However, there also occurs the diagrams shown in Fig. 3.8(b) where the single bubble occurs twice. In fact, there are three identical terms in W_4 which have the pair of bubbles. The factor of 3 occurs because if one chooses τ_1 as one of the vertices of a bubble, the other τ variable for the same bubble is either τ_2, τ_3, or τ_4, and these three choices each give the same contribution. Thus this factor is

$$\tfrac{3}{24} \int_0^\beta d\tau_1 \int_0^\beta d\tau_2 \langle T\tau \hat{V}(\tau_1)\hat{V}(\tau_2)\rangle \int_0^\beta d\tau_3 \int_0^\beta d\tau_4 \langle T_\tau \hat{V}(\tau_3)\hat{V}(\tau_4)\rangle = \tfrac{1}{2}(U_2)^2$$

Similarly, one can show that each of the connected diagrams in Fig. 3.8(d) occurs six times, where the factor of 6 comes from different ways to shift the dummy variables around. Thus we have

$$W_4 = \tfrac{1}{2}(U_2)^2 + U_4$$

Next we consider terms with $n = 6$. There will be many connected diagrams which contain six electron lines and three phonon lines; try to draw them all. These give U_6. There are also disconnected diagrams. There will be diagrams containing three bubbles, as shown in Fig. 3.8(c)—in fact there will be 15 of these, all equal to each other. There will also be disconnected diagrams which contain one four-line term from Fig. 3.8(d) and one from Fig. 3.8(a). Thus we get

$$W_6 = \frac{1}{3!}(U_2)^3 + U_2 U_4 + U_6$$

The terms in U_2 appear to be generating the series

$$1 + U_2 + \frac{1}{2!}(U_2)^2 + \frac{1}{3!}(U_2)^3 + \cdots = e^{U_2} \tag{3.6.10}$$

There is a simple argument which shows that the nth term is really appropriate for the exponential series. Since n is even, let $n = 2m$, and there are m bubbles of the type shown in Fig. 3.8(a). We must determine how many different ways one can get m bubbles. Take the first variable, τ_1. It may be paired in the bubble,

$$U_2 = \tfrac{1}{2}\sum_{\mathbf{q},\mathbf{p}} M_{\mathbf{q}}^2 \int_0^\beta d\tau_1 \int_0^\beta d\tau_j \mathscr{G}^{(0)}(\mathbf{p}, \tau_1 - \tau_j)\mathscr{G}^{(0)}(\mathbf{p} + \mathbf{q}, \tau_j - \tau_1) \times \mathscr{D}^{(0)}(\mathbf{q}, \tau_1 - \tau_j) \tag{3.6.11}$$

with the variable τ_j being any of the other $2m - 1$ variables. Next, take any of the $2m - 2$ variables left after pairing τ_1 and τ_j, say τ_s, and it may be paired with any of the $2m - 3$ left. Thus the number of different arrangements is

$$(2m - 1)(2m - 3)(2m - 5) \cdots 5 \cdot 3 \cdot 1 = \frac{(2m)!}{m!\, 2^m}$$

Thus we have that this factor in W_n is

$$W_n = \frac{(2m)!}{m!\, 2^m} \frac{1}{(2m)!} \left[\int_0^\beta d\tau_1 \int_0^\beta d\tau_j \langle T_\tau \hat{V}(\tau_1)\hat{V}(\tau_j)\rangle\right]^m = \frac{1}{m!}(U_2)^m$$

which is the form desired in (3.6.10). Thus we have shown that, at least, this one term sums correctly. One must now convince oneself that any other contribution also sums correctly. This is not too hard. It is more subtle to show that all the cross products such as U_2U_4 occur with just the right counting.

First we must prove another identity:

$$U_l = \frac{(-1)^l}{l!} \int_0^\beta d\tau_1 \cdots \int_0^\beta d\tau_l \langle T_\tau \hat{V}(\tau_1) \cdots \hat{V}(\tau_l)\rangle_{\text{connected}}$$

The same U_l is in (3.6.8). Thus we are really asserting that for each term l there are $(l - 1)!$ terms alike, so that the prefactor in front of the number of different connected diagrams is

$$\frac{(l-1)!}{l!} = \frac{1}{l}$$

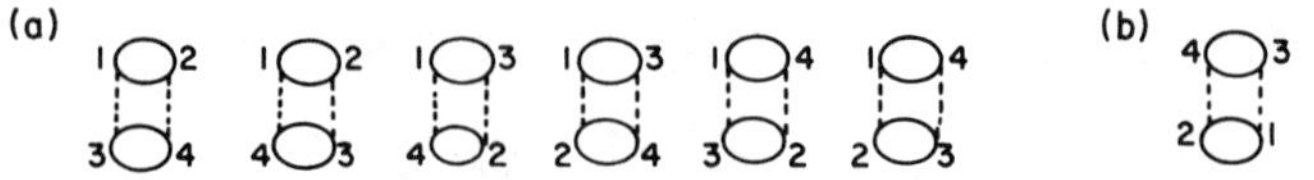

FIGURE 3.9

Again the proof of this is just an exercise in combinatorial counting. We are going to expand our S matrix in terms of Green's functions and have expressions such as

$$U_l = \frac{1}{l!} \int_0^\beta d\tau_1 \cdots \int_0^\beta d\tau_l \mathscr{G}(p, \tau_1 - \tau_j)\mathscr{G}(p', \tau_2 - \tau_k) \cdots$$

The combinatorial question is, How many different combinations are really alike if we just relabel the dummy variables τ_1, τ_j, τ_2, etc.? The answer is $(l - 1)!$. The reader should try to work some examples to convince oneself that it works. The trivial case of $l = 2$ so that $(l - 1)! = 1$ was done in (3.6.11). A more complex example is shown in Fig. 3.9(a), where the $(4 - 1)! = 6$ different terms are shown for one of the different diagrams with $l = 4$. The numbers 1, 2, 3, 4 refer to the τ variables: τ_1, τ_2, τ_3, τ_4. These six are the only different arrangements. One might ask why the figures are drawn such that τ_1 is always the upper left-hand corner. Why not the labeling shown in Fig. 3.9(b)? The answer is that this is exactly the same term as the first one in Fig. 3.9(a). They are topologically identical but just drawn upside down and backwards, or rotated by 180°. However, and most importantly, both diagrams yield identical results when written down with Green's functions,

$$\begin{aligned} \frac{1}{4\nu^2} \sum_{\mathbf{q}} \sum_{\mathbf{p}_1\mathbf{p}_2} M_q^4 \int_0^\beta d\tau_1 \int_0^\beta d\tau_2 \int_0^\beta d\tau_3 \int_0^\beta d\tau_4 \mathscr{D}^{(0)}(\mathbf{q}, \tau_1 - \tau_3)\mathscr{D}^{(0)}(\mathbf{q}, \tau_2 - \tau_4) \\ \times \mathscr{G}^{(0)}(\mathbf{p}_1, \tau_1 - \tau_2)\mathscr{G}^{(0)}(\mathbf{p}_1 + \mathbf{q}, \tau_2 - \tau_1)\mathscr{G}^{(0)}(\mathbf{p}_2, \tau_3 - \tau_4) \\ \times \mathscr{G}^{(0)}(\mathbf{p}_2 + \mathbf{q}, \tau_4 - \tau_3) \end{aligned} \tag{3.6.12}$$

which is the ultimate test of whether they are topologically identical.

In the prior section, the Green's function $\mathscr{G}(\mathbf{p}, \tau)$ was evaluated by a similar expansion technique. There the connected diagrams obeyed the counting rules that

$$\begin{aligned} \frac{1}{l!} \int_0^\beta d\tau_1 \cdots \int_0^\beta d\tau_l \langle T\hat{C}_{\mathbf{p}}(\tau)\hat{V}(\tau_1) \cdots \hat{V}(\tau_l)\hat{C}_{\mathbf{p}}^\dagger\rangle_{\text{connected}} \\ = \int_0^\beta d\tau_1 \cdots \int_0^\beta d\tau_l \langle T_\tau \cdots \rangle_{\substack{\text{different} \\ \text{connected}}} \end{aligned}$$

Here there were $l!$ arrangements which are identical. In the linked cluster method there are only $(l-1)!$. The difference is that in $\mathscr{G}(\mathbf{p}, \tau)$ we have a reference variable τ, so that we consider arrangements with respect to this variable, and there turn out to be more of them—by a factor of l.

The general proof of our theorem proceeds along the following lines. Let $U_{m_1}, U_{m_2}, \ldots, U_{m_j}, \ldots$ denote the set of possible connected diagrams, with $m_1, m_2, \ldots$ denoting the order of the contribution; i.e., the number of factors V, or powers of λ, with that contribution. We assume that these numbers are all different: $m_1 \neq m_2 \neq m_j \cdots$. Thus for each value of m_j, U_{m_j} contains all the different connected diagrams. However, when expanding a general term of order n, each U_{m_j} may occur more than once or not at all. Let p_j denote the number of times U_{m_j} does occur. Then one has that

$$\sum_j p_j m_j = n \tag{3.6.13}$$

This rule is that the n powers of V are distributed among the various U_m but that the powers of V still must add up to n. The important combinatorial question is to determine how many different arrangements of the n objects there are among the set of boxes (m_j, p_j), where p_j is the number of boxes with m_j. The answer is

$$\frac{n!}{\prod_i [(m_j!)^{p_j} p_j!]}$$

Thus the general term is

$$\begin{aligned} W_n = \sum_{\substack{m_1, m_2, \ldots \\ p_1, p_2, \ldots}} (-1)^n & \frac{1}{(m_1!)^{p_1}(m_2!)^{p_2} \cdots p_1! p_2! \cdots} \\ & \times \left[\int_0^\beta d\tau_1^{(1)} \cdots \int_0^\beta d\tau_{m_1}^{(2)} \langle T_\tau \hat{V}(\tau_1^{(1)}) \cdots \hat{V}(\tau_{m_1}^{(1)}) \rangle \right]^{p_1} \\ & \times \left[\int_0^\beta d\tau_1^{(2)} \cdots \int_0^\beta d\tau_{m_2}^{(2)} \langle T_\tau \hat{V}(\tau_1^{(2)}) \cdots \hat{V}(\tau_{m_2}^{(2)}) \rangle \right]^{p_2} \cdots \end{aligned}$$

The summation is over all possible combinations of (p_j, m_j) for each n, subject to the condition that (3.6.13) is valid. Our series is

$$\sum_n \lambda^n W_n = \prod_{m_j} \left(\sum_{p_j=0}^{\infty} \lambda^{m_j p_j} \frac{U_{m_j}^{p_j}}{p_j!} \right) = \exp\left(\sum_{m_j} \lambda^{m_j} U_{m_j} \right)$$

which proves the theorem (3.6.7).

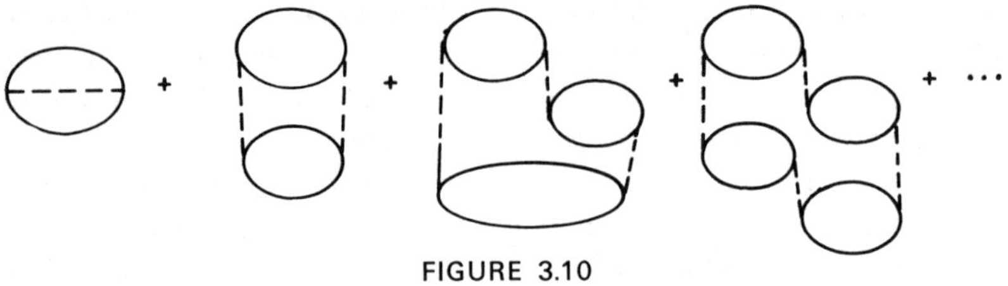

FIGURE 3.10

Now we shall consider the evaluation of this thermodynamic potential. As an example, we shall evaluate the series of diagrams shown in Fig. 3.10 for the electron–phonon interaction. They can be summed to give a simple answer. Of course, this series is not the exact answer, since there are additional terms in which the bubbles may have internal phonon lines also. The rules for constructing diagrams for the terms in the thermodynamic potential are similar to those in Sec. 3.5. One follows all those rules and then multiplies the result by β/l. The $1/l$ factor is just the one which occurs in (3.6.8), which we have explained already. The factor β comes from the extra τ integration. For example, in (3.6.11), which is the first bubble, the argument is only a function of $\tau_1 - \tau_j$:

$$\int_0^\beta d\tau_1 \int_0^\beta d\tau_j f(\tau_1 - \tau_j)$$

Thus if we expand

$$f(\tau_1 - \tau_j) = \frac{1}{\beta} \sum_{iq_n} e^{-iq_n(\tau_1 - \tau_j)} f(iq_n)$$

the first integration gives

$$\frac{1}{\beta} \int_0^\beta d\tau_j e^{-iq_n\tau_j} = \delta_{n=0}$$

while the second just gives β. Each term has this extra τ integration, which gives the one extra factor of β. Thus for the first bubble in Fig. 3.10 we get

$$U_2 = \frac{\beta}{2} \lambda^2 \frac{1}{\beta} \sum_{\mathbf{q}, iq_n} M_q{}^2 \mathscr{D}^{(0)}(\mathbf{q}, iq_n) P^{(1)}(\mathbf{q}, iq_n) \tag{3.6.14}$$

where the polarization diagram $P^{(1)}(\mathbf{q}, iq_n)$ is for the single bubble. This was already evaluated in (3.4.15):

$$P^{(1)}(\mathbf{q}, iq_n) = \frac{2}{\beta\nu} \sum_{\mathbf{p}, ip_n} \mathscr{G}^{(0)}(\mathbf{p}, ip_n) \mathscr{G}^{(0)}(\mathbf{p} + \mathbf{q}, ip_n + iq_n) \tag{3.4.15}$$

The second term in Fig. 3.10 has two bubbles connected by two phonon lines. This was described earlier in (3.6.12) in τ space. In Fourier transform space, it is simply

$$U_4 = \frac{\beta}{4} \frac{\lambda^4}{\beta} \sum_{iq,\mathbf{q}} [M_q{}^2 \mathscr{D}^{(0)}(\mathbf{q}, iq_n) P^{(1)}(\mathbf{q}, iq_n)]^2$$

Momentum and frequency conservation requires both phonons to have the same variables $(\mathbf{q}, iq)$, and both polarization diagrams are also just functions of this combination. The term with n bubbles and n phonon lines is just

$$U_{2n} = \frac{\beta}{2n} \frac{\lambda^{2n}}{\beta} \sum_{\mathbf{q}, iq_n} [M_q{}^2 \mathscr{D}^{(0)}(\mathbf{q}, iq_n) P^{(1)}(\mathbf{q}, iq_n)]^n$$

Thus it is simple to sum the series:

$$\Omega - \Omega_0 = -\frac{1}{\beta} \sum_{n=1}^{\infty} U_{2n} = \frac{1}{2\beta} \sum_{\mathbf{q}, iq_n} \ln[1 - \lambda^2 M_q{}^2 \mathscr{D}^{(0)}(\mathbf{q}, iq_n) P^{(1)}(\mathbf{q}, iq_n)] \tag{3.6.15}$$

The right-hand side of (3.6.15) is a correction to the thermodynamic potential. It is not exact, as explained above, since only part of the diagrams are included. The right-hand side is proportional to the volume ν, as is obvious when the q summation is changed to an integration:

$$\sum_q = \nu \int \frac{d^3q}{(2\pi)^3}$$

Similarly, the p summation in $P^{(1)}(q, iq_n)$ can be changed to an integration:

$$\frac{1}{\nu} \sum_p = \int \frac{d^3p}{(2\pi)^3}$$

Thus the argument of the logarithm is not dependent on ν, and ν only enters by multiplying the result.

The answer is a logarithm. We still need to do the summation over iq_n, and this summation is inconvenient (but still possible) when the important functions are inside the argument of the logarithm. There is a standard trick for eliminating the logarithm, although it does not eliminate it but instead just disguises it. If we treat

$$\eta = \lambda^2$$

as a variable of integration,

$$\int_0^{\lambda^2} d\eta \frac{M_q^2 \mathscr{D}^{(0)} P^{(1)}}{1 - \eta \mathscr{D}^{(0)} P^{(1)} M_q^2} = -\ln(1 - \lambda^2 \mathscr{D}^{(0)} P^{(1)} M_q^2)$$

then for $\lambda = 1$ we get

$$\Omega - \Omega_0 = -\frac{\nu}{2\beta} \sum_{iq_n} \int \frac{d^3q}{(2\pi)^3} \int_0^1 d\eta \frac{M_q^2 \mathscr{D}^{(0)} P^{(1)}}{1 - \eta M_q^2 \mathscr{D}^{(0)} P^{(1)}} \tag{3.6.16}$$

The logarithm has been eliminated, but we have introduced another problem. Now the right-hand side must be evaluated for each value of η and then the integral taken. Since η enters in the same way as a coupling constant, this is called a coupling constant integration. This integration can be taken outside of the summation over iq_n, which makes this summation easier in some cases. There are some other manipulations on this formula, which put it into a form which appears simpler. First, define the phonon self-energy as

$$\pi^{(1)}(\eta, \mathbf{q}, iq_n) = \eta M_q^2 P^{(1)}(\mathbf{q}, iq_n) \tag{3.6.17}$$

where we have added the coupling constant η. The superscript (1) on $\pi^{(1)}$ indicates that this is an approximate self-energy, with only the single-electron bubble included. The phonon Green's function is

$$\mathscr{D}^{(1)}(\eta, q, iq_n) = \frac{\mathscr{D}^{(0)}}{1 - \mathscr{D}^{(0)} \pi^{(1)}}$$

The superscript (1) on $\mathscr{D}^{(1)}$ indicates that it is an approximate formula for the Green's function, since the self-energy is approximate. It is also a function of the coupling constant strength η. Thus (3.6.16) is just

$$\Omega - \Omega_0 = -\frac{\nu}{2\beta} \sum_{iq_n} \int \frac{d^3q}{(2\pi)^3} \int_0^1 \frac{d\eta}{\eta} \pi^{(1)}(\eta, \mathbf{q}, iq_n) \mathscr{D}^{(1)}(\eta, \mathbf{q}, iq_n)$$

There is a theorem which we are going to state but not prove. The exact correction to the thermodynamic potential from electron–phonon interactions is

$$\Omega - \Omega_0 = -\frac{\nu}{2\beta} \sum_{iq_n} \int \frac{d^3q}{(2\pi)^3} \int_0^1 \frac{d\eta}{\eta} \pi(\eta, \mathbf{q}, iq_n) \mathscr{D}(\eta, \mathbf{q}, iq_n) \tag{3.6.18}$$

where $\pi(\eta, \mathbf{q}, iq_n)$ is the exact phonon self-energy and $\mathscr{D}(\eta, \mathbf{q}, iq_n)$ is the exact phonon Green's function. Both are found as a function of η, and then one has to perform the integration over η. Proofs of this theorem are given

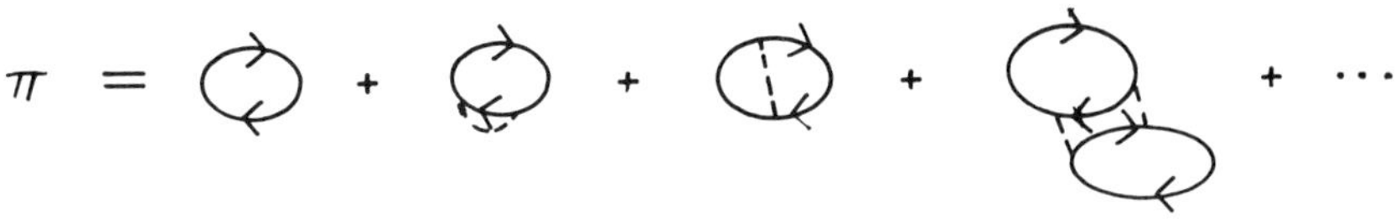

FIGURE 3.11

in Abrikosov *et al.* (1963). The exact phonon self-energy $\pi(\eta, \mathbf{q}, iq_n)$ is the sum of an infinite number of self-energy diagrams, some of which are shown in Fig. 3.11. There are bubbles with internal phonon lines and also bubbles connected by more than one phonon line.

Of course, if we also account for electron–electron interactions, then the phonon self-energy can also contain internal Coulomb interactions, etc. Of course, we should also consider the effect of Coulomb interaction on the thermodynamic potential. The general theorem (3.6.8) and (3.6.9) is true for all interactions—Coulomb, phonon, or others.

It is possible to combine the effects of the phonons and Coulomb interaction in a simple way. If we consider electrons interacting by either or both interactions, then the single-bubble diagram appears twice. As shown in Fig. 3.12, the two electron vertices can be connected by a single Coulomb line or a single phonon line. Thus the sum of these contributions is

$$U_2 = \frac{\eta}{2}\beta\frac{1}{\beta}\sum_{\mathbf{q},iq_n} P^{(1)}(\mathbf{q}, iq_n)[v_\mathbf{q} + M_\mathbf{q}^2\mathscr{D}^{(0)}(\mathbf{q}, iq_n)]$$

$$v_q = \frac{4\pi e^2}{q^2}$$

Similarly, there are four diagrams with two bubbles. The two bubbles interact by two Coulomb lines, two phonon lines, or a mix with one of each. This contribution is

$$U_4 = \frac{\eta^2}{4}\beta\frac{1}{\beta}\sum_{\mathbf{q},iq_n} \{P^{(1)}(\mathbf{q}, iq_n)[v_q + M_q^2\mathscr{D}^{(0)}(\mathbf{q}, iq_n)]\}^2$$

It seems desirable to introduce a combined interaction propagator, Coulomb

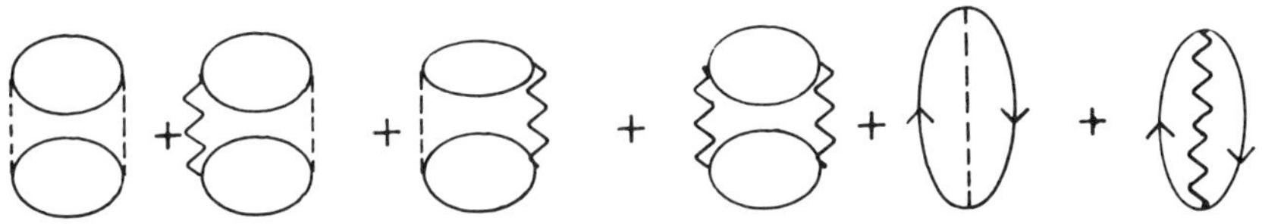

FIGURE 3.12

plus phonon, which is

$$W^{(0)}(\mathbf{q}, iq_n) = v_{\mathbf{q}} + M_q^2 \mathscr{D}^{(0)}(\mathbf{q}, iq_n) \tag{3.6.19}$$

It obeys a Dyson equation, which is

$$W(\mathbf{q}, iq_n) = \frac{W^{(0)}}{1 - W^{(0)}\pi} \tag{3.6.20}$$

where $\pi(\mathbf{q}, iq_n)$ is the total self-energy diagram, which is the sum of the diagrams shown in Fig. 3.11. Of course, now the internal lines in Fig. 3.11 refer to $W^{(0)}$, since they may be either Coulomb or phonon. The generalization of (3.6.18) to include both Coulomb and phonon effects is

$$\begin{aligned} \Omega - \Omega_0 = &- \frac{\nu}{2\beta} \sum_{iq_n} \int \frac{d^3q}{(2\pi)^3} \int_0^1 \frac{d\eta}{\eta}\, \pi(\eta, \mathbf{q}, iq_n) W(\eta, \mathbf{q}, iq_n) \\ &- \frac{n}{2} \int \frac{d^3q}{(2\pi)^3}\, v_q \end{aligned} \tag{3.6.21}$$

The last term contains the Coulomb self-energy of an electron interacting with itself, and this unwanted contribution is subtracted out in the second term. The two interactions in (3.6.19) do not complete the possibilities. If the Hamiltonian has additional interactions due to impurities, ions, etc., then one has additional contributions which may contribute to the thermodynamic potential.

The expression (3.6.18) seems to favor the phonon self-energies. It is reasonable to ask what happened to the electron self-energies. The answer is that they are in there too. The diagrams we have been considering can also be viewed as corrections to the electron self-energy. This viewpoint does not introduce an additional contribution but is another equivalent way of expressing the answer we just obtained. For example, the basic bubble diagram, shown in Fig. 3.8(a) and expressed in (3.6.11), may also be written as

$$U_2 = \frac{\beta}{2} \frac{\lambda}{\beta} \sum_{\mathbf{p}, ip, \sigma} \mathscr{G}^{(0)}(\mathbf{p}, ip) \Sigma^{(1)}(\mathbf{p}, ip)$$

where $\Sigma^{(1)}$ is the electron self-energy from one-phonon processes, which was given earlier in (3.4.14):

$$\Sigma^{(1)}(\mathbf{p}, ip) = \frac{1}{\beta\nu} \sum_{\mathbf{q}, iq} M_q^2 \mathscr{G}^{(0)}(\mathbf{p} + \mathbf{q}, ip + iq) \mathscr{D}^{(0)}(\mathbf{q}, iq) \tag{3.4.14}$$

We repeat, for emphasis, that the above U_2 is not a different result from

(3.6.14). It is the same result, expressed with different symbols; the same Green's functions have been grouped differently. Similarly, the exact answer is

$$\Omega - \Omega_0 = \frac{1}{\beta} \sum_{\mathbf{p}, ip_n} \int_0^1 \frac{d\eta}{\eta} \mathscr{G}(\eta, \mathbf{p}, ip)\Sigma(\eta, \mathbf{p}, ip) \qquad (3.6.22)$$

where $\Sigma(\eta, \mathbf{p}, ip_n)$ is the exact self-energy of the electron and $\mathscr{G}$ is the exact Green's function obtained from Dyson's equation with that exact self-energy. If we use the propagator $W^{(0)}$ in (3.6.19) for all internal lines so that the electrons can interact via phonons or Coulomb interactions, then (3.6.22) is exactly equal to (3.6.21). The two results represent the same set of corrections. They are not different contributions but are the same contribution expressed in different notation.

These formulas are often used to calculate the ground state energy in the limit of $T \to 0$. There are two limits which are then taken: $\nu \to \infty$ and $T \to 0$. In studying the electron gas, Kohn and Luttinger (1960) observed that the right answer was obtained if they were taken first in the limit of $\nu \to \infty$ and then $T \to 0$. The reverse order omits important terms. These are of the form

$$\beta n_F(\xi_{\mathbf{p}})[1 - n_F(\xi_{\mathbf{p}})]$$

If $\nu \neq \infty$, then all levels are discrete in the finite volume, and as $T \to 0$, these terms give zero since either n_F or $1 - n_F$ is zero. However, if one first takes $\nu \to \infty$, so that the levels are continuous, then the limit $T \to 0$ gives

$$\lim_{T\to 0} \beta n_F(1 - n_F) = \lim_{T\to 0} \frac{d}{d\xi_p} [-n_F(\xi_{\mathbf{p}})] = \delta(\xi_{\mathbf{p}})$$

There are many terms of this kind in the perturbation expansion for the ground state energy of the electron gas.

It is illustrative to show a simple example of evaluating the thermodynamic potential. Let us assume that the only effect of the phonon self-energy is to change the unperturbed frequencies $\omega_{\mathbf{q}}^2$ to a new set of renormalized frequencies $\Omega_{\mathbf{q}}^2$. Since the Green's function is

$$\mathscr{D} = \frac{2\omega_{\mathbf{q}}}{(i\omega_n)^2 - \omega_{\mathbf{q}}^2 - 2\omega_{\mathbf{q}}\pi(\mathbf{q}, i\omega_n)}$$

then we assume that

$$\mathscr{D}(\mathbf{q}, i\omega_n) = \frac{2\omega_{\mathbf{q}}}{(i\omega_n)^2 - \Omega_q^2}$$

This form for $\mathscr{D}(q, i\omega_n)$ can be accomplished by making the choice

$$2\omega_q\pi(\eta, \mathbf{q}, i\omega_n) = \eta(\Omega_\mathbf{q}^2 - \omega_\mathbf{q}^2) \tag{3.6.23}$$

or

$$\mathscr{D}(\eta, \mathbf{q}, iq_n) = \frac{2\omega_q}{(iq_n)^2 - \omega_\mathbf{q}^2 - \eta(\Omega_\mathbf{q}^2 - \omega_\mathbf{q}^2)}$$

The choice (3.6.23) is not the only one possible to renormalize the frequencies. It is the choice one would get if it is assumed that the change from $\omega_\mathbf{q}$ to $\Omega_\mathbf{q}$ is accomplished by the one-bubble polarization diagram given in (3.6.17). Then the coupling constant η enters as just a multiplicative factor, as shown in (3.6.23). If further self-energy diagrams are needed to get a good phonon self-energy π, then η would enter in a more complicated fashion. We shall show in Chapter 6 that the single-bubble approximation is often adequate, so that the present derivation applies to many systems.

We now consider the steps necessary to evaluate (3.6.18), or perhaps just (3.6.16). Thus the expression to be evaluated is

$$\Omega - \Omega_0 = -\frac{1}{2}\frac{1}{\beta}\sum_{\mathbf{q}, iq_n}\int_0^1 d\eta\, \frac{\Omega_\mathbf{q}^2 - \omega_\mathbf{q}^2}{(iq_n)^2 - \omega_\mathbf{q}^2 - \eta(\Omega_\mathbf{q}^2 - \omega_\mathbf{q}^2)} \tag{3.6.24}$$

First, introduce a frequency which depends on a coupling constant:

$$\Omega_\eta^2 = \omega_\mathbf{q}^2 + \eta(\Omega_\mathbf{q}^2 - \omega_\mathbf{q}^2) > 0$$

The summation over Matsubara frequencies may be done by the techniques suggested in Sec. 3.5. Alternately, one might recognize the answer as

$$\frac{1}{\beta}\sum_{iq_n}\frac{1}{(iq_n)^2 - \Omega_\eta^2} = \frac{1}{2\Omega_\eta}\bar{\mathscr{D}}(\Omega_\eta, \tau = 0) = -\frac{1}{2\Omega_\eta}\,[2n_B(\Omega_\eta) + 1]$$

where $\bar{\mathscr{D}}(\Omega_\eta, \tau)$ is the Green's function which is the Fourier transform of

$$\bar{\mathscr{D}}(\Omega_\eta, iq_n) = \frac{2\Omega_\eta}{(iq_n)^2 - \Omega_\eta^2}$$

Thus our evaluation of (3.6.24) is now just the coupling constant integral,

$$\Omega - \Omega_0 = +\frac{\beta}{2}\sum_\mathbf{q}(\Omega_\mathbf{q}^2 - \omega_\mathbf{q}^2)\int_0^1 \frac{d\eta}{2\Omega_\eta}\left(1 + \frac{2}{e^{\beta\Omega_\eta} - 1}\right)$$

which can be done exactly:

$$\Omega - \Omega_0 = \frac{1}{2}\sum_q\left[\Omega_\eta + \frac{2}{\beta}\ln(1 - e^{-\beta\Omega_\eta})\right]_{\eta=0}^{\eta=1}$$

The answer is

$$\Omega - \Omega_0 = +\sum_q \left[\frac{1}{2}(\Omega_{\mathbf{q}} - \omega_{\mathbf{q}}) + \frac{1}{\beta}\ln(1 - e^{-\beta\Omega_{\mathbf{q}}}) - \frac{1}{\beta}\ln(1 - e^{-\beta\omega_{\mathbf{q}}})\right]$$

The right-hand side is the thermodynamic potential from the phonons at the new frequency $\Omega_{\mathbf{q}}$ minus the contribution from the phonons at the old frequencies $\omega_{\mathbf{q}}$. When this is combined with the result (3.6.5) for Ω_0, the terms from $\omega_{\mathbf{q}}$ all cancel. The final answer is

$$\Omega = -\frac{2\nu}{\beta}\int \frac{d^3p}{(2\pi)^3}\ln(1 + e^{-\xi_{\mathbf{p}}\beta}) + \frac{\nu}{\beta}\int \frac{d^3q}{(2\pi)^3} \times [\tfrac{1}{2}\beta\Omega_{\mathbf{q}} + \ln(1 - e^{-\beta\Omega_{\mathbf{q}}})] \tag{3.6.25}$$

Thus the thermodynamic potential is just the summation of the unperturbed electron contribution plus the contribution from the phonons at the new frequencies $\Omega_{\mathbf{q}}$. Of course, this form is what one would expect from simple considerations. If the only effect of the electron–phonon interactions is to change the phonon frequencies to the new values $\Omega_{\mathbf{q}}$, then one must be able to solve the Hamiltonian exactly and write it as

$$H = \sum_{\mathbf{p}\sigma} \xi_{\mathbf{p}} C^{\dagger}_{\mathbf{p}\sigma} C_{\mathbf{p}\sigma} + \sum_{\mathbf{q}} \Omega_{\mathbf{q}}(a_{\mathbf{q}}^{\dagger} a_{\mathbf{q}} + \tfrac{1}{2})$$

The thermodynamic potential for this simple Hamiltonian is just (3.6.25). Unfortunately, most corrections to the thermodynamic potential are not as easy to evaluate as this simple example.

B. Green's Functions

So far we have used only the linked cluster, or cumulant, expansion to evaluate the thermodynamic potential. One could consider the possibility of using similar expansion techniques to evaluate other types of correlation functions, such as Green's functions (Brout and Carruthers, 1963). This method has been used with success on two problems: One is the polaron problem for a single particle in a band, studied by Dunn (1975). The other has been in treating X-ray processes, where the hole dynamics have been studied by a linked cluster method (Mahan, 1975). Here the hole has also been treated as a single-hole problem. It has not been applied to a many-polaron, or many-hole, problem. The procedure begins with the S-matrix expansion for a particle Green's function, given in (3.4.12),

$$\mathscr{G}(\mathbf{p}, \tau) = \sum_{n=0}^{\infty} \lambda^n W_n(\mathbf{p}, \tau)$$
$$W_n(\mathbf{p}, \tau) = -\frac{(-1)^n}{n!} \int_0^\beta d\tau_1 \cdots \int_0^\beta d\tau_n \, {}_0\langle T_\tau \hat{C}_{\mathbf{p}}(\tau)\hat{V}(\tau_1) \cdots \hat{V}(\tau_n)\hat{C}_{\mathbf{p}}^\dagger(0)\rangle \tag{3.6.26}$$

where we have added the coupling parameter λ to the expansion. In the cumulant expansion, we assume that the series can be regrouped as an exponential power series in λ:

$$\mathscr{G}(\mathbf{p}, \tau) = \mathscr{G}^{(0)}(\mathbf{p}, \tau) \exp\left[\sum_{n=1}^{\infty} \lambda^n F_n(\mathbf{p}, \tau)\right] \tag{3.6.27}$$

The functions $F(\mathbf{p}, \tau)$ are found by equating terms with like powers of λ in the series expansions. That is, we write

$$\exp\left(\sum_{j=1} \lambda^j F_j\right) = 1 + \lambda F_1 + \lambda^2\left(F_2 + \frac{1}{2!} F_1^2\right) + \lambda^3\left(F_3 + F_2F_1 + \frac{1}{3!} F_1^3\right) + \cdots$$

so that, term by term, we have the equalities

$$W_1 = F_1\mathscr{G}^{(0)}$$
$$W_2 = \left(F_2 + \frac{1}{2!} F_1^2\right)\mathscr{G}^{(0)}$$
$$W_3 = \left(F_3 + F_1F_2 + \frac{1}{3!} F_1^3\right)\mathscr{G}^{(0)}$$

The functions $W_n(\mathbf{p}, \tau)$ are scalar functions which we know how to calculate using Wick's theorems. The unknown functions $F_n(\mathbf{p}, \tau)$ can be evaluated in terms of W_n by solving the above equations (Brout and Carruthers, 1963):

$$F_1 = \frac{W_1}{\mathscr{G}^{(0)}}$$
$$F_2 = \frac{W_2}{\mathscr{G}^{(0)}} - \frac{1}{2} F_1^2 \tag{3.6.28}$$
$$F_3 = \frac{W_3}{\mathscr{G}^{(0)}} - F_1F_2 - \frac{1}{3!} F_1^3$$
$$\vdots$$

The basic idea is very simple. It is just a different way of regrouping the series which results from the S matrix. It provides an alternative to using Dyson equations. Presumably, one gets the exact answer if all terms are taken and evaluated. However, since one can only get the exact answer in trivial problems, the method is really useful in problems where one can calculate a few terms in the perturbation expansion.

There are two possible difficulties with this procedure. The first is that the regrouping may not make mathematical sense. Nonsensical results are found in cases where $\mathscr{G}(\mathbf{p}, \tau)$ describes a Green's function of a particle in a system with N like particles, where $N \propto \nu$. In that case the terms F_n do not converge when put into the exponent. The offending terms are ones which contribute to the total energy of the system—terms which are proportional to ν and hence change the chemical potential. The chemical potential is not changed by V in one-particle or one-hole problems, since the one particle does not contribute any energy term proportional to ν. The series in $F_n(\tau)$ shows excellent convergence in these cases. These results will be discussed in Chapters 6 and 8. The other problem with this method is that when it works, one has some fairly complicated functions of τ in the exponent. One is invariably forced to use a computer to perform the Fourier transforms in frequency space. This step is not really a difficulty—just some work!

The physical model implied by this regrouping will be discussed in Chapter 6. It is a systematic development of the Tomonaga (1947) model of pion emission in particle theory. Tomonaga assumed that all pion emissions from a particle were statistically independent—there was no correlation between successive emissions. His model corresponds to the approximation of just using F_1 alone. However, if one uses $F_1 + F_2$, then F_2 puts in the correlation between pairs of emissions. Similarly, F_3 puts in emissions between three-particle events, etc. The advantage of the method is that the Green's function includes many-particle emission processes, even when one keeps only a few terms in the series for F_n. For our problems we do not have pions but other boson excitations—such as phonons or the boson-like excitations of the electron gas.

3.7. REAL TIME GREEN'S FUNCTIONS

Six different Green's function of time were introduced in Chapter 2. Two of them, the retarded and advanced functions, have been discussed at nonzero temperature. Here we wish to discuss the other four. The ultimate usefulness of the real time functions is in the treatment of nonequilibrium phenomena. Nonequilibrium transport theory using real time Green's function is treated in Chapter 7. The present section discusses their properties at finite temperature.

The Matsubara method is unsuitable for nonequilibrium since there is no thermodynamic basis for temperature for a system out of equilibrium. The entire Matsubara method is based upon temperature, and no method has been found so far for extending it to nonequilibrium processes.

The real time functions at finite temperature have formal definitions very similar to those at zero temperature. Comparing with (2.9.2), the Green's functions for electrons of momentum **p** are

$$\begin{aligned} G^{<}(\mathbf{p}, t_1, t_2) &= i\langle C_{\mathbf{p}}^{\dagger}(t_2) C_{\mathbf{p}}(t_1)\rangle \\ G^{>}(\mathbf{p}, t_1, t_2) &= -i\langle C_{\mathbf{p}}(t_1) C_{\mathbf{p}}^{\dagger}(t_2)\rangle \\ G_t(\mathbf{p}, t_1, t_2) &= \theta(t_1 - t_2) G^{>}(\mathbf{p}, t_1, t_2) + \theta(t_2 - t_1) G^{<}(\mathbf{p}, t_1, t_2) \\ G_{\bar{t}}(\mathbf{p}, t_1, t_2) &= \theta(t_2 - t_1) G^{>}(\mathbf{p}, t_1, t_2) + \theta(t_1 - t_2) G^{<}(\mathbf{p}, t_1, t_2) \end{aligned} \tag{3.7.1}$$

These definitions appear to be identical to those at zero temperature. The important difference at nonzero temperatures is that the brackets $\langle \cdots \rangle$ have a different meaning. They have several different meanings, depending upon the circumstance:

Case 1. At zero temperature, and in equilibrium, the brackets $\langle \cdots \rangle$ denote the ground state of the interacting system.

Case 2. At finite temperature, and in equilibrium, the brackets denote the thermodynamic average as given in (3.2.2) or (3.3.1).

Case 3. When not in equilibrium, the brackets denote an average over the accessible phase space. However, the available phase space depends upon the recent history of the system, and kinetic constraints such as energy conservation. The meaning of the brackets is poorly understood for systems out of equilibrium.

Another feature of (3.7.1) is that the Green's functions are not expressed on the left as functions of $(t_1 - t_2)$ but of (t_1, t_2) separately. In equilibrium, they are functions of $(t_1 - t_2)$, but that may not be true for systems out of equilibrium.

In thermal equilibrium, the real time functions each have a simple relation to the retarded function. Each is found easily from G_{ret}. Since G_{ret} is found easily from the Matsubara function, the easiest way to find the real time functions is by first finding the Matsubara function of complex frequency. These statements apply only in thermal equilibrium.

Again using the states $|n\rangle$ and $|m\rangle$, which are exact eigenstates of the Hamiltonian, two of the real time functions in (3.7.1) can be written as

$$G^<(\mathbf{p}, t_1 - t_2) = i \sum_{n,m} e^{-\beta E_m} |\langle n | C_\mathbf{p} | m\rangle|^2 e^{i(E_n - E_m)(t_1 - t_2)}$$

$$G^>(\mathbf{p}, t_1 - t_2) = -i \sum_{n,m} e^{-\beta E_n} |\langle n | C_\mathbf{p} | m\rangle|^2 e^{i(E_n - E_m)(t_1 - t_2)}$$

The time dependence is Fourier transformed to ω, which produces the delta function $2\pi\delta(\omega + E_n - E_m)$ in each term. These terms have many of the same factors that occur in the spectral function $A(\mathbf{p}, \omega)$ in (3.3.15). They differ mainly in the thermal factors $\exp(-\beta E_{n,m})$. They can be made to be identical by utilizing some algebraic relationships such as

$$\begin{aligned} e^{-\beta E_m}\delta(\omega + E_n - E_m) &= e^{-\beta E_m}[1 + e^{\beta\omega}]n_F(\omega)\delta(\omega + E_n - E_m) \\ &= [e^{-\beta E_m} + e^{-\beta E_n}]n_F(\omega)\delta(\omega + E_n - E_m) \end{aligned}$$

In the first line the factor of $n_F(\omega)$ has been added to numerator and denominator. In the second line, one of these factors has been changed to the sum of two exponential factors by using $E_m - \omega = E_n$. The part in brackets is exactly the combination that appears in the spectral function. So we have derived

$$G^<(\mathbf{p}, \omega) = i n_F(\omega) A(\mathbf{p}, \omega) \tag{3.7.2}$$

The other real time functions can be derived using the same technique:

$$\begin{aligned} G^>(\mathbf{p}, \omega) &= -i[1 - n_F(\omega)]A(\mathbf{p}, \omega) \\ G_t(\mathbf{p}, \omega) &= [1 - n_F(\omega)]G_{\text{ret}}(\mathbf{p}, \omega) + n_F(\omega)G_{\text{adv}}(\mathbf{p}, \omega) \\ G_{\bar{t}}(\mathbf{p}, \omega) &= -[1 - n_F(\omega)]G_{\text{adv}}(\mathbf{p}, \omega) - n_F(\omega)G_{\text{ret}}(\mathbf{p}, \omega) \end{aligned} \tag{3.7.3}$$

These expressions are identical to (2.9.15). In equilibrium the real time Green's functions are found easily from the retarded and advanced functions. The latter two are found easily from the Matsubara functions. In equilibrium there is no need to set up a separate formalism for the real time functions.

The primary usefulness of the real time Green's functions is in the theory of nonequilibrium phenomena. Some of these ideas and techniques are introduced here. The first step is to derive Dyson's equation. In fact the result (2.9.13) at zero temperature is also correct for nonzero temperature. The matrices have the definitions that are given in (2.9.12). There is no reason to rederive the same equation. Instead it seems appropriate to comment on the differences in the derivation between zero and nonzero temperatures.

Equation (2.9.13) is generally regarded as being the correct form for Dyson's equation even for systems out of equilibrium. However, we have never seen a satisfactory proof for nonequilibrium systems. For systems in equilibrium, one can derive (2.9.13) at finite temperature by starting from Dyson's equation for the Matsubara Green's functions. Treating τ as a complex variable, one can deform the contour of integration, and end up with (2.9.13) (Langreth, 1973).

For nonequilibrium, the Matsubara functions are obviously an invalid starting point. Instead, one can try to expand the S matrix in real time. Then one encounters the problems mentioned in Sec. 3.1 regarding expanding two S matrices: one for $\exp(-\beta H)$ and another one for $\exp(\pm itH)$. All derivations ignore the former. Including only the S matrix of time, one again finds (2.9.13). It appears correct to omit the S matrix expansion of $\exp(-\beta H)$. This point is discussed in the next section. Equation (2.9.13) is used for Dyson's equation for nonzero temperature, for equilibrium and nonequilibrium, since there is nothing else available.

Nonequilibrium theory usually proceeds by first deriving an equation of motion for the Green's function, which is similar to a Boltzmann equation. This equation is usually derived for systems in equilibrium, or slightly out of equilibrium. Then the equation is applied to systems far from equilibrium.

The first step is to find an equation of motion for the interacting Green's function when it is not in equilibrium. Such an equation can be found from (2.9.13) by operating by $(i\partial/\partial t - H_0)$ on both sides of the equation. First one needs to know the time evolution of the noninteracting Green's functions. This behavior is deduced easily from (2.9.6) and (2.9.2). The time derivatives can be expressed compactly in the matrix notation. Remember that $\partial\,\theta(t)/\partial t = \delta(t)$.

$$\left(i\frac{\partial}{\partial t} - \varepsilon_k\right)\tilde{G}_0(\mathbf{k}, t) = \delta(t)\tilde{I}$$

$$\left[i\frac{\partial}{\partial t} - H_0(x)\right]\tilde{G}_0(x) = \delta^4(x)\tilde{I} \tag{3.7.4}$$

On the right-hand side of (2.9.13), this operator only acts upon $\tilde{G}_0$, for which we use the above result to find

$$\left[i\frac{\partial}{\partial t_1} - H_0(x_1)\right]\tilde{G}(x_1, x_2) = \delta^4(x_1 - x_2)\tilde{I} + \int dx_3\, \Sigma(x_1, x_3)\tilde{G}(x_3, x_2) \tag{3.7.5}$$

This formula provides the equation of motion for the interacting Green's function. It is the basis for the nonequilibrium theory of interacting systems.

The structure of this equation is interesting. On the left-hand side, only the noninteracting terms are contained in the Hamiltonian H_0. The contribution from the interactions is provided by the self-energy functions on the right-hand side.

So far, the equation of motion has been derived for the variable x_1 in $\tilde{G}(x_1, x_2)$. It is useful to have an equation of motion for the other variable x_2. Since the definition (2.9.2) of the Green's functions contains the conjugate wave function $\psi^\dagger(x_2)$, the equation of motion on this variable is the complex conjugate of Schrödinger's equation. Furthermore, Dyson's equation can be written in an alternate form. Instead of (2.9.13), it is correct to have $\tilde{G}_0$ on the right in the interaction terms:

$$\tilde{G}(x_1, x_2) = \tilde{G}_0(x_1 - x_2) + \int dx_3 \int dx_4 \tilde{G}(x_1, x_3)\tilde{\Sigma}(x_3, x_4)\tilde{G}_0(x_4 - x_2)$$

The equation of motion on the x_2 variable, when using this form for Dyson's equation, still acts only upon $\tilde{G}_0$ on the right and produces delta functions. These steps produce the alternate equation of motion for the Green's function:

$$\left[-i\frac{\partial}{\partial t_2} - H_0(\mathbf{r}_2, -\mathbf{p}_2)\right]\tilde{G}(x_1, x_2) = \delta^4(x_1 - x_2)\tilde{I} + \int dx_3 \tilde{G}(x_1, x_3)\tilde{\Sigma}(x_3, x_2) \tag{3.7.6}$$

The sign change $\mathbf{p}_2$ to $-\mathbf{p}_2$ comes from the fact that the left-hand side of the above equation is the complex conjugate of Schrödinger's equation. The operator $\mathbf{p} = -i\nabla$ changes sign under complex conjugation. This behavior is different from the Hermitian conjugate, where H_0 acts to the left on $\psi^\dagger(x_2)$ and $\mathbf{p}$ does not change sign. These two equations of motion will be used in Chapter 7 to develop a quantum Boltzmann equation for nonequilibrium phenomena. The quantum Boltzmann equation is based upon a many-body distribution function first suggested by Wigner (1932).

Wigner Distribution Function

The traditional Boltzmann equation is expressed in terms of the distribution function $f(\mathbf{r}, \mathbf{v}, t)$. The three variables are position, $\mathbf{r}$, velocity, $\mathbf{v}$, and time, t. The point of view is semiclassical since it is assumed that the position and velocity (momentum) of the particle can be defined simultaneously. In order to use this distribution for quantum systems, it is necessary to perform some type of averaging in order to remove effects due to the uncertainty principle.

If quantum effects are important, it is necessary to introduce another variable into the distribution function. This variable could either be energy E or equivalently a frequency $\omega = E$. The resulting distribution function is often called a *Wigner distribution function* $f(\mathbf{k}, \omega; \mathbf{r}, t)$. The velocity $\mathbf{v}$ has been changed to the wave vector $\mathbf{k} = m\mathbf{v}$. The distribution function is derived from the Green's function $G^<(x_1, x_2)$ defined in Eq. (2.9.2). The Wigner distribution function (WDF) is derived using the following series of steps. First one goes to a center-of-mass coordinate system:

$$(\mathbf{R}, T) = \tfrac{1}{2}(x_1 + x_2)$$
$$(\mathbf{r}, t) = x_1 - x_2$$

Note that T means center-of-mass time, rather than temperature or time ordering. The symbol β is used for the inverse of Boltzmann's constant times temperature. The notation on the Green's function is altered to these center-of-mass coordinates:

$$G^<(\mathbf{r}, t; \mathbf{R}, T) = i\langle\psi^+(\mathbf{R} - \tfrac{1}{2}\mathbf{r}, T - \tfrac{1}{2}t)\psi(\mathbf{R} + \tfrac{1}{2}\mathbf{r}, T + \tfrac{1}{2}t)\rangle$$

The next step is to Fourier transform the relative variables $(\mathbf{r}, t)$ into $(\mathbf{k}, \omega)$:

$$G^<(\mathbf{k}, \omega; \mathbf{R}, T) = \int d^3r e^{-i\mathbf{k}\cdot\mathbf{r}} \int dt\, e^{i\omega t} G^<(\mathbf{r}, t; \mathbf{R}, T) \tag{3.7.7}$$

The relation to the WDF is quite simple:

$$f(\mathbf{k}, \omega; \mathbf{R}, T) = -iG^<(\mathbf{k}, \omega; \mathbf{R}, T)$$

Regard this assertion as the definition of $f(\mathbf{k}, \omega; \mathbf{R}, T)$. This choice is made reasonable by showing that various moments of f provide the macroscopic quantities of particle density $n(\mathbf{R}, T)$, particle current $j_\mu(\mathbf{R}, T)$, and energy density $n_E(\mathbf{R}, T)$:

$$n(\mathbf{R}, T) = \int \frac{d^3k}{(2\pi)^3} \int \frac{d\omega}{2\pi} f(\mathbf{k}, \omega; \mathbf{R}, T) = \langle\psi^+(\mathbf{R}, T)\psi(\mathbf{R}, T)\rangle$$

$$\mathbf{j}(\mathbf{R}, T) = \int \frac{d^3k}{(2\pi)^3} \frac{\mathbf{k}}{m} \int \frac{d\omega}{2\pi} f(\mathbf{k}, \omega; \mathbf{R}, T)$$

$$n_E(\mathbf{R}, T) = \int \frac{d^3k}{(2\pi)^3} \int \frac{d\omega}{2\pi} \omega f(\mathbf{k}, \omega; \mathbf{R}, T) \tag{3.7.8}$$

$$= i\left[\frac{\partial}{\partial t} \langle \psi^+(\mathbf{R}, T - \tfrac{1}{2}t)\psi(\mathbf{R}, T + \tfrac{1}{2}t)\rangle\right]_{t=0}$$

$$= \langle \psi^+(\mathbf{R}, T) H \psi(\mathbf{R}, T)\rangle$$

The technique for solving nonequilibrium problems is very simple. The equation of motion in (3.7.5) or (3.7.6) for $G^<(\mathbf{k}, \omega; \mathbf{R}, T)$ is just the quantum Boltzmann equation (QBE). This equation is then solved, which yields directly the Wigner distribution function $f(\mathbf{k}, \omega; \mathbf{R}, T)$. Macroscopic variables such as current and density are found by taking the above integrals over d^3k and $d\omega$. This technique is also useful in systems that are not homogeneous because there is a slowly varying potential $V(\mathbf{R})$.

The above expressions are useful for nonequilibrium situations. For example, the answer is zero if the equilibrium function (3.7.2) for $G^<$ is used to calculate the current j_μ in (3.7.8). That is the right answer. A system carrying current can be in steady state but not in equilibrium. In order to obtain a finite current it is necessary to solve for the nonequilibrium Green's function $G^<$ when the Hamiltonian contains an electric field. This procedure is described in Chapter 7.

The semiclassical Boltzmann distribution function $f(\mathbf{r}, \mathbf{v}, t)$ is found by taking the frequency integral of the WDF:

$$f(\mathbf{r}, \mathbf{v}, t) = \int_{-\infty}^{\infty} \frac{d\omega}{2\pi} f(m\mathbf{v}, \omega; \mathbf{r}, t) \tag{3.7.9}$$

This method of deriving the semiclassical Boltzmann equation is an alternative to the usual technique of coarse grain averaging.

The Green's function $G^<$ plays a central role in the nonequilibrium theory. In order to solve it one also needs to know the retarded function. These two Green's functions are the most important. Here we untangle the matrix equation in (3.7.5) and (3.7.6) in order to present the individual equation of motion for these separate Green's functions:

$$\left[i\frac{\partial}{\partial t_1} - H_0(x_1)\right]G^<(x_1, x_2) = \int dx_3[\Sigma_t(x_1, x_3)G^<(x_3, x_2) - \Sigma^<(x_1, x_3)G_{\bar{t}}(x_3, x_2)]$$

$$\left[i\frac{\partial}{\partial t_1} - H_0(x_1)\right]G_{\mathrm{ret}}(x_1, x_2) = \delta^4(x_1 - x_2) + \int dx_3 \Sigma_{\mathrm{ret}}(x_1, x_3)G_{\mathrm{ret}}(x_3, x_2)$$

$$\left[-i\frac{\partial}{\partial t_2} - H_0(\mathbf{r}_2, -\mathbf{p}_2)\right]G^<(x_1, x_2) = \int dx_3[G_t(x_1, x_3)\Sigma^<(x_3, x_2) - G^<(x_1, x_3)\Sigma_{\bar{t}}(x_3, x_2)] \quad (3.7.10)$$

$$\left[-i\frac{\partial}{\partial t_2} - H_0(\mathbf{r}_2, -\mathbf{p}_2)\right]G_{\text{ret}}(x_1, x_2) = \delta^4(x_1 - x_2) + \int dx_3\, G_{\text{ret}}(x_1, x_3)\Sigma_{\text{ret}}(x_3, x_2)$$

The equations are the starting point for the derivation of the quantum Boltzmann equation.

The WDF is an elegant and useful formalism. It does have some liabilities. If one constructs the WDF for any nontrivial Hamiltonian, then $f(\mathbf{r}, \omega; \mathbf{R}, T)$ has regions where it is negative. The function is not positive definite, which means that it cannot be interpreted as a probability density. This feature can be shown by a simple example. Consider in one dimension a particle in a box of length L. The eigenvalues and eigenfunctions are

$$\phi_n(x) = \sqrt{\frac{2}{L}}\sin(k_n x), \qquad \varepsilon_n = \varepsilon n^2$$

$$k_n = \pi n/L, \qquad \varepsilon = k_1^2/2m$$

Using the representation in (2.9.5), and setting $t = t_1 - t_2$, gives

$$G^<(x_1, x_2, t) = \frac{2i}{L}\sum_n n_F(\varepsilon_n)\sin(k_n x_1)\sin(k_n x_2)e^{-it\varepsilon n^2}$$

Here $x_{1,2}$ are not four-vectors but one-dimensional coordinates of position. The product of two sine functions can be written as the difference of two cosine functions. Taking the Fourier transform of time gives the function of frequency. Since we are in equilibrium, we can write $G^< = in_F(\omega)A$. In the center-of-mass notation we find

$$f(x, \omega; X) = n_F(\omega)A(x, \omega; X)$$

$$A(x, \omega; X) = \frac{2\pi}{L}\sum_n [\cos(k_n x) - \cos(2k_n X)]\delta(\omega - \varepsilon n^2)$$

where $x = x_1 - x_2$ is relative position and $X = (x_1 + x_2)/2$ is center-of-mass position. The function $A(x, \omega; X)$ is a sum of delta functions. The amplitude of the delta functions have variable sign, depending upon the values of x and X. Neither A nor f are positive at all points.

For many Hamiltonians of interest the function f is not positive definite. That creates philosophical problems if f is interpreted as a probability function. Do not interpret f this way. Instead, f is regarded as one step in the calculation. It is never the last step, since f is used to calculate other quantities that can be measured. We always find physically sensible results for measurable quantities such as the particle density and current. No problems are encountered as long as one avoids interpreting f as a probability density.

3.8. KUBO FORMULA FOR ELECTRICAL CONDUCTIVITY

Many experiments in condensed matter physics measure the linear response to an external perturbation on the material. The experimentalists may put the sample in a magnetic field, electric field, optical field, temperature gradient, or pressure field and measure the magnetization, electrical current, light absorption, or whatever. Linear response means that the signal is directly proportional to the intensity of the external perturbation. Usually the assumption of linear response is valid at low magnitude of perturbing field.

Kubo formulas are the name applied to the correlation function which describes the linear response. There are many of them, since there are many possible perturbations and many linear responses for each perturbation. Formulas of this type were first proposed by Green (1952, 1954) for transport in liquids. Kubo (1959) first derived the equations for electrical conductivity in solids and in the form which we shall use here. His derivation and result will be given below. Other Kubo formulas will be derived in this section. In fact, we shall give several derivations of the formula for the electrical conductivity in order to familiarize the reader with different types of derivations in the literature. One finds formulas which appear to be quite different but are really identical. We also wish to demonstrate that the longitudinal and transverse conductivities are identical at long wavelength.

In electrical conduction, a time-dependent external electric field,

$$E_\alpha^{(\mathrm{ext})}(\mathbf{r}, t) = \Xi_\alpha^{(\mathrm{ext})} e^{i\mathbf{q}\cdot\mathbf{r} - i\omega t} \tag{3.8.1}$$

is applied to the solid, where $\alpha = x, y$, or z are the directions in space.

In linear response, the induced current is proportional to the applied electric field:

$$J_\alpha(\mathbf{r}, t) = \sum_\beta \sigma'_{\alpha\beta}(\mathbf{q}, \omega) \Xi_\beta^{(\text{ext})} e^{i(\mathbf{q}\cdot\mathbf{r} - \omega t)}$$

Although the symbol σ' has the appearance and dimensions of a conductivity, it is not the conductivity we want. Instead, we want the conductivity which is the response to the total electric field in the solid. The applied or external field $E^{(\text{ext})}$ induces currents which in turn make other electric fields. The summation of all these fields is the total electric field, which is called $E(\mathbf{r}, t)$. The conductivity we want is the one which responds to the actual electric field in the solid:

$$\begin{aligned} J_\alpha(\mathbf{r}, t) &= \sum_\beta \sigma_{\alpha\beta}(\mathbf{q}, \omega) E_\beta(\mathbf{r}, t) \\ E_\alpha(\mathbf{r}, t) &= \Xi_\alpha \exp[i(\mathbf{q}\cdot\mathbf{r} - \omega t)] \\ \sigma_{\alpha\beta} &= \operatorname{Re}(\sigma_{\alpha\beta}) + i \operatorname{Im}(\sigma_{\alpha\beta}) \end{aligned} \tag{3.8.2}$$

We take Eq. (3.8.2) as the fundamental definition of the microscopic conductivity. It also introduces the coordinated problem of determining the proportionality between the external electric field $E_\alpha^{(\text{ext})}$ and the total internal electric field $E_\alpha(\mathbf{r}, t)$. For a static, longitudinal electric field, the two electric fields are related by the continuity of the normal component of $\mathbf{D}$ at the surface. We shall later derive the general method for transverse fields. Equation (3.8.2) is correct for a homogeneous material. Actual solids are not homogeneous, although crystals are periodic. However, (3.8.2) implies that we can write the space-time response as

$$J_\alpha(\mathbf{r}, t) = \int d^3r' \int_{-\infty}^{t} dt' \sigma_{\alpha\beta}(\mathbf{r} - \mathbf{r}'; t - t') E_\beta(\mathbf{r}', t')$$

where repeated indices imply summation. This equation is just the Fourier transform of (3.8.2). It assumes that the current response of a material at $\mathbf{r}$ is only a function of the separation $\mathbf{r} - \mathbf{r}'$ from the external electric field at $\mathbf{r}'$. This assumption is incorrect on an atomic scale. It is certainly not valid in the atomic-like core states of a solid or in any case where the separation $\mathbf{r} - \mathbf{r}'$ is a few angstroms. A rigorous formulation would have the conductivity as a function of $\mathbf{r}$ and $\mathbf{r}'$ separately, $\sigma(\mathbf{r}, \mathbf{r}'; t - t')$. In solids, it is permissible to use (3.8.2) only when it is understood that the current is to be averaged over many unit cells of the solid. Usually it is applied when $\mathbf{q}$ is small and long-wavelength excitations are being studied.

Quite often our interest is in the dc conductivity, which is obtained by taking the limits $\mathbf{q} \to 0$ and $\omega \to 0$ in that order. Then the conductivity is only real. We shall derive the Kubo formula assuming that only a single frequency is perturbing the system and that $\sigma_{\alpha\beta}(\mathbf{q}, \omega)$ is the response to this single frequency. Actually, using a single frequency is not a restriction. Since we assume that the system is linear and that perturbations at different frequencies act independently, the total current is the summation of the responses at different frequencies.

The Hamiltonian for the system is taken to have the form

$$H + H'$$

The term H' contains the interaction between the total electric field and the particles of the system. We use (2.10.1) as the basic form of the interaction between electromagnetic fields and charges. The electric field is expressed as a vector potential, so that

$$\begin{aligned} H' &= -\frac{1}{c}\int d^3r\, j_\alpha(\mathbf{r}) A_\alpha(\mathbf{r}, t) \\ \frac{1}{c} A_\alpha(\mathbf{r}, t) &= \frac{-i}{\omega} E_\alpha(\mathbf{r}, t) \end{aligned} \tag{3.8.3}$$

We are using the Coulomb gauge, where $\nabla \cdot \mathbf{A} = 0$, as explained in Sec. 2.10. Also, the electric and vector potentials are taken to be transverse, so that the scalar potential φ is set equal to zero. The terms in (2.10.1) with $(A)^2$ are dropped, since their effects are nonlinear in the electric field. The term H contains all the other terms and interactions in the solid or liquid. There are interactions such as electron–electron, electron–phonon, spin–spin, with impurities, etc. Our goal is to calculate the electrical conductivity when all these other interactions are present.

The current operator in (3.8.3) was discussed in Chapter 1. It has the form of

$$j_\alpha(\mathbf{r}) = \frac{1}{2m} \sum_i e_i [\mathbf{p}_i \delta(\mathbf{r} - \mathbf{r}_i) + \delta(\mathbf{r} - \mathbf{r}_i)\mathbf{p}_i]_\alpha$$

The notation is more compact if we do the r integral in (3.8.3) and obtain H' in terms of the Fourier transform of the current operator:

$$\begin{aligned} H' &= \frac{i}{\omega} j_\alpha(\mathbf{q}) \Xi_\alpha e^{-i\omega t} \\ j_\alpha(\mathbf{q}) &= \frac{1}{2m} \sum_i e_i (\mathbf{p}_i e^{i\mathbf{q}\cdot\mathbf{r}_i} + e^{i\mathbf{q}\cdot\mathbf{r}_i} \mathbf{p}_i)_\alpha \end{aligned} \tag{3.8.4}$$

In terms of creation and destruction operators, the current operator is conventionally written as

$$j_\alpha(\mathbf{q}) = \sum_{\lambda\delta} p_\alpha^{\lambda\delta}(\mathbf{q}) C_\lambda^\dagger C_\delta$$

$$p_\alpha^{\lambda\delta}(\mathbf{q}) = \frac{e}{2m_i} \int d^3 r e^{i\mathbf{q}\cdot\mathbf{r}} [\psi_\lambda(\mathbf{r})^* \nabla_\alpha \psi_\delta(\mathbf{r}) - \psi_\delta(\mathbf{r}) \nabla_\alpha \psi_\lambda(\mathbf{r})^*]$$

where λ, δ are the states associated with some unperturbed Hamiltonian H_0 which is chosen as the basis for the perturbation expansion. Several of these possibilities were discussed in Chapter 1. A distinction is made between the current operator j_α in (3.8.4) and the induced current J_α in (3.8.2). The operator j_α is used in the Hamiltonian, while J_α is the actual current measured by the experimentalist. The measured value of the current is the average value for the velocity of the particles in the system, which we take as the summation over all the particle velocities divided by the volume:

$$J_\alpha(\mathbf{r}, t) = \frac{e}{\nu} \left\langle \sum_i v_{i\alpha} \delta(\mathbf{r} - \mathbf{r}_i) \right\rangle = \frac{e}{\nu} \sum_i \langle v_{i\alpha} \rangle$$

When we quantize the particle velocity in (1.5.15), the velocity is the momentum minus the vector potential:

$$\mathbf{v}_i = \frac{1}{m} \left[\mathbf{p}_i - \frac{e}{c} \mathbf{A}(\mathbf{r}_i) \right]$$

$$J_\alpha(\mathbf{r}, t) = \frac{e}{m\nu} \sum_i \langle p_{i\alpha} \rangle - \frac{e^2}{mc\nu} \sum_i A_\alpha(\mathbf{r}_i, t)$$

The momentum operator $\mathbf{p}_i$ is proportional to the current operator, $\mathbf{j} = e\mathbf{p}_i/m$. In the last term, we use the relationship (3.8.3) between the vector potential and the electric field. For this term, the summation over particles divided by volume is replaced by the density, since we again assume long-wavelength disturbance:

$$J_\alpha(\mathbf{r}, t) = \langle j_\alpha(\mathbf{r}, t) \rangle + i \frac{n_0 e^2}{m\omega} E_\alpha(\mathbf{r}, t)$$

There is one term in the current proportional to the electric field and another term given by the expectation value of the local current operator $\langle \mathbf{j}(\mathbf{r}, t) \rangle$. We shall show that the latter term is also proportional to the electric field and that the constant of proportionality is given by the Kubo formula. These two terms we call $\mathbf{J}^{(1)}$ and $\mathbf{J}^{(2)}$:

$$\mathbf{J} = \mathbf{J}^{(1)} + \mathbf{J}^{(2)}$$

$$\mathbf{J}^{(2)} = \langle \mathbf{j}(\mathbf{r}, t) \rangle$$

$$\mathbf{J}^{(1)} = \frac{in_0 e^2}{m\omega} \mathbf{E}(\mathbf{r}, t)$$

Now we must derive $\mathbf{J}^{(2)}$, which is the derivation of the Kubo formula.

A. Transverse Fields, Zero Temperature

The following derivation of the Kubo formula is valid at zero temperature. The first step is to consider the expectation value of the current operator as a function of time:

$$J_\alpha^{(2)}(\mathbf{r}, t) = \langle \psi' \mid e^{i(H+H')t} j_\alpha(\mathbf{r}) e^{-it(H+H')} \mid \psi' \rangle \tag{3.8.5}$$

The Heisenberg representation is used, as explained in Sec. 2.2. Next we go into the interaction representation, where H' is treated as the perturbation. Thus we write

$$e^{-it(H+H')} = e^{-itH} U(t)$$

$$U(t) = e^{itH} e^{-it(H+H')}$$

and

$$J_\alpha^{(2)}(\mathbf{r}, t) = \langle \psi' \mid U^\dagger(t) e^{itH} j_\alpha(\mathbf{r}) e^{-iHt} U(t) \mid \psi' \rangle$$

The operator $U(t)$ was defined earlier in (2.1.4), and a formal solution was derived in (2.1.7):

$$U(t) = T \exp\left[-i \int_0^t dt' H'(t')\right]$$

We are also using the definitions

$$H'(t) = e^{iHt} H' e^{-iHt}$$

$$j(t) = e^{iHt} j e^{-iHt}, \qquad \text{etc.}$$

The wave function $\mid \psi' \rangle$ in (3.8.5) is the Schrödinger wave function at $t = 0$

for an interacting system with both $H + H'$ as the Hamiltonian. As in Sec. 3.1, we want the wave function $|\psi\rangle$ to be appropriate when H' is absent from the system. The relationship between the two wave functions is [see (2.2.2)]

$$|\psi'\rangle = T\exp\left[-i\int_{-\infty}^{0} dt' H'(t')\right]|\psi\rangle$$

The time development of the system is given by combining these results:

$$\begin{aligned} U(t)\,|\psi'\rangle &= T\exp\left[-i\int_{-\infty}^{t} dt' H'(t')\right]|\psi\rangle \\ &\equiv S(t, -\infty)\,|\psi\rangle \end{aligned}$$

The expression for the expectation value of the current is now

$$J_\alpha^{(2)}(\mathbf{r}, t) = \langle\psi\,|\,S^\dagger(t, -\infty) j_\alpha(\mathbf{r}, t) S(t, -\infty)\,|\,\psi\rangle \tag{3.8.6}$$

In the present case, we do not need the exact form of the solution. Only terms linear in the electric field E_α are desired, so we need to keep terms linear only in H'. It is sufficient, for linear response, to keep the terms

$$S(t, -\infty)\,|\psi\rangle = \left[1 - i\int_{-\infty}^{t} dt' H'(t')\right]|\psi\rangle + O(H')^2$$

The Hermitian conjugate of this wave function is

$$\langle\psi\,|\,S(t, -\infty)^\dagger = \langle\psi\,|\left[1 + i\int_{-\infty}^{t} dt' H'(t')\right] + O(H')^2$$

From (3.8.6), the expectation value of the current operator is

$$\begin{aligned} J_\alpha^{(2)}(\mathbf{r}, t) &= \langle\psi\,|\left[1 + i\int_{-\infty}^{t} dt' H'(t')\right] j_\alpha(\mathbf{r}, t)\left[1 - i\int_{-\infty}^{t} dt' H'(t')\right]|\,\psi\rangle \\ &= \langle\psi\,|\left\{j_\alpha(\mathbf{r}, t) - i\int_{-\infty}^{t} dt' [j_\alpha(\mathbf{r}, t)H'(t') - H'(t')j_\alpha(\mathbf{r}, t)]\right\}|\,\psi\rangle \end{aligned}$$

The first term is assumed to vanish,

$$\langle\psi\,|\,j_\alpha(\mathbf{r}, t)\,|\,\psi\rangle = 0$$

since there is usually no current in the solid in the absence of an electric field or something equivalent such as a time-varying magnetic field. The first nonzero term is the one linear in H', and this is the term we want. There are two terms, which can be expressed as a commutator:

$$J_\alpha^{(2)}(\mathbf{r}, t) = -i \int_{-\infty}^{t} dt' \langle \psi \mid [j_\alpha(\mathbf{r}, t), H'(t')] \mid \psi \rangle \tag{3.8.7}$$

Our goal is to derive Eq. (3.8.2). Thus we need to remove a factor $E(\mathbf{r}, t)$ from the integrand of the above equation. With H' given in (3.8.4), the integrand has the factors

$$\begin{aligned}[j_\alpha(\mathbf{r}, t), H'(t')] &= \frac{i}{\omega} \Xi_\beta e^{-i\omega t'} [j_\alpha(\mathbf{r}, t), j_\beta(\mathbf{q}, t')] \\ &= \frac{i}{\omega} E_\beta(\mathbf{r}, t) e^{-i\mathbf{q}\cdot\mathbf{r}} e^{i\omega(t-t')} [j_\alpha(\mathbf{r}, t), j_\beta(\mathbf{q}, t')]\end{aligned}$$

which we regroup in the following way:

$$J_\alpha^{(2)}(\mathbf{r}, t) = +\frac{1}{\omega} E_\beta(\mathbf{r}, \mathbf{t}) e^{-i\mathbf{q}\cdot\mathbf{r}} \int_{-\infty}^{t} dt' e^{i\omega(t-t')} \langle \psi \mid [j_\alpha(\mathbf{r}, t), j_\beta(\mathbf{q}, t')] \mid \psi \rangle$$

Thus we just put a factor $\exp[i(\mathbf{q} \cdot \mathbf{r} - \omega t)]$ outside of the integral and, to compensate, leave the inverse of this factor inside the integral. If this result is compared with (3.8.2), we find they have the same form. The current $J_\alpha^{(2)}$ is proportional to the electric field. The constant of proportionality is the conductivity:

$$\begin{aligned}\sigma_{\alpha\beta}(\mathbf{q}, \omega) = &+\frac{1}{\omega} e^{-i\mathbf{q}\cdot\mathbf{r}} \int_{-\infty}^{t} dt' e^{i\omega(t-t')} \langle \psi \mid [j_\alpha(\mathbf{r}, t), j_\beta(\mathbf{q}, t')] \mid \psi \rangle \\ &+ \frac{n_0 e^2}{m\omega} i\delta_{\alpha\beta}\end{aligned}$$

However, this result is not quite right. We need to average over the space variable $\mathbf{r}$ in order to eliminate atomic fluctuations. To take this average, we just integrate over all volume $\int d^3r$ and then divide by ν. However, the only r dependence of the expression are the factors

$$\int d^3r e^{-i\mathbf{q}\cdot\mathbf{r}} j_\alpha(\mathbf{r}, t) = j_\alpha(-\mathbf{q}, t) = j_\alpha^{\dagger}(\mathbf{q}, t)$$

Thus we obtain the final result for the conductivity:

$$\sigma_{\alpha\beta}(\mathbf{q}, \omega) = \frac{1}{\omega\nu} \int_{-\infty}^{t} dt' e^{i\omega(t-t')} \langle \psi \mid [j_\alpha{}^{\dagger}(\mathbf{q}, t), j_\beta(\mathbf{q}, t')] \mid \psi \rangle$$

$$+ \frac{n_0 e^2}{m\omega} i\delta_{\alpha\beta} \tag{3.8.8}$$

Equation (3.8.8) is the Kubo formula.

The correlation function in (3.8.8) is only a function of the time difference $t - t'$. If this difference is made the variable of time integration $t - t' \rightarrow t$, the formula can be expressed as

$$\sigma_{\alpha\beta}(\mathbf{q}, \omega) = + \frac{1}{\omega\nu} \int_{0}^{\infty} dt e^{i\omega t} \langle \psi \mid [j_\alpha{}^{\dagger}(\mathbf{q}, t), j_\alpha(\mathbf{q}, 0)] \mid \psi \rangle + \frac{n_0 e^2}{m\omega} i\delta_{\alpha\beta};$$

which is the way it is usually written. The right-hand side is definitely not a function of t.

The wave function $\mid \psi \rangle$ in (3.8.8) is the ground state of the many-body Hamiltonian H. This Hamiltonian contains all the possible interactions in the solid except the interaction with the vector potential H'. The conductivity is calculated using (3.8.8), which has no mention of photon field. Thus the conductivity is an intrinsic property of the ground state of the system. Equation (3.8.2) can be viewed as a Taylor series in the applied electric field $\Xi_\beta{}^{(\text{ext})}$:

$$J_\alpha(\Xi_\beta) = J_\alpha(0) + \left(\frac{\partial J_\alpha}{\partial \Xi_\beta} \right)_{\Xi_\beta = 0} \Xi_\beta{}^{(\text{ext})} + O(\Xi_\beta{}^{(\text{ext})})^2$$

The conductivity $\sigma'_{\alpha\beta} = (\partial J_\alpha / \partial \Xi_\beta)$ is calculated for $\Xi_\alpha{}^{(\text{ext})} = 0$. It is a characteristic of all linear response correlation functions that they are ground state properties. A major difficulty is that we do not know the ground state ψ of most many-body systems. Then one must employ the S-matrix expansion and Green's function analysis, which has been described in Chapter 2 and this chapter. Thus Eq. (3.8.8) is often the starting point for a many-body calculation, as we shall demonstrate in later chapters.

The Kubo formulas contain a retarded, two-particle Green's function. From the definition of the retarded Green's function in (3.3.3), we can define the retarded correlation function of the current operator as

$$\Pi_{\alpha\beta}(\mathbf{q}, t - t') = - \frac{i\theta(t - t')}{\nu} \langle \psi \mid [j_\alpha{}^{\dagger}(\mathbf{q}, t), j_\beta(\mathbf{q}, t')] \mid \psi \rangle$$

Its Fourier transform is

$$\Pi_{\alpha\beta}(\mathbf{q}, \omega) = -\frac{i}{\nu} \int_{-\infty}^{\infty} dt\theta(t - t')e^{i\omega(t-t')}\langle \psi \mid [j_\alpha^\dagger(\mathbf{q}, t), j_\beta(\mathbf{q}, t')] \mid \psi \rangle$$

By comparing this definition with (3.7.8), we find that

$$\sigma_{\alpha\beta}(\mathbf{q}, \omega) = +\frac{i}{\omega}\left[\Pi_{\alpha\beta}(\mathbf{q}, \omega) + \frac{n_0 e^2}{m}\delta_{\alpha\beta}\right] \tag{3.8.9}$$

The conductivity is the retarded correlation function of the current multiplied by i and divided by ω. The correlation function $\Pi_{\alpha\beta}(\mathbf{q}, \omega)$ is usually called the *current–current* correlation function.

It is usually easiest to calculate the retarded correlation function in the Matsubara formalism. Thus one follows the prescription outlined in (3.3.8) and (3.3.10). First one defines the equivalent current–current correlation function in the Matsubara formalism:

$$\begin{aligned} \Pi_{\alpha\beta}(\mathbf{q}, \tau) &= -\frac{1}{\nu}\langle T_\tau j_\alpha^\dagger(\mathbf{q}, \tau)\, j_\beta(\mathbf{q}, 0)\rangle \\ \Pi_{\alpha\beta}(\mathbf{q}, i\omega) &= \int_0^\beta d\tau e^{i\omega\tau}\Pi_{\alpha\beta}(\mathbf{q}, \tau) \end{aligned} \tag{3.8.10}$$

The Matsubara function is evaluated as best one can—perhaps using the diagrammatic techniques described in Sec. 3.4. Then the desired retarded function is obtained from

$$\underset{i\omega\to\omega+i\delta}{\text{change}}\ \Pi_{\alpha\beta}(\mathbf{q}, i\omega) \to \Pi_{\alpha\beta}(\mathbf{q}, \omega) \tag{3.8.11}$$

and the conductivity from (3.8.9). These are the steps we shall always use in calculating the conductivity.

The dc conductivity is obtained by taking the limit $\mathbf{q} \to 0$ and then the limit $\omega \to 0$. The wrong answer may be obtained if the order of these limits is reversed, which may be understood on physical grounds. The limit $\omega = 0$, $\mathbf{q} \neq 0$ describes a static electric field, which is periodic in space. Here the charge will seek a new equilibrium, after which no current will flow. Thus it is usually important to first take the limit of $\mathbf{q} \to 0$:

$$\lim_{\mathbf{q}\to 0}\begin{cases} \sigma_{\alpha\beta}(\mathbf{q}, \omega) = \sigma_{\alpha\beta}(\omega) \\ \Pi_{\alpha\beta}(\mathbf{q}, i\omega) = \Pi_{\alpha\beta}(i\omega) \\ \Pi_{\alpha\beta}(\mathbf{q}, \omega) = \Pi_{\alpha\beta}(\omega) \\ j_\alpha(\mathbf{q}, \tau) = j_\alpha(\tau) \end{cases}$$

The limit $\mathbf{q} \to 0$ presents no problem. The current operator is well behaved in the limit of $\mathbf{q} \to 0$, so that the correlation functions are well behaved. In fact, when only the dc conductivity is needed, it simplifies the derivation to set $\mathbf{q} = 0$ at the beginning of the calculation.

The limit of $\omega \to 0$ is more delicate. Here the conductivity is real:

$$\operatorname{Re} \sigma_{\alpha\beta} = -\lim_{\omega\to 0} \frac{1}{\omega} \operatorname{Im}[\Pi_{\alpha\beta}(\omega)] \tag{3.8.12}$$

The right-hand side contains the imaginary part of the retarded correlation function. The combination—$\operatorname{Im}[\Pi_{\alpha\beta}(\omega)]$—is just the spectral function of that operator; call it $R_{\alpha\beta}(\omega)$. By using the general formula (3.3.13) for the spectral function, we can write a formal solution for $\operatorname{Im}(\Pi)$:

$$-\operatorname{Im}[\Pi_{\alpha\beta}(\omega)] = \tfrac{1}{2} R_{\alpha\beta}(\omega)$$

$$= \frac{\pi}{\nu} (1 - e^{-\beta\omega}) e^{-\beta\Omega} \sum_{n,m} e^{-\beta E_n} \langle n \mid j_\alpha^\dagger \mid m \rangle \langle m \mid j_\beta \mid n \rangle \delta(\omega + E_n - E_m)$$

Now it is straightforward to take the limit $\omega \to 0$, since the prefactor is

$$\lim_{\omega\to 0} \frac{1}{\omega} (1 - e^{-\beta\omega}) = \beta$$

and

$$\operatorname{Re}(\sigma_{\alpha\beta}) = \frac{\pi\beta}{\nu} e^{\beta\Omega} \sum_{n,m} e^{-\beta E_n} \mid \langle n \mid j_\alpha^\dagger \mid m \rangle \langle m \mid j_\beta \mid n \rangle \mid \delta(E_n - E_m) \tag{3.8.13}$$

The conductivity is finite as $\omega \to 0$. There is no divergence in this limit, even though there is the ω^{-1} factor in (3.8.12).

It is a curious feature of the Kubo formula that in order to calculate the dc conductivity, it is necessary to calculate the ac conductivity and then to take the limit $\omega \to 0$. At least this is the easiest way to do it. Later another Kubo formula will be derived strictly at $\omega = 0$. However, the latter formula is cumbersome to evaluate, and it is faster to evaluate $\sigma(\omega)$ by the procedure we just described—by taking the ac conductivity and the limit that $\omega \to 0$. We also find it easier, at zero temperature, to use the Matsubara formalism and then to take the limit of $T \to 0$ at the end. We shall soon demonstrate that (3.8.8) is the right Kubo formula, even at finite temperatures.

The current–current correlation function is a *two-particle* correlation function. The current operator contains a product of one creation and one destruction operator, so the correlation function (3.8.10) contains at least four such operators:

$$\Pi_{\alpha\beta}(\mathbf{q}, \tau) = -\frac{1}{\nu} \sum_{\substack{\lambda\delta \\ \mu\nu}} p_\alpha{}^{\delta\lambda}(\mathbf{q}) p_\beta{}^{\mu\nu}(\mathbf{q}) \langle T_\tau C_\lambda{}^\dagger(\tau) C_\delta(\tau) C_\mu{}^\dagger(0) C_\nu(0)\rangle$$

The correlation function describes how two particles are created and destroyed. The conductivity arises from correlations between these two events.

The conductivity is a measurable quantity. The conductivity we are describing is the same as that measured in a circuit or by an optical probe. We shall find that measureable quantities always involve retarded correlation functions with at least two particles. The one-particle Green's function can never be measured—at least not in the rigorous sense. The one-particle Green's function describes a Gedanken experiment which can never happen in practice. One creates a particle at time τ',

$$\mathscr{G}(\mathbf{p}, \tau - \tau') = -\langle T C_\mathbf{p}(\tau) C_\mathbf{p}{}^\dagger(\tau')\rangle$$

in state $\mathbf{p}$ and then destroys it at another time τ. One asks for the correlation between these events. This experiment is impossible, since real particles cannot be created or destroyed. In a two-particle Green's function, one describes a sequence of events which is realistic. A particle

$$\langle T_\tau C_\nu{}^\dagger(\tau) C_\mu(\tau) C_\mu{}^\dagger(0) C_\nu(0)\rangle$$

is changed from one state to another, say from ν to μ, and then at a later time it is changed back. The two-particle correlation function, wherein a particle changes its state, is always found in the correlation functions of linear response. In elementary particle physics, a particle can be absolutely destroyed or created. But this event always happens in conjunction with some other event which involves other particles. For example, an electron and positron can mutually annihilate and make several photons. Then one would have terms in the current operator involving the creation or destruction of two particles:

$$j_\alpha(\mathbf{q}) = \sum [p_\alpha{}^{\lambda\delta}(\mathbf{q}) C_\lambda d_\delta + \text{h.c.}] \tag{3.8.14}$$

Even in this case, the current–current correlation function involves four operators: two for one particle and two for another. If the particles interact, i.e., by a Coulomb interaction, then the correlation function may *not* be divided into two independent Green's functions:

$$\langle T_\tau C_\lambda(\tau)\, d_\delta(\tau)\, d_\delta^\dagger(0) C_\lambda^\dagger(0)\rangle \neq \langle T_\tau C_\lambda(\tau) C_\lambda^\dagger(0)\rangle \langle T\, d_\delta(\tau)\, d_\delta^\dagger(0)\rangle$$

We shall encounter current operators of the form (3.8.14). First, positron annihilation is an important experiment in metal physics (West, 1974). The form of (3.8.14) is also used in other contexts. If an electron is destroyed in a filled band, the resulting excitation is called a *hole*. This terminology is used in the filled valence band of a semiconductor and also in the core levels of all solids. Thus we have the identity

$$C_\lambda \left| \begin{matrix} \text{filled} \\ \text{electron} \\ \text{band} \end{matrix} \right\rangle \equiv d_\lambda^\dagger \left| \begin{matrix} \text{empty} \\ \text{hole} \\ \text{band} \end{matrix} \right\rangle$$

where $d_\lambda^\dagger$ is the creation operator for the hole. Thus if one takes an electron from a filled band and moves it to an empty or partially filled band, then one has a current operator of the form (3.8.14), which is used to describe these electron-hole excitation processes.

B. Finite Temperatures

The preceding derivation of the Kubo formula has several restrictions. It was limited to zero temperature and transverse electric fields. Both of these restrictions are unnecessary, since (3.8.8) is correct for both finite temperatures and longitudinal electric fields. At finite temperatures, the bracket is interpreted as a thermodynamic average. Our restriction to transverse electric fields was actually unnecessary. One could use a gauge wherein the scalar potential φ was zero but not the longitudinal vector potential. Then the longitudinal electric field can be expressed in terms of the longitudinal vector potential. The derivation is the same, step by step, with the same answer.

We shall now present another derivation which is valid at finite temperatures. A time-varying perturbation $H'(t)$ is put into the system at finite temperatures. The central question concerns the degree to which the thermodynamic averaging is influenced by the time-varying interaction: do we put H' into the thermodynamic weighting factor $\exp[\beta(\Omega - H - H' + \mu N)]$ as well as into the time development of the operators?

As an example, do we take for the time development of the current operator an expression of the type

$$J^{(2)}(t) \overset{?}{=} \mathrm{Tr}\{e^{\beta[\Omega-H-H'(t)+\mu N]}e^{it[H+H'(t)]}je^{-it[H+H'(t)]}\}$$

Although this expression is reasonable, it is believed to be wrong. It is incorrect to just put H' into the thermodynamic weighting factor. If the time oscillation is fast enough, the heat bath will not follow the oscillations. Thus it is not clear to which degree one should put the time perturbations into the thermodynamic weighting factor. For example, the above formula may be manipulated by moving the right-hand operator in the trace to the left, and by expressing the time operators as a time-dependent density operator:

$$\begin{aligned} J^{(2)}(t) &= \mathrm{Tr}[\varrho(t)j] \\ \varrho(t) &\overset{?}{=} e^{-it[H+H'(t)]}e^{\beta[\Omega-H-H'(t)+\mu N]}e^{it[H+H'(t)]} \end{aligned} \tag{3.8.15}$$

However, since the middle exponential expression also contains the operator combination $H + H'$, it commutes with the time operators, so that the above predicts

$$\varrho(t) \overset{?}{=} e^{\beta[\Omega-H-H'(t)+\mu N]}$$

Thus the time dependence of the current operator plays no role if we also assume that H' goes into the thermodynamic weighting factors. Since our intuition is that the time development of the current operator should not cancel, the above ansatz is incorrect.

Kubo's derivation of the time-dependent density matrix proceeds in the following fashion. The density matrix $\varrho_0 = \exp[\beta(\Omega - H + \mu N)]$ applies to the equilibrium system in the absence of $H'(t)$. It is assumed that the system is described by this density matrix at the initial point in time, which is taken at $t = -\infty$. The perturbation H' is adiabatically switched on as the system is brought forward in time to the present. Now the time-dependent density matrix, for the interacting system including $H'(t)$, is called $\varrho(t)$. It obeys a Heisenberg equation of motion, which when solved yields an expression for the electrical current:

$$\begin{aligned} \frac{d}{dt}\varrho(t) &= -i[H + H'(t), \varrho(t)] \\ J_\alpha^{(2)} &= \mathrm{Tr}[\varrho(t)j_\alpha] \end{aligned}$$

Kubo's starting point is tantamount to not including $H'(t)$ in the thermodynamic weighting factor. The density matrix is defined as $\varrho(t) = \varrho_0 + f(t)$, where ϱ_0 is the density matrix in the absence of H'. The equilibrium density matrix ϱ_0 is time-independent so that the above equations simplify to

$$i\frac{d}{dt}f(t) = [H, \varrho_0] + [H, f] + [H', \varrho_0] + [H', f]$$
$$[H, \varrho_0] = 0$$

The objective is to solve for the term in $J^{(2)}$ which is proportional to H', which is treated as infinitesimally small. Since f is proportional to H', it follows that f is small, and we can neglect terms proportional to $(H')^2$ such as $[H', f]$.

$$i\frac{\partial}{\partial t}f = [H, f] + [H', \varrho_0] + O([H', f])$$

This equation is solved by moving the first term on the right to the left,

$$i\frac{\partial}{\partial t}f - [H, f] = [H', \varrho_0]$$

and then expressing the left-hand side as

$$e^{-iHt}\left\{i\frac{\partial}{\partial t}(e^{iHt}fe^{-iHt})\right\}e^{iHt} = i\frac{\partial f}{\partial t} - [H, f]$$

The linear differential equation may be integrated to give

$$\begin{aligned} i\frac{\partial}{\partial t}(e^{iHt}fe^{-iHt}) &= e^{iHt}[H', \varrho_0]e^{-iHt} = [H'(t), \varrho_0] \\ f(t) &= f(-\infty) - ie^{-iHt}\left\{\int_{-\infty}^{t} dt'[H'(t'), \varrho_0]\right\}e^{iHt} \end{aligned} \tag{3.8.16}$$

There is a temptation in (3.8.16) to write $e^{-iHt}H'(t')e^{iHt} = H'(t' - t)$, which would normally be correct. It is wrong in the present context because H' contains the external field whose time development is not governed by H. We assume that the interaction is switched on slowly in time, so that at $t = -\infty$ there is no interaction, and $H' = 0$. Of course, then $f(-\infty) = 0$ since this term only exists when H' exists. Thus we have obtained a solution of f which is proportional to H', which is adequate for a description of linear response.

The evaluation of (3.8.15) is

$$J_\alpha^{(2)}(\mathbf{r}, t) = \mathrm{Tr}[\varrho_0 j_\alpha(r)] + \mathrm{Tr}[f(t) j_\alpha(\mathbf{r})]$$

where the first term on the right,

$$\mathrm{Tr}[\varrho_0 j_\alpha(r)] = 0$$

vanishes since there is no current if there is no external applied field. The expectation value of the current, proportional to H', is

$$J_\alpha^{(2)}(\mathbf{r}, t) = -i\,\mathrm{Tr}\left\{e^{-iHt}\left[\int_{-\infty}^{t} dt'[H'(t'), \varrho_0]\right]e^{iHt} j_\alpha(\mathbf{r})\right\}$$

The quantity in curly braces can be rearranged by using the cyclic properties of the trace:

$$J_\alpha^{(2)}(\mathbf{r}, t) = -i\,\mathrm{Tr}\left\{\int_{-\infty}^{t} dt'[H'(t'), \varrho_0]e^{iHt} j_\alpha(\mathbf{r}) e^{-iHt}\right\}$$

The factors on the right are just the time development of the current operator:

$$J_\alpha^{(2)}(\mathbf{r}, t) = -i\,\mathrm{Tr}\left\{\int_{-\infty}^{t} dt'[H(t'), \varrho_0] j_\alpha(\mathbf{r}, t)\right\}$$

By using the cyclic properties of the trace, the three operators can be rearranged:

$$\begin{aligned}\mathrm{Tr}\{[H'(t'), \varrho_0] j_\alpha(\mathbf{r}, t)\} &= \mathrm{Tr}[H'(t')\varrho_0 j_\alpha(\mathbf{r}, t) - \varrho_0 H'(t') j_\alpha(\mathbf{r}, t)] \\ &= \mathrm{Tr}\{\varrho_0[j_\alpha(\mathbf{r}, t)H'(t') - H'(t') j_\alpha(\mathbf{r}, t)]\}\end{aligned}$$

The term in curly braces is just a commutator:

$$\mathrm{Tr}\{\varrho_0[j_\alpha(\mathbf{r}, t), H'(t')]\}$$

Thus we finally obtain for the expectation value of the current, the thermodynamic average of the commutator of $H'(t')$ and $j(\mathbf{r}, t)$:

$$J_\alpha^{(2)}(\mathbf{r}, t) = -i\int_{-\infty}^{t} dt'\langle[j_\alpha(\mathbf{r}, t), H'(t')]\rangle \tag{3.8.7'}$$

Equation (3.8.7′) is precisely the zero temperature equation which was derived earlier in (3.8.7). It leads, in the same way, to the Kubo formula (3.8.8). Previously the angle brackets meant to take the expectation value in the ground state at zero temperature, while now it means a trace over the thermal distribution at finite temperatures. Equation (3.8.8) is valid at finite temperatures.

C. Zero Frequency

The Kubo formula (3.8.8) has been shown to be valid at finite temperatures and for transverse electric fields. We also indicated a proof that it is valid for longitudinal electric fields. However, the purist might still object. All the derivations introduce a frequency ω, and the dc conductivity is obtained by letting $\omega \to 0$ at the end of the derivation. It is possible to provide a derivation of the dc conductivity which assumes a constant electric field throughout the derivation? Of course, the answer is affirmative. This provides the motivation for the next derivation, which follows a method suggested by Luttinger (1964).

We begin by assuming an electric field $\mathbf{E}_0$ which is static in time and constant in space. It causes a scalar potential

$$\phi(\mathbf{r}) = -\mathbf{E}_0 \cdot \mathbf{r}$$

and the external vector potential is zero. This scalar potential introduces a perturbation term into the Hamiltonian which is

$$e^{st}F = e^{st} \int d^3r \varrho(\mathbf{r}) \phi(\mathbf{r})$$

where $\varrho(\mathbf{r})$ is the charge density operator of the system. We have added a factor $\exp(st)$ to this term so that it vanishes as $t \to -\infty$. This factor represents the switching on of the potential as time develops from the past. We shall assume that the perturbation on the density matrix,

$$\varrho = \varrho_0 + fe^{st}$$

also has this term. The first term $\varrho_0 = \exp[\beta(\Omega - H)]$ is the density matrix when F is absent. The perturbation fe^{st} is due to F. The starting point is the equation of motion for the density operator,

$$i\frac{\partial \varrho}{\partial t} = i\frac{\partial}{\partial t}(fe^{st}) = (H + e^{st}F, \varrho + fe^{st})$$

which is solved as before. The term proportional to the perturbation F is given in (3.8.16):

$$fe^{st} = -i\int_{-\infty}^{t} dt' e^{-iHt} e^{st'}[F(t'), \varrho_0]e^{iHt} = -i\int_{-\infty}^{t} dt' e^{st'}[F(t'-t), \varrho_0]$$

It is convenient to change integration variables to $t' - t \Rightarrow -t$:

$$f = -i\int_0^{\infty} dt e^{-st}[F(-t), \varrho_0]$$

The crucial trick is to write the following identity for the commutator:

$$[F(-t), \varrho_0] = -i\varrho_0 \int_0^{\beta} d\beta' \frac{\partial}{\partial t} F(-t - i\beta')$$

The identity is easy to prove,

$$\int_0^{\beta} d\beta' \frac{\partial F(-t - i\beta')}{\partial t} = i\int_0^{\beta} d\beta' \frac{\partial F(-t - i\beta')}{\partial \beta'} = i[F(-t - i\beta) - F(-t)]$$

so that

$$\begin{aligned} -i^2\varrho_0[F(-t - i\beta) - F(-t)] &= \varrho_0[e^{+\beta H}F(-t)e^{-\beta H} - F(-t)] \\ &= [F(-t), \varrho_0] \end{aligned}$$

Thus for the expectation value of the current we get the expression

$$J_\alpha(\mathbf{r}) = \mathrm{Tr}[fj_\alpha(\mathbf{r})] = -\int_0^{\infty} dt e^{-st} \int_0^{\beta} d\beta' \,\mathrm{Tr}\left[\varrho_0 \frac{\partial F(-t - i\beta')}{\partial t} j_\alpha(\mathbf{r})\right] \tag{3.8.17}$$

The reason that the trick identity is introduced is that one can now do the following manipulations to produce the electric field. First, the time development of $F(t)$ is given by

$$F(-t) = \int d^3r \varrho(\mathbf{r}, -t)\phi(\mathbf{r})$$

since the scalar potential is time independent. The time derivative of the density is related to the current through the equation of continuity:

$$\frac{\partial F(-t)}{\partial t} = \int d^3\mathbf{r}\phi(\mathbf{r}) \frac{\partial \varrho(-t)}{\partial t} = -\int d^3\mathbf{r}\phi(\mathbf{r})\nabla \cdot \mathbf{j}(\mathbf{r}, -t)$$

We integrate by parts on the space variable,

$$\frac{\partial F(-t)}{\partial t} = \int d^3\mathbf{r}\mathbf{j}_\alpha(\mathbf{r}, -t) \cdot \nabla\phi(\mathbf{r}) = -\int d^3\mathbf{r}\mathbf{j}(\mathbf{r}, -t) \cdot \mathbf{E}_0$$

and thereby obtain the electric field E_0:

$$J_\alpha(r) = \int_0^\infty dt e^{-st} \int_0^\beta d\beta' \operatorname{Tr}\left[\varrho_0 \int d^3\mathbf{r}' j_\beta(\mathbf{r}', -t - i\beta') E_{0\beta} j_\alpha(\mathbf{r})\right]$$

Since the electric field is a constant, it can be removed from the integrals, and the conductivity is given by

$$\operatorname{Re}(\sigma_{\alpha\beta}) = \int_0^\infty dt e^{-st} \int_0^\beta d\beta' \int d^3\mathbf{r}' \operatorname{Tr}[\varrho_0 j_\beta(\mathbf{r}', -t - i\beta) j_\alpha(\mathbf{r})]$$

The right-hand side appears to depend on the spacial variable $\mathbf{r}$. This dependence is eliminated, as before, by averaging over the volume ν of the sample. This manipulation just introduces the $\mathbf{q} \to 0$ limit of the current operator $j_\alpha(\mathbf{q})$:

$$\lim_{\mathbf{q}\to 0} j_\alpha(\mathbf{q}, t) = j_\alpha(t) = \int d^3 r j_\alpha(\mathbf{r}, t)$$

Thus the Kubo formula for the dc electrical conductivity is

$$\operatorname{Re}(\sigma_{\alpha\beta}) = \frac{1}{\nu} \int_0^\infty dt e^{-st} \int_0^\beta d\beta' \operatorname{Tr}[\varrho_0 j_\beta(-t - i\beta') j_\alpha] \qquad (3.8.18)$$

The result does not contain the frequency. The elimination of ω has been achieved by paying a penalty: There is a second integration, this time over the temperature-like variable β'. This second integration makes the results appear different from our prior result (3.8.8). The two results are identical for $\omega = 0$. To prove this, again introduce the representation $|n\rangle$, $|m\rangle$, which are exact eigenstates of the Hamiltonian H. The important matrix element is

$$\begin{aligned}\langle n \mid j_\alpha(-t - i\beta') \mid m\rangle &= \langle n \mid e^{-iH(t+i\beta')} j_\alpha e^{iH(t+i\beta')} \mid m\rangle \\ &= \langle n \mid j_\alpha \mid m\rangle e^{-it(E_n - E_m)} e^{\beta'(E_n - E_m)}\end{aligned}$$

so that

$$\operatorname{Re}(\sigma_{\alpha\beta}) = \frac{\operatorname{Re}}{\nu}\left[\sum_{n,m} e^{-\beta E_n}\langle n \mid j_\beta \mid m\rangle\langle m \mid j_\alpha \mid n\rangle\right]$$

$$\times \int_0^\infty ds e^{-it(E_n - E_m - is)} \int_0^\beta d\beta' e^{\beta'(E_n - E_m)}$$

Both integrals can be done easily. We want the real part of the conductivity, and the only complex term is the time integral:

$$\operatorname{Re}\left[\int_0^\infty dt e^{-it(E_n - E_m - is)}\right] = \operatorname{Re}\left[\frac{1}{i(E_n - E_m - is)}\right] = \pi\delta(E_n - E_m)$$

The β' integral is now easy,

$$\delta(E_n - E_m)\int_0^\beta d\beta' e^{\beta'(E_n - E_m)} = \beta\delta(E_n - E_m)$$

so that the final result is

$$\operatorname{Re}(\sigma_{\alpha\beta}) = \frac{\pi\beta}{\nu}\sum_{n,m} e^{-\beta E_n}\langle n \mid j_\beta \mid m\rangle\langle m \mid j_\alpha \mid n\rangle\delta(E_n - E_m) \qquad (3.8.13)$$

This equation is identical to the earlier result (3.8.13), since $j_\alpha^\dagger = j_\alpha$ for $\mathbf{q} = 0$. We have proved that (3.8.18) is identical to the $\omega \to 0$ limit of the usual Kubo formula (3.8.8). In the limit of $\omega \to 0$, the longitudinal and transverse conductivities are identical.

D. Photon Self-Energy

A formula which is identical to the Kubo formula can also be derived from the photon self-energy. We consider the photon Green's function (3.2.17),

$$\mathscr{D}_{\mu\nu}(\mathbf{k}, \tau) = -\langle T_\tau A_\mu(\mathbf{k}, \tau) A_\nu(-\mathbf{k}, 0)\rangle$$

and treat as the perturbation

$$V = -\frac{1}{c}\int d^3r\left[j_\alpha(\mathbf{r})A_\alpha(\mathbf{r}) + \frac{e^2 n(\mathbf{r})}{2mc}\mathbf{A}(\mathbf{r})^2\right]$$

$$= -\frac{1}{\sqrt{\nu}}\sum_{\mathbf{k}} j_\alpha(\mathbf{k})A_\alpha(\mathbf{k}) + \frac{n_0 e^2}{2m}\sum_{\mathbf{k}}\mathbf{A}(\mathbf{k})\cdot\mathbf{A}(-\mathbf{k})$$

The $A(\mathbf{k})^2$ term contributes a self-energy of $\Pi^{(1)} = (n_0e^2)/m$. The other term in the self-energy comes from the $n = 2$ term in the expansion for the S matrix for the $\mathbf{j} \cdot \mathbf{A}$ interaction,

$$-\sum_{\mathbf{k}',\mathbf{k}''} \frac{1}{2!\nu} \int_0^\beta d\tau_1 \int_0^\beta d\tau_2 \langle T_\tau j_l(\mathbf{k}', \tau_1)\, j_m(\mathbf{k}'', \tau_2)\rangle$$

$$\times \langle T_\tau A_\mu(\mathbf{k}, \tau) A_l(\mathbf{k}', \tau_1) A_m(\mathbf{k}'', \tau_2) A_\nu(-\mathbf{k}, 0)\rangle$$

$$= \sum_{\mathbf{k}} \int_0^\beta d\tau_1 \int_0^\beta d\tau_2 \Pi_{lm}(\mathbf{k}, \tau_1 - \tau_2) \mathscr{D}^{(0)}_{\mu l}(\mathbf{k}, \tau - \tau_1) \mathscr{D}^{(0)}_{m\nu}(\mathbf{k}, \tau_2)$$

where

$$\Pi_{lm}(\mathbf{k}, \tau_1 - \tau_2) = -\frac{1}{\nu} \langle T_\tau j_l(-\mathbf{k}, \tau_1) j_m(\mathbf{k}, \tau_2)\rangle = -\frac{1}{\nu} \langle T_\tau j_l^\dagger(\mathbf{k}, \tau_1) j_m(\mathbf{k}, \tau_2)\rangle$$

By taking the Fourier transform of this expression, we find that the self-energy contribution is

$$\Pi_{lm}(\mathbf{k}, i\omega) = -\frac{1}{\nu} \int_0^\beta d\tau e^{i\omega_n \tau} \langle T_\tau j_l^\dagger(\mathbf{k}, \tau)\, j_m(\mathbf{k}, 0)\rangle \qquad (3.8.19)$$

If we recall (2.10.11) for the dielectric function, in retarded or Matsubara notation,

$$\varepsilon(\mathbf{k}, i\omega) = 1 - \frac{4\pi}{(i\omega)^2}\left[\Pi(\mathbf{k}, i\omega) + \frac{n_0e^2}{m}\right]$$

$$\varepsilon_{\text{ret}}(\mathbf{k}, \omega) = 1 - \frac{4\pi}{\omega^2}\left[\Pi(\mathbf{k}, \omega) + \frac{n_0e^2}{m}\right] \qquad (3.8.20)$$

where Π is the scalar part of the self-energy $\Pi_{\alpha\beta}$ for homogeneous, isotopic systems:

$$\Pi(\mathbf{k}, i\omega) = \tfrac{1}{3} \sum_\alpha \Pi_{\alpha\alpha}(\mathbf{k}, i\omega)$$

Now it is often written that the long-wavelength dielectric function can be given by the conductivity

$$\varepsilon(\omega) = \lim_{\mathbf{k}\to 0} \varepsilon_{\text{ret}}(\mathbf{k}, \omega) = 1 - \frac{4\pi\sigma(\mathbf{k}, \omega)}{i\omega} = 1 - \frac{4\pi\sigma(\omega)}{i\omega} \qquad (3.8.21)$$

If we compare (3.8.20) with (3.8.21), we obtain the expression for the conductivity $\sigma(\omega)$ in terms of the retarded correlation function $\Pi(\omega)$:

$$\sigma(\omega) = \frac{i}{\omega}\left[\Pi(\omega) + \frac{n_0e^2}{m}\right]$$

Since Π is the current-current correlation function, we have just rederived (3.8.9).

We restate the important point that in evaluating the Kubo formula for the current–current correlation function, do not include the self-energy terms which arise from the vector potential $\mathbf{A}(\mathbf{r}, t)$. These terms have already been included in the formalism, indirectly, by changing the external field $\mathbf{E}^{(\text{ext})}$ to the total field $\mathbf{E}$. They do not reenter the formalism again except as internal interactions within the polarization bubble.

3.9. OTHER KUBO FORMULAS

There are other measurements besides electrical conductivity. These other measurements require additional correlation functions or Kubo formulas. Some of these will be derived here. The first is magnetic susceptibility. Next comes the derivation of transport coefficients such as thermal conductivity or thermoelectric power.

A. Pauli Paramagnetic Susceptibility

The magnetic susceptibility of a solid depends rather strikingly on whether the system is already spontaneously magnetized. If it is, then an additional weak, external magnetic field may not change the magnetization appreciably—there may be no linear response. We shall not consider this case but only the situation when the solid is not in a magnetic state; it is not ferromagnetic, antiferromagnetic, etc. A magnetic state has long-range order, although there is usually short-range ordering, from the interactions, even in the absence of long-range order.

The perturbation term in the Hamiltonian for Pauli paramagnetism has just the interaction with the electron spin,

$$V = -\mathbf{m} \cdot \mathbf{H}_0 e^{-i\omega t}$$

where the ac magnetic field is

$$H(t) = \mathbf{H}_0 e^{-i\omega t}$$

and the magnetization is

$$m_\alpha = g\mu_0 \sum_i S_{i,\alpha}$$

the sum over all spins $S_{i,\alpha}$ at position i with vector direction α. The other factors are the Bohr magneton $\mu_0 = |e| \hbar/2mc$ and gyromagnetic ratio $g \simeq 2$. Another contribution to the magnetization is the orbital, or Landau, contribution, which is not discussed here. In the Pauli paramagnetism the objective is to evaluate the magnetization,

$$M_\alpha(t) = \langle m_\alpha(t) \rangle = X_{\alpha\beta}(\omega) H_\beta(t)$$

for terms linear in the magnetic field. The derivation proceeds exactly as in the preceding section. In analogy with (3.8.7), we obtain

$$M_\alpha(t) = -i \int_{-\infty}^{t} dt' \langle [m_\alpha(t), V(t')] \rangle$$

so that the magnetic susceptibility is

$$\chi_{\alpha\beta}(\omega) = +i \int_{-\infty}^{t} dt' e^{i\omega(t-t')} \langle [m_\alpha(t), m_\beta(t')] \rangle \qquad (3.9.1)$$

Equation (3.9.1) is just the Fourier transform of a retarded correlation function. Thus it may also be evaluated conveniently in the Matsubara formalism. Define a correlation function

$$\chi_{\alpha\beta}(i\omega) = + \int_0^\beta d\tau \langle T_\tau m_\alpha(\tau) m_\beta(0) \rangle e^{i\omega\tau} \qquad (3.9.2)$$

and (3.9.1) is just the retarded function

$$\chi_{\alpha\beta}(i\omega) \xrightarrow[i\omega \to \omega + i\delta]{} \chi_{\alpha\beta}(\omega)$$

The spin operators are often written in terms of raising and lowering operators,

$$S_\alpha^{(\pm)} = S_x \pm iS_y$$
$$S^{(+)} = \sum_{\mathbf{k}} C_{\mathbf{k}\uparrow}^\dagger C_{\mathbf{k}\downarrow}$$

so that the three operators $S^{(+)}, S^{(-)}, S_z$ are used instead of S_x, S_y, S_z. Of course, one can construct susceptibility functions from these operators also.

The correlation function will be evaluated for a simple example. It is the Pauli susceptibility of a free-electron gas at zero temperature. First we shall derive the answer from simple considerations. In a magnetic field H_0 in the z direction, the spin up and spin down electrons have the following energies and occupation functions

$$\xi_\uparrow(p) = \xi_p - \mu_0 H_0, \qquad n_\uparrow = n_F(\xi_p - \mu_0 H_0)$$
$$\xi_\downarrow(p) = \xi_p + \mu_0 H_0, \qquad n_\downarrow = n_F(\xi_p + \mu_0 H_0)$$

The net magnetization will be the difference between the number of spin up and spin down electrons. If we expand the occupation numbers in a power series in the magnetic field and keep the first term, we obtain

$$M = +\mu_0 \sum_{\mathbf{p}} [n_\uparrow(\mathbf{p}) - n_\downarrow(\mathbf{p})] = -2\mu_0{}^2 H_0 \sum_{\mathbf{p}} \frac{\partial n_F}{\partial \xi_{\mathbf{p}}} (\xi_{\mathbf{p}})$$
$$\lim_{T\to 0} \frac{M}{H_0} = 2\mu_0{}^2 \sum_{\mathbf{p}} \delta(\xi_{\mathbf{p}}) = \mu_0{}^2 N_F = \chi_P$$
$$N_F = \frac{mk_F}{\pi^2}$$

The Pauli susceptibility χ_p is given by the square of the Bohr magneton μ_0 times the density of states N_F at the Fermi energy. This density of states is the total for both spin components. The notation is confusing, since often the same symbol is used as the density of states for each spin component, so that the value is only one-half as large.

The same result is obtained from the correlation function. The operator m_α is given in terms of creation and destruction operators as

$$m_z = \mu_0 \sum_{\substack{\mathbf{k} \\ \sigma=\pm 1}} \sigma C^\dagger_{\mathbf{k}\sigma} C_{\mathbf{k}\sigma}$$

We use $\sigma = \pm 1$ to denote electron spin up or down. The Kubo formula for the correlation function for the static susceptibility ($i\omega = 0$) is

$$\chi_{ZZ} = \mu_0{}^2 \int_0^\beta d\tau \sum_{\substack{kp \\ \sigma\sigma'=\pm 1}} \sigma\sigma' \langle T_\tau C^\dagger_{\mathbf{k}\sigma}(\tau) C_{\mathbf{k}\sigma}(\tau) C^\dagger_{\mathbf{p}\sigma'} C_{\mathbf{p}\sigma'} \rangle$$

For the free-particle system, the Hamiltonian is just the kinetic energy term, so that $C^\dagger_{\mathbf{k}\sigma}(\tau) = e^{\tau\xi_k} C^\dagger_{\mathbf{k}\sigma}$ and $C_{\mathbf{k}\sigma}(\tau) = e^{-\tau\xi_k} C_{\mathbf{k}\sigma}$. Then the operator $C^\dagger_{k\sigma} C_{k\sigma}$ has no τ dependence, so the τ integral just gives β. The combination of operators can be averaged directly. Some care must be taken with the terms which have $\sigma = \sigma'$ and $p = k$. Using the fact that $\langle n_{\mathbf{k}}^2 \rangle = \langle n_{\mathbf{k}} \rangle$ if both refer to the same spin state, we can derive

$$\chi_{ZZ} = \mu_0^2 \beta \Big(\sum_{\sigma \mathbf{p}} \sigma^2 n_{\mathbf{p}} + \sum_{\substack{\mathbf{p} \neq \mathbf{k} \\ \sigma, \sigma'}} \sigma \sigma' n_k n_{\mathbf{p}} + \sum_{\substack{\sigma \neq \sigma' \\ \mathbf{p}}} \sigma \sigma' n_{\mathbf{p}}^2 \Big)$$

$$= \mu_0^2 \beta \Big[\sum_{\sigma \mathbf{p}} \left(\sigma^2 n_{\mathbf{p}} - n_{\mathbf{p}}^2 \right) + \Big(\sum_{\sigma \mathbf{p}} \sigma n_{\mathbf{p}} \Big)^2 \Big]$$

The second term on the right vanishes, since there are as many up spins as down spins in equilibrium. The correlation function is evaluated, of course, in the absence of any external magnetic field. Thus we are left with the combination of operators

$$\chi_{ZZ} = \mu_0^2 \beta \sum_{\sigma \mathbf{p}} n_{\mathbf{p}}(1 - n_{\mathbf{p}}) \equiv -\mu_0^2 \sum_{\sigma \mathbf{p}} \left(\frac{\partial n_F}{\partial \xi_{\mathbf{p}}} \right) = 2\mu_0^2 \sum_{\mathbf{p}} \delta(\xi_{\mathbf{p}})$$

which is identical to $\partial n_{\mathbf{p}}/\partial \xi_{\mathbf{p}}$. Thus we obtain exactly the same combination of factors as previously, so we derive the same result as the simple derivation.

Another method of evaluating the correlation function is to define the wave-vector-dependent susceptibility $\chi(\mathbf{q}, i\omega)$ by

$$\chi_{\alpha\beta}(\mathbf{q}, i\omega) = \int_0^\beta d\tau e^{i\omega\tau} \langle m_\alpha(\mathbf{q}, \tau) m_\beta(-\mathbf{q}, 0) \rangle$$

$$m_z(\mathbf{q}, \tau) = \mu_0 \sum_{\substack{\mathbf{p} \\ \sigma = \pm 1}} \sigma C^\dagger_{\mathbf{p}+\mathbf{q},\sigma} C_{\mathbf{p}\sigma}$$

$$m_x(\mathbf{q}, \tau) = \frac{\mu_0}{2^{1/2}} \sum_{\mathbf{p}} (C^\dagger_{\mathbf{p}+\mathbf{q}\uparrow} C_{\mathbf{p}\downarrow} + C^\dagger_{\mathbf{p}+\mathbf{q}\downarrow} C_{\mathbf{p}\uparrow}) \tag{3.9.3}$$

$$m_y(\mathbf{q}, \tau) = \frac{i\mu_0}{2^{1/2}} \sum_{\mathbf{p}} (C^\dagger_{\mathbf{p}+\mathbf{q}\uparrow} C_{\mathbf{p}\downarrow} - C^\dagger_{\mathbf{p}+\mathbf{q}\downarrow} C_{\mathbf{p}\uparrow})$$

We evaluate χ_{ZZ} for finite values of $\mathbf{q}$ and $i\omega = 0$ and formally take the limit $\mathbf{q} \to 0$ at the end of the calculation:

$$\chi_{ZZ}(\mathbf{q}, 0) = \mu_0^2 \sum_{ss'pk} ss' \int_0^\beta d\tau \langle T_\tau C^\dagger_{\mathbf{p}+\mathbf{q}s}(\tau) C_{\mathbf{p}s}(\tau) C^\dagger_{\mathbf{k}-\mathbf{q}s'}(0) C_{\mathbf{k}s'}(0) \rangle$$

For free electrons, this correlation function is finite only if $s = s'$, and then it is identical to the polarization operator $P^{(1)}(\mathbf{q}, i\omega)$ which we have evaluated previously. For $i\omega = 0$ this is

$$\chi_{ZZ}(\mathbf{q}, 0) = -\mu_0^2 P^{(1)}(\mathbf{q}, 0) = -2\mu_0^2 \sum_p \frac{n_\mathbf{p} - n_{\mathbf{p}+\mathbf{q}}}{\xi_\mathbf{p} - \xi_{\mathbf{p}+\mathbf{q}}}$$

If we now take the limit of $\mathbf{q} \to 0$ on the right-hand side, we obtain our previous result $\partial n_p / \partial \xi_\mathbf{p}$ in the summation over p. Thus all methods give the same result for the free-electron system.

B. Thermal Currents and Onsager Relations

There are two conceptual difficulties in deriving a formula for the thermal conductivity. The first is that one must put a temperature gradient on the solid. However, all our thermal averaging assumes a constant temperature. Of course, the thermal conductivity is still defined in the limit that the temperature difference ΔT goes to zero, so that the correlation function can be evaluated at a single temperature. But the problem lies in formally deriving the correlation function without assuming that ΔT exists somewhere in the formalism. Another way to express this difficulty is to consider how one raises the temperature by heating the sample. The heating changes the energy of the system, so that one might try to calculate the time rate of change of the energy. But the energy operator is the Hamiltonian, and its rate of change is $\dot{H} = i[H, H] = 0$, which does not help much. This problem has been well studied in nonequilibrium statistical mechanics (de Groot, 1952, 1962), and it has been found that the important quantity is the rate of change of the entropy.

Another problem with the thermal conductivity is that there are several definitions of heat current. We shall mention three. We show that the same definition of experimental quantities is obtained—thermal conductivity, thermoelectrical power, etc.—regardless of which definition is adopted. But one must be careful to do the right calculation for each definition of heat current.

In linear response, there are current $\mathbf{J}_i$ which flow as a result of "forces" $\mathbf{X}_i$ on the system. These forces might be temperature gradients ∇T, or electric fields $\mathbf{E} = -\nabla V$, or concentration gradients which are expressed as gradients of the chemical potential $\nabla \mu$. Linear response assumes these are proportional:

$$\mathbf{J}_i = \sum_j \mathbf{Z}^{(ji)} \mathbf{X}_j \qquad (3.9.4)$$

The coefficients $\mathbf{Z}^{(ij)}$ are the measurable constants for which we are trying to deduce the proper correlation function. There are also the Onsager relationships, which specify that $\mathbf{Z}^{(ij)} = \mathbf{Z}^{(ji)}$. A moment's reflection will show that the Onsager relationship is not valid for any arbitrary choice of currents and forces. For example, if they are valid for a force $\nabla(1/T) = -(\nabla T/T^2)$, then changing the force to ∇T, so that $1/T^2$ is absorbed in $\mathbf{Z}$, means the Onsager relation is no longer valid. There must be a criterion for choosing the forces and currents, which has been given in nonequilibrium statistical mechanics by de Groot (1952). In a nonequilibrium process, there is a net generation of entropy, so that $\partial S/\partial t > 0$. Here S is just the part of the entropy which is generated nonreversibly. If one requires that this be expressed as

$$\frac{\partial S}{\partial t} = \sum_i \mathbf{J}_i \cdot \mathbf{X}_i \tag{3.9.5}$$

then the Onsager relations are valid. There is not a unique set of currents and forces, and many possibilities satisfy (3.9.5). Each choice defines a different set of coefficients $\mathbf{Z}^{(ij)}$, but each set obeys the relationship $\mathbf{Z}^{(ij)} = \mathbf{Z}^{(ji)}$.

The symbol $\mathbf{J}$ will refer to particle current. There are, as we mentioned, several definitions of energy current. Our discussion will follow Barnard (1972). The first energy current which comes to mind is the one we shall call the *energy current* $\mathbf{J}_E$. For a free-particle system it is

$$\begin{aligned} \mathbf{j}_E &= \sum_p \mathbf{v}_p \varepsilon_p n_p \\ \mathbf{J}_E &= \langle \mathbf{j}_E \rangle \end{aligned} \tag{3.9.6}$$

just the velocity of the particles times the energy times the number. Other definitions were given in Sec. 1.2 for particles which were not free. For a system with only these two currents, their forces are (Luttinger, 1964)

$$\begin{aligned} \mathbf{J}_1 &= \mathbf{J}, \qquad & \mathbf{X}_1 &= -\frac{e}{T}\nabla \mathbf{V} - \nabla\left(\frac{\mu}{T}\right) \\ \mathbf{J}_2 &= \mathbf{J}_E, \qquad & \mathbf{X}_2 &= \nabla\left(\frac{1}{T}\right) \end{aligned} \tag{3.9.7}$$

where e is the charge of the particle. The rate of entropy production is

$$\frac{\partial S}{\partial t} = -\mathbf{J} \cdot \left[\frac{e}{T}\nabla V + \nabla\left(\frac{\mu}{T}\right)\right] + \mathbf{J}_E \cdot \nabla\left(\frac{1}{T}\right) \tag{3.9.8}$$

and the linear response equations (3.9.4) are

$$J_\alpha = -M_{\alpha\delta}^{(11)}\left[\frac{e}{T}\nabla_\delta V + \nabla_\delta\left(\frac{\mu}{T}\right)\right] + M_{\alpha\delta}^{(12)}\nabla_\delta\left(\frac{1}{T}\right)$$
$$J_{E,\alpha} = -M_{\alpha\delta}^{(21)}\left[\frac{e}{T}\nabla_\delta V + \nabla_\delta\left(\frac{\mu}{T}\right)\right] + M_{\alpha\delta}^{(22)}\nabla_\delta\left(\frac{1}{T}\right) \tag{3.9.9}$$

With these definitions, one has the relationship $M_{\alpha\delta}^{(12)} = M_{\alpha\delta}^{(21)}$.

The trouble with (3.9.6) is that the energy current is not the heat current. As discussed by Taylor (1970), the heat current should be

$$\mathbf{J}_Q = \mathbf{J}_E - \mu\mathbf{J} \tag{3.9.10}$$

which for free particles has the form

$$\mathbf{j}_Q = \sum_p n_p \mathbf{v}_p(\varepsilon_p - \mu)$$
$$\mathbf{J}_Q = \langle \mathbf{j}_Q \rangle$$

so that the reference energy is the chemical potential μ. The latter definition makes sense only for particles with a positive chemical potential—e.g., electrons in metals. If one takes an electron below the chemical potential and moves it down to the other end of the sample, one has moved energy. But when the electron arrives at the new location, it finds locally a filled Fermi distribution, so that the only states available to it are above the Fermi energy. But the energy gain required to increase its energy, to get it above the Fermi energy, must come from the surroundings. The energy gain from the surroundings must cool the locality, so that the electron has brought coldness with it. Only electrons which arrive with energy above the Fermi energy, and give energy to the surroundings, bring heat. With this choice of thermal current, the forces are

$$\mathbf{J}_1 = \mathbf{J}, \qquad \mathbf{X}_1 = -\frac{1}{T}\boldsymbol{\nabla}\bar{\mu}$$
$$\mathbf{J}_2 = \mathbf{J}_Q, \qquad \mathbf{X}_2 = \boldsymbol{\nabla}\left(\frac{1}{T}\right) \tag{3.9.11}$$

where

$$\bar{\mu} = \mu + eV$$

With this choice, the rate of entropy production is

$$\frac{\partial S}{\partial t} = \frac{-\mathbf{J}}{T}\cdot\boldsymbol{\nabla}\bar{\mu} + \mathbf{J}_Q\cdot\boldsymbol{\nabla}\left(\frac{1}{T}\right) \tag{3.9.12}$$

and the linear response equations are

$$\begin{aligned} J_\alpha &= -\frac{1}{T} L^{(11)}_{\alpha\delta} \nabla_\delta \bar{\mu} + L^{(12)}_{\alpha\delta} \nabla_\delta \left(\frac{1}{T} \right) \\ J_{Q\alpha} &= -\frac{1}{T} L^{(21)}_{\alpha\delta} \nabla_\delta \bar{\mu} + L^{(22)}_{\alpha\delta} \nabla_\delta \left(\frac{1}{T} \right) \end{aligned} \tag{3.9.13}$$

where $L^{(12)} = L^{(21)}$. It is easy to show that these equations are consistent with the set (3.9.7) and (3.9.9). In (3.9.7) we regroup the terms,

$$\frac{e}{T} \mathbf{\nabla} V + \mathbf{\nabla}\left(\frac{\mu}{T} \right) = \frac{e}{T} \mathbf{\nabla} V + \frac{1}{T} \mathbf{\nabla}\mu + \mu \mathbf{\nabla}\left(\frac{1}{T} \right) = \frac{1}{T} \mathbf{\nabla}\bar{\mu} + \mu \mathbf{\nabla}\left(\frac{1}{T} \right)$$

so that $\partial S/\partial t$ becomes

$$\begin{aligned} \frac{\partial S}{\partial t} &= -\mathbf{J} \cdot \left[\frac{1}{T} \mathbf{\nabla}\bar{\mu} + \mu \nabla\left(\frac{1}{T} \right) \right] + \mathbf{J}_E \cdot \mathbf{\nabla}\left(\frac{1}{T} \right) \\ &= -\frac{\mathbf{J}}{T} \cdot \mathbf{\nabla}\bar{\mu} + (\mathbf{J}_E - \mu \mathbf{J}) \cdot \mathbf{\nabla}\left(\frac{1}{T} \right) \end{aligned}$$

which is identical to (3.9.12). Similarly, we can operate on the current equations (3.9.9) in order to find the relationship between the coefficients $M^{(ij)}$ and $L^{(ij)}$. The same regrouping on (3.9.9) gives

$$\begin{aligned} J_\alpha &= -\frac{1}{T} M^{(11)}_{\alpha\delta} \nabla_\delta \bar{\mu} + (M^{(12)}_{\alpha\delta} - \mu M^{(11)}_{\alpha\delta}) \nabla_\delta \left(\frac{1}{T} \right) \\ J_{E,\alpha} &= -\frac{1}{T} M^{(21)}_{\alpha\delta} \nabla_\delta \bar{\mu} + (M^{(22)}_{\alpha\delta} - \mu M^{(21)}_{\alpha\delta}) \nabla_\delta \left(\frac{1}{T} \right) \end{aligned}$$

Next, consider the following combination of these equations:

$$\begin{aligned} J_{Q,\alpha} = J_{E,\alpha} - \mu J_\alpha = &-\frac{1}{T} (M^{(21)}_{\alpha\delta} - \mu M^{(11)}_{\alpha\delta}) \nabla_\delta (\bar{\mu}) \\ &+ (M^{(22)}_{\alpha\delta} - \mu M^{(21)}_{\alpha\delta} - \mu M^{(12)}_{\alpha\delta} + \mu^2 M^{(11)}_{\alpha\delta}) \nabla_\delta \left(\frac{1}{T} \right) \end{aligned}$$

This equation for $J_{\alpha,Q}$ now has the same form as the currents in (3.9.13). Thus we have the relationships

$$\begin{aligned} L^{(11)} &= M^{(11)} \\ L^{(21)} &= L^{(12)} = M^{(12)} - \mu M^{(11)} \\ L^{(22)} &= M^{(22)} - 2\mu M^{(12)} + \mu^2 M^{(11)} \end{aligned} \tag{3.9.14}$$

A third possible definition of heat or energy current is the choice

$$\mathbf{J}_W = \mathbf{J}_E + eV\mathbf{J} = \mathbf{J}_Q + \bar{\mu}\mathbf{J} \tag{3.9.15}$$

Here one includes the fact that the potential $V(r)$ may not be constant in the sample, so that a charged particle which moves will acquire a change in potential energy. We leave as a homework assignment the problem of determining the forces which go with this choice of energy current.

The thermal conductivity is usually measured under conditions of no particle current $\mathbf{J} = 0$. From (3.9.9), this condition leads to

$$\frac{e}{T}\boldsymbol{\nabla}V + \boldsymbol{\nabla}\left(\frac{\mu}{T}\right) = \frac{M^{(12)}}{M^{(11)}}\boldsymbol{\nabla}\left(\frac{1}{T}\right)$$

which for the energy current gives

$$\mathbf{J}_E = \boldsymbol{\nabla}\left(\frac{1}{T}\right)\left[M^{(22)} - \frac{(M^{(12)})^2}{M^{(11)}}\right]$$

We are, for the moment, treating the M as scalar quantities, which is permissible if all the forces are in the same symmetry direction. Since the thermal conductivity K is usually written as

$$\mathbf{J}_E = -K\boldsymbol{\nabla}T$$

this gives

$$K = \frac{1}{T^2}\left[M^{(22)} - \frac{(M^{(12)})^2}{M^{(11)}}\right] \tag{3.9.16}$$

Equations (3.9.4) and (3.9.5) and the constraint $\partial S/\partial t > 0$ can be used to show that the right-hand side is always positive (Kubo, 1959). If there is no particle flow, then $\mathbf{J}_Q = \mathbf{J}_E$. Thus we should be able to obtain the thermal conductivity from (3.9.13) in a similar fashion. The restriction that no current flows gives

$$\frac{1}{T}\boldsymbol{\nabla}\bar{\mu} = \frac{L^{(12)}}{L^{(11)}}\boldsymbol{\nabla}\left(\frac{1}{T}\right) \tag{3.9.17}$$

and the heat current is

$$\mathbf{J}_Q = \boldsymbol{\nabla}\left(\frac{1}{T}\right)\left[L^{(22)} - \frac{(L^{(12)})^2}{L^{(11)}}\right]$$

The thermal conductivity is

$$K = \frac{1}{T^2}\left[L^{(22)} - \frac{(L^{(12)})^2}{L^{(11)}}\right]$$

It is easy to use the identities (3.9.14) to show that this definition is consistent with (3.9.16).

The electrical conductivity is usually defined when there is no temperature gradient $\nabla T = 0$ and no concentration gradient $\nabla\mu = 0$. This gives

$$\sigma = \frac{e^2}{T} L^{(11)}$$

The coefficient $L^{(12)}$ may also be measured, since it is the thermoelectric coefficient. If a solid has a temperature gradient ΔT and no particle currents $\mathbf{J} = 0$ and no concentration gradients are allowed ($\nabla\mu = 0$), then a voltage difference ΔV is measured which is proportional to ΔT and is given by (3.9.17):

$$\Delta V = -\frac{1}{eT}\frac{L^{(12)}}{L^{(11)}}\Delta T$$

This equation is the definition of the thermoelectric coefficient S, which is also called the thermopower:

$$S = \frac{\Delta V}{\Delta T} = -\frac{1}{eT}\frac{L^{(12)}}{L^{(11)}}$$

The thermopower S may have either sign. In ion diffusion models in solids, the sign of the thermopower provides a direct determination of the sign of the charge on the diffusing ion (Girvin, 1978). The correlation function is also related to the Peltier coefficient (Barnard, 1972).

C. Correlation Functions

The coefficients $M^{(ij)}$ and $L^{(ij)}$ are correlation functions of current operators. They are expressed by formulas similar to those of the preceding section. One wishes to obtain the measurable currents $\mathbf{J}_i$ by evaluating the expectation value of the operator $\mathbf{J}_i$ of the same quantity:

$$\mathbf{J}_i = \langle \mathbf{j}_i \rangle$$

The derivation is identical to the steps used to get (3.8.17). One introduces a term F into the Hamiltonian, and this term causes the change in the density matrix. In direct analogy with (3.8.17), we have

$$\mathbf{J}_i = -\int_0^\infty dte^{-st} \int_0^\beta d\beta' \operatorname{Tr}\left[\varrho_0 \frac{\partial F}{\partial t}(-t - i\beta')\mathbf{j}_i(\mathbf{r})\right] \tag{3.9.18}$$

The only remaining step is to make the proper choice for $\partial F/\partial t$. It is the time rate of change of the energy of the system. At least it is the dissipative part brought on by the transport and forces. We identify this with the heat production or equivalently with the irreversible rate of change of entropy S:

$$\frac{\partial F}{\partial t} = \frac{dQ}{dt} = T\frac{\partial S}{\partial t} = \frac{1}{\beta}\sum_l \mathbf{j}_l \cdot \mathbf{X}_l$$

The small j_i symbols for current operators are used here, since they are operators, not their averages. If this is used in (3.9.18),

$$J_{i\beta} = -\frac{1}{\beta}\int_0^\infty dte^{-st} \int_0^\beta d\beta' \operatorname{Tr}\left[\varrho_0 \sum_l j_{l,\alpha}(-t - i\beta')X_{l,\alpha} j_{i,\beta}\right]$$

Since the forces $\mathbf{X}_l$ are constants, they may be removed from the integrals. Thus we have that $\mathbf{J}_i$ is proportional to $\mathbf{X}_l$, and the constants of proportionality are just the transport coefficient:

$$Z_{\beta\alpha}^{(il)} = -\frac{1}{\beta}\int_0^\infty dte^{-st} \int_0^\beta d\beta' \operatorname{Tr}[\varrho_0 j_{l,\alpha}(-t - i\beta')j_{i,\beta}(r)] \tag{3.9.19}$$

Equation (3.9.19) is a very simple result. The transport coefficient between force $\mathbf{X}_l$ and current $\mathbf{J}_i$ is just the current-current correlation function of j_l and j_i. For example, the coefficients in (3.9.13) are

$$L^{(11)} = -\frac{1}{\beta}\int_0^\infty dte^{-st} \int_0^\beta d\beta' \operatorname{Tr}[\varrho_0 j_\alpha(-t - i\beta')j_\beta]$$

$$L^{(12)} = -\frac{1}{\beta}\int_0^\infty dte^{-st} \int_0^\beta d\beta' \operatorname{Tr}[\varrho_0 j_{Q,\alpha}(-t - i\beta')j_\beta]$$

$$L^{(22)} = -\frac{1}{\beta}\int_0^\infty dte^{-st} \int_0^\beta d\beta' \operatorname{Tr}[\varrho_0 j_{Q,\alpha}(-t - i\beta')j_{Q,\beta}]$$

The $M^{(il)}$ coefficients in (3.9.9) are similar; the only difference is that $\mathbf{j}_E$ is used instead of $\mathbf{j}_Q$. Since $\mathbf{j}_Q = \mathbf{j}_E - \mu\mathbf{j}$, it is easy to show that the relations (3.9.14) are automatically satisfied. It is also easy to prove the Onsager relations. First one writes the correlation function in the $|n\rangle$ and $|m\rangle$ notation of (3.8.13):

$$Z^{(il)} = \pi \sum_{n,m} e^{-\beta E_n} \langle n \mid j_l \mid m\rangle\langle m \mid j_i \mid n\rangle \delta(E_n - E_m)$$

$$Z^{(li)} = \pi \sum_{n,m} e^{-\beta E_n} \langle n \mid j_i \mid m\rangle\langle m \mid j_l \mid n\rangle \delta(E_n - E_m)$$

These two expressions can be shown to be identical by interchanging dummy summation variables $|n\rangle$ and $|m\rangle$ and using $E_n = E_m$.

These correlation functions can also be evaluated by using the Matsubara formalism. One just evaluates the correlation function

$$Z^{(il)} = \frac{i}{i\omega\beta} \int_0^\beta d\tau e^{i\omega\tau} \langle T_\tau j_l(\tau) j_i(0)\rangle$$

by the usual S-matrix techniques. After a suitable result is obtained, one gets the retarded function by letting $i\omega \to \omega + i\delta$ and then zero frequency by letting $\omega \to 0$. It was shown in Sec. 3.7 that these steps are identical to the evaluation of (3.9.19).

PROBLEMS

1. Prove the following moments of the spectral functions:

$$\int_{-\infty}^{\infty} \frac{d\omega}{2\pi} n_F(\omega)\omega^2 A(\mathbf{p}, \omega) = -\langle [H, C_\mathbf{p}^\dagger][H, C_\mathbf{p}]\rangle$$

$$\int_{-\infty}^{\infty} \frac{d\omega}{2\pi} \omega n_B(\omega) B(\mathbf{q}, \omega) = \langle A(\mathbf{q})[H, A(-\mathbf{q})]\rangle$$

Evaluate these moments for noninteracting particles and phonons.

2. Take the Fourier transform of the correlation function (3.1.1). Define its spectral function as $C(\mathbf{p}, \omega)$. Show that

$$A(\mathbf{p}, \omega) = (1 + e^{-\beta\omega}) C(\mathbf{p}, \omega)$$

What is the result when (3.1.1) is time-ordered?

3. Let $U(t)$ be $\varrho_{\mathbf{q}}(t)$ = density operator of wave vector $\mathbf{q}$. Prove that

$$S(\mathbf{k}) = \int_{-\infty}^{\infty} \frac{d\omega}{2\pi} n_B(\omega) S(\mathbf{k}, \omega)$$

where $S(\mathbf{k}) = \langle \varrho_{\mathbf{k}}^{\dagger} \varrho_{\mathbf{k}} \rangle$ is the pair distribution function and $S(\mathbf{k}, \omega)$ is the spectral function for the density–density correlation function.

4. At zero temperature,

a. Show that $n_{\mathbf{p}} = \int_{-\infty}^{0} (d\omega/2\pi) A(\mathbf{p}, \omega)$.

b. Evaluate this for a Lorentzian: $A(\mathbf{p}, \omega) = 2\Gamma_{\mathbf{p}}/[(\omega - \xi_{\mathbf{p}})^2 + \Gamma_{\mathbf{p}}^2]$.

c. Plot n_p vs. $\xi_{\mathbf{p}}$ for cases (1) $\Gamma_{\mathbf{p}} = \Gamma_0$ = constant and (2) $\Gamma_{\mathbf{p}} = [\Gamma_0 \mid \xi_{\mathbf{p}} \mid /(\Gamma_0^2 + \xi_{\mathbf{p}}^2)^{1/2}]$.

5. Let $C_{\mathbf{p}}^{\dagger}$ be a creation operator describing a spinless boson whose number is conserved; this model is sometimes applied to ^{4}He atoms if one can ignore the internal structure (excitations, degrees of freedom). The particles will have a chemical potential μ. Define

$$\mathscr{G}(\mathbf{p}, \tau) = -\langle T_\tau e^{\beta(\Omega - K)} C_{\mathbf{p}}(\tau) C_{\mathbf{p}}^{\dagger}(0) \rangle$$

$$K = H - \mu N$$

a. Show that $\mathscr{G}^{(0)}(\mathbf{p}, \omega) = 1/(i\omega - \xi_{\mathbf{p}})$.

b. What is the definition of the retarded function?

6. Prove the following general theorem for particles which are fermions or bosons:

$$G_{\text{ret}}(\mathbf{p}, t) = -i\theta(t) \int_{-\infty}^{\infty} \frac{d\omega}{2\pi} A(\mathbf{p}, \omega) e^{-i\omega t}$$

7. Evaluate the thermodynamic average of

$$\left\langle \sum_k a_{\mathbf{k}}^{\dagger}(\tau) a_{\mathbf{k}}(\tau) \sum_{\mathbf{p}} a_{\mathbf{p}}^{\dagger}(\tau') a_{\mathbf{p}}(\tau') \right\rangle$$

by doing it (a) directly and (b) using Wick's theorem. Both should give the same answer. Terms with $\mathbf{p} = \mathbf{k}$ have to be treated carefully. Do this problem for the case where the operators represent bosons.

8. Use the rules for constructing diagrams to write down the two two-phonon self-energy diagrams of the electron: $\Sigma^{(2)}(\mathbf{p}, ip_n)$.

9. The two diagrams in Fig. 3.13 contribute to the electron self-energy and arise from the electron-electron interaction. Use the rules for constructing dia-

FIGURE 3.13
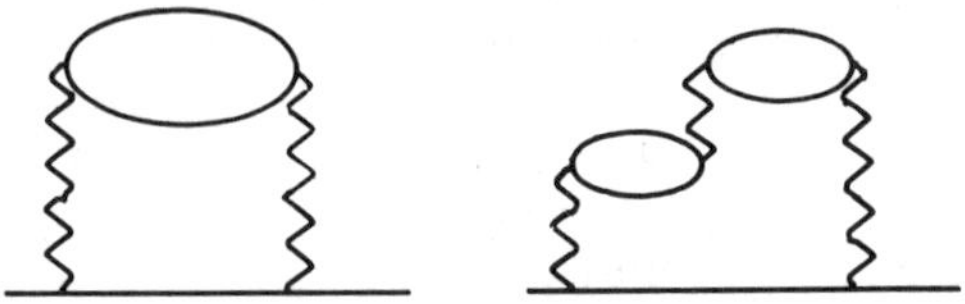

grams, and write down these contributions in the Matsubara formalism. However, do not evaluate them—i.e., do not do either the frequency summations or the wave vector integrals.

10. How is (3.4.16) altered if the internal line is a Coulomb interaction?

11. Derive the summation formulas (3.5.2) and (3.5.3). Also evaluate

$$\frac{1}{\beta}\sum_{i\omega_n} \mathscr{D}^{(0)}(\mathbf{q}, i\omega_n)\mathscr{D}^{(0)}(\mathbf{k}, i\omega_n + iq_n)$$

12. Evaluate the Matsubara frequency summation for

$$\frac{1}{\beta}\sum_{iq_n} \mathscr{G}^{(0)}(\mathbf{p}, ip_n + iq_n)\mathscr{D}(\mathbf{q}, iq_n)$$

13. Prove that

$$\Omega_0 = \frac{2}{\beta}\sum_{\mathbf{p}, ip_n} e^{ip_n\tau} \ln \mathscr{G}^{(0)}(\mathbf{p}, ip_n), \qquad \tau \to 0^+$$

where $\mathscr{G}^{(0)}$ is for electrons and Ω_0 is the free-electron part of the thermodynamic potential (Luttinger and Ward, 1960).

14. Consider the terms W_8 for the electron–phonon interaction. How many times do each of the terms appear: U_4^2, U_2^4, U_6U_2, $U_4U_2^2$?

15. Evaluate $W_1(\mathbf{p}, \tau)$ in (3.6.26) using $\hat{V}(\tau_1)$ as the electron–electron interaction. Which terms would not make sense if one tried to write this as $\exp(F_1)$?

16. Show that at finite temperatures the real part of the conductivity can also be written as

$$\mathrm{Re}[\sigma_{\alpha\beta}(\omega)] = \frac{1 - e^{-\beta\omega}}{2\omega\nu}\int_{-\infty}^{\infty} dt e^{i\omega t}\langle j_\alpha^\dagger(\mathbf{q}, t) j_\beta(\mathbf{q}, 0)\rangle$$

17. Find the forces $\mathbf{X}_i$ which go with the currents $\mathbf{J}$, $\mathbf{J}_w$ in (3.9.15). What are the Kubo formulas for the transport coefficients?

18. Show that the relations (3.9.14) are consistent with (3.9.19).

19. Show that the second-order phonon self-energy, from the electron–phonon

interaction, may be written as $\Pi(\mathbf{q}, \omega) = M_q^2 P^{(1)}(\mathbf{q}, \omega)$:

$$P^{(1)}(\mathbf{q}, \omega) = 2 \int \frac{d^3k}{(2\pi)^3} n_{\mathbf{k}} \left(\frac{1}{\omega - \varepsilon_{\mathbf{k}} + \varepsilon_{\mathbf{k}+\mathbf{q}} + i\delta} - \frac{1}{\omega + \varepsilon_{\mathbf{k}} - \varepsilon_{\mathbf{k}+\mathbf{q}} + i\delta} \right)$$

Evaluate this expression by doing the wave vector integrals. Assume a free-electron gas of Fermi momentum k_F at zero temperature. Give both the real and imaginary parts of $P^{(1)}$.

20. Derive the change in the thermodynamic potential when the electron self-energy is $\Sigma(\eta) = \eta \Sigma_0$.

21. Derive the change in the thermodynamic potential when the electron self-energy is $\Sigma(\eta) = \eta^2 \Sigma_0$.

22. The thermodynamic potential Ω, internal energy U, and entropy S are related by $\Omega = U - TS - \mu N$, where

$$S = -\left(\frac{d\Omega}{dT} \right)_{\mu,\nu}$$

$$C_V = T\left(\frac{dS}{dT} \right)_{\mu,\nu}$$

Derive the definitions of S, U, and C_V for the free-electron gas. Show that the specific heat is proportional to T at low T, and find the constant of proportionality.

23. Derive the relationship between the total internal and external electric fields by solving Maxwell's equations.

24. Show that Wick's theorem is invalid when H_0 contains particle–particle interactions. As an example, evaluate the correlation function W below for a system of two states. Do it exactly and also by Wick's theorem.

$$H_0 = \sum_{\alpha=1}^{2} \xi_\alpha C_\alpha^\dagger C_\alpha + U n_1 n_2$$

$$W = \langle T_\tau \sum_{\alpha=1}^{2} C_\alpha^\dagger(\tau_1) C_\alpha(\tau_1) \sum_{\delta=1}^{2} C_\delta^\dagger(\tau_2) C_\delta(\tau_2) \rangle$$

25. Derive the equilibrium real time Green's functions in (3.7.3) starting from (3.7.1)

26. In one dimension find the Wigner distribution function for the simple harmonic oscillator of mass m and frequency ω_0. In center-of-mass coordinates show that its spectral function is given in terms of Laguerre polynomials

$$A(k, \omega; X) = 4\pi e^{-\Delta} \sum_n (-1)^n L_n(2\Delta) \delta[\omega - \omega_0(n + 1/2)]$$

$$\Delta = X^2/l^2 + k^2 l^2, \qquad l^2 = 1/m\omega_0$$

27. Prove the following identity:

$$\Sigma_{\text{ret}}(\mathbf{k}, \omega; T) = i \int_{-\infty}^{\infty} \frac{d\omega'}{\omega - \omega' + i\delta} [\Sigma^{>}(\mathbf{k}, \omega'; T) - \Sigma^{<}(\mathbf{k}, \omega'; T)]$$

This result is easy to prove in equilibrium. Can you also prove it for systems out of equilibrium?

Chapter 4

Exactly Solvable Models

Every many-body theorist should be knowledgeable about the available exactly solvable models. First, there are not many of them. Second, they are useful for gaining insight into many-particle systems. If the problem to be solved can be related to an exactly solvable one, however vaguely, one can usually gain some insight.

Several models are presented here to introduce the discussion of Green's functions. In each case, we shall solve the model without Green's functions and again with them. Models which are exactly solvable may be solved by a variety of techniques.

The models we shall present are only a small sampling of the possible exactly solvable ones. Others have been presented by Schweber (1961). Many one-dimensional models may be solved exactly (Lieb and Mattis, 1966).

4.1. POTENTIAL SCATTERING

A very simple problem is an impurity potential $V(r)$ in an otherwise free-particle system. All other interactions are ignored except that of the free particles with the impurity, which is assumed to be at the origin. The potential is assumed to have no internal structure—spin, excited states, etc. It is a simple function of position r of the particle from the origin and is spherically symmetric.

The wave functions ψ_λ and energy ε_λ for each particle may be obtained by solving a one-electron Schrödinger equation:

$$H\psi_\lambda = \left[-\frac{\hbar^2}{2m} \nabla^2 + V(r) \right] \psi_\lambda = \varepsilon_\lambda \psi_\lambda \tag{4.1.1}$$

This may be accomplished by numerical means if the Schrödinger equation is not solvable analytically.

In many-body theory, the impurity problem is usually encountered as a scattering center. If the free-particle states are plane waves,

$$\psi(\mathbf{r}) = \frac{1}{\nu^{1/2}} \sum_{\mathbf{k}} C_{\mathbf{k}} e^{i\mathbf{k}\cdot\mathbf{r}}$$

the Hamiltonian is expressed as operators

$$\begin{aligned} H &= \sum_{\mathbf{k}} \varepsilon_k C_{\mathbf{k}}^{\dagger} C_{\mathbf{k}} + \frac{1}{\nu} \sum_{\mathbf{kk}'} V_{\mathbf{kk}'} C_{\mathbf{k}}^{\dagger} C_{\mathbf{k}'} \\ V_{\mathbf{kk}'} &= \int d^3r V(r) e^{-i\mathbf{r}\cdot(\mathbf{k}-\mathbf{k}')} = V(\mathbf{k}-\mathbf{k}') \end{aligned} \tag{4.1.2}$$

The last term is the potential scattering of the free particles. The object is to diagonalize the Hamiltonian (4.1.2). Of course, we already know that the solutions are given by (4.1.1). However, the problem is not entirely solved. Since the equation is second order, it has two solutions. These may be chosen as the ingoing and outgoing waves. Alternately, these two may be combined to give standing waves. We shall explain how these choices are related to the scattering problem implied in (4.1.2).

Our discussion will benefit from the fact that we already know that the solution is of the form (4.1.1). Thus we next need to know the boundary conditions on the wave function. Thus we consider the integral equation for the wave function:

$$\begin{aligned} \psi_{\mathbf{k}}(\mathbf{r}) &= \phi_{\mathbf{k}}(\mathbf{r}) + \sum_{\mathbf{k}'} \frac{\phi_{\mathbf{k}'}(\mathbf{r})}{\varepsilon_k - \varepsilon_{k'}} \int d^3r' \phi_{\mathbf{k}'}^{*}(\mathbf{r}') V(r') \psi_{\mathbf{k}}(\mathbf{r}') \\ \phi_{\mathbf{k}}(\mathbf{r}) &= \frac{1}{\nu^{1/2}} e^{i\mathbf{k}\cdot\mathbf{r}} \\ \varepsilon_k &= \frac{k^2}{2m} \end{aligned} \tag{4.1.3}$$

This form of the integral equation is valid for the free-particle states with energy $\varepsilon_k = k^2/2m$. The Schrödinger equation (4.1.1) may also have bound states. We shall not discuss these explicitly. They obey an equation similar to (4.1.3), but the energy is changed to the binding energy $\varepsilon_k \to -\varepsilon_B$ ($\varepsilon_B > 0$), and the $\phi_{\mathbf{k}}(\mathbf{r})$ term on the right is absent.

First we must prove that (4.1.3) is equivalent to (4.1.1). We operate on both sides of the equation by $H_0 - \varepsilon_k$,

$$(H_0 - \varepsilon_k)\psi_{\mathbf{k}} = (H_0 - \varepsilon_k)\phi_{\mathbf{k}} + \sum_{k'} \frac{(H_0 - \varepsilon_k)\phi_{\mathbf{k}'}(\mathbf{r})}{\varepsilon_k - \varepsilon_{k'}} \int d^3r' \phi^*_{\mathbf{k}'}(\mathbf{r}')V(r')\psi_{\mathbf{k}}(\mathbf{r}')$$

and use the fact that

$$(H_0 - \varepsilon_k)\phi_{\mathbf{k}} = -\left(\frac{\hbar^2}{2m}\nabla^2 + \varepsilon_k\right)\phi_{\mathbf{k}} = 0$$

to get

$$(H_0 - \varepsilon_k)\psi_{\mathbf{k}} = -\sum_{\mathbf{k}'} \phi_{\mathbf{k}'}(\mathbf{r}) \int d^3r' \phi^*_{\mathbf{k}'}(\mathbf{r}')V(r')\psi_{\mathbf{k}}(\mathbf{r}')$$

If we use the completeness relation for the summation over the set of states

$$\sum_{\mathbf{k}'} \phi_{\mathbf{k}'}(\mathbf{r})\phi^*_{\mathbf{k}'}(\mathbf{r}') = \delta^3(\mathbf{r} - \mathbf{r}')$$

then the right-hand side is

$$-V(r)\psi_{\mathbf{k}}(\mathbf{r})$$

If this is transferred to the left, we have the final equation

$$(H_0 + V - \varepsilon_k)\psi_{\mathbf{k}}(\mathbf{r}) = 0$$

which is the desired answer. The above equation is precisely (4.1.1). Thus we have shown that (4.1.3) is equivalent to the usual Schrödinger equation. Next we must discuss boundary conditions. The differential equation (4.1.1) is second order, so there are two independent solutions. As a mathematical problem, we may combine these two solutions in any possible way, depending on the choice of boundary conditions. However, as a physics problem, we usually choose the combination of the two solutions to correspond to a desirable physical situation. One possible choice is to have the wave function a standing wave. This leads to the reaction matrix equation. Other possible choices are to have the wave function an incoming wave or an outgoing wave, which leads to T-matrix theory. From scattering theory, and causality, we often wish to take the outgoing wave. This has the asymptotic form $\exp(ikr)/r$ at large distance. This choice of phase [rather than $\exp(-ikr)/r$] arises from the physics convention that the time development is $\exp(-i\omega t)$. The integral equation (4.1.3) is a convenient starting point for this discussion of boundary conditions, since the various choices of standing, outgoing, or ingoing waves are determined only by the complex part of the energy denominator. The factor is $\varepsilon_k - \varepsilon_{k'}$ for standing waves, so that the principal part is chosen for the denominator. The factor is $\varepsilon_k - \varepsilon_{k'} + i\delta$ for outgoing waves, and $\varepsilon_k - \varepsilon_{k'} - i\delta$ for ingoing waves.

A. Reaction Matrix

Here one chooses the energy denominator to be real and given by the principal part. The free-particle Green's function is then defined as

$$G_0(\mathbf{k}, \mathbf{r} - \mathbf{r}') = P \sum_{\mathbf{k}'} \frac{\phi_{\mathbf{k}'}(\mathbf{r})\phi^*_{\mathbf{k}'}(\mathbf{r}')}{\varepsilon_k - \varepsilon_{k'}} = \frac{P}{\nu} \sum_{k'} \frac{e^{i\mathbf{k}'\cdot(\mathbf{r}-\mathbf{r}')}}{\varepsilon_k - \varepsilon_{k'}}$$
$$= P \int \frac{d^3k'}{(2\pi)^3} \frac{e^{i\mathbf{k}'\cdot(\mathbf{r}-\mathbf{r}')}}{\varepsilon_k - \varepsilon_{k'}}$$

The last step on the right takes $\nu \to \infty$ and changes the summation into an integration over wave vectors. The integral is standard and gives

$$G_0(\mathbf{k}, \mathbf{r} - \mathbf{r}') = -\pi\varrho(k) \frac{\cos k \mid \mathbf{r} - \mathbf{r}' \mid}{k \mid \mathbf{r} - \mathbf{r}' \mid}$$

We have introduced a factor $\varrho(k)$ which is the density of states of the particles:

$$\varrho(k) = \int \frac{d^3k'}{(2\pi)^3} \delta(\varepsilon_k - \varepsilon_{k'}) = \frac{km}{2\hbar^2\pi^2}$$

For particles with spin, this is the density of states per spin configuration. In electron systems, for example, the net density of states is twice this when spin degeneracy is considered. The Green's function may be expanded as a function of $\mathbf{r}$ and $\mathbf{r}'$:

$$G_0(\mathbf{k}, \mathbf{r} - \mathbf{r}') = +\pi\varrho(k) \sum_l (2l + 1)P_l(\hat{r} \cdot \hat{r}')j_l(kr)\eta_l(kr'), \qquad r < r'$$
$$= +\pi\varrho(k) \sum_l (2l + 1)P_l(\hat{r} \cdot \hat{r}')j_l(kr_<)\eta_l(kr_>) \tag{4.1.4}$$

where the notation $r_<$ and $r_>$ means the lesser and greater of r and r', respectively. The $j_l(kr)$ and $\eta_l(kr)$ are spherical Bessel functions of the first and second kind, and the $P_l(\cos\theta)$ are Legendre functions. These results may be put into the integral equation (4.1.3):

$$\psi_{\mathbf{k}}(\mathbf{r}) = \phi_{\mathbf{k}}(\mathbf{r}) + \int d^3r' G_0(\mathbf{k}, \mathbf{r} - \mathbf{r}')V(r')\psi_k(\mathbf{r}') \tag{4.1.4a}$$

At this point it is convenient to reduce the integral equation to angular momentum components l. A plane wave may be expanded as

$$e^{i\mathbf{k}\cdot\mathbf{r}} = \sum_l (2l+1) i^l P_l(\hat{k}\cdot\hat{r}) j_l(kr)$$

A similar expansion is used for the actual wave function:

$$\psi_{\mathbf{k}}(\mathbf{r}) = \sum_l (2l+1) i^l P_l(\hat{k}\cdot\hat{r}) R_l(kr)$$

The radial function $R_l(kr)$ is the quantity we are trying to determine. Of course, it satisfies a radial Schrödinger equation of the form

$$-\frac{\hbar^2}{2m}\left[\frac{1}{r^2}\frac{\partial}{\partial r} r^2 \frac{\partial R}{\partial r} - \frac{l(l+1)}{r^2} R\right] + [V(r) - \varepsilon_k]R = 0 \qquad (4.1.5)$$

but this does not yet determine the boundary conditions. These are obtained by substituting these forms into (4.1.4a). Each angular momentum component is selected by multiplying the equation by $P_l(\hat{k}\cdot\hat{r})$ and then integrating over all spatial angles. One uses the fact that

$$\int d\Omega_r P_l(\hat{k}\cdot\hat{r}) P_m(\hat{r}\cdot\hat{p}) = \frac{4\pi}{2l+1}\,\delta_{lm} P_l(\hat{k}\cdot\hat{p})$$

to reduce the equation down to one which involves only the same angular momentum component:

$$\begin{aligned} R_l(kr) &= j_l(kr) + 4\pi^2\varrho(k)\int_0^\infty r'^2\,dr'\,j_l(kr_<)\eta_l(kr_>)V(r')R_l(kr') \\ &= j_l(kr) + 4\pi^2\varrho(k)\Big[\eta_l(kr)\int_0^r r'^2\,dr'\,j_l(kr')V(r')R_l(kr') \\ &\quad + j_l(kr)\int_r^\infty r'^2\,dr'\eta_l(kr')V(r')R_l(kr')\Big] \end{aligned} \qquad (4.1.6)$$

It is important that we assumed the potential was spherically symmetric. Otherwise the scattering term would mix angular momentum components, which usually makes the equation much harder to solve.

The solution is examined in the limit as $kr \to \infty$. From the differential equation (4.1.5) we can show that the radial wave function must asymptotically approach the value

$$\lim_{kr\to\infty} R_l(kr) \to \frac{C_l(k)}{kr}\sin\left[kr + \delta_l(k) - \frac{l\pi}{2}\right] \qquad (4.1.7)$$

The prefactor C_l is yet to be determined. The asymptotic limit of the integral equation (4.1.6) is

$$\lim R_l(kr) \rightarrow j_l(kr) + D_l(k)\eta_l(kr)$$
$$\rightarrow \frac{\sin(kr - l\pi/2)}{kr} - D_l \frac{\cos(kr - l\pi/2)}{kr} \tag{4.1.8}$$

where

$$D_l(k) = 4\pi^2 \varrho(k) \int_0^\infty r'^2\, dr' j_l(kr')V(r')R_l(kr') \tag{4.1.9}$$

The potential $V(r)$ is assumed to be of short range: It falls off faster than r^{-2} at large distances. This is necessary for the integral in (4.1.9) to be well defined. The scattering is then described by a phase shift $\delta_l(k)$ which depends on angular momentum and wave vector. The two asymptotic expansions (4.1.7) and (4.1.8) must be identical, which is accomplished by setting

$$D_l(k) = -\tan \delta_l = 4\pi^2 \varrho(k) \int_0^\infty r'^2\, dr'\, j_l(kr')V(r')R_l(kr') \tag{4.1.10}$$

so that (4.1.8) becomes

$$\lim_{kr\to\infty} R_l(kr) \rightarrow \frac{1}{kr \cos \delta}\left[\cos \delta \sin\left(kr - \frac{l\pi}{2}\right) + \sin \delta \cos\left(kr - \frac{l\pi}{2}\right)\right]$$
$$= \frac{\sin[kr + \delta_l - (l\pi/2)]}{kr \cos \delta}$$

Thus we have shown that the normalization coefficient in (4.1.7) is

$$C_l = \frac{1}{\cos \delta_l}$$

When solving the radial wave function (4.1.5), the solution obtained is well behaved at the origin. This solution is followed outward in r until the region is reached where $V(r) \simeq 0$ and the centrifugal barrier $\hbar l(l+1)/(2mr^2)$ is small. Then the solution has the form (4.1.7) with $C_l = \{\cos[\delta_l(k)]\}^{-1}$. This provides the proper normalization of the wave function. The wave function is now properly normalized at large r, and by following it back toward the origin it is normalized everywhere. These steps provide the proper solution to the scattering equation:

$$\psi_{\mathbf{k}}(\mathbf{r}) = \phi_{\mathbf{k}}(\mathbf{r}) + P \sum_{k'} \frac{\phi_{k'}(r)}{\varepsilon_k - \varepsilon_{k'}} \int d^3r' \phi^*_{\mathbf{k}'}(\mathbf{r}')V(r')\psi_{\mathbf{k}}(\mathbf{r}')$$

The *reaction matrix* is defined as the quantity

$$R_{\mathbf{k}'\mathbf{k}} = \int d^3r \phi^*_{\mathbf{k}'}(r) V(r) \psi_{\mathbf{k}}(r) \tag{4.1.11}$$

It may be expanded in angular momentum states by using the expansions for the wave function and the plane wave. This gives

$$R_{\mathbf{k}'\mathbf{k}} = 4\pi \sum_l (2l+1) P_l(\hat{k} \cdot \hat{k}') R_l(k', k')$$

where the radial integral is

$$R_l(k', k) = \int_0^\infty r^2 \, dr j_l(k'r) V(r) R_l(kr) \tag{4.1.12}$$

It is defined for the general case where $k' \neq k$. Of course, if they happen to be equal, then the answer is just (4.1.10):

$$R_l(k, k) = -\frac{\tan \delta_l}{4\pi^2 \varrho(k)} = -\frac{\hbar^2 \tan \delta_l}{2mk}$$

where we have used the expression for the density of states $\varrho(k)$. We repeat that the reaction matrix is only proportional to $\tan \delta_l$ for the diagonal terms with $k = k'$. Otherwise, one must do the integral in (4.1.12).

The reaction matrix obeys an integral equation which is deduced by putting (4.1.11) into the integral equation (4.1.3) for $\psi_{\mathbf{k}}(\mathbf{r})$:

$$R_{\mathbf{k}'\mathbf{k}} = V_{\mathbf{k}'\mathbf{k}} + P \sum_{k_1} \frac{V_{\mathbf{k}'\mathbf{k}_1} R_{\mathbf{k}_1\mathbf{k}}}{\varepsilon_k - \varepsilon_{k_1}} \tag{4.1.13}$$

This integral equation is frequently found in scattering problems. Now we know how to solve it. For each angular momentum state, the one-particle radial equation (4.1.5) is solved with the boundary condition that the asymptotic limit go to (4.1.7). This wave function is used in (4.1.12), which determines the answer. It is a simple procedure to solve (4.1.13). The important point is that the wave functions must be normalized correctly. The designation of principal parts on the energy denominator in (4.1.13) specifies the unique way this is to be done.

B. *T* Matrix

The other common choice of boundary conditions uses outgoing waves. which is accomplished by adding an infinitesimal complex part $i\delta$ to the energy denominator in (4.1.3). Thus we have the integral equation

$$\psi_{\mathbf{k}}(\mathbf{r}) = \phi_{\mathbf{k}}(\mathbf{r}) + \int d^3r' G_0(\mathbf{k}, \mathbf{r} - \mathbf{r}')V(r')\psi_{\mathbf{k}}(\mathbf{r}') \tag{4.1.14}$$

where the Green's function for outgoing waves is

$$G_0(k, r - r') = \frac{1}{\nu} \sum_{\mathbf{k}'} \frac{e^{-i\mathbf{k}' \cdot (\mathbf{r}-\mathbf{r}')}}{\varepsilon_k - \varepsilon_{k'} + i\delta}$$

This integral is evaluated to give

$$G_0(k, R) = -\pi\varrho(k)\frac{e^{ikR}}{kR} = -\pi\varrho(k)\left(\frac{\cos kR}{kR} + i\frac{\sin kR}{kR}\right)$$

The cosine term was evaluated earlier in (4.1.4). The sine term is

$$\frac{\sin(k\,|\,\mathbf{r} - \mathbf{r}'\,|)}{k|\mathbf{r} - \mathbf{r}'|} = \sum_l (2l + 1)P_l(\hat{r} \cdot \hat{r}')j_l(kr)j_l(kr')$$

These results can be put into the integral equation (4.1.14) to give an equation for the wave function. This may be reduced to each angular momentum state:

$$\tilde{R}_l(kr) = j_l(kr) + 4\pi^2\varrho(k) \int_0^\infty r'^2\, dr' [j_l(kr_<)\eta_l(kr_>) - ij_l(kr)j_l(kr')] \times V(r')\tilde{R}_l(kr')$$

The radial part of the wave function is denoted by $\tilde{R}_l$. It will be a different wave function from that found for the standing-wave boundary conditions. The difference between the two is only a different choice of normalization coefficient $C_l(k)$ in (4.1.7). This different choice is what is to be determined. The above integral equation may be rewritten as

$$\begin{aligned}\tilde{R}_l(kr) = j_l(kr)\Big[1 &+ 4\pi^2\varrho(k) \int_r^\infty r'^2\, dr' \eta_l(kr')V(r')\tilde{R}_l(kr') \\ &- 4\pi^2 i\varrho(k) \int_0^\infty r'^2\, dr' j_l(kr')V(r')\tilde{R}_l(kr')\Big] + 4\pi^2\varrho(k)\eta_l(kr) \\ &\times \int_0^r dr' r'^2 j_l(kr')V(r')\tilde{R}_l(kr')\end{aligned}$$

Then one can take the limit $kr \to \infty$ and obtain

$$\begin{aligned}\lim_{kr\to\infty} \tilde{R}_l(kr) &= j_l(kr)[1 - i\tilde{D}_l(k)] + \tilde{D}_l(k)\eta_l(kr) \\ \tilde{D}_l(k) &= 4\pi^2\varrho(k) \int_0^\infty r'^2\, dr' j_l(kr')V(r')\tilde{R}_l(k\; r')\end{aligned} \tag{4.1.15}$$

This has to go to the form (4.1.7) because the radial wave function $\tilde{R}_l(kr)$ also obeys the differential equation (4.1.5). Our result (4.1.15) has that form if we set

$$\tilde{D}_l(k) = -e^{+i\delta_l} \sin \delta_l$$

so that the factor multiplying $j_l(kr)$ is

$$\begin{aligned} 1 - i\tilde{D}_l &= 1 + ie^{+i\delta_l} \sin \delta_l = 1 + \frac{e^{+i\delta_l}}{2} (e^{i\delta_l} - e^{-i\delta_l}) \\ &= e^{+i\delta_l} \cos \delta_l \end{aligned}$$

Thus for the asymptotic limit we get

$$\begin{aligned} \lim_{kr\to\infty} \tilde{R}_l(kr) &= e^{+i\delta_l}[j_l(kr) \cos \delta_l - \eta_l(kr) \sin \delta_l] \\ &= \frac{e^{+i\delta_l}}{kr} \left[\sin\left(kr - \frac{l\pi}{2}\right) \cos \delta + \cos\left(kr - \frac{l\pi}{2}\right) \sin \delta \right] \\ &= e^{+i\delta_l} \frac{\sin}{kr} \left(kr + \delta - \frac{l\pi}{2}\right) \end{aligned}$$

which is indeed the right form, (4.1.7). Thus for outgoing wave boundary conditions, the proper choice of normalization coefficient is

$$C(k) = e^{+i\delta_l}$$

Thus one would proceed by solving (4.1.5) and insisting that it have this asymptotic limit. The easiest way is to solve it first for a real wave function and make this real wave function go to $\sin[(kr + \delta - l\pi/2)/kr]$ (which determines δ_l). Then one can multiply the wave function everywhere by the phase factor $e^{+i\delta_l}$.

The *T matrix* is defined as

$$T_{\mathbf{k}'\mathbf{k}} = \int d^3r \phi_{\mathbf{k}'}{}^*(\mathbf{r}) V(r) \psi_{\mathbf{k}}(\mathbf{r}) = 4\pi \sum_l (2l + 1) P_l(\hat{k} \cdot \hat{k}') T_l(k', k) \tag{4.1.16}$$

$$T_l(k', k) = \int_0^\infty r^2 \, dr j_l(k'r) V(r) \tilde{R}_l(kr)$$

The original integral equation for the wave function (4.1.14) may be used to generate the T-matrix equation

$$T_{\mathbf{k}'\mathbf{k}} = V_{\mathbf{k}'\mathbf{k}} + \sum_{\mathbf{k}_1} \frac{V_{\mathbf{k}'\mathbf{k}_1} T_{\mathbf{k}_1\mathbf{k}}}{\varepsilon_k - \varepsilon_{k_1} + i\delta} \tag{4.1.17}$$

This equation is often encountered in scattering problems. The solution is easy. One solves Schrödinger's equation (4.1.1) for the wave function $\psi_k(r)$. Only the radial part of the wave function is difficult, since one must solve (4.1.5). The solution is normalized by insisting that the radial wave function have the form (4.1.7) at long distance from the potential. The coefficient is $C = e^{+i\delta_l}$. These radial solutions are used in (4.1.16), which then gives an exact solution for the T-matrix equation. This result differs from the reaction matrix result only in the choice of the coefficient C which multiplies the radial wave function. Thus the angular momentum components of the two scattering functions are related by the ratio of these normalization coefficients:

$$\tilde{R}_l(kr) = e^{+i\delta_l} \cos(\delta_l) R_l(kr)$$
$$T_l(k', k) = e^{+i\delta_l(k)} \cos[\delta_l(k)] R_l(k', k)$$

The T matrix is a complex quantity since it was defined with a complex $e^{+i\delta_l}$ phase factor. The diagonal T matrix is

$$T(k, k) = \frac{-1}{4\pi^2 \varrho(k)} e^{+i\delta_l} \sin[\delta_l(k)] = -\frac{1}{2mk} e^{+i\delta_l} \sin \delta_l$$

Again we repeat that this simple form is valid only when $k = k'$. The imaginary part of the diagonal T matrix is

$$-\text{Im}\, T_l(k, k) = \frac{1}{4\pi^2 \varrho(k)} \sin^2(\delta_l)$$
$$-\text{Im}\, T_{\mathbf{kk}} = \frac{1}{\pi \varrho(k)} \sum_l (2l + 1) \sin^2(\delta_l)$$

which is related to the total scattering cross section for the potential:

$$\sigma = \frac{4\pi}{k^2} \sum_l (2l + 1) \sin^2(\delta_l) = -\frac{2}{v_k} \text{Im}\, T_{\mathbf{kk}} \tag{4.1.18}$$

One can also prove the *optical theorem*

$$-2\, \text{Im}\, T_{\mathbf{kk}} = 2\pi \int \frac{d^3k}{(2\pi)^3} \mid T_{\mathbf{k'k}} \mid^2 \delta(\varepsilon_k - \varepsilon_{k'}) \tag{4.1.19}$$

It is important to realize that the reaction matrix is *not* identical to the real part of the T matrix. Either of these two different quantities may be selected as the energy shift of a particle interacting with a potential.

C. Friedel's Theorem

There are several important theorems involving phase shifts. These are conveniently proved by letting the solid have a spherical shape of radius R, with the impurity located at the center of the sphere. The distance R is very large, and eventually we shall take the limit as $R \to \infty$. The reason that we take a finite size sample is that we are going to count nodes in the wave function, so we must start from some point and move inward. We assume, for convenience, that the wave functions vanish at the surface of the sphere.

In the absence of the potential, the solutions which are regular at the origin have a radial part $j_l(kr)$. The condition that they vanish at the surface of the sphere is

$$j_l(k_\alpha R) = 0$$

Since we are usually concerned with small values of l, say $l < 5$, for most applications and since R is very large, we may adequately use the approximation that

$$j_l(k_\alpha R) \to \frac{1}{k_\alpha R} \sin\left(k_\alpha R - \frac{l\pi}{2}\right)$$

so that the condition for the vanishing of the wave function at the surface is

$$k_\alpha R = \left(n + \frac{l}{2}\right)\pi$$

Thus for each value of l, there is a solution for each additional integer n, where each additional integer has a solution with an extra node in the wave function. Similarly, we expand the continuum wave functions (i.e., for $\varepsilon_k > 0$) in the presence of the impurity at large r,

$$R_l(kr) \to \frac{C_l}{kr} \sin\left(kr + \delta_l - \frac{l\pi}{2}\right)$$

and observe that the solution is obtained for each k_n which satisfies the equation

$$k_n R + \delta_l(k_n) = (n + \tfrac{1}{2}l)\pi$$

We are trying to count the additional particle states in the presence of the impurity. The number of states dn between k and $k + dk$ is

$$\frac{dn}{dk} = \frac{R}{\pi} + \frac{1}{\pi}\frac{d\delta_l}{dk}$$

The first term R/π is just what one would have without the impurity. Thus the extra states δn from the impurity are given by the formula

$$\frac{d(\delta n)}{dk} = \frac{1}{\pi} \frac{d\delta_l}{dk} \tag{4.1.20}$$

The quantity on the right $(d/dk)(\delta_l(k)/\pi)$ is interpreted as the change in the number of particle states caused by the impurity. For example, if the potential is repulsive so that particles are pushed away from the impurity region, then $d\delta_l/dk$ will be negative. Similarly, an attractive potential will draw particles inward, so that $d\delta/dk$ is positive. Many potentials have $d\delta/dk$ positive for some wave vectors and negative for others. The quantity $d\delta/dk$ is the change in the density of states for each angular momentum state, each magnetic quantum number m_l, and each spin quantum number m_s. The total change in the density of states is obtained by summing over all these quantum numbers:

$$\frac{d}{dk} N = \frac{d}{dk} \sum_{m_s m_l l} \left(\frac{\delta_{l,m_l,m_s}}{\pi} \right) \tag{4.1.21}$$

In a metal, the electron states are occupied up to the Fermi wave vector k_F. The *Friedel sum rule* (Friedel, 1952) is obtained by integrating up to the Fermi wave vector,

$$\int_0^{k_F} dk \left(\frac{dN}{dk} \right) = Z = \sum_{m_s m_l l} \left(\frac{\delta_{l,m_l,m_s}(k_F)}{\pi} \right) \tag{4.1.22}$$

where Z is the charge on the impurity. The Friedel sum rule is a statement of charge neutrality. In a static electron gas, there are no long-range Coulomb potentials of the form r^{-1}. Instead, the electron gas is displaced in the vicinity of an impurity charge. The displaced electronic charge exactly cancels the impurity charge. For example, an impurity of positive valance Z has electrons drawn in, until an extra Z of electrons, or $-eZ$ of charge, surrounds the impurity. This extra charge is called a screening charge. It is shown in Fig. 4.1(a). The phase shifts are calculated for the screened potential of impurity plus screening charge. If the potential is repulsive, electrons are depleted around the impurity, so that charge neutrality is maintained. This depletion is shown in Fig. 4.1(b). The Friedel sum rule is a statement of charge neutrality: The change in electron charge around an impurity is exactly equal in magnitude and opposite in sign to the charge of the impurity. A neutral impurity would have a Friedel sum of zero. This does not mean that all the phase shifts vanish. Generally, some are plus,

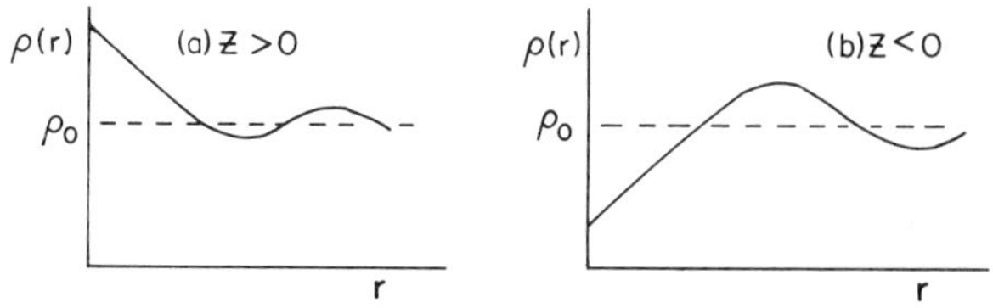

FIGURE 4.1. Electron density $\varrho(r)$ when an impurity charge of Z is put at the origin. (a) For $Z > 0$, (b) for $Z < 0$.

some are minus, and all vary with the wave vector—only the summation at the Fermi energy is zero. Another way to express that the screening charge equals the impurity charge is

$$Z = 4\pi \int_0^\infty r^2 \, dr [\varrho(r) - \varrho_0]$$

where ϱ_0 is the equilibrium charge density in the metal.

The charge densities in Fig. 4.1 are schematic. In Chapter 5 we shall discuss the method of actually calculating these curves. All such curves, when calculated correctly, have the oscillations in the charge density at large distance. These are called *Friedel oscillations.* Their magnitude has been exaggerated in Fig. 4.1, since they are small in amplitude. But they occur in real solids and have been observed by several techniques. The Friedel oscillations may be derived by taking the asymptotic limit of the change in charge density,

$$\varrho(r) - \varrho_0 = 2 \int_{k<k_F} \frac{d^3k}{(2\pi)^3} \, [| \psi_{\mathbf{k}}(\mathbf{r}) |^2 - | \phi_{\mathbf{k}}(\mathbf{r}) |^2]$$

which is

$$\lim_{r\to\infty} [\varrho(r) - \varrho_0] = \frac{8\pi}{(2\pi)^3 r^2} \sum_l (2l + 1) \int_0^{k_F} dk \left[\sin^2\left(kr + \delta_l - \frac{l\pi}{2}\right) - \sin^2\left(kr - \frac{l\pi}{2}\right) \right]$$

and

$$\sin^2\left(kr + \delta_l - \frac{l\pi}{2}\right) - \sin^2\left(kr - \frac{l\pi}{2}\right) = \tfrac{1}{2}[\cos(2kr - l\pi) - \cos(2kr + 2\delta_l - l\pi)]$$

The wave vector integral is difficult because the phase shifts depend on k. We can derive an approximate answer by writing this dependence as

$\delta_l(k) = \delta_l(k_F) + (k - k_F)(d\delta/dk)$. The k integral is then elementary and the r-dependent part is

$$\int_0^{k_F} dk(-1)^l \left\{\cos(2kr) - \cos\left[2k\left(r + \frac{d\delta}{dk}\right) + 2\delta_l - 2k_F \frac{d\delta}{dk}\right]\right\}$$
$$\to \frac{1}{2r}\{-\sin(2k_F r) + \sin[2k_F r + 2\delta_l(k_F)]\}$$

and the change in density is

$$\lim_{r\to\infty}[\varrho(r) - \varrho_0] = \frac{1}{4\pi^2 r^3}\sum_l (2l+1)(-1)^l \sin[\delta_l(k_F)] \times \cos[2k_F r + \delta_l(k_F)] + O\left(\frac{1}{r^4}\right)$$

At large distances, the changes in charge density oscillate with a period of $2k_F$ and decrease in amplitude as r^{-3}. This asymptotic equation is independent of the nature of the impurity. The impurity determines only the values for the phase shift $\delta_l(k_F)$.

The Friedel sum rule is believed to be exact in real systems. Langer and Ambegaokar (1961) have shown it to be valid even in an interacting many-particle system. If one knows the exact impurity potential and the exact screening charge profile which it caused, one should find that the Friedel sum rule is valid. In practice, one does a calculation by constructing an impurity potential by pseudopotential or other means and screens it with a good dielectric function. One numerically evaluates the phase shifts and typically finds that the Friedel sum rule errs by a small percentage. This error is no fault of the rule but the choice of potential or screening function. In fact, the theoretical system is not charge neutral. Usually one adjusts the potential slightly, by altering a screening length, for example, to force the Friedel sum rule to be satisfied. Then one has a consistent model of the neutral system, and the phase shifts are probably reasonably accurate.

Quite often the phase shifts do not depend on m_l or m_s, but they strongly depend on l. Then one can represent the answer for electrons as

$$Z = \frac{2}{\pi}\sum_l (2l+1)\delta_l(k_F) \tag{4.1.22a}$$

which is the way that it is usually presented. The factor of 2 is spin degeneracy and $2l + 1$ is orbital degeneracy. The quantity

$$\frac{2}{\pi}(2l+1)\delta_l(k_F)$$

is interpreted as the amount of screening charge in the angular momentum channel l.

In real solids, the impurity may occupy the site ordinarily occupied by a host atom. Then the normal host atom would have its own set of phase shifts $\delta_l^{(h)}$ which characterize the normal metal. For example, if Mg^{2+} is a substitutional impurity in sodium metal, it could be where a Na^+ usually is placed. The Na^+ ion has its own phase shifts $\delta_l^{(h)}$, and the Mg^{2+} ion has its own phase shifts $\delta_l^{(i)}$. If one repeats the arguments leading to (4.1.22a), the Friedel sum rule is given by the difference of the two sets of phase shifts:

$$Z = \frac{2}{\pi} \sum_l (2l+1)[\delta_l^{(i)}(k_F) - \delta_l^{(h)}(k_F)]$$

Note that the host phase shifts do not obey the Friedel sum rule.

Another exact result may be obtained in terms of phase shifts. It is the total energy of the impurity as caused by its interactions with the surrounding electrons in a metal. The theorem is due to Fumi (1955) and relates the total energy E_i to an energy integral over the phase shifts:

$$\begin{aligned} E_i &= -\int_0^{E_F} dE \sum_{lm_lm_s} \frac{\delta_l(E)}{\pi} \\ &= -\frac{\hbar^2}{m} \int_0^{k_F} k\, dk \sum_{lm_lm_s} \frac{\delta_l(k)}{\pi} \end{aligned} \tag{4.1.23}$$

The energy integral starts at the bottom of the conduction band. It is assumed there are no bound states. The presence of bound states will change the answer. One can also use the second form of (4.1.23). This includes just the sum over wave vector states for the conduction electrons.

Fumi's theorem is proved in the following way. The Hamiltonian is the kinetic energy plus the potential energy. Well outside the impurity, or its screening charge, the potential is zero, and the Hamiltonian only has the kinetic energy term. Here the energy must just be the summation of the kinetic energies of all the particles, which is a discrete summation over the states allowed in the sphere of radius R:

$$E = \sum_{l,m,\alpha} \frac{\hbar^2}{2m} k_\alpha^2 = \frac{\hbar^2}{2mR^2} \sum_{\alpha,l,m} \left[\left(n + \frac{l}{2}\right)\pi - \delta_l\right]^2$$

The quantity we desire is actually the change in kinetic energy, so the result without the impurity is subtracted:

$$E_i = \frac{\hbar^2}{2mR^2} \sum_{\substack{\alpha,l\\ m_l,m_s}} \left\{ \left[\left(n + \frac{l}{2} \right)\pi - \delta_l \right]^2 - \left(n + \frac{l}{2} \right)^2 \pi^2 \right\}$$

$$\simeq -\frac{\hbar^2}{mR} \sum_{\substack{\alpha,l\\ m_s,m_l}} k_\alpha \delta_l(k_\alpha)$$

Next we let the radius of the sphere go to infinity, which changes the wave vector summation to a continuous integration. This change is

$$\frac{1}{R} \sum_\alpha k_\alpha \delta_l(k_\alpha) \to \frac{1}{\pi} \int_0^{k_F} dk k \delta_l(k)$$

and the result is

$$E_i = -\frac{\hbar^2}{m\pi} \sum_{\substack{l m_l\\ m_s}} \int_0^{k_F} k\, dk \delta_l(k)$$

which proves the theorem for the conduction band states. The theorem may be expressed another way. Let $Z(k)$ be the charge alteration around the impurity for wave vectors up to k:

$$Z(k) = \frac{1}{\pi} \sum_{\substack{l,m_l\\ m_s}} \delta_l(k)$$

$$E_i = -\frac{\hbar^2}{m} \int_0^{k_F} k\, dk Z(k)$$

An integration by parts changes this into

$$E_i = -\frac{\hbar^2}{m} \left[\frac{k_F{}^2}{2} Z(k_F) - \int_0^k dk \left(\frac{k^2}{2} \right) \frac{dZ}{dk} \right]$$

$$= -E_F Z(k_F) + \int_0^{k_F} dk \varepsilon_k \frac{dZ}{dk}$$

The factor dZ/dk is now the change in charge density per unit wave vector. The energy of the impurity is two terms. The first, $-E_F Z$, states that Z electrons of energy E have been removed ($Z < 0$) or added ($Z > 0$) to the system. The second term states that they have been redistributed at lower energies. This is illustrated in Fig. 4.2. The excess impurity charge is indicated at the origin in position space, whereas actually it is distributed

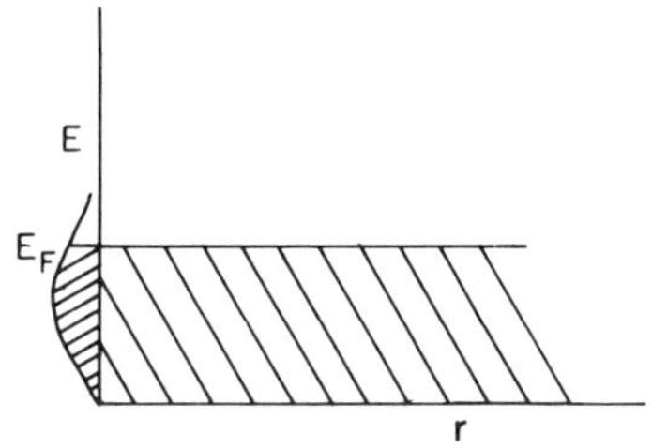

FIGURE 4.2. A schematic drawing which shows the energy distribution of the screening charge around an impurity. This charge must come from the Fermi surface.

over position as shown in Fig. 4.1. For attractive potentials, the charge is removed from the Fermi level and occupies the lower energy state around the impurity. This lowering of the energy accounts for the impurity energy E_i.

D. Phase Shifts

Several examples of phase-shift calculations will be given to illustrate their properties. The first is the hard sphere potential. This potential has the property

$$V(r) = \begin{cases} \infty, & r < a \\ 0, & r > a \end{cases}$$

The infinite potential forces the wave function $\psi_{\mathbf{k}}(\mathbf{r})$ to vanish at $r < a$. Outside the hard sphere, the radial wave functions must have the general form

$$R_l(kr) = j_l(kr) - \tan(\delta_l)\eta_l(kr)$$

where we have selected the reaction matrix form of normalization (4.1.8). These wave functions vanish at the hard sphere $r = a$ with the choice

$$\tan(\delta_l) = \frac{j_l(ka)}{\eta_l(ka)}$$

which determines the phase shifts. The first three phase shifts are plotted in Fig. 4.3 as a function of ka. The s-wave result is particularly simple. For $l = 0$, the radial wave function is written as

$$R_0(kr) = \frac{1}{kr}\sin[k(r-a)] = \frac{1}{kr}\sin(kr + \delta_0)$$

so that

$$\delta_0 = -ka$$

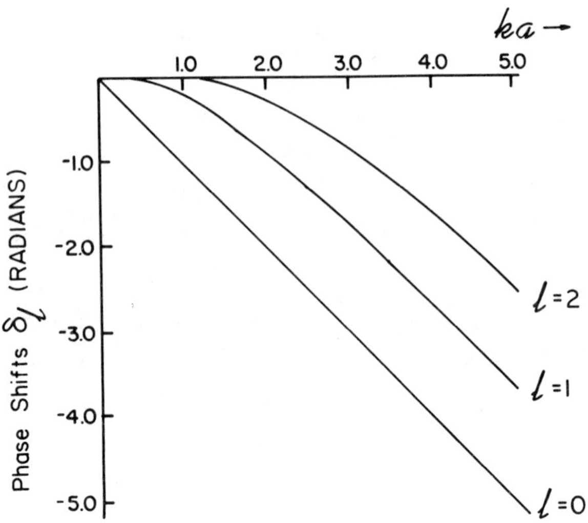

FIGURE 4.3. Phase shifts for a hard sphere potential of radius a. Results are given only for $l = 0, 1, 2$; higher ones exist but are not shown.

The other phase shifts start at the origin with the dependence $\delta_l \approx (ka)^{2l+1}$ and asymptotically go to $\delta_l \approx (l\pi/2) - ka$ at large values of ka.

The second example of phase shifts is for a Lennard-Jones potential. This potential, which is also called the 6-12 potential, has the spatial dependence

$$V(r) = 4\varepsilon\left[\left(\frac{\sigma}{r}\right)^{12} - \left(\frac{\sigma}{r}\right)^{6}\right]$$

It is frequently used to represent the interaction potential between neutral molecules. Its popularity is caused by its mathematical simplicity, and our ignorance of the real potential, and not by its accuracy for real systems. The potential has a depth ε and a "hard sphere" distance σ. We need to solve the radial part of Schrödinger's equation with this potential:

$$\left\{-\frac{\hbar^2}{2\mu}\frac{1}{r^2}\left[\frac{\partial}{\partial r}r^2\frac{\partial}{\partial r} - l(l+1)\right] + V(r) - \frac{\hbar^2 k^2}{2\mu}\right\}R_l(kr) = 0$$

It is convenient to use dimensionless units. Distance is measured in units of σ, $x = r/\sigma$:

$$\left[-\frac{1}{x^2}\frac{\partial}{\partial x}x^2\frac{\partial}{\partial x} + \frac{l(l+1)}{x^2} + 4g\left(\frac{1}{x^{12}} - \frac{1}{x^6}\right) - (k\sigma)^2\right]R_l(kr) = 0$$

where the important coupling constant is the dimensionless number

$$g = \frac{2\mu\varepsilon\sigma^2}{\hbar^2}$$

This potential has a finite number of bound states, which is true of all potentials with a short-ranged interaction; only r^{-1} potentials have an infinite number of bound states. For values of g sufficiently small, there are no bound states at all. The critical value of coupling constant for the 6-12 potential is (Goldberg *et al.*, 1966)

$$g_c = 5.596$$

There are no bound states for $g < g_c$. For $g > g_c$, one bound state appears in the s wave, and for higher values of g additional bound states may appear.

The potential between two atoms of ^{4}He is usually characterized by a Lennard-Jones potential. The best parameters are (Feltgen *et al.*, 1982)

$$\varepsilon = 1.484 \times 10^{-15} \text{ erg}$$

$$\sigma = 2.648 \text{ Å}$$

$$m_{\text{He}} = 6.648 \times 10^{-24} \text{ g}$$

With these numbers, the coupling constant $g = 6.22$, where one must remember that $2\mu = m_{\text{He}}$ for two helium atoms. This number is remarkably close to the value of critical binding. In fact, whether two helium atoms bind and form the dimer He_2 is debated among chemists. The present calculation suggests binding.

Whether or not a bound state does exist between two helium atoms, it is certain that the phase shifts will have a remarkable low-energy behavior. They are shown in Fig. 4.4(a). The bound state, or nearly bound state, produces a resonance in the scattering cross section. The s-wave phase shift is shown for the case of no binding, $g = 5.5$, and the case of binding, $g = 5.7$. For no binding, the phase shift rises rapidly from zero at low energy and levels off near $\pi/2$. Then it gradually decreases linearly with increasing k. Except for the initial onset, it is well represented by the formula

$$\delta_0 \simeq \frac{\pi}{2} - k\sigma$$

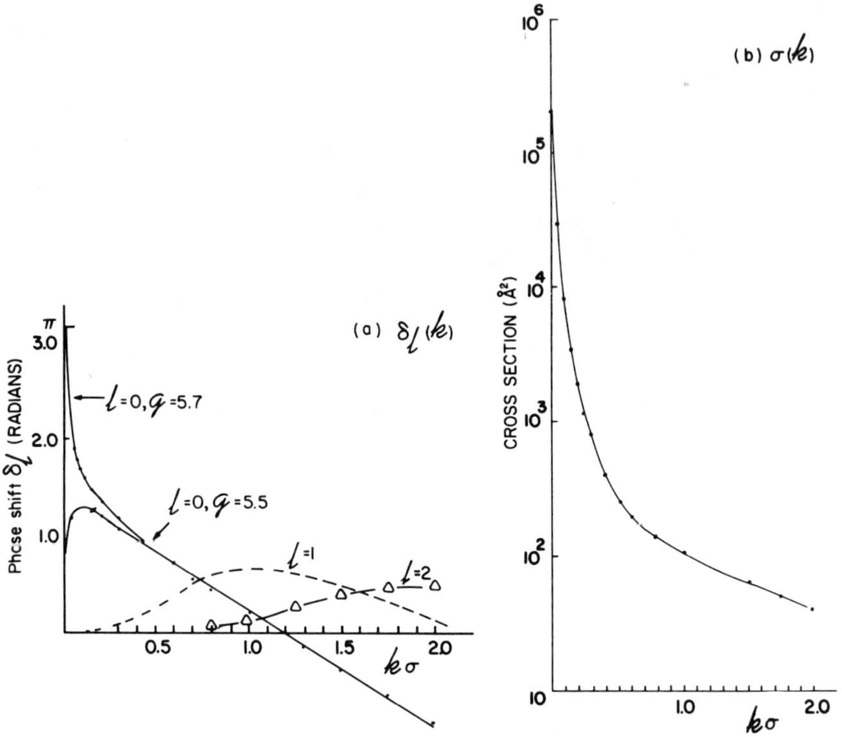

FIGURE 4.4. (a) Phase shifts for the scattering of two atoms of ^{4}He. The two curves for $l = 0$ have dimensionless coupling constants $g = 5.5$ and $g = 5.7$. (b) The total scattering cross section found from the phase shifts in (a). The low-energy resonance is apparent.

The factor $-k\sigma$ is hard-sphere behavior, while $\pi/2$ is the effect of the attractive potential. The bound case, $g = 5.7$, has the phase shift starting at π in agreement with Levinson's theorem. It falls rapidly to $\pi/2$ and then is indistinguishable from the other case. Both results produce a cross section as shown in Fig. 4.4(b), with a large resonance at zero energy. The p-wave phase shift shows only slight effects of the potential. Because of the symmetry of the two-particle wave function, the p-wave case is unobservable; see Sec. 10.1A.

These helium phase shifts will be useful when we discuss theories of superfluidity of ^{4}He in Chapter 10. Further examples of phase shifts are given in the problems at the end of the chapter.

E. Impurity Scattering

How are Green's functions used to describe impurity scattering? They do not work very well for describing the scattering of a particle from a single impurity. We calculate, in Dyson's equation, the change in energy of the particle from the interactions. One impurity in a large system of volume ν changes the energy of a particle only by order $1/\nu$. This assumes, of course, that the particle does not get bound by the impurity. A free-particle state has a density $|\phi_k|^2 = 1/\nu$, so that a local impurity changes the energy only by $1/\nu$. To alter the energy, one must add N_i impurities, where $N_i/\nu \to n_i$ as $\nu \to \infty$. Then the particle energy gets changed as a function of the concentration n_i of impurities. The isolated impurity case should be studied by the following procedure: One solves the self-energy as a function of n_i and then takes the limit $n_i \to 0$.

The potential which scatters the electron is taken as a summation of impurity potentials:

$$\bar{V}(\mathbf{r}) = \sum_i V(\mathbf{r} - \mathbf{R}_i)$$

The Fourier transform is $\bar{V}(\mathbf{q})$:

$$\bar{V}(\mathbf{q}) = \varrho_i(\mathbf{q}) V(\mathbf{q}) \varrho(\mathbf{q})$$

where $\varrho(-\mathbf{q})$ and $\varrho_i(\mathbf{q})$ are the particle densities of electrons and impurities, respectively:

$$\varrho(\mathbf{q}) = \sum_{\mathbf{k}\sigma} C^\dagger_{\mathbf{k}+\mathbf{q}\sigma} C_{\mathbf{k}\sigma}$$

$$\varrho_i(\mathbf{q}) = \sum_i e^{i\mathbf{q}\cdot\mathbf{R}_i}$$

When the S matrix is expanded, we shall encounter products of the impurity density operator:

$$f_n(\mathbf{q}_1 \cdots \mathbf{q}_n) = \langle \varrho_i(\mathbf{q}_1)\varrho_i(\mathbf{q}_2) \cdots \varrho_i(\mathbf{q}_n)\rangle_{\text{av}}$$

These must be averaged. This average is not taken over temperature. Instead, it is an average over the possible positions which the impurities may have in the solid. Usually it is assumed that the impurities are randomly located and that there is no correlation between their positions. This method of impurity averaging was suggested by Kohn and Luttinger (1957). Their result will be derived by first examining the results for a small number of operators. The first term is $n = 1$:

$$f_1(\mathbf{q}) = \left\langle \sum_i e^{i\mathbf{q}\cdot\mathbf{R}_i} \right\rangle = N_i \delta_{\mathbf{q}=0}$$

If the R_i are located randomly, this sum is zero unless $\mathbf{q} = 0$, and then the sum gives the number of impurities N_i. The second case is $n = 2$:

$$\left\langle \sum_{ij} e^{i\mathbf{q}_1\cdot\mathbf{R}_i} e^{i\mathbf{q}_2\cdot\mathbf{R}_j} \right\rangle = \left\langle \sum_{i=j} e^{i\mathbf{R}_i\cdot(\mathbf{q}_1+\mathbf{q}_2)} + \sum_{i\neq j} e^{i(\mathbf{q}_1\cdot\mathbf{R}_i+\mathbf{q}_2\cdot\mathbf{R}_j)} \right\rangle$$

The first term is zero unless $\mathbf{q}_1 + \mathbf{q}_2 = 0$, and the second term is zero unless both $\mathbf{q}_1$ and $\mathbf{q}_2$ equal zero:

$$f_2(\mathbf{q}_1, \mathbf{q}_2) = N_i \delta_{\mathbf{q}_1+\mathbf{q}_2=0} + N_i(N_i - 1)\delta_{\mathbf{q}_1=0}\delta_{\mathbf{q}_2=0}$$

The general result found by Kohn and Luttinger is

$$f_n(\mathbf{q}_1 \cdots \mathbf{q}_n) = N_i \delta_{\Sigma_i^n \mathbf{q}_j=0} + N_i(N_i - 1) \sum_m \delta_{\Sigma_{j=1}^m \mathbf{q}_j=0}\delta_{\Sigma_{j=m+1}^n \mathbf{q}_j=0}$$
$$+ N_i(N_i - 1)(N_i - 2)\delta_{\Sigma\mathbf{q}=0}\delta_{\Sigma\mathbf{q}=0}\delta_{\Sigma\mathbf{q}=0} + \cdots$$

In terms with products of several delta functions $\delta_{\Sigma\mathbf{q}=0}$, one must take all possible combinations of distributing the $\mathbf{q}_j$s among the delta functions. This general result is proved by generalizing the low-order examples which we have already provided. For example, the result for $n = 3$ is

$$f_3(\mathbf{q}_1, \mathbf{q}_2, \mathbf{q}_3) = N_i \delta_{\mathbf{q}_1+\mathbf{q}_2+\mathbf{q}_3} + N_i(N_i - 1)(\delta_{\mathbf{q}_1}\delta_{\mathbf{q}_2+\mathbf{q}_3} + \delta_{\mathbf{q}_2}\delta_{\mathbf{q}_1+\mathbf{q}_3} + \delta_{\mathbf{q}_3}\delta_{\mathbf{q}_1+\mathbf{q}_2})$$
$$+ N_i(N_i - 1)(N_i - 2)\delta_{\mathbf{q}_1}\delta_{\mathbf{q}_2}\delta_{\mathbf{q}_3}$$

The next step is to learn to draw Feynman diagrams with these impurity averages. Since $N_i > 10^{10}$ for real systems, we can approximate this by

$$f(\mathbf{q}_1 \cdots \mathbf{q}_n) = N_i \delta_{\Sigma\mathbf{q}_j} + N_i^2 \delta_{\Sigma\mathbf{q}}\delta_{\Sigma\mathbf{q}} + N_i^3 \delta_{\Sigma\mathbf{q}}\delta_{\Sigma\mathbf{q}}\delta_{\Sigma\mathbf{q}} + \cdots \qquad (4.1.24)$$

Thus there occur the combinations of the function

$$N_i \delta_{\Sigma_j \mathbf{q}_j=0} \qquad (4.1.25)$$

This factor is interpreted as the particle scattering from a single impurity.

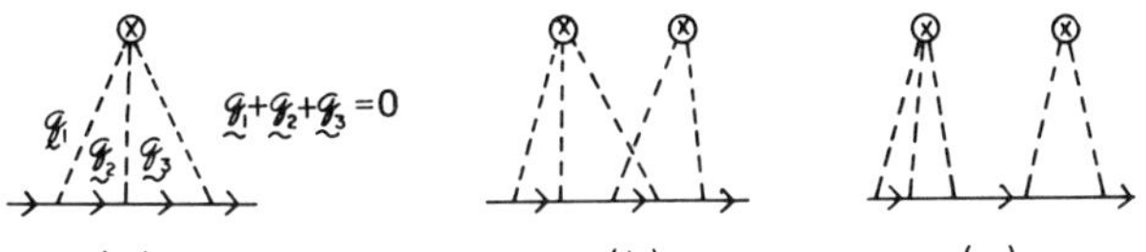

FIGURE 4.5

The $\delta_{\Sigma q=0}$ states that momentum is conserved for the particle while scattering from the impurity. In Fig. 4.5(a), the impurity is represented by an X, and the solid line is the particle. The dashed lines represent impurity interactions $V(\mathbf{q}_j)$. Momentum conservation requires the momenta from a single impurity to sum to zero. Diagrams with two factors of (4.1.25) involve the scattering from two impurities, as shown in Figs. 4.5(b) and 4.5(c).

In Fig. 4.5(b), the scattering from the two impurities interfere—the momentum lines cross each other. Of course, this could happen if the impurities were nearby, so that such diagrams are important when the concentration of impurities is large. In Fig. 4.5(c), the two scattering events are *disconnected*, since the interaction lines do not overlap. Each connected diagram gives a separate self-energy contribution. Edwards (1958) has shown that one has the Dyson equation

$$G(\mathbf{p}, ip) = \frac{1}{ip - \xi_p - \Sigma(\mathbf{p}, ip)}$$

where the self-energy $\Sigma(\mathbf{p}, ip)$ from impurity scattering contains all the connected contributions. The diagram in Fig. 4.5(b) is connected, so it contributes a term to Σ. The self-energy diagrams for scattering from a single impurity are shown in Fig. 4.6. These are the terms which are important when the concentration of impurities is small.

The first self-energy diagram in Fig. 4.6 gives a contribution

$$\Sigma_{(1)}(\mathbf{p}, ip) = \frac{N_i}{\nu} \sum_{\mathbf{q}} \delta_{\mathbf{q}=0} V(\mathbf{q}) = n_i V(0)$$

The second diagram in the series has a particle line as an intermediate state.

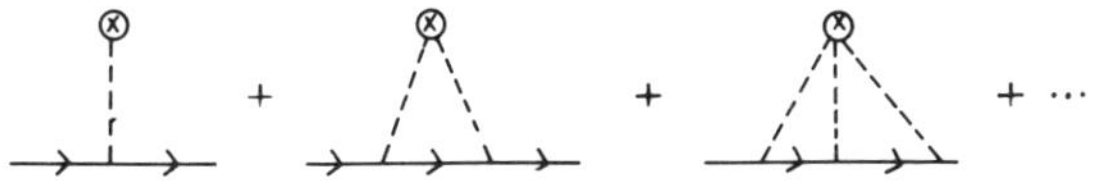

FIGURE 4.6

This self-energy term is

$$\Sigma_{(2)}(\mathbf{p}, ip) = \frac{N_i}{\nu^2} \sum_{\mathbf{q}_1\mathbf{q}_2} \delta_{\mathbf{q}_1+\mathbf{q}_2} V(\mathbf{q}_1)V(\mathbf{q}_2)G^{(0)}(\mathbf{p} + \mathbf{q}_1, ip)$$

which may also be written as $(\nu \to \infty)$

$$\Sigma_{(2)}(\mathbf{p}, ip) = n_i \int \frac{d^3p'}{(2\pi)^3} \frac{V(\mathbf{p} - \mathbf{p}')V(\mathbf{p}' - \mathbf{p})}{ip - \xi_{p'}}$$

There is no summation over internal frequency variables. The impurity is considered rigid and cannot absorb or transfer energy to the particle which is scattering. Consequently, all the internal lines have the same energy ip_n as the initial particle. This result is not assumed but instead is derived directly from the Green's function expansion. For example, the above self-energy term arose from second-order terms and had the τ integrals

$$\int_0^\beta d\tau e^{ip_n\tau} \int_0^\beta d\tau_1 \int_0^\beta d\tau_2 G^{(0)}(\mathbf{p}, \tau - \tau_1)G^{(0)}(\mathbf{p} + \mathbf{q}_1, \tau_1 - \tau_2)G^{(0)}(\mathbf{p}, \tau_2)$$
$$= G^{(0)}(\mathbf{p}, ip_n)^2 G^{(0)}(\mathbf{p} + \mathbf{q}_1, ip_n)$$

The fact that all electron lines have the same energy comes directly from the τ integrals. It is true for an arbitrary order of diagram—all particle Green's functions have the same energy ip_n. As stated before, there is no other excitation in the model to which energy may be transferred.

The third term in the series in Fig. 4.6 is

$$\Sigma_{(3)}(\mathbf{p}, ip_n) = n_i \int \frac{d^3p_1}{(2\pi)^3} \int \frac{d^3p_2}{(2\pi)^3} \frac{V(\mathbf{p} - \mathbf{p}_1)V(\mathbf{p}_1 - \mathbf{p}_2)V(\mathbf{p}_2 - \mathbf{p})}{(ip - \xi_{p_1})(ip - \xi_{p_2})}$$

At this point, one may deduce the nth term by inspection. It is

$$\Sigma_{(n)}(\mathbf{p}, ip) = n_i \int \frac{d^3p_1}{(2\pi)^3} \cdots \frac{d^3p_n}{(2\pi)^3} V(\mathbf{p} - \mathbf{p}_1)V(\mathbf{p}_1 - \mathbf{p}_2) \cdots V(\mathbf{p}_n - \mathbf{p})$$
$$\times \prod_{j=1}^{n} \frac{1}{ip - \xi_{pj}} \qquad (4.1.26)$$

Now that the nth term is evident, it is also possible to sum the series of terms and get the total self-energy which is proportional to the concentration n_i. The first step in this summation is to write an integral equation for the self-energy. This integral equation is in terms of a *vertex function*

$\Gamma(\mathbf{p}, \mathbf{p}')$ which has the equation

$$\Gamma(\mathbf{p}', \mathbf{p}) = V(\mathbf{p}' - \mathbf{p}) + \int \frac{d^3p_1}{(2\pi)^3} \frac{V(\mathbf{p}' - \mathbf{p}_1)\Gamma(\mathbf{p}_1, \mathbf{p})}{ip - \xi_{p_1}} \tag{4.1.27}$$

$$\begin{aligned} \Sigma(\mathbf{p}, ip) &= \sum_{n=1}^{\infty} \Sigma_{(n)}(\mathbf{p}, ip) = n_i V(0) + n_i \int \frac{d^3p'}{(2\pi)^3} \frac{V(\mathbf{p} - \mathbf{p}')\Gamma(\mathbf{p}', \mathbf{p})}{ip - \xi_{p'}} \\ &= n_i \Gamma(\mathbf{p}, \mathbf{p}) \end{aligned} \tag{4.1.28}$$

Repeated iteration of Eq. (4.1.27) will generate the series of terms which give the successive diagrams in Fig. 4.6, of which (4.1.26) is the nth term. The series is summed by solving the integral equation. To this end, define the function

$$\pi(\mathbf{r}, \mathbf{p}) = \int \frac{d^3p'}{(2\pi)^3} \frac{e^{i\mathbf{p}'\cdot\mathbf{r}}\Gamma(\mathbf{p}', \mathbf{p})}{ip - \xi_{p'}}$$

This quantity and $\Gamma(\mathbf{p}', \mathbf{p})$ are both functions of ip_n, but this dependence is not explicitly added to the notation. In terms of this new function, the self-energy (4.1.28) is

$$\Sigma(\mathbf{p}, ip) = n_i V(0) + n_i \int d^3r e^{-i\mathbf{p}\cdot\mathbf{r}} V(r)\pi(\mathbf{r}, \mathbf{p}) \tag{4.1.29}$$

To solve for the function $\pi(\mathbf{r}, \mathbf{p})$, consider the effects of the differential operator acting on it:

$$\left[-\frac{\hbar^2}{2m}\nabla^2 - \mu - ip\right]\pi(\mathbf{r}, \mathbf{p}) = \int \frac{d^3p'}{(2\pi)^3} \frac{\xi_{\mathbf{p}'} - ip}{ip - \xi_{p'}} e^{i\mathbf{p}'\cdot\mathbf{r}}\Gamma(\mathbf{p}', \mathbf{p})$$

$$\left[-\frac{\hbar^2\nabla^2}{2m} - \mu - ip\right]\pi(\mathbf{r}, \mathbf{p}) = -\int \frac{d^3p'}{(2\pi)^3} \Gamma(\mathbf{p}', \mathbf{p})e^{i\mathbf{p}'\cdot\mathbf{r}}$$

The right-hand side of this equation may be simplified. First, note that the vertex function may be written as

$$\Gamma(\mathbf{p}', \mathbf{p}) = \int d^3r e^{-i\mathbf{p}'\cdot\mathbf{r}} V(r)[e^{i\mathbf{p}\cdot\mathbf{r}} + \pi(\mathbf{r}, \mathbf{p})]$$

which follows directly from its definition (4.1.27). From this relation, it is easy to show that

$$\int \frac{d^3p'}{(2\pi)^3} \Gamma(\mathbf{p}', \mathbf{p})e^{i\mathbf{p}'\cdot\mathbf{r}} = V(r)[e^{i\mathbf{p}\cdot\mathbf{r}} + \pi(\mathbf{r}, \mathbf{p})]$$

Thus we deduce that the function $\pi(\mathbf{r}, \mathbf{p})$ obeys an inhomogeneous differential equation:

$$\left[-\frac{\hbar^2\nabla^2}{2m} - \mu - ip + V(r)\right]\pi(\mathbf{r}, \mathbf{p}) = -V(r)e^{i\mathbf{p}\cdot\mathbf{r}} \tag{4.1.30}$$

This equation is solved in the following fashion. First we find the solution to the following Schrödinger equation:

$$H\psi_\lambda = \left[-\frac{\hbar^2\nabla^2}{2m} + V(r)\right]\psi_\lambda = E_\lambda\psi_\lambda$$

This differential equation was discussed extensively at the beginning of this section. We take the outgoing wave boundary conditions. The reason for this choice will be clear below. In terms of these wave functions, the special function $\pi(\mathbf{r}, \mathbf{p})$ has the following solution to the inhomogeneous equation (4.1.30),

$$\pi(\mathbf{r}, \mathbf{p}) = \int d^3r' e^{i\mathbf{p}\cdot\mathbf{r}'} V(r') \sum_\lambda \frac{\psi_\lambda^*(\mathbf{r}')\psi_\lambda(\mathbf{r})}{ip_n + \mu - E_\lambda} = \sum_\lambda \frac{\psi_\lambda(\mathbf{r})T^*_{\lambda\mathbf{p}}}{ip + \mu - E_\lambda}$$

where we have used the T matrix defined in (4.1.16). This result for $\pi(\mathbf{r}, \mathbf{p})$ can be directly verified by operating on both sides by $H - \mu - ip_n$ and then using the completeness property:

$$\sum_\lambda \psi_\lambda^*(\mathbf{r}')\psi_\lambda(\mathbf{r}) = \delta^3(\mathbf{r} - \mathbf{r}')$$

The summation λ is taken over all the eigenstates of H, which include the bound states as well as the continuum states. When this result is used in (4.1.29), we obtain the final result for the self-energy in terms of the T matrices which were defined in (4.1.16):

$$\Sigma(\mathbf{p}, ip) = n_i V(0) + n_i \sum_\lambda \frac{|T_{\mathbf{p}\lambda}|^2}{ip + \mu - E_\lambda} \tag{4.1.31}$$

Equation (4.1.31) is the exact term in the self-energy which is proportional to the concentration n_i. It has been obtained by summing the set of diagrams shown in Fig. 4.6, where the particle has multiple scattering events from the same impurity. There are other terms in the self-energy which are proportional to higher powers of the concentration $(n_i)^m$. For example, one term which contributes to order n_i^2 is shown in Fig. 4.5(b). However, in the limit of dilute impurities, as $n_i \to 0$, the result (4.1.31) is the most important.

Our interest is usually in the retarded self-energy, which we call $\Sigma(\mathbf{p}, \omega)$, omitting the "ret" subscript:

$$\Sigma(\mathbf{p}, \omega) = n_i\left(V_{\mathbf{pp}} + \sum_\lambda \frac{|T_{\mathbf{p}\lambda}|^2}{\omega - \xi_\lambda + i\delta}\right) \tag{4.1.32}$$

There is one particular case which is very important: when the energy ω is set equal to ξ_p. Then the self-energy (4.1.32) is just n_i times the T matrix:

$$\Sigma(\mathbf{p}, \xi_p) = n_i T_{\mathbf{pp}} = -\frac{2\pi n_i}{mp}\sum_l (2l+1)e^{+i\delta_l(p)}\sin[\delta_l(p)]$$

This identification is made by noting that the T-matrix equation (4.1.17) is identical to the vertex equation (4.1.27) when $ip \to \xi_p + i\delta$. This result is frequently used in representing the self-energy of the particle. The imaginary part of the self-energy has a simple formula:

$$-2\,\mathrm{Im}[\Sigma(\mathbf{p}, \omega)] = 2\pi n_i \sum_\lambda |T_{\mathbf{p}\lambda}|^2\,\delta(\omega - \xi_\lambda)$$

The right-hand side may be further reduced in the special case $\omega = \xi_p$. If we combine the two prior results (4.1.18) and (4.1.19), this may be shown to be proportional to the scattering cross section:

$$-2\,\mathrm{Im}[\Sigma(\mathbf{p}, \xi_p)] = -2n_i\,\mathrm{Im}(T_{\mathbf{pp}}) = n_i v_p \sigma(p)$$

The imaginary part of the self-energy for $\omega = \xi_p$ is the cross section times $v_p n_i$. The factor $v_p n_i$ is just the rate at which the particle encounters the impurities. Thus the lifetime of the particle may be identified as

$$\frac{1}{\tau(p)} = n_i v_p \sigma(p)$$

Similarly, the factor $n_i\sigma$ is just the inverse mean free path of the particle:

$$l_p = \frac{1}{n_i\sigma(p)} = v_p\tau(p)$$

This provides an example of the earlier assertion that the imaginary part of the retarded self-energy is related to the damping of the particle.

The energy-dependent self-energy may also be defined in terms of a generalized, energy-dependent T-matrix equation:

$$\begin{aligned} \Sigma(\mathbf{p}, \omega) &= n_i T_{\mathbf{pp}}(\omega) \\ T_{\mathbf{pp}'}(\omega) &= V_{\mathbf{pp}'} + \int \frac{d^3p''}{(2\pi)^3}\frac{V_{\mathbf{pp}''}T_{\mathbf{p}''\mathbf{p}'}(\omega)}{\omega - \xi_{p''} + i\delta} \end{aligned} \tag{4.1.33}$$

This equation is the same one which is satisfied by the retarded vertex function (4.1.27). This vertex function is just the energy-dependent T matrix.

An interesting question is the relationship between these self-energy expressions and Fumi's theorem (4.1.23). The two results are quite compatible. The self-energy expressions describe the effect of the impurities on single-particle states. Fumi's theorem gives the energy change of the *system.* We can derive Fumi's theorem by averaging the single-particle properties. This is shown in the next section.

F. Ground State Energy

Here we wish to show that the self-energy expression (4.1.32), when averaged correctly, will give Fumi's theorem (4.1.23) for the energy per impurity. What we intend to prove, say at zero temperature, is that the impurities cause a change in the ground state energy of

$$\Delta\Omega = -N_i \frac{2}{\pi m} \int_0^{k_F} p\, dp \sum_l (2l+1)\delta_l(p) + O(N_i^2) \qquad (4.1.34)$$

Terms proportional to N_i^2 give energy terms arising from interactions between impurities. The term proportional to a single power N_i is the average energy per impurity from interactions with the electrons.

The result (4.1.34) has an interesting history. Most of our references have been to solid state physics work such as that of Friedel and Fumi. A parallel development was occurring in nuclear theory. Brueckner *et al.* (1954) proposed that the ground state energy of the system was an average over the energy of the single-particle states. The energy per particle was taken to be the reaction matrix, so they took the energy change to be

$$\Delta\Omega = 2N_i \int \frac{d^3p}{(2\pi)^3} n_F(\xi_p) R_{\mathbf{pp}} = -\frac{2N_i}{\pi m} \int_0^{k_F} p\, dp \sum_l (2l+1) \tan\delta_l$$

This has $\tan\delta_l$ instead of δ_l and so is incorrect. Fukuda and Newton (1956) showed that the correct result should be δ_l rather than $\tan\delta_l$. In fact, they give results nearly identical to Fumi. Fukuda and Newton and also DeWitt (1956) proved some important theorems regarding the energy (4.1.34). The result is not dependent on the spherical box which was used in the derivation—the same result was demonstrated for a cube and other shapes. Similarly, it does not depend on the assumption that the wave function vanished at the surface of the box—the same result is obtained for other boundary conditions. Fumi's theorem appears to be a general result.

The ground state energy is calculated using the formula (3.6.22) derived earlier, which contains a coupling constant integration

$$\Delta\Omega = \frac{\nu}{\beta} \sum_{ip,\sigma} \int \frac{d^3p}{(2\pi)^3} \int_0^1 \frac{d\eta}{\eta} \mathscr{G}(\mathbf{p}, ip, \eta)\Sigma(\mathbf{p}, ip; \eta)$$

Since we only wish to find the term proportional to n_i, and since the self energy Σ is already proportional to n_i, we can replace $\mathscr{G}$ by $\mathscr{G}^{(0)}$. The difference is just self-energy terms which give higher powers in $(n_i)^m$, which we are going to neglect. The coupling constant integration is important for weighting each factor V of impurity potential (and not each factor of the T matrix). From (4.1.26), this self-energy is

$$\Sigma(\mathbf{p}, ip; \eta) = n_i\Big[\eta V_{pp} + \eta^2 \int \frac{d^3p'}{(2\pi)^3} \frac{V_{\mathbf{pp}'}V_{\mathbf{p}'\mathbf{p}}}{ip - \xi_{p'}}$$
$$+ \eta^3 \int \frac{d^3p'}{(2\pi)^3} \int \frac{d^3p''}{(2\pi)^3} \frac{V_{\mathbf{pp}'}V_{\mathbf{p}'\mathbf{p}''}V_{\mathbf{p}''\mathbf{p}}}{(ip - \xi_{p'})(ip - \xi_{p''})} + \cdots\Big] \tag{4.1.35}$$

$$\Delta\Omega = 2N_i \int \frac{d^3p}{(2\pi)^3} \frac{1}{\beta} \sum_{ip} \frac{1}{ip - \xi_p} \Big[V_{\mathbf{pp}} + \frac{1}{2} \int \frac{d^3p'}{(2\pi)^3} \frac{V_{\mathbf{pp}'}V_{\mathbf{p}'\mathbf{p}}}{ip - \xi_{p'}}$$
$$+ \frac{1}{3} \int \frac{d^3p'}{(2\pi)^3} \int \frac{d^3p''}{(2\pi)^3} \frac{V_{\mathbf{pp}'}V_{\mathbf{p}'\mathbf{p}''}V_{\mathbf{p}''\mathbf{p}}}{(ip - \xi_{p'})(ip - \xi_{p''})} + \cdots\Big]$$
$$+ O(N_i^2)$$

where we have done the coupling constant integration. There are several ways to evaluate this expression.

Our first derivation will give tan δ, instead of δ, so that it gives the wrong answer. This is included, not just for fun, but to show how simple it can be to get the wrong result. While reading this derivation, try to find the incorrect step, which leads to this wrong result. The incorrect derivation proceeds by doing the summation on Matsubara frequencies in (4.1.35) term by term:

$$\frac{1}{\beta} \sum_{ip} \mathscr{G}^{(0)}(\mathbf{p}, ip) = n_F(\xi_p)$$

$$\frac{1}{\beta} \sum_{ip} \mathscr{G}^{(0)}(\mathbf{p}, ip)\mathscr{G}^{(0)}(\mathbf{p}', ip) = \frac{n_F(\xi_{\mathbf{p}}) - n_F(\xi_{\mathbf{p}'})}{\xi_p - \xi_{p'}}$$

The nth term is found from the contour integral,

$$\oint \frac{dZ}{2\pi i} n_F(Z) \prod_{j=1}^{n} \frac{1}{Z - \xi_j} = 0 \tag{4.1.36}$$

where the contour is taken at infinity so that it includes all the poles. This leads to the identity

$$\frac{1}{\beta}\sum_{ip}\prod_{j=1}^{n}\left(\frac{1}{ip-\xi_j}\right)=\sum_{l=1}^{n} n_F(\xi_l)\prod_{\substack{j=1\\ j\neq l}}^{n}\frac{1}{\xi_l-\xi_j} \tag{4.1.37}$$

The expression for the ground state energy becomes

$$\begin{aligned}
\Delta\Omega = 2N_i\Bigg\{&\int\frac{d^3p}{(2\pi)^3}\,n_F(\xi_p)V_{\mathbf{pp}}+\frac{1}{2}\int\frac{d^3p}{(2\pi)^3}\int\frac{d^3p'}{(2\pi)^3}\\
&\times V^2_{\mathbf{pp'}}\left[\frac{n_F(\xi_p)}{\xi_p-\xi_{p'}}+\frac{n_F(\xi_{p'})}{\xi_{p'}-\xi_p}\right]+\frac{1}{3}\int\frac{d^3p}{(2\pi)^3}\int\frac{d^3p'}{(2\pi)^3}\int\frac{d^3p''}{(2\pi)^3}\\
&\times V_{\mathbf{pp'}}V_{\mathbf{p'p''}}V_{\mathbf{p''p}}\left[\frac{n_F(\xi_p)}{(\xi_p-\xi_{p'})(\xi_p-\xi_{p''})}+\frac{n_F(\xi_{p'})}{(\xi_{p'}-\xi_p)(\xi_{p'}-\xi_{p''})}\right.\\
&\left.+\frac{n_F(\xi_{p''})}{(\xi_{p''}-\xi_p)(\xi_{p''}-\xi_{p'})}\right]+\cdots\Bigg\}
\end{aligned} \tag{4.1.38}$$

There are two terms proportional to V^2. They are identical, after changing around the dummy variables of integration, so this cancels the $\frac{1}{2}$ factor. Similarly, the three terms with three powers of V are all equal, which cancels the $\frac{1}{3}$ factor. The $1/n$ factor is canceled in the nth term. We get the result

$$\begin{aligned}
\Delta\Omega = 2N_i\int\frac{d^3p}{(2\pi)^3}\,n_F(\xi_p)\bigg[&V_{\mathbf{pp}}+\int\frac{d^3p'}{(2\pi)^3}\,\frac{V_{\mathbf{pp'}}V_{\mathbf{p'p}}}{\xi_p-\xi_p}\\
&+\int\frac{d^3p'}{(2\pi)^3}\int\frac{d^3p''}{(2\pi)^3}\,\frac{V_{\mathbf{pp'}}V_{\mathbf{p'p''}}V_{\mathbf{p''p}}}{(\xi_p-\xi_{p'})(\xi_p-\xi_{p''})}+\cdots\bigg]
\end{aligned}$$

The series in brackets is just the reaction matrix equation (4.1.13), so that we can write this as

$$\begin{aligned}
R_{\mathbf{pp}} &= V_{\mathbf{pp}}+P\int\frac{d^3p'}{(2\pi)^3}\,\frac{V_{\mathbf{pp'}}V_{\mathbf{p'p}}}{\xi_p-\xi_{p'}}\\
&\quad+P\int\frac{d^3p}{(2\pi)^3}\int\frac{d^3p''}{(2\pi)^3}\,\frac{V_{\mathbf{pp'}}V_{\mathbf{p'p''}}V_{\mathbf{p''p}}}{(\xi_p-\xi_{p'})(\xi_p-\xi_{p''})}
\end{aligned} \tag{4.1.39}$$

$$\Delta\Omega = 2N_i\int\frac{d^3p}{(2\pi)^3}\,n_F(\xi_p)R_{\mathbf{pp}} = -\frac{2N_i}{\pi m}\int_0^{n_F} p\,dp\sum_l(2l+1)\tan[\delta_l(p)]$$

This gives $\tan\delta_l$, which is the wrong answer.

Did you find the wrong step? The reaction matrix equation (4.1.39) contains a P for principle parts. It is an instruction to omit the term $p = p^{(n)}$ where the energy denominator vanishes. The T-matrix equation does not have this instruction, because one is supposed to include the terms with $p = p^{(n)}$. Our derivation mishandled these terms. Our "theorem" (4.1.37) is correct only as long as the ξ_j are different; it is incorrect if they are alike. The correct result, valid for all cases, is obtained by evaluating the contour integral (4.1.36) by drawing a branch cut along the real axis. The summation over Matsubara frequencies is the difference between the integration along the top and bottom of the branch cut (see Sec. 3.5):

$$\frac{1}{\beta}\sum_{ip}\prod_{j=1}\frac{1}{ip-\xi_j}$$
$$= -\int_{-\infty}^{\infty}\frac{d\varepsilon}{2\pi i}\,n_F(\varepsilon)\,\mathrm{Im}\left[\prod_j\left(\frac{1}{\varepsilon-\xi_j+i\delta}\right)-\prod_j\left(\frac{1}{\varepsilon-\xi_j-i\delta}\right)\right]$$
$$= -\int_{-\infty}^{\infty}\frac{d\varepsilon}{2\pi i}\,n_F(\varepsilon)\,\mathrm{Im}\left\{\prod_j\left[P\frac{1}{\varepsilon-\varepsilon_j}-i\pi\delta(\varepsilon-\varepsilon_j)\right]\right.$$
$$\left.-\prod_j\left[P\frac{1}{\varepsilon-\xi_j}+i\pi\delta(\varepsilon-\xi_j)\right]\right\}$$

When taking the imaginary part of the right-hand side, it is possible to get products of odd numbers of delta functions. For $n = 3$ the result is

$$\frac{1}{\beta}\sum_{ip}\frac{1}{(ip-\xi_1)(ip-\xi_2)(ip-\xi_3)}$$
$$=\frac{n_F(\xi_1)}{(\xi_1-\xi_2)(\xi_1-\xi_3)}+\frac{n_F(\xi_2)}{(\xi_2-\xi_1)(\xi_2-\xi_3)}+\frac{n_F(\xi^3)}{(\xi_3-\xi_1)(\xi_3-\xi_2)}$$
$$-\pi^2 n_F(\xi_1)\delta(\xi_1-\xi_2)\delta(\xi_1-\xi_3)$$

The third term in (4.1.38) should really be written as

$$\frac{1}{3}\int\frac{d^3p}{(2\pi)^3}\int\frac{d^3p'}{(2\pi)^3}\int\frac{d^3p''}{(2\pi)^3}\,V_{\mathbf{pp}'}V_{\mathbf{p}'\mathbf{p}''}V_{\mathbf{p}''\mathbf{p}}\left[\frac{n_F(\xi_p)}{(\xi_p-\xi_{p''})(\xi_p-\xi_{p'})}\right.$$
$$+\frac{n_F(\xi_{p'})}{(\xi_{p'}-\xi_p)(\xi_{p'}-\xi_{p''})}+\frac{n_F(\xi_{p''})}{(\xi_{p''}-\xi_p)(\xi_{p''}-\xi_{p'})}$$
$$\left.-\pi^2\delta(\xi_p-\xi_{p'})\delta(\xi_p-\xi_{p''})\right]$$

Similar correction terms, with odd powers of delta functions, occur in the fourth and all higher terms in the series (4.1.38). An examination of these terms shows that we are generating a series of reaction matrices:

$$\begin{aligned}\Delta\Omega = 2N_i \int \frac{d^3p}{(2\pi)^3}\, n_F(\xi_p)\Big\{R_{\mathbf{pp}} - \frac{1}{3}\int \frac{d^3p_1\, d^3p_2}{(2\pi)^6} \\ \times R_{\mathbf{pp}_1}R_{\mathbf{p}_1\mathbf{p}_2}R_{\mathbf{p}_2\mathbf{p}}\pi^2\delta(\xi_p - \xi_{p_1})\delta(\xi_p - \xi_{p_2}) \\ + \frac{1}{5}\int \frac{d^3p_1\, d^3p_2\, d^3p_3\, d^3p_4}{(2\pi)^{12}} R_{\mathbf{pp}_1}R_{\mathbf{p}_1\mathbf{p}_2}R_{\mathbf{p}_2\mathbf{p}_3}R_{\mathbf{p}_3\mathbf{p}_4}R_{\mathbf{p}_4\mathbf{p}} \\ \times \prod_{j=1}^{4} [\pi\delta(\xi_p - \xi_p{}^{(j)})] \cdots\Big\}\end{aligned}$$

The wave vector integrals may be done term by term. The $dp^{(n)}$ integral eliminates the delta function $\delta(\xi_{\mathbf{p}} - \xi_{\mathbf{p}}{}^{(n)})$ and makes $p^{(n)} = p$. The angular integrations force all l components to be the same. Thus we derive the series

$$\begin{aligned}\Delta\Omega = \frac{2N_i}{\pi m}\int_0^\infty p\, dp n_F(\xi_p) \sum_l (2l+1) \\ \times \{[4\pi^2\varrho R_l(p, p)] - \tfrac{1}{3}[4\pi^2\varrho R_l(p, p)]^3 + \tfrac{1}{5}(4\pi^2\varrho R_l)^5 \cdots\}\end{aligned}$$

The factor $4\pi^2\varrho R_l(p, p) = \tan[\delta_l(p)]$, where $\varrho = mp/2\pi^2$ and $R_l = [\tan(\delta_l)]/2mp$. The series is just that for the arctan $x = x - \frac{1}{3}x^3 + \frac{1}{5}x^5 \cdots$, where $x = \tan\delta$. Thus we derive the ground state energy per particle as given by Fumi's theorem:

$$\Delta\Omega = -\frac{2N_i}{\pi m}\int_0^\infty p\, dp n_F(\xi_p) \sum_l (2l+1)\delta_l(p)$$

This derivation contains two messages. The first is that Fumi's theorem with δ_l is correct and may be obtained by using the conventional formulas for the ground state energy of the system. The second message is that the energy of the system is *not* just a simple average of the single-particle energies, such as $\sum_{\mathbf{p}} n_F(\xi_p)R_{\mathbf{pp}}$ or $\sum_{\mathbf{p}} n_F(\xi_p)T_{\mathbf{pp}}$. Both of the latter guesses are incorrect.

A more elegant derivation of Fumi's theorem is possible by following the techniques of Langer and Ambegaokar (1961). This avoids evaluating the series (4.1.35) on a term-by-term basis. It may be summed, at least formally, by recognizing that it is the power series expansion of the operator function,

$$\Delta\Omega = -2N_i \int \frac{d^3p}{(2\pi)^3}\, \frac{1}{\beta}\sum_{ip} \langle \mathbf{p} \,|\, \ln\Big(1 - \frac{1}{ip - H_0} V\Big) \,|\, \mathbf{p}\rangle$$

where the operator expansion is

$$-\langle \mathbf{p} \mid \ln\left(1 - \frac{1}{ip - H_0} V\right) \mid \mathbf{p}\rangle$$

$$= \frac{1}{ip - \xi_p}\left[V_{\mathbf{pp}} + \frac{1}{2}\int \frac{d^3p'}{(2\pi)^3}\frac{V_{\mathbf{pp'}}V_{\mathbf{p'p}}}{ip - \xi_{p'}} + \frac{1}{3}\int \frac{d^3p'\,d^3p''}{(2\pi)^3}\frac{V_{\mathbf{pp'}}V_{\mathbf{p'p''}}V_{\mathbf{p''p}}}{(ip - \xi_{p'})(ip - \xi_{p''})} + \cdots\right]$$

where $\langle \mathbf{p} \mid$ and $\mid \mathbf{p}\rangle$ are the plane-wave states which are eigenstates of the free-particle Hamiltonian H_0 and V is the impurity potential. The summation over Matsubara frequencies is evaluated next by introducing the contour integral:

$$\oint \frac{dz}{2\pi i} n_F(z) \ln\left(1 - \frac{1}{z - H_0} V\right)$$

It has a branch cut along the real axis, and the summation over ip is equal to the contribution from integrating along this branch cut:

$$\frac{1}{\beta}\sum_{ip} \ln(\cdots) = -\int_{-\infty}^{\infty} \frac{d\varepsilon}{2\pi} n_F(\varepsilon) \operatorname{Im}\left(\ln\left\{\frac{1 - [1/(\varepsilon + i\delta - H_0)]V}{1 - [1/(\varepsilon - i\delta - H_0)]V}\right\}\right)$$

The next step is to recognize that the operator equation for the T matrix is

$$T_{pp'}(\varepsilon \pm i\delta) = \langle p \mid V \frac{1}{1 - [1/(\varepsilon \pm i\delta - H_0)]V} \mid p'\rangle$$

The argument of the logarithm can be cast into T matrices by inserting the operator V in both the numerator and denominator. Langer and Ambegaokar show that these operator manipulations can be justified. Thus the ground state energy is

$$\Delta\Omega = 2N_i \int \frac{d^3p}{(2\pi)^3} \int_{-\infty}^{\infty} \frac{d\varepsilon}{2\pi} n_F(\varepsilon) \langle \mathbf{p} \mid \operatorname{Im}\left\{\ln\left[\frac{T(\varepsilon - i\delta)}{T(\varepsilon + i\delta)}\right]\right\} \mid \mathbf{p}\rangle$$

The T matrix has an amplitude and a phase:

$$T(\varepsilon \pm i\delta) = \mid T \mid e^{\pm i\phi(\varepsilon)}$$

$$\Delta\Omega = -2N_i \int \frac{d^3p}{(2\pi)^3} \int_{-\infty}^{\infty} \frac{d\varepsilon}{\pi} n_F(\varepsilon) \langle p \mid \phi(\varepsilon) \mid p\rangle$$

The ground state energy is just the phase of the T matrix. For spherically symmetric scattering centers, this is $\phi(\varepsilon) = \delta(\varepsilon - \varepsilon_p)\delta_l(\varepsilon_p)$, which proves Fumi's theorem again.

4.2. LOCALIZED STATE IN THE CONTINUUM

In this section we shall solve exactly the Hamiltonian

$$H = \varepsilon_c b^\dagger b + \sum_{\mathbf{k}} \varepsilon_{\mathbf{k}} c_{\mathbf{k}}^\dagger c_{\mathbf{k}} + \sum_{\mathbf{k}} A_{\mathbf{k}}(c_{\mathbf{k}}^\dagger b + b^\dagger c_{\mathbf{k}}) \tag{4.2.1}$$

It describes a localized state of fixed energy ε_c and operators b and $b^\dagger$. This localized state will be called the impurity, and we assume only one exists. There is a continuous set of states of energy $\varepsilon_{\mathbf{k}}$ with operators $c_{\mathbf{k}}$ and $c_{\mathbf{k}}^\dagger$. This set of states could have a finite bandwidth, as often occurs in tight binding models in solids, or else it could be a free-particle model. The last term in the Hamiltonian includes the mixing between these two kinds of states. It contains processes whereby the continuum particle hops onto the impurity ($b^\dagger c_{\mathbf{k}}$) and where the particle hops off the impurity into the continuum ($c_{\mathbf{k}}^\dagger b$). If the particles have spin, it is assumed that this hopping on and off the impurity preserves the particle spin state. Thus the spin never changes and is unimportant, so its dependence is suppressed in all subscripts and labels. If the hopping particle could change its spin orientation, the problem would become harder.

Since the Hamiltonian is quadratic in operators, its solution is equivalent to diagonalizing a matrix. The solution may be obtained in this fashion, although we shall not do it this way. Since the Hamiltonian is quadratic, the statistics are irrelevant—the same eigenvalues are obtained for fermions or bosons. It is also irrelevant whether there are one or many particles in the system. For a Hamiltonian which contains only quadratic operators of fermions or bosons, one just diagonalizes the Hamiltonian to find the eigenstates, and then all eigenstates behave independently.

This model Hamiltonian was introduced simultaneously by Anderson (1961) and Fano (1961). Anderson applied it to solid-state physics, while Fano used it in atomic spectra. It tends to be called the Anderson model or the Fano model depending on whether the speaker is a solid-state or atomic physicist. We shall call it the *Fano–Anderson model*. As explained in Chapter 1, it is related to the famous Anderson model, which is (4.2.1) plus another term.

The Hamiltonian will first be solved without using Green's functions.

Afterwards, the Green's function solution will be given. The nature of the solution depends critically on whether the energy ε_c is within the band of states ε_k. Thus if the continuous band of states $\varepsilon_{\mathbf{k}}$ is confined to the range

$$w_1 < \varepsilon_{\mathbf{k}} < w_2$$

then the solution depends on whether ε_c is also within this range. Actually, this statement is incorrect. Because of the interactions with the continuous band of states, the energy of the localized state is altered to a new value $\bar{\varepsilon}_c$. Of course, since we have not yet solved the problem, we do not yet know how to find this renormalized energy $\bar{\varepsilon}_c$. We shall jump ahead and give the result that the new energy is

$$\bar{\varepsilon}_c = \varepsilon_c + \sum_{\mathbf{k}} \frac{A_{\mathbf{k}}^2}{\bar{\varepsilon}_c - \varepsilon_{\mathbf{k}}} \tag{4.2.2}$$

If this value is within the range

$$w_1 \leq \bar{\varepsilon}_c \leq w_2$$

then the solution has an important property: There are no localized states in the system. A continuum particle may hop onto the impurity, but after a while it may hop off again. Thus, particles spend only part time on the impurity, so this is not a well-defined eigenstate. The impurity state has become a scattering resonance.

Of course, if the new energy $\bar{\varepsilon}_c$ is outside of the band of continuum states, then a true localized state will exist. Then the solution has a distinct form—a real bound state.

We shall first solve for the case where $\bar{\varepsilon}_c$ is within the continuous band and no bound state exists. All states are continuum states. We define a new set of operators $\alpha_{\mathbf{k}}$ and $\alpha_{\mathbf{k}}^\dagger$ which refer to the eigenstates of (4.2.1). The old operators b and $c_{\mathbf{k}}$ can be expanded in terms of this new set:

$$\begin{aligned} b &= \sum_{\mathbf{k}} \nu_{\mathbf{k}} \alpha_{\mathbf{k}} \\ c_{\mathbf{k}} &= \sum_{\mathbf{k}'} \eta_{\mathbf{k},\mathbf{k}'} \alpha_{\mathbf{k}'} \end{aligned} \tag{4.2.3}$$

One impurity in the presence of $N \approx 10^{23}$ particles changes their energy by a negligible amount—remember the assumption of no bound states—so that these new operators still have the energy $\varepsilon_{\mathbf{k}}$. All that is needed is the vector $\nu_{\mathbf{k}}$ and the matrix $\eta_{\mathbf{k},\mathbf{k}'}$ for a complete solution. We shall also solve the model now in one dimension. The extension to two and three dimensions will be easy to describe at the end. It is also assumed that the dispersion $\varepsilon_{\mathbf{k}}$ does not permit two states $\mathbf{k}$ and $\mathbf{k}'$ to have the same energy $\varepsilon_{\mathbf{k}} = \varepsilon_{\mathbf{k}'}$. Of course, this is never realized in practice, since states of $\mathbf{k}$ and $-\mathbf{k}$ usually have the same energy. The problem is that the hopping on and off the impurity preserves energy but not wave vector information. Thus the hopping will mix the states which exist at the same energy. This is a nuisance to describe, so it will be omitted for the moment, and $\mathbf{k}$ becomes a scalar.

Since the new operators α_k and $\alpha_k^\dagger$ describe eigenstates, the Hamiltonian may be written as

$$H = \sum_k \varepsilon_k \alpha_k^\dagger \alpha_k$$

The commutator $[b, H]$ is evaluated using both the new operators and the old:

$$[b, H] = b\varepsilon_c + \sum_k A_k c_k = \sum_k \varepsilon_k \nu_k \alpha_k$$

Then the old operators are expressed as (4.2.3), and this equation is

$$\varepsilon_c \sum_k \nu_k \alpha_k + \sum_{kk'} A_k \eta_{k,k'} \alpha_{k'} = \sum_k \varepsilon_k \nu_k \alpha_k$$

Each coefficient of α_k is independent; i.e., take the commutator of this equation with $\alpha_k^\dagger$ and get

$$\nu_k(\varepsilon_k - \varepsilon_c) = \sum_{k'} A_{k'} \eta_{k',k} \tag{4.2.4}$$

The same procedure is used with the commutator $[c_k, H]$, which gives

$$[c_k, H] = \varepsilon_k c_k + A_k b = \sum_{k'} \eta_{k,k'} \varepsilon_{k'} \alpha_{k'}$$

or expressing b and c_k as $\alpha_{k'}$ gives another equation:

$$\eta_{k,k'}(\varepsilon_k - \varepsilon_{k'}) = -A_k \nu_{k'}$$

The last equation gives a result for $\eta_{k,k'}$ when $\varepsilon_k \neq \varepsilon_{k'}$, which means that $k \neq k'$, because of our assumptions. It provides no information about $\eta_{k,k'}$ for the case where $k = k'$. It is necessary to introduce another unknown function Z_k which is proportional to the value of $\eta_{k,k'}$ when $k = k'$:

$$\eta_{k,k'} = -\frac{A_k \nu_{k'}}{\varepsilon_k - \varepsilon_{k'}} + \delta_{kk'} Z_k \nu_k A_k \tag{4.2.5}$$

The other factors $A_k \nu_k$ are added to the last term for convenience. The energy denominator in the first term is taken as a principal part; the term is omitted when $\varepsilon_k = \varepsilon_{k'}$. This will be the case for all such energy denominators, and the conventional symbol P will be omitted. If this expression for $\eta_{k,k'}$ is used in (4.2.4), the equation for ν_k, it becomes

$$\nu_k(\varepsilon_k - \varepsilon_c) = \sum_{k'} A_{k'}\left(-\frac{A_{k'}\nu_k}{\varepsilon_{k'} - \varepsilon_k} + \delta_{kk'}Z_k\nu_k A_k\right)$$

The ν_k dependence factors out completely. This leaves us with an equation in which the only unknown quantity is Z_k,

$$\varepsilon_k - \varepsilon_c = -\sum_{k'} \frac{A_{k'}^2}{\varepsilon_{k'} - \varepsilon_k} + Z_k A_k^2$$

so that this quantity is now determined. It also simplifies the notation to introduce the self-energy function

$$\Sigma(\varepsilon_k) = \sum_{k'} \frac{A_{k'}^2}{\varepsilon_k - \varepsilon_{k'}} \tag{4.2.6}$$

so that Z_k is written as

$$Z_k = \frac{1}{A_k^2}\,[\varepsilon_k - \varepsilon_c - \Sigma(\varepsilon_k)] \tag{4.2.7}$$

We repeat that all quantities on the right-hand side of this result are known, so that Z_k is now known. To make further progress toward obtaining $\eta_{k,k'}$ and ν_k, we need more equations. These are obtained from the commutation relations for the old operators, which must still be obeyed when they are expressed in terms of the α_ks:

$$[b, b^\dagger] = 1 = \sum_k \nu_k^2 \tag{4.2.8}$$

$$[c_k, c_{k'}^\dagger] = \delta_{kk'} = \sum_{k''} \eta_{k,k''}\eta_{k,k'} \tag{4.2.9}$$

$$[b, c_k^\dagger] = 0 = \sum_{k'} \eta_{k,k'}\nu_{k'} \tag{4.2.10}$$

The last equation is used first. If the result (4.2.5) for $\eta_{k,k'}$ is used, (4.2.10) becomes

$$0 = \sum_{k'} \nu_{k'}\left(-\frac{A_k\nu_{k'}}{\varepsilon_k - \varepsilon_{k'}} + \delta_{kk'}Z_k\nu_k A_k\right)$$

or

$$0 = A_k\left(Z_k\nu_k^2 + \sum_{k'} \frac{\nu_{k'}^2}{\varepsilon_{k'} - \varepsilon_k}\right) \tag{4.2.11}$$

Since A_k is not zero, the quantity in parentheses must vanish. This result will be used below. Next we take the second commutator (4.2.9), insert the result for $\eta_{k,k'}$, and find

$$\delta_{kk'} = \sum_{k'} \left(\frac{A_k \nu_{k''}}{\varepsilon_{k''} - \varepsilon_k} + \delta_{kk''} Z_k \nu_k A_k \right) \left(\frac{A_{k'} \nu_{k''}}{\varepsilon_{k''} - \varepsilon_{k'}} + \delta_{k'k''} Z_{k'} \nu_{k'} A_{k'} \right)$$

This equals

$$\delta_{kk'} = \delta_{kk'} Z_k^2 \nu_k^2 A_k^2 + \frac{A_k A_{k'}}{\varepsilon_k - \varepsilon_{k'}} (Z_k \nu_k^2 - Z_{k'} \nu_{k'}^2) + A_k A_{k'} \sum_{k''} \times \frac{\nu_{k''}^2}{(\varepsilon_{k''} - \varepsilon_k)(\varepsilon_{k''} - \varepsilon_{k'})} \tag{4.2.12}$$

The last term must be rearranged. Since the energy denominators are principal parts, we must use *Poincaré's theorem*:

$$P \frac{1}{\varepsilon_{k''} - \varepsilon_k} \frac{1}{\varepsilon_{k''} - \varepsilon_{k'}} = P \frac{1}{\varepsilon_k - \varepsilon_{k'}} \left(\frac{1}{\varepsilon_{k''} - \varepsilon_k} - \frac{1}{\varepsilon_{k''} - \varepsilon_{k'}} \right) + \pi^2 \delta(\varepsilon_{k''} - \varepsilon_k) \delta(\varepsilon_{k''} - \varepsilon_{k'})$$

The delta functions of energy must be changed back to Kronecker deltas, since we are using box normalization, in a box of length L. This alteration is

$$\delta(\varepsilon_k - \varepsilon_{k'}) = \frac{L}{2\pi v_k} \delta_{k,k'} \tag{4.2.13}$$

where v_k is the velocity of the particle:

$$v_k = \frac{\partial \varepsilon_k}{\partial k} \tag{4.2.14}$$

Thus we can regroup the terms in (4.2.12):

$$\delta_{kk'} = \delta_{kk'} \nu_k^2 A_k^2 \left[Z_k^2 + \left(\frac{L}{2v_k} \right)^2 \right] + \frac{A_k A_{k'}}{\varepsilon_k - \varepsilon_{k'}} \left[\left(Z_k \nu_k^2 + \sum_{k''} \frac{\nu_{k''}^2}{\varepsilon_{k''} - \varepsilon_k} \right) - \left(Z_{k'} \nu_{k'}^2 + \sum_{k''} \frac{\nu_{k''}^2}{\varepsilon_{k''} - \varepsilon_{k'}} \right) \right]$$

The last two quantities in parentheses vanish, since each is identical to (4.2.11). All the remaining terms are proportional to $\delta_{kk'}$ and so exist only when $k = k'$. This, finally, gives the equation for ν_k:

$$\nu_k^2 = \left\{ A_k^2 \left[Z_k^2 + \left(\frac{L}{2v_k} \right)^2 \right] \right\}^{-1} \tag{4.2.15}$$

Of course, once ν_k is known, then $\eta_{k,k'}$ is obtained easily from (4.2.5). The result for ν_k may be reworked into a form in which the physics is more transparent. The earlier result (4.2.7) for Z_k is used to rewrite (4.2.15) as

$$\nu_k^2 = \frac{A_k^2}{[\varepsilon_k - \varepsilon_c - \Sigma(\varepsilon_k)]^2 + (LA_k^2/2v_k)^2} \tag{4.2.16}$$

Furthermore, the self energy $\Sigma(\varepsilon_k)$ will be interpreted as the real part of the retarded self-energy:

$$\Sigma_{\text{ret}}(\varepsilon) = \sum_{k'} \frac{A_{k'}^2}{\varepsilon - \varepsilon_{k'} + i\delta}$$

$$\text{Re}[\Sigma_{\text{ret}}(\varepsilon_k)] = P \sum_{k'} \frac{A_{k'}^2}{\varepsilon_k - \varepsilon_{k'}} \equiv \Sigma(\varepsilon_k)$$

Similarly, the imaginary part of the retarded self-energy is

$$\text{Im}[\Sigma(\varepsilon)] = -\pi \sum_{k'} A_{k'}^2 \delta(\varepsilon - \varepsilon_{k'})$$

By using the previous identity (4.2.13) for the delta function, the imaginary part of the retarded self-energy becomes

$$\text{Im}[\Sigma(\varepsilon_k)] = -\pi \sum_{k'} A_{k'}^2 \left(\frac{L}{2\pi v_k}\right) \delta_{k,k'} = -\frac{L}{2v_k} A_k^2$$

The expression on the right is recognized as the same factor which occurs in the denominator of (4.2.16) for ν_k^2. In fact, this quantity may be re-written as

$$\nu_k^2 = -\left(\frac{2v_k}{L}\right) \frac{\text{Im}[\Sigma(\varepsilon_k)]}{\{\varepsilon_k - \varepsilon_c - \text{Re}[\Sigma(\varepsilon_k)]\}^2 + \{\text{Im}[\Sigma(\varepsilon_k)]\}^2}$$

$$= -\frac{2v_k}{L} \text{Im}\{[\varepsilon_k - \varepsilon_c - \Sigma_{\text{ret}}(\varepsilon_k)]^{-1}\}$$

so that ν_k^2 is proportional to the imaginary part of a retarded Green's function. It was shown in Sec. 3.3 that the imaginary part of a retarded Green's function is proportional to the spectral function:

$$A(\varepsilon) = -2\,\text{Im}\{[\varepsilon - \varepsilon_c - \Sigma_{\text{ret}}(\varepsilon)]^{-1}\}$$

$$\nu_k^2 = +\frac{v_k}{L} A(\varepsilon_k)$$

The remaining equation (4.2.8) which must be satisfied may now be considered:

$$1 = \sum_k \nu_k^2$$

Since we now know ν_k^2, this is readily substituted, and we obtain the integral:

$$1 = \sum_k \frac{v_k}{L} A(\varepsilon_k) = \frac{1}{2\pi} \int dk v_k A(\varepsilon_k)$$

The integration variable is altered to the energy of the particle, so that the integral is now

$$1 = \int \frac{d\varepsilon_k}{2\pi} A(\varepsilon_k)$$

This integral is just the sum rule for spectral functions which was proved in (3.3.16). The sum rule applies to any spectral function and must also apply to the particular one we have just derived. Thus (4.2.8) is satisfied. Thus, the transformation (4.2.3) to the α_k has been achieved. The form we have derived for ν_k satisfies all the commutation relations (4.2.8)–(4.2.10) as well as the commutation relations with the original Hamiltonian. The problem has been solved exactly.

Since we have identified the quantity

$$G(\varepsilon) = \frac{1}{\varepsilon - \varepsilon_c - \Sigma_{\text{ret}}(\varepsilon)} \tag{4.2.17}$$

as a retarded Green's function, we might inquire for which particle it is the Green's function. It is the Green's function of the localized particle, or what became of the localized particle after it became delocalized. This assertion is somewhat evident if we consider the definition of the localized retarded Green's function for fermions:

$$\begin{aligned} G(t) &= -i\langle b(t)b^\dagger(0) + b^\dagger(0)b(t)\rangle\theta(t) \\ &= -i\sum_k \nu_k^2\langle\alpha_k(t)\alpha_k^\dagger(0) + \alpha_k^\dagger(0)\alpha_k(t)\rangle\theta(t) \\ &= -i\sum_k \nu_k^2 e^{-i\varepsilon_k t}\theta(t) \\ &= \frac{-i}{L}\theta(t)\sum_k v_k A(\varepsilon_k)e^{-i\varepsilon_k t} \\ &= -i\theta(t)\int_{-\infty}^{\infty}\frac{d\varepsilon_k}{2\pi} A(\varepsilon_k)e^{-i\varepsilon_k t} \end{aligned} \tag{4.2.18}$$

In the last step we have changed the summation to an integral over ε_k. The last line is the identity for the retarded Green's function assigned in Problem 6 of Chapter 3.

The Fano–Anderson model is just a description of a localized scattering resonance. The continuum particles come to the impurity, spend some time in the resonant state, and then depart in another continuum state. This model is really no different from that of the prior section, on impurity scattering, if the impurity potential is made to have a resonance. The phase shifts for the present model are defined as

$$\tan[\delta(\varepsilon)] = \frac{\mathrm{Im}[\Sigma_{\mathrm{ret}}(\varepsilon)]}{\varepsilon - \varepsilon_c - \mathrm{Re}[\Sigma_{\mathrm{ret}}(\varepsilon)]}$$

The phase shift is all that is needed to use the results of the prior section to describe the Fano–Anderson model. The resonant behavior comes from approximating $-\mathrm{Im}(\Sigma) \approx \Gamma$ as a constant, or at least as a slowly varying function of energy. Then the spectral function $A(\varepsilon)$ is Lorentzian. If the width Γ is small, this describes a resonance.

As an example of a scattering potential, consider the potential $V(r)$ shown in Fig. 4.7(a). The resonance behavior is chosen by first solving for the bound states of Fig. 4.7(b). For the latter case, bound states exist when

$$\tan ka = -\frac{k}{\alpha}$$

$$\alpha = \left(\frac{2mV_0}{\hbar^2} - k^2\right)^{1/2}$$

We can make a bound state at half of the well depth by choosing $ka = \alpha a = 3\pi/4$, where k^2 is the particle energy relative to the bottom of the well. The well shape of Fig. 4.7(a) will have a resonance at nearly the same energy. Now the finite extent of the repulsive part will allow the particle to leave and impart a width to the state—and to the scattering resonance. The phase shift is shown in Fig. 4.7(c) for the potential of part (a). The steep rise in the phase shift, of about π, occurs at the value $ka \approx 3\pi/4$.

The change in phase shift by π across the resonance may be understood from the Friedel sum rule. If one fills up the eigenstates of the system, then one particle will reside in the vicinity of the impurity and contribute unity to the Friedel sum over δ/π.

The initial Hamiltonian was written as (4.3.1) because we wrote it in terms of eigenstates which were not orthogonal. All our effort was merely an exercise in orthogonalization. This could have been avoided by

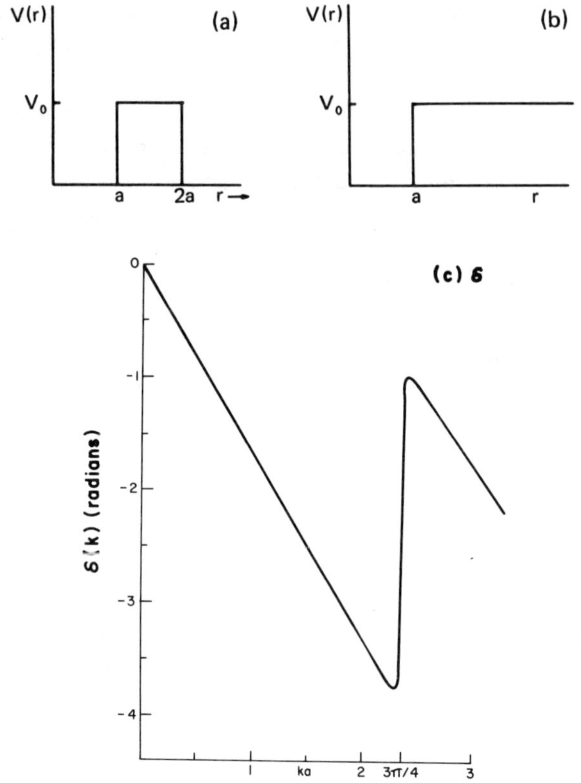

FIGURE 4.7. (a) The potential whose phase shifts can show a resonance behavior. The resonance occurs at the bound state of the potential shown in (b). (c) The phase shift jumps in value by almost π at the wave vector of the resonance.

writing H in an orthogonal basis. This basis would have the particles with a scattering resonance from the impurity. Thus the Hamiltonian describes one-particle behavior.

The model is very important in a number of applications. The d states of transition metals seem well described as scattering resonances of the *sp* electrons (see Anderson and McMillan, 1967). Another example is in surface physics. When atoms come from a gas and are absorbed on the surface of a metal, the conduction electrons of the metal may hop onto the atomic-like states of the absorbed atom. This charging effect on the absorbed atoms seems well described by the model (Schönhammer and Gunnarsson, 1977). It has also found numerous applications in atomic physics.

The model will now be solved by using Green's functions. This solution is much quicker than the prior method, since there is only one self-energy diagram. We write the Hamiltonian as $H_0 + V$, where V is the last term in (4.2.1) and H_0 is the first two terms. We start expanding the Green's function for the localized state:

$$\mathscr{G}(ip) = -\int_0^\beta d\tau e^{ip\tau}\langle Tb(\tau)b^\dagger(0)\rangle$$

The first self-energy term comes from the $n = 2$ term in the S-matrix expansion:

$$-\frac{1}{2}\int_0^\beta d\tau e^{ip\tau}\int_0^\beta d\tau_1\int_0^\beta d\tau_2 \sum_{k_1k_2} A_{k_1}A_{k_2}\langle Tb(\tau)[c_{k_1}(\tau_1)b^\dagger(\tau_1)$$
$$+ b(\tau_1)c_{k_1}^\dagger(\tau_1)][c_{k_2}(\tau_2)b^\dagger(\tau_2) + b(\tau_2)c_{k_2}^\dagger(\tau_2)]b^\dagger(0)\rangle$$

The correlation function is easily evaluated in terms of the unperturbed Green's functions of the localized state $\mathscr{G}^{(0)}(\tau)$ and the continuum states $\mathscr{G}^{(0)}(k, \tau)$ to give

$$+\sum_k A_k^2\int_0^\beta d\tau e^{ip\tau}\int_0^\beta d\tau_1\int_0^\beta d\tau_2 \mathscr{G}^{(0)}(\tau - \tau_1)\mathscr{G}^{(0)}(k, \tau_1 - \tau_2)\mathscr{G}^{(0)}(\tau_2)$$
$$= +\sum_k A_k^2\mathscr{G}^{(0)}(ip)^2\mathscr{G}^{(0)}(k, ip) = \mathscr{G}^{(0)}(p)^2\Sigma(ip)$$

where

$$\Sigma(ip) = \sum_k A_k^2\mathscr{G}^{(0)}(k, ip) = \sum_k \frac{A_k^2}{ip - \xi_k}$$

This turns out to be the only self-energy diagram. The higher terms in the S matrix only produce higher powers of this self-energy contribution:

$$\mathscr{G}(ip) = \mathscr{G}^{(0)}(ip)[1 + \mathscr{G}^{(0)}\Sigma + (\mathscr{G}^{(0)}\Sigma)^2 + \cdots]$$

This series may be summed to give the Dyson equation for the Matsubara Green's function:

$$\mathscr{G}(ip) = \frac{1}{ip - \varepsilon_c - \Sigma(ip)}$$

Changing $ip \to \varepsilon + \mu + i\delta$ gives the retarded Green's function,

$$G(\varepsilon) = \frac{1}{\varepsilon - \varepsilon_c - \Sigma_{\text{ret}}(\varepsilon)}$$

which is the same as noted earlier in (4.2.17). This proves that v_k^2 is really proportional to the spectral function of the $G(\varepsilon)$. The equivalence of this result to (4.2.18) is just an example of the general theorem proved in Problem 6 of Chapter 3:

$$G_{\text{ret}}(t) = -i\theta(t) \int_{-\infty}^{\infty} \frac{d\varepsilon}{2\pi} A(\varepsilon) e^{-i\varepsilon t}$$

The result may be generalized to higher dimension and other energy bands. This extension is best done using the Green's function technique, since the derivation is the easiest. The derivation is the same as in one dimension. The Matsubara form of the Green's function of the b operators (4.2.18) is

$$\mathscr{G}(ip) = \frac{1}{ip - \varepsilon_c - \Sigma(ip)}$$

where the self-energy operator is summed over all states in the system:

$$\Sigma(ip) = \sum_{\mathbf{k},\lambda} \frac{A_{\mathbf{k},\lambda}^2}{ip - \varepsilon_{\mathbf{k},\lambda}}$$

The real part of the denominator of the retarded Green's function is

$$\varepsilon - \varepsilon_c - \text{Re}[\Sigma_{\text{ret}}(\varepsilon)]$$

The Green's function has poles at the points where this denominator vanishes. The poles of the Green's function corresponds to excitations of the system. The poles in this case are at the energy $\bar{\varepsilon}_c$ which satisfies

$$\bar{\varepsilon}_c = \varepsilon_c + \text{Re}[\Sigma_{\text{ret}}(\bar{\varepsilon}_c)]$$

This expression for $\bar{\varepsilon}_c$ is just the result which was asserted earlier in (4.2.2). If this pole lies within the continuum of states, then the resonance occurs, and there is no bound state. But if the pole occurs outside of the band of continuum states, then the system has a real bound state. A pole occurs when $\text{Im}\,\Sigma = 0$ at the same point that the real part of the denominator vanishes. Generally, $\text{Im}\,\Sigma$ is not zero throughout the continuum band, so that $\text{Im}\,\Sigma = 0$ only outside of this band.

If we assume that the bound state occurs outside of the band, the spectral function has the form

$$A(\varepsilon) = 2\pi\delta\{\varepsilon - \varepsilon_c - \text{Re}[\Sigma(\varepsilon)]\} + \frac{-2\,\text{Im}[\Sigma(\varepsilon)]}{[\varepsilon - \varepsilon_c - \text{Re}(\Sigma)]^2 + [\text{Im}(\Sigma)]^2}$$

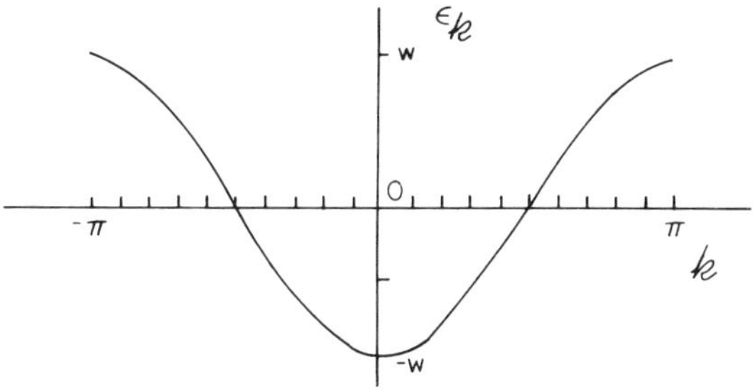

FIGURE 4.8. Tight binding energy bands in one dimension.

The first term comes from the pole of the Green's function, while the second term is from those parts where Im $\Sigma \neq 0$. The first term may be rewritten as a simple delta function plus a renormalization factor. This procedure was described earlier in (3.3.23):

$$A(\varepsilon) = 2\pi Z\delta(\varepsilon - \bar{\varepsilon}_c) - \frac{2\,\mathrm{Im}(\Sigma)}{[\varepsilon - \varepsilon_c - \mathrm{Re}(\Sigma)]^2 + [\mathrm{Im}(\Sigma)]^2}$$

$$Z = \left(1 - \frac{d\,\mathrm{Re}(\Sigma)}{d\varepsilon}\right)^{-1}_{\bar{\varepsilon}_c}$$

It may be possible that there is more than one bound state, and then the first term becomes a series of delta functions.

These points are well illustrated by an example. Take a one-dimensional tight binding model on a solid of unit separation $a = 1$ and length L with a constant coupling constant:

$$\varepsilon_k = -w \cos k$$

$$A_k^2 = \frac{C}{L}$$

Also, take the initial bound state energy ε_c in the middle of the band $\varepsilon_c = 0$. The band structure is shown in Fig. 4.8. The Brillouin zone is defined by $-\pi \leq k \leq \pi$. The self-energy function Σ_{ret} is now elementary to evaluate:

$$\Sigma_{\mathrm{ret}}(\varepsilon) = \frac{C}{L}\sum_k \frac{1}{\varepsilon + w\cos k + i\delta} = \frac{C}{2\pi}\int_{-\pi}^{\pi} \frac{dk}{\varepsilon + w\cos k + i\delta}$$

The real part is

$$\mathrm{Re}(\Sigma) = \frac{C}{2\pi}\int_{-\pi}^{\pi} \frac{dk}{\varepsilon + w\cos k} = \frac{C\,\mathrm{sgn}\,\varepsilon}{(\varepsilon^2 - w^2)^{1/2}} \qquad \text{if } \varepsilon^2 > w^2$$

$$= 0 \qquad \text{if } \varepsilon^2 < w^2$$

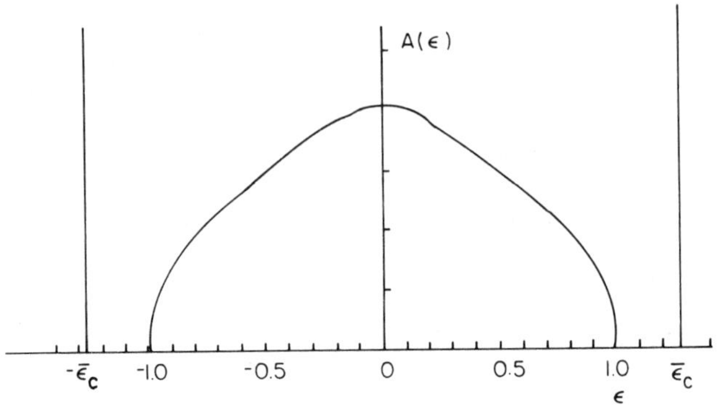

FIGURE 4.9. Spectral function from Eq. (4.2.19). $C = 1,\ w = 1.0$

where sgn ε is the sign of ε. The imaginary part is

$$-\mathrm{Im}[\Sigma(\varepsilon)] = \frac{C}{2\pi}\int_{-\pi}^{\pi} dk\pi\delta(\varepsilon + w\cos k)$$

$$= \frac{C}{w\,|\sin k\,|} = \frac{C}{(w^2 - \varepsilon^2)^{1/2}} \qquad \text{if } w^2 > \varepsilon^2$$

Thus the spectral function of the localized state is

$$A(\varepsilon) = \left\{2\pi\delta\left[\varepsilon - \frac{C}{(\varepsilon^2 - w^2)^{1/2}}\right] + 2\pi\delta\left[\varepsilon + \frac{C}{(\varepsilon^2 - w^2)^{1/2}}\right]\right\}\theta(\varepsilon^2 - w^2)$$
$$+ \frac{2C\theta(w^2 - \varepsilon^2)}{(w^2 - \varepsilon^2)^{1/2}\{\varepsilon^2 + [C^2/(w^2 - \varepsilon^2)]\}} \tag{4.2.19}$$

The step function $\theta(x)$ is unity if the argument is positive and zero if negative. The first term is finite only outside of the band of continuum states, while the last term is finite only inside the band. Figure 4.9 shows a plot of the entire spectral function. There are two sharp bound states, one below the band and one above. The continuous contribution throughout the band is due to the last term in (4.2.19). According to the sum rule, the total area under all the contributions must be 2π. The solution to the equation

$$\bar{\varepsilon}_c = \frac{C}{(\bar{\varepsilon}_c^{\,2} - w^2)^{1/2}}$$

is

$$\bar{\varepsilon}_c = \pm \frac{1}{(2)^{1/2}}\,[w^2 + (w^4 + 4C^2)^{1/2}]^{1/2}$$

and the renormalization factor is

$$Z_c = \frac{\bar{\varepsilon}_c^{\,2} - w^2}{2\bar{\varepsilon}_c^{\,2} - w^2}$$

and is the same at both poles. For strong coupling, where C becomes very large, $\bar{\varepsilon}_c \gg w$. In this case the renormalization factors approach $\frac{1}{2}$, so that one-half the spectral weight is in each pole.

4.3. INDEPENDENT BOSON MODELS

The independent boson model is very important in many-body physics. It is an exactly solvable model which describes some relaxation phenomena. It has become very useful for describing a wide variety of effects in solid-state physics. In this section we shall deal with it at great length. Two different derivations of the basic mathematical result will be provided. In addition, several variations on the model will also be briefly described. An exact solution may be obtained by a variety of techniques. We shall follow our usual procedure and solve it first by ordinary operator algebra. The second derivation will employ Green's functions.

The first Hamiltonian which will be solved is

$$H = c^\dagger c\left[\varepsilon_c + \sum_{\mathbf{q}} M_{\mathbf{q}}(a_{\mathbf{q}} + a_{\mathbf{q}}^\dagger)\right] + \sum_{\mathbf{q}} \omega_{\mathbf{q}} a_{\mathbf{q}}^\dagger a_{\mathbf{q}} \tag{4.3.1}$$

The Hamiltonian describes a fixed particle of energy ε_c interacting with a set of phonons with frequencies $\omega_{\mathbf{q}}$. The interaction occurs only when the state is occupied and $c^\dagger c = 1$. The phonons are the independent bosons. An alert reader will note that this was solved exactly in Chapter 1. By making a canonical transformation, the Hamiltonian may be rewritten as [see (1.3.3)]

$$\bar{H} = c^\dagger c(\varepsilon_c - \Delta) + \sum_{\mathbf{q}} \omega_{\mathbf{q}} a_{\mathbf{q}}^\dagger a_{\mathbf{q}}$$

where the self-energy is

$$\Delta = \sum_{\mathbf{q}} \frac{M_{\mathbf{q}}^2}{\omega_{\mathbf{q}}}$$

The solution to this problem is identical with the problem of a charge on a harmonic spring in a uniform electric field. The electric field causes a displacement of the charge to a new equilibrium position, about which it vibrates with the same frequency as before; see (1.1.11).

The present objective is to obtain a better description of the fluctuations about equilibrium. The self-energy Δ represents the zero temperature ground state configuration of the system. To study relaxation effects, we shall also need to understand the fluctuations. This is obtained from the following Green's function,

$$G(t) = -i\langle Tc(t)c^{\dagger}(0)\rangle \tag{4.3.2}$$

where a full description of the time variation is needed. We shall solve this for the real-time Green's function. This is permissible in the present case because the single impurity state $c^{\dagger}c$ will not alter the phonon energies. When we do the thermodynamic averaging over the phonon states, the perturbation V has no effect, so we do not need to worry about the perturbation expansion for the exp $(-\beta H)$ part. Thus we begin to solve to the Green's function of time at finite temperature.

A. Solution by Canonical Transformation

The Hamiltonian is first solved by a canonical transformation. This solution is nearly identical to the procedure described in Chapter 1, but here it is done with more rigor. A new Hamiltonian is desired by a transformation of the type

$$\bar{H} = e^{s}He^{-s} = c^{\dagger}c(\varepsilon_c - \Delta) + \sum_{\mathbf{q}} \omega_{\mathbf{q}}a_{\mathbf{q}}^{\dagger}a_{\mathbf{q}} \tag{4.3.3}$$

The transformation must be done so that $s^{\dagger} = -s$. The transformation on any product of operators is done by taking the product of the transformed operators. The last theorem is shown by inserting $1 = e^{-s}e^{s}$ between each operator:

$$e^{s}ABCD \cdots Ze^{-s} = (e^{s}Ae^{-s})(e^{s}Be^{-s})(e^{s}Ce^{-s}) \cdots e^{s}Ze^{-s} = \bar{A}\bar{B}\bar{C} \cdots \bar{Z}$$

If we assume that any function of operators may be expressed as a power series, then the transformation on a function of operators is just the function of the transformed operators:

$$e^{s}f(A)e^{-s} = e^{s} \sum_{n=0}^{\infty} a_n A^n e^{-s} = \sum_{n=0}^{\infty} a_n(\bar{A})^n = f(\bar{A}) \tag{4.3.4}$$

Thus we need to consider only the transformation on each operator separately, and the transformed Hamiltonian is the old one with the new operators. These are evaluated using

$$\bar{A} = e^{s}Ae^{-s} = A + [s, A] + \frac{1}{2!}[s, [s, A]] + \cdots$$

$$s = c^{\dagger}c\sum_{\mathbf{q}}\frac{M_{\mathbf{q}}}{\omega_{\mathbf{q}}}(a_{\mathbf{q}}{}^{\dagger} - a_{\mathbf{q}})$$

which gives

$$\begin{aligned}
\bar{c} &= cX \\
\bar{c}^{\dagger} &= c^{\dagger}X^{\dagger} \\
\bar{a} &= a - \frac{M_{\mathbf{q}}}{\omega_{\mathbf{q}}}c^{\dagger}c \\
\bar{a}^{\dagger} &= a^{\dagger} - \frac{M_{\mathbf{q}}}{\omega_{\mathbf{q}}}c^{\dagger}c
\end{aligned} \tag{4.3.5}$$

We have introduced the operator

$$X = \exp\left[-\sum_{\mathbf{q}}\frac{M_{\mathbf{q}}}{\omega_{\mathbf{q}}}(a_{\mathbf{q}}{}^{\dagger} - a_{\mathbf{q}})\right] \tag{4.3.6}$$

Since this commutes with the c operator, then the number operator is the same in the new representation,

$$\bar{c}^{\dagger}\bar{c} = c^{\dagger}cX^{\dagger}X = c^{\dagger}c$$

since

$$X^{\dagger} = X^{-1}$$

The transformed Hamiltonian is

$$\begin{aligned}
\bar{H} = \varepsilon_{c}\bar{c}^{\dagger}\bar{c} &+ \sum_{\mathbf{q}}\omega_{\mathbf{q}}\left(a_{\mathbf{q}}{}^{\dagger} - \frac{M_{\mathbf{q}}}{\omega_{\mathbf{q}}}c^{\dagger}c\right)\left(a_{\mathbf{q}} - \frac{M_{\mathbf{q}}}{\omega_{\mathbf{q}}}c^{\dagger}c\right) \\
&+ \sum_{\mathbf{q}}M_{\mathbf{q}}\left(a_{\mathbf{q}} + a_{\mathbf{q}}{}^{\dagger} - 2\frac{M_{\mathbf{q}}}{\omega_{\mathbf{q}}}c^{\dagger}c\right)c^{\dagger}c
\end{aligned}$$

which is

$$\bar{H} = c^{\dagger}c(\varepsilon_{c} - \Delta) + \sum_{q}\omega_{\mathbf{q}}a_{\mathbf{q}}{}^{\dagger}a_{\mathbf{q}}$$

This transformed Hamiltonian is precisely the form which was our objective in (4.3.3). It is also the form which was derived earlier in Chapter 1. The reason for repeating the analysis was that we needed the important factors X in (4.3.6).

This transformation is now applied to the Green's function (4.3.2). The factor $1 = e^{-s}e^{s}$ is inserted into the trace, say for $t > 0$:

$$\mathrm{Tr}(e^{-\beta H}e^{iHt}ce^{-itH}c^{\dagger}e^{-s}e^{s})$$

Using the cyclic properties of the trace, we may alter this to

$$\mathrm{Tr}(e^{s}e^{-\beta H}e^{iHt}ce^{-itH}c^{\dagger}e^{-s}) = \mathrm{Tr}(e^{-\beta \bar{H}}e^{i\bar{H}t}\bar{c}e^{-it\bar{H}}\bar{c}^{\dagger})$$

By using our previous theorems, we see that everything in the trace is now changed to the transformed representation. Thus the Green's function may be written as $(t > 0)$

$$G(t) = -ie^{\beta\Omega}\,\mathrm{Tr}(e^{-\beta\bar{H}}e^{i\bar{H}t}\bar{c}e^{-i\bar{H}t}\bar{c}^{\dagger})$$

It should be emphasized that this $G(t)$ will be exactly equal to the earlier definition (4.3.2). The new equation for $G(t)$ is just another way of evaluating the same thing. At first glance it appears that this evaluation is now trivial, since the Hamiltonian is diagonal in the $\bar{c}^{\dagger}\bar{c}$ operators. But this is untrue and misleading, since $\bar{c}$ and $\bar{c}^{\dagger}$ do not commute with a or $a^{\dagger}$ because of the X factor in (4.3.5). Thus it is necessary to stick with the c and a representation and to put in X explicitly. Thus the Green's function becomes

$$G(t) = -ie^{\beta\Omega}\,\mathrm{Tr}(e^{-\beta\bar{H}}e^{i\bar{H}t}cXe^{-i\bar{H}t}c^{\dagger}X^{\dagger})$$

However, it is now possible to achieve a great simplification. Since $\bar{H}$ is diagonal in c and a, it is easy to commute it through cX and obtain the time development of the operators:

$$e^{i\bar{H}t}cXe^{-i\bar{H}t} = e^{-it(\varepsilon_c-\Delta)}cX(t)$$

$$X(t) = \exp\left[-\sum_{\mathbf{q}}\frac{M_{\mathbf{q}}}{\omega_{\mathbf{q}}}(a_{\mathbf{q}}{}^{\dagger}e^{i\omega_q t} - a_{\mathbf{q}}e^{-i\omega_q t})\right]$$

The phonon and electron parts of the trace may now be completely separated:

$$G(t) = -ie^{\beta\Omega}\,\mathrm{Tr}(e^{-\beta\bar{\varepsilon}_c n}cc^{\dagger})\,\mathrm{Tr}[e^{-\beta\Sigma\omega_q n_q}X(t)X^{\dagger}(0)]$$

$$\bar{\varepsilon}_c = \varepsilon_c - \Delta$$

The result for $G(t)$ has the great simplification mentioned above. The particle part is trivial (assuming they are fermions),

$$e^{\beta\Omega_{\mathrm{el}}}\,\mathrm{Tr}(e^{-\beta n\bar{\varepsilon}_c}cc^{\dagger}) = 1 - n_F(\bar{\varepsilon}_c)$$

so that there remains only the problem of evaluating the phonon part of the trace. This evaluation is nontrivial, although it may be done exactly. The method we shall use was introduced by Feynman (1951).

B. Feynman Disentangling of Operators

The objective now is to evaluate the trace over the phonon distributions of the operator:

$$F(t) = e^{\beta\Omega_{\text{ph}}}\,\text{Tr}[e^{-\Sigma_{\mathbf{q}}\beta n_{\mathbf{q}}\omega_{\mathbf{q}}}X(t)X^{\dagger}(0)]$$

Each wave vector state q is averaged independently, and the final result is the product over $\mathbf{q}$ states:

$$F(t) = \prod_{\mathbf{q}} \mathscr{F}_{\mathbf{q}}(t) \tag{4.3.7}$$

$$\mathscr{F}_{\mathbf{q}}(t) = e^{\beta\Omega_{\mathbf{q}}}\,\text{Tr}\{e^{-\beta n_{\mathbf{q}}\omega_{\mathbf{q}}}\exp[(-M_{\mathbf{q}}/\omega_{\mathbf{q}})(a_{\mathbf{q}}{}^{\dagger}e^{i\omega_{\mathbf{q}}t} - a_{\mathbf{q}}e^{-i\omega_{\mathbf{q}}t})]e^{(M_{\mathbf{q}}/\omega_{\mathbf{q}})(a_{\mathbf{q}}{}^{\dagger}-a_{\mathbf{q}})}\}$$

For each state $\mathbf{q}$, the trace is merely a summation over all possible integer values of n_q between zero and infinity,

$$e^{\beta\Omega_{\mathbf{q}}}\,\text{Tr}(\quad) = (1 - e^{-\beta\omega_{\mathbf{q}}})\sum_{n_{\mathbf{q}}=0}^{\infty}\langle n_q \mid (\quad) \mid n_q\rangle$$

where the prefactor is the normalization:

$$e^{\beta\Omega_q} = \left(\sum_{n_q=0}^{\infty} e^{-\beta n_{\mathbf{q}}\omega_{\mathbf{q}}}\right)^{-1} = 1 - e^{-\beta\omega_{\mathbf{q}}}$$

We shall simplify the present notation by first dropping all $\mathbf{q}$ subscripts. We shall also use $\lambda = M_{\mathbf{q}}/\omega_{\mathbf{q}}$, so that we need to find

$$\mathscr{F} = (1 - e^{-\beta\omega})\sum_{n=0}^{\infty}\langle n \mid e^{-\beta n\omega}\exp[-\lambda(a^{\dagger}e^{i\omega t} - ae^{-i\omega t})]e^{\lambda(a^{\dagger}-a)} \mid n\rangle \tag{4.3.8}$$

The state $\mid n\rangle$ is the state with n excitations and is given in terms of the operators as

$$\mid n\rangle = \frac{(a^{\dagger})^n}{(n!)^{1/2}}\mid 0\rangle$$

The first step is to separate the operators in the exponential. This step is where we use Feynman's theorem on the disentangling of operators, which is as follows.

Theorem: If the operators A and B have the property that their commutator $C = [A, B]$ commutes with both A and B, then

$$e^{A+B} = e^{A}e^{B}e^{-1/2[A,B]} \tag{4.3.9}$$

This theorem was proved in Problem 5 of Chapter 2. It is used to separate the exponents in $X(t)$ and $X^{\dagger}(0)$. To evaluate $X(t)$, set

$$A = -\lambda a^{\dagger}e^{i\omega t}$$

$$B = \lambda a e^{-i\omega t}$$

so that

$$[A, B] = \lambda^2$$

and we obtain

$$\begin{aligned} X(t) &= e^{A+B} \\ &= e^{-1/2\lambda^2} \exp(-\lambda a^{\dagger}e^{i\omega t}) \exp(\lambda a e^{-i\omega t}) \end{aligned}$$

The result for $X^{\dagger}(0)$ is just the Hermitian conjugate at $t = 0$:

$$X^{\dagger}(0) = e^{-1/2\lambda^2}e^{\lambda a^{\dagger}}e^{-\lambda a}$$

$$X(t)X^{\dagger}(0) = e^{-\lambda^2}e^{-\lambda a^{\dagger}(t)}e^{\lambda a(t)}e^{\lambda a^{\dagger}}e^{-\lambda a}$$

The next step is to get all the destruction operators on the right and the creation operators on the left. Thus we need to exchange the center two operators. Since they do not commute, this exchange will produce another complex phase factor. These two operators are written as

$$e^{\lambda a(t)}e^{\lambda a^{\dagger}} = e^{\lambda a^{\dagger}}[e^{-\lambda a^{\dagger}}e^{\lambda a(t)}e^{\lambda a^{\dagger}}]$$

The factor in brackets has exactly the form derived earlier in (4.3.4). Thus if we evaluate

$$\begin{aligned} e^{-\lambda a^{\dagger}}ae^{-i\omega t}e^{\lambda a^{\dagger}} &= e^{-i\omega t}\left(a - \lambda[a^{\dagger}, a] + \frac{\lambda^2}{2}[a^{\dagger}[a^{\dagger}, a]] \cdots\right) \\ &= e^{-i\omega t}(a + \lambda) \end{aligned}$$

then

$$e^{-\lambda a^{\dagger}}e^{\lambda a(t)}e^{\lambda a^{\dagger}} = \exp[\lambda e^{-i\omega t}(a + \lambda)] = \exp(\lambda^2 e^{-i\omega t})e^{\lambda a(t)}$$

$$e^{\lambda a(t)}e^{\lambda a^{\dagger}} = \exp(\lambda^2 e^{-i\omega t})e^{\lambda a^{\dagger}}e^{\lambda a(t)}$$

so (4.3.8) is finally arranged into the desired form:

$$\mathscr{F}(t) = (1 - e^{-\beta\omega}) \exp[-\lambda^2(1 - e^{-i\omega t})]$$
$$\times \sum_{n=0}^{\infty} \langle n \mid e^{-\beta n\omega} \exp[\lambda a^\dagger(1 - e^{i\omega t})] \exp[-\lambda a(1 - e^{-i\omega t})] \mid n\rangle$$

All the terms with a can be collected together in exponentials since all these terms commute—and likewise for all the terms with $a^\dagger$. Next we wish to prove that

$$(1 - e^{-\beta\omega}) \sum_{n=0}^{\infty} e^{-\beta n\omega} \langle n \mid e^{u^* a^\dagger} e^{-ua} \mid n\rangle = e^{-|u|^2 N}$$
$$N = \frac{1}{e^{\beta\omega} - 1} \tag{4.3.10}$$

where, for our case, $u = \lambda(1 - e^{-i\omega t})$. Equation (4.3.10) is proved by expanding the exponents in a power series:

$$e^{-ua} \mid n\rangle = \sum_{l=0}^{\infty} \frac{(-u)^l}{l!} a^l \mid n\rangle$$

Now recall the properties of destruction operators acting on a state:

$$a \mid n\rangle = n^{1/2} \mid n - 1\rangle$$
$$a^2 \mid n\rangle = [n(n-1)]^{1/2} \mid n - 2\rangle$$
$$a^l \mid n\rangle = \left[\frac{n!}{(n-l)!}\right]^{1/2} \mid n - l\rangle$$

The feature that $a^l \mid n\rangle = 0$ for $l > n$ is very useful: It terminates the power series after n terms:

$$e^{-ua} \mid n\rangle = \sum_{l=0}^{n} \frac{(-u)^l}{l!} \left[\frac{n!}{(n-l)!}\right]^{1/2} \mid n - l\rangle$$

Of course, this is why the destruction operators were arranged to the right. The other operator may be taken to operate the left and just produces the Hermitian conjugate of the above result:

$$\langle n \mid e^{u^* a^\dagger} = \sum_{m=0}^{n} \frac{(u^*)^m}{m!} \left[\frac{n!}{(n-m)!}\right]^{1/2} \langle n - m \mid$$

These two results must be multiplied together. Using the basic orthogonality of the states,

$$\langle n - m \mid n - l \rangle = \delta_{n-m,n-l} \equiv \delta_{m,l}$$

produces the compact result

$$\langle n \mid e^{u^* a^\dagger} e^{-ua} \mid n \rangle = \sum_{l=0}^{n} \frac{(-\mid u \mid^2)^l}{(l!)^2} \frac{n!}{(n-l)!}$$

This power series should be familiar to every student of physics: It is just the Laguerre polynomial of order n (remember hydrogen wave functions):

$$\langle n \mid e^{u^* a^\dagger} e^{-ua} \mid u \rangle = L_n(\mid u \mid^2)$$

The final step is to sum the series over n. The last series is just the generating function of Laguerre polynomials:

$$(1 - z)^{-1} e^{[\mid u \mid^2 z/(z-1)]} = \sum_{n=0}^{\infty} L_n(\mid u \mid^2) z^n$$

In our case, to prove the theorem (4.3.10), we identify

$$z = e^{-\beta\omega}$$

$$\frac{z}{z-1} = -N$$

When these factors are collected, this does prove (4.3.10). The result for $\mathscr{F}(t)$ is

$$\mathscr{F}(t) = e^{-\phi(t)}$$

$$\phi(t) = \lambda^2[(1 - e^{-i\omega t}) + N \mid 1 - e^{i\omega t} \mid^2]$$

The factor $\phi(t)$ may also be written as

$$\phi(t) = \lambda^2[(N+1)(1 - e^{-i\omega t}) + N(1 - e^{i\omega t})]$$

We return to (4.3.7) and reintroduce the product over all q states. The function $F(t)$ contains a summation of the exponential factor:

$$F(t) = \prod_q \mathscr{F}_q(t) = \exp\left[-\sum_{\mathbf{q}} \phi_{\mathbf{q}}(t)\right] \equiv \exp[-\Phi(t)]$$

$$\Phi(t) = \sum_{\mathbf{q}} \left(\frac{M_{\mathbf{q}}}{\omega_{\mathbf{q}}}\right)^2 [N_{\mathbf{q}}(1 - e^{i\omega_{\mathbf{q}} t}) + (N_{\mathbf{q}} + 1)(1 - e^{-i\omega_{\mathbf{q}} t})] \qquad (4.3.11)$$

$$N_{\mathbf{q}} = (e^{\beta\omega_{\mathbf{q}}} - 1)^{-1}$$

The final result for the particle Green's function for $t > 0$ is

$$G(t) = -ie^{-it(\varepsilon_c - \Delta)}e^{-\Phi(t)}(1 - n_F) \tag{4.3.12}$$

This is the exact result and in the form we shall use it. We shall now proceed to a description of the physics.

C. Einstein Model

The physics is best understood by examining a simple application of the model. All the phonons are taken to have the same energy ω_0, which is called the Einstein model. The case of zero temperature will be discussed first, and the finite temperature modifications will follow afterwards. For zero temperature, all the phonon occupation factors are zero:

$$N_q = N_0 = 0$$

Furthermore, the summation over wave vector just produces a coupling constant g:

$$g = \sum_{\mathbf{q}} \frac{M_{\mathbf{q}}^2}{\omega_{\mathbf{q}}^2}$$

The Green's function still has the form (4.3.12), but now the factors are quite simple:

$$\Delta = g\omega_0$$

$$\Phi(t) = g(1 - e^{-i\omega_0 t})$$

The particle Green's function will be evaluated for the case of a single particle, so set $n_F = 0$. Thus we wish to evaluate

$$G(t) = -i\theta(t)\exp[-it\varepsilon_c - g(1 - i\omega_0 t - e^{-i\omega_0 t})]$$

The same result is obtained for the retarded function. The spectral function is then the imaginary part of the retarded Green's function of frequency:

$$\begin{aligned} A(\omega) &= -2\,\mathrm{Im}(-i)\int_0^\infty dt e^{i\omega t}\exp[-it\varepsilon_c - g(1 - i\omega_0 t - e^{-i\omega_0 t})] \\ &= 2\,\mathrm{Re}\left\{\int_0^\infty dt \exp[it(\omega - \varepsilon_c + \Delta) - g + ge^{-i\omega_0 t}]\right\} \end{aligned}$$

The time integral may be evaluated by expanding the $ge^{-i\omega_0 t}$ part of the exponent in a power series,

$$\exp(ge^{-i\omega_0 t}) = \sum_l \frac{g^l}{l!} e^{-i\omega_0 lt}$$

so that the time integral contains terms such as

$$\int_0^\infty dt \exp[it(\omega - \varepsilon_c + \Delta - \omega_0 l)] = \frac{i}{\omega - \varepsilon_c + \Delta - \omega_0 l + i\delta}$$

The factor $i\delta$ is added to force the convergence of the oscillating integrand at large values of time. Then take the limit $\delta \to 0$ and obtain

$$\frac{i}{\omega - \varepsilon_c + \Delta - \omega_0 l + i\delta} = P \frac{i}{\omega - \varepsilon_c + \Delta - \omega_0 l} + \pi\delta(\omega - \varepsilon_c + \Delta - \omega_0 l)$$

The spectral function is the real part of this time integral, which is just the delta function:

$$A(\omega) = 2\pi e^{-g} \sum_{\lambda=0}^{\infty} \frac{g^l}{l!} \delta(\omega - \varepsilon_c + \Delta - \omega_0 l) \tag{4.3.13}$$

The spectral function is a series of delta functions, spaced exactly ω_0 apart. The distribution of peak heights follows a Poisson distribution.

Remember that the spectral function is the probability that the particle has frequency ω. If there were no interactions, the particle would always have energy ε_c, and we would get a single delta function at $\omega = \varepsilon_c$. This limit is obtained from (4.3.13) by setting $g = 0$. For $g \neq 0$ the particle has a finite probability of occupying other states which have l phonons with it. These are not excited states. Since we have set the temperature equal to zero, we must be describing the ground state. In this ground state of the coupled system of particle and phonon, some probability exists that the system will have the different sets of frequencies $\varepsilon_c - \Delta + l\omega_0 \equiv \omega_l$. The different values of ω_l obviously correspond to the particle being coupled to some phonons, which is certainly to be expected in this system.

The spectral function (4.3.13) is shown in Fig. 4.10 for two different values of the coupling constant $g = 0.5$ and 5.5. These correspond to the weak and strong coupling cases, respectively. Weak coupling is $g < 1$. Here the $l = 1$ peak is smaller than $l = 0$, and higher l peaks get smaller very rapidly. For strong coupling, the peak strength increases with l up to l values of approximately $l \approx g$, and then the peaks decrease again.

It is useful to test the sum rules. The first one is [see (3.3.16)]

$$1 = \int_{-\infty}^{\infty} \frac{d\omega}{2\pi} A(\omega) = e^{-g} \sum_l \frac{g^l}{l!} \int_{-\infty}^{\infty} d\omega \delta(\omega - \omega_l) = e^{-g} \sum \frac{g^l}{l!} = 1$$

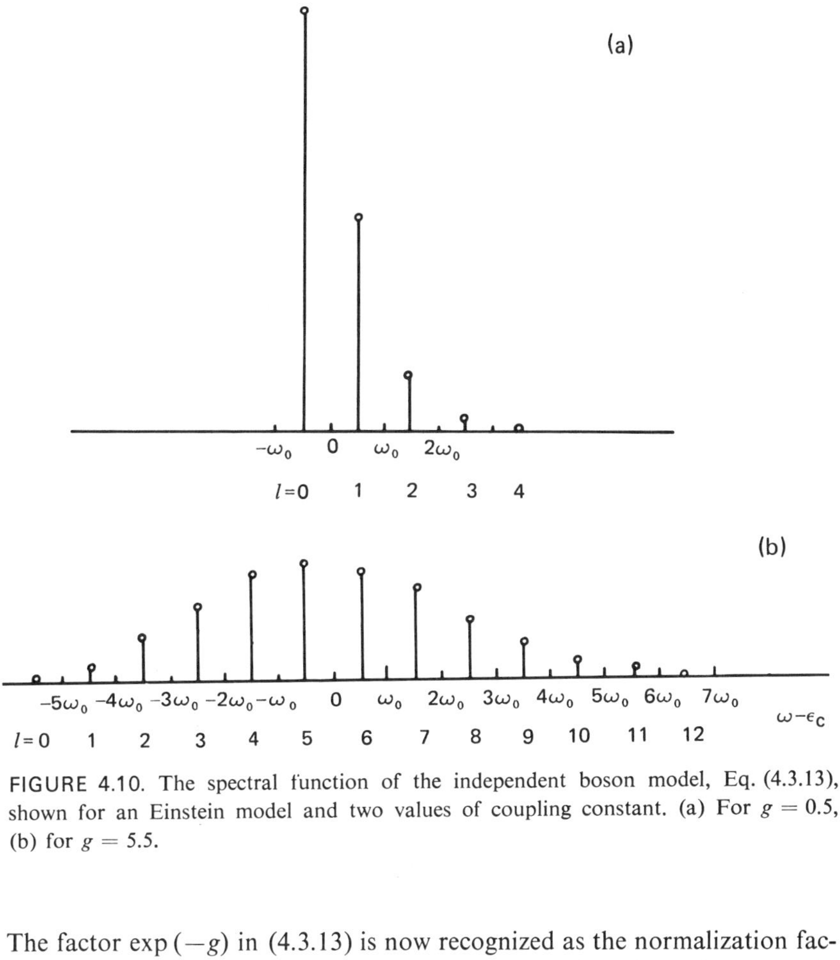

FIGURE 4.10. The spectral function of the independent boson model, Eq. (4.3.13), shown for an Einstein model and two values of coupling constant. (a) For $g = 0.5$, (b) for $g = 5.5$.

The factor $\exp(-g)$ in (4.3.13) is now recognized as the normalization factor which maintains the sum rule. These spectral functions also have the property that

$$\varepsilon_c = \int_{-\infty}^{\infty} \frac{d\omega}{2\pi}\, \omega A(\omega) = \langle \omega \rangle \tag{4.3.14}$$

This moment is easily proved by direct evaluation,

$$\langle \omega \rangle = \int_{-\infty}^{\infty} \frac{d\omega}{2\pi}\, \omega A(\omega) = e^{-g} \sum_{l=0}^{\infty} \frac{g^l}{l!} (\varepsilon_c - \Delta + \omega_0 l) = \varepsilon_c - \Delta + \omega_0 g = \varepsilon_c$$

which does give (4.3.14). This last integral is called the *first moment*. Higher moments $\langle \omega^n \rangle$ may also be evaluated, and they will depend on coupling constant. By using the relation

$$G_{\text{ret}}(t) = -i\theta(t) \int_{-\infty}^{\infty} \frac{d\omega}{2\pi} A(\omega) e^{-i\omega t}$$

the moments may also be shown to equal derivatives of the retarded Green's function:

$$\langle \omega^n \rangle = i \left[\frac{d^n}{dt^n} G_{\text{ret}}(t) \right]_{t \to 0^+}$$

Since the retarded Green's function is finite only at $t > 0$, one takes the time derivatives first and the limit of $t \to 0^+$ second.

An inspection of Fig. 4.10(b) verifies that the first moment is independent of coupling strength g. The delta function peaks have an intensity envelope which does appear to have a maximum near $\omega \approx \varepsilon_c$. Thus as one increases the coupling constant g, the self-energy $\Delta = g\omega_0$ becomes larger in magnitude, so that the lowest energy peak shifts downward in energy. But its intensity lowers also because of the factor $\exp(-g)$, and the envelope of the delta function peaks becomes a Gaussian. A Poisson distribution becomes a Gaussian for large g, which may be shown in the following way. For large g, we need large values of l, so that Stirling's approximation may be used for the factorial l:

$$l! = (2\pi)^{1/2} l^{l+1/2} e^{-l}$$

In this approximation the Poisson intensity becomes

$$\begin{aligned}
e^{-g} \frac{g^l}{l!} &\simeq \frac{1}{(2\pi l)^{1/2}} \exp\left[-g + l + l \ln\left(\frac{g}{l}\right)\right] \\
&= \frac{1}{(2\pi l)^{1/2}} \exp\left[-(g-l) + l \ln\left(1 + \frac{g-l}{l}\right)\right] \\
&\simeq \frac{1}{(2\pi l)^{1/2}} \exp\left[-(g-l) + l\left(\frac{g-l}{l}\right) + \frac{l}{2}\left(\frac{g-l}{l}\right)^2 + \cdots\right] \\
&\simeq \frac{1}{(2\pi l)^{1/2}} \exp\left[-\frac{(g-l)^2}{2l}\right]
\end{aligned}$$

which is a Gaussian.

The zero temperature result was described first, with an Einstein model, to emphasize that these additional peaks in the spectral function are really ground state properties of the system. For acoustical phonons, one gets a continuous distribution of possible phonon energies up to the bandwidth.

This causes the spectral function $A(\omega)$ to become a continuous function of ω. The spectral shapes are dependent on whether the coupling is piezoelectric or deformation potential to the acoustical phonons (Duke and Mahan, 1965).

Now we shall do the Einstein model at finite temperature. The phonon contribution $\Phi(t)$ in the Green's function is

$$\Phi(t) = g[(N+1)(1 - e^{-i\omega_0 t}) + N(1 - e^{i\omega_0 t})]$$

$$N = \frac{1}{e^{\beta\omega_0} - 1}$$

Because of the relationship

$$\frac{N+1}{N} = e^{\beta\omega_0}, \qquad \left(\frac{N+1}{N}\right)^{1/2} = e^{\beta\omega_0/2}$$

this exponent may have its terms grouped in the following fashion:

$$\Phi(t) = g\{2N + 1 - [N(N+1)]^{1/2}(e^{-i\omega_0(t+i\beta/2)} + e^{i\omega_0(t+i\beta/2)})\}$$

The reason for this grouping will now become apparent. Our objective is to expand $\exp[-\Phi(t)]$ in a power series in $\exp(i\omega_0 t)$, so that one can evaluate the spectral function. Now recall the series which generates the Bessel functions of complex argument, $I_n(z)$:

$$e^{+z\cos\theta} = \sum_{l=-\infty}^{\infty} I_l(z) e^{il\theta}$$

Our function

$$e^{-\Phi} = e^{-g(2N+1)} \exp\{g[N(N+1)]^{1/2}(e^{-i\omega_0(t+i\beta/2)} + e^{i\omega_0(t+i\beta/2)})\}$$

has this form if we identify

$$z = 2g[N(N+1)]^{1/2}$$

$$\theta = \omega_0(t + i\beta/2)$$

Thus the retarded Green's function at finite temperatures may be expanded as

$$\begin{aligned} G_{\text{ret}}(t) &= -i\theta(t)\exp[-it(\varepsilon_c - \Delta) - \Phi(t)] \\ &= -i\theta(t)\exp[-it(\varepsilon_c - \Delta) - g(2N+1)] \sum_{l=-\infty}^{\infty} I_l\{2g[N(N+1)]^{1/2}\} \\ &\quad \times e^{(l\omega_0/2)\beta} e^{-it\omega_0 l} \end{aligned}$$

and its spectral function is

$$A(\omega) = 2\pi \exp[-g(2N+1)] \sum_{l=-\infty}^{\infty} I_l\{2g[N(N+1)]^{1/2}\}e^{l(\beta\omega_0/2)}$$
$$\times \delta(\omega - \varepsilon_c + \Delta - \omega_0 l) \tag{4.3.15}$$

The spectral function contains a summation over the frequencies $\omega_l = \varepsilon_c - \Delta + \omega_0 l$ and is a delta function at these values. In this sense it is similar to the zero temperature result (4.3.13). Now the coefficient is far more complicated. Another important difference is that negative values of l are now permitted. Although $I_n = I_{-n}$, the factor $\exp[\beta(\omega_0/2)l]$ skews the envelope of intensities to the positive side.

D. Optical Absorption and Emission

The effects which were described above can, in some cases, actually be measured. Of course, one can never measure the properties of a one-particle Green's function, as we stated in Sec. 3.7. Linear response theory always gives a two-particle Green's function, which describes how the system responds when an external probe causes the system to change its state. However, in the many-boson model, there are some two-particle Green's functions which have properties nearly identical to the one-particle properties which have just been derived.

One important model is a localized defect with several possible localized electronic states. Each of these states may have a different matrix element for coupling to the phonon field:

$$\begin{aligned} H &= \sum_{\mathbf{q}} \omega_{\mathbf{q}} a_{\mathbf{q}}^{\dagger} a_{\mathbf{q}} + \sum_i H_i \\ H_i &= C_i^{\dagger} C_i \Big[\varepsilon_i + \sum_{\mathbf{q}} M_{\mathbf{q},i}(a_{\mathbf{q}} + a_{\mathbf{q}}^{\dagger}) \Big] = h_i C_i^{\dagger} C_i \end{aligned} \tag{4.3.16}$$

This Hamiltonian may also be exactly diagonalized. A canonical transformation of the previous form,

$$\bar{H} = e^s H e^{-s}$$
$$s = \sum_i s_i = \sum_i C_i^{\dagger} C_i \sum_{\mathbf{q}} \frac{M_{\mathbf{q},i}}{\omega_{\mathbf{q}}} (a_{\mathbf{q}}^{\dagger} - a_{\mathbf{q}})$$

gives

$$\begin{aligned} \bar{H} &= \sum_{\mathbf{q}} \omega_{\mathbf{q}} a_{\mathbf{q}}^{\dagger} a_{\mathbf{q}} + \sum_i (\varepsilon_i - \Delta_i) C_i^{\dagger} C_i \\ \Delta_i &= \sum_{\mathbf{q}} \frac{M_{\mathbf{q},i}^2}{\omega_{\mathbf{q}}} \end{aligned} \tag{4.3.17}$$

The terms such as $n_i n_j$ will be set equal to zero, since we shall assume that there is only one electron on the impurity. It may be in different states but not in two different electronic states at once. The Hamiltonian (4.3.16) was written with the electronic states not interacting with each other, except through the phonons. Any terms which permit a direct interaction between the states usually render the Hamiltonian unsolvable, at least exactly. For example, we do not include terms such as

$$(C_i^\dagger C_j + C_j^\dagger C_i) \sum_{\mathbf{q}} M_{\mathbf{q}i,j}(a_{\mathbf{q}} + a_{\mathbf{q}}^\dagger)$$

which permit the particle to change its state by emitting a phonon. These probably exist in real systems, but their addition to the Hamiltonian makes the problem much more difficult. Thus they are customarily omitted.

In an optical absorption process, an electron may change its electronic state, say from i to j, by the absorption of a photon of frequency ω. This process is described by the Kubo formula, using the current–current correlation function. As in Sec. 4.3C, the real-time correlation function may be employed. Thus it is convenient to use the version of the Kubo formula given in Problem 16 of Chapter 3. For optical frequencies such that $\beta\omega \gg 1$, this is

$$\mathrm{Re}(\sigma_{\alpha\beta}) = \frac{1}{2}\,\frac{1}{\omega}\int_{-\infty}^{\infty} dt e^{i\omega t}\langle j_\alpha(t) j_\beta(0)\rangle \tag{4.3.18}$$

The relation $\beta\omega \gg 1$ is easily satisfied, since typically $\hbar_i\omega \approx 2$–3 eV is in the visible spectrum, while at room temperature $\beta = 40\ \mathrm{eV}^{-1}$. For the transition between two localized states, the current operator is [see (1.2.17)]

$$j_\alpha = \sum_{ij} p_{ij,\alpha} C_i^\dagger C_j$$

The matrix element p_{ij} is treated as a constant in this problem. It plays no role in the many-body physics which follows. Thus we need to evaluate the correlation function:

$$\begin{aligned} \langle j_\alpha(t) j_\beta(0)\rangle &= \sum_{ijkl} p_{ij,\alpha} p_{kl,\beta} \langle C_i^\dagger(t) C_j(t) C_k^\dagger C_l\rangle \\ U = \langle C_i^\dagger(t) C_j(t) C_k^\dagger C_l\rangle &= \mathrm{Tr}(e^{-\beta H} e^{iHt} C_i^\dagger C_j e^{-iHt} C_k^\dagger C_l) e^{\beta\Omega} \end{aligned} \tag{4.3.19}$$

The Hamiltonian H is that in (4.3.16). This correlation function may be solved exactly by using the same steps that were used to solve the Green's function (4.3.2) to get (4.3.12). First, the unit operator $1 = e^{-s}e^{s}$ is inserted into the trace:

$$U = \mathrm{Tr}(e^{-\beta H} e^{iHt} C_i^\dagger C_j e^{-iHt} C_k^\dagger C_l e^{-s} e^{s})$$

Using the cyclic property of the trace, we obtain

$$U = \mathrm{Tr}(e^{s}e^{-\beta H}e^{iHt}C_i^{\dagger}C_je^{-iHt}C_k^{\dagger}C_le^{-s})$$

The canonical transformation is taken on each term inside of the trace, which gives

$$U = \mathrm{Tr}(e^{-\beta\bar{H}}e^{i\bar{H}t}X_i^{\dagger}X_jC_i^{\dagger}C_je^{-i\bar{H}t}X_k^{\dagger}X_lC_k^{\dagger}C_l)$$

The factors X_i result from the transformation of the particle operators:

$$\bar{C}_i = e^{s}C_ie^{-s} = e^{s_i}C_ie^{-s_i} = C_iX_i$$

$$X_i = \exp\left[-\sum_{\mathbf{q}}\frac{M_{\mathbf{q},i}}{\omega_{\mathbf{q}}}(a_{\mathbf{q}}^{\dagger} - a_{\mathbf{q}})\right]$$

The transformed Hamiltonian $\bar{H}$ in (4.3.17) is diagonal in the operators C and a. Thus the time development of the correlation function may be found at once:

$$e^{i\bar{H}t}X_i^{\dagger}X_jC_i^{\dagger}C_je^{-i\bar{H}t} = X_i^{\dagger}(t)X_j(t)C_i^{\dagger}C_je^{it(\varepsilon_i-\varepsilon_j-\Delta_i+\Delta_j)}$$

$$U = e^{it(\varepsilon_i-\varepsilon_j-\Delta_i+\Delta_j)}\,\mathrm{Tr}[e^{-\beta\bar{H}}X_i^{\dagger}(t)X_j(t)X_k^{\dagger}(0)X_l(0)C_i^{\dagger}C_jC_k^{\dagger}C_l]$$

The electron and phonon parts of the trace may be separated, which is permissible because the X_i operators do not depend on particle states:

$$U = e^{it(\varepsilon_i-\varepsilon_j-\Delta_i+\Delta_j)}U_{\mathrm{el}}U_{\mathrm{ph}}(t)$$

$$U_{\mathrm{el}} = \mathrm{Tr}(e^{-\beta\bar{H}}C_i^{\dagger}C_jC_k^{\dagger}C_l)$$

$$U_{\mathrm{ph}}(t) = \mathrm{Tr}[e^{-\beta\bar{H}}X_i^{\dagger}(t)X_j(t)X_k^{\dagger}(0)X_l(0)]$$

The electron part is quickly evaluated. Using Wick's theorem, the subscripts referring to particle states must be paired. In fact, one must have that $j = k$ and $i = l$:

$$U_{\mathrm{el}} = \mathrm{Tr}(e^{-\beta\bar{H}}C_l^{\dagger}C_kC_k^{\dagger}C_l) = n_l(1 - n_k)$$

$$U_{\mathrm{ph}}(t) = \mathrm{Tr}[e^{-\beta\bar{H}}X_l^{\dagger}(t)X_k(t)X_k^{\dagger}(0)X_l(0)]$$

This result is useful when we consider the evaluation of the phonon part of the trace. It is similar in form to the earlier trace for the one-particle Green's function. The only difference is that there are now four factors instead of two. The evaluation procedure is the same, since the four operators can be paired and combined into two. This happens because operators which are at the same time can be combined, since their exponents commute:

$$X_l^\dagger(t)X_k(t) = e^{iHt}\left(\prod_{\mathbf{q}} e^{\lambda_l(a_\mathbf{q}^\dagger - a_\mathbf{q})}e^{-\lambda_k(a_\mathbf{q}^\dagger - a_\mathbf{q})}\right)e^{-iHt}$$
$$= e^{iHt}\prod_{\mathbf{q}} e^{(\lambda_l-\lambda_k)(a_\mathbf{q}^\dagger - a_\mathbf{q})}e^{-iHt}$$
$$= \exp\sum_{\mathbf{q}} \frac{M_{\mathbf{q},l} - M_{\mathbf{q},k}}{\omega_\mathbf{q}}(a_\mathbf{q}^\dagger e^{+i\omega_\mathbf{q}t} - a_\mathbf{q}e^{-i\omega_\mathbf{q}t})$$

It is only possible to add two exponential operators in this fashion when they commute. They do in this case when taken at the same time. Thus our correlation function $U_{\text{ph}}(t)$ above is immediately simplified to

$$U_{\text{ph}}(t) = \prod_{\mathbf{q}} \text{Tr}\{e^{-\beta H_\mathbf{q}}\exp[-\lambda(a_\mathbf{q}^\dagger e^{i\omega_\mathbf{q}t} - a_\mathbf{q}e^{-i\omega_\mathbf{q}t})]e^{\lambda(a_\mathbf{q}^\dagger - a_\mathbf{q})}\}$$
$$\lambda = \frac{M_{\mathbf{q},k} - M_{\mathbf{q},l}}{\omega_\mathbf{q}}$$

This equation for $U_{\text{ph}}(t)$ is precisely the form (4.3.7) which was untangled previously, and now the effective coupling constant is $\lambda = (M_{\mathbf{q},k} - M_{\mathbf{q},l})/\omega_\mathbf{q}$. Thus we can use the previous untangling result (4.3.11) to obtain the evaluation of $U_{\text{ph}}(t)$:

$$U_{\text{ph}}(t) = \exp\left\{-\sum_{\mathbf{q}} \frac{(M_{\mathbf{q},k} - M_{\mathbf{q},l})^2}{\omega_\mathbf{q}^2}[(N_\mathbf{q}+1)(1-e^{-i\omega_\mathbf{q}t}) + N_\mathbf{q}(1-e^{i\omega_\mathbf{q}t})]\right\}$$

These various results are collected, and the result for the correlation function for the conductivity is

$$\text{Re}(\sigma_{\alpha\beta}) = \frac{1}{2}\frac{1}{\omega}\sum_{l,k} n_l(1-n_k)p_{kl,\alpha}p_{kl,\beta}$$
$$\times\int_{-\infty}^{\infty} dt\,\exp\{it[\omega + \varepsilon_l - \varepsilon_k - \Delta_l + \Delta_k - \Phi_{kl}(t)]\}$$
$$\Phi_{kl}(t) = \sum_{\mathbf{q}} \frac{(M_{\mathbf{q},k} - M_{\mathbf{q},l})^2}{\omega_\mathbf{q}^2}[(N_\mathbf{q}+1)(1-e^{-i\omega_\mathbf{q}t}) + N_\mathbf{q}(1-e^{i\omega_\mathbf{q}t})] \tag{4.3.20}$$

The function Φ_{kl} in the exponent is similar to the one for the particle Green's function. The only difference is that the effective matrix element is the difference between the two single-state matrix elements:

$$M_{\text{eff},\mathbf{q}} = M_{\mathbf{q},k} - M_{\mathbf{q},l}$$

For example, this means that if the two matrix elements M_k and M_l happen to be equal, then the effective matrix element vanishes. In this unlikely circumstance the spectral function is just a delta function at the frequency $\omega = \varepsilon_k - \varepsilon_l$. Usually M_l and M_k are not equal, at least not for all different wave vectors, so that phonon effects are present in the transition.

The model we have just solved describes dynamic relaxation. In the initial state of the system, the electron is in a state l, and the phonons are relaxed about their equilibrium configuration for the state l. Recall that if the phonon part of the Hamiltonian is written in terms of harmonic oscillator coordinates, the phonons relax to an equilibrium configuration given by [see (1.3.3)]

$$Q_{\mathbf{q}}^{(l)} = -2\left(\frac{\hbar}{2\varrho\omega_{\mathbf{q}}\nu}\right)^{1/2} \frac{M_{\mathbf{q},l}}{\omega_{\mathbf{q}}}$$

In the optical absorption, the electron starts in state l and ends in state k. The phonons start with an equilibrium configuration about the point $Q_{\mathbf{q}}^{(l)}$ but end the optical transition with the equilibrium configuration about the point $Q_{\mathbf{q}}^{(k)}$. The phonon system must alter its equilibrium configuration during the transition. This is a relaxation process, since it must relax to the new equilibrium configuration during the optical step. The physics problem is to determine how the phonon relaxation process affects the absorption spectra. This information is contained in the result (4.3.20).

The process is indicated schematically in Fig. 4.11, which shows a potential energy diagram for each oscillator coordinate $Q_{\mathbf{q}}$. There are two parabolic curves, with parabolicity $\omega_{\mathbf{q}}^2$. The lower curve describes the ground state of the system. The electronic energy is ε_l. If there were no coupling to the particle in state l, the phonon parabola would be a minimum at this point, $Q_q = 0$. However, because of the coupling $M_{q,l}$ to this particle, the potential minimum is at $Q_q^{(l)}$. The upper curve describes the final state potential energy curves of the phonons plus particle. The particle energy is ε_k, and the curve crosses here because the phonon system has this energy when $Q_q = 0$. This potential energy curve has the minimum at $Q_q^{(k)}$. This minimum has been drawn on the other side of the axis to emphasize that it is usually a point different from the ground state minimum. Figure 4.11 is called a *configurational coordinate diagram* (Williams and Hebb, 1951). Such diagrams were originally constructed with a single-particle coordinate Q, which represented the real atom displacement of the first shell of atoms about the impurity. However, the use of configurational coordinate diagrams is much more rigorous when interpreted as the potential energy of the individual phonon modes.

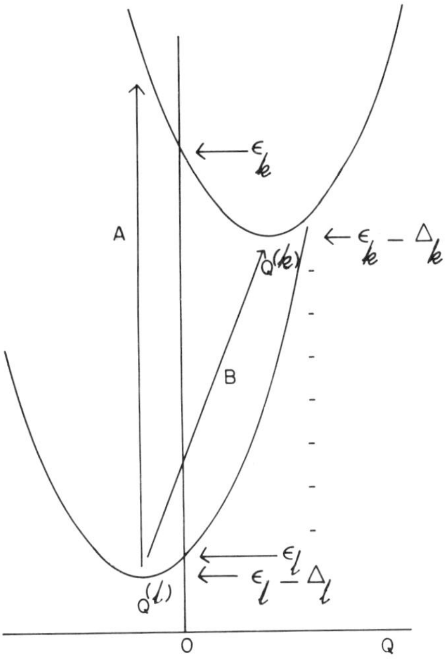

FIGURE 4.11. Configurational coordinate drawing of the independent boson model. The two parabolas represent the phonon potential energy of the initial and final electronic states in the transition. Transition path A is most likely, while path B is less likely but gives the zero-phonon probability.

The physical question is whether the optical transition happens vertically on this diagram or along some other trajectory. For example, the arrow B is from one potential minimum to the other. This transition is called the *zero-phonon line*. In reality, all these transitions are possible, and each has a probability of occurrence. These probabilities are given by (4.3.20). The physics is well illustrated by again using the Einstein model for phonons. We expand this in the same manner which we used to derive the one-particle Green's function (4.3.15):

$$\operatorname{Re} \sigma_{\alpha\beta}(\omega) = \frac{\pi}{\omega} \sum_{lk} n_l(1 - n_k) p_{kl,\alpha} p^*_{kl,\beta} e^{-g_{kl}(T)} \sum_{m=-\infty}^{\infty} \delta(\omega - \omega_m) e^{m\beta\omega/2} I_m(\gamma_{kl})$$

$$g_{kl}(T) = (2N + 1) g_{kl}$$

$$g_{kl} = \sum_{\mathbf{q}} \frac{(M_{\mathbf{q},k} - M_{\mathbf{q},l})^2}{\omega_{\mathbf{q}}^{\ 2}} \tag{4.3.21}$$

$$\omega_m = \varepsilon_k - \varepsilon_l - \Delta_k + \Delta_l + \omega_0 m$$

$$\gamma_{kl} = 2[N_0(N_0 + 1)]^{1/2} g_{kl}$$

This formula for $\operatorname{Re} \sigma(\omega)$ has exactly the same form as the one-particle spectral function (4.3.15), as an expansion in Bessel functions of complex argument. Now the coupling constant to the phonons is determined by $M_{\mathbf{q},k} - M_{\mathbf{q},l}$. If $M_{\mathbf{q},k} = M_{\mathbf{q},l}$, then $g = 0$, $\gamma = 0$, and the conductivity is a single delta function at $\omega = \varepsilon_k - \varepsilon_l$.

The zero-phonon line is given by the term with $m = 0$. It has the energy $\omega = \varepsilon_k - \varepsilon_l - \Delta_k + \Delta_l$, which corresponds to the B arrow in Fig. 4.11. The probability of this occurring is proportional to

$$e^{-g_{kl}(T)} I_0(\gamma_{kl})$$

The envelope of delta function heights is determined by the m dependence of the factor

$$I_m(\gamma_{kl}) e^{m\beta\omega_0/2}$$

For coupling constants g_{kl} greater than unity, the delta function intensifies with m and then falls off at higher values. Thus the strongest delta function occurs at positive finite values of m, which is illustrated in Fig. 4.12. The circles show the peak heights calculated for $g_{kl} = 4$ and $N_0 = \frac{1}{2}$. The points are the emission spectra, which will be derived below.

The usual interpretation is that terms with $m > 0$ correspond to the creation of m phonons during the optical transition. The zero-phonon line is just that—the transition where no phonons are created or destroyed.

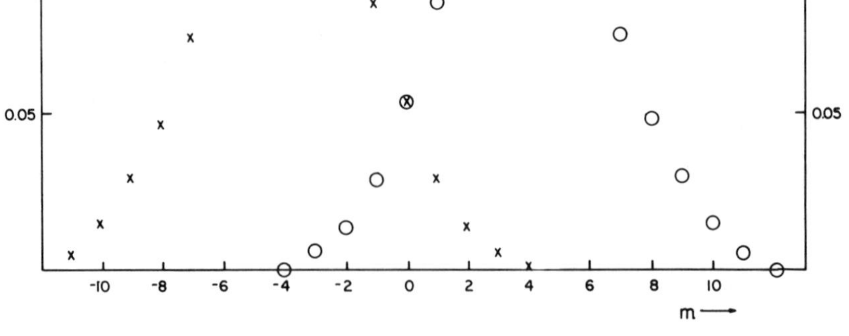

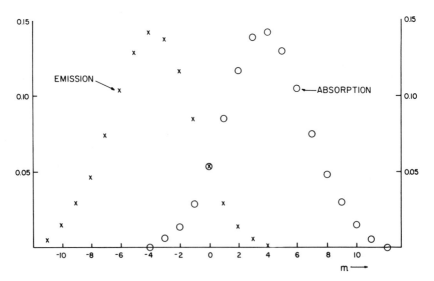

FIGURE 4.12. The emission and absorption spectra for an Einstein model, in the independent boson approximation, from Eq. (4.3.21). The O's and **X**'s mark the peak intensities of each phonon sideband.

Transitions in which $m > 0$ are phonon emission, and transitions in which $m < 0$ are phonon absorption. The phonon absorption process is possible at finite temperature, when phonons are thermally present in the initial system. This interpretation ignores the fact that both the ground and excited states are an admixture of phonon states. The spectral functions (4.3.15) of both initial and final states contain admixtures of different numbers of phonons. These differing admixtures are due to fluctuations in the system. Sometimes the particle has one phonon around it, while other times it has three or four, etc. The probability of different fluctuations is given by the Bessel function coefficients in (4.3.15).

Thus the other interpretation is that the absorption spectra of Fig. 4.12 correspond to transitions between states of different fluctuations. The initial state may be in a state with a fluctuation of plus four phonons and the final state with plus six. Thus it takes a net of plus two phonon energies to complete the transition with energy conservation. This model is not really different from the above interpretation. The phonons in the fluctuation still have to dissipate away. However, one state may have a different number of average phonons, in its fluctuating cloud, than the other. The number of phonons which must be created or destroyed to make up this average is not really creating a net amount to dissipate into the system. They will, instead, stay at the impurity site and take part in the fluctuations.

The emission spectra will now be considered. It is assumed that the particle has been elevated to an excited state at an earlier time, perhaps by an optical absorption process. However, experimentally it is often done by electron bombardment. The phonons in the excited state relax around the particle and eventually attain the equilibrium configuration plus fluctuations. This assumption is valid only for long-lived excited states. Then a photon is emitted while the particle drops to a lower energy state. Of course, the phonon system must adjust during the emission process just as it did during the absorption. The same relaxation processes are encountered again.

If we are sufficiently clever, it will be possible to deduce the emission result without doing any more work. It may be obtained from the absorption result by a simple argument. To calculate the emission, we need to calculate the rate per unit time that photons are produced in the solid. Of course, the absorption is the rate at which they disappear. That is, we write a simple rate equation for the average number of protons N of frequency ω,

$$\frac{dN}{dt} = -wN \tag{4.3.22}$$

which has the solution

$$N(t) = N_0 e^{-tw}$$

where w is the rate of absorption. Since the conductivity $\sigma(\omega)$ also has the units of $\sec^{-1}$, one might expect that $w \propto \sigma$. In fact, the exact relationship is

$$w = \frac{4\pi\sigma}{\varepsilon_1} \tag{4.3.23}$$

This may be derived in the following way. A beam of photons, in a material, are absorbed according to *Beer's law*,

$$N = N_0 e^{-\alpha x}$$

where α is the absorption coefficient and x is the distance traveled. This distance is given by the group velocity c/n times the time,

$$x = t\frac{c}{n}$$

$$\alpha = 2\frac{\omega}{c}k$$

and so

$$\alpha x = 2t\omega\frac{k}{n}$$

$$w = 2\omega\frac{k}{n}$$

where k is the extinction coefficient and n is the refractive index:

$$n + ik = (\varepsilon_1 + i\varepsilon_2)^{1/2}$$

The square of this equation gives the imaginary part of the dielectric function,

$$\varepsilon_2 = 2nk$$

$$w = \frac{\omega\varepsilon_2}{n^2}$$

which is related to the real part of the conductivity according to (3.7.21):

$$\varepsilon_2 = \frac{4\pi\sigma}{\omega}, \qquad \varepsilon_1 \simeq n^2$$

$$w = \frac{4\pi\sigma}{\varepsilon_1}$$

We have proved the assertion (4.3.23). The rate of photon absorption is just proportional to the real part of the conductivity, as given by the Kubo formula. Another way to understand this is to consider rewriting the Kubo formula with some additional factors:

$$\sigma(\omega) \to (1 - e^{-\beta\omega}) \int_{-\infty}^{\infty} dt \langle j(t) b^{\dagger}(t) j(0) b(0) \rangle$$

The two additional factors are the creation and destruction operators for photons b and $b^{\dagger}$. The right-hand part of the correlation factor,

$$j_{\alpha}(0) b(0) = \sum_{kl} p_{kl,\alpha} c_k{}^{\dagger} c_l b$$

describes the process whereby a photon is destroyed and an electron has its state changed from l to k. This term arises from the $p \cdot A$ interaction. The left-hand part of the correlation function contains the factors

$$j_{\beta}(t) b^{\dagger}(t) = \sum_{kl} p_{kl,\beta}^{\alpha} c_l{}^{\dagger}(t) c_k(t) b(t)$$

which describe the inverse process at another time t. The correlation function between these two events is the Kubo formula. The photon parts may be factored out of the expression, $b^{\dagger}(t) = b^{\dagger} e^{i\omega t}$ and $\langle b^{\dagger} b \rangle = N$:

$$(1 - e^{-\beta\omega}) \langle b^{\dagger} b \rangle \int_{-\infty}^{\infty} dt e^{i\omega t} \langle j_{\beta}(t) j_{\alpha}(0) \rangle = N\sigma(\omega)$$

This quantity is, except for the factor of 4π, just the right-hand side of Eq. (4.3.22). It is the rate that photons are destroyed in the sample by optical absorption.

For emission, we wish to calculate the rate at which photons are created. Now the first step in the correlation function whould describe a photon being created while the electron changes its state. This correlation function should be

$$\int_{-\infty}^{\infty} dt \langle j_{\alpha}(t) b(t) j_{\beta}(0) b^{\dagger} \rangle$$

where the prefactor $1 - e^{-\beta\omega}$ is dropped. If we take these photon factors out of the integrand, using $b(t) = b e^{-i\omega t}$ and $\langle b b^{\dagger} \rangle = N + 1$, we obtain

$$\langle b b^{\dagger} \rangle \int_{-\infty}^{\infty} dt e^{-i\omega t} \langle j_{\alpha}(t) j_{\beta}(0) \rangle = (N + 1) \frac{I(\omega)}{4\pi}$$

This, multiplied by 4π, must be the rate that photons are being made in the solid:

$$\frac{dN}{dt} = I(N+1)$$

$$I(\omega) = 4\pi \int_{-\infty}^{\infty} dt e^{-i\omega t} \langle j_\alpha(t) j_\beta(0) \rangle$$

The formula for $I(\omega)$ is identical with the Kubo formula except that the frequency has changed sign. If we compare with the Einstein model result (4.3.21), for example, we only change $\delta(\omega + \omega_m)$:

$$I(\omega) \propto e^{-g(2N_0+1)} \sum_{m=-\infty}^{\infty} \delta(\omega + \omega_m) e^{1/2m\beta\omega_0} I_m(\gamma)$$

The outcome of this is very simple. The emission spectra are just the mirror image of the absorption spectra—reflected across the zero-phonon line. The emission is illustrated by the points marked **X** in Fig. 4.12. The emission spectra are mostly on the low-frequency side of the zero-phonon line. In general, the emission spectra have a lower average frequency than the corresponding absorption spectra for the same processes. The difference is a consequence of relaxation, which is illustrated using the configurational coordinate diagram of Fig. 4.13. The vertical arrow "abs" shows the most likely absorption event. Of course, other absorption events are possible, but the peak of the envelope of delta functions is at this vertical transition.

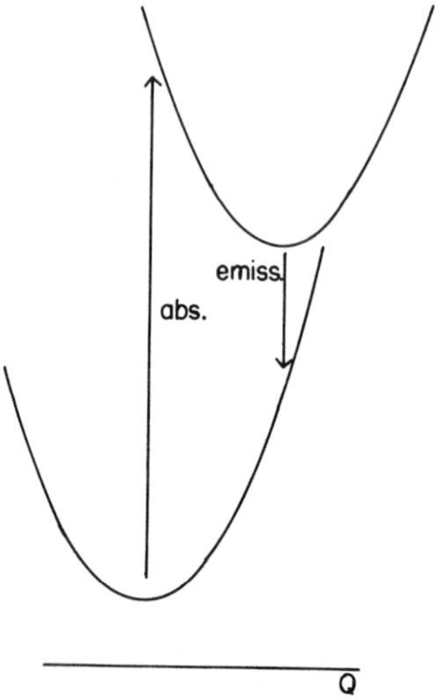

FIGURE 4.13. Configurational coordinate drawing shows that the average absorption frequency is higher than the average emission frequency.

Similarly, the most likely emission event is the downward arrow marked "emiss." It is shorter than the absorption arrow, which indicates that the average emission takes less energy. No matter how one draws those two parabolas with respect to each other, the downward emission arrow is always shorter than (or equal to) the upward arrow. This fact can also be demonstrated by using the various theoretical formulas which have been derived. However, the simple diagram of Fig. 4.13 is probably the clearest proof that the emission has an average lower energy than absorption.

The lower average energy in emission, compared to absorption, is understood using the first interpretation mentioned after (4.3.21). During absorption, one usually makes phonons. Thus the photon energy is given by the change in particle energy plus the phonon energy:

$$\omega = \Delta(\text{particle energy}) + \text{phonons}$$

This is higher than the particle transition energy. In the emission event, the energy comes from the particle transition, but here, too, one makes phonons, on the average. Thus energy conservation is

$$\Delta(\text{particle energy}) = \omega + \text{phonons}$$

and here the average photon energy ω is less than the particle transition energy.

Figure 4.14 shows an experimental result for absorption and emission at an impurity center in ZnTe, reported by Dietz *et al.* (1962). The emission and absorption are mirror images about the zero-phonon lines. The arrows at the top show the separation of the optical phonon lines, which are very clear in the emission spectra. The two other peaks between successive LO phonons are sidebands due to LA and TA phonons. These data were taken at 20 K. It is an unusually good example of the relation between emission and absorption.

E. Sudden Switching

There is another way to derive the result for optical emission and absorption. Of course it obtains the same answer, since both derivations are exact. This other derivation emphasizes the switching aspects of the problem. It is just an alternate method of obtaining the same answer but perhaps provides a slightly different physical insight. The many-body problem is presented as the system response when a potential is suddenly switched on. We return to the Kubo formula and consider the time cor-

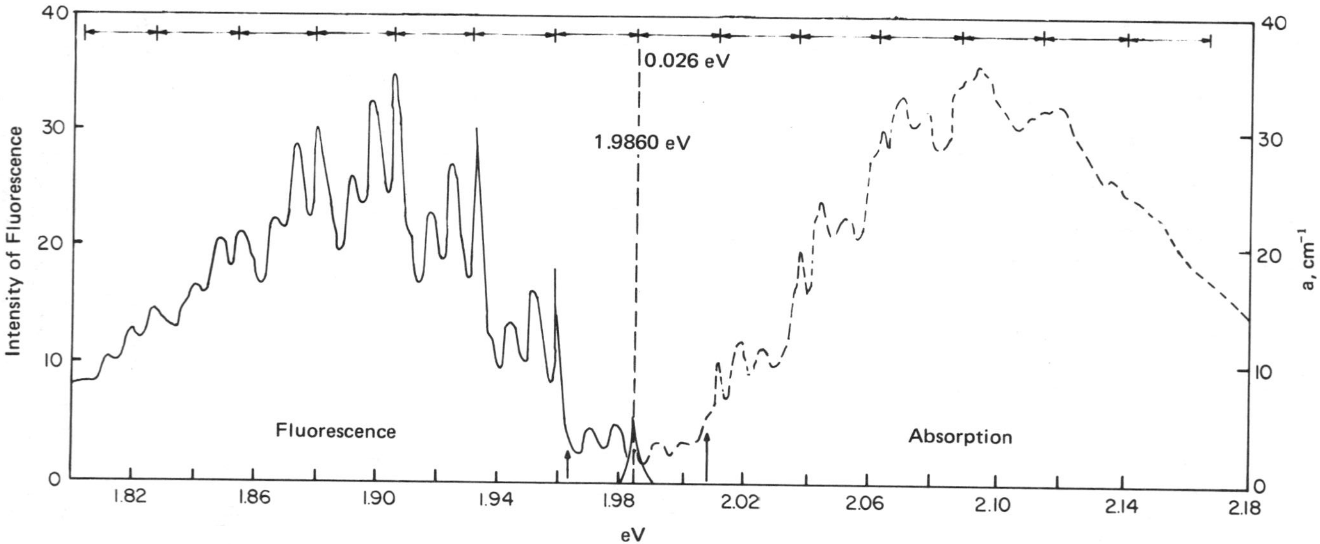

FIGURE 4.14. The fluorescence and absorption at 20 K from an impurity center in ZnTe. The positions of the peak are mirrored about a central no-phonon line common to both spectra. The LO phonon energy intervals are indicated. The other two peaks in each interval are from TA and LA phonons. *Source*: Dietz *et al.* (1962) (used with permission).

relation function, which is evaluated:

$$U = \mathrm{Tr}(e^{-\beta H} e^{iHt} C_l^\dagger C_k e^{-iHt} C_k^\dagger C_l) \tag{4.3.24}$$

In writing this expression, we have utilized the fact that the particle operators were paired up. Now the Hamiltonian H occurs three places in this expression. We shall show that it has two different forms: one for each different place it occurs in the sequence of operators. The factor on the right is

$$e^{-iHt} C_k^\dagger C_l$$

To its right the operators $C_k^\dagger C_l$ destroy the particle in state l and create it in state k. Thus H operates on a system with a particle in k but not in l. According to the form (4.3.16) for the Hamiltonian,

$$H = H_p + \sum_i h_i C_i^\dagger C_i$$

$$H_p = \sum_{\mathbf{q}} \omega_{\mathbf{q}} a_{\mathbf{q}}^\dagger a_{\mathbf{q}}$$

$$h_i = \varepsilon_i + \sum_{\mathbf{q}} M_{\mathbf{q},i}(a_{\mathbf{q}} + a_{\mathbf{q}}^\dagger) \equiv \varepsilon_i + v_i$$

H on a state $|k\rangle = C_k^\dagger |0\rangle$ gives

$$H|k\rangle = (H_p + h_k)|k\rangle$$

Thus, $\exp(-iHt)C_k^\dagger C_l$ may be replaced in the operator sequence by

$$e^{-iHt} C_k^\dagger C_l = e^{-i(H_p+h_k)t} C_k^\dagger C_l$$

This result is exact, and no approximation is involved in the replacement. The other two factors of H are at the left of the operator sequence. On their right is the immediate sequence of particle operators,

$$e^{-\beta H} e^{iHt} C_l^\dagger C_k \cdots$$

which return the particle back to the state l. Thus these Hamiltonians operate on the system with a particle in the state l and therefore produce

$$e^{-\beta(H_p+h_l)} e^{it(H_p+h_l)} C_l^\dagger C_k$$

The correlation function $U(t)$ in (4.3.24) is exactly equal to

$$U = \mathrm{Tr}(e^{-\beta(H_p+h_l)} e^{it(H_p+h_l)} C_l^\dagger C_k e^{-it(H_p+h_k)} C_k^\dagger C_l)$$

The particle operators may now be removed from the trace. The time development of the correlation function involves only the operators H_p and h_j, and these contain only phonon operators. The particle operators have done their job, in the correlation function, by determining which effective Hamiltonians operate in the sequence. Thus we now need to evaluate the phonon part of the trace:

$$U(t) = e^{it(\varepsilon_l - \varepsilon_k)} n_l(1 - n_k)\bar{U}_{\text{ph}}(t)$$
$$\bar{U}_{\text{ph}}(t) = \text{Tr}(e^{-\beta(H_p+v_l)} e^{it(H_p+v_l)} e^{-it(H_p+v_k)})$$

This correlation function could be evaluated by Feynman disentangling. However, let us develop another method.

When a Hamiltonian is solved in the interaction representation, one writes $H = H_0 + V$, where H_0 is a part one can solve and V is the perturbation. Now the procedure is slightly different. We shall call H_0 the Hamiltonian for the state with a particle in l,

$$H_0 \equiv H_p + h_l$$

and $H_0 + V$ the Hamiltonian for the particle in the state k,

$$H_0 + V = H_p + h_k$$

Obviously, the perturbation V is the difference between these two Hamiltonians:

$$V = h_k - h_l = \sum_{\mathbf{q}} (M_{\mathbf{q},k} - M_{\mathbf{q},l})(a_{\mathbf{q}} + a_{\mathbf{q}}^\dagger)$$

The phonon part of the correlation function may be written as

$$\bar{U}_{\text{ph}}(t) = \text{Tr}(e^{-\beta H_0} e^{itH_0} e^{-it(H_0+V)}) \tag{4.3.25}$$

The next step is to make H_0 a diagonal Hamiltonian, which is not difficult, since the unitary transformation is given earlier in (4.3.17):

$$\begin{aligned}
\bar{H}_0 &= e^{s_l} H_0 e^{-s_l} = \varepsilon_l - \Delta_l + \sum_{\mathbf{q}} \omega_{\mathbf{q}} a_{\mathbf{q}}^\dagger a_{\mathbf{q}} \\
(\bar{H}_0 + \bar{V}) &= e^{s_l}(H_0 + V)e^{-s_l} = \varepsilon_k - \Delta_k + \sum_{\mathbf{q}} \omega_{\mathbf{q}} a_q^\dagger a_q + \sum_{\mathbf{q}} \frac{1}{\omega_{\mathbf{q}}} (M_{\mathbf{q},l} - M_{\mathbf{q},k})^2 \\
&\quad + \sum_{\mathbf{q}} (M_{\mathbf{q},k} - M_{\mathbf{q},l})(a_{\mathbf{q}} + a_{\mathbf{q}}^\dagger) \\
s_l &= \sum_{\mathbf{q}} \frac{M_{\mathbf{q},l}}{\omega_{\mathbf{q}}} (a_{\mathbf{q}}^\dagger - a_{\mathbf{q}}) \\
\Delta_{kl} &= \sum_{q} \frac{1}{\omega_{\mathbf{q}}} (M_{\mathbf{q},k} - M_{\mathbf{q},l})^2
\end{aligned} \tag{4.3.26}$$

Thus the transformed potential $\bar{V}$ is obtained by subtracting $\bar{H}_0$ from $\bar{H}_0 + \bar{V}$:

$$\bar{V} = \varepsilon_k - \varepsilon_l - \Delta_k + \Delta_l + \Delta_{kl} + \overline{\delta V}$$

$$\overline{\delta V} = \sum_{\mathbf{q}} (M_{\mathbf{q},k} - M_{\mathbf{q},l})(a_{\mathbf{q}} + a_{\mathbf{q}}^{\dagger})$$

The unitary transformation in (4.3.26) is applied to all the operators in the correlation function (4.3.25). It is done following the same steps used to transform (4.3.19), except that here we use only $\exp(s_l)$,

$$U_{\text{ph}}(t) = \text{Tr}(e^{-\beta\bar{H}_0}e^{it\bar{H}_0}e^{-it(\bar{H}_0+\bar{V})}) \tag{4.3.27}$$

where $\bar{H}_0$ now has the diagonal form in (4.3.26).

This correlation function has the form of the switching phenomena, as mentioned at the beginning of this subsection. It describes the response of the system to suddenly switching on the potential $\bar{V}$ at time $t = 0$. This switching on causes numerous transients in the phonon system, and these transients are the cause of the phonon sidebands observed in absorption.

To illustrate this switching, consider a system which has a Hamiltonian given by

$$H = \bar{H}_0 + \bar{V}d^{\dagger}d$$

where $d^{\dagger}$ and d are creation and destruction operators for some particle. The Green's function of the d particle is given by $(t > 0)$

$$G_d(t) = -i\langle 0 \mid e^{iHt}de^{-iHt}d^{\dagger} \mid 0\rangle$$

The particle d is created at time $t = 0$. Thus e^{-iHt} operates on the state with a particle in d and $Hd^{\dagger} \mid 0\rangle = (\bar{H}_0 + \bar{V})d^{\dagger} \mid 0\rangle$. Then at a later time t the particle is destroyed by the destruction operator d, and e^{iHt} operates on a state with no d particle. Thus this d-particle Green's function is

$$G_d(t) = -i\langle dd^{\dagger}\rangle\langle e^{i\bar{H}_0t}e^{-i(\bar{H}_0+\bar{V})t}\rangle$$

which is exactly the form of the correlation function (4.3.27). All the d particle did for us was to switch on the potential $\bar{V}$ at $t = 0$ and switch it off at time t. This switching on causes the phonons to respond and adjust to the new potential $\bar{V}$. These transients are the same relaxation processes which were discussed earlier.

The correlation function (4.3.27) is evaluated by first recalling the interaction representation result (2.1.7):

$$e^{i\bar{H}_0 t}e^{-it(\bar{H}_0+\bar{V})} = T\exp\left[-i\int_0^t dt_1 \bar{V}(t_1)\right]$$

In this case, the operator $\bar{V}(t)$ is

$$\bar{V}(t) = e^{i\bar{H}_0 t}\bar{V}e^{-i\bar{H}_0 t}$$

$$\bar{V}(t) = \varepsilon_k - \varepsilon_l - \Delta_k + \Delta_l + \delta\bar{V}(t)$$

$$\delta\bar{V}(t) = \sum (M_{q,l} - M_{q,k})(a_q e^{-i\omega_q t} + a_q^{\dagger} e^{+i\omega_q t})$$

The constant terms may be immediately removed from the correlation function. Only the operator part $\bar{\delta}V(t)$ needs to be time-ordered:

$$e^{i\bar{H}_0 t}e^{-it(\bar{H}_0+\bar{V})}$$
$$= \exp[-it(\varepsilon_j - \varepsilon_l + \Delta_k - \Delta_l + \Delta_{kl})]T\exp\left[-i\int_0^t dt_1\, \overline{\delta V}(t_1)\right]$$

Now $\delta V(t)$ contains only two terms, which are, respectively, proportional to $a^{\dagger}$ and a. Since the commutator of these results is a constant, one immediately considers evaluating these time-ordered exponents by using Feynman's theorem (4.3.9). It is not correct to use precisely the form of (4.3.9). The exponent operates at different times, and this fact must be considered. The proper procedure is done in Problem 5 of Chapter 2, which is to separate the two terms in the time-ordered exponent by the equations

$$T\exp\left[-i\int_0^t dt_1\, \overline{\delta V}(t_1)\right] = e^{a^{\dagger}\phi(t)}T\exp\left[-i\lambda\int_0^t dt_1 e^{-a^{\dagger}\phi(t_1)}ae^{+a^{\dagger}\phi(t_1)}e^{-i\omega t_1}\right]$$

$$\phi(t) = -i\lambda\int_0^t dt_1 e^{i\omega t_1} = \frac{\lambda}{\omega}(1 - e^{i\omega t})$$

$$\lambda = (M_l - M_k)$$

where we have written the result for one **q** state and have left off the **q** subscripts. In the exponent, the time dependence of the integrand is

$$e^{-i\omega t_1}e^{-a^{\dagger}\phi(t_1)}ae^{+a^{\dagger}\phi(t_1)} = e^{-i\omega t_1}[a + \phi(t_1)]$$

This time integral may now be evaluated and gives

$$-i\lambda \int_0^t dt_1 e^{-i\omega t_1}\left[a + \frac{\lambda}{\omega}(1 - e^{i\omega t_1})\right]$$
$$= -\frac{\lambda}{\omega} a(1 - e^{-i\omega t}) + i\frac{\lambda^2}{\omega} t - \frac{\lambda^2}{\omega^2}(1 - e^{-i\omega t})$$

The phonon correlation function is now

$$U_{\text{ph}}(t) = e^{-it(\varepsilon_k - \varepsilon_l - \Delta_k + \Delta_l)} \exp\left[-\sum_{\mathbf{q}} \frac{\lambda_{\mathbf{q}}^2}{\omega_{\mathbf{q}}^2}(1 - e^{-i\omega_{\mathbf{q}} t})\right] \text{Tr}(e^{-\beta H_0} e^{a^\dagger \phi} e^{-a\phi})$$

Note that the factor Δ_{kl} has canceled out of this expression. The trace is evaluated using the previous result (4.3.10). The normalization factor $1 - e^{-\beta\omega} = [\text{Tr}(e^{-\beta H_0})]^{-1}$ has not been written explicitly in these equations but obviously should be included. Thus for the final result we obtain

$$U_{\text{ph}}(t) = \exp[-it(\varepsilon_k - \varepsilon_l - \Delta_k + \Delta_l) - \Phi_{kl}(t)]$$

$$\Phi_{kl}(t) = \sum_{\mathbf{q}} \frac{(M_{\mathbf{q},k} - M_{\mathbf{q},l})^2}{\omega_{\mathbf{q}}^2} [(N_{\mathbf{q}} + 1)(1 - e^{-i\omega_{\mathbf{q}} t}) + N_{\mathbf{q}}(1 - e^{i\omega_{\mathbf{q}} t})]$$

which is precisely what had been found earlier in (4.3.20).

The following physical picture emerges from this derivation. The coupled system of particle and phonon has been sitting in a state, say l, where the phonons have been fluctuating around the equilibrium position. Suddenly the particle is moved to state k by the optical absorption process. From the point of view of the phonon system, it appears that the potential

$$\delta V = \sum_{\mathbf{q}} (M_{\mathbf{q},k} - M_{\mathbf{q},l})(a_{\mathbf{q}} + a_{\mathbf{q}}^\dagger) \tag{4.3.28}$$

has been suddenly switched on at that time—which we take to be $t = 0$. The transient response in the phonon system is measured by switching off the potential at a later time and measuring the correlations which result. The resulting function of time describes the temporal evolution of the phonon system. Its Fourier transform gives the phonon sideband structure which is observed in the optical spectra. This picture explains the mirror relation between emission and absorption. In absorption the potential which is switched on is (4.3.28), while in emission, if the particle transition is in the opposite direction, the potential which is suddenly switched on is just the negative of this.

F. Linked Cluster Expansion

Another way to evaluate the time-ordered exponential operator

$$U(t) = T\left\langle \exp\left[-i\int_0^t dt_1 V(t_1)\right]\right\rangle \tag{4.3.29}$$

is by the linked cluster expansion, which was discussed in Sec. 3.6. It is sufficient to derive the result for a single-phonon state of wave vector $\mathbf{q}$. The summation over the $\mathbf{q}$ vector in the exponent is easy to do at the end. Thus we consider the case where we can write $V(t)$ as

$$V(t) = \lambda A(t)$$

$$A(t) = ae^{-i\omega t} + a^{\dagger}e^{i\omega t}$$

The S matrix is expanded directly, and each term is evaluated. The series is presumed to be an exponential series. In the present example, as we shall see, there is only one term which remains after the resummation. The nth-order term is

$$U(t) = \sum_{n=0}^{\infty} (-i)^n U_n(t)$$

$$U_n(t) = \frac{\lambda^n}{n!}\int_0^t dt_1 \cdots \int_0^t dt_n \langle TA(t_1) \cdots A(t_n)\rangle$$

Since $A(t)$ describes the creation or destruction of a phonon, the A operators always exist in pairs. This means that only terms with n even are finite; the others are zero. Since n is even, define $n = 2m$:

$$U(t) = \sum_{m=0}^{\infty} (-1)^m U_{2m}(t)$$

$$U_{2m}(t) = \frac{\lambda^{2m}}{(2m)!}\int_0^t dt_1 \cdots \int_0^t dt_{2m} \langle TA(t_1) \cdots A(t_{2m})\rangle$$

According to Wick's theorem, the A operators pair up to form phonon Green's functions [see (2.3.7)]:

$$\langle TA(t_1)A(t_2)\rangle = iD(t_1 - t_2) = (N + 1)e^{-i\omega|t_1-t_2|} + Ne^{i\omega|t_1-t_2|}$$

$$N = \frac{1}{e^{\beta\omega} - 1}$$

Thus the mth term has m phonons:

$$\langle TA(t_1) \cdots A(t_{2m}) \rangle = i^m \sum_{\substack{\text{all} \\ \text{combinations}}} D(t_i - t_j)D(t_2 - t_s) \cdots D(t_u - t_{2m})$$

Each phonon Green's function depends on two time variables $D(t_i - t_j)$. The time integrals over those two variables define a function

$$\phi(t) = i \int_0^t dt_1 \int_0^t dt_2 D(t_1 - t_2)$$

Doing the integrals gives

$$\phi(t) = \frac{2}{\omega^2} [(N + 1)(1 - e^{-i\omega t}) + N(1 - e^{i\omega t}) - it\omega]$$

The quantity $U_{2m}(t)$ is just proportional to $\phi(t)^m$, since each combination of time-ordered products gives this same result. The only remaining question is the combinatorial determination of the number of different arrangements. The number of such combinations is

$$\frac{(2m)!}{2^m m!}$$

which is obtained in the following way. The first variable, t_1, may be paired with any of the $2m - 1$ other variables. The next time variable, which is t_2 if it was not paired with t_1, is paired with any of the other $2m - 3$ variables. Thus the number of combinations is

$$(2m - 1)(2m - 3) \cdots 3 \cdot 1 = \frac{(2m)!}{2^m m!}$$

Thus we obtain the result

$$U_{2m}(t) = \frac{\lambda^{2m}}{m!} \left[\frac{\phi(t)}{2} \right]^m$$

which is summed to obtain the correlation function:

$$U(t) = \exp\left[-\frac{\lambda^2}{2} \phi(t) \right]$$

The function $\frac{1}{2}\lambda^2\phi(t)$ is now recognized as the exponential function of time, which has already been derived several different ways. It is just the double integral over the phonon Green's function. The linked cluster expansion is obviously simple, since there is only one distinct linked cluster, which is $\phi(t)$.

Another way to evaluate the Green's function is to leave the particle operators in the S-matrix expansion. There are occasions when it is not immediately apparent that they can be factored out, as was done in (4.3.24), to give the starting point for (4.3.27). There are several exactly solvable models which can be evaluated this way. Thus it is useful to learn the technique. Consider the Green's function for a single particle in a band, which is coupled to phonons:

$$\begin{aligned}
H &= H_0 + V \\
H_0 &= \varepsilon_0 C^\dagger C + \omega a^\dagger a \\
V &= \lambda C^\dagger C(a + a^\dagger) \equiv \lambda C^\dagger C A \\
A &= a + a^\dagger \\
G(t) &= -i\langle TC(t)C^\dagger(0)\rangle = -i\theta(t)\langle e^{iH_0 t} C e^{-iHt} C^\dagger\rangle \\
G^{(0)}(t) &= -i\theta(t)e^{-i\varepsilon_0 t}
\end{aligned}$$

Since there is only one particle in the band, the creation operator must be to the right, so $t > 0$. The unperturbed Green's function $G^{(0)}(t)$ has a simple form. The interacting Green's function is written, following Chapter 2, as

$$\begin{aligned}
G(t) &= -i\theta(t)\langle T\hat{C}(t)U(t)\hat{C}^\dagger(0)\rangle \\
\hat{C}(t) &= e^{iH_0 t} C e^{-iH_0 t} = e^{-it\varepsilon_0} C \\
U(t) &= e^{iH_0 t} e^{-iHt} = T \exp\left[-i\int_0^t dt_1 \hat{V}(t_1)\right]
\end{aligned}$$

The $U(t)$ matrix, or S matrix, is now expanded in an infinite series of terms. This infinite series will generate the linked cluster expansion. However, now the particle operators $C^\dagger$ and C are included, along with the phonon operators a and $a^\dagger$. Again only terms with n even are nonzero, and the $2m$ term is

$$\begin{aligned}
G_{2m}(t) = -i\theta(t)\frac{(-i)^{2m}}{(2m)!}\lambda^{2m}\int_0^t dt_1 \cdots \int_0^t dt_{2m}\langle T\hat{A}(t_1)\cdots\hat{A}(t_{2m})\rangle \\
\times\langle T\hat{C}^\dagger(t)\hat{C}^\dagger(t_1)\hat{C}(t_1)\cdots\hat{C}^\dagger(t_{2m})\hat{C}(t_{2m})C^\dagger(0)\rangle \qquad (4.3.30)
\end{aligned}$$

Now there is a time-ordered product of particle operators. At first this appears to complicate the evaluation procedure. After all, this time-ordered product of operators,

$$W(t, t_1, \ldots, t_{2m}) = \langle T\hat{C}(t)\hat{C}^\dagger(t_1)\hat{C}(t_1)\cdots\hat{C}^\dagger(t_{2m})\hat{C}(t_{2m})\hat{C}^\dagger(0)\rangle \qquad (4.3.31)$$

must be expanded, according to Wick's theorem, into all possible pairings. However, a careful inspection shows that this time-ordered product has a trivial evaluation. The number operator $C^\dagger C$ is time independent in the interaction representation since H_0 commutes with the number operator $C^\dagger C$,

$$\hat{C}^\dagger(t_j)\hat{C}(t_j) = e^{iH_0t_j}C^\dagger Ce^{-iH_0t_j} = C^\dagger C = n$$

so that the time-ordered correlation function just contains a product of number operators:

$$W(t, t_1, \ldots, t_{2m}) = \langle T\hat{C}(t)n^{2m}\hat{C}^\dagger(0)\rangle = \langle T\hat{C}(t)C^\dagger(0)\rangle \qquad (4.3.32)$$

The number operator gives unity when operating upon the state with one particle. The W function does not actually depend on any time variable except t. The term remaining, on the right-hand side, is just proportional to $G^{(0)}(t)$. Since this same factor occurs in each term in the S-matrix expansion, it may be factored out of the series:

$$G(t) = \sum_{m=0}^{\infty} G_{2m}(t)$$

$$G_{2m}(t) = G^{(0)}(t)\frac{(-i)^{2m}\lambda^{2m}}{(2m)!}\int_0^t dt_1 \cdots \int_0^t dt_{2m}\langle T\hat{A}(t_1)\cdots\hat{A}(t_{2m})\rangle$$

What remains in the series is just the same linked cluster expansion which was evaluated above. Thus the exact Green's function is, again, our result which is now very familiar:

$$G(t) = G^{(0)}(t)e^{-\lambda^2\phi(t)/2}$$

Thus the presence of the C operators in the linked cluster expansion did not, in this case, change the result. Usually that does not happen. For most Hamiltonians, the presence of the C operators changes the evaluation of each term in the S-matrix expansion. These cases are often *not* exactly solvable and certainly are not by the method we are discussing.

The usual polaron problem, for one free particle in a band, is described by the Hamiltonian derived in Sec. 1.3:

$$H = \sum_{\mathbf{q}} \omega_{\mathbf{q}} a_{\mathbf{q}}^\dagger a_{\mathbf{q}} + \sum_{\mathbf{k}} \varepsilon_{\mathbf{k}} C_{\mathbf{k}}^\dagger C_{\mathbf{k}} + \sum_{\mathbf{q}} M_{\mathbf{q}} A_{\mathbf{q}} \varrho_{\mathbf{q}} = H_0 + V$$

$$A_{\mathbf{q}} = a_{\mathbf{q}} + a_{-\mathbf{q}}^\dagger$$

$$\varrho_{\mathbf{q}} = \sum_{\mathbf{k}} C_{\mathbf{k}+\mathbf{q}}^\dagger C_{\mathbf{k}}$$

The coupling between the particle and phonons depends on the particle density operator,

$$\varrho_{\mathbf{q}}(t) = \sum_{\mathbf{k}} C^{\dagger}_{\mathbf{k}+\mathbf{q}} C_{\mathbf{k}} e^{it(\varepsilon_{\mathbf{k}+\mathbf{q}} - \varepsilon_{\mathbf{k}})}$$

and this is time dependent. Thus if we were to attempt to solve for the particle Green's function by a linked cluster expansion, the particle part of the correlation function becomes far more complicated. This Green's function,

$$G(\mathbf{k}, t) = -i\theta(t)\langle e^{iHt} C_{\mathbf{k}} e^{-iHt} C_{\mathbf{k}}{}^{\dagger}\rangle = -i\theta(t)\langle e^{iH_0 t} C_{\mathbf{k}} e^{-it(H_0+V)} C_{\mathbf{k}}{}^{\dagger}\rangle$$

$$V = \sum_{\mathbf{q}} M_{\mathbf{q}} A_{\mathbf{q}} \varrho_{\mathbf{q}} \tag{4.3.33}$$

when evaluated by expanding the S matrix, produces the $2m$th term of the same general form as (4.3.30):

$$G(\mathbf{k}, t) = -i\theta(t) \sum_{m=0}^{\infty} G_{2m}(\mathbf{k}, t)$$

$$G_{2m}(\mathbf{k}, t) = \frac{(-i)^m}{(2m)!} \int_0^t dt_1 \cdots \int_0^t dt_{2m} \sum_{q_1} \cdots \sum_{q_{2m}} M_{\mathbf{q}_1} \cdots M_{\mathbf{q}_{2m}}$$
$$\times \langle T A_{\mathbf{q}_1}(t_1) \cdots A_{\mathbf{q}_{2m}}(t_{2m})\rangle \langle T C_{\mathbf{k}}(t) \varrho_{\mathbf{q}_1}(t_1) \cdots \varrho_{\mathbf{q}_{2m}}(t_{2m}) C_{\mathbf{k}}{}^{\dagger}(0)\rangle \tag{4.3.34}$$

Now the particle part of the correlation function depends on products of the density operators at different times. Usually this does not have a simple time dependence, and the polaron problem is not exactly solvable. One may still evaluate it by a linked cluster expansion. But this now becomes an approximate procedure, whereby one evaluates only a few terms in an infinite series, and the remaining terms are omitted. This approximation will be discussed in Chapter 6.

Van Haeringen (1965) proved the theorem which specifies the most general possible conditions for which (4.3.33) may be solved exactly by the linked cluster method. An exact solution is obtained whenever $\varrho_{\mathbf{q}}(t)$ has the form

$$\varrho_{\mathbf{q}}(t) = e^{itf(\mathbf{q})} \varrho_{\mathbf{q}}$$

$$\varrho_{\mathbf{q}} \equiv \varrho_{\mathbf{q}}(0) = \sum_{\mathbf{k}} C^{\dagger}_{\mathbf{k}+\mathbf{q}} C_{\mathbf{k}}$$

Since the general time dependence is given by the kinetic energy difference

$f(\mathbf{k}, \mathbf{q}) = \varepsilon_{\mathbf{k}+\mathbf{q}} - \varepsilon_{\mathbf{k}}$, his condition is that this difference is independent of $\mathbf{k}$:

$$f(\mathbf{q}) = \varepsilon_{\mathbf{k}+\mathbf{q}} - \varepsilon_{\mathbf{k}} \tag{4.3.35}$$

If these conditions are met, the exact solution to the Green's function may still be obtained from the linked cluster method. Since the factor $\exp[itf(\mathbf{q})]$ is a c number, and not an operator, it may be removed from the time-ordered product of particle operators:

$$\begin{aligned} &\langle TC_{\mathbf{k}}(t)\varrho_{\mathbf{q}_1}(t_1) \cdots \varrho_{\mathbf{q}_{2m}}(t_{2m})C_{\mathbf{k}}^{\dagger}(0)\rangle \\ &\quad = \exp\left[\sum_{j=1}^{2m} it_j f(\mathbf{q}_j)\right]\langle TC_{\mathbf{k}}(t)\varrho_{\mathbf{q}_1} \cdots \varrho_{\mathbf{q}_{2m}}C_{\mathbf{k}}^{\dagger}(0)\rangle \end{aligned}$$

This leaves, in the time-ordered product, just the time-independent density operators $\varrho_{\mathbf{q}_j}$. These have the same effect that the number operators did previously. They do not change the value of the correlation function. It is exactly equal to

$$\exp\left[i\sum_{j=1}^{2m} t_j f(\mathbf{q}_j)\right]\langle TC_{\mathbf{k}}(t)C_{\mathbf{k}}^{\dagger}(0)\rangle = iG^{(0)}(\mathbf{k}, t)\exp\left[i\sum_{j=1}^{2m} t_j f(\mathbf{q}_j)\right] \tag{4.3.36}$$

The validity of (4.3.36) may be shown in the following way. The effect of $\varrho_{\mathbf{q}}$ on $C_{\mathbf{k}}^{\dagger}\mid 0\rangle$ is just to change the particle from the state $\mathbf{k}$ to $\mathbf{k}+\mathbf{q}$:

$$\varrho_{\mathbf{q}}C_{\mathbf{k}}^{\dagger}\mid 0\rangle = \sum_{\mathbf{k}'} C_{\mathbf{k}+\mathbf{q}}^{\dagger}C_{\mathbf{k}'}C_{\mathbf{k}}^{\dagger}\mid 0\rangle = C_{\mathbf{k}+\mathbf{q}}^{\dagger}\mid 0\rangle$$

The effect of a product of such density operators is to change $(C_{\mathbf{k}}^{\dagger}\mid 0\rangle$ by a summation of all their $\mathbf{q}$ vectors:

$$\varrho_{q_1}\varrho_{q_2} \cdots \varrho_{q_{2m}}C_{\mathbf{k}}^{\dagger}\mid 0\rangle = C_{\mathbf{Q}}^{\dagger}\mid 0\rangle$$

$$\mathbf{Q} = \mathbf{k} + \sum_{j=1}^{2m} \mathbf{q}_j$$

Thus the particle correlation function,

$$\langle 0 \mid C_{\mathbf{k}}(t)\varrho_{\mathbf{q}_1}\varrho_{\mathbf{q}_2} \cdots \varrho_{\mathbf{q}_{2m}}C_{\mathbf{k}}^{\dagger}(0)\mid 0\rangle = \langle 0 \mid C_{\mathbf{k}}(t)C_{\mathbf{k}}^{\dagger}(0)\mid 0\rangle$$

does give $iG^{(0)}(\mathbf{k}, t)$ if the summation of all the q vectors is zero, which is, in fact, the case. The $\sum \mathbf{q}_j = 0$ is required by the phonon part of the correlation function. When this correlation function is evaluated, by pairing according to Wick's theorem, the pairing forces the $\mathbf{q}$ values to be equal and opposite in sign:

$$\langle TA_{\mathbf{q}_i}(t_i)A_{\mathbf{q}_j}(t_j)\rangle = \delta_{\mathbf{q}_i+\mathbf{q}_j=0}\langle TA_{\mathbf{q}_i}(t_i)A_{-\mathbf{q}_i}(t_j)\rangle = i\delta_{\mathbf{q}_i+\mathbf{q}_j}D(\mathbf{q}_i, t_i - t_j)$$

Thus the summation of all the $\mathbf{q}$ values is certainly zero, since they are paired in sets of $\mathbf{q}$ and $-\mathbf{q}$. Thus we have proved the assertion (4.3.36).

The extra time factors $\exp(itf)$ go into the evaluation of the linked cluster expansion. Thus we define a new function:

$$\bar{\phi}(t) = \tfrac{1}{2}\sum_{\mathbf{q}} M_{\mathbf{q}}^2 \int_0^t dt_1 \int_0^t dt_2 iD(\mathbf{q}, t_1 - t_2)e^{i[f(\mathbf{q})t_1 + f(-\mathbf{q})t_2]}$$

It is evaluated for the case $f(-\mathbf{q}) = -f(\mathbf{q})$, which is the interesting physical case discussed below:

$$\begin{aligned}\bar{\phi}(t) = \sum_{\mathbf{q}} M_{\mathbf{q}}^2 \bigg\{ & \frac{-it\omega_{\mathbf{q}}}{\omega_{\mathbf{q}}^2 - f^2} + (N_{\mathbf{q}} + 1)\bigg[\frac{1 - e^{-it(\omega_{\mathbf{q}} - f)}}{(\omega_{\mathbf{q}} - f)^2}\bigg] \\ & + N_{\mathbf{q}}\bigg[\frac{1 - e^{it(\omega_{\mathbf{q}} + f)}}{(\omega_{\mathbf{q}} + f)^2}\bigg]\bigg\}\end{aligned}$$

Terms with $\omega_{\mathbf{q}} - f$ are identical to those with $\omega_{\mathbf{q}} + f$ by the variable change: $\mathbf{q} \to -\mathbf{q}$. The exact result for the Green's function (4.3.33) is

$$G(\mathbf{k}, t) = -i\theta(t)e^{-i\varepsilon_{\mathbf{k}}t}e^{-\bar{\phi}(t)} \tag{4.3.37}$$

Equation (4.3.37) is the most general form for $G(\mathbf{k}, t)$, which may be exactly solvable, according to van Haeringen.

What kind of systems can we have which obey his condition that $\varepsilon_{\mathbf{k}+\mathbf{q}} - \varepsilon_{\mathbf{k}} = f(\mathbf{q})$? One possibility is that the energy is a constant,

$$\varepsilon_{\mathbf{k}} = \varepsilon_0$$

so that $f = 0$. It is this example which has been repeatedly worked throughout this section. However, there is another case which is now recognized as an exactly solvable model. It has the particle moving with a constant velocity $\mathbf{v}$, so that its energy is

$$\varepsilon_{\mathbf{k}} = \mathbf{k} \cdot \mathbf{v}$$

$$f(\mathbf{q}) = \mathbf{q} \cdot \mathbf{v} = -f(-\mathbf{q})$$

We should remark that the system of equations is not Galilean invariant. Even though the particle is moving with constant velocity, one cannot transform the equations into a system where it is standing still, in a stationary system of phonons. The phonons, particularly the acoustical phonons, have their own velocity and are not invariant under a Galilean transformation. Instead, the phonon frequencies appear Doppler shifted to the particle. Thus the particle Green's function could be evaluated for the case that it was going faster or slower than the acoustical phonons system. Presumably the phonon relaxation around the particle will vary considerably between these two cases. This model is illustrated in the assigned problems. The one-dimensional constant velocity model for polarons was solved by Engelsberg and Varga (1964).

In the polaron problem the particle–phonon coupling produces fluctuations in the number of phonons surrounding the particle. This number fluctuates from time to time and from particle to particle. In the independent boson model, each phonon, in this fluctuating cloud around the particle, is assumed to exist with a probability which is independent of whether other phonons are also simultaneously present. Each phonon is fluctuating independently of the others. If a particle is fixed, after emitting a phonon it is still fixed. Thus the probability of emitting the second phonon is the same as the probability of emitting the first. When the probability of each emission or absorption is the same, then one has the independent boson model. Of course, a particle moving with a constant velocity **v** also has its motion unaffected by how many phonons have been emitted. Again the probability of emitting each phonon is independent of others which may have been emitted, so that all phonon emissions are independent.

A free particle with kinetic energy $\varepsilon_k = k^2/2m$ is not described by the independent boson model. After emission of one phonon with wave vector $\mathbf{q}$, the particle goes to an intermediate state with wave vector $\mathbf{k} + \mathbf{q}$ and energy $(\mathbf{k} + \mathbf{q})^2/2m$. The intermediate state is different from the starting one, so that the emission of the second phonon has a probability that is different from the emission of the first. Each emission has a different probability, so that the probability of n phonon emissions is not just a Poisson distribution,

$$P_n = e^{-\alpha} \frac{\alpha^n}{n!}$$

where α is the probability of a single emission. The Poisson distribution is the characteristic zero temperature distribution of phonon sidebands only when each emission has the same probability, which is independent of the number of other phonons emitted.

The independent boson model has been used widely, with a number of variations. Almbladh and Minnhagen (1978) solved the Fano-Anderson model with phonon coupling to the localized level. A number of related models have been reviewed by Cini and D'Andrea (1988).

4.4. TOMONAGA MODEL

The Tomonaga model (Tomonaga, 1950) describes a one-dimensional electron gas. The procedure is to examine the Hamiltonian of the one-dimensional electron gas and make some approximations on it. As a consequence of these approximations, the Hamiltonian becomes exactly solvable. Thus the one-dimensional electron gas is not exactly solvable but only an approximate version of it.

The important physics is the recognition that the excitations of the electron gas are approximate bosons, although the elementary particles—electrons—are fermions. The excitations involve two-particle states, for example, moving an electron from one state to another. The wave function of the two fermion states has boson properties. The Tomonaga model assumes that the excitations are exactly bosons, which is the important approximation.

The model has been useful in several kinds of problems. First, there are organic solids such as TTF–TCNQ whose conductivity is thought to be largely one-dimensional (see Heeger, 1974). The Tomonaga model has played a role in the interpretation of electrical conductivity in these materials (see Luther and Emery, 1974). Second, in impurity problems, or X-ray absorption problems, the response of the electron gas to the central impulse can be factored into spherical harmonics associated with different angular momentum states l. Each angular momentum channel l then becomes a one-dimensional electron gas (see Schotte and Schotte, 1969) to which one may apply the Tomonaga model.

A. Tomonaga Model

The original model of Tomonaga (1950) discusses the Hamiltonian for the one-dimensional interacting electron gas:

$$\begin{aligned} H &= v_F \sum_{ks} | k | a_{ks}^\dagger a_{ks} + \frac{1}{2L} \sum_k V_k \varrho(k) \varrho(-k) \\ \varrho(k) &= \sum_{p,s} a_{p-k/2,s}^\dagger a_{p+k/2,s} \end{aligned} \tag{4.4.1}$$

The system has length L, and v_F is the Fermi velocity of the particles, which are assumed to have a linear dispersion relation. The summation over

k states may be turned into integrals by the usual transformation as $L \to \infty$:

$$\sum_k f(k) = \frac{L}{2\pi} \int dk f(k)$$

The spin index is $s = \pm 1$, and $\varrho(k)$ is the electron density operator. The V_k is an electron–electron interaction term, whose form will be specified below. It will not be $4\pi e^2/k^2$, which is dimensionally incorrect in one dimension, since V_k has units of erg-cm. Dimensional analysis then suggests the form $V_k \propto e^2(k_F/k)^n$, where n is any exponent.

The basic step in the Tomonaga model is to divide the density operator into two terms:

$$\begin{aligned} \varrho_1(k) &= \sum_{s,p>0} a^\dagger_{p-k/2,s} a_{p+k/2,s} \\ \varrho_2(k) &= \sum_{s,p<0} a^\dagger_{p-k/2,s} a_{p+k/2,s} \\ \varrho(k) &= \varrho_1(k) + \varrho_2(k) \end{aligned} \tag{4.4.2}$$

The density operator $\varrho(k)$ commutes with any other density operator $\varrho(k')$. However, the two parts ϱ_1 and ϱ_2 do not commute with the same parts for other wave vectors. Thus, examine the commutation relations:

$$\begin{aligned} [\varrho_1(k), \varrho_1(k')] &= \sum_{\substack{s,s' \\ p,p'>0}} [a^\dagger_{p-k/2,s} a_{p+k/2,s}, a^\dagger_{p'-k'/2,s'} a_{p'+k'/2,s'}] \\ &= \sum_{s,p>0} [a^\dagger_{p-k/2,s} a_{p+k'+k/2,s} \theta(p + k/2 + k'/2) \\ &\quad - a^\dagger_{p-k'-k/2,s} a_{p+k/2,s} \theta(p - k/2 - k'/2)] \end{aligned} \tag{4.4.3}$$

An important special case is

$$[\varrho_1(k), \varrho_1(-k)] = \sum_{s,p>0} (n_{p-k/2,s} - n_{p+k/2,s}) = \sum_{\substack{-k/2 \le \bar{p} \le k/2 \\ s}} n_{\bar{p},s} \tag{4.4.4}$$

The right-hand side shows that the commutation relations depend on the operator $n_{\bar{p},s}$ over a range of p values. The operator n is replaced by its average in the ground state of the free-particle system. Thus, set

$$\sum_{\bar{p}} n_{\bar{p},s} = \sum_{\bar{p},s} \langle n_{\bar{p},s} \rangle = 2 \sum_{-k/2 \le \bar{p} \le k/2} \theta(k_F - |\bar{p}|) = \begin{cases} 2\left(\dfrac{kL}{2\pi}\right), & k < 2k_F \\ \dfrac{2k_F L}{\pi}, & k > 2k_F \end{cases}$$

and the commutation relations (4.4.4) can be written for $k < 2k_F$ as

$$[\varrho_1(k), \varrho_1(-k)] = \frac{kL}{\pi}$$

$$[\varrho_2(k), \varrho_2(-k)] = -\left(\frac{kL}{\pi}\right)$$

$$[\varrho_1(k), \varrho_2(-k)] = 0$$

We have also included the analogous result for $[\varrho_2(k), \varrho_2(-k)]$ and $[\varrho_1, \varrho_2]$, which can be derived in the same fashion. The Tomonaga model assumes that these density operators obey the exact commutation relations of

$$\begin{aligned}
[\varrho_1(k), \varrho_1(-k')] &= \delta_{kk'} \frac{Lk}{\pi} \\
[\varrho_2(k), \varrho_2(-k')] &= -\delta_{kk'} \frac{Lk}{\pi} \\
[\varrho_1(k), \varrho_2(-k')] &= 0
\end{aligned} \tag{4.4.5}$$

These relations are the central approximation of the Tomonaga model. The commutation relations are not exact, since the commutators give operators, as in (4.4.3). However, when we take the expectation value of the exact commutation relations, we do get these results. For example, in (4.4.3) we have

$$\langle[\varrho_1(k), \varrho_1(k')]\rangle = \sum_{s,p>0} [\langle a^\dagger_{p-k/2,s} a_{p+k'+k/2,s}\rangle - \langle a^\dagger_{p+k'-k/2,s} a_{p+1/2k,s}\rangle]$$

In the right-hand side, the average of $\langle a^\dagger_{p-1/2k} a_{p+k'+1/2k}\rangle$ is zero unless $k' = -k$, so that we have

$$\begin{aligned}
\langle[\varrho_1(k), \varrho_1(-k')]\rangle &= \delta_{kk'} \sum_{s,p>0} [\langle n_{p-k/2,s}\rangle - \langle n_{p+k/2,s}\rangle] \\
&= 2\delta_{kk'} \sum_{-k/2 \le \bar{p} \le k/2} \langle n_{\bar{p}}\rangle
\end{aligned}$$

Thus, although the commutation relations (4.4.5) are not exact, the expectation values of these commutators are given exactly. Thus we have made an approximation, but not a very bad one.

It is convenient to express the density operators $\varrho_j(\pm k)$ in terms of creation and destruction operators. This step will be done so that the creation operators are dimensionless and the commutation relations (4.4.5) are obeyed. The creation–destruction operators are for bosons. These definitions are given below, where the symbol k is always positive:

$$\begin{aligned}
\varrho_1(k) &= b_k\left(\frac{kL}{\pi}\right)^{1/2} \\
\varrho_1(-k) &= b_k^{\dagger}\left(\frac{kL}{\pi}\right)^{1/2} \\
\varrho_2(k) &= b_{-k}^{\dagger}\left(\frac{kL}{\pi}\right)^{1/2} \\
\varrho_2(-k) &= b_{-k}\left(\frac{kL}{\pi}\right)^{1/2} \\
[b_k, b_{k'}^{\dagger}] &= \delta_{kk'}
\end{aligned} \tag{4.4.6}$$

Thus, when k is positive, we have $\varrho_1(k) \propto b_k$, and when k is negative, we have $\varrho_1(k) \propto b_{-k}^{\dagger}$. The operators ϱ_1 always commute with ϱ_2. For example, $[\varrho_1(k), \varrho_2(k)] = (kL/\pi)[b_k, b_{-k}^{\dagger}] = 0$. Thus the choice (4.4.6) does satisfy the approximate commutation relations (4.4.5).

In the Hamiltonian (4.4.1), the second term may now be written in terms of these boson operators:

$$\begin{aligned}
\frac{1}{2L}\sum_k V_k\varrho(k)\varrho(-k) &= \sum_k \bar{V}_k(b_k + b_{-k}^{\dagger})(b_k^{\dagger} + b_{-k}) \\
\bar{V}_k &= \frac{|k| V_k}{2\pi}
\end{aligned} \tag{4.4.7}$$

The electron–electron interaction term has been recast into an interaction between the boson excitations of the electron gas.

The first term in (4.4.1), which is the particle kinetic energy, requires some additional work before it is expressed in terms of boson coordinates. It is not immediately obvious how to express $\sum_k |k| a_k^{\dagger}a_k$ in terms of the new boson operators. When faced with this predicament, we instead examine the commutation relations of this operator. The objective is to find a boson representation of the kinetic energy operator which reproduces the commutation relations. If this cannot be done exactly, at least try to find a good approximation. The commutator algebra completely specifies the excitation spectrum of the system, so that the excitations are adequately described by operators with accurate commutation relations.

Call the kinetic energy term H_0. Its commutator with $\varrho_1(k)$ is

$$[\varrho_1(k), H_0] = v_F \sum_{\substack{s,p>0 \\ s'k'}} |k'| [a^\dagger_{p-k/2,s} a_{p+k/2,s}, a^\dagger_{k's'} a_{k's'}]$$

$$[\varrho_1(k), H_0] = v_F \sum_{s,p>0} a^\dagger_{p-k/2,s} a_{p+k/2,s} (|p + k/2| - |p - k/2|)$$

$$|p + k/2| - |p - k/2| = \begin{cases} k & \text{if } p > \dfrac{k}{2} \\[2ex] 2p & \text{if } p < \dfrac{k}{2} \end{cases}$$

For small values of k, we have that $p > k/2$ over most of the p summation. Thus for small k values, the above commutator is approximately given by

$$[\varrho_1(k), H_0] = v_F k \sum_{s,p} a^\dagger_{p-1/2k,s} a_{p+1/2k,s} = k v_F \varrho_1(k) \tag{4.4.8}$$

The above is a desirable form for the commutator, since the right-hand side is also proportional to $\varrho_1(k)$. With the boson representations (4.4.5) and (4.4.6) for $\varrho_1(k)$, the approximate commutation relation (4.4.8) is

$$[b_k, H_0] = k v_F b_k \equiv \omega_k b_k$$

$$\omega_k = v_F |k|$$

Of course, the same result would be given by the choice of $H_0 = \sum_{k>0} \times \omega_k b_k^\dagger b_k$. We must also consider the commutator of H_0 with ϱ_2. The same approximation in this case leads to

$$[\varrho_2(k), H_0] = -\omega_k \varrho_2(k)$$

Both of these approximate commutators are satisfied with the following choice for H_0:

$$H_0 = \frac{v_F \pi}{L} \sum_{k>0} [\varrho_1(-k)\varrho_1(+k) + \varrho_2(+k)\varrho_2(-k)]$$

$$H_0 = \sum_k \omega_k b_k^\dagger b_k \tag{4.4.9}$$

$$H = \sum_k \omega_k b_k^\dagger b_k + \sum_k \bar{V}_k (b_k + b^\dagger_{-k})(b_k^\dagger + b_{-k})$$

The one-dimensional electron gas (4.4.1) has been recast into the boson Hamiltonian (4.4.9). The latter is exactly solvable, as will soon be shown.

The Tomonaga model (4.4.9) has been derived from (4.4.1) with several key approximations on commutation relations. The model form (4.4.9) is a description of the electron gas as due to boson excitations.

Equation (4.4.9) may be solved exactly by a variety of techniques. Probably the easiest method is to change to a coordinate representation for the boson operators:

$$Q_k = \frac{1}{(2\omega_k)^{1/2}} (b_k + b^\dagger_{-k})$$

$$P_k = i\left(\frac{\omega_k}{2}\right)^{1/2} (b_k^\dagger - b_{-k})$$

$$[Q_k, P_{k'}] = i\delta_{kk'}$$

In this representation the Hamiltonian is written as

$$\begin{aligned} H_0 &= \tfrac{1}{2} \sum_k (P_{-k}P_k + \omega_k^2 Q_k Q_{-k}) \\ H &= \tfrac{1}{2} \sum_k [P_k P_{-k} + Q_k Q_{-k}(\omega_k^2 + 4\omega_k \bar{V}_k)] \end{aligned} \tag{4.4.10}$$

where we have added the zero-point motion term $\frac{1}{2} \sum_k \omega_k$ to H_0. Equation (4.4.10) shows that the new eigenfrequencies are

$$E_k = (\omega_k^2 + 4\omega_k \bar{V}_k)^{1/2} = |\, k \,| \left(v_F^2 + \frac{2}{\pi} V_k v_F\right)^{1/2} \tag{4.4.11}$$

If we now change back to a new set of boson normal mode operators, which are normalized to the new eigenfrequencies, we have diagonalized the Hamiltonian:

$$\begin{aligned} Q_k &= \frac{1}{(2E_k)^{1/2}} (\alpha_k + \alpha^\dagger_{-k}) \\ P_k &= i\left(\frac{E_k}{2}\right)^{1/2} (\alpha_k^\dagger - \alpha_{-k}) \\ H &= \sum_k E_k(\alpha_k^\dagger \alpha_k + \tfrac{1}{2}) \\ [\alpha_k, \alpha_{k'}^\dagger] &= \delta_{kk'} \end{aligned} \tag{4.4.12}$$

These series of steps may be summarized by the observation that we really just changed boson operators in the following way:

$$b_k + b_{-k}^\dagger = \left(\frac{\omega_k}{E_k}\right)^{1/2} (\alpha_k + \alpha_{-k}^\dagger)$$
$$b_k^\dagger - b_{-k} = \left(\frac{E_k}{\omega_k}\right)^{1/2} (\alpha_k^\dagger - \alpha_{-k}) \tag{4.4.13}$$

These transformations will be useful for other problems.

The Hamiltonian of the one-dimensional electron gas (4.4.1) has been solved approximately. It should be emphasized that we have only been solving for the excitation spectrum of the electron gas. Some of these excitations are fluctuations in the density operator $\varrho(k)$. Very similar results to the Tomonaga model are obtained by writing an equation of motion for the density operator and solving it approximately. This approach is used in Chapter 5.

So far we have not specified the form of the interaction potential V_k. In fact, physicists often choose a variety of forms for this interaction in order to suit their problem. The units of V_k are the same as v_F: erg-cm ($\hbar v_F$ is in erg-cm). One possible choice is to take $V_k \propto e^2 = \text{constant} = V_0$. The energy spectrum $E_k = |k| [v_F^2 + (2V_0/\pi)v_F]^{1/2}$ is just altered by having the Fermi velocity increased:

$$E_k = |k| \bar{v}_F$$
$$\bar{v}_F = v_F\left(1 + \frac{2}{\pi}\frac{V_0}{v_F}\right)^{1/2}$$

The constant V_0 is assumed to be positive, since it describes interactions between electrons.

Another possible choice is to take $V_k = (2/3)(e^2 k_F^2/k^2)$. This leads to the long-wavelength modes having a constant frequency, which is the plasma frequency:

$$E_k = (k^2 v_F^2 + \omega_p^2)^{1/2}$$
$$\omega_p^2 = 4\omega_k \bar{V}_k = \frac{2}{\pi} k^2 v_F \left(\frac{2}{3}\frac{e^2 k_F^2}{k^2}\right) = \frac{4\pi e^2 n_0}{m}$$
$$n_0 = \frac{k_F^3}{3\pi^2}$$

In the electron gas, there are two different types of excitations. One is the plasma modes at long wavelength, and the other is the electron–hole

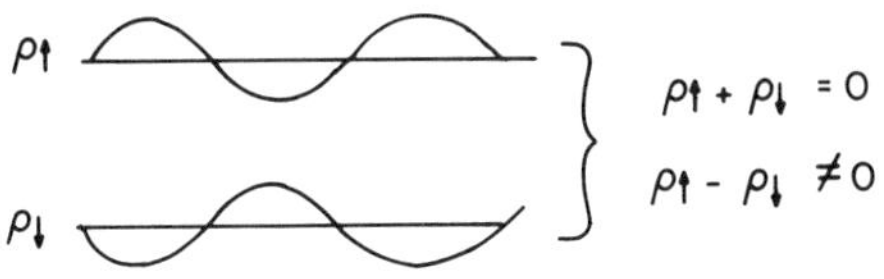

FIGURE 4.15. Spin up and spin down charge densities.

excitations at shorter wavelength. The latter are probably best described by the choice $V_k = V_0$.

B. Spin Waves

The Hamiltonian (4.4.1) of the one-dimensional electron gas has other collective excitations besides the density oscillations which were discussed above. These other excitations have the character of spin waves, or magnons. Overhauser (1965) has shown that the excitation spectrum is completely described by the sum of these two types of excitations: density oscillations and spin waves. This feature of one dimension does not apply to three dimensions. The density oscillations are the excitations which occur when there are external perturbations such as electric fields. The spin waves respond to magnetic perturbations and contribute to the spin susceptibility.

The spin waves are described by the operators

$$\begin{aligned}
\sigma(k) &= \sigma_1(k) + \sigma_2(k) \\
\sigma_1(k) &= \sum_{s,p>0} s a^\dagger_{p-k/2,s} a_{p+k/2,s} \qquad (4.4.14) \\
\sigma_2(k) &= \sum_{s,p<0} s a^\dagger_{p-k/2,s} a_{p+k/2,s}
\end{aligned}$$

where the spin index is $s = \pm$ for $\uparrow, \downarrow$. The nature of the spin wave excitations is shown in Fig. 4.15. The spin up and spin down densities have opposite variations, so that there is no net change in the particle density. There is a variation in $\varrho_\uparrow - \varrho_\downarrow$ which we call σ:

$$\begin{aligned}
\varrho_s &= \sum_p a^\dagger_{p-k/2,s} a_{p+k/2,s} \\
\varrho &= \varrho_\uparrow + \varrho_\downarrow \\
\sigma &= \varrho_\uparrow - \varrho_\downarrow
\end{aligned}$$

The properties of these operators are examined in the same fashion that we used for the density operators. The commutation relations are found among these operators and between them and the density operators. Some typical results are

$$[\sigma_1(k), \sigma_1(-k')] = \sum_{s,p>0} s^2(a^\dagger_{p-k/2,s}a_{p-k'+k/2,s} - a^\dagger_{p+k'-k/2,s}a_{p+k/2,s})$$

$$[\sigma_1(k), \varrho_1(-k')] = \sum_{s,p>0} s(a^\dagger_{p-k/2,s}a_{p-k'+k/2,s} - a^\dagger_{p+k'-k/2,s}a_{p+k/2,s})$$

Since $s^2 = 1$, the commutator $[\sigma_1(k), \sigma_1(-k)]$ is equal to $[\varrho_1(k), \varrho_1(-k)]$ in (4.4.4) and is given the same value. The commutator $[\sigma_1(k), \varrho_1(-k)]$ contains one factor of s, and the two terms $s = 1$ and $s = -1$ will cancel. This cancellation occurs when the two spin states are occupied with equal probability—so that the system is not magnetic. We have been making this assumption, since otherwise the excitation spectrum is quite different. In the same spirit which was used to derive the approximate commutation relations (4.4.5) for the density operators, we deduce a similar approximate set of commutators for the spin operators:

$$\begin{aligned}
[\sigma_1(k), \sigma_1(-k')] &= \delta_{kk'} \frac{kL}{\pi} \\
[\sigma_2(k), \sigma_2(-k')] &= -\delta_{kk'} \frac{kL}{\pi} \\
[\sigma_1(k), \sigma_2(k')] &= 0 \\
[\sigma_i(k), \varrho_j(k')] &= 0 \qquad (i, j = 1, 2)
\end{aligned} \tag{4.4.15}$$

The spin operators commute with the density operators and so describe an independent set of excitations. These excitations can be represented by a new set of creation and destruction operators, which for $k > 0$ are

$$\begin{aligned}
\sigma_1(k) &= c_k\left(\frac{kL}{\pi}\right)^{1/2} \\
\sigma_1(-k) &= c_k^\dagger\left(\frac{kL}{\pi}\right)^{1/2} \\
\sigma_2(k) &= c^\dagger_{-k}\left(\frac{kL}{\pi}\right)^{1/2} \\
\sigma_2(-k) &= c_{-k}\left(\frac{kL}{\pi}\right)^{1/2} \\
[c_{k'}c^\dagger_{k'}] &= \delta_{kk'} \\
[c_{k'}b^\dagger_{k'}] &= 0
\end{aligned} \tag{4.4.16}$$

The next step is to examine the commutation relation of $\sigma_j(k)$ with the Hamiltonian (4.4.1), which will establish the energy spectrum of these spin wave operators. Of course they commute with the second term in (4.4.1), from electron–electron interactions, since they commute with the density operators. Thus we need to consider only the commutator with the kinetic energy term H_0:

$$[\sigma_1(k), H_0] = v_F \sum_{s,p>0} s a^{\dagger}_{p-k/2,s} a_{p+k/2,s}(|p + k/2| - |p - k/2|)$$

$$\simeq v_F k \sigma_1(k)$$

The commutator is evaluated in the same approximation which was used to get (4.4.8). Thus the commutator of $\sigma_1(k)$ with the Hamiltonian just gives $\omega_k \sigma_1(k)$. Exactly the same result is obtained by representing H_0 by $\sum_{k>0} (\pi V_F/L)\sigma_1(-k)\sigma_1(k)$ and using the commutation relations (4.4.15). Thus the spin wave part of the Hamiltonian, from both σ_1 and σ_2 parts, is

$$H_{\text{sw}} = \frac{v_F \pi}{L} \sum_{k>0} [\sigma_1(-k)\sigma_1(k) + \sigma_2(k)\sigma_2(-k)] = \sum_k \omega_k c_k{}^{\dagger} c_k \tag{4.4.17}$$

$$H_T = H + H_{\text{sw}} = \sum_k \omega_k (b_k{}^{\dagger} b_k + c_k{}^{\dagger} c_k) + \sum_k \bar{V}_k (b_k + b^{\dagger}_{-k})(b_k{}^{\dagger} + b_{-k})$$

$$H_T = \sum_k (E_k \alpha_k{}^{\dagger} \alpha_k + \omega_k c_k{}^{\dagger} c_k) \tag{4.4.18}$$

We have combined H_{sw} with the density operator parts, in (4.4.9), (4.4.18), and (4.4.12), to give the total Hamiltonian H_T for the excitation spectra of the one-dimensional electron gas. The original model of Tomonaga actually described a spinless electron gas. For spin one-half systems, the two possible spin orientations lead to another type of independent excitation which we have called spin waves. The total Hamiltonian (4.4.17) has the density and spin wave excitations decoupled.

The original Hamiltonian (4.4.1) did not contain any terms which would cause interactions between spin waves; there were no terms of the type $\sigma(k)\sigma(-k)$. Thus the spin wave excitation spectrum ω_k is the same as for the free particles, at least in the approximations we have made to derive the Tomonaga model. There are often situations where a fermion will interact selectively with other fermions of the same spin state, so that there might be terms in the Hamiltonian such as

$$U = \frac{1}{L}\sum_{k,s} U_k\Big(\sum_p a^\dagger_{p-k/2,s}a_{p+k/2,s}\Big)\Big(\sum_{p'} a^\dagger_{p'+k/2,s}a_{p'-k/2,s}\Big)$$

$$= \frac{1}{L}\sum_{k,s} U_k\varrho_s(k)\varrho_s(-k) = \frac{1}{L}\sum_k U_k[\varrho_\uparrow(k)\varrho_\uparrow(-k) + \varrho_\downarrow(k)\varrho_\downarrow(-k)] \tag{4.4.19}$$

This term can be algebraically factored into density and spin wave components:

$$U = \frac{1}{2L}\sum_k U_k\{[\varrho_\uparrow(k) + \varrho_\downarrow(k)][\varrho_\uparrow(-k) + \varrho_\downarrow(-k)]$$

$$+ [\varrho_\uparrow(k) - \varrho_\downarrow(k)][\varrho_\uparrow(-k) - \varrho_\downarrow(-k)]\}$$

$$= \frac{1}{2L}\sum_k U_k[\varrho(k)\varrho(-k) + \sigma(k)\sigma(-k)] \tag{4.4.20}$$

The first term contributes to the interaction among the density excitations and just adds to the second term in (4.4.1). The second term of U provides an effective interaction between the spin waves. We can solve the spin wave Hamiltonian with this new interaction, since it has the same mathematical form as (4.4.9), whose solution is (4.4.11) and (4.4.12). Thus spin-selective forces, such as given in (4.4.19), are easily incorporated into the formalism.

The spin wave part of the excitation spectrum can be used to derive the Pauli spin susceptibility. The starting point for this calculation is (3.8.3):

$$\chi(k, i\omega) = -\int_0^\beta d\tau \langle T_\tau \sigma(k, \tau)\sigma(-k, 0)\rangle e^{i\omega\tau} \tag{4.4.21}$$

In the Tomonaga model, the correlation function may be evaluated exactly by using the operator representation (4.4.16):

$$\sigma(-k, 0) = \sigma_1(-k) + \sigma_2(-k) = \left(\frac{kL}{\pi}\right)^{1/2}(c_k^\dagger + c_{-k})$$

$$\sigma(k, \tau) = e^{\tau H_{\rm sw}}\sigma(k)e^{-\tau H_{\rm sw}} = \left(\frac{kL}{\pi}\right)^{1/2}(c_k e^{-\tau\omega_k} + c^\dagger_{-k}e^{\tau\omega_k})$$

The τ independence is determined by $H_{sw} = \sum_k \omega_k c_k^{\dagger} c_k$. The further steps in the evaluation of the correlation function are identical to the derivation of the unperturbed phonon Green's function in (3.2.16). Thus we are immediately led to

$$\chi(k, i\omega) = \left(\frac{|k| L}{\pi}\right) \frac{2\omega_k}{(i\omega)^2 - \omega_k^2}$$

The retarded correlation function is found from the analytical continuation $i\omega \to \omega + i\delta$ so that we have the final answer

$$\chi(k, \omega) = \left(\frac{|k| L}{\pi}\right) \frac{2\omega_k}{\omega^2 - \omega_k^2 + i\omega\delta} \tag{4.4.22}$$

The susceptibility is found to be proportional to the length L of the electron gas. This dependence on L is correct, since the susceptibility is the total magnetization M divided by the magnetic field, and the total magnetization is indeed proportional to the size of the system. A more meaningful quantity would be the magnetization per unit volume, which is the above result divided by L. The susceptibility demonstrates a resonance phenomenon, so that it is singular whenever the external perturbations (k, ω) exactly match those of the excitation spectrum $\omega_k = kv_F$.

C. Luttinger Model

A model proposed by Luttinger (1963) is a slight variation on the Tomonaga model. It has the advantage of being exactly solvable, with fewer approximations, yet is identical to the Tomonaga model in some of its essential properties. The basic feature of the Luttinger model is that the system has two types of fermions. One has an energy spectrum given by $\varepsilon_k = kv_F$, while the other has an energy spectrum given by $\varepsilon_k = -kv_F$. They are shown by the solid and dashed lines in Fig. 4.16(a). There is an infinite number of each kind of particle, since the occupied energy states stretch to negative infinity.

In the Tomonaga model (4.4.1) we assumed that the energy spectrum is that shown in Fig. 4.16(b). The particles have a linear dispersion rela-

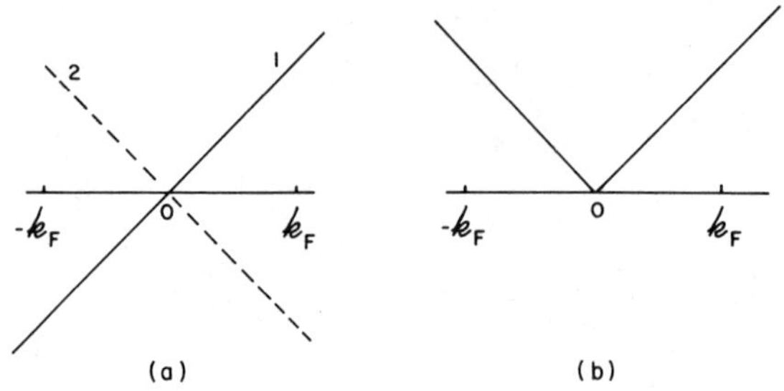

FIGURE 4.16. (a) The Luttinger model has two distinct particles, with separate energy bands. (b) The Tomonaga model has one particle, whose energy band $v_F \mid k \mid$.

tion, but the same kind of particle is represented throughout the band of states.

The two kinds of fermions in the Luttinger model are denoted by the operators $a_{1,k,s}$ and $a_{2,k,s}$, where the subscript 1 or 2 designates the particle. The two bands are quite independent, so the two fermion operators anticommute:

$$\{a_{i,k',s'}, a^\dagger_{j,k,s}\} = \delta_{ij}\delta_{kk'}\delta_{ss'}$$

It is conventional to define operators $\varrho_i(k)$ and $\sigma_i(k)$ in a manner analogous to the Tomonaga model ($p > 0$):

$$\varrho_i(p) = \sum_{\substack{\text{all } k,\\ s=\pm 1}} a^\dagger_{i,k+p,s} a_{i,k,s}$$

$$\varrho_i(-p) = \sum_{\text{all } k,s} a^\dagger_{i,k,s} a_{i,k+p,s} = \varrho_i(p)^\dagger$$

$$\sigma_i(p) = \sum_{\text{all } k,s} s a^\dagger_{i,k+p,s} a_{i,k,s}$$

$$\sigma_i(-p) = \sum_{\text{all } k,s} s a^\dagger_{i,k,s} a_{i,k+p,s} = \sigma_i(p)^\dagger$$

The advantage of the Luttinger model is that it has the same kind of commutation relations as found for the Tomonaga model. However, now they are valid for all p, whereas they were valid only for $p < 2k_F$ in the

Tomonaga model:

$$[\varrho_1(-p), \varrho_1(+p')] = \delta_{pp'}\left(\frac{pL}{\pi}\right)$$

$$[\varrho_2(p), \varrho_2(-p')] = \delta_{pp'}\left(\frac{pL}{\pi}\right)$$

$$[\varrho_1(p), \varrho_2(-p')] = 0$$

$$[\sigma_1(-p), \sigma_1(p')] = \delta_{pp'}\left(\frac{pL}{\pi}\right)$$

$$[\sigma_2(-p), \sigma_2(p')] = \delta_{pp'}\left(\frac{pL}{\pi}\right)$$

$$[\sigma_1(p), \sigma_2(p')] = 0$$

$$[\sigma_i(p), \varrho_j(p')] = 0$$

These commutation relations depend, in an important way, on the assumption that there is an infinite number of negative energy particles. For example, the first commutator is

$$[\varrho_1(-p), \varrho_1(p)] = 2\sum_k (n_{1,k} - n_{1,k+p}) = \frac{pL}{\pi}$$

The factor of 2 comes from the summation over the two spin configurations $s = \pm 1$. If there were a finite number of particles, each summation over particle number would just give the number of 1-particles N_1,

$$N_1 = 2\sum_k n_{1,k} = 2\sum_k n_{1,k+p}$$

and the commutator would be zero. However, when there is an infinite number of particles in negative energy states, a finite result is obtained, which is illustrated in Fig. 4.17. For a finite band, the difference

$$\sum_k (n_{1,k} - n_{1,k+p})$$

equals $pL/2\pi$ at the top end of the band, but it equals the negative of this at the bottom end of the band, so that there is no net difference. For a semiinfinite band, there is no bottom part, so only the top difference is counted.

The kinetic energy term in the Luttinger model is

$$H_0 = v_F \sum_{sk} k(a^\dagger_{1,k,s}a_{1,k,s} - a^\dagger_{2,k,s}a_{2,k,s}) \tag{4.4.23}$$

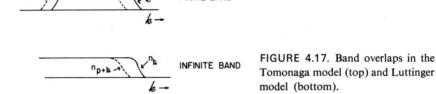

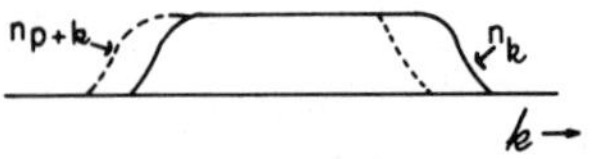

FINITE BAND

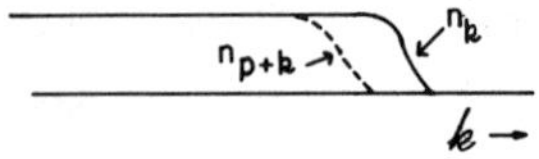

INFINITE BAND

FIGURE 4.17. Band overlaps in the Tomonaga model (top) and Luttinger model (bottom).

H_0 has the exact commutation relations with the operators ($p \geq 0$)

$$[H_0, \varrho_1(p)] = v_F p \varrho_1(p), \qquad [H_0, \varrho_2(p)] = -v_F p \varrho_2(p)$$
$$[H_0, \sigma_1(p)] = v_F p \sigma_1(p), \qquad [H_0, \sigma_2(p)] = -v_F p \sigma_2(p)$$

Therefore the kinetic energy term is exactly represented by the operator

$$H_0 = \frac{\pi}{L} v_F \sum_{p>0} [\varrho_1(p)\varrho_1(-p) + \varrho_2(-p)\varrho_2(p) + \sigma_1(p)\sigma_1(-p) + \sigma_2(-p)\sigma_2(p)] \tag{4.4.24}$$

In the Tomonaga model, the boson approximation applies only for excitation with small k. This restriction is removed in the Luttinger model. The transformation to boson operators is

$$\varrho_1(-p) = b_{1p}\left(\frac{pL}{\pi}\right)^{1/2}, \qquad \varrho_1(p) = b_{1p}^\dagger\left(\frac{pL}{\pi}\right)^{1/2}$$
$$\varrho_2(-p) = b_{2,-p}^\dagger\left(\frac{pL}{\pi}\right)^{1/2}, \qquad \varrho_2(p) = b_{2,-p}\left(\frac{pL}{\pi}\right)^{1/2}$$
$$\sigma_1(-p) = c_{1p}\left(\frac{pL}{\pi}\right)^{1/2}, \qquad \sigma_1(p) = c_{1p}^\dagger\left(\frac{pL}{\pi}\right)^{1/2}$$
$$\sigma_2(-p) = c_{2,-p}^\dagger\left(\frac{pL}{\pi}\right)^{1/2}, \qquad \sigma_2(p) = c_{2,-p}\left(\frac{pL}{\pi}\right)^{1/2}$$
$$H_0 = \sum_{p>0} p v_F (b_{1p}^\dagger b_{1p} + b_{2,-p}^\dagger b_{2,-p} + c_{1p}^\dagger c_{1p} + c_{2,-p}^\dagger c_{2,-p})$$

The operator $\varrho_1(p)$ for $p > 0$ takes a particle from state k and puts it into $p + k$. This operation will make an electron–hole pair when $k < k_F$ and $p + k > k_F$. The summation over all such electron–hole pairs is represented by the boson creation operator $b_{1p}^\dagger$. For particle 2, the Fermi "surface" is at the negative wave vector $-k_F$. Electron–hole pairs are made mostly at negative wave vectors. Thus the operator $\varrho_2(-p) = \sum_{ks} a_{2ks}^\dagger a_{2k+p,s}$ for $p > 0$ creates the bosons, since it takes an electron from the occupied state $-k_F < k + p$ to the unoccupied state $k < -k_F$, where k is negative.

Various kinds of interaction terms may be added to the Luttinger model. Those which arise from electron–electron interactions are expressed as the product of four fermion operators, or two density operators. They are usually taken to have the form

$$V_1 = \frac{1}{2L} \sum_{p>0} V_{1p}[\varrho_1(p)\varrho_1(-p) + \varrho_2(-p)\varrho_2(+p)]$$

$$= \frac{1}{2\pi} \sum_{p>0} V_{1p} p(b_{1p}^{\dagger} b_{1p} + b_{2,-p}^{\dagger} b_{2,-p})$$

$$V_2 = \frac{1}{2L} \sum_{\text{all } p} V_{2p} \varrho_1(p)\varrho_2(-p)$$

$$= \frac{1}{2\pi} \sum_{p>0} V_{2p} p(b_{1p} b_{2,-p} + b_{1p}^{\dagger} b_{2,-p}^{\dagger})$$

The first term V_1 describes processes whereby one electron–hole pair from the j particles ($j = 1, 2$) is created while another is destroyed. Momentum conservation prevents the spontaneous creation of two electron–hole pairs of the same particle type. The second interaction V_2 describes processes whereby two electron–hole pairs are created, or both destroyed, by coupling between particles 1 and 2. These processes are similar to the interactions in the Tomonaga model (4.4.1), which has $V_{1p} = V_{2p} = V_p$. The Hamiltonian

$$H = H_0 + V_1 + V_2 \tag{4.4.25}$$

is exactly solvable, since it describes linear coupling between two harmonic oscillator systems.

The Luttinger model has the advantage of being exactly solvable, at least insofar as we have included the terms in (4.4.25). Of course, one could add other terms which might render it no longer exactly solvable. The disadvantage of the model is that it is unphysical, since it contains the infinite reservoir of negative energy particles.

D. Single-Particle Properties

Some of the most interesting applications of the Tomonaga–Luttinger models are concerned with single-particle properties of the electron gas. One would like, for example, to calculate the occupation number $n_{i,k,s}$

$= \langle a^\dagger_{i,k,s} a_{i,k,s} \rangle$ in the interacting system. A more ambitious calculation would be the one-particle Green's function

$$G_{i,s}(k, t) = -i\langle T a_{i,k,s}(t) a^\dagger_{i,k,s}(0)\rangle$$

To obtain these quantities, we need to have a representation of the single-fermion operator $a_{i,k,s}$ in terms of our boson operators. Our discussion of this procedure will follow Mattis and Lieb (1965) and Luther and Peschel (1974).

The representation of the single-fermion operator $a_{i,k,s}$ in terms of bosons is found, as always, by examining the commutation relations. A representation of $a_{i,k,s}$ is satisfactory if it obeys all the proper commutation relations with the other operators. The first step is to Fourier-transform $a_{i,k,s}$ into a real-space representation:

$$\begin{aligned} \Psi_{i,s}(x) &= \frac{1}{L^{1/2}} \sum_k e^{ikx} a_{i,k,s} \\ \Psi^\dagger_{i,s}(x) &= \frac{1}{L^{1/2}} \sum_k e^{-ikx} a^\dagger_{i,k,s} \end{aligned} \tag{4.4.26}$$

The advantage of this representation becomes clear when we consider the commutator of $\Psi_{i,s}(x)$ with the density operators. We shall use the Luttinger form of the Tomonaga model for this discussion. Typical commutators are

$$\begin{aligned} [\Psi_{is}(x), \varrho_j(p)] &= \delta_{ij} e^{ipx} \Psi_{is}(x) \\ [\Psi_{is}(x), \sigma_j(p)] &= \delta_{ij} s e^{ipx} \Psi_{is}(x) \end{aligned} \tag{4.4.27}$$

which may be derived, for example, in the following way:

$$\begin{aligned} [\Psi_{1s}(x), \varrho_i(p)] &= \frac{1}{L^{1/2}} \sum_{kk's'} e^{ikx} [a_{1ks}, a^\dagger_{1k'+p,s'} a_{1k's'}] \\ &= \frac{1}{L^{1/2}} \sum_{kk's'} e^{ikx} a_{1k's'} \delta_{ss'} \delta_{k=k'+p} \\ &= e^{ixp} \frac{1}{L^{1/2}} \sum_{k'} e^{ik'x} a_{1k's} = e^{ixp} \Psi_{1s}(x) \end{aligned}$$

The commutator of $[\Psi_{1s}(x), \varrho_i(p)]$ has a simple form, since it is just proportional to $\Psi_{1s}(x)$. The solution would be simpler if the commutator were a constant or even proportional to a density operator. It is not, so the solution of (4.4.27) is more complicated. One *possible* solution has the form

$$\Psi_{1s}(x) = F_1(x)\exp[J(x)]$$

$$J(x) = \frac{-\pi}{L}\sum_{p>0}\frac{1}{p}\{e^{-ipx}[\varrho_1(+p) + s\sigma_1(+p)] - e^{+ipx}[\varrho_1(-p) + s\sigma_1(-p)]\} \tag{4.4.28}$$

The prefactor $F_1(x)$ can be a function of x but is a c number in the sense that it must commute with both $\sigma_1(p)$ and $\varrho_1(p)$. First, we show that this does satisfy the equation by considering the commutator

$$\begin{aligned}[\Psi_{1s}, \varrho_1] &= F_1(e^J\varrho_1 - \varrho_1 e^J) = F_1(e^J\varrho_1 e^{-J} - \varrho_1)e^J \\ &= F_1[J, \varrho_1]e^J = [J, \varrho_1]\Psi_{1s}\end{aligned}$$

The last line is valid only for the case where the commutator $[J, \varrho_1]$ is a c number which commutes with the operator J. We shall soon see that this is true for the $J(x)$ in (4.4.28):

$$[J, \varrho_1(p)] = \frac{\pi}{L}\sum_{p'>0}\frac{1}{p'}e^{ip'x}[\varrho_1(-p'), \varrho_1(p)] = e^{ipx}$$

$$[J, \varrho_1(-p)] = -\frac{\pi}{L}\sum_{p'>0}\frac{1}{p'}e^{-ip'x}[\varrho_1(p'), \varrho_1(-p)] = e^{-ipx}$$

The next observation is that the factor

$$\varrho_1(p) + s\sigma_1(p) = \sum_{k,s'}(1 + ss')a^\dagger_{1,k+p,s'}a_{1,k,s'} = 2\sum_k a^\dagger_{1,k+p,s}a_{1,k,s}$$

since the factor $1 + ss' = 0$ unless $s = s'$, and then it is 2. The equations can be condensed by introducing the notation of a spin-dependent density operator ($p > 0$):

$$\begin{aligned}\varrho_{is}(p) &= \sum_{\text{all } k} a^\dagger_{i,k+p,s}a_{i,k,s} \\ \varrho_{is}(-p) &= \sum_{\text{all } k} a^\dagger_{i,k,s}a_{i,k+p,s} \\ \varrho_i(p) &= \sum_s \varrho_{is}(p) \\ \sigma_i(p) &= \sum_s s\varrho_{is}(p)\end{aligned} \tag{4.4.29}$$

$$J_{1s}(x) = -\frac{2\pi}{L}\sum_{\text{all } p}\frac{e^{-ipx}}{p}\varrho_{1s}(p) \tag{4.4.30}$$

$$\Psi_{1s}(x) = F_1(x)\exp[J_{1s}(x)]$$

The spin-dependent density operators obey the commutation relations

$$[\varrho_{1s}(-p), \varrho_{1s}(p')] = \delta_{pp'}\left(\frac{pL}{2\pi}\right)$$

$$[\varrho_{2s}(p), \varrho_{2s}(-p')] = \delta_{pp'}\left(\frac{pL}{2\pi}\right)$$

and thus are represented by boson operators of the form

$$\begin{aligned} \varrho_{1s}(-p) &= b_{1s,p}\left(\frac{pL}{2\pi}\right)^{1/2}, \qquad \varrho_{1s}(p) = b_{1s,p}^{\dagger}\left(\frac{pL}{2\pi}\right)^{1/2} \\ \varrho_{2s}(-p) &= b_{2s,-p}^{\dagger}\left(\frac{pL}{2\pi}\right)^{1/2}, \qquad \varrho_{2s}(p) = b_{2s,-p}\left(\frac{pL}{2\pi}\right)^{1/2} \\ H_0 &= \frac{2\pi v_F}{L}\sum_{\substack{p>0\\ s}} [\varrho_{1s}(p)\varrho_{1s}(-p) + \varrho_{2s}(-p)\varrho_{2s}(p)] \end{aligned} \tag{4.4.31}$$

The form of $\Psi_{1s}(x)$ in (4.4.28) is indeed a solution to the commutator equation (4.4.27). Unfortunately this solution has some undesirable properties which will force a modification. The need for changes in $\Psi_{1s}(x)$ may be understood by examining the form of the operator $\langle\Psi_{1s}^{\dagger}(x)\Psi_{1s}(x')\rangle_0$ for the noninteracting electron system, which is the Luttinger model with just the Hamiltonian H_0 in (4.4.23). At zero temperature, the noninteracting system has the feature that the momentum distributions for particles 1 and 2 have the form

$$n_{1,k,s} = \theta(k_F - k)$$

$$n_{2,k,s} = \theta(k_F + k)$$

This fact can be used to evaluate the correlation function $\langle\Psi_{1s}^{\dagger}(x)\Psi_{1s}(x')\rangle_0$ by using the inverse of the transformation (4.4.26):

$$\begin{aligned} \langle\Psi_{1s}^{\dagger}(x)\Psi_{1s}(x')\rangle &= \frac{1}{L}\sum_{kk'} e^{-i(kx-k'x')}\langle a_{1ks}^{\dagger}a_{1k's}\rangle \\ &= \frac{1}{L}\sum_{kk'} e^{-i(kx-k'x')}\delta_{kk'}n_{1,k,s} \\ &= \int_{-\infty}^{\infty}\frac{dk}{2\pi}e^{-ik(x-x')}\theta(k_F - k) \\ \langle\Psi_{1s}^{\dagger}(x)\Psi_{1s}(x')\rangle &= -\frac{e^{-ik_F(x-x')}}{2\pi i(x - x' + i\alpha)} \\ \langle\Psi_{2s}^{\dagger}(x)\Psi_{2s}(x')\rangle &= +\frac{e^{+ik_F(x-x')}}{2\pi i(x - x' - i\alpha)} \end{aligned} \tag{4.4.32}$$

where the factors of $\pm i\alpha$ are added to aid convergence at infinity. The objective in choosing the representation $\Psi_{1s}(x) = F_1(x)\exp[J_{1s}(x)]$ is to make the result for $\langle \Psi_{1s}^\dagger(x)\Psi_{1s}(x')\rangle$ be like (4.4.32) for the noninteracting Luttinger model. A method of doing this was suggested by Luther and Peschel (1974). It uses a limiting process, where the wave function $\Psi_{1s}(k)$ contains a parameter α, and the limit $\alpha \to 0$ is taken at the end of the calculation. Using this parameter α, we can represent the position space operators as

$$\begin{aligned}
\Psi_{1s}(x) &= \frac{1}{(2\pi\alpha)^{1/2}} \exp[ik_F x + J_{1s}(\alpha, x)] \\
\Psi_{2s}(x) &= \frac{1}{(2\pi\alpha)^{1/2}} \exp[-ik_F x - J_{2s}(\alpha, x)] \\
J_{is}(\alpha, x) &= -\frac{2\pi}{L} \sum_{k>0} \frac{e^{-\alpha k/2}}{k} [e^{-ikx}\varrho_{is}(k) - e^{ikx}\varrho_{is}(-k)] \\
&= -J_{is}(\alpha, x)^\dagger
\end{aligned} \tag{4.4.33}$$

This new form for J_{is} may be expressed in terms of the boson operators defined in (4.4.31):

$$\begin{aligned}
J_{1s}(\alpha, x) &= \sum_{p>0} e^{-\alpha p/2}\left(\frac{2\pi}{pL}\right)^{1/2} (b_{1s,p}e^{ipx} - b_{1s,p}^\dagger e^{-ipx}) \\
J_{2s}(\alpha, x) &= \sum_{p>0} e^{-\alpha p/2}\left(\frac{2\pi}{pL}\right)^{1/2} (b_{2s,-p}^\dagger e^{ipx} - b_{2s,-p}e^{-ipx})
\end{aligned} \tag{4.4.34}$$

The exponential factor $J_{1s}(\alpha, x)$ has the same form as (4.4.28) in the limit where $\alpha \to 0$, so the commutator (4.4.27) is obeyed in this limit. The prefactor is $(2\pi\alpha)^{-1/2}$, which will be explained below.

Now consider the evaluation of the quantity

$$\begin{aligned}
\langle \Psi_{1s}^\dagger(x)\Psi_{1s}(x')\rangle &= \frac{1}{2\pi\alpha} e^{-ik_F(x-x')}\langle e^{-J_{1s}(\alpha,x)}e^{J_{1s}(\alpha,x')}\rangle \\
&= \frac{1}{2\pi\alpha} e^{-ik_F(x-x')} \\
&\quad \times \prod_{k>0} \left\langle \exp\left[e^{-\alpha k/2}\left(\frac{2\pi}{kL}\right)^{1/2} (e^{-ikx}b_k^\dagger - e^{ikx}b_k)\right]\right. \\
&\quad \times \left.\exp\left[e^{-\alpha k/2}\left(\frac{2\pi}{kL}\right)^{1/2} (e^{ikx'}b_k - e^{-ikx'}b_k^\dagger)\right]\right\rangle
\end{aligned}$$

using this new representation. The right-hand side of this expression is an average of exponential functions of boson operators. These are exactly the same forms which were evaluated in Sec. 4.4. Each exponent is separated by using the Feynman theorem $\exp(A+B) = \exp(A)\exp(B)\exp(-\frac{1}{2}[A, B])$. Then the factors are commuted until all the destruction operators are on the right:

$$\begin{aligned}\langle \Psi_{1s}^{\dagger}(x)\Psi_{1s}(x')\rangle &= \frac{1}{2\pi\alpha}\exp\left[-ik_F(x-x') - \frac{2\pi}{L}\sum_{k>0}\frac{e^{-\alpha k}}{k}(1-e^{ik(x-x')})\right] \\ &\times \prod_{k>0}\left\langle \exp\left[\left(\frac{2\pi}{kL}\right)^{1/2} e^{-\alpha k/2} b_k^{\dagger}(e^{-ikx} - e^{-ikx'})\right]\right. \\ &\times \exp\left.\left[\left(\frac{2\pi}{kL}\right)^{1/2} e^{-\alpha k/2} b_k(e^{ikx'} - e^{+ikx})\right]\right\rangle \end{aligned} \tag{4.4.35}$$

At zero temperature, the quantity in the final brackets gives unity, so we have the following prediction for the noninteracting electron gas at zero temperature:

$$\begin{aligned}\langle \Psi_{1s}^{\dagger}(x)\Psi_{1s}(x')\rangle &= \frac{1}{2\pi\alpha}\exp[-ik_F(x-x') - \phi_0(x-x')] \\ \phi_0(x) &= \frac{2\pi}{L}\sum_{k>0}\frac{e^{-\alpha k}}{k}(1-e^{ikx}) = \int_0^{\infty}\frac{dk}{k}e^{-\alpha k}(1-e^{ikx})\end{aligned} \tag{4.4.36}$$

The next step is to do the integral for $\int dk$. We expand the exponential $\exp(ikx)$ and integrate term by term. The factor $e^{-\alpha k}$ ensures the convergence of these integrals, which is the primary role played by α:

$$\begin{aligned}\phi_0(x) &= -\sum_{l=1}^{\infty}\frac{(ix)^l}{l!}\int_0^{\infty} dk\, k^{l-1}e^{-\alpha k} = -\sum_{l=1}^{\infty}\left(\frac{ix}{\alpha}\right)^l\frac{1}{l} \\ \phi_0(x) &= \ln\left(1-\frac{ix}{\alpha}\right) \\ e^{-\phi_0(x)} &= \frac{1}{1-ix/\alpha}\end{aligned} \tag{4.4.37}$$

The series for $\phi_0(x)$ is recognized as a logarithm, so we finally obtain

$$\begin{aligned}\langle \Psi_{1s}^{\dagger}(x)\Psi_{1s}(x')\rangle &= \frac{1}{2\pi\alpha}\frac{1}{1-i(x-x')/\alpha}\exp[-ik_F(x-x')] \\ &= \frac{1}{2\pi}\frac{1}{\alpha-i(x-x')}\exp[-ik_F(x-x')]\end{aligned}$$

Now the limit $\alpha \to 0$ does indeed recover the noninteracting value (4.4.32).

It is easy to check that the factor $\langle \Psi_{2s}^\dagger(x)\Psi_{2s}(x')\rangle$ is also given correctly. Thus the representation (4.4.33) for $\Psi_{1s}(x)$ reproduces the commutation relation with the density operators and also gives the correct ground state momentum distribution for the noninteracting system. All these results, of course, apply in the limit where $\alpha \to 0$. The commutator of $\Psi_{1s}(x)$ with H_0 is also given correctly, since the latter is expressed in terms of the density operators, which have the correct commutators. Thus the representation (4.4.33) is suitable for the single-fermion operators.

This representation can be used to calculate many interesting properties of the Luttinger model. For example, the electron Green's function is

$$\begin{aligned} G_{1s}(x - x', t) &= -i\langle T\Psi_{1s}(x, t)\Psi_{1s}^\dagger(x', 0)\rangle \\ &= -i\theta(t)\langle e^{iH_0t}\Psi_{1s}(x)e^{-iH_0t}\Psi_{1s}^\dagger(x')\rangle \\ &\quad + i\theta(-t)\langle \Psi_{1s}^\dagger(x')e^{iH_0t}\Psi_{1s}(x)e^{-iH_0t}\rangle \end{aligned} \tag{4.4.38}$$

The correlation functions can be evaluated at zero temperature, using the same steps which led to (4.4.35). The time dependence of $\Psi_{1s}(x)$ for the noninteracting Hamiltonian is

$$\Psi_{1s}(x, t) = \frac{1}{(2\pi\alpha)^{1/2}} \exp[ik_F x + J_{1s}(\alpha, x, t)]$$

$$\begin{aligned} J_{1s}(\alpha, x, t) &= e^{iH_0t}J_{1s}(\alpha, x)e^{-iH_0t} \\ &= \sum_{k>0} e^{-\alpha k/2}\left(\frac{2\pi}{Lk}\right)^{1/2}(e^{ikx-i\omega_k t}b_{1s,k} - e^{-ikx}e^{i\omega_k t}b_{1s,k}^\dagger) \end{aligned}$$

$$J_{1s}(\alpha, x, t) = J_{1s}(\alpha, x - v_F t)$$

The time dependence of $\Psi_{1s}(x)$ merely changes the factor x in $J_{1,s}(\alpha, x)$ to $x - v_F t$. This rather trivial change makes it possible to use the previous result for $\langle \Psi_{1s}^\dagger(x)\Psi_{1s}(x')\rangle$ to evaluate $\langle \Psi_{1s}^\dagger(x')\Psi_{1s}(x, t)\rangle$:

$$\begin{aligned} \langle \Psi_{1s}^\dagger(x')\Psi_{1s}(x, t)\rangle &= \frac{1}{2\pi\alpha} \exp[ik_F(x - x') - \phi_0(x' - x + v_F t)] \\ &= \frac{e^{ik_F(x-x')}}{2\pi[\alpha + i(x - x' - v_F t)]} \end{aligned}$$

$$\begin{aligned} \langle \Psi_{1s}(x, t)\Psi_{1s}^\dagger(x')\rangle &= \frac{1}{2\pi\alpha} \exp[-ik_F(x - x') - \phi_0{}^*(x' - x + v_F t)] \\ &= \frac{e^{ik_F(x-x')}}{2\pi[\alpha - i(x - x' - v_F t)]} \end{aligned}$$

Similarly, the factor $\langle \Psi_{1s}(x, t)\Psi_{1s}^\dagger(x')\rangle$ has just the Hermitian conjugate of ϕ_0. Thus the Green's function for the noninteracting system is easily obtained ($\alpha \to 0$):

$$G_{1s}(x, t) = \frac{e^{ik_F x}}{2\pi}\left[\frac{\theta(t)}{x - v_F t + i\alpha} + \frac{\theta(-t)}{x - v_F t - i\alpha}\right] \tag{4.4.39}$$

This equation can be Fourier-transformed to obtain the Green's function in the wave vector representation:

$$\begin{aligned} G_{1s}(k, t) &= \int_{-\infty}^{\infty} dx e^{-ikx} G_{1s}(x, t) \\ &= \int_{-\infty}^{\infty} \frac{dx}{2\pi} e^{ix(k_F - k)}\left[\frac{\theta(t)}{x - v_F t + i\alpha} + \frac{\theta(-t)}{x - v_F t - i\alpha}\right] \\ &= [-i\theta(t)\theta(k - k_F)e^{-iv_F t(k-k_F)} + i\theta(-t)\theta(k_F - k)e^{-iv_F t(k-k_F)}] \end{aligned}$$

The integrals are done by contour integration. For example, the term with $\theta(-t)$ has a pole in the upper half-plane at $x = v_F t + i\alpha$. If $k_F > k$, the contour is closed in the upper half-plane and picks up this pole. If $k_F < k$, the contour is closed in the lower half-plane where there is no pole. The preceding is precisely the same result for $G(k, t)$ that we obtain in the fermion representation:

$$\begin{aligned} G_{1s}(k, t) &= -i\langle Ta_{1ks}(t)a^{\dagger}_{1ks}(0)\rangle \\ &= -ie^{-itv_F(k-k_F)}[\theta(t)\theta(k - k_F) - \theta(-t)\theta(k_F - k)] \end{aligned}$$

where the energy has been normalized to the Fermi energy.

Our success in calculating $G_{1s}(k, t)$ again illustrates that the boson representation (4.4.33) for the single-particle operators will faithfully reproduce the results obtained directly from the fermion representation. The virtue of the boson representation is that we can now proceed to solve more difficult problems. In particular, we can add interaction terms such as V_1 or V_2 in (4.4.25) and continue to solve for the one-particle properties. Exact expressions can be found for Green's functions, or other correlation functions, although they are usually difficult to evaluate analytically.

E. Interacting System of Spinless Fermions

An exact solution can be obtained for various correlation functions, even for the interacting electron gas in one dimension. We shall first solve for the occupation number. This solution relies upon the representation of the single-particle operators which was developed in the prior subsection. The Hamiltonian in this part is taken to be the Luttinger model given earlier in (4.4.25). We shall initially follow Mattis and Lieb (1965) and discuss a spinless fermion system:

$$\begin{aligned}
H &= \sum_{p>0} [\bar{\omega}_p(b^\dagger_{1,p}b_{1,p} + b^\dagger_{2,-p}b_{2,-p}) + \bar{V}_p(b^\dagger_{1,p}b^\dagger_{2,-p} + b_{2,-p}b_{1,p})] \\
\bar{\omega}_p &= pv_F + p\,\frac{V_{1p}}{2\pi} \qquad (4.4.40)\\
\bar{V}_p &= p\,\frac{V_{2p}}{2\pi}
\end{aligned}$$

The first step in the solution is to learn the method of diagonalizing this Hamiltonian. There are several ways to do this, and all give the same result. We shall use a canonical transformation to a new set of boson operators β_p, α_p, which are defined as

$$\begin{aligned}
b_{1,p} &= \beta_p \cosh(\lambda_p) - \alpha_p{}^\dagger \sinh(\lambda_p) \\
b^\dagger_{1,p} &= \beta_p{}^\dagger \cosh(\lambda_p) - \alpha_p \sinh(\lambda_p) \\
b_{2,-p} &= \alpha_p \cosh(\lambda_p) - \beta_p{}^\dagger \sinh(\lambda_p) \\
b^\dagger_{2,-p} &= \alpha_p{}^\dagger \cosh(\lambda_p) - \beta_p \sinh(\lambda_p) \qquad (4.4.41)\\
[b_{1,p}, b^\dagger_{1,p}] &= [\beta_p, \beta_p{}^\dagger] \cosh^2(\lambda_p) + [\alpha^\dagger_{-p}, \alpha_{-p}] \sinh^2(\lambda_p) \\
&= \cosh^2(\lambda_p) - \sinh^2(\lambda_p) = 1 \\
[\alpha_p, \beta_p{}^\dagger] &= 0 \\
[\alpha_p, \alpha_p{}^\dagger] &= 1
\end{aligned}$$

One can check that the various commutation relations are still obeyed in this new representation. The parameter λ_p is chosen so that the Hamiltonian (4.4.40) is diagonalized. It is first written out in terms of the transformed operators:

$$\begin{aligned}
H = \sum_{s,p} \big((\beta_p{}^\dagger\beta_p + \alpha_p{}^\dagger\alpha_p)\{[\cosh^2(\lambda_p) + \sinh^2(\lambda_p)]\bar{\omega}_p - 2\bar{V}_p \sinh(\lambda_p)\cosh(\lambda_p)\} \\
+ (\beta_p{}^\dagger\alpha_p{}^\dagger + \alpha_p\beta_p)\{[\cosh^2(\lambda_p) + \sinh^2(\lambda_p)]\bar{V}_p - 2\bar{\omega}_p \sinh(\lambda_p)\cosh(\lambda_p)\}\big)
\end{aligned}$$

Since these are boson operators, the ordering of terms such as $\alpha\beta = \beta\alpha$ does not matter. We also have ignored the zero-point motion terms which result from the rearrangements such as $\alpha\alpha^\dagger = \alpha^\dagger\alpha + 1$. Two combinations of hyperbolic functions seem to occur:

$$\cosh^2 \lambda + \sinh^2 \lambda = \cosh 2\lambda$$
$$2 \sinh \lambda \cosh \lambda = \sinh 2\lambda$$

The Hamiltonian is diagonalized by setting to zero the coefficient of the term $\beta^\dagger\alpha^\dagger + \alpha\beta$. This step gives $\tanh(2\lambda_p) = \bar{V}_p/\bar{\omega}_p$, so that the diagonalized Hamiltonian is

$$\begin{aligned} H &= \sum_{p>0} E_p(\beta_p{}^\dagger\beta_p + \alpha_p{}^\dagger\alpha_p) \\ E_p &= (\bar{\omega}_p{}^2 - \bar{V}_p{}^2)^{1/2} \\ \cosh(2\lambda_p) &= \frac{\bar{\omega}_p}{E_p} \\ \sinh(2\lambda_p) &= \frac{\bar{V}_p}{E_p} \end{aligned} \tag{4.4.42}$$

The transformation to the new operators will be used whenever we are evaluating the properties of the system. The α and β operators refer to the actual boson normal modes in the interacting system. The ground state of the system is thus the state of the vacuum of α and β particles; i.e., one has $\alpha \mid 0\rangle = 0$, $\beta \mid 0\rangle = 0$. These are the same set of normal modes we found for the Tomonaga model (4.4.12) when $V_{1p} = V_{2p} = V_p$.

First consider the evaluation of the fermion occupation number. As shown previously in (4.4.35), we need to consider the ground state expectation value of the operator combination:

$$\langle \Psi_1{}^\dagger(x)\Psi_1(x')\rangle = \frac{1}{2\pi\alpha} \exp[-ik_F(x - x') - \phi(x - x')]$$
$$e^{-\phi(x)} = \langle e^{-J_1(\alpha,x)} e^{J_1(\alpha,0)}\rangle$$

where $J_1(\alpha, x)$ is given in (4.4.34). The ground state of the system must be the particle vacuum of the bosons with excitation energy E_p in (4.4.42), since these are the normal modes. Thus we must change the $J_1(\alpha, x)$ operator to the α- and β-particle representation in order to use the ground state property that $\alpha \mid 0\rangle = 0$ and $\beta \mid 0\rangle = 0$. The transformation (4.4.41) produces a redefined operator form for $J_{1s}(\alpha, x)$:

$$\begin{aligned} J_1(\alpha, x) = \sum_{p>0} e^{-1\alpha p/2}\left(\frac{2\pi}{pL}\right)^{1/2} &\{e^{ipx}[\beta_p \cosh(\lambda_p) - \alpha_p{}^\dagger \sinh(\lambda_p)] \\ &- e^{-ipx}[\beta_p{}^\dagger \cosh(\lambda_p) - \alpha_p \sinh(\lambda_p)]\} \end{aligned}$$

It contains operators of both the α and β types. These operators are independent, since they describe an independent boson system; our two independent systems $b_{1,p}$ and $b_{2,-p}$ have been recombined into the two new modes with operators α_p and β_p. Thus we must average each of these boson systems independently. The ground state average of $\langle e^{-J(x)}e^{J(0)}\rangle$ is

$$\begin{aligned}
\phi(x) &= \phi_a(x) + \phi_b(x) \\
e^{-\phi_a(x)} &= \langle e^{-J_a(x)}e^{J_a(0)}\rangle \\
e^{-\phi_b(x)} &= \langle e^{-J_b(x)}e^{J_b(0)}\rangle \\
J_a &= \sum_{p>0} e^{-\alpha p/2}\left(\frac{2\pi}{pL}\right)^{1/2} \sinh(\lambda_p)(e^{-ipx}\alpha_p - e^{ipx}\alpha_p{}^\dagger) \\
J_b &= \sum_{p>0} e^{-\alpha p/2}\left(\frac{2\pi}{pL}\right)^{1/2} \cosh(\lambda_p)(e^{ipx}\beta_p - e^{-ipx}\beta_p{}^\dagger)
\end{aligned}$$

The separate averages for $\phi_a(x)$ and $\phi_b(x)$ are similar to the one found earlier in (4.4.36). The average for ϕ_b is the same as the earlier average for $\phi_0(x)$, except for the extra kernal $\cosh(\lambda_p)$. The average for $\phi_a(x)$ also contains a unique kernal $\sinh(\lambda_p)$ and has $x \to -x$. Thus by analogy with (4.4.36), at zero temperature we obtain

$$\begin{aligned}
\phi_b(x) &= \frac{2\pi}{L}\sum_{p>0}\frac{e^{-\alpha p}}{p}\cosh^2(\lambda_p)(1 - e^{ipx}) \\
\phi_a(x) &= \frac{2\pi}{L}\sum_{p>0}\frac{e^{-\alpha p}}{p}\sinh^2(\lambda_p)(1 - e^{-ipx})
\end{aligned} \tag{4.4.43}$$

The result for ϕ_b is manipulated by replacing $\cosh^2(\lambda_p)$ by its equivalent $1 + \sinh^2 \lambda$. The term with "1" is identical to the $\phi_0(x)$ we evaluated previously in the noninteracting electron gas,

$$\begin{aligned}
\phi_b(x) &= \phi_0(x) + \frac{2\pi}{L}\sum_{p>0}\frac{e^{-\alpha p}}{p}\sinh^2(\lambda_p)(1 - e^{ipx}) \\
\phi_0(x) &= \frac{2\pi}{L}\sum_{p>0}\frac{e^{-\alpha p}}{p}(1 - e^{ipx}) = \ln\left(1 - \frac{ix}{\alpha}\right)
\end{aligned}$$

so that

$$\begin{aligned}
\phi_a + \phi_b &= \phi_0(x) + \phi_s(x) \\
\phi_s(x) &= \frac{2\pi}{L}\sum_{p>0}\frac{e^{-\alpha p}}{p}\sinh^2(\lambda_p)[(1-e^{ipx}) + (1-e^{-ipx})] \qquad (4.4.44) \\
\langle \Psi_1{}^\dagger(x)\Psi_1(x')\rangle &= \frac{e^{-ik_F(x-x')}}{2\pi[\alpha - i(x - x')]}\, e^{-\phi_s(x-x')}
\end{aligned}$$

The effect of the interactions on the electron gas is contained in the exponential factor $\exp(-\phi_s)$. If this term were absent, we would have the same result as for the noninteracting electron gas.

Any evaluation of the factor $\phi_s(x)$ must assume some specific form of the potential between electrons. By using the relation $\sinh^2 \lambda = \frac{1}{2}[\cosh(2\lambda) - 1]$ and our prior result (4.4.42) for $\cosh 2\lambda = \bar{\omega}_p/E_p$, we can express this as

$$\phi_s(x) = \int_0^\infty \frac{dp}{p} e^{-\alpha p}\left(\frac{\bar{\omega}_p}{E_p} - 1\right)(1 - \cos px)$$

One possible model is to take $V_{1p} = V_{2p} = \text{constant} = V_0$. A constant potential, in momentum space, is obtained from a delta function interaction in real space $V(x) = V_0\delta(x)$. Thus this model assumes that the particles interact only when they directly collide. In this case we find that

$$\bar{\omega}_p = p\left(v_F + \frac{V_0}{2\pi}\right)$$

$$E_p = pv_F\left(1 + \frac{V_0}{\pi v_F}\right)^{1/2}$$

$$\sinh^2(\lambda_p) = \frac{1}{2}\left(\frac{1 + V_0/2\pi v_F}{(1 + V_0/\pi v_F)^{1/2}} - 1\right)$$

and the factor $\sinh^2(\lambda_p)$ is a constant, which we shall call g:

$$g = \sinh^2(\lambda_p)$$

$$\phi_s(x) = 2g \int_0^\infty \frac{dp}{p} e^{-\alpha p}(1 - \cos px)$$

The integral for the exponential factor $\phi_s(x)$ is now simple to evaluate, since it has the same form as earlier for $\phi_0(x)$ in (4.4.36):

$$\phi_s(x) = g[\phi_0(x) + \phi_0(x)^*] = g \ln\left(1 + \frac{x^2}{\alpha^2}\right)$$

$$\langle \Psi_1^\dagger(x)\Psi_1(x')\rangle = \frac{e^{-ik_F(x-x')}}{-2\pi i(x - x' + i\alpha)} \frac{1}{[1 + (x - x')^2/\alpha^2]^g}$$

Thus in the delta function model, we obtain the following prediction for the momentum distribution of the 1-particles:

$$n_{1,k} = \int_{-\infty}^{\infty} dx e^{ik(x-x')}\langle \Psi_1^\dagger(x)\Psi_1(x')\rangle = -\int_{-\infty}^{\infty} \frac{dx}{2\pi i} \frac{e^{ix(k-k_F)}}{x + i\alpha} \frac{\alpha^{2g}}{(x^2 + \alpha^2)^g}$$

If we set $g = 0$, then we recover the noninteracting case $n_{1,k} = \theta(k_F - k)$, which is obtained by closing the contour of integration in the UHP (upper half-plane) when $k_F < k$ and in the LHP (lower half-plane) when $k_F > k$. The pole at $x = -i\alpha$ is circled only in the latter case.

Mattis and Lieb (1965) showed that a more interesting result is found for the case where the coupling constant g is finite. Here we find that $n_{1,k}$ = constant, independent of k, and that the Fermi distribution is totally destroyed. This happens even in the limit where g is infinitesimally small. As $g \to 0$, the constant is $\frac{1}{2}$.

To show this, first change the integration variables to $y = x/\alpha$:

$$n_{1,k} = \int_{-\infty}^{\infty} \frac{dy}{2\pi i} \frac{e^{iy\alpha(k-k_F)}}{y + i} \frac{1}{(y^2 + 1)^g}$$

The only α dependence is in the exponential factor. For $g = 0$, we need this exponential factor, since it tells us whether to close the integration contour in the upper or lower half plane. However, for a finite value of g the integral converges even without the exponential factor, since at large values of y we have that $\int dy y^{-1-2g} \approx y^{-2g} \to 0$, which is convergent. Thus we can set $\alpha = 0$ before doing the integral and consider

$$n_{1,k} = -\int_{-\infty}^{\infty} \frac{dy}{2\pi i} \frac{1}{y + i} \frac{1}{(1 + y^2)^g}$$

We now have the result that the right-hand side is no longer a function of $k - k_F$. The integral for $y > 0$ is added to that for $y < 0$ by changing the variable $y \to -y$ in the latter to give the real integral:

$$\begin{aligned} n_{1,k} &= -\int_{0}^{\infty} \frac{dy}{2\pi i} \frac{1}{(1 + y^2)^g} \left(\frac{1}{y + i} + \frac{1}{-y + i} \right) \\ &= \int_{0}^{\infty} \frac{dy}{\pi} \frac{1}{(1 + y^2)^{1+g}} \end{aligned}$$

This integral is in a standard form, which is given in tables [G&R, 3.194(3) after changing $y^2 = x$] in terms of gamma functions:

$$n_{1,k} = \frac{1}{2(\pi)^{1/2}} \frac{\Gamma(\frac{1}{2} + g)}{\Gamma(1 + g)}$$

In the limit where $g \to 0$ we have that $\Gamma(\frac{1}{2}) = \pi^{1/2}$ and that $\Gamma(1) = 1$, so

$$\lim_{g \to 0} n_{1,k} = \tfrac{1}{2}$$

Thus the distribution function is a nonanalytic function of the coupling constant g. For $g = 0$, we get the usual noninteracting distribution function. But when we introduce an arbitrarily weak delta function potential, the Fermi distribution is destroyed, and each wave vector state is occupied with an equal probability. For the case where $g \rightarrow 0$, this probability approaches $\frac{1}{2}$. Thus the $g = 0$ result is not obtained in the $g \rightarrow 0$ limit. This result would be difficult to prove by perturbation theory and shows the value of an exact solution. These results pertain only to the one-dimensional electron gas.

F. Electron Exchange

When a calculation is completed, one should always try to use simple arguments to ascertain whether the result is sensible. Such a discussion is very revealing for the present problem. For an N-particle spinless fermion system in one dimension, the many-particle wave function $\Psi(x_1, x_2, \ldots, x_n)$ has the property that it must change sign if the positions of two particles are interchanged, e.g., particles x_1 and x_2:

$$\Psi(x_2, x_1, x_3, \ldots, x_n) = -\Psi(x_1, x_2, x_3, \ldots, x_n) \tag{4.4.45}$$

The wave function must vanish if two particles occupy the same position $x_i = x_j$, since Eq. (4.4.45) becomes a statement that the wave function equals its negative and so is zero. This antisymmetric property is correctly maintained if we represent the wave function as a Slater determinant:

$$\Psi(x_1, x_2, \ldots, x_N) = \frac{1}{(N!)^{1/2}} \begin{vmatrix} \Psi_1(x_1) & \Psi_1(x_2) & \cdots & \Psi_1(x_N) \\ \Psi_2(x_1) & \Psi_2(x_2) & \cdots & \Psi_2(x_N) \\ \vdots & & & \\ \Psi_N(x_1) & & \cdots & \Psi_N(x_N) \end{vmatrix}$$

The Slater determinant vanishes when $x_i = x_j$. However, the vanishing of the wave function when two particles are at the same position in space is a general property of the many-particle system, which is true even when the wave functions are not determinants. The point we are making is that the expectation value of the pair potential between two particles is

$$\langle V \rangle = \int \prod_{j=1}^{N} dx_j V(x_1 - x_2) \mid \Psi(x_1 \cdots x_N) \mid^2$$

This expectation value $\langle V \rangle$ is zero for a delta function potential. This potential acts only when two particles are at the same point, and we have shown that the probability of this is zero since the wave function vanishes whenever this might happen. Thus the delta function potential, between spinless fermions, should have no effect on the system whatsoever. A correct solution to this problem would show that the system behaves exactly as a noninteracting system.

This conclusion is correct, but so was our solution to the Luttinger model. In the Luttinger model, there are two kinds of fermions, particles 1 and 2. The Fermi statistics only prevent particle 1 from being at the same point as another particle 1 and particle 2 from being at the same point as another particle 2. Particles 1 can be at the same point as a particle 2 and hence feel the effects of a delta function interaction. The drastic effects of the delta function potential were caused by the interaction between these different kinds of fermions. However, the Tomonaga model, if solved correctly, should not show any effects of a delta function interaction. To obtain this correct result, one has to include the effects of *electron exchange*.

First, we wish to show that the instability in the interacting electron gas was caused by the terms V_2 in the Hamiltonian (4.4.40). If we set $\bar{V}_p = 0 = V_{2p}$, then sinh $\lambda^2 = 0$, and the term ϕ_s is also zero. This shows that the instability is caused by the interactions between particles 1 and 2, i.e., by scattering events between the two sides of the Fermi "surface." The term V_{1p} changes the dispersion relation for the boson excitations but not the result that $n_{1,k} = \theta(k_F - k)$. In fact, V_{1p} should not have any effect on the spectra at all, since it is a delta function potential between like fermions. This error is made by our neglect of exchange events.

Electron–electron interaction terms such as

$$\tfrac{1}{2} \sum_q V_q \sum_{kk'} a^\dagger_{k+q} a^\dagger_{k'-q} a_{k'} a_k$$

describe scattering events whereby two particles in states k and k' are scattered to $k + q$ and $k' - q$. In the boson approximation, we have been grouping these operators in a particular way,

$$\tfrac{1}{2} \sum_q V_q \Big(\sum_k a^\dagger_{k+q} a_k\Big)\Big(\sum_{k'} a^\dagger_{k'-q} a_{k'}\Big)$$

so that we identify the two transitions $k \to k + q$ and $k' \to k' - q$, which are shown by the solid lines in Fig. 4.18(a). However, since the electrons are indistinguishable, one gets the same net transition when the two par-

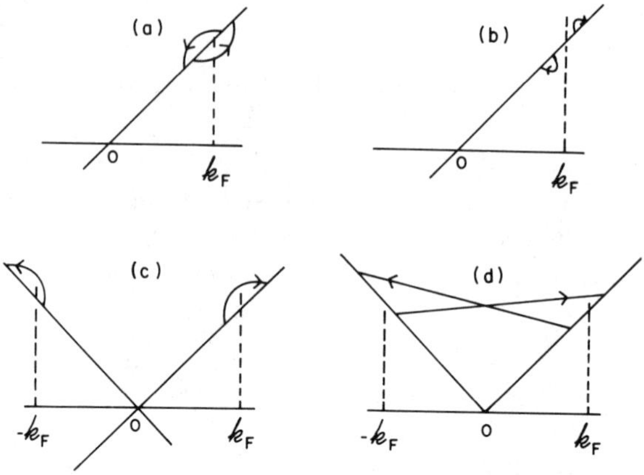

FIGURE 4.18. Energy loss processes. Event (b) is the exchange process of (a). Event (c) is important in both the Luttinger and Tomonaga models, but only in the Tomonaga model does it have the exchange process in (d).

ticles have the scatterings $k \to k' - q$ and $k' \to k + q$ shown in Fig. 4.18(b). In both (a) and (b), the two initial and final states are the same. The second transition (b) is called an exchange transition or exchange event. It is equivalent to the following grouping of our operators:

$$\begin{aligned} &\text{Direct:} \qquad \sum_q V_q\Big(\sum_k a^\dagger_{k+q}a_k\Big)\Big(\sum_{k'} a^\dagger_{k'-q}a_{k'}\Big) \\ &\text{Exchange:} \qquad -\sum_{qkk'} V_q(a^\dagger_{k+q}a_{k'})(a^\dagger_{k'-q}a_k) \end{aligned} \tag{4.4.46}$$

where parentheses have been put around operators which refer to a solid line in Figs. 4.18(a) and 4.18(b). The exchange term has a negative sign because the grouping of operators in (4.4.46) is obtained after an odd number of anticommutations of fermion operators. Our earlier mistake is now clear. The exchange events also are excitations, which can be described by boson operators. They should also be included in the dynamical description of the boson system. The sign change in front of the exchange term will cause them to cancel the effects of the direct term. This cancellation is the reason the delta function has no effect, as will be shown below.

The exchange term (4.4.46) is manipulated by first changing variables $k' = k + Q + q$ and then changing $q = p - k$, so that it is rewritten as

$$-\sum_{Qpk} V_{p-k}(a_p^\dagger a_{p+Q})(a^\dagger_{k+Q}a_k)$$

The operator groupings have the form of density operators, which we have been turning into boson operators. The potential term V_{p-k} prevents this assignment in most cases. However, for the delta function potential, we have $V_q = V_0 =$ constant, in which case the direct and exchange terms are

$$\text{Direct:} \qquad V_0 \sum_q \left(\sum_k a_{k+q}^\dagger a_k\right)\left(\sum_{k'} a_{k'-q}^\dagger a_{k'}\right)$$

$$\text{Exchange:} \qquad -V_0 \sum_Q \left(\sum_p a_p{}^\dagger a_{p+Q}\right)\left(\sum_k a_{k+Q}^\dagger a_k\right)$$

They are equal and opposite, so that the net interaction is zero. Thus the exchange events exactly cancel the direct events for the spinless fermion problem. The delta function interaction has no effect, and this is true in one or any dimension.

In the Luttinger model, the term which causes the instability in the electron gas is V_2, whose scatterings are shown in Fig. 4.18(c). This interaction makes an electron–hole pair on each particle line. The Tomonaga model has the exchange event shown in Fig. 4.18(d), which will cancel this interaction for the delta function potential. This exchange event is not permitted in the Luttinger model, since the two lines are different particles. For the delta function potential the Luttinger model has an instability because it lacks the exchange event shown in Fig. 4.18(d), while the Tomonaga model does not have the instability.

4.5. POLARITONS

A. Semiclassical Discussion

The word *polariton* was coined by Hopfield (1958) to describe the normal modes in solids which propagate as electromagnetic waves. The word is a combination of *polarization* and *photon*, because these modes are combinations of free photons and the polarization modes of the solid. A new word was needed, because a new view was then emerging about the optical properties of solids. Hopfield deserves credit for the popularization of this new physics, although similar ideas had been discussed earlier by Fano (1956, 1960) and by Born and Huang (1954).

In the old view of electromagnetic wave propagation in solids, the light shone upon the surface of a sample and went into it. The polarization modes of the solid, e.g., TO phonons, could absorb some of this light.

The new view is that the light and the polarization modes in the solid are coupled into a new set of normal modes. These new modes are called polaritons. When light is shone upon the surface, polaritons may be created which propagate inward. The mathematics is trivial; since both the photons and the polarization modes are usually described by harmonic oscillator equations, the new modes are obtained by solving coupled harmonic oscillator equations. The physical effect is semiclassical and need not involve quantum mechanics. The photon Green's function $\mathscr{D}_{\mu\nu}(q, \omega)$ in a system with dielectric function $\varepsilon_{\mu\nu}(q, \omega)$ was derived in (2.10.12). A transverse wave will propagate with $\hat{\varepsilon} \perp q$, and the normal modes are the poles of the Green's functions. Assuming they exist, these poles are at

$$\omega^2 \varepsilon_t(q, \omega) = \omega_q{}^2 \equiv (qc)^2 \tag{4.5.1}$$

where $\varepsilon_t(q, \omega)$ is the transverse dielectric function.

The normal modes are the ω which satisfy this equation; call them Ω_q. For example, consider the phenomenological equation from optical phonons in ionic crystals for the transverse dielectric function, which is the same as the longitudinal dielectric function at long wavelength:

$$\varepsilon_t(\omega) = \varepsilon_\infty + \frac{\varepsilon_0 - \varepsilon_\infty}{1 - \omega^2/\omega_{\mathrm{TO}}^2} \tag{4.5.2}$$

When this form of the dielectric function is used in (4.5.1), there results a quadratic equation for ω. It may be solved in order to obtain the following two solutions:

$$\begin{aligned} \Omega_q{}^2 &= \tfrac{1}{2}(\tilde{\omega}_q{}^2 + \omega_{\mathrm{LO}}^2) \pm \tfrac{1}{2}[(\omega_{\mathrm{LO}}^2 + \tilde{\omega}_q{}^2)^2 - 4\tilde{\omega}_q{}^2\omega_{\mathrm{TO}}^2]^{1/2} \\ \tilde{\omega}_q{}^2 &= \frac{c^2q^2}{\varepsilon_\infty} \end{aligned} \tag{4.5.3}$$

These solutions are plotted in Fig. 4.19. The solid lines are the solutions to (4.5.3) and are the actual normal modes in the solid. The dashed lines show the unperturbed photon mode $\tilde{\omega}_q$ and phonon mode ω_{TO}. It is these two which are coupled to form the new set of normal modes.

The picture of energy propagation is illustrated in Fig. 4.20. It shows a slab of polar material upon which light of frequency ω is incident I from the left. Some of the light is reflected R, while other parts may be trans-

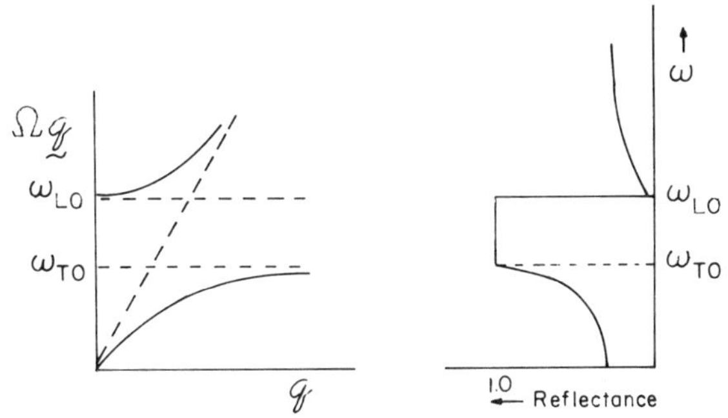

FIGURE 4.19. Polariton energy bands are shown at the left. There are no bands allowed for $\omega_{TO} < \omega < \omega_{LO}$, so that the solid must have a reflectivity of unity in this interval.

mitted through the slab and out the other side T. However, inside of the slab, the normal modes which propagate are those given by Eq. (4.5.3). The wave vectors of the polariton modes are determined by $\omega = \Omega_q$ and not by the free-photon value $q = (\omega/c)(\varepsilon_\infty)^{1/2}$. This point is further illustrated by the observation that there are no polariton modes of real frequency in the range $\omega_{LO} > \omega > \omega_{TO}$, as shown in Fig. 4.19. Since there are no polariton modes in this gap, no energy can be transmitted through the slab (actually modes in the range $\omega_{LO} > \omega > \omega_{TO}$ have complex wave vector $q_R + iq_I = (\omega/c)[\varepsilon(\omega)]^{1/2}$ so that energy can be transmitted through the slab of thickness d with the probability $\exp(-2dq_I)$, which we neglect). Thus all the energy is reflected from the slab, as shown in Fig. 4.19. This complete reflectivity for $\omega_{LO} > \omega > \omega_{TO}$ was known earlier for a simple oscillator in a solid, but the physics is clarified when cast into the form of an energy gap in the spectrum of normal modes.

The equations in the system we have been using have no damping in them, so that light is not absorbed in this model. All the electromagnetic

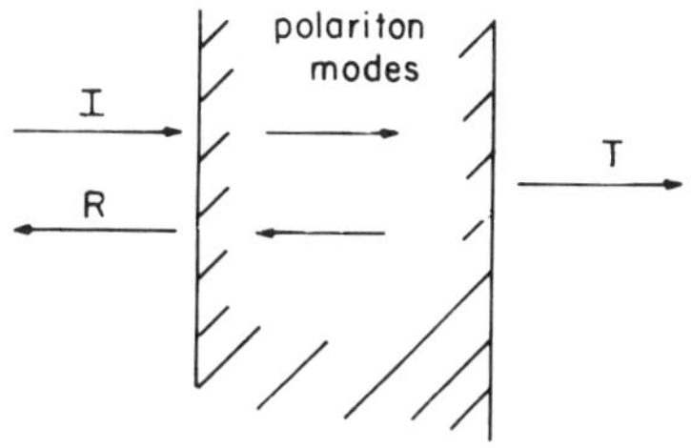

FIGURE 4.20. When light has normal incidence I upon a slab, there are the usual reflected R and transmitted components T. Inside the slab the normal modes are polaritons.

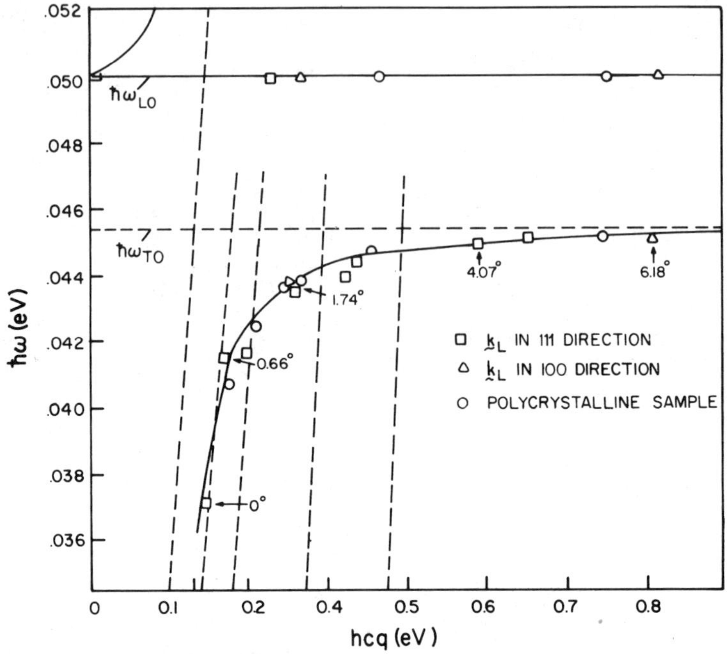

FIGURE 4.21. A plot of the observed energies and wave vectors of the polaritons and of the LO phonons in GaP. The theoretical dispersion curves are shown by the solid lines. The dispersion curves for the uncoupled dispersion curve are shown by the dashed lines. *Source*: Henry and Hopfield (1965) (used with permission).

energy, incident upon the sample, is either reflected or transmitted. The transmitted parts get carried through the slab by the polariton modes. The absorption of electromagnetic energy can be introduced into the equations by introducing damping into the phonons or additional scattering into the photons—such as due to defects. However, if the TO-phonon system has no damping, then it does not cause energy absorption. The polariton picture is a rather different physical model than the older one in which the phonons caused the absorption of energy in the solid, and this absorption in turn led to reflection.

The most direct experimental verification of the theory is to measure the polariton dispersion curves, which has been done for the polaritons associated both with optical phonons and excitons. The first measurement was by Henry and Hopfield (1965), who measured the polaritons associated

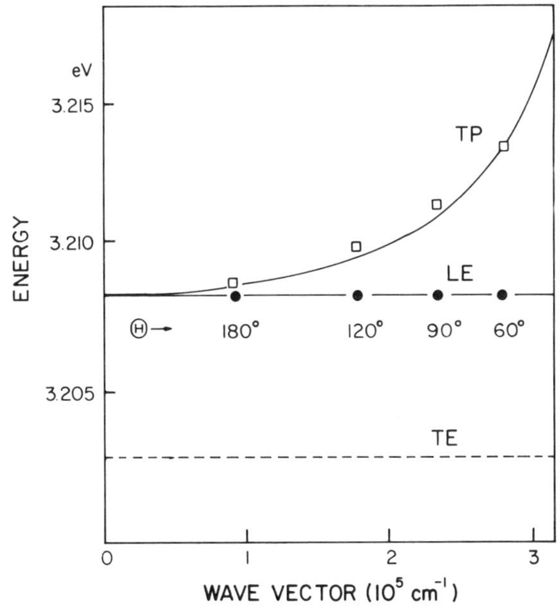

FIGURE 4.22. Polariton diagram for the first exciton of CuCl. ●, longitudinal exciton (LE); □, upper branch of transverse exciton polariton (TP); ——, dispersion curve calculated from reflectivity data; – – –, transverse exciton (TE). *Source*: Fröhlich *et al.* (1971) (used with permission).

with the optical phonons in GaP. Their results are shown in Fig. 4.21, where the solid line is the theoretical curve (4.5.3), which compares well with their experimental points. They measured these modes by Raman scattering. The frequency of the Raman shift gave the frequency of the excited mode, while the directional change during the Raman scattering gives the information needed to deduce the wave vector of the polariton. There is obviously excellent agreement between theory and experiment. Experimental results are reviewed by Borstal and Folge (1977).

The polaritons associated with the exciton in CuCl was measured by Fröhlich *et al.* (1971), whose data is shown in Fig. 4.22. They also used a two-photon experiment, but in this case it is two-photon absorption. This permits them to measure the upper branch of the transverse dispersion curve (labeled TP), which has the shape predicted by the theory. They also measured the longitudinal exciton (LE), which does not show the mixing with the photon. The polariton energy is found experimentally as the sum of the individual photon frequencies, and the polariton wave vector is the vector sum of the two-photon wave vectors. The curve marked TE shows where the transverse exciton would be had it not mixed with the photon and formed the polariton mode.

B. Phonon–Photon Coupling

The semiclassical theory will be calculated using Green's functions. The self-energy operators for the photon, at least the parts due to the phonons, may be evaluated exactly, which will leave us with the semiclassical theory. An interesting aspect to this derivation is that there are contributions from both the first- and second-order self-energy. Fortunately that is all, and the exact self-energy is found after the first two orders of perturbation theory.

The Hamiltonian has the form

$$H = H_0 + H', \qquad H_0 = \omega_{\mathrm{LO}} \sum_{\mathbf{q}} b_{\mathbf{q}}^{\dagger} b_{\mathbf{q}} + \sum_{\mathbf{q}\lambda} \omega_{\mathbf{q}} a_{\mathbf{q}\lambda}^{\dagger} a_{\mathbf{q}\lambda}$$

$$H' = -\frac{\bar{q}}{Mc} \sum_j \mathbf{P}_j \cdot \mathbf{A}(R_j) + \frac{\bar{q}^2}{2Mc^2} \sum_j \mathbf{A}(\mathbf{R}_j) \cdot \mathbf{A}(\mathbf{R}_j) \tag{4.5.4}$$

The operators b_q, $b_q^{\dagger}$ are for phonons and $a_{\mathbf{q}\lambda}$, $a_{\mathbf{q}\lambda}^{\dagger}$ for photons. The interaction between the photons and optical phonons is contained in the term H'. The ions in the solid are treated as spherical balls with charge $\pm\bar{q}$, equilibrium position $\mathbf{R}_j$, and momentum $\mathbf{P}_j$. The reduced mass M is used for the relative motions of positive and negative ions in the optical mode of oscillation. In terms of these coordinates, the current $\mathbf{J}_j$ and vector potential $\mathbf{A}(\mathbf{R}_j)$ are (see Secs. 1.1 and 2.10) expressed in terms of phonon and photon operators:

$$\mathbf{J}_j = \frac{\bar{q}}{M} \mathbf{P}_j = i\bar{q}\Sigma_{\mathbf{q}} \left(\frac{\omega_0}{2NM}\right)^{1/2} \hat{\xi}_{\mathbf{q}} (b_{\mathbf{q}}^{\dagger} - b_{-\mathbf{q}}) e^{-i\mathbf{q}\cdot\mathbf{R}_j} = \Sigma_{\mathbf{q}} \frac{1}{N^{1/2}} \mathbf{J}(\mathbf{q}) e^{-i\mathbf{q}\cdot\mathbf{R}_j}$$

$$A_\mu(\mathbf{R}_j) = \sum_{\mathbf{q}\lambda} \left(\frac{2\pi c^2}{\omega_{\mathbf{q}} \nu}\right)^{1/2} \xi_{\mu\lambda} e^{i\mathbf{q}\cdot\mathbf{R}_j} (a_{\mathbf{q}\lambda} + a_{-\mathbf{q}\lambda}^{\dagger}) = \frac{c}{\nu^{1/2}} \Sigma_{\mathbf{q}} A_\mu(\mathbf{q}) e^{i\mathbf{q}\cdot\mathbf{R}_j}$$

$$J_\mu(\mathbf{q}) = i\bar{q} \left(\frac{\omega_0}{2M}\right)^{1/2} \hat{\xi}_{\mu\lambda} (b_{\mathbf{q}}^{\dagger} - b_{-\mathbf{q}}), \qquad A_\mu(\mathbf{q}) = \sum_\lambda \xi_{\mu\lambda} \left(\frac{2\pi}{\omega_q}\right)^{1/2} (a_{\mathbf{q}\lambda} + a_{-\mathbf{q}\lambda}^{\dagger})$$

The first self-energy will be from the photon self-energy, which arises from the term in A^2. The interaction in the second term in (4.5.4) is

$$H_{A^2} = \frac{\bar{q}^2}{2Mv_0} \sum_{\mathbf{q},\mu} A_\mu(\mathbf{q}) A_\mu(-\mathbf{q})$$

where $v_0 = v/N$ is the volume of the unit cell in a binary crystal of positive and negative ions; i.e., v_0 is the volume per phonon mode. The self-energy comes from the first-order term in the expansion of the S matrix:

$$\begin{aligned}
\mathscr{D}_{\mu\nu}(\mathbf{k}, i\omega) &= -\int_0^\beta d\tau e^{i\omega\tau} \langle T A_\mu(\mathbf{k}, \tau) A_\nu(-\mathbf{k}, 0)\rangle \\
&= \mathscr{D}^{(0)}_{\mu\nu}(\mathbf{k}, i\omega) + \frac{\bar{q}^2}{2Mv_0} \int_0^\beta d\tau e^{i\omega\tau} \int_0^\beta d\tau_1 \sum_{\mathbf{q},\delta} \\
&\quad \times \langle T_\tau \tilde{A}_\mu(\mathbf{k}, \tau) \tilde{A}_\delta(\mathbf{q}, \tau_1) \tilde{A}_\delta(-\mathbf{q}, \tau_1) \tilde{A}_\nu(-\mathbf{k}, 0)\rangle + \cdots \\
A_\mu(\mathbf{k}, \tau) &= e^{\tau H} A_\mu e^{-\tau H} \\
\tilde{A}_\mu(\mathbf{k}, \tau) &= e^{\tau H_0} A_\mu e^{-\tau H_0}
\end{aligned} \tag{4.5.5}$$

The four-operator sequence in the interaction term is equal to

$$\begin{aligned}
&\int_0^\beta d\tau \int_0^\beta d\tau_1 e^{i\omega\tau} \sum_{\mathbf{q}\delta} \langle \tilde{A}_\mu(\mathbf{k}, \tau) \tilde{A}_\delta(\mathbf{q}, \tau_1) \tilde{A}_\delta(-\mathbf{q}, \tau_1) \tilde{A}_\nu(-\mathbf{k}, 0)\rangle \\
&\quad = 2 \sum_\delta \mathscr{D}^{(0)}_{\mu\delta}(\mathbf{k}, i\omega) \mathscr{D}^{(0)}_{\delta\nu}(\mathbf{k}, i\omega)
\end{aligned}$$

The factor of 2 arises because of the two choices of pairing: $\mathbf{k} = \mathbf{q}$ or $\mathbf{k} = -\mathbf{q}$. Thus the series in (4.5.5) is

$$\begin{aligned}
\mathscr{D}_{\mu\nu}(\mathbf{k}, i\omega) &= \mathscr{D}^{(0)}_{\mu\nu}(\mathbf{k}, i\omega) + \Pi^{(1)} \sum_\delta \mathscr{D}^{(0)}_{\mu\delta}(\mathbf{k}, i\omega) \mathscr{D}^{(0)}_{\delta\nu}(\mathbf{k}, i\omega) + \cdots \\
\Pi^{(1)} &= \frac{\bar{q}^2}{Mv_0}
\end{aligned} \tag{4.5.6}$$

Now according to (2.10.11) the self-energy $\Pi^{(1)}(\mathbf{k}, i\omega)$ enters into the denominator of the photon Green's function with an additional factor of 4π:

$$\begin{aligned}
\mathscr{D}_{\mu\nu}(\mathbf{k}, i\omega) &= \frac{4\pi(\delta_{\mu\nu} - k_\mu k_\nu / k^2)}{(i\omega)^2 - \omega_{\mathbf{k}}{}^2 - 4\pi\Pi^{(1)}(\mathbf{k}, i\omega)} \\
4\pi\Pi^{(1)} &= \frac{4\pi\bar{q}^2}{Mv_0} = \omega_p{}^2
\end{aligned} \tag{4.5.7}$$

The quantity $4\pi\Pi^{(1)} = 4\pi\bar{q}^2/Mv_0$ is just a plasma frequency. It would be the longitudinal frequency $\omega_L = \omega_p$ of the ion vibrational oscillation if there were no other restoring force. Of course, there is a restoring force, which leads to the transverse frequency $\omega_{\rm TO} \equiv \omega_0$. The longitudinal frequency is then determined by the combination of the short-range restoring force $\omega_{\rm TO}^2 = \varkappa/M$ and the long-range Coulomb force ω_p^2, $\omega_{\rm LO}^2 = \omega_{\rm TO}^2 + \omega_p^2$. All this discussion is slightly irrelevant because we are solving for transverse modes, not longitudinal modes. For the transverse case, the modes are at the poles of the Green's function in (4.5.7). The poles are at the frequencies $\omega^2 = \omega_k^2 + 4\pi\Pi_{\rm ret}(\mathbf{k}, \omega)$, which would be at $c^2k^2 + \omega_p^2$ if there were no other contributions to the self-energy. However, there is another contribution to $\Pi(\mathbf{k}, \omega)$ from the $\mathbf{P} \cdot \mathbf{A}$ term in H', which cancels ω_p^2.

The A^2 term in the interaction H' does not contribute any further terms to the self-energy of the photon. The self-energy contribution $\Pi^{(1)}$ provides an exact evaluation of this term in the Hamiltonian. This exact result could have been deduced without the aid of Green's functions. The Hamiltonian for free photons can be written in terms of harmonic oscillator coordinates by identifying the displacement $q_\mu(\mathbf{k}) = A_\mu(\mathbf{k})/(4\pi)^{1/2}$. Of course, the Hamiltonian has the form $H_{0,\rm photon} = \frac{1}{2}\sum_k [p(\mathbf{k})p(-\mathbf{k}) + \omega_k^2 q(\mathbf{k})q(-\mathbf{k})]$. When the A^2 term is written in the same notation, it is $\frac{1}{2}\omega_p^2 \sum_k q(\mathbf{k})q(-\mathbf{k})$. Thus when we add the A^2 term to H_0, we have the combination $\frac{1}{2}[p(\mathbf{k})p(-\mathbf{k}) + \Omega_k^2 q(\mathbf{k})q(-\mathbf{k})]$, where the new modes are $\Omega_k^2 = \omega_k^2 + \omega_p^2$. The same result was obtained using Green's functions.

The next step is to evaluate the self-energy arising from the $\mathbf{J} \cdot \mathbf{A}$ term in the interaction H' in (4.5.4). The $\mathbf{J}_j$ means the current of the vibrating ion pair. The $\mathbf{J} \cdot \mathbf{A}$ term is rewritten in wave vector space as

$$H_{p \cdot A} = -\frac{1}{(\nu_0)^{1/2}} \sum_{\mathbf{k}\delta} J_\delta(\mathbf{k}) A_\delta(-\mathbf{k})$$

One factor which will occur in the self-energy expressions for the photon is the correlation function of the ion momentum $J_\mu(\mathbf{k})$ with itself. Thus define a correlation function

$$\begin{aligned}
\Phi_{\mu\nu}(\mathbf{k}, i\omega) &= -\int_0^\beta d\tau \langle T_\tau[e^{\tau H_0} J_\mu(\mathbf{k}) e^{-\tau H_0}] J_\nu(-\mathbf{k}) \rangle e^{i\omega\tau} \\
&= \delta_{\mu\nu} \frac{\bar{q}^2 \omega_0}{2M} \int_0^\beta d\tau [e^{\tau\omega_0} \langle b_\mathbf{k}^\dagger b_\mathbf{k} \rangle + e^{-\tau\omega_0} \langle b_\mathbf{k} b_\mathbf{k}^\dagger \rangle] \\
&= \delta_{\mu\nu} \frac{\omega_0^2}{(i\omega)^2 - \omega_0^2} \frac{\bar{q}^2}{M}
\end{aligned} \tag{4.5.8}$$

which is easily evaluated since the properties of the phonon operators are known.

The $J \cdot A$ interaction does not contribute any term to the first-order self-energy expression $\Pi^{(1)}$, because the first-order term has only a single factor of the operator $J_\mu(q)$, and the average of this operator is zero. The second-order term in the expansion for the S matrix is

$$\frac{1}{2!}\frac{1}{\nu_0}\int_0^\beta d\tau \int_0^\beta d\tau_1 \int_0^\beta d\tau_2 e^{i\omega\tau} \sum_{\substack{\delta\delta' \\ \mathbf{k}_1\mathbf{k}_2}} \langle T_\tau \bar{J}_\delta(\mathbf{k}_1, \tau_1)\bar{J}_{\delta'}(\mathbf{k}_2, \tau_2)\rangle$$
$$\times \langle T_\tau \bar{A}_\mu(\mathbf{k}, \tau)\bar{A}_\delta(\mathbf{k}_1, \tau_1)\bar{A}_{\delta'}(\mathbf{k}_2, \tau_2)\bar{A}_\nu(-\mathbf{k}, 0)\rangle$$

which is evaluated by the usual rules to give

$$\frac{1}{\nu_0}\sum_{\delta\delta'} \mathscr{D}^{(0)}_{\mu\delta}(\mathbf{k}, i\omega)\Phi_{\delta\delta'}(\mathbf{k}, i\omega)\mathscr{D}^{(0)}_{\delta'\nu}(\mathbf{k}, i\omega)$$

where $\Phi_{\delta\delta'}$ is given in (4.5.8). A factor of 2 comes from the two possible pairings $\mathbf{k} = \mathbf{k}_1$ or $\mathbf{k} = \mathbf{k}_2$. This self-energy expression is

$$\Pi^{(2)}_{\mu\nu}(\mathbf{k}, i\omega) = \frac{1}{\nu_0}\Phi_{\mu\nu}(\mathbf{k}, i\omega) = \frac{1}{4\pi}\frac{\omega_p{}^2\omega_0{}^2\delta_{\mu\nu}}{(i\omega)^2 - \omega_0{}^2}$$

This self-energy term is the only one from this interaction. Thus the total self-energy is obtained by adding the two contributions:

$$4\pi\Pi(\mathbf{k}, i\omega) = 4\pi(\Pi^{(1)} + \Pi^{(2)}) = \omega_p{}^2\left[1 + \frac{\omega_0{}^2}{(i\omega)^2 - \omega_0{}^2}\right] = \frac{(i\omega)^2\omega_p{}^2}{(i\omega)^2 - \omega_0{}^2}$$

The photon self-energy for the Hamiltonian (4.5.4) has been found exactly. The energy denominator in the photon Green's function (4.5.7) can now be written in terms of the dielectric function:

$$(i\omega)^2 - \omega_k{}^2 - 4\pi\Pi(\mathbf{k}, i\omega) = (i\omega)^2\varepsilon(i\omega) - \omega_k{}^2$$
$$\varepsilon(i\omega) = 1 + \frac{\omega_p{}^2}{\omega_0{}^2 - (i\omega)^2} \tag{4.5.9}$$
$$\mathscr{D}_{\mu\nu}(\mathbf{k}, i\omega) = \frac{4\pi(\delta_{\mu\nu} - k_\mu k_\nu)k^2}{(i\omega)^2\varepsilon(i\omega) - \omega_k{}^2}$$

This dielectric function should be compared with (4.5.2) from optical phonons. In the present discussion, we have been assuming that $\varepsilon_\infty = 1$. Otherwise, a value of ε_∞ other than unity will modify the vector potential with factors of ε_∞. Similarly, we identify the static dielectric constant as $\varepsilon_0 = 1 + \omega_p{}^2/\omega_0{}^2$. Another feature of the dielectric function (4.5.9) is that the frequency of the longitudinal mode oscillation is determined by the condition that $\varepsilon(\omega_L) = 0$, which has the solution $\omega_L^2 = \omega_0^2 + \omega_p^2$. The transverse frequency is $\omega_{\mathrm{TO}} \equiv \omega_0$. With these identifications, the two expressions (4.5.2) and (4.5.9) for the dielectric functions are identical.

Thus the Green's function solution of the Hamiltonian for the coupled phonon–photon modes yields exactly the same equation $\omega^2\varepsilon(\omega)^2 = \omega_k^2$ as found from the semiclassical arguments in Sec. 4.5A. The two modes, photon and phonon, are coupled and form two new modes which are polaritons. In fact, we could just solve the original Hamiltonian (4.5.4) by a canonical transformation to a new set of operators, say α_k and β_k, which diagonalize the Hamiltonian. These new creation and destruction operators are for the polariton modes. This exercise is assigned as a problem.

C. Exciton–Photon Coupling

The polariton concept applies to the mixing of any polarization waves with the photons. Although the phonons are an obvious example, the physics is the same for waves which are associated with electronic polarization. Here we shall analyze a simple case of this. The solid is treated as a collection of atoms which are only weakly interacting, say by van der Waals forces. The electrons in each atom can occupy a set of discrete energy levels ε_m. This model is applied to molecular solids, in which molecules are weakly bound together, and the optical properties of the solid are often only weak perturbations on the optical properties of the free molecules (see Davydov, 1971, or Mahan, 1975).

A light wave, which is traveling through the solid, can induce the electrons to change their electronic state from n to m. This excitation process can really happen if the photon energy ω matches the energy of electronic excitation $\hbar\omega = \varepsilon_m - \varepsilon_n$. However, for photon frequencies ω which do not match any of the excitation frequencies $\omega \neq \omega_\alpha = \varepsilon_m - \varepsilon_n$, real excitations cannot occur. Then the light wave will induce a *virtual* transition to this excited state: A virtual process is an excitation which starts to occur but violates energy conservation, so the excitation process ceases. Nevertheless, the electronic system is *polarized* during this virtual excitation process. The atom will have dipole moment $\boldsymbol{\mu}$ whose magnitude is proportional to the amplitude of the electric field of the light which is acting upon the atom:

$$\mu_\mu = \sum_\nu \alpha_{\mu\nu}(\omega) E_\nu e^{-i\omega t}$$

An expression for the atomic polarizability $\alpha_{\mu\nu}(\omega)$ will be derived below.

The Hamiltonian for each atom of atomic number Z will be taken to have the nonrelativistic form

$$H = \frac{1}{2m}\sum_i p_i^2 - Ze^2\sum_i \frac{1}{r_i} + \frac{e^2}{2}\sum_{i,j}\frac{1}{|\mathbf{r}_i - \mathbf{r}_j|} \tag{4.5.10}$$

This is assumed to have solutions which are single-particle orbitals $\phi_n(r)$ with energy ε_n. In the Hartree–Fock approximation, the state of the atom is a Slater determinant of the occupied orbital states. The ground state is the Slater determinant of the set of orbitals with the lowest energy. An excited electronic state is obtained by raising an electron from one of the ground state orbitals to another one of higher energy, which is unoccupied in the ground state. Denote by Greek subscripts α or β the possible excitation energies $\omega_\alpha = \varepsilon_m - \varepsilon_n$. Each excitation α corresponds to moving one electron from a ground state to an excited state orbital. We shall introduce the operator $b_\alpha^\dagger$ to describe this excitation, which can be defined in terms of the operators a_n of the electronic states:

$$b_\alpha^\dagger = a_m^\dagger a_n$$

The a_n operators are fermions. The b_α operators are the product of two fermion operators and thus have boson properties. However, they are not bosons since the number in each state is limited: One has that $(b_\alpha^\dagger)^N = 0$, where the value N depends on the maximum number of electrons which are in state n or may be in state m. However, we shall approximate the b_α operators as pure bosons. It is an adequate approximation as long as the density of such bosons is low, so that the average number of bosons on each atomic site is much less than unity. Then the restriction on the site occupancy of the bosons is irrelevant, and the b_α operators obey good boson statistics. Thus we can approximate the Hamiltonian of a single atom as

$$\begin{aligned} H &= E_g + \sum_\alpha \omega_\alpha b_\alpha^\dagger b_\alpha \\ [b_\alpha, b_\beta^\dagger] &= \delta_{\alpha,\beta} \end{aligned} \tag{4.5.11}$$

where E_g is the ground state energy and $b_\alpha^\dagger b_\alpha$ is the number of excitations.

Our model of an atom is far too simple by today's theoretical standards. The Hartree–Fock (HF) approximation is seldom adequate to describe an atom, and usually one has to take a linear combination of HF states (configurational interaction, or CI in the chemist's notation). The single-particle orbitals $\phi_n(r)$ also vary with the excited state configuration, which requires a more complicated Hamiltonian than (4.5.11) in order to describe excitations. We shall omit these complications, which hopefully are only a side issue for the discussion of excitons.

An important quantity for our discussion is the dipole matrix element $\mathbf{X}(\alpha)$ for the states ϕ_n and ϕ_m which comprise the transition

$$\mathbf{X}(\alpha) = \int d^3r \phi_m^*(\mathbf{r})\mathbf{x}\phi_n(\mathbf{r})$$

This matrix element will be zero in most cases, since it is finite only when the two electronic orbitals ϕ_n and ϕ_m differ by one unit of angular momentum. This matrix element is important because we are treating the case wherein the excitations occur by the interaction with light. The rate for this is determined by the dipole matrix element.

We shall now introduce an operator P_μ which is the *polarization* operator for the atom, where μ denotes vector direction and α denotes excitation:

$$P_\mu = e \sum_\alpha X_\mu(\alpha)(b_\alpha + b_\alpha^\dagger)$$

It should not be confused with the momentum, which has the same symbol. This operator has the units of polarization, which is charge times length. One can think of the quantity $X_\mu(b_\alpha + b_\alpha^\dagger)$ as a typical displacement operator, which in this case refers to the displacement of electronic charge in the atom. It is quite analogous to the displacement operator for the harmonic oscillator system, which is $X_\mathbf{q}(a_q + a_q^\dagger)$, where $X_\mathbf{q} = (\hbar/2M\omega_\mathbf{q})^{1/2}$ is the unit of length.

We shall use the symbol $\mathscr{P}_{\mu\nu}(i\omega)$ to denote the correlation function of the polarization operator P_μ with itself:

$$\mathscr{P}_{\mu\nu}(i\omega) = -\int_0^\beta d\tau e^{i\omega\tau}\langle T_\tau P_\mu(\tau)P_\nu(0)\rangle \qquad (4.5.12)$$

It is easily evaluated for the noninteracting Hamiltonian (4.5.11):

$$\mathscr{P}_{\mu\nu}^{(0)}(i\omega) = -\int_0^\beta d\tau e^{i\omega\tau} \sum_\alpha e^2 X_\mu(\alpha)X_\nu(\alpha)(e^{-\tau\omega_\alpha}\langle b_\alpha b_\alpha^\dagger\rangle + e^{\tau\omega_\alpha}\langle b_\alpha^\dagger b_\alpha\rangle)$$

$$\mathscr{P}_{\mu\nu}^{(0)}(i\omega) = \sum_\alpha \frac{2e^2 X_\mu X_\nu \omega_\alpha}{(i\omega)^2 - \omega_\alpha^2} \qquad (4.5.13)$$

The quantity $-\mathscr{P}^{(0)}$ is actually the polarizability $\alpha_{\mu\nu}(i\omega)$ of the atom.

To keep this discussion simple, we shall assume that this quantity is isotropic:

$$-\mathscr{F}^{(0)}_{\mu\nu}(i\omega) = \alpha_{\mu\nu}(i\omega) = \delta_{\mu\nu}\alpha(i\omega)$$

$$\alpha(i\omega) = 2e^2 \sum_{\alpha} \frac{X_\mu(\alpha)^2 \omega_\alpha}{\omega_\alpha{}^2 - (i\omega)^2}$$

So far our discussion has been confined to the properties of a single atom. Next we shall consider the additional properties which result when the atoms are collected together in the solid. The discussion about photons is deferred to later, and we shall now consider only the original Hamiltonian (4.5.10) as applied to a collection of atoms. The electrons on one atom will interact with the nucleus and electrons on adjacent atoms. Denote the position of an electron $\mathbf{r}_i = \mathbf{R}_i + \mathbf{x}_i$, where $\mathbf{R}_i$ is the position of the atom and $\mathbf{x}_i$ is the displacement of the electron from the center of the atom. For two electrons on different atoms, the interaction is expanded assuming that $|\mathbf{R}_j - \mathbf{R}_i| > |\mathbf{x}_j - \mathbf{x}_i|$:

$$\frac{1}{|\mathbf{r}_i - \mathbf{r}_j|} = \frac{1}{|\mathbf{R}_i - \mathbf{R}_j|} + \frac{(\mathbf{x}_i - \mathbf{x}_j)(\mathbf{R}_i - \mathbf{R}_j)}{|\mathbf{R}_i - \mathbf{R}_j|}$$
$$+ \tfrac{1}{2} \sum (\mathbf{x}_i - \mathbf{x}_j)_\mu \phi_{\mu\nu}(\mathbf{R}_i - \mathbf{R}_j)(\mathbf{x}_i - \mathbf{x}_j)_\nu + \cdots$$

$$\phi_{\mu\nu}(\mathbf{R}) = \frac{\delta_{\mu\nu}}{R^3} - \frac{3R_\mu R_\nu}{R^5}$$

The first two terms can usually be neglected, since the first is a constant, and the second will vanish for crystals with inversion symmetry. Thus the first important term, resulting from the interaction between electrons on different atoms, is the dipole–dipole interaction:

$$\mathbf{x}_i \cdot \boldsymbol{\phi}(\mathbf{R}_i - \mathbf{R}_j) \cdot \mathbf{x}_j \tag{4.5.14}$$

We shall follow the standard practice and retain only this term in the theory, which is called the dipole approximation. Sometimes it is improved by also including the next terms, which involve quadrupoles and octopoles. For a treatment of exciton theory without making a monopole expansion but instead retaining all the terms in the electron–electron interaction, see Agranovitch (1960, 1961) for the formalism and Mahan (1975) for a sample calculation.

The notation needs to be expanded to include the site l of the atom in the solid. Thus we shall refer to exciton operators as $b_{\mu l}$ and polarization operators as $P_\mu(l)$. For one atom per unit cell, each quantity may also be expanded in wave vector space:

$$P_\mu(l) = e \sum_\alpha X_\mu(\alpha)(b_{\alpha,l} + b^\dagger_{\alpha,l})$$

$$P_\mu(\mathbf{k}) = \frac{1}{N^{1/2}} \sum_l e^{i\mathbf{k}\cdot\mathbf{R}_l} P_\mu(l) = e \sum_\alpha X_\mu(\alpha)(b_{\alpha,\mathbf{k}} + b^\dagger_{\alpha,-\mathbf{k}})$$

$$b_{\alpha,\mathbf{k}} = \frac{1}{N^{1/2}} \sum_l e^{i\mathbf{k}\cdot\mathbf{R}_l} b_{\alpha,l}$$

In the dipole–dipole interaction (4.5.14), the position vector of the electron $e\mathbf{x}_i$ is replaced by the equivalent displacement operator $P_\mu(i)$. The dipole–dipole interaction between excitations on different atoms becomes:

$$\begin{aligned}
\frac{e^2}{2} \sum_{ij} \mathbf{x}_i \cdot \boldsymbol{\phi}(\mathbf{R}_i - \mathbf{R}_j) \cdot \mathbf{x}_j &\to \frac{1}{2} \sum_{ij,\mu\nu} P_\mu(i)\ \phi_{\mu\nu}(\mathbf{R}_i - \mathbf{R}_j) P_\nu(j) \\
&= \frac{1}{2} \sum_{\mathbf{k},\mu\nu} P_\mu(\mathbf{k}) P_\nu(-\mathbf{k}) \sum_{\mathbf{R}}{}' e^{i\mathbf{k}\cdot\mathbf{R}} \phi_{\mu\nu}(\mathbf{R}) \\
&= \frac{1}{2} \sum_{\substack{\mathbf{k} \\ \mu\nu}} P_\mu(\mathbf{k}) P_\nu(-\mathbf{k}) \frac{4\pi}{v_0} T_{\mu\nu}(\mathbf{k})
\end{aligned}$$

$$T_{\mu\nu}(\mathbf{k}) = \frac{v_0}{4\pi} \sum_l{}' e^{i\mathbf{k}\cdot\mathbf{R}_l} \phi_{\mu\nu}(\mathbf{R}_l)$$

The interaction between excitations on the same atom is omitted, since this contribution is already included implicitly in the free-atom Hamiltonian (4.5.10). The last step above gives the interaction in wave vector notation. The Fourier transform of $\phi_{\mu\nu}(\mathbf{R})$ is called $T_{\mu\nu}(\mathbf{k})$. We shall need to know this only at small wave vectors, which for cubic crystals is (Cohen and Keffer, 1955)

$$\lim_{k\to 0} T_{\mu\nu}(\mathbf{k}) = \frac{k_\mu k_\nu}{k^2} - \frac{1}{3} \delta_{\mu\nu} + O(ka) \tag{4.5.15}$$

where a is a lattice constant. The Hamiltonian we consider for the moment is the original atomic part (4.5.11) plus the dipole–dipole interaction term. The atomic part can also be expressed in wave vector notation, so that we come to consider

$$\begin{aligned}
H &= \sum_{\alpha,\mathbf{k}} \omega_\alpha b^\dagger_{\alpha,\mathbf{k}} b_{\alpha,\mathbf{k}} + \tfrac{1}{2} \sum_{\mathbf{k},\alpha\beta} V_{\alpha\beta}(\mathbf{k})(b_{\alpha,\mathbf{k}} + b^\dagger_{\alpha,-\mathbf{k}})(b_{\beta,-\mathbf{k}} + b^\dagger_{\beta,\mathbf{k}}) \\
V_{\alpha\beta}(\mathbf{k}) &= \frac{4\pi e^2}{v_0} \sum_{\mu\nu} X_\mu(\alpha) T_{\mu\nu}(\mathbf{k}) X_\nu(\beta)
\end{aligned} \tag{4.5.16}$$

This form of a Hamiltonian has been diagonalized before—see (4.4.9). However, the present case is more complicated, because of the summation over the different excitations α and β. Thus we solve (4.5.16) by the use of a Green's function. The correlation function (4.5.12) is redefined to include wave vectors:

$$\mathscr{F}_{\mu\nu}(\mathbf{k}, i\omega) = -\int_0^\beta d\tau e^{i\omega\tau} \langle TP_\mu(\mathbf{k}, \tau) P_\nu(-\mathbf{k}, 0)\rangle \qquad (4.5.17)$$

This correlation function is evaluated by a diagrammatic expansion. The first term in the Hamiltonian (4.5.16) is treated as H_0, and the second term is the interaction V. For this H_0, the unperturbed Green's function was evaluated earlier in (4.5.13) and is the negative of the atomic polarizability:

$$\mathscr{F}^{(0)}_{\mu\nu}(\mathbf{k}, i\omega) = \sum_\alpha \frac{2e^2 X_\mu(\alpha) X_\nu(\alpha)\omega_\alpha}{(i\omega)^2 - \omega_\alpha{}^2} = -\alpha_{\mu\nu}(i\omega)$$

The self-energy term for the perturbation V occurs in first order. An important feature of the evaluation of this term is the separability of the potential $V_{\alpha\beta}$,

$$V = \frac{1}{2}\,\frac{4\pi}{\nu_0} \sum_{\mu\nu,\mathbf{k}} P_\mu(\mathbf{k}) T_{\mu\nu}(\mathbf{k}) P_\nu(-\mathbf{k})$$

so that the different α and β summations just go into the factors $P_\mu(\mathbf{k})$ and $P_\nu(-\mathbf{k})$. For example, the first-order term in the expansion of the S matrix is

$$\begin{aligned}\mathscr{F}_{\mu\nu}(\mathbf{k}, i\omega) &= \mathscr{F}^{(0)}_{\mu\nu} + \int_0^\beta d\omega e^{i\omega\tau} \int_0^\beta d\tau_1 \sum_{\mathbf{k}'\mu'\nu'} \frac{2\pi}{\nu_0} T_{\mu'\nu'}(\mathbf{k}') \\ &\quad \times \langle T\bar{P}_\mu(\mathbf{k}, \tau)\bar{P}_{\mu'}(-\mathbf{k}', \tau_1)\bar{P}_{\nu'}(\mathbf{k}', \tau_1)\bar{P}_\nu(-\mathbf{k}, \tau)\rangle + \cdots \\ &= \mathscr{F}^{(0)}_{\mu\nu}(\mathbf{k}, i\omega) + \frac{4\pi}{\nu_0} \sum_{\mu'\nu'} \mathscr{F}^{(0)}_{\mu\mu'}(\mathbf{k}, i\omega) T_{\mu'\nu'}(\mathbf{k}) \\ &\quad \times \mathscr{F}^{(0)}_{\nu'\nu}(\mathbf{k}, i\omega) + \cdots\end{aligned}$$

The self-energy is $4\pi T_{\mu\nu}/\nu_0$. It is the only self-energy term, and higher-order terms in the S-matrix expansion just produce multiples of this term. Thus the Dyson equation for the correlation function is the matrix equation

$$\mathscr{F}_{\mu\nu}(\mathbf{k}, i\omega) = \mathscr{F}^{(0)}_{\mu\nu}(\mathbf{k}, i\omega) + \frac{4\pi}{\nu_0} \sum_{\mu'\nu'} \mathscr{F}^{(0)}_{\mu\mu'}(\mathbf{k}, i\omega) T_{\mu'\nu'}(\mathbf{k}) \mathscr{F}_{\nu'\nu}(\mathbf{k}, i\omega)$$

This equation is not too hard to solve in the general case, since it is only a

3 by 3 matrix in the direction variables $(\mu, \nu) = (x, y, z)$. We shall only consider the simplest possible case, which has an isotropic polarizability $-\mathscr{F}^{(0)} = \delta_{\mu\nu}\alpha$ and the long-wavelength form of the interaction $T_{\mu\nu}$ as given in (4.5.15). Then the equation to be solved simplifies to

$$\mathscr{F}_{\mu\nu} = -\delta_{\mu\nu}\alpha - \frac{4\pi\alpha}{v_0}\sum_\lambda \left(\frac{k_\mu k_\lambda}{k^2} - \frac{\delta_{\mu\lambda}}{3}\right)\mathscr{F}_{\lambda\nu} \tag{4.5.18}$$

The matrix $\mathscr{F}_{\mu\nu}$ depends only on the scalar functions C and D,

$$\mathscr{F}_{\mu\nu} = C\delta_{\mu\nu} + D\frac{k_\mu k_\nu}{k^2}$$

$$\sum_\lambda \left(\frac{k_\mu k_\lambda}{k^2} - \frac{\delta_{\mu\lambda}}{3}\right)\mathscr{F}_{\lambda\nu} = C\left(\frac{k_\mu k_\nu}{k^2} - \frac{\delta_{\mu\nu}}{3}\right) + \frac{2D}{3}\frac{k_\mu k_\nu}{k^2}$$

where the definition of $\mathscr{F}_{\mu\nu}(\mathbf{k})$ contains the most general dependence of an isotropic system on the indices μ and ν. We can generate equations in only C and D by inserting the above results in (4.5.18) and then equating coefficients of the terms $\delta_{\mu\nu}$ and $k_\mu k_\nu/k^2$:

$$C = -\alpha + \frac{4\pi\alpha}{3v_0}C$$

$$D = -\frac{4\pi\alpha}{v_0}\left(C + \frac{2D}{3}\right)$$

These two equations for C and D can be solved algebraically:

$$C = -\frac{\alpha}{1 - 4\pi\alpha/3v_0}$$

$$D = \frac{\alpha(4\pi\alpha/v_0)}{(1 - 4\pi\alpha_0/3v_0)(1 + 8\pi\alpha/3v_0)} \tag{4.5.19}$$

$$\mathscr{F}_{\mu\nu} = -\frac{\alpha}{1 - 4\pi\alpha/3v_0}\left[\delta_{\mu\nu} - \frac{4\pi\alpha/v_0}{1 + 8\pi\alpha/3v_0}\frac{k_\mu k_\nu}{k^2}\right]$$

Equation (4.5.19) is the long-wavelength form of $\mathscr{F}_{\mu\nu}(\mathbf{k}, i\omega)$.

The quantity $-\mathscr{F}_{\mu\nu}$ is the polarizability of the solid, just as $-\mathscr{F}^{(0)}_{\mu\nu}$ is the polarizability of the atom. The interactions in V modify the atomic polarizability. These modifications are just the local field corrections. In a cubic dielectric, the dielectric function at long wavelength has the Lorenz–Lorentz form:

$$\varepsilon(\omega) = 1 + \frac{4\pi\alpha/v_0}{1 - 4\pi\alpha/3v_0} \tag{4.5.20}$$

The factor $4\pi\alpha/3v_0$ in the denominator is a local field correction. We have derived exactly the same factors in the first term in $\mathscr{P}_{\mu\nu}$ in (4.5.19). The atomic polarizability α has dimensions of cm^3, so the polarizability per unit volume α/v_0 is dimensionless. Often this combination itself is written as α. In any case, we shall soon show that the dielectric function (4.5.20) may be derived from the correlation function $\mathscr{P}_{\mu\nu}$ in (4.5.19).

We are now in a position to solve the Hamiltonian for the interaction between excitons and photons:

$$H = \frac{1}{2m} \sum_i \left[\mathbf{p}_i - \frac{e}{c} \mathbf{A}(r_i) \right]^2 - Ze^2 \sum_i \frac{1}{r_i} + \frac{e^2}{2} \sum_{ij} \frac{1}{|\mathbf{r}_i - \mathbf{r}_j|} + \sum_{\mathbf{k}\lambda} \omega_{\mathbf{k}} a^\dagger_{\mathbf{k}\lambda} a_{\mathbf{k}\lambda} \tag{4.5.21}$$

This Hamiltonian can be rewritten in terms of the operators for excitons and photons:

$$\begin{aligned} H &= \sum_{\mathbf{k}\lambda} \omega_{\mathbf{k}} a^\dagger_{\mathbf{k}\lambda} a_{\mathbf{k}\lambda} + \sum_{\mathbf{k}\alpha} \omega_\alpha b^\dagger_{\alpha\mathbf{k}} b_{\alpha,\mathbf{k}} + \frac{2\pi}{v_0} \sum_{\mu\nu\mathbf{k}} P_\mu(\mathbf{k}) T_{\mu\nu}(\mathbf{k}) P_\nu(\mathbf{k}) \\ &\quad - \frac{1}{(v_0)^{1/2}} \sum_{\mathbf{k}} J_\mu(\mathbf{k}) A_\mu(\mathbf{k}) + \frac{e^2 Z}{2mv_0} \sum_{\mathbf{k}\mu} A_\mu(\mathbf{k}) A_\mu(-\mathbf{k}) \\ J_\mu(\mathbf{k}) &= \frac{e}{m} \sum_\alpha p_\mu(\alpha)(b^\dagger_{\alpha,\mathbf{k}} - b_{\alpha,\mathbf{k}}) \\ p_\mu(\alpha) &= \frac{1}{i} \int d^3 r \phi_m(r)^* \frac{\partial}{\partial x_\mu} \phi_n(r) \end{aligned} \tag{4.5.22}$$

We have also introduced the operator $J_\mu(\mathbf{k})$, which is the current operator for the exciton system. The two operators J_μ and P_μ are simply related. The classical expression has the current as the time derivative of the polarization:

$$\begin{aligned} \mathbf{J} &= \frac{\partial}{\partial t} \mathbf{P} \\ J_\mu(\mathbf{k}, \tau) &= i \frac{\partial}{\partial \tau} P_\mu(\mathbf{k}, \tau) \end{aligned} \tag{4.5.23}$$

This expression is also correct for our operators, which can be derived from the relation $\mathbf{p} = -im[\mathbf{r}, H]$ between momentum and position.

The photon Green's function is now obtained by the same procedure as used in Sec. 4.5B. The two self-energy terms are evaluated as before. The first-order self-energy, which comes from the A^2 term, is evaluated with the same result [see (4.5.6)]:

$$\Pi_{\mu\nu}^{(1)} = \frac{e^2 Z}{m v_0} \delta_{\mu\nu}$$

The factor of Z is the number of electrons in each atom in unit volume ν_0. In enters into the A^2 term because the Hamiltonian in (4.5.21) contains the summation over i, which is the summation over all electrons, i.e., the summation over each atom and then the summation over each electron on each atom. The plasma frequency in this system is then $4\pi\Pi^{(1)} = 4\pi e^2 Z/(m\nu_0) \equiv \omega_p{}^2$.

The other self-energy contribution comes from the $J \cdot A$ term in the Hamiltonian. In analogy with (4.5.8), this term in the self-energy arises from the correlation function:

$$\Pi_{\mu\nu}^{(2)} = \Phi_{\mu\nu}(\mathbf{k}, i\omega) = \frac{1}{\nu_0} \int_0^\beta d\tau e^{i\omega\tau} \langle T J_\mu(\mathbf{k}, \tau) J_\nu(-\mathbf{k}, 0) \rangle$$

We need to evaluate this correlation function. It will be found by relating it to the correlation function for polarizability $\mathscr{P}_{\mu\nu}$ which is defined in (4.5.17) and solved in (4.5.19). This solution is the appropriate one for our case, since $\mathscr{P}_{\mu\nu}$ has been evaluated with the full exciton Hamiltonian, which includes all terms in the exciton–photon Hamiltonian (4.5.22) except the $J \cdot A$ and A^2 terms. Thus we shall obtain the exact solution to the photon self-energy, for the Hamiltonian (4.5.22), if we can relate $\Phi_{\mu\nu}$ to $\mathscr{P}_{\mu\nu}$. They are related by using the operator identity (4.5.23),

$$\Phi_{\mu\nu}(\mathbf{k}, i\omega) = -\frac{1}{\nu_0} \int_0^\beta d\tau e^{i\omega\tau} \left\langle T \left[\frac{\partial}{\partial\tau} P_\mu(\mathbf{k}, \tau) \right] \left[\frac{\partial}{\partial\tau'} P(-\mathbf{k}, \tau') \right]_{\tau'=0} \right\rangle$$

and then integrating by parts on the τ variable. The first integration by parts brings

$$\begin{aligned} \Phi_{\mu\nu}(\mathbf{k}, i\omega) = &-\frac{1}{\nu_0} \left\langle [P_\mu(\mathbf{k}, \beta) - P_\mu(\mathbf{k}, 0)] \left(\frac{\partial P_\mu}{\partial\tau'} \right)_{\tau'} \right\rangle \\ &- (i\omega) \int_0^\beta d\tau e^{i\omega\tau} \left\langle TP(\mathbf{k}, \tau) \left[\frac{\partial}{\partial\tau'} P(-\mathbf{k}, \tau') \right]_{\tau'=0} \right\rangle \end{aligned}$$

The argument of the correlation function is rewritten as

$$-\left\langle TP(\mathbf{k}, 0) \frac{\partial}{\partial\tau} P(-\mathbf{k}, -\tau) \right\rangle$$

so that we can further integrate by parts to bring us to the expression

$$\Phi_{\mu\nu}(\mathbf{k}, i\omega) = -\frac{1}{v_0}\left\langle [P_\mu(\mathbf{k}, \beta) - P_\mu(\mathbf{k}, 0)]\left(\frac{\partial P_\mu}{d\tau'}\right)_{\tau'=0}\right\rangle$$
$$- (i\omega)\langle P_\mu(\mathbf{k}, 0)[P_\nu(-\mathbf{k}, -\beta) - P_\nu(-\mathbf{k}, 0)]\rangle$$
$$+ (i\omega)^2 \int_0^\beta d\tau e^{i\omega\tau}\langle TP_\mu(\mathbf{k}, \tau)P_\nu(-\mathbf{k}, 0)\rangle \qquad (4.5.24)$$

The last term has the desired form of $(i\omega)^2\mathscr{F}_{\mu\nu}$. The first two terms, which are constants of integration, must be evaluated next. The second term is easy, since it is zero:

$$\langle P_\mu(\mathbf{k}, 0)[P_\nu(-\mathbf{k}, -\beta) - P_\mu(-\mathbf{k}, 0)]\rangle$$
$$= \mathrm{Tr}\{e^{-\beta H}[P_\mu(\mathbf{k})e^{-\beta H}P_\nu(-\mathbf{k})e^{\beta H} - P_\mu(\mathbf{k})P_\nu(-\mathbf{k})]\}$$
$$= -\mathrm{Tr}\{e^{-\beta H}[P_\mu(\mathbf{k}), P_\nu(-\mathbf{k})]\} = 0$$

where we have used the cyclic properties of the trace to rearrange the first term. The first term in $\Phi_{\mu\nu}$ requires more care, since it is finite and makes an important contribution to the result:

$$R = -\frac{1}{v_0}\left\langle [P_\mu(\mathbf{k}, \beta) - P_\mu(\mathbf{k}, 0)]\left(\frac{\partial P_\nu}{\partial \tau}\right)\right\rangle = -\frac{1}{v_0}\left\langle \frac{\partial P_\nu}{\partial \tau}P_\mu - P_\mu\frac{\partial P}{\partial \tau}\right\rangle$$
$$= -\frac{i}{v_0}[P_\mu, J_\nu]$$

The commutators of J and P can be evaluated from their definitions:

$$R = -\frac{i}{v_0}[P_\mu, J_\nu] = -\frac{e^2}{imv_0}\sum_{\alpha,\beta} X_\mu(\alpha)p_\nu(\beta)[b_{\alpha,\mathbf{k}} + b^\dagger_{\alpha,-\mathbf{k}}, b^\dagger_{\alpha,\mathbf{k}} - b_{\alpha,-\mathbf{k}}]$$
$$= -\frac{2e^2}{imv_0}\sum_\alpha X_\mu(\alpha)p_\nu(\alpha)$$

To evaluate the remaining summation over α, we rewrite the momentum matrix element as $ip/m = [x, H]$, so that this term is

$$\frac{1}{m}p_\nu(\alpha) = i\omega_\alpha X_\nu(\alpha)$$
$$R = -\frac{2e^2}{v_0}\sum_\alpha \omega_\alpha X_\mu(\alpha)X_\nu(\alpha) = -\frac{e^2}{v_0 m}\delta_{\mu\nu}\sum_\alpha f_\alpha$$
$$f_\alpha = \frac{2\omega_\alpha X_\mu(\alpha)^2 m}{\hbar}$$

The summation over oscillator strength f_α can be trivially evaluated by using the *f-sum rule*, which states that the summation

$$\sum_\alpha f_\alpha = Z$$

where Z is the number of electrons on each atom. Thus we finally derive that the first term in (4.5.24) is just

$$R = -\frac{e^2 Z}{m v_0}\,\delta_{\mu\nu}$$

$$\Pi^{(2)}_{\mu\nu} = \Phi_{\mu\nu} = -\frac{{\omega_p}^2}{4\pi}\,\delta_{\mu\nu} + (i\omega)^2 \mathscr{F}_{\mu\nu}$$

The two self-energy expressions may be added,

$$4\pi(\Pi^{(1)}_{\mu\nu} + \Pi^{(2)}_{\mu\nu}) = 4\pi(i\omega)^2 \mathscr{F}_{\mu\nu}(\mathbf{k}, i\omega)$$

and the plasma frequency cancels from the result. The photon self-energy function is just proportional to the correlation function $\mathscr{F}_{\mu\nu}$ of the polarization operator. It was in anticipation of this result that we earlier called $\mathscr{F}_{\mu\nu}$ the polarizability of the solid. It is given in (4.5.19). Next we solve the Dyson equation for the photon Green's function:

$$\mathscr{D}_{\mu\nu} = \mathscr{D}^{(0)}_{\mu\nu} + \sum_{\lambda\alpha} \mathscr{D}^{(0)}_{\mu\lambda} \Pi_{\lambda\alpha} \mathscr{D}_{\alpha\nu} \tag{4.5.25}$$

By using the fact that $\mathscr{D}_{\mu\nu}$ and $\mathscr{D}^{(0)}_{\mu\nu}$ both have the factor $\delta_{\mu\nu} - k_\mu k_\nu/k^2$, it is easy to show that the term $D(k_\mu k_\nu/k^2)$ in $\mathscr{F}_{\mu\nu}$ makes no contribution, since it acts along the direction of $\mathbf{k}$, while the factor $\delta_{\mu\nu} - k_\mu k_\nu/k^2$ is designed to select the parts of the self-energy operator which are perpendicular to $\mathbf{k}$; i.e.,

$$\sum_\lambda \left(\delta_{\mu\lambda} - \frac{k_\mu k_\lambda}{k^2}\right)\left(C\delta_{\lambda\gamma} + D\,\frac{k_\lambda k_\gamma}{k^2}\right) = C\left(\delta_{\mu\gamma} - \frac{k_\mu k_\gamma}{k^2}\right)$$

The solution to (4.5.25) is

$$\mathscr{D}_{\mu\nu}(\mathbf{k}, i\omega) = \frac{4\pi(\delta_{\mu\nu} - k_\mu k_\nu/k^2)}{(i\omega)^2 \varepsilon(i\omega) - c^2 k^2}$$

$$\varepsilon(i\omega) = 1 + \frac{4\pi\alpha(i\omega)/v_0}{1 - (4\pi\alpha/v_0)/3}$$

Thus we derive that the photon Green's function is governed by the transverse dielectric function $\varepsilon(i\omega)$, which has the Lorenz–Lorentz local field correction. Again we find that the propagating normal modes are governed by the relationship $\omega^2\varepsilon(\omega) = \omega_k^2$, which gives the polariton modes. Thus we should view $\mathscr{D}_{\mu\nu}$ as actually the Green's function for the polariton, since it has the correct dispersion relation. This is the proper interpretation, since it has the true eigenfrequencies of the exciton–photon system.

Since both the exciton and photon system are bosons, we could just as easily have solved the problem from the exciton viewpoint. Then the photons merely give rise to self-energy corrections for the exciton system. Thus when we calculate the self-energy of the exciton, with all photon effects included, we can again derive the same dispersion relation for polaritons. Since the photon and exciton systems have equal footing, the problem can be treated successfully from either starting point. The solution from the exciton system is assigned as a problem.

The only real difference between the phonon and exciton solutions, in Sec. 4.5B and this section, is the inclusion of the local field corrections for the excitons. They arise from the term in the Hamiltonian describing the dipole–dipole interaction between polarization modes on different atoms. Of course, the phonon system also has this term in the Hamiltonian, which should be included in the analysis.

PROBLEMS

1. For the hard sphere potential, show that the diagonal reaction matrix is

$$R_l(k, k) = \frac{\hbar^2}{2mk} \frac{j_l(ka)}{\eta_l(ka)}$$

Next show that the exact off-diagonal reaction matrix is

$$R_l(k', k) = \frac{\hbar^2}{2mk} \frac{j_l(k'a)}{\eta_l(ka)}$$

(*Hint*: Let the hard sphere be a finite step with height V_0, and solve for $V_0 \to \infty$.) What is the T matrix for the hard sphere?

2. Derive the formula for the phase shift shown in Fig. 4.7(c). Does your plot give the same result?

3. Consider a one-dimensional harmonic chain of infinite length. Let one atom have a mass m different from the others of mass M. Solve exactly for the vibrational normal modes, and obtain the condition for the formation of a local mode of vibration. Does this happen when the impurity mass m is lighter or heavier than M?

4. Consider the vibrational modes of a three-dimensional solid. Let all atoms be alike except for one. Obtain the equation which determines the frequency of the normal mode. (a) Isotope effect: Let the one atom have a different mass but the same force constants. (b) Impurity: Let both mass and force constants be different.

5. Let two particles move with the same constant velocity $\mathbf{v}$ in a phonon field. The Hamiltonian is

$$H = \mathbf{v} \cdot (\mathbf{p}_1 + \mathbf{p}_2) + \sum_{\mathbf{q}} \omega_{\mathbf{q}} a_{\mathbf{q}}^{\dagger} a_{\mathbf{q}} + \sum_{\mathbf{q}} \frac{M_{\mathbf{q}}}{\nu^{1/2}} A_{\mathbf{q}} (e^{i\mathbf{q} \cdot \mathbf{R}_1} + e^{i\mathbf{q} \cdot \mathbf{R}_2})$$

a. Show that the interaction energy due to phonon exchange is ($r = \mathbf{R}_1 - \mathbf{R}_2$)

$$V(\mathbf{r}, \mathbf{v}) = -2 \sum_{\mathbf{q}} \frac{M_{\mathbf{q}}^2}{\nu} \frac{\cos(\mathbf{q} \cdot \mathbf{r})}{\omega_{\mathbf{q}} - \mathbf{q} \cdot \mathbf{v}}$$

b. For a Debye model with piezoelectric coupling, this may be approximated as

$$V(\mathbf{r}, \mathbf{v}) = -M_0 \int_{-\infty}^{\infty} \frac{dq}{2\pi} \int \frac{d\Omega_q}{4\pi} \frac{e^{i\mathbf{q} \cdot \mathbf{r}}}{c_s - \hat{\mathbf{q}} \cdot \mathbf{v}}$$

Evaluate this as a function of $\mathbf{r}$ and $\mathbf{v}$.

6. Let the following Hamiltonian describe the phonon interaction with a collection of constant velocity particles: $H = H_0 + V$, where

$$H_0 = \sum_j \mathbf{v}_j \cdot \mathbf{p}_j + \sum_q \omega_{\mathbf{q}} a_{\mathbf{q}}^{\dagger} a_{\mathbf{q}}$$

$$V = \frac{1}{\nu^{1/2}} \sum_{\mathbf{q}} M_{\mathbf{q}} A_{\mathbf{q}} \sum_j e^{i\mathbf{q} \cdot \mathbf{R}_j}$$

where $\mathbf{R}_j$ and $\mathbf{p}_j$ are the position and momentum of the particles. Evaluate

$$U(t) = \langle e^{iH_0 t} e^{-iHt} \rangle$$

where the average is over the phonon thermodynamics.

7. The coherent neutron scattering from the atoms in a solid is described by the correlation function

$$S(\mathbf{q}, t) = \frac{1}{N} \sum_{j,l} \langle e^{i\mathbf{q} \cdot \mathbf{R}_j(t)} e^{-i\mathbf{q} \cdot \mathbf{R}_l(0)} \rangle$$

For a crystal, write $\mathbf{R}_j = \mathbf{R}_j^{(0)} + \mathbf{Q}_j$, where $\mathbf{Q}_j$ is the displacement due to phonons. Find the exact result for this correlation function.

8. Derive a result for the second moment

$$f(T) = \langle \omega^2 \rangle - \langle \omega \rangle^2$$

for the spectral function (4.3.15). See (4.3.14) for definitions of $\langle \omega^n \rangle$.

9. The 6-10 Lennard-Jones potential may be written as

$$V_{6-10}(r) = a\varepsilon\left[\left(\frac{\sigma}{r}\right)^{10} - \left(\frac{\sigma}{r}\right)^{6}\right]$$

a. Find the value of a such that ε is the maximum well depth.

b. Schrödinger's equation with $E = 0$ may be solved exactly for this potential. Do this, and show that the value of critical binding is $g_c = 2(3)^{3/2}/(5)^{1/2} = 4.65$.

10. Consider the bound states of the Kondo Hamiltonian for a localized spin one-half particle interacting with a single particle in a band of continuum states:

$$H = \sum_{\sigma\mathbf{k}} \varepsilon_k c^{\dagger}_{\mathbf{k}\sigma} c_{\mathbf{k}\sigma} - \sum_{\mathbf{k}\mathbf{k}'} [J_z S^{(z)}(c^{\dagger}_{\mathbf{k}\uparrow} c_{\mathbf{k}'\uparrow} - c^{\dagger}_{\mathbf{k}\downarrow} c_{\mathbf{k}'\downarrow}) + J_{\perp}(c^{\dagger}_{\mathbf{k}\uparrow} c_{\mathbf{k}'\downarrow} S^{(-)} + c^{\dagger}_{\mathbf{k}\downarrow} c_{\mathbf{k}'\uparrow} S^{(+)})]$$

where $S^{(z)}$, $S^{(-)}$, and $S^{(+)}$ refer to the localized spins. Show that the following are exact bound states of H:

$$\varphi_1 = \sum_{\mathbf{k}} a_{\mathbf{k}} c^{\dagger}_{\mathbf{k}\uparrow} \mid \uparrow\rangle$$

$$\varphi_2 = \sum_{\mathbf{k}} b_{\mathbf{k}} c^{\dagger}_{\mathbf{k}\downarrow} \mid \downarrow\rangle$$

$$\varphi_3 = \sum_{\mathbf{k}} d_{\mathbf{k}} (c^{\dagger}_{\mathbf{k}\uparrow} \mid \downarrow\rangle - c^{\dagger}_{\mathbf{k}\downarrow} \mid \uparrow\rangle)$$

$$\varphi_4 = \sum_{\mathbf{k}} e_{\mathbf{k}} (c^{\dagger}_{\mathbf{k}\uparrow} \mid \downarrow\rangle + c^{\dagger}_{\mathbf{k}\downarrow} \mid \uparrow\rangle)$$

where a_k, b_k, d_k, and e_k are coefficients and $\mid\uparrow\rangle$, $\mid\downarrow\rangle$ are configurations of the localized spins. Find the eigenvalue equations for each case, and identify the triplet and singlet states.

11. Solve the Tomonaga model when the interaction U in (4.4.19) is added to the original Hamiltonian (4.4.1). Diagonalize this new Hamiltonian, and use it to evaluate the Pauli spin susceptibility (4.4.21).

12. Consider the static density–density correlation function

$$c(\xi) = \frac{1}{L}\int_0^L dx \varrho(x)\varrho(x+\xi) = \frac{1}{L^2}\sum_k \cos(k\xi)\langle\varrho(k)\varrho(-k)\rangle$$

a. Evaluate this for the noninteracting electron system at $T = 0$.

b. Use the Tomonaga model to evaluate this for the interacting electron system.

13. In the spinless Luttinger model, compare the commutator $[\Psi_1(x), H_0]$ as obtained in the fermion representation and the boson representation.

14. Show that the free energy of the Fano–Anderson model can be written exactly as $F = F_0 + \delta F$ where

$$\beta F_0 = 2 \sum_{\mathbf{k}} \ln[1 + e^{-\beta\varepsilon_k}] + \ln[1 + e^{-\beta\varepsilon_c}]$$

$$\delta F = \int_0^1 d\lambda \sum_{ip} \frac{\Sigma(ip)}{ip - \varepsilon_c - \lambda\Sigma(ip)}$$

Evaluate this expression at zero temperature, and show that it agrees with Fumi's theorem.

15. Calculate the retarded density–density correlation function for the spinless Luttinger model:

$$\chi(p, i\omega) = \int_0^\beta d\tau e^{i\omega\tau} \langle T_\tau \varrho(p, \tau)\varrho(-p, 0)\rangle$$

$$\varrho(p) = [\varrho_1(p) + \varrho_2(p)]$$

16. Interesting effects happen to the susceptibilities $\chi(q, \omega)$ when $q \approx 2k_F$. This wave vector region may be evaluated in the Luttinger model by taking operators which are the product of particle-1 and -2 fermions. For example, evaluate the correlation function $\chi(q, \omega)$ in the noninteracting spinless Luttinger model, where

$$\Lambda(q) = \int dx e^{-iqx} \Psi_{2s}^\dagger(x) \Psi_{1s}(x)$$

$$\chi(q, t) = \langle \Lambda(q, t) \Lambda^\dagger(q, 0)\rangle$$

17. Find the eigenvalue equation of the boson modes of a one-dimensional electron gas described by the spinless Luttinger model, which interacts with phonons. The *ds* are phonon operators:

$$H = v_F \sum_k k(a_{1k}^\dagger a_{1k} - a_{2k}^\dagger a_{2k}) + \sum_k \Omega_k d_k^\dagger d_k + \sum_{\text{all } q} \frac{M_q}{L^{1/2}} (d_q + d_{-q}^\dagger)[\varrho_1(q) + \varrho_2(q)]$$

18. Write the Hamiltonian (4.5.4) in terms of harmonic oscillator coordinates for both the photons and the phonons. Solve it as a coupled oscillator problem.

19. Solve the polariton Hamiltonian (4.5.22) from the exciton viewpoint. Calculate the self-energies of the exciton, including those arising from the interaction with photons. You may simplify the problem by neglecting the local field term. Your answer should have the poles of the exciton Green's function at the polariton modes.

Chapter 5

Electron Gas

The use of diagram techniques in many-particle physics began in the early 1950s, soon after their introduction into field theory. Although these methods were applied to a variety of problems, some areas of work were more successful than others. The two areas which enjoyed the early success were the homogeneous electron gas and the polaron problem. Later there were other successes such as the theories of superconductivity and superfluidity. However, the theory of the homogeneous electron gas, as it is presently understood, was worked out by many contributors during the period 1957–1958. They brought a variety of theoretical approaches to this problem, but all used diagrammatic techniques in some form.

5.1. EXCHANGE AND CORRELATION

The homogeneous electron gas is described by the Hamiltonian

$$\begin{aligned} H &= \sum_{\mathbf{p}\sigma} \varepsilon_p c^\dagger_{\mathbf{p}\sigma} c_{\mathbf{p}\sigma} + \frac{1}{2\nu} \sum_{\substack{\mathbf{k}\mathbf{k}'\sigma\sigma' \\ \mathbf{q}\neq 0}} v_q c^\dagger_{\mathbf{k}+\mathbf{q}\sigma} c^\dagger_{\mathbf{k}'-\mathbf{q}\sigma'} c_{\mathbf{k}'\sigma'} c_{\mathbf{k}\sigma} \\ \varepsilon_p &= \frac{p^2}{2m} \\ v_q &= \frac{4\pi e^2}{q^2} \end{aligned} \tag{5.1.1}$$

which was derived in (1.2.13). The electrons are free particles, which mutually interact by Coulomb's law e^2/r. There are N electrons in a large volume ν, with an average density $n_0 = N/\nu$. A positive charge of density n_0 is spread uniformly through the volume ν. This maintains the overall charge neutrality of the system. The homogeneous electron gas is also called the *jellium model* of a solid.

There are two electronic properties which will be evaluated in this section. First is the self-energy of an electron of momentum $\mathbf{p}$. Second, the properties of all the electrons will be averaged to obtain the total energy of the system. This will be done at zero temperature, so that we obtain the ground state energy. The quantity we want is the ground state energy per particle E_g which can depend only on the particle density $E_g(n_0)$. The total energy of the N-particle system is just $NE_g = E_T$, since we ignore surface effects.

The parameter r_s is universally used to describe the density of an electron gas,

$$\frac{4\pi}{3} r_s^3 = \frac{1}{n_0 a_0^3}$$

where a_0 is the Bohr radius. In an electron gas with uniform density n_0, r_s is the radius in atomic units of the sphere which encloses one unit of electron charge. Thus r_s is small for a high-density electron gas and large for a low-density one. Other properties of an electron gas may be expressed in terms of this parameter. The density may be related to the Fermi wave vector,

$$n_0 = 2\int \frac{d^3k}{(2\pi)^3} n_k = \frac{1}{\pi^2}\int_0^{k_F} k^2\, dk = \frac{k_F^3}{3\pi^2}$$

so that the Fermi wave vector and energy are related to r_s,

$$k_F a_0 = (3\pi^2 n_0)^{1/3} a_0 = \left(\frac{9\pi}{4}\right)^{1/3}\left(\frac{4\pi n_0 a_0^3}{3}\right)^{1/3} = \left(\frac{9\pi}{4}\right)^{1/3}\frac{1}{r_s} = \frac{1.9192}{r_s}$$

$$E_F = \frac{k_F^2}{2m} = (k_F a_0)^2\left(\frac{\hbar^2}{2ma_0^2}\right) = \frac{3.6832}{r_s^2} E_{ry}$$

where $E_{ry} = 13.6$ eV will be the standard unit of energy. Similarly, the plasma frequency is

$$\hbar\omega_p = \hbar\left(\frac{4\pi n_0 e^2}{m}\right)^{1/2} = \left(\frac{12}{r_s^3}\right)^{1/2} E_{ry} = \frac{3.4641}{r_s^{3/2}} E_{ry}$$

In the homogeneous electron gas, the average kinetic energy of the electrons is going to be proportional to $E_F \approx \langle\text{K.E.}\rangle \approx k_F^2$, and by dimensional analysis this is inversely proportional to the square of the characteristic length of the system, which is r_s. Thus we deduce that $\langle\text{K.E.}\rangle \propto$

$1/r_s^2$. Similarly, dimensional analysis suggests that the average Coulomb energy per particle will be e^2 divided by the characteristic length, or $\langle \text{P.E.} \rangle \propto 1/r_s$. When the electron gas has sufficiently high density, which is small r_s, the kinetic energy term will be larger than the potential energy term. In this case, the electrons will behave like free particles, since the potential energy is a perturbation on the dominant kinetic energy. In the high-density limit, the free-particle picture is expected to be valid. This limiting case will be investigated below. The kinetic energy and potential energy contributions will be calculated. The potential energy terms cannot be found exactly, but the result can be expressed as a power series—with logarithm terms—in the parameter r_s. This series should be accurate at small values of r_s. Below we shall give a term-by-term derivation of this series. Later, in Sec. 5.2, the other limit of low density is considered where the potential energy is larger than the kinetic energy.

A. Kinetic Energy

The first energy term is the kinetic energy. For a single particle this is

$$\varepsilon_k = \frac{k^2}{2m}$$

The contribution to the ground state energy is obtained by summing over all the particles in the ground state:

$$\begin{aligned} E_{T,KE} &= \sum_{p,\sigma} \varepsilon_k n_k = 2\nu \int \frac{d^3k}{(2\pi)^3} \frac{k^2}{2m} n_k = \left(\frac{N}{n_0}\right) \frac{1}{\pi^2 2m} \int_0^{k_F} dk k^4 \\ &= \frac{3}{10} \frac{k_F{}^2}{m} N \end{aligned}$$

The average kinetic energy is $\frac{3}{5}E_F$, which may be given in terms of r_s:

$$\frac{3}{5} E_F = \frac{2.2099}{r_s{}^2} \tag{5.1.2}$$

B. Direct Coulomb

All the remaining terms in the energy come from the Coulomb interaction between the particles. This contribution has not been evaluated exactly. Instead, approximate expressions are obtained by a variety of means. We shall follow the derivation of Gell-Mann and Brueckner (1957) and start by examining the terms generated by ordinary perturbation theory.

The first term which occurs is the Coulomb interaction between the electrons and the uniform positive background. In the model of the homogeneous electron gas, the time-averaged electron density is uniform throughout the system, as is the positive background. These equal and opposite charge densities exactly cancel, so that the net system is charge neutral. The direct Coulomb energy is zero. That is, this energy is given by the equation

$$NE_0 = \frac{e^2}{2} \int dr_1\, dr_2 \frac{[\varrho_e(r_1) - \varrho_i(r_1)][\varrho_e(r_2) - \varrho_i(r_2)]}{|\mathbf{r}_1 - \mathbf{r}_2|} \tag{5.1.3}$$

but the ion and electron particle densities are $\varrho_i = \varrho_e = n_0$, and the contribution is zero. This fact has already been used in writing Eq. (5.1) by the omission of the $\mathbf{q} = 0$ term from the Coulomb interaction. This $\mathbf{q} = 0$ term is the direct Coulomb interaction among the electrons. This term is omitted because it is canceled by the direct interaction with the positive background.

C. Exchange

The Coulomb interactions in (5.1.1) provide other energy contributions in addition to the direct term. The direct term corresponds to the following pairing of the operators in (5.1.1) when one evaluates the average value of H:

$$\sum_{\substack{\mathbf{k}\mathbf{k}' \\ \sigma\sigma'}} c^\dagger_{\mathbf{k}+\mathbf{q},\sigma} c^\dagger_{\mathbf{k}'-\mathbf{q}\sigma'} c_{\mathbf{k}'\sigma'} c_{\mathbf{k}\sigma} = \delta_{\mathbf{q}=0} \left(\sum_{k\sigma} c^\dagger_{\mathbf{k},\sigma} c_{\mathbf{k},\sigma} \right)^2$$

Another way to pair the same operators is

$$\begin{aligned} \sum_{\substack{\mathbf{k}\mathbf{k}' \\ \sigma\sigma'}} c^\dagger_{\mathbf{k}+\mathbf{q},\sigma} c^\dagger_{\mathbf{k}'-\mathbf{q},\sigma'} c_{\mathbf{k}'\sigma'} c_{\mathbf{k}\sigma} &= -\sum_{\substack{k\sigma \\ \mathbf{k}'\sigma'}} n_{\mathbf{k}\sigma} n_{\mathbf{k}+\mathbf{q}\sigma} \delta_{\sigma\sigma'} \delta_{\mathbf{k}'=\mathbf{k}+\mathbf{q}} \\ &= -\sum_{\mathbf{k}\sigma} n_{\mathbf{k}\sigma} n_{\mathbf{k}+\mathbf{q}\sigma} \end{aligned}$$

This leads to the exchange energy of the electron, which was derived previously in (2.8.5), and also to the exchange contribution to the ground state energy:

$$\begin{aligned} \Sigma_x(k) &= -\frac{1}{\nu} \sum_{\mathbf{q}} v_{\mathbf{q}} n_{\mathbf{k}+\mathbf{q}} \\ E_{g,x} &= \frac{1}{2} \frac{1}{N} \sum_{\mathbf{k}\sigma} n_{\mathbf{k}} \Sigma_x(k) \end{aligned} \tag{5.1.4}$$

The self-energy $\Sigma_x(k)$ depends only on the wave vector k of the particle and not upon its energy variable $i\omega$ or ω. The wave vector integrals are elementary ($\mathbf{k}' = \mathbf{k} + \mathbf{q}$, $\nu = \cos\theta$):

$$\Sigma_x(k) = -\int \frac{d^3k'}{(2\pi)^3} \frac{4\pi e^2}{|\mathbf{k} - \mathbf{k}'|^2} n_{\mathbf{k}'}$$

$$= -\frac{e^2}{\pi}\int_0^{k_F} k'^2\, dk' \int_{-1}^{1} \frac{d\nu}{k^2 + k'^2 - 2kk'\nu}$$

$$= -\frac{e^2}{\pi k}\int_0^{k_F} k'\, dk' \ln\left|\frac{k + k'}{k - k'}\right|$$

$$\Sigma_x(k) = -\frac{e^2 k_F}{\pi}\left(1 + \frac{1 - y^2}{2y}\ln\left|\frac{1 + y}{1 - y}\right|\right)$$

$$y = \frac{k}{k_F}$$

A particle at the Fermi energy $k = k_F$ has $y = 1$ and

$$\Sigma_x(k_F) = -\frac{e^2 k_F}{\pi} \tag{5.1.5}$$

Thus it is convenient to write

$$\Sigma_x(k) = +\frac{e^2 k_F}{\pi} S(y)$$

$$S(y) = -\left(1 + \frac{1 - y^2}{2y}\ln\left|\frac{1 + y}{1 - y}\right|\right)$$

where $S(y)$ is a function which gives the wave vector dependence of the exchange energy. This function is shown in Fig. 5.1. Its value at $y = 0$ is $S = -2$. In the vicinity of $y = 1$ it rises steeply and approaches zero at large values of y.

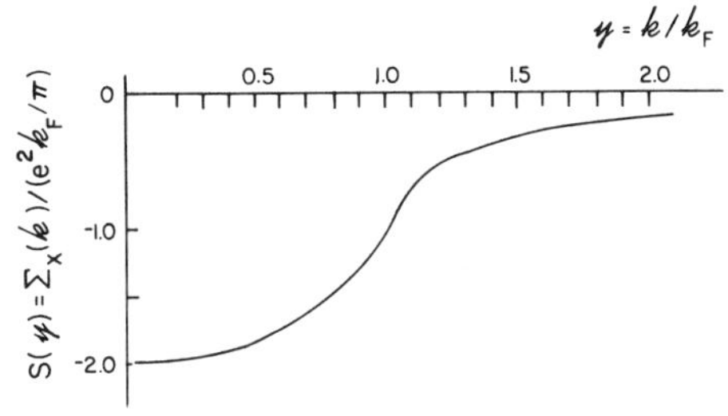

FIGURE 5.1. The unscreened exchange energy of the electron gas, according to Eq. (5.1.5).

At the Fermi energy the value is $S(1) = -1$. This point is interesting because the function $S(y)$ has an infinite slope. The derivative of $S(y)$ is

$$\frac{d}{dy} S(y) = \frac{1}{2y}\left(\frac{1+y^2}{y} \ln\left|\frac{1+y}{1-y}\right| - 2\right)$$

This has a logarithmic divergence as $y \to 1$. This fact is interesting because it predicts that the effective mass is zero. The effective mass of a particle was defined in (3.3.25). Since the exchange self-energy is not frequency dependent, the effective mass is given by

$$\left(\frac{m}{m^*}\right) = 1 + \frac{\partial}{\partial \varepsilon_k} \Sigma_x(k) = 1 + \frac{m}{k} \frac{\partial}{\partial k} \Sigma_x(k)$$

For the exchange energy, this is

$$\left(\frac{m}{m^*}\right) = \frac{e^2 m}{\pi k} \frac{\partial}{\partial y} S(y) = \frac{e^2 m}{2\pi k_F} \frac{1}{y^2}\left(\frac{1+y^2}{y} \ln\left|\frac{1+y}{1-y}\right| - 2\right)$$

which diverges at the Fermi energy $y \to 1$. If the inverse effective mass really diverged at the Fermi surface, it would have several observable consequences. The electron gas would be unstable at low temperatures, and the specific heat would diverge. Many metals have unusual properties at low temperatures; e.g., some become superconducting. But this is not due to electron–electron interactions but to electron–phonon interactions. The alkali metals are stable at low temperatures, so that this exchange instability is regarded as being absent. In fact, an examination of further terms in the perturbation theory produces another divergence in the effective mass which exactly cancels the one due to exchange. Thus the effective mass and specific heat are not divergent. This subject is pursued further in Sec. 5.8.

The exchange energy contribution to the ground state energy is obtained from (5.1.4). Summing over spins gives

$$E_{g,x} = \frac{1}{N} \sum_k n_k \Sigma_x(k) = \frac{1}{n_0} \int \frac{d^3k}{(2\pi)^3} n_k \Sigma_x(k)$$

The integral is straightforward and gives

$$\begin{aligned} E_{g,x} &= -\frac{1}{n_0} \frac{k_F{}^3}{2\pi^2}\left(\frac{e^2 k_F}{\pi}\right) \int_0^1 y^2\, dy\left(1 + \frac{1-y^2}{2y} \ln\left|\frac{1+y}{1-y}\right|\right) \\ E_{g,x} &= -\frac{3}{4} \frac{e^2 k_F}{\pi} \end{aligned} \qquad (5.1.6)$$

The average exchange energy per electron is $\frac{3}{2}$ that at the Fermi energy. The additional factor of $\frac{1}{2}$ enters the ground state energy to account for the fact that the exchange energy is a pair interaction. In terms of the parameter r_s the total ground state exchange energy per electron is

$$E_{g,x} = -\frac{3}{2\pi}(k_F a_0)\left(\frac{e^2}{2a_0}\right) = -\frac{0.9163}{r_s} E_{ry} \tag{5.1.7}$$

where we have used the earlier result for $k_F a_0$.

So far we have found two terms for the energy of the particle,

$$E(k) = \frac{\hbar^2 k^2}{2m} + \Sigma_x(k) + \cdots \tag{5.1.8}$$

and the corresponding two terms for the ground state energy per particle,

$$E_g = \frac{2.2099}{r_s^2} - \frac{0.9163}{r_s} + \cdots \tag{5.1.9}$$

The ground state energy has the appearance of a power series, in increasing powers of r_s. Although it is usually unsafe to extrapolate from just two terms, in fact this is a series in r_s. The next term will be of order $(r_s)^0$. The zeroth power could be interpreted as either a constant or as $\ln(r_s)$. In fact, both of these terms are present. Thus the series has the form

$$E_g = \frac{2.2099}{r_s^2} - \frac{0.9163}{r_s} + A + B\ln(r_s) + Cr_s + \cdots$$

Here we shall derive only the constant and the logarithm term. This will yield the answer first given by Gell-Mann and Brueckner (1957):

$$E_g = \frac{2.2099}{r_s^2} - \frac{0.9163}{r_s} - 0.094 + 0.0622\ln(r_s) + \cdots \tag{5.1.10}$$

This result has since been obtained by a variety of perturbation techniques. The Gell-Mann and Brueckner derivation was by Rayleigh–Schrödinger perturbation theory. Alternate derivations have been presented by Sawada *et al.* (1957), Hubbard (1957), Nozieres and Pines (1958), and Quinn and Ferrell (1958). Our derivation will follow Quinn and Ferrell.

The energy terms we have evaluated comprise the Hartree–Fock theory. This is defined to be the kinetic energy, the direct Coulomb energy which is zero, and the exchange energy. Wigner and Seitz (1933, 1934) suggested the name *correlation energy* to mean the energy terms beyond Hartree–

Fock. This is usually applied to the ground state energy. Thus we write the total ground state energy per particle as

$$E_g = \frac{2.2099}{r_s^2} - \frac{0.9163}{r_s} + E_c$$

where the correlation energy E_c needs to be determined. Of course, we have already given some terms in its power series in r_s:

$$E_c = -0.094 + 0.0622 \ln(r_s) + O(r_s) \tag{5.1.11}$$

This result is accurate in the limit of $r_s \to 0$. There is some uncertainty regarding the maximum value of r_s for which these few terms provide an accurate description, but it is around $r_s \equiv 1$. Actual metals have values of r_s up to about 6, and this series does not give sensible numbers at these low densities.

The term correlation energy is often applied to other quantities besides the total ground state energy. For example, the correlation energy of a particle of wave vector k is those terms beyond Hartree–Fock:

$$E(k) = \frac{k^2}{2m} + \Sigma_x(k) + \Sigma_c(k) \tag{5.1.12}$$

If we compute the correlation energy for a particle, we can average this to obtain the average correlation energy per particle—which is the ground state correlation energy.

D. Seitz' Theorem

The derivation of Quinn and Ferrell (1958) does calculate the correlation energy per particle; they then average this to obtain the correlation contribution to the ground state energy. This averaging is based on a theorem of Seitz (1940) which relates the ground state energy to the chemical potential. The chemical potential is defined as the energy it takes to add or remove an electron from the material. It is the energy which divides the empty from the occupied states at zero temperature. Of course, this is just the Fermi energy of the metal. Thus the chemical potential is the energy of an electron of momentum k_F:

$$\mu = E(k_F) = \frac{k_F^2}{2m} - \frac{e^2 k_F}{\pi} + E_c(k_F)$$

The chemical potential μ is thus only a function of the density of the electron gas n_0. The theorem of Seitz is that

$$\mu(n_0) = \frac{\partial}{\partial n_0}\,[n_0 E_g(n_0)] = E_g + n_0 \frac{\partial}{\partial n_0} E_g \tag{5.1.13}$$

The proof is based on the definition of the chemical potential. It is the energy difference between the system with N particles and that with $N + 1$:

$$\mu = E_T(N + 1) - E_T(N)$$

For an N-particle system, we have for the total energy

$$E_T = NE_g$$

An $(N + 1)$-particle system, with volume ν, has a density $(1/\nu)(N + 1) = n_0 + (1/\nu)$. Thus the total energy of the $(N + 1)$-particle system is

$$\begin{aligned} E_T(N + 1) &= (N + 1)E_g\left(n_0 + \frac{1}{\nu}\right) = (N + 1)\left(E_g(n_0) + \frac{1}{\nu}\frac{\partial E_g}{\partial n_0}\right) \\ &= NE_g + E_g(n_0) + \frac{n_0\,\partial E_g}{\partial n_0} + O\left(\frac{1}{\nu}\right) \end{aligned}$$

The difference between this expression and NE_g is the chemical potential and gives the assertion in (5.1.13). In proving this theorem, we keep the volume ν fixed and also fix the amount of positive charge. The $(N + 1)$-particle system has a slight charge imbalance, since it has one more unit of electron charge than positive charge. However, this is a negligible contribution to the energy. For example, if a body of average dimension L is uniformly charged with one unit of charge, then its Coulomb energy is of order e^2/L. But if L for a macroscopic body is $L \approx 1$ cm, then e^2/L is $\approx 10^{-7}$ eV, which is negligible.

The chemical potential is the negative of the work function. This is the energy required to remove an electron from the solid and take it to infinity with zero kinetic energy. If this vacuum level is called zero, then the measured work function just gives the negative of the value of the chemical potential. However, there is a surface correction to the work function, or chemical potential, but not to the volume part of the ground state energy per particle. That is, we may write the total energy as

$$E_T = NE_g + AE_S$$

where A is the total surface area and E_S is the energy per unit surface area.

For macroscopic bodies, E_g does not depend on the surface area. However, μ does have a term which depends on the surface—actually on the surface dipole layer. We shall write

$$\mu = \mu_B + \Delta V$$

where ΔV is the contribution from the surface dipole layer and the μ_B are the bulk terms which we are trying to calculate. The surface contribution to the work function was first recognized by Bardeen (1936) and calculated accurately by Lang and Kohn (1971). The theorem of Seitz actually just refers to μ_B:

$$\mu_B = E_g + n_0 \frac{\partial E_g}{\partial n_0}$$

This relationship is valid term by term in the energy series. The Hartree–Fock terms which have been derived so far may be expressed as powers of n_0:

$$E_{g,HF} = \frac{2.2099}{r_s^2} - \frac{0.9163}{r_s} = \frac{3}{5} X n_0^{2/3} - \frac{3}{4} Y n_0^{1/3}$$

$$\frac{3}{5} X n_0^{2/3} = \frac{2.2099}{r_s^2} = \frac{3}{5} \frac{E_F}{E_{ry}} = \frac{3}{5} (3\pi^2 n_0 a_0^3)^{2/3}$$

$$\frac{3}{4} Y n_0^{1/3} = \frac{0.9163}{r_s} = \frac{3}{2\pi} (3\pi^2 n_0 a_0^3)^{1/3}$$

Applying the theorem, for the chemical potential we obtain

$$\mu_{B,HF} = \frac{d}{dn_0} (n_0 E_{g,HF}) = X n_0^{2/3} - Y n_0^{1/3} = E_F - (e^2/\pi) k_F$$

The theorem correctly gives that the average kinetic energy is $\frac{3}{5}$ of the value at the Fermi energy, and the average exchange energy is $\frac{3}{4}$ the value at the Fermi energy. These results follow from their respective powers of n_0: $\frac{2}{3}$ and $\frac{1}{3}$.

The correlation energy for the ground state is a series in r_s or $n_0^{1/3}$. The theorem may be applied to this series in a term-by-term fashion:

$$\begin{aligned} E_c &= A + B \ln(r_s) + C r_s + \cdots \\ E_c &= A - \frac{1}{3} B \ln\left(\frac{4\pi n_0 a_0^3}{3}\right) + C n_0^{-1/3} \left(\frac{4\pi a_0^3}{3}\right)^{-1/3} + \cdots \end{aligned} \tag{5.1.14}$$

$$\mu_c = \frac{d}{dn_0} (n_0 E_c) = A - \frac{1}{3} B + B \ln(r_s) + \frac{2}{3} C r_s + \cdots \tag{5.1.15}$$

Quinn and Ferrell used this theorem to make a backward deduction. They first found μ_c by calculating the self-energy terms of an electron at the Fermi energy. Then the above relationships may be used to find the correlation energy for the ground state energy. The logarithm term has the same coefficient B for the two energy terms. Once this is determined, the constant term for the ground state energy is obtained by adding $B/3$ to the constant term found for the chemical potential.

The self-energy of an electron, from Coulomb interactions, may be calculated by the methods described in Chapters 2 and 3. The self-energy function has an infinite number of terms. The basic procedure is to start examining some low-order terms and to deduce which terms contribute to the constant and $\ln(r_s)$ terms. The term *low-order* refers to the order of the self-energy diagram, which is the number of internal Coulomb lines.

E. $\Sigma^{(2a)}$

The exchange energy graph is shown as the first diagram in Fig. 5.2. It is the only contribution with one Coulomb line. The correlation energy is the sum of all contributions with two or more Coulomb lines. There are three diagrams, or self-energy terms, with two Coulomb lines. One of these is shown in Fig. 5.2(a). By using the rules for constructing diagrams in Sec. 3.4, this self-energy contribution is

$$\Sigma^{(2a)}(k) = \frac{1}{\nu^2} \sum_{\mathbf{q}\mathbf{q}'} v_q v_{q'} \frac{1}{\beta^2} \sum_{iq_n iq_n'} \mathscr{G}^{(0)}(k+q)\mathscr{G}^{(0)}(k+q')\mathscr{G}^{(0)}(k+q+q')$$

The first summation over frequencies is given in (3.5.2):

$$\frac{1}{\beta} \sum_{iq_n} \mathscr{G}^{(0)}(k+q)\mathscr{G}^{(0)}(k+q+q') = \frac{n_F(\xi_{\mathbf{k}+\mathbf{q}}) - n_F(\xi_{\mathbf{k}+\mathbf{q}+\mathbf{q}'})}{iq_n' + \xi_{\mathbf{k}+\mathbf{q}} - \xi_{\mathbf{k}+\mathbf{q}+\mathbf{q}'}}$$

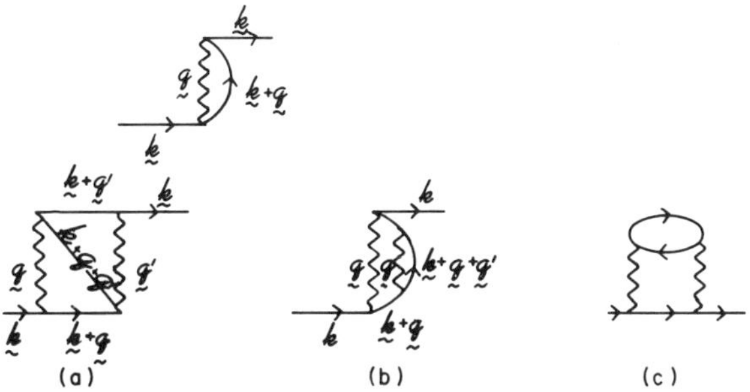

FIGURE 5.2

The second summation requires a new derivation, since one energy denominator has odd multiples of $i\pi/\beta$, and the other has even multiples. The answer is

$$\frac{1}{\beta}\sum_{iq_n'}\frac{\mathscr{G}^{(0)}(k+q')}{iq_n'+\xi_{\mathbf{k}+\mathbf{q}}-\xi_{\mathbf{k}+\mathbf{q}+\mathbf{q}'}}=-\frac{n_B(\xi_{\mathbf{k}+\mathbf{q}+\mathbf{q}'}-\xi_{\mathbf{k}+\mathbf{q}})+n_F(\xi_{\mathbf{k}+\mathbf{q}'})}{ik_n+\xi_{\mathbf{k}+\mathbf{q}+\mathbf{q}'}-\xi_{\mathbf{k}+\mathbf{q}}-\xi_{\mathbf{k}+\mathbf{q}'}}$$

where one occupation function is a boson distribution and the other fermion. The boson distribution may be eliminated by using the identity

$$[n_F(\xi_{\mathbf{k}+\mathbf{q}})-n_F(\xi_{\mathbf{k}+\mathbf{q}+\mathbf{q}'})]n_B(\xi_{\mathbf{k}+\mathbf{q}+\mathbf{q}'}-\xi_{\mathbf{k}+\mathbf{q}})=n_F(\xi_{\mathbf{k}+\mathbf{q}+\mathbf{q}'})[1-n_F(\xi_{\mathbf{k}+\mathbf{q}})]$$

which gives the final self-energy term:

$$\begin{aligned}\Sigma^{(2a)}(k)=-\frac{1}{\nu^2}\sum_{\mathbf{q}\mathbf{q}'}&\frac{v_q v_{q'}}{ik_n+\xi_{\mathbf{k}+\mathbf{q}+\mathbf{q}'}-\xi_{\mathbf{k}+\mathbf{q}}-\xi_{\mathbf{k}+\mathbf{q}'}}\\&\times\{n_F(\xi_{\mathbf{k}+\mathbf{q}'})[n_F(\xi_{\mathbf{k}+\mathbf{q}})-n_F(\xi_{\mathbf{k}+\mathbf{q}+\mathbf{q}'})]\\&+n_F(\xi_{\mathbf{k}+\mathbf{q}+\mathbf{q}'})[1-n_F(\xi_{\mathbf{k}+\mathbf{q}})]\}\end{aligned}$$

The result we need is the self-energy of a particle on the Fermi surface. Thus, set $k=k_F$ and $ik_n=\xi_k=k^2/2m-\mu$. The terms in the energy denominator largely cancel,

$$\xi_{\mathbf{k}}+\xi_{\mathbf{k}+\mathbf{q}+\mathbf{q}'}-\xi_{\mathbf{k}+\mathbf{q}}-\xi_{\mathbf{k}+\mathbf{q}'}=\frac{\mathbf{q}\cdot\mathbf{q}'}{m}$$

so that we need to evaluate the expression $(\nu\to\infty)$

$$\begin{aligned}\Sigma^{(2a)}(k_F,\xi_F)=-\frac{(4\pi e^2)^2m}{(2\pi)^6}\int\frac{d^3q}{q^2}\int\frac{d^3q'}{(q')^2}\frac{1}{\mathbf{q}\cdot\mathbf{q}'}\\\times\{n_F(\xi_{\mathbf{k}+\mathbf{q}'})[n_F(\xi_{\mathbf{k}+\mathbf{q}})-n_F(\xi_{\mathbf{k}+\mathbf{q}+\mathbf{q}'})]+\cdots\}\end{aligned}$$

It is evaluated by using dimensionless units. All wave vectors are normalized to the Fermi wave vector $\mathbf{x}=\mathbf{q}/k_F$, $\mathbf{y}=\mathbf{q}'/k_F$, $\hat{z}=\mathbf{k}_F/k_F$:

$$\begin{aligned}\Sigma^{(2a)}=-\frac{E_{ry}}{2\pi^4}\int\frac{d^3x}{x^2}\int\frac{d^3y}{y^2}\frac{1}{\mathbf{x}\cdot\mathbf{y}}\{n_F(\hat{z}+\mathbf{y})[n_F(\hat{z}+\mathbf{x})-n_F(\hat{z}+\mathbf{x}+\mathbf{y})]\\+n_F(\hat{z}+\mathbf{x}+\mathbf{y})[1-n_F(\hat{z}+\mathbf{x})]\}\end{aligned}\tag{5.1.16}$$

The quantity on the right is independent of electron density. The integral is convergent and gives a finite result. It contributes to the constant term A in (5.1.15). Since Seitz's theorem is true for each term, this must equal the same constant as found for the equivalent term for E_c. This was evaluated by Onsager *et al.* (1966) and found to be $\Sigma^{(2a)} = \frac{1}{3}\ln(2) - (3/2\pi^2)\zeta(3) = 0.04836$.

F. $\Sigma^{(2b)}$

The second self-energy term involving two Coulomb lines is that shown in Fig. 5.2(b). This contribution is omitted because it is part of a sum of diagrams which gives zero. To show this, consider the summation shown in Fig. 5.3(a). Each term has one more exchange diagram. All terms may be summed by evaluating the exchange energy with an electron Green's function in the self-energy which includes the exchange energy. Thus this summation is given by the self-energy

$$\Sigma_x'(k) = \frac{-1}{\nu}\sum_{\mathbf{q}} v_q \frac{1}{\beta}\sum_{ik_n} \mathscr{G}(\mathbf{k}+\mathbf{q}, ik_n)$$

where the Green's function now has a self-energy

$$\mathscr{G}(\mathbf{k}+\mathbf{q}, ik_n) = \frac{1}{ik_n - \xi_{\mathbf{k}+\mathbf{q}} - \Sigma_x(\mathbf{k}+\mathbf{q})}$$

The fact that the self-energy $\Sigma_x(k)$ does not depend on frequency means that the frequency summation of the Green's function yields the simple number operator as in (3.5.4),

$$\frac{1}{\beta}\sum_{ik_n} \mathscr{G}(\mathbf{k}+\mathbf{q}, ik_n) = n_F[\xi_{\mathbf{k}+\mathbf{q}} + \Sigma_x(\mathbf{k}+\mathbf{q})] \equiv \frac{1}{e^{\beta(\xi+\Sigma)}+1}$$

and the self-energy is

$$\Sigma_x'(k) = \frac{-1}{\nu}\sum_{\mathbf{q}} v_q n_F[\xi_{\mathbf{k}+\mathbf{q}} + \Sigma_x(\mathbf{k}+\mathbf{q})] = \frac{-1}{\nu}\sum_{\mathbf{k}'} v_{\mathbf{k}-\mathbf{k}'} n_F[\xi_{k'} + \Sigma_x(k')] \tag{5.1.17}$$

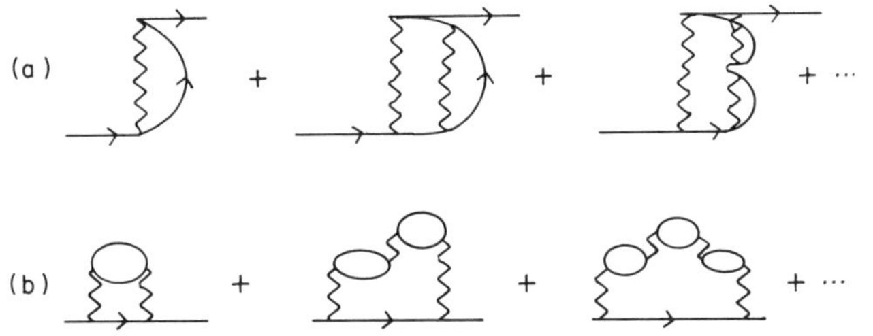

FIGURE 5.3

At zero temperature the electron distribution function $n_F[\xi_{k'} + \Sigma_x(k')]$ is just a step function which equals unity for electrons beneath the chemical potential and zero for those above. This step function must also be normalized so that the electron density is still n_0, as in

$$n_0 = 2 \int \frac{d^3k}{(2\pi)^3} n_F[\xi_k + \Sigma_x(k)]$$

The effect of the exchange energy $\Sigma_x(k)$ in the argument of n_F is just to change the chemical potential. Of course, we know that, since it is precisely the effect we are trying to calculate. This means the Fermi wave vector k_F is the same, even after we have included the exchange energy in the occupation function. The addition of the exchange energy is just canceled by an equal change in chemical potential. The result is that at zero temperature,

$$\lim_{T\to 0} n_F[\xi_k + \Sigma_x(k)] = \lim_{T\to 0} n_F(\xi_k) = \theta(k_F - k)$$

the new Fermi distribution is unchanged from the unperturbed one. This also means that the exchange energy $\Sigma_x'(k)$ in (5.1.17) is exactly equal to the one calculated earlier. If we reexamine Fig. 5.3(a), then the summation of these terms just yields the value of the first term alone. All the subsequent terms in that series sum to zero. This comes about because of a shift in the chemical potential.

The evaluation of these self-energy terms has become more subtle. It is not just a straightforward evaluation of terms in a series. One must understand each term in order to gauge its actual contribution. This is further emphasized when we examine the other second-order term.

The exchange self-energy has a remarkable effect upon the zero temperature electron distribution. It leaves k_F unchanged. This simple result is a consequence of the feature that $\Sigma_x(k)$ is independent of frequency. Most self-energy terms depend on frequency and so have a larger effect upon the wave vector distributions: Recall Problem 4 in Chapter 3.

G. $\Sigma^{(2c)}$

The third self-energy with two Coulomb lines is shown in Fig. 5.2(c). This self-energy diagram may be written as

$$\Sigma^{(2c)}(k) = -\frac{1}{\nu} \sum_{\mathbf{q}} v_q^2 \frac{1}{\beta} \sum_{iq_n} P^{(1)}(\mathbf{q}, iq_n) \mathscr{G}^{(0)}(\mathbf{k} + \mathbf{q}, ik_n + iq_n)$$

$$P^{(1)}(q) = \frac{2}{\nu} \sum_{\mathbf{k}} \frac{n_F(\xi_{\mathbf{k}}) - n_F(\xi_{\mathbf{k}+\mathbf{q}})}{iq_n + \xi_{\mathbf{k}} - \xi_{\mathbf{k}+\mathbf{q}}}$$

The closed fermion loop gives the polarization diagram $P^{(1)}(\mathbf{q}, iq_n)$ which was evaluated in Chapter 3. Although one could proceed with the evaluation of this self-energy, it obviously has one drawback. The wave vector integral appears to diverge at small values of q ($\nu \to \infty$):

$$\int \frac{d^3q}{q^4} = \int_0 \frac{dq}{q^2} = \lim_{\eta \to 0} \frac{1}{\eta}$$

Indeed it does diverge, so this self-energy term is infinite. The divergence is removed by summing a series of self-energy diagrams. This series is shown in Fig. 5.3(b). Each term has one more polarization bubble and one more Coulomb line—hence an additional factor of $v_q P^{(1)}$. The summation of these terms gives the simple series

$$\begin{aligned} \Sigma^{(3b)}(k) &= -\frac{1}{\nu} \sum_{\mathbf{q}} v_q \frac{1}{\beta} \sum_{iq_n} \mathscr{G}^{(0)}(k+q) \\ &\quad \times \{v_q P^{(1)}(q) + [v_q P^{(1)}(q)]^2 + [v_q P^{(1)}(q)]^3 + \cdots\} \\ &= -\frac{1}{\nu} \sum_{\mathbf{q}} v_q \frac{1}{\beta} \sum_{iq} \mathscr{G}^{(0)}(k+q) \frac{v_q P^{(1)}(q)}{1 - v_q P^{(1)}(q)} \end{aligned} \tag{5.1.18}$$

The summation of terms shown in Fig. 5.3 is called the *random phase approximation* (RPA). The "approximation" in RPA is that we use this series of terms to represent the entire answer. For example, the RPA result for the electron correlation energy is that from Fig. 5.3(b) or (5.1.18). This ignores other terms, such as (5.1.16), which was found from the diagram in Fig. 5.2(a). The RPA ignores many more terms in each higher order of perturbation series.

The denominator in (5.1.18) is a very important quantity because it is very often used in calculations. It is the RPA approximation to the dielectric function and is defined as

$$\varepsilon_{\text{RPA}}(q) = 1 - v_q P^{(1)}(q) \tag{5.1.19}$$

The self-energy series may be written as

$$\Sigma^{(3b)}(k) = -\frac{1}{\nu} \sum_{\mathbf{q}} v_q \frac{1}{\beta} \sum_{iq_n} \mathscr{G}^{(0)}(k+q) \left[\frac{1}{\varepsilon_{\text{RPA}}(q)} - 1\right] \tag{5.1.20}$$

The "-1" term in the brackets is easy to evaluate. It is

$$\frac{1}{\nu} \sum_{\mathbf{q}} v_q \frac{1}{\beta} \sum_{iq_n} \mathscr{G}^{(0)}(k+q) = \frac{1}{\nu} \sum_{\mathbf{q}} v_q n_{\mathbf{k}+\mathbf{q}} = -\Sigma_x(k)$$

which is just the negative of the exchange energy. Thus if we add the exchange energy to the above result, we are left with just a term containing the RPA dielectric function:

$$\Sigma_{\text{RPA}}(k) = \Sigma^{(3b)}(k) + \Sigma_x(k) = -\frac{1}{\nu}\sum_{\mathbf{q}} v_q \frac{1}{\beta}\sum_{iq_n} \frac{\mathscr{G}^{(0)}(k+q)}{\varepsilon_{\text{RPA}}(q)}$$

The summation of these two term is defined as the RPA self-energy. This is frequently used to evaluate the self-energy of the electron (see Lundqvist, 1969). We shall discuss this result in Sec. 5.8. The exchange energy is logically included as the first term in the series shown in Fig. 5.3(b). The RPA self-energy, defined above, is also called the *screened exchange energy*. This self-energy has one feature which is worth mentioning now, although we shall show it much later in Sec. 5.8: The inverse effective mass no longer diverges at the Fermi surface.

Since the exchange energy has been calculated previously, we shall just evaluate $\Sigma^{(3b)}$ in (5.1.18). The frequency summation is evaluated, at zero temperature, by changing it into a continuous integral, as described in Sec. 3.5. At $T \to 0$ the points $\omega_n = 2\pi n k_B T$ become closer together, and in the limit the summation becomes an integral:

$$\lim_{T\to 0} \frac{1}{\beta}\sum_{iq_n} \frac{\mathscr{G}^{(0)}(\mathbf{k}+\mathbf{q}, ik_n + iq_n)}{\varepsilon_{\text{RPA}}(q, iq_n)} = \int_{-\infty}^{\infty} \frac{d\omega}{2\pi} \frac{\mathscr{G}^{(0)}(\mathbf{k}+\mathbf{q}, ik_n + i\omega)}{\varepsilon_{\text{RPA}}(q, i\omega)}$$

$$\Sigma^{(3b)}(\mathbf{k}, ik_n) = -\frac{(4\pi e^2)^2}{(2\pi)^4} \int \frac{d^3q}{q^4} \times \int \frac{d\omega}{\varepsilon_{\text{RPA}}(\mathbf{q}, i\omega)} \frac{P^{(1)}(\mathbf{q}, i\omega)}{ik_n + i\omega - \xi_{\mathbf{k}+\mathbf{q}}} \tag{5.1.21}$$

It is very convenient to evaluate the integral this way, since $\varepsilon_{\text{RPA}}(q, i\omega)$ is a real function of its arguments q and ω.

The polarization operator ($\nu \to \infty$)

$$P^{(1)}(\mathbf{q}, i\omega) = \frac{2}{(2\pi)^3} \int d^3k \frac{n_F(\xi_{\mathbf{k}}) - n_F(\xi_{\mathbf{k}+\mathbf{q}})}{i\omega + \xi_{\mathbf{k}} - \xi_{\mathbf{k}+\mathbf{q}}}$$

is evaluated by a direct integration. The variable change $\mathbf{k} + \mathbf{q} \to \mathbf{k}$ in the second term reduces the integral to the form at $T = 0$ ($\nu = \cos\theta$):

$$P^{(1)}(\mathbf{q}, i\omega) = -\frac{2}{(2\pi)^2}\int_0^{k_F} k^2\,dk \int_{-1}^{1} d\nu \times\left[\frac{1}{\varepsilon_q + (qk\nu/m) + i\omega} + \frac{1}{\varepsilon_q + (qk\nu/m) - i\omega}\right]$$

$$= -\frac{m}{q2\pi^2}\int_0^{k_F} k\,dk \times\left\{\ln\left[\frac{\varepsilon_q + (qk/m) + i\omega}{\varepsilon_q - (qk/m) + i\omega}\right] + \ln\left[\frac{\varepsilon_q + (qk/m) - i\omega}{\varepsilon_q - (qk/m) - i\omega}\right]\right\}$$

$$= -\frac{mk_F}{2\pi^2}\left\{1 + \frac{m^2}{2k_Fq^3}[4E_F\varepsilon_q - (\varepsilon_q + i\omega)^2]\ln\left[\frac{\varepsilon_q + qv_F + i\omega}{\varepsilon_q - qv_F + i\omega}\right] + \frac{m^2}{2k_Fq^3}[4E_F\varepsilon_q - (\varepsilon_q - i\omega)^2]\ln\left[\frac{\varepsilon_q + qv_F - i\omega}{\varepsilon_q - qv_F - i\omega}\right]\right\} \quad (5.1.22)$$

This result for $P^{(1)}(\mathbf{q}, i\omega)$ appears to be complex. Actually, it is a real function of $\mathbf{q}$ and ω. This can be seen immediately by taking the complex conjugate of this result, which yields the same expression. It is also useful to write out the expression in terms of real quantities:

$$P^{(1)}(\mathbf{q}, i\omega) = -\frac{mk_F}{2\pi^2}\left\{1 + \frac{m^2}{2k_Fq^3}(4E_F\varepsilon_q - \varepsilon_q^{\,2} + \omega^2)\ln\left[\frac{(\varepsilon_q + qv_F)^2 + \omega^2}{(\varepsilon_q - qv_F)^2 + \omega^2}\right] + \frac{\omega}{qv_F}\left[\tan^{-1}\left(\frac{\omega}{\varepsilon_q + qv_F}\right) - \tan^{-1}\left(\frac{\omega}{\varepsilon_q - qv_F}\right)\right]\right\} \quad (5.1.23)$$

Another feature of this result is that it is an even function of frequency. The RPA dielectric function $\varepsilon_{\text{RPA}}(q, i\omega) = 1 - v_q P^{(1)}(\mathbf{q}, i\omega)$ also has these features: It is real and a symmetric function of frequency.

We usually do not wish to evaluate (5.1.21) but rather the retarded function obtained by setting $ik_n \to \xi_k + i\delta$. Here we must exercise care in the order in which we do operations. In doing the summations over Matsubara frequencies in Sec. 3.5, it was important to evaluate all these summations *before* performing any analytical continuations such as replacing $ik_n \to \xi + i\delta$. In the present calculation, the frequency integral is the same as the summation over Matsubara frequencies—it is just being done at zero temperatures. Thus it should be done before the analytical continuation. Unfortunately the $d\omega$ integral has an argument which is complicated, so that the integral will need to be done numerically. Thus we shall learn how to interchange the orders of these two operations: the analytical continuation and the integration over frequencies. This is done by expressing the final retarded function as two terms:

$$\Sigma^{(3b)}(k, \xi_k) = \Sigma^{(\text{line})}(k, \xi_k) + \Sigma^{(\text{res})}(k, \xi_k) \tag{5.1.24}$$

The notations $\Sigma^{(\text{line})}$ and $\Sigma^{(\text{res})}$ follow that of Quinn and Ferrell (1958). The first term $\Sigma^{(\text{line})}$ is what is obtained if one could interchange the two steps of analytical continuation and contour integration. It is just

$$\Sigma^{(\text{line})}(k, \xi_k) = -\frac{(4\pi e^2)^2}{(2\pi)^4} \int \frac{d^3q}{q^4} \int_{-\infty}^{\infty} \frac{d\omega}{\varepsilon_{\text{RPA}}(q, i\omega)} \frac{P^{(1)}(q, i\omega)}{i\omega + \xi_k - \xi_{\mathbf{k}+\mathbf{q}}} \tag{5.1.25}$$

which is what one gets by taking (5.1.21) and making the replacement $ik_n \to \xi_k + i\delta$. This function $\Sigma^{(\text{line})}$ will be shown below to be totally real. It could not be the entire self-energy, or else the electron would suffer no damping. The second term $\Sigma^{(\text{res})}$ is defined as the difference between the exact retarded self-energy $\Sigma_k^{(3b)}(k, \xi_k)$ and $\Sigma^{(\text{line})}$. A simple definition of $\Sigma^{(\text{res})}$ is that it is the error one makes in interchanging the orders of the two steps: analytical continuation and frequency summations. To determine its value, examine the frequency integral:

$$I(ik_n) = \int_{-\infty}^{\infty} \frac{d\omega}{2\pi} f(i\omega)\left(\frac{1}{i\omega + \xi_{\mathbf{k}} - \xi_{\mathbf{k}+\mathbf{q}}} - \frac{1}{i\omega + ik_n - \xi_{\mathbf{k}+\mathbf{q}}}\right)$$

$$f(i\omega) = \frac{P^{(1)}}{\varepsilon_{\text{RPA}}}$$

This integral is similar to the argument of (5.1.21) and (5.1.25). The integral $I(ik_n)$ is just what is needed to find $\Sigma^{(\text{res})}$. If the $d\omega$ integral is done and then the analytic continuation $ik_n \to \xi_k$, then in terms of the retarded function I_R we have

$$\Sigma^{(\text{res})}(k, \xi_k) = +\frac{(4\pi e^2)^2}{(2\pi)^4} \int \frac{d^3q}{q^4} I_R(\xi_k)$$

The integral $I(ik_n)$ is evaluated by a contour integration at which one uses only the residues at the poles of the factor $[(i\omega + \xi_{\mathbf{k}} - \xi_{\mathbf{k}+\mathbf{q}})^{-1} - (i\omega + ik_n - \xi_{\mathbf{k}+\mathbf{q}})^{-1}]$. The poles of the function $f(i\omega)$, or its branch cuts, give no contribution to $I_R(\xi_k)$. For example, consider that $f(i\omega)$ had a simple pole at $\omega = \omega_j$ of residue R_j:

$$f(i\omega) = \frac{R_j}{\omega - \omega_j}$$

$$I(ik_n) = \int_{-\infty}^{\infty} \frac{d\omega}{2\pi i} \frac{R_j}{\omega - \omega_j} \left[\frac{1}{\omega - i(\xi_{\mathbf{k}} - \xi_{\mathbf{k+q}})} - \frac{1}{\omega + k_n + i\xi_{\mathbf{k+q}}} \right]$$

Then the $d\omega$ integral can be done by completing the contour as a semicircle in the upper half-plane at infinity. If ω_j is in the upper half-plane, the integral is

$$I(ik_n) = R_j \left[\frac{1}{\omega_j - i(\xi_{\mathbf{k}} - \xi_{\mathbf{k+q}})} - \frac{1}{\omega_j + k_n + i\xi_{\mathbf{k+q}}} \right] + \frac{\theta(\xi_{\mathbf{k}} - \xi_{\mathbf{k+q}}) R_j}{i(\xi_{\mathbf{k}} - \xi_{\mathbf{k+q}}) - \omega_j} - \frac{\theta(-\xi_{\mathbf{k+q}}) R_j}{-k_n - i\xi_{\mathbf{k+q}} - \omega_j}$$

However, the term from the pole of $f(i\omega)$ vanishes when we make the analytical continuation $ik_n \to \xi_{\mathbf{k}} + i\delta$:

$$I_R(\xi_k) = \frac{R_j}{i(\xi_{\mathbf{k}} - \xi_{\mathbf{k+q}}) - \omega_j} [\theta(\xi_{\mathbf{k}} - \xi_{\mathbf{k+q}}) - \theta(-\xi_{\mathbf{k+q}})]$$

The factors $\theta(x)$ are step functions which are unity if $x > 0$. They arise in the contour integration because the poles at $i(\xi_{\mathbf{k}} - \xi_{\mathbf{k+q}})$ and $-k_n - i\xi_{\mathbf{k+q}}$ are only in the upper half-plane if they are satisfied. The same result is obtained if the contour is closed in the lower half plane. One can generalize this derivation to any form of $f(i\omega)$, with the following result:

$$\begin{aligned} I_R(\xi_k) &= f(\xi_{\mathbf{k+q}} - \xi_{\mathbf{k}})[\theta(\xi_{\mathbf{k}} - \xi_{\mathbf{k+q}}) - \theta(-\xi_{\mathbf{k+q}})] \\ \Sigma^{(\mathrm{res})}(k, \xi_k) &= + \frac{(4\pi e^2)^2}{(2\pi)^3} \int \frac{d^3q}{q^4} [\theta(\xi_{\mathbf{k}} - \xi_{\mathbf{k+q}}) - \theta(-\xi_{\mathbf{k+q}})] \\ &\quad \times \frac{P^{(1)}(q, \xi_{\mathbf{k+q}} - \xi_{\mathbf{k}})}{\varepsilon_{\mathrm{RPA}}(q, \xi_{\mathbf{k+q}} - \xi_{\mathbf{k}})} \end{aligned} \tag{5.1.26}$$

This is the desired result for $\Sigma^{(\mathrm{res})}$. The combination of step functions $\theta(\xi_{\mathbf{k}} - \xi_{\mathbf{k+q}}) - \theta(-\xi_{\mathbf{k+q}})$ equals the following: plus one whenever $\xi_{\mathbf{k}} > \xi_{\mathbf{k+q}} > 0$, minus one whenever $\xi_{\mathbf{k}} < \xi_{\mathbf{k+q}} < 0$, and zero otherwise. Our present interest is in evaluating this on the Fermi surface, which is $\xi_{\mathbf{k}} = 0$. The term $\Sigma^{(\mathrm{res})}$ is zero in this case, so that we can ignore this term for the further discussion of the correlation energy or ground state energy. This term is important in the later discussions of other properties, such as the effective mass.

For the energy of an electron on the Fermi surface, we need to evaluate $\Sigma^{(\text{line})}$ in (5.1.25). The first integral to be evaluated is the angular integral,

$$\int_{-1}^{1} d\nu \frac{1}{i\omega - \varepsilon_q - \nu q v_F} = \frac{1}{q v_F} \ln\left(\frac{i\omega - \varepsilon_q + q v_F}{i\omega - \varepsilon_q - q v_F}\right)$$

which gives

$$\Sigma^{(3b)}(k_F) = -\frac{2e^2 m}{\pi k_F} \int_0^\infty \frac{dq}{q^3} \int_{-\infty}^{\infty} \frac{d\omega}{\varepsilon_{\text{RPA}}(q, i\omega)} P^{(1)}(q, i\omega)$$
$$\times \ln\left(\frac{i\omega - \varepsilon_q + q v_F}{i\omega - \varepsilon_q - q v_F}\right)$$

The last logarithm may be written as

$$\ln\left(\frac{i\omega - \varepsilon_q + q v_F}{i\omega - \varepsilon_q - q v_F}\right) = \frac{1}{2} \ln\left(\frac{\omega^2 + (\varepsilon_q - q v_F)^2}{\omega^2 + (\varepsilon_q + q v_F)^2}\right) + i \tan^{-1}\left(\frac{\omega}{\varepsilon_q - q v_F}\right)$$
$$- i \tan^{-1}\left(\frac{\omega}{\varepsilon_q + q v_F}\right)$$

The complex terms vanish since they are odd functions of frequency, while the rest of the integrand is an even function of frequency:

$$\Sigma^{(3b)}(k_F) = -\frac{e^4 m}{\pi k_F} \int_0^\infty \frac{dq}{q^3} \int_{-\infty}^{\infty} d\omega \frac{P^{(1)}(q, i\omega)}{\varepsilon_{\text{RPA}}(q, i\omega)} \ln\left(\frac{\omega^2 + (\varepsilon_q - q v_F)^2}{\omega^2 + (\varepsilon_q + q v_F)^2}\right) \tag{5.1.27}$$

The right-hand side of this expression is now just a real function: The factor $\varepsilon_{\text{RPA}}(q, i\omega)$ is purely real, as mentioned earlier. The two integrals which remain in this expression are hard to do analytically. However, there is one result which may be deduced with little effort. This is the coefficient of the $\ln r_s$ term in the correlation energy. This term comes from integrating the region of small wave vector q. The integral is not divergent as $q \to 0$. However, this integration region must be handled carefully. As $q \to 0$, the polarization operator is

$$\lim_{q \to 0} P^{(1)}(q, i\omega) = -\frac{m k_F}{\pi^2} \left[1 - \left(\frac{\omega}{q v_F}\right) \tan^{-1}\left(\frac{q v_F}{\omega}\right)\right]$$

and the logarithm term is

$$\lim_{q \to 0} \ln\left[\frac{\omega^2 + (\varepsilon_q - q v_F)^2}{\omega^2 + (\varepsilon_q + q v_F)^2}\right] = -\frac{4(q/2k_F)}{(\omega/q v_F)^2 + 1} + O(q^3)$$

We redefine the variables $Z = q/2k_F$ and $\nu = \omega/qv_F$, and the integration over small wave vectors may be written as

$$\Sigma^{(3b)}(k_F) = -\frac{4}{\pi^3} E_{ry} \int_{-\infty}^{\infty} \frac{d\nu A(\nu)}{\nu^2 + 1} \int_0 \frac{dZ}{Z(1 + A\gamma/Z^2)}$$

$$A = 1 - \nu \tan^{-1}\left(\frac{1}{\nu}\right)$$

$$\gamma = \frac{1}{\pi(k_F a_0)} = \frac{r_s}{6.0293}$$

At small values of Z, this integral is

$$\int_0^{?} \frac{dZZ}{Z^2 + \gamma A} = \frac{1}{2} \ln(Z^2 + \gamma A)\Big|_0^{?} = -\frac{1}{2} \ln \gamma A = -\frac{1}{2} \ln r_s + \cdots$$

This produces the term $\ln r_s$. It arises from the long-wavelength part of the integration region. The diagram with just two Coulomb lines, in Fig. 5.2(c), is divergent in this limit. We summed a series of terms, which made the divergence vanish, and replaced it by this $\ln r_s$ contribution. Thus there is a nonanalytic dependence on r_s, which can be obtained only by summing an infinite set of diagrams. The coefficient of the $\ln r_s$ term is given by the integral

$$B = +\frac{2}{\pi^3} \int_{-\infty}^{\infty} \frac{d\nu}{\nu^2 + 1} \left[1 - \nu \tan^{-1}\left(\frac{1}{\nu}\right)\right] = +\frac{2}{\pi^2} (1 - \ln 2)$$

$$= +0.0622$$

The coefficient B was first found by Macke (1950). The constant term from the integrals in (5.1.27) may also be obtained.

H. High-Density Limit

The first two terms in the series for the ground state energy in the RPA are

$$E_{c,\mathrm{RPA}} = -0.142 + 0.0622 \ln r_s + O(r_s, r_s \ln r_s)$$

If we add the Onsager result for the second-order diagram $\Sigma^{(2a)}$ [see (5.1.16)], we obtain the result of Gell-Mann and Brueckner:

$$E_c = -0.094 + 0.0622 \ln r_s + \cdots$$

Another term in the series has been obtained by Carr and Maradudin (1964):

$$E_c = -0.094 + 0.0622 \ln r_s + 0.018 r_s \ln r_s + a r_s + O(r_s^2) \quad (5.1.28)$$

They were unable to obtain all the terms contributing to the coefficient a of the linear term in r_s. They obtained all its contributions except one, and they were able to put limits on its value.

Carr and Maradudin also discussed the convergence of this series at large values of r_s. Because of the term $\ln r_s$, this series is not analytic as $r_s \to 0$, which is the limit of very high densities. But the series is accurate in this limit. Metallic densities are roughly $1.8 < r_s < 6$. It is an important question whether the formula (5.1.28) is valid for these values of r_s. Figure 5.4 shows a plot of the correlation energy found by Carr and Maradudin. Only the terms shown in (5.1.28) have been plotted ($a = 0$). It diverges at small r_s, which is just the effect of the $\ln(r_s)$ term. It is negative up to values about $r_s \approx 2.5$, where it crosses the axis and becomes positive. Now there are simple arguments which indicate that the correlation energy must be negative. These will be given below in great detail. These arguments show that the perturbation formula for E_c is not valid in the range of metallic densities.

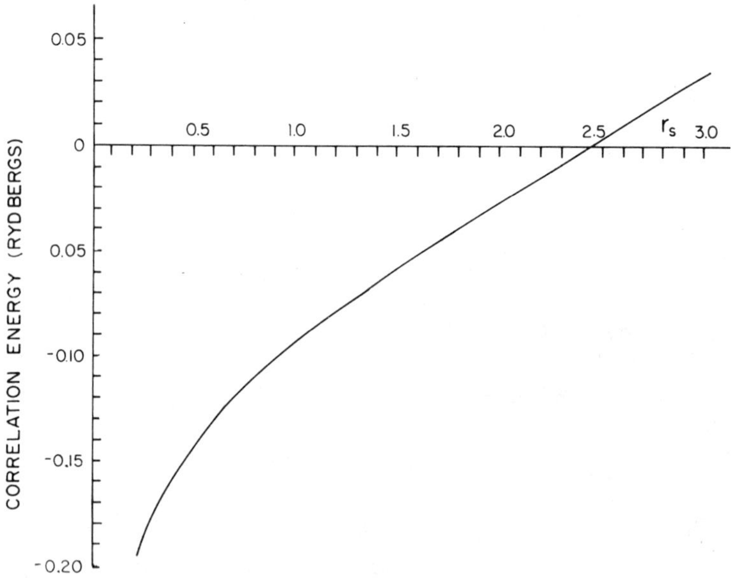

FIGURE 5.4. The correlation energy of the electron gas as given by the expansion in r_s in Eq. (5.1.28).

Large values of r_s correspond to the low-density electron gas. Wigner (1934) showed that the low-density electron gas behaved very differently from the high-density electron gas. At low density, the electrons become localized, and the calculation of the correlation energy becomes quite different. In the next section we shall find this low-density limit of Wigner and then shall discuss the extrapolation procedures between the low- and high-density limits. By these extrapolation procedures one can obtain a formula for the correlation energy which is valid in the region of metallic densities.

I. Pair Distribution Function

The correlation energy is the improvement in the ground state energy beyond the Hartree–Fock approximation. The correlation energy may be understood, at least conceptually, only after understanding Hartree–Fock. Thus first we shall examine an important property of the Hartree–Fock theory: the *pair distribution functions* $g(r)$, which was introduced in Sec. 1.6. The pair distribution function $g(r)$ is the probability that an electron is at $\mathbf{r}$ if there is already one at $r = 0$. It depends on spin, so there are two different pair distribution functions in an unmagnetized electron gas: $g_{\uparrow\uparrow}(r) = g_{\downarrow\downarrow}(r)$ and $g_{\uparrow\downarrow}(r)$. In the notation $g_{ss'}(r)$ the first spin index s is for the central electron, while the second s' is for the electrons at $\mathbf{r}$. Thus $g_{\downarrow\uparrow}(r)$ is the probability that a spin up electron is at $\mathbf{r}$ if a spin down electron is at $r = 0$. Of course, the electrons are not fixed, but they are usually moving rapidly. Thus these pair distribution functions are averages for the moving particles. Even the electron at $r = 0$ is not fixed, so that this reference point moves with the electron.

The pair distribution functions are rather easy to calculate in the Hartree–Fock approximation. Here the N-particle wave function is a Slater determinant,

$$\Psi_{\lambda 1 \cdots \lambda N}(r_1 \cdots r_N) = \frac{1}{(N!)^{1/2}} \begin{vmatrix} \phi_{\lambda 1}(r_1) & \phi_{\lambda 1}(r_2) & \cdots & \phi_{\lambda 1}(r_N) \\ \phi_{\lambda 2}(r_1) & \phi_{\lambda 2}(r_2) & \cdots & \phi_{\lambda 2}(r_N) \\ \vdots & & & \\ \phi_{\lambda N}(r_1) & & \cdots & \phi_{\lambda N}(r_N) \end{vmatrix} \tag{5.1.29}$$

where the λ_j are the quantum numbers which describe the states, and there is one wave function for every occupied electron state. The square of this many-particle wave function is the N-particle density matrix. The pair distribution function is given by the two-particle density matrix. This is obtained by integrating the N-particle density over all but two space coordinates and taking the similar expectation value over the $N - 2$ spin variables:

$$g(\mathbf{r}_1, \mathbf{r}_2) = \nu^2 \int d^3r_3 \cdots d^3r_N \mid \Psi_{\lambda 1 \cdots \lambda N}(r_1 \cdots r_N) \mid^2$$

If the one-electron orbitals $\phi_\lambda(r)$ are assumed orthogonal, then this integration just yields the sum over all possible pair wave functions:

$$g_{ss'}(\mathbf{r}_1, \mathbf{r}_2) = \frac{\nu^2}{N(N-1)} \sum_{\lambda_i \lambda_j} \begin{vmatrix} \phi_{\lambda i}(\mathbf{r}_1) & \phi_{\lambda i}(\mathbf{r}_2) \\ \phi_{\lambda j}(\mathbf{r}_1) & \phi_{\lambda j}(\mathbf{r}_2) \end{vmatrix}^2$$

The sum over $\lambda_i \lambda_j$ is over all occupied states, so each pair is summed twice. For the homogeneous electron gas, the orbitals must describe plane waves,

$$\phi_\lambda(\mathbf{r}) = \chi_s \frac{e^{i\mathbf{k} \cdot \mathbf{r}}}{\nu^{1/2}}$$

where the χ_s are the spin functions. The pair distribution function just describes the orbital part, so that one averages over spin functions. These spin averages are $\langle \chi_\uparrow \chi_\uparrow \rangle = 1$, $\langle \chi_\uparrow \chi_\downarrow \rangle = 0$, etc. Thus the two pair distribution functions are

$$g_{\uparrow\downarrow}(\mathbf{r}_1 - \mathbf{r}_2) = \frac{1}{n_0^2} \sum_{\mathbf{k}_1, \mathbf{k}_2} \left(\left| \frac{e^{i(\mathbf{k}_1 \cdot \mathbf{r}_1 + \mathbf{k}_2 \cdot \mathbf{r}_2)}}{\nu} \right|^2 + \left| \frac{e^{i(\mathbf{k}_1 \cdot \mathbf{r}_2 + \mathbf{k}_2 \cdot \mathbf{r}_1)}}{\nu} \right|^2 \right)$$

$$= \frac{2}{N^2} \left(\frac{N}{2} \right)^2 = \frac{1}{2}$$

$$g_{\uparrow\uparrow}(\mathbf{r}_1 - \mathbf{r}_2) = \frac{1}{n_0^2} \sum_{\mathbf{k}_1, \mathbf{k}_2} \left| \frac{e^{i(\mathbf{k}_1 \cdot \mathbf{r}_1 + \mathbf{k}_2 \cdot \mathbf{r}_2)}}{\nu} - \frac{e^{i(\mathbf{k}_1 \cdot \mathbf{r}_2 + \mathbf{k}_2 \cdot \mathbf{r}_1)}}{\nu} \right|^2$$

$$= \frac{2}{N^2} \sum_{\mathbf{k}_1, \mathbf{k}_2} (1 - e^{i(\mathbf{k}_1 - \mathbf{k}_2) \cdot (\mathbf{r}_1 - \mathbf{r}_2)}) = \frac{1}{2} [1 - \phi(\mathbf{r}_1 - \mathbf{r}_2)^2]$$

$$\phi(\mathbf{r}) = \frac{2}{N} \sum_{\mathbf{k}_1} e^{i\mathbf{k}_1 \cdot \mathbf{r}}$$

The antiparallel spin distribution function $g_{\uparrow\downarrow} = g_{\downarrow\uparrow} = \frac{1}{2}$ in the Hartree–Fock approximation. The cross term which results from the determinant is zero because of the orthogonality of the two spin functions. This means that there is no correlation in the position of electrons of opposite spin. The parallel spin $g_{\uparrow\uparrow} = g_{\downarrow\downarrow}$ has a definite spatial dependence. This results from the cross term which is retained since the spins are parallel and the

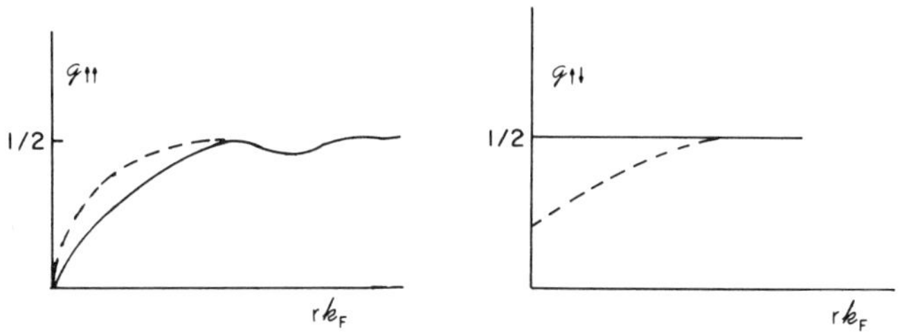

FIGURE 5.5. The pair distribution functions for parallel (left) and antiparallel (right) spins. The solid line is the Hartree–Fock approximation, while the dashed line includes correlation.

spin averages are unity. The function $\phi(r)$ is calculated to be

$$\phi(r) = \frac{2}{n_0}\int \frac{d^3k}{(2\pi)^3} n_k e^{i\mathbf{k}\cdot\mathbf{r}} = \frac{3}{rk_F{}^3}\int_0^{k_F} k\, dk \sin kr$$
$$= \frac{3}{(rk_F)^3}[\sin(rk_F) - (rk_F)\cos(rk_F)] = \frac{3}{rk_F} j_1(rk_F)$$

where $j_1(x)$ is the spherical Bessel function. These two pair distribution functions are shown as the solid lines in Fig. 5.5. The parallel spin $g_{\uparrow\uparrow}(r)$ vanishes at $r = 0$ and approaches $\frac{1}{2}$ at large distances. The pair distribution function $g_{\uparrow\uparrow}(r)$ must vanish at $r \rightarrow 0$ as a consequence of the exclusion principle for fermions. This feature is built into the many-particle wave function (5.1.29): If any pair of orbitals have the same spin and position, these two columns of the determinant are identical, and it must vanish. This is an important feature of Hartree–Fock wave functions.

The total pair distribution function for a particle is the combination of the result for parallel and antiparallel spin distributions:

$$g(r) = g_{\uparrow\uparrow}(r) + g_{\uparrow\downarrow}(r)$$

In a nonmagnetic system, the same result is obtained if the central particle has spin down. This pair distribution function has two features which are important for our discussion. First, there is the normalization integral:

$$n_0 \int d^3r[g(r) - 1] = -1 \tag{5.1.30}$$

The second is a statement about the ground state energy of the electron gas. We know that this has kinetic energy plus Coulomb parts. The Coulomb energy around the hole in the Hartree–Fock approximation is

$$E_{\text{Coul}} = e^2 n_0 \int d^3r \, \frac{g(r) - 1}{r} \tag{5.1.31}$$

The integral E_{Coul} is related to the exchange which is discussed in Sec. 5.6C.

The above two relations for $g(r)$ can be checked in the Hartree–Fock approximation. In this case $g(r) = \frac{1}{2} + g_{\uparrow\uparrow}(r)$. The normalization integral yields ($x = rk_F$)

$$n_0 \int d^3r \left[g_{\uparrow\uparrow}(r) - \frac{1}{2} \right] = -\frac{1}{2} n_0 \int d^3r \phi(r)^2 = -\frac{6}{\pi} \int_0^\infty dx j_1(x)^2 = -1$$

while the Coulomb integral yields

$$E_{\text{Coul}} = e^2 n_0 \int \frac{d^3r}{r} \left[g_{\uparrow\uparrow}(r) - \frac{1}{2} \right] = \frac{e^2 n_0}{2} \int d^3r \, \frac{\phi(r)^2}{r}$$

$$= -\frac{6}{\pi} e^2 k_F \int_0^\infty dx \, \frac{j_1(x)^2}{x} = -\frac{3}{4} \frac{e^2 k_F}{\pi}$$

These are the correct results for Hartree–Fock. This is the same exchange energy per particle which was derived by diagrammatic means.

The pair distribution function $g(r)$ is the distribution of electrons, on the average, about any electron. It goes to unity at large distances, which is a result of the uniform distribution n_0 of electron charge. That is, the electron charge density is $-en_0 g(r)$. Since $g(r)$ is less than unity near $r \approx 0$, the electron charge is depleted in the vicinity of the electron. This reduction may be viewed as a hole in the electron density. Wigner suggested the name *exchange* and *correlation hole*. This hole moves with the electron.

According to the normalization (5.1.30), the total charge missing from this hole is one electron charge. The factor $-n_0(g - 1)$ is the density of hole charge, and this integrates to 1. This is really a statement of charge neutrality. The hole has a positive charge density, since it is the absence of electrons. The integral of this charge density must be e. Thus each electron of charge $-e$ has in its hole the amount of charge e, and the system is neutral. The homogeneous electron gas has a uniform charge density only on the average. To a particular electron, the system is not uniform at all, since other electrons are not as likely to venture near to it as they are to other points.

The Coulomb integral gives the potential energy of the system. The electron interacts with its own hole charge. It is not influenced by the electrons or uniform charge density farther away, since they cancel on the average. Near the electron the positive background charge is not canceled by the other electrons, since the other electrons are not as likely to be nearby. In the Hartree–Fock approximation, only electrons of parallel spin make the exchange and correlation hole. Since $g_{\uparrow\downarrow} = \frac{1}{2}$, the antiparallel spins are not affected.

It is possible to arrange $g(r)$ so that the sum rule (5.1.30) is still obeyed, yet the potential energy is lower (i.e., a larger negative value) than found in Hartree–Fock. This is done by permitting $g_{\uparrow\downarrow}(r)$ to become less than $\frac{1}{2}$ near the point $r \approx 0$, as shown by the dashed line in Fig. 5.5. There is no particular reason it must go to zero. The only fixed rule is that it may not be negative, since it is a probability. Since the total hole charge must be unity, this change in $g_{\uparrow\downarrow}(r)$ must be accompanied by a change in $g_{\uparrow\uparrow}(r)$ which causes it to rise more steeply with r, as shown by the dashed line in Fig. 5.5. It is these changes which give rise to the correlation energy. Changes in $g_{\uparrow\downarrow}$ and $g_{\uparrow\uparrow}$ will usually cost some kinetic energy, so that one cannot just adjust $g_{\uparrow\downarrow}$ and $g_{\uparrow\uparrow}$ to maximize the potential energy alone. The system will seek the lowest energy state at $T = 0$, and this is the state with some "correlation" between the motion of electrons with antiparallel spins. This "correlation" between the motion of pairs of electrons is what gives rise to the "correlation energy."

5.2. WIGNER LATTICE AND METALLIC HYDROGEN

Actual electron systems exist over a range of densities. The range in metals is roughly $1.8 < r_s < 6$. But electron densities inside atoms are also quite variable. Thus it is necessary to have a description of the correlation energy which is valid for a wide range of densities. In the previous section, a formula was derived for the high-density limit of an homogeneous electron gas. Next one must derive the result for the low-density limit. The final formulas are found by interpolating between these extremes. This procedure was first carried through by Wigner (1934), and his formula‡

$$E_c = -\frac{0.88}{r_s + 7.8} E_{ry} \tag{5.2.1}$$

is still widely used. Recent interpolation formulas are not obviously better.

‡ Wigner's original paper contained an error. The corrected result is given.

At low densities, Wigner speculated that the electrons would become localized and form a regular lattice. We are still using the model that the positive charge is uniformly spread through the system. The electron lattice would presumably be a close-packed structure such as bcc, fcc, or hcp, in which electrons would vibrate around their equilibrium positions. There would be vibrational modes of the electrons, and these would be at the plasmon frequency. At very low density the potential energy becomes more important than the kinetic energy. The kinetic energy is in the zero-point motion of the vibrational modes. Localization cannot occur until this is less than the potential energy.

Wigner calculated the potential energy in the following fashion. A Wigner–Seitz model was taken for the unit cell of the lattice. This is a sphere of radius $r_s a_0$, with the electron at the center. Each sphere has overall neutrality, since the one-electron charge at the center is canceled by the positive charge inside the volume of the sphere. For a sphere of radius $r_s a_0$, this is one unit of charge, since that was the definition of r_s. Outside of each sphere, the electric field is zero. If all unit cells are spheres, then they exert no electric fields on each other. The electric fields inside a sphere arise only from the electron and positive charge within that sphere. In the Wigner–Seitz model, the spheres exert no electrical forces on each other. Of course this is only an approximate model, since the unit cells are not truly spheres—spheres cannot be packed together to cover all volume. We shall show that the error made by this approximation is remarkably small.

The first term is the potential energy between the electron and the uniform positive background. This we write as

$$E_{\text{ep}} = \int d^3r\left(-\frac{e^2}{r}\right)n_0 = -\frac{3e^2}{r_s^3 a_0^3}\int_0^{r_s a_0} r\,dr = -\frac{3e^2}{2a_0 r_s} = -\frac{3}{r_s}\left(\frac{e^2}{2a_0}\right)$$

where $-e^2/r$ is the potential energy and $n_0 = 3/(4\pi r_s^3 a_0^3)$ is the density of positive charge at each distance r. This gives $-3/r_s$ in atomic units, which is a large energy term.

The second term in the potential energy is the interaction of the positive charge with itself. The potential energy $V(r)$ from the positive charge at a distance r from the center is obtained by solving first for its equivalent electric field,

$$-\frac{\partial V}{\partial r} = +eE(r) = +\frac{e^2}{r^2}n_0\left(\frac{4\pi r^3}{3}\right) = +\frac{e^2 r}{r_s^3 a_0^3}$$

which is e^2/r^2 times the total charge within the sphere of radius r. When this is integrated to obtain $V(r)$, we get a constant of integration. This constant is determined by the condition that the total potential, from electron and positive charge, must vanish at the surface of the sphere. Thus the constant is chosen to make $V(r)$ be $e^2/r_s a_0$ at the surface. The result is

$$V(r) = \frac{e^2}{2r_s a_0}\left[3 - \left(\frac{r}{a_0 r_s}\right)^2\right] = \frac{1}{r_s}\left[3 - \left(\frac{r}{a_0 r_s}\right)^2\right]\left(\frac{e^2}{2a_0}\right)$$

The potential energy of the positive charge interacting with itself is found using

$$E_{\text{pp}} = \frac{1}{2}\int d^3r\, V(r) n_0 = \frac{3}{4}\,\frac{e^2}{(r_s a_0)^4}\int_0^{r_s a_0} r^2\, dr\left[3 - \left(\frac{r}{a_0 r_s}\right)^2\right]$$
$$= \frac{3}{5}\,\frac{e^2}{r_s a_0} = \frac{6}{5r_s}\left(\frac{e^2}{2a_0}\right)$$

where we multiply by $\frac{1}{2}$ because this is a self-energy. In atomic units this is $1.2/r_s$. These are the only two potential energy terms. The interaction of the electron with itself is not included. Aside from the fact that it is infinity, it does not change in the metal and so does not contribute to the cohesive energy of the system.

The sum of these two terms is $-1.8/r_s$. This is the total potential energy in the Wigner lattice in the Wigner–Seitz approximation. This is a large term; e.g., it has a larger coefficient than the exchange and correlation energies which were found for the free-particle system. The system apparently has gained energy by the localization of the electrons.

The actual energy of several Wigner lattices was calculated by Sholl (1967) using Madelung summation methods. This is the energy of a lattice of point charges in a uniform positive background. He expressed his results as $-A/r_s$, where the parameter A was found for several lattices:

Lattice	A
sc	1.760
fcc	1.79175
bcc	1.79186
hcp	1.79168

The sc lattice has a unit cell which is not very spherical, and its A is different. But the other lattices are more close-packed, and the coefficient 1.792 is remarkably close to the Wigner–Seitz value of 1.8. The latter approximation errs by less than $\frac{1}{2}\%$.

Is the Wigner lattice stable? To answer that question, first consider the potential energy on an individual electron in the vicinity of its equilibrium point. In the Wigner–Seitz model, this is just the negative of the potential energy from the positive charge:

$$V_{ec}(r) = -V(r) = -\frac{1}{r_s}\left[3 - \left(\frac{r}{a_0 r_s}\right)^2\right]E_{ry}$$

At $r = 0$ this is our prior result $-3/r_s$. Away from $r = 0$ the potentials increase quadratically. If each electron moved independently of the others, then the electron would have harmonic vibrations about the equilibrium point. The zero-point motion of these vibrations is another term in the ground state energy, and includes the kinetic energy as well as some potential energy. The zero point energy was evaluated by Wigner, who found that its contribution toward the ground state energy was proportional to $r_s^{-3/2}$. At large r_s this term becomes smaller than the potential energy term, which falls off as r_s^{-1}. He concluded the lattice was stable at sufficiently large values of r_s. Of course the particles do not move independently. Their movements form collective vibrational modes. A better test of lattice stability is to calculate all the phonon modes and show that their frequencies are all real—so the lattice is stable for collective motions. This was done by Carr (1961). A summary of the theory of the Wigner lattice was given by Care and March (1975).

In the limit of low density, as $r_s \to \infty$, the potential energy of the Wigner lattice is proportional to $-1.792/r_s$. This is the total Coulomb energy, which is correlation plus exchange. If the exchange energy $-0.9163/r_s$ is subtracted from this result, we are left with the low-density limit of the correlation energy:

$$\lim_{r_s \to \infty} E_c = -\frac{0.8757}{r_s}$$

This is rounded off to 0.88. Wigner also estimated that the high-density limit of the correlation energy was

$$\lim_{r_s \to 0} E_c = -0.113$$

This estimate is now known to be incorrect, because of the $\ln r_s$ term which causes a divergence in E_c at small values of r_s. The final formula (5.2.1) is a simple interpolation scheme which satisfies these two limits. This formula is plotted in Fig. 5.6 as the line marked W.

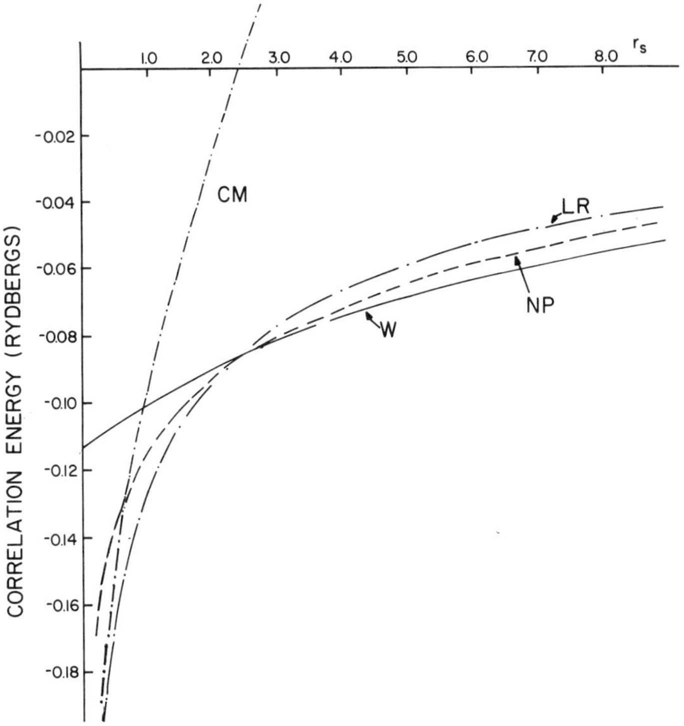

FIGURE 5.6. The correlation energy of the electron gas. The line CM (Carr–Maradudin) is the same small r_s expansion given in Fig. 5.4. The other curves are interpolation schemes according to Wigner (W), Nozieres and Pines (NP), and Lindgren and Rosen (LR).

The interpolation procedure assumes that the correlation energy is a smooth function between the limits of high and low density. This would be reasonable if the system were in the same state in both limits. But the two limits describe systems in different phases. The high-density system has free-electron states, and the low-density system has localized electrons. The smooth interpolation through a phase boundary is not obviously correct—or incorrect.

Other interpolation schemes have been proposed. One by Nozieres and Pines (1958) is

$$E_c = -0.115 + 0.031 \ln r_s$$

This is shown in Fig. 5.6 as the curve marked NP. It is quite similar to Wigner's result in the range of metallic densities.

Another interpolation formula has been proposed by Lindgren and Rosen (1970) (LR),

$$E_c = -\left[r_s + 3 + 4(r_s)^{1/2} - \frac{0.08}{(r_s)^{1/2}}\right]^{-1}$$

which is also shown in Fig. 5.6. It was chosen to agree with Carr and Maradudin at low values of r_s. Although the LR formula does not have an $\ln r_s$ term, it does agree well with the exact high-density expansion at low values of r_s. In the range of metallic densities, it has a lower (in magnitude) correlation energy than the previous two formulas.

There have been other calculations of the correlation energies at metallic densities. These are usually based on (5.1.21) or the equivalent (5.1.27). The quantity $\varepsilon_{\text{RPA}}(q, i\omega)$ is replaced by a better dielectric function —one which is more accurate at low densities. This leads to an improved correlation energy. Even a direct evaluation of (5.1.27) leads to a reasonable correlation energy for metallic densities (Lundqvist, 1969). There is no need to evaluate only the low r_s expansion of this result, as we did before. One can numerically solve it for all values of r_s. Some of these improved dielectric functions are discussed in Sec. 5.4.

Metallic Hydrogen

It has not been possible to study the three-dimensional Wigner lattice in a laboratory. There is another system, which is superficially quite similar, for which one could consider attempting experiments. This is metallic hydrogen. If one takes the Wigner lattice and changes the sign of all the charges, one has the model for metallic hydrogen. The protons are well approximated as point charges which are embedded in a free-electron gas. Since all other monovalent atoms form metals, perhaps hydrogen may also be made metallic.

Of course, at ordinary pressures, hydrogen does not form a metal. It is the gas H_2 at room temperatures. The lowering of temperature turns it into a liquid and then a solid. In both condensed phases the molecule H_2 retains its identity, so that the solid is a molecular crystal and an insulator. It may be possible to make hydrogen a metal by the application of pressure. Reports of success at high pressures are by Grigor'ev *et al.* (1972) and by Hawke *et al.* (1978). MacDonald and Burgess (1982) suggest that the high–pressure phase is a liquid metal.

Our discussion of metallic hydrogen will adopt the approximation that the electrons are a uniform electron gas. Better calculations show that the electron density is slightly nonuniform, since the electron density is higher near the proton than at the edge of the unit cell. Nevertheless, the assumption of uniformity makes only a small error in the analysis of the ground state energy. For a homogeneous electron gas, the electronic contribution to the ground state energy is our familiar result

$$E_g = \frac{2.2099}{r_s^2} - \frac{0.9163}{r_s} + E_c(r_s) + E_H$$

where we should use one of the formulas for the correlation energy E_c. The first three terms for the ground state energy are for the case where the positive charge is spread uniformly throughout the system. In metallic hydrogen, the positive charge is a regular lattice of points. The extra Coulomb energy which results from the localization of the positive charge was calculated above for the Wigner lattice. Changing the sign of all the charges yields the same result: $-1.792/r_s = E_H$. The quantity E_H is the Coulomb energy calculated in the Hartree approximation, which neglects exchange or the existence of the correlation hole.

There are, however, some subtle aspects to this calculation worth mentioning. The interaction energy between the protons and the uniform electron gas is calculated as before, e.g., also includes the interactions among the protons, but omits the proton interacting with itself. The electron–electron interactions are evaluated assuming a constant charge density n_0. Consider the Wigner–Seitz model of a spherical unit cell. Although there is one unit of electron charge in this unit cell, this is not treated as a single electron—then the electron–electron interactions would be an electron interacting with itself, which should not be included. Instead, each electron is spread uniformly throughout the solid. In the free-electron model, the unit cell has one charge because 10^{23} electrons each contribute 10^{-23} of their charge. Then the calculation of electron–electron interactions has a negligible contribution from one electron interacting with itself. This picture is changed by the concept of the exchange and correlation hole. The electrons are not charges which are uniformly spread throughout the material. They are points which, on the average, may be found with uniform probability anywhere. But if one electron is at a point, the others are not, on the average, within its exchange and correlation hole. However, we already include this energy effect in the correlation energy. Once it is included, the lattice energy from the protons is found by assuming the electron density is uniform.

The net result of this is that the energy from electron–electron interactions is small. In the Wigner–Seitz spherical model, the total energy from electron–electron interactions is

$$E_{\mathrm{ee}} = E_x + E_c + \frac{1.2}{r_s} = \frac{0.2837}{r_s} + E_c(r_s) \tag{5.2.2}$$

This is the sum of exchange, correlation, and the $1.2/r_s$ which comes from the uniform density of electrons interacting with themselves in the spherical unit cell. The term $0.28/r_s$ is small and positive, while E_c is small and negative. They largely cancel, so the net electron–electron energy is small. This means that the concept of the exchange–correlation hole is correct. If we put one electron in a unit cell, then others will not likely be there because of this hole. And if other electrons are not in the cell, there are no electron–electron interactions. Thus it is easy to understand why the net electron–electron interaction is small.

The ground state energy for metallic hydrogen is

$$\begin{aligned} E_g &= \frac{2.2099}{r_s^{\,2}} - \frac{1}{r_s}(0.916 + 1.792) + E_c(r_s) \\ &= \frac{2.210}{r_s^{\,2}} - \frac{2.708}{r_s} + E_c(r_s) \end{aligned}$$

$E_g(r_s)$ must be minimized, with respect to r_s, to obtain the predicted density of metallic hydrogen; the predicted value of r_s is defined as r_{s0}. The minimization will be done in the Hartree–Fock approximation by ignoring the correlation energy E_c, because the variation of $E_c(r_s)$ with respect to r_s is small at metallic densities, so the correlation energy would have little influence upon the choice of r_s. Thus we get the simple equation to minimize:

$$0 = \frac{d}{dr_s} E_{\mathrm{HF}} = \frac{d}{dr_s}\left(\frac{2.210}{r_s^{\,2}} - \frac{2.708}{r_s}\right) = \frac{-4.420}{r_{s0}^3} + \frac{2.708}{r_{s0}^2}$$

The minimum occurs at $r_{s0} = 1.632$, and the Hartree–Fock energy at this value is

$$E_{\mathrm{HF}}(r_{s0}) = \frac{2.210}{r_{s0}^2} - \frac{2.708}{r_{s0}} = -0.830 E_{ry}$$

The Wigner correlation energy at this value of r_{s0} is -0.093, so that the predicted ground state energy is $-0.923 E_{ry}$. This is the energy per electron. This binding energy is not even as large as atomic hydrogen, which is 1 Rydberg per electron. So far we have predicted that metallic hydrogen is not as bound as atomic hydrogen. Of course, it is possible to do a better calculation of E_g for metallic hydrogen, which includes the nonuniformity of the electron gas. The best result so far is $E_g = -1.048E_{ry}$ per atom at $r_{s0} = 1.60$ by Hammerberg and Ashcroft (1974). Although this is now more bound than atomic hydrogen, it is still not as bound as molecular hydrogen or solid molecular hydrogen which has $r_s = 3.3$. The conclusion is that hydrogen would rather be a molecule than a metal, and this conclusion is in accord with experiments at atmospheric pressure.

5.3. COHESIVE ENERGY OF METALS

The ground state energy of an electron gas is the binding energy of that ideal system. One experimental test of the theory is to compare with the binding energy of real metals. This comparison was first done carefully by Brooks (1958a, 1958b, 1963), who showed that the cohesive energy of simple metals could be explained by this theory. "Simple" metals are those without d electrons in the conduction band. The transition metals are not simple, since many of their properties are dominated by d electrons, and these are not described by our free-electron model. The quantity which is usually calculated is the cohesive energy, which is the difference in binding energy between the atom and the metal. For the atom, its relevant binding energy is the sum of the ionization potentials of the valence electrons which become part of the conduction band of the metal. This we define as E_I to be a positive number. The cohesive energy is

$$E_{\text{coh}} = -(E_g + E_I) \tag{5.3.1}$$

where we define E_{coh} to be positive, while E_g is negative. The ground state energy of the metal is calculated using the same type of energy terms we found for metallic hydrogen. For $Z = 1$ these are

$$\begin{aligned} E_g &= \frac{2.2099}{m_B r_s^2} + E_{\text{ee}} + E_{\text{i}} \\ E_{\text{ee}} &= \frac{0.2837}{r_s} + E_c(r_s) \end{aligned} \tag{5.3.2}$$

in which E_g includes a kinetic energy with an effective mass m_B, the electron–electron interactions defined in (5.2.2), and the electron–ion term E_{i}. The introductory discussion will be kept simple, so that we shall calculate the latter two quantities in the Wigner–Seitz model of a spherical unit cell. We shall also quote some of the results of Brooks, and the reader is encouraged to read his careful treatment.

The formulas in (5.3.2) apply to monovalent ions ($Z = 1$). For a higher valence, they must be altered to reflect the increased ion charge, which increases the Wigner–Seitz radius to $r_{\mathrm{WS}} = Z^{1/3}r_s$. For a Wigner–Seitz sphere with a point nucleus at the center, the total electrostatic energy from electron–electron interactions E_{ee} and electron–nucleus interactions E_{en} is obtained following the same arguments as in Sec. 5.2:

$$E_{\mathrm{ee}} = \frac{6}{5}\frac{Z^2}{r_{\mathrm{WS}}} E_{ry} = 1.2\frac{Z^{5/3}}{r_s} E_{ry}$$

$$E_{\mathrm{en}} = -\frac{3Z^2}{r_{\mathrm{WS}}} E_{ry}$$

These must be divided by Z to obtain the contribution per electron. The total electron–electron interaction energy, which includes the influence of correlation and exchange, is altered from (5.3.2) to the result

$$E_{\mathrm{ee}} = E_{\mathrm{ee}} + E_x + E_c = \frac{1}{r_s}(1.2Z^{2/3} - 0.9163) + E_c(r_s)$$

For the point nucleus, the quantity $E_{\mathrm{i}} = E_{\mathrm{en}}/Z$. For a finite nucleus, this is amended to

$$E_{\mathrm{i}} = -\frac{3Z}{r_{\mathrm{WS}}}(1 - \delta)E_{ry}$$

where the quantity δ accounts for the finite size of the nucleus.

The electron–ion term is the only contribution which poses difficulties. The other contributions are now familiar and can be evaluated using formulas which have already been derived. However, to discuss the electron–ion interaction, we need a description of the ionic potential which acts upon the electron. This has been calculated a variety of ways. For simple metals, one easy procedure is to construct a pseudopotential which gives the correct electron behavior for motion outside of the ion. This is usually done by fitting the potential to atomic energy levels. For example, the model potential of Heine and Abarenkov (1964, 1965) is shown in Fig. 5.7. The potential in atomic units (distances in Bohr radii, energies in Rydbergs) is

$$V_{\mathrm{i}}(r) = \begin{cases} A_l, & r < r_{\mathrm{i}} \\ -\dfrac{2Z}{r}, & r > r_{\mathrm{i}} \end{cases} \tag{5.3.3}$$

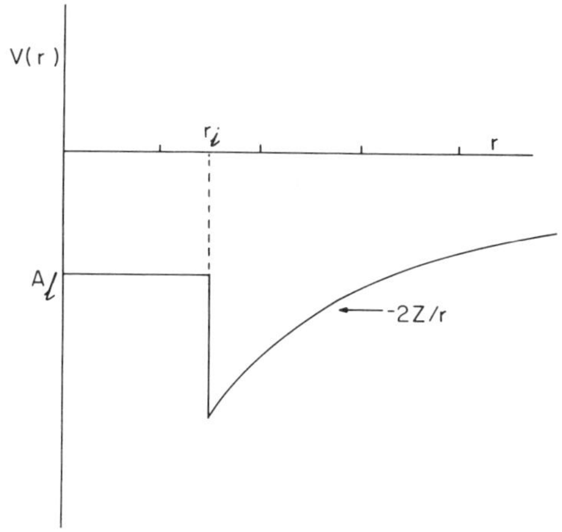

FIGURE 5.7. The ion pseudopotential according to Heine and Abarenkov.

The distance r_i is interpreted as the radius of the ion core. This is not a precise quantity, so that one may select r_i from a range of reasonable values. After a value of r_i is chosen, it remains fixed throughout the rest of the calculations. The charge of the ion is Z. The core parameter A_l is then varied to give the proper atomic energy states. It is found to depend on the angular momentum state l and also on the energy of the atomic state. For example, one finds $A_l(E_{ns})$ for a succession of states such as Mg^{2+}: $3S$, $4S$, $5S$, etc. For each eigenvalue, $A_l(E_{ns})$ has a different value. The reason the potential is useful is that the energy dependence is quite linear,

$$A_l(E) = A_l + (E - E_l)\left(\frac{dA}{dE}\right) \tag{5.3.4}$$

where A_l and dA/dE are both constants and E_l is the lowest atomic eigenvalue. This is shown in Fig. 5.8 for two different values of r_i. The ion core potential is described by just a small set of constants: r_i and the $[A_l, (dA/dE)_l]$ for each value of l. Some numerical results for S waves ($l = 0$) are shown in Table 5.1 (Mahan, 1974). Similar tabulations have been made by Animalu (1965), Harrison (1966), and Ashcroft (1968). The linear energy dependence is expected of pseudopotentials. Heine and Abarenkov tried several potential shapes before deciding that the simple constant core was superior. It permits an easy computation of the potential for an electron of any energy and orbital momentum.

The electron–ion interaction energy E_i has a simple physical interpretation. It is the energy of a particle at the bottom of the conduction band when calculated ignoring the effects of correlation and electron–

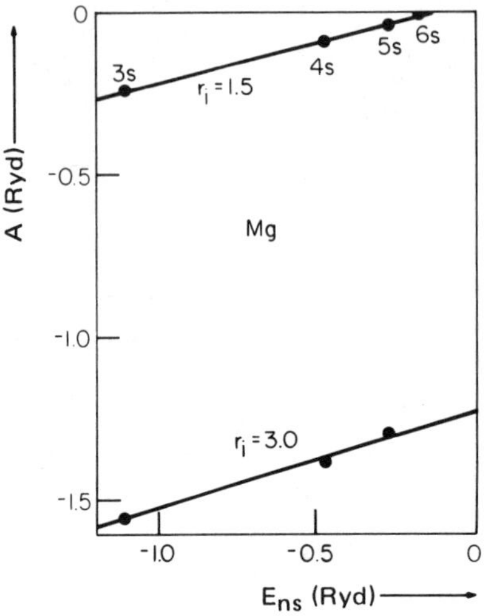

FIGURE 5.8. Pseudopotential core parameter A_0 for magnesium nS atomic states for two values of r_i. The straight-line fit is apparent. *Source*: Mahan (1974).

TABLE 5.1. Pseudopotentials

Z	Metal	E_{ns}	r_s	r_{WS}	r_j	A_0	dA/dE
1	Li	−0.3963	3.24	3.24	1.50	0.8648	0.0065
1	Na	−0.3777	3.96	3.96	1.80	+0.1357	0.386
1	K	−0.3190	4.96	4.96	2.30	+0.1328	0.701
					2.50	−0.1235	0.568
1	Rb	−0.3090	5.23	5.23	2.50	+0.0291	0.877
1	Cs	−0.2892	5.63	5.63	3.00	−0.1632	0.815
2	Be	−1.3385	1.87	2.356	1.00	2.0236	−0.169
2	Mg	−1.1051	2.66	3.351	1.50	−0.2453	0.238
2	Ca	−0.8725	3.26	4.107	1.70	3.2704	0.0505
					1.87	0.8607	0.27
2	Zn	−1.3203	2.30	2.898	1.20	−0.5952	0.952
2	Sr	−0.8107	3.54	4.460	2.14	+0.3263	0.464
					2.20	+0.0586	0.459
2	Cd	−1.2427	2.60	3.276	1.20	+0.4485	2.26
2	Ba	−0.7353	3.70	4.660	2.40	0.4310	0.614
					2.53	−0.0479	0.552
3	Al	−2.0909	2.07	2.985	1.20	0.6363	0.0213
3	Ga	−2.2572	2.18	3.144	1.10	+0.2789	0.701
					1.20	−1.2859	0.574
3	In	−2.0604	2.41	3.476	1.20	1.1593	1.45
					1.30	−0.5873	0.985
4	Sn	−2.9040	2.22	3.524	1.20	0.7907	0.840
4	Pb	−3.2775	2.30	3.651	1.20	−0.7386	1.242

TABLE 5.2. Electron–Ion Interactions

Metal	r_i	r_s	$A(E_1)$	δ	E_i'	$E_i{}^a$
Li	1.50	3.24	0.863	0.307	−0.642	−0.687
Na	1.80	3.96	0.046	0.212	−0.597	−0.609
K	2.30	4.96	0.016	0.218	−0.473	−0.486
Rb	2.50	5.23	−0.102	0.209	−0.454	−0.459
Cs	3.00	5.63	−0.271	0.207	−0.423	−0.421

[a] H. Brooks (1963).

electron interactions, i.e., all the terms we call E_{ee}. One may take an energy band calculation, such as that of Ham (1962) for the alkali metals, and just read off the band energy at the Γ point. Similar values of Brooks are shown in the last column of Table 5.2. These are the values we shall use in the cohesive energy calculation. Nearly the identical values are obtained if one solves the Wigner–Seitz model with the potential (5.3.3). Schrödinger's equation is solved in a sphere of radius r_s, with the condition that the wave function be normal to the surface of the sphere $(d\psi/dr)_{r_s} = 0$. Although this sounds crude compared to a full energy band calculation, the results are almost identical (Mahan, 1974). A model which is even less accurate is to assume that the electron density is spread uniformly throughout the sphere. This produces an approximate (classical) value which we call E_i':

$$E_i' = \frac{1}{Z}\int d^3 r n_0 V_i(r)$$

The integral is easily done for the model potential (5.3.3) and for $Z = 1$ gives

$$E_i' = \frac{3}{r_s^3}\left(A\int_0^{r_i} r^2\, dr - 2\int_{r_i}^{r_s} r\, dr\right) = \frac{-3}{r_s}(1-\delta)$$

$$\delta = \left(\frac{r_i}{r_s}\right)^2\left(1 + \frac{Ar_i}{3}\right)$$

The parameter δ expresses the effect of the finite size of the ion. It would be zero for a point ion, as in our calculation of metallic hydrogen, but is not for real ions in metals. Table 5.2 shows the calculated values of δ and E_i'. In each case the value of A was found self-consistently from

(5.3.4) for $E = E_i$. The E_i' are quite similar to the values found by Brooks. One concludes that the quantity E_i is mostly potential energy. For simple atomic states one has the virial theorem which states that the kinetic energy equals the binding energy, so that the potential energy is twice the binding energy. The theorem does not apply for Wigner–Seitz boundary conditions. Here the potential energy is nearly equal to the binding energy, so that the kinetic energy of electrons at the bottom of the conduction band is small in the pseudopotential model. It would obviously be much larger for real wave functions which included the oscillations in the core region.

For real metals, the kinetic energy term must reflect the fact that the effective electron mass may not be unity. The effective mass m_B is defined as one-half the inverse coefficient of the k^2 term in the electron energy, averaged over the Fermi sphere. It is influenced by band structure effects and particularly by the nonsphericity of the Fermi surface. The sodium Fermi surface is very spherical, but those for lithium and cesium are not. There are additional contributions to the effective mass of a particle from electron–electron interactions and also electron–phonon interactions. Neither of these enter the cohesive energy calculation. The electron–electron effects are included in the term E_{ee}. For example, we computed the wave vector dependence of the exchange energy and then averaged over the Fermi sphere to obtain (5.1.6). We obviously do not need to include the same wave vector dependence elsewhere in the calculation. For electron–phonon interactions, the effective mass is influenced only by electron states within a Debye energy of the Fermi surface, which is only a small fraction of the total electrons in the Fermi sphere. It has a negligible effect upon cohesion. Our values for the m_B were taken from Pines and Nozieres (1966).

Table 5.3 shows the theoretical cohesive energies compared with the experimental values. The agreement is not good for lithium but is fair for the other alkali metals. The numbers are very small, so that the metals are not strongly bound. Indeed, they are mechanically quite soft and at room temperature have the consistency of warm butter. The cohesive energy calculation is difficult because the final result is small. One must compute several large numbers and subtract them to get a small residue. Each large number must be found accurately for this procedure to work well. Of the four numbers which we have determined, certainly E_I is the most accurate, and E_i is probably second. The kinetic energy term is difficult because of the uncertainties in the band mass. The term E_{ee} is small, but it has an uncertainty larger than its value. We used Wigner's correlation energy, but choosing that of Nozieres and Pines will change the result $0.005E_{ry}$, which is the level of disagreement between theory and experiment.

TABLE 5.3. Cohesive Energies of Alkali Metals

Metal	r_s	E_1	m_B	$E_{\text{K.E.}}$	E_{ee}	E_I	E_{coh} (theory)	E_{coh} (exp.)
Li	3.24	−0.687	1.45	0.145	0.008	0.396	0.138	0.117
Na	3.96	−0.609	0.98	0.144	−0.003	0.378	0.090	0.084
K	4.86	−0.486	0.93	0.101	−0.011	0.319	0.077	0.073
Rb	5.23	−0.459	0.89	0.091	−0.013	0.307	0.075	0.061
Cs	5.63	−0.421	0.83	0.084	−0.015	0.286	0.066	0.061

$$E_{\text{K.E.}} = \frac{2.210}{r_s^2} \frac{1}{m_B}$$

$$E_{ee} = \frac{-0.916}{r_s} + \frac{1.2}{r_s} + E_c(r_s) \qquad \text{(Wigner)}$$

$$-E_{\text{coh}}\,(\text{theory}) = E_1 + E_{\text{K.E.}} + E_{ee} + E_I$$

The level of agreement between theory and experiment must be considered satisfactory, considering the uncertainties in the last two terms.

The smallness of the electron–electron interaction term confirms the notion about the exchange and correlation hole. The interactions are small because the electrons tend to avoid each other.

5.4. LINEAR SCREENING

Screening is one of the most important concepts in many-body theory. Charges, which are able, will move in response to an electric field. This charge movement will stabilize into a new distribution of charge around the electric field. This new distribution is just the right amount of charge to cancel the electric field at large distances. The proof of this is rather trivial. If the electric field is not canceled at large distances, more charge will still be attracted until it is sufficient for cancellation. If the electric field is caused by an impurity charge distribution $\varrho_i(\mathbf{r})$, with net charge $Q_i = \int d^3r\varrho_i(\mathbf{r})$, the amount of mobile charge attracted to the surroundings is exactly $-Q_i$. This fact was already used in discussing the Friedel sum rule in Sec. 4.1. The name *screening charge* is applied to the mobile charge attracted by the impurity electric field. It will also have its own distribution in space $\varrho_s(\mathbf{r})$. The *screened potential* from the impurity charge and the screening charge is given by

$$\varphi(\mathbf{r}) = \int d^3r' \, \frac{\varrho_i(\mathbf{r}') + \varrho_s(\mathbf{r}')}{|\mathbf{r} - \mathbf{r}'|} \tag{5.4.1}$$

This is an exact result, as long as $\varrho_s(\mathbf{r})$ is found exactly.

The screening charge is not necessarily in bound states due to the electric field from the impurity. Of course, this could happen if the electric field from the impurity is strong enough. But quite often the screening charge is from the unbound conduction electrons of the metal or semiconductor. In their motion through the crystal, they spend a little more time near the impurity potential, if it is attractive, than they do elsewhere in the solid. When these motions are averaged, there is more electron density near the impurity than elsewhere, and this is the screening charge. If the impurity potential is repulsive for electrons, they tend to spend less time near the impurity, so the average charge is depleted near the impurity. Here the screening charge is positive, since it signifies a reduction in the average density of electrons, which have negative charge.

The classical macroscopic theory is quite familiar. The electric field $\mathbf{E}$ and displacement field $\mathbf{D}$ obey the equations

$$\begin{aligned} \nabla \cdot \mathbf{D}(\mathbf{r}) &= 4\pi\varrho_i(\mathbf{r}) \\ \nabla \cdot \mathbf{E}(\mathbf{r}) &= 4\pi[\varrho_i(\mathbf{r}) + \varrho_s(\mathbf{r})] \end{aligned} \tag{5.4.2}$$

All equations are Fourier-transformed to give

$$\begin{aligned} i\mathbf{q} \cdot \mathbf{D}(\mathbf{q}) &= 4\pi\varrho_i(\mathbf{q}) \\ i\mathbf{q} \cdot \mathbf{E}(\mathbf{q}) &= 4\pi[\varrho_i(\mathbf{q}) + \varrho_s(\mathbf{q})] \end{aligned}$$

The components of $\mathbf{D}(\mathbf{q})$ and $\mathbf{E}(\mathbf{q})$ along the direction $\mathbf{q}$ are the longitudinal fields $D_l(\mathbf{q})$ and $E_l(\mathbf{q})$. The longitudinal electric field is related to the scalar potential $E_l(\mathbf{r}) = -\nabla\varphi(\mathbf{r})$ or its transform $\varphi(q) = iE_l(q)/q$:

$$\begin{aligned} D_l(q) &= \frac{4\pi}{iq} \varrho_i(q) \\ E_l(q) &= \frac{4\pi}{iq} [\varrho_i(q) + \varrho_s(q)] \\ \varphi(q) &= \frac{4\pi}{q^2} [\varrho_i(q) + \varrho_s(q)] \end{aligned}$$

The *dielectric response function* is defined as the ratio $D_l(q)/E_l(q)$ in the limit where $\varrho_i \to 0$:

$$\varepsilon(q) = \lim_{\varrho_i \to 0} \frac{D_l(q)}{E_l(q)} = \lim_{\varrho_i \to 0} \left[\frac{\varrho_i(q)}{\varrho_i(q) + \varrho_s(q)} \right] \tag{5.4.3}$$

In this limit $\varepsilon(q)$ becomes a property of the material and is independent of the charge distribution. One of our goals is to calculate this dielectric function. The linear screening model assumes this definition is true for finite $\varrho_i(q)$, which gives for the potential

$$\begin{aligned} \varphi'(q) &= \frac{4\pi}{q^2} \frac{\varrho_i(q)}{\varepsilon(q)} \\ \varphi'(\mathbf{r}) &= \int \frac{d^3q}{(2\pi)^3} \frac{4\pi}{q^2} \frac{\varrho_i(q)}{\varepsilon(q)} e^{i\mathbf{q}\cdot\mathbf{r}} \end{aligned} \tag{5.4.4}$$

The potential φ' is the total potential from screening charge and impurity charge. That is obvious from its definition, since $\mathbf{E}(\mathbf{r})$ is the electric field in (5.4.2) from both screening and impurity charges. The potential φ' should be similar to the exact screened potential φ in (5.4.1). They are identical in the limit where ϱ_i is small. The linear screening approximation is to calculate $\varphi'(r)$ in place of $\varphi(r)$. This is a much easier function to calculate, at least after $\varepsilon(q)$ has been determined. Another feature of the linear screening model is that the screening charge $\varrho_s(q)$ density is proportional, in q space, to the impurity charge density $\varrho_i(q)$. Linear screening models assume that

$$\varepsilon(q) = \frac{\varrho_i(q)}{\varrho_s(q) + \varrho_i(q)}$$

$$\frac{\varrho_s(q)}{\varrho_i(q)} = \frac{1}{\varepsilon(q)} - 1$$

are valid for finite values of ϱ_i, rather than infinitesimal ones.

A simple example will illustrate when linear screening does not apply. Consider a point charge $Z = -1$ in the homogeneous electron gas with high density ϱ_0. Here the screening charge is positive, which requires a depletion of the electron density around the impurity. This is sketched in Fig. 5.9(a). We suppose that the linear screening model works well for $Z = -1$, although that may not be a good assumption in metals. We shall now consider the response to a large charge of $Z = -10$. In the linear response theory, the screening charge is multiplied 10-fold, so that

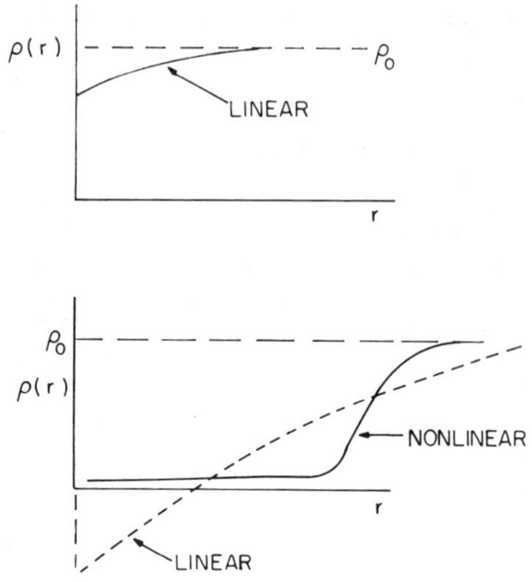

FIGURE 5.9. (Top) Change in electron charge density from linear screening (neglecting Friedel oscillations). (Bottom) A nonlinear screening model must be used for a much larger impurity charge.

it predicts the electron density $\varrho(r)$ to have the dashed line in Fig. 5.9(b). Of course, this is nonsense, since it predicts that the electron particle density is negative for small values of r. The actual electron density has the form shown by the solid line. In this case the screening charge is not just 10 times that for $Z = -1$.

The macroscopic theory defined the dielectric function $\varepsilon(q)$. But it did not provide a clue to the technique of finding it. This is the role of a microscopic theory. Our immediate goal is to derive a rigorous definition of $\varepsilon(q)$ in terms of microscopic operators. In the following section we shall show how this exact equation is solved, approximately, to give model dielectric functions which are used in calculations.

Our derivation of the exact equation for $\varepsilon(q)$ is done by considering the interaction between two impurity charges Z_1e and Z_2e in the homogeneous electron gas. The interaction potential between these two charges, in the linear screening model, is proportional to the product Z_1Z_2. We evaluate the ground state energy of the system and extract all the energy terms proportional to Z_1Z_2. The summation of all these terms is what is defined as the interaction potential between the two charges. Of course, there will also be terms proportional to Z_1^n or Z_2^n ($n = 1, 2, \ldots$), which are the energies needed to put each separate charge in by itself. The terms $Z_1^nZ_2^m$ for $m, n \geq 2$ are contributions to the nonlinear interaction. We shall ignore all terms except Z_1Z_2, since our immediate interest is the derivation of linear screening.

The Hamiltonian of the homogeneous electron gas, with two impurity charges Z_1e and Z_2e at $\mathbf{R}_1$ and $\mathbf{R}_2$, is written as

$$H = H_0 + \frac{Z_1Z_2e^2}{|\mathbf{R}_1 - \mathbf{R}_2|} - \frac{1}{\nu}\sum_{\mathbf{q}} v_q \varrho(\mathbf{q}) \sum_{j=1}^{2} Z_j e^{i\mathbf{q}\cdot\mathbf{R}_j}$$
$$\varrho(\mathbf{q}) = \sum_{p\sigma} C^\dagger_{\mathbf{p}+\mathbf{q},\sigma} C_{\mathbf{p},\sigma} \tag{5.4.5}$$

The first term H_0 is the Hamiltonian for the homogeneous electron gas (5.1.1). It includes the electron-electron interactions, as well as the kinetic energy. The term $e^2Z_1Z_2/|\mathbf{R}_1 - \mathbf{R}_2|$ is the direct interaction between the two charges. The last term in (5.4.5) is the interaction potential between each impurity charge Z_je and the electrons of the homogeneous electron gas. These are represented by their density operator $\varrho(\mathbf{q})$.

The ground state energy is calculated from the thermodynamic potential, which is found from the linked cluster theorems of (3.6.7)–(3.6.9):

$$\Omega = \Omega_0 - \frac{1}{\beta}\sum_{l=1}^{\infty} U_l$$
$$U_l = \frac{(-1)^l}{l}\int_0^\beta d\tau_1 \cdots \int_0^\beta d\tau_l \langle T_\tau V(\tau_1) \cdots V(\tau_l)\rangle_{\substack{\text{different}\\ \text{connected}}}$$

where the U_l are the different connected diagrams. We shall evaluate only the terms in this series which are proportional to Z_1Z_2. The unperturbed thermodynamic potential Ω_0 comes from H_0. At zero temperature, this is the ground state energy of the homogeneous electron gas, which was evaluated in Secs. 5.1 and 5.2. The last two terms in (5.4.5) are the interaction potential V which enters the perturbation expansion for the thermodynamic potential.

The term U_1 in the expansion has only one power of V. In this order, the only contribution proportional to Z_1Z_2 is from the direct interaction

$$U_1 = -\beta e^2 \frac{Z_1Z_2}{|\mathbf{R}_1 - \mathbf{R}_2|} = -\beta Z_1 Z_2 \int \frac{d^3q}{(2\pi)^3} v_q e^{i\mathbf{q}\cdot(\mathbf{R}_1 - \mathbf{R}_2)}$$

The other first-order term is zero, since the average of $\varrho(\mathbf{q})$ is zero in the electron gas unless $\mathbf{q} = 0$. Of course, these averages are taken with the Hamiltonian H_0, which is without the impurities present.

The term U_2 has two powers of V, and two powers of the last term in (5.4.5) give a contribution

$$U_2 = \frac{1}{2}\int_0^\beta d\tau_1 \int_0^\beta d\tau_2 \frac{1}{\nu^2}\sum_{\mathbf{q}\mathbf{q}'} v_q v_{q'} \langle T_\tau \varrho(\mathbf{q}, \tau_1)\varrho(\mathbf{q}', \tau_2)\rangle$$
$$\times (Z_1 e^{i\mathbf{q}\cdot\mathbf{R}_1} + Z_2 e^{i\mathbf{q}\cdot\mathbf{R}_2})(Z_1 e^{i\mathbf{q}'\cdot\mathbf{R}_1} + Z_2 e^{i\mathbf{q}'\cdot\mathbf{R}_2})$$

This is the only term in U_2; the other perturbation $e^2Z_1Z_2/|\mathbf{R}_1 - \mathbf{R}_2|$ does not enter again since it has only the one connected diagram. In U_2 there is a term proportional to Z_1^2 and another proportional to Z_2^2. They are part of the energy needed to put each charge, separately, into the homogeneous electron gas. This is not the total energy, since there are further terms proportional to $Z_j^n (n \geq 3)$ which also contribute. The term U_2 also has a cross term, which is proportional to Z_1Z_2:

$$U_2 = \frac{Z_1Z_2}{\nu^2} \sum_{\mathbf{qq}'} e^{i(\mathbf{q}\cdot\mathbf{R}_1+\mathbf{q}'\cdot\mathbf{R}_2)} v_q v_{q'} \int_0^\beta d\tau_1 \int_0^\beta d\tau_2 \langle T_\tau \varrho(\mathbf{q}, \tau_1)\varrho(\mathbf{q}', \tau_2)\rangle + \cdots$$

There are no other terms proportional to Z_1Z_2. The higher linked cluster terms U_l have only higher powers of the charges. Thus our derivation is completed. The net interaction between the two charges is the sum of the terms from U_1 and U_2. The U_2 term is simplified by the fact that in a homogeneous system it is finite only when $\mathbf{q} + \mathbf{q}' = 0$, since the density–density correlation function of the electron operators is nonzero only in this circumstance:

$$\Delta\Omega = Z_1Z_2 \int \frac{d^3q}{(2\pi)^3} v_q e^{i\mathbf{q}\cdot(\mathbf{R}_1-\mathbf{R}_2)} \times\left[1 - \frac{v_q}{\nu\beta}\int_0^\beta d\tau_1 \int_0^\beta d\tau_2 \langle T_\tau \varrho(\mathbf{q}, \tau_1)\varrho(-\mathbf{q}, \tau_2)\rangle\right]$$

This formula is compared with the linear screening model (5.4.4) for the potential from a charge distribution. One charge, say Z_1, is the impurity $\varrho_1(\mathbf{q}) = Z_1$. The other charge Z_2 is the test charge which measures the strength of the screened potential. Thus, the net interaction between the charges can be written as a screened coulomb interaction of the form

$$V'(\mathbf{R}_1 - \mathbf{R}_2) \equiv \Delta\Omega = Z_1Z_2 \int \frac{d^3q}{(2\pi)^3} \frac{v_q}{\varepsilon(q)} e^{i\mathbf{q}\cdot(\mathbf{R}_1-\mathbf{R}_2)}$$

This provides the rigorous definition of the dielectric function:

$$\frac{1}{\varepsilon(\mathbf{q})} = 1 - \frac{v_q}{\nu\beta}\int_0^\beta d\tau_1 \int_0^\beta d\tau_2 \langle T_\tau \varrho(\mathbf{q}, \tau_1)\varrho(-\mathbf{q}, \tau_2)\rangle$$

The correlation term can be further simplified. The correlation function depends only on the difference of the two arguments $\tau_1 - \tau_2$. This fact, along with the periodicity (3.1.6) of the argument, permits one of the τ

integrals to be eliminated to give

$$\int_0^\beta d\tau_1 \int_0^\beta d\tau_2 \langle T_\tau \varrho(\mathbf{q}, \tau_1)\varrho(-\mathbf{q}, \tau_2)\rangle = \beta \int_0^\beta d\tau \langle T_\tau \varrho(\mathbf{q}, \tau)\varrho(-\mathbf{q}, 0)\rangle$$

so the inverse dielectric function is

$$\frac{1}{\varepsilon(\mathbf{q})} = 1 - \frac{v_q}{\nu} \int_0^\beta d\tau \langle T_\tau \varrho(\mathbf{q}, \tau)\varrho(-\mathbf{q}, 0)\rangle \tag{5.4.6}$$

The exact result (5.4.6) is very important. It relates the dielectric function to the density–density correlation function. The time variation of the operators $\varrho(\mathbf{q}, \tau) = e^{\tau H_0}\varrho(\mathbf{q})e^{-\tau H_0}$ is governed by H_0, which is the full Hamiltonian for the homogeneous electron gas but without the potential of the impurities.

The static density–density correlation function $\langle \varrho(\mathbf{q})\varrho(-\mathbf{q})\rangle$ is related to the static structure factor $S(\mathbf{q})$, which was defined in Sec. 1.6:

$$\frac{1}{N} \langle \varrho(\mathbf{q})\varrho(-\mathbf{q})\rangle = N\delta_{\mathbf{q}=0} + S(\mathbf{q}) \tag{5.4.7}$$

From a knowledge of $S(\mathbf{q})$ we can obtain the pair distribution function $g(\mathbf{r})$ of the electron gas. The latter quantity is important in describing correlation in the electron gas. This raises the question of whether there is any relationship between $S(\mathbf{q})$ and $1/\varepsilon(\mathbf{q})$. In fact, there is. To show this, first we generalize (5.4.6) to finite values of frequency:

$$\frac{1}{\varepsilon(\mathbf{q}, i\omega)} = 1 - \frac{v_q}{\nu} \int_0^\beta d\tau e^{i\omega\tau} \langle T_\tau \varrho(\mathbf{q}, \tau)\varrho(-\mathbf{q}, 0)\rangle \tag{5.4.8}$$

This result is true, although we have not proved it here. Actually, we could just regard it as the definition of the longitudinal dielectric function $\varepsilon(q, i\omega)$. We wish to use (5.4.8) to prove the following important theorem:

$$S(\mathbf{q}) + N\delta_{\mathbf{q}=0} = -\frac{1}{n_0 v_q} \int_{-\infty}^{\infty} \frac{d\omega}{\pi} \frac{1}{1 - e^{-\beta\omega}} \operatorname{Im}\left[\frac{1}{\varepsilon(\mathbf{q}, \omega)}\right] \tag{5.4.9}$$

The function $\varepsilon(\mathbf{q}, \omega)$ is the retarded function obtained from $\varepsilon(\mathbf{q}, i\omega_n)$ by $i\omega_n \to \omega + i\delta$. The subscript "ret" will be omitted, since we always mean the retarded function when writing $\varepsilon(\mathbf{q}, \omega)$. The factor $n_0 = k_F^3/3\pi^2$ is the particle density, and $v_q = 4\pi e^2/q^2$.

To prove this theorem, we shall introduce the sets of states $| n\rangle$ and $| m\rangle$, which are exact eigenstates of the Hamiltonian. These states were used in Sec. 3.3 for proving several theorems. The Hamiltonian is for the homogeneous electron gas. These complete sets of states are introduced into (5.4.8):

$$\frac{1}{\varepsilon(\mathbf{q}, i\omega)} = 1 - \frac{v_q}{\nu} \int_0^\beta d\tau \sum_{n,m} e^{-\beta E_n} \,|\, \langle n \,|\, \varrho(\mathbf{q}) \,|\, m \rangle \,|^2\, e^{\tau(i\omega + E_n - E_m)}$$

This permits the τ integral to be done and the retarded function to be found in a formal manner:

$$\begin{aligned} \frac{1}{\varepsilon(\mathbf{q}, i\omega)} &= 1 - \frac{v_q}{\nu} \sum_{n,m} e^{-\beta E_n} \,|\, \langle n \,|\, \varrho(\mathbf{q}) \,|\, m \rangle \,|^2\, \frac{e^{\beta(E_n - E_m)} - 1}{i\omega + E_n - E_m} \\ \frac{1}{\varepsilon(\mathbf{q}, \omega)} &= 1 - \frac{v_q}{\nu} \sum_{n,m} e^{-\beta E_n} \,|\, \langle n \,|\, \varrho(\mathbf{q}) \,|\, m \rangle \,|^2\, \frac{e^{\beta(E_n - E_m)} - 1}{\omega + E_n - E_m + i\delta} \end{aligned} \tag{5.4.10}$$

The quantity on the right is real except for the $i\delta$ factor. Hence the imaginary part of this term just gives a delta function:

$$\begin{aligned} \operatorname{Im}\left[\frac{1}{\varepsilon(\mathbf{q}, \omega)}\right] &= \pi \frac{v_q}{\nu} \sum_{n,m} e^{-\beta E_n} \,|\, \langle n \,|\, \varrho(\mathbf{q}) \,|\, m \rangle \,|^2\, \delta(\omega + E_n - E_m) \\ &\quad \times (e^{\beta(E_n - E_m)} - 1) \\ &= -\pi(1 - e^{-\beta\omega}) \frac{v_q}{\nu} \sum_{n,m} e^{-\beta E_n} \,|\, \langle n \,|\, \varrho(\mathbf{q}) \,|\, m \rangle \,|^2 \\ &\quad \times \delta(\omega + E_n - E_m) \end{aligned} \tag{5.4.11}$$

We divide by $1 - e^{-\beta\omega}$ and integrate over all ω:

$$\begin{aligned} \int_{-\infty}^{\infty} \frac{d\omega}{\pi} \frac{1}{1 - e^{-\beta\omega}} \operatorname{Im}\left[\frac{1}{\varepsilon(\mathbf{q}, \omega)}\right] &= -\frac{v_q}{\nu} \sum_{n,m} e^{-\beta E_n} \,|\, \langle n \,|\, \varrho(\mathbf{q}) \,|\, m \rangle \,|^2 \\ &\equiv -\frac{v_q}{\nu} \langle \varrho(\mathbf{q}) \varrho(-\mathbf{q}) \rangle \end{aligned}$$

This term on the far right is the static density–density correlation function. This gives the static structure factor $S(\mathbf{q})$, as shown in (5.4.7). This completes the proof of (5.4.9). At zero temperature, this becomes ($\mathbf{q} \neq 0$)

$$S(\mathbf{q}) = -\frac{1}{n_0 v_q} \int_0^\infty \frac{d\omega}{\pi} \operatorname{Im}\left[\frac{1}{\varepsilon(\mathbf{q}, \omega)}\right] \tag{5.4.12}$$

which is the way it is often written. The pair distribution function $g(r)$

or the static structure factor $S(\mathbf{q})$ is obtained only from a knowledge of the frequency-dependent dielectric function $\varepsilon(\mathbf{q}, \omega)$. The latter theorem is not dependent on any assumptions regarding linear screening. It is exact. It arises because both $S(\mathbf{q})$ and $\varepsilon(q, \omega)$ are related to the density–density correlation function. The assumption of linear screening is merely using (5.4.4) to calculate the screened potential from the impurity charge distribution $\varrho_i(\mathbf{q})$.

The density–density correlation function

$$-\int_0^\beta d\tau e^{i\omega_n \tau} \langle T_\tau \varrho(\mathbf{q}, \tau) \varrho(-\mathbf{q}, 0) \rangle$$

has the appearance of a Green's function in the Matsubara representation. One has an operator $\varrho(-\mathbf{q}, 0)$ acting at $\tau = 0$ and its inverse at τ. Since the operator is the density, this provides the response of the system to a density fluctuation. To develop the analogy further, the function

$$S(\mathbf{q}, \omega) = -\frac{1}{n_0 v_q} \operatorname{Im}\left[\frac{1}{\varepsilon(\mathbf{q}, \omega)}\right] \tag{5.4.13}$$

is the spectral function of this operator, since it is proportional to the imaginary part of the retarded Green's functions associated with this correlation function. This is an important observation, since the spectral functions provide direct physical information. For the electron or the phonon, peaks in their spectral functions are interpreted as excitations of these operators. In a similar way, we interpret the peaks in $S(\mathbf{q}, \omega)$ as longitudinal excitations of the electron gas. These are two-particle excitations, since the density operator itself contains two operators—one creation and one destruction. In fact, some of the excitations we find, such as plasmons, are collective excitations of many particles.

The density operator has boson properties, since it is the product of two fermion operators. Thus $S(\mathbf{q}, \omega)$ is a spectral function for boson operators. Consequently, it has many features in common with other spectral functions for bosons; i.e., compare (5.4.9) with the similar phonon result (3.3.20):

$$2N_{\mathbf{q}} + 1 = \int_{-\infty}^{\infty} \frac{d\omega}{2\pi} n_B(\omega) B(\mathbf{q}, \omega)$$

Another feature of $S(\mathbf{q}, \omega)$ is that it is positive for $\omega > 0$ and negative for $\omega < 0$, with $S(\mathbf{q}, -\omega) = -S(\mathbf{q}, \omega)$. This may be shown directly from (5.4.11).

5.5. MODEL DIELECTRIC FUNCTIONS

No one has yet derived the exact dielectric function of the homogeneous electron gas. Instead, approximate solutions have been obtained to (5.4.6). Some of these have been very successful, perhaps because they are simple or perhaps because they are accurate. They acquire the name of their inventor(s). We shall describe four of them, which is only a small subset of the vast number which are available.

A. Thomas–Fermi

The Thomas (1927)–Fermi (1928) theory is also called the Fermi–Thomas theory about half the time. The derivation begins with the exact equation for the screened potential energy from an impurity charge distribution $\varrho_i(r)$,

$$\nabla^2 V(r) = 4\pi e[\varrho_i(r) + \varrho_s(r)]$$

where $\varrho_s(r)$ is the screening charge. We use the symbol ϱ for charge density, and $n(r)$ for particle density $\varrho = -en$, and $\varrho(q)$ for the density operator $\varrho(q) = \sum_{k\sigma} c^{\dagger}_{\mathbf{k}+\mathbf{q}\sigma} c_{\mathbf{k}\sigma}$. This equation can be obtained from (5.4.1) by the application of $-e\nabla^2$. The factor $-e$ arises because this is the potential energy for electrons. Now we start to make approximations. In Thomas–Fermi theory, the electron density $n(r)$ is represented locally as a free-particle system. Thus we write the screening charge as the difference between $n(r)$ and the equilibrium charge density n_0,

$$\varrho_s(r) = -e[n(r) - n_0]$$

where the electron charge is $-e$. For a free-particle system, the local density is $n(r) = k_F{}^3(r)/3\pi^2$, where the Fermi wave vector is now a local quantity. This, in turn, is determined by the condition that the chemical potential E_F is independent of position:

$$\frac{k_F{}^2(r)}{2m} = E_F(r) = E_F - V(r)$$

The physics is illustrated in Fig. 5.10. Assume the potential $V(r)$ is slowly varying in space. In each part of space we can draw a little box—as the shaded region—and treat it as a gas of fermions. If the absolute Fermi level is at E_F, then the effective Fermi level is reduced or raised by the value of the local potential V. If these approximations are collected, there

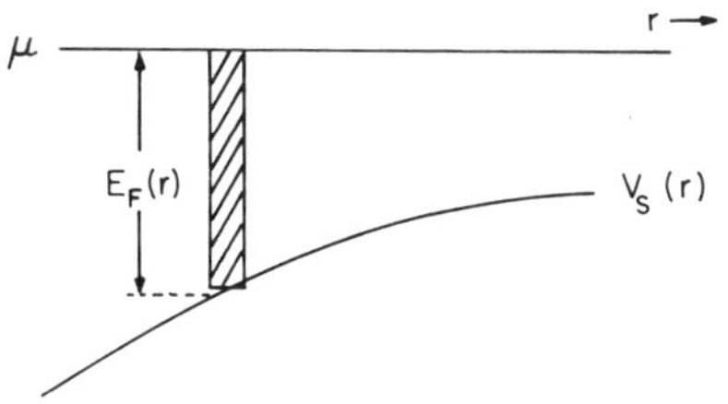

FIGURE 5.10. The Thomas–Fermi model has a Fermi degeneracy energy E_F which varies with position according to the variation in the potential energy.

results the equation

$$\nabla^2 V = +4\pi e\left\{\varrho_i(r) + en_0 - en_0\left[1 - \frac{V(r)}{E_F}\right]^{3/2}\right\}$$

For atoms, this approximate equation is solved exactly with $E_F = 0$ and $\varrho_i = Ze\delta^3(r)$ to give a good description of atomic potentials and charge distributions (Landau and Lifshitz, 1958). Thus the assumption that $V(r)$ is slowly varying does not seem unduly restrictive. To get a linear screening model, and hence a dielectric function, we need to make a further assumption. This is that $V/E_F \ll 1$, so we may expand the root $(1 - V/E_F)^{3/2} \simeq 1 - \frac{3}{2}V/E_F$ to obtain the equation

$$\nabla^2 V = +4\pi e\varrho_i(r) + \frac{6\pi e^2 n_0}{E_F} V(r)$$

The term on the right, proportional to V, is now moved to the left. Its coefficient is defined as the square of the Thomas–Fermi screening wave vector $q_{\rm TF}$:

$$\begin{aligned}(\nabla^2 - q_{\rm TF}^2)V(r) &= +4\pi e\varrho_i(r) \\ q_{\rm TF}^2 &= \frac{6\pi e^2 n_0}{E_F}\end{aligned} \qquad (5.5.1)$$

This equation may be solved in Fourier transform space to give

$$V(r) = -4\pi e \int \frac{d^3q}{(2\pi)^3} \frac{\varrho_i(q)}{q^2 + q_{\rm TF}^2} e^{i\mathbf{q}\cdot\mathbf{r}}$$

If we compare this with (5.4.4), we conclude that the Thomas–Fermi dielectric function is

$$\varepsilon(q) = 1 + \frac{q_{\rm TF}^2}{q^2} \qquad (5.5.2)$$

It has a simple form. This makes it easy to use in a variety of calculations. This is the chief explanation for its popularity.

For example, an analytical result can be obtained when the impurity is a point charge $\varrho_i(q) = Q_i$. Then we need to evaluate the integrals

$$\begin{aligned} V(r) &= -\frac{eQ_i}{\pi} \int_0^\infty \frac{q^2\,dq}{q^2 + q_{\mathrm{TF}}^2} \int_{-1}^{1} d\nu e^{iqr\nu} \\ &= -2\frac{eQ_i}{\pi r} \int_0^\infty \frac{q\,dq}{q^2 + q_{\mathrm{TF}}^2} \sin qr \\ &= -e\frac{Q_i}{i\pi r} \int_{-\infty}^{\infty} \frac{q\,dq e^{iqr}}{q^2 + q_{\mathrm{TF}}^2} \\ &= -\frac{eQ_i}{r} e^{-q_{\mathrm{TF}} r} \end{aligned}$$

The last integral is done by closing the integration contour in the upper half plane and picking up the pole at iq_{TF}. The screened interaction has the form of a Yukawa potential. The interaction declines rapidly at large distances because of the exponential dependence $\exp(-q_{\mathrm{TF}} r)$. In metals, the Thomas–Fermi wave vector has a typical value of 1 Å^{-1}. Thus the screened coulomb potential declines rapidly on the scale of a unit cell. The screening wave vector may be expressed in atomic units as

$$a_0 q_{\mathrm{TF}} = \left(\frac{4}{\pi} k_F a_0\right)^{1/2} = \frac{1.5632}{r_s^{1/2}}$$

For example, at sodium density $r_s = 3.96$, one finds $q_{\mathrm{TF}} = 1.48$ Å^{-1}.

Thomas–Fermi theory provides only a static model for $\varepsilon(q)$. It is not usually used to describe the dynamic response $\varepsilon(q, \omega)$.

B. Lindhard, or RPA

The Lindhard (1954) dielectric function is more commonly called the RPA, for random phase approximation. It is a model for a static $\varepsilon(q)$ or dynamic $\varepsilon(q, \omega)$ dielectric function. It was already introduced in Sec. 5.1 for the discussion of correlation energies. It is rather easy to derive and has a simple conceptual basis. It also predicts correctly a number of properties of the electron gas such as plasmons. In the early days of electron gas theory, it was *the* dielectric function. Nowadays there is a tendency to use one of the recent models, which are better for describing the response of the electron gas. Two derivations of the RPA will be presented: one from equations of motion and the second using Green's functions and diagrams.

The derivation by equations of motion is also called the method of *self-consistent field* (Ehrenreich and Cohen, 1959). One introduces an impurity charge density $\varrho_i(r, t)$ or its equivalent Fourier transform $\varrho_i(q, \omega)$. The equivalent impurity potential is $V_i(q, \omega) = -e\varphi_i(q, \omega)$. There are three potential energies in the problem. The first is from the impurity $V_i(q, \omega)$, the second is from the screening charge $V_s(q, \omega)$, and the third is the total potential, which is the sum of these two:

$$V(q, \omega) = V_i(q, \omega) + V_s(q, \omega)$$
$$\nabla^2 V_s(r, t) = 4\pi e \varrho_s(r, t) \quad \text{or} \quad V_s(q, \omega) = -\frac{4\pi e}{q^2}\varrho_s(q, \omega) \tag{5.5.3}$$
$$\nabla^2 V_i(r, t) = 4\pi e \varrho_i(r, t) \quad \text{or} \quad V_i(q, \omega) = -\frac{4\pi e}{q^2}\varrho_i(q, \omega)$$

The major assumption in the derivation is that the electrons respond to the total energy V. Thus when we solve the equations for the motion of the electron, we use the potential $V(r, t)$ or its transform $V(q, \omega)$. This presents a minor problem, since initially we do not know $V(q, \omega)$. That is the object of the calculation, and $V_i(q, \omega)$ is assumed known. Once $V(q, \omega)$ is known, the dielectric function in the linear screening model is

$$\varepsilon(q, \omega) = \frac{V_i(q, \omega)}{V(q, \omega)}$$

and the calculation is completed. In the method of self-consistent field, we assume that the electrons respond to V and try to determine this function self-consistently. Thus we write as the effective Hamiltonian for the electrons

$$H = \sum_{p\sigma} \varepsilon_p c^\dagger_{\mathbf{p}\sigma} c_{\mathbf{p}\sigma} + \frac{1}{\nu}\sum_{\mathbf{q}} V(\mathbf{q}, t)\varrho(\mathbf{q})$$
$$\varrho(\mathbf{q}) = \sum_{p,\sigma} c^\dagger_{\mathbf{p}+\mathbf{q},\sigma} c_{\mathbf{p},\sigma} \tag{5.5.4}$$

The time dependence $V(\mathbf{q}, t)$ is put directly into the Hamiltonian (5.5.4). The impurity charge is regarded as a classical system which is oscillating, and we are trying to find the quantum response of the electron gas to this classical oscillation. Furthermore, the impurity is assumed to oscillate at a single frequency: $\varrho_i(\mathbf{r}, t) = \varrho_i(\mathbf{r})e^{-i\omega t}$ and $V_i(\mathbf{r}, t) = V_i(\mathbf{r})e^{-i\omega t}$. But the average response of the system will depend on ω, so that we write the average of $\varrho(\mathbf{q}, t)$ as $\langle \varrho(\mathbf{q}, t)\rangle = \varrho(\mathbf{q}, \omega)e^{-i\omega t}$ and the average of $V(\mathbf{q}, t)$ as $\langle V(\mathbf{q}, t)\rangle = V(\mathbf{q}, \omega)e^{-i\omega t}$. For the homogeneous electron gas, the density operator $\varrho(\mathbf{q})$ has an expectation value of zero for $\mathbf{q} \neq 0$. When the impurity is present, the expectation value is nonzero and is proportional to the average for the screening charge:

$$\begin{aligned}\langle \varrho_s(\mathbf{q}, t)\rangle &= -e\langle \varrho(\mathbf{q}, t)\rangle = -e \sum_{p\sigma} \langle c^{\dagger}_{\mathbf{p}+\mathbf{q},\sigma} c_{\mathbf{p}\sigma}\rangle \\ &= -e\varrho(\mathbf{q}, \omega)e^{-i\omega t}\end{aligned}$$

Since the averages for ϱ_s and ϱ are proportional, it simplifies the discussion to use only one symbol, which we choose to be ϱ. For example, in terms of the average $\langle \varrho_s\rangle = -e\langle \varrho\rangle$, Eq. (5.5.3) is

$$V_s(\mathbf{q}, \omega) = \frac{4\pi e^2}{q^2} \varrho(\mathbf{q}, \omega) \tag{5.5.3a}$$

The dielectric response function is defined as the ratio

$$\varepsilon(\mathbf{q}, \omega) = \frac{V_i(\mathbf{q})}{V(\mathbf{q}, \omega)}$$

The first term in (5.5.4) is the kinetic energy of the electrons, and the second is their interaction with the self-consistent potential $V(\mathbf{q}, t)$. Note that there are no explicit electron–electron interactions. They are included, indirectly, in the interaction term: The part of $V(\mathbf{q}, t)$ from the screening $V_s(\mathbf{q}, t)$ is caused by electron–electron interactions. This is a rather crude way to include these interactions, since it neglects all effects of correlation and exchange. This is the major defect of the RPA, which is remedied by the subsequent models discussed below. Indeed, the main objective of other models has been to improve the RPA by including the effects of the exchange and correlation hole.

To obtain the screened potential V_s in (5.5.3), we need to derive an expression for the screening particle density $\varrho_s(\mathbf{q}, t)$. This is obtained by writing an equation of motion for this operator and then solving it approximately. In the homogeneous electron gas, the particle density operator $n(\mathbf{q}, t)$ has an expectation value of zero unless there is a perturbation of the system. A perturbation on the system of $(\mathbf{q}, \omega)$ will cause polarization of the electron system, so that the average $\langle \varrho(\mathbf{q}, t)\rangle$ will now have a finite value. In the linear screening model, we assume that this average is proportional to the potential causing the perturbation, so $\langle \varrho(\mathbf{q}, t)\rangle \propto \langle V(\mathbf{q}, t)\rangle$. Our goal is to determine the constant of proportionality.

The equation of motion of the density operator comes from

$$i \frac{d}{dt} \varrho(\mathbf{q}, t) = [H, \varrho(\mathbf{q}, t)]$$

Actually, it is more convenient to evaluate the equation of motion for the operator:

$$i \frac{d}{dt} c^{\dagger}_{\mathbf{p}+\mathbf{q},\sigma} c_{\mathbf{p},\sigma} = [H, c^{\dagger}_{\mathbf{p}+\mathbf{q},\sigma} c_{\mathbf{p},\sigma}]$$

After this has been solved, the result is summed over $(\mathbf{p}, \sigma)$ to obtain $n(\mathbf{q}, t)$. The impurity potential $V(\mathbf{q}, t)$ is assumed to be oscillating at a single frequency $\exp(-i\omega t)$ so that the time derivative on the left of (5.4.5) gives $-i\omega c^{\dagger}_{\mathbf{p}+\mathbf{q},\sigma} c_{\mathbf{p},\sigma}$. The commutators on the right are evaluated for the Hamiltonian (5.5.4),

$$\sum_{\mathbf{k},s} \varepsilon_k [c^{\dagger}_{\mathbf{k}s} c_{\mathbf{k}s}, c^{\dagger}_{\mathbf{p}+\mathbf{q}\sigma} c_{\mathbf{p}\sigma}] = (\varepsilon_{\mathbf{p}+\mathbf{q}} - \varepsilon_{\mathbf{p}}) c^{\dagger}_{\mathbf{p}+\mathbf{q}\sigma} c_{\mathbf{p}\sigma}$$

$$\frac{1}{\nu} \sum_{\mathbf{q}'\mathbf{k}s} V(\mathbf{q}', t) [c^{\dagger}_{\mathbf{k}+\mathbf{q}'s} c_{\mathbf{k}s}, c^{\dagger}_{\mathbf{p}+\mathbf{q}\sigma} c_{\mathbf{p}\sigma}]$$
$$= \frac{1}{\nu} \sum_{\mathbf{q}'} V(\mathbf{q}', t) (c^{\dagger}_{\mathbf{p}+\mathbf{q}+\mathbf{q}'\sigma} c_{\mathbf{p}\sigma} - c^{\dagger}_{\mathbf{p}+\mathbf{q},\sigma} c_{\mathbf{p}-\mathbf{q}',\sigma})$$

so that we obtain the equation

$$(\varepsilon_p - \varepsilon_{\mathbf{p}+\mathbf{q}} + \omega) c^{\dagger}_{\mathbf{p}+\mathbf{q}\sigma} c_{\mathbf{p}\sigma} = \frac{1}{\nu} \sum_{\mathbf{q}'} V(\mathbf{q}', t) (c^{\dagger}_{\mathbf{p}+\mathbf{q}+\mathbf{q}',\sigma} c_{\mathbf{p}\sigma} - c^{\dagger}_{\mathbf{p}+\mathbf{q},\sigma} c_{\mathbf{p}-\mathbf{q}',\sigma})$$
$$\simeq \frac{V(\mathbf{q}, t)}{\nu} (c^{\dagger}_{\mathbf{p}\sigma} c_{\mathbf{p}\sigma} - c^{\dagger}_{\mathbf{p}+\mathbf{q}\sigma} c_{\mathbf{p}+\mathbf{q}\sigma})$$

The last term on the right is approximated by taking only the term in the summation which has $\mathbf{q}' = -\mathbf{q}$. The terms with other values of $\mathbf{q}'$ are neglected. It is assumed they average out to zero. This is what is meant by the random phase approximation. The approximate equation can now be solved:

$$c^{\dagger}_{\mathbf{p}+\mathbf{q},\sigma} c_{\mathbf{p}\sigma} = \frac{V(\mathbf{q}, t)}{\nu} \left(\frac{c^{\dagger}_{\mathbf{p}\sigma} c_{\mathbf{p},\sigma} - c^{\dagger}_{\mathbf{p}+\mathbf{q}\sigma} c_{\mathbf{p}+\mathbf{q}\sigma}}{\varepsilon_{\mathbf{p}} - \varepsilon_{\mathbf{p}+\mathbf{q}} + \omega} \right)$$

The above equation is now summed over $(\mathbf{p}, \sigma)$ to give

$$\varrho(q, t) = \sum_{\mathbf{p}\sigma} c^{\dagger}_{\mathbf{p}+\mathbf{q},\sigma} c_{\mathbf{p},\sigma} = \frac{V(q, t)}{\nu} \sum_{\mathbf{p},\sigma} \frac{c^{\dagger}_{\mathbf{p}\sigma} c_{\mathbf{p}\sigma} - c^{\dagger}_{\mathbf{p}+\mathbf{q}\sigma} c_{\mathbf{p}+\mathbf{q}\sigma}}{\varepsilon_{\mathbf{p}} - \varepsilon_{\mathbf{p}+\mathbf{q}} + \omega}$$

This equation shows that the operator on the left, $\varrho(\mathbf{q}, t)$, is proportional to the operator on the right, $V(\mathbf{q}, t)$. The average is taken of this equation, so that $\langle \varrho \rangle$ and $\langle V \rangle$ are replaced by $\varrho(\mathbf{q}, \omega) e^{-i\omega t}$ and $V(\mathbf{q}, \omega) e^{-i\omega t}$. In addition, the number operators $c^\dagger_{\mathbf{p}\sigma}c_{\mathbf{p}\sigma}$ and $c^\dagger_{\mathbf{p}+\mathbf{q}\sigma}c^\dagger_{\mathbf{p}+\mathbf{q}\sigma}$ are replaced by their averages $n_F(\xi_\mathbf{p})$ and $n_F(\xi_{\mathbf{p}+\mathbf{q}})$. These steps give an equation which relates the average of these operators:

$$\varrho(\mathbf{q}, \omega) = \frac{V(\mathbf{q}, \omega)}{\nu} \sum_{\mathbf{p},\sigma} \frac{n_F(\xi_\mathbf{p}) - n_F(\xi_{\mathbf{p}+\mathbf{q}})}{\varepsilon_\mathbf{p} - \varepsilon_{\mathbf{p}+\mathbf{q}} + \omega} \equiv V(q, \omega)P^{(1)}(q, \omega)$$

This result can now be used in (5.5.3a):

$$V_s(\mathbf{q}, \omega) = \frac{4\pi e^2}{q^2} \varrho(\mathbf{q}, \omega) = V(\mathbf{q}, \omega)v_q P^{(1)}(\mathbf{q}, \omega)$$

The result has the screening particle density $\varrho(\mathbf{q}, \omega)$ proportional to the self-consistent potential $V(\mathbf{q}, \omega)$. The constant of proportionality is $P^{(1)}(\mathbf{q}, \omega)$, which was evaluated earlier in (5.1.22).

The equations may now be solved to obtain the dielectric function. In the equation for $V(\mathbf{q}, \omega)$ we substitute our new result for $V_s(q, \omega)$ and then solve for $V(\mathbf{q}, \omega)$ in terms of $V_i(\mathbf{q}, \omega)$:

$$\begin{aligned} V(\mathbf{q}, \omega) &= V_i(\mathbf{q}, \omega) + V_s(\mathbf{q}, \omega) = V_i(\mathbf{q}, \omega) + v_q P^{(1)} V(\mathbf{q}, \omega) \\ &= \frac{V_i(\mathbf{q}, \omega)}{1 - v_q P^{(1)}(\mathbf{q}, \omega)} \end{aligned}$$

The ratio of these two quantities is just the RPA dielectric function:

$$\begin{aligned} \varepsilon_{\mathrm{RPA}}(\mathbf{q}, \omega) &= 1 - v_q P^{(1)}(\mathbf{q}, \omega) \\ P^{(1)}(\mathbf{q}, \omega) &= \frac{1}{\nu} \sum_{\mathbf{q},\sigma} \frac{n_F(\xi_\mathbf{p}) - n_F(\xi_{\mathbf{p}+\mathbf{q}})}{\varepsilon_\mathbf{p} - \varepsilon_{\mathbf{p}+\mathbf{q}} - \omega - i\delta} \end{aligned}$$

This completes the derivation from the method of self-consistent fields. In Sec. 5.5D we shall return to this technique when deriving the Singwi–Sjölander dielectric function.

The second method of deriving $\varepsilon_{\mathrm{RPA}}$ is a diagrammatic analysis using Green's functions. The basic definition of $1/\varepsilon(q, i\omega)$ in (5.4.8) is rewritten in the interaction representation:

$$\frac{1}{\varepsilon(q, i\omega)} = 1 - \frac{v_q}{\nu} \int_0^\beta d\tau e^{i\omega\tau} \frac{\langle T_\tau \hat{S}(\beta)\hat{\varrho}(\mathbf{q}, \tau)\hat{\varrho}(-\mathbf{q}, 0)\rangle}{\langle S(\beta)\rangle}$$

$$V = \frac{1}{\nu} \sum_{\substack{\mathbf{kpq}' \\ \sigma\sigma'}} v_{\mathbf{q}'} c^\dagger_{\mathbf{p}+\mathbf{q}'\sigma} c^\dagger_{\mathbf{k}-\mathbf{q}'\sigma'} c_{\mathbf{k}\sigma'} c_{\mathbf{p}\sigma}$$

where H_0 is the kinetic energy term in the homogeneous electron gas and

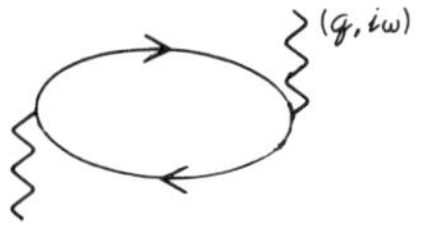

FIGURE 5.11 (q, iω)

V is the electron–electron interactions. The operator $\hat{\varrho}$ has its time dependence determined by H_0 in the interaction representation. The S matrix will be expanded term by term to see what sort of terms develop. One particular subset of these will be summed and will yield the RPA.

The first term in the expansion for the S matrix is just $P^{(1)}(\mathbf{q}, i\omega)$:

$$\begin{aligned}
P^{(1)}(\mathbf{q}, i\omega) &= -\frac{1}{\nu}\int_0^\beta d\tau e^{i\omega\tau}\langle T_\tau \hat{\varrho}(\mathbf{q}, \tau)\hat{\varrho}(-\mathbf{q}, 0)\rangle \\
&= -\frac{1}{\nu}\sum_{\substack{\mathbf{kp}\\ \sigma\sigma'}}\int_0^\beta d\tau e^{i\omega\tau}\langle T_\tau \hat{c}^\dagger_{\mathbf{p}+\mathbf{q}\sigma}(\tau)\hat{c}_{\mathbf{p}\sigma}(\tau)\hat{c}^\dagger_{\mathbf{k}-\mathbf{q}\sigma'}\hat{c}_{\mathbf{k}\sigma'}\rangle \\
&= \frac{1}{\nu}\sum_{\mathbf{p}\sigma}\int_0^\beta d\tau e^{i\omega\tau}\mathscr{G}^{(0)}(\mathbf{p}, \tau)\mathscr{G}^{(0)}(\mathbf{p}+\mathbf{q}, -\tau) \\
&= \frac{1}{\nu}\sum_{\mathbf{p}\sigma}\frac{n_F(\xi_\mathbf{p}) - n_F(\xi_{\mathbf{p}+\mathbf{q}})}{\varepsilon_\mathbf{p} - \varepsilon_{\mathbf{p}+\mathbf{q}} - i\omega}
\end{aligned}$$

This is shown as a diagram with a single-fermion closed loop in Fig. 5.11. The wiggly lines at each end are just added to define the two vertices of the polarization diagram. They could indicate that the polarization term is in response to an excitation with wave vector $\mathbf{q}$ and frequency $i\omega$. The calculation for ε_{RPA} is not terminated at this point, since so far we have shown that $1/\varepsilon = 1 + v_q P^{(1)} + \cdots$ rather than $\varepsilon_{\text{RPA}} = 1 - v_q P^{(1)}$. Obviously, more terms are needed to get to the RPA.

The next term in the S-matrix expansion is

$$\begin{aligned}
&\frac{1}{\nu}\sum_{\substack{\mathbf{kpq}'\\ \sigma\sigma'}}\int_0^\beta d\tau e^{i\omega\tau}\int_0^\beta d\tau_1 v_{q'} \\
&\qquad\times\langle T_\tau \hat{\varrho}(\mathbf{q}, \tau)\hat{c}^\dagger_{\mathbf{p}+\mathbf{q}'\sigma'}(\tau_1)\hat{c}^\dagger_{\mathbf{k}-\mathbf{q}'\sigma'}(\tau_1)\hat{c}_{\mathbf{k}\sigma'}(\tau_1)\hat{c}_{\mathbf{p}\sigma}(\tau_1)\hat{\varrho}(-\mathbf{q}, 0)\rangle
\end{aligned}$$

There are four terms which result when Wick's theorem is applied to this correlation function. All contributions have four electron Green's functions and one Coulomb interaction $v_{q'}$. Their diagrams are shown in Fig. 5.12. The first one is a vertex correction to the basic bubble diagram. The next two are exchange energy diagrams for the Green's functions in the bubble;

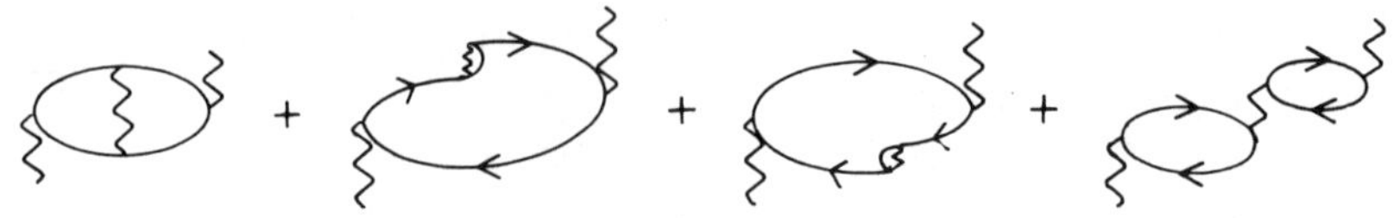

FIGURE 5.12

they contribute to the self-energy of these Green's functions. The last diagram contains two bubbles which are connected by the Coulomb line v_q.

We remarked earlier that the density–density correlation function had the appearance of a Green's function. It also has a Dyson equation. The exact evaluation of this correlation function may be written as

$$-\frac{1}{\nu}\int_0^\beta d\tau e^{i\omega\tau}\langle T_\tau \varrho(\mathbf{q},\tau)\varrho(-\mathbf{q},0)\rangle = \frac{P(\mathbf{q}, i\omega)}{1 - v_q P(\mathbf{q}, i\omega)}$$

Here the density operators $\varrho(q, \tau)$ have the τ dependence governed by $H = H_0 + V$, instead of only H_0 as in $P^{(1)}$. The polarization diagram $P(q, \omega)$ is the *summation* of all "different" polarization terms. Polarization diagrams are not "different" if any of their parts are linked by a single Coulomb line v_q. For example, the last diagram in Fig. 5.12 is not a different polarization diagram. This term arises from the expansion

$$\frac{P^{(1)} + P^{(2)} + \cdots}{1 - v_q(P^{(1)} + \cdots)} = (P^{(1)} + \cdots)[1 + v_q(P^{(1)} + \cdots)\cdots]$$
$$= P^{(1)} + v_q(P^{(1)})^2 + \cdots$$

where it is the term $v_q(P^{(1)})^2$. There are terms in $P(\mathbf{q}, i\omega)$ which have more than one bubble, but they must be connected by more than one Coulomb line.

The random phase approximation is approximating the exact polarization diagram $P(\mathbf{q}, \omega)$ by its first term, which is $P^{(1)}(\mathbf{q}, i\omega)$. Thus we write

$$\frac{1}{\varepsilon_{\text{RPA}}} = 1 + \frac{v_q P^{(1)}}{1 - v_q P^{(1)}} = \frac{1}{1 - v_q P^{(1)}}$$

which does give $\varepsilon_{\text{RPA}} = 1 - v_q P^{(1)}(\mathbf{q}, i\omega)$. This derivation makes clear the approximate nature of the RPA. The total polarization operator $P(\mathbf{q}, i\omega)$ has an infinite number of terms, while the RPA retains one. The exact dielectric function is easily shown to be

$$\varepsilon(\mathbf{q}, i\omega) = 1 - v_q P(\mathbf{q}, i\omega)$$

An obvious way to improve the dielectric function is to include more terms in the summation of polarization contributions (Geldart and Taylor, 1970). This is not as simple as it sounds. There are an infinite number of possibilities, so some physics must be used to guide the choice. Neither does it help that the obvious possibilities, such as the first diagram in Fig. 5.12, are not simple to evaluate analytically. In fact, most progress has been made by nondiagrammatic means, as will be discussed below.

The RPA dielectric function is evaluated once $P^{(1)}(\mathbf{q}, i\omega)$ is obtained. The wave vector integrals were evaluated earlier in (5.1.22) for the Matsubara function. The retarded function is obtained by taking the analytical continuation $i\omega \rightarrow \omega + i\delta$. The retarded dielectric function is complex, and its real and imaginary parts are called ε_1 and ε_2:

$$\varepsilon_{\text{RPA}}(\mathbf{q}, \omega) = \varepsilon_1(\mathbf{q}, \omega) + i\varepsilon_2(\mathbf{q}, \omega)$$

$$\varepsilon_1(\mathbf{q}, \omega) = 1 + \frac{1}{2}\frac{q_{\text{TF}}^2}{q^2}\left\{1 + \frac{m^2}{2k_F q^3}\,[4E_F\varepsilon_q - (\varepsilon_q + \omega)^2]\ln\left|\frac{\varepsilon_q + qv_F + \omega}{\varepsilon_q - qv_F + \omega}\right| \right.$$
$$\left. + \frac{m^2}{2k_F q^3}\,[4E_F\varepsilon_q - (\varepsilon_q - \omega)^2]\ln\left|\frac{\varepsilon_q + qv_F - \omega}{\varepsilon_q + qv_F + \omega}\right|\right\} \qquad (5.5.5)$$

For $q < 2k_F$

$$\begin{aligned} \varepsilon_2(\mathbf{q}, \omega) &= +2\omega\left(\frac{e^2 m^2}{q^3}\right), & qv_F - \varepsilon_q > \omega > 0 \\ &= +\left(\frac{e^2 m}{q^3}\right)\left[k_F^2 - \left(\frac{m}{q}\right)^2(\omega - \varepsilon_q)^2\right], & \varepsilon_q + qv_F > \omega > qv_F - \varepsilon_q \\ &= 0, & \omega > \varepsilon_q + qv_F \end{aligned} \qquad (5.5.6)$$

For $q > 2k_F$

$$\varepsilon_2(\mathbf{q}, \omega) = +\left(\frac{e^2 m}{q^3}\right)\left[k_F^2 - \left(\frac{m}{q}\right)^2(\omega - \varepsilon_q)^2\right], \qquad \varepsilon_q + qv_F \geq \omega \geq \varepsilon_q - qv_F$$

The real part $\varepsilon_1(\mathbf{q}, \omega)$ may be represented by a single formula. The imaginary part $\varepsilon_2(\mathbf{q}, \omega)$ has a variety of functional relations for different values of (q, ω).

Figure 5.13 shows graphs of ε_1, ε_2, and $-\text{Im}(1/\varepsilon_{\text{RPA}})$ for $r_s = 3$ and all relevant frequencies. For large values of q, say $q \gtrsim 2k_F$, one finds that $\varepsilon_1 \simeq 1$, ε_2 is small, so that $\varepsilon_2 \simeq -\text{Im}(1/\varepsilon)$. Thus there is only a single line drawn for these cases.

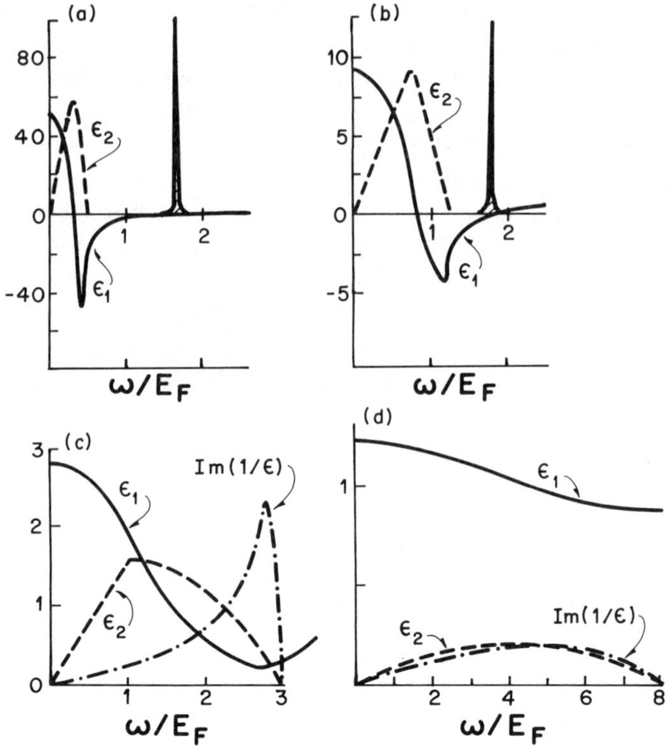

FIGURE 5.13. The RPA predictions regarding ε_1, ε_2, and $\mathrm{Im}(1/\varepsilon)$ as a function of q and ω. Results shown for $r_s = 3.0$.

The real part ε_1 always approaches 1 at large ω. At $\omega \to 0$ we have $\varepsilon_2 \to 0$, and the static $\varepsilon_{\mathrm{RPA}}$ is just $\varepsilon_1(\mathbf{q}, 0)(x = q/2k_F)$,

$$\begin{aligned}\varepsilon_{\mathrm{RPA}}(\mathbf{q}) &\equiv \varepsilon_{\mathrm{RPA}}(\mathbf{q}, \omega = 0)\\ &= 1 + \frac{1}{2}\,\frac{q_{\mathrm{TF}}^2}{q^2}\left[1 + \frac{1}{2x}(1 - x^2)\ln\left|\frac{1+x}{1-x}\right|\right]\end{aligned} \tag{5.5.7}$$

which is always positive. The prefactor to the bracket includes the square of the Thomas–Fermi wave vector q_{TF}^2. However, for values of $q \lesssim k_F$, $\varepsilon_1(q, \omega)$ becomes negative for intermediate values of ω. This requires two crossings of the $\varepsilon_1 = 0$ axis. The low-frequency crossing always happens when ε_2 is large, so that $-\mathrm{Im}(1/\varepsilon) = \varepsilon_2/(\varepsilon_1^2 + \varepsilon_2^2)$ is well behaved when $\varepsilon_1 \to 0$. However, the high-frequency point where $\varepsilon_1 = 0$ occurs has $\varepsilon_2 = 0$. In that case we interpret $\lim_{\varepsilon_2 \to 0} \varepsilon_2/(\varepsilon_1^2 + \varepsilon_2^2) = \pi\delta(\varepsilon_1)$, so that a delta function is obtained. This delta function is the plasmon peak which is the sharp singularity on the right of the graph. It is given a finite width in Fig. 5.13 to aid the eye. Remember that we interpret peaks in $S(\mathbf{q}, \omega)$ as excitations of the system

$$S(\mathbf{q}, \omega) = -\frac{1}{n_0 v_q}\,\mathrm{Im}\left[\frac{1}{\varepsilon(\mathbf{q}, \omega)}\right] = \frac{1}{n_0 v_q}\,\frac{\varepsilon_2}{\varepsilon_1^2 + \varepsilon_2^2} \tag{5.5.8}$$

Plasmons are excitations which exist in real metals and in any electron gas. They were discovered in ionized gases by Langmuir. They occur at small values of q, as is evident in Fig. 5.13. Thus we examine the limit of (5.5.5) when $q \to 0$ while ω remains finite. In particular, we expand $\varepsilon_1(\mathbf{q}, \omega)$ while assuming $\varepsilon_q/\omega \ll 1$, $qv_F/\omega \ll 1$ to get

$$\begin{aligned} \lim_{q\to 0} \varepsilon_1(\mathbf{q}, \omega) &= 1 - \frac{\omega_p^2}{\omega^2}\left\{1 + \frac{1}{\omega^2}\left[\frac{3}{5}(qv_F)^2 - \varepsilon_q^2\right] + O\left(\frac{1}{\omega^4}\right)\right\} \\ \omega_p^2 &= \frac{4\pi n_0 e^2}{m} \end{aligned} \tag{5.5.9}$$

In fact this calculation is tedious, since quite a few terms in the expansion of the logarithm must be retained to cancel the q^{-3} prefactors of the logarithm and the q^{-2} prefactor in the curly braces of (5.5.5). For $\varepsilon_2 = 0$ the condition that $\varepsilon_1 = 0$ predicts that the plasmon peak occurs at the frequency

$$\omega = \omega_p\left(1 + \frac{3}{10}\,\frac{q^2 v_F^2}{\omega_p^2} + \cdots\right) \tag{5.5.10}$$

At $q = 0$ the prediction is quite simple, $\omega = \omega_p$. The quantity ω_p is called the *plasma frequency* of the electron gas. It depends only on the electron density n_0 and mass m, which is actually the effective band mass, although these are close to the free-electron value for many metals. Table 5.4 shows some actual plasma frequencies measured in metals, which are compared with this simple formula. The prediction $\omega = \omega_p$ is found to be amazingly accurate. One great reason for the popularity of the RPA dielectric function is that it obviously describes plasmons very well. We shall show later that it is exact in the limit where $q \to 0$, so that this success is understandable.

TABLE 5.4. Plasma Frequencies (eV)

Metal	Experimental	ω_p	ω_p'	Reference
Li	7.12	8.03	7.95	*b*
Na	5.71	5.90	5.77	*b*
	5.85			*c*
K	3.72	4.36	4.02	*b*
	3.87			*d*
Mg	10.6	10.88	10.70	*e*
Al	15.3	15.77	15.55	*f*

[a] $\omega_p = \left(\dfrac{4\pi e^2 n_0}{m}\right)^{1/2}, \quad \omega_p' = \left(\dfrac{4\pi e^2 n_0}{\varepsilon_{\text{core}} m}\right)^{1/2}.$

[b] C. Kunz, *Phys. Lett.* **15**, 312 (1965).

[c] J. B. Swan, *Phys. Rev.* **135**, A1467 (1964).

[d] J. L. Robins and F. E. Best, *Proc. Phys. Soc. London* **79**, 110 (1962).

[e] C. J. Powell and J. B. Swan, *Phys. Rev.* **116**, 81 (1959).

[f] C. J. Powell and J. B. Swan, *Phys. Rev.* **115**, 869 (1959).

The formulas for $\varepsilon_2(q, \omega)$ in (5.5.6) are complicated. They may be derived in the following fashion. First, begin with the definition of the imaginary part of the retarded function:

$$\begin{aligned}
\varepsilon_2(\mathbf{q}, \omega) &= -v_q \operatorname{Im}[P^{(1)}_{\text{ret}}(q, \omega)] \\
&= 2\pi v_q \int \frac{d^3p}{(2\pi)^3} \delta(\varepsilon_\mathbf{p} - \varepsilon_{\mathbf{p+q}} + \omega)[n_F(\xi_\mathbf{p}) - n_F(\xi_{\mathbf{p+q}})] \qquad (5.5.11)
\end{aligned}$$

A variable change is made in the term with $n_F(\xi_{\mathbf{p+q}})$ by replacing $\mathbf{p} \rightarrow -\mathbf{p} - \mathbf{q}$, so that this becomes

$$\begin{aligned}
\varepsilon_2(\mathbf{q}, \omega) &= \frac{v_q}{(2\pi)^2} \int d^3p n_F(\xi_p)[\delta(\varepsilon_\mathbf{p} - \varepsilon_{\mathbf{p+q}} + \omega) - \delta(\varepsilon_\mathbf{p} - \varepsilon_{\mathbf{p+q}} - \omega)] \\
&= \frac{v_q}{2\pi} \int_0^{k_F} p^2\, dp \int_{-1}^{1} d\nu \left[\delta\left(-\omega + \varepsilon_q + \frac{qp\nu}{m}\right) \right. \\
&\quad \left. - \delta\left(+\omega + \varepsilon_q + \frac{pq\nu}{m}\right)\right] \qquad (5.5.12)
\end{aligned}$$

This formula shows that $\varepsilon_2(\mathbf{q}, \omega)$ is antisymmetric in frequency $\varepsilon_2(q, -\omega) = -\varepsilon_2(q, \omega)$. We shall evaluate it only for $\omega > 0$. The angular integrations

($\nu = \cos\theta$) are done first, and they eliminate the delta function:

$$\int_{-1}^{1} d\nu\, \delta\left(\varepsilon_q + \frac{qp\nu}{m} \pm \omega\right) = \frac{m}{pq}\, \theta(p - p_{1,2})$$

$$\varepsilon_2(q, \omega) = \frac{2me^2}{q^3}\left(\int_{p_1}^{k_F} p\, dp - \int_{p_2}^{k_F} p\, dp\right)$$

The p integrals are elementary and give $\frac{1}{2}(k_F^2 - p_1^2)$ and $\frac{1}{2}(k_F^2 - p_2^2)$, respectively. The complicated aspect comes from the lower limits of integration p_1 and p_2. These limits are imposed by the angular integral over the delta function. For example, in the first integral the delta function forces $\nu = \cos\theta$ to equal

$$-1 \le \nu = (\omega - \varepsilon_q)\frac{m}{pq} \le 1$$

and ν must have values between ± 1. The latter conditions restrict the value of p to

$$-p \le \frac{m}{q}(\omega - \varepsilon_q) \le p$$

which may be summarized as

$$p > \frac{m}{q}\,|\,\omega - \varepsilon_q\,| \equiv p_1$$

Of course, the integral over p contributes only when $k_F^2 > p_1^2$. In the second integral, the argument is similar—only the sign of ω is changed, so $p_2 = (m/q)|\,\omega + \varepsilon_q\,|$. The result for ε_2 may be written as

$$\varepsilon_2(\mathbf{q}, \omega) = \frac{e^2 m}{q^3}\,[\theta(k_F - p_1)(k_F^2 - p_1^2) - \theta(k_F - p_2)(k_F^2 - p_2^2)] \quad (5.5.13)$$

This result is identical with (5.5.6), although there we simplified the result. For example, $k_F > p_2$ may only be satisfied when $q < 2k_F$, so the second term may be eliminated when $q > 2k_F$.

At small values of q, $\varepsilon_2(q, \omega)$ is proportional to ω at small values of ω. This is shown in (5.5.6) and is evident from the graphs in Fig. 5.13. The proportionality to ω is an important feature of ε_2. It arises in (5.5.13) whenever both inequalities $k_F > p_1, p_2$ are satisfied, since $p_1^2 - p_2^2 = 2m\omega$. The linear dependence of ε_2 on ω must also occur for the exact dielectric function, which may be shown by a simple argument. The physical process under consideration is the rate at which electron–hole pairs are made in the electron gas. A *hole* is a state which has an electron removed from the filled Fermi sea. An initial electron of momentum p and energy ξ_p is excited by a perturbation with $(\mathbf{q}, \omega)$. Thus it is excited to a new state with momentum $\mathbf{p} + \mathbf{q}$ and energy $\xi_{\mathbf{p}+\mathbf{q}} = \xi_{\mathbf{p}} + \omega$. The electron can only be scattered into states which are previously unoccupied, so $\xi_{\mathbf{p}+\mathbf{q}}$ must be above the occupied Fermi sea. Thus the basic process takes an electron from below to above the Fermi level. It leaves a vacancy in the Fermi sea, which is the hole. Thus the excitation process makes electron–hole pairs. The net rate of making such pairs is

$$R_{\text{pairs}} = 2(2\pi) \int \frac{d^3p}{(2\pi)^3} \, \delta(\xi_{\mathbf{p}} + \omega - \xi_{\mathbf{p}+\mathbf{q}})$$
$$\times \{n_F(\xi_{\mathbf{p}})[1 - n_F(\xi_{\mathbf{p}+\mathbf{q}})] - n_F(\xi_{\mathbf{p}+\mathbf{q}})[1 - n_F(\xi_{\mathbf{p}})]\}$$

where $n_F(\xi_{\mathbf{p}})[1 - n_F(\xi_{\mathbf{p}+\mathbf{q}})]$ is the rate of making pairs, while $n_F(\xi_{\mathbf{p}+\mathbf{q}}) \times [1 - n_F(\xi_{\mathbf{p}})]$ is the return rate. We derive a formula similar to (5.5.11), except that neither ξ_p nor $n_F(\xi_p)$ need to be their free-particle values. They could be interpreted as the energy and occupation number for the fully interacting system. If we now change integration variables to $p\, dp \to d\xi p = d\xi$ and $p\, dv = (m/q)\, d\xi_{\mathbf{p}+\mathbf{q}} \equiv (m/q)\, d\xi'$, the above integral may be written as

$$R_{\text{pairs}} \simeq \frac{m}{\pi q} \int d\xi \int d\xi' \, \delta(\xi + \omega - \xi')[n_F(\xi) - n_F(\xi')]$$

where we ignore the limits of integration in this simple argument. The integral over ξ' may be done to eliminate the delta function:

$$R_{\text{pairs}} \simeq \frac{m}{\pi q} \int d\xi [n_F(\xi) - n_F(\xi + \omega)]$$

The remaining integral is over a distribution $n_F(\xi)$ minus the same integral over the displaced distribution $n_F(\xi + \omega)$. The difference between these two must be proportional to ω, at small ω, since we can write $n_F(\xi + \omega) = n_F(\xi) + \omega[dn_F(\xi)/d\xi]$:

$$R_{\text{pairs}} = \frac{m\omega}{\pi q} \int d\xi \left[-\frac{dn_F(\xi)}{d\xi} \right]$$

Thus the rate of making electron–hole pairs is proportional to ω. The derivation applies only to small values of ω, since at large values the limits of integration change the result—as they did in the case of the RPA.

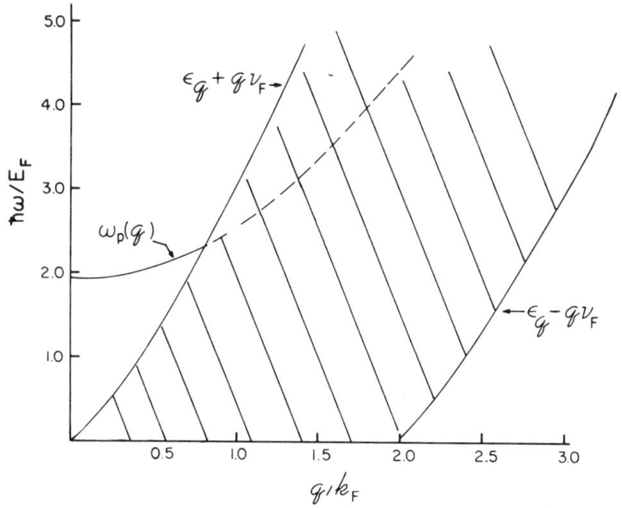

FIGURE 5.14. The excitation region of the electron gas in $(\mathbf{q}, \omega)$ space. The plasmon line ω_p becomes highly damped in the region of electron–hole pairs, which is shown hatched.

The limits in $(\mathbf{q}, \omega)$ space over which $\varepsilon_2(\mathbf{q}, \omega)$ is finite are shown in Fig. 5.14 by the shaded region. These are bounded by the two lines $\omega = \varepsilon_q \pm qv_F$. The excitation spectrum of the electron gas is given by $S(\mathbf{q}, \omega)$ in (5.5.8). It is finite when ε_2 is finite, so that it also exists in the shaded area of Fig. 5.14. In addition, $S(\mathbf{q}, \omega)$ has the plasmon peak which exists where $\varepsilon_2 = 0$. This is also shown in Fig. 5.14. Thus this figure actually describes the excitation spectrum of the electron gas to density fluctuations. At large values of q/k_F, $S(\mathbf{q}, \omega)$ is given accurately by the approximation $S(\mathbf{q}, \omega) \sim \varepsilon_2/n_0 v_q$, so that it describes the process whereby single electron–hole pairs are made. At intermediate values of q/k_F, the excitation spectrum has the complex shape shown in Fig. 5.13. These curves can be measured in metals by the inelastic scattering of energetic electrons through large momentum transfers. However, we defer a comparison with experiments until later sections, after several other dielectric models have been introduced.

The RPA form of $\varepsilon_2(\mathbf{q}, \omega)$ in (5.5.12) can also be evaluated analytically at finite temperatures. Here the fermion occupation number is $n_F(\xi_p) = (e^{\beta\xi} + 1)^{-1}$:

$$\varepsilon_2(\mathbf{q}, \omega) = \frac{v_q}{(2\pi)} \int_0^\infty \frac{p^2\,dp}{e^{\beta(\varepsilon_p - \mu)} + 1} \times \int_{-1}^{1} d\nu \left[\delta\left(\varepsilon_q + \frac{pq\nu}{m} - \omega\right) - \delta\left(\varepsilon_q + \frac{pq\nu}{m} + \omega\right) \right]$$

The angular integral $d\nu$ eliminates the delta function, and the requirement that $|\nu| \le 1$ is equivalent to $p > (m/q)\,|\varepsilon_q \pm \omega|$, or

$$\int_{-1}^{1} d\nu\,\delta\left(\varepsilon_q + \frac{pq\nu}{m} \pm \omega\right) = \frac{m}{pq}\,\theta\left(p - \frac{m}{q}\,|\varepsilon_q \pm \omega|\right)$$

$$p_{1,2} = \frac{m}{q}\,|\varepsilon_q \pm \omega|$$

Thus we have the momentum integral

$$\begin{aligned}
\varepsilon_2(\mathbf{q}, \omega) &= \frac{2e^2 m}{q^3}\int_{p_1}^{p_2} p\,dp\,\frac{1}{e^{\beta(\varepsilon_p - \mu)} + 1} = -\frac{2e^2m^2}{q^3\beta}\ln(1 + e^{-\beta(\varepsilon_p - \mu)})_{p_1}^{p_2} \\
&= \frac{2e^2m^2}{q^3\beta}\ln\left(\frac{1 + e^{-\beta(\varepsilon^{(-)} - \mu)}}{1 + e^{-\beta(\varepsilon^{(+)} - \mu)}}\right) \qquad (5.5.14) \\
\varepsilon^{(\pm)} &= \frac{(\varepsilon_q \pm \omega)^2}{4\varepsilon_q}
\end{aligned}$$

This analytical result is valid at finite temperatures and for any density of particles. Our previous zero temperature result (5.5.13) is the limit where $\beta \to \infty$. Another interesting situation is when Maxwell–Boltzmann statistics apply to the fermion gas. This classical distribution is obtained by letting the chemical potential μ become negative, so that $\beta\mu \to -\infty$ and $\exp(\beta\mu) = \frac{1}{2}n_0(2\pi\beta/m)^{1/2}$. In this limit, the exponential factors in the logarithm (5.5.14) are small, so we expand the argument of the logarithm, and the leading terms give

$$\begin{aligned}
\varepsilon_2(\mathbf{q}, \omega) &= \frac{2e^2m^2}{q^3\beta}\,e^{\beta\mu}(e^{-\beta\varepsilon^{(-)}} - e^{-\beta\varepsilon^{(+)}}) \\
&= \frac{n_0 e^2 m^2}{q^3\beta}\left(\frac{2\pi\beta}{m}\right)^{3/2}(e^{-\beta(\varepsilon_q - \omega)^2/4\varepsilon_q} - e^{-\beta(\varepsilon_q + \omega)^2/4\varepsilon_q}) \qquad (5.5.15)
\end{aligned}$$

This can also be obtained directly from (5.5.12) by using the Maxwell–Boltzmann form of $n_F(\xi_p) \sim \exp(-\beta\varepsilon_p)$.

C. Hubbard

Hubbard (1957) introduced a correction factor to the RPA of the form

$$\varepsilon_H(\mathbf{q}, \omega) = 1 - \frac{v_q P^{(1)}(\mathbf{q}, \omega)}{1 + v_q G(q) P^{(1)}(\mathbf{q}, \omega)} \qquad (5.5.16)$$

$$G(q) = \frac{1}{2}\,\frac{q^2}{q^2 + k_F{}^2} \qquad (5.5.17)$$

This curious formula was regarded as an improvement to the RPA in many properties. Yet its true worth was unappreciated, because it could not be compared with the better dielectric functions which only became available later. The best dielectric functions today are written in precisely the form (5.5.16), with a $G(q)$ which is only slightly different from the simple form proposed by Hubbard. Thus his result, which was somewhat of a stab in the dark, is now regarded as being well ahead of its time.

The factor $G(q)$ is introduced to account for the existence of the exchange and correlation hole around the electron. The dielectric function describes how the potential V_i is affected by the screening charge from the conduction electrons of the metal. Because of the exchange and correlation hole around each electron, when one electron is participating in the dielectric screening, others are less likely to be found nearby. This should have some affect upon the nature of the dielectric screening.

This may be treated by explicitly putting electron–electron interactions into the problem. Thus we rewrite the Hamiltonian (5.5.4) in a more fundamental form, with explicit reference to the impurity potential V_i and the electron–electron interactions:

$$H = \sum_{\mathbf{p}\sigma} \xi_p c_{\mathbf{p}\sigma}^\dagger c_{\mathbf{p}\sigma} + \frac{1}{2\nu} \sum_{\substack{\mathbf{q},\mathbf{pk} \\ \sigma\sigma'}} v_q c_{\mathbf{p}+\mathbf{q}\sigma}^\dagger c_{\mathbf{k}-\mathbf{q}\sigma'}^\dagger c_{\mathbf{k}\sigma'} c_{\mathbf{p}\sigma} + \frac{1}{\nu} \sum_{\mathbf{q}} V_i(\mathbf{q}, t)\varrho(\mathbf{q})$$

The total potential on the electrons $V(\mathbf{q}, t)$ is the sum of the impurity potential $V_i(\mathbf{q}, t)$ and the contribution from the screening charge. The latter quantity is $V_s(\mathbf{q}, t) = v_q\varrho(\mathbf{q}, t)$, where $\varrho(\mathbf{q}, t)$ is the particle density operator used earlier in the section. In this notation, the average value of the total potential is now dependent on the average of the particle density operator:

$$V(\mathbf{q}, \omega) = V_i(\mathbf{q}) + v_q\varrho(\mathbf{q}, \omega)$$

The basic procedure for the derivation is the same as used earlier for the RPA. We write an equation of motion for the particle density operator $\varrho(\mathbf{q}, t)$. This equation is manipulated until we obtain an equation which has $\varrho(\mathbf{q}, t)$ given in terms of V_i or $\varrho(\mathbf{q}, t)$ itself. Then this equation may be solved to get the self-consistent solution.

The equation we examine has the form

$$\varrho(\mathbf{q}, t) = \sum_{\mathbf{p}\sigma} c_{\mathbf{p}+\mathbf{q}\sigma}^\dagger(t) c_{\mathbf{p}\sigma}(t)$$

$$i\frac{d}{dt} c_{\mathbf{p}+\mathbf{q}\sigma}^\dagger c_{\mathbf{p}\sigma} = [H, c_{\mathbf{p}+\mathbf{q}\sigma}^\dagger c_{\mathbf{p}\sigma}]$$

The three terms in H give three terms in the commutator with $c^\dagger_{\mathbf{p}+\mathbf{q}\sigma}c_{\mathbf{p}\sigma}$. The kinetic energy term and the potential term V_i have the same form which was used earlier for the discussion of the RPA. Thus these results are again

$$\left[\sum_{ks} \xi_{\mathbf{k}} c^\dagger_{\mathbf{k}s} c_{\mathbf{k}s},\, c^\dagger_{\mathbf{p}+\mathbf{q}\sigma}c_{\mathbf{p}\sigma}\right] = (\varepsilon_{\mathbf{p}+\mathbf{q}} - \varepsilon_{\mathbf{p}}) c^\dagger_{\mathbf{p}+\mathbf{q},\sigma}c_{\mathbf{p}\sigma}$$

$$\frac{1}{\nu}\sum_{\mathbf{q}'} V_i(q')[\varrho(q', t),\, c^\dagger_{\mathbf{p}+\mathbf{q}\sigma}c_{\mathbf{p}\sigma}] \simeq \frac{1}{\nu} V_i(\mathbf{q})[c^\dagger_{\mathbf{p}\sigma}c_{\mathbf{p}\sigma} - c^\dagger_{\mathbf{p}+\mathbf{q}\sigma}c_{\mathbf{p}+\mathbf{q}\sigma}]$$

The interesting term is the commutator with the electron–electron interaction. This commutator gives two terms, each with four operators:

$$[H_{\text{ee}},\, c^\dagger_{\mathbf{p}+\mathbf{q}\sigma}c_{\mathbf{p}\sigma}] = \frac{1}{\nu}\sum_{\mathbf{k}\mathbf{q}'s} v_{q'}(c^\dagger_{\mathbf{p}+\mathbf{q}'+\mathbf{q},\sigma} c^\dagger_{\mathbf{k}s} c_{\mathbf{k}+\mathbf{q}'s} c_{\mathbf{p}\sigma} - c^\dagger_{\mathbf{p}+\mathbf{q}\sigma} c^\dagger_{\mathbf{k}-\mathbf{q}'s} c_{\mathbf{k}s} c_{\mathbf{p}-\mathbf{q}'\sigma})$$

This result is approximated by pairing up the operators. We select values of the summation variables which produce operators of the type we are seeking. These are either combinations such as $c_{\mathbf{p}}{}^\dagger c_{\mathbf{p}}$ or $c^\dagger_{\mathbf{p}+\mathbf{q}}c_{\mathbf{p}}$. There are three combinations which are interesting: $\mathbf{q}' = -\mathbf{q}$; $\mathbf{k} = \mathbf{p}$, $s = \sigma$; and $\mathbf{k} = \mathbf{p} + \mathbf{q}$, $s = \sigma$. These give the combinations

$$\begin{aligned}[H_{\text{ee}},\, c^\dagger_{\mathbf{p}+\mathbf{q}\sigma}c_{\mathbf{p}\sigma}] &= \frac{1}{\nu}[c^\dagger_{\mathbf{p}\sigma}c_{\mathbf{p}\sigma} - c^\dagger_{\mathbf{p}+\mathbf{q}\sigma}c_{\mathbf{p}+\mathbf{q}\sigma}] \\ &\quad \times \left[v_q \sum_{s\mathbf{k}} c^\dagger_{\mathbf{k}+\mathbf{q}s}c_{\mathbf{k}s} - \sum_{\mathbf{q}'} v_{q'} c^\dagger_{\mathbf{p}+\mathbf{q}+\mathbf{q}'\sigma}c_{\mathbf{p}+\mathbf{q}'\sigma}\right] \\ &\quad - c^\dagger_{\mathbf{p}+\mathbf{q}\sigma}c_{\mathbf{p}\sigma}\frac{1}{\nu}\sum_{\mathbf{q}'} v_{q'}(c^\dagger_{\mathbf{p}+\mathbf{q}+\mathbf{q}'\sigma}c_{\mathbf{p}+\mathbf{q}+\mathbf{q}'\sigma} - c^\dagger_{\mathbf{p}+\mathbf{q}'\sigma}c_{\mathbf{p}+\mathbf{q}'\sigma})\end{aligned}$$

The last two of these terms we recognize as the difference of the exchange energies $\Sigma_x(p) = -(1/\nu)\sum_{\mathbf{q}'} v_q n_F(\xi_{\mathbf{p}+\mathbf{q}})$ [see (5.1.4)] of the electron in the initial and final state. The other four terms provide the combination, after averaging, of

$$\frac{1}{\nu}[n_F(\xi_{\mathbf{p}}) - n_F(\xi_{\mathbf{p}+\mathbf{q}})]\left[v_q\varrho(\mathbf{q}, \omega)e^{-i\omega t} - \sum_{q'} v_{q'}\langle c^\dagger_{\mathbf{p}+\mathbf{q}+\mathbf{q}'\sigma}c_{\mathbf{p}+\mathbf{q}'\sigma}\rangle\right]$$

The problem is in the last term. It is neither of the combinations which are desirable, such as $n_F(\xi_p)$ or $c^\dagger_{\mathbf{p}+\mathbf{q}\sigma}c_{\mathbf{p}\sigma}$. In fact, for a linear screening model, this term must be proportional to $v_q\langle\varrho(q, t)\rangle$, which is the screening charge. The term we are evaluating is the contribution of electron–electron interactions to the density variation. In a linear screening model, this must be proportional to the screening potential $V_s(\mathbf{q}, t) = v_q\varrho(q, t)$. Thus Hubbard assumed that it was and that the constant of proportionality is called $G(q)$:

$$\frac{1}{\nu}\sum_{\mathbf{q}'} v_{\mathbf{q}'}\langle c^{\dagger}_{\mathbf{p}+\mathbf{q}+\mathbf{q}'\sigma}c_{\mathbf{p}+\mathbf{q}'\sigma}\rangle = G(q)v_q\langle\varrho(\mathbf{q}, t)\rangle$$

The function $G(q)$ is found in the following way. First, rewrite the summation by changing the dummy variable $\mathbf{q}'$ to $\mathbf{p}' = \mathbf{p} + \mathbf{q}'$, so that this term is

$$\sum_{\mathbf{p}'} v_{\mathbf{p}-\mathbf{p}'}\langle c^{\dagger}_{\mathbf{p}'+\mathbf{q}\sigma}c_{\mathbf{p}'\sigma}\rangle$$

If the factor $v_{\mathbf{p}-\mathbf{p}'}$ were not in the summation, this would be the particle density operator. Thus Hubbard took it out: He replaced $v_{\mathbf{p}-\mathbf{p}'}$ by $4\pi e^2/(q^2 + k_F{}^2)$. The particle density operator also has the summation over the two spin directions σ, which multiplies the result by 2. Thus this average may now be written approximately as

$$\sum_{\mathbf{p}'} v_{\mathbf{p}-\mathbf{p}'}\langle c^{\dagger}_{\mathbf{p}'+\mathbf{q}\sigma}c_{\mathbf{p}'\sigma}\rangle \to \frac{4\pi e^2}{q^2 + k_F{}^2}\,\frac{1}{2}\sum_{\mathbf{p}'\sigma}\langle c^{\dagger}_{\mathbf{p}'+\mathbf{q}\sigma}c_{\mathbf{p}'\sigma}\rangle$$

This term is indeed proportional to $v_q\langle\varrho(\mathbf{q}, t)\rangle$ when we set $G(q) = \frac{1}{2}q^2/(q^2 + k_F{}^2)$. These are the main elements of the Hubbard approximation.

Of course, we still have to derive (5.5.16), which shows how $G(q)$ enters into the dielectric function. First we solve the equation of motion for $c^{\dagger}_{\mathbf{p}+\mathbf{q}\sigma}c_{\mathbf{p}\sigma}$, which we have been so laboriously deriving. This equation is

$$\begin{aligned}[\omega + \varepsilon_p - \varepsilon_{\mathbf{p}+\mathbf{q}} + \Sigma_x(p) - \Sigma_x(\mathbf{p}+\mathbf{q})]\langle c^{\dagger}_{\mathbf{p}+\mathbf{q}\sigma}c_{\mathbf{p}\sigma}\rangle \\ = \frac{1}{\nu}[n_F(\xi_{\mathbf{p}}) - n_F(\xi_{\mathbf{p}+\mathbf{q}})]\{v_q\langle\varrho(\mathbf{q}, t)\rangle[1 - G(q)] + V_i(\mathbf{q}, t)\}\end{aligned}$$

which may be solved for the particle density operator:

$$\langle\varrho(\mathbf{q}, t)\rangle = \bar{P}(\mathbf{q}, \omega)\{V_i(q, t) + v_q\langle\varrho(\mathbf{q}, t)\rangle[1 - G(q)]\} \qquad (5.5.18)$$

$$\bar{P}(\mathbf{q}, \omega) = \frac{1}{\nu}\sum_{\mathbf{p}\sigma}\frac{n_F(\xi_{\mathbf{p}}) - n_F(\xi_{\mathbf{p}+\mathbf{q}})}{\omega + \varepsilon_{\mathbf{p}} - \varepsilon_{\mathbf{p}+\mathbf{q}} + \Sigma_x(\mathbf{p}) - \Sigma_x(\mathbf{p}+\mathbf{q})} \qquad (5.5.19)$$

The polarization diagram $\bar{P}(\mathbf{q}, \omega)$ is not the same as used in the RPA. The electron exchange energies are in the energy denominator for both states $\mathbf{p}$ and $\mathbf{p} + \mathbf{q}$. This should be expected, since these terms belong in the energy denominator. They are really just a consequence of Dyson's equation. One gets the same polarization contribution by evaluating the basic bubble diagram using a Green's function which has the exchange energy as a self-energy. As indicated in Fig. 5.13, such self-energy terms are to be expected in the polarization diagram. Thus if we take the polarization diagram to be

$$\bar{P}(q, i\omega) = \frac{1}{\nu} \sum_{\mathbf{p},\sigma} \frac{1}{\beta} \sum_{ip} \mathscr{G}^{(x)}(\mathbf{p}, ip) \mathscr{G}^{(x)}(\mathbf{p} + \mathbf{q}, ip + i\omega)$$

$$\mathscr{G}^{(x)}(p) = \frac{1}{ip - \xi_{\mathbf{p}} - \Sigma_x(\mathbf{p})}$$

the Matsubara summations can all be done since Σ_x does not depend on ip and gives the Matsubara function

$$\bar{P}(q, i\omega) = \frac{2}{\nu} \sum_{p} \frac{n_F[\xi_{\mathbf{p}} + \Sigma_x(\mathbf{p})] - n_F[\xi_{\mathbf{p}+\mathbf{q}} + \Sigma_x(\mathbf{p} + \mathbf{q})]}{i\omega + \xi_{\mathbf{p}} + \Sigma_x(\mathbf{p}) - \xi_{\mathbf{p}+\mathbf{q}} - \Sigma_x(\mathbf{p} + \mathbf{q})}$$

The retarded function is obtained from $i\omega \rightarrow \omega + i\delta$ and indeed gives (5.5.19). One consequence of this derivation is the observation that the exchange energies should also go into the occupation numbers $n_F[\xi_p + \Sigma_x(p)]$. At zero temperature, this substitution just renormalizes the chemical potential μ. In practice, these exchange effects are usually left out of the polarization term, and $\bar{P}(q, \omega)$ is replaced by the less accurate RPA result $P^{(1)}(q, \omega)$. The reason is simple: $\bar{P}(q, \omega)$ may not be evaluated in an analytical fashion, so it is more cumbersome to use than $P^{(1)}$.

The dielectric function $\varepsilon_H(q, \omega)$ is derived by combining (5.5.18) with the definition of the total potential:

$$V(q, \omega) = V_i(q) + v_q \varrho(q, \omega) \tag{5.5.20}$$

These two equations are solved to eliminate $\varrho(q, \omega)$. The remaining equation gives a proportionality between V and V_i, which is just ε_H. For example, from (5.5.18) we find $[\langle \varrho(q, t) \rangle = \varrho(q, \omega) e^{-i\omega t};\ V_i(q, t) = V_i(q) e^{-i\omega t}]$

$$\varrho(q, \omega) = \frac{\bar{P}(q, \omega) V_i(q)}{1 - v_q \bar{P}(1 - G)}$$

which when put into (5.5.20) gives

$$V(q, \omega) = V_i(q) \left[1 + \frac{v_q \bar{P}(q, \omega)}{1 - v_q \bar{P}(1 - G)} \right] = \frac{V_i(q)}{\varepsilon_H(q, \omega)}$$

The two terms in the brackets are combined to give

$$\frac{1}{\varepsilon_H} = \frac{1 + v_q \bar{P} G}{1 - v_q \bar{P}(1 - G)}$$

which is inverted to give the dielectric function:

$$\varepsilon_H(q, \omega) = 1 - \frac{v_q \bar{P}(q, \omega)}{1 + v_q \bar{P}(q, \omega) G(q)}$$

The actual Hubbard form (5.5.16) uses $P^{(1)}$ in place of $\bar{P}$.

The effect of $G(q)$ is to reduce the electron–electron interactions in the dielectric screening. This is apparent from (5.5.18), where the factor $1 - G$ multiplies the screening potential $v_q\langle\varrho(q, t)\rangle$. If $G = 0$, we have the RPA result. A finite G is interpreted as arising from the exchange hole around the electron. We use the word *exchange* rather than the term *exchange and correlation* because the Hubbard model includes only the exchange hole. The subtracted term $-\sum v_{q'}\langle c^\dagger_{\mathbf{p}+\mathbf{q}+\mathbf{q}'\sigma} c_{\mathbf{p}+\mathbf{q}'\sigma}\rangle$ acts only on excitations of the same spin σ as the initial disturbance, so it does not include $g_{\uparrow\downarrow}(r)$ effects.

The word *exchange* has a particular meaning when applied to the solution by equation of motion. When one has a term such as $c^\dagger_{\mathbf{p}+\mathbf{q}+\mathbf{q}',\sigma} c^\dagger_{\mathbf{k}s} c_{\mathbf{k}+\mathbf{q}'s} c_{\mathbf{p}\sigma}$ which is being decoupled, the *direct* terms are those which pair subscripts with both p or both k. For example, the term $\mathbf{q} + \mathbf{q}' = 0$ gives the term $\langle c^\dagger_{\mathbf{p}\sigma} c_{\mathbf{p}\sigma}\rangle c^\dagger_{\mathbf{k}s} c_{\mathbf{k}-\mathbf{q}s}$, which is a direct coulomb term. The *exchange* terms are those which pair together subscripts with a $\mathbf{p}$ and a $\mathbf{k}$. For example, the choice $\mathbf{k} = \mathbf{p} + \mathbf{q}$ gives the exchange energy term $-\langle c^\dagger_{\mathbf{p}+\mathbf{q}+\mathbf{q}'\sigma} c_{\mathbf{p}+\mathbf{q}+\mathbf{q}'\sigma}\rangle \times c^\dagger_{\mathbf{p}+\mathbf{q}\sigma} c_{\mathbf{p}\sigma}$, while the choice $\mathbf{k} = \mathbf{p}$ gives the term $-\langle c^\dagger_{\mathbf{p}\sigma} c_{\mathbf{p}\sigma}\rangle c^\dagger_{\mathbf{p}+\mathbf{q}+\mathbf{q}'\sigma} c_{\mathbf{p}+\mathbf{q}'\sigma}$. The exchange energy arose from the coulomb interaction of a particle with its exchange hole. Similarly, the last term—which contributes to $G(q)$—reduces the coulomb interactions in the screening term. This is also caused by the exchange hole.

The function $G(q)$ is usually taken to be frequency independent. It would be better to obtain a $G(q)$ which also had frequency dependence, since presumably the exchange–correlation hole is frequency dependent.

D. Singwi–Sjölander

The 1967 Singwi–Sjölander dielectric function has the same form [(5.5.16)] as Hubbard's but with a different choice of $G(q)$. Singwi has collaborated with a variety of authors to develop improvements in the method of choosing $G(q)$. Once could remark, in a humorous way, that $G(q)$ is time dependent because of its improvements over the years. This makes our discussion difficult, since there are several possible choices to describe. We shall derive the original version of Singwi–Sjölander and refer to the literature of the others (see Vashishta and Singwi, 1972; Singwi and Tosi, 1981).

Their original derivation is very attractive, because it explicitly includes the exchange–correlation hole. They derive an equation of motion for the screening charge and then insert the pair distribution function $g(r)$ into the Coulomb integral between particles. They solved these equations and obtained

$$\varepsilon(q, \omega) = 1 - \frac{v_q P^{(1)}(q, \omega)}{1 + v_q P^{(1)}(q, \omega) G(q)} \tag{5.5.16}$$

$$G(q) = -\frac{1}{n_0} \int \frac{d^3q'}{(2\pi)^3} \frac{\mathbf{q} \cdot \mathbf{q}'}{\mathbf{q}'^2} [S(\mathbf{q} - \mathbf{q}') - 1] \tag{5.5.21}$$

where $S(q)$ is the static structure factor associated with $g(r)$. Their derivation employs Wigner distribution functions, which will not be introduced here for lack of space. Instead, we shall derive their $G(q)$ by another method given by Singwi *et al.* (1968). It uses equations of motion. They find the second time derivative of the particle density operator $\varrho(q, t)$. The first time derivative is a single commutator with H, so the second time derivative is a double commutator,

$$\ddot{\varrho}(\mathbf{q}, t) = -[H, [H, \varrho(\mathbf{q}, t)]]$$

$$\varrho(\mathbf{q}, t) = \sum_{\mathbf{p}\sigma} c^{\dagger}_{\mathbf{p}+\mathbf{q}\sigma} c_{\mathbf{p}\sigma}$$

$$H = \sum_{\mathbf{p}\sigma} \xi_p c^{\dagger}_{\mathbf{p}\sigma} c_{\mathbf{p}\sigma} + \frac{1}{2\nu} \sum_{\substack{\mathbf{pkq} \\ \sigma s}} v_q c^{\dagger}_{\mathbf{p}+\mathbf{q}\sigma} c^{\dagger}_{\mathbf{k}-\mathbf{q}s} c_{\mathbf{k}s} c_{\mathbf{p}\sigma}$$

where H is the Hamiltonian for the homogeneous electron gas. We are trying to determine the effect of the electron–electron interactions, so the other terms can be conveniently ignored for this discussion. When we consider the first commutator,

$$[H, \varrho(\mathbf{q}, t)] = \sum_{\mathbf{p}\sigma} (\varepsilon_{\mathbf{p}+\mathbf{q}} - \varepsilon_{\mathbf{p}}) c^{\dagger}_{\mathbf{p}+\mathbf{q},\sigma} c_{\mathbf{p},\sigma}$$

we find that $\varrho(\mathbf{q}, t)$ commutes with the electron–electron term in the Hamiltonian. This is not surprising, since we can write it as $\sum v_q \varrho(\mathbf{q})\varrho(\mathbf{q})$ except for self-interaction effects. An operator commutes with itself, and the electron–electron term contains just density operators, which is why it commutes with $\varrho(\mathbf{q}, t)$. The second commutator provides the interesting result. The commutator with the kinetic term in H just gives a repetition of the energy difference factor:

$$\left[\sum_{\mathbf{k}\sigma} \xi_{\mathbf{k}\sigma} c^{\dagger}_{\mathbf{k}\sigma} c_{\mathbf{k}\sigma}, \sum_{\mathbf{p}\sigma} (\varepsilon_{\mathbf{p}+\mathbf{q}} - \varepsilon_{\mathbf{p}}) c^{\dagger}_{\mathbf{p}+\mathbf{q}\sigma} c_{\mathbf{p}\sigma}\right] = \sum_{\mathbf{p}\sigma} (\varepsilon_{\mathbf{p}+\mathbf{q}} - \varepsilon_{\mathbf{p}})^2 c^{\dagger}_{\mathbf{p}+\mathbf{q}\sigma} c_{\mathbf{p}\sigma}$$

The commutator with the interaction gives four terms, which can be combined to give two different ones by rearranging the order of operators. They are

$$\frac{1}{2\nu}\sum_{\substack{\mathbf{p}'\mathbf{kq}'\mathbf{p}\\ ss'}} v_{q'}(\varepsilon_{\mathbf{p}+\mathbf{q}}-\varepsilon_{\mathbf{p}})[c^{\dagger}_{\mathbf{p}'+\mathbf{q}'s}c^{\dagger}_{\mathbf{k}-\mathbf{q}'s'}c_{\mathbf{k}s'}c_{\mathbf{p}'s},\ c^{\dagger}_{\mathbf{p}+\mathbf{q}\sigma}c_{\mathbf{p}\sigma}]$$

$$=\frac{1}{\nu}\sum_{\substack{\mathbf{kq}'\mathbf{p}\\ s\sigma}}(\varepsilon_{\mathbf{p}+\mathbf{q}}-\varepsilon_{\mathbf{p}})v_{q'}[c^{\dagger}_{\mathbf{p}+\mathbf{q}+\mathbf{q}'\sigma}c^{\dagger}_{\mathbf{k}-\mathbf{q}'s}c_{\mathbf{k}s}c_{\mathbf{p}\sigma}-c^{\dagger}_{\mathbf{p}+\mathbf{q}\sigma}c^{\dagger}_{\mathbf{k}-\mathbf{q}'s}c_{\mathbf{k}s}c_{\mathbf{p}-\mathbf{q}'\sigma}]$$

The change in dummy variables $\mathbf{p}\to\mathbf{p}+\mathbf{q}'$ in the last term makes the operator sequence the same as the first term:

$$=\frac{1}{\nu}\sum_{\substack{\mathbf{kpq}'\\ s\sigma}} v_{q'}c^{\dagger}_{\mathbf{p}+\mathbf{q}+\mathbf{q}'\sigma}c^{\dagger}_{\mathbf{k}-\mathbf{q}'s}c_{\mathbf{k}s}c_{\mathbf{p}\sigma}[(\varepsilon_{\mathbf{p}+\mathbf{q}}-\varepsilon_{\mathbf{p}})-(\varepsilon_{\mathbf{p}+\mathbf{q}+\mathbf{q}'}-\varepsilon_{\mathbf{p}+\mathbf{q}'})]$$

The kinetic energies largely cancel in this expression, yielding only $-\mathbf{q}\cdot\mathbf{q}'/m$. When we write this term, the operator sequence $c^{\dagger}_{\mathbf{p}+\mathbf{q}+\mathbf{q}'\sigma}c^{\dagger}_{\mathbf{k}-\mathbf{q}'s}c_{\mathbf{k}s}c_{\mathbf{p}\sigma}$ will be rearranged into $\sum_{p\sigma} c^{\dagger}_{\mathbf{p}+\mathbf{q}+\mathbf{q}'\sigma}c_{\mathbf{p}\sigma}\sum_{ks} c^{\dagger}_{\mathbf{k}-\mathbf{q}s}c_{\mathbf{k}s}=\varrho(\mathbf{q}+\mathbf{q}',t)\varrho(-\mathbf{q}',t)$. The rearrangement is permissible as long as we do not count the energy of a particle acting on itself; a self-interaction is included in $\varrho\varrho$ unless we take care to leave it out. The various terms are collected into the equation of motion for the density operator:

$$-\ddot{\varrho}(\mathbf{q},t)=+\sum_{\mathbf{p}\sigma}(\varepsilon_{\mathbf{p}+\mathbf{q}}-\varepsilon_{\mathbf{p}})^2c^{\dagger}_{\mathbf{p}+\mathbf{q}\sigma}c_{\mathbf{p}\sigma}$$
$$+\frac{1}{\nu}\sum_{\mathbf{q}'}v_{q'}\left(\frac{-\mathbf{q}\cdot\mathbf{q}'}{m}\right)\varrho(\mathbf{q}+\mathbf{q}',t)\varrho(-\mathbf{q}',t)$$

One obvious way to evaluate the last term is to take $\mathbf{q}'=-\mathbf{q}$ in the summation. This term alone gives the equation $[\varrho(0)=\sum_{\mathbf{p}\sigma}c^{\dagger}_{\mathbf{p}\sigma}c_{\mathbf{p}\sigma}=N]$

$$-\ddot{\varrho}(\mathbf{q},t)=\frac{v_q q^2}{m}\,\frac{N}{\nu}\,\varrho(\mathbf{q},t)=\omega_p{}^2\varrho(\mathbf{q},t)$$

which describes the plasma oscillations in the electron gas. Since the other terms all vanish when $\mathbf{q}\to 0$, we have shown that the long-wavelength density oscillations are the same plasma oscillations as predicted in the RPA. Thus the RPA is exact in this limit.

The Singwi–Sjölander result is obtained from treating $\langle\varrho(\mathbf{q}+\mathbf{q}')\varrho(-\mathbf{q}')\rangle$ more carefully. They wrote it in terms of a direct summation over particle positions, as was done in Sec. 1.6:

$$\varrho(\mathbf{q} + \mathbf{q}')\varrho(-\mathbf{q}') = \sum_i e^{i\mathbf{r}_i \cdot (\mathbf{q}+\mathbf{q}')} \sum_{\substack{j \\ j \neq i}} e^{-i\mathbf{r}_j \cdot \mathbf{q}'}$$

The term $i \neq j$ is excluded, since this is a particle interacting with itself. The factors in the exponent are rearranged so that we have a summation over a particle $\mathbf{r}_j$ and the other summation over the particle separations $\mathbf{r}_i - \mathbf{r}_j$:

$$\varrho(\mathbf{q} + \mathbf{q}')\varrho(-\mathbf{q}') = \sum_j e^{i\mathbf{r}_j \cdot \mathbf{q}} \sum_{i \neq j} e^{i(\mathbf{r}_i - \mathbf{r}_j) \cdot (\mathbf{q}+\mathbf{q}')}$$

By following the definitions in Sec. 1.6, the summation over $\mathbf{r}_i - \mathbf{r}_j$ can be replaced by its average in the electron gas:

$$\sum_{i \neq j} e^{i(r_i - r_j) \cdot (\mathbf{q}+\mathbf{q}')} \Rightarrow n_0 \int d^3r g(r) e^{i\mathbf{r} \cdot (\mathbf{q}+\mathbf{q}')} = N\delta_{\mathbf{q}+\mathbf{q}'} + S(\mathbf{q} + \mathbf{q}') - 1$$

The other summation $\sum e^{i\mathbf{q} \cdot \mathbf{r}_j}$ is just $\varrho(\mathbf{q}, t)$, so that we have derived

$$\langle \varrho(\mathbf{q} + \mathbf{q}')\varrho(-\mathbf{q}')\rangle' = \langle \varrho(\mathbf{q}, t)\rangle [N\delta_{\mathbf{q}+\mathbf{q}'=0} + S(\mathbf{q} + \mathbf{q}') - 1] \quad (5.5.22)$$

The prime on the bracket means one is to omit the interaction of a particle with itself in doing the averages. This omission produces the -1 term on the right. The term $N\delta_{\mathbf{q}+\mathbf{q}'}$ is the same one which we took above, which gave plasmons. The other term gives $G(q)$. It is already proportional to $\varrho(\mathbf{q}, t)$, so that $G(q)$ is obtained without difficulty:

$$\frac{1}{m\nu} \sum_{\mathbf{q}'} v_{q'}(-\mathbf{q} \cdot \mathbf{q}') \langle \varrho(\mathbf{q}, t)\rangle [N\delta_{\mathbf{q}+\mathbf{q}'=0} + S(\mathbf{q} + \mathbf{q}') - 1]$$
$$= n_0 \frac{v_q q^2}{m} \langle \varrho(\mathbf{q}, t)\rangle [1 - G(q)]$$

$$G(q) = -\frac{1}{n_0} \frac{1}{\nu} \sum_{\mathbf{q}'} \frac{(-\mathbf{q} \cdot \mathbf{q}')}{q'^2} [S(\mathbf{q} + \mathbf{q}') - 1]$$

$$\omega^2 \langle \varrho(\mathbf{q}, t)\rangle = \sum_{p\sigma} (\varepsilon_{\mathbf{p}+\mathbf{q}} - \varepsilon_{\mathbf{p}})^2 \langle C^\dagger_{\mathbf{p}+\mathbf{q}\sigma} C_{\mathbf{p}}\rangle + \omega_p^2 \langle \varrho(\mathbf{q}, t)\rangle [1 - G(q)]$$

The factor of $G(q)$ provides a modification of the plasmon dispersion at finite q due to electron correlations. The formula for $G(q)$ is the same one which was announced in (5.5.21) after changing dummy variables $\mathbf{q}' \to -\mathbf{q}'$. The derivation of the dielectric function (5.5.16) is the same as the Hubbard theory: In the theory of self-consistent fields, the screening term has the factor $1 - G$ because of the role of correlations.

The Singwi–Sjölander formula for $G(q)$ can only be evaluated with a knowledge of $S(\mathbf{q})$. This, in turn, is obtained from a knowledge of $\varepsilon(\mathbf{q}, \omega)$ through the relationship (5.4.12) derived earlier:

$$S(\mathbf{q}) = -\frac{1}{n_0 v_q} \int_0^\infty \frac{d\omega}{\pi} \operatorname{Im}\left[\frac{1}{\varepsilon(\mathbf{q}, \omega)}\right] \tag{5.4.12}$$

But $\varepsilon(\mathbf{q}, \omega)$ depends on $G(q)$. Thus the three equations (5.5.21), (5.5.16), and (5.4.12) form a triad which link the three functions $G(q)$, $S(\mathbf{q})$, and $\varepsilon(\mathbf{q}, \omega)$. They were solved self-consistently on the computer. This must be done for each value of r_s, since the results depend on density. In discussing some later versions of $G(q)$, they (Singwi *et al.*, 1970; Vashishta and Singwi, 1972) remarked that their results could be adequately fitted by the simple expression

$$G(q) = A(1 - e^{-B(q/k_F)^2})$$

The constants A and B, both dimensionless, are given in Table 5.5 for different values of r_s. Since they depend smoothly on density, the parameters A and B may be obtained for other values of r_s by interpolation. This form for $G(q)$ fits their computed one well at small and intermediate values of q/k_F but not larger values. However, G is relatively unimportant at large q, so this drawback is not serious.

Sometimes $g(r)$ is known, rather than $S(q)$. In this case, their formula can be rearranged so that $G(q)$ is given directly from $g(r)$. We use the

TABLE 5.5. $G(q) = A(1 - e^{-B(q/k_F)^2})$

r_s	Vashishta and Singwi (1972)		Singwi *et al.* (1970)	
	A	B	A	B
1	0.70853	0.36940	0.7756	0.4307
2	0.85509	0.33117	0.8994	0.3401
3	0.97805	0.30440	0.9629	0.2924
4	1.08482	0.28430	0.9959	0.2612
5	1.17987	0.26850	1.0138	0.2377
6	1.26569	0.25561	1.0216	0.2189

Fourier transform relations (1.6.2) to write $G(q)$ as

$$G(q) = -\int d^3r[1 - g(r)]I(\mathbf{r}, \mathbf{q})$$
$$I(\mathbf{r}, \mathbf{q}) = \int \frac{d^3q'}{(2\pi)^3} \frac{\mathbf{q} \cdot \mathbf{q}'}{(q')^2} e^{i\mathbf{r} \cdot (\mathbf{q}+\mathbf{q}')} \tag{5.5.23}$$

The $I(\mathbf{r}, \mathbf{q})$ integral is evaluated in the following way. Set $\mathbf{r} = \hat{z}r$ so that $\mathbf{q}' \cdot \mathbf{r} = q'r\nu$ and $\mathbf{q} \cdot \mathbf{r} = qr\nu_0$. The law of cosines gives

$$\mathbf{q} \cdot \mathbf{q}' = qq'[\nu\nu_0 + (1 - \nu^2)^{1/2}(1 - \nu_0^2)^{1/2} \cos \varphi]$$

The integral over $d\varphi$ eliminates the $\cos \varphi$ term, so we have

$$I(\mathbf{r}, \mathbf{q}) = e^{i\mathbf{r} \cdot \mathbf{q}} \frac{q\nu_0}{(2\pi)^2} \int_0^\infty q' \, dq' \int_{-1}^1 d\nu \nu e^{iq'r\nu}$$
$$= e^{i\mathbf{r} \cdot \mathbf{q}} \frac{q\nu_0 i}{2\pi^2} \int_0^\infty q' \, dq' j_1(rq')$$

The integral over the spherical Bessel function yields $\pi/2r^2$:

$$I(\mathbf{r}, \mathbf{q}) = \frac{i\mathbf{r} \cdot \mathbf{q} e^{i\mathbf{r} \cdot \mathbf{q}}}{4\pi r^3}$$

This is put into the integral (5.5.23), and the angular integrals are done:

$$G(q) = -\frac{qi}{2} \int_0^\infty dr[1 - g(r)] \int_{-1}^1 d\nu_0 \nu_0 e^{iqr\nu_0}$$
$$= q \int_0^\infty dr[1 - g(r)] j_1(qr)$$

The angular integral gives another spherical Bessel function. The factor $1 - g$ describes the full exchange–correlation hole around the electron. The electron correlation is included for other electrons of both spin configurations, and the net is $1 - g$. This is the physically reasonable result and is why the Singwi–Sjölander result is so attractive.

In the next section we shall calculate a number of properties of the electron gas and use all four dielectric functions. In general the accuracy increases in the order we presented them: Thomas–Fermi is least accurate, while Singwi–Sjölander is the best.

5.6. PROPERTIES OF THE ELECTRON GAS

There are a number of properties of the electron gas which may be predicted with a knowledge of the dielectric function $\varepsilon(q, \omega)$. These properties may also be measured for metals. However, one should be careful in applying predictions of the homogeneous electron gas to real metals, because the ion cores often have a significant influence. Thus a favorable comparison to experiment may not, in this case, be the signature of the best theory. We shall discuss four properties: the pair distribution function, the screening charge, compressibility, and correlation energy.

A. Pair Distribution Function

The pair distribution function $g(r)$ may be obtained from a knowledge of the static structure factor $S(q)$ through the transform relation (1.6.2):

$$g(r) = 1 + \frac{3}{2rk_F{}^3} \int_0^\infty k\,dk \sin(kr)[S(k) - 1]$$

The static structure factor $S(k)$ is obtained from the dielectric function through the relation (5.4.12). The Thomas–Fermi theory is a static model of dielectric screening, so that it has no prediction regarding $g(r)$. Hence we shall only compare the theories of RPA, Hubbard, and Singwi–Sjölander. These are obtained only after a numerical computation.

The three results for $g(r)$ are shown in Fig. 5.15 for six values of r_s. The RPA and Hubbard theories are seen to have negative values at small values of r. This is viewed as a great deficiency, since $g(r)$ is—by definition—strictly a positive function. The pair distribution function is defined as the probability that another electron is a distance r away from the first; as a probability, it must be positive. The negative values predicted by the RPA and Hubbard theories are regarded as significant failures of these theories. The Singwi–Sjölander theory is slightly negative at large r_s but otherwise positive, so it is much better in this respect. For most metallic densities, it predicts $g(r)$ to be positive for all r. Thus the Singwi–Sjölander theory is a significant improvement over the earlier theories.

B. Screening Charge

The screened potential $V(r)$ about a point charge $Ze\delta(\mathbf{r})$ may be calculated and plotted. This is slightly inconvenient, since $V(r)$ diverges as r^{-1} at small r, as do all Coulomb potentials. Instead, it is customary to plot the density of screening charge $n_s(r)$ about the point impurity. This quantity is finite at the origin $r = 0$ and also has the Friedel oscillations at large distance. In the linear screening model, derived in Sec. 5.4, the potential energy from an impurity charge of Ze is

$$V(r) = Z \int \frac{d^3q}{(2\pi)^3} \frac{v_q}{\varepsilon(q)} e^{i\mathbf{q}\cdot\mathbf{r}}$$

We operate on both sides of this equation with ∇^2. From Poisson's equation, we know that $\nabla^2 V$ is proportional to the total charge density in the system. This must be the central impurity $Ze\delta(\mathbf{r})$ plus the screening particle density $n_s(r)$. Thus we derive the equation for this screening charge:

$$\nabla^2 V(r) = -4\pi e^2[Z\delta(\mathbf{r}) - n_s(r)] = -4\pi e^2 Z \int \frac{d^3q}{(2\pi)^3} \frac{e^{i\mathbf{q}\cdot\mathbf{r}}}{\varepsilon(q)}$$

$$n_s(r) = -Z \int \frac{d^3q}{(2\pi)^3} e^{i\mathbf{q}\cdot\mathbf{r}}\left[\frac{1}{\varepsilon(q)} - 1\right]$$

The distribution is spherically symmetric in the homogeneous electron gas, so the angular integrals may be performed at once:

$$n_s(r) = \frac{Z}{2\pi^2 r} \int_0^\infty q\, dq \sin qr\left[1 - \frac{1}{\varepsilon(q)}\right] \tag{5.6.1}$$

In this case a knowledge of the static dielectric function $\varepsilon(q)$ is sufficient to determine the density of screening charge. Thus the Thomas–Fermi model may also be used to make predictions. In Sec. 5.5A we showed that it predicted the screened potential energy to be $Ze^2 \exp(-q_{\mathrm{TF}}r)/r$, so that the screening particle density is predicted to be $Zq_{\mathrm{TF}}^2 \exp(-q_{\mathrm{TF}}r)/r$. This is not plotted in Fig. 5.16. The rest, which are shown in Fig. 5.16, have to be obtained by numerical Fourier transform, which was first done by Langer and Vosko (1960) for the RPA. All curves are very similar, except for Thomas–Fermi, which is the only one to diverge at $r \to 0$. The RPA (1), Hubbard (2), Singwi–Sjölander (3), and Singwi *et al.* (4) results all show Friedel oscillations which are similar. These are lacking in the Thomas–Fermi theory. These Friedel oscillations are real features of impurities in metals, so that the Thomas–Fermi theory is deficient in several respects. However, the other theories predict remarkably similar results, which should not be surprising. Screening is basically a one-body property, and little correlation is evident in one-body amplitudes. The effects of $G(q)$ are much more apparent in properties involving two-body correlations.

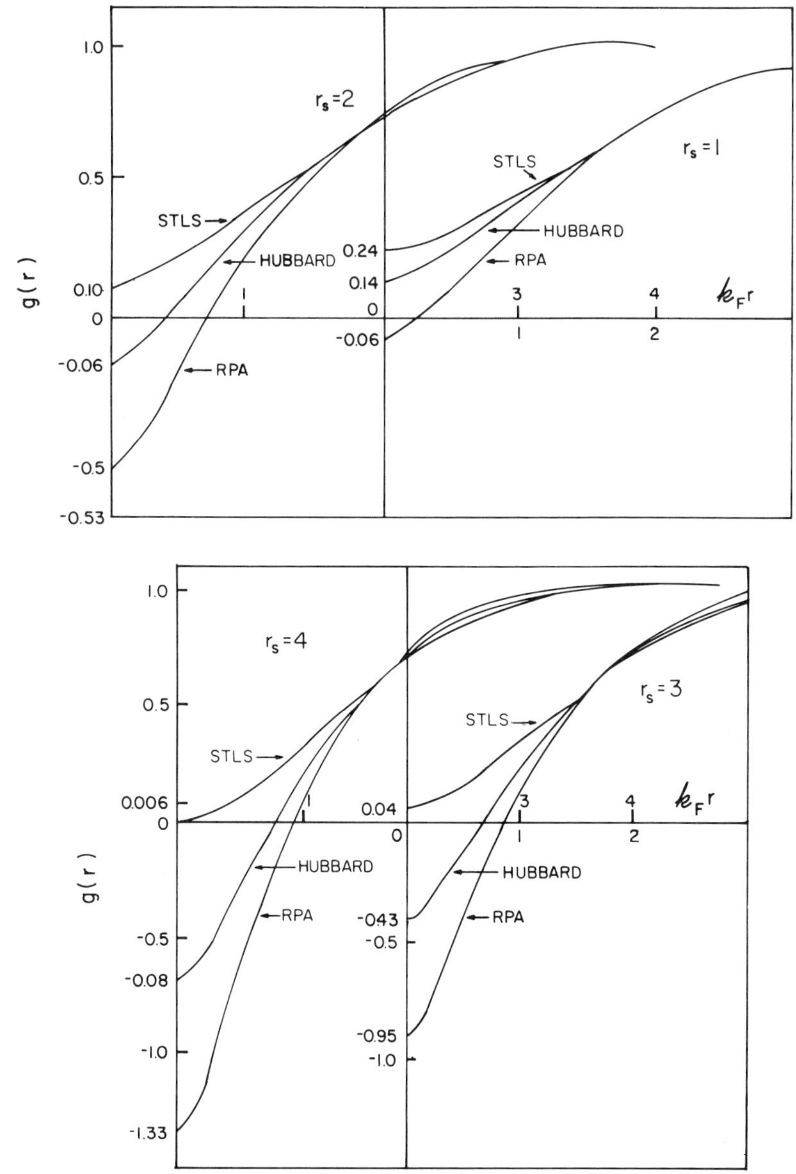

FIGURE 5.15. The pair distribution function $g(r)$ for the electron gas at various densities for several model dielectric functions. *Source*: Singwi *et al.* (1968) (used with permission).

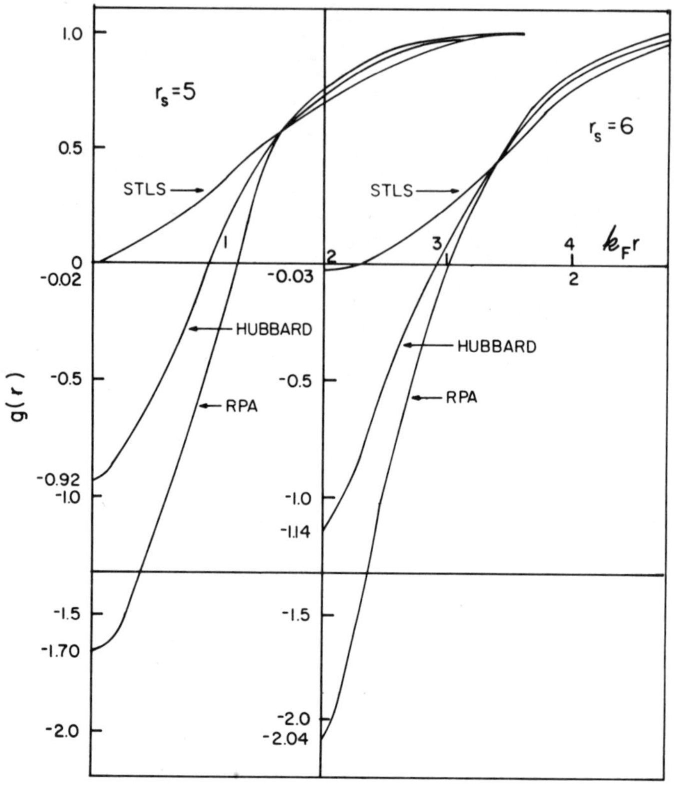

FIGURE 5.15 *(continued)*

C. Correlation Energies

A knowledge of $\varepsilon(q, \omega)$ is also sufficient for predicting the correlation energy of the electron gas. This should be expected, since the correlation energy arises from the correlation hole around the electron. The exchange–correlation hole was put into the derivation of $G(q)$. Presumably one could invert this argument and obtain $g(r)$ from $G(q)$. Now we shall show how $\varepsilon(q, \omega)$ may be manipulated to obtain the correlation energy. Our discussion follows Singwi *et al.* (1968).

The coulomb interaction energy of the electron gas may be written as

$$\frac{1}{2}\frac{1}{\nu}\sum_{\substack{\mathbf{q}\\ \mathbf{q}\neq 0}} v_q \varrho(\mathbf{q})\varrho(-\mathbf{q})$$

as long as we remember to avoid all terms which have an electron interacting

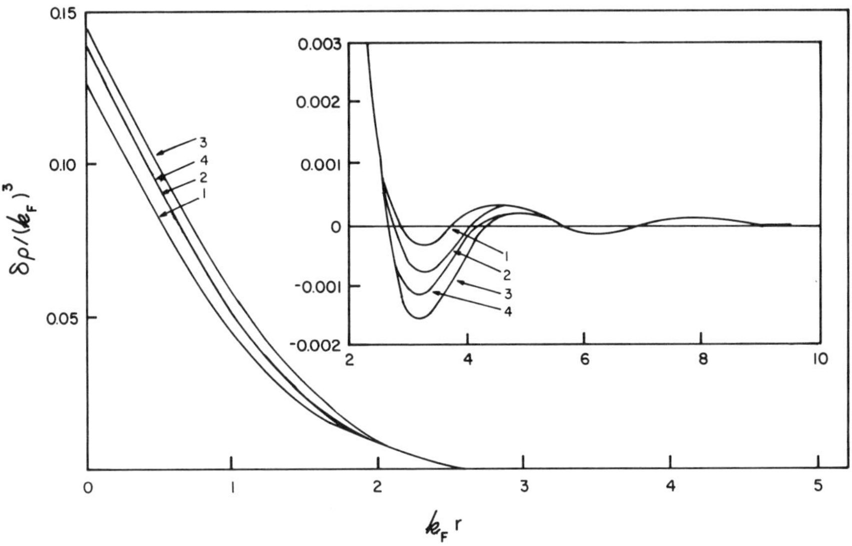

FIGURE 5.16. Screening charge density near a point impurity in the electron gas with $r_s = 3$ according to several model dielectric functions: (1) RPA, (2) Hubbard, (3) Singwi–Sjölander, and (4) Singwi *et al.* *Source*: Singwi *et al.* (1970) (used with permission).

with itself. The average value of $\varrho(\mathbf{q})\varrho(-\mathbf{q})$ in the electron gas is

$$\langle \varrho(\mathbf{q})\varrho(-\mathbf{q})\rangle' = N[N\delta_{\mathbf{q}=0} + S(q) - 1]$$

The prime on the bracket means one is to omit the particle interacting with itself. This explains why this differs by -1 from the similar average in (5.4.7). This may be derived by following the same steps used to get (5.5.22). Thus the interaction energy E_{int} per electron of the electron gas is

$$NE_{\text{int}}(e^2) = \frac{1}{2}\,\frac{N}{\nu}\sum_{\mathbf{q}} v_q[S(q) - 1] = 2\pi e^2 N \int \frac{d^3q}{(2\pi)^3}\,\frac{1}{q^2}\,[S(q) - 1] \tag{5.6.2}$$

The summation has been converted to an integration over q. In doing this step, remember that the $\mathbf{q} = 0$ term is not in the summation, so that the term $N\delta_{\mathbf{q}=0}$ has no effect. The quantity $E_{\text{int}}(e^2)$ is not the Coulomb contribution to the ground state energy. To obtain the ground state energy per particle E_g, one must do a coupling constant integration of the type discussed earlier in Sec. 3.6:

$$E_g = \frac{3}{5}\,E_{F0} + \int_0^{e^2} \frac{d\lambda}{\lambda}\,E_{\text{int}}(\lambda) \tag{5.6.3}$$

This may be derived starting from (3.6.22), which is assigned as a problem. Equation (5.6.3) is an exact result for the ground state energy of the homogeneous electron gas. One must calculate the Coulomb interaction energy for each value of λ up to e^2. In practice, this is the same as finding $S(q)$ as a function of density. The first term $\frac{3}{5}E_{F0}$ is the average kinetic energy, which is calculated assuming that $e^2 = 0$.

The angular integrals in the definition (5.6.2) of E_{int} may be done, yielding

$$E_{\text{int}}(e^2) = \frac{e^2}{\pi} \int_0^\infty dq[S(q) - 1]$$

This depends on e^2 through the prefactor on the right. It also has a dependence through $S(q)$, which is decidedly dependent on electron–electron interactions. It is conventional to introduce a dimensionless function

$$\gamma = -\frac{1}{2k_F} \int_0^\infty dq[S(q) - 1] \tag{5.6.4}$$

which depends on density. In our coupling constant integration (5.6.3), γ also depends on λ. The interaction energy may be written as

$$\begin{aligned} E_{\text{int}} &= -\frac{2}{\pi} e^2 k_F \gamma \\ &= -\frac{4}{\pi r_s}\left(\frac{9\pi}{4}\right)^{1/3} \gamma \end{aligned}$$

This has been put into Rydberg units in the second line by using $k_F a_0 = (9\pi/4)^{1/3}/r_s$. Next we wish to investigate the coupling constant (e^2) dependence of these terms. The Rydberg energy E_{Ryd} is proportional to e^4 or λ^2. Similarly, $r_s \propto \lambda$ since $r_s \propto a_0^{-1}$ and the Bohr radius is inversely proportional to e^2. Thus if we write these dependences as $r_s \to \lambda r_s$ or $e^4 \to \lambda^2 e^4$, where $0 \to \lambda \to 1$, we have

$$E_{\text{int}}(\lambda) = -\lambda \frac{4}{\pi r_s}\left(\frac{9\pi}{4}\right)^{1/3} \gamma(\lambda r_s)$$

The factor γ is also dependent on density, which we write as $\gamma(\lambda r_s)$ or as $\gamma(r_s)$ when $\lambda = 1$. The coupling constant integration may be done, at least formally, and the ground state energy per particle may be obtained from (5.6.3):

$$E_g = \frac{2.2099}{r_s^2} - \frac{4}{\pi r_s}\left(\frac{9\pi}{4}\right)^{1/3} \int_0^1 d\lambda \gamma(\lambda r_s)$$

The coupling constant integration involves only the factor γ. All the other λ's have factored out of the expression. Of course, we chose γ so this would happen. The correlation energy consists of all contributions to the ground state energy except kinetic and exchange. Thus the correlation energy is

$$E_c = \frac{0.9163}{r_s} - \frac{4}{\pi r_s}\left(\frac{9\pi}{4}\right)^{1/3}\int_0^1 d\lambda\gamma(\lambda r_s) \tag{5.6.5}$$

One can verify that this vanishes when $S(q)$ and $\gamma(r_s)$ are calculated in the Hartree–Fock approximation. This expression is evaluated numerically after obtaining $S(q)$ and $\gamma(r_s)$ as functions of density. Another way to write this expression is to change integration variables to $r_s' = \lambda r_s$, which gives

$$E_{\text{corr}} = \frac{0.9163}{r_s} - \frac{4}{\pi r_s^2}\left(\frac{9\pi}{4}\right)^{1/3}\int_0^{r_s} dr_s'\gamma(r_s')$$

Some results are given in Fig. 5.17. Four curves are shown, corresponding to RPA, Hubbard, Singwi–Sjölander, and Nozieres–Pines. The last is included to provide a comparison with the earlier plot of the correlation energy in Fig. 5.6. The RPA curve predicts a correlation energy

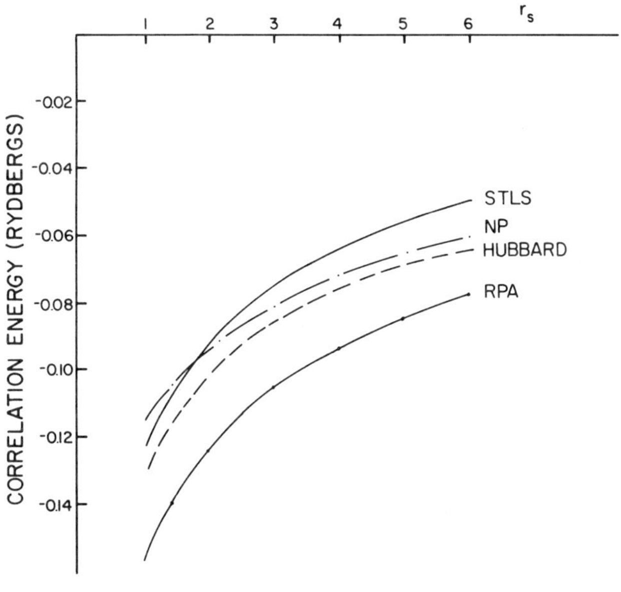

FIGURE 5.17. Correlation energy of the electron gas according to different model dielectric functions: RPA, Hubbard, and Singwi *et al.* Also shown is the interpolation formula of Nozieres and Pines to facilitate comparison with Fig. 5.6.

with a magnitude larger than the others. The other three are similar and also similar to that of Wigner. Thus the correlation energy may be obtained with a knowledge of the static structure factor $S(q)$—if it is known for all densities. We presume the Singwi–Sjölander correlation energy is best, because of its success in predicting a positive $g(r)$. The rather direct relationship between $g(r)$ and $S(q)$ implies that the best $g(r)$ is the best $S(q)$, which in turn is the best correlation energy.

D. Compressibility

The compressibility of the electron gas is defined as follows. Let E be the total energy of the system, which was calculated in Sec. 5.1. This is an extrinsic quantity; i.e., it is proportional to the number of particles N in the system of volume ν. The pressure P is defined as the rate of change in E with volume at constant N. The inverse compressibility is the rate of change of P under the same conditions:

$$P = -\left(\frac{dE}{d\nu}\right)_N$$

$$\frac{1}{K} = -\nu\left(\frac{dP}{d\nu}\right)_N$$

The definition of K^{-1} is multiplied by ν so that K is not dependent on the size of the system—it is not dependent on N. The compressibility is usually compared to that of a free-particle Fermi gas at zero temperature. There the only energy term is the kinetic energy, which we showed in Sec. 5.1 to be

$$E = \frac{3}{5} E_F N = \frac{3}{10} \frac{\hbar^2}{m} N\left(\frac{3\pi^2 N}{\nu}\right)^{2/3}$$

We differentiate this twice with respect to ν, while keeping N fixed, and find

$$P = +\frac{2}{5} E_F n_0 = +\frac{1}{5} \frac{\hbar^2}{m} (3\pi^2)^{2/3}\left(\frac{N}{\nu}\right)^{5/3}$$

$$\frac{1}{K_f} = \frac{2}{3} E_F n_0 = \frac{1}{3} \frac{\hbar^2}{m} \frac{N}{\nu} \left(\frac{3\pi^2 N}{\nu}\right)^{2/3}$$

so that we have shown the free-particle compressibility to be

$$K_f = \frac{3}{2E_F n_0}$$

Most of the numerical results are presented as the ratio K_f/K, where K_f is the above result. An effective mass of unity is assumed.

This procedure for finding the compressibility may be generalized. Ignoring surface effects, the total energy of the system is N times the exact ground state energy per particle $E = NE_g$. For a fixed N, the only volume dependence is through the density dependence of $E_g(n_0)$, where $n_0 = N/v$. Thus twice differentiating the total energy yields

$$P = -N \frac{dE_g}{dv} = n_0^2 \frac{dE_g}{dn_0}$$

$$\frac{1}{K} = -v\left(\frac{dP}{dv}\right) = n_0 \frac{d}{dn_0} P(n_0) = 2n_0^2 E_g' + n_0^3 E_g''$$

$$= n_0^2 \frac{d^2}{dn_0^2} (n_0 E_g)$$

The last line shows that this may be also represented as the second derivative of the quantity $f(n_0) = n_0 E_g$. Since $E = NE_g = vf$, $f(n_0)$ is just the energy per unit volume. The compressibility is easily found as the second derivative of this quantity multiplied by n_0^2. The calculation of the ground state energy E_g was described in Sec. 5.6C. Using this method, one may find $f(n_0)$ numerically. This permits a computation of the compressibility from the previous calculation.

Another relation for the compressibility may be obtained by using the theorem of Seitz, which was given in (5.1.13):

$$\mu = \frac{d}{dn_0} f(n_0)$$

Here μ is the chemical potential of the electrons, evaluated ignoring surface effects. The compressibility is related to the change in chemical potential with density:

$$\frac{1}{K} = n_0^2 \frac{d}{dn_0} \mu \tag{5.6.6}$$

In some cases it may be easier to calculate the properties of a particle on the Fermi surface rather than the total energy. In particular, only a single derivative is needed to evaluate (5.6.6).

Another method of evaluating K is through the *compressibility sum rule.* This is an exact relationship between the compressibility and the long-wavelength limit of the static dielectric function (see Nozieres, 1964). Take the longitudinal dielectric function $\varepsilon(q, \omega)$, set $\omega = 0$, and then take the limit $q \to 0$. In this limit, one obtains

$$\lim_{q\to 0} \varepsilon(q, 0) = 1 + \frac{4\pi e^2}{q^2} n_0{}^2 K \tag{5.6.7}$$

By substituting the free-particle compressibility $K_f = 3/(2n_0 E_F)$, the compressibility sum rule may also be written as

$$\lim_{q\to 0} \varepsilon(q, 0) = 1 + \frac{q_{\text{TF}}^2}{q^2}\left(\frac{K}{K_f}\right)$$

This shows immediately that the RPA predicts $K/K_f = 1$, since $\varepsilon_{\text{RPA}} \to 1 + q_{\text{TF}}^2/q^2$. It also provides a method of relating K to $G(q)$. If one takes the limit $q \to 0$ for the dielectric function (5.5.16) and remembers that $v_q P^{(1)} \to -q_{\text{TF}}^2/q^2$,

$$\begin{aligned}\lim_{q\to 0} \varepsilon(q, 0) &= 1 - \frac{v_q P^{(1)}}{1 + G(q) v_q P^{(1)}} \\ &= 1 + \frac{q_{\text{TF}}^2/q^2}{1 - [q_{\text{TF}}^2 G(q)/q^2]}\end{aligned}$$

The function $G(q)$ vanishes as q^2, so that the limit $G(q)/q^2$ yields a constant, which we shall call $a/k_F{}^2$. The constant a is dimensionless. The Hubbard theory predicts that $a = \frac{1}{2}$, so that the compressibility is

$$\begin{aligned}\frac{K_f}{K} &= \lim_{q\to 0}\left[1 - q_{\text{TF}}^2 \frac{G(q)}{q^2}\right] = 1 - a\left(\frac{q_{\text{TF}}}{k_F}\right)^2 \\ \left(\frac{K_f}{K}\right)_{\text{Hubbard}} &= 1 - \frac{1}{2}\left(\frac{q_{\text{TF}}}{k_F}\right)^2 = 1 - \frac{r_s}{3.01}\end{aligned} \tag{5.6.8}$$

The constant a may also be obtained from the Singwi–Sjölander theory, where $G(q)$ is given by (5.5.21):

$$G(q) = -\frac{1}{n_0}\int \frac{d^3q'}{(2\pi)^3} \frac{\mathbf{q}\cdot\mathbf{q}'}{q'^2} [S(\mathbf{q} - \mathbf{q}') - 1]$$

Change variables of integration to $\mathbf{Q} = \mathbf{q} - \mathbf{q}'$, and evaluate this in the limit where $q \to 0$:

$$\begin{aligned}\lim_{q\to 0} G(q) &= -\frac{3\pi^2}{k_F{}^3}\frac{1}{(2\pi)^2}\int_0^\infty Q^2\, dQ[S(Q) - 1]\int_{-1}^{1} d\nu \frac{q(q - Q\nu)}{Q^2 + q^2 - 2qQ\nu} \\ &= \frac{-3}{4k_F{}^3}\int_0^\infty Q^2\, dQ[S(Q) - 1]\int_{-1}^{\prime} d\nu\left[\frac{q^2}{Q^2}(1 - 2\nu^2) + O(q^4)\right] \\ &= -\frac{1}{2}\frac{q^2}{k_F{}^3}\int_0^\infty dQ[S(Q) - 1]\end{aligned}$$

The combination of factors on the right is the same as in the definition of γ in (5.6.4). Thus we conclude that the Singwi–Sjölander theory predicts that $a = \gamma$ or that

$$\frac{K_f}{K} = 1 - \gamma\left(\frac{q_{\text{TF}}}{k_F}\right)^2$$

The value of a may be deduced from Table 5.5, since the limit $q \to 0$ yields $a = AB$, where A and B are the constants listed. This value of a varies with density, but it is generally smaller than the Hubbard value of $\frac{1}{2}$ and nearer to $\frac{1}{4}$.

We have described two ways to evaluate the compressibility. The first is by finding the ground state energy per particle and differentiating twice. The second is from the long-wavelength limit of the dielectric function. These would give the same result if all quantities could be found exactly. But when we use approximate theories for $\varepsilon(q, \omega)$, the two methods of finding the compressibility will give different results. For example, consider the compressibility in the Hartree–Fock approximation. The ground state energy now includes the exchange term, and a second derivative of the ground state energy gives $a = \frac{1}{4}$ as the Hartree–Fock prediction. But when the exchange hole is used to get $g(r)$, as at the end of Sec. 5.1, and then $S(q)$, and then $G(q)$, and finally a, one obtains the different result $a = \frac{3}{8}$. The Hartree–Fock approximation predicts different results in the two methods of calculation. The form of $G(r)$ suggested by Vashishta and Singwi (1972) and Schneider *et al.* (1970) has the virtue that it gives almost the same compressibility by both methods of calculation. If we denote their function as $G_{\text{VS}}(q)$, while $G(q)$ is the function in (5.5.21), they are related by

$$G_{\text{VS}}(q) = \left(1 + \frac{2}{3} n_0 \frac{d}{dn_0}\right) G(q)$$

The derivative is taken with respect to density. This form is regarded as being particularly accurate for the compressibility.

Figure 5.18 shows some theoretical compressibilities as a function of r_s. The solid lines are results obtained by differentiating the ground state energy. All the theories except Hartree–Fock predict identical compressibilities when evaluated in this fashion. And the Hartree–Fock result is not very different. Presumably, the curve labeled RPA, SS (for Singwi–Sjölander), Hubbard, VS (for Vashishta–Singwi) is the proper result, since it is given by so many theories. The dashed lines are the results obtained from the compressibility sum rule—from the limit of $\varepsilon(q)$ or $G(q)$ as $q \to 0$.

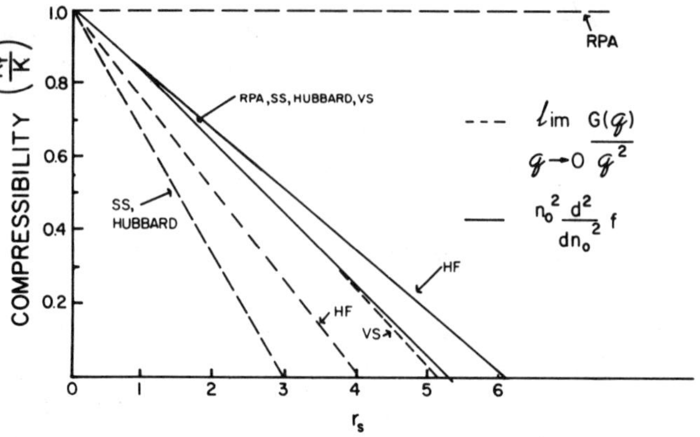

FIGURE 5.18. The compressibility K of the electron gas compared to that of the non-interacting gas K_f. The five theories shown are Hartree–Fock, RPA, Singwi–Sjölander, Hubbard, and Vashishta–Singwi. Each theory makes two predictions: the solid line from the derivative of the ground state energy and the dashed line from the compressibility sum rule.

These vary widely among the theories, except for the coincidence of Hubbard with the original Singwi–Sjölander result (see Singwi *et al.*, 1968). The Vashishta–Singwi curve is nearly identical to that calculated by the other method.

The compressibility is predicted to go negative about $r_s = 5$. This suggests that the electron gas is unstable at densities lower than this critical value. Real metals exist with larger values than this—i.e., Cs has $r_s = 5.63$—but they are not homogeneous electron gases. However, metals do not exist with r_s values greater than Cs, which implies that there may be a fundamental limit beyond which the density may no longer be reduced. Unfortunately, this point cannot be tested. One may put solids under pressure but not under tension.

5.7. SUM RULES

The longitudinal dielectric function $\varepsilon(q, \omega)$ of the homogeneous electron gas has some exact moments which are useful for checking results. Some of these have already been used and mentioned. For example, the long-wavelength limit is $\lim_{q \to 0} \varepsilon(q, \omega) = 1 + v_q n_0^2 K$, where K is the isothermal compressibility. Similarly, the high-frequency limit is

$$\lim_{\omega \to \infty} \varepsilon(q, \omega) = 1 - \frac{\omega_p^2}{\omega^2} + O(\omega^{-4}) \tag{5.7.1}$$

where ω_p is the plasma frequency. The frequency ω is large when it is larger than other energies in the system, such as the plasma frequency or the Fermi energy.

Another class of exact results are called *sum rules*. The "sums" are actually integrals over frequency. Several of them are

$$\int_0^\infty d\omega\omega\varepsilon_2(q, \omega) = \frac{\pi}{2} \omega_p^2 \tag{5.7.2}$$

$$\lim_{q\to 0} \int_0^\infty \frac{d\omega}{\omega} \varepsilon_2(q, \omega) = \frac{\pi}{2} (v_q n_0^2 K) \tag{5.7.3}$$

$$\int_0^\infty d\omega\omega \, \mathrm{Im}\left[\frac{1}{\varepsilon(q, \omega)}\right] = -\frac{\pi}{2} \omega_p^2 \tag{5.7.4}$$

$$\lim_{q\to 0} \int_0^\infty \frac{d\omega}{\omega} \, \mathrm{Im}\left[\frac{1}{\varepsilon(q, \omega)}\right] = -\frac{\pi}{2} \tag{5.7.5}$$

These are exact results. Approximate formulas for $\varepsilon(q, \omega)$ may not automatically obey these relationships. Indeed, approximate formulas are considered virtuous according to whether they satisfy these identities. Our sign convention has $\varepsilon_2(q, \omega) > 0$ for $\omega > 0$. This implies that $\mathrm{Im}(1/\varepsilon) = -\varepsilon_2/|\varepsilon|^2$ is a negative quantity for $\omega > 0$. The last two sum rules may also be written as

$$\int_0^\infty d\omega\omega \frac{\varepsilon_2(q, \omega)}{\varepsilon_1^2 + \varepsilon_2^2} = +\frac{\pi}{2} \omega_p^2$$

$$\lim_{q\to 0} \int_0^\infty \frac{d\omega}{\omega} \frac{\varepsilon_2(q, \omega)}{\varepsilon_1^2 + \varepsilon_2^2} = \frac{\pi}{2}$$

These exact results may be proved by a variety of methods. The third relation (5.7.4) is often called the *longitudinal f-sum rule*. It may be proved from the double commutator

$$C = \langle[[H, \varrho(\mathbf{q})], \varrho(-\mathbf{q})]\rangle$$

where the average is taken, at finite temperature, over the thermodynamic states of the system. The Hamiltonian H is for the full homogeneous electron gas. However, the density operator $\varrho(\mathbf{q}) = \sum_{\mathbf{k}s} c_{\mathbf{k}+\mathbf{q}s} c_{\mathbf{k}s}$ commutes with all terms in this Hamiltonian except the kinetic energy. Thus the first commutator is

$$\begin{aligned}
[H, \varrho(\mathbf{q})] &= \sum_{\substack{\mathbf{p}\mathbf{k}' \\ s\sigma}} \xi_p [c_{\mathbf{p}s}^\dagger c_{\mathbf{p}s}, c_{\mathbf{k}+\mathbf{q}\sigma} c_{\mathbf{k}\sigma}] = \sum_{\mathbf{p}s} \xi_{\mathbf{p}} (c_{\mathbf{p}s}^\dagger c_{\mathbf{p}-\mathbf{q}s} - c_{\mathbf{p}+\mathbf{q}s}^\dagger c_{\mathbf{p}s}) \\
&= \sum_{\mathbf{p}s} c_{\mathbf{p}+\mathbf{q}s}^\dagger c_{\mathbf{p}s} (\varepsilon_{\mathbf{p}+\mathbf{q}} - \varepsilon_{\mathbf{p}}) = \frac{q^2}{2m} \varrho(q) + \frac{1}{m} \sum_{\mathbf{p}s} \mathbf{q} \cdot \mathbf{p} c_{\mathbf{p}+\mathbf{q}s}^\dagger c_{\mathbf{p}s}
\end{aligned}$$

The term proportional to $\varrho(\mathbf{q})$ will commute with $\varrho(-\mathbf{q})$. Only the other term needs to be further evaluated,

$$
\begin{aligned}
C = \langle[[H, \varrho(\mathbf{q})], \varrho(-\mathbf{q})]\rangle &= \left\langle\left[\frac{1}{m}\sum \mathbf{q}\cdot\mathbf{p}c^{\dagger}_{\mathbf{p}+\mathbf{q}s}c_{\mathbf{p}s}, \varrho(-\mathbf{q})\right]\right\rangle \\
&= \frac{1}{m}\sum_{\mathbf{p}\mathbf{q}'} \mathbf{q}\cdot\mathbf{p}\langle(c^{\dagger}_{\mathbf{p}+\mathbf{q}s}c_{\mathbf{p}+\mathbf{q}s} - c^{\dagger}_{\mathbf{p}s}c_{\mathbf{p}s})\rangle \\
&= \frac{1}{m}\sum_{\mathbf{p}\mathbf{q}} \langle c^{\dagger}_{\mathbf{p}s}c_{\mathbf{p}s}\rangle[(\mathbf{p}-\mathbf{q})\cdot\mathbf{q} - \mathbf{p}\cdot\mathbf{q}] = -\frac{q^2}{m}\sum_{\mathbf{p}s} n_{\mathbf{p}} = -\frac{q^2N}{m}
\end{aligned}
$$

and yields the simple result that this double commutator is just $-Nq^2/m$. Next, this double commutator will be shown to be proportional to the sum in (5.7.4). It is evaluated by inserting the complete sets of states $|\,n\rangle$ and $|\,m\rangle$, which are exact eigenstates of the Hamiltonian, and then collecting terms:

$$
\begin{aligned}
C = \langle[[H, \varrho(\mathbf{q})], \varrho(-\mathbf{q})]\rangle &= \sum_{n,m}\{e^{-\beta E_n}\langle n\,|\,[H, \varrho(\mathbf{q})]\,|\,m\rangle\langle m\,|\,\varrho(-\mathbf{q})\,|\,n\rangle \\
&\quad - e^{-\beta E_m}\langle m\,|\,\varrho(-\mathbf{q})\,|\,n\rangle\langle n\,|\,[H, \varrho(\mathbf{q})]\,|\,m\rangle\} \\
&= \sum_{n,m} |\langle n\,|\,\varrho(\mathbf{q})\,|\,m\rangle|^2 (e^{-\beta E_n} - e^{-\beta E_m})(E_n - E_m)
\end{aligned}
$$

Earlier in (5.4.11) it was shown that

$$
\mathrm{Im}\left[\frac{1}{\varepsilon(q,\omega)}\right] = -\pi(1-e^{-\beta\omega})\frac{v_q}{\nu}\sum_{n,m} e^{-\beta E_n}|\langle n\,|\,\varrho(\mathbf{q})\,|\,m\rangle|^2\,\delta(\omega + E_n - E_m)
$$

so that the sum rule is

$$
\begin{aligned}
&\int_{-\infty}^{\infty} d\omega\omega\,\mathrm{Im}\left[\frac{1}{\varepsilon(q,\omega)}\right] \\
&\qquad = -\pi\frac{v_q}{\nu}\sum_{n,m}|\langle n\,|\,\varrho(\mathbf{q})\,|\,m\rangle|^2 (E_m - E_n)(e^{-\beta E_n} - e^{-\beta E_m})
\end{aligned}
$$

When we compare the sum rule with the double commutator, we find that they are proportional, and this provides the derivation of the sum rule (5.7.4):

$$
\int_{-\infty}^{\infty} d\omega\omega\,\mathrm{Im}\left[\frac{1}{\varepsilon(q,\omega)}\right] = \frac{\pi v_q}{\nu}C = -\frac{4\pi^2 e^2 n_0}{m} = -\pi\omega_p^{\,2}
$$

The integral is from $-\infty$ to ∞, but the negative and positive parts contribute equally, so the left-hand side is twice the integral from 0 to ∞.

The other three sum rules may be derived by using the *Kramers–Kronig relations* (Kronig, 1926; Kramers, 1927). Let $B(\omega)$ be a function of complex variable ω, which is analytic in the upper half plane.

We also need to assume that $B(\omega) \to 0$ as $|\omega| \to \infty$. This means that it has no poles or branch cuts above the real axis. We shall prove that

$$B(\omega) = \frac{1}{i\pi} \int_{-\infty}^{\infty} d\omega' B(\omega') P \frac{1}{\omega' - \omega} \tag{5.7.6}$$

This integral is along the real axis. On the real axis, the function $B(\omega)$ has real and imaginary parts, which are called B_1 and B_2, so that $B(\omega) = B_1(\omega) + iB_2(\omega)$. They are related by taking the real and imaginary parts of (5.7.6):

$$\begin{aligned} B_1(\omega) &= \frac{1}{\pi} \int_{-\infty}^{\infty} d\omega' B_2(\omega') P \frac{1}{\omega' - \omega} \\ B_2(\omega) &= \frac{-1}{\pi} \int_{-\infty}^{\infty} d\omega' B_1(\omega') P \frac{1}{\omega' - \omega} \end{aligned} \tag{5.7.7}$$

These two identities are the Kramers–Kronig relations. They are useful throughout all areas of physics. Our own interest is in applying them to the real and imaginary parts of the dielectric function $\varepsilon(\omega) = \varepsilon_1 + i\varepsilon_2$. Another application is to define the refractive index $n(\omega)$ and extinction coefficient $k(\omega)$, where $\tilde{n} = n + ik = \varepsilon^{1/2}$. The complex refractive index $\tilde{n}$ obeys the conditions of the theorem and hence the preceding identities. There are other applications in optics, scattering theory, and virtually all branches of physics.

The theorem is proved by examining the contour integral

$$0 = \int \frac{d\omega' B(\omega')}{\omega' - \omega}$$

where the contour is shown in Fig. 5.19. The contour integral is zero because there are no poles or branch cuts within the contour. The singular point $\omega' = \omega$ is avoided by a local semicircle around it. The contour is closed by a semicircle with a large radius R. Take the limit $R \to \infty$, and this contribution will vanish, since it is assumed that $B(\omega')$ vanishes sufficiently rapidly at large argument. There only remains the contour integral along the real axis plus the small semicircle about the point $\omega' = \omega$. These two contributions are

$$0 = \int_{-\infty}^{\infty} d\omega' B(\omega') P \frac{1}{\omega' - \omega} + i \int_{\pi}^{0} d\theta B(\omega + \delta e^{i\theta})$$

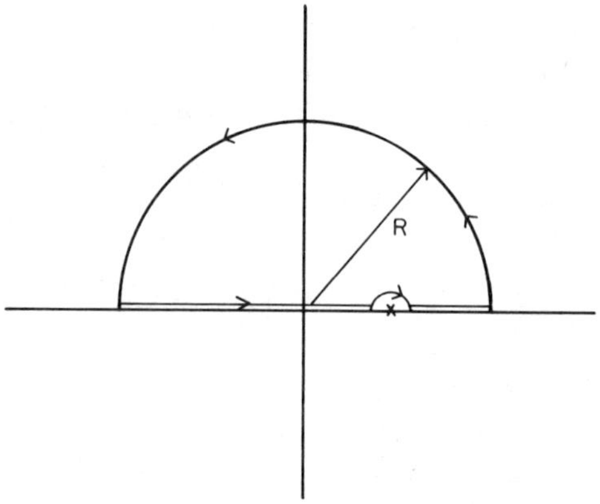

FIGURE 5.19

The second term on the right, from the small semicircle ($\delta \to 0$), gives $-i\pi B(\omega)$. This term is moved to the left of the equal sign, and the basic theorem (5.7.6) is thereby demonstrated.

The most significant hypothesis in the theorem is that $B(\omega)$ is analytic in the upper half-plane. This is a property of causal functions. Any physical function of time $b(t)$, which describes the response of a system, has the property that $b(t) = 0$ for $t < 0$. To show this, relate the signal $S(t)$ to a force $F(t)$ by the relation

$$S(t) = \int_{-\infty}^{\infty} dt' b(t - t') F(t')$$

where $b(t - t')$ is the response at t to the driving term at t'. A causal function has the property that there is no signal before the response, so that $b(t - t')$ is nonzero only when $t \geq t'$. Thus causal functions have $b(t) = 0$ for $t < 0$. We write this as a Fourier transform of $B(\omega)$:

$$b(t) = \int_{-\infty}^{\infty} d\omega e^{-i\omega t} B(\omega)$$

For $t < 0$ we can evaluate the integral on the right by closing the contour of integration in the upper half plane. If $B(\omega)$ is analytic there, with no poles or cuts, the contour integral is zero. Any poles or cuts would make $b(t)$ have a finite value for $t < 0$, which is unphysical. Thus the analyticity in the upper half-plane, of $B(\omega)$, is a consequence of causality for physical functions. Our retarded correlation functions are causal and are analytic in the upper half-plane. All the retarded correlation functions obey these Kramers–Kronig relations.

The advanced Green's functions, which are not analytic in the upper half-plane, do *not* obey the relationships (5.7.7). The advanced Green's functions are zero for $t > 0$, which makes them analytic in the lower half-plane of frequency space. They obey the complex conjugate of (5.7.6).

The other sum rules are easy to prove as simple applications of the Kramers–Kronig relations. First take the case where $B(\omega) = 1/\varepsilon(q, \omega) - 1$. The -1 term is included to make B vanish as $|\omega| \to \infty$. Using the first Kramers–Kronig relation gives

$$\mathrm{Re}\left[\frac{1}{\varepsilon(q, \omega)}\right] - 1 = \frac{1}{\pi}\int_{-\infty}^{\infty} d\omega' \, \mathrm{Im}\left[\frac{1}{\varepsilon(q, \omega')}\right] P\,\frac{1}{\omega' - \omega}$$

The right-hand side may be simplified by using the fact that $\mathrm{Im}[1/\varepsilon(q, \omega)]$ is asymmetric in frequency $\mathrm{Im}[1/\varepsilon(q, -\omega)] = -\mathrm{Im}[1/\varepsilon(q, \omega)]$. The integrand $(-\infty, 0)$ is evaluated by changing $\omega \to -\omega$ and combining this with the $(0, \infty)$ part to give

$$\mathrm{Re}\left[\frac{1}{\varepsilon(q, \omega)}\right] = 1 + \frac{2}{\pi}\int_{0}^{\infty} d\omega' \, \mathrm{Im}\left[\frac{1}{\varepsilon(q, \omega')}\right] P\,\frac{\omega'}{\omega'^2 - \omega^2} \tag{5.7.8}$$

This identity is also a sum rule but not one of the simple ones we are trying to prove. However, they are just simple limits of this result. First take the limit where $\omega \to 0$:

$$\frac{1}{\varepsilon(q)} = \frac{1}{\varepsilon(q, 0)} = 1 + \frac{2}{\pi}\int_{0}^{\infty} \frac{d\omega'}{\omega'} \, \mathrm{Im}\left[\frac{1}{\varepsilon(q, \omega')}\right] \tag{5.7.9}$$

The quantity $\varepsilon(q, 0)$ is real, so (5.7.9) expresses the static dielectric function as an integral over the loss function. The sum rule (5.7.5) is obtained by taking the limit where $q \to 0$. The quantity $\varepsilon(q)$ is evaluated from the compressibility sum rule (5.6.7),

$$\begin{aligned} &\lim_{q\to 0} \varepsilon(q) = 1 + v_q n_0^2 K \to O\left(\frac{1}{q^2}\right) \to \infty \\ &\lim_{q\to 0} \frac{1}{\varepsilon(q)} = \lim_{q\to 0} \frac{q^2}{q^2 + 4\pi e^2 n_0^2 K} = 0 \end{aligned} \tag{5.7.10}$$

which shows that the left-hand side is zero. Thus we obtain

$$\lim_{q\to 0}\int_{0}^{\infty} \frac{d\omega'}{\omega'} \, \mathrm{Im}\left[\frac{1}{\varepsilon(q, \omega')}\right] = \frac{-\pi}{2}$$

which is (5.7.5).

The longitudinal f-sum rule, which was proved by taking commutators, may be demonstrated by another method. Take the limit $\omega \to \infty$ in (5.7.8). For very large values of ω, the factor $\omega'^2 - \omega^2$ may be replaced by $-\omega^2$, so that (5.7.8) becomes

$$\lim_{\omega\to\infty} \mathrm{Re}\left[\frac{1}{\varepsilon(q, \omega)}\right] = 1 - \frac{1}{\omega^2}\frac{2}{\pi}\int_0^\infty d\omega'\omega' \,\mathrm{Im}\left[\frac{1}{\varepsilon(q, \omega')}\right] + O(\omega^{-4}) \tag{5.7.11}$$

From (5.7.1), the left-hand side becomes

$$\lim_{\omega\to\infty} \mathrm{Re}\left[\frac{1}{\varepsilon(q, \omega)}\right] = \frac{1}{1 - \dfrac{\omega_p^2}{\omega^2}} = 1 + \frac{\omega_p^2}{\omega^2} + O(\omega^{-4})$$

If we equate the two terms of order ω^{-2}, we derive the sum rule (5.7.4):

$$\int_0^\infty d\omega'\omega' \,\mathrm{Im}\left[\frac{1}{\varepsilon(q, \omega')}\right] = \frac{-\pi}{2}\omega_p^2$$

An alert reader will note that we never actually proved that (5.7.1) was a rigorous result. Thus our use of it in the present demonstration is not conclusive. However, we can turn the argument around. Since we have demonstrated the longitudinal f-sum rule by another method, using commutators, it is certainly rigorous. This sum rule, and (5.7.11), serves to prove that

$$\lim_{\omega\to\infty} \mathrm{Re}\left[\frac{1}{\varepsilon(q, \omega)}\right] = 1 + \frac{\omega_p^2}{\omega^2} + O\left(\frac{1}{\omega^4}\right)$$

which is thus a proof of the validity of (5.7.1).

The remaining two sum rules, (5.7.2) and (5.7.3), are obtained from the Kramers–Kronig relations using $\varepsilon - 1$ for $B(\omega)$. Again the -1 term is included so that $\varepsilon - 1$ vanishes at large values of $|\omega|$. From (5.7.7) we obtain

$$\begin{aligned} \varepsilon_1(q, \omega) - 1 &= \frac{1}{\pi}\int_{-\infty}^\infty d\omega'\varepsilon_2(q, \omega')P\frac{1}{\omega' - \omega} \\ \varepsilon_1(q, \omega) &= 1 + \frac{2}{\pi}\int_0^\infty d\omega'\varepsilon_2(q, \omega')P\frac{\omega'}{\omega'^2 - \omega^2} \end{aligned} \tag{5.7.12}$$

The second equation here is derived using the fact that $\varepsilon_2(q, -\omega) = -\varepsilon_2(q, \omega)$. This equation is evaluated first for $\omega = 0$:

$$\varepsilon_1(q) = 1 + \frac{2}{\pi}\int_0^\infty \frac{d\omega'}{\omega'}\varepsilon_2(q, \omega') \tag{5.7.13}$$

Equation (5.7.13) is an exact relation between the imaginary part of the dielectric function $\varepsilon_2(q, \omega')$ and the static part $\varepsilon_1(q)$. In the limit where $q \to 0$, the compressibility sum rule (5.7.10) gives $\varepsilon_1 \to 1 + v_q n_0^2 K$, so that

$$\lim_{q\to 0} \int_0^\infty \frac{d\omega'}{\omega'} \varepsilon_2(q, \omega') = + \frac{\pi}{2} v_q n_0^2 K$$

which establishes (5.7.3).

The remaining sum rule (5.7.2) is found from the limit $\omega \to \infty$ in (5.7.12):

$$\lim_{\omega\to\infty} \varepsilon_1(q, \omega) = 1 - \frac{2}{\omega^2 \pi} \int_0^\infty d\omega' \omega' \varepsilon_2(q, \omega') + O\left(\frac{1}{\omega^4}\right)$$

By using (5.7.1), the left-hand side becomes $1 - \omega_p^2/\omega^2$, and equating terms of order ω^{-2} produces the result

$$\int_0^\infty d\omega' \omega' \varepsilon_2(q, \omega') = \frac{\pi}{2} \omega_p^2$$

Thus we have demonstrated the four sum rules. These are not the only sum rules which exist. Others are sometimes found useful, in particular the third moment (Mihara and Puff, 1968; Hopfield, 1970).

It is curious that the first moment sum rules of both $\varepsilon_2(q, \omega)$ and $-\mathrm{Im}[1/\varepsilon(q, \omega)]$ yield the same result $(\pi/2)\omega_p^2$. And this result is independent of q. One should not conclude that ε_2 and $-\mathrm{Im}(1/\varepsilon)$ are identical. Earlier, in Fig. 5.13, it was shown that these two functions are similar at large q but quite different at small values of q/q_F.

Plasmons play a particularly important role in these sum rules at small values of q. In the limit where $q \to 0$, from (5.5.9) we have

$$\lim_{q\to 0} \varepsilon(q, \omega) = 1 - \frac{\omega_p^2}{\omega^2} + O(\omega^4)$$

$$\lim_{q\to 0} \left[\frac{1}{\varepsilon(q, \omega)} - 1\right] = \frac{\omega^2}{\omega^2 - \omega_p^2} - 1 = \frac{\omega_p^2}{\omega^2 - \omega_p^2}$$
$$= \frac{\omega_p}{2}\left(\frac{1}{\omega - \omega_p} - \frac{1}{\omega + \omega_p}\right)$$

For a retarded function, we let $\omega \to \omega + i\delta$, so that the imaginary part of this expression gives

$$\lim_{q\to 0} \mathrm{Im}\left[\frac{1}{\varepsilon(q, \omega)}\right] = -\frac{\pi}{2} \omega_p[\delta(\omega - \omega_p) - \delta(\omega + \omega_p)]$$

This relation is rigorously valid only in the limit where $q \to 0$. Nevertheless, it is curious that it satisfies the last two sum rules exactly:

$$\lim_{q\to 0}\int_0^\infty d\omega\omega \operatorname{Im}\left[\frac{1}{\varepsilon(q,\omega)}\right] = \int_0^\infty d\omega\omega\left[-\frac{\pi}{2}\omega_p\delta(\omega-\omega_p)\right] = -\frac{\pi}{2}\omega_p^2$$

$$\lim_{q\to 0}\int_0^\infty \frac{d\omega}{\omega}\operatorname{Im}\left[\frac{1}{\varepsilon(q,\omega)}\right] = \int_0^\infty \frac{d\omega}{\omega}\left[\frac{-\pi}{2}\omega_p\delta(\omega-\omega_p)\right] = \frac{-\pi}{2}$$

This shows that plasmons provide all the contributions to $\operatorname{Im}[1/\varepsilon(q,\omega)]$ in the limit where $q \to 0$. Since the interpretation of $\operatorname{Im}[1/\varepsilon(q,\omega)]$ is that it describes longitudinal excitations of the electron gas, plasmons are the *only excitation* in the limit where $q \to 0$. Any other excitation would make a contribution to the sum rule, which is impossible since plasmons provide the entire result. Thus any other excitation has zero strength in the limit where $q \to 0$.

5.8. ONE-ELECTRON PROPERTIES

In this section we shall discuss single-particle properties of electrons in the homogeneous electron gas. These include the effective mass, spin susceptibility, and mean free path. The present section is put at the end of this chapter deliberately to make a point: The one-electron properties are not very important for discussions of most features of the electron gas. Our discussions of ground state energies and pair correlations did not rely significantly on the properties of the individual electrons. The excitation spectra of the electron gas are also not very dependent on the properties of single electrons. In fact, the converse is true: We need to know the collective properties of the electron gas in order to discuss the behavior of a single electron. This single electron interacts with the other electrons, which are represented as collective excitations. Thus our procedure has been to describe the collective excitations first. This was done assuming that the particles were nearly free (i.e., RPA) or had some correlation (i.e., Hubbard, Singwi–Sjölander). But we ignored the finite mean free path or possible effective mass changes of the electron. They do not change these collective properties to a significant degree.

This is true in the electron gas because our original assumption is that the electrons are free plane waves, and the final description is not very different. Thus the self-energy effects we find do not alter the basic picture. If our calculations were to show that something drastic happened to the individual electrons—that they were magnetic or superconducting or semiconducting—obviously the collective properties would be altered in a significant fashion. Thus it is important that the initial eigenstates are an approximate description of the actual physical system. It is only in this circumstance that one finds that the one-particle self-energies have little influence on collective properties.

The one-electron properties are obtained from a calculation of the electron self-energy due to electron–electron interactions. This self-energy was discussed earlier, in Sec. 5.1, in the discussion of electron correlation. If we had chosen to follow the original discussion of Gell-Mann and Brueckner (1957), the correlation energy could be derived without first discussing the electron self-energies. Again we make the point that the collective properties are not significantly dependent on first deriving the one-electron properties. For the discussion of electron correlation we followed Quinn and Ferrell (1958) and first considered the electron self-energy. The results of that discussion can be used as the starting point for the present derivations. The best calculations which have been reported of single-electron properties calculate only the screened exchange interaction.

The electron self–energy from screened exchange energy is

$$\Sigma_{sx}(\mathbf{k}, ik_n) = -\frac{1}{\beta}\sum_m \int \frac{d^3q}{(2\pi)^3} v_q \frac{\Gamma(\mathbf{q}, iq_m)}{\varepsilon(\mathbf{q}, iq_m)} \mathscr{G}_0(\mathbf{k}+\mathbf{q}, ik_n + iq_m) \tag{5.8.1}$$

$$\varepsilon(\mathbf{q}, iq_m) = 1 - v_q P(\mathbf{q}, iq_m)\Gamma(\mathbf{q}, iq_m)$$

$$\Gamma(\mathbf{q}, iq_m) = 1/[1 + v_q G(q) P(q, iq_m)]$$

The integrand contains a vertex function $\Gamma(q, iq)$. The vertex function should depend upon the k variable. In four-vector notation, it is a function of $\Gamma(k, k+q)$. Neglecting the k dependence is an approximation. All dielectric functions of the Hubbard-type, with a $G(q)$ factor, make this approximation of neglecting the k dependence of the vertex function.

For dielectric functions of the Hubbard type, Ting, Lee, and Quinn (1975) showed that the same function Γ enters the numerator as enters the dielectric function. This formula is derived below. The RPA result in Sec. 5.1G is obtained by setting $\Gamma = 1$ in the above expression.

Ting, Lee, and Quinn (1975) derived (5.8.1) by starting from the definition of the ground-state energy of the interacting electron system. We use the formula in Sec. 5.6, which gives E_g as a coupling constant integral over $[S(q) - 1]$. Then we use (5.4.8) to express $S(q)$ as a frequency summation over the inverse dielectric function:

$$E_g = E_{g0} - \frac{\nu}{2\beta_0} \int_0^1 \frac{d\eta}{\eta} \int \frac{d^3q}{(2\pi)^3} \sum_{iq} \left[\frac{1}{\varepsilon_\eta(\mathbf{q}, iq)} - 1 \right] \tag{5.8.2}$$

$$\varepsilon_\eta(\mathbf{q}, iq) = 1 - \eta v_q P(\mathbf{q}, iq)/[1 + \eta v_q G(q) P(\mathbf{q}, iq)]$$

The coupling constant η has been made dimensionless by having $\eta = 0$ at $e^2 = 0$ and $\eta = 1$ at e^2. $G(q)$ may also be a functional of the product $\eta\chi$, where $\chi = v_q P$.

The self-energy can be obtained as the functional derivative of the ground-state energy with respect to the occupation number n_k. For example, the kinetic energy term can be written as

$$E_{g0} = \sum_{\mathbf{k}} n_{\mathbf{k}} \varepsilon_{\mathbf{k}}$$

$$\frac{\delta E_{g0}}{\delta n_{\mathbf{k}}} = \varepsilon_{\mathbf{k}} \tag{5.8.3}$$

$$\Sigma_{sx}(\mathbf{k}, ik) = \delta E_g / \delta n_{\mathbf{k}} - \varepsilon_{\mathbf{k}}$$

The screened exchange energy is the functional derivative of the interaction term in (5.8.2). We assume the only dependence upon $n_{\mathbf{k}}$ in the integrand is in the susceptibility χ. The functional derivative has the form for $f = 1/\varepsilon$

$$\frac{\delta f}{\delta n_{\mathbf{k}}} = v_q \frac{\delta f}{\delta \chi} \frac{\delta P}{\delta n_{\mathbf{k}}}$$

where

$$P = \frac{1}{\nu} \sum_{\mathbf{k}\sigma} n_{\mathbf{k}} \left(\frac{1}{ik - \xi_{\mathbf{k}+\mathbf{q}} + iq} \quad \frac{1}{ik - \xi_{\mathbf{k}+\mathbf{q}} - iq} \right)$$

$$\frac{\delta P}{\delta n_{\mathbf{k}}} = \frac{1}{\nu} [\mathscr{G}_0(\mathbf{k} + \mathbf{q}, iq_m + ik) + \mathscr{G}_0(\mathbf{k} + \mathbf{q}, -iq_m + ik)]$$

The two Green's function terms in $\delta P/\delta n$ contribute equally, since $\varepsilon(\mathbf{q}, iq_m)$ is a symmetric function of complex frequency iq_m. This removes the factor of 2 in front of the interaction term in (5.8.2). The two steps of coupling constant integral and functional derivative can be arranged to cancel. Introduce the variable $y = \eta\chi$, which becomes the variable of both integration and differentiation:

$$\int_0^1 \frac{d\eta}{\eta} \frac{\partial f(\eta\chi)}{\partial \chi} = \frac{1}{\chi} \int_0^\chi dy \frac{\partial f(y)}{\partial y} = \frac{1}{\chi} [f(\chi) - f(0)]$$

When $f(\chi) = 1/\varepsilon = 1/(1 - \Gamma\chi)$ then $f(\chi) - f(0) = \chi\Gamma/\varepsilon$. Collecting all of these results produces (5.8.1). This derivation is independent of the particular form for the vertex function Γ. It is also valid for any form for $G(q)$ as long as one assumes that $G(q)$ is a functional of the susceptibility χ.

Equation (5.8.1) is most easily evaluated following the method discussed in Sec. 5.1G. One first separates the exchange contribution, so that one is left with the correlation part of the electron self-energy. Zero temperature is assumed, so the frequency summation over m can be changed to a continuous integral over $d\omega$. The analytic continuation $ik_n => \xi_k$ produces two terms, which are called the line and residue parts:

$$\Sigma_{sx}(\mathbf{k}, \xi_k) = \Sigma_x(k) + \Sigma^{(\text{line})}(\mathbf{k}, \xi_k) + \Sigma^{(\text{res})}(\mathbf{k}, \xi_k) \tag{5.8.4}$$

$$\Sigma^{(\text{line})}(\mathbf{k}, \xi_k) = -\int \frac{d\omega}{2\pi} \int \frac{d^3q}{(2\pi)^3} v_q \left[\frac{\Gamma(\mathbf{q}, i\omega)}{\varepsilon(\mathbf{q}, i\omega)} - 1 \right] \tag{5.8.5}$$

$$\Sigma^{(\text{res})}(\mathbf{k}, \xi_k) = \int \frac{d^3q}{(2\pi)^3} v_q \left[\frac{\Gamma(q, \xi_{\mathbf{k}+q} - \xi_{\mathbf{k}})}{\varepsilon(q, \xi_{\mathbf{k}+\mathbf{q}} - \xi_{\mathbf{k}})} - 1 \right] [\Theta(\xi_{\mathbf{k}} - \xi_{\mathbf{k}+\mathbf{q}}) - \Theta(-\xi_{\mathbf{k}+\mathbf{q}})] \tag{5.8.6}$$

Both Σ_x and $\Sigma^{(\text{line})}$ are real.

Calculations using these formulas are shown in Fig. 5.20. Figure 5.20a shows the real part of the self-energy Σ_{sx} as a function of k for r_s values appropriate to common metals. All of the curves have a similar shape. The strong dip in the curve is due to the coupling of the electron to the plasmon. The imaginary self-energy Im Σ has a big jump in magnitude when the electron has enough kinetic energy to emit a plasmon. The Kramers–Kronig relation between Re Σ and Im Σ guarantees that Re Σ has a dip when Im Σ has a jump.

A similar dip in the self-energy is obtained whenever the electron couples strongly to an oscillator. The self-energy of electrons coupled to LO phonons in insulators has the same shape, as discussed in Problem 6.7. Of course, there the dip happens at the energy characteristic of the phonon, while here it occurs at the energy of the plasmon.

The electron self-energy also contributes to the energy width of the occupied band of electrons. This change in width can be crudely estimated as the difference between the self-energy calculated at $k = 0$ and at $k = k_F$.

$$\Delta W = \Sigma_{sx}(k_F, 0) - \text{Re}\, \Sigma_{sx}(0, -E_F)$$

The energy $\xi = 0$ at the Fermi energy, and equals $-E_F$ at the bottom of

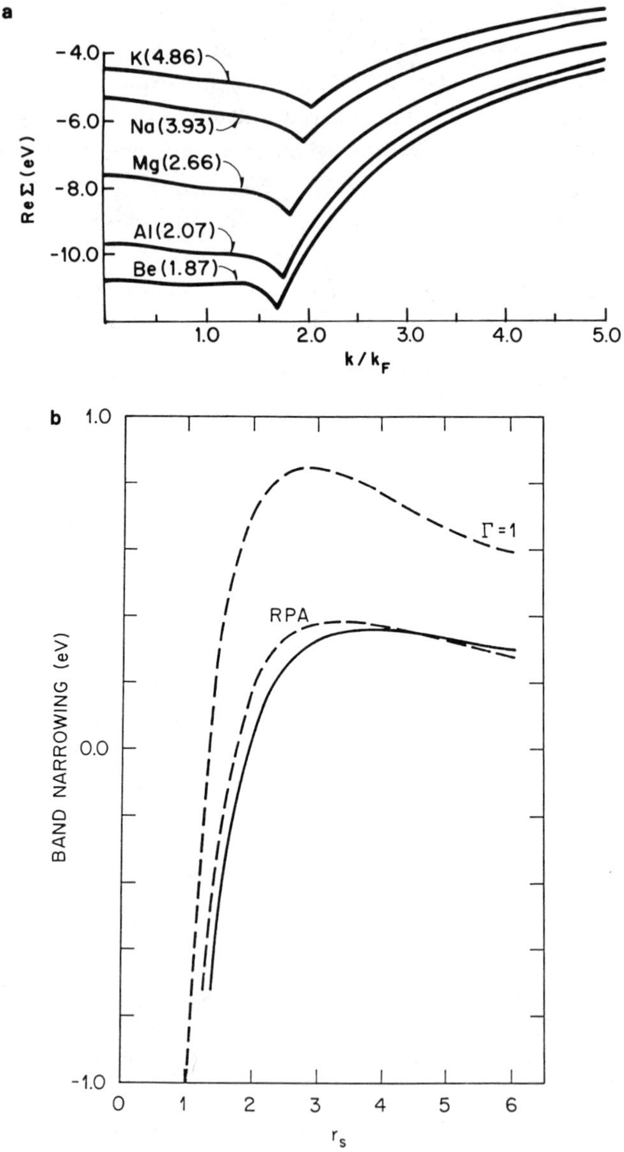

FIGURE 5.20. (a) The real part of the screened exchange energy $\Sigma_{sx}(k, \xi_k)$ as a function of k for value of r_s appropriate for simple metals. The dip is caused by plasmon emission. *Source*: Shung, Sernelius, and Mahan (1987). (b) The change in energy band width for electrons interacting with electron–electron interactions. The RPA result ($G = 0$) is compared to the $G(q)$ of Vashishta and Singwi. The results are quite similar. *Source*: Mahan and Sernelius (1989).

the band. Figure 5.20b compares ΔW calculated using RPA and the Vashishta–Singwi (VS) form for $G(q)$ given in Table 5.5. The results are quite similar. The local field factor of $G(q)$ has only a small influence upon band width of the electrons. The curve labeled $\Gamma = 1$ uses $G(q)$ in $\epsilon(q, \omega)$ but not in Γ. This is called the 'GW' approximation.

A. Renormalization Constant Z_F

The renormalization coefficient of the one-electron Green's function is defined for a point $(k, \varepsilon = \xi_k)$ as [see (3.3.23)]

$$Z(k) = \frac{1}{(1 - \partial \Sigma_{sx}/\partial \varepsilon)_{\varepsilon = \xi_k}}$$

Usually the retarded self-energy Σ_{sx} is complex, as is the renormalization coefficient. At the particular point, $k = k_F$ and $\xi = 0$, the self-energy Σ_{sx} is exactly real, since $\text{Im}(\Sigma^{(\text{res})})$ vanishes. Electrons on the Fermi energy have an infinitely long mean free path, or lifetime. This occurs because they have no unoccupied electron states into which they might scatter. Thus the renormalization coefficient is real at this point. This particular value is called $Z_F = Z(k_F)$:

$$Z_F = \frac{1}{(1 - \partial \Sigma_{sx}/\partial \varepsilon)_{\substack{k = k_F \\ \varepsilon = 0}}}$$

The renormalization coefficient is interpreted as the amount of single-particle behavior of the particle-like excitation in the electron gas. Since the excitations behave like particles, they are called *quasiparticles*. Generally the spectral function $A(k, \varepsilon)$ has a central peak plus some smooth background wings. This is illustrated schematically in Fig. 5.21. The renormalization coefficient is defined as the area under the central peak of the spectral function. If the peak is broad, then the definition is ambiguous, since the limits of the central peak are ill-defined. Precisely at the Fermi surface the

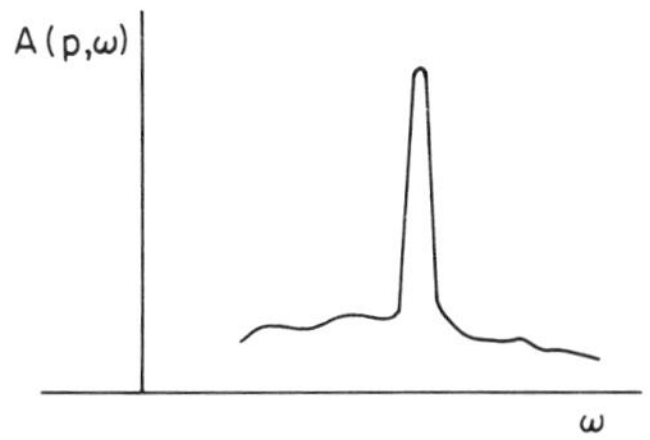

FIGURE 5.21. Spectral function A.

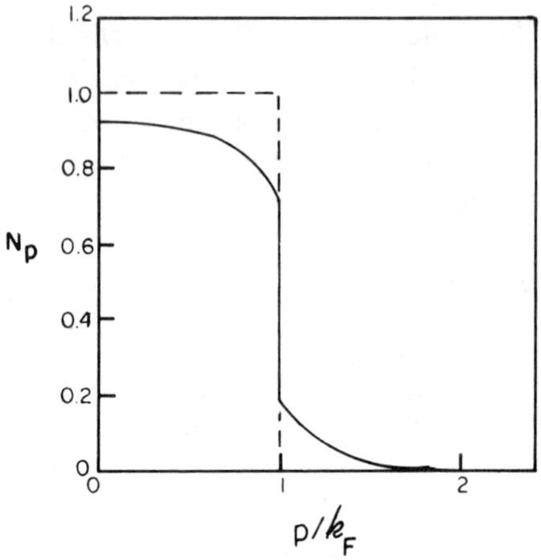

FIGURE 5.22. Momentum density of the homogeneous electron gas for $r_s = 3.97$. *Source*: Daniel and Vosko (1960) (used with permission).

quasiparticle central peak is a sharp delta function, because $\mathrm{Im}\,\Sigma = 0$. Here the renormalization coefficient has a precise meaning: the spectral weight under the delta function peak. As discussed in Sec. 3.3, the renormalization coefficient Z_F must be less than or equal to unity since the total area under $A(k, \varepsilon)$ is unity.

The quantity Z_F has an additional interpretation, which can be measured experimentally. The momentum distribution of the electron gas is given by [see (3.3.19)]

$$n_k = \int \frac{d\varepsilon}{2\pi} A(k, \varepsilon) n_F(\varepsilon)$$

Actual results for an interacting electron gas have the form illustrated by the solid line in Fig. 5.22 (Daniel and Vosko, 1960). The dashed line is for a noninteracting Fermi system, which has all the particles below $k = k_F$ at $T = 0$. For the interacting Fermi system, even at $T = 0$, the momentum distribution is the solid line, with some components for $k > k_F$. The quantity $n_{\mathbf{k}}$ should be distinguished from the energy distribution $n_F(\varepsilon) = (e^{\beta\varepsilon} + 1)^{-1}$, which is a sharp step at zero temperature. The renormalization coefficient Z_F is the magnitude of the step at $k = k_F$ in the momentum distribution n_k. This is easily shown by using the definition of n_k and the fact that

$$A(k_F, \varepsilon) = 2\pi Z_F \delta(\varepsilon) + f(\varepsilon)$$

where $f(\varepsilon)$ is a smooth function. The step in the momentum distribution in the Fermi gas may be measured in actual metals by Compton scattering.

First we discuss the theoretical calculations. The derivative of (5.8.4) is evaluated at the point $\varepsilon = 0$. First consider $\Sigma^{(\text{res})}$ in (5.8.6). The factor $\theta(\varepsilon - \xi_{\mathbf{k}+\mathbf{q}}) - \theta(-\xi_{\mathbf{k}+\mathbf{q}})$ will vanish at $\varepsilon = 0$, so the only finite term is when the derivative acts upon this factor:

$$\left(\frac{\partial \Sigma^{(\text{res})}}{\partial \varepsilon}\right)_{\varepsilon=0} = \int \frac{d^3q}{(2\pi)^3} \frac{v_q}{\varepsilon(q, 0)} \delta(\xi_{\mathbf{k}+\mathbf{q}})$$

At $k = k_F$ the integral of the delta function gives

$$\int_{-1}^{1} d\nu\, \delta\left(\varepsilon_q + \frac{qk_F\nu}{m}\right) = \frac{m}{qk_F} \theta(2k_F - q)$$

so that the final contribution is

$$\frac{\partial \Sigma^{(\text{res})}}{\partial \varepsilon} = \frac{e^2 m}{\pi k_F} \int_0^{2k_F} \frac{dq}{q\varepsilon(q, 0)} \tag{5.8.7}$$

This integral is done numerically. The evaluation of $\Sigma^{(\text{line})}$ is similar to that discussed earlier, as in deriving (5.1.27). The factor $\varepsilon(\mathbf{q}, i\omega)$ is a real and symmetric function of q and ω. The angular integrals are done on the electron Green's function:

$$\int_{-1}^{1} d\nu\, \mathscr{G}^{(0)}(\mathbf{k} + \mathbf{q}, i\omega + \varepsilon) = \int_{-1}^{1} \frac{d\nu}{i\omega + \varepsilon - \xi_k - \varepsilon_q - (kq\nu/m)}$$

$$= -\frac{m}{kq} \ln\left[\frac{i\omega + \varepsilon + \mu - (k + q)^2/2m}{i\omega + \varepsilon + \mu - (k - q)^2/2m}\right]$$

Using the fact that the rest of the integrand is symmetric in ω, we only need the symmetric part, which gives

$$\Sigma^{(\text{line})}(k, \varepsilon) = +\frac{e^2 m}{4\pi^2 k} \int_0^\infty \frac{dq}{q} \int_{-\infty}^{\infty} \frac{d\omega}{\varepsilon(q, i\omega)}$$

$$\times \ln\left\{\frac{\omega^2 + [\varepsilon + \mu - (k + q)^2/2m]^2}{\omega^2 + [\varepsilon + \mu - (k - q)^2/2m]^2}\right\}$$

This double integral is evaluated numerically. It can be differentiated with respect to either k or ε and is well behaved.

TABLE 5.6. Z_F

r_s	RPA[a]	Hubbard[b]
0	1	1
1	0.859	0.87
2	0.768	0.77
3	0.700	0.70
4	0.646	0.63
5	0.602	
6	0.568	

[a] Hedin (1965).
[b] Rice (1965).

There have been two major types of calculations with these formulas. The first is the RPA, where we use Hedin's (1965) results. The other is the revised Hubbard model work of Rice (1965). The two calculations of Z_F are shown in Table 5.6 for different values of r_s. Rice only reported values of $r_s \leq 4$. The two calculations are in very good agreement. The two models of the dielectric function make the same prediction regarding this quantity.

Eisenberger *et al.* (1972) reported Compton scattering experiments in metallic sodium and lithium which demonstrated the existence of the high-momentum tail in n_k. Compton scattering is the inelastic scattering of photons by the electrons in the solid. The photons are scattered by all the electrons, both in core states and the conduction band. The conduction band profiles, which are our present interest, are obtained after subtracting the core scattering. The best results are obtained for metals with low atomic number, since they have fewer core levels to subtract. Eisenberger *et al.* analyzed their data in terms of the sudden approximation. This theory assumes that the scattering rate of photons is proportional to

$$\frac{d^2\sigma}{d\omega\, d\Omega} \propto \int d^3p n_p \delta(\omega + \varepsilon_{\mathbf{p}} - \varepsilon_{\mathbf{p}+\mathbf{q}})$$

where ω is the energy transferred by the photon and $\mathbf{q}$ is the momentum transferred. The electrons are scattered from $\mathbf{p}$ to $\mathbf{p} + \mathbf{q}$, and the average is taken over the initial distributions n_p of the interacting electron system. When the angular integral is done to eliminate the delta function, it is found that the experimental quantity is only a function of the combination of

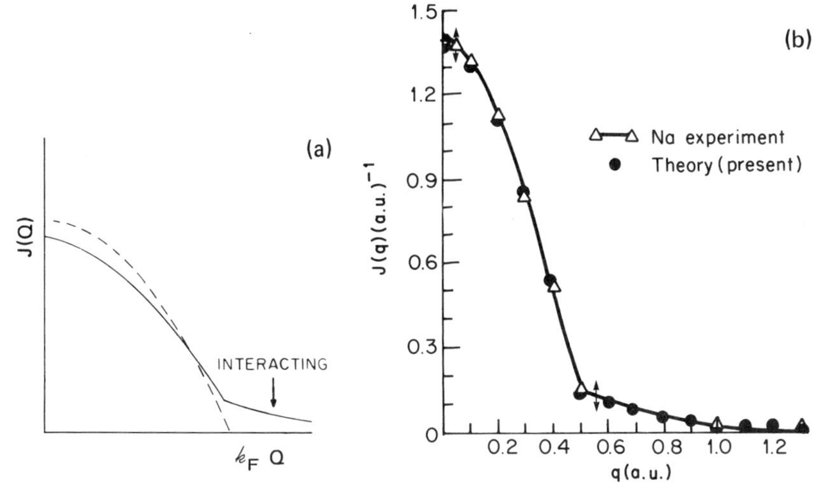

FIGURE 5.23. (a) The dashed line shows the momentum distribution in Compton scattering from a noninteracting electron gas, and the solid line is the prediction for an interacting system. (b) The experimental results in metallic sodium (solid points) compared with the theory (open triangles) for this metal. *Source*: Pandry and Lam (1973) (with permission).

variables called $Q = (m/q)(\varepsilon_q - \omega)$:

$$\int_{-1}^{1} d\nu\,\delta\left(\omega - \varepsilon_q - \frac{pq\nu}{m}\right) = \frac{m}{pq}\,\theta(p - |\,Q\,|)$$

$$\frac{d^2\sigma}{d\omega\, d\Omega} \propto J(Q)$$

$$J(Q) = \int_{|Q|}^{\infty} p\; dpn_p$$

For a noninteracting electron gas, one has that $n_p = \theta(k_F - p)$ and that the scattering function $J(Q)$ is an inverted parabola:

$$J(Q) = \tfrac{1}{2}(k_F{}^2 - Q^2)\theta(k_F - Q)$$

$$Q = \frac{m}{q}\,(\varepsilon_q - \omega)$$

This is shown as the dashed line in Fig. 5.23(a). The solid line shows the type of behavior expected for an interacting electron gas. There is a long tail at high values of Q, which arises from the similar high tail in n_p for values of $p > k_F$.

The experimental and theoretical results of Pandrey and Lam (1973) are shown in Fig. 5.23(b). The triangular points are the experiment, and they are joined by the solid line. The circular points are the theory. The agreement is obviously excellent. Sodium is the ideal metal for testing the distribution n_k since its Fermi surface is quite spherical. Other experimentalists (Eisenberger *et al.*, 1972) find that the Compton scattering depends only upon Q, which can be tested by different choices of q and ω.

B. Effective Mass

The formula for the effective mass was given previously in (3.3.25). It will be evaluated for electrons on the Fermi surface of the metal:

$$\left(\frac{m}{m^*}\right) = \left|\frac{1 + \partial\Sigma(k, \varepsilon)/\partial\varepsilon_k}{1 - \partial\Sigma(k, \varepsilon)/\partial\varepsilon}\right|_{\substack{k=k_F \\ \varepsilon=0}} = Z_F\left[1 + \frac{\partial\Sigma(k, \varepsilon)}{\partial\varepsilon_k}\right]_{\substack{k=k_F \\ \varepsilon=0}}$$

This assumes the Brillouin–Wigner form of perturbation theory. Another form of perturbation theory uses the self-energy to determine the quasiparticle energy by the approximation

$$E(k) = \xi_k + \mathrm{Re}[\Sigma(k, \xi_k)]$$

$$\left(\frac{m}{m^*}\right) = 1 + \left[\frac{\partial\Sigma(k, \varepsilon)}{\partial\varepsilon_k} + \frac{\partial\Sigma(k, \varepsilon)}{\partial\varepsilon}\right]_{\substack{\varepsilon=0 \\ k=k_F}}$$

and the effective mass formula is different. The differences are of order $(\partial\Sigma/\partial\varepsilon)^2(\partial\Sigma/\partial\varepsilon + \partial\Sigma/\partial\varepsilon_k)$, which turns out to be small since $\partial\Sigma/\partial\varepsilon + \partial\Sigma/\partial\varepsilon_k \sim 0$. Both terms are large but have opposite signs and largely cancel. The values for m/m^* turn out to be close to unity. We shall quote the results of Rice, who used the approximate formula for m/m^*. They are shown in Table 5.7. The ratio m/m^* is less than unity for values of r_s less than 2, while it is slightly larger than unity for larger values of r_s. The effective masses of electrons on the Fermi surface of the metal may be measured quite accurately by many types of cyclotron resonance experiments. The effective mass values for real metals are determined by three major contributions. The first is the electron–electron interactions, and we have just found that they make only a small contribution to the mass. The second is due to band structure, and this is quite variable. It is small in sodium but large in lithium and many other metals. The third contribution is due to electron–phonon interactions, and this is quite sizable. Thus we must

TABLE 5.7. m/m^*

r_s	Hubbard[a]
0	1
1	0.96
2	0.99
3	1.02
4	1.06

[a] Rice (1965).

defer any comparison with experiment until this phonon contribution is calculated in Sec. 6.4.

Another application of the effective mass theory is in the low-temperature specific heat, or heat capacity. The free-electron theory of metals predicts that the specific heat of a metal, at low temperature, is linear in temperature. A free-electron gas of N particles has the result

$$C_{v0} = \pi^2 N k_B{}^2 T \frac{m}{(3\pi^2 n_0)^{2/3}}$$

For an interacting system, the prediction is that the free-electron mass is replaced by the effective mass at the Fermi energy m^*. Thus the ratio of the linear term in the specific heat, at low temperature, is just the ratio of the effective masses:

$$\frac{C_v}{C_{v0}} = \frac{m^*}{m}$$

The quantity C_{v0}/C_v is calculated exactly in the same manner as m/m^*. Again this ratio is influenced by phonon effects.

C. Pauli Paramagnetic Susceptibility

The Pauli susceptibility is another one of the many parameters of the electron gas which are calculated using Green's functions. The susceptibility of a free-electron metal such as sodium or aluminum has a number of contributions. One is that of the ion cores, and another is the orbital diamagnetism, which is called Landau susceptibility. These two effects must be subtracted from the experimental value in order to deduce the Pauli part, which is due to the spin of the electron.

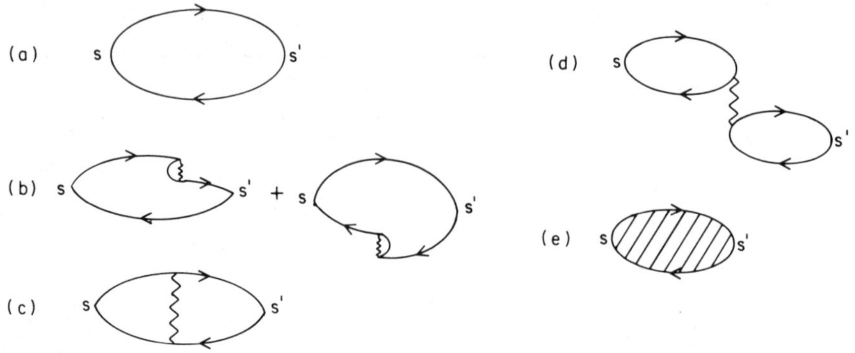

FIGURE 5.24

One way to calculate the Pauli susceptibility is by the use of the Kubo formula derived in Sec. 3.8A,

$$\chi_{\alpha\beta}(\mathbf{q}, i\omega) = \int_0^\beta d\tau e^{i\omega\tau}\langle T_\tau M_\alpha(\mathbf{q}, \tau) M_\beta(-\mathbf{q}, 0)\rangle$$

$$M_\alpha(\mathbf{q}) = \mu_0 \sum_{p,s,s'} \sigma_{ss'}^{(\alpha)} c_{\mathbf{p}+\mathbf{q}s}^\dagger c_{\mathbf{p}s'}$$

where $\sigma^{(\alpha)}$ are the Pauli spin matrices. We use $\sigma = \pm 1$ to signify spin up or down for the $s = \frac{1}{2}$ system. Let us evaluate the static susceptibility, which has $i\omega = 0$. Our Hamiltonian is for the homogeneous electron gas. Let us consider the type of terms which might occur by expanding the S matrix and examining the lowest-order terms. The first term, the one which has no interactions, was evaluated in Sec. 3.8A. It gave the result for the free-electron gas, which is called $\chi_F = \mu_0^2 N_F$. The first interaction term is of the form

$$\lim_{q\to 0} \chi_{zz}(q) = \chi_F - \int_0^\beta d\tau \int_0^\beta d\tau_1 \frac{1}{2\nu} \sum_{\substack{\mathbf{q}'\mathbf{p}\mathbf{k} \\ \sigma\sigma'}} v_{q'}$$
$$\times \langle T_\tau M_z(\mathbf{q}, \tau) c_{\mathbf{p}+\mathbf{q}'\sigma}^\dagger(\tau_1) c_{\mathbf{k}-\mathbf{q}'\sigma'}^\dagger(\tau_1) c_{\mathbf{k}\sigma'}(\tau_1) c_{\mathbf{p}\sigma}(\tau_1) M_z(-\mathbf{q}, 0)\rangle$$

$$M_z(\mathbf{q}) = \mu_0 \sum_{\substack{\mathbf{p} \\ \sigma=\pm 1}} \sigma c_{\mathbf{p}+\mathbf{q}\sigma}^\dagger c_{\mathbf{p}\sigma}$$

We shall not evaluate the several connected terms which result from this contribution but instead shall refer to Fig. 5.24, where they are drawn. The free-electron term is given by the bubble diagram shown in Fig. 5.24(a).

The vertices are labeled s to indicate that they contain a spin operator and so are not the same vertex as found for the density or current operators. Figure 5.24(b) contains the exchange self-energy diagrams for the electron Green's function. These can be incorporated by using a $\mathscr{G}$ instead of $\mathscr{G}^{(0)}$ for this propagator, where $\mathscr{G}$ contains the self-energy terms of the electron gas. Figure 5.24(c) has a Coulomb line connecting the two propagators. It is called a *vertex correction*. It is actually a type of exchange scattering, which the Hubbard and Singwi–Sjölander terms $G(q)$ are trying to simulate. Figure 5.24(d) has two closed loops, connected by a Coulomb line. This diagram is zero. It vanishes because each bubble has the correlation function of an M_α operator and a density operator

$$\langle T_\tau M_\alpha(\mathbf{q}, \tau)\varrho(-\mathbf{q}, 0)\rangle = 0$$

This combination is zero because of the averaging over $\sigma = \pm 1$. One can show that all diagrams with more than one bubble are zero, because the Coulomb interaction is spin independent. Therefore the correlation function is limited to terms with a single bubble, which has an s vertex on each end. This is indicated schematically in Figure 5.24(e). We shall call this bubble diagram $\bar{P}$. It is not the same as the exact P which enters into the definition of the longitudinal dielectric function, since the latter may have contributions from more than one bubble if connected by two or more Coulomb lines.

One possible approximation for $\bar{P}$ is to use that of the Hubbard or Singwi–Sjölander theories:

$$P(q, i\omega) = \frac{P^{(1)}(q, i\omega)}{1 + v_q G(q) P^{(1)}(q, i\omega)}$$

For the static susceptibility $\chi_{zz} \equiv \chi$, we take the limit $i\omega = 0$ and then the limit $q \to 0$. Recall that in this limit $P^{(1)}$ goes to minus the density of states $N_F = mk_F/\pi^2$, $\chi = \mu_0{}^2 N_F$:

$$\chi = -\mu_0{}^2 \lim_{q\to 0} P(q) = \frac{+\mu_0{}^2 N_F}{1 - v_q G(q) N_F} = \frac{\chi_F}{1 - a(q_{\mathrm{TF}}/k_F)^2}$$

$$\frac{\chi}{\chi_F} = \frac{1}{1 - a(q_{\mathrm{TF}}/k_F)^2} = \frac{K}{K_f}$$

In this approximation, the ratio χ/χ_F is identical to the ratio of the compressibilities K/K_f which was found in Sec. 5.6D. The constant a was discussed there. It varies with density, and among the various theories, but is approximately equal to $\frac{1}{4}$.

Our discussion of the Pauli susceptibility is similar to one by Wolff (1960). Better methods of calculating the susceptibility have been given by Silverstein (1963) and Rice (1965).

D. Mean Free Path

The mean free path l_k is the average distance the electron travels between scattering events. Similarly, the lifetime τ_k is the average time between scattering events and is related by $l_k = v_k \tau_k$. In the electron gas, most scattering processes are inelastic, and the electron loses energy each time it scatters. These quantities are defined in terms of the imaginary part of the self-energy:

$$\frac{\hbar}{\tau_k} = -2\,\mathrm{Im}[\Sigma(k, \xi_k)]$$

$$\frac{1}{l_k} = -\frac{2}{v_k}\,\mathrm{Im}[\Sigma(k, \xi_k)]$$

Most calculations have been done with the screened exchange energy in (5.8.3). Its imaginary part is given by the imaginary part of $\Sigma^{(\mathrm{res})}$ in (5.8.6), since the other term, $\Sigma^{(\mathrm{line})}$, is totally real. Thus we are led to consider the expression

$$\frac{1}{l_k} = -\frac{2}{v_k}\int \frac{d^3q}{(2\pi)^3}\, v_q[\theta(\xi_{\mathbf{k}} - \xi_{\mathbf{k+q}}) - \theta(-\xi_{\mathbf{k+q}})]\,\mathrm{Im}\left[\frac{1}{\varepsilon(q, \xi_{\mathbf{k+q}} - \xi_{\mathbf{k}})}\right] \tag{5.8.8}$$

This has formed the basis for most treatments of electron scattering in metals (Ferrell, 1956, 1957). The expression is expected once we identify $\mathrm{Im}[1/\varepsilon(q, \omega)]$ as the rate of making excitations (q, ω) in the solid. The excitation scatters the electron from its initial energy ξ_k to its final energy $\xi_{\mathbf{k+q}}$. Thus the scattering rate may be derived from the probability rate,

$$\int_0^\infty d\omega P(q, \omega)\delta(\omega + \xi_{\mathbf{k}} - \xi_{\mathbf{k+q}})$$

$$P(\mathbf{q}, \omega) = -\mathrm{Im}\left[\frac{1}{\varepsilon(q, \omega)}\right]$$

where we have summed over all energy transfers $d\omega$. The delta function eliminates the $d\omega$ integral, and we are left with an expression of the type in (5.8.8).

For small values of initial energy ξ_k we would need to numerically integrate this expression. It is a double integral, since one must do both the dq and dv ($v = \cos\theta$) integrations. This has been done by Lundqvist (1969). For more energetic particles, the excitation spectrum of $\mathrm{Im}[1/\varepsilon(q, \omega)]$ is dominated by long-wavelength plasmons. One way to evaluate this expression approximately is to assume that the plasmons totally dominate the excitation spectrum. At small q we can use the limiting form

$$\lim_{q\to 0} \varepsilon(q, \omega) = 1 - \frac{\omega_p{}^2}{\omega^2}$$

$$\lim_{q\to 0} \frac{1}{\varepsilon(q, \omega)} = \frac{1}{1 - \omega_p{}^2/\omega^2} = 1 + \frac{\omega_p}{2}\left(\frac{1}{\omega - \omega_p + i\delta} - \frac{1}{\omega + \omega_p + i\delta}\right)$$

to obtain

$$-\mathrm{Im}\left[\frac{1}{\varepsilon(q, \omega)}\right] = \frac{\pi}{2}\,\omega_p[\delta(\omega - \omega_p) - \delta(\omega + \omega_p)]$$

Now the analytical result may be derived for the mean free path. We insert the delta function for $\mathrm{Im}(1/\varepsilon)$ into the integrand and change integration variables to $\mathbf{k}' = \mathbf{k} + \mathbf{q}$:

$$\frac{1}{l_k} = \frac{e^2\omega_p}{v_k}\,\theta(\xi_k - \omega_p)\int_{k_F}^{\infty} k'^2\,dk'\,\delta\!\left(\frac{k^2}{2m} - \omega_p - \frac{k'^2}{2m}\right) \times \int_{-1}^{1} dv\,\frac{1}{k^2 + k'^2 - 2kk'v} \tag{5.8.9}$$

The factor $\theta(\xi_{\mathbf{k}} - \omega_p)$ arises from the original factors in the integrand of $\theta(\xi_{\mathbf{k}} - \xi_{\mathbf{k}+\mathbf{q}}) - \theta(-\xi_{\mathbf{k}+\mathbf{q}})$. They reduce to the condition that $\xi_{\mathbf{k}} > \omega_p > 0$, $\xi_{\mathbf{k}'} > 0$. The integrals may all be evaluated:

$$\frac{1}{l_k} = \frac{1}{2a_0}\left(\frac{\omega_p}{\varepsilon_k}\right)\theta(\xi_{\mathbf{k}} - \omega_p)\ln\left[\frac{(\varepsilon_k)^{1/2} + (\varepsilon_k - \omega_p)^{1/2}}{(\varepsilon_k)^{1/2} - (\varepsilon_k - \omega_p)^{1/2}}\right] \tag{5.8.10}$$

The formula is given in terms of the kinetic energy ε_k measured from the bottom of the conduction band, although plasmons are emitted only when $\xi_k > \omega_p$ or $\varepsilon_k > \omega_p + E_F$. Equation (5.8.10) is almost identical to an old formula of Bethe (1930) for the rate of energy loss of charged particles in solids.

Figure 5.25 shows a comparison of these theories with the experiments in aluminum. The metal aluminum is chosen as a basis of comparison because the best experimental results are available of any free-electron

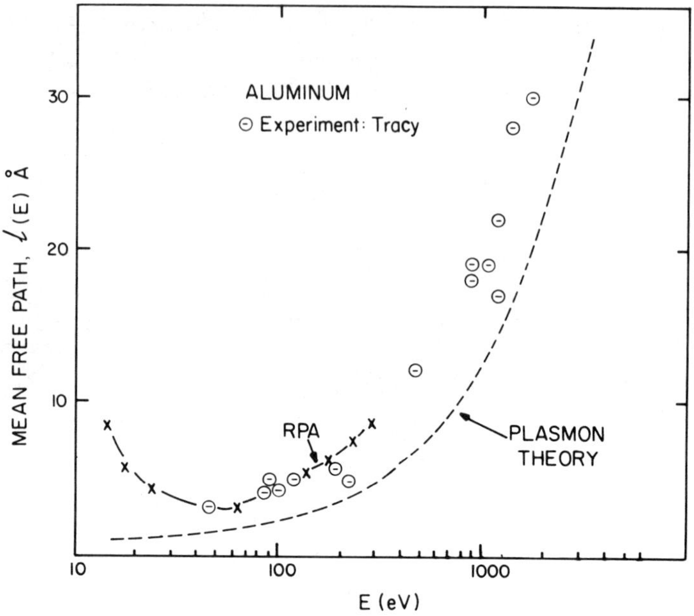

FIGURE 5.25. The mean free path of electrons in aluminum as a function of energy above the Fermi surface. Circles are data points from Tracy (1974). The dashed line with X points is the RPA calculation of Lundqvist (1969). The other dashed line is the plasmon theory which ignores pair production, and its predictions are inaccurate.

metal. The data is due to Tracy (1974). The plasmon formula (5.8.10) only needs the aluminum plasmon energy of $\hbar\omega_p = 15.5$ eV to determine the dashed line. One sees that the plasmon theory curve in Fig. 5.25 is too low by about 30–40%. The electron mean free path is longer than would be predicted by the simple plasmon formula. The plasmon theory is unreliable at lower energies. Experiments show that the rate of making plasmons vanishes below 40 eV (Flodstrom *et al.*, 1977). The other theoretical curve is the RPA result, where we have used the calculations of Lundqvist (1969) for $r_s = 2$, which is close to the aluminum value of $r_s = 2.07$. He numerically evaluated (5.8.8). His curve is in excellent agreement with the experimental results. Similar measurements have been done on other metals, although few on free-electron metals. See the reviews by Powell (1974) and Brundle (1975).

Another experiment is to measure the inelastic scattering of electrons by thin metal foils. The electrons are shot through the foil, and their cross section for inelastic scattering is measured as a function of angle and energy. The electrons must be fairly energetic to penetrate the foil. The experiments show that the electrons mostly lose energy by exciting long-wavelength plasmons (Marton *et al.*, 1962). This is what we assumed in the calculation of the mean free path.

First let us calculate the angular dependence of the scattering when the electron emits one long-wavelength plasmon. The process is that the energetic electron—usually in the kilovolt range—loses one plasmon of energy and is scattered through an angle θ. The energetic electron scatters from $\mathbf{k}$ to $\mathbf{k}'$. The scattering rate may be obtained from (5.8.9) by eliminating the angular integrals. This gives us the differential scattering rate as a function of angle:

$$\frac{dw}{d\Omega} = \frac{e^2\omega_p}{2\pi} \int_{k_F}^{\infty} dk' k'^2 \frac{\delta(\varepsilon_k - \omega_p - \varepsilon_{k'})}{k^2 + k'^2 - 2kk' \cos\theta}$$

The delta function determines the value of k'. For energetic electrons, this is

$$k' = (k^2 - 2m\omega_p)^{1/2} \simeq k - \frac{\omega_p}{v}$$

It turns out that the typical scattering angle is very small. Thus it is possible to use the small-angle approximation for $\cos\theta = 1 - \frac{1}{2}\theta^2$. The denominator can be simplified to

$$k^2 + k'^2 - 2kk' \cos\theta \simeq (k - k')^2 + kk'\theta^2 \simeq k^2(\theta_0{}^2 + \theta^2)$$

$$\theta_0 = \frac{\omega_p}{2E}$$

and the cross section has the simple form derived by Ritchie (1957):

$$\frac{dw}{d\Omega} = \frac{e^2\omega_p}{2\pi\hbar v} \frac{1}{\theta^2 + \theta_0{}^2}$$

The scattering distribution is Lorentzian. The angular width θ_0 is typically a fraction of a degree when E is a kilovolt. Thus the angular distribution is sharply peaked in the forward direction. This is exactly the behavior found experimentally (Sueoka, 1965).

Another aspect of electron energy loss through metal foils is shown in Fig. 5.26, which is the data of Marton *et al.* (1962) for aluminum. The peak on the left of Fig. 5.26(a) is from the electrons which go through the foil with no energy loss. The next peak is from those which excite one plasmon, and the nth is from electrons which excite $n - 1$ plasmons. The angular dependence of these peaks was measured and found to be Lorentzian. Another feature of these peaks is shown in Fig. 5.26(b), which is the peak intensity, or the area under each curve. The background is subtracted

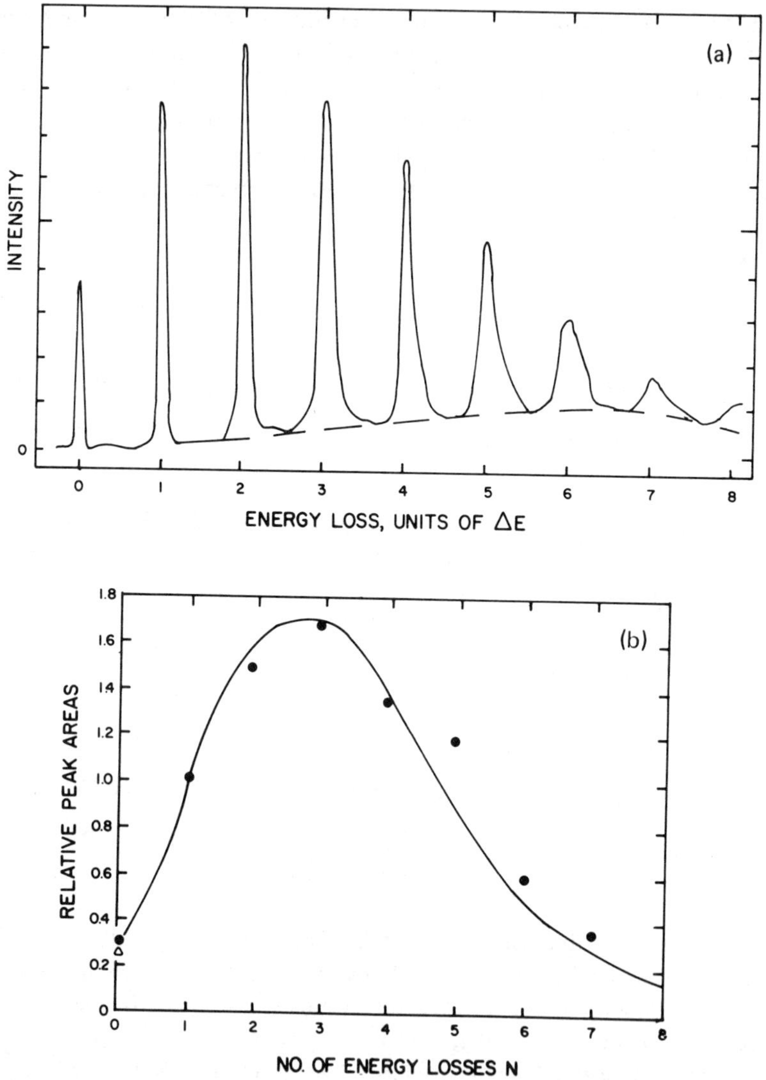

FIGURE 5.26. Energy loss spectrum at $\theta = 0$ for electrons going through aluminum films. The relative peak heights are shown in (b) and are well fit by a Poisson distribution shown by the solid line. *Source*: Marton *et al.* (1962) (used with permission).

according to the dashed line in Fig. 5.26(a). The peak intensities follow a Poisson distribution (solid line), so that the intensity of the nth peak is given by

$$P_n = e^{-\alpha} \frac{\alpha^n}{n!}$$

$$\alpha = \frac{t}{l}$$

The parameter α is the thickness of the film divided by the mean free path for plasmon emission. The electron energy was 20 keV. The Poisson distribution is a characteristic of the independent boson model of Sec. 4.3. A Poisson distribution occurs whenever the bosons, in this case plasmons, are emitted independently of previous emissions. This is a reasonable model for these fast electrons. An electron of 20 keV does not lose a significant fraction of its energy when it emits a plasmon of 15 eV. Similarly, we have shown that its angle does not change appreciably, since the scattering is mostly in the forward direction. Thus the electron trajectory is largely unchanged by the plasmon emissions. All emission events have the same likelihood and are independent of previous emissions. It makes sense that the independent boson model should be applicable here. Of course we use the zero temperature version of the theory, since even at room temperatures, the number of plasmons thermally present is $(e^{\beta\omega_p} - 1)^{-1} \sim 0$, which is negligible.

PROBLEMS

1. Consider the ground state of the ferromagnetic electron gas. The particles are plane waves, but all spins point in the same direction.

   a. Find the ground state energy of the system in the Hartree–Fock approximation.
   b. Compare this with the paramagnetic state (equal numbers of up and down spins) calculated in the same approximation. Are there values of r_s for which the ferromagnetic state is lower in energy?

2. In the Wigner lattice the electron feels a harmonic potential $V(r) = -3/r_s + r^2/r_s^3$. Find the zero-point energy from oscillations in this well by assuming that each electron moves independently of the others.

3. The correlation energy may also be evaluated by Rayleigh–Schrödinger perturbation theory (Gell-Mann and Brueckner, 1957). The second-order ground

state energy is

$$E_g^{(2)} = \sum_I \frac{|\langle I \mid V \mid i\rangle|^2}{E_i - E_I}$$

Obtain the ground state energy terms which contribute in second order, and write them down in terms of wave vector summations.

4. Calculate the cohesive energy of metallic helium assuming it is a uniform electron gas, with α particles spaced on a crystal lattice. Compare your result to atomic helium.

5. Calculate the cohesive energy of Be and Mg, which have $Z = 2$. Assume $m_B = 1$, and estimate E_1 by assuming a constant electron density with the pseudopotential of Table 5.1. The experimental values are 0.123 and $0.058 E_{ry}$ per electron.

6. Consider putting a single point charge Q into a homogeneous electron gas. Show that to order Q^2 the energy required to do this is

$$\Delta\Omega = \frac{1}{2} \int \frac{d^3q}{(2\pi)^3} \frac{4\pi Q^2}{q^2} \left[\frac{1}{\varepsilon(q)} - 1\right] + O(Q^3)$$

Estimate this result for the Thomas–Fermi model.

7. Use (5.4.12) to estimate $S(q)$ at large q in the RPA. At large q, one has that $\text{Im}(1/\varepsilon) \simeq -\varepsilon_2$.

8. Use (5.4.12) to find the behavior of $S(q)$ as $q \to 0$.

9. Set $1 - g(r) = A \exp(-rk_F)$.

   a. Determine A by the normalization condition (5.1.30).
   b. Determine $S(q)$.
   c. Determine $G(q)$. Compare this on a piece of graph paper with Hubbard and Singwi–Sjölander for $r_s = 3$.

10. Calculate the compressibility of the electron gas from the ground state energy:

   a. In the Hartree–Fock approximation.
   b. Including both HF and Wigner correlation energy.

Plot both results on a piece of graph paper as K_f/K vs. r_s for metallic densities.

11. Calculate $S(q)$ in the Hartree–Fock approximation. First show that the starting point is $S(q) = 1 - (2/N) \sum_{\mathbf{k}} n_{\mathbf{k}} n_{\mathbf{k}+\mathbf{q}}$.

12. Calculate $\gamma(r_s)$ in the Hartree–Fock approximation using the results of Problem 11. Verify that the correlation energy (5.6.5) is zero with this $\gamma(r_s)$.

13. Use the form of the dielectric function (5.5.16) to show that the plasma dispersion relation is

$$\omega_q = \omega_p + \alpha\left(\frac{\hbar q^2}{m}\right) + O(q^4)$$

$$\alpha = \frac{3}{10}\,\frac{\hbar k_F^2}{m\omega_p}\left[1 - \frac{5}{9}a\left(\frac{q_{TF}}{k_F}\right)^2\right]$$

where a is the parameter which enters into the compressibility (5.6.8).

14. Prove the ground state energy theorem (5.6.3) starting from (3.6.21) using just the electron–electron interaction.

15. In the Vashishta–Singwi theory, the constant a in the compressibility is no longer equal to γ as given by (5.6.4). Find the new relationship to γ.

16. Does the RPA satisfy the sum rule (5.7.13)? Show by explicitly doing the integrations.

17. Derive the sum rule for the third moment for a Hamiltonian of particles in a potential field $V(r)$. Do this by evaluating the expectation of the triple commutator with H:

$$H = \sum_{\mathbf{k}} \xi_k c_{\mathbf{k}}^{\dagger} c_{\mathbf{k}} + \sum_{\mathbf{k}} \varrho(\mathbf{k}) V(\mathbf{k})$$

$$\langle \omega^3 \rangle = \langle [H, [H, [H, \varrho(\mathbf{q})]]], \varrho(-\mathbf{q}) \rangle$$

18. At long wavelength the dielectric function $\varepsilon(q, i\omega) \to 1 + \omega_p^2/\omega^2$. Calculate $\Sigma^{(3b)}$ in (5.1.21) for this case, and verify the results for $\Sigma^{(\mathrm{res})}$ and $\Sigma^{(\mathrm{line})}$.

19. Prove that the step in n_k at $k = k_F$ is Z_F.

20. The screened exchange energy (5.8.3) may be calculated exactly in the Thomas–Fermi approximation. Do this, and use your answer to obtain analytical expressions for

a. The self-energy at $p = 0$.
b. The self-energy at $p = k_F$.
c. The effective mass at the Fermi energy.

Plot all three results as a function of r_s on a piece of graph paper.

21. Calculate the average kinetic and exchange energies per electron for a two-dimensional electron gas with Coulomb interactions e^2/r. Define r_s by $\pi a_0^2 n_0 = 1/r_s^2$, and your answer should be $E_g = r_s^{-2} - 1.20/r_s \ldots$. Find the analytical form of the coefficient of the exchange energy (it is *not* $\frac{6}{5}$). For a review of correlation calculations, see Jonson (1976).

22. Draw the connected Feynman diagrams for the electron self-energy from electron–electron interactions that contain two or three internal Coulomb lines. Show that all terms can be written as $v_q \Gamma/\varepsilon$ in (5.8.1). Which diagrams contribute to Γ and which to ε?

23. One contribution to $\Gamma(q)$ and hence $G(q)$ is the screened ladder diagrams inside the bubble $P(q)$. In the model where $\Gamma(p, p + q) \approx \Gamma(q)$, show that this contribution gives $G(q) = 1/[2\varepsilon(q)]$. Since $\varepsilon(q)$ depends upon $G(q)$, one has a self-consistent equation for $G(q)$. Solve it. Plot on a piece of graph paper $G(q)$ vs. q/k_F for $r_s = 4$, and compare with the results of Hubbard and with Vashishta and Singwi.

Chapter 6

Electron–Phonon Interaction

6.1. FRÖHLICH HAMILTONIAN

The Fröhlich Hamiltonian describes the interaction between a single electron in a solid and LO (longitudinal optical) phonons:

$$H = \sum_{\mathbf{p}} \frac{p^2}{2m} c_{\mathbf{p}}^{\dagger} c_{\mathbf{p}} + \omega_0 \sum_{\mathbf{q}} a_{\mathbf{q}}^{\dagger} a_{\mathbf{q}} + \sum_{\mathbf{q}\mathbf{p}} \frac{M_0}{\nu^{1/2}} \frac{1}{|q|} c_{\mathbf{p}+\mathbf{q}}^{\dagger} c_{\mathbf{p}} (a_{\mathbf{q}} + a_{\mathbf{q}}^{\dagger})$$

$$M_0^2 = \frac{4\pi\alpha\hbar(\hbar\omega_0)^{3/2}}{(2m)^{1/2}} \tag{6.1.1}$$

$$\alpha = \frac{e^2}{\hbar}\left(\frac{m}{2\hbar\omega_0}\right)^{1/2}\left(\frac{1}{\varepsilon_\infty} - \frac{1}{\varepsilon_0}\right)$$

This Hamiltonian was derived in Chapter 1, with the form of the interaction given in Sec. 1.3E. The LO phonons are usually represented by an Einstein model; i.e., the phonon frequency $\omega_0 \equiv \omega_{\text{LO}}$ is taken to be a constant. Since there is a single electron, the Hamiltonian may also be written as

$$H = \frac{p^2}{2m} + \omega_0 \sum_{\mathbf{q}} a_{\mathbf{q}}^{\dagger} a_{\mathbf{q}} + \sum_{\mathbf{q}} \frac{M_0}{\nu^{1/2}} \frac{e^{i\mathbf{q}\cdot\mathbf{r}}}{|q|} (a_{\mathbf{q}} + a_{\mathbf{q}}^{\dagger}) \tag{6.1.2}$$

where $\mathbf{r}$ and $\mathbf{p}$ are the conjugate coordinates of the electron. The unperturbed electron is taken to have free-particle motion with an effective mass m. Since there is only one electron in the problem, the results are independent of the statistics of the particle. The same results are obtained for any fermion or boson in the solid, such as holes, positions, etc., as long as they are free to move. The phonon modes are unaffected by the one electron in the solid, so the phonon self-energy is zero (actually of order ν^{-1}), and the phonon Green's function $\mathscr{D}$ is always $\mathscr{D}^{(0)}$. The model also assumes that the motion is isotropic in direction and that the energy bands of the solid are nondegenerate. These rather restricted conditions describe what is called the *Fröhlich polaron problem.* The model actually applies in some cases to conduction bands of semiconductors and ionic solids which have their minimum at the Γ point and have an isotropic effective mass. For other symmetry points in the solid, one usually has to improve the model by taking into account anisotropy in the effective mass or degeneracy of the bands. The latter is important for holes in semiconductors.

The Fröhlich polaron problem was an important problem in mathematical physics during the 1950s. Numerous mathematical techniques were tried out on this problem. We shall describe several of them here: Brillouin–Wigner perturbation theory, Rayleigh–Schrödinger perturbation theory, strong coupling theory, and linked cluster theory. The Green's function method is, as we shall show, equivalent at zero temperature to Brillouin–Wigner theory. Several other methods were tried, including that of Low *et al.* (1953). The problem was most accurately solved by Feynman (1955). He introduced a variational method based on path integrals. After lengthy algebra, he obtained a result which even today is the best available. We shall not present his discussion, since his theory is very lengthy. The Feynman results will, however, be used as the standard to which other theories are compared. The methods we fail to discuss, such as that of Low *et al.*, give poor results when compared with Feynman's method.

The polaron Hamiltonian (6.1.1) describes the motion of a particle while it is linearly coupled to a system of boson particles. In the Fröhlich Hamiltonian, the bosons are optical phonons in a polar solid. The classical picture has the particle exerting forces upon the ions, which respond and move. The ion motion creates new forces which act back upon the particle. The finite ion frequency ω_0 makes the reaction forces of the ions on the particle retarded in time. The quantum nature of the phonons makes these forces occur in discrete units. In both the classical and quantum pictures, the ion motion is pictured as a polarization of the surrounding medium by the particle. The particle must drag this polarization with it during its motion through the solid, which affects its energy and effective mass.

A. Brillouin–Wigner Perturbation Theory

The Brillouin (1932, 1933)–Wigner (1935) perturbation theory method is a historical predecessor of the modern Green's functions method. It is equivalent to solving the equation for the energy spectrum E_p of a particle of momentum p,

$$E_p = \varepsilon_p + \mathrm{Re}[\Sigma_{\mathrm{ret}}(p, E_p)]$$
$$\varepsilon_p = \frac{p^2}{2m} \tag{6.1.3}$$

where $\Sigma_{\mathrm{ret}}(p, E)$ is the retarded self-energy. The method ignores the imaginary part of the self-energy of the electron. That is, we know that the particle properties are actually described by the spectral function:

$$A(p, E) = -2\,\mathrm{Im}\left[\frac{1}{E - \varepsilon_p - \Sigma_{\mathrm{ret}}(p, E)}\right]$$

If the imaginary part of the retarded self-energy is zero, then we replace it by an infinitesimal value $-i\delta$, and the spectral function just becomes a delta function, whose argument is (6.1.3). Thus the Brillouin–Wigner method is exact if the imaginary part of the retarded self-energy is actually zero. Then if the retarded self-energy function Σ_{ret} is found exactly, the exact result is obtained for the particle motion.

The self-energy function cannot be obtained exactly unless the Hamiltonian can be solved exactly. This has not yet been achieved for (6.1.1). In practice, one usually evaluates a few terms in the perturbation expansion and thereby obtains an approximate $\Sigma_{\mathrm{ret}}(p, E)$. Equation (6.1.3) is solved with this approximate self-energy, and one has an approximate solution. This procedure is Brillouin–Wigner perturbation theory.

For one particle interacting with a set of optical phonons, the imaginary self-energy does vanish at zero temperature. This statement is only true for particles whose kinetic energy is less than the phonon energies ω_0. The argument for this is the following. For a particle of energy E_i to emit a phonon, it must go to a final state $E_f = E_i - \omega_0$. However, if E_i is less than ω_0, this equation cannot be satisfied since the energy E_f cannot be below the bottom of the band. Thus the process is forbidden by energy conservation. Of course, a particle can absorb a phonon and increase its energy to $E_f = E_i + \omega_0$. This step is always possible as long as there are phonons in the system. But the number of phonons is proportional to the thermal occupation factor,

$$N_0 = \frac{1}{e^{\beta\omega_0} - 1}$$

and this vanishes as $T \to 0$. Thus as $T = 0$, the low-energy particle can neither absorb nor emit phonons. Since these are the only two loss mechanisms, the particle cannot lose or gain energy. Its mean free path is infinite, and the imaginary part of the retarded self-energy is zero, which means that Brillouin–Wigner perturbation theory is exact at $T = 0$. Of course, it is exact only if the retarded self-energy is found exactly. In practice, this does not happen, and Brillouin–Wigner perturbation theory is usually a poor approximation.

The self-consistent energy function E_p is a smooth function of momentum at small values of p. Thus we can expand it in a power series:

$$E_p = E_0 + \frac{\hbar^2}{2m^*} p^2 + O(p^4) \tag{6.1.4}$$

The quantity E_0 is the downward shift of the band minimum from polaron effects. E_0 is negative and gives the amount an electron with zero momentum lowers its energy by interacting with phonons. In Sec. 4.3 the self-energy of a fixed particle is $-\sum_{\mathbf{q}} M_{\mathbf{q}}{}^2/\omega_{\mathbf{q}}$, but it will be different now that the particle can move. The quantity m^* is just the coefficient of the p^2 term in the momentum expansion. It is called the *effective mass*. In solids there are several effective masses. Each energy band will have its own curvature at the band minimum, which defines the effective band mass m_b. The band mass should be used for m in (6.1.1) and (6.1.2). We leave off the b subscript, but m is identical to m_b. The effective mass m^* in (6.1.4) is that resulting from the band curvature and polarons. It expresses the way that m (i.e., m_b) is changed by the polaron interactions. A formula for m/m^* was given previously in (3.3.25).

The first term in the perturbation series for the self-energy has one phonon in the self-energy diagram. The Feynman diagram is shown in Fig. 6.1(a). From the rules for constructing diagrams, the self-energy is

$$\Sigma^{(1)}(\mathbf{p}, ip) = \sum_{\mathbf{q}} \frac{M_0{}^2}{\nu} \frac{1}{q^2} \frac{1}{\beta} \sum_{iq_n} \mathscr{G}^{(0)}(\mathbf{p}+\mathbf{q}, ip + iq)\mathscr{D}^{(0)}(\mathbf{q}, iq_n)$$

The Matsubara summation was given in Sec. 3.5:

$$\Sigma^{(1)}(\mathbf{p}, ip) = \frac{M_0{}^2}{\nu} \sum_{\mathbf{q}} \frac{1}{q^2} \left(\frac{N_0 + n_F}{ip_n + \omega_0 - \xi_{\mathbf{p}+\mathbf{q}}} + \frac{N_0 + 1 - n_F}{ip - \omega_0 - \xi_{\mathbf{p}+\mathbf{q}}} \right) \tag{6.1.5}$$

At zero temperature, $N_0 = 0$. The fermion occupation factors n_F are all zero if there is only one particle in a band. Set $ip = E - \mu + i\delta$ so that the real part of the retarded self-energy is ($\nu \rightarrow \infty$)

$$\text{Re}[\Sigma_{\text{ret}}^{(1)}(\mathbf{p}, E)] = \frac{M_0{}^2}{(2\pi)^3} \int \frac{d^3q}{q^2} \frac{1}{E - \omega_0 - (\mathbf{p}+\mathbf{q})^2/2m} \tag{6.1.6}$$

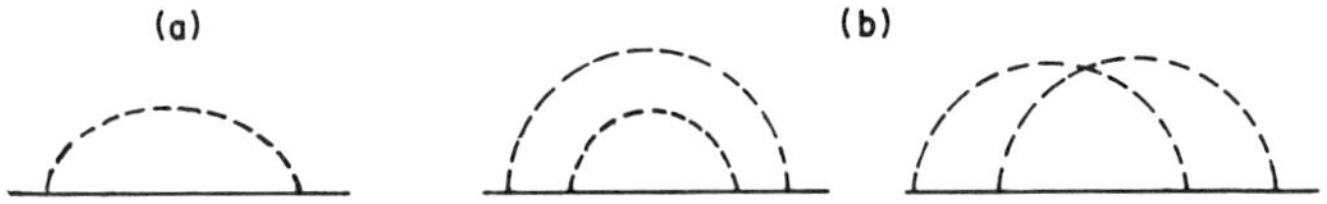

FIGURE 6.1

This result is also easy to obtain from the zero temperature techniques of Chapter 2. The energy E is now measured from the bottom of the conduction band. A free particle has $E = 0$ when it has no kinetic energy. The wave vector integrals may be done exactly. Here we shall only do the case where $E < \omega_0$. The finite temperature results, for all values of E, are given elsewhere (Mahan, 1972). Let $\nu = \cos\theta = \hat{p}\cdot\hat{q}$ be the angular variable, so that we need to do two integrals:

$$\text{Re}[\Sigma_{\text{ret}}^{(1)}(\mathbf{p}, \varepsilon)] = -\frac{M_0^2}{(2\pi)^2}\int_0^\infty dq \int_{-1}^{1} d\nu \frac{1}{-E + \omega_0 + (1/2m)(p^2 + q^2 + 2pq\nu)}$$

The integrand remains unchanged if we change q to $-q$ and ν to $-\nu$. This means that the answer is unchanged if we alter the q limits to $(-\infty, \infty)$ and divide the result by 2; the two intervals $(-\infty, 0)$ and $(0, \infty)$ contribute equally. The integral can now be simplified by changing the q variable to

$$x = \frac{q + p\nu}{(2m)^{1/2}}$$
$$\text{Re}[\Sigma_{\text{ret}}^{(1)}(\mathbf{p}, E)] = -\frac{M_0^2(2m)^{1/2}}{8\pi^2}\int_{-\infty}^{\infty} dx \int_{-1}^{1} d\nu \frac{1}{\omega_0 - E + \varepsilon_p(1-\nu^2) + x^2} \tag{6.1.7}$$

The x integral can be done first. The integral is of the form

$$\int_{-\infty}^{\infty} \frac{dx}{x^2 + A} = \frac{\pi\theta(A)}{A^{1/2}}$$

where A represents everything else in the denominator of (6.1.7). The integral is finite only when $A > 0$. This is always the case when $\omega_0 > E$, as we have assumed. The remaining integral is just an arcsine:

$$\int_{-1}^{1} \frac{d\nu}{[\omega_0 - E + \varepsilon_p(1 - \nu^2)]^{1/2}} = \frac{2}{(\varepsilon_p)^{1/2}} \sin^{-1}\left(\frac{\varepsilon_p}{\omega_0 - E + \varepsilon_p}\right)^{1/2}$$
$$\text{Re}[\Sigma_{\text{ret}}^{(1)}(p, E)] = -\frac{\alpha\omega_0^{3/2}}{(\varepsilon_p)^{1/2}} \sin^{-1}\left(\frac{\varepsilon_p}{\omega_0 - E + \varepsilon_p}\right)^{1/2} \tag{6.1.8}$$

Equation (6.1.8) is the result for the one-phonon part of the retarded self-energy when $E < \omega_0$. To obtain the parameter E_0, we need the self-energy at $p = 0$. By using the expansion for the arcsine, $\sin^{-1} x = x + x^3/6 + O(x^5)$, the $p \to 0$ limit may be shown to be

$$\text{Re}[\Sigma_{\text{ret}}^{(1)}(0, E_0)] = -\frac{\alpha\omega_0^{3/2}}{(\omega_0 - E_0)^{1/2}} \qquad (6.1.9)$$

Equation (6.1.9) is the result for a zero-momentum particle. The result for a fixed particle is quite different. In Sec. 4.3 it is shown to be

$$\Delta = -\frac{M_0^2}{\nu}\sum_{\mathbf{q}}\frac{1}{q^2} = -\frac{M_0^2}{(2\pi)^3}4\pi\int_0^\infty dq$$

The integral does not converge unless some cutoff is used at the high-momentum values. Of course, the cutoff could be the size of the Brillouin zone or alternately the spatial extent of the localized particle wave function. No cutoff is needed in the free-polaron problem because of the recoil energy of the electron. The zero-momentum result is different from the fixed-particle result, a fact which is easily understood from the classical picture. The ions in the vicinity of the electron get either repulsed or attracted to the electron. Their motion changes the potential field felt by the electron. If the electron is fixed, it does not respond to this change. The ions seek their new equilibrium position without any change in the potential exerted upon them by the electron. A free electron can respond to the changing potential of the ions. Thus when the ions move in response to the potential exerted upon them by the electron, the electron in turn begins to move in response to the changing potential of the ions. The final solution describes the coordinated motion of the electron and ions. For a free electron, the recoil of the electron makes polaron theory harder. Thus a free polaron of zero momentum does not mean the electron is fixed. It is constantly moving, in response to the continual interplay with the force field of the ion. For $p = 0$, the motion has, on the average, zero momentum. But the electron is moving in a random, stochastic fashion. When it has finite momentum, it still has this random part of its motion, even as it is drifting. The total polaron momentum

$$\mathbf{P} = \mathbf{p} + \sum_{\mathbf{q}} \mathbf{q} a_{\mathbf{q}}^{\dagger} a_{\mathbf{q}} \qquad (6.1.10)$$

commutes with the Hamiltonian and therefore is a constant of motion. Thus momentum is an acceptable eigenvalue to assign to the polaron.

The imaginary part of the one-phonon self-energy may also be obtained from (6.1.5). With $ip_n \to E - \mu$ and $n_F = 0$, the imaginary part is

$$-\mathrm{Im}[\Sigma_{\mathrm{ret}}^{(1)}(\mathbf{p}, E)] = \frac{\pi M_0^2}{(2\pi)^3} \int \frac{d^3q}{q^2} \times [N_0 \delta(E + \omega_0 - \varepsilon_{\mathbf{p}+\mathbf{q}}) + (N_0 + 1)\delta(E - \omega_0 - \varepsilon_{\mathbf{p}+\mathbf{q}})] \tag{6.1.11}$$

There are two terms in the brackets. The first term corresponds to the absorption of a phonon by the electron. This term is proportional to the phonon number density N_0, which vanishes at zero temperature. The second term comes from the emission of a phonon by the electron. It is proportional to the factor $1 + N_0$, which is finite even at $T = 0$. However, this term is finite only if $E > \omega_0$, so that the particle has more energy than the phonon, and is zero when $E < \omega_0$. Thus $\mathrm{Im}[\Sigma_{\mathrm{ret}}^{(1)}] = 0$ at $T = 0$, as asserted above.

The *Tamm–Dancoff* (TD) *approximation* constitutes solving Brillouin–Wigner perturbation theory with only the one-phonon self-energy:

$$E_p = \varepsilon_p + \mathrm{Re}[\Sigma_{\mathrm{ret}}^{(1)}(\mathbf{p}, E_p)]$$

For zero momentum, the TD approximation gives the particularly simple equation

$$E_0 = -\frac{\alpha \omega_0^{3/2}}{(\omega_0 - E_0)^{1/2}}$$

This is a cubic equation for E_0, which is easily solved. These results are summarized in Table 6.1 for several values of α. They are labeled E_{TD}.

The other columns are the Feynman results E_F and also the results of Rayleigh–Schrödinger perturbation theory $E_{\mathrm{RS}}^{(1,2)}$. $E_{\mathrm{RS}}^{(1)}$ is also evaluated in the one-phonon approximation to provide a fair comparison. This

TABLE 6.1

α	E_F/ω_0	E_{TD}/ω_0	$E_{\mathrm{RS}}^{(1)}/\omega_0$	$E_{\mathrm{RS}}^{(2)}/\omega_0$
1	−1.013	−0.76	−1.00	−1.016
2	−2.055	−1.31	−2.00	−2.064
3	−3.133	−1.80	−3.00	−3.143

equation is very simple, $E_{\rm RS}^{(1)} = -\alpha\omega_0$, as will be shown below. $E_{\rm RS}^{(2)}$ also includes the two-phonon terms. The table shows that the Tamm–Dancoff approximation is a poorer approximation than Rayleigh–Schrödinger. It provides the larger error when compared with the Feynman result, particularly for values of $\alpha \sim 1$, which is called the intermediate coupling regime.

The Tamm–Dancoff gives poor results in the intermediate coupling regime. This is easy to understand, since we have introduced a gap in the excitation spectrum. The energy denominator in (6.1.6) has the difference between the initial energy of the particle E and the value in the intermediate state $\varepsilon_{\mathbf{p}+\mathbf{q}}$. However, E has the minimum value E_0, while $\varepsilon_{\mathbf{p}+\mathbf{q}}$ has the minimum value of zero. Thus to get to the excited state in this approximation, one has an excitation energy of $-E_0$. Tamm–Dancoff is a poor approximation because there really is no excitation energy in the spectrum. The Tamm–Dancoff approximation just happens to insert a gap, which explains why it gives poor results for intermediate coupling.

The above results are suitable for a single particle in a band. A different approach must be used if there is a small but nonzero density n_0 of particles in the band. Then one must consider the change in the chemical potential $\delta\mu$. In insulators, the chemical potential μ has a negative value if the band minimum E_m is defined as zero energy. If n_0 is the density of particles, and λ is the deBroglie wavelength, then for the noninteracting system with Maxwell–Boltzmann statistics one has that $\mu - E_m = k_BT\ln(n_0\lambda^3)$. $\mu - E_m$ is negative if the argument of the logarithm is less than unity, which happens when n_0 is small. If polaron interactions cause a change δE_m in the energy of the band minimum, and if n_0 is unchanged, then the chemical potential must change according to $\delta\mu = \delta E_m$.

In the polaron problem, $E_m = 0$ and $\delta E_m = E_0$. How does $\delta\mu$ enter into the calculation? In going from (6.1.5) to (6.1.6) we must set $ip = E - \mu - \delta\mu + i\delta$. Effectively E is replaced by $E - \delta\mu$ in all of the following equations. Since $\delta\mu = E_0$, (6.1.9) is replaced by

$$\mathrm{Re}[\Sigma_{\rm ret}^{(1)}(0, E_0] = -\alpha\omega_0$$

The right-hand side no longer depends upon E_0. The change in band minimum is the $E_0 = -\alpha\omega_0$. This result now agrees with the Rayleigh–Schrödinger formula, which is given below. It is in much better agreement with the Feynman result than is the Tamm–Dancoff approximation.

Brillouin–Wigner perturbation theory does better when more terms are included in the self-energy expression. The anticipated improvement is difficult to test, since the next terms are quite formidable.

There are two self-energy terms involving two phonons. They have not yet been evaluated completely in that their wave vector integrals are difficult to do analytically. We shall be content to write these terms merely to inspect the difficulties. The two self-energy diagrams are shown in Fig. 6.1(b). Both have two phonons and three intermediate electron states. The self-energy may be written from ordinary perturbation theory. Here one puts in an energy denominator for each electron intermediate state and sums over both phonons. Each energy denominator contains the difference between the initial energy E and the intermediate-state energy. At zero temperature, the real part of the retarded self-energy is

$$\begin{aligned}
\mathrm{Re}[\Sigma_{\mathrm{ret}}^{(2)}(\mathbf{p}, E)] &= \frac{M_0^3}{\nu^2} \sum_{\mathbf{q}_1\mathbf{q}_2} \frac{1}{q_1^2} \frac{1}{q_2^2} \left[\frac{1}{(E - \varepsilon_{\mathbf{p}+\mathbf{q}_1} - \omega_0)^2} \frac{1}{E - \varepsilon_{\mathbf{p}+\mathbf{q}_1+\mathbf{q}_2} - 2\omega_0} \right. \\
&\left. + \frac{1}{E - \varepsilon_{\mathbf{p}+\mathbf{q}_1} - \omega_0} \frac{1}{E - \varepsilon_{\mathbf{p}+\mathbf{q}_1+\mathbf{q}_2} - 2\omega_0} \frac{1}{E - \varepsilon_{\mathbf{p}+\mathbf{q}_2} - \omega_0} \right]
\end{aligned} \tag{6.1.12}$$

Remember that E must be renormalized by the change in the chemical potential $\delta\mu$. The first term corresponds to the first diagram in Fig. 6.1(b) and the second to the last diagram. The same result could also be obtained using the rules for constructing diagrams. In the Matsubara notation, these self-energy terms are

$$\begin{aligned}
\Sigma^{(2)}(\mathbf{p}, ip) = \frac{1}{\beta^2} \frac{M_0^4}{\nu^2} \sum_{\mathbf{q}_1\mathbf{q}_2} \sum_{iq_n i\omega_n} & \mathscr{D}^{(0)}(\mathbf{q}_1, iq)\mathscr{D}^{(0)}(\mathbf{q}_2, i\omega)(1/q_1^2)(1/q_2^2) \\
& \times \mathscr{G}^{(0)}(\mathbf{p} + \mathbf{q}_1, ip + iq)\mathscr{G}^{(0)}(\mathbf{p} + \mathbf{q}_1 + \mathbf{q}_2, ip + iq_1 + i\omega) \\
& \times [\mathscr{G}^{(0)}(\mathbf{p} + \mathbf{q}_1, ip + iq) + \mathscr{G}^{(0)}(\mathbf{p} + \mathbf{q}_2, ip + i\omega)]
\end{aligned} \tag{6.1.13}$$

One may do the two Matsubara summations in each term and obtain the most general finite temperature result. This is left as an exercise. At zero temperature, the real part of the retarded self-energy does become (6.1.12). We are not aware of any evaluation of this expression.

B. Rayleigh–Schrödinger Perturbation Theory

The Rayleigh–Schrödinger form of perturbation theory is the standard kind which is described in quantum mechanical textbooks such as that of Schiff (1955). It is also called *on the mass shell* perturbation theory. Energy and momentum are no longer separate variables. In evaluating $\Sigma(p, E)$ the energy E is set equal to ε_p, so the self-energy is just a function of one variable p—or, equivalently, ε_p. Of course, if the imaginary part of the self-energy is actually zero, then energy and momentum are uniquely related. Thus Rayleigh–Schrödinger is an exact procedure under the same conditions that Brillouin–Wigner perturbation theory is valid—a zero value for the imaginary self-energy. We shall restrict our discussion to the case of zero temperature and $\varepsilon_p < \omega_0$, for which this condition is valid.

The two perturbation theories have a direct link to the potential scattering theories of Sec. 4.1. The Rayleigh–Schrödinger perturbation theory is analogous to the reaction matrix theory of potential scattering, while Brillouin–Wigner is analogous to T-matrix theory. In the former, one calculates real quantities on the mass shells i.e., with $E = \xi_p$. In the latter, one calculates a complex quantity for a general value of E. Recall that the reaction matrix result was not equal to the real part of the T-matrix result. Similarly, the Rayleigh–Schrödinger electron–phonon self-energy of the electron is *not* found from the Brillouin–Wigner form by just setting $E = \xi_p$ and taking the real part. This happens to work for the one-phonon self-energy but not when higher-order terms are included.

We shall call successive terms in the self-energy function $\Sigma_{\text{RS}}^{(n)}(p)$. The superscript denotes the number of phonons in the self-energy term. The self-energy is real, so we shall omit the signature "Re" before the self-energy expression. The one-phonon self-energy is

$$\Sigma_{\text{RS}}^{(1)}(\mathbf{p}) = \frac{M_0^2}{(2\pi)^3} \int \frac{d^3q}{q^2} \frac{1}{\varepsilon_p - \omega_0 - \varepsilon_{\mathbf{p}+\mathbf{q}}} \tag{6.1.14}$$

The energy denominator contains the difference between the initial state energy ε_p and the intermediate state energy $\varepsilon_{\mathbf{p}+\mathbf{q}} + \omega_0$; the latter has one phonon excited. The summation over wave vectors is a summation over all intermediate states subject to momentum conservation. The self-energy is the same as the Brillouin–Wigner result (6.1.6) after replacing E by ε_p. This identity is true only for the one-phonon self-energies. For higher orders, we must not only replace E by ε_p but also add some terms to the Brillouin–Wigner result in order to get Rayleigh–Schrödinger.

The wave vector integrals in (6.1.14) are elementary. Indeed, we just need to take (6.1.8) and replace E by ε_p, with the result for $\varepsilon_p < \omega_0$

$$\Sigma_{\text{RS}}^{(1)}(\mathbf{p}) = -\alpha \frac{\omega_0^{3/2}}{(\varepsilon_p)^{1/2}} \sin^{-1}\left(\frac{\varepsilon_p}{\omega_0}\right)^{1/2}$$

In the limit where $\varepsilon_p \to 0$, we obtain the simple result mentioned earlier:

$$\Sigma_{\mathrm{RS}}^{(1)}(0) = -\alpha\omega_0$$

It is no accident that the self-energy is just $-\alpha\omega_0$. The constant α was defined so that this happens.

The polaron effective mass is derived by expanding the arcsine in a power series in $(\varepsilon_p/\omega_0)^{1/2}$:

$$\Sigma_{\mathrm{RS}}^{(1)}(\mathbf{p}) = -\alpha\omega_0\left[1 + \frac{1}{6}\frac{\varepsilon_p}{\omega_0} + O\left(\frac{\varepsilon_p}{\omega_0}\right)^2\right]$$

$$\begin{aligned} E_p &= \varepsilon_p - \alpha\omega_0\left(1 + \frac{1}{6}\frac{\varepsilon_p}{\omega_0}\right) + O(p^4) \qquad (6.1.15) \\ &= -\alpha\omega_0 + \frac{p^2}{2m}(1 - \alpha/6) + O(p^4) \end{aligned}$$

Thus we deduce that the effective mass is $m^* = m/(1 - \alpha/6)$. The polaron effects make the particle appear to be heavier than the band mass m. The extra mass arises because of the interaction between the electron and the ions. The electron causes a change in the equilibrium positions of the ions in its vicinity. The change in equilibrium position arises from the mutual interaction between the electron and ion, which was mentioned earlier. When the electron moves with a momentum p, it must drag this ionic deformation with it. It takes energy to move this deformation. The drag is what causes the mass to increase as the polaron coupling constant is increased.

In the Tamm–Dancoff approximation, the formula (3.3.24) gives the effective mass $m^* = m(1 + \alpha/2)/(1 + \alpha/3)$. The details of this calculation are assigned as a problem. This agrees with the Rayleigh–Schrödinger result at small values of α, since both formulas are proportional to $m^* = m[1 + \alpha/6 + O(\alpha^2)]$. However, they behave quite differently in the intermediate coupling regime. The Rayleigh–Schrödinger result predicts that something calamitous happens at $\alpha \sim 6$. A similar catastrophe is not implied by the Brillouin–Wigner formula, which is well behaved for all values of α. We shall see that the Rayleigh–Schrödinger result is better. Something quite important does indeed happen at $\alpha \sim 6$: The particle becomes localized. This will be shown in the next subsection on strong coupling theory. We have already remarked, in connection with Table 6.1, that the zero-momentum values are much better in the Rayleigh–Schrödinger picture.

At zero temperature, the two-phonon self-energy in the Rayleigh–Schrödinger theory is

$$\Sigma_{\mathrm{RS}}^{(2)}(\mathbf{p}) = \frac{M_0^4}{\nu^2} \sum_{\mathbf{q}_1\mathbf{q}_2} \frac{1}{q_1^2 q_2^2} \left(\frac{1}{\varepsilon_p - \varepsilon_{\mathbf{p}+\mathbf{q}_1} - \omega_0} \right) \left[\frac{1}{\varepsilon_p - \varepsilon_{\mathbf{p}+\mathbf{q}_1+\mathbf{q}_2} - 2\omega_0} \right.$$
$$\times \left(\frac{1}{\varepsilon_p - \varepsilon_{\mathbf{p}+\mathbf{q}_1} - \omega_0} + \frac{1}{\varepsilon_p - \varepsilon_{\mathbf{p}+\mathbf{q}_2} - \omega_0} \right)$$
$$\left. - \frac{1}{(\varepsilon_p - \varepsilon_{\mathbf{p}+\mathbf{q}_2} - \omega_0)^2} \right] \tag{6.1.16}$$

The wave vector integrals have been evaluated analytically for zero momentum. The integrals are complicated, so that we shall only give the result:

$$\Sigma_{\mathrm{RS}}^{(2)}(0) = -\alpha^2\omega_0 \left[\ln\left(1 + \frac{3}{2(2)^{1/2}}\right) - \frac{1}{2^{1/2}} \right] = -0.0159\alpha^2\omega_0$$

We combine $\Sigma_{\mathrm{RS}}^{(2)}$ with the earlier result for the one-phonon self-energy. The three-phonon self-energy was evaluated numerically by Sheng and Dow (1971), who computed the next terms in the Rayleigh–Schrödinger series for E_0 and m^*:

$$E_0 \equiv E_{\mathrm{RS}} \equiv -\omega_0[\alpha + 0.0159\alpha^2 + 0.008765\alpha^3 + O(\alpha^4)]$$

$$\frac{m}{m^*} = 1 - \frac{1}{6}\alpha + 0.02263\alpha^2 + O(\alpha^3)$$

Some values are tabulated in the Table 6.1 and agree well with the Feynman values. The exact ground state energy is a power series in α. We have just the first three terms in this series. It is remarkable that the coefficients of the α^2 and α^3 terms are so small. Even for intermediate coupling values $1 \leq \alpha \leq 6$, the term $0.0159\alpha^2$ makes less than a 10% contribution to the total value of E_0. Another way to say this is that first-order perturbation theory is a good approximation even for intermediate coupling strengths. This is further evidence for the superiority of Rayleigh–Schrödinger over Brillouin–Wigner for the zero temperature polaron problem. We assume that higher-order terms in α^n also have a small coefficient and will also not contribute much toward E_0 for intermediate values of coupling constant.

It is natural to ask why the Rayleigh–Schrödinger method is so good. This point will be explained in detail in the next section. But it seems worthwhile to summarize some of these findings here. In Chapter 4 we solved exactly the independent boson model for a model system of electrons and phonons. The polaron model may not be solved exactly, but the Rayleigh–Schrödinger perturbation theory is related to the independent boson model. The self-energy for the interacting system of electrons and phonons arises from terms in which various numbers of phonons are virtually emitted into intermediate states. The one-phonon self-energy in the Rayleigh–Schrödinger picture corresponds to the assumption that all the phonons are virtually emitted independently of the others—as in the zero-boson model. The two-phonon self-energy describes the correlations between the virtual emission of pairs of phonons. That is, $\Sigma_{\rm RS}^{(1)}(p)$ contains the basic contribution from emitting one phonon, two phonons—and all numbers of phonons. The next term $\Sigma_{\rm RS}^{(2)}$ describes the correlations between pairs of phonon emissions. The following term $\Sigma_{\rm RS}^{(3)}$ describes three-phonon correlations, which are not just pairwise correlations. The term $0.0159\alpha^2$ is small, apparently, because correlations are not important at intermediate coupling. For this reason, one expects the α^3 terms to be similarly small.

At first it seems surprising that $\Sigma_{\rm RS}^{(1)}$ has the self-energy from emissions of all different numbers of phonons—but without correlations. However, this is similar to our findings for the independent boson model. Here the exact self-energy was $-\sum_{\mathbf{q}} M_{\mathbf{q}}^2/\omega_q$, yet this described a ground state which had a mixture of large numbers of phonons. If we take the Rayleigh–Schrödinger theory and set all kinetic energy terms equal to the same constant ε_0, we immediately recover the independent boson model—in which the particle had constant energy. The one-phonon self-energy becomes

$$\Sigma_{\rm RS}^{(1)} = -\frac{M_0^2}{\nu}\sum_{\mathbf{q}}\frac{1}{q^2}\frac{1}{\omega_0} \equiv -\sum_{\mathbf{q}}\frac{M_{\mathbf{q}}^2}{\omega_{\mathbf{q}}}$$

$\Sigma_{\rm RS}^{(1)}$ is the exact self-energy when the kinetic energy is constant, which means that all higher self-energy terms are zero. An inspection of $\Sigma_{\rm RS}^{(2)}$ in (6.1.16) shows that it vanishes when all kinetic energies are a constant. However, we already know it must vanish, since the exact self-energy for the independent boson model is just a linear function of α.

In Sec. 4.3 we solved the independent boson model by a variety of methods. But we never did solve it by using Dyson's equation and trying to write all the self-energy diagrams. At zero temperature, this would correspond to trying to solve it by Brillouin–Wigner perturbation theory. In fact, the simple exact self-energy cannot be obtained this way—at least no one has ever succeeded in doing it. One apparently has to evaluate all the terms in all orders of perturbation theory. It is not surprising that this approach does not work well for the polaron problem either.

A lover of antiques will enjoy that, for the polaron problem, the old-fashioned perturbation theory works better than Dyson's equations. One might ask whether this is a general feature. If so, why bother to learn about Green's functions? A general rule cannot be given. In some Hamiltonians the Rayleigh–Schrödinger method is best, and in others Dyson's equation is better. Coupled mode problems, such as found in the Tomonaga model of Sec. 4.4 or in polaritons in Sec. 4.5, need to be described by Brillouin–Wigner perturbation theory. Another example: Whenever the imaginary part of the self-energy is significant, it seems necessary to use a Green's function approach. The old-fashioned techniques do not allow for this in a systematic manner. The theory of strongly coupled superconductors, for example, could not be done without Green's functions. The actual message is that Green's functions are not the best way to solve all problems. Which problems are best solved by Green's functions? Unfortunately, at the moment, the only way to tell is by trial and error. No general rules are available which predict when one perturbation method is better than another.

The word *polaron* describes the coupled system of electrons and ions. The self-energy we calculate is for the mutually interacting system of electrons and phonons. Although we call it the electron self-energy, part of the energy resides in the ions themselves. The ions are moved from their equilibrium position by the presence of the electron. The displacement takes some vibrational energy, which is part of the electron self-energy, because it follows the electron in its journey through the crystal. A simple analogy is the charge on a spring, which was discussed in Sec. 1.1. The application of the electric field causes a deformation of the spring. The final self-energy $-(eF)^2/2K$ contains a part which is the energy needed to compress the spring to the new equilibrium. In the polaron motion, this happens locally wherever the polaron is at the moment.

The classical picture we have been using is that the ions in the polar lattice deform around the electron. The quantum picture is the same, except that the motion of the ions is quantized. The number of phonons is discrete. The amplitude of the ion displacements may not have a continuous range of values but only discrete amounts, which are phonons. If the average number of phonons around the electron is large, this makes little difference. A laboratory spring does not appear to be quantized, although it surely is, but the displacements which usually are observed are so large, with such large quantum numbers, that the quantum nature is irrelevant. The same is true with our polaron if it has a large average number of phonons. These phonons are called the *phonon cloud*.

One may try to calculate the average number of phonons in the polaron. As with all things in polaron theory, we may only calculate it approximately. Since the Rayleigh–Schrödinger method is the best, we shall use it again. From first-order perturbation theory, the wave function of the electron is

$$\Psi_{\mathbf{p}}(r) = |\,\mathbf{p}\rangle + \sum_{\mathbf{p}'} \frac{|\,\mathbf{p}'\rangle\langle\mathbf{p}'\,|\,V\,|\,\mathbf{p}\rangle}{\varepsilon_p - \varepsilon_{p'} + \omega_0}$$

$$= \frac{1}{\nu^{1/2}}\left[e^{i\mathbf{p}\cdot\mathbf{r}}\,|\,0\rangle + \frac{M_0}{\nu^{1/2}}\sum_{\mathbf{q}} \frac{1}{q}\,\frac{e^{i(\mathbf{p}+\mathbf{q})\cdot\mathbf{r}} a_{\mathbf{q}}^{\dagger}\,|\,0\rangle}{\varepsilon_{\mathbf{p}} - \varepsilon_{\mathbf{p}+\mathbf{q}} - \omega_0}\right] \tag{6.1.17}$$

where $|\,0\rangle$ is the phonon vacuum, so $a_{\mathbf{q}}^{\dagger}\,|\,0\rangle$ has one phonon with wave vector $\mathbf{q}$. The total number of phonons is found by taking the expectation of the phonon number operator $\sum_{\mathbf{k}} a_{\mathbf{k}}^{\dagger}a_{\mathbf{k}}$ with this state:

$$\mathrm{Nu}(p) = \int d^3r\,\Psi_{\mathbf{p}}^{\dagger}(r)\sum_{\mathbf{k}} a_{\mathbf{k}}^{\dagger}a_{\mathbf{k}}\Psi_{\mathbf{p}}(r)$$

We use the symbol Nu for number, to prevent confusion with other N symbols such as the thermal average number of phonons. The number operator on the first term in (6.1.17) is zero, and on the second term it is unity. Thus the first-order Rayleigh–Schrödinger prediction is

$$\mathrm{Nu}(p) = \frac{M_0^2}{\nu}\sum_{\mathbf{q}} \frac{1}{q^2}\,\frac{1}{(\varepsilon_{\mathbf{p}} - \varepsilon_{\mathbf{p}+\mathbf{q}} - \omega_0)^2}$$

This wave vector integral is done by making the same variable changes that were used to obtain (6.1.7) [$\nu = \cos\theta$, $x = (q + p\nu)/(2m)^{1/2}$]:

$$\mathrm{Nu}(p) = \frac{M_0^2(2m)^{1/2}}{8\pi^2}\int_{-\infty}^{\infty} dx \int_{-1}^{1} d\nu\,\frac{1}{(\omega_0 - \nu^2\varepsilon_p + x^2)^2}$$

The two integrals give, in turn, the values (for $\omega_0 > \varepsilon_p$)

$$\int_{-\infty}^{\infty} dx\,\frac{1}{(\omega_0 - \nu^2\varepsilon_p + x^2)^2} = \frac{\pi}{2}\,\frac{1}{(\omega_0 - \nu^2\varepsilon_p)^{3/2}}$$

$$\int_{-1}^{1} d\nu\,\frac{1}{(\omega_0 - \varepsilon_p\nu^2)^{3/2}} = \frac{1}{\omega_0}\,\frac{2}{(\omega_0 - \varepsilon_p)^{1/2}}$$

so the final result is ($\omega_0 > \varepsilon_p$)

$$\mathrm{Nu}(p) = \frac{\alpha}{2}\left(\frac{\omega_0}{\omega_0 - \varepsilon_p}\right)^{1/2}$$

For a zero-momentum particle, the result is $\mathrm{Nu} = \frac{1}{2}\alpha$. For $\alpha > 2$, the average number of phonons is greater than 1. This shows, again, that the "one-phonon" self-energy term actually describes many phonons coupled to the electron—only without correlation. It should be remembered that this is an *average number* of phonons. The actual number fluctuates from time to time about this average number.

These results apply for $\varepsilon_p < \omega_0$. An attempt to calculate the formula for $\varepsilon_p > \omega_0$ gives the result infinity. This is a nonsensical answer, but the question was also nonsensical. A particle with kinetic energy larger than ω_0 will eventually emit a real phonon and lower its energy to $\varepsilon_p - \omega_0$. Thus the number of phonons about the electron is not a stationary quantity when $\varepsilon_p > \omega_0$. Thus it is not reasonable to try to evaluate a stationary matrix element such as Nu.

Another quantity to calculate is the mean free path of the electron. At zero temperature, this is infinite when $\varepsilon_p < \omega_0$. However, at finite temperatures, there is a finite probability N_0 that some thermally excited phonons do exist. One of them may be absorbed by the electron, thereby changing its energy and momentum. According to the golden rule (Schiff, 1955), the transition probability is

$$w = \frac{2\pi}{\hbar} \int \frac{d^3q}{(2\pi)^3} \frac{M_0^2}{q^2} N_0 \delta(\varepsilon_{\mathbf{p}} + \omega_0 - \varepsilon_{\mathbf{p+q}})$$

This is nearly the same formula as twice the imaginary part of the retarded self-energy given in (6.1.11). The main difference is the replacement of E by ε_p. We have also left off the second term, proportional to $N_0 + 1$, which gives the rate of phonon emission; it is zero if $\varepsilon_p < \omega_0$. The integrals are evaluated by changing variables in the manner used in (6.1.7) to give $[\nu = \cos\theta,\ x = (q + p\nu)/(2m)^{1/2}]$

$$w = \frac{M_0^2(2m)^{1/2}}{4\pi} N_0 \int_{-1}^{1} d\nu \int_{\infty}^{\infty} dx\, \delta(\omega_0 + \varepsilon_p \nu^2 - x^2)$$

The two integrals may be evaluated in turn,

$$\int_{-\infty}^{\infty} dx\, \delta(\omega_0 + \varepsilon_p \nu^2 - x^2) = \frac{1}{(\omega_0 + \varepsilon_p \nu^2)^{1/2}}$$

$$\int_{-1}^{1} d\nu \frac{1}{(\omega_0 + \varepsilon_p \nu^2)^{1/2}} = \frac{1}{(\varepsilon_p)^{1/2}} \ln\left[\frac{(\omega_0 + \varepsilon_p)^{1/2} + (\varepsilon_p)^{1/2}}{(\omega_0 + \varepsilon_p)^{1/2} - (\varepsilon_p)^{1/2}}\right]$$

to give the scattering rate for phonon absorption. The scattering rate is the same as the inverse lifetime of the particle $w = 1/\tau$:

$$\frac{1}{\tau_p} = \alpha N_0 \frac{\omega_0^{3/2}}{(\varepsilon_p)^{1/2}} \ln\left[\frac{(\omega_0 + \varepsilon_p)^{1/2} + (\varepsilon_p)^{1/2}}{(\omega_0 + \varepsilon_p)^{1/2} - (\varepsilon_p)^{1/2}}\right]$$

The mean free path l_p is found from the classical expression $l_p = \tau_p v_p$, so that one multiplies τ_p by p/m. At zero momentum, the lifetime is evaluated as

$$\frac{1}{\tau_0} = 2\alpha N_0 \omega_0 \tag{6.1.18}$$

C. Strong Coupling Theory

The strong coupling theory for polarons was invented by Landau and Pekar (1946). Their theory was the first work on polarons, which even preceded the word *polaron*. Their theory, and its subsequent improvements, is now known to be valid at large values of α—hence the current name of strong coupling theory. The method of calculation is radically quite different from the prior perturbation theories. It is basically a variational calculation on a Gaussian wave function. We shall do the calculation first and discuss the physics afterward.

The Hamiltonian (6.1.2) is rewritten so that the phonon operators appear as displacements $Q_{\mathbf{q}}$ and their conjugate momenta P_q, which are chosen so they are dimensionless:

$$Q_{\mathbf{q}} = \frac{1}{2^{1/2}} (a_{\mathbf{q}} + a^{\dagger}_{-\mathbf{q}}) \qquad a_{\mathbf{q}} = \frac{1}{2^{1/2}} (Q_{\mathbf{q}} + iP_{\mathbf{q}})$$

$$P_{\mathbf{q}} = \frac{-i}{2^{1/2}} (a_{\mathbf{q}} - a^{\dagger}_{-\mathbf{q}}) \qquad a_{\mathbf{q}}^{\dagger} = \frac{1}{2^{1/2}} (Q_{\mathbf{q}} - iP_{\mathbf{q}})$$

$$H = \frac{p^2}{2m} + \frac{\omega_0}{2} \sum_{\mathbf{q}} (P_{\mathbf{q}}^2 + Q_{\mathbf{q}}^2) + \frac{2^{1/2} M_0}{\nu^{1/2}} \sum_{\mathbf{q}} \frac{Q_{\mathbf{q}}}{q} e^{i\mathbf{q}\cdot\mathbf{r}}$$

The wave function of the many-particle system $\Phi(\mathbf{r}; Q_{\mathbf{q}})$ must contain the coordinates of the electron $\mathbf{r}$ and the ion displacements $Q_{\mathbf{q}}$. We make what appears to be a drastic assumption: that the electron is localized with a Gaussian wave function. Later we shall soften this interpretation and show that the particle is not quite localized. But for the moment we assume that the total wave function is a simple product of electron and phonon coordinates:

$$\Phi(\mathbf{r}; Q_q) = \varphi(\mathbf{r})\psi_n(Q_\mathbf{q} + \delta Q_\mathbf{q})$$
$$\varphi(\mathbf{r}) = \frac{\beta^{3/2}}{\pi^{3/4}} \exp\left(\frac{-\beta^2 r^2}{2}\right) \tag{6.1.19}$$

where β is a variational parameter. The phonon wave functions ψ_n are the usual harmonic oscillator wave functions, except that they are centered about an equilibrium displacement $\delta Q_\mathbf{q}$, which also needs to be determined. The first step in the calculation is to take the expectation

$$\mathscr{H}(Q_q) = \int d^3r \varphi(\mathbf{r}) H \varphi(\mathbf{r})$$

of the Hamiltonian over the electron part of the coordinates. For the two r-dependent terms the Gaussian wave functions give

$$\int d^3r \varphi \frac{p^2}{2m} \varphi = \frac{3}{4} \frac{\beta^2}{m}$$

$$\int d^3r \varphi(\mathbf{r})^2 e^{i\mathbf{q}\cdot\mathbf{r}} = e^{-q^2/4\beta^2}$$

The second integral is most easily evaluated in x, y, z coordinates. The expectation value of the Hamiltonian is

$$\mathscr{H}(Q_q) = \frac{3}{4} \frac{\beta^2}{m} + \frac{\omega_0}{2} \sum_\mathbf{q} (P_\mathbf{q}{}^2 + Q_\mathbf{q}{}^2) + \sum_\mathbf{q} L_\mathbf{q} Q_\mathbf{q}$$

$$L_\mathbf{q} = \frac{2^{1/2} M_0}{\nu^{1/2}} \frac{e^{-q^2/4\beta^2}}{q}$$

The next step is to choose the equilibrium displacement $\delta Q_\mathbf{q}$ so that the term linear in Q_q is eliminated:

$$\delta Q_\mathbf{q} = \frac{L_\mathbf{q}}{\omega_0}$$

$$\mathscr{H}(Q_q) = \frac{\omega_0}{2} \sum_\mathbf{q} [P_\mathbf{q}{}^2 + (Q_\mathbf{q} + \delta Q_\mathbf{q})^2] + \frac{3}{4} \frac{\beta^2}{m} - \frac{1}{2\omega_0} \sum_\mathbf{q} L_\mathbf{q}{}^2$$
$$= \frac{\omega_0}{2} \sum_\mathbf{q} [P_q{}^2 + (Q_q + \delta Q_q)^2] + E(\beta)$$

The first term on the right describes the harmonic vibrations of the phonons about their new equilibrium positions $-\delta Q_{\mathbf{q}}$. These have the harmonic wave functions in (6.1.19) and the eigenvalues $(n + \frac{1}{2})\omega$. The second term $3\beta^2/4m$ is the kinetic energy of the electron in the Gaussian wave function. The last term is the potential energy of interaction between the phonons and electron. The parameter β is varied to give the lowest energy for these last two terms. First the potential term must be evaluated:

$$\frac{1}{2\omega_0} \sum_{\mathbf{q}} L_{\mathbf{q}}^2 = \frac{M_0^2}{\omega_0} \int \frac{d^3q}{(2\pi)^3} \frac{e^{-q^2/2\beta^2}}{q^2} = \alpha\left(\frac{\beta^2\omega_0}{m\pi}\right)^{1/2}$$

$$E(\beta) = \frac{3}{4}\frac{\beta^2}{m} - \alpha\left(\frac{\beta^2\omega_0}{m\pi}\right)^{1/2}$$

If we set $B = \beta^2/2m$, the last two terms are

$$E(B) = \frac{3}{2} B - B^{1/2}\alpha\left(\frac{2\omega_0}{\pi}\right)^{1/2}$$

$$\frac{dE}{dB} = \frac{3}{2} - \frac{1}{2B^{1/2}}\alpha\left(\frac{2\omega_0}{\pi}\right)^{1/2}$$

The variation shows that the minimum value is $B_0^{1/2} = (\alpha/3)(2\omega_0/\pi)^{1/2}$, so that the minimum energy is

$$E(B_0) = -\frac{\alpha^2\omega_0}{3\pi} = -0.106\alpha^2\omega_0$$

As usual with bound state variational calculations, the potential energy is twice the kinetic energy, in agreement with the virial theorem. This is the result obtained by Landau and Pekar. The energy is proportional to α^2 rather than α.

In strong coupling theory, one assumes that α is very large and tries to develop the energy as a power series in $1/\alpha$. The philosophy is very similar to that for deducing the correlation energy of the homogeneous electron gas in Secs. 5.1 and 5.2: One perturbation expansion was developed for small r_s and another for high r_s. For the polaron we shall do the same. The Rayleigh–Schrödinger expansion is valid at small α, and the strong coupling is valid at large α. We shall see that the interpolation between these two limits is remarkably easy. The best available result for the strong coupling limit (Miyake, 1976) is

$$\lim_{\alpha\to\infty} E_0(\alpha) = -\omega_0[0.1085\alpha^2 + 2.836 + O(1/\alpha^2)] \qquad (6.1.20)$$

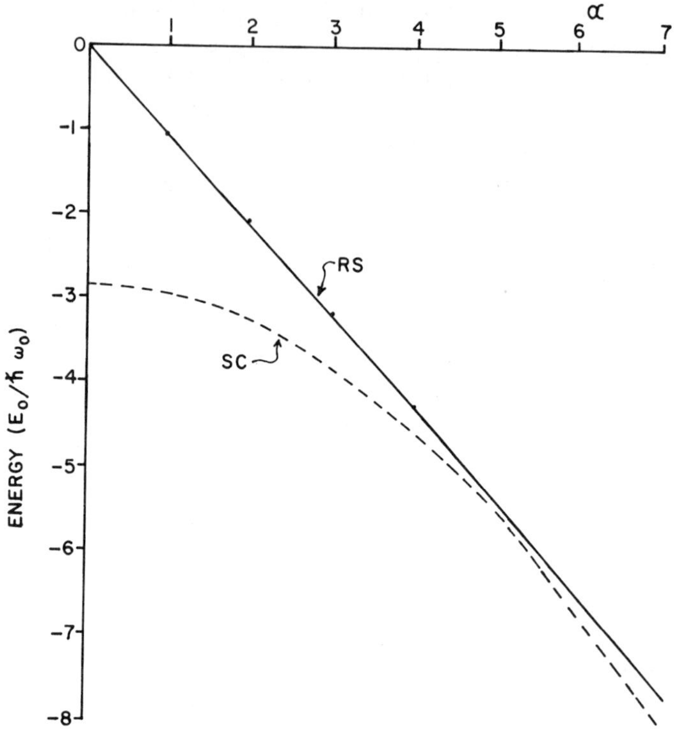

FIGURE 6.2. Energy of a polaron with zero momentum as a function of coupling constant. The solid line is the Rayleigh–Schrödinger theory, and the dashed line is the strong coupling theory.

The coefficient of the α^2 term is 0.1085, which is remarkably close to the Landau–Pekar result of 0.1061. This agreement shows that the Gaussian assumption is very accurate.

Figure 6.2 shows a plot of the ground state energy for Rayleigh–Schrödinger perturbation theory—the solid curve labeled RS; also shown by the dashed line is the strong coupling theory using (6.1.20). The strong coupling theory appears to predict a lower energy state for small values of α. This is deceptive, since we have omitted terms in the asymptotic series of order α^{-2}, and these obviously diverge at small α. Thus we should really believe the strong coupling theory results only for values of α above, say, about 5 or 6. The two curves almost touch for values of $\alpha \sim 5$. The value $\alpha \sim 5$ is believed to be the crossover region between the two theories. Thus the correct theory is the Rayleigh–Schrödinger result for values up to $\alpha \leq 5$ and the strong coupling theory for $\alpha \geq 5$. This is precisely the behavior of the Feynman theory.

Good insight into the strong coupling theory has been provided by Evrard (1972). The first question concerns the actual size of the Gaussian wave functions. From the minimal value of B_0, we deduce that $\beta_0 = (\alpha/3)[(4/\pi)\omega_0 m]^{1/2}$. β_0 has the dimensions of cm^{-1}, so its inverse is roughly the size of the localized wave function. If we restore $\hbar$, the quantity

$$l_0 = \left(\frac{\hbar}{2m\omega_0}\right)^{1/2} \tag{6.1.21}$$

is the characteristic polaron length and is roughly 10 Å for $\hbar\omega_0 \simeq 0.03$ eV and m equal to the electron mass. Of course, m is actually the band mass, which is usually less than an electron mass. Anyway, our basic polaron size is $\beta_0^{-1} \simeq 40$ Å$/\alpha$. For α values around 5 to 6, this is 7 Å, or about the size of the atomic unit cell. The localized electron will be influenced by the fact that the ions are atoms. Our initial Hamiltonian (6.1.2) assumes a continuum theory for the ions, which is probably a reasonable approximation for small values of α but surely fails when the polaron size is of atomic dimension. Thus strong coupling theory cannot be applied to real solids without additional modifications to account for the atomic nature of the phonons. From now on we must consider it a mathematical model with interesting properties. The useful piece of physics we have deduced is that polarons become localized on atomic sites for α values stronger than 5 to 6. The Fröhlich Hamiltonian does not describe their behavior after they are localized.

The theory of polaron localization has a catch. One needs to know the band mass m to calculate α. Although one could deduce m from a very good energy band calculation, these seldom have the required accuracy. Instead, one usually measures the polaron mobility and the effective mass, and the combined results will yield m, m^*, and α (see Hodby, 1972). The catch is that this procedure works only if the polaron is mobile. After it becomes localized, its mobility drops precipitously. Thus one can no longer measure the effective mass by cyclotron resonance. It is difficult to deduce what the polaron constant α should have been after the particle is localized. One can only deduce it accurately if the particle is not localized. Some of Hodby's results are shown in Table 6.2. An interesting feature is that there are no values of α reported above 4. Perhaps this is the cutoff for localization of the polaron.

TABLE 6.2. Measured Properties of Certain Alkali, Silver, and Thallium Halides[a]

Material	ε_s	ε_i	ε_∞	$\hbar\omega_{LO}$ (cm^{-1})	Carrier	M_{band}	α	$M_{polaron}$ (measured)
KCl	4.49[b]	—	2.20[b]	212[c]	Electron	0.489	3.69	0.922 ± 0.04[d]
KBr	4.52[b]	—	2.39[b]	166[c]	Electron	0.404	3.22	0.700 ± 0.03[d]
KI	4.68[b]	—	2.68[b]	143[c]	Electron	0.345	2.60	0.536 ± 0.03[d]
RbCl	4.53[b]	—	2.20[b]	182[c]	Electron	0.507	4.09	1.03 ± 0.10[d]
RbI	4.55[b]	—	2.61[b]	105.5[c]	Electron	0.405	3.35	0.72 ± 0.07[d]
AgCl	9.50[b]	—	3.97[b]	196[e]	Electron	0.299	1.90	0.411 ± 0.02_f
					Hole	Self-traps		
AgBr	10.60[b]	—	4.68[b]	140[e]	Electron	0.222	1.58	0.2897 ± 0.004[g]
					Hole	1.22	—	1.71 ± 0.15(m_1)[h]
					(111 Ellipsoid)	0.48	2.31	0.79 ± 0.05$(m_t)_h$
TlCl	37.6[b]	—	5.00[b]	175[e]	Electron	0.355	2.59	0.551 ± 0.03[i]
					Hole			0.58 ± 0.03(m_1)[i]
					(100 Ellipsoid)			0.98 ± 0.04(m_t)[i]
TlBr	35.1[b]	—	5.64[b]	117[e]	Electron	0.336	2.64	0.525 ± 0.03[i]
					Hole			0.55 ± 0.03(m_1)[i]
					(100 Ellipsoid)			0.74 ± 0.03(m_t)[i]
Cu_2O	7.5[j]	7.11[k]	6.46[k]	660[l], 150[l]	Electron	0.93	0.18, 0.19	0.99 ± 0.03[m]
					Hole	0.56	0.14, 0.15	0.58 ± 0.03[m]
					Hole	0.66	0.15, 0.16	0.69 ± 0.04[m]
HgI_2	8.5 $\parallel c$	—	6.8 $\parallel c$	140 $\parallel c$	Electron	0.31		0.39 ± 0.02$(m \parallel c)$[p]
	25.9 $\perp c$[n]	—	5.15 $\perp c$[n]	116, 32 $\perp c$[o]	(Ellipsoid)	0.28		0.36 ± 0.015$(m \perp c)$[p]
					Hole	1.29		1.55 ± 0.1$(m \parallel c)$[p]
					(Ellipsoid)	0.56		1.025 ± 0.1$(m \perp c)$[p]

[a] Compiled by J. Hodby (used with permission). All tabulated properties are measured at 5 K or lower temperature. ε_s and ε_∞ are, respectively, the static and optical dielectric constants. ε_i is the dielectric constant for Cu_2O between 150 and 660 cm^{-1}. $\hbar\omega_{LO}$ is the energy of the longitudinal optical phonon at the zone center as measured by direct optical techniques. M_{band} (referred to the mass of the free electron) is deduced from the measured polaron mass ($M_{polaron}$) by the relation $M_{band} = M_{polaron}[(1 - \alpha/12)/(1 + \alpha/12)]$. All polaron masses (except those labeled *g*) have been measured at 137 GHz. No correction has been applied for the nonzero value of this frequency. The error bars define the points of half-maximum height on the observed resonance curves.

[b] R. P. Lowndes and D. H. Martin, *Proc. R. Soc. London Ser. A* **308**, 473 (1969).

[c] R. P. Lowndes, *Phys. Rev. B* **1**, 2754 (1970).

[d] J. W. Hodby, *J. Phys. C* **4**, 19 (1971).

[e] R. P. Lowndes, *Phys. Rev. B* **6**, 1490 (1972).

[f] J. W. Hodby, A. Hirano, S. Komiyama, and T. Masumi, to be published.

[g] J. W. Hodby, J. G. Crowder, and C. C. Bradley, *J. Phys. C* **7**, 3033 (1974).

[h] H. Tamura and T. Masumi, *Solid State Commun.* **12**, 1183 (1973).

[i] J. W. Hodby, G. T. Jenkin, K. Kobayashi, and H. Tamura, *Solid State Commun.* **10**, 1017 (1972).

[j] M. O'Keefe, *J. Chem. Phys.* **39**, 1789 (1963).

[k] C. Carabatos, A. Diffine, and M. Sieskind, *J. Phys. Paris* **29**, 529 (1968).

[l] A. Compaan and H. Z. Cummins, *Phys. Rev. B* **6**, 4753 (1972); P. T. Yu, T. R. Shen, and Y. Petroff, *Solid State Commun.* **12**, 973 (1973).

[m] J. W. Hodby, T. E. Jenkins, C. Schwab, H. Tamura, and D. Trivich, *J. Phys. C* **9**, 1429 (1976).

[n] H. Burkhard, private communication.

[o] Y. Ogawa, I. Harada, H. Matsuura, T. Shimanouchi, and J. Hiraishi, *Spectrochim. Acta Part A* **32**, 49 (1976).

[p] P. D. Bloch, J. W. Hodby, C. Schwab, and D. W. Stacey, *J. Phys. C* **11**, 2579 (1978).

We wish to examine further the process of localization: not the process whereby an electron gets trapped, but rather how the electron wave function changes with increasing α. The point is that this process is not all that dramatic. Consider the Rayleigh–Schrödinger wave function (6.1.17), which is written as

$$\psi_{\mathbf{p}}(\mathbf{r}) = \frac{e^{i\mathbf{p}\cdot\mathbf{r}}}{\nu^{1/2}}\,\eta_{\mathbf{p}}(\mathbf{r})\,|\,0\rangle$$

$$\eta_{\mathbf{p}}(\mathbf{r}) = 1 + \frac{M_0}{\nu^{1/2}}\sum_{\mathbf{q}}\frac{e^{i\mathbf{q}\cdot\mathbf{r}}}{q}\,\frac{a_{\mathbf{q}}^{\dagger}}{\varepsilon_{\mathbf{p}} - \varepsilon_{\mathbf{p}+\mathbf{q}} - \omega_0}$$

We wish to evaluate $\eta_{\mathbf{p}}(\mathbf{r})$ as a function of r. Of course, it depends on the operator $a_{\mathbf{q}}^{\dagger}$. This could be replaced by the equivalent operator $2^{1/2}Q_{\mathbf{q}} = (a_{\mathbf{q}} + a_{\mathbf{q}}^{\dagger})$, since $a_{\mathbf{q}}$ gives zero when operating on the vacuum. Thus $\eta_{\mathbf{p}}(\mathbf{r})$ may also be written as a function of the electron coordinate $\mathbf{r}$ and the phonon coordinate $\mathbf{Q}_{\mathbf{q}}$. Here the parts are not separable. The operator $\mathbf{Q}_{\mathbf{q}}$ describes the amount of ion displacement around the electron, which varies by discrete amounts. However, to get an approximate wave function, let us replace $\mathbf{Q}_{\mathbf{q}}$ by its average value, which is approximately

$$\langle \mathbf{Q}_{\mathbf{q}} \rangle \simeq \frac{M_0}{\nu^{1/2}}\,\frac{1}{q}\,\frac{1}{\varepsilon_{\mathbf{p}} - \varepsilon_{\mathbf{p}+\mathbf{q}} - \omega_0}$$

$$\eta_{\mathbf{p}}(\mathbf{r}) \simeq 1 + \frac{M_0^2}{\nu}\sum_{\mathbf{q}}\frac{e^{i\mathbf{q}\cdot\mathbf{r}}}{q^2}\,\frac{1}{(\varepsilon_{\mathbf{p}} - \varepsilon_{\mathbf{p}+\mathbf{q}} - \omega_0)^2}$$

The second term in the wave function may now be integrated. It is simple to do this when $\mathbf{p} = 0$ $[q_0 = (2m\omega_0)^{1/2}]$,

$$\eta_0(\mathbf{r}) = 1 + \frac{(2m)^2 M_0^2}{4\pi^2}\int_0^{\infty} dq \int_{-1}^{1} d\nu\,\frac{e^{iqr\nu}}{(q^2 + q_0^2)^2}$$

$$= 1 + \frac{(2m)^2 M_0^2}{4\pi^2}\,\frac{2}{r}\int_0^{\infty}\frac{dq}{q}\,\frac{\sin qr}{(q^2 + q_0^2)^2}$$

$$= 1 + \frac{(2m)^2 M_0^2}{4\pi^2}\,\frac{2}{r}\,\frac{\pi}{2q_0^4}\left[1 - \frac{1}{2}\,e^{-q_0 r}(2 + q_0 r)\right]$$

so that for the wave function we obtain

$$\eta_0(\mathbf{r}) = 1 + \frac{\alpha}{x}\,[1 - e^{-x}(1 + x/2)]$$

$$x = q_0 r$$

At $\mathbf{r} = 0$ it gives the simple result $\eta_0(0) = 1 + (\alpha/2)$. It goes to unity at large values of r. The scale of length is $q_0^{-1} = (2m\omega_0/\hbar)^{-1/2}$, which is just the l_0 which was introduced earlier in (6.1.21). Thus polaron theory predicts that the electron wave function is larger near the origin. The origin is the point we have chosen as the center of the ion displacement. One could easily choose another reference point $\mathbf{R}_0$ by having the average $\langle Q_\mathbf{q} \rangle$ multiplied by $\exp(-i\mathbf{q} \cdot \mathbf{R}_0)$. Thus the electron is more likely to be in the region of space where the ions are displaced. Conversely, the phonons are more likely to be displaced where the electron is located. This bootstrap relationship combines to produce a maximum at some point in space where they are both more likely to be found. Curves for $\alpha = 1$ and $\alpha = 4$ are shown in Figs. 6.3(a) and 6.3(b). As α increases, the hump in the center is larger. In each case, there is still a finite probability that the electron is not at this location and is far away. The strong coupling wave function for $\alpha = 6$ is shown in Fig. 6.3(c). This is just a plot of the Gaussian wave function. Of course it has a hump in the center. Now the wave function at large $\mathbf{r}$ goes to zero rather than unity. It no longer has the possibility of being somewhere else. It has become trapped in the potential well formed by the ion displacements. These ion displacements are caused by the presence of the electron. The two features—potential well and electron wave function—must be determined self-consistently.

There are several other properties worth mentioning for polarons in strong coupling theory. The first is that the localization may occur anywhere. If the phonons were really a continuum, then the Gaussian packet would drift about and have an effective mass. The effective mass is quite large: Allcock (1956) has estimated it to be $m^* = 0.0208m\alpha^4$. For example, at $\alpha = 5$ it is $m^*/m = 13$. A similar estimate, $m^* = 0.0227m\alpha^4$, is given by Miyake (1976). The polaron is heavy because it has to drag with it the potential well of the phonons. Recall the prediction from Rayleigh–Schrödinger perturbation theory that the effective mass m^*/m was proportional to $(1 - \alpha/6)^{-1}$ so that something important seemed to happen at $\alpha \sim 6$. This is now seen to be the localization, which makes the polaron much heavier. The heavy mass is not measurable because, as we said earlier, strongly coupled polarons are a model which we cannot apply to real systems without modification.

Another feature of strongly coupled polarons is that they have excited states which are also localized. For example, one could try to construct a polaron state which had p-wave symmetry. A suitable trial wave function might be

$$\varphi(r) = \left(\frac{2\beta_1^5}{\pi^{3/2}}\right)^{1/2} z e^{(-\beta_1^2 r^2)/2}$$

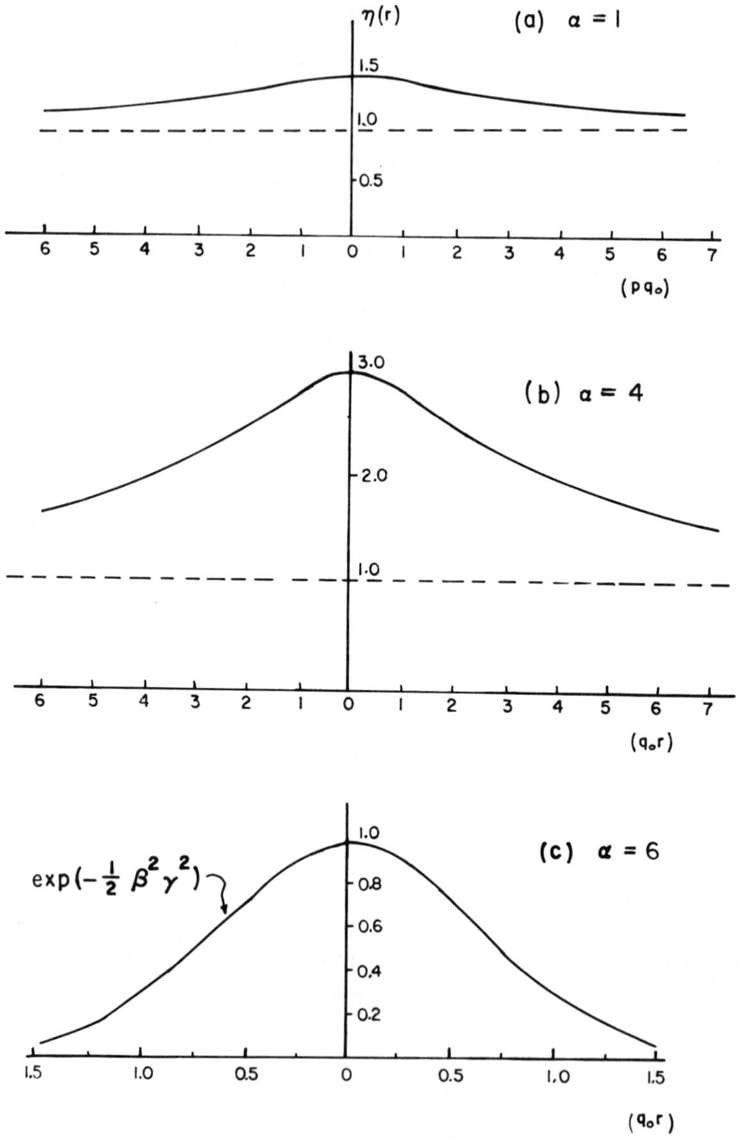

FIGURE 6.3. The polaron wave function in real space for different coupling constants. Results in (a) and (b) are free polarons in Rayleigh–Schrödinger theory, while (c) is from strong coupling theory.

One could repeat the variational procedure with β as a variational parameter (see Problem 10 at the end of this chapter). This will be different from the s-wave value β_0, and it will be smaller. The wave function is more spread out, and the electron is less bound. This has been called the *relaxed excited state* (see Kartheuser *et al.*, 1969). If the excited state were stable, one could observe actual optical transitions to this state from the ground state of s symmetry. The excited state is not stable, since it is degenerate in energy with the ground s state plus L phonons, where L is whatever number is necessary to make up the energy difference. Nevertheless, theoretical calculations show a sharp line in the optical spectra, which has been interpreted as this s- to p-wave transition. Thus the strongly coupled polaron can create its own internal structure.

The strong coupling limit is calculated in the adiabatic approximation (see Allcock, 1962). The electron has sufficient binding energy that its oscillatory motion in the potential well is much faster than the vibrational frequency of the phonons. Thus the phonons do not have time to adjust to the individual oscillations of the electron. Instead they adjust to the average motion of the electrons. Thus we treat them as a rigid potential well, in which the electron adiabatically oscillates. A quite different picture applies to the weak coupling limit. There the phonon energy is larger than that of the electrons. The picture is that the phonons, or ion polarization, follow the electron during its motion.

D. Linked Cluster Theory

The linked cluster theory was advocated by Brout and Carruthers (1963) as a general method to attack many-body problems. It is surprising that the method waited so long to be applied to the Fröhlich polaron. It was not until the work of Dunn (1975) that this was done, although some earlier work of Mahan (1966) was similar. The surprise is even greater when it is realized that this is an excellent method of obtaining the Green's function $G(\mathbf{p}, t)$ for the polaron in the weak and intermediate coupling regimes. Since we have learned, in the prior section, that the Fröhlich polaron Hamiltonian is useful only for these coupling values, the linked cluster theory is applicable for all relevant coupling strength. One disagreeable aspect of the technique is that the final numbers have to be generated on the computer, since the formulas are too complicated to permit analytical evaluation. This is the price one pays for good Green's functions.

The linked cluster methods were described in Sec. 3.6. Since we are concerned with a one-electron problem, we are permitted the luxury of working in real time. Although we shall work at finite temperatures, the real-time formalism is valid. We do not need the Matsubara technique since the interactions do not change the thermodynamic averages. That is, if we write the Hamiltonian as $H = H_0 + V$, then we can use H_0 in $\exp(-\beta H)$ rather than H, since V does not affect the thermodynamic averages. The only averaging we shall do is over the phonons, and one electron in a macroscopic solid does not alter the phonon energies. This substitution, of using H_0 for H in the thermodynamic averaging, is not appropriate in a many-particle system where the particles influence the phonon modes—as in a metal or a heavily doped semiconductor. But it is acceptable in the one-electron Fröhlich Hamiltonian. Thus for the phonon Green's function at finite temperature we can take

$$\begin{aligned} D^{(0)}(t) &= -i[(N_0 + 1)e^{-i\omega_0|t|} + N_0 e^{i\omega_0|t|}] \\ N_0 &= [\exp(\beta\omega_0) - 1]^{-1} \end{aligned} \tag{6.1.22}$$

$D^{(0)}(t)$ has no q dependence, since the phonons are assumed to have no dispersion. In this case $D^{(0)}(t) = D(t)$ since all the phonon self-energies are zero. Thus our only concern is to evaluate the electron Green's function. If we adapt the formulas of Sec. 3.6 to real time, the Green's function for the electron may be written as an exponential function of momentum and time:

$$\begin{aligned} G(\mathbf{p}, t) &= G^{(0)}(\mathbf{p}, t) \exp[F(\mathbf{p}, t)] \\ G^{(0)}(\mathbf{p}, t) &= -i\theta(t)e^{-i\varepsilon_p t} \end{aligned}$$

The function $F(\mathbf{p}, t)$ is generated as a series of terms, which are obtained by the following systematic procedure. The electron Green's function

$$G(\mathbf{p}, t) = -i\langle 0 \mid Te^{iHt}c_{\mathbf{p}}e^{-iHt}c_{\mathbf{p}}^{\dagger} \mid 0\rangle = -i\theta(t)\langle T\hat{c}_{\mathbf{p}}(t)U(t)\hat{c}_{\mathbf{p}}^{\dagger}\rangle$$

is evaluated in the interaction representation:

$$G(\mathbf{p}, t) = -i\theta(t)\sum_n W_n(\mathbf{p}, t)$$

$$W_n(\mathbf{p}, t) = \frac{(-i)^{2n}}{(2n)!}\int_0^t dt \cdots \int_0^t dt_{2n}\langle T\hat{c}_{\mathbf{p}}(t)\hat{V}(t_1) \cdots \hat{V}(t_{2n})\hat{c}_{\mathbf{p}}^{\dagger}(0)\rangle$$

The S-matrix expansion generates a series of scalar functions $W_n(\mathbf{p}, t)$. These are resumed as an exponential series of terms $F_n(\mathbf{p}, t)$:

$$G(\mathbf{p}, t) = -i\theta(t)\sum_n W_n(\mathbf{p}, t) \equiv -i\theta(t)e^{-i\varepsilon_p t}\exp\left[\sum_{n=1}^{\infty} F_n(\mathbf{p}, t)\right]$$

$$F(\mathbf{p}, t) = \sum_{n=1}^{\infty} F_n(\mathbf{p}, t)$$

These two series may be equated term by term by assuming that both W_n and F_n are proportional to the coupling constant α^n. That is, we define W_n to be all the terms in the S-matrix expansion which are proportional to α^n. Similarly, when we resum the series as an exponential, F_n contains all the terms proportional to α^n. For the polaron problem, the terms in the S matrix are zero when l is odd, since they contain an odd number of phonon creation or destruction operators. Only the terms with l even contribute, and we equate $n = 2l$. With this convention, we get the relationship between the two series:

$$F_1 = e^{i\varepsilon_p t} W_1(\mathbf{p}, t)$$

$$F_2 = e^{i\varepsilon_p t} W_2(\mathbf{p}, t) - \frac{1}{2!} F_1^2$$

$$F_3 = e^{i\varepsilon_p t} W_3(\mathbf{p}, t) - F_1 F_2 - \frac{1}{3!} F_1^3$$

The terms are listed only up to $n = 3$, since that is already beyond what has been evaluated for the polaron problem. So far only the first two terms have been computed. However, these appear to be adequate to describe the polaron Green's function for low and intermediate coupling. The term W_1, or equivalently F_1, was first evaluated by Mahan (1966). This term has the form

$$W_1(\mathbf{p}, t) = \frac{(-i)^2}{2!} \int_0^t dt_1 \int_0^t dt_2 \langle T\hat{c}_{\mathbf{p}}(t) \hat{V}(t_1) \hat{V}(t_2) \hat{c}_{\mathbf{p}}^{\dagger}(0) \rangle$$

If we write the Fröhlich Hamiltonian as $H = H_0 + V$,

$$H_0 = \sum_{\mathbf{p}} \varepsilon_p c_{\mathbf{p}}^{\dagger} c_{\mathbf{p}} + \omega_0 \sum_{\mathbf{q}} a_{\mathbf{q}}^{\dagger} a_{\mathbf{q}}$$

$$V = \sum_{\mathbf{qp}} \frac{M_0}{q(\nu)^{1/2}} c_{\mathbf{p}+\mathbf{q}}^{\dagger} c_{\mathbf{p}} (a_{\mathbf{q}} + a_{-\mathbf{q}}^{\dagger})$$

then the interaction potential term is

$$\hat{V}(t) = \frac{M_0}{\nu^{1/2}} \sum_{\mathbf{q}} \frac{e^{-it(\varepsilon_{\mathbf{p}} - \varepsilon_{\mathbf{p}+\mathbf{q}})}}{q} c_{\mathbf{p}+\mathbf{q}}^{\dagger} c_{\mathbf{p}} \hat{A}_{\mathbf{q}}(t)$$

$$\hat{A}_{\mathbf{q}}(t) = a_{\mathbf{q}} e^{-i\omega_0 t} + a_{-\mathbf{q}}^{\dagger} e^{i\omega_0 t}$$

The two phonon operators just give the phonon Green's function (6.1.22), and the electron correlation function can be written as a product of three

$G^{(0)}$ by using Wick's theorem,

$$\langle T\hat{A}_{\mathbf{q}}(t_1)\hat{A}_{\mathbf{q}'}(t_2)\rangle = i\delta_{\mathbf{q}+\mathbf{q}'=0}D(t_1 - t_2)$$

$$\langle T\hat{c}_{\mathbf{p}}(t)\hat{c}^{\dagger}_{\mathbf{p}_1-\mathbf{q}}(t_1)\hat{c}_{\mathbf{p}_1}(t_1)\hat{c}^{\dagger}_{\mathbf{p}_2+\mathbf{q}}(t_2)\hat{c}_{\mathbf{p}_2}(t_2)\hat{c}_{\mathbf{p}}{}^{\dagger}(t)\rangle$$
$$= 2i^3G^{(0)}(\mathbf{p}, t - t_1)G^{(0)}(\mathbf{p} + \mathbf{q}, t_1 - t_2)G^{(0)}(\mathbf{p}, t_2)$$

so that we have to evaluate

$$W_1(\mathbf{p}, t) = \frac{i^2}{\nu}\sum_{\mathbf{q}}\frac{M_0{}^2}{q^2}\int_0^t dt_1\int_0^t dt_2 D(t_1 - t_2)G^{(0)}(\mathbf{p}, t - t_1)$$
$$\times G^{(0)}(\mathbf{p} + \mathbf{q}, t_1 - t_2)G^{(0)}(\mathbf{p}, t_2)$$

For one particle in a band, the electron Green's function $G^{(0)}(\mathbf{p}, t)$ is zero unless $t > 0$, since the other time ordering for $t < 0$ has the destruction operator operating on the vacuum. Thus by writing all electron Green's functions as $G^{(0)}(\mathbf{p}, t) = -i\theta(t) \exp -i\varepsilon_p t$, one may simplify the expression to

$$W_1(\mathbf{p}, t) = -\frac{e^{-i\varepsilon_p t}}{\nu}\sum_{\mathbf{q}}\frac{M_0{}^2}{q^2}\int_0^t dt_1\int_0^{t_1} dt_2 e^{i(\varepsilon_{\mathbf{p}} - \varepsilon_{\mathbf{p}+\mathbf{q}})(t_1 - t_2)}$$
$$\times [(N_0 + 1)e^{-i\omega_0(t_1 - t_2)} + N_0 e^{i\omega_0(t_1 - t_2)}] \tag{6.1.23}$$

The two time integrals may be done without difficulty:

$$e^{i\varepsilon_p t}W_1(\mathbf{p}, t) = F_1(\mathbf{p}, t)$$
$$= -\frac{M_0{}^2}{\nu}\sum_{\mathbf{q}}\frac{1}{q^2}\left[it\left(\frac{N_0 + 1}{\varepsilon_{\mathbf{p}} - \varepsilon_{\mathbf{p}+\mathbf{q}} - \omega_0} + \frac{N_0}{\varepsilon_{\mathbf{p}} - \varepsilon_{\mathbf{p}+\mathbf{q}} + \omega_0}\right)\right.$$
$$\left. + (N_0 + 1)\frac{1 - e^{it(\varepsilon_{\mathbf{p}} - \varepsilon_{\mathbf{p}+\mathbf{q}} - \omega_0)}}{(\varepsilon_{\mathbf{p}} - \varepsilon_{\mathbf{p}+\mathbf{q}} - \omega_0)^2} + \frac{N_0(1 - e^{it(\varepsilon_{\mathbf{p}} - \varepsilon_{\mathbf{p}+\mathbf{q}} + \omega_0)})}{(\varepsilon_{\mathbf{p}} - \varepsilon_{\mathbf{p}+\mathbf{q}} + \omega_0)^2}\right] \tag{6.1.24}$$

Thus we obtain the function $F_1(\mathbf{p}, t)$. It has several features which are immediately interesting. There is a linear term in t. Its coefficient is just $-i\Sigma^{(1)}_{\text{RS}}(p)$, where

$$\Sigma^{(1)}_{\text{RS}}(p) = \frac{M_0{}^2}{\nu}\sum_{\mathbf{q}}\frac{1}{q^2}\left(\frac{N_0 + 1}{\varepsilon_{\mathbf{p}} - \varepsilon_{\mathbf{p}+\mathbf{q}} - \omega_0} + \frac{N_0}{\varepsilon_{\mathbf{p}} - \varepsilon_{\mathbf{p}+\mathbf{q}} + \omega_0}\right)$$

$\Sigma^{(1)}_{\text{RS}}$ is the one-phonon self-energy from Rayleigh–Schrödinger theory. It is just the real part of the self-energy, since the principal part of the energy denominator is taken. The formula is the result appropriate for finite temperatures. Our result (6.1.14), for zero temperature, is obtained by setting $N_0 \rightarrow 0$. The first term describes processes whereby an electron of momentum $\mathbf{p}$ emits a phonon of wave vector $-\mathbf{q}$ and goes to the state $\mathbf{p}+\mathbf{q}$. The second term describes processes where an electron in state $\mathbf{p}$ absorbs a phonon of $\mathbf{q}$ and goes to $\mathbf{p}+\mathbf{q}$. The energy denominator contains $\pm\omega_0$ depending on whether the intermediate state $\mathbf{p}+\mathbf{q}$ has one more or one less phonon than the initial state. The evaluation of these wave vector integrals is assigned as a problem.

At large values of t, the exponential factors $\exp[it(\varepsilon_{\mathbf{p}} - \varepsilon_{\mathbf{p}+\mathbf{q}} \pm \omega_0)]$ oscillate rapidly and average to zero. Thus at first it appears that the large time limit is

$$\lim_{t\rightarrow\infty} F_1(\mathbf{p}, t) \overset{?}{\rightarrow} -it\Sigma_{\mathrm{RS}}^{(1)}(p) - L^{(1)}(p)$$

$$L^{(1)}(p) = \frac{M_0^2}{\nu}\sum_{\mathbf{q}}\frac{1}{q^2}\left[\frac{N_0+1}{(\varepsilon_{\mathbf{p}} - \varepsilon_{\mathbf{p}+\mathbf{q}} - \omega_0)^2} + \frac{N_0}{(\varepsilon_{\mathbf{p}} - \varepsilon_{\mathbf{p}+\mathbf{q}} + \omega_0)^2}\right]$$

The quantity $L^{(1)}(p)$ may also be evaluated as a function of p. The first term resembles the operator $\mathrm{Nu}(p)$ for the number of phonons in the polaron cloud. The second term in $L^{(1)}(p)$ is more interesting because it equals infinity. This occurs because the energy denominator $\Omega_+ = \varepsilon_{\mathbf{p}} - \varepsilon_{\mathbf{p}+\mathbf{q}} + \omega_0$ goes to zero during the q integration, and it enters as $(\Omega_+)^{-2}$, which gives a nonintegrable divergence. However, there is no real divergence in F_1. If we collect all the terms in Ω_+, including those from the self-energy, we have

$$\lim_{\Omega_+\rightarrow 0}\frac{1}{(\Omega_+)^2}(1 + it\Omega_+ - e^{it\Omega_+}) \rightarrow -\frac{1}{2}t^2$$

which is well behaved when $\Omega_+ \rightarrow 0$. Thus there is no actual divergence in the wave vector integration, but we must proceed more carefully with the evaluation. One approach is to do all the wave vector integrals in $F_1(p, t)$ and then take the limit where $t \rightarrow \infty$. Another method is to add an infinitesimal convergence factor $i\delta$ to Ω_+ wherever it occurs. This makes the functions $L^{(1)}(p)$ and $\Sigma_{\mathrm{RS}}^{(1)}(p)$ both complex and ensures the convergence at large times of the factors $\exp[it(\Omega_+ + i\delta)]$. Both of these two methods give identical results, although the second is usually easier. Thus at large times, for $\varepsilon_p < \omega_0$ one finds

$$\lim_{t\rightarrow\infty} F_1(p, t) = -it\Sigma_{\mathrm{RS}}^{(1)} - L^{(1)}(p)$$

$$\Sigma_{\mathrm{RS}}^{(1)}(p) = -\alpha\left(\frac{\omega_0^3}{\varepsilon_p}\right)^{1/2}\left[(N_0+1)\sin^{-1}\left(\frac{\varepsilon_p}{\omega_0}\right)^{1/2} + iN_0\sinh^{-1}\left(\frac{\varepsilon_p}{\omega_0}\right)^{1/2}\right]$$

$$L^{(1)}(p) = \frac{1}{2}\alpha(\omega_0)^{1/2}\left[\frac{N_0+1}{(\omega_0 - \varepsilon_p)^{1/2}} - i\frac{N_0}{(\omega_0 + \varepsilon_p)^{1/2}}\right]$$

Thus if the Fourier transform is taken to obtain the Green's function of energy $G(p, E)$, its imaginary part, the spectral function, should have a central peak which is an asymmetric Lorentzian:

$$\begin{aligned} A(p, E) &= -2\,\mathrm{Im}\left\{\int_{-\infty}^{\infty} dt e^{it(E-\varepsilon_p)}[-i\theta(t)]e^{F(p,t)}\right\} \\ &\simeq 2\,\mathrm{Re}\left(\int_0^{\infty} dt e^{it(E-\varepsilon_p-\Sigma)}e^{-L}\right) \\ &\simeq -2\,\mathrm{Im}\left(\frac{e^{-L}}{E-\varepsilon_p-\Sigma}\right) \end{aligned}$$

The factor $\mathrm{Im}[\Sigma_{\mathrm{RS}}(p)]$ enters as the width of the central peak, in agreement with our notions that the spectral width gives the lifetime uncertainty of the state. The factor

$$Z_p \equiv e^{-\mathrm{Re}[L(p)]}$$

is the renormalization factor $Z(p)$. This renormalization factor was discussed earlier in (3.3.22). There it came from a Dyson's equation approach and was related to the amplitude of the delta function. Here it is also a peak amplitude. Again we find that it must be less than unity. Thus the linked cluster theory produces the same features in the spectral function as found from Dyson's equation. However, the linked cluster method appears to give them more accurately.

Most numerical work has been done for the momentum state $\mathbf{p} = 0$. In this case $F_1(0, t)$ can be obtained analytically in terms of the Fresnel integrals (Mahan, 1966) ($s = t\omega_0$):

$$c_2(x) + is_2(x) = \int_0^x \frac{dt e^{it}}{(2\pi t)^{1/2}} \equiv E_2(x)$$

$$\begin{aligned} F_1(0, t) = -\frac{\alpha}{(i\pi)^{1/2}}\Big\{(N_0+1)\Big[-is^{1/2}e^{-is} + (2s+i)\left(\frac{\pi}{2}\right)^{1/2} E_2(s)^*\Big] \\ + N_0\Big[is^{1/2}e^{is} + (2s-i)\left(\frac{1}{2\pi}\right)^{1/2} E_2(s)\Big]\Big\} \end{aligned} \tag{6.1.25}$$

The easiest way to derive this result is to follow the method of Dunn (1975). His suggestion is to go back to (6.1.23) ($\nu \to \infty$),

$$F_1(0, t) = -M_0^2 \int \frac{d^3q}{(2\pi)^3} \frac{1}{q^2} \int_0^t dt_1 \int_0^{t_1} dt_2 e^{-i\varepsilon_q(t_1-t_2)}$$
$$\times [(N_0 + 1)e^{-i\omega_0(t_1-t_2)} + N_0 e^{i\omega_0(t_1-t_2)}]$$

and to do the wave vector integrals before the time integrals t_1 and t_2. The wave vector integrals are all of the form

$$4\pi \int_0^\infty dq e^{-iq^2(t/2m)} = (2\pi)^{3/2}\left(\frac{2m}{it}\right)^{1/2} \qquad \text{for } t > 0$$

Since the integrand is only a function of $t_1 - t_2$, we can change the integration variable to $t' = t_1 - t_2$ and then change the order of integrations to get

$$\int_0^t dt_1 \int_0^{t_1} dt_2 f(t_1 - t_2) = \int_0^t dt_1 \int_0^{t_1} dt' f(t')$$
$$= \int_0^t dt' f(t')(t - t')$$

or

$$F_1(0, t) = -\frac{M_0^2}{(2\pi)^{3/2}} (2m)^{1/2} \int_0^t \frac{dt'}{(it')^{1/2}} (t - t')[(N_0 + 1)e^{-i\omega_0 t'} + N_0 e^{i\omega_0 t'}]$$

The time integrals are now recognized as just Fresnel integrals or integrals related to them. Thus one easily derives the result (6.1.25). An approximate Green's function is obtained when we approximate $F(p, t)$ by its first term $F_1(p, t)$. The approximate expression can be numerically Fourier-transformed to obtain the spectral function

$$A^{(1)}(E) = -2 \operatorname{Im}\left[(-i) \int_0^\infty dt e^{iEt} e^{F_1(0,t)}\right]$$

In this way, one gets an approximation to the imaginary part of the Green's function. The real part may also be obtained.

The accuracy of this approximate Green's function is not obvious. About the only way to judge is to also evaluate the next term $F_2(p, t)$, which was done by Dunn. This analysis begins with an examination of $W_2(p, t)$:

$$W_2(p, t) = \frac{(-i)^4}{4!} \int_0^t dt_1 \cdots \int_0^t dt_4 \langle T\hat{c}_p(t)\hat{V}(t_1)\hat{V}(t_2)\hat{V}(t_3)\hat{V}(t_4)\hat{c}_p^\dagger(0)\rangle$$

The correlation function is evaluated in the usual way, according to Wick's theorem. There are three different terms, and each occurs 4! times. They are

$$W_2(p,t) = W_2^{(a)}(p,t) + W_2^{(b)}(p,t) + W_2^{(c)}(p,t)$$

$$W_2^{(a)}(p,t) = -\sum_{\mathbf{q}\mathbf{q}'} \frac{M_0^4}{\nu^2 q^2 q'^2} \int_0^t dt_1 \cdots \int_0^t dt_4 D(t_1-t_4) D(t_2-t_3) G^{(0)}(\mathbf{p}, t-t_1)$$
$$\times G^{(0)}(\mathbf{p}+\mathbf{q}, t_1-t_2) G^{(0)}(\mathbf{p}+\mathbf{q}+\mathbf{q}', t_2-t_3)$$
$$\times G^{(0)}(\mathbf{p}+\mathbf{q}, t_3-t_4) G^{(0)}(\mathbf{p}, t_4)$$

$$W_2^{(b)}(p,t) = -\sum_{\mathbf{q}\mathbf{q}'} \frac{M_0^4}{\nu^2 q^2 q'^2} \int_0^t dt_1 \cdots \int_0^t dt_4 D(t_1-t_3) D(t_2-t_4) G^{(0)}(\mathbf{p}, t-t_1)$$
$$\times G^{(0)}(\mathbf{p}+\mathbf{q}, t_1-t_2) G^{(0)}(\mathbf{p}+\mathbf{q}+\mathbf{q}', t_2-t_3)$$
$$\times G^{(0)}(\mathbf{p}+\mathbf{q}', t_3-t_4) G^{(0)}(\mathbf{p}, t_4)$$

$$W_2^{(c)}(p,t) = -\sum_{\mathbf{q}\mathbf{q}'} \frac{M_0^4}{\nu^2 q^2 q'^2} \int_0^t dt_1 \cdots \int_0^t dt_4 D(t_1-t_2) D(t_3-t_4) G^{(0)}(\mathbf{p}, t-t_1)$$
$$\times G^{(0)}(\mathbf{p}+\mathbf{q}, t_1-t_2) G^{(0)}(\mathbf{p}, t_2-t_3) G^{(0)}(\mathbf{p}+\mathbf{q}', t_3-t_4) G^{(0)}(\mathbf{p}, t_4)$$

These three terms can be represented by the Feynman diagrams in Fig. 6.4. Of course, usually such Feynman diagrams are associated with Dyson's equation, which is not the case here. Such diagrams are convenient to use, but here they have a different interpretation. For example, the diagram in Fig. 6.4(c) is not evaluated in Dyson's theory, since it is just the square of the one-phonon term. This is not true in the linked cluster theory, so it must now be evaluated. In fact the theory is not one of "linked clusters" at all, so the name is inappropriate. All diagrams are evaluated. A more descriptive name would be something like *exponential resummation*, which lacks pizzazz.

In the long time limit, one can show that

$$\lim_{t\to\infty} F_2(p,t) = -it\Sigma_{\mathrm{RS}}^{(2)}(p) - L^{(2)}(p)$$

where $\Sigma_{\mathrm{RS}}^{(2)}(p)$ is the two-phonon self-energy from Rayleigh–Schrödinger theory. It is the result (6.1.16) obtained earlier, except with two modifica-

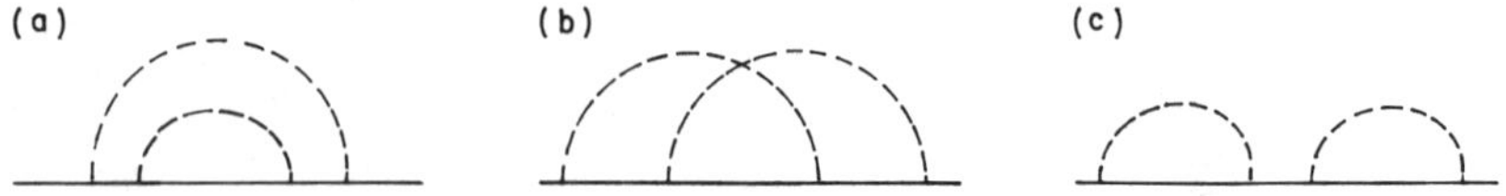

FIGURE 6.4

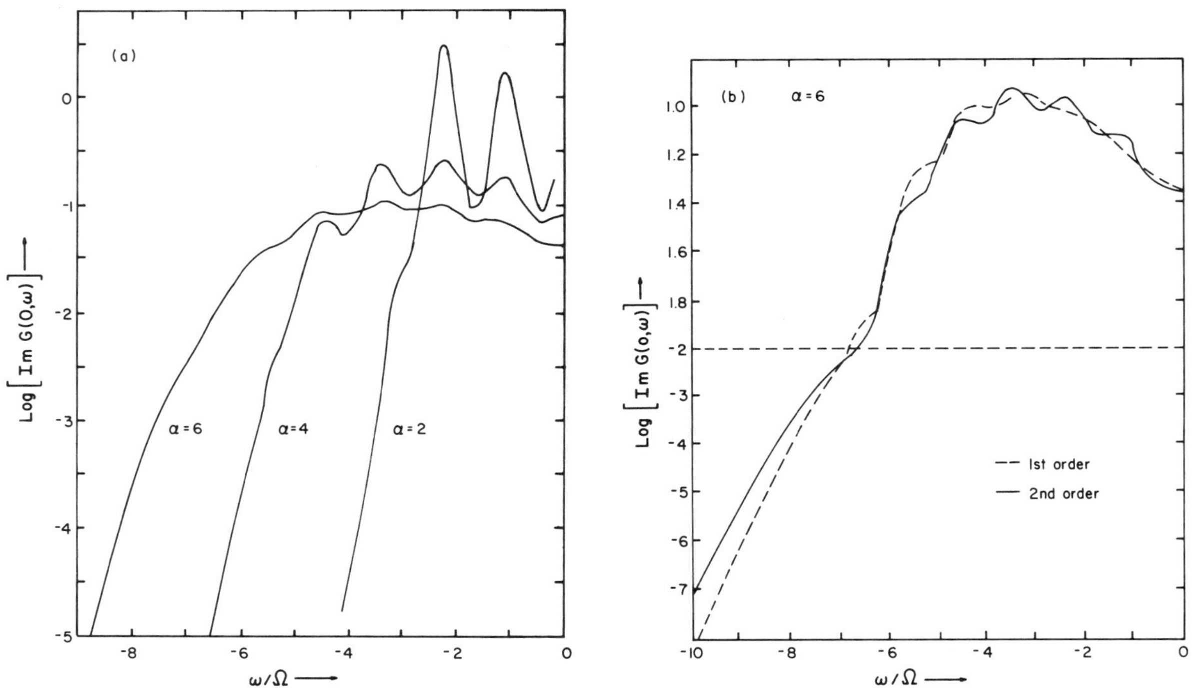

FIGURE 6.5. The spectral function of the polaron with zero momentum and $\beta\omega_0 = 2.5$. (a) Results for three coupling strengths, calculated with one- and two-phonon clusters. (b) A comparison of the results for one-phonon (dashed line) and one-phonon plus two-phonon (solid line) clusters. *Source*: Dunn (1975) (used with permission).

tions: It is valid for finite temperature, and it is complex because the infinitesimal factor $i\delta$ is added to all energy denominators. These results suggest that the exact function $F(p, t)$ should have as its long time limit

$$\lim_{t\to\infty} F(p, t) = -it\Sigma_{\text{RS}}(p) - L(p)$$

where Σ_{RS} is the exact Rayleigh–Schrödinger self-energy. This result, however, has not been proved. But if it is correct, it shows that the main peak in the spectral function is shifted to a new maximum at $E_1 = \varepsilon_p + \text{Re}[\Sigma_{\text{RS}}(p)]$. This seems to be a reasonable result. Of course, this supposes that the renormalization factor $\exp\{-\text{Re}[L(p)]\} = Z(p)$ is not small. Otherwise, the peak in the spectral function will be somewhere else.

Although the large time limit of F_2 has only linear powers of t, this is not true of W_2. It has a term proportional to t^2, which arises from $W_2{}^{(c)}$. But when we set $F_2 = e^{i\varepsilon_p t} W_2 - \frac{1}{2}F_1{}^2$, the t^2 term is canceled by the t^2 term which comes from $F_1{}^2$. The resummation into an exponential is apparently the proper choice because the higher powers of t^n cancel. For the three-phonon term, W_3 contains terms with both t^3 and t^2, and these should all cancel so that F_3 would be linear only in t. It is this feature which makes the exponential the proper function to choose for the resummation. For example, why not choose some other function such as $\ln(\sum F_n) = \sum W_n$ or $\cosh(\sum F_n) = \sum W_n$? For the polaron problem, the exponential resummation appears to be the appropriate choice because of the cancellation of all terms in t^n. By no means is this always the proper choice for other many-body problems. A similar cancellation does not occur in the Kondo problem, for example.

Some numerical results of Dunn (1975) are shown in Fig. 6.5. In Fig. 6.5(a) the spectral functions for $p = 0$ and $k_B T = 0.4\omega_0$ are shown for three values of α. Dunn found that for $\alpha \leq 1$ the two-phonon results did not change the spectral function at all. The curves calculated with $F = F_1 + F_2$ were identical with Mahan's, which used $F = F_1$. For higher values of α there were changes, but they were not dramatic. Figure 6.5(b) shows a comparison of the one-phonon and two-phonon spectra functions for $\alpha = 6$. The theory is not applicable at this high α value, because one should be using strong coupling theory with its internal excited states. However, one can conclude that the linked cluster theory converges rapidly for low and intermediate coupling strengths.

The spectral functions shown in Fig. 6.5 describe the Green's functions for all frequencies ω. The low-energy exponential tail is for states below the conduction band minimum. The peaks at multiples of ω_0 are states with different numbers of phonons associated with the polaron. These are similar to those found for the independent boson model, except that here the peaks are broadened by the recoil of the electron. For $\alpha = 2$ the lowest two peaks are the main ones. But for $\alpha = 4$ and 6 the lowest peaks have small amplitude, which shows the influence of the renormalization factor.

The one-phonon term F_1 is very similar in form to the independent boson model. Compare (6.1.24) with (4.3.11) and (4.3.12). The only difference is the substitution of $\Omega_{\pm} = \varepsilon_{\mathbf{p}} - \varepsilon_{\mathbf{p}+\mathbf{q}} \pm \omega_0$ in the independent boson model. Thus we give F_1 the same interpretation. It describes the Green's function when successive emissions of phonons are independent. The approximate Green's function, with only F_1, does describe an electron coupled to many different numbers of phonons—but all uncorrelated. The addition of F_2 included correlations between pairs of phonon emissions. F_2 has little effect on either the self-energy or the spectral function itself at $p = 0$.

The linked cluster theory is quite different from the Dyson's equation approach. There the one-phonon self-energy $\Sigma^{(1)}(p, E)$ adds only a one-phonon peak to the self-energy. The two-phonon peak comes from the two-phonon self-energy, etc. Thus to get a full Green's function, with many phonon peaks, one needs to evaluate many self-energy diagrams. These higher-order diagrams are difficult to evaluate and have never been calculated. Thus the Dyson's equation method has not been solved to the same accuracy as the linked cluster theory. Of course, one can also obtain $G(p, t)$ from Feynman theory.

6.2. SMALL POLARON THEORY

Polarons become "small" when they become localized, as in the strong coupling theory. In contrast, Fröhlich polarons are sometimes called "large." Small-polaron theory assumes that the size of the polaron corresponds with atomic dimensions. It recognizes the periodicity of the solid and thereby assumes that the motion of the particle is no longer translationally continuous. Instead, it assumes that the particle, usually an electron, may occupy an orbital state $\phi(\mathbf{r} - \mathbf{R}_j)$ centered on atomic site $\mathbf{R}_j$. The orbital states are identical on each site, so there is periodicity. The particle may move from site to site, exactly as in the tight binding model. The motion from site to site may be caused by the overlap, or nonorthogonality, of the orbitals on adjacent sites. The phonons are coupled to the particle at whichever site it is on. Thus we consider the following Hamiltonian for small polarons:

$$H = J\sum_{j\delta} C_{j+\delta}^{\dagger}C_j + \sum_{\mathbf{q}} \omega_q a_{\mathbf{q}}^{\dagger} a_{\mathbf{q}} + \sum_{j\mathbf{q}} C_j^{\dagger}C_j e^{i\mathbf{q}\cdot\mathbf{R}_j} M_q(a_{\mathbf{q}} + a_{-\mathbf{q}}^{\dagger}) \qquad (6.2.1)$$

where C_j is the destruction operator for a particle on site R_j. Spin indices are not important in this problem and so are omitted. If (6.2.1) is compared with the Fröhlich Hamiltonian (6.1.1), only the first term is different. The present Hamiltonian (6.2.1), which includes the periodicity of the solid, is much more realistic.

Our discussion will follow the classic treatment of Holstein (1959), which followed the pioneering work of Tiablikov (1952) and Yamashita and Kurosawa (1958). The small-polaron Hamiltonian exhibits two types of behavior which are very different. First we shall describe each type of behavior in detail. Afterwards, we shall discuss the interesting problem of the transition region between these limiting cases.

A. Large Polarons

The first class of behavior is "large" polaron motion, of the Fröhlich type. This happens whenever the bandwidth zJ is large, where z is the coordination number. The condition on the bandwidth will be stated later with more precision. When the bandwidth is large, the Hamiltonian is solved in wave vector space. Thus we transform to collective coordinates:

$$C_{\mathbf{k}} = \frac{1}{N^{1/2}} \sum_j C_j e^{i\mathbf{k}\cdot\mathbf{R}_j}$$

$$H = zJ\sum_{\mathbf{k}} \gamma_{\mathbf{k}} C_{\mathbf{k}}^{\dagger}C_{\mathbf{k}} + \sum_{\mathbf{q}} \omega_{\mathbf{q}} a_{\mathbf{q}}^{\dagger} a_{\mathbf{q}} + \sum_{\mathbf{kq}} C_{\mathbf{k}+\mathbf{q}}^{\dagger} C_{\mathbf{k}} M_{\mathbf{q}}(a_{\mathbf{q}} + a_{-\mathbf{q}}^{\dagger})$$

$$\gamma_{\mathbf{k}} = \frac{1}{z} \sum_{\boldsymbol{\delta}} e^{i\mathbf{k}\cdot\boldsymbol{\delta}}$$

Now the small-polaron Hamiltonian has an even closer resemblance to the Fröhlich polaron Hamiltonian. The only difference is the replacement of the free-particle energy $\varepsilon_k = k^2/2m$ by the tight binding form $\varepsilon_{\mathbf{k}} = zJ\gamma_{\mathbf{k}}$. The wave vector summation for particles extends only over the Brillouin zone. In a realistic model of a solid, there would be many bands to be summed. For a large bandwidth, the particle will confine its motion to states near the bottom of the band. For J negative, this occurs at $\mathbf{k} = 0$. If we expand $\gamma_{\mathbf{k}}$ about the point $\mathbf{k} = 0$, in cubic crystals we find $\varepsilon_{\mathbf{k}} = zJ - (zJ\delta^2/6)k^2$, so that the particle has an effective mass of $m^{-1} = zJ\delta^2/3$.

If the polaron coupling strength M_q is small, this can be described by weak coupling theory. There is a slight change in the particle's energy and effective mass because of polaron effects. These changes must correspond, in the tight binding model, to a change in bandwidth. The polaron self-energy, in first-order Rayleigh–Schrödinger perturbation theory, is

$$\Sigma_{\mathrm{RS}}^{(1)}(\mathbf{k}) = \sum_{\mathbf{q}} M_{\mathbf{q}}^2 \left[\frac{N_q + 1 - n_F(\varepsilon_{\mathbf{k}})}{\varepsilon_{\mathbf{k}} - \varepsilon_{\mathbf{k}+\mathbf{q}} - \omega_{\mathbf{q}}} + \frac{N_{\mathbf{q}} + n_F(\varepsilon_{\mathbf{k}})}{\varepsilon_{\mathbf{k}} - \varepsilon_{\mathbf{k}+\mathbf{q}} + \omega_{\mathbf{q}}} \right] \tag{6.2.2}$$

The wave vector integrals are rather hard to evaluate for any form of M_q^2, because the energy denominators have the inconvenient expressions $\varepsilon_{\mathbf{k}+\mathbf{q}} = zJ\gamma_{\mathbf{k}+\mathbf{q}}$. Some results have been obtained in one dimension (Suna, 1964). Fröhlich-type polaron effects are obtained when the bandwidth is large and the polaron effects are small.

B. Small Polarons

Next we consider the other limiting case of the small-polaron Hamiltonian (6.2.1). It is assumed that the polaron effects are dominant and that the bandwidth is small. The physical picture is that the polaron effects localize the particle on a site and that hopping occurs infrequently from site to site. Thus the tight binding term $J \sum C_{j+\delta}^\dagger C_j$ is the perturbation, while the particle–phonon term is large. The Hamiltonian is solved in position space—for the particles—without resorting to collective coordinates. The first step is to apply a canonical transformation which diagonalizes the last two terms in the Hamiltonian. These last two terms are the same as found in the exactly solvable models of Sec. 4.3 for the many-boson model. Thus the canonical transformation has the same form:

$$\begin{aligned}
\bar{H} &= e^S H e^{-S} \\
S &= -\sum_{j\mathbf{q}} n_j e^{i\mathbf{q}\cdot\mathbf{R}_j} \frac{M_q}{\omega_q} (a_{\mathbf{q}} - a_{-\mathbf{q}}^\dagger) \\
\bar{H} &= J \sum_{j\delta} C_{j+\delta}^\dagger C_j X_{j+\delta}^\dagger X_j + \sum_{\mathbf{q}} \omega_q a_{\mathbf{q}}^\dagger a_{\mathbf{q}} - \sum_j n_j \Delta \\
\Delta &= \sum_{\mathbf{q}} \frac{M_q^2}{\omega_q} \\
X_j &= \exp\left[\sum_{\mathbf{q}} e^{i\mathbf{q}\cdot\mathbf{R}_j} \frac{M_q}{\omega_q} (a_{\mathbf{q}} - a_{-\mathbf{q}}^\dagger) \right]
\end{aligned} \tag{6.2.3}$$

The polaron self-energy is Δ. The factors X_j were encountered previously in Sec. 4.3D. They arise from the canonical transformation of the particle operators $e^S C_j e^{-S} = C_j X_j$. The number operator $n_j = C_j^{\dagger} C_j$ commutes with S and so is unaffected by the transformation. But the tight binding term $JC_{j+\delta}^{\dagger} C_j$ produces factor, $X_{j+\delta}^{\dagger} X_j$. The first term is not solvable exactly, so the canonical transformation did not diagonalize this Hamiltonian. However, so far we have not made any approximations. The eigenstates and energy levels of $\bar{H}$ are the same as H. Thus it is fruitful to investigate the solutions of $\bar{H}$. In the interaction representation this is set equal to $\bar{H} = H_0 + V$:

$$H_0 = \sum_{\mathbf{q}} \omega_q a_{\mathbf{q}} a_{\mathbf{q}}^{\dagger} - \sum_j n_j \Delta$$

$$V = J \sum_{j\delta} C_{j+\delta}^{\dagger} C_j X_{j+\delta}^{\dagger} X_j \tag{6.2.4}$$

The exponents in the factor $X_{j+\delta}^{\dagger} X_j$ can be combined since they commute:

$$X_{j+\delta}^{\dagger} X_j = \exp\left[\sum_{\mathbf{q}} e^{i\mathbf{q}\cdot\mathbf{R}_j}(1 - e^{i\mathbf{q}\cdot\boldsymbol{\delta}})\left(\frac{M_q}{\omega_q}\right)(a_{\mathbf{q}} - a_{-\mathbf{q}}^{\dagger})\right]$$

This simplifies the form of this operator for taking the expectation of phonon operators. The perturbation V describes the hopping of the polaron from one site j to the neighboring site $j + \delta$. The amplitude of this process is

$$J\langle f \mid X_{j+\delta}^{\dagger} X_j \mid i\rangle$$

where $\mid i\rangle$ and $\mid f\rangle$ describe the phonon occupation numbers in the initial and final states of the transition. Since the factor $X_{j+\delta}^{\dagger} X_j$ permits phonons to be made or destroyed, the states $\mid i\rangle$ and $\mid f\rangle$ may have different numbers of phonons.

Holstein (1959) made a distinction between events in which the number of phonons were changed during the hop and events where they did not change. If the number of phonons is changed in the hop, this is an inelastic scattering process. The particle loses its phase coherence by this emission or absorption of phonons. Each hop becomes a statistically independent event. The particle motion is diffusive, since it has Brownian motion by randomly hopping from site to site. These are called *nondiagonal transitions*.

The other situation is a *diagonal transition* in which all the phonon occupation numbers remain the same during the hop. The number of phonons $n_{\mathbf{q}}$ in each state $\mathbf{q}$ remains unaltered. This is a rather strict condition. For example, if the number of phonons is unchanged but one state gains a phonon while another loses one, this is a nondiagonal transition. The phase coherence of a particle is maintained during a diagonal transition. If this is the most likely hopping event, the particle will hop from site to site while retaining phase coherence. Then it is a Bloch particle and forms energy bands. The diagonal part of the hopping probability is defined as $J\exp(-S_T)$, where

$$e^{-S_T} = \langle i \mid X_{j+\delta}^\dagger X_j \mid i \rangle \tag{6.2.5}$$

If the nondiagonal transition probability is small, the particle will form energy bands with a band energy

$$\bar{\varepsilon}_{\mathbf{k}} = zJ\gamma_{\mathbf{k}} e^{-S_T}$$

The effect of the polarons is to reduce the bandwidth by a factor $\exp(-S_T)$. The effective mass is increased by the same factor. In the Holstein picture, we can think of the diagonal hops as contributing to the real part of the self-energy and the nondiagonal transitions to the imaginary part of the self-energy.

Thus the large-polaron behavior, of the Fröhlich type, occurs when the diagonal transitions dominate. The small-polaron behavior, with diffusive hopping, occurs whenever the nondiagonal transitions dominate.

C. Diagonal Transitions

The diagonal and nondiagonal transition rates are obtained by using the same type of operator techniques introduced in Sec. 4.3. First we shall calculate the diagonal transition rate from (6.2.5). The factor $X_{j+\delta}^\dagger X_j$ has two operators in the exponential which are Feynman disentangled, with the destruction operators to the right. Using the techniques introduced into Sec. 4.3 $\{\lambda = \exp(i\mathbf{q}\cdot\mathbf{R}_j)[1 - \exp(i\mathbf{q}\cdot\boldsymbol{\delta})]M_q/\omega_q\}$, we obtain

$$\begin{aligned} e^{-S_T} &= \prod_{\mathbf{q}} \langle i \mid e^{\lambda a - \lambda^* a^\dagger} \mid i \rangle \\ &= \prod_{\mathbf{q}} \exp(-\tfrac{1}{2} \mid \lambda_{\mathbf{q}} \mid^2) \langle i \mid e^{-\lambda^* a^\dagger} e^{\lambda a} \mid i \rangle \end{aligned}$$

We wish to take events with the same number of phonons in each state q for both $\langle i \mid$ and $\mid i \rangle$. We shall assume that the number of phonons in each state is determined by a thermal average. Thus the summation is taken over each possible number n_q of phonons in each state, with the thermal probability $\exp(-\beta\omega_q n_q)$ for this occurrence [see (4.3.10)]:

$$e^{-S_T} = \prod_{\mathbf{q}} \exp\left(\frac{|\lambda_{\mathbf{q}}|^2}{2}\right)(1 - e^{-\beta\omega_q}) \sum_{n=0}^{\infty} e^{-\beta n\omega_q} \langle n \mid e^{\lambda^\dagger a^\dagger} e^{-\lambda a} \mid n\rangle$$
$$= \prod_{\mathbf{q}} \exp\left(-\frac{|\lambda_{\mathbf{q}}|^2}{2} - N_q |\lambda_{\mathbf{q}}|^2\right)$$

Thus we conclude that the thermal factor S_T has the form

$$S_T = \sum_{\mathbf{q}} \left(\frac{M_q}{\omega_q}\right)^2 [1 - \cos(\mathbf{q}\cdot\boldsymbol{\delta})](2N_q + 1)$$
$$N_q = (e^{\beta\omega_q} - 1)^{-1} \tag{6.2.6}$$

The factor S_T is temperature dependent and increases with increasing temperature. For temperatures larger than the Debye temperature, one may often use the expansion $N_q \simeq K_B T/\omega_q - \frac{1}{2} + O(\beta\omega_q)$ to write

$$S_T \simeq 2K_B T \sum_{\mathbf{q}} \frac{M_q^2}{\omega_q^3} [1 - \cos(\mathbf{q}\cdot\boldsymbol{\delta})] + O\left(\frac{1}{T}\right) \tag{6.2.7}$$

so that S_T increases linearly with temperature at high temperature. The factor $\exp(-S_T)$ determines the rate of diagonal transitions and decreases with increasing temperature. Thus the Fröhlich-type polaron effects, with coherent motion, are most likely to be found at lower temperatures. As the temperature is increased, the polaron bandwidth $zJ\exp(-S_T)$ becomes smaller and smaller. It is reasonable to expect that the type of motion will change at some temperature. In this model, the change is to a hopping-type motion.

The factor δ is the distance the polaron hops in a single motion. It is presumably a lattice constant, or at least a distance associated with some fundamental unit of length on the scale of the crystal unit cell. The factor $1 - \cos(\mathbf{q}\cdot\boldsymbol{\delta})$ suggests that phonons of short wavelength are important, since this factor vanishes for $\mathbf{q}\to 0$. Thus the integrations in (6.2.6) and (6.2.7) probably have significant contributions from phonons near the edge of the Brillouin zone. These integrals must be done carefully and probably numerically. Simple models, such as the Debye model, are probably very inaccurate. Of course, the matrix elements M_q must also be found accurately for zone edge phonons. The Fröhlich result $M_q \simeq q^{-1}$ is a valid approximation only at long wavelength and may not be adequate for the q integrations in (6.2.6) or (6.2.7).

D. Nondiagonal Transitions

Nondiagonal transitions are hops, from site $\mathbf{R}_j$ to $\mathbf{R}_j + \boldsymbol{\delta}$, in which the number of phonons is not conserved. It is an inelastic scattering process. Eventually we shall only consider events in which the total energy is conserved in the transition. But this may happen when 10 phonons are being made in one state and 5 lost in another, so that energy can be conserved but not the phonon number. The matrix element $\langle f \mid X_{j+\delta}^{\dagger} X_j \mid i \rangle$ describes the transition from one quantum state to another. By itself, it does not convey meaningful information. Instead, one must evaluate the transition rate w, which is defined as the rate per unit time at which the hop occurs. It is calculated by first finding the correlation function $W(t)$, where we sum over all final states except the state $|f\rangle = |i\rangle$,

$$\begin{aligned} W(t) &= J^2 \sum_{f \neq i} \langle i \mid X_j^{\dagger}(t) X_{j+\delta}(t) \mid f \rangle \langle f \mid X_{j+\delta}^{\dagger} X_j \mid i \rangle \\ &= J^2 [\langle i \mid X_j^{\dagger}(t) X_{j+\delta}(t) X_{j+\delta}^{\dagger}(0) X_j(0) \mid i \rangle - | \langle i \mid X_{j+\delta}^{\dagger} X_j \mid i \rangle |^2] \end{aligned}$$

and then evaluating the Fourier transform:

$$w = \frac{1}{\hbar^2} \int_{-\infty}^{\infty} dt W(t) \tag{6.2.8}$$

This is just the Fermi golden rule (Schiff, 1955). The transition rate is directly proportional to the dc conductivity calculated from the Kubo formula. The latter relationship is $\sigma = \frac{1}{3}\beta e^2 \delta^2 n_0 w$. The derivation of σ will be provided later, after the evaluation of w. The first term in $W(t)$ is a time-dependent correlation function, which is identical in mathematical form to that found previously for the optical absorption in the many-boson model [see (4.3.20)]. The previous result may be utilized, by direct analogy, to give the following result:

$$\langle i \mid X_j^{\dagger}(t) X_{j+\delta}(t) X_{j+\delta}^{\dagger}(0) X_j(0) \mid i \rangle = \exp[-\Phi(t)]$$

$$\Phi(t) = \sum_{\mathbf{q}} \frac{M_q^2}{\omega_q^2} \mid 1 - e^{i\mathbf{q}\cdot\boldsymbol{\delta}} \mid^2 [(N_q + 1)(1 - e^{-i\omega_q t}) + N_q(1 - e^{i\omega_q t})]$$

The second term in $W(t)$ arises from the diagonal transition rate. One's first impulse is to set this equal to the square of the amplitude, or $J^2 \exp(-2S_T)$. This step is too hasty. When we thermally average $|\langle i \mid X_{j+\delta}^{\dagger} X_j \mid i \rangle|^2$, we do not usually find that the average of a square of a quantity is equal to the square of the average. Recall that the diagonal transition matrix element, for a state with n phonons, is

$$\langle n \mid X_{j+\delta}^{\dagger} X_j \mid n\rangle = e^{-|u_{\mathbf{q}}|^2/2}\langle n \mid e^{u_{\mathbf{q}}^* a_{\mathbf{q}}^{\dagger}} e^{-u_{\mathbf{q}} a_{\mathbf{q}}} \mid n\rangle = e^{-|u_{\mathbf{q}}|^2/2} L_n(\mid u_{\mathbf{q}} \mid^2)$$

$$u_{\mathbf{q}} = \frac{M_q}{\omega_q}(1 - e^{i\mathbf{q}\cdot\boldsymbol{\delta}})$$

where $L_n(x)$ is the Laguerre polynominal. The quantity $\exp(-S_T)$ is obtained by averaging this amplitude over all possible values of n. A slightly different result is obtained if we average the square of the amplitude over all possible values of n:

$$\begin{aligned} e^{-S_T} &= \prod_{\mathbf{q}} e^{-|u|^2/2}(1 - e^{-\beta\omega_q}) \sum_{n=0}^{\infty} e^{-\beta n\omega_q} L_n(\mid u_q^2 \mid) \\ &= \exp\Big[-\sum_{\mathbf{q}} \mid u_{\mathbf{q}} \mid^2 (N_q + \tfrac{1}{2})\Big] \\ \mid \langle i \mid X_{j+\delta}^{\dagger} X_j \mid i\rangle \mid_{\mathrm{av}}^2 &= \prod_{q} e^{-|u_{\mathbf{q}}|^2}(1 - e^{-\beta\omega_q}) \sum_{n=0}^{\infty} e^{-\beta n\omega_q} L_n(\mid u_q \mid^2)^2 \\ &= \prod_{\mathbf{q}} \exp(-2S_T) I_0(2 \mid u_{\mathbf{q}} \mid^2 [N_q(N_q + 1)]^{1/2}) \end{aligned} \tag{6.2.9}$$

The average of the hopping probability contains another factor, which is a Bessel function, which is obtained for each state $\mathbf{q}$. The final diagonal hopping probability is obtained by taking the product of this result over all q states. Thus for the function $W(t)$ we obtain the result

$$\begin{aligned} W(t) = J^2 \exp(-2S_T)\Big\{&\exp\Big\{\sum_{\mathbf{q}} \mid u_{\mathbf{q}} \mid^2 [(N_q + 1)e^{-i\omega_q t} + N_q e^{i\omega_q t}]\Big\} \\ &- \exp\Big[\sum_{\mathbf{q}} \ln(I_0\{2 \mid u_{\mathbf{q}} \mid^2 [N_q(N_q + 1)]^{1/2}\})\Big]\Big\} \end{aligned} \tag{6.2.10}$$

The time integral must be done next to obtain the final transition rate in (6.2.8). However, there are some additional simplifications of this expression which can be made. We shall discuss separately the case of an Einstein model and phonons with dispersion. The second case is easier, at least conceptually, and is discussed first.

E. Dispersive Phonons

Dispersive phonons have the frequency ω_q depending on $\mathbf{q}$. We shall first treat this case. The basic assumption is that there are not large numbers of phonons, in different states $\mathbf{q}$, which have the same energy. Of course, in a solid the states in the star of the wave vector have the same energy, so there are always several $\mathbf{q}$ states with the same energy. But we are not talking about a few states but large sections of the Brillouin zone. The precise physical restrictions will be made clearer in the next section dealing with the Einstein model. The restriction to dispersive phonons occurs because we do not wish to consider events in which energy can be conserved by phonons being destroyed in one state, say $\mathbf{q}_1$, while being created in another, $\mathbf{q}_2$. This can happen easily if $\omega_{q_1} = \omega_{q_2}$ but not otherwise.

We now take the limit where the volume of the solid ν goes to infinity. The particle–phonon matrix element M_q actually has a factor of $\nu^{-1/2}$. Thus we write this matrix element as $M_q = \bar{M}_q/\nu^{1/2}$, where $\bar{M}_q$ does not have any volume dependence and is only a function of q. In the limit $\nu \to \infty$, the summations which determine S_T and $\Phi(t)$ become integrals which are well behaved:

$$\begin{aligned} 2S_T &= \sum_{\mathbf{q}} \frac{M_q^2}{\omega_q^2}(2N_q+1)\,|1-e^{i\mathbf{q}\cdot\boldsymbol{\delta}}|^2 = \int \frac{d^3q}{(2\pi)^3}\frac{\bar{M}_q^2}{\omega_q^2}|1-e^{i\mathbf{q}\cdot\boldsymbol{\delta}}|^2(2N_q+1) \\ \Phi(t) &= 2S_T - \varphi(t) \\ \varphi(t) &= \sum_{\mathbf{q}} \frac{M_q^2}{\omega_q}\,|1-e^{i\mathbf{q}\cdot\boldsymbol{\delta}}|^2[(N_q+1)e^{-i\omega_q t} + N_q e^{i\omega_q t}] \\ &= \int \frac{d^3q}{(2\pi)^3}\frac{\bar{M}_q^2}{\omega_q^2}\,|1-e^{i\mathbf{q}\cdot\boldsymbol{\delta}}|^2[(N_q+1)e^{-i\omega_q t} + N_q e^{i\omega_q t}] \end{aligned} \tag{6.2.11}$$

However, the Bessel function I_0 in (6.2.9) or (6.2.10) has an argument which vanishes in this limit. When $z \to 0$, we obtain the following limits:

$$\lim_{Z\to 0} I_0(Z) = 1 + \frac{Z^2}{4} + O(Z^4)$$

$$\ln[I_0(Z)] = \frac{Z^2}{4} + O(Z^4)$$

$$\begin{aligned} \lim_{\nu\to\infty}\sum_{\mathbf{q}} \ln\left\{I_0\left[\frac{\theta(q)}{\nu}\right]\right\} &= \frac{\nu}{(2\pi)^3}\int d^3q\,\ln\left\{I_0\left[\frac{\theta(q)}{\nu}\right]\right\} \\ &= \frac{1}{4}\,\frac{1}{\nu}\int \frac{d^3q}{(2\pi)^3}\,\theta(q)^2 \to 0 \end{aligned}$$

The last term in (6.2.10) has an exponent which goes to zero as $1/\nu$, so this term becomes unity. Thus we obtain

$$W(t) = J^2 e^{-2S_T}(e^{+\varphi(t)} - 1)$$
$$w = \frac{J^2}{\hbar^2} e^{-2S_T} \int_{-\infty}^{\infty} dt(e^{\varphi(t)} - 1) \tag{6.2.12}$$

The time integral must be done next in order to obtain the final transition rate w. For a general spectrum of phonons, containing both acoustical and optical types, all dispersive, the exponential function $\varphi(t)$ has a complicated time dependence, and the time integrals should be done numerically. An approximate result may be obtained by a saddle-point integration, which will be done later. First, it is useful to discuss the convergence and properties of the time integral.

At very large values of t, $\varphi(t)$ approaches zero. Thus the integrand goes to zero at large values of time. This is fortunate, since otherwise the integral would diverge. However, in one dimension DeWit (1968) has shown that the approach to unity is too slow. He showed that $\lim_{t\to\infty} \varphi(t) \to O(1/t^{1/2})$, so that the integral in (6.2.12) does not converge. Since there are no one-dimensional solids, perhaps this is no disaster. However, from the pedagogical viewpoint, this is unfortunate since one-dimensional problems are nice for homework. In three dimensions the convergence is fast enough, so that w has a definite value.

The physics is understood by investigating the frequency spectrum of just the first factor in $W(t)$. We shall call its Fourier transform $U(\omega)$:

$$U(\omega) = \int_{-\infty}^{\infty} dte^{i\omega t}e^{-\Phi(t)} = e^{-2S_T} \int_{-\infty}^{\infty} dte^{i\omega t}e^{+\varphi(t)}$$

A typical spectrum is shown in Fig. 6.6. There is a delta function at $\omega = 0$ plus a smooth spectrum. The delta function is the *no-phonon* transition. It is the probability that the transition takes place without any phonons

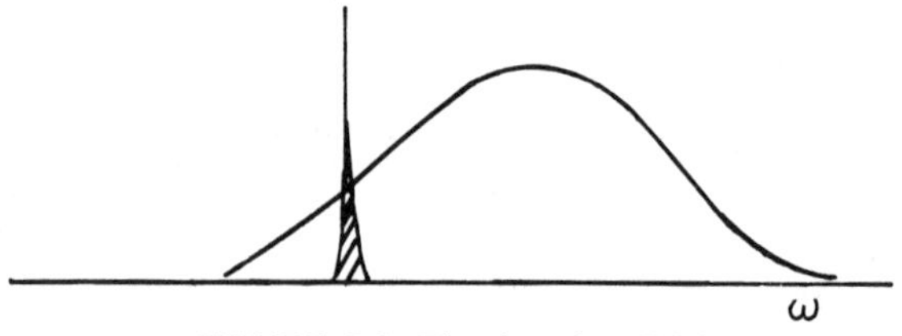

FIGURE 6.6. The function $U(\omega)$.

being emitted or absorbed. As long as this has a finite probability, one will see a delta function in the spectrum. The intensity of the delta function is just proportional to $\exp(-2S_T)$, which is called the *Debye–Waller* factor. In the hopping transition, the equivalent of the no-phonon line is hopping by a diagonal transition. We wish to eliminate these terms. The subtraction of the -1 factor from the time integrand in the definition of w is just what is needed to eliminate the delta function. Thus the value of w is just the smooth part of the curve in Fig. 6.6 evaluated at $\omega = 0$.

An approximate result for w may be obtained by a saddle-point integration (Schotte, 1966). This approximate result is very accurate in the limit where the particle–phonon coupling is large, so the polaron effects are significant.

First we rewrite $\varphi(t)$ in (6.2.11) as

$$w = \frac{J^2}{\hbar^2} e^{-2S_T} \int_{-\infty}^{\infty} dt(e^{\varphi(t)} - 1)$$

$$\varphi(t) = \sum_{\mathbf{q}} | u_{\mathbf{q}} |^2 [N_q(N_q + 1)]^{1/2} 2 \cos\left[\omega_q\left(t + \frac{i\beta}{2}\right)\right]$$

$$u_{\mathbf{q}} = \frac{M_q}{\omega_q}(1 - e^{i\mathbf{q}\cdot\boldsymbol{\delta}})$$

The t integration is treated as a contour integral in a complex space. The path of integration is deformed to go over the saddle point. The saddle point t_s is located at the point $t_s = -i\beta/2$. The deformed contour is shown in Fig. 6.7, with the saddle point at $\oplus$. The function $\varphi(t) \equiv \bar{\varphi}(z)$ is expanded as a power series about the saddle point. Let the distance from the saddle point be $z = t + i\beta/2$, so that

$$\bar{\varphi}(z) = 2 \sum_{\mathbf{q}} | u_{\mathbf{q}} |^2 [N_q(N_q + 1)]^{1/2} \cos(\omega_q z)$$

$$\bar{\varphi}(z) \simeq \bar{\varphi}(0) - \gamma z^2 + O(z^4)$$

$$\bar{\varphi}(0) = 2 \sum_{\mathbf{q}} | u_{\mathbf{q}} |^2 [N_q(N_q + 1)]^{1/2}$$

$$\gamma = \sum_{\mathbf{q}} \omega_q^2 | u_{\mathbf{q}} |^2 [N_q(N_q + 1)]^{1/2}$$

An approximate result is obtained by neglecting all terms except the first two. The time integration, in the vicinity of the saddle point, is now just a

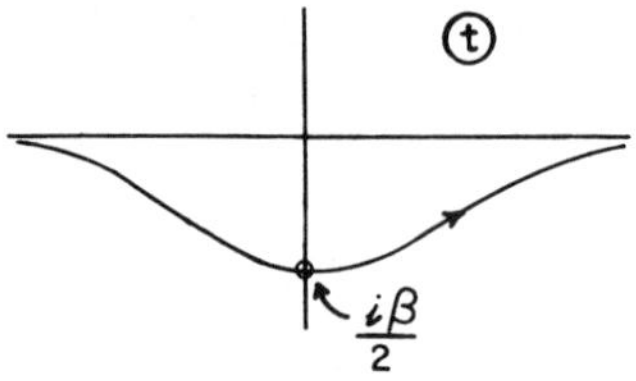

FIGURE 6.7. Path of integration in complex t plane.

Gaussian integral:

$$\int dz e^{\bar{\varphi}(z)} \simeq e^{\bar{\varphi}(0)} \int_{-\infty}^{\infty} dz e^{-\gamma z^2} = \left(\frac{\pi}{\gamma}\right)^{1/2} e^{\bar{\varphi}(0)}$$

$$w \simeq \frac{J^2}{\hbar^2}\left(\frac{\pi}{\gamma}\right)^{1/2} \exp[-2S_T + \bar{\varphi}(0)] \tag{6.2.13}$$

The saddle-point integral just does the $\exp[\varphi(t)]$ part of the integrand and not the -1 term. However, the latter contribution just eliminates the delta function corresponding to the no-phonon transitions. The saddle-point integration does not have this contribution. It provides the smooth background part of $U(\omega)$ at $\omega = 0$, which is precisely the contribution which is needed for w.

The two exponential terms in (6.2.13) can be combined,

$$2S_T - \bar{\varphi}(0) = \sum_{\mathbf{q}} | u_{\mathbf{q}} |^2[(N_q + 1)^{1/2} - (N_q)^{1/2}]^2$$

which is a positive definite form. The exponent in (6.2.13) is always negative. At high temperatures, the phonon occupation numbers may be expanded: $N_q \to (K_BT/\omega_q) - \frac{1}{2} + O(\omega_q/K_BT)$. If we define $x = \beta\omega_q$, the factors in the exponent give

$$\lim_{x\to 0} N_q = \frac{1}{e^x - 1} = \frac{1}{x} - \frac{1}{2} + O(x)$$

$$\lim_{x\to 0} [(N_q + 1)^{1/2} - (N_q)^{1/2}]^2 = \frac{2}{x} - 2\left(\frac{1}{x^2} - \frac{1}{4}\right)^{1/2}$$

$$= \frac{2}{x}\left[1 - \left(1 - \frac{x^2}{4}\right)^{1/2}\right] = \frac{x}{4} + O(x^3)$$

$$2S_T - \bar{\varphi}(0) \simeq \frac{1}{4K_BT} \sum_{\mathbf{q}} | u_{\mathbf{q}} |^2 \omega_q$$

The high-temperature expansion defines an energy $\bar{\Delta}$,

$$\begin{aligned}
\bar{\Delta} &= \frac{1}{4}\sum_{\mathbf{q}} |u_{\mathbf{q}}|^2\omega_q = \frac{1}{4}\sum_{\mathbf{q}} \frac{M_q^2}{\omega_q} |1 - e^{i\mathbf{q}\cdot\boldsymbol{\delta}}|^2 \\
&= \frac{1}{2}\int \frac{d^3q}{(2\pi)^3} \frac{\bar{M}_q^2}{\omega_q} [1 - \cos(\mathbf{q}\cdot\boldsymbol{\delta})] \\
w &= \frac{J^2}{\hbar^2}\left(\frac{\pi}{\gamma}\right)^{1/2} \exp\left(-\frac{\bar{\Delta}}{K_BT}\right)
\end{aligned} \tag{6.2.14}$$

which enters the hopping rate w as an activation energy. The same factor $\bar{\Delta}$ defines $\gamma = 4\bar{\Delta}K_BT$ at high temperature.

Small-polaron theory predicts that the hopping rate is thermally activated. The rate of hopping increases at higher temperatures. This behavior is in direct contrast to the rate $J^2 \exp(-2S_T)$ of diagonal transitions, which decreases with increasing temperatures. Thus as one increases the temperature, the diagonal transitions become less likely, while the nondiagonal transitions become more likely. The band-type polaron motion, or large polarons of the Fröhlich type, will exist only at small temperatures. Holstein estimated the transition temperature between band motion and hopping motion to occur around 40% of the Debye temperature. The estimate is remarkably insensitive to the magnitude of J. In the Holstein model, the low-temperature motion should be band-like, while the high-temperature motion is hopping. There have been many experimental systems with these characteristics which have been ascribed to small-polaron theory. One example is TiO_2 (Bogomolov *et al.*, 1967). Bogomolov *et al.* observe the transition from band to hopping conductivity at about 300°C. The conduction bands of transition metal oxides are often d bands, and conduction in them seems well described by small-polaron theory. See the reviews by Adler (1967), Appel (1967), and Böttger and Bryksin (1976).

There are many other systems in which the electron conductivity is thermally activated. These may often be otherwise explained, e.g., by an activation energy for freeing bound electrons from defects or alternately by a static barrier to the electron motion. However, it is thought that in most systems in which the electron motion is by hopping there is some polarization of the electron by its immediate surroundings. The hopping electron must carry its local polarization along, which gives the activation energy. The small-polaron picture applies when this polarization is due to phonons—or atomic realignment near the electron. Of course, there could also be polarization of the electronic states, for example, the dielectric screening by the material of the electron charge. Electronic polarization can be described by an *electronic polaron model*, which has the same mathematical form as small-polaron theory. Here the bosons are not phonons but density operators representing electron–hole pair excitations of the system. This concept applies even in insulators, where the pair oscillations have an energy gap.

Another application of the model is to the diffusion of ions in solids. Here the diffusing particle is an ion. When hopping, it polarizes the ions around it, and this polarization must move with it. Flynn (1972) has argued that polarization leads to an activation energy for hopping, just as in small-polaron theory. The application of small-polaron theory to ions must be done delicately. The term J obviously does not represent tunneling amplitudes between sites or wave function overlap, since these are negligible for ions. Instead, J must be a jump probability, which acknowledges that the ion occasionally does hop because of thermal fluctuations. Thus J already contains some thermal fluctuations—which are phonons—and additional degrees of freedom must be added carefully. This can be done, as shown by Flynn.

The important message in small-polaron theory is that the motion is thermally activated. If the particle polarizes its surroundings, then it can hop only by moving this polarization along. The greater the polarization, the less likely this occurs. Each jump may occur only when the polarization arrangements on initial and final sites are the same. Since the system is in a state of continual fluctuation, this coincidence sometimes does happen. It is less likely when the polarization is most severe. This is the physical reason for the appearance of thermal activation.

The degree of polarization of the medium is given by the polaron binding energy $\Delta = \sum_{\mathbf{q}} M_q^2/\omega_q$, which is not the same as the activation energy $\bar{\Delta}$. The activation energy is smaller, although both will have similar magnitudes.

F. Einstein Model

The Einstein model assumes that all the phonons have the same energy. Although this is never realized in solids, often there are many phonon states with similar frequencies. It is this situation which we wish to discuss. The Einstein model is a simple, albeit extreme, limit of this behavior.

There is an important distinction between a *zero-energy transition* and a *zero-phonon transition*. The zero-phonon transition is defined as a diagonal transition in which the number of phonons $n_{\mathbf{q}}$ in each state $\mathbf{q}$ does not change.

In the limit of infinite volume ($\nu \to \infty$), it was shown that the average amplitude and probability for these diagonal transitions are

$$\langle i \mid X_{j+\delta}^{\dagger} X_j \mid i \rangle_{\text{av}} = \exp(-S_T)$$

$$\mid \langle i \mid X_{j+\delta}^{\dagger} X_j \mid i \rangle \mid_{\text{av}}^2 = \exp(-2S_T)$$

A zero-energy transition is defined as one where the energy is conserved. In an Einstein model, this means that the net number of phonons is not changed. However, this can occur by having different $\mathbf{q}$ states gain or lose phonons as long as the total number is unchanged. A state $\mathbf{q}_1$ may lose three phonons if they are absorbed by other states, say two in $\mathbf{q}_2$ and one in $\mathbf{q}_3$. If there are $3N$ phonon modes in the solid, the number of possible combinations which conserve total phonon number is quite large. Thus the probability of a zero-energy transition is much larger than the probability of a zero-phonon transition.

The probability P_{ze} of a zero-energy transition may be calculated several different ways. The easiest is to evaluate the coefficient of the delta function at zero energy in $U(\omega)$. When all the phonons have the same frequency, the exponential factor is

$$\Phi(t) = 2S_T - \varphi(t)$$

$$2S_T = \frac{2N_0 + 1}{\omega_0^2} \sum_{\mathbf{q}} M_q^2 \mid 1 - e^{i\mathbf{q}\cdot\boldsymbol{\delta}} \mid^2 \equiv g(2N_0 + 1)$$

$$\varphi(t) = g[(N_0 + 1)e^{-i\omega_0 t} + N_0 e^{i\omega_0 t}] = 2g[N_0(N_0+1)]^{1/2} \cos\left[\omega_0\left(t + \frac{i\beta}{2}\right)\right]$$

$$g = \frac{1}{\omega_0^2} \sum_{\mathbf{q}} M_q^2 \mid 1 - e^{i\mathbf{q}\cdot\boldsymbol{\delta}} \mid^2$$

The term $U(t)$ may be expressed as a summation over Bessel functions, exactly as in the derivation of (4.3.15):

$$\begin{aligned} U(t) &= J^2 \exp[-\Phi(t)] \\ &= J^2 \exp(-2S_T) \exp\left[\varepsilon \cos \omega_0\left(t + \frac{i\beta}{2}\right)\right] \\ &= J^2 \exp(-2S_T) \sum_{l=-\infty}^{\infty} I_0(\varepsilon) e^{i\omega_0 l(t+i\beta/2)} \end{aligned}$$

$$\varepsilon = 2g[N_0(N_0 + 1)]^{1/2}$$

The frequency spectrum $U(\omega)$ is just a summation of delta functions at the points $\omega = l\omega_0$. The zero-energy transition is the term with $l = 0$. It has the probability

$$P_{ze} = J^2 \exp(-2S_T) I_0(\varepsilon) \tag{6.2.15}$$

This is larger than the zero–phonon probability by the factor $I_0(\varepsilon)$. The argument of the Bessel function does not vanish as $\nu \to \infty$. Now the Bessel function provides an actual enhancement to the probability. When the coupling is strong and the argument of the Bessel function is large, we can use the asymptotic expansion $I_0(z) \to (2\pi z)^{-1/2} \exp(z)$ to get

$$P_{ze} \simeq \frac{J^2}{(2\pi\varepsilon)^{1/2}} \exp(-2S_T + \varepsilon)$$

The zero-energy probability P_{ze} has an exponent of the same form found in w:

$$2S_T - \varepsilon = g[(N_0 + 1)^{1/2} - (N_0)^{1/2}]^2$$

$$\lim_{\omega_0/K_B T \ll 1} (2S_T - \varepsilon) = \frac{\bar{\Delta}}{K_B T}$$

$$\bar{\Delta} = \tfrac{1}{4}\omega_0 g$$

It also may be represented as an activation energy $\bar{\Delta}$ at higher temperature.

As first derived by Nagaev (1963), the zero-energy probability P_{ze} increases with increasing temperature. The zero-phonon probability $J^2 \exp(-2S_T)$ decreases with increasing temperature. Thus they have quite different behavior.

The same result can also be obtained by a direct summation over each phonon mode. By following the method used to derive (6.2.9), one can show that the thermally averaged probability of a transition which makes m phonons in the state $\mathbf{q}$ is proportional to the Bessel function I_m:

$$|\langle i + m \mid X_{j+\delta}^{\dagger} X_j \mid i\rangle|^2 = e^{-2s_q} e^{\beta\omega_0 m/2} I_m(\varepsilon_q)$$

$$2s_q = (2N_0 + 1) \frac{M_q^2}{\omega_q^2} |1 - e^{i\mathbf{q}\cdot\boldsymbol{\delta}}|^2$$

$$\varepsilon_q = 2[N_0(N_0 + 1)]^{1/2} \frac{M_q^2}{\omega_q^2} |1 - e^{i\mathbf{q}\cdot\boldsymbol{\delta}}|^2$$

The final probability is obtained by taking the product over all q states and the summation over all possible values for each m_q, consistent with the condition that the total summation over all m_q values is zero:

$$P_{\text{ze}} = J^2 \sum_{m_1} \sum_{m_2} \cdots \sum_{m_N} \delta_{m_1+m_2+\cdots+m_N=0} \prod_q e^{-2S_q} I_m(\varepsilon_q) e^{\beta\omega_0 m_i/2}$$

The summation over m_q can be done by successively using the rule

$$\sum_m I_m(\varepsilon_1) I_{n-m}(\varepsilon_2) = I_n(\varepsilon_1 + \varepsilon_2)$$

which leads to the final probability:

$$P_{\text{ze}} = J^2 \Big(\prod_q e^{-2s_q}\Big) I_0\Big(\sum_q \varepsilon_q\Big) = J^2 e^{-2S_T} I_0(\varepsilon)$$

$$\sum_q s_q = S_T$$

$$\sum_q \varepsilon_q = \varepsilon$$

The result for P_{ze} is the same as in (6.2.15). In this derivation, the preceding summations must be done before one takes the limit $\nu \to \infty$.

When we try to evaluate the nondiagonal hopping rate using (6.2.12),

$$w = J^2 e^{-2S_T} \int_{-\infty}^{\infty} dt (e^{\varphi(t)} - 1)$$

there is a delta function with a coefficient

$$w = J^2 e^{-2S_T} [I_0(\varepsilon) - 1] \int_{-\infty}^{\infty} dt$$

which is the difference between the zero-energy probability and the zero-phonon probability. Thus the hopping rate for Einstein phonons is ill-defined without some modification of the formula. For dispersive phonons, the zero-energy probability was equal to the zero-phonon probability, so no contradiction arose. These probabilities are different in the Einstein model.

The resolution to these difficulties is not entirely clear. If one reexamines the Holstein arguments, it seems that in calculating the nondiagonal transitions one probably should subtract the zero-energy probability rather than the zero-phonon probability. Thus the proper form for w may be

$$w = J^2 e^{-2S_T} \int_{-\infty}^{\infty} dt [e^{\varphi(t)} - I_0(\varepsilon)]$$

This choice occurs because one really wishes to subtract all terms with no net time dependence, which is P_{ze}. In this case, there is no longer a delta function at zero energy, and w is now well defined.

Nagaev made the additional argument that one should use $P_{ze}^{1/2}$ as the estimate for the probability of diagonal transitions. His suggestion implies, as already noted, that the bandwidth increases, rather than decreases, with increasing temperature. His suggestion is not widely accepted.

G. Kubo Formula

The Kubo formula for small polarons was evaluated by Lang and Firsov (1963, 1964). The conductivity $\sigma(\omega)$ is evaluated from the current–current correlation function. The limit of $\omega \to 0$ provides the dc conductivity, while the optical absorption is given by the result for finite frequencies. Both results are interesting for small polarons. Our discussion will begin with the form of the Kubo formula given in Problem 16 of Chapter 3:

$$\mathrm{Re}(\sigma) = \frac{1 - e^{-\beta\omega}}{2\omega} \int_{-\infty}^{\infty} dt e^{i\omega t} \langle \mathscr{J}_\alpha^\dagger(t) \mathscr{J}_\alpha(0) \rangle$$

$$\mathscr{J} = i \frac{Je}{\hbar} \sum_{j\delta} \boldsymbol{\delta} C_{j+\delta}^\dagger C_j X_{j+\delta}^\dagger X_j$$

The electrical current operator is the hopping form (1.2.18), which has been subsequently altered by the canonical transformation (6.2.3). The spatial subscripts α are replaced by the diagonal sum in cubic crystals:

$$\mathrm{Re}[\sigma(\omega)] = \frac{(1 - e^{-\beta\omega})J^2}{6\omega} \frac{e^2}{\hbar^2} \sum_{\substack{\delta\delta' \\ jj'}} \boldsymbol{\delta} \cdot \boldsymbol{\delta}' \int_{-\infty}^{\infty} dt e^{i\omega t}$$
$$\times \langle C_j^\dagger(t) C_{j+\delta}(t) C_{j'+\delta'}^\dagger C_{j'} \rangle \langle X_j^\dagger(t) X_{j+\delta}(t) X_{j'+\delta'}^\dagger X_{j'} \rangle$$

In the electron correlation function, the operators have a time development governed by $e^{i\bar{H}t} C e^{-i\bar{H}t}$, where $\bar{H}$ also contains the hopping term V in (6.2.4). In the interaction representation, the expansion of the S matrix will result in terms of higher power in J. These are seldom considered, and it is customary to evaluate only the term of order J^2. This is certainly the leading term. However, the neglect of higher terms in J is mostly expediency, since they are hard to calculate. There is generally no proof available that these higher-order terms are smaller.

In the interaction representation, the zeroth-order term in the S-matrix expansion just replaces the electron part of the correlation function by its value obtained using the time development governed by H_0 rather than $\bar{H}$. This leading term in the conductivity will be called $\sigma^{(0)}$. In this approximation, the electron correlation function is easy to evaluate since there is no time dependence:

$$\langle \hat{C}_j^\dagger(t)\hat{C}_{j+\delta}(t)C_{j'+\delta'}^\dagger C_{j'}\rangle = \delta_{jj'}\delta_{\delta=\delta'}\langle n_j(1 - n_{j+\delta})\rangle$$

The correlation function is just equal to $c(1 - c)$, which is the probability c that the initial site is occupied times the probability $1 - c$ that the final site is empty. This is the correct average for this correlation function whenever there is no correlation between the site occupations of neighboring particles. Such correlations exist, for example, whenever the forces between particles on different sites are included. Most model calculations omit such forces.

The electron correlation function is nonzero only when $j = j'$ and $\delta = \delta'$. The calculation of the real part of the conductivity $\sigma^{(0)}$ is now reduced to just the evaluation of the correlation function for phonon coordinates, which is the nondiagonal transition rate for each site, summed over sites

$$\text{Re}[\sigma^{(0)}(\omega)] = \frac{1}{6}\,\frac{J^2e^2}{\hbar^2}\,\frac{1}{\omega}\,(1 - e^{-\beta\omega})c(1 - c)\sum_{j\delta}\delta^2\int_{-\infty}^{\infty} dte^{i\omega t}$$
$$\times\langle X_j^\dagger(t)X_{j+\delta}(t)X_{j+\delta}^\dagger(0)X_j(0)\rangle$$

We assume that the solid has sufficient symmetry that the rate for each pair of sites is the same. The summation over sites j just yields the number of sites, which is called N_0. The summation over δ yields the coordination number z. The number of particles is $N = N_0c$, where c is the concentration. For dimensional reasons, we must divide by the volume of the solid, so that for the real part of the conductivity we finally obtain

$$\text{Re}[\sigma^{(0)}(\omega)] = \frac{1}{6}\left(\frac{N_0}{\nu}\right)z\,\frac{J^2e^2}{\hbar^2}\,c(1 - c)\,\frac{1}{\omega}\,(1 - e^{-\beta\omega})U(\omega)$$

$$U(\omega) = \int_{-\infty}^{\infty} dte^{i\omega t}\langle X_j^\dagger(t)X_{j+\delta}(t)X_{j+\delta}(0)X_j^\dagger(0)\rangle$$

The conductivity is proportional to the function $U(\omega)$, which has been mentioned above. This function has a delta function at zero frequency, which should also be eliminated from $\text{Re}[\sigma(\omega)]$. Thus the real part of the conductivity $\sigma(\omega)$ is just proportional to the Fourier transform of the hopping correlation function $W(t)$. The dc result $\omega \to 0$ was evaluated earlier in (6.2.1) by a saddle-point integration, which for the dc conductivity gives

$$\sigma_{\rm dc}^{(0)} = \frac{1}{6}\left(\frac{N_0}{\nu}\right)zc(1-c)e^2\beta w$$

$$w \simeq \frac{J^2}{\hbar^2}\left(\frac{\pi}{\gamma}\right)^{1/2}\exp[-S_T+\bar{\varphi}(0)] \simeq \frac{J^2}{\hbar^2}\left(\frac{\pi}{\gamma}\right)^{1/2}\exp\left(-\frac{\bar{\Delta}}{K_BT}\right)$$

The same saddle-point integration also provides an estimate of the frequency-dependent conductivity (Reik, 1972). Again the integration variable is changed to $z = t + \frac{1}{2}i\beta$, and $\bar{\varphi}(z)$ is expanded about the point $z = 0$:

$$\begin{aligned} U(\omega) &= e^{-2S_T}\int_{-\infty}^{\infty} dt e^{i\omega t}e^{\varphi(t)} \\ &= e^{-2S_T+(1/2)\beta\omega}\int dz e^{i\omega z}e^{\bar{\varphi}(z)} \\ &\simeq e^{-2S_T}e^{\beta\omega/2}e^{\bar{\varphi}(0)}\int_{-\infty}^{\infty} dz e^{i\omega z}e^{-\gamma z^2} \end{aligned}$$

The integral is still of the Gaussian form and yields a Gaussian function of frequency. The prefactor $\exp(\frac{1}{2}\beta\omega)$ is combined with $1 - e^{-\beta\omega}$ to give

$$U(\omega) \simeq \left(\frac{\pi}{\gamma}\right)^{1/2}\exp\left[-2S_T+\bar{\varphi}(0)+\frac{1}{2}\beta\omega\right]\exp\left(-\frac{\omega^2}{4\gamma}\right)$$

$$\text{Re}[\sigma^{(0)}(\omega)] = \sigma_{\rm dc}^{(0)}\frac{\sinh(\frac{1}{2}\beta\omega)}{\frac{1}{2}\beta\omega}e^{-\omega^2/4\gamma} \tag{6.2.16}$$

The result for $\text{Re}[\sigma(\omega)]$ is important for relating theory and experiment. It provides a prediction of proportionality between the dc conductivity and the optical absorption. For example, it predicts that the optical absorption is Gaussian, with a width given by $4\gamma = 16\bar{\Delta}K_BT$, where $\bar{\Delta}$ is the activation energy observed in the dc conductivity. This may be tested experimentally. It also predicts that the magnitude of the absorption is proportional to the magnitude of the dc conductivity at each temperature. These predictions are confirmed in TiO_2. Figure 6.8 shows the optical data of Kudinov *et al.* (1970). The absorption data is in satisfactory agreement with the theoretical curve, which is calculated using parameters which also fit the dc conductivity.

The physics of the absorption process may be understood using a configurational coordinate picture suggested by Polder (unpublished; see Reik, 1972). The coordinates are illustrated in Fig. 6.9. A particle sitting at site j has a parabolic potential energy curve which represents the potential

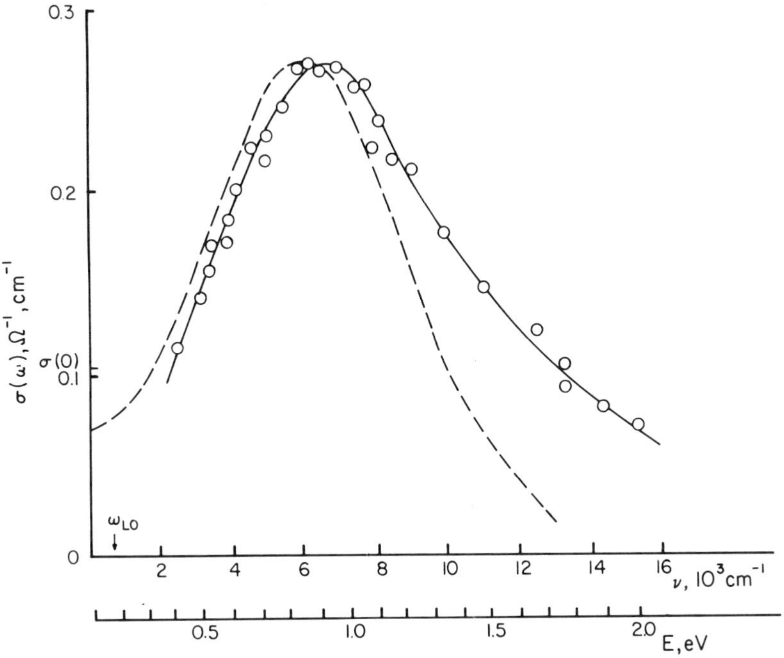

FIGURE 6.8. Optical absorption by small polarons in TiO_2 for $E \parallel c$ at $T = 600$ K given by points. Dashed curve is theory with $\hbar\omega_0 = 800\ \text{cm}^{-1} = 0.1$ eV. *Source*: Kudinov *et al.* (1970) (used with permission).

energy of the phonons for a displacement Q about the equilibrium point. The neighboring site $j + \delta$ has an identical parabola, displaced to the new equilibrium point, at a distance Q_0. The dc conductivity occurs by thermal activation over the intervening potential barrier $\bar{\Delta}$, which is midway between the two parabolas. If the parabolas have a curvature given by α so that

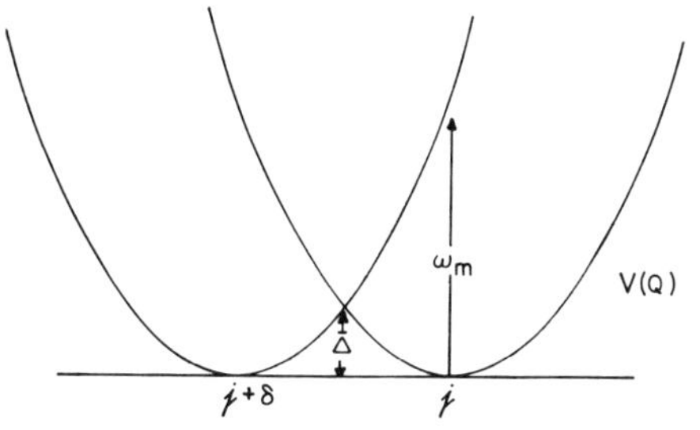

FIGURE 6.9. Configurational coordinate drawing of small-polaron motion. The probable optical absorption is the transition ω_m, while the hopping barrier is $\bar{\Delta}$.

the potential energy curves are $V(Q) = \alpha Q^2$, then we can deduce that the activation energy is $\bar{\Delta} = V(Q_0/2) = \frac{1}{4}\alpha Q_0^2$. Similarly, the most probable optical transition is the vertical arrow in Fig. 6.9. It puts the particle on the neighboring parabola, and later the particle relaxes down to the minimum at the neighboring point. The length of this arrow is the frequency $\omega_m = V(Q_0) = \alpha Q_0^2$, which is four times the activation energy. The proportionality $\omega_m/\bar{\Delta} = 4$ agrees with the result (6.2.16), which may be rewritten as

$$\mathrm{Re}[\sigma^{(0)}(\omega)] = \sigma_{\mathrm{dc}}^{(0)} \frac{1 - e^{-\beta\omega}}{\beta\omega} \exp\left[-\frac{(\omega - \beta\gamma)^2}{4\gamma} + \frac{\bar{\Delta}}{K_B T} \right]$$

which shows that the maximum frequency is at $\omega_m = \beta\gamma = 4\bar{\Delta}$. The configurational coordinate model may be used, in fact, to derive (6.2.16) directly.

6.3. HEAVILY DOPED SEMICONDUCTORS

An interesting application of polaron theory is in many-particle systems. If the solid has many electrons, one has to consider electron–electron interactions along with the electron–phonon interactions. This is very important in metals, which will be discussed in the next section. Other systems, which mix the two theories of polarons and electron–electron interactions, are semiconductors which are heavily doped with impurities. Many III–V semiconductors are polar and may also be doped sufficiently with impurities to form a degenerate electron gas. Then at low temperatures the electron gas has a well-defined Fermi energy, and the electrical resistivity is finite.

These systems are interesting from several viewpoints. As an electron gas, they may have an effective density which is very high. The actual density is quite low, perhaps only 10^{18}–10^{19} cm^{-3}. However, when we calculate the value of r_s, we must use the effective Bohr radius,

$$a_0^* = \frac{\hbar^2 \varepsilon_0}{e^2 m^*}$$

$$r_s = \left(\frac{4\pi}{3} n_0 a_0^{*3} \right)^{-1/3}$$

with the effective mass m^* and static dielectric function ε_0 for the semiconductor. Because typical values are $m^*/m \simeq 0.1$–0.2 and $\varepsilon_0 \simeq 10$, the effective Bohr radius may be as large as 50 Å. Even for an electron concentration of $n_0 \simeq 10^{18}$ cm^{-3}, for a Bohr radius of $a_0^* \simeq 50$ Å we deduce that

$r_s \simeq 1$, which is a smaller value than found in metals. Even smaller values of r_s are obtainable in many semiconductors. Thus semiconductors are a suitable environment in which to study the high-density electron gas.

Many semiconductors are polar. The interaction between electrons in the conduction band and LO phonons is well described by the Fröhlich Hamiltonian. Many of these semiconductors may also be doped to high electron concentration, so that experimentally one may study polaron theory in a many-particle system. However, nature does not seem to like these two conditions—strong polaron interactions and high doping levels—to occur in the same materials. Materials which are very polar, such as alkali halides or II–VI semiconductors, can seldom be doped to where the electron gas is degenerate. Group IV (Si, Ge) and III–V semiconductors may be doped readily to high levels but are either weakly polar or not polar. Thus the theory of polarons, in many-particle semiconductors, only needs to treat systems with weak coupling between electrons and phonons. An advantage of these experimental systems is that all the parameters of the Hamiltonian are known from other measurements. The only parameters of the theory are m^*, n_0, ε_0, and ε_∞, which all can be obtained by simple measurements. Thus the many-body theory may be tested in situations where there are no adjustable parameters.

The self-energy of an electron from electron–phonon interactions is dramatically altered by the presence of other electrons. We shall describe several effects. First, the phonon energies themselves are altered by the screening properties of the electron gas. Second, the electron–phonon interaction is screened by the electron gas. This is a significant feature, so that even intermediate values of α are reduced to weak coupling. Third, the nature of the electron's self-energy is changed dramatically by the Pauli exclusion principle. When electrons interact with phonons, they can only scatter into states not occupied by other electrons. The sum of these three effects is that polaron theory is entirely changed from the simple one-particle theory of the previous sections.

A. Screened Interaction

A little bit of thought shows we have a complicated system. The phonons and the electron gas are mutually interacting and affecting each other's properties. Should the first step in the calculation be to calculate the effect of the phonons on the electrons, or vice versa? The answer is neither. Instead, one should calculate the total longitudinal dielectric function of the system, which includes parts from both electrons and phonons. This function will be sufficient to describe all the relevant physics.

The effective interaction between electrons may be written as

$$V_{\text{eff}}(q, \omega) = \frac{4\pi e^2}{q^2 \varepsilon_{\text{total}}(q, \omega)} = \frac{v_q}{\varepsilon_{\text{total}}} \tag{6.3.1}$$

where $\varepsilon_{\text{total}}(q, \omega)$ is the total dielectric function from all sources. This total dielectric function will be derived in three ways. The first will be a phenomenological derivation. The second and third derivations use Green's functions.

The simple derivation is given first. The dielectric function is the summation of the three contributions: ε_∞, the electron–electron interactions, and electron–phonon interactions. The first two we already know how to obtain, so that we can immediately write

$$\varepsilon_{\text{total}}(q, \omega) = \varepsilon_\infty - v_q P(q, \omega) + \text{phonon term}$$

where $P(q, \omega)$ is the polarization of the electron gas, which was discussed in Sec. 5.5. The phonon term will be calculated assuming that there are no electrons present. For optical phonons in polar crystals, the ions will vibrate in response to an oscillating electric field. If X is the distance between a pair of plus and minus ions with reduced mass M, the classical equations of motion are

$$M\ddot{X} + KX = eE_0 e^{-i\omega t} = eE(t)$$

The force constant is $K = M\omega_{\text{TO}}^2$. The oscillatory solution has a periodic displacement of

$$X = \frac{-eE_0}{M} \frac{e^{-i\omega t}}{\omega^2 - \omega_{\text{TO}}^2}$$

The net polarization of the system is $P = en_0 X$, where e and n_0 are the charge and density of the ion pairs. The contribution to the dielectric function is

$$\Delta\varepsilon = 4\pi\alpha = \frac{4\pi P}{E(t)} = \frac{4\pi e^2 n_0}{M} \frac{1}{\omega_{\text{TO}}^2 - \omega^2}$$

The constants $4\pi e^2 n_0/M$ may be determined phenomenologically by realizing that in the absence of the electron–electron term, the long-wavelength dielectric function of a polar crystal may be written as

$$\varepsilon(\omega) = \varepsilon_\infty + \frac{\varepsilon_0 - \varepsilon_\infty}{1 - \omega^2/\omega_{\text{TO}}^2}$$

$$\varepsilon_0 - \varepsilon_\infty = \frac{4\pi e^2 n_0}{M\omega_{\text{TO}}^2}$$

Thus at zero frequency one finds $\varepsilon(0) = \varepsilon_0$, while at frequencies above the reststrahl $\omega \gg \omega_{\rm TO}$ we have $\varepsilon(\omega) = \varepsilon_\infty$. These are the definitions of the two quantities ε_0 and ε_∞. The total dielectric function is

$$\varepsilon_{\rm total}(q, \omega) = \varepsilon_\infty + \frac{\varepsilon_0 - \varepsilon_\infty}{1 - \omega^2/\omega_{\rm TO}^2} - v_q P(q, \omega) \tag{6.3.2}$$

In this derivation, we have treated the three contributions to the dielectric function as independent and additive. This is not quite right. The rigorous derivation, which will be given later, shows that the phonons affect the electron–electron contribution. When calculating the polarization diagrams for $P(q, \omega)$, we must also include diagrams in which the basic electron bubble has internal phonon lines. However, this has never been done in practice. Instead, the term $v_q P(q, \omega)$ is approximated by either the RPA or the Thomas–Fermi model. In this approximation, the three terms are then independent.

The rigorous derivation of the total screening function proceeds by summing diagrams. The Hamiltonian has the form

$$\begin{aligned}
H &= H_0 + H_{ee} + H_{e\text{-ph}} + H_{\text{ph-ph}} \\
H_0 &= \sum_{\mathbf{p}\sigma} \xi_p c^\dagger_{\mathbf{p}\sigma} c_{\mathbf{p}\sigma} + \sum_{\mathbf{q}\lambda} \omega_{q\lambda} a^\dagger_{\mathbf{q}\lambda} a_{\mathbf{q}\lambda} \\
H_{ee} &= \frac{1}{2\nu} \sum_{\mathbf{q}} v_q{}^\infty \varrho(\mathbf{q}) \varrho^\dagger(\mathbf{q}) \\
H_{e\text{-ph}} &= \frac{1}{\sqrt{\nu}} \sum_{\mathbf{q}\lambda} M_{q\lambda} \varrho(\mathbf{q}) A_\lambda{}^\dagger(\mathbf{q}) \\
H_{\text{ph-ph}} &= \tfrac{1}{2} \sum_{\mathbf{q}\lambda} V_\lambda(\mathbf{q}) A_\lambda(\mathbf{q})^\dagger A_\lambda(\mathbf{q}) \\
v_q{}^\infty &= \frac{v_q}{\varepsilon_\infty} = \frac{4\pi e^2}{\varepsilon_\infty q^2}, \qquad A_\lambda(\mathbf{q}) = a_{\mathbf{q}\lambda} + a^\dagger_{-\mathbf{q}\lambda}, \qquad \varrho(\mathbf{q}) = \sum_{p\sigma} c^\dagger_{\mathbf{p}+\mathbf{q}\sigma} c_{\mathbf{p}\sigma}
\end{aligned} \tag{6.3.3}$$

The electron density operator is $\varrho(\mathbf{q})$. The electron–phonon matrix element is $M_{q\lambda}$. The phonon–phonon matrix element is $V_\lambda(\mathbf{q})$. The electron–electron interaction has been written as the product of two density operators. This form has the liability that it permits an electron to interact with itself. These terms are meant to be absent and are ignored. The shorthand notation is convenient for the following discussion.

The term H_0 describes the noninteracting systems of electrons and phonons. The electron's energy $\xi_p = \varepsilon_p - \mu$ is measured from its chemical potential. We shall usually assume that the kinetic energy $\varepsilon_p = p^2/2m^*$ is described by an effective mass m^* which is ordinarily smaller than an electron mass. The electron–phonon part of the Hamiltonian has the usual form. Our applications will mostly be to polar coupling, where $M_{q\lambda}$ has the form in (6.1.1). The electron–electron interaction has an interaction $v_q^\infty = v_q/\varepsilon_\infty$. The factor ε_∞ is included to account for the dielectric screening of the material. There are three sources of dielectric screening in the system. The first is the electron–electron interactions from the mobile electrons in the conduction band; their contribution to the screening is included explicitly when we include H_{ee}. The second is from the optical phonons, and these are included through the term $H_{e\text{-ph}}$. The third is from the high-energy electronic excitations—across the band gap of the semiconductor. They give rise to ε_∞. We could also include them as an additional term in the Hamiltonian, which would be solved to give ε_∞. However, for the low-frequency excitations of interest in the present discussion, these high-energy excitations are uninteresting because they give a constant contribution to $\varepsilon(q, \omega)$. Thus it is simpler to include these effects through the constant ε_∞, which is introduced in the Hamiltonian at the beginning.

There are two different ways to solve these equations. Both methods give the same answer. The easiest method is done first. In order to keep the discussion simple, we specialize the equations to longitudinal optical (LO) phonons. The ions are treated as point masses with an effective charge eZ. Then the various interaction terms are approximated as

$$\omega_{q\lambda} = \omega_{\mathrm{TO}}, \qquad M_{q\lambda} = v_q^\infty ZqX(\mathbf{q})$$
$$V_\lambda(\mathbf{q}) = v_q^\infty Z^2q^2X(\mathbf{q})^2, \qquad X(\mathbf{q})^2 = \frac{n_0}{2M\omega_{\mathrm{TO}}}$$

The phonon frequency $\omega_{q\lambda}$ is chosen to be the TO frequency, rather than the LO frequency. The difference between these frequencies arises from the long-range Coulomb interactions in $H_{\mathrm{ph-ph}}$. As long as the phonon–phonon interactions are explicitly included in the interaction term of the Hamiltonian, then we must have the TO frequency in H_0. Later it is shown how the term $H_{\mathrm{ph-ph}}$ changes ω_{TO} to ω_{LO}.

All of the interactions are due to long-range Coulomb interactions: electron–electron, electron–phonon, and phonon–phonon. All of the interaction terms can be combined into one term

$$\begin{aligned} H_{\mathrm{int}} &= H_{ee} + H_{e\text{-ph}} + H_{\mathrm{ph-ph}} \\ &= \tfrac{1}{2} \sum_{\mathbf{q}} v_q^\infty \varrho_T^\dagger(\mathbf{q}) \varrho_T(\mathbf{q}) \end{aligned}$$

where

$$\varrho_T(\mathbf{q}) = \frac{1}{\sqrt{\nu}}\, \varrho(\mathbf{q}) + ZqX(\mathbf{q})A(\mathbf{q})$$

The total density operator ϱ_T is the summation of the density operators for electron and phonons.

The dielectric function is calculated for this interaction by following the same steps used in Sec. 5.5B to obtain RPA. The difference between that case and the present one is the phonons. By following the same steps, one can show that

$$\varepsilon(\mathbf{q}, i\omega) = \varepsilon_\infty[1 - v_q{}^\infty P_T{}^{(1)}(\mathbf{q}, i\omega)]$$

$$P_T{}^{(1)}(\mathbf{q}, i\omega) = -\int_0^\beta d\tau\, e^{i\omega\tau} \langle T_\tau \varrho_T(\mathbf{q}, \tau)\varrho_T{}^\dagger(\mathbf{q}, 0)\rangle$$

The electron and phonon operators act independently. The most important terms in $P^{(1)}$ are the summation of the separate susceptibilities for electrons and phonons:

$$P_T{}^{(1)}(\mathbf{q}, i\omega) = P_e{}^{(1)}(\mathbf{q}, i\omega) + P_{\text{ph}}^{(1)}(\mathbf{q}, i\omega)$$

$$P_{\text{ph}}^{(1)}(\mathbf{q}, i\omega) = Z^2q^2X(q)^2\, \mathscr{D}\, (\mathbf{q}, i\omega)$$

$$-\, v_q P_{\text{ph}}^{(1)}(\mathbf{q}, \omega) = \frac{4\pi e^2 n_0 Z^2}{M}\, \frac{1}{\omega_{\text{TO}}^2 - \omega^2}$$

The electron polarization $P_e{}^{(1)}$ is identical to the RPA result in Sec. 5.5B. This dielectric function is identical to the one we found in (6.3.2). It is the RPA result for a system of electrons *and* LO phonons. This derivation is rather simple because all long-range interactions were included in H_{int}.

The third derivation of this same result is the one presented in most reviews. It is also the most complicated derivation, because it takes two steps. The first step is to solve the Hamiltonian in the absence of conduction electrons. This subset of terms is called H_0':

$$H_0' = H_0 + H_{\text{ph-ph}}$$

This Hamiltonian can be solved exactly. The phonon–phonon interaction term just changes ω_{TO} to ω_{LO}.

$$H_0' = \sum_{\mathbf{q}\sigma} \xi_p c_{\mathbf{p}\sigma}^\dagger c_{\mathbf{p}\sigma} + \sum_{\mathbf{q}} \omega_{\text{LO}} a_{\mathbf{q}}{}^\dagger a_{\mathbf{q}}$$

An easy way to show this is by calculating the phonon self-energy. In first-order perturbation theory, the term $H_{\text{ph-ph}}$ gives a self-energy of

$$\Pi^{(1)}(q, i\omega) = V_\lambda(q) = 4\pi e^2 Z^2 n_0/(2M\omega_{\text{TO}}\varepsilon_\infty)$$

and the exact interacting Green's function is

$$D(q, \omega) = \frac{-2\omega_{\text{TO}}}{\omega^2 - \omega_{\text{TO}}^2 - 2\omega_{\text{TO}}\Pi^{(1)}} = \frac{-2\omega_{\text{TO}}}{\omega^2 - \omega_{\text{LO}}^2}$$

$$\omega_{\text{LO}}^2 = \omega_{\text{TO}}^2 + \frac{4\pi e^2 n_0 Z^2}{M\varepsilon_\infty}$$

The denominator of the Green's function contains the LO phonon frequency rather than the LO frequency. The long-range interactions between the phonons cause the energy difference between LO and TO phonons.

The next step is to include the electrons. The Hamiltonian is written as

$$H = H_0' + H_{ee} + H_{e\text{-ph}}$$

The interactions contain an electron–electron term and an electron–phonon term. The summation of diagrams that comprise the screening will be different from the RPA form, since the phonon–phonon interaction term is no longer included in the interaction term. It is present, but in H_0'. Because there is no longer an explicit phonon–phonon interaction, the only polarization diagrams that can occur are those for the electrons. The phonon polarization diagrams have already been summed to turn H_0 into H_0'.

We consider the scattering of two electrons. Fig. 6.10(a) shows two electrons which are being scattered by the unscreened electron–electron interaction. Each such diagram will have a factor v_q^∞. Similarly, Fig. 6.10(b) shows two electrons scattering by sending a phonon from one to the other. Each such interaction has a factor $V_{\text{ph}}(q, \omega)$ which is proportional to the

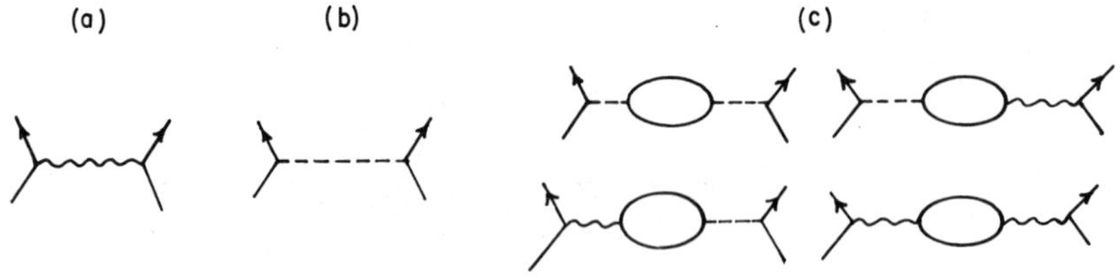

FIGURE 6.10

square of the electron–phonon matrix element and the phonon Green's function:

$$V_{\rm ph}(q, \omega) = M_q^{\,2} D^{(0)}(q, \omega) = -M_q^{\,2} \frac{2\omega_q}{\omega_q^{\,2} - \omega^2}$$

The phonon frequencies ω_q are those which include phonon–phonon interactions. These two ways of scattering electrons—by electron–electron interactions and by electron–phonon interactions—are treated on an equal footing. Thus when we consider diagrams which have a single polarization bubble $P(q, \omega)$, there are four possible diagrams, as shown in Fig. 6.10(c). Both the interaction to the right and left of the bubble may be either a Coulomb line or a phonon line. Diagrams with n bubbles in a chain can be shown to have 2^{n+1} possible arrangements. These may all be summed exactly by realizing that the effective propagator of the system is the sum of these two contributions. This was defined earlier in (3.6.19) as

$$W^{(0)}(q, \omega) = v_q^{\,\infty} + V_{\rm ph}(q, \omega)$$

Thus we can write the effective interaction as the following summation of terms (see Sec. 5.5B):

$$\begin{aligned} V_{\rm eff}(q, \omega) &= W^{(0)} + P(W^{(0)})^2 + P^2(W^{(0)})^3 + P^3(W^{(0)})^4 + \cdots \\ &= \frac{W^{(0)}}{1 - W^{(0)}P} \end{aligned}$$

The factors on the right are Dyson's equation for W, which is a suitable definition of the effective interaction. $V_{\rm eff}$ is set equal to $v_q/\varepsilon_{\rm total}(q, \omega)$ as in (6.3.1), which provides a definition of the total dielectric function:

$$\frac{1}{\varepsilon_{\rm total}(q, \omega)} = \frac{q^2}{4\pi e^2} \frac{v_q^{\,\infty} + V_{\rm ph}(q, \omega)}{1 - [v_q^{\,\infty} + V_{\rm ph}(q, \omega)]P(q, \omega)} \tag{6.3.4}$$

The formula (6.3.4) for $1/\varepsilon_{\rm total}(q, \omega)$ is valid for any kind of phonons—LA, TA, etc. The only assumption in the derivation is that the electrons had electron–phonon and electron–electron interactions. For systems where more than one type of phonon interacts with the electron, $V_{\rm ph}$ is interpreted as the summation over all the phonon modes of the product $| M_\lambda |^2 \mathscr{D}_\lambda(q, \omega)$. In the present section, we shall apply this formula only to optical phonons in polar crystals. In the next section, it will be applied to all types of phonons in metals.

The next step must be to show that this is identical to the earlier result (6.3.2). At the moment the two forms do not seem much alike.

The first step is to invert Eq. (6.3.4):

$$\varepsilon_{\text{total}} = v_q\left[\frac{1}{W^{(0)}} - P(q, \omega)\right] = \frac{v_q}{W^{(0)}} - v_q P \tag{6.3.5}$$

This splits off the last term, which is the electron–electron contribution. The first term must give the other two terms in (6.3.2). As an example, consider the Fröhlich interaction between electrons and optical phonons; the terms are

$$W^{(0)} = v_q^{\infty} + V_{\text{ph}}$$

$$v_q^{\infty} = \frac{v_q}{\varepsilon_\infty} = \frac{4\pi e^2}{q^2 \varepsilon_\infty}$$

$$V_{\text{ph}} = -\frac{4\pi e^2}{q^2}\,\frac{\omega_{\text{LO}}^2}{\omega_{\text{LO}}^2 - \omega^2}\left(\frac{1}{\varepsilon_\infty} - \frac{1}{\varepsilon_0}\right)$$

Thus the first term of (6.3.5) is

$$\frac{v_q}{W^{(0)}} = \frac{v_q}{v_q^{\infty} + V_{\text{ph}}} = \left[\frac{1}{\varepsilon_\infty} - \frac{\omega_{\text{LO}}^2}{\omega_{\text{LO}}^2 - \omega^2}\left(\frac{1}{\varepsilon_\infty} - \frac{1}{\varepsilon_0}\right)\right]^{-1}$$

$$= \frac{\varepsilon_\infty}{1 - [1/(\omega_{\text{LO}}^2 - \omega^2)][\omega_{\text{LO}}^2 - (\varepsilon_\infty/\varepsilon_0)\omega_{\text{LO}}^2]}$$

Next we use the Lyddane–Sachs–Teller (1941) relationship

$$\frac{\omega_{\text{LO}}^2}{\omega_{\text{TO}}^2} = \frac{\varepsilon_0}{\varepsilon_\infty}$$

to rewrite the last term in the denominator:

$$\frac{v_q}{W^{(0)}} = \frac{\varepsilon_\infty}{1 - [(\omega_{\text{LO}}^2 - \omega_{\text{TO}}^2)/(\omega_{\text{LO}}^2 - \omega^2)]}$$

If we finally split off the term ε_∞, we get the desired result:

$$\frac{v_q}{W^{(0)}} = \varepsilon_\infty + \frac{\varepsilon_\infty[(\omega_{\mathrm{LO}}^2 - \omega_{\mathrm{TO}}^2)/(\omega_{\mathrm{LO}}^2 - \omega^2)]}{1 - [(\omega_{\mathrm{LO}}^2 - \omega_{\mathrm{TO}}^2)/(\omega_{\mathrm{LO}}^2 - \omega^2)]}$$

$$= \varepsilon_\infty + \frac{\varepsilon_\infty(\omega_{\mathrm{LO}}^2 - \omega_{\mathrm{TO}}^2)}{\omega_{\mathrm{TO}}^2 - \omega^2}$$

$$= \varepsilon_\infty + \frac{\varepsilon_0 - \varepsilon_\infty}{1 - \omega^2/\omega_{\mathrm{TO}}^2} \tag{6.3.6}$$

where the Lyddane–Sachs–Teller relationship was used again for rearranging the last numerator. Thus from (6.3.5) we have shown that the total dielectric function (6.3.4) is in agreement with (6.3.2):

$$\varepsilon_{\mathrm{total}}(q, \omega) = \varepsilon_\infty + \frac{\varepsilon_0 - \varepsilon_\infty}{1 - \omega^2/\omega_{\mathrm{TO}}^2} - v_q P(q, \omega) \tag{6.3.2}$$

The first method of derivation is obviously the easiest. It also has a message which should be remembered: The total dielectric function is often well approximated by the sum of the independent contributions. For example, the piezoelectric contribution from acoustical phonons provides another term in the above series (Mahan, 1972) for crystals which are piezoelectric. The starting point of a many-body calculation for any system with several contributions should be to find the dielectric function. Often a satisfactory approximation is obtained by just adding the separate contributions. It will be shown later that this does provide a description of the mixing and interference between these modes. The mixing occurs because all these various modes—LO phonons, plasmons, piezoelectric phonons—have longitudinal electric fields. The longitudinal electric fields of the various modes cause the mutual coupling. These are well described by the dielectric function.

A curious feature of the optical phonon part of the dielectric function $(\varepsilon_0 - \varepsilon_\infty)/(1 - \omega^2/\omega_{\mathrm{TO}}^2)$ is that the resonance frequency is ω_{TO}. The resonance is the frequency where the energy denominator vanishes. This position of resonance is in contrast to the Green's function for optical phonons, which has its pole or resonance frequency at the LO phonon frequency ω_{LO}. There is no disagreement between these two results. The longitudinal excitations of the system are not at the poles of $\varepsilon(\omega)$ but are at the poles of $1/\varepsilon(\omega)$. Thus if the polar solid had no conduction electrons and was an insulator, the total dielectric function would just be the first two terms of (6.3.2) or

$$\varepsilon(\omega) = \varepsilon_\infty - \frac{\varepsilon_0 - \varepsilon_\infty}{1 - \omega^2/\omega_{\text{TO}}^2}$$
$$\frac{1}{\varepsilon(\omega)} = \frac{1}{\varepsilon_\infty} - \frac{\omega_{\text{LO}}^2}{\omega_{\text{LO}}^2 - \omega^2}\left(\frac{1}{\varepsilon_\infty} - \frac{1}{\varepsilon_0}\right) \tag{6.3.7}$$

The derivation of $1/\varepsilon(\omega)$ is simple, since from (6.3.6) we observe that it is just the quantity $W^{(0)}/v_q$. The poles of $1/\varepsilon(\omega)$ are indeed at ω_{LO}. Of course, in a polar solid there are both LO and TO phonons. However, only the LO phonons make long-range electric fields, so only they respond to a longitudinal electric field.

The combined system of electrons and phonons has the dielectric function $/\varepsilon_{\text{total}}(q, \omega)$given in (6.3.2). The excitations of this system must be given by the poles in $1/\varepsilon_{\text{total}}(q, \omega)$. Let $Z_i(q)$ denote the complex values of ω at which these poles are found. According to the theorem of residues, the complex function may be exactly expressed as the summation over these poles and residues:

$$\frac{1}{\varepsilon_{\text{total}}(q, Z)} = \sum_i \frac{R_i(q)}{Z^2 - Z_i(q)^2}$$

The poles always occur in the pairs $\pm Z_i$ since $\varepsilon_{\text{total}}$ is an even function of ω. Each pole is then interpreted as an excitation of the system, with a coupling strength given by the residue. The expansion in poles and residues would be an exact way to describe the excitations, and their couplings, if $\varepsilon_{\text{total}}^{-1}$ were only a sum of poles. However, our previous experience indicates that there are branch cuts in the analytic function $1/\varepsilon_{\text{total}}$. In the electron gas, these described the pair excitations, and the same excitations should occur here. Thus the exact description of the excitation spectrum would be all contributions where there is a nonzero contribution from

$$S(q, \omega) = -\frac{1}{n_0 v_q} \operatorname{Im}\left[\frac{1}{\varepsilon_{\text{total}}(q, \omega + i\delta)}\right] \tag{6.3.8}$$

which would include poles as well as branch cuts.

There are two important frequencies in the coupled system: the LO phonon frequency ω_{LO} and the plasma frequency of the electron gas:

$$\omega_p{}^2 = \frac{4\pi e^2 n_0}{\varepsilon_\infty m^*} \tag{6.3.9}$$

The coupled modes have a quite different character depending on whether $\omega_{\text{LO}} > \omega_p$ or $\omega_p > \omega_{\text{LO}}$. The usual case is $\omega_p \gg \omega_{\text{LO}}$, which will be discussed here. The other case $\omega_{\text{LO}} > \omega_p$ may also occur and will be treated in the next section. The experimentalist may vary n_0, and hence ω_p, so the two situations $\omega_p < \omega_{\text{LO}}$ or $\omega_p > \omega_{\text{LO}}$ may be achieved in the same material.

The total dielectric function is being denoted as $\varepsilon_{\text{total}}$. Perhaps this should just be called $\varepsilon(q, \omega)$. However, it is conventional to reserve the latter name for the electron–electron parts of the dielectric function. Thus we define

$$\varepsilon(q, \omega) = 1 - v_q^{\infty} P(q, \omega) \tag{6.3.10}$$

In the case where $\omega_p > \omega_{\text{LO}}$, it is customary—although perhaps not necessary—to explicitly separate the electron–electron part of the effective interaction:

$$V_{\text{eff}}(q, \omega) = \frac{v_q}{\varepsilon_{\text{total}}} = \frac{v_q^{\infty}}{\varepsilon(q, \omega)} + V_{\text{sc-ph}}(q, \omega)$$

The second term on the right is the screened electron–phonon interaction. It is defined by this equation, so $V_{\text{sc-ph}}$ is the difference between V_{eff} and the screened Coulomb interaction. This definition may be manipulated by combining these two terms:

$$V_{\text{sc-ph}} = \frac{v_q^{\infty} + V_{\text{ph}}}{1 - (v_q^{\infty} + V_{\text{ph}})P} - \frac{v_q^{\infty}}{1 - v_q^{\infty} P}$$

$$= \frac{V_{\text{ph}}}{\varepsilon[1 - (v_q^{\infty} + V_{\text{ph}})P]}$$

The last denominator may be regrouped as

$$\varepsilon[1 - (v_q^{\infty} + V_{\text{ph}})P] = \varepsilon(\varepsilon - V_{\text{ph}} P) = \varepsilon^2\left(1 - \frac{V_{\text{ph}} P}{\varepsilon}\right)$$

$$V_{\text{sc-ph}} = \frac{V_{\text{ph}}}{\varepsilon^2(1 - V_{\text{ph}} P/\varepsilon)}$$

From the definition of $V_{\text{ph}} = M_q^2 D^{(0)}$, we can rewrite the effective interaction as

$$V_{\text{eff}} = \frac{v_q^{\infty}}{\varepsilon} + \frac{M_q^2}{\varepsilon^2} D$$

$$D(q, \omega) = \frac{D^{(0)}}{1 - M_q^2 D^{(0)} P/\varepsilon} = \frac{2\omega_q}{\omega^2 - \omega_q^2 - 2\omega_q (M_q^2 P/\varepsilon)} \tag{6.3.11}$$

The screened electron–phonon interaction $V_{\text{sc-ph}}$ is expressed as the product of the screened matrix element $M_q^2/\varepsilon(q, \omega)^2$ and the phonon Green's function $D(q, \omega)$. The phonon Green's function contains self-energy terms arising from the polarization of the electron gas. The quantity $P(q, \omega)$ describes the polarization of the electron gas, while M_q^2 is the coupling between the electrons and phonons. It is important to note that this phonon Green's function is different from the one which would be derived from Dyson's equation by ignoring the electron–electron interactions. The latter result would be Dyson's equation with no electron–electron interactions:

$$D' = \frac{D^{(0)}}{1 - M_q^2 D^{(0)} P}$$

The difference is the additional factor of $1/\varepsilon$ in the self-energy term. The derivation of D includes electron–electron interactions properly and show that the factor of $1/\varepsilon$ actually belongs. Since ε is usually greater than unity, including the $1/\varepsilon$ factor will considerably weaken the self-energy effects.

Another interesting feature of the screened electron–phonon interaction is that the squared matrix element M_q^2 is divided by the square of the dielectric function. Since M_q is something in the nature of the electron–ion potential, one might naively expect it to be divided by the dielectric function. The derivation shows that M_q is divided by ε or M_q^2 by ε^2.

There is a good reason for separating the effective interaction into the Coulomb and screened phonon parts. When $\omega_p \gg \omega_{\text{LO}}$, the dielectric function $\varepsilon(q, \omega)$ which occurs in the phonon part needs to be evaluated at $\omega \simeq \omega_{\text{LO}}$. For $\omega_p \gg \omega_{\text{LO}}$, $\varepsilon(q, \omega_{\text{LO}})$ is well approximated by the static limit of $\varepsilon(q)$. Thus it is a good approximation to rewrite the effective interaction as

$$\begin{aligned} V_{\text{eff}}(q, \omega) &\simeq \frac{v_q^\infty}{\varepsilon(q, \omega)} + \frac{M_q^2}{\varepsilon(q)^2} D(q, \omega) \\ D(q, \omega) &= \frac{2\omega_q}{\omega^2 - \omega_q^2 - 2\omega_q M_q^2 P(q)/\varepsilon(q)} \end{aligned} \tag{6.3.12}$$

In evaluating the screened Coulomb term, the full frequency dependence must be retained in $\varepsilon(q, \omega)$. But the static limits $\varepsilon(q)$ and $P(q)$ are satisfactory in the phonon term. Usually treatments have used either Thomas–Fermi theory or the RPA. Although it would be more accurate to use a Singwi–Sjölander dielectric function, the additional labor may not be worth the incremental increase in accuracy.

B. Experimental Verifications

First-order Raman scattering is permitted in crystals which lack an inversion center. This is the case for zincblende and wurtzite, which are the crystal structure of most III–V and II–VI semiconductors. The Raman experiment measures the frequency of excitations which have the same magnitude of wave vector as the incident light. For optical frequencies, the wave vector is typically 10^5 cm^{-1}, which is essentially the limit of $q \rightarrow 0$ for $P(q, \omega)$. In this limit, the electron-gas term $-v_q{}^{\infty}P(q, \omega)$ becomes $-\omega_p{}^2/\omega^2$, so the total dielectric function is

$$\lim_{q\to 0} \varepsilon_{\text{total}}(q, \omega) = \varepsilon_{\text{total}}(\omega) = \varepsilon_\infty + \frac{\varepsilon_0 - \varepsilon_\infty}{1 - \omega^2/\omega_{\text{TO}}^2} - \frac{\omega_p{}^2}{\omega^2} \tag{6.3.13}$$

The poles of $1/\varepsilon_{\text{total}}(\omega)$ occur where $\varepsilon_{\text{total}}(\omega) = 0$. The equation

$$\varepsilon_{\text{total}}(\omega) = 0 = \varepsilon_\infty + \frac{\varepsilon_0 - \varepsilon_\infty}{1 - \omega^2/\omega_{\text{TO}}^2} - \frac{\omega_p{}^2}{\omega^2}$$

is a quadratic equation for ω^2, which is easily solved to give

$$\omega_\pm{}^2 = \tfrac{1}{2}(\omega_p{}^2 + \omega_{\text{LO}}^2) \pm \tfrac{1}{2}[(\omega_p{}^2 + \omega_{\text{LO}}^2)^2 - 4\omega_p{}^2\omega_{\text{TO}}^2]^{1/2} \tag{6.3.14}$$

These two solutions are plotted as the solid lines in Fig. 6.11 as $\omega_p{}^2$ is changed by increasing n_0. The parameters are appropriate for GaAs, which has $m^* = 0.072$ and $\varepsilon_\infty = 11.3$, $\omega_{\text{TO}} = 268$ cm^{-1}, and $\omega_{\text{LO}} = 291$ cm^{-1}. Also shown is the experimental data of Mooradian and Wright (1966) for GaAs; they measured the frequencies by Raman scattering. There is excellent agreement between theory and experiment.

The LO phonons and the plasmons are two modes which are mutually

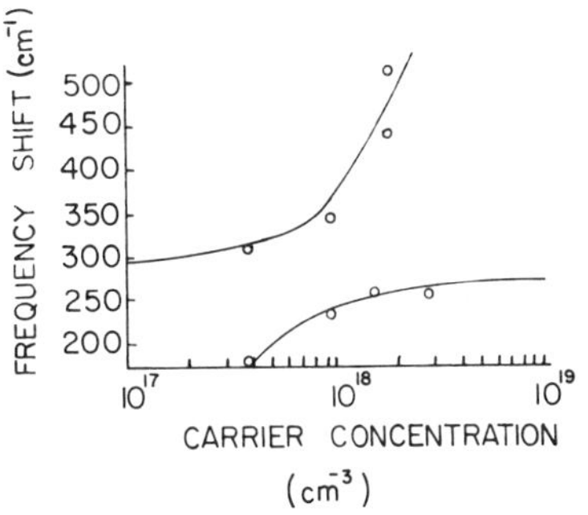

FIGURE 6.11. Plasma frequency in GaAs as a function of electron density n_0, as determined by Raman scattering. Solid line is theory using $\omega_{\text{TO}} = 268$ cm^{-1}, $\omega_{\text{LO}} = 291$ cm^{-1}, $\varepsilon_\infty = 11.3$, and $m^* = 0.07$. *Source*: Mooradian and Wright (1966) (used with permission).

coupled. Figure 6.11 represents a typical crossing phenomenon of two coupled modes. This is a particularly clear example, since both modes are completely classical. Even if we restored $\hbar$ to our equations, it would not appear in (6.3.14). It is also useful to understand the asymptotic limits of very high and low density. At high values of n_0, the analytical solution of (6.3.14) gives the roots

$$\omega_+^2 \to \omega_p^2$$

$$\omega_-^2 \to \omega_{\mathrm{TO}}^2$$

The phonon-like mode has a frequency ω_{TO}, not ω_{LO}. The choice ω_{LO} occurs because the frequency difference between ω_{LO} and ω_{TO} is caused by long-ranged Coulomb interactions. The electron gas screens these long-ranged Coulomb effects whenever $\omega_p^2 > \omega_{\mathrm{LO}}^2$. In this case, the electron gas can respond with a higher frequency than the phonons and can follow the motion of the phonons. Thus the electron gas will screen the phonons and prevent long-range electric fields, the frequency difference between ω_{LO} and ω_{TO} vanishes at long wavelengths, and the longitudinal phonon excitation has frequency ω_{TO}. Similarly, the phonons cannot follow the plasma oscillations of the electron gas, so the phonons do not contribute to the screening of the electron–electron interactions. Thus the plasma frequency in (6.3.9) has a dielectric constant ε_∞, not ε_0. The terms which contribute to ε_∞ are high-frequency interband transitions, which can certainly follow these low-frequency plasma oscillations.

The other limit of density has low values of n_0. When $\omega_p^2 \ll \omega_{\mathrm{LO}}^2$, the two roots of (6.3.14) are

$$\omega_+^2 \to \omega_{\mathrm{LO}}^2$$

$$\omega_-^2 \to \frac{\omega_p^2 \omega_{\mathrm{TO}}^2}{\omega_{\mathrm{TO}}^2} = \frac{4\pi e^2 n_0}{\varepsilon_0 m^*}$$

The Lyddane–Sachs–Teller relationship has been used in the last identity. Now the phonon-like mode has a frequency ω_{LO}. The electron gas cannot oscillate as fast as the phonons and does not screen the long-ranged coulomb fields of the phonons. Thus the phonons have a longitudinal frequency given by ω_{LO}. Of course, this must be the correct limit as $n_0 \to 0$. Since the phonons can now follow the oscillations of the electron gas, they do contribute to the screening of the plasma oscillations. Hence the plasma frequency now contains the screening factor ε_0 rather than ε_∞.

C. Electron Self-Energies

The motion of the electrons are affected by the other electrons and by the phonons. These interactions determine the self-energy of the electron. The self-energy will have contributions from electron–electron as well as electron–phonon interactions. These will combine into the screened interaction, described in Sec. 6.3A and summarized in (6.3.1) and (6.3.2). It is a rather cumbersome expression in its general form. Actual evaluation, to get analytical results, will require making some approximations at some point in the calculation. The only calculations which have been reported, and verified experimentally, are those for the limit that $\omega_p \gg \omega_{\mathrm{LO}}$. In this case it is permissable to separate the self-energy effects into an electron part and a screened phonon part. The electron–electron part may be evaluated by the procedures for the homogeneous electron gas given in Chapter 5. Nothing further needs to be mentioned about this term.

Our present objective is to calculate the electron self-energy from electron–phonon interactions. The first step is to calculate the one-phonon self-energy term. For the screened electron–phonon interaction in (6.3.11), this is

$$\Sigma(p, ip) = -\frac{1}{\beta}\sum_{i\omega_n}\int \frac{d^3q}{(2\pi)^3}\,\frac{M_q^{\,2}}{\varepsilon(q, i\omega)^2}\,\mathscr{D}(q, i\omega)\mathscr{G}^{(0)}(\mathbf{p}+\mathbf{q}, ip+i\omega) \tag{6.3.15}$$

There should also be a summation over the different phonon modes λ, which we shall omit, since our discussion will be confined to the polar coupling. Later we shall need to decide whether this one-phonon term is sufficient. Perhaps more terms will be needed, with higher numbers of phonons. It will turn out that the one-phonon term is adequate, since the coupling is weak. The screening weakens the effective coupling constant. Materials which have intermediate values of α when they are insulators become weak coupling when the electron gas screens. This will be shown later in great detail. For the moment, we only wish to make the point that the one-phonon self-energy is going to be sufficient. We also defer, for the moment, the question of whether we shall use the Rayleigh–Schrödinger or Brillouin–Wigner form of the self-energy. We shall designate the energy as ip, which will be ξ_p for the Rayleigh–Schrödinger form of perturbation theory and ω for the other.

The one-phonon self-energy (6.3.15) will be evaluated for the degenerate semiconductor. By degenerate, we mean a semiconductor which has been doped with a sufficient number of carriers that they become an electron gas, with a well-defined Fermi surface, in the limit of zero temperature. The degenerate limit will simplify the calculation in a number of ways, since we can use the homogeneous electron-gas results to describe the dielectric function $\varepsilon(q, i\omega)$.

In writing the self-energy in (6.3.15), we have used a free-particle electron Green's function $\mathscr{G}^{(0)}$ but a renormalized phonon propagator $\mathscr{D}$. Our procedure is to solve the phonon system assuming that the electrons are a free-particle gas. The self-energy terms calculated for the phonons have already been discussed. The phonons change their frequency at long wavelength, because of the screening of the electron gas. Thus we could use these newly found modes in calculating the properties of the electrons; the phonon Green's function in (6.3.12) could be used in (6.3.15). Although this would be a reasonable, and systematic, procedure, it makes the analytical calculation too difficult. Thus instead, we shall use an unperturbed phonon Green's function $\mathscr{D}^{(0)}$ in (6.3.15). For optical phonons, this assumes an Einstein model for the phonon frequencies:

$$\mathscr{D}^{(0)}(q, i\omega) = \mathscr{D}^{(0)}(i\omega) = \frac{-2\omega_0}{\omega_0^{\,2} + \omega^2}$$

The basic problem with (6.3.15) is that there is too much physics in it. It describes how the electron gas screens and modifies the phonons and their interactions with electrons. It does this sufficiently well that it is too cumbersome to evaluate without making some approximations. One may as well make these approximations at the beginning of the calculation, since this saves a lot of work. The first approximation we have already described, which is to replace the phonon Green's function $\mathscr{D}$ by the unperturbed propagator $\mathscr{D}^{(0)}$.

The next approximation is much more drastic. The dielectric function $\varepsilon(q, i\omega)$ will be approximated by its static value $\varepsilon(q)$. We shall ignore the frequency dependence of the dielectric function. This is reasonable as long as we are interested in frequencies much less than the plasma frequency ω_p. Thus the approximation $\varepsilon(q, \omega) \approx \varepsilon(q)$ is allowable whenever $\omega_{\text{LO}} \ll \omega_p$. In fact, the dielectric function may now be combined with the matrix element to give a screened interaction:

$$V_s(q) = \frac{M_q^{\,2}}{\varepsilon(q)^2}$$

The self-energy function in (6.3.15) now has the form

$$\Sigma(\mathbf{p}, ip) = -\frac{1}{\beta}\sum_{i\omega}\int\frac{d^3q}{(2\pi)^3}\, V_s(q)\mathscr{D}^{(0)}(i\omega)\mathscr{G}^{(0)}(\mathbf{p}+\mathbf{q}, ip+i\omega)$$

The summation over Matsubara frequencies now has the simple form which was evaluated in Sec. 3.5:

$$\Sigma(p) = \int\frac{d^3q}{(2\pi)^3}\, V_s(q)\left[\frac{1+N_0-n_F(\xi_{\mathbf{p}+\mathbf{q}})}{ip-\xi_{\mathbf{p}+\mathbf{q}}-\omega_0} + \frac{N_0+n_F(\xi_{\mathbf{p}+\mathbf{q}})}{ip-\xi_{\mathbf{p}+\mathbf{q}}+\omega_0}\right]$$

A much more complicated result is obtained when the more accurate forms of $\varepsilon(q, i\omega)$ and $\mathscr{D}(q, i\omega)$ are retained in doing the summations. The preceding simple result will be the basis for our discussion.

Our previous discussions of polaron theory always occurred in one-electron systems in which the Fermi occupation factors n_F could be set equal to zero. In the present discussion, they must be retained, and they play an important role in the nature of the final result. We shall only discuss the case for the zero temperature limit, where we can neglect the phonon thermal occupation factors $N_0 = (e^{\beta\omega_0}-1)^{-1}$. Thus we are led to consider the wave vector integrals, where we change integration variables to $\mathbf{p}' = \mathbf{p}+\mathbf{q}$:

$$\Sigma(p, ip) = \int\frac{d^3p'}{(2\pi)^3}\, V_s(\mathbf{p}-\mathbf{p}')\left[\frac{1-n_F(\xi_{p'})}{ip-\xi_{p'}-\omega_0} + \frac{n_F(\xi_{p'})}{ip-\xi_{p'}+\omega_0}\right] \tag{6.3.16}$$

A model form must be chosen for the dielectric function $\varepsilon(q)$. Of course, it would be super-accurate to use the Singwi–Sjölander form, but this is sufficiently complicated that we would need the computer to find any results. Instead, we shall chose the Thomas–Fermi form. Although less accurate, it permits an analytical answer. Then we shall argue that the answer is insensitive to the screening, so that the Thomas–Fermi approximation is probably adequate. However, this assertion has never been thoroughly checked by doing the self-energy calculation with the several alternate dielectric models of Chapter 5.

The retarded self-energy has both real and imaginary parts. The imaginary part will be done first, since it can be evaluated exactly—now that we have made enough approximations. It is given by $(ip \rightarrow \omega + i\delta)$

$$\text{Im}[\Sigma(\mathbf{p}, \omega)] = -\frac{\pi}{(2\pi)^3} \int p'^2\, dp' \{[1 - n_F(\xi_{p'})]\delta(\omega - \xi_{p'} - \omega_0) + n_F(\xi_{p'})\delta(\omega - \xi_{p'} + \omega_0)\} \int d\Omega' V_s(\mathbf{p} - \mathbf{p}')$$

The delta functions take out the dp' integrals. There remains just the angular integrals. Let $\nu = \cos\theta$ give the angle between $\mathbf{p}$ and $\mathbf{p}'$. For polar coupling, with Thomas–Fermi screening, the angular integrals are

$$\int d\Omega' V_s(\mathbf{p} - \mathbf{p}') = \frac{8\pi^2 \alpha \omega_0{}^{3/2}}{(2m)^{1/2}} \int_{-1}^{1} d\nu \frac{(\mathbf{p} - \mathbf{p}')^2}{[q_{\text{TF}}^2 + (\mathbf{p} - \mathbf{p}')^2]^2}$$

The integrals are straightforward and give a lengthy result (Mahan and Duke, 1966):

$$\text{Im}[\Sigma(p, \omega)] = -\frac{\alpha \omega_0{}^{3/2}}{4(\varepsilon_p)^{1/2}} [\theta(\omega - \omega_0) g_p(\omega - \omega_0) + \theta(-\omega - \omega_0) g_p(\omega + \omega_0)] \tag{6.3.17}$$

$$g_p(Z) = \varepsilon_s \left\{ \frac{1}{[(\varepsilon_p)^{1/2} + (Z + E_F)^{1/2}]^2 + \varepsilon_s} - \frac{1}{[(\varepsilon_p)^{1/2} - (Z + E_F)^{1/2}]^2 + \varepsilon_s} \right\} + \ln \frac{\varepsilon_s + [(\varepsilon_p)^{1/2} + (Z + E_F)^{1/2}]^2}{\varepsilon_s + [(\varepsilon_p)^{1/2} - (Z + E_F)^{1/2}]^2} \tag{6.3.18}$$

$$\varepsilon_s = \frac{\hbar^2 q_{\text{TF}}^2}{2m}$$

The imaginary self-energy is shown in Fig. 6.12. It is plotted as a function of the energy ω relative to the Fermi energy for some value of p. The striking feature in the result is that $\text{Im}(\Sigma)$ is zero for energies within ω_0 of the Fermi surface. Those electrons outside of this range can lose energy by emitting or absorbing phonons. But those electrons within ω_0 of the Fermi energy cannot and have a zero $\text{Im}(\Sigma)$.

The gap in the allowed values of $\text{Im}(\Sigma)$ can be understood in a simple way. The excitations with energy $\omega > \omega_0$ above the Fermi sea are electrons, or electron-like quasiparticles. They can decay by emitting an LO phonon ω_0. If they initially have an energy ω, then after emitting the optical phonon, they have a final energy $\omega - \omega_0$. The electron states up to the energy $E_F(\omega = 0)$ are all occupied, so that the final state must have an energy larger than this. Otherwise the transition cannot occur since the final states are all occupied. Thus we conclude that $\omega - \omega_0 > 0$. This shows that phonon emission by electrons can occur only for initial energies at least ω_0 above the Fermi energy. These processes are illustrated qualitatively

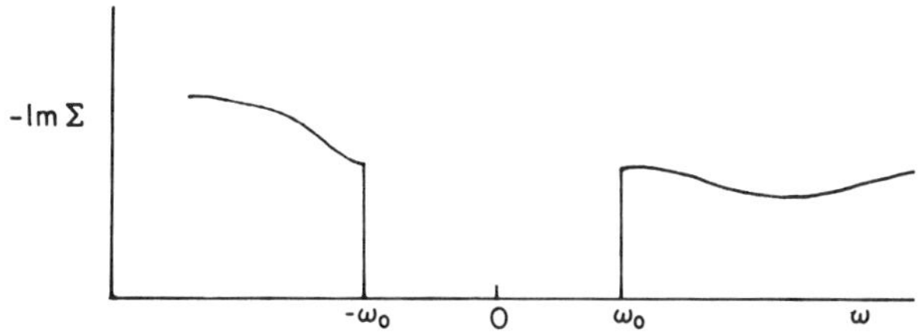

FIGURE 6.12. The imaginary part of the electron self-energy from scattering by optical phonons. The energy ω is measured from the Fermi surface.

in Fig. 6.13(a). The allowed transition has enough initial energy that the final state is above the occupied levels. The forbidden transition would go to a state already occupied. This explains why $\text{Im}(\Sigma)$ is zero for $0 \leq \omega \leq \omega_0$.

The behavior for $\omega < 0$ may be understood by realizing that the excitations are not electrons. Instead, the excitations below the Fermi energy are states where an electron is missing. These are called *Holes* in a metal. Unfortunately, the word *hole* in a semiconductor usually means an excitation of the valence band. Our word *Hole*, with a capital "H," is an excitation of the conduction band. It has an electron missing from the otherwise filled Fermi sea in the conduction band. These Holes decay by having an electron jump into this empty state, which is illustrated in Fig. 6.13(b). The decay of a Hole is allowed whenever the Hole can be filled by an electron already present. Thus if the Hole has initial energy $\omega < -\omega_0$, then an electron in the Fermi sea can emit an optical phonon

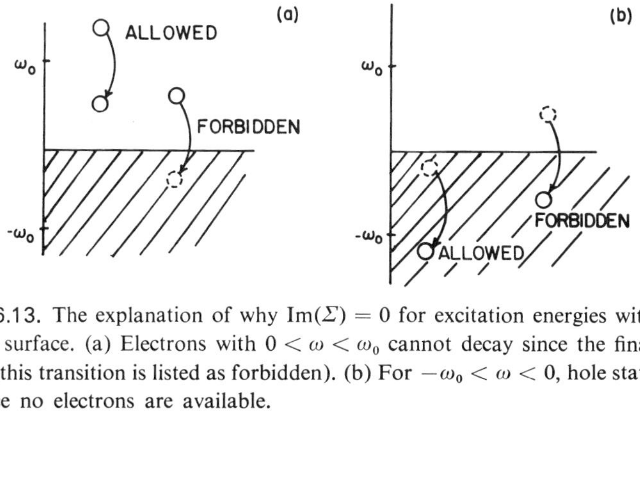

FIGURE 6.13. The explanation of why $\text{Im}(\Sigma) = 0$ for excitation energies within ω_0 of the Fermi surface. (a) Electrons with $0 < \omega < \omega_0$ cannot decay since the final state is occupied (this transition is listed as forbidden). (b) For $-\omega_0 < \omega < 0$, hole states cannot decay since no electrons are available.

and jump into this Hole state. This process is shown as the "allowed" transition in Fig. 6.13(b). The process is equivalent to the Hole jumping to lower energy by an amount ω_0 since the zero energy for a Hole is the Fermi energy. The "forbidden" process of Hole decay in Fig. 6.13 cannot occur because it requires the initial electron to be above the Fermi sea. This is not possible at zero temperature if the system is in equilibrium. The forbidden hole decay explains why $\mathrm{Im}(\Sigma)$ is zero for $-\omega_0 < \omega < 0$. Thus we have explained the energy gap in $\mathrm{Im}(\Sigma)$ for $-\omega_0 \leq \omega \leq \omega_0$. Of course, this gap exists only for the self-energy contribution from optical phonons. The electron will have finite values of $\mathrm{Im}(\Sigma)$ in this energy region arising from scattering by acoustical phonons and from electron–electron interactions.

The self-energies we are calculating describe the excitation spectrum of the interacting system. They are for electron-like quasiparticles for energies above the Fermi energy and Holes below the Fermi energy.

The function $g_p(z)$ in (6.3.18) is a smooth function of both ε_p and z. This lack of structure is due to the screening of the electron gas, which eliminates all sharp features. Since $g_p(z)$ is sufficiently dull, it is a good approximation to treat it as a constant. A simple place to evaluate this constant is at $z = 0$ and $p = 0$. This limit defines a coupling constant we call f_0:

$$f_0 = \lim_{\substack{p\to 0\\ z\to 0}} \frac{\alpha\omega_0^{3/2}}{4(\varepsilon_p)^{1/2}}\, g_p(z) = \frac{\alpha\omega_0^{3/2}E_F^{3/2}}{(E_F+\varepsilon_s)^2}$$
$$\mathrm{Im}[\Sigma(\omega)] = -f_0[\theta(\omega-\omega_0)+\theta(-\omega-\omega_0)] \tag{6.3.19}$$

The real part of the self-energy may now be obtained by a simple argument. The retarded self-energy function $\Sigma(p, \omega)$ is causal and hence obeys a Kramers–Kronig relation (see Sec. 5.7). Thus the real part may be derived from the imaginary part by

$$\mathrm{Re}[\Sigma(\omega)] = \pi \int_{-\infty}^{\infty} d\omega' \, \frac{\mathrm{Im}[\Sigma(\omega')]}{\omega'-\omega}$$
$$= \frac{f_0}{\pi}\left(\int_{-\infty}^{-\omega_0} \frac{d\omega'}{\omega'-\omega} + \int_{\omega_0}^{\infty} \frac{d\omega'}{\omega'-\omega}\right)$$

For the simple function in (6.3.19), the integrals are just logarithms:

$$\mathrm{Re}[\Sigma(\omega)] = -\frac{f_0}{\pi}\ln\left|\frac{\omega+\omega_0}{\omega-\omega_0}\right| \tag{6.3.20}$$

The result (6.3.20) was first obtained by Englesberg and Schrieffer (1963). They were actually discussing metals with optical phonons, but the result also applies to doped semiconductors. The total effect of the screened polar interaction is contained in the coupling constant f_0. The same type of answer is obtained for any electron gas with an Einstein phonon, but the detailed definition of f_0 is altered. This is a consequence of electron screening, which makes the details of the coupling constant not terribly important.

There have been several attempts to verify this theory, in degenerate semiconductors, through electron tunneling experiments. Usually the tunneling is done through the Schottky barrier at a metal–semiconductor interface. The metal effects are negligible, and all structure in the electron tunneling conductance can be attributed to the semiconductor—unless the metal is superconducting, and then the Bardeen–Cooper–Schrieffer (BCS) gap is apparent. But for normal metals, even if ferromagnetic, the metal–semiconductor junction is determined by the properties of the semiconductor (Conley and Mahan, 1967).

The most direct confirmation of the theory was obtained by Tsui (1974), who measured the constant f_0 in InAs as the magnitude of the step increase in the scattering rate of the electrons. He was measuring metal–semiconductor junctions in a magnetic field. The differencial conductance d^2/Id^2V showed oscillations, similar to the deHaas–van Alphen (dHvA) oscillations in a metal. Tsui's data is shown in Fig. 6.14. The oscillations change their amplitude at voltages greater than 30 meV, which is the energy of the LO phonon in InAs. The tunneling voltage is the difference of the two chemical potentials and is the energy in the semiconductor which the electrons have after they tunnel from the metal.

In the theory of the dHvA effect, the amplitude is given by $A \simeq \exp(-H_0/H)$, where H_0 is a constant related to the Dingle relaxation time:

$$\tau_0 = \frac{\pi m^* e}{e H_0}$$

The ratio of the amplitudes above and below the phonon onset gives the increased scattering rate, which gives $\mathrm{Im}(\Sigma)$:

$$\tau_0 = -\frac{\hbar}{2\,\mathrm{Im}(\Sigma)} = +\frac{\hbar}{2 f_0} = \frac{(E_F + E_s)^2}{2\alpha(\omega_0 + E_F)^{3/2}}$$

Using parameters for InAs, Tsui used this formula to predict a value for

τ_0 and compared it with his experimental number:

$$\text{Theory:} \qquad \tau_0 = 5.3 \times 10^{-13}\ \text{sec}$$

$$\text{Experiment:} \qquad \tau_0 = (5.1 \pm 0.3) \times 10^{-13}\ \text{sec}$$

This is excellent agreement between theory and experiment.

In many respects, the degenerate semiconductor is the ideal environment in which to study the electron gas and polaron effects in an electron gas. The electron gas can be made to have an effective high density at which the theory is supposed to be valid. All parameters of the theory, such as α or m^*, can be determined by independent experiments in the

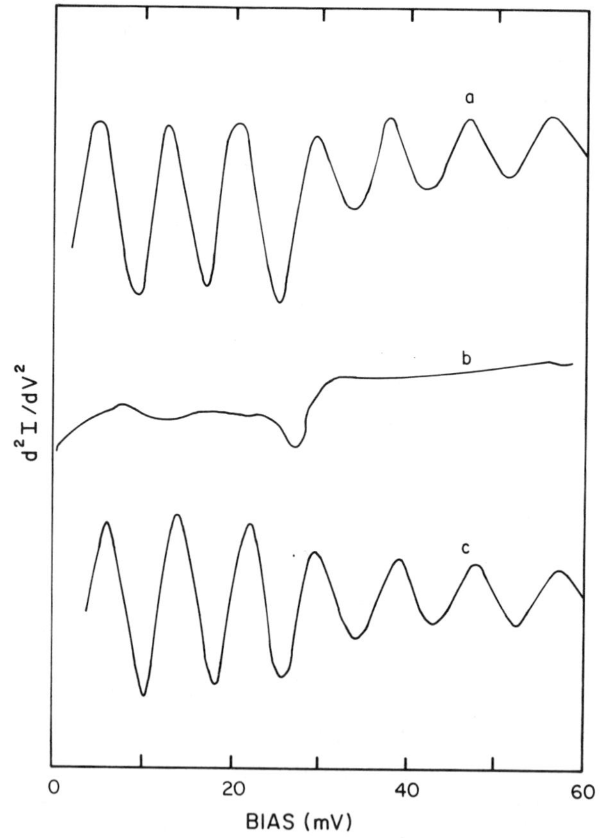

FIGURE 6.14. d^2I/dV^2 vs. V data from an InAs–oxide–Pb junction. The InAs sample has $n_0 = 5.5 \times 10^{17}\ \text{cm}^{-3}$ and tunneling data taken at $T = 4.2$ K. (a) Taken with $H =$ 22 kG, (b) taken with $H = 2$ kG, and (c) the difference between (a) and (b). *Source*: Tsui (1974) (used with permission).

insulating semiconductor. Thus one can do theory with no adjustable parameters. The experimentalists have systems which are relatively easy to measure and which have an electron density which may be varied continuously.

6.4. METALS

Electron–phonon interactions in metals have been studied intensely for several reasons. The phonons in metals are significantly affected by the electron gas, so any study of the phonon system requires an understanding of the electron–phonon interaction. The electrons near the Fermi surface have their motions influenced by the phonons, so that many experiments which measure these electrons—such as transport or cyclotron resonance—are affected by electron–phonon interactions. The interaction also causes superconductivity in many metals. Thus there are many experimental methods which have been applied to this topic, and ample reasons to do so.

There is an initial problem facing any theoretical treatment. Since the two systems of electrons and phonons affect each other very much, it is not obvious where to begin the many-body calculations. Does one first solve the phonons assuming the electrons have no phonon influences, or solve for the electron properties using the free propagators for phonons? That is, does one first solve for the phonon or the electron self-energies? Neither approach may work if the two systems are strongly coupled. Actually, there is a right way to begin, as discussed by Migdal (1958). The justification of the procedure requires, to some extent, knowing the solution to the problem. One should start by first discussing the phonons using electron states which do not contain phonon effects. Although the phonons affect the electrons, they alter electron states only within a Debye energy of the Fermi surface. These are a small fraction of the electrons in the system. However, when we solve for the electrons' influence upon the phonons, we shall average over all the occupied states of the electron gas. This average is influenced to a negligible fashion by the few electrons at the Fermi surface, which themselves are influenced by phonons. Thus a good answer is obtained by first solving for the phonon states using electron states which have no phonons included. Then one can solve for the electron states using these renormalized phonon states. Thus we shall follow the standard procedure and first solve for the properties of the phonons. We shall employ an electron Green's function which does not have a self-energy contribution from phonons. But it does contain the effects of electron–electron interactions, which are very important for obtaining the correct physical description.

A. Phonons in Metals

For simplicity we shall discuss metals, such as sodium or aluminum, which have only one atom per unit cell. The many-body system is a collection of ions and electrons. It is charge neutral, so the ions and electrons have the same average charge density $|e|n_0 = eZn_i$, where n_0 and n_i are the particle densities of electrons and ions. The ions of valence Z will be considered as rigid objects. The excitations of the inner core electrons can be neglected, since these energetic events are not induced by the motion of the ions. These core excitations will contribute a small amount of dielectric screening, which can be included as a phenomenological high-frequency dielectric constant ε_{i}. The Hamiltonian was discussed earlier in Sec. 1.3:

$$H = \sum_i \frac{p_i^2}{2m} + \sum_j \frac{P_j^2}{2M_j} + \frac{1}{2} e^2 \sum_{ij} \frac{e^2}{|\mathbf{r}_i - \mathbf{r}_j|} + \sum_{i\alpha} V_{\mathrm{ei}}(\mathbf{r}_i - \mathbf{R}_\alpha) + \sum_{\alpha\beta} V_{\mathrm{ii}}(\mathbf{R}_\alpha - \mathbf{R}_\beta) \tag{6.4.1}$$

The potential $V_{\mathrm{ei}}(\mathbf{r}_i - \mathbf{R}_\alpha)$ is between an electron at $\mathbf{r}_i$ and an ion at $\mathbf{R}_\alpha$. It was called V_A in Sec. 1.3. Similarly, the potential $V_{\mathrm{ii}}(\mathbf{R}_\alpha - \mathbf{R}_\beta)$ is between two ions. Both of these potentials are unscreened, so they behave at large distances as

$$\begin{aligned} \lim_{r\to\infty} V_{\mathrm{ei}}(r) &= -\frac{Ze^2}{\varepsilon_{\mathrm{i}} r} \\ \lim_{r\to\infty} V_{\mathrm{ii}}(R) &= \frac{Z^2e^2}{\varepsilon_{\mathrm{i}} R} \end{aligned} \tag{6.4.2}$$

In a free-electron metal such as sodium or aluminum, the ions are sufficiently far apart that the Coulomb form (6.4.2) of V_{ii} is probably valid for all ion pairs. However, the electron–ion interaction V_{ei} must always be treated realistically at small distances. The customary procedure is to use a pseudopotential [see (5.3.3) or Harrison, 1966]. The electron–electron interactions are also included in the Hamiltonian. They will cause the potentials to be screened.

Our first discussion will be at an introductory level, approximately the same as Schrieffer (1964). In this model the ions are point charges, and the electron gas is jellium. Later we shall discuss real metal effects at a more advanced level. In the introductory treatment we shall try to explain the physics and avoid the bewildering notation of the correct treatment.

The first step is to solve the Hamiltonian of the electrons in the absence of phonons. The ions are fixed in their equilibrium positions $\mathbf{R}_\alpha^{(0)}$. Then (6.4.1) becomes the Hamiltonian of electrons in a periodic, neutral system:

$$H_{0e} = \sum_i \frac{p_i^2}{2m} + \sum_{i\alpha} V_{ei}(\mathbf{r}_i - \mathbf{R}_\alpha^{(0)}) + \frac{1}{2}\sum_{ij} \frac{e^2}{|\mathbf{r}_i - \mathbf{r}_j|} + \frac{1}{2}\sum_{\alpha\beta} V_{ii}(\mathbf{R}_\alpha^{(0)} - \mathbf{R}_\beta^{(0)})$$

The solution of this problem is quite formidable. The electrons states are Bloch waves in the periodic potential. They must be calculated with the full effects of correlation and exchange in the metallic environment, which has only been done approximately. The problem is quite difficult, but it is not the problem at hand. For the discussion of electron–phonon effects, we shall assume that the electron part of the Hamiltonian has been solved. Furthermore, in this introductory treatment, the electron states of our quasi-jellium model of a solid are approximated as plane waves.

The Hamiltonian of the phonons is first solved without reference to the electrons. In the harmonic approximation (see Sec. 1.1) the ions are assumed to have small displacements $\mathbf{Q}_\alpha$ about their equilibrium positions $\mathbf{R}_\alpha^{(0)}$ so that the Hamiltonian can be written as

$$\mathbf{R}_\alpha = \mathbf{R}_\alpha^{(0)} + \mathbf{Q}_\alpha$$

$$H = H_{0e} + H_{0p} + H_{ep}$$

$$H_{0p} = \sum_\alpha \frac{P_\alpha^2}{2M_\alpha} + \frac{1}{4}\sum_{\alpha\beta} (\mathbf{Q}_\alpha - \mathbf{Q}_\beta)_\mu (\mathbf{Q}_\alpha - \mathbf{Q}_\beta)_\nu \Phi_{\mu\nu}(\mathbf{R}_\alpha^{(0)} - \mathbf{R}_\beta^{(0)}) \tag{6.4.3}$$

$$H_{ep} = \sum_{j\alpha} \mathbf{Q}_\alpha \cdot \nabla V_{ei}(\mathbf{r}_j - \mathbf{R}_\alpha^{(0)}) \tag{6.4.4}$$

$$\Phi_{\mu\nu}(\mathbf{R}) = \nabla_\mu \nabla_\nu V_{ii}(\mathbf{R})$$

The term H_{0p} is called the *bare-phonon* Hamiltonian. These are the phonons calculated by ignoring the electron–electron and electron–phonon interactions. The ion motions are calculated using only the direct ion–ion interaction potential V_{ii}. The bare-phonon Hamiltonian, once solved, describes a set of eigenfrequencies and normal modes which are quite unrealistic. The poor results are obtained because the phonon Hamiltonian is for a system which is not charge neutral. When the ions vibrate, the bare-phonon

Hamiltonian H_{0p} treats the electron gas as a rigid background, which does not follow the ion motion. Hence the ion vibrations cause long-range Coulomb fields. These are just plasma oscillations, of the ions, so the long-wavelength phonons will have the frequency

$$\omega_{\mathrm{ip}}^2 = \frac{4\pi e^2 Z^2 n_i}{M\varepsilon_{\mathrm{i}}} \tag{6.4.5}$$

In usual solids, and metals, the long-wavelength oscillations of the atoms or ions are described by acoustical phonons, which have the dispersion relation $\omega(q) = c_s q$ proportional to the wave vector, where c_s is the speed of sound. In solids with more than one atom per unit cell there may also be optical phonons, which have a constant frequency at long wavelength. However, the above plasma modes of H_{0p} are not from optical modes of vibration, since we explicitly chose to solve for solids with only one atom per cell and no optical modes. In fact, the bare-phonon Hamiltonian H_{0p} is sufficiently unrealistic that it does not have acoustical mode solutions, which we know are necessary for real metals. The ion plasma modes are the correct long-wavelength normal modes of H_{0p} but do not realistically approximate the actual modes in a metal.

In normal metals, the electrons and ions respond differently because of the great difference in their masses. When the ions vibrate, the electrons follow the ion motion. Thus there are no long-range Coulomb fields and no ion plasma waves. When the electrons oscillate at their natural frequency ω_p, the ions are too heavy to follow, so the electron motion is not screened. The electrons can freely oscillate at their plasma frequency, while the ions cannot. Our bare-phonon Hamiltonian gives the wrong answer because we did not let the electrons follow the ion motion. Permitting the electrons to follow the ion motion is another set of words to describe the screening by the electrons of the motion of the ions.

The method of solving (6.4.3) for the bare phonons was described in Sec. 1.1. The ion displacements $\mathbf{Q}_\alpha$ and conjugate momenta $\mathbf{P}_\alpha$ are expanded in a set of normal modes,

$$\mathbf{Q}_\alpha = \frac{1}{(N_{\mathrm{i}})^{1/2}} \sum_{\mathbf{k}} \mathbf{Q}_{\mathbf{k}} e^{i\mathbf{k}\cdot\mathbf{R}_\alpha^{(0)}}$$

$$\mathbf{P}_\alpha = \frac{1}{(N_{\mathrm{i}})^{1/2}} \sum_{\mathbf{k}} \mathbf{P}_{\mathbf{k}} e^{-i\mathbf{k}\cdot\mathbf{R}_\alpha^{(0)}}$$

$$H_{0p} = \sum_{\mathbf{k}} \left[\frac{1}{2m} \mathbf{P}_{\mathbf{k}} \cdot \mathbf{P}_{-\mathbf{k}} + \frac{1}{2} Q_{\mathbf{k}\mu} Q_{-\mathbf{k}\nu} \phi_{\mu\nu}(\mathbf{k}) \right]$$

where the interaction term is

$$\phi_{\mu\nu}(\mathbf{k}) = \frac{1}{2} \sum_{\alpha,\beta} (e^{i\mathbf{k}\cdot\mathbf{R}_\alpha{}^{(0)}} - e^{-i\mathbf{k}\cdot\mathbf{R}_\beta{}^{(0)}})(e^{-i\mathbf{k}\cdot\mathbf{R}_\alpha{}^{(0)}} - e^{i\mathbf{k}\cdot\mathbf{R}_\beta{}^{(0)}})$$
$$\times \Phi_{\mu\nu}(\mathbf{R}_\alpha{}^{(0)} - \mathbf{R}_\beta{}^{(0)})$$
$$= -\frac{1}{\nu_0} \sum_{\mathbf{G}} [\bar{\Phi}_{\mu\nu}(\mathbf{G} + \mathbf{k}) - \bar{\Phi}_{\mu\nu}(\mathbf{G})] \tag{6.4.6}$$

$$\bar{\Phi}_{\mu\nu}(\mathbf{q}) = \int d^3R \Phi_{\mu\nu}(\mathbf{R}) e^{i\mathbf{q}\cdot\mathbf{R}}$$

The summation $\mathbf{G}$ is over the reciprocal lattice vectors of the solid, ν_0 is the volume of the unit cell of the crystal $\nu_0 = \nu/N_i = n_i^{-1}$, and N_i is the number of ions in the solid $ZN_i = N$. If the ions were point charges, these terms would be

$$V_{\text{ii}}(R) = \frac{Z^2e^2}{\varepsilon_{\text{i}} R}$$

$$\Phi_{\mu\nu} = -\frac{Z^2e^2}{\varepsilon_{\text{i}}} \left(\frac{\delta_{\mu\nu}}{R^3} - \frac{3R_\mu R_\nu}{R^5} \right)$$

$$\bar{\Phi}_{\mu\nu}(q) = \frac{-4\pi Z^2 e^2 q_\mu q_\nu}{\varepsilon_{\text{i}} q^2}$$

The summation over $\mathbf{G}$ converges slowly, which may be helped by using Ewald techniques (see Joshi and Rajagopal, 1968). The normal modes of the bare-phonon system are found as the eigenfrequencies $\Omega_{\mathbf{k}\lambda}$ and eigenstates $\hat{\xi}_{k\lambda}$ of the equation

$$\det[M\Omega_{\mathbf{k}\lambda}^2 \delta_{\mu\nu} - \phi_{\mu\nu}(\mathbf{k})] = 0 \tag{6.4.7}$$

These frequencies and eigenstates are used to define a set of creation and destruction operators through the expansion (1.1.16):

$$\mathbf{Q}_{\mathbf{k}} = \sum_\lambda \left(\frac{\hbar}{2\varrho v \Omega_{\mathbf{k}\lambda}} \right)^{1/2} \hat{\xi}_{\mathbf{k}\lambda} (a_{\mathbf{k}\lambda} + a^\dagger_{-\mathbf{k}\lambda})$$
$$H_{0\text{p}} = \sum_{\mathbf{k}\lambda} \Omega_{\mathbf{k}\lambda} (a^\dagger_{\mathbf{k}\lambda} a_{\mathbf{k}\lambda} + \tfrac{1}{2}) \tag{1.1.16}$$

We have completed a formal description of the solution of the bare-phonon system. The modes which are obtained are conceptually useful, but we do not need to calculate them. We use the set of eigenfrequencies and eigenstates only as the basis for solving the electron–ion interaction H_{ep}. When we now solve for this interacting system, we shall eventually obtain a new equation for the phonon modes. These new equations contain electron screening and provide a realistic description of the atomic motions. These are worth solving numerically. The eigenstates of the bare-phonon system are useful only as a concept upon which to build our solutions in terms of Green's functions. The interaction term between the electron and phonons is written as

$$H_{\text{ep}} = \frac{1}{\nu^{1/2}} \sum_{\mathbf{q}\lambda,\mathbf{G}} M_\lambda(\mathbf{q}+\mathbf{G}) e^{i(\mathbf{q}+\mathbf{G})\cdot r}(a_{\mathbf{q}\lambda} + a^\dagger_{-\mathbf{q}\lambda})$$
$$M_\lambda(\mathbf{q}+\mathbf{G}) = \left(\frac{\hbar}{2\varrho\Omega_{\mathbf{q}\lambda}}\right)^{1/2} \hat{\xi}_{\mathbf{q}\lambda}\cdot(\mathbf{G}+\mathbf{q}) V_{\text{ei}}(\mathbf{G}+\mathbf{q}) \tag{6.4.8}$$

The phonon states in this basis are the bare phonons. It is an unrealistic basis but serves only as the starting point for the Green's function calculation. The electron states are put into the second quantized notation, so that the effective Hamiltonian has the form

$$H = \sum_{\mathbf{k}} \xi_k c_{\mathbf{k}}^\dagger c_{\mathbf{k}} + \frac{1}{2}\sum_{\substack{\mathbf{qkp}\\ \sigma\sigma'}} v_q c^\dagger_{\mathbf{k}+\mathbf{q}\sigma} c^\dagger_{\mathbf{p}-\mathbf{q}\sigma'} c_{\mathbf{p}\sigma'} c_{\mathbf{k}\sigma} + \sum_{\mathbf{q}\lambda} \Omega_{q\lambda} a^\dagger_{\mathbf{q}\lambda} a_{\mathbf{q}\lambda}$$
$$+ \sum_{n\mathbf{q}\lambda\mathbf{k}\sigma} \frac{M_\lambda(q)}{\nu^{1/2}} c^\dagger_{\mathbf{k}+\mathbf{q}\sigma} c_{\mathbf{k}\sigma}(a_{\mathbf{q}\lambda} + a^\dagger_{-\mathbf{q}\lambda})$$

The solution to this Hamiltonian was discussed in Sec. 6.3. The present problem corresponds to the case where the electron plasma frequency ω_p is very much larger than the phonon frequencies. In this case the effective interaction between two electrons may be written as (6.3.11) with a screened Coulomb interaction and a screened phonon interaction:

$$V_{\text{eff}}(\mathbf{q}, i\omega) = \frac{v_q}{\varepsilon_{\text{i}}\varepsilon(\mathbf{q}, i\omega)} + \sum_\lambda \frac{M_\lambda^2(\mathbf{q})}{\varepsilon(\mathbf{q}, i\omega)^2} \mathscr{D}_\lambda(\mathbf{q}, i\omega)$$
$$\varepsilon(\mathbf{q}, i\omega) = 1 - \frac{v_q}{\varepsilon_{\text{i}}} P(\mathbf{q}, i\omega)$$
$$\mathscr{D}_\lambda(\mathbf{q}, i\omega) = \frac{\mathscr{D}_\lambda^{(0)}}{1 - M_\lambda^2 \mathscr{D}_\lambda^{(0)} P(\mathbf{q}, i\omega)/\varepsilon(q, i\omega)}$$

The quantity $\mathscr{D}_\lambda$ is the phonon Green's function. Its denominator contains the factor $M_\lambda^2 \mathscr{D}^{(0)} P/\varepsilon$. The term $1/\varepsilon(q, i\omega)$ comes from the screening of the electron gas. It describes how the electrons can follow the motion of the ions and hence screen out the long-range Coulomb forces. The screening eliminates the ion–plasma solutions and makes the modes acoustical. To show this, first rewrite the phonon Green's function as $(i\omega \to \omega + i\delta)$

$$D_{\text{ret}}(q, \omega) = \frac{2\Omega_{q\lambda}}{\omega^2 - \Omega_{q\lambda}^2 - 2\Omega_{q\lambda}(M_\lambda{}^2 P/\varepsilon)}$$

At long wavelength, we have the following limits for the factors which enter into the denominator of this Green's function:

$$\lim_{q\to 0}\begin{cases} \Omega_{q\lambda} \to \omega_{\text{ip}} \\ M_\lambda \to q\left(\dfrac{\hbar}{2\varrho\omega_{\text{ip}}}\right)^{1/2} \dfrac{4\pi e^2 Z}{q^2 \varepsilon_{\text{i}} \nu_0} \\ 2\Omega_{q\lambda} M_\lambda{}^2 \to \dfrac{v_q}{\varepsilon_{\text{i}}} \dfrac{4\pi e^2 Z^2}{\varrho \nu_0{}^2 \varepsilon_{\text{i}}} \equiv \dfrac{v_q}{\varepsilon_{\text{i}}} \omega_{\text{ip}}^2 \\ \omega^2 - \Omega_{q\lambda}^2 - 2\Omega_{q\lambda} \dfrac{M_\lambda{}^2 P}{\varepsilon} \to \omega^2 - \omega_{\text{ip}}^2 - \omega_{\text{ip}}^2\left(\dfrac{1}{\varepsilon} - 1\right) = \omega^2 - \dfrac{\omega_{\text{ip}}^2}{\varepsilon} \end{cases}$$

The last term on the right shows that the phonon energy denominator becomes the simple expression $\omega^2 - \omega_{\text{ip}}^2/\varepsilon$. For the small frequencies of phonons, the static dielectric function $\varepsilon(q, 0)$ may be used. In the limit $q \to 0$ it is given by the compressibility sum rule (5.6.7). The zero of the phonar energy denominator defines the new phonon modes $\omega_\lambda(q)^2 = \omega_{\text{ip}}^2/\varepsilon(q)$:

$$\lim_{q\to 0} \omega_\lambda(q) = \frac{q}{q_{\text{TF}}}\, \omega_{\text{ip}}\left(\frac{K_f}{K}\right)^{1/2}$$

Thus the long-wavelength excitation of the coupled system of ions and electrons are now acoustical phonons. The speed of sound is given by $c_s = \omega_{\text{ip}}(K/K_f)^{1/2} q_{\text{TF}}^{-1}$, where the factors are ω_{ip}, ion–plasma frequency; q_{TF}, Thomas–Fermi screening length; and K/K_f, ratio of compressibility to that of the free-electron gas. The formula for c_s is similar to the result first obtained by Bohm and Staver (1951), who found that $c_s = \omega_{\text{ip}}/q_{\text{TF}} = v_F(Zm/3M)^{1/3}$. These ordinary sound waves are obtained because the electron gas follows the vibrations of the ions. When the ion of charge Ze moves, an amount of screening charge nearly $-eZ$ follows it, so that the vibrating entity is nearly neutral. No long-range Coulomb fields are present, and one does not obtain waves at the plasma frequency of the ions. The screening charge has contributions from all electrons in the Fermi sea.

The situations in metals is quite analogous to that of semiconductors.

There we found that the electron–electron interactions caused the LO phonon modes to be given by

$$\omega^2 = \omega_{\rm TO}^2 + \frac{\omega_{\rm LO}^2 - \omega_{\rm TO}^2}{\varepsilon(q)}$$

To apply this formula to metals, set $\omega_{\rm LO}^2 = \omega_{\rm ip}^2$ and $\omega_{\rm TO} = 0$. The TO frequency is zero in metals since there is no restoring force for transverse oscillations. These steps give the same formula $\omega^2 = \omega_{\rm ip}^2/\varepsilon(q)$.

The phonon Green's function can be written approximately as

$$D(q, \omega) = \frac{2\Omega_{q\lambda}}{\omega^2 - \omega_\lambda(q)^2}$$

This approximate form is valid only when the damping of the phonon states can be ignored. The phonon eigenfrequencies ω_λ are the ones found with the inclusion of electron–electron interactions and represent the actual phonon modes measured in the solid. The electron–phonon matrix element $M_\lambda{}^2$ in (6.4.8) contains the factor $\Omega_{q\lambda}^{-1}$. Thus in the screened phonon interaction, the unphysical frequency $\Omega_{q\lambda}$ completely cancels from the product $M_\lambda{}^2 D(q, \omega)$, and the final formula contains no reference to $\Omega_{q\lambda}$ whatsoever.

The preceding discussion was much too simple to find phonon modes in actual metals. Our arguments applied only to longitudinal modes of vibration and only at long wavelength. A better description is needed to calculate transverse modes. Dyson's equation for phonons is really a matrix equation, so that the phonon energy denominator is not just a scalar quantity but is derived from the determinant of a matrix.

The reader is referred to Joshi and Rajagopal (1968) for a thorough treatment of phonons in metals. They suggest one method of using the dielectric function for the homogeneous electron gas to calculate realistic phonon modes for metals. If the Fourier transform of $V_{\rm ii}(r)$ and $V_{\rm ei}(r)$ are $\bar{V}_{\rm ii}(q)$ and $\bar{V}_{\rm ei}(q)$, the screened interaction $\bar{V}_{\rm si}(q)$ potential between ions is given by the effective potential:

$$\bar{V}_{\rm si}(q) = \bar{V}_{\rm ii}(q) + \frac{\bar{V}_{\rm ei}(q)^2 P(q)}{\varepsilon(q)} \tag{6.4.9}$$

The first term is from the direct interaction between ions. The second term is from the interaction mediated by the electrons. The latter term arises when the ions polarize the electrons, and this polarization acts upon other ions. The motivation for choosing this interaction comes from the form of the denominator of the phonon's Green's function, whose poles predict that the phonon frequencies are given by

$$\omega^2 = \Omega_{q\lambda}^2 + 2\Omega_{q\lambda} \frac{M_\lambda{}^2 P(q)}{\varepsilon(q)}$$

The first term on the right, $\Omega_{q\lambda}^2$, is roughly given by $\phi_{\mu\nu}(q)/M \sim q_\mu q_\nu \times \bar{V}_{\text{ii}}(q)/\varrho$. The second term on the right is

$$2\Omega_{q\lambda} M_\lambda{}^2 \frac{P}{\varepsilon} = \frac{(\hat{\varepsilon} \cdot \mathbf{q})^2}{\varrho} \bar{V}_{\text{ei}}^2 \frac{P(q)}{\varepsilon(q)}$$

Thus the two terms combined have the form

$$\omega^2 \sim \frac{q^2}{M} \left[\bar{V}_{\text{ii}}(q) + \bar{V}_{\text{ei}}(q)^2 \frac{P(q)}{\varepsilon(q)} \right]$$

The term in brackets is just the effective interaction in (6.4.9). Our Green's function derivation has indicated this form to be the appropriate one for the potential between two ions. The effective interaction will be used to solve for the phonon modes in the metal. These modes are found by solving a determinantal equation.

For a system of point charges—metallic hydrogen—we would have that all potentials are pure Coulomb, so the expression can be simplified ($Z = 1$):

$$\text{Point ions} \left\{ \begin{array}{l} \bar{V}_{\text{ii}} = \bar{V}_{\text{ei}} = v_q \\ \bar{V}_{\text{si}} = \dfrac{v_q}{\varepsilon(q)} \end{array} \right.$$

For real metals it is *not* a good approximation to express the screened ion potential as $\bar{V}_{\text{ii}}/\varepsilon(q)$. It would be a good approximation if we had the identity $\bar{V}_{\text{ei}}^2 = v_q \bar{V}_{\text{ii}}/\varepsilon_{\text{i}}$, but this is not usually valid. The electron–ion potential $\bar{V}_{\text{ei}}$ is quite unlike a pure Coulomb potential at large wave vector. Thus in general, it is preferable to use the expression (6.4.9).

The phonon modes $\omega_\lambda(q)$ may be obtained in the same fashion as described in (6.4.6) and (6.4.7). A dynamical matrix is constructed from the screened ion potential,

$$\phi_{\mu\nu}(\mathbf{k}) = \frac{1}{\nu_0} \sum_{\mathbf{G}} [(\mathbf{k} + \mathbf{G})_\mu (\mathbf{k} + \mathbf{G})_\nu \bar{V}_{\text{si}}(\mathbf{k} + \mathbf{G}) - G_\mu G_\nu \bar{V}_{\text{si}}(\mathbf{G})]$$

which is used in the determinantal equation (6.4.7). That was the procedure followed by Woll and Kohn (1962) to calculate phonon modes in aluminum. Other calculations are described by Ziman (1960), Sham and Ziman (1963), and Harrison (1966). It is a very active field. Examples of recent calculations are Na and K by Shukla and Taylor (1976); K, Sn, and Sb by Kay and Reissland (1976); In by Garrett and Swihart (1976); and Na, K, and Rb by Srivastava and Singh (1976).

A significant feature of phonon modes in metals is the *Kohn* (1959) *anomaly*. In the RPA, the electron polarization operator is

$$P^{(1)}(q) = \frac{mk_F}{-2\pi^2}\left[1 + \frac{1}{2x}(1 - x^2)\ln\left|\frac{1+x}{1-x}\right|\right]$$

$$x = \frac{q}{2k_F}$$

In the equations for $\omega_\lambda^2(q)$, from $P^{(1)}(q)$ there appears the factor

$$(q^2 - 4k_F^2)\ln(q - 2k_F)$$

which is finite for $q = 2k_F$ but has an infinite slope. *This* means that the phonon modes have a dispersion $(\partial/\partial q)\omega_\lambda(q)$ which is logarithmically divergent at $q = 2k_F$.

B. Electron Self-Energies

The electron–phonon interactions in metals have an important influence upon electron states near the Fermi energy. The energy scale is set by the phonon energies themselves and so is roughly a Debye energy. This is only 20–30 meV in many metals, compared to Fermi degeneracies E_F of several electron volts. The actual self-energy $\Sigma(k, u)$ of electrons from electron–phonon interactions is a small energy, which is negligible. However, its derivative $\partial\Sigma/\partial u$ is large, so that it makes a large contribution to the electron effective mass. Exactly the opposite behavior was found earlier in Chapter 5 for the electron self-energies from electron–electron interactions: There the energies are large, but the contributions to the effective mass are small.

For the electron–phonon interaction, we may now assume that the phonon states have been determined satisfactorily, perhaps by the methods already discussed. Having started with a set of bare phonons with frequency $\Omega_{q\lambda}$ and matrix element M_λ, the actual phonon states in the metal have been determined by numerical means. Thus there is available a new set of phonon frequencies $\omega_\lambda(\mathbf{q})$, eigenvectors, and matrix element $\bar{M}_\lambda(\mathbf{q})$ for electron–phonon interaction, where λ is the mode index: TA, LA, etc. The matrix element is obtained as follows: the screened electron–phonon interaction (6.3.11) is rewritten in terms of the new matrix elements and frequencies:

$$V_{\text{sp}} = \sum_\lambda \frac{M_\lambda{}^2}{\varepsilon(\mathbf{q}, \omega)^2} \frac{2\Omega_{q\lambda}}{\omega^2 - \Omega_{\mathbf{q}\lambda}^2 - 2\Omega_{\mathbf{q}\lambda}P(\mathbf{q}, \omega)M_\lambda{}^2/\varepsilon(\mathbf{q}, \omega)}$$

$$= \bar{M}_\lambda{}^2(\mathbf{q}) \frac{2\omega_\lambda(\mathbf{q})}{\omega^2 - \omega_\lambda(\mathbf{q})} = \bar{M}_\lambda{}^2(\mathbf{q})\bar{D}^{(0)}(\mathbf{q}, \omega)$$

where we ignore the frequency dependence in $P(q, \omega)$ and $\varepsilon(q, \omega)$. The preceding equation serves to define the renormalized matrix element from (6.4.8),

$$\bar{M}_\lambda(\mathbf{q}) = \frac{M_\lambda(\mathbf{q})}{\varepsilon(\mathbf{q})} \left(\frac{\Omega_{\mathbf{q}\lambda}}{\omega_\lambda(\mathbf{q})}\right)^{1/2} = \left(\frac{\hbar}{2\varrho\omega_\lambda(\mathbf{q})}\right)^{1/2} \mathbf{q} \cdot \hat{\xi}_{q\lambda} \frac{V_{\text{ei}}(\mathbf{q})}{\varepsilon(\mathbf{q})} \tag{6.4.10}$$

which is expressed in terms of the renormalized frequencies $\omega_\lambda(\mathbf{q})$. This renormalized version of the screened electron–phonon interaction is exactly what we would obtain from a Hamiltonian of the form

$$H = \sum_{s\mathbf{k}} \xi_k c_{\mathbf{k}s}^\dagger c_{\mathbf{k}s} + \sum_{\mathbf{q}\lambda} \left[\omega_\lambda(\mathbf{q}) a_{\mathbf{q}\lambda}^\dagger a_{\mathbf{q}\lambda} + \frac{\bar{M}_\lambda(\mathbf{q})}{\nu^{1/2}} (a_{\mathbf{q}\lambda} + a_{-\mathbf{q}\lambda}^\dagger) \sum_{\mathbf{k}s} c_{\mathbf{k}+\mathbf{q}s}^\dagger c_{\mathbf{k}s}\right] \tag{6.4.11}$$

Equation (6.4.11) is just a simple electron–phonon Hamiltonian using the renormalized phonon frequencies and matrix elements. Its significant feature is that electron–electron interactions are omitted, for when they are included, they serve to screen the electron–phonon interactions. This phenomenon was used to derive the renormalized frequencies and matrix elements. Were we to again add electron–electron interactions, we would again renormalize the quantities which have been renormalized once. A second renormalization is incorrect. Thus one can use the effective Hamiltonian (6.4.11), with renormalized frequencies, only as long as the screening of the interaction is not included again.

The self-energy of the electron, from electron–phonon interactions, is calculated using the renormalized phonons and matrix elements. The only diagram which is ever calculated is the one-phonon basic bubble diagram:

$$\Sigma(\mathbf{k}, ik_n) = -\frac{1}{\beta} \sum_{\lambda\mathbf{q}, i\omega} \bar{M}_\lambda^2(\mathbf{q}) \mathscr{G}(\mathbf{k}+\mathbf{q}, ik+i\omega) \bar{\mathscr{D}}^{(0)}(\mathbf{q}, i\omega)$$

$$= +\sum_\lambda \int \frac{d^3q}{(2\pi)^3} \bar{M}_\lambda^2(\mathbf{q}) \frac{1}{\beta} \sum_{i\omega} \frac{2\omega_\lambda(\mathbf{q})}{\omega^2 + \omega_\lambda(\mathbf{q})^2} \frac{1}{ik + i\omega - \xi_{\mathbf{k}+\mathbf{q}}}$$

This one-phonon self-energy makes only a small contribution to the energy of the electron. Higher-order terms, involving two or more phonons, are neglected on the basis that their contribution is even smaller, as first shown by Migdal (1958). The one-phonon self-energy diagram is evaluated by making several approximations. The free-particle propagator $\mathscr{G}^{(0)}$ is used for the electron Green's function. The electron–phonon matrix elements $\bar{M}_\lambda(\mathbf{q})$ are treated as functions only of $\mathbf{q}$, not of $i\omega$. Thus we ignore the frequency dependence in the longitudinal dielectric function $\varepsilon(\mathbf{q}, i\omega)$. With these approximations, the Matsubara summation is the same one which has been evaluated previously in Sec. 3.5 for unperturbed electron and phonon Green's functions:

$$\Sigma(\mathbf{k}, ik_n) = \int \frac{d^3q}{(2\pi)^3} \sum_\lambda \bar{M}_\lambda^2(\mathbf{q}) \times \left[\frac{n_B(\omega_\lambda) + 1 - n_F(\xi_{\mathbf{k}+\mathbf{q}})}{ik_n - \xi_{\mathbf{k}+\mathbf{q}} - \omega_\lambda(\mathbf{q})} + \frac{n_B(\omega_\lambda) + n_F(\xi_{\mathbf{k}+\mathbf{q}})}{ik_n - \xi_{\mathbf{k}+\mathbf{q}} + \omega_\lambda(\mathbf{q})} \right] \quad (6.4.12)$$

The retarded function is obtained by $ik_n \to u + i\delta$. Our discussion will be confined to zero temperature properties, so set $n_B = 0$. Thus we consider the expression

$$\Sigma(\mathbf{k}, u) = \int \frac{d^3q}{(2\pi)^3} \sum_\lambda \bar{M}_\lambda^2(\mathbf{q}) \left[\frac{1 - n_F(\xi_{\mathbf{k}+\mathbf{q}})}{u - \xi_{\mathbf{k}+\mathbf{q}} - \omega_\lambda + i\delta} + \frac{n_F(\xi_{\mathbf{k}+\mathbf{q}})}{u - \xi_{\mathbf{k}+\mathbf{q}} + \omega_\lambda + i\delta} \right] \quad (6.4.13)$$

This self-energy is customarily evaluated using the Brillouin–Wigner form of perturbation theory. The energy u is retained as a separate parameter, and the self-energy is calculated as a function of both u and k.

The variation of $\Sigma(\mathbf{k}, u)$ with respect to u is far more important than with respect to k. The variation with respect to u is on the scale of a phonon energy, while the variation with respect to k is on the scale of k_F. An order of magnitude estimate of these derivatives is

$$\frac{\partial \Sigma}{\partial u} \simeq \frac{\Sigma}{\omega_D}$$

$$\frac{\partial \Sigma}{\partial \xi_k} \simeq \frac{\Sigma}{E_F}$$

where ω_D is the Debye energy of the solid. The derivative of $\Sigma(\mathbf{k}, u)$ with respect to u is larger than the derivative with respect to ξ_k by the factor $(\omega_D/E_F)^{-1} \simeq 10^{+2}$. It is usually possible to neglect $\partial\Sigma/\partial\xi_k$, so that the effective mass and renormalization factors are given by the same expression:

$$\left(\frac{m}{m^*}\right) = Z_F = \left(1 - \frac{\partial \Sigma}{\partial u}\right)^{-1}_{\substack{u=0\\ k=k_F}} \tag{6.4.14}$$

In fact the electron self-energy $\Sigma(\mathbf{k}, u)$ from electron–phonon interactions has appreciable value only for u within a Debye energy of the Fermi surface, or $-\omega_D \lesssim u \lesssim \omega_D$. In this narrow energy range, k hardly changes from k_F. For metals with a spherical Fermi surface, the k dependence of $\Sigma(\mathbf{k}, u)$ is unimportant, and k dependence is often even suppressed in the notation:

$$\Sigma(u) \equiv \Sigma(k_F, u)$$

When the metal has its Fermi sphere cut by Bragg planes, so that the Fermi surface is divided into pockets, the self-energy may depend on the position on the Fermi surface in the region near these Bragg planes. In that case one should retain the $\mathbf{k}$ dependence.

The two terms in the integrand of (6.4.13) which contain the factor n_F are far more important than the factor which contains the 1. Thus it is natural to separate Σ into two terms:

$$\Sigma = \Sigma^{(a)} + \Sigma^{(b)}$$

$$\Sigma^{(a)}(\mathbf{k}, u) = \sum_\lambda \int \frac{d^3q}{(2\pi)^3} \bar{M}_\lambda^2(\mathbf{q}) \frac{1}{u - \xi_{\mathbf{k}+\mathbf{q}} - \omega_\lambda(\mathbf{q}) + i\delta} \tag{6.4.15}$$

$$\Sigma^{(b)}(\mathbf{k}, u) = \sum_\lambda \int \frac{d^3q}{(2\pi)^3} n_F(\xi_{\mathbf{k}+\mathbf{q}}) \sum_\lambda \bar{M}_\lambda^2(\mathbf{q}) \left[\frac{1}{u - \xi_{\mathbf{k}+\mathbf{q}} + \omega_\lambda(\mathbf{q}) + i\delta} - \frac{1}{u - \xi_{\mathbf{k}+\mathbf{q}} - \omega_\lambda(\mathbf{q}) + i\delta}\right]$$

The first term $\Sigma^{(a)}$ is a rather dull function of its two arguments. Not only is this self-energy term small in magnitude but so are its derivatives—with respect to k and u. Hence it makes a negligible contribution to the energy, effective mass, or other properties of the electron. An estimate of this term can be obtained in the following way. The integration over q extends throughout the Brillouin zone. The screened Coulomb interaction will have appreciable value until $q \simeq q_{\text{TF}}$. Thus the maximum effective wave vector is either the zone boundary $\sim\pi/a$ or q_{TF}, which are comparable in magnitude. We estimate all quantities in the integrand by setting $q \simeq q_{\text{TF}}$. The phonon energy can be neglected in the energy denominator. Thus the size of the self-energy is estimated to be

$$\Sigma^{(a)} \simeq q_{\text{TF}}^3 \frac{\bar{M}_\lambda^2}{\varepsilon_{q\text{TF}}} = q_{\text{TF}}^3 \left(\frac{4\pi e^2 n_0}{q_{\text{TF}}^2}\right)^2 \frac{q_{\text{TF}}^2}{2\varrho\omega_q} \frac{2m}{q_{\text{TF}}^2}$$

$$\simeq \frac{m}{M} \frac{E_F^2}{\omega_D}$$

where $4\pi e^2 n_0/q_{\text{TF}}^2 \simeq E_F$, $\omega_q \simeq \omega_D$ is the phonon Debye frequency, and $q_{\text{TF}}^3/\varrho \simeq 1/M$, where M is the ion mass. Thus when we estimate the self-energy of the electron, we encounter the two dimensionless factors

$$\frac{\Sigma^{(a)}}{E_F} \simeq \frac{m}{M}\left(\frac{E_F}{\omega_D}\right) \simeq 10^{-2}$$

$$\frac{E_F}{\omega_D} \simeq 10^2$$

$$\frac{m}{M} \simeq 10^{-4}$$

Because of the inverse phonon energy, the large factor $E_F/\omega_D \simeq 10^2$. But this is amply reduced by the ratio of electron to ion masses, $m/M \simeq 10^{-4}$. It is the latter ratio that makes the self-energy term small. The self-energy term, when evaluated, turns out to be several millielectron volts or about the size of a phonon energy. This small self-energy is consistent with the results in the previous sections of this chapter. The polaron self-energies are always about a phonon energy in magnitude. For semiconductors, this may be a large number compared to the reference energy, which is k_BT. Thus polaron energies in semiconductors or ionic solids are important because they are large compared to the reference energy, which is small. In metals, the reference energy is the Fermi degeneracy E_F. Our phonon self-energies are very small compared with this number and may be neglected.

The other self-energy term, $\Sigma^{(b)}$ in (6.4.15), is about the same magnitude. Its importance comes from its contribution to the electron effective mass, which is determined by $\partial\Sigma/\partial u$ and $\partial\Sigma/\partial\varepsilon_k$. First consider the real part of $\partial\Sigma^{(b)}/\partial u$:

$$\operatorname{Re}\left(\frac{\partial\Sigma^{(b)}}{\partial u}\right)_{u=0} = -\sum_\lambda \int \frac{d^3q}{(2\pi)^3}\, n_F(\xi_{\mathbf{k}+\mathbf{q}})\bar{M}_\lambda^2(q) \times\left\{\frac{1}{[\xi_{\mathbf{k}+\mathbf{q}} - \omega_\lambda(q)]^2} - \frac{1}{[\xi_{\mathbf{k}+\mathbf{q}} + \omega_\lambda(q)]^2}\right\}$$

The integral is evaluated by integrating by parts on the angular variable $\nu = \cos\theta = \hat{p}\cdot\hat{q}$, so that we have the two factors

$$\int u\, dv = uv - \int v\, du$$

$$v = -\frac{m}{kq}(\xi_{\mathbf{k}+\mathbf{q}} \pm \omega_\lambda)^{-1}, \qquad dv = d\nu(\xi_{\mathbf{k}+\mathbf{q}} \pm \omega_\lambda)^{-2}$$

$$u = n_F(\xi_{\mathbf{k}+\mathbf{q}}), \qquad du = \frac{kq}{m}\, d\nu\left[\frac{\partial}{\partial\xi}\, n_F(\xi_{\mathbf{k}+\mathbf{q}})\right]$$

The integration by parts gives

$$\begin{aligned}\operatorname{Re}\left(\frac{\partial\Sigma^{(b)}}{\partial u}\right)_{u=0} &= \sum_\lambda \int \frac{q^2\, dq}{(2\pi)^2}\, \bar{M}_\lambda^2(q)\left[n_F(\xi_{\mathbf{k}+\mathbf{q}})\left(\frac{1}{\xi_{\mathbf{k}+\mathbf{q}} \pm \omega_\lambda} - \frac{1}{\xi_{\mathbf{k}+\mathbf{q}} + \omega_\lambda}\right)\Bigg|_{\nu=-1}^{\nu=1}\right. \\ &\quad \left. - \int_{-1}^{1} d\nu\, \frac{\partial n_F(\xi_{\mathbf{k}+\mathbf{q}})}{\partial\xi_{\mathbf{k}+\mathbf{q}}}\left(\frac{1}{\xi_{\mathbf{k}+\mathbf{q}} - \omega_\lambda} - \frac{1}{\xi_{\mathbf{k}+\mathbf{q}} + \omega_\lambda}\right)\right]\end{aligned} \tag{6.4.16}$$

The first term in (6.4.16) is smaller than the second by a factor $\omega_{\lambda q}/E_F$ and may be neglected. The second term is the important one. The factor $\partial n_F(\xi)/\partial\xi = -\delta(\xi)$ at zero temperature. Changing the integration variable to $\mathbf{k}' = \mathbf{k} + \mathbf{q}$ gives, for this last term alone,

$$\begin{aligned}\operatorname{Re}\left(\frac{\partial\Sigma^{(b)}}{\partial u}\right)_{u=0} &= -\sum_\lambda \int \frac{d^3k'}{(2\pi)^3}\, \delta(\xi_{k'})\, \frac{2}{\omega_\lambda(\mathbf{k}-\mathbf{k}')}\, \bar{M}_\lambda^2(\mathbf{k}-\mathbf{k}') \\ &= -\frac{2}{(2\pi)^3}\sum_\lambda \int \frac{d^2k'}{v_F'}\, \frac{\bar{M}_\lambda^2(\mathbf{k}-\mathbf{k}')}{\omega_\lambda(\mathbf{k}-\mathbf{k}')}\end{aligned}$$

The remaining integration is over the two-dimensional area of the Fermi surface. For spherical Fermi surfaces, this factor is $d^2k/v_F = mk_F\delta\Omega_{k'}$, but most Fermi surfaces are not spherical. The term $[\partial\, \mathrm{Re}(\Sigma)^{(b)}/\partial u]$ is quite large because of the factor ω_λ^{-1}, since phonon energies are small. This important expression was first derived by Nakajima and Watabe (1963). It is called λ,

$$\mathrm{Re}\left(-\frac{\partial\Sigma^{(b)}}{\partial u}\right)_{u=0} \equiv \lambda(\hat{k}) = \frac{2}{(2\pi)^3}\sum_\lambda \int \frac{d^2k'}{v_F'}\,\frac{\bar{M}_\lambda(\mathbf{k}-\mathbf{k}')^2}{\omega_\lambda(\mathbf{k}-\mathbf{k}')} \tag{6.4.17}$$

which is somewhat unfortunate, since λ is also used to designate the summation over the polarization modes of the phonon system. No confusion should arise over these quite different parameters.

The parameter $\lambda(\hat{k})$, as defined in (6.4.17), is a function of the position on the Fermi surface of the metal. The point $\mathbf{k}$ is the reference point, and the average is taken over other points $\mathbf{k}'$ on the Fermi surface. Quite often average value of λ is given for a metal, which is obtained by averaging $\lambda(\hat{k})$ over the Fermi surface:

$$\lambda = \frac{\int (d^2k/v_F)\lambda(\hat{k})}{\int d^2k/v_F} = \frac{2}{(2\pi)^3}\sum_\lambda \int \frac{d^2k}{v_F}\int \frac{d^2k'}{v_F'}\,\frac{\bar{M}_\lambda(\mathbf{k}-\mathbf{k}')^2}{\omega_\lambda(\mathbf{k}-\mathbf{k}')}\bigg/\int \frac{d^2k}{v_F} \tag{6.4.18}$$

The other contribution to the effective mass of a particle at the Fermi surface is from the derivative of $\Sigma(\mathbf{k}, u)$ with respect to ξ_k. However, $\partial\Sigma/\partial\xi_k$ is much smaller than λ by a factor $\sim \omega_D/E_F$, so that the most significant term is λ. The other terms are usually entirely neglected, and the effective mass of electrons at the Fermi energy, from electron–phonon contributions, is given by

$$\left(\frac{m^*}{m}\right) = 1 + \lambda \tag{6.4.19}$$

The first calculations of this quantity were by Swihart, Scalapino, and Wada (1965), and by Ashcroft and Wilkins (1965), and many have been done since. The quantity λ can be determined from electron tunneling experiments, so that experimental values are available. Both theory and experiment have been reviewed by Grimvall (1981). His suggested values are shown in Table 6.3 for some nontransition metals. Some of the values are quite large—they are 1.6 and 1.5 for mercury and lead. These large values show that the electron–phonon coupling is large for these metals. This observation is significant for the theory of superconductivity, which is due to

TABLE 6.3. λ: Electron–Phonon Mass Enhancement[a]

Li	Be		
0.41 ± 0.15	0.24 ± 0.05		
Na	Mg	Al	
0.16 ± 0.04	0.35 ± 0.04	0.43 ± 0.05	
K	Zn	Ga	
0.13 ± 0.03	0.37 ± 0.05	0.97 ± 0.05	
Rb	Cd	In	Sn
0.16 ± 0.04	0.40 ± 0.05	0.8	0.72
Cs	Hg	Tl	Pb
0.15 ± 0.04	1.6	0.8	1.5

[a] From Grimvall (1981).

electron–phonon coupling—as described in Chapter 9. Metals which have a low value of λ, such as the alkali metals, are not superconducting.

The phonon density of states is defined to be

$$F(\omega) = \sum_{\lambda} \int \frac{d^3q}{(2\pi)^3} \, \delta[\omega - \omega_\lambda(q)]$$

where the summation λ is over different modes, and the integration is over the Brillouin zone. McMillan (1968) has introduced another function, which is called α^2F. It is treated as a single function, although the notation implies incorrectly that it is the product of two functions. It is defined as

$$\begin{aligned} \alpha_{\mathbf{k}}{}^2(\omega)F_{\mathbf{k}}(\omega) &= \frac{1}{(2\pi)^3} \int \frac{d^2k'}{v_F{}'} \sum_{\lambda} | \bar{M}_\lambda(\mathbf{k} - \mathbf{k}') |^2 \delta[\omega - \omega_\lambda(\mathbf{k} - \mathbf{k}')] \\ \lambda(\hat{k}) &= 2 \int_0^\infty \frac{d\omega}{\omega} \, \alpha_{\mathbf{k}}{}^2(\omega)F_{\mathbf{k}}(\omega) \end{aligned} \tag{6.4.20}$$

$\alpha_{\mathbf{k}}{}^2F_{\mathbf{k}}$ is the frequency spectrum one obtains by starting at a point on the Fermi surface $\mathbf{k}$ and integrating over all other points on the Fermi surface $\mathbf{k}'$. It will vary from point to point on the Fermi surface of a metal. The same quantity without the subscript $\mathbf{k}$ is the average over all different points on the Fermi surface:

$$\alpha^2(\omega)F(\omega) = \int \frac{d^2k}{v_F} \, \alpha_{\mathbf{k}}{}^2(\omega)F_{\mathbf{k}}(\omega) \Big/ \int \frac{d^2k}{v_F}$$

Nowadays it is calculated with a computer, using realistic electron wavefunctions, realistic phonon eigenstates and frequencies, and realistic matrix elements. The quantity $\alpha^2(\omega)F(\omega)$ can be obtained experimentally, as will be explained in Chapter 9, from electron tunneling experiments. As an example, $\alpha^2(\omega)F(\omega)$ for Pb and Sn are given in Fig. 6.15. These results are from the experimental data of Rowell *et al.* (1969a,b, 1971). The dashed line in Fig. 6.15(a) is $F(\omega)$ for Pb, which is similar to α^2F because the matrix elements vary only by about a factor of 2 over the various phonon states.

Other physical quantities may be calculated from α^2F. For example, the lifetime of electrons on the Fermi surface is

$$\frac{1}{\tau(\mathbf{k})} = 4\pi \int_0^\infty d\omega \alpha_{\mathbf{k}}^2(\omega) F_{\mathbf{k}}(\omega) \left(\frac{1}{e^{\beta\omega}+1} + \frac{1}{e^{\beta\omega}-1} \right) \tag{6.4.21}$$

$1/\tau(\mathbf{k})$ is zero at zero temperature but is finite for a finite temperature. This lifetime is between scatterings of the electron by phonon emission or absorption. It is not the same lifetime which enters a calculation of the resistivity, for example. These differences are treated in Chapter 7. The electron lifetime in (6.4.21) is derived from the imaginary part of the retarded self-energy in (6.4.13):

$$\frac{1}{\tau(\mathbf{k})} = -2\,\mathrm{Im}[\Sigma(\mathbf{k}, \xi_k = 0)] = 2\pi \sum_\lambda \int \frac{d^3q}{(2\pi)^3} \mid \bar{M}_\lambda(\mathbf{q}) \mid^2 \{[n_B(\omega_\lambda) + 1 - n_F(\xi_{\mathbf{k}+\mathbf{q}})]\delta(\xi_{\mathbf{k}+\mathbf{q}} + \omega_\lambda) + [n_B(\omega_\lambda) + n_F(\xi_{\mathbf{k}+\mathbf{q}})]\delta(\xi_{\mathbf{k}+\mathbf{q}} - \omega_\lambda)\}$$

If we change variables to $\mathbf{k}' = \mathbf{k} + \mathbf{q}$, the integral is

$$\begin{aligned} \frac{1}{\tau(\mathbf{k})} &= 2\pi \sum_\lambda \int \frac{d^3k'}{(2\pi)^3} \mid \bar{M}_\lambda(\mathbf{k}-\mathbf{k}') \mid^2 \{n_B[\omega_\lambda(\mathbf{k}-\mathbf{k}')] + n_F[\omega_\lambda(\mathbf{k}-\mathbf{k}')]\} \\ &\quad \times \{\delta[\xi_{\mathbf{k}'} + \omega_\lambda(\mathbf{k}-\mathbf{k}')] + \delta[\xi_{\mathbf{k}'} - \omega_\lambda(\mathbf{k}-\mathbf{k}')]\} \\ &= \frac{4\pi}{(2\pi)^3} \int \frac{d^2k'}{v_F'} \sum_\lambda \mid M_\lambda(\mathbf{k}-\mathbf{k}') \mid^2 [n_B(\omega_\lambda) + n_F(\omega_\lambda)] \end{aligned}$$

which is the same as (6.4.21). Results for aluminum and lead have been reported by Tomlinson and Carbotte (1976). The same calculation for finite u gives

$$\begin{aligned} \frac{1}{\tau(\mathbf{k}, u)} &= -2\,\mathrm{Im}[\Sigma(\mathbf{k}, u)] \\ &= 2\pi \int_0^\infty d\omega \alpha_{\mathbf{k}}^2(\omega) F_{\mathbf{k}}(\omega)[2n_B(\omega) + n_F(\omega + u) + n_F(\omega - u)] \end{aligned} \tag{6.4.22}$$

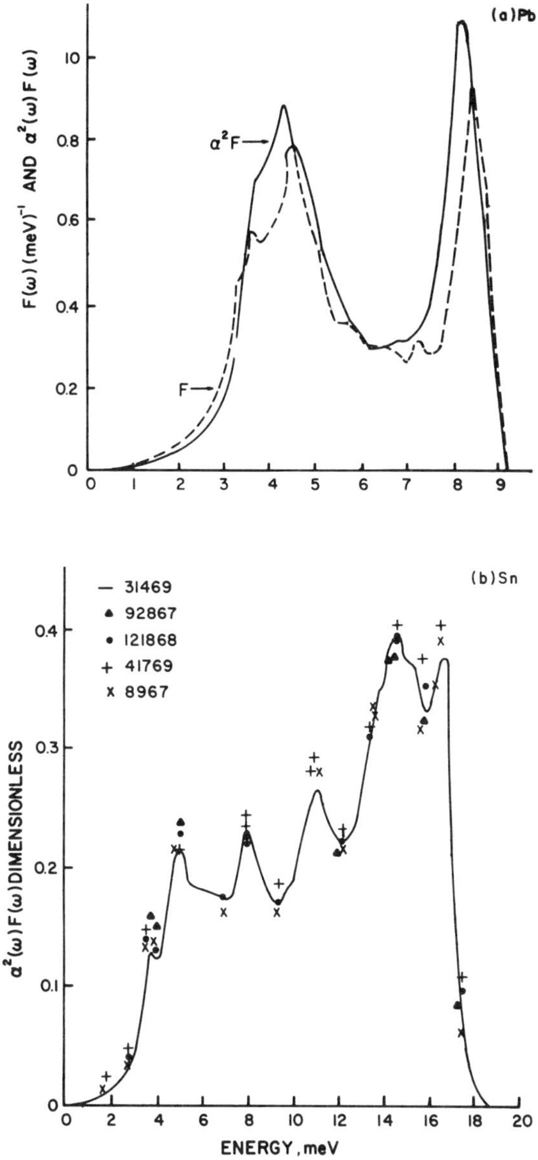

FIGURE 6.15. The energy dependence of the electron–phonon coupling $\alpha^2(\omega)F(\omega)$ in (a) lead and (b) tin as determined by electron tunneling in superconductors. In (a) the dashed line is the phonon density of state $F(\omega)$ alone. In (b) the points indicate the spread of values for different tunnel junctions. *Source*: Rowell *et al.* (1969a, 1971) (used with permission).

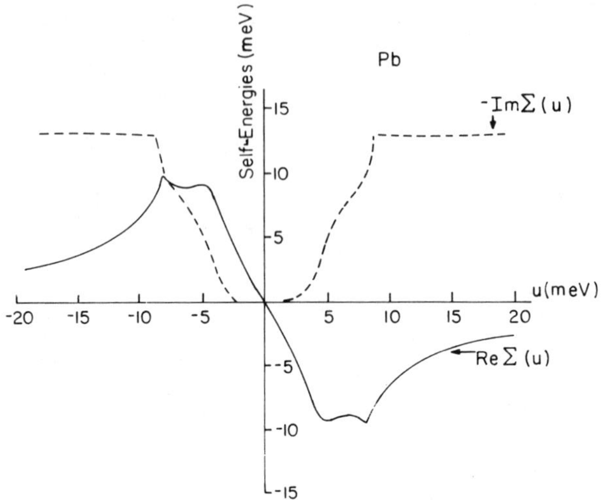

FIGURE 6.16. The real and imaginary self-energies at zero temperature of an electron in lead, from the electron–phonon interaction. *Source*: Grimvall (1976) (used with permission).

Equations (6.4.21) and (6.4.22) are the formulas for calculating Im(Σ).

Next a formula is needed for Re[$\Sigma(\mathbf{k}, u)$]. From (6.4.15), we integrate by parts on the $\nu = \cos\theta$ variable in the same fashion as was done in deriving (6.4.16):

$$\int_{-1}^{1} d\nu n_F(\xi_{\mathbf{k+q}})\left[\frac{1}{u - \xi_{\mathbf{k+q}} + \omega_\lambda(q)} - \frac{1}{u - \xi_{\mathbf{k+q}} - \omega_\lambda(q)}\right]$$

$$= -\frac{m}{pq} n_F(\xi_{\mathbf{k+q}}) \ln\left|\frac{u - \xi_{\mathbf{k+q}} + \omega_\lambda(q)}{u - \xi_{\mathbf{k+q}} - \omega_\lambda(q)}\right|_{\nu=-1}^{\nu=1}$$

$$+ \int_{-1}^{1} d\nu \frac{\partial}{\partial \xi_{\mathbf{k+q}}} n_F(\xi_{\mathbf{k+q}}) \ln\left|\frac{u - \xi_{\mathbf{k+q}} + \omega_\lambda(q)}{u - \xi_{\mathbf{k+q}} - \omega_\lambda(q)}\right|$$

$$\simeq -\ln\left|\frac{u + \omega_\lambda(q)}{u - \omega_\lambda(q)}\right| \int_{-1}^{1} d\nu \delta(\xi_{\mathbf{k+q}})$$

The term $(\partial/\partial\xi)n_F(\xi) = -\delta(\xi)$ is the important one, and the other may be neglected in comparison: A careful analysis shows it to be smaller by the factor ω_D/E_F. Hence we obtain the result for zero temperature:

$$\text{Re}[\Sigma(\mathbf{k}, u)] = -\sum_\lambda \int \frac{d^3k'}{(2\pi)^3} \delta(\xi_{\mathbf{k}'}) \bar{M}_\lambda^2(\mathbf{k} - \mathbf{k}')^2 \ln\left|\frac{u + \omega_\lambda(\mathbf{k} - \mathbf{k}')}{u - \omega_\lambda(\mathbf{k} - \mathbf{k}')}\right|$$

$$= -\int_0^\infty d\omega \alpha_{\mathbf{k}}^2(\omega) F_{\mathbf{k}}(\omega) \ln\left|\frac{\omega + u}{\omega - u}\right| \tag{6.4.23}$$

This vanishes at $u = 0$ and is positive for $u < 0$ and negative for $u > 0$. The usual procedure is to find $\alpha^2 F$ from superconducting tunneling experiments. $\alpha_{\mathbf{k}}{}^2(\omega)F_{\mathbf{k}}(\omega)$, which is a property of both the normal metal and the superconductor, is used in (6.4.22) and (6.4.23) to find the electron self-energy from electron–phonon interactions. Corresponding experimental information may be obtained from electron tunneling in normal metals, as shown by Rowell *et al.* (1969b). Their results for Pb are shown in Fig. 6.16. The self-energy has no **k** dependence since they use an averaged value of $\alpha^2(\omega)F(\omega)$. $\mathrm{Re}[\Sigma(u)]$ is quite small—on the order of millielectron volts. However, it has a very steep slope at point $u = 0$, which gives $\partial\Sigma/\partial u$ its large value.

PROBLEMS

1. The polaron self-energy $-\alpha\omega_0$ contains some kinetic energy of the electron and some potential energy. Use the Rayleigh–Schrödinger wave function to show these have the ratio of 1 : 3 at $p = 0$.

2. Show that the total momentum of the polaron in (6.1.10) commutes with the Hamiltonian in (6.1.1).

3. Calculate m/m^* for the Fröhlich polaron in the Tamm–Dancoff approximation.

4. Do the summations over Matsubara frequencies in the two-phonon self-energy (6.1.13). Show that this reduces to (6.1.12) at zero temperature.

5. Use the two-phonon wave function, in Rayleigh–Schrödinger perturbation theory, to find the α^2 term for the number of phonons in the polaron cloud at $p = 0$. Be sure your wave function is normalized to unity. You should get $-0.1036\alpha^2$, so correlations reduce the number of phonons in the cloud.

6. Calculate $\Sigma_{\mathrm{RS}}^{(1)}(p)$ at finite temperature for all values of ε_p. Also calculate its imaginary part, which is the imaginary part of the retarded self-energy with $E = \varepsilon_p$, which is also half the mean free path for one-phonon emissions or absorptions.

7. Use the answer of Problem 6 to plot, on a piece of graph paper, $\Sigma_{\mathrm{RS}}^{(1)}(\varepsilon_p)$ for $\alpha = 1$ and $0 < \varepsilon_p < \omega_0$. Evaluate the self-energy in the Tamm–Dancoff approximation for the same parameters, and plot this on the same set of graph paper.

8. Consider the one-phonon self-energy in Rayleigh–Schrödinger perturbation theory at finite temperatures:

$$\Sigma_{\mathrm{RS}}^{(1)}(p) = \sum_q \frac{M_q^2}{\nu}\left(\frac{N_q+1}{\varepsilon_{\mathbf{p}} - \varepsilon_{\mathbf{p+q}} - \omega_q} + \frac{N_q}{\varepsilon_{\mathbf{p}} - \varepsilon_{\mathbf{p+q}} + \omega_q}\right)$$

At high temperatures, make the approximation $N_q \simeq (k_B T/\omega_q)$, and keep only the term proportional to kT. Evaluate this by assuming a Debye spectrum and a piezoelectric matrix element (1.3.7). You should find that the self-energy vanishes when v_p is less than the speed of sound.

9. Minimize the ground state energy, in strong coupling theory, with the variational wave function $\varphi(r) = A\exp(-\beta r)$. Show that the energy is $-0.098\alpha^2\omega_0$, which is not as low as found with a Gaussian wave function.

10. In strong coupling theory, calculate the energy of the lowest p state by using the variational wave function $\varphi(r) = AZ\exp(-\frac{1}{2}\beta_1^2 r^2)$, where β_1 is a variational parameter.

11. In the limit of small wave vectors, the interaction between an electron and the homogeneous electron gas has the form

$$v_q\left(\frac{1}{\varepsilon} - 1\right) \to \frac{v_q\omega_p^2}{\omega^2 - \omega_p^2}$$

The electron–plasmon interaction is identical in form to the Fröhlich polaron Hamiltonian. What is the equivalent α for the electron gas? Find the numerical values for metallic sodium and aluminum.

12. Consider a small-polaron Hamiltonian which has the phonons localized on each site. This model could apply to molecular crystals, where the molecular vibrations are highly localized:

$$H = J\sum_{i\delta} c_{i+\delta}^{\dagger} c_i + \omega_0 \sum_i a_i^{\dagger} a_i + \lambda\omega_0 \sum_i c_i^{\dagger} c_i (a_i + a_i^{\dagger})$$

Show that this may be derived from (6.2.1) with the choice M_q = constant. Find the canonical transformation which diagonalizes the last two terms in the Hamiltonian. Use the transformed Hamiltonian to calculate

a. The thermally averaged amplitude and probability that a hop occurs without any change in the number of phonons at each site.

b. The thermally averaged amplitude and probability that the hop occurs without any change in the number of phonons but that they may be exchanged between the two sites.

13. Show that the effective interaction (6.3.2) with the dielectric function (6.3.3) may be written as

$$V_{\text{eff}} = v_q\left[\frac{1}{\varepsilon_\infty - v_qP} - \frac{\Omega(q)^2}{\Omega(q)^2 - \omega^2}\left(\frac{1}{\varepsilon_\infty - v_qP} - \frac{1}{\varepsilon_0 - v_qP}\right)\right]$$

where $\Omega(q)$ is the renormalized phonon modes of (6.3.15). At $\omega = 0$ V_{eff} is $v_q/(\varepsilon_0 - v_qP)$, and at $\omega \to \infty$ it is $v_q/(\varepsilon_\infty - v_qP)$ (Appel, 1966).

14. The Raman scattering of Fig. 6.13 occurs because the light couples non-linearly to the phonons but not to the plasmons. Assume a term in the Hamiltonian of the form

$$I \sum_{\mathbf{kk}'} (b_{\mathbf{k}}^{\dagger} b_{\mathbf{k}'} A_{\mathbf{k}'-\mathbf{k}} + \text{h.c.})$$

which describes the scattering of light from $\mathbf{k}' \to \mathbf{k}$ with photon operators $b_{\mathbf{k}}^{\dagger} b_{\mathbf{k}'}$ and phonon operators $A_{\mathbf{q}} = a_{\mathbf{q}} + a_{-\mathbf{q}}^{\dagger}$.

a. Derive a correlation function which describes the Raman scattering to order I^2.
b. Solve it at long wavelengths to get the Raman intensity of the modes $\omega_\pm$ in (6.3.13) as a function of n_0.

15. Compare Rayleigh–Schrödinger and Brillouin–Wigner perturbation theory by plotting both theories using the self-energy in (6.3.21):

$$\text{RS:} \qquad E(p) = \xi_p + \text{Re}[\Sigma(\xi_p)]$$

$$\text{BW:} \qquad E(p) = \xi_p + \text{Re}\{\Sigma[E(p)]\}$$

Use the parameters $f_0 = 10$ meV and $\omega_0 = 30$ meV. Which theory appears more reasonable?

16. Find the form of $\bar{V}_{\text{ei}}(q)$ for the Heine–Abarankov pseudopotential (5.3.3), and give the form of (6.4.9).

17. Use the RPA for $\varepsilon_2(q, \omega)$ to estimate the damping of phonons from electron–hole creation in a metal.

18. Prove that the sum rule $\int_0^\infty d\omega\omega\alpha^2(\omega)F(\omega)$ is independent of phonon properties (McMillan, 1968). Evaluate the sum rule analytically for metallic hydrogen, and estimate the formula numerically.

19. Derive Eq. (6.4.22) for $\tau(k, u)^{-1}$. Show that it goes to a constant at $u \to \infty$.

20. Use (6.4.22) and (6.4.23) to show that at zero temperature $\text{Re}[\Sigma(u)]$ and $\text{Im}[\Sigma(u)]$ obey a Kramers–Kronig relation. Calculate by direct integration of (5.7.7).

21. Show that for a jellium model of a metal with point ions, λ is exactly equal and opposite (in sign) to the electron–electron contribution (5.8.7).

22. Discuss how one could evaluate the electron effective mass directly from the effective interaction (6.3.2) *without* separating it into Coulomb and phonon parts. Evaluate the electron self-energy contribution,

$$\Sigma(p) = -\int \frac{d^3q}{(2\pi)^3} \frac{1}{\beta} \sum_{iq} V_{\text{eff}}(q)\mathscr{G}^{(0)}(p+q)$$

by following the procedure beginning at (5.8.3). Derive the term $(\partial\Sigma^{(\text{res})}/\partial u)_{u=0}$ for this case.

23. Show that in the jellium model of a metal, with the ions as point charges, the longitudinal modes at long wavelength are described by a total dielectric function [see (6.3.3)]:

$$\varepsilon_{\text{total}} = \varepsilon_1(1 - \omega_{\text{ip}}^2/\omega^2) - v_q P(q, \omega)$$

With this choice, one can write the effective interaction in (6.3.2) as the correct electron–electron and electron–phonon parts.

24. If $\Sigma(T, k, \varepsilon)$ is the retarded self-energy at temperature T, show that the electron self-energy in a metal from electron–phonon interactions obeys

$$\text{Re}[\Sigma(T, k, \varepsilon)] = \int_{-\infty}^{\infty} d\varepsilon' \left[-\frac{\partial n_F(\varepsilon')}{\partial \varepsilon'} \right] \text{Re}[\Sigma(0, k, \varepsilon - \varepsilon')]$$

25. Use (6.4.21) to show that at high temperatures

$$1/\tau = 2\pi\lambda k_B T$$

Chapter 7

dc Conductivities

7.1. ELECTRON SCATTERING BY IMPURITIES

This chapter is mostly concerned with the methods for calculating the electrical conductivity. Four different methods are discussed: (1) solving the Boltzmann equation, (2) evaluating the Kubo formula for the current–current correlation function, (3) evaluating the force–force correlation function, and (4) solving the quantum Boltzmann equation. For scattering from fixed impurities they all give the same answer. For scattering by phonons two different answers are obtained. One we call the Ziman (1960) formula, and the other the Holstein (1964) formula. Two criteria are important in comparing these methods: Which is easiest to use, and which gives the most accurate answer?

The electrical resistivity, or conductivity, from impurity scattering is an important topic. From the experimental viewpoint, all solids have impurities which make a contribution to the total resistivity. In many metals and semiconductors, the low-temperature resistivity is dominated by impurities, since all other contributions are temperature dependent and vanish at low temperature. In metals, as we shall see, the resistivity from impurity scattering is largely temperature independent, except for temperature variations on the scale of the Fermi temperature $T_F = E_F/k_B$. The subject is important, on the theoretical side, because it was one of the earliest evaluations of the Kubo formula. The importance of vertex corrections became apparent. Indeed, the derivation showed that vertex corrections are usually very important and should be assumed important until shown otherwise. This conclusion, and message, continues to be relevant even for calculations of other quantities. The present chapter is really about vertex corrections.

A. Boltzmann Equation

The electrical resistivity from impurity scattering is easily derived by using the Boltzmann equation. This derivation is presented for several reasons. First, it is probably the easiest way to get the answer. Second, the resistivity was first found this way; the Green's function evaluation of the Kubo formula only confirmed the result known earlier from transport theory. The Green's function derivation is complicated and subtle, and it is useful to know and believe the right answer in order to recognize it when it is finally derived.

Our objective is to derive a formula for the electrical resistivity with the least possible fuss. Thus we shall take the simplest possible model for the solid. It is a homogeneous system except for randomly located impurities. The electron states are plane waves except for occasional scattering from isolated impurities. The impurities are very dilute, so that interference between successive scatterings can be neglected. In the Boltzmann theory, the electrons are described by a classical distribution function $f(\mathbf{r}, \mathbf{k}, t)$. The time rate of change of this distribution function is governed by the equation

$$\frac{df}{dt} = \frac{\partial f}{\partial t} + \mathbf{v} \cdot \nabla f + \frac{\partial \mathbf{k}}{\partial t} \cdot \nabla_k f = \left(\frac{df}{dt}\right)_{\text{collisions}}$$

The right-hand side is the time rate of change due to collisions with the impurities. In our model system, there is no $\mathbf{r}$ dependence in $f(\mathbf{r}, \mathbf{k}, t)$ since we have assumed it to be homogeneous. Also, for the dc conductivity, there is no time dependence. We assume that the system has a weak external electric field and that the current flows in a steady-state fashion. Thus the distribution function is only a function of wave vector $f(\mathbf{k})$ and obeys the equation

$$\frac{\partial \mathbf{k}}{\partial t} \cdot \nabla_k f(\mathbf{k}) = \left(\frac{df}{dt}\right)_{\text{collisions}}$$

In a solid, the factor $\partial \mathbf{k}/\partial t$ is equivalent to an acceleration which is equal to the forces on the electron (see Kittel, 1963):

$$\frac{\partial \mathbf{k}}{\partial t} = -e\mathbf{E} - \frac{e}{c} \mathbf{v} \times \mathbf{H}_0$$

In the present problem, we assume only an electric field **E** and no magnetic field $\mathbf{H}_0 = 0$, so that

$$-e\mathbf{E} \cdot \nabla_k f(\mathbf{k}) = \left(\frac{df}{dt}\right)_{\text{collisions}} \tag{7.1.1}$$

The collision term is the most interesting. It is evaluated in the *relaxation time* approximation. This assumes that collisions seek to return the system to the equilibrium configuration $f_0(k)$, which is the configuration the system would have in the absence of an electric field. It is assumed that the rate of change of $f(\mathbf{k})$ due to collisions is proportional to the degree that $f(\mathbf{k})$ is different from $f_0(k)$:

$$\left(\frac{df}{dt}\right)_{\text{collisions}} = \frac{-[f(\mathbf{k}) - f_0(k)]}{\tau_1(k)}$$

This equation defines the relaxation time $\tau_1(k)$. A more detailed derivation can be found in Ziman (1960). Our job will be to find and solve an equation for $\tau_1(k)$. Once this is done, we have an equation for the distribution function $f(\mathbf{k})$:

$$f(\mathbf{k}) = f_0(k) + e\tau_1(k)\mathbf{E} \cdot \nabla_k f(\mathbf{k})$$

When the electric field is small, only a small amount of current flows. The system is only slightly out of equilibrium. Thus the distribution function $f(\mathbf{k}) = f_0(k) + f_1(\mathbf{k})$, where the change $f_1(\mathbf{k})$ is small. Thus we can replace f by f_0 on the right-hand side, so that $f(\mathbf{k})$ is finally evaluated:

$$\begin{aligned} f(\mathbf{k}) &= f_0(k) + e\tau_1(k)\mathbf{E} \cdot \nabla_k f_0(k) \\ f(\mathbf{k}) &= f_0(k) + e\tau_1(k)\frac{\mathbf{E} \cdot \mathbf{k}}{m}\frac{\partial f_0(k)}{\partial \varepsilon_k} \end{aligned} \tag{7.1.2}$$

The electrical current density **J** is the product of the electron charge $-e$, the electron's density n_0, and the average velocity $\langle \mathbf{v} \rangle$, which is obtained by averaging over the electron distribution:

$$\mathbf{J} = -en_0\langle v \rangle = -en_0 \int \frac{d^3k}{(2\pi)^3} f(\mathbf{k}) \frac{\hbar\mathbf{k}}{m}$$

$$1 = \int \frac{d^3k}{(2\pi)^3} f(\mathbf{k}) = \int \frac{d^3k}{(2\pi)^3} f_0(k)$$

The distribution function $f(\mathbf{k})$ is normalized to give unity. By using the result (7.1.2) for $f(\mathbf{k})$, the term f_0 gives an average $\langle v \rangle$ of zero, since presumably no current flows when there is no electric field; as many electrons are going one way as another. Thus the current is proportional to the second term, and it is proportional to the electric field:

$$\mathbf{J} = -e^2 n_0 \int \frac{d^3k}{(2\pi)^3} \tau_1(k) \mathbf{v}_k (\mathbf{E} \cdot \mathbf{v}_k) \frac{\partial}{\partial \varepsilon_k} f_0(k)$$

In a homogeneous, isotropic system, the current $\mathbf{J}$ flows in the direction of $\mathbf{E}$. The quantity $f_0(k)$ is independent of k direction. The only angular factors are $\mathbf{v}_k(\mathbf{v}_k \cdot \mathbf{E})$. The angular integrals will average this to $\frac{1}{3} v_k^2 \mathbf{E}$. The conductivity σ is the ratio of J to E, which gives the final formula for the electrical conductivity:

$$\sigma = -\frac{1}{3} e^2 n_0 \int \frac{d^3k}{(2\pi)^3} v_k^2 \tau_1(k) \frac{\partial}{\partial \varepsilon_k} f_0(k) \tag{7.1.3}$$

It is a positive quantity since $\partial f_0 / \partial \varepsilon_k$ is always negative. Equation (7.1.3) will be used as the basis of all the calculations. There remains the important task of deriving a formula for the relaxation time $\tau_1(k)$. It is *not* just the time between scattering events, which we found was given by the imaginary part of the retarded self-energy. This is an important point, since it makes our life difficult. The relaxation time in the Boltzmann equation is a special quantity.

The impurities have been assumed to be static, fixed objects with a spherically symmetric potential. It has no internal excitations, so the electron scatters from it elastically. The impurity causes the electron in state $\mathbf{k}$ to scatter to $\mathbf{k}'$, which has the same energy, so that $|\mathbf{k}'| = |\mathbf{k}|$ or $\varepsilon_k = \varepsilon_{k'}$. The net rate of scattering out of the state $\mathbf{k}$ is the rate of going from $\mathbf{k}$ to $\mathbf{k}'$, which is proportional to $f(\mathbf{k})[1 - f(\mathbf{k}')]$ minus the rate from $\mathbf{k}'$ to $\mathbf{k}$, which is proportional to $f(\mathbf{k}')[1 - f(\mathbf{k})]$:

$$-\left(\frac{df}{dt}\right)_{\text{collisions}} = \frac{f(\mathbf{k}) - f_0(k)}{\tau_1(k)} = 2\pi n_i \int \frac{d^3k'}{(2\pi)^3} \delta(\varepsilon_k - \varepsilon_{k'})$$
$$\times \{|T_{\mathbf{kk}'}|^2 f(\mathbf{k})[1 - f(\mathbf{k}')] - |T_{\mathbf{k}'\mathbf{k}}|^2 f(\mathbf{k}')[1 - f(\mathbf{k})]\}$$

where n_i is the concentration of impurities. The quantity $T_{\mathbf{kk}'}$ is the matrix element for scattering from $\mathbf{k}$ to $\mathbf{k}'$. It is just the T matrix which was defined in Sec. 4.1. It is symmetric in its indices $T_{\mathbf{kk}'} = T_{\mathbf{k}'\mathbf{k}}$, so that we can simplify this to

$$\frac{f(\mathbf{k}) - f_0(k)}{\tau_1(k)} = 2\pi n_i \int \frac{d^3k'}{(2\pi)^3} \delta(\varepsilon_k - \varepsilon_{k'}) \mid T_{\mathbf{kk'}} \mid^2 [f(\mathbf{k}) - f(\mathbf{k}')] \qquad (7.1.4)$$

The integrand contains the factor $f(\mathbf{k}) - f(\mathbf{k}')$. The integral is evaluated by assuming the form in (7.1.2), which is written as

$$f(\mathbf{k}) = f_0(k) + \mathbf{k} \cdot \mathbf{E}C(k)$$

$$f(\mathbf{k}') = f_0(k) + \mathbf{k}' \cdot \mathbf{E}C(k)$$

Since $\mid \mathbf{k} \mid = \mid \mathbf{k}' \mid$ and $f_0(k') = f_0(k)$, the quantities $f(\mathbf{k})$ and $f(\mathbf{k}')$ differ only in the angular part of the second term. The angular part is treated as follows: Define a coordinate system in which the Z direction is $\hat{k}$, so that

$$\hat{k} \cdot \hat{E} = \cos\theta$$

$$\hat{k}' \cdot \hat{k} = \cos\theta'$$

$$\hat{k}' \cdot \hat{E} = \cos\theta \cos\theta' + \sin\theta \sin\theta' \cos\varphi$$

where we have used the law of cosines in the last identity. The difference of the two distribution functions may now be written out in terms of these angular variables:

$$f(\mathbf{k}) - f(\mathbf{k}') = f(\mathbf{k}) - f_0(k) - [f(\mathbf{k}') - f_0(k)]$$

$$= kEC(k)[\cos\theta(1 - \cos\theta') - \sin\theta \sin\theta' \cos\varphi]$$

The last term on the right, which contains the factor $\cos\varphi$, will vanish when we do the integral $\int^{2\pi} d\varphi$. There is no other φ dependence in the integrand of (7.1.4), and the average of $\cos\varphi$ is zero. The remaining term may be written as

$$\int d\Omega_{k'}[f(\mathbf{k}) - f(\mathbf{k}')] = kEC(k)\cos\theta \int d\Omega_{k'}(1 - \cos\theta')$$

$$= [f(\mathbf{k}) - f_0(k)] \int d\Omega_{k'}(1 - \cos\theta')$$

The term $f(\mathbf{k}) - f_0(k)$ is factored from both sides of (7.1.4), which leaves the definition for the reciprocal of the relaxation time:

$$\frac{1}{\tau_1(k)} = 2\pi n_i \int \frac{d^3k'}{(2\pi)^3} \delta(\varepsilon_k - \varepsilon_{k'}) \mid T_{\mathbf{kk}'} \mid^2 (1 - \cos\theta') \tag{7.1.5}$$

The important factor in the integrand is $1 - \cos\theta' = 1 - \mathbf{k} \cdot \mathbf{k}'/k^2$. It makes the relaxation time in the Boltzmann equation different from the usual relaxation time $\tau(k)$, which is the time between scattering events. The latter quantity is simply

$$\frac{1}{\tau(k)} = 2\pi n_i \int \frac{d^3k'}{(2\pi)^3} \delta(\varepsilon_k - \varepsilon_{k'}) \mid T_{\mathbf{kk}'} \mid^2 = \frac{4\pi n_i}{mk} \sum_l (2l+1) \sin^2 \delta_l(k)$$

$$= v_k \sigma(k) n_i$$

$$\sigma(k) = \frac{4\pi}{k^2} \sum_l (2l+1) \sin^2 \delta_l(k)$$

The usual inverse relaxation time is just the scattering cross section $\sigma(k)$ times the particle velocity v_k times the impurity density n_i. An equivalent result may be obtained for the relaxation time $\tau_1(k)$ of the Boltzmann equation. The T matrix is expanded as in (4.1.16) for the case where $\mid \mathbf{k} \mid = \mid \mathbf{k}' \mid$:

$$T_{\mathbf{kk}'} = \frac{-2\pi}{mk} \sum_l (2l+1) P_l(\cos\theta') e^{+i\delta_l(k)} \sin \delta_l(k)$$

where θ' is the angle between $\mathbf{k}$ and $\mathbf{k}'$. The angular integrals in (7.1.5) are straightforward but cumbersome, so we shall only quote the result:

$$\frac{1}{\tau_1(k)} = \frac{4\pi n_i}{mk} \sum_{l=1}^{\infty} l \sin^2[\delta_l(k) - \delta_{l-1}(k)]$$

This formula for $1/\tau_1$ has a combination of phase shifts different from that found in the formula for the cross section.

The factor $1 - \cos\theta'$ weights the amount of scattering of the electron by the impurity. Small-angle scattering, where $\cos\theta' \simeq 1$, is relatively unimportant in contributing to τ_1^{-1}. These events do little to impede the flow of electrons and so contribute little to the resistivity. The factor $1 - \cos\theta'$ obviously favors large-angle scattering events, which are more important for the electrical resistivity. The relaxation time in the transport equation is not identical to the average scattering rate because there is an additional factor to weight the amount of scattering in each event.

We shall give one example of evaluating the conductivity in (7.1.3).

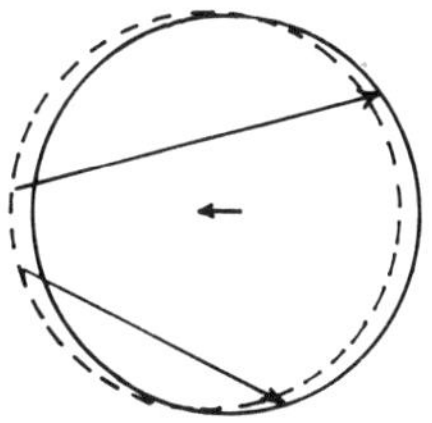

FIGURE 7.1. The circle represents the Fermi sea, which begins to move in response to an applied electric field. Steady state is maintained by relaxation back to other points on the Fermi surface.

For a free-electron metal, the distribution function is

$$f_0(k) = \frac{2}{n_0} \frac{1}{e^{\beta \xi_k} + 1}$$

$$\lim_{T\to 0} \frac{\partial}{\partial \xi_k} f_0 = \frac{-2}{n_0} \delta(\xi_k)$$

where the factor of 2 is for spin degeneracy. At zero temperature, the derivative $\partial f_0/\partial \xi_k$ becomes a delta function which fixes $k = k_F$. The angular integrals have already been done, so that we obtain

$$\sigma = \frac{2}{3} e^2 \frac{4\pi}{(2\pi)^3} v_F{}^2 m k_F \tau_1(k_F)$$

$$= \frac{e^2 n_0 \tau_1(k_F)}{m} \tag{7.1.6}$$

where we have used $n_0 = k_F{}^3/3\pi^2$. The relevant relaxation time is for electrons at the Fermi surface. Yet the conductivity is proportional to the density n_0 of all conduction electrons and not just to those at the Fermi surface. This surprising result is quite reasonable once the physics is understood. When the electric field is first imposed, the equation $\dot{\mathbf{k}} = -e\mathbf{E}$ shows that all electrons in the Fermi sea start accelerating equally. The Fermi sea is translationally shifted in wave vector space. The scattering tends to relax the Fermi distribution back to its undisturbed configuration. As shown in Fig. 7.1, electrons in the leading edge of the displaced Fermi distribution are scattered back to the rear regions. Only those electrons at the Fermi surface can scatter. The electrons well below the Fermi surface cannot elastically scatter, since all states with the same energy are already occupied with other electrons. Above the Fermi surface there are no thermally excited electrons. Thus only electrons at the Fermi surface are available to elastically scatter to other points on the Fermi surface.

The conductivity is relatively insensitive to temperature as long as the density of states of the metal is a smooth function of energy near the Fermi surface. The resistivity $\varrho = \sigma^{-1}$, which is the inverse of the conductivity, is proportional to the concentration of impurities. This proportionality is experimentally verified.

Since impurity scattering is elastic, it does not change the energy of the electron. As the current of electrons moves through the solid, each electron gains energy from the electric field. How does the electron lose this energy, if it only scatters elastically? The next few paragraphs will answer this question.

Let $\nu = \cos\theta$, where θ is the angle between **k** and **E**. It order to keep the discussion simple, it is assumed that the distribution $f(\mathbf{k})$ is isotropic in the absence of a field. When the field is present, the distribution function can be expanded in a Legendre series in ν. The first few terms are

$$f(\mathbf{k}) = f_s(k) + \nu f_p(k) + P_2(\nu) f_d(k) + \cdots$$

where $f_s(k)$ is the isotropic part of the distribution while $f_p(k)$ is the $l = 1$ part of the distribution. Note that $f_s(k)$ is *not* the equilibrium part of the distribution, which we call $f_0(k)$. The electrical current is determined by the p distribution f_p.

The conduction process can be viewed as having the following steps:

1. For $t < 0$, $E = 0$ and the initial distribution is f_0.
2. At $t = 0$ the field E is switched on. It accelerates the particles and creates the p distribution f_p.
3. The elastic scattering takes the particles from the p distribution f_p to the s distribution f_s. This step has a time constant τ_1.
4. The isotropic distribution f_s relaxes back to the equilibrium distribution f_0. This step has a different time constant.

The energy relaxation occurs in the last step where f_s is brought to equilibrium. The electrons can lose energy to their heat bath, usually phonons. This process has a very different time constant than τ_1 and is usually much slower. The relaxation time τ_1 from elastic scattering determines the rate at which particles scatter out of the p distribution f_p into other distributions such as f_s and f_d. The current is determined by τ_1 since it gives the steady state amplitude f_p. The energy relaxation occurs elsewhere in the chain of events. The following discussion amplifies these remarks.

For **E** in the z direction the Boltzmann equation is

$$\frac{\partial}{\partial t} f(\mathbf{k}) - eE\left[\nu \frac{\partial}{\partial k} + \frac{1}{k}(1 - \nu^2)\frac{\partial}{\partial \nu}\right] f(\mathbf{k}) = -[f(\mathbf{k}) - f_0(k)]/\tau_1$$

The Legendre expansion is used for $f(\mathbf{k})$. Then one takes the first two angular moments. That is, one first integrates the above equation over all 4π of solid angle. This step removes all of the angular factors. Next, one multiplies the above equation by v and integrates again over all of the solid angle. These steps produce the two equations

$$\frac{\partial f_s}{\partial t} - (eE/3k^2)\frac{\partial}{\partial k}(k^2 f_p) = -(f_s - f_0)/\tau_1$$

$$\frac{\partial f_p}{\partial t} - eE\frac{\partial f_s}{\partial k} = -f_p/\tau_1$$

These first-order differential equations are of the form

$$\frac{\partial a(t)}{\partial t} + a/\tau = b(t)$$

which have the solution

$$a(t) = a(0)e^{-t/\tau} + \int_0^t dt'\, e^{-(t-t')/\tau} b(t')$$

Define $\delta f = f_s - f_0$. Assume that the field is turned on at $t = 0$, so f_p and δf vanish at $t = 0$. The solution to the above equations can be written as

$$f_p(k, t) = eE\tau_1[1 - e^{-t/\tau_1}]\frac{\partial f_0}{\partial k} + eE\int_0^t dt'\, e^{-(t-t')/\tau_1}\frac{\partial}{\partial k}\delta f\,(k, t')$$

$$\delta f\,(k, t) = (eE/3k^2)\int_0^t dt'\, e^{-(t-t')/\tau_1}\frac{\partial}{\partial k}[k^2 f_p(k, t')]$$

The current is determined by the distribution f_p. After the field E is switched on at $t = 0$, the distribution f_p reaches its steady-state value of $eE\tau_1 \partial f_0/\partial k$. The change in the isotropic distribution δf grows in time according to E^2. The Joule heating of $O(E^2)$ affects the isotropic part of the distribution. The effect on f_p is $O(E^3)$. The current is only affected by Joule heating through nonlinear terms giving $j \alpha E^3$.

The elastic scattering by impurities determines the rate by which electrons in the p distribution are scattered to the s distribution. This determines the current, since it gives the steady-state occupation of the p distribution. Energy relaxation occurs in the s distribution, as f_s tries to relax towards f_0. This process has a very different rate. It does not affect the current unless the s distribution heats up, and the temperature is changed.

B. Kubo Formula: Approximate Solution

The electrical conductivity can be calculated from the Kubo formula by using the technique described in (3.7.10)–(3.7.12). The correlation function is evaluated for finite temperatures and frequencies:

$$\pi(i\omega) = -\frac{1}{3\nu}\int_0^\beta d\tau e^{i\omega\tau}\langle T_\tau \mathbf{j}(\tau)\cdot\mathbf{j}(0)\rangle \tag{7.1.7}$$

The retarded function, $\pi_{\text{ret}}(\omega)$, is obtained by letting $i\omega \to \omega + i\delta$, and the dc conductivity is given by the limit of $\omega \to 0$:

$$\sigma = -\lim_{\omega\to 0}\left\{\frac{\text{Im}[\pi_{\text{ret}}(\omega)]}{\omega}\right\} \tag{7.1.8}$$

This correlation function for σ will be evaluated for the same model system described in Sec. 7.1A. There is a free-particle system with a dilute concentration of simple scattering centers. Thus we evaluate (7.1.7) for the following Hamiltonian and current operator:

$$H = \sum_{\mathbf{k}\sigma} \xi_k C^\dagger_{\mathbf{k}\sigma} C_{\mathbf{k}\sigma} + V$$

$$V = \frac{1}{\nu}\sum_{\mathbf{q}} V(q)\varrho_i(\mathbf{q})\varrho(\mathbf{q})$$

$$\varrho_i(\mathbf{q}) = \sum_j e^{i\mathbf{q}\cdot\mathbf{R}_j}$$

$$\varrho(\mathbf{q}) = \sum_{\mathbf{k}\sigma} C^\dagger_{\mathbf{k}+\mathbf{q}\sigma} C_{\mathbf{k}\sigma}$$

$$\mathbf{j} = -\frac{e}{m}\sum_{\mathbf{k}\sigma} \mathbf{k} C^\dagger_{\mathbf{k}\sigma} C_{\mathbf{k}\sigma}$$

The impurities are at positions $\mathbf{R}_j$, and an average will be taken over the possible distributions of impurity positions. This averaging technique was described earlier in Sec. 4.1E.

The theoretical calculation is divided into two parts. The first part, in this section, is simply to reproduce the Boltzmann result which was just derived. We wish to derive a conductivity which is proportional to the relaxation time τ_1 and where $1/\tau_1(k)$ is defined as a scattering probability weighted by the $1 - \cos\theta'$ factor. The derivation will entail a summation over a set of vertex diagrams. The treatment will be kept as introductory as possible, since it is one of the first summation over vertex diagrams which we have encountered. In Sec. 7.1C, we shall present a formally exact solution to the correlation function due to Langer (1960).

The logical way to evaluate the correlation function (7.1.7) is as a power series in the concentration of impurities. We know that averages over the impurity density operators in (4.1.24),

$$f_n(\mathbf{q}_1 \cdots \mathbf{q}_n) = \langle \varrho_i(\mathbf{q}_1)\varrho_i(\mathbf{q}_2) \cdots \varrho_i(\mathbf{q}_n)\rangle$$
$$= N_i \delta_{\Sigma \mathbf{q}_j} + N_i^2 \delta_{\Sigma \mathbf{q}_j} \delta_{\Sigma \mathbf{q}_j} + N_i^3 \delta_{\Sigma \mathbf{q}} \delta_{\Sigma \mathbf{q}} \delta_{\Sigma \mathbf{q}} + \cdots$$

are expressed as a power series in the number of impurities N_i. At first sight it appears possible to evaluate (7.1.7) by just expanding the S matrix and collecting all terms proportional to N_i, then N_i^2, etc. A simple expansion in powers of N_i does not work, as is apparent from the answer we wish to obtain. It has $\sigma \propto \tau_1$ and $\tau_1 \propto 1/n_i$, so that $\sigma \simeq 1/n_i$. Thus the first term has the conductivity inversely proportional to N_i, which is impossible to obtain as a simple series in N_i except by summing a set of diagrams. Thus we recognize, from the outset, that we are going to have to sum series of diagrams in order to obtain the expected answer.

The correlation function (7.1.7) is written in the interaction representation as

$$\pi(i\omega) = -\frac{1}{3\nu}\int_0^\beta d\tau e^{i\omega\tau}\langle T_\tau S(\beta)e^{\tau H_0}\mathbf{j}e^{-\tau H_0}\cdot \mathbf{j}\rangle$$

The logical way to evaluate this expression is to expand the S matrix and consider each term. The first term has $S = 1$, and this correlation function is called π_0:

$$\pi_0(i\omega) = \frac{-e^2}{3m^2\nu}\int_0^\beta d\tau e^{i\omega\tau} \sum_{\mathbf{p}\mathbf{p}'\sigma\sigma'} \mathbf{p}\cdot\mathbf{p}'\langle T_\tau C^\dagger_{\mathbf{p}\sigma}(\tau)C_{\mathbf{p}\sigma}(\tau)C^\dagger_{\mathbf{p}'\sigma}C_{\mathbf{p}'\sigma}\rangle$$

$$= +\frac{2e^2}{3m^2\nu}\sum_{\mathbf{p}} p^2 \int_0^\beta d\tau \mathscr{G}^{(0)}(\mathbf{p}, \tau)\mathscr{G}^{(0)}(\mathbf{p}, -\tau)e^{i\omega\tau}$$

This expression is zero unless $i\omega = 0$, since the number operators $C_{\mathbf{p}}^{\dagger}C_{\mathbf{p}}$ are τ independent, and the τ integral is just $\int_0^\beta d\tau \exp i\omega\tau = \beta\delta_{\omega=0}$. The term π_0 gives a conductivity of zero. This is not surprising, since it is the correlation function of free particles, and they have zero resistivity. Thus we should have found that the correlation function for the conductivity was infinity, but the $\pi_0 = 0$ is sufficient to tell us we asked a nonsensical question and got a nonsensical answer. We are going to have a resistive system only by putting damping into the particle motion.

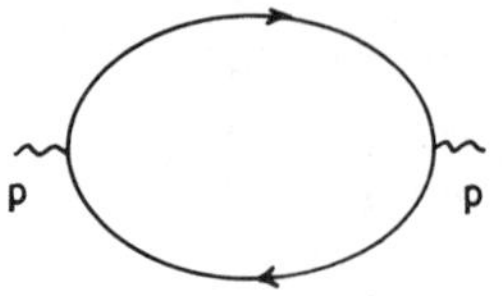

FIGURE 7.2

The next logical step is to replace all $\mathscr{G}^{(0)}$ by $\mathscr{G}$. The self-energy of the particles, from impurity scattering, is included in all the particle Green's functions. Of course, $\mathscr{G}$ is obtained by summing a series of diagrams, which is Dyson's equation. The self-energy $\Sigma(\mathbf{p}, ip)$ from impurity scattering was evaluated in Sec. 4.1 in the limit of low n_i. It is a retarded function with real and imaginary parts, where the imaginary parts are due to the damping of the particle motion. Thus the step of replacing $\mathscr{G}^{(0)}$ by $\mathscr{G}$ does put in damping of the particle motion. The first correlation function which will be evaluated is shown in Fig. 7.2. It is a simple bubble diagram, with the smooth lines denoting $\mathscr{G}$ and the two vertices having the vector vertex $\mathbf{p}$. This correlation function is called $\pi^{(0)}(i\omega)$:

$$\pi^{(0)}(i\omega) = +\frac{2e^2}{3m^2\nu}\sum_{\mathbf{p}} p^2 \frac{1}{\beta}\sum_{ip_n} \mathscr{G}(\mathbf{p}, ip + i\omega)\mathscr{G}(\mathbf{p}, ip) \tag{7.1.9}$$

The wiggly lines at the two ends of the bubble, which are connected to the vertices, represent the incoming frequency $i\omega$. The first step is to evaluate the summation over Matsubara frequencies:

$$\begin{aligned} S &= \frac{1}{\beta}\sum_{ip} \mathscr{G}(\mathbf{p}, ip)\mathscr{G}(\mathbf{p}, ip + i\omega) \\ &= \frac{1}{\beta}\sum_{ip} \frac{1}{ip - \xi_p - \Sigma(\mathbf{p}, ip)} \, \frac{1}{ip + i\omega - \xi_p - \Sigma(\mathbf{p}, ip + i\omega)} \end{aligned}$$

The procedure for doing this was described in Sec. 3.5. A contour integral

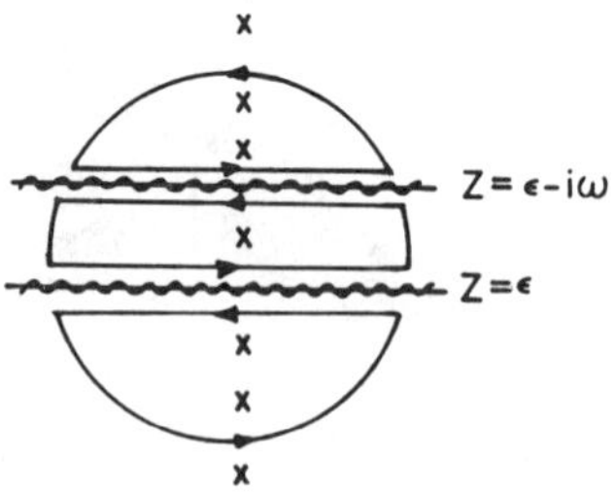

FIGURE 7.3

is evaluated as follows:

$$\int \frac{dZ}{2\pi i} n_F(Z)\mathscr{G}(\mathbf{p}, Z)\mathscr{G}(\mathbf{p}, Z + i\omega)$$

This has branch cuts on the two lines $Z = \varepsilon$ and $Z = \varepsilon - i\omega$, where ε is real. The contour is shown in Fig. 7.3. There are also the poles of $n_F(Z)$ at $Z = (2n + 1)i\pi/\beta$. The summation is equal to the integrals along the branch cuts, where the contributions above and below the cuts are subtracted:

$$S = -\int_{-\infty}^{\infty} \frac{d\varepsilon}{2\pi i} n_F(\varepsilon)\{\mathscr{G}(\mathbf{p}, \varepsilon + i\omega)[\mathscr{G}(\mathbf{p}, \varepsilon + i\delta) - \mathscr{G}(\mathbf{p}, \varepsilon - i\delta)]$$
$$+ \mathscr{G}(\mathbf{p}, \varepsilon - i\omega)[\mathscr{G}(\mathbf{p}, \varepsilon + i\delta) - \mathscr{G}(\mathbf{p}, \varepsilon - i\delta)]\}$$

The factor

$$\mathscr{G}(\mathbf{p}, \varepsilon + i\delta) - \mathscr{G}(\mathbf{p}, \varepsilon - i\delta) = G_{\text{ret}}(\mathbf{p}, \varepsilon) - G_{\text{adv}}(\mathbf{p}, \varepsilon)$$
$$= +2i \operatorname{Im}[G_{\text{ret}}(\mathbf{p}, \varepsilon)] = -iA(\mathbf{p}, \varepsilon)$$

is just the spectral function of the retarded Green's function, so that the summation over Matsubara frequencies is

$$S = +\int_{-\infty}^{\infty} \frac{d\varepsilon}{2\pi} n_F(\varepsilon)A(\mathbf{p}, \varepsilon)[\mathscr{G}(\mathbf{p}, \varepsilon + i\omega) + \mathscr{G}(\mathbf{p}, \varepsilon - i\omega)]$$

$$\pi^{(0)}(i\omega) = \frac{2e^2}{3m^2\nu} \sum_{\mathbf{p}} p^2 \int_{-\infty}^{\infty} \frac{d\varepsilon}{2\pi} n_F(\varepsilon)A(\mathbf{p}, \varepsilon)[\mathscr{G}(\mathbf{p}, \varepsilon + i\omega) + \mathscr{G}(\mathbf{p}, \varepsilon - i\omega)]$$

It is now the proper time to make the analytical continuation $i\omega \to \omega + i\delta$. This step must only be performed after the summation over Matsubara frequencies. The analytical continuation changes the $\mathscr{G}(\mathbf{p}, \varepsilon \pm i\omega)$ into $G(\mathbf{p}, \varepsilon \pm \omega \pm i\delta)$. The next step, indicated in (7.1.8), is to take the imaginary part of this expression. The only imaginary factors are in the retarded $G(\mathbf{p}, \varepsilon + \omega + i\delta)$ and advanced $G(\mathbf{p}, \varepsilon - \omega - i\delta)$ Green's functions, and their imaginary parts are just proportional to their spectral function $A(\mathbf{p}, \varepsilon \pm \omega)$:

$$\operatorname{Im}[G_{\text{ret}}(\mathbf{p}, \varepsilon + \omega) + G_{\text{adv}}(\mathbf{p}, \varepsilon - \omega)] = -\tfrac{1}{2}[A(\mathbf{p}, \varepsilon + \omega) - A(\mathbf{p}, \varepsilon - \omega)]$$

$$\operatorname{Im}[\pi_{\text{ret}}^{(0)}(\omega)] = -\frac{e^2}{3m^2\nu} \sum_{\mathbf{p}} p^2 \int_{-\infty}^{\infty} \frac{d\varepsilon}{2\pi} n_F(\varepsilon)A(\mathbf{p}, \varepsilon)$$
$$\times [A(\mathbf{p}, \varepsilon + \omega) - A(\mathbf{p}, \varepsilon - \omega)]$$

The change of variables $\varepsilon \to \varepsilon + \omega$ in the second term allows this expression to be rewritten as

$$\mathrm{Im}[\pi_{\mathrm{ret}}^{(0)}(\omega)] = -\frac{e^2}{3\nu m^2} \sum_{\mathbf{p}} p^2 \int_{-\infty}^{\infty} \frac{d\varepsilon}{2\pi} A(\mathbf{p}, \varepsilon) A(\mathbf{p}, \varepsilon+\omega)[n_F(\varepsilon) - n_F(\varepsilon+\omega)]$$

The next step in the derivation (7.1.8) is to divide by ω and then to take the limit $\omega \to 0$. The important frequency dependence is in the last factor:

$$\lim_{\omega \to 0} \frac{1}{\omega} [n_F(\varepsilon) - n_F(\varepsilon + \omega)] = -\frac{\partial}{\partial \varepsilon} n_F(\varepsilon)$$

Thus we obtain the conductivity, which is given by the correlation function $\pi^{(0)}$. This term in the conductivity is called $\sigma^{(0)}$:

$$\sigma^{(0)} = -\frac{e^2}{3m^2} \int \frac{d^3p p^2}{(2\pi)^3} \int \frac{d\varepsilon}{2\pi} A(\mathbf{p}, \varepsilon)^2 \left[\frac{\partial}{\partial \varepsilon} n_F(\varepsilon)\right] \tag{7.1.10}$$

In the last step, the limit $\nu \to \infty$ changed the summation over $\mathbf{p}$ into a continuous integral. The right-hand side is positive since $\partial n_F/\partial \varepsilon$ is negative. Before discussing this result, it is useful to review the order of the steps in the derivation. They will be used in all discussions of evaluating the Kubo formula:

1. Do all summations over Matsubara frequencies ip.
2. Analytically continue $i\omega \to \omega + i\delta$ to get the retarded function $\pi_{\mathrm{ret}}(\omega)$.
3. Take the imaginary part $\mathrm{Im}\, \pi_{\mathrm{ret}}$.
4. Divide by ω, and then take the limit $\omega \to 0$.

Equation (7.1.10) for $\sigma^{(0)}$ has several interesting features. The factor $(\partial/\partial\varepsilon)n_F(\varepsilon)$ equals $-\delta(\varepsilon)$ at zero temperature, which is rather convenient, since it serves to eliminate the integral over $d\varepsilon$. The sharp step in $n_F(\varepsilon)$ is in contrast to the momentum distribution (3.3.19),

$$n_{\mathbf{p}} = \int \frac{d\varepsilon}{2\pi} n_F(\varepsilon) A(\mathbf{p}, \varepsilon)$$

which no longer has a discontinuous step at $p = k_F$ since the impurity scattering causes a general smearing of this distribution. However, the energy distribution $n_F(\varepsilon)$ is always a sharp function of ε at zero temperature regardless of the interactions.

Another crucial feature of (7.1.10) is that the spectral function is squared, or $A(\mathbf{p}, \varepsilon)^2$. That this is important may be shown by examining the limit as the impurity concentration n_i vanishes. Since the self-energy is proportional to n_i, it will vanish in this limit. If we define Δ as $-\mathrm{Im}(\Sigma)$, then we have

$$\lim_{n_i \to 0} A(p, \varepsilon) = \lim_{\Delta \to 0} \frac{2\Delta}{[\varepsilon - \xi_p - \mathrm{Re}(\Sigma)]^2 + \Delta^2} = 2\pi\delta(\varepsilon - \xi_p)$$

The spectral function becomes a delta function when $n_i \to 0$. This limiting behavior is reasonable, since in the absence of self-energy effects, the particles are all free, and the spectral function is indeed a delta function. The question at hand is what happens to A^2 as $n_i \to 0$, since it appears to go as the square of a delta function. In fact, A^2 does diverge as $n_i \to 0$, which makes the conductivity diverge to infinity when $n_i \to 0$. We need to find a method of handling this divergence. The answer is provided by considering the integrals

$$\int_{-\infty}^{\infty} \frac{dx}{2\pi} \frac{2\Delta}{x^2 + \Delta^2} = 1$$

$$\int_{-\infty}^{\infty} \frac{dx}{2\pi} \left(\frac{2\Delta}{x^2 + \Delta^2} \right)^2 = \frac{1}{\Delta}$$

where $x = \varepsilon - \xi_p - \mathrm{Re}(\Sigma)$. The first integral has the right behavior as $\Delta \to 0$, since it gives the same result as does $2\Delta/(x^2 + \Delta^2) = A = 2\pi\delta(x)$. The second integral shows that we can make the replacement

$$A^2 = \lim_{\Delta \to 0} \left(\frac{2\Delta}{x^2 + \Delta^2} \right)^2 = \lim_{\Delta \to 0} \frac{2\pi\delta(x)}{\Delta} \tag{7.1.11}$$

which will give the right behavior as $\Delta \to 0$. The replacement $A^2 \to 2\pi\delta(x)/\Delta$ will be made in the limit as $n_i \to 0$. Furthermore, the quantity 2Δ is recognized as the inverse mean free path of electrons on the Fermi surface,

$$\Delta(p_F, \varepsilon = 0) = \frac{1}{2\tau(p_F)} = -\mathrm{Im}(\Sigma)$$

so that our conductivity formula may now be written as

$$\sigma^{(0)} = \frac{2e^2}{3m^2} \int \frac{d^3p}{(2\pi)^3} \delta(\xi_p) p^2 \tau(p) \tag{7.1.12}$$

This equation looks like the right answer for σ, since it seems to have exactly the same combination of factors as (7.1.3). But there is a very important difference between (7.1.12) and the Boltzmann result—in the relaxation time. The formula (7.1.12) has a relaxation time *without* the $1 - \cos\theta'$ factor, since the relaxation time in (7.1.12) is from $-\mathrm{Im}(\Sigma)$, which is just the average time between scattering events. The $1 - \cos\theta'$ factor was important for weighting the large-angle scattering processes, which were important for the resistivity. Thus the preliminary answer (7.1.12) is not the Boltzmann result and has serious deficiencies.

At least we should congratulate ourselves for one achievement. We have succeeded in deriving a term in σ which is inversely proportional to n_i. The relaxation time $\tau(p)$, although the wrong one for the conductivity, at least has the virtue that it is inversely proportional to n_i, which makes $\sigma^{(0)}$ also inversely proportional to n_i.

Since our calculation has not yet yielded the right result, we must press on. We have not yet evaluated all the possible diagrams. Among those remaining, there must be a subset which, when evaluated, will lead us to the right answer. Since we are expanding the S matrix for impurity scattering, the remaining terms contain higher powers in the impurity interaction V. At first it appears that higher powers in the impurity interaction must imply that the additional terms are higher powers in the impurity concentration n_i. This is not the case, as is obvious from the answer. The σ we want is proportional to n_i^{-1}, while our preliminary term $\sigma^{(0)}$ is also proportional to n_i^{-1}. Thus the important correction terms we shall sum must also yield terms in the conductivity which are proportional to n_i^{-1}. That is, we must search for those terms in the S-matrix expansion which cause σ to diverge as n_i^{-1} when $n_i \rightarrow 0$, although these terms must come from higher terms in the S-matrix expansion. Higher-order terms in S can go proportionally with n_i^{-1} if they also contain higher powers of the spectral function A^n.

The correlation function $\sigma^{(0)}$ contained Green's functions $\mathscr{G}$ which include all self-energy effects. The remaining diagrams are called vertex corrections. They are defined as diagrams in which the impurity scattering links the Green's functions on both sides of the bubble. Some vertex diagrams are shown in Fig. 7.4(a). There is a single impurity with a varying number of scattering events from the electron line on either side of the bubble. If there were no scattering line connecting one side of the bubble, the diagram would be a self-energy term on the other side. A diagram in which the two electron lines, on both sides of the bubble, scatter from the same impurity cannot be a self-energy diagram of either one and so is called

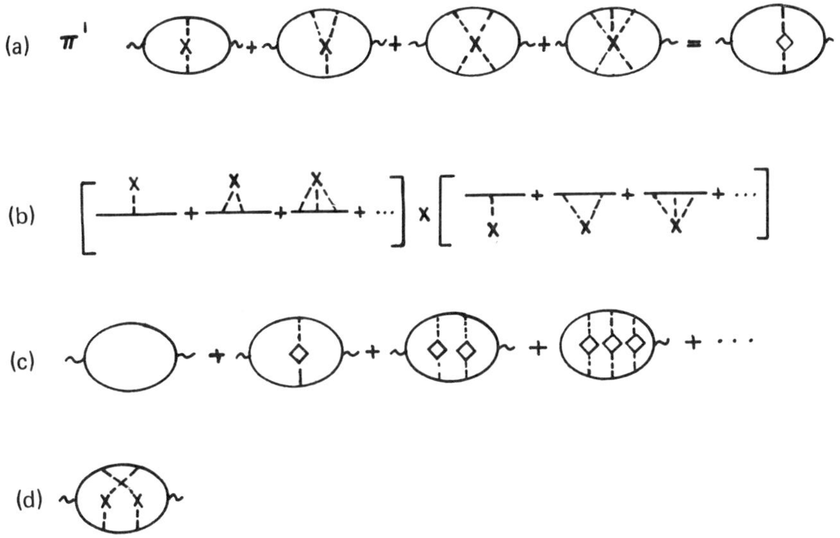

FIGURE 7.4

a vertex diagram. Figure 7.4(a) only shows vertex diagrams with a single impurity participating in the scattering. Vertex diagrams can occur with scattering from several impurities. These are equally important and will be considered later.

The sum of all the diagrams in Fig. 7.4(a) can be evaluated in a simple way and gives the answer $n_i T_{\mathbf{pp}'}(ip) T_{\mathbf{p}'\mathbf{p}}(ip + i\omega)$, so that the correlation function is

$$\pi^{(1)} = \frac{2e^2}{\nu^2 3m^2} \sum_{\mathbf{p}} \frac{1}{\beta} \sum_{ip} \mathbf{p} \cdot \mathbf{p}' \mathscr{G}(\mathbf{p}, ip) \mathscr{G}(\mathbf{p}, ip + i\omega) \mathscr{G}(\mathbf{p}', ip)$$
$$\times \mathscr{G}(\mathbf{p}', ip + i\omega) n_i T_{\mathbf{pp}'}(ip) T_{\mathbf{p}'\mathbf{p}}(ip + i\omega) \tag{7.1.13}$$

Equation (7.1.13) is easy to prove once we remember the rules about all scatterings from a single impurity: (1) Total momentum is conserved, and (2) energy is not changed from ip or $ip + i\omega$. The second rule says that all electron lines in one string, say the top, are at the same energy ip, while the ones on the bottom are at $ip + i\omega$. The momentum conservation requires that the momentum transfer on the top electron line be $\mathbf{p} - \mathbf{p}'$, which is exactly the opposite of the other, $\mathbf{p}' - \mathbf{p}$. The evaluation is easy, since the two sides decouple, as is illustrated schematically in Fig. 7.4(b). The lower sum of diagrams is

$$\mathscr{G}(\mathbf{p}, ip + i\omega)\mathscr{G}(\mathbf{p}', ip + i\omega)\Big[V_{\mathbf{pp}'} + \int \frac{d^3p_1}{(2\pi)^3} V_{\mathbf{pp}_1} V_{\mathbf{p}_1\mathbf{p}'} \mathscr{G}(\mathbf{p}_1, ip + i\omega)$$
$$+ \int \frac{d^3p_1\, d^3p_2}{(2\pi)^3} V_{\mathbf{pp}_1} \mathscr{G}(\mathbf{p}_1, ip + i\omega) V_{\mathbf{p}_1\mathbf{p}_2} \mathscr{G}(\mathbf{p}_2, ip + i\omega) V_{\mathbf{p}_2\mathbf{p}'} + \cdots \Big]$$
$$= \mathscr{G}(\mathbf{p}, ip + i\omega)\mathscr{G}(\mathbf{p}', ip + i\omega) T_{\mathbf{pp}'}(ip + i\omega)$$

The part in brackets is just the equation for the energy-dependent T matrix which was derived in (4.1.33), except the T-matrix equation has $\mathscr{G}^{(0)} = (ip - \xi_p)^{-1}$ instead of $\mathscr{G}$ in the terms in the summation. That is, self-energy effects are not usually included in the internal lines of the scattering process. We shall also make the approximation of replacing $\mathscr{G}$ by $\mathscr{G}^{(0)}$ in these internal lines, so that we recover the T-matrix equation. The top electron line in Fig. 7.4(b) is evaluated exactly the same way, except it goes from $\mathbf{p}'$ to $\mathbf{p}$, with energy ip:

$$\mathscr{G}(\mathbf{p}, ip)\mathscr{G}(\mathbf{p}', ip) T_{\mathbf{p}'\mathbf{p}}(ip)$$

The product of these two results gives (7.1.13).

The sum of the series in Fig. 7.4(a) is represented by a diamond in the figures,

$$= n_i T_{\mathbf{pp}'}(ip) T_{\mathbf{p}'\mathbf{p}}(ip + i\omega) \equiv W^{(1)}_{\mathbf{pp}'}(ip, ip + i\omega) \tag{7.1.14}$$

which is illustrated to the right of Fig. 7.4(a). The diamond is the total vertex scattering from a single impurity, which is abbreviated as $W^{(1)}_{\mathbf{pp}'}(ip, ip + i\omega)$.

Figure 7.4(c) shows the sum of correlation functions which have increasing numbers of diamonds in them. These are called *ladder diagrams.* When they are summed, we shall obtain our objective of a relaxation time with the factor $1 - \cos\theta'$. It is important to realize that the sum of diagrams in Fig. 7.4(c) is not the only contribution with scattering from several impurity sites. An example of a nonladder diagram is shown in Fig. 7.4(d), and this contribution is not included in the series shown in Fig. 7.4(c). The sum of ladder diagrams omits many terms. However, the omitted terms are not as important in the limit where $n_i \to 0$.

The first two terms in the sum of ladder diagrams have already been derived; they are $\pi^{(0)}$ and $\pi^{(1)}$ in (7.1.9) and (7.1.13). The other terms are

$$\pi^{(L)}(i\omega) = \frac{2e^2}{3m^2\nu} \sum_{\mathbf{p}} \frac{1}{\beta} \sum_{ip} \mathscr{G}(\mathbf{p}, ip)\mathscr{G}(\mathbf{p}, ip + i\omega)\mathbf{p}$$
$$\times\Big[\mathbf{p} + \frac{1}{\nu}\sum_{\mathbf{p}'} W^{(1)}_{\mathbf{pp}'}\mathscr{G}(\mathbf{p}', ip)\mathscr{G}(\mathbf{p}', ip + i\omega)\mathbf{p}' + \frac{1}{\nu^2}\sum_{\mathbf{p}'\mathbf{p}''} W^{(1)}_{\mathbf{pp}'}$$
$$\times\mathscr{G}(\mathbf{p}', ip)\mathscr{G}(\mathbf{p}', ip+i\omega)W^{(1)}_{\mathbf{p}'\mathbf{p}''}\mathscr{G}(\mathbf{p}'', ip)\mathscr{G}(\mathbf{p}'', ip+i\omega)\mathbf{p}''+\cdots\Big]$$

The superscript (L) denotes ladder sum. The series of terms in the ladder sum can be generated by representing it as a vector vertex function $\mathbf{\Gamma}^{(L)}(\mathbf{p}, ip, ip + i\omega)$:

$$\pi^{(L)}(i\omega) = \frac{2}{3}\frac{e^2}{m^2\nu} \sum_{\mathbf{p}} \frac{1}{\beta} \sum_{ip} \mathscr{G}(\mathbf{p}, ip)\mathscr{G}(\mathbf{p}, ip + i\omega)$$
$$\times\mathbf{p}\cdot\mathbf{\Gamma}^{(L)}(\mathbf{p}, ip, ip + i\omega) \tag{7.1.15}$$

$$\mathbf{\Gamma}^{(L)}(\mathbf{p}, ip, ip + i\omega) = \mathbf{p} + \frac{1}{\nu}\sum_{\mathbf{p}'} \mathbf{\Gamma}^{(L)}(\mathbf{p}', ip, ip + i\omega)W^{(1)}_{\mathbf{pp}'}(ip, ip + i\omega)$$
$$\times\mathscr{G}(\mathbf{p}', ip)\mathscr{G}(\mathbf{p}', ip + i\omega) \tag{7.1.16}$$

Repeated iteration of (7.1.16) will generate the series of terms in the ladder summation shown in Fig. 7.4(c). The ladder summation will be evaluated, approximately, in order to obtain the $1 - \cos\theta'$ factor in the relaxation time. The next section contains a systematic discussion of all the vertex terms.

The first term in the ladder summation, after $\pi^{(0)}$, is $\pi^{(1)}$, which is given in (7.1.13). It will be evaluated using the four steps summarized after Eq. (7.1.10). The first step is to evaluate the summation over ip in the combination of terms:

$$\frac{1}{\beta}\sum_{ip} \mathscr{G}(\mathbf{p}, ip)\mathscr{G}(\mathbf{p}, ip + i\omega)\mathscr{G}(\mathbf{p}', ip)\mathscr{G}(\mathbf{p}', ip + i\omega)W^{(1)}_{\mathbf{pp}'}(ip, ip + i\omega)$$
$$= -2n_i \int_{-\infty}^{\infty} \frac{d\varepsilon}{2\pi}\, n_F(\varepsilon)\{\mathrm{Im}[G_{\mathrm{ret}}(\mathbf{p}, \varepsilon)\mathscr{G}(\mathbf{p}', \varepsilon)T_{\mathbf{pp}'}(\varepsilon)]\mathscr{G}(\mathbf{p}, i\omega + \varepsilon)$$
$$\times\mathscr{G}(\mathbf{p}', ip' + \varepsilon)T_{\mathbf{p}'\mathbf{p}}(i\omega + \varepsilon) + \mathrm{Im}[G_{\mathrm{ret}}(\mathbf{p}, \varepsilon)G_{\mathrm{ret}}(\mathbf{p}', \varepsilon)T_{\mathbf{p}'\mathbf{p}}(\varepsilon)]$$
$$\times\mathscr{G}(\mathbf{p}, \varepsilon - i\omega)\mathscr{G}(\mathbf{p}', \varepsilon - i\omega)T_{\mathbf{pp}'}(\varepsilon - i\omega)\}$$

The summation result, on the right-hand side, is found by constructing the usual contour integral, which has cuts along the real axis $Z = \varepsilon$ and also $Z = \varepsilon - i\omega$. The next two steps are the analytical continuation $i\omega \rightarrow \omega + i\delta$, followed by taking the imaginary part:

$$\begin{aligned}\mathrm{Im}[\pi^{(1)}_{\mathrm{ret}}(\omega)] = \frac{-4e^2n_i}{m^2\nu^2}\sum_{\mathbf{pp}'}\mathbf{p}\cdot\mathbf{p}'\int_{-\infty}^{\infty}\frac{d\varepsilon}{2\pi}\,n_F(\varepsilon)\\ \times\{\mathrm{Im}[G_{\mathrm{ret}}(\mathbf{p},\varepsilon)G_{\mathrm{ret}}(\mathbf{p}',\varepsilon)T_{\mathbf{pp}'}(\varepsilon)]\\ \times\mathrm{Im}[G_{\mathrm{ret}}(\mathbf{p},\varepsilon+\omega)G_{\mathrm{ret}}(\mathbf{p}',\varepsilon+\omega)T_{\mathbf{p}'\mathbf{p}}(\varepsilon+\omega)]\\ +\mathrm{Im}[G_{\mathrm{ret}}(\mathbf{p},\varepsilon)G_{\mathrm{ret}}(\mathbf{p}',\varepsilon)T_{\mathbf{p}'\mathbf{p}}(\varepsilon)]\\ \times\mathrm{Im}[G_{\mathrm{adv}}(\mathbf{p},\varepsilon-\omega)G_{\mathrm{adv}}(\mathbf{p}',\varepsilon-\omega)T_{\mathbf{pp}'}(\varepsilon-\omega-i\delta)]\}\end{aligned}$$

Next we make the variable change $\varepsilon \to \varepsilon + \omega$ in the second term in the curly braces. Then change the advanced functions to retarded, which are the complex conjugate functions, so that it changes the sign of the imaginary part. These rearrangements lead to the expression

$$\begin{aligned}\mathrm{Im}[\pi^{(1)}_{\mathrm{ret}}(\omega)] = \frac{-4e^2n_i}{m^2\nu^2}\sum_{\mathbf{pp}'}\mathbf{p}\cdot\mathbf{p}'\int_{-\infty}^{\infty}\frac{d\varepsilon}{2\pi}\,[n_F(\varepsilon)-n_F(\varepsilon+\omega)]\\ \times\mathrm{Im}[G_{\mathrm{ret}}(\mathbf{p},\varepsilon)G_{\mathrm{ret}}(\mathbf{p}',\varepsilon)T_{pp'}(\varepsilon)]\\ \times\mathrm{Im}[G_{\mathrm{ret}}(\mathbf{p},\varepsilon+\omega)G_{\mathrm{ret}}(\mathbf{p}',\varepsilon+\omega)T_{\mathbf{p}'\mathbf{p}}(\varepsilon+\omega)]\end{aligned}$$

Finally, we divide by ω and take the limit where $\omega \to 0$. This derives the conductivity contribution $\sigma^{(1)}$ from the first ladder contribution:

$$\begin{aligned}\sigma^{(1)} = \frac{4e^2n_i}{3m^2}\int\frac{d^3p}{(2\pi)^6}\,d^3p'\,\mathbf{p}\cdot\mathbf{p}'\int\frac{d\varepsilon}{2\pi}\left[-\frac{\partial n_F(\varepsilon)}{\partial\varepsilon}\right]\\ \times\mathrm{Im}[G_{\mathrm{ret}}(\mathbf{p},\varepsilon)G_{\mathrm{ret}}(\mathbf{p}',\varepsilon)T_{\mathbf{pp}'}(\varepsilon)]\,\mathrm{Im}[G_{\mathrm{ret}}(\mathbf{p},\varepsilon)G_{\mathrm{ret}}(\mathbf{p}',\varepsilon)T_{\mathbf{p}'\mathbf{p}}(\varepsilon)]\end{aligned}$$

In the last step, the limit $\nu \to \infty$ changes the wave vector summations into integrations. To keep the argument simple, several approximations will be introduced at this stage. The next section contains a more careful justification of the final answer without these approximations. First, the T matrices $T_{\mathbf{pp}'}$ are taken to be real and removed from the imaginary operator—of course they are not strictly real; this is an approximation. The ε dependence is removed from the T matrices by setting $\varepsilon = \xi_p$ in $T_{\mathbf{pp}'}$ and $T_{\mathbf{p}'\mathbf{p}}$. The remaining factors are

$$\{\mathrm{Im}[G_{\mathrm{ret}}(\mathbf{p},\varepsilon)G_{\mathrm{ret}}(\mathbf{p}',\varepsilon)]\}^2$$

If we write

$$G_{\mathrm{ret}}(\mathbf{p},\varepsilon) = \mathrm{Re}(G) - \frac{i}{2}A$$

$$\{\mathrm{Im}[G_{\mathrm{ret}}(\mathbf{p}, \varepsilon)G_{\mathrm{ret}}(\mathbf{p}', \varepsilon)]\}^2 = \tfrac{1}{4}\{\mathrm{Re}[G(\mathbf{p}, \varepsilon)]A(\mathbf{p}', \varepsilon) + \mathrm{Re}[G(\mathbf{p}', \varepsilon)]A(\mathbf{p}, \varepsilon)\}^2$$
$$= \tfrac{1}{4}\{\mathrm{Re}[G(\mathbf{p}, \varepsilon)]^2 A(\mathbf{p}', \varepsilon)^2 + \mathrm{Re}[G(\mathbf{p}', \varepsilon)]^2 A(\mathbf{p}, \varepsilon)^2$$
$$+ 2\,\mathrm{Re}[G(\mathbf{p}, \varepsilon)]\,\mathrm{Re}[G(\mathbf{p}', \varepsilon)]A(\mathbf{p}, \varepsilon)A(\mathbf{p}', \varepsilon)\}$$

The first two terms on the right-hand side are equal after interchanging dummy variables $\mathbf{p}$ and $\mathbf{p}'$. The other term is neglected as being less important when $n_i \to 0$. Thus we can collect terms and express $\sigma^{(1)}$ as

$$\sigma^{(1)} = \frac{e^2}{3m^2} \int \frac{d^3p}{(2\pi)^3}\, p^2 \int_{-\infty}^{\infty} \frac{d\varepsilon}{2\pi}\, A(\mathbf{p}, \varepsilon)^2 \left[\frac{-\partial n_F(\varepsilon)}{\partial \varepsilon}\right] \Lambda(\mathbf{p}, \varepsilon)$$

$$\Lambda(\mathbf{p}, \varepsilon) = \frac{2n_i}{p^2} \int \frac{d^3p'}{(2\pi)^3}\, \mathbf{p} \cdot \mathbf{p}' \mid T_{\mathbf{pp}'} \mid^2 \{\mathrm{Re}[G_{\mathrm{ret}}(\mathbf{p}', \varepsilon)]\}^2$$

Equation (7.1.10) for $\sigma^{(0)}$ is the same as the preceding except that $\Lambda = 1$. If we write

$$\sigma^{(0)} + \sigma^{(1)} = \frac{e^2}{3m^2} \int \frac{d^3p}{(2\pi)^3}\, p^2 \int_{-\infty}^{\infty} \frac{d\varepsilon}{2\pi}\, A(\mathbf{p}, \varepsilon)^2 \left[\frac{-\partial n_F(\varepsilon)}{\partial \varepsilon}\right] [1 + \Lambda(\mathbf{p}, \varepsilon)] \tag{7.1.17}$$

then $\Lambda(\mathbf{p}, \varepsilon)$ is the vertex correction in the integrand which arises from the ladder diagram $\pi^{(1)}$. To evaluate $\Lambda(\mathbf{p}, \varepsilon)$ when $n_i \to 0$, we need to know how to handle the expression $[\mathrm{Re}(G_{\mathrm{ret}})]^2$. We shall proceed in the same spirit which was earlier used to get $A(\mathbf{p}, \varepsilon)^2$. Consider the following:

$$\mathrm{Re}(G) = \frac{x}{x^2 + \Delta^2}, \qquad x = \varepsilon - \xi_p - \mathrm{Re}(\Sigma)$$

$$\lim_{\Delta \to 0} [\mathrm{Re}(G)]^2 = \frac{x^2}{(x^2 + \Delta^2)^2} = \frac{1}{x^2 + \Delta^2} - \frac{\Delta^2}{(x^2 + \Delta^2)^2}$$
$$= \frac{1}{2\Delta} A - \frac{1}{4} A^2$$
$$= 2\pi\delta(x)\left(\frac{1}{2\Delta} - \frac{1}{4\Delta}\right) \simeq \frac{2\pi\delta(x)}{4\Delta}$$

In the limit where $\Delta \to 0$, we find that $[\mathrm{Re}(G_{\mathrm{ret}})]^2$ diverges and behaves roughly as $A(\mathbf{p}, \varepsilon)/4\Delta$. Thus we obtain

$$\lim_{n_i \to 0} \Lambda(\mathbf{p}, \varepsilon) = 2\pi n_i \int \frac{d^3p'}{(2\pi)^3}\, \frac{\mathbf{p} \cdot \mathbf{p}'}{p^2} \mid T_{\mathbf{pp}'} \mid^2 \frac{\delta(\varepsilon - \xi_{p'})}{\Delta(\mathbf{p}', \varepsilon)}$$

We use the fact that $\partial n_F/\partial \varepsilon$ in (7.1.17) will force $\varepsilon = 0$, while $\delta(-\xi_{p'})$

in the preceding will force $p' = k_F$. We define the transport scattering rate as

$$\frac{1}{\tau_1(k)} = 2\Delta_T = 2\pi n_i \int \frac{d^3p'}{(2\pi)^3} \mid T_{\mathbf{pp}'} \mid^2 \delta(-\xi_{p'})\left(1 - \frac{\mathbf{p} \cdot \mathbf{p}'}{p^2}\right)$$

$$\Lambda = \frac{1}{\Delta}(\Delta - \Delta_T)$$

where $\cos\theta' = (\mathbf{p} \cdot \mathbf{p}')/p^2$. We see that Λ is given by the ratio of $(\Delta - \Delta_T)/\Delta$. Thus Λ is independent of n_i in the limit where $n_i \to 0$. The ladder diagram does make a contribution to the conductivity which is of order n_i^{-1}. Although the term started out as a scattering term which was proportional to n_i, its evaluation introduced enough factors of A^2 and $[\mathrm{Re}(G)]^2$ to give extra factors of n_i^{-1}. The vertex corrections do cause contributions to the conductivity which are the same order in n_i^{-1} as the basic bubble diagram.

The higher terms in the series of ladder diagrams just cause additional powers of Λ^n. Thus the leading terms in n_i^{-1} in this summation of vertex corrections just generates the series

$$\sigma^{(L)} = \sum_{n=0}^{\infty} \sigma(n) = \frac{e^2}{3m^2} \int \frac{d^3p}{(2\pi)^4} p^2 A(p, 0)^2 [1 + \Lambda + \Lambda^2 + \Lambda^3 + \cdots]$$

$$= \frac{e^2}{3m^2} \int \frac{d^3p}{(2\pi)^4} p^2 \frac{A(p, 0)^2}{1 - \Lambda}$$

where the term Λ^n is given by n ladders. The factor in brackets is a series which is easily summed,

$$\frac{1}{1 - \Lambda} = \frac{1}{1 - (\Delta - \Delta_T)/\Delta} = \frac{\Delta}{\Delta_T}$$

to give the ratio of Δ/Δ_T. Thus the result for the conductivity is $(A^2 \to A/\Delta)$

$$\lim_{n_i \to 0} \sigma^{(L)} = \frac{e^2}{3m^2} \int \frac{d^3p}{(2\pi)^3} p^2 \frac{\delta(\xi_p)}{\Delta_T} = \frac{2e^2}{3m^2} \int \frac{d^3p}{(2\pi)^3} p^2 \tau_1(p) \times \left[\frac{-\partial n_F(\xi_p)}{\partial \xi_p}\right] \tag{7.1.18}$$

Equation (7.1.18) is the desired result for the conductivity, since the relaxation time $\tau_1(p)$ now has the factor $1 - \cos\theta'$ which is known from classical transport theory. Equation (7.1.18) is identical to the Boltzmann equation result (7.1.3).

The Green's function derivation of this result is certainly harder than the transport equation of Sec. 7.1A. The one advantage of the Green's function method is that we do not have to make the approximation that $n_i \to 0$. The $n_i \to 0$ limit is implicitly made in the Boltzmann equation result when we assume the particles are plane waves except for occasional scattering events from isolated impurities. The Green's function derivation of (7.1.18) emphasizes the importance of this limit $n_i \to 0$, since it was continually invoked to pick out the divergent terms to the neglect of others. In Sec. 7.1C we shall discuss how to evaluate the Kubo formula without having to assume that $n_i \to 0$.

There are several lessons to be learned from this summation of ladder diagrams. The ladder diagrams, although they appear in a higher order of perturbation theory, do not lead to terms in the final answer which are smaller than lower-order terms. Thus a term is not necessarily small if it occurs in a higher order of perturbation theory. It is this feature of vertex corrections which means they can never be dismissed without an investigation. They may be small, but one should always check. It would be nice to have some rules of thumb which would establish whether vertex corrections are important. The way to tell is to examine the scattering process which causes the vertex functions. If the two particle states, on each side of the bubble, can scatter quasielastically so that their relative energy changes little, then vertex corrections are large. The vertex correction is basically a potential divided by an energy denominator $\Lambda \simeq V/\Delta E$, where ΔE is the change in energy of the two particles. If ΔE is small, the small denominator will compensate for the small potential V, so that vertex corrections become sizable. Repeated scatterings, as in a series of ladder diagrams, just cause additional powers of the factor $(V/\Delta E)^n$. Thus vertex corrections are large when the scattering by the potential causes only a small change in the relative energy of the two particles.

C. Kubo Formula: Rigorous Solution

The evaluation of the Kubo formula in the previous section was mainly intended to rederive the result known from the Boltzmann equation. Another objective was to emphasize the importance of vertex corrections in the evaluation of two-particle correlation functions. It is seldom a good approximation to evaluate a two-particle correlation function as the product of two single-particle Green's functions. The previous derivation assumed the limit $n_i \to 0$ and retained only the most divergent terms.

The same correlation function is evaluated much more rigorously in the present section. We follow the treatment of Langer (1960), who found a formally exact solution. The basic model Hamiltonian is the same: The electrons scatter from randomly located impurities which are treated as spherically symmetric scattering potentials. Since the potentials are fixed, with no degrees of freedom, the electron does not change its energy ip upon scattering. In a Feynman diagram, all electron Green's functions which fall on a continuous line between vertices of a bubble have the same energy ip or $ip + i\omega$. It is this feature which is useful for obtaining a formal solution to the problem. Since the energy is specified for each Green's function, the scattering processes affect only the momentum distribution of the particles, and the integral equations are in terms of momentum variables.

The first step in a formal solution to the correlation function is to define the *irreducible scattering vertex* $W_{\mathbf{pp}'}(ip, ip + i\omega)$, which is the summation of all *different* scattering terms. The word *different* has a carefully prescribed meaning. The vertex occurs as a scattering between two electron lines, as shown in Fig. 7.5, such that one can draw a box around it which only cuts the electron lines. Three different scattering vertices have light boxes drawn around them in Fig. 7.5(a). The irreducible vertex $W_{\mathbf{pp}'}(ip, ip + i\omega)$ is the summation of all different vertex terms. These are evaluated using Green's functions $\mathscr{G}$ with exact self-energies. The two vertices in Fig. 7.5(b) are different since they have different numbers of potential lines to each electron line. Figure 7.5(c) has two identical vertex terms. Only one is included in the irreducible vertex $W_{\mathbf{pp}'}$, and the double vertex in Fig. 7.5(c) will be included in the final answer by terms with more than one $W_{\mathbf{pp}'}$. One contribution to the irreducible vertex is the summation of all scattering terms from a single impurity center. This sum of diagrams is shown in Fig. 7.4(a) and was evaluated in (7.1.14) as

$$W^{(1)}_{\mathbf{pp}'}(ip, ip + i\omega) = n_i T_{\mathbf{pp}'}(ip) T_{\mathbf{p}'\mathbf{p}}(ip + i\omega)$$

Other terms in the irreducible vertex $W^{(j)}$ arise from terms which have simultaneous scattering from two or more impurities.

In Feynman diagrams, the irreducible vertex is represented as a shaded rectangle, as shown in Fig. 7.6(a). The exact correlation function $\pi(i\omega)$ for the Kubo formula is the summation of bubble diagrams shown in Fig. 7.6(b). Each term in the series of bubbles has one more irreducible

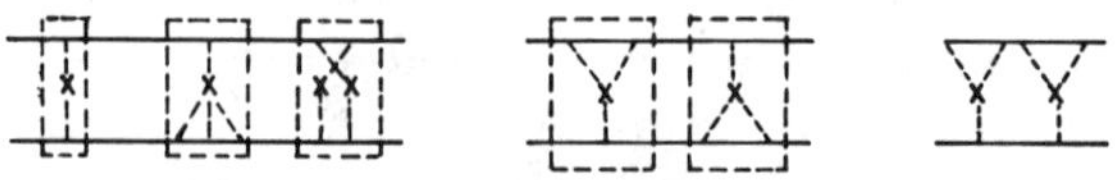

(a) (b) (c) FIGURE 7.5

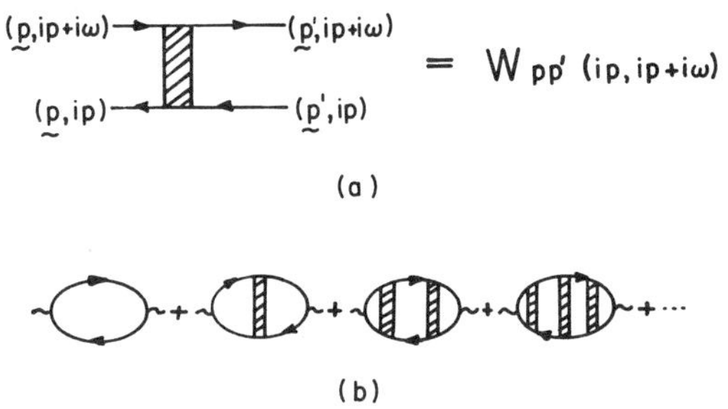

FIGURE 7.6

vertex and two more Green's functions $\mathscr{G}(\mathbf{p}_j, ip)$, $\mathscr{G}(\mathbf{p}_j, ip + i\omega)$:

$$\begin{aligned}\pi(i\omega) = \frac{2e^2}{3m^2\nu\beta} \sum_{\mathbf{p},ip} \Big\{ &\mathscr{G}(\mathbf{p}, ip)\mathscr{G}(\mathbf{p}, ip + i\omega) \\ &\times \Big[p^2 + \frac{1}{\nu} \sum_{\mathbf{p}'} \mathbf{p} \cdot \mathbf{p}' W_{\mathbf{pp}'}(ip, ip + i\omega)\mathscr{G}(\mathbf{p}', ip + i\omega)\mathscr{G}(\mathbf{p}', ip) \\ &+ \frac{1}{\nu^2} \sum_{\mathbf{p}'\mathbf{p}''} \mathbf{p} \cdot \mathbf{p}'' W_{\mathbf{pp}'} W_{\mathbf{p}'\mathbf{p}''} \mathscr{G}(\mathbf{p}', ip)\mathscr{G}(\mathbf{p}', ip + i\omega) \\ &\times \mathscr{G}(\mathbf{p}'', ip)\mathscr{G}(\mathbf{p}'', ip + i\omega) + \cdots \Big\}\end{aligned}$$

This series of terms may be summed in a formal manner by defining a vector vertex function $\mathbf{\Gamma}(\mathbf{p}, ip, ip + i\omega)$ which obeys the equation

$$\begin{aligned}\mathbf{\Gamma}(\mathbf{p}, ip + i\omega) &= \mathbf{p} + \frac{1}{\nu} \sum_{\mathbf{p}'} W_{\mathbf{pp}'}(ip, ip + i\omega)\mathscr{G}(\mathbf{p}', ip)\mathscr{G}(\mathbf{p}', ip + i\omega) \\ &\quad \times \mathbf{\Gamma}(\mathbf{p}', ip, ip + i\omega) \\ \pi(i\omega) &= \frac{2e^2}{3m^2\nu} \frac{1}{\beta} \sum_{\mathbf{p},ip} \mathscr{G}(\mathbf{p}, ip)\mathscr{G}(\mathbf{p}, ip + i\omega)\mathbf{p} \cdot \mathbf{\Gamma}(\mathbf{p}, ip, ip + i\omega)\end{aligned} \tag{7.1.19}$$

This set of equations, using Green's functions $\mathscr{G}$ with the exact self-energy function $\Sigma(\mathbf{p}, ip)$, includes all terms in the correlation function. They are written in terms of the self-energy Σ and irreducible vertex W, which themselves are each a summation of an infinite number of terms. In practice, these two functions cannot be found exactly but only approximately, so that one never obtains the exact solution. However, we can continue with the formal description of the evaluation of the correlation function assuming that we can obtain these functions.

The correlation function is a function of ip and $i\omega$—in fact, in the combination ip and $ip + i\omega$. Thus let us define the quantity

$$P(ip, ip + i\omega) = \frac{2e^2}{3m^2} \frac{1}{\nu} \sum_{\mathbf{p}} \mathscr{G}(\mathbf{p}, ip)\mathscr{G}(\mathbf{p}, ip + i\omega)\mathbf{p} \cdot \mathbf{\Gamma}(\mathbf{p}, ip, ip + i\omega)$$
$$\pi(i\omega) = \frac{1}{\beta} \sum_{ip} P(ip, ip + i\omega) \tag{7.1.20}$$

The summation over Matsubara frequencies ip is evaluated, as usual, by examining the contour integral $\int dz n_F(z) P(z, z + i\omega)$. The integrand has the poles of $n_F(z)$, which give the summation over ip, and also branch cuts along the two axes $z = \varepsilon$ and $z = \varepsilon - i\omega$, where ε is real:

$$\pi(i\omega) = \int_{-\infty}^{\infty} \frac{d\varepsilon}{2\pi i} n_F(\varepsilon)[P(\varepsilon + i\delta, \varepsilon + i\omega) - P(\varepsilon - i\delta, \varepsilon + i\omega) + P(\varepsilon - i\omega, \varepsilon + i\delta) - P(\varepsilon - i\omega, \varepsilon - i\delta)]$$

Next we find the retarded function from the analytical continuation $i\omega \to \omega + i\delta$. A variable change $\varepsilon \to \varepsilon + \omega$ in the last two terms brings us to the point

$$\pi_{\text{ret}}(\omega) = -\int \frac{d\varepsilon}{2\pi i} \{n_F(\varepsilon)[P(\varepsilon + i\delta, \varepsilon + \omega + i\delta) - P(\varepsilon - i\delta, \varepsilon + \omega + i\delta)] + n_F(\varepsilon + \omega)[P(\varepsilon - i\delta, \varepsilon + \omega + i\delta) - P(\varepsilon - i\delta, \varepsilon + \omega - i\delta)]\}$$

The second and third terms have the same factor $P(\varepsilon - i\delta, \varepsilon + \omega + i\delta)$. The first and fourth terms have the factors $P(\varepsilon \pm i\delta, \varepsilon + \omega \pm i\delta)$, which are complex conjugates of each other:

$$\begin{aligned} n_F(\varepsilon)P(\varepsilon + i\delta, \varepsilon + \omega + i\delta) - n_F(\varepsilon + \omega)P(\varepsilon - i\delta, \varepsilon + \omega - i\delta) \\ = [n_F(\varepsilon) - n_F(\varepsilon + \omega)] \operatorname{Re}[P(\varepsilon + i\delta, \varepsilon + \omega + i\delta)] \\ + i[n_F(\varepsilon) + n_F(\varepsilon + \omega)] \operatorname{Im}[P(\varepsilon + i\delta, \varepsilon + \omega + i\delta)] \end{aligned}$$

The next step is to take the imaginary part of this expression, which eliminates the term $\operatorname{Im}[P(\varepsilon + i\delta, \varepsilon + \omega + i\delta)]$. The limit of $\omega \to 0$ brings us to a formal expression for the dc conductivity:

$$\begin{aligned} \sigma &= -\lim_{\omega \to 0} \frac{\operatorname{Im}(\pi_{\text{ret}})}{\omega} \\ &= \int_{-\infty}^{\infty} \frac{d\varepsilon}{2\pi} \left[\frac{-\partial n_F(\varepsilon)}{\partial \varepsilon}\right] \{P(\varepsilon - i\delta, \varepsilon + i\delta) - \operatorname{Re}[P(\varepsilon + i\delta, \varepsilon + i\delta)]\} \end{aligned} \tag{7.1.21}$$

The term $P(\varepsilon - i\delta, \varepsilon + i\delta)$ is real, since the function $P(z, z')$ is symmetric in its two arguments, and a complex conjugation turns $P(\varepsilon - i\delta, \varepsilon + i\delta)$ into itself.

There are only two functions which need to be found, $P(\varepsilon - i\delta, \varepsilon + i\delta)$ and $P(\varepsilon + i\delta, \varepsilon + i\delta)$. At zero temperature, where $-\partial n_F/\partial\varepsilon = \delta(\varepsilon)$, they need to be found only at $\varepsilon = 0$. These two functions have quite different behavior and are obtained by different methods. Both are usually important, but the most singular is $P(\varepsilon - i\delta, \varepsilon + i\delta)$, and this term leads to the form $1 - \cos\theta'$ in the limit $n_i \to 0$. To see this, first consider the definition of P in (7.1.20):

$$P(\varepsilon-i\delta, \varepsilon+i\delta) = \frac{2e^2}{3m^2}\frac{1}{\nu}\sum_{\mathbf{p}} \mathscr{G}(\mathbf{p}, \varepsilon+i\delta)\mathscr{G}(\mathbf{p}, \varepsilon-i\delta)\mathbf{p}\cdot\mathbf{\Gamma}(\mathbf{p}, \varepsilon-i\delta, \varepsilon+i\delta)$$

The product of Green's functions are entirely real:

$$\begin{aligned}\mathscr{G}(\mathbf{p}, \varepsilon+i\delta)\mathscr{G}(\mathbf{p}, \varepsilon-i\delta) &= G_{\text{ret}}(\mathbf{p}, \varepsilon)G_{\text{adv}}(\mathbf{p}, \varepsilon) = \frac{1}{[\varepsilon-\xi_{\mathbf{p}}-\text{Re}(\Sigma)]^2+[\text{Im}(\Sigma)]^2}\\ &= -\frac{A(\mathbf{p}, \varepsilon)}{-2\,\text{Im}[\Sigma(\mathbf{p}, \varepsilon)]} \equiv \frac{A(\mathbf{p}, \varepsilon)}{2\Delta(\mathbf{p}, \varepsilon)}\\ P(\varepsilon - i\delta, \varepsilon + i\delta) &= \frac{e^2}{3m^2}\int\frac{d^3p}{(2\pi)^3}\frac{A(\mathbf{p}, \varepsilon)}{\Delta(\mathbf{p}', \varepsilon)}\mathbf{p}\cdot\mathbf{\Gamma}(\mathbf{p}, \varepsilon - i\delta, \varepsilon + i\delta)\end{aligned} \tag{7.1.22}$$

The combination $G_{\text{ret}}G_{\text{adv}}$ is rigorously defined as the spectral function $A(p, \varepsilon)$ divided by $2\Delta = -2\,\text{Im}(\Sigma)$. Note that we have not assumed that $A^2 \to A/2\Delta$ but have in fact rigorously derived $A/2\Delta$ for this combination of Green's functions. The same combination is found in the equation for the vertex function:

$$\begin{aligned}\mathbf{\Gamma}(\mathbf{p}, \varepsilon - i\delta, \varepsilon + i\delta) = \mathbf{p} &+ \int\frac{d^3p'}{(2\pi)^3}\mathbf{\Gamma}(\mathbf{p}', \varepsilon - i\delta, \varepsilon + i\delta)\\ &\times W_{\mathbf{pp}'}(\varepsilon - i\delta, \varepsilon + i\delta)G_{\text{ret}}(\mathbf{p}, \varepsilon)G_{\text{adv}}(\mathbf{p}, \varepsilon)\end{aligned}$$

The vector function $\mathbf{\Gamma}$ must point in the direction of $\mathbf{p}$ since that is the only vector in its function arguments. Thus it is convenient to define an integral equation for the scalar function:

$$\begin{aligned}\mathbf{\Gamma}(\mathbf{p}, \varepsilon - i\delta, \varepsilon + i\delta) &= \mathbf{p}\gamma_1(\mathbf{p}, \varepsilon)\\ \gamma_1(\mathbf{p}, \varepsilon) &= 1 + \int\frac{d^3p'}{(2\pi)^3}\frac{A(\mathbf{p}', \varepsilon)}{2\Delta(\mathbf{p}', \varepsilon)}\frac{\mathbf{p}\cdot\mathbf{p}'}{p^2}\\ &\quad\times W_{\mathbf{pp}'}(\varepsilon - i\delta, \varepsilon + i\delta)\gamma_1(\mathbf{p}', \varepsilon)\\ P(\varepsilon - i\delta, \varepsilon + i\delta) &= \frac{e^2}{3m^2}\int\frac{d^3p}{(2\pi)^3}p^2\frac{A(\mathbf{p}, \varepsilon)}{\Delta(\mathbf{p}, \varepsilon)}\gamma_1(\mathbf{p}, \varepsilon)\end{aligned} \tag{7.1.23}$$

Equation (7.1.23) is a one-dimensional integral equation for the scalar function $\gamma_1(\mathbf{p}, \varepsilon)$, where p is the integration variable. The angular integrals $\int d\Omega_{p'}$ just average the quantity $\mathbf{p} \cdot \mathbf{p}' W_{pp'}$ over angles to provide the kernal for the integral equation. The integral equation should not be difficult to solve with modern computers for realistic self-energies and irreducible vertices.

The Boltzmann result is obtained in the twin limits $T \to 0$ $(\varepsilon \to 0)$ and $n_i \to 0$, where the quantities become

$$A(\mathbf{p}', 0) = 2\pi\delta(\xi_{\mathbf{p}'})$$

$$W_{\mathbf{pp}'}(-i\delta, i\delta) \to W^{(1)}_{\mathbf{pp}'}(-i\delta, i\delta) = n_i \mid T_{\mathbf{pp}'} \mid^2$$

Equation (7.1.23) then reduces to the integral equation

$$\gamma_1(\mathbf{p}) = 1 + \int \frac{d^3p'}{(2\pi)^3} \frac{2\pi\delta(\xi_{p'})}{2\Delta(\mathbf{p}')} n_i \mid T_{\mathbf{pp}'} \mid^2 \frac{\mathbf{p} \cdot \mathbf{p}'}{p^2} \gamma_1(\mathbf{p}')$$

$$\gamma_1(p_F) = 1 + \gamma_1(p_F) \frac{\Delta - \Delta_T}{\Delta}$$

$$2\Delta_T = 2\pi \int \frac{d^3p'}{(2\pi)^3} n_i \mid T_{\mathbf{pp}'} \mid \delta(\xi_{\mathbf{p}'})(1 - \mathbf{p} \cdot \mathbf{p}'/p^2)$$

which is easily solved to give

$$\gamma_1(p_F) = \frac{\Delta}{\Delta_T}$$

The factor $(1 - \mathbf{p} \cdot \mathbf{p}'/p^2) = 1 - \cos\theta'$ since $\mid \mathbf{p}' \mid = \mid \mathbf{p} \mid$. The solution $\gamma_1 = \Delta/\Delta_T$ is put into (7.1.22) to give

$$P(-i\delta, +i\delta) = \frac{e^2}{3m^2} \int \frac{d^3p}{(2\pi)^3} 2\pi \frac{\delta(\xi_p)p^2}{\Delta_T(\mathbf{p})}$$

which gives the same conductivity as (7.1.3) or (7.1.18) when put into (7.1.21). The term $P(\varepsilon - i\delta, \varepsilon + i\delta)$ leads to the important contribution as $T \to 0$ and $n_i \to 0$. This does not mean that the other term can be neglected.

The other term $P(\varepsilon + i\delta, \varepsilon + i\delta)$ is complex but not nearly as singular in the limit where $n_i \to 0$. As is evident from the definition (7.1.20), the singular parts should arise from the Green's function product:

$$G(\mathbf{p}, \varepsilon + i\delta)^2 = G_{\text{ret}}(\mathbf{p}, \varepsilon)^2 = \frac{1}{[\varepsilon - \xi_p - \text{Re}(\Sigma) - i\,\text{Im}(\Sigma)]^2}$$

In the limit where $n_i \to 0$ or $\Delta \to 0$, this becomes $[x = \varepsilon - \xi_{\mathbf{p}} - \mathrm{Re}(\Sigma)$, $\Delta = -\mathrm{Im}(\xi)]$

$$G_{\mathrm{ret}}(\mathbf{p}, \varepsilon)^2 \to \frac{x^2 - \Delta^2 - 2ix\Delta}{(x^2 + \Delta^2)^2} \to \frac{1}{x^2 + \Delta^2} - \frac{2\Delta^2}{x^2 + \Delta^2} - \frac{2ix\Delta}{(x^2 + \Delta^2)^2}$$

$$\underset{\Delta \to 0}{\to} \frac{A}{2\Delta} - \frac{1}{2} A^2 - \frac{ix}{\Delta} A^2$$

As $\Delta \to 0$ we can replace $A^2 \to A/\Delta$ and find that the real part of this expression vanishes. The imaginary part becomes $x\delta(x)$, which is also zero. Thus the singular parts of this expression vanish as $\Delta \to 0$ or $n_i \to 0$. The vertex corrections to $P(\varepsilon + i\delta, \varepsilon + i\delta)$ are not of order unity but actually of order n_i. Thus in this case, the vertex corrections may actually be a series of terms which are successively smaller, so that the vertex corrections may be obtained by just evaluating the first few. The situation is quite different from that for $P(\varepsilon - i\delta, \varepsilon + i\delta)$, where one has to solve the vertex equation and sum all the ladder diagrams. Of course, one can evaluate $P(\varepsilon + i\delta, \varepsilon + i\delta)$ by solving a vertex equation similar to (7.1.23). If we define the scalar vertex $\gamma_2(p, \varepsilon)$ by $\mathbf{\Gamma}(\mathbf{p}, \varepsilon + i\delta, \varepsilon + i\delta) = \mathbf{p}\gamma_2(\mathbf{p}, \varepsilon)$, then we have

$$P(\varepsilon + i\delta, \varepsilon + i\delta) = \frac{2}{3} \frac{e^2}{m^2} \int \frac{d^3p}{(2\pi)^3} p^2 \gamma_2(\mathbf{p}, \varepsilon) G_{\mathrm{ret}}(\mathbf{p}, \varepsilon)^2$$

$$\gamma_2(\mathbf{p}, \varepsilon) = 1 + \int \frac{d^3p'}{(2\pi)^3} \frac{\mathbf{p} \cdot \mathbf{p}}{p^2} \gamma_2(\mathbf{p}', \varepsilon) W_{\mathbf{pp}'}(\varepsilon + i\delta, \varepsilon + i\delta) \times G_{\mathrm{ret}}(\mathbf{p}, \varepsilon)^2$$

The vertex function $\gamma_2(p, \varepsilon)$ is complex, as are the irreducible vertex $W_{\mathbf{pp}'}(\varepsilon + i\delta, \varepsilon + i\delta)$ and the product $G_{\mathrm{ret}}(\mathbf{p}, \varepsilon)^2$. Thus this is a two-coupled equation for the real and imaginary parts. It may be obtained from the *Ward* (1950) *identity*,

$$\mathbf{\Gamma}(\mathbf{p}, \varepsilon + i\delta, \varepsilon + i\delta) = \mathbf{p} + m\nabla_p \Sigma(\mathbf{p}, \varepsilon + i\delta)$$

which is an exact identity between the exact vertex function $\mathbf{\Gamma}(\mathbf{p}, \varepsilon + i\delta, \varepsilon + i\delta)$ and the exact retarded self-energy $\Sigma(\mathbf{p}, \varepsilon + i\delta)$ from impurity scattering. This self-energy is not the one we found in Sec. 4.1, since that was only from scattering from a single impurity and only had unperturbed Green's functions $G^{(0)}$ as the internal lines in the scattering equation. The exact self-energy is found from scattering from all numbers of impurities and using exact Green's functions as internal lines of diagrams—although this procedure must be done carefully in order not to count the same contribution twice. The Ward identity is very convenient, since it permits the vertex function to be obtained from the self-energy function by a simple operation. It will be proved in the next section.

D. Ward Identities

The evaluation of a two-particle correlation function, such as the Kubo formula, often requires an evaluation of a vertex function. The Ward (1950) identity is an exact relationship between vertex functions and the self-energies in the problem. As an example, two types of Ward identities permit the evaluation of the scalar vertex function $\Gamma(\mathbf{p}, ip)$ or the vector vertex function $\mathbf{\Gamma}(\mathbf{p}, ip)$, which satisfy the equations

$$\Gamma(\mathbf{p}, ip) = 1 + \int \frac{d^3p'}{(2\pi)^3} \Gamma(\mathbf{p}', ip)\mathscr{G}(\mathbf{p}', ip)^2 W_{\mathbf{pp}'}(ip, ip) \tag{7.1.24}$$

$$\mathbf{\Gamma}(\mathbf{p}, ip) = \mathbf{p} + \int \frac{d^3p'}{(2\pi)^3} \mathbf{\Gamma}(\mathbf{p}', ip)\mathscr{G}(\mathbf{p}', ip)^2 W_{\mathbf{pp}'}(ip, ip) \tag{7.1.25}$$

The Ward identity states that these two functions are given by

$$\Gamma(\mathbf{p}, ip) = 1 - \left[\frac{\partial \Sigma(\mathbf{p}, z)}{\partial z}\right]_{z=ip} \tag{7.1.26}$$

$$\mathbf{\Gamma}(\mathbf{p}, ip) = \mathbf{p} + m\nabla_p \Sigma(\mathbf{p}, ip) \tag{7.1.27}$$

Thus an evaluation of the self-energy function $\Sigma(\mathbf{p}, ip)$ permits an easy evaluation of these two vertex functions. We shall prove these relationships later.

An important point regarding the Ward identities is that they are not useful for evaluating all the vertex functions which are encountered. An example is provided in the last section, where the Ward identities were useful for finding $\mathbf{\Gamma}(\mathbf{p}, \varepsilon + i\delta, \varepsilon + i\delta)$ but not $\mathbf{\Gamma}(\mathbf{p}, \varepsilon - i\delta, \varepsilon + i\delta)$. The Ward identities cannot be applied blindly; one must use them only when appropriate. These circumstances are delineated after the identities are proved.

The Ward identities for impurity scattering were proved by Langer (1961). Similar theorems for electron–phonon interactions were derived by Engelsberg and Schrieffer (1963). We shall prove only the result for the ladder diagrams obtained by scattering from a single impurity. In this case, the self-energy diagram is that for scattering from a single impurity:

$$\Sigma(\mathbf{p}, ip) = n_i\left\{V_{\mathbf{pp}} + \int \frac{d^3p'}{(2\pi)^3} V_{\mathbf{pp}'}V_{\mathbf{p}'\mathbf{p}}\, \mathscr{G}(\mathbf{p}', ip)\right.$$
$$\left. + \int \frac{d^3p'\, d^3p''}{(2\pi)^2} V_{\mathbf{pp}'}V_{\mathbf{p}'\mathbf{p}''}V_{\mathbf{p}''\mathbf{p}}\mathscr{G}(\mathbf{p}', ip)\mathscr{G}(\mathbf{p}'', ip) + \cdots\right\} \tag{7.1.28}$$

An important condition is that the Green's functions in this self-energy diagram are those calculated with the self-energy

$$\mathscr{G} = (ip - \xi_{\mathbf{p}} - \Sigma)^{-1}$$

Thus (7.1.28) is really a self-consistent equation for the self-energy Σ, since it depends functionally on itself. Unfortunately, this means that the Ward identities do not really permit us to avoid solving an integral equation. Instead, we only exchange one integral equation for another. In this sense the Ward identities are not very useful in practice.

Rather than prove the two separate results (7.1.26) and (7.1.27), we shall prove a general theorem for which these are the limiting cases. The general theorem is obtained by subtracting the expressions (7.1.28) for $\Sigma(\mathbf{p}, ip) \equiv \Sigma(p)$ by the same result for $\Sigma(\mathbf{p} + \mathbf{q}, ip + i\omega) \equiv \Sigma(p + q)$:

$$\Sigma(p + q) - \Sigma(p)$$
$$= n_i \int \frac{d^3p_1}{(2\pi)^3} V_{\mathbf{pp}_1}V_{\mathbf{p}_1\mathbf{p}}[\mathscr{G}(p_1 + q) - \mathscr{G}(p_1)] + n_i \int \frac{d^3p_1\, d^3p_2}{(2\pi)^3}$$
$$\times V_{\mathbf{pp}_1}V_{\mathbf{p}_1\mathbf{p}_2}V_{\mathbf{p}_2\mathbf{p}}\, [\mathscr{G}(p_1 + q)\mathscr{G}(p_2 + q) - \mathscr{G}(p_1)\mathscr{G}(p_2)]$$
$$+ n_i \int \frac{d^3p_1\, d^3p_2\, d^3p_3}{(2\pi)^3} V_{\mathbf{pp}_1}V_{\mathbf{p}_1\mathbf{p}_2}V_{\mathbf{p}_2\mathbf{p}_3}V_{\mathbf{p}_3\mathbf{p}}$$
$$\times [\mathscr{G}(p_1 + q)\mathscr{G}(p_2 + q)\mathscr{G}(p_3 + q) - \mathscr{G}(p_1)\mathscr{G}(p_2)\mathscr{G}(p_3)] + \cdots \tag{7.1.29}$$

By purely algebraic manipulations, this series can be shown to be identical to

$$\Sigma(p + q) - \Sigma(p) = n_i \int \frac{d^3p'}{(2\pi)^3} \mathscr{G}(p')\mathscr{G}(p' + q)T_{\mathbf{pp}'}(ip)T_{\mathbf{p}'+\mathbf{q},\mathbf{p}+\mathbf{q}}(ip + i\omega)$$
$$\times [\Sigma(p' + q) - \Sigma(p') + \xi_{\mathbf{p}'+\mathbf{q}} - \xi_{\mathbf{p}'} - i\omega]$$
$$= n_i \int \frac{d^3p'}{(2\pi)^3} T_{\mathbf{pp}'}(ip)T_{\mathbf{p}'+\mathbf{q},\mathbf{p}+\mathbf{q}}(ip + i\omega)$$
$$\times [\mathscr{G}(p' + q) - \mathscr{G}(p')] \tag{7.1.30}$$

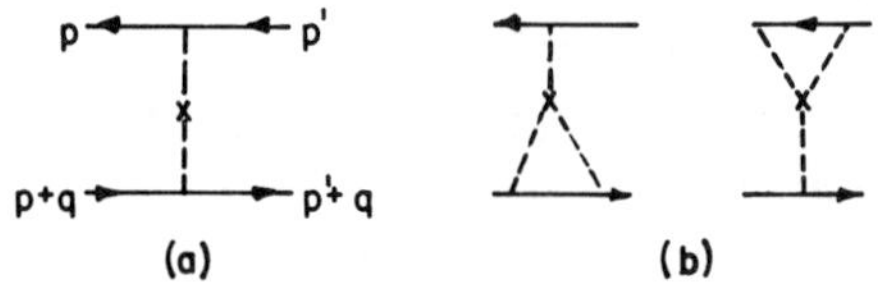

FIGURE 7.7

This rather startling result may be demonstrated term by term. The first nonvanishing term is

$$n_i \int \frac{d^3p'}{(2\pi)^3} V_{\mathbf{pp'}} V_{\mathbf{p'p}} [\mathscr{G}(p'+q) - \mathscr{G}(p')]$$
$$= n_i \int \frac{d^3p'}{(2\pi)^3} V_{\mathbf{pp'}} V_{\mathbf{p'p}} \mathscr{G}(p') \mathscr{G}(p'+q)$$
$$\times [\Sigma(p'+q) - \Sigma(p') + \xi_{\mathbf{p'+q}} - \xi_{\mathbf{p'}} - i\omega]$$

which is just the vertex diagram shown in Fig. 7.7(a). The next term in the series (7.1.29) is

$$n_i \int \frac{d^3p_1\, d^3p_2}{(2\pi)^3} V_{\mathbf{pp}_1} V_{\mathbf{p}_1\mathbf{p}_2} V_{\mathbf{p}_2\mathbf{p}} [\mathscr{G}(p_1+q)\mathscr{G}(p_2+q) - \mathscr{G}(p_1)\mathscr{G}(p_2)]$$

The Green's function factors in brackets may be rearranged into

$$\mathscr{G}(p_1+q)\mathscr{G}(p_2+q) - \mathscr{G}(p_1)\mathscr{G}(p_2) = \mathscr{G}(p_1+q)[\mathscr{G}(p_2+q) - \mathscr{G}(p_2)]$$
$$+ \mathscr{G}(p_2)[\mathscr{G}(p_1+q) - \mathscr{G}(p_1)] \qquad (7.1.31)$$

The first bracket on the right is the first diagram in Fig. 7.7(b). The factor

$$n_i \int \frac{d^3p_1}{(2\pi)^3} V_{\mathbf{pp}_1} V_{\mathbf{p}_1\mathbf{p}_2} \mathscr{G}(p_1+q)$$

is just the multiple scattering from the impurity by the electron on the lower line. Similarly, the second term in (7.1.31) corresponds to the second diagram in Fig. 7.7(b), where the factor $\int d^3p_2 V_{\mathbf{p}_1\mathbf{p}_2} V_{\mathbf{p}_2\mathbf{p}} \mathscr{G}(p_2)$ is the multiple scattering of the top electron line. These are just the series of terms which generate the factor $T_{\mathbf{pp'}}(ip)T_{\mathbf{p'+q,p+q}}(ip+i\omega)$, as illustrated earlier in Figs. 7.4(a) and 7.4(b). The further terms in (7.1.29) serve to provide the remaining terms in the series for $T_{\mathbf{pp'}}T_{\mathbf{p'+q,p+q}}$. In this manner, one can establish the validity of (7.1.30).

From (7.1.30), we see that the quantity

$$\Lambda(p, p+q) = \Sigma(p+q) - \Sigma(p) + \xi_{\mathbf{p+q}} - \xi_{\mathbf{p}} - i\omega \qquad (7.1.32)$$

obeys the vertex equation

$$\Lambda(p, p+q) = \xi_{\mathbf{p}+\mathbf{q}} - \xi_{\mathbf{p}} - i\omega + n_i \int \frac{d^3p'}{(2\pi)^3} \mathscr{G}(p')\mathscr{G}(p'+q) \times T_{\mathbf{pp}'} T_{\mathbf{p}'+\mathbf{q},\mathbf{p}+\mathbf{q}} \Lambda(p', p'+q) \tag{7.1.33}$$

The two equations (7.1.32) and (7.1.33) provide the most general type of Ward identity. They are useful, since any equation which can be cast into the form of (7.1.33) has the solution (7.1.32). Langer (1961) and Engelsberg and Schrieffer (1963) show that this equation is related to the equation of continuity $\nabla \cdot \mathbf{j} + \dot{\varrho} = 0$.

The first Ward identity (7.1.26) is shown by taking the limit $\mathbf{q} = 0$ and then dividing Eq. (7.1.33) by $-i\omega$ with the result

$$\frac{\Lambda(\mathbf{p}, ip, ip+i\omega)}{-i\omega} = 1 + \int \frac{d^3p'}{(2\pi)^3} \mathscr{G}(\mathbf{p}', ip)\mathscr{G}(\mathbf{p}', ip+i\omega) \times W^{(1)}_{\mathbf{pp}'}(ip, ip+i\omega)\left[\frac{\Lambda(\mathbf{p}', ip, ip+i\omega)}{-i\omega}\right] \tag{7.1.34}$$

The quantity $\Gamma(\mathbf{p}, ip)$ in (7.1.24) obeys the same equation as $\Lambda/(-i\omega)$ in the limit $i\omega \to 0$, so they are equal. Of course, we have shown it only for the case where $W_{\mathbf{pp}'} = W^{(1)}_{\mathbf{pp}'}$. From (7.1.32) one has the solution

$$\Gamma(\mathbf{p}, ip) = \lim_{i\omega\to 0}\left[\frac{\Lambda(\mathbf{p}, ip, ip+i\omega)}{-i\omega}\right] = 1 - \left[\frac{\partial\Sigma(\mathbf{p}, Z)}{\partial Z}\right]_{Z=ip}$$

which proves the Ward identity (7.1.26).

The other Ward identity is found as the limit $i\omega \to 0$, followed by letting $\mathbf{q} \to 0$. The latter limit is taken slowly, so that one can retain terms proportional to $\mathbf{q}$. In this limit Eqs. (7.1.32) and (7.1.33) become

$$\lim_{\mathbf{q}\to 0} \Lambda = \frac{1}{m}\left[\mathbf{q}\cdot\mathbf{p} + m\mathbf{q}\cdot\nabla_p\Sigma(\mathbf{p}, ip)\right] \tag{7.1.35}$$

$$\lim_{\mathbf{q}\to 0} \Lambda = \frac{\mathbf{q}\cdot\mathbf{p}}{m} + n_i \int \frac{d^3p'}{(2\pi)^3} \Lambda(p', p'+q)\mathscr{G}(p')\mathscr{G}(p'+q) T_{\mathbf{pp}'} T_{\mathbf{p}'+\mathbf{q},\mathbf{p}+\mathbf{q}} \tag{7.1.36}$$

The vertex function Λ is proportional to $\mathbf{q}$, so define the vector vertex function by the limit ($i\omega = 0$)

$$\lim_{\mathbf{q}\to 0} \Lambda(p, p+q) = \frac{1}{m}\mathbf{q}\cdot\mathbf{\Gamma}(\mathbf{p}, ip)$$

Then the preceding two equations can be expressed in terms of this vector vertex function:

$$\mathbf{\Gamma}(p, ip) = \mathbf{p} + m\nabla_p\Sigma(\mathbf{p}, ip)$$

$$\mathbf{\Gamma}(p) = \mathbf{p} + \int \frac{d^3p'}{(2\pi)^3} \mathbf{\Gamma}(p')\mathscr{G}(p')^2 n_i \mid T_{\mathbf{pp}'}(ip) \mid^2$$

This equation is the same as (7.1.25), so we have proved the other Ward identity (7.1.27). These are both now understood to be limiting cases of the general result (7.1.33). The Ward identities are useful anytime one can cast the vertex equation into the form (7.1.33).

The factor $[1 - (\partial\Sigma/\partial Z)]$ is recognized as the inverse of the renormalization Z defined earlier and discussed, for example, in Sec. 5.8A. This quantity is sometimes called the effective charge. Similarly, the vector vertex is part of the factors which give the effective mass of the particle. The Ward identities relate the vertex corrections to a change in the effective charge and mass of the particle, which is why it is related to the equation of continuity.

Our last example is a vertex function which is *not* given by a Ward identity. If we wish to sum the set of ladder diagrams in Fig. 7.6(b) for a correlation evaluated at finite $\mathbf{q}$ and $i\omega$, then the ladder sum for a scalar vertex is

$$\Gamma(p, p + q) = 1 + n_i \int \frac{d^3p'}{(2\pi)^3} \mathscr{G}(p')\mathscr{G}(p' + q)T_{\mathbf{pp}'}(ip) \times T_{\mathbf{p}'+\mathbf{q},\mathbf{p}+\mathbf{q}}(ip + i\omega)\Gamma(p', p' + q)$$

This integral equation *cannot* be cast into the form of (7.1.33), so it *cannot* be evaluated by a Ward identity. This example is provided to emphasize again the limited usefulness of Ward identities. They do not help evaluate the vertex function encountered in a wide variety of problems. Generally they are useful only in the case where $\mathbf{q} \to 0$. However, they are useful even for finite $i\omega$, since (7.1.34) is certainly a scalar vertex equation which was derived from Ward identities.

7.2. MOBILITY OF FRÖHLICH POLARONS

The Fröhlich Hamiltonian between electrons and Einstein phonons

($\omega_0 \equiv \omega_{LO}$) was defined and discussed in Sec. 6.1:

$$\begin{aligned} H &= \sum_{\mathbf{p}} \varepsilon_{\mathbf{p}} c_{\mathbf{p}}^{\dagger} c_{\mathbf{p}} + \omega_0 \sum_{\mathbf{q}} a_{\mathbf{q}}^{\dagger} a_{\mathbf{q}} + \frac{M_0}{\nu^{1/2}} \sum_{\mathbf{qp}} \frac{1}{|\mathbf{q}|} c_{\mathbf{p}+\mathbf{q}}^{\dagger} c_{\mathbf{p}} (a_{\mathbf{q}} + a_{-\mathbf{q}}^{\dagger}) \\ M_0^2 &= \frac{4\pi\alpha\hbar(\hbar\omega_0)^{3/2}}{(2m_B)^{1/2}} \\ \alpha &= \frac{e^2}{\hbar}\left(\frac{m_B}{2\hbar\omega_0}\right)^{1/2}\left(\frac{1}{\varepsilon_\infty} - \frac{1}{\varepsilon_0}\right) \\ \varepsilon_p &= \frac{p^2}{2m_B} \end{aligned} \tag{7.2.1}$$

There we calculated several important quantities such as the effective mass m^* and the ground state energy E_0.

The effective mass of a particle can, in principle, be measured by cyclotron resonance. Such experiments have been done for polarons (Hodby, 1972) but only recently for solids with large values of α. The measurement provides the experimental value of the effective mass m^*, which is a function of the band mass m_B and the polar coupling constant α. A separate measurement of the two dielectric functions ε_0 and ε_∞, as well as ω_{LO} itself, will permit a determination of the band mass m_B and α from m^*. Of course, this analysis takes a theory of the polaron mass $m^*(m_B, \alpha)$, but that was provided in Sec. 6.1. However, the situation was not as clear a decade ago when there were numerous competing theories of polarons. Then it was not obvious which theory was the best, so a primary objective of experiments was to test the polaron theories. This test took another measurement to provide another experimental number which depended on m_B and α. The ground state energy E_0 was not suitable, since it only renormalizes the band minimum, which is not useful since the unrenormalized band minimum cannot be determined independently.

The other important measurement was usually the mobility of electrons in polar crystals. Typical experimental results are shown in Figs. 7.8 and 7.9. Figure 7.8 shows the Hall mobility of CdTe measured by Segall *et al.* (1963). The steep rise around 200 K is due to optical mode scattering. At lower temperatures the mobility saturates because of the scattering from impurities in the crystals. Impurity scattering varies from sample to sample, depending on the concentration and type of impurity. Figure 7.9 shows the mobility in the semiconductor CdS, which behaves much differently from most other materials. It is one of the most piezoelectric of materials,

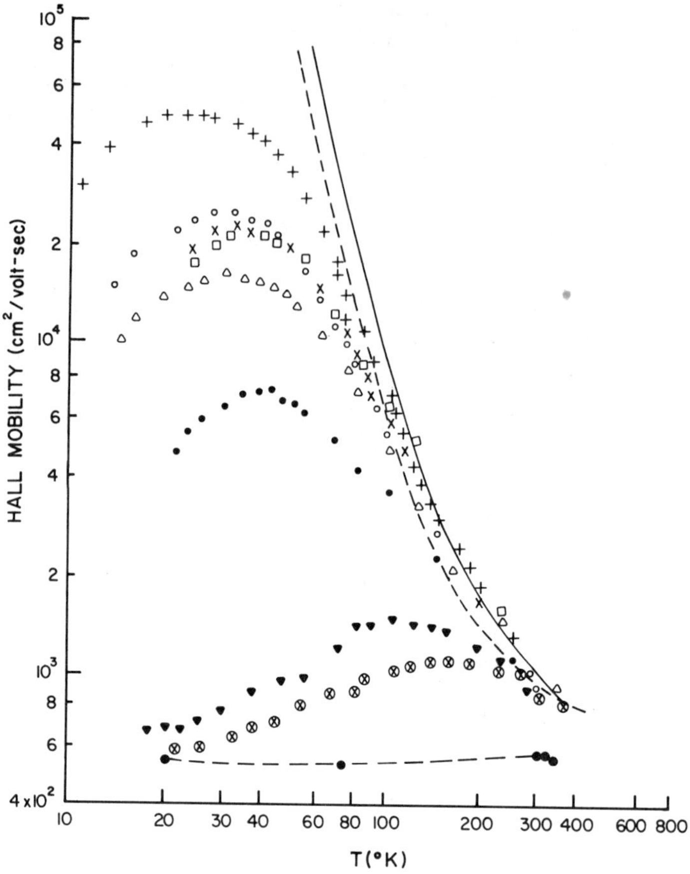

FIGURE 7.8. The temperature dependence of the electron mobility of several samples of *n*-type CdTe. Different samples have different kinds and concentrations of defects or impurities. Solid line is the theoretical mobility from optical mode scattering, including the temperature dependence of the static dielectric constant. The dashed line neglects this change. *Source*: Segall *et al.* (1963) (used with permission).

and its low-temperature mobility in very pure sample is limited by the piezoelectric scattering of acoustical phonons. The $T^{-1/2}$ prediction of piezoelectric scattering is clearly confirmed by the results of Fuijita *et al.* (1965) in Fig. 7.9(b). The higher-temperature mobilities found by Devlin (1967) are shown in Fig. 7.9(a), and they are limited by optical phonon scattering.

The average value of the current operator is the particle density n_0

times the charge e times the average velocity $\langle v \rangle$. The average velocity $\langle v \rangle$ is proportional to the applied electric field F, and the constant of proportionality is the mobility:

$$\langle v \rangle = \mu F$$

$$J = -en_0\mu F$$

The mobility μ is the average velocity of each electron per unit applied electric field. Of course, it is strictly defined in the limit of vanishing electric field. Since the electrical conductivity $\sigma = n_0 e^2 \tau/m = -n_0 e \mu$ is the ratio of the current J to the field F, we also have $\mu = -e\tau/m$. Later we shall explain the distinction between the mobility and the Hall mobility.

Different techniques are required for different materials in order to measure the mobility. The main problem is to get enough mobile electrons in the sample to do a measurement. Ordinary semiconductors have enough carriers due to defects, as is the case in Fig. 7.8. But in very pure semiconductors and in ionic solids with wide band gaps, the carriers must be introduced by optical excitation. Sometimes infrared light is used, and the carriers come from defect levels in the band gap. In other cases, ultraviolet light is used, and the carriers are excited across the energy gap of the solid. These techniques are reviewed by Hodby (1972). In most cases a separate measurement is done to also determine the carrier concentration n_0, which is a strong function of temperature when the electrons are thermally excited from defect levels. Often the concentration n_0 is found from a Hall measurement (see Ziman, 1960). The term *Hall mobility* is used for the result obtained by measuring the conductivity σ and then dividing it by the concentration n_0 found by the Hall effect. The mobility and the Hall mobility differ by small factors, which are discussed by Devlin (1967) and Ziman (1960).

The theories of the electron mobility in insulating materials, such as alkali halides and II–VI semiconductors, treat it as a property of a single electron. The electron lifetime is calculated for the scattering from impurities and by acoustical and optical phonons. The electron–electron interactions can be ignored in the limit where the concentration of electrons is very low.

There are just as many different ways to calculate the polaron mobility as there are to calculate the effective mass m^* or ground state energy E_0. Each theoretical technique was applied to the mobility as well as to the other quantities. As summarized by Langreth (1967), these various methods usually agree in the limit of weak coupling ($\alpha \ll 1$) and low temperature

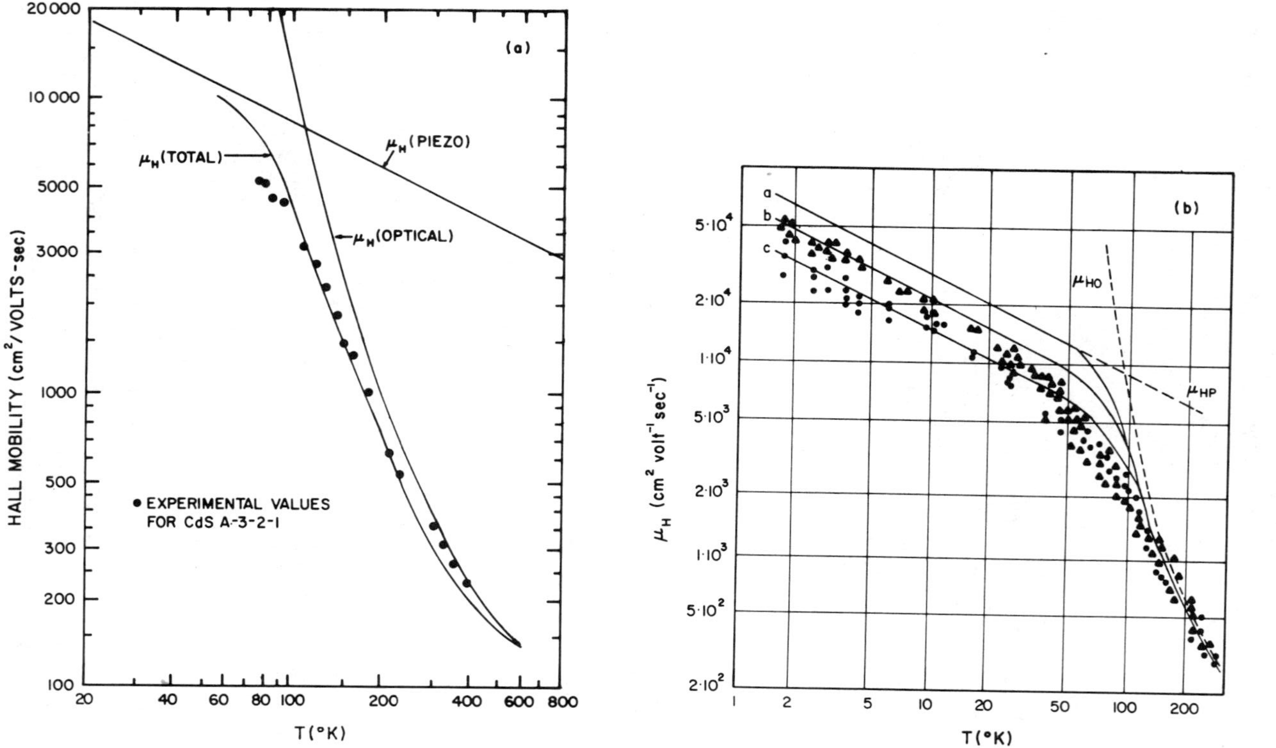

FIGURE 7.9. The temperature dependence of electron mobility in undoped *n*-type CdS, which is very piezoelectric. (a) High-temperature results from Devlin (1967). (b) Low-temperature results from Fujita *et al.* (1965) (used with permission).

($\beta\omega_0 \gg 1$). The theories predict

$$\lim_{\substack{T\to 0\\ \alpha\to 0}} \mu = \frac{-e\tau_0}{m_B} = -\frac{e}{2\alpha m_B \omega_0}(e^{\beta\omega_0} - 1) \equiv \mu_0 \tag{7.2.2}$$

$$\frac{1}{\tau_0} = 2\alpha N_0 \omega_0$$

$$N_0 = \frac{1}{e^{\beta\omega_0} - 1}$$

The lifetime τ_0 is the result obtained in (6.1.18) as $1/\tau_0 = -2\,\mathrm{Im}[\Sigma^{(1)}(p, \varepsilon)]$ in the limit where $p \to 0$ and $\varepsilon \to 0$, where $\Sigma^{(1)}$ is the one-phonon self-energy. This limit is appropriate, since at very low temperatures the electrons are in states within $k_B T$ of the bottom of the band. These low-energy particles cannot emit phonons, since this event is prevented by energy conservation. They can only absorb them, and the rate of absorption is proportional to the thermal average density of phonons N_0. The factor N_0 makes the mobility increase exponentially with decreasing temperature, since the electron scattering becomes less likely as the number density of phonons declines. The exponential increase in the mobility is evident in the experimental data of Figs. 7.8 and 7.9. The behavior of large polarons is opposite to that of small polarons, whose mobility increases with increasing temperature.

The asymptotic result (7.2.2) is useful for checking theories but not useful for comparing with experiments. The mobility is mostly limited to optical phonon scattering in the temperature region $\beta\omega_0 \simeq \frac{1}{2}$, so we need to calculate the mobility for higher temperature. Second, many of the interesting materials have polaron constants in the intermediate coupling range of $1 \leq \alpha \leq 3$, so we need a theory which is valid in this range. As usual, the most successful theories employ the Feynman path integral method (see Thornber, 1972). The Green's function method has not done as well, so we shall present only the calculations for low temperature and small α.

One feature of the mobility formula (7.2.2) is that it is proportional to the inverse of α. Our starting point for the theoretical calculation will again be the Kubo formula, which will be evaluated for the potential of electron–phonon interactions. The expansion of the S matrix for this potential will generate a series in the parameter α. To obtain a leading term in the inverse power of α, we shall have to sum a subset of diagrams. The situation is similar to the mobility from impurity scattering, where we needed to sum diagrams to get the conductivity inversely proportional to the impurity concentration n_i. We do not sum the same subset of diagrams for the polaron mobility. This conclusion is already evident from the result presented in (7.2.2): Here the relaxation time is *not* calculated with the $1 - \cos\theta'$ factor in the angular average. Thus the polaron mobility is calculated in a different way than the scattering from impurities. Actually it is calculated in the same way, but we obtain a different result. The reason for this difference will be discussed in great detail. It arises from the inelastic nature of the polaron scattering—as first shown by Howarth and Sondheimer (1953).

Our general approach will be that adopted by Langreth and Kadanoff (1964). They used the Green's function method, starting from the Kubo formula, to derive (7.2.2). Next they tried to obtain all the correction terms of order α^0. The answer for the mobility μ is apparently a power series in α, with the leading term in (7.2.2) of order α^{-1}:

$$\mu = \frac{a_{-1}}{\alpha} + a_0\alpha^0 + a_1\alpha + a_2\alpha^2 + \cdots \tag{7.2.3}$$

They tried to calculate the next coefficient a_0 in this series in the low-temperature limit. This objective is simple but has slippery aspects. When we write down a term in the S-matrix expansion, it is a subtle procedure to determine its leading term in α. The situation is similar to that which was just discussed for impurity scattering. Each term in the series (7.2.3) is obtained by summing subsets of diagrams.

We shall derive only the first term of Langreth and Kadanoff. Their answer is quite simple, and it is probably worthwhile to state it in advance. Define as μ_0 the result (7.2.2) for the limits $\alpha \to 0$ and $T \to 0$. They found

$$\begin{aligned} \frac{\mu}{\mu_0} &= 1 - \frac{\alpha}{6} + O(\alpha^2) \\ \mu_0 &= -\frac{e}{2\alpha m_B \omega_0 N_0} \end{aligned} \tag{7.2.4}$$

They observed that μ/μ_0 is precisely the expansion in α given by the equations

$$\begin{aligned} \mu &= -\frac{e\tau}{m^*} \\ \frac{\tau}{\tau_0} &= 1 + O(\alpha^2) \\ \frac{m^*}{m_B} &= 1 + \frac{\alpha}{6} + O(\alpha^2) \end{aligned} \tag{7.2.5}$$

where we have also given the separate expansions for $m^*(\alpha)$ and $\tau(\alpha)$. The ballistic formula $\mu = -e\tau/m^*$ supports the quasiparticle picture—that the particle acts as if it has an effective mass m^* and lifetime τ. They speculate that the quasiparticle picture would be valid for all values of α and that the inclusion of all terms in α would just reproduce the product series of m^* and τ.

A. Single-Particle Properties

Before launching into the derivation of the electron mobility, it is necessary to derive the expressions for some single-particle properties. These will be needed in the limit of zero temperature. Thus we drop all terms of order N_0 compared to unity.Many of these single-particle properties were derived in Sec. 6.1. For example, the first self-energy term, which is proportional to α, is the one-phonon result in (6.1.8):

$$\mathrm{Re}[\Sigma^{(1)}(p, \omega)] = -\alpha \frac{\omega_0^{3/2}}{(\varepsilon_p)^{1/2}} \sin^{-1}\left(\frac{\varepsilon_p}{\omega_0 - \omega + \varepsilon_p}\right)^{1/2} + O(\alpha N_0, \alpha^2)$$

At zero temperature, we need this self-energy evaluated at small ε_p and small ω, which is

$$\mathrm{Re}[\Sigma^{(1)}(p, \omega)] = -\alpha[\omega_0 + \tfrac{1}{2}\omega - \tfrac{1}{3}\varepsilon_p + O(\omega^2, \varepsilon_p{}^2, \omega\varepsilon_p)] \tag{7.2.6}$$

This expansion permits a quick derivation of the effective mass m^* and the renormalization coefficient Z:

$$\begin{aligned} Z_0 &= \left(1 - \frac{\partial \Sigma}{\partial \omega}\right)^{-1}_{\substack{p=0\\ \omega=0}} = (1 + \tfrac{1}{2}\alpha)^{-1} = 1 - \tfrac{1}{2}\alpha + O(\alpha^2) \\ \left(\frac{m_B}{m^*}\right)_{\substack{p=0\\ \omega=0}} &= Z_0\left(1 + \frac{\partial \Sigma}{\partial \varepsilon_p}\right) = \frac{1 + \frac{1}{3}\alpha}{1 + \frac{1}{2}\alpha} \simeq 1 - \tfrac{1}{6}\alpha \end{aligned} \tag{7.2.7}$$

These two results will be used extensively. Another quantity we shall need is the lifetime τ. The electron lifetime is defined as

$$\begin{aligned} \frac{1}{\tau(p)} &= Z(p)\{-2\,\mathrm{Im}[\Sigma(p, E_p)]\} \\ Z(p) &= \left(1 - \frac{\partial \Sigma}{\partial \omega}\right)^{-1}_{\omega=E_p} \end{aligned} \tag{7.2.8}$$

where $Z(p)$ is the renormalization coefficient. The imaginary part of the self-energy Σ is not evaluated at the energy $\omega \to 0$ but at the ground state energy E_p. Of course, this choice is made because that is the particle energy, which comes from the self-consistent solution to the equations

$$E_p = \varepsilon_p + \mathrm{Re}[\Sigma(p, E_p)]$$
$$E_0 = \lim_{p\to 0} E_p = -\alpha\omega_0 + O(\alpha^2)$$

This ground state energy is only needed to order α, which is the simple result $E_p = -\alpha\omega_0 + p^2/2m^* + O(p^4)$.

Equation (7.2.8) shows that the renormalization coefficient $Z(p)$ enters into the definition of the lifetime. The argument for this is as follows. The spectral function is defined as

$$\begin{aligned} A(p, \omega) &= -2\,\mathrm{Im}[G_{\mathrm{ret}}(p, E)] \\ &= \frac{-2\,\mathrm{Im}[\Sigma(p, \omega)]}{\{\omega - \varepsilon_p - \mathrm{Re}[\Sigma(p, \omega)]\} + [\mathrm{Im}(\Sigma)]^2} \end{aligned} \qquad (7.2.9)$$

In the limit where $\mathrm{Im}\,\Sigma \to 0$, the spectral function goes into a delta function which expresses energy conservation multiplied by the renormalization coefficient:

$$\begin{aligned} \lim_{\mathrm{Im}\,\Sigma\to 0} A(p, \omega) &= 2\pi\delta\{\omega - \varepsilon_p - \mathrm{Re}[\Sigma(p, \omega)]\} \\ &= Z(p)\times 2\pi\delta(\omega - E_p) \end{aligned}$$

A suitable definition of $\tau(p)$ is obtained by examining this limit more carefully when $\mathrm{Im}(\Sigma)$ is small but not infinitesimal. From the earlier definition of the retarded Green's function (Problem 6 in Chapter 3),

$$G_{\mathrm{ret}}(p, t) = \theta(t) \int_{-\infty}^{\infty} \frac{d\omega}{2\pi i}\, e^{-i\omega t} A(p, \omega) \qquad (7.2.10)$$

we shall define the relaxation time from the decay of the Green's function. In the vicinity of the peak $\omega \simeq E_p$ of the spectral function (7.2.9) we can write

$$\begin{aligned} \omega - \varepsilon_p - \mathrm{Re}[\Sigma(p, \omega)] &\simeq \omega - \varepsilon_p - \mathrm{Re}[\Sigma(p, E_p)] - (\omega - E_p)\frac{\partial\,\mathrm{Re}(\Sigma)}{\partial\omega} \\ &\simeq (\omega - E_p)\left[1 - \frac{\partial\,\mathrm{Re}(\Sigma)}{\partial\omega}\right] \simeq \frac{\omega - E_p}{Z(p)} \end{aligned}$$

so that the spectral function is approximately

$$A(p, \omega) \simeq \frac{-2\,\mathrm{Im}(\Sigma)}{(\omega - E_p)^2/Z^2 + [\mathrm{Im}(\Sigma)]^2}$$

The definition of the relaxation time (7.2.8) allows this to be rewritten as

$$A(p, \omega) \simeq Z(p)\left(\frac{1/\tau(p)}{(\omega - E_p)^2 + (\frac{1}{2}\tau)^2}\right)$$

If $\tau(p)$ is treated as a function of p but not ω, for the retarded Green's function the Fourier transform in (7.2.10) gives

$$G_{\text{ret}}(p, t) = -iZ(p)e^{-itE_p}e^{-t/2\tau}\theta(t)$$

The Green's function has the desired form, with the relaxation time $\tau(p)$ determining the decay of the excitation.

There is another way to understand the factor Z in the definition of $\tau(p)$. The quantity $-2\,\text{Im}(\Sigma)$ is the rate of decay of a state (p, ω). The factor Z is the fraction of the quasiparticle strength at the value (p, ω). The rest of the quasiparticle strength is usually dispersed throughout the spectrum.

We now are in a position to evaluate the contribution of the one-phonon self-energy to the quasiparticle lifetime. The imaginary self-energy is evaluated at the quasiparticle energy $\omega = E_0 = -\alpha\omega_0$ and is multiplied by the factor of Z given in (7.2.7). The imaginary self-energy is calculated from the expression

$$\begin{aligned}
-2\,\text{Im}[\Sigma(p, \omega)] &= 2\pi \int \frac{d^3q}{(2\pi)^3} \frac{M_0^{\,2}}{q^2} \\
&\quad \times [N_0\delta(\omega + \omega_0 - \varepsilon_{\mathbf{p}+\mathbf{q}}) + (N_0 + 1)\delta(\omega - \omega_0 - \varepsilon_{\mathbf{p}+\mathbf{q}})] \\
&= \alpha \frac{\omega_0^{\,3/2}}{(\varepsilon_p)^{1/2}} \left[N_0\theta(\omega + \omega_0) \ln\left| \frac{(\omega + \omega_0)^{1/2} + \varepsilon_p}{(\omega + \omega_0)^{1/2} - \varepsilon_p} \right| \right. \\
&\quad \left. + (N_0 + 1)\theta(\omega - \omega_0) \ln\left| \frac{(\omega - \omega_0)^{1/2} + (\varepsilon_p)^{1/2}}{(\omega - \omega_0)^{1/2} - (\varepsilon_p)^{1/2}} \right| \right]
\end{aligned} \tag{7.2.11}$$

to be, for $\omega \simeq -\alpha\omega_0 \lesssim 0$, $p \to 0$,

$$-2\,\text{Im}[\Sigma(0, \omega)] = \frac{2\alpha N_0\omega_0^{\,3/2}}{(\omega_0 + \omega)^{1/2}} \simeq 2\alpha N_0\left[\omega_0 - \frac{1}{2}\omega + O\left(\frac{\omega^2}{\omega_0}\right)\right]$$

$$-2\,\text{Im}[\Sigma(0, -\alpha\omega_0)] = 2\alpha N_0\omega_0[1 + \tfrac{1}{2}\alpha + O(\alpha^2)]$$

$$\begin{aligned}
\frac{1}{\tau} &= Z[-2\,\text{Im}(\Sigma)] = \frac{1}{\tau_0}\left(1 - \frac{1}{2}\alpha\right)\left(1 + \frac{1}{2}\alpha\right) \\
&= \frac{1}{\tau_0}[1 + O(\alpha^2)]
\end{aligned}$$

The first correction terms in α from $Z = 1 - \frac{1}{2}\alpha$ and $\text{Im}[\Sigma(0)] \propto 1 + \frac{1}{2}\alpha$ cancel to order α^2. Thus the one-phonon term provides no correction term to τ of α^2.

B. α^{-1} Term in the Mobility

Our general procedure for the calculation of the electron mobility will be the same which was used for impurity scattering. We shall evaluate the Kubo formula for the electrical conductivity:

$$\pi(i\omega) = -\frac{1}{3\nu}\int_0^\beta d\tau e^{i\omega\tau}\langle T_\tau \mathbf{j}(\tau) \cdot \mathbf{j}(0)\rangle$$

$$\sigma = -\lim_{\omega\to 0}\left\{\text{Im}\left[\frac{\pi_{\text{ret}}(\omega)}{\omega}\right]\right\}$$

After it is evaluated, the mobility is found by dividing σ by $-en_0$, where $-e$ and n_0 are the charge and concentration of electrons. The Fröhlich Hamiltonian (7.2.1) is used, where the electron–phonon interaction provides the basic scattering mechanism for the electrons.

The first term which is evaluated is the basic bubble diagram, shown in Fig. 7.10(a). The solid lines are total Green's functions $\mathscr{G}(p, ip)$ which include the self-energies. The evaluation of this term is identical to that used for the same bubble diagram for impurity scattering. The contribution to the Kubo formula contains the two Green's functions as in Eq. (7.1.9):

$$\pi^{(0)}(i\omega) = \frac{2e^2}{3m_B^2}\int\frac{d^3p}{(2\pi)^3}p^2\frac{1}{\beta}\sum_{ip_n}\mathscr{G}(\mathbf{p}, ip_n)\mathscr{G}(\mathbf{p}, ip_n + i\omega)$$

which are evaluated by the standard series of steps to yield Eq. (7.1.10):

$$\sigma^{(0)} = \frac{e^2}{3m_B^2}\int\frac{d^3p}{(2\pi)^3}p^2\int\frac{d\varepsilon}{2\pi}A(p, \varepsilon)^2\left[-\frac{\partial n_F(\varepsilon)}{\partial\varepsilon}\right] \qquad (7.2.13)$$

Of course, in the present application, the self-energies in the Green's func-

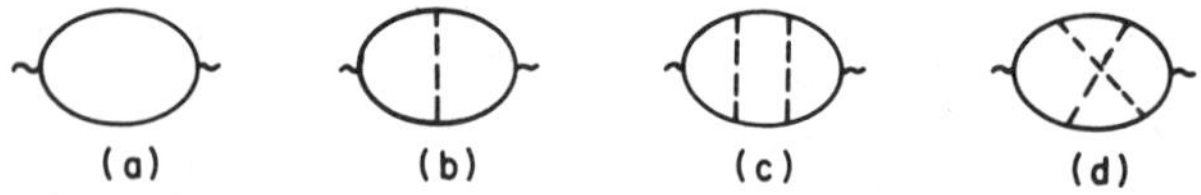

FIGURE 7.10

tions are evaluated from the Fröhlich Hamiltonian rather than impurity scattering. In a more realistic model of a solid, we should probably include both self-energy expressions—from phonons *and* impurity scattering.

An approximate evaluation of this contribution to the conductivity is obtained in the limit where $\alpha \to 0$. The electron distribution $n_F(\varepsilon)$ is assumed to be Maxwell–Boltzmann:

$$n_F(\varepsilon) = \frac{1}{2} n_0 \left(\frac{2\pi\beta}{m^*} \right)^{3/2} e^{-\beta(\varepsilon - E_0)}$$

where $n_F(\varepsilon)$ is the energy distribution for each spin state, while n_0 is the total concentration for both spin states. This term in the conductivity is divided by $-en_0$ to get the corresponding term in the mobility of each electron:

$$\mu^{(0)} = -\frac{e\beta}{6m_B^2} \left(\frac{2\pi\beta}{m^*} \right)^{3/2} \int \frac{d^3p}{(2\pi)^3} p^2 \int \frac{d\varepsilon}{2\pi} e^{-\beta(\varepsilon - E_0)} A(p, \varepsilon)^2 \qquad (7.2.14)$$

The square of the spectral function is first expressed as

$$A(p, \varepsilon)^2 = \frac{\{-2\,\mathrm{Im}[\Sigma(p, \varepsilon)]\}^2}{\{[\varepsilon - \varepsilon_p - \mathrm{Re}(\Sigma)]^2 + [\mathrm{Im}(\Sigma)]^2\}^2}$$
$$= \frac{Z(p)^2 \tau(p)^{-2}}{\{(\varepsilon - E_p)^2 + [2\tau(p)]^{-2}\}^2}$$

In the limit where $\mathrm{Im}(\Sigma) \to 0$ [or $\tau(p) \to \infty$], this can be approximated by

$$A(p, \varepsilon)^2 \simeq 4\pi\tau(p) Z(p)^2 \delta(\varepsilon - E_p)$$

$$\mu^{(0)} \simeq -\frac{e\beta}{3m_B^2} \left(\frac{2\pi\beta}{m^*} \right)^{3/2} \int \frac{d^3p}{(2\pi)^3} p^2 Z(p)^2 \tau(p) e^{-\beta(E_p - E_0)}$$

The electron lifetime $\tau(p)$ is inversely proportional to α, so that the mobility $\mu^{(0)}$ is inversely proportional to α. The preceding integral is evaluated in the limit where the temperature $T \to 0$. The new variable of integration is $x = p(\beta/2m^*)^{1/2}$, so the energy exponent is

$$\beta(E_p - E_0) = \beta \frac{p^2}{2m^*} \left[1 + O\left(\frac{\varepsilon_p}{\omega_0} \right) \right]$$
$$= x^2 \left[1 + O\left(\frac{x^2}{\beta\omega_0} \right) \right]$$

Similarly, the p dependence of $Z(p)$ and $\tau(p)$ can be neglected in the limit where $T \to 0$, since the variable $p = x(2m^*/\beta)^{1/2}$ goes to zero. The mobility integral becomes a simple Gaussian ($d^3p = 4\pi p^2\, dp$):

$$\mu^{(0)} = \left[\frac{-e\tau(0)Z(0)^2 m^*}{m_B{}^2}\right] \frac{8}{3(\pi)^{1/2}} \int_0^\infty dx x^4 e^{-x^2}$$

$$= -e\tau(0)Z(0)^2 \frac{m^*}{m_B{}^2} \tag{7.2.15}$$

The result (7.2.15) is the contribution from the simple bubble diagram of Fig. 7.10(a). This bubble result has factors which depend on α as follows: $Z(0) \equiv Z_0 = 1 - \frac{1}{2}\alpha$, $\tau(0) \equiv \tau = \tau_0$, $m^*/m_B = 1 + \frac{1}{2}\alpha$. Thus we derive the α dependence of $\mu^{(0)}$:

$$\mu^{(0)} = -\frac{e\tau_0}{m_B}\left(1 - \frac{5}{6}\alpha\right) \tag{7.2.16}$$

Other contributions can be derived from the other diagrams, which are the vertex corrections. In the limits $T \to 0$ and $\alpha \to 0$ the vertex corrections do *not* contribute to the mobility a term which goes as α^{-1}. Thus the simple bubble result is the final answer at low temperature and weak coupling. This conclusion is quite different from the situation we found for impurity scattering. There we had to sum a series of ladder diagrams in order to derive the final answer, and each ladder diagram gave a term which was the same inverse power of the coupling constant—the impurity concentration n_i. This does not happen for optical phonon scattering because of the inelastic nature of the phonon scattering. The ladder diagrams are not as important at low temperature. Mahan (1966) showed that the two phonon ladder diagrams provide corrections to σ of $O(T/\omega_0)$.

7.3. ELECTRON–PHONON INTERACTIONS IN METALS

A. Force-Force Correlation Function

In pure metals the electrical resistivity has two components. There is usually a constant resistivity from electron scattering by impurities, which is the largest part of the resistance at small temperatures. There is also a resistivity from electron scattering by phonons, which is temperature dependent and becomes large at high temperature. "Matthiessen's rule" (1862) is that these two contributions to the resistance are additive. It should be regarded as a rule of thumb, rather than an ironclad rule. There are enough "deviations from Matthiessen's rule" to make the abbreviation DMR a familiar acronym (Bass, 1982).

Electron scattering by acoustical phonons presents a hard problem in transport theory. The scattering is slightly inelastic. We can apply to this problem neither the elastic scattering theory of Sec. 7.1 nor the inelastic scattering theory of Sec. 7.2. Instead we must derive and solve an integral equation for the energy dependence of the scattering process. The slightly inelastic nature of the scattering process makes this calculation much harder than the previous cases.

Two methods for obtaining the electrical conductivity are emphasized in this book. One uses equilibrium methods and evaluates the Kubo formula for the current–current correlation function. The resistivity from phonons will be found using this method, which follows the original derivation by Holstein (1964). The second method utilizes the quantum Boltzmann equation (QBE), which is a nonequilibrium theory, Mahan and Hänsch (1983) used the QBE to derive the Holstein formula. Both of these derivations are complicated. They end by deriving the same integral equation for the scattering function, which must be solved by further work. Their virtue is that they are formally exact starting points, although approximations are made in obtaining the solution.

Other methods for obtaining the resistivity have been proposed, partly to avoid all of the work associated with the exact methods. These other methods are approximate. However, often the theories are both simple and accurate, which make them useful approximations. One of them is the *force–force* correlation function. If $F(t)$ is the fluctuating force that acts on the electron, then define $R(i\omega)$ as the force–force correlation function:

$$R(i\omega) = -\frac{1}{3}\int_0^\beta d\tau\, e^{i\omega\tau}\langle T_\tau \mathbf{F}(\tau)\cdot\mathbf{F}(0)\rangle$$

$$\varrho = \frac{1}{e^2 n_0^2}\lim_{\omega\to 0}[\mathrm{Im}\, R_{\mathrm{ret}}(\omega)/\omega]$$

This formula is just the quantum analogy of the Nyqvist theorem (Kubo *et al.*, 1985). After calculating this correlation function, the retarded function is obtained by letting $i\omega \to \omega + i\delta$. The resistance ϱ is found by dividing by ω and taking the limit of $\omega \to 0$. For example, assume that the force on the electron has two terms: $\mathbf{F}_i$ from impurities and $\mathbf{F}_{\mathrm{ph}}$ from phonons. If they are uncorrelated, then the correlation function has no cross terms. Symbolically we can write

$$R = \langle(\mathbf{F}_i + \mathbf{F}_{\mathrm{ph}})\cdot(\mathbf{F}_i + \mathbf{F}_{\mathrm{ph}})\rangle = \langle\mathbf{F}_i\cdot\mathbf{F}_i\rangle + \langle\mathbf{F}_{\mathrm{ph}}\cdot\mathbf{F}_{\mathrm{ph}}\rangle$$

In this case the resistivities from impurities and phonons are additive, in agreement with Matthiessen's rule.

As an example, the resistivity is calculated from impurity scattering. The potential energy of the electron scattering from the impurities at R_i is discussed in Sec. 4.1E

$$V(\mathbf{r}) = \sum_i V_{ei}(\mathbf{r} - \mathbf{R}_i) = \frac{1}{\nu} \sum_{i\mathbf{q}} V(\mathbf{q}) \exp[i\mathbf{q} \cdot (\mathbf{r} - \mathbf{R}_i)]$$

The force $\mathbf{F}$ is the gradient of the potential. The factor of $\exp(-i\mathbf{q} \cdot \mathbf{r})$ can also be written as the electron density operator $\varrho(\mathbf{q})$. The factor of $\exp(i\mathbf{q} \cdot \mathbf{R}_i)$ can be written as the impurity density operator $\varrho_i(\mathbf{q})$.

$$\mathbf{F}(\mathbf{r}) = -\frac{i}{\nu} \sum_{\mathbf{q}i} \mathbf{q} V(\mathbf{q}) \exp[i\mathbf{q} \cdot (\mathbf{r} - \mathbf{R}_i)] = -\frac{i}{\nu} \sum_{\mathbf{q}} \mathbf{q} V(\mathbf{q}) \varrho(\mathbf{q}) \varrho_i(\mathbf{q})$$

The next step is to evaluate the force–force correlation function $R(i\omega)$. In correlating $\mathbf{F}$ with itself, there are two separate factors. One is $\langle \varrho_i \varrho_i \rangle$, which equals the number of impurities N_i if $\mathbf{q} = -\mathbf{q}'$. The other is the electron density–density correlation function, which is given exactly in (5.4.8) in terms of the inverse dielectric function

$$R(i\omega) = -\frac{N_i}{3\nu^2} \sum_{\mathbf{q}} q^2 \frac{V(q)^2}{v_q} \left[\frac{1}{\varepsilon(\mathbf{q}, i\omega)} - 1 \right]$$

The next step is to take the imaginary part of the retarded function. The only retarded function on the right-hand side of the above equation is the inverse dielectric function. Its imaginary part is $\varepsilon_2/\varepsilon_1^2 + \varepsilon_2^2)$. At low frequency $\varepsilon_2 = 2\omega e^2 m^2/q^3$. The frequency factor is eliminated when we divide by ω. The formula for the resistivity from impurity scattering is

$$\varrho = \frac{n_i m^2}{6\pi n_0^2 e^2} \int \frac{d^3q}{(2\pi)^2} q \left| \frac{V(q)}{\varepsilon(q)} \right|^2$$

This formula is the exact result for the zero-temperature resistivity from impurity scattering, when the scattering is calculated in the second Born approximation. If $V(q)/\varepsilon(q)$ is replaced by the T matrix for scattering, then it is the exact result, period. It is the formula $\varrho = m/(n_0 e^2 \tau_1)$, where τ_1 is defined in (7.1.5). It even includes the $(1 - \cos\theta)$ factor, although this assertion is not immediately obvious. One has to perform the angular integral in (7.1.5), which eliminates the delta function, in order to show its equivalence with the above formula for the resistivity.

The force–force correlation function gives the right resistivity for impurity scattering. No vertex equation or integral equation was needed in the derivation. The ease of derivation has made this approach popular.

Several caveats are needed. One is that impurity scattering is the only known example where the force–force correlation function gives the correct answer. In other cases it give an approximate answer. The second caveat is that *the right answer is obtained by a wrong derivation*. The derivation contains two important limits. One is setting the volume $\nu \to \infty$, while the second is $\omega \to 0$. The wrong answer is obtained by taking these limits in the wrong order. In fact, we did take them in the wrong order. If they had been done correctly, we would have found that $R_{\text{ret}}(\omega) \sim \omega^3$ and the force–force correlation function would give a zero result as $\omega \to 0$. These points are discussed by Argyres and Sigel (1974), Huberman and Chester (1975), Kubo *et al.* (1985), and Fishman (1989).

The force–force correlation function may also be evaluated for the electron scattering by phonons. The result is

$$\varrho(T) = C' \sum_{\lambda} \int q\, d^3q \mid W(q) \mid^2 (\xi_\lambda \cdot \mathbf{q})^2 \left[-\frac{\partial}{\partial \omega} n_B(\omega) \right]_{\omega=\omega_\lambda(\mathbf{q})} \tag{7.3.1}$$

$$C' = \frac{3\hbar \nu_0}{Me^2 16 v_F{}^2 k_F{}^4}$$

here $W(q)$ is the screened electron–ion interaction. This formula was first derived by Ziman (1960) as a variational solution to the Boltzmann equation. It is the formula that is most often evaluated when calculating the temperature dependence of the resistivity of metals. Figure 7.11 shows a theoretical calculation of Dynes and Carbotte (1968) compared with experiments for Na, K, and Al. An important feature of these calculations is numerically integrating over the Brillouin zone for all the phonon states, while employing accurate values for the phonon frequencies $\omega_\lambda(\mathbf{q})$ and polarization vectors ξ_λ. Another calculation is by Shukla and Taylor (1976), who used the dielectric function of Geldart and Taylor (1970) in screening the electron–ion interaction.

B. Kubo Formula

We shall present a theory of the dc conductivity based on an evaluation of the Kubo formula by Holstein (1964). This theory sums the ladder diagrams for phonons and reduces the vertex function to an integral equa-

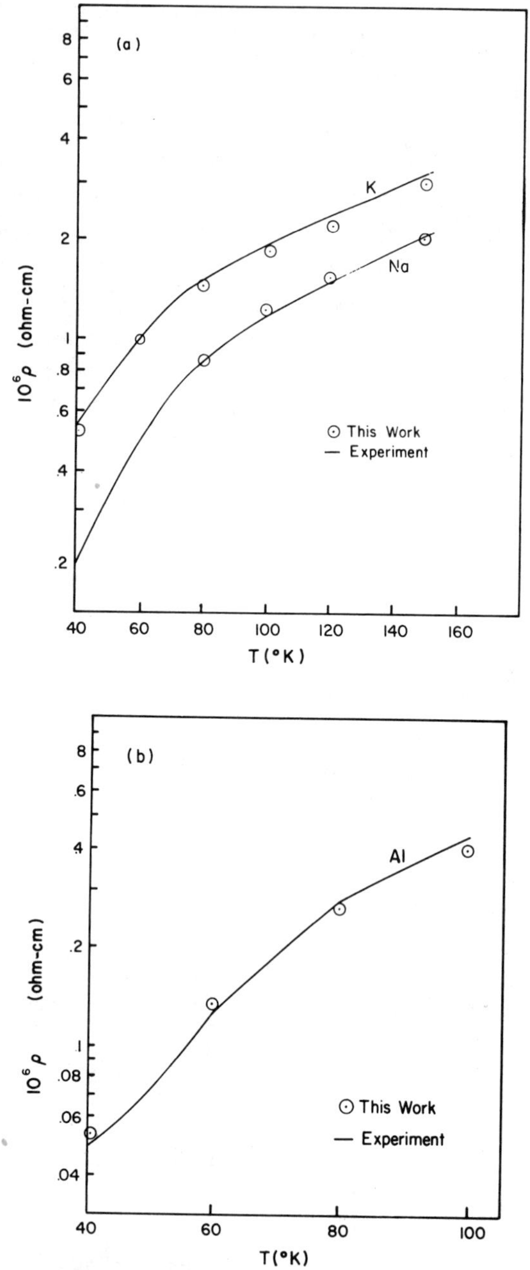

FIGURE 7.11. Resistivity as a function of temperature for (a) Na and K and (b) Al. Solid line is experiment, and points are theory. *Source*: Dynes and Carbotte (1968) (used with permission).

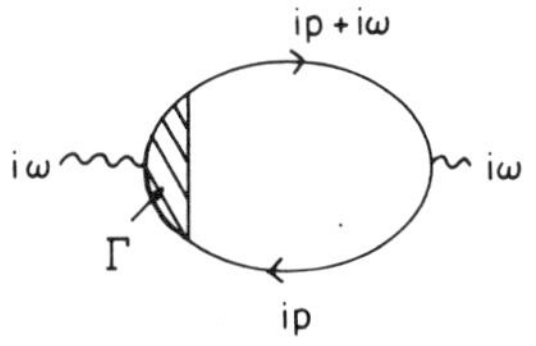

FIGURE 7.12

tion which is solved numerically. So far a solution is available only for a spherical Fermi surface, so that the result is the Kubo formula analogy of (7.3.1). The results are expressed in terms of the McMillan function $\alpha^2(\omega)F(\omega)$ and a similar function $\alpha^2_{\rm tr}(\omega)F(\omega)$, which is used in transport theory.

Our goal is to evaluate the current–current correlation function in the presence of the electron–phonon interaction. This correlation function can always be expressed as a product of two Green's functions and the vertex function. The Green's function $\mathscr{G}(\mathbf{p}, ip)$ used here represent fully interacting particles, with a self-energy found from the electron–phonon interaction plus any additional interactions of interest. The Feynman diagram for the correlation function is shown in Fig. 7.12 where the vertex function is put only at one end of the bubble in order not to overcount the vertex terms:

$$\begin{aligned}\pi(i\omega) &= -\frac{1}{3\nu}\int_0^\beta d\tau e^{i\omega\tau}\langle T_\tau \mathbf{j}(\tau)\cdot\mathbf{j}(0)\rangle \\ &= -\frac{e^2}{3m^2\nu}\sum_{\substack{\mathbf{pp}'\\ \sigma\sigma'}}\mathbf{p}\cdot\mathbf{p}'\int_0^\beta d\tau e^{i\omega\tau}\langle T_\tau c^\dagger_{\mathbf{p}\sigma}(\tau)c_{\mathbf{p}\sigma}(\tau)c^\dagger_{\mathbf{p}'\sigma'}c_{\mathbf{p}'\sigma'}\rangle \\ &= \frac{2e^2}{3m^2}\int\frac{d^3p}{(2\pi)^3}\frac{1}{\beta}\sum_{ip}\mathscr{G}(\mathbf{p}, ip)\mathscr{G}(\mathbf{p}, ip+i\omega)\mathbf{p}\cdot\mathbf{\Gamma}(\mathbf{p}, ip, ip+i\omega)\end{aligned} \tag{7.3.2}$$

The dc conductivity is found by the same steps used in Secs. 7.1 and 7.2. One evaluates the correlation function for values of $i\omega$ and analytically continues this, $i\omega \to \omega + i\delta$, to find the retarded function. The dc conductivity is the imaginary part of the retarded function divided by ω in the limit where $\omega \to 0$:

$$\sigma = -\lim_{\omega\to 0}\frac{\operatorname{Im}[\pi_{\rm ret}(\omega)]}{\omega}$$

The vertex function $\mathbf{\Gamma}(\mathbf{p}, ip, ip+i\omega)$ will be evaluated later. It depends on both the frequency variables ip and $i\omega$, which we write in the combination $(ip, ip + i\omega)$. The reason for this choice will become clearer later, when we actually solve for the vertex function, but the two arguments ip

and $ip + i\omega$ come from the two-electron Green's functions which have the same frequency arguments. In a homogeneous electron gas, the vector vertex function $\mathbf{\Gamma}(\mathbf{p}, ip, ip + i\omega)$ must point in the vector direction $\mathbf{p}$, although in real metals the crystalline potential defines other possible directions. However, our approximation of treating the Fermi surface as strictly spherical is equivalent to neglecting crystal directions, so we can assume $\mathbf{\Gamma}$ points in the direction $\mathbf{p}$. Therefore it is convenient to introduce the scalar function $\gamma(\mathbf{p}, ip, ip + i\omega)$, which is the amplitude of the vector vertex function:

$$\mathbf{\Gamma}(\mathbf{p}, ip, ip + i\omega) = \mathbf{p}\gamma(\mathbf{p}, ip, ip + i\omega)$$

$$\pi(i\omega) = \frac{2e^2}{3m^2} \int \frac{d^3p}{(2\pi)^3} p^2 S(\mathbf{p}, i\omega)$$

$$S(\mathbf{p}, i\omega) = \frac{1}{\beta} \sum_{ip} \mathscr{G}(\mathbf{p}, ip)\mathscr{G}(\mathbf{p}, ip + i\omega)\gamma(\mathbf{p}, ip, ip + i\omega)$$

The scalar function γ is not the same as the scalar vertex function in (7.1.24). The first step is to evaluate the summation over Matsubara frequencies ip to obtain $S(\mathbf{p}, i\omega)$. To this end, we construct the usual contour integral which has cuts along the axes where $ip \rightarrow$ real and also $ip + i\omega \rightarrow$ real. It gives

$$\begin{aligned} S(\mathbf{p}, i\omega) = \int_{-\infty}^{\infty} \frac{d\varepsilon}{2\pi i} n_F(\varepsilon)\{&\mathscr{G}(\mathbf{p}, \varepsilon + i\omega) \\ &\times [G_{\text{ret}}(\mathbf{p}, \varepsilon)\gamma(\mathbf{p}, \varepsilon + i\delta, \varepsilon + i\omega) - G_{\text{adv}}(\mathbf{p}, \varepsilon)\gamma(\mathbf{p}, \varepsilon - i\delta, \varepsilon + i\omega)] \\ &+ \mathscr{G}(\mathbf{p}, \varepsilon - i\omega)[G_{\text{ret}}(\mathbf{p}, \varepsilon)\gamma(\mathbf{p}, \varepsilon - i\omega, \varepsilon + i\delta) \\ &- G_{\text{adv}}(\mathbf{p}, \varepsilon)\gamma(\mathbf{p}, \varepsilon - i\omega, \varepsilon - i\delta)]\} \end{aligned}$$

The next step is to analytically continue $i\omega \rightarrow \omega + i\delta$ and then take the imaginary part. The advanced and retarded Green's functions G_{adv} and G_{ret} are selected according to whether the analytical continuation approaches the real axes from above or below. It is also important for the vertex function to know whether its energy variables fall above or below the branch cuts, so we keep track of its $i\delta$ parts:

$$\begin{aligned} \text{Im}[S_{\text{ret}}(\mathbf{p}, \omega)] = -\text{Re}\Big[\int_{-\infty}^{\infty} \frac{d\varepsilon}{2\pi} n_F(\varepsilon)\{&G_{\text{ret}}(\mathbf{p}, \varepsilon + \omega)[G_{\text{ret}}(\mathbf{p}, \varepsilon) \\ &\times \gamma(\mathbf{p}, \varepsilon + i\delta, \varepsilon + \omega + i\delta) - G_{\text{adv}}(\mathbf{p}, \varepsilon)\gamma(\mathbf{p}, \varepsilon - i\delta, \varepsilon + \omega + i\delta)] \\ &+ G_{\text{adv}}(\mathbf{p}, \varepsilon - \omega)[G_{\text{ret}}(\mathbf{p}, \varepsilon)\gamma(\mathbf{p}, \varepsilon - \omega - i\delta, \varepsilon + i\delta) \\ &- G_{\text{adv}}(\mathbf{p}, \varepsilon)\gamma(\mathbf{p}, \varepsilon - \omega - i\delta, \varepsilon - i\delta)]\}\Big] \end{aligned}$$

Next we change integration variables $\varepsilon \to \varepsilon + \omega$ in the last two terms. This change makes the second and third terms become identical except for the factors of n_F. Similarly, except for the factors of n_F, the first and last terms become complex conjugates of each other; since we want their real part, they are identical:

$$\begin{aligned}\operatorname{Im}[S_{\text{ret}}(\mathbf{p}, \omega)] = \operatorname{Re}\Big\{\int_{-\infty}^{\infty} \frac{d\varepsilon}{2\pi} & [n_F(\varepsilon + \omega) - n_F(\varepsilon)] \\ & \times [G_{\text{adv}}(\mathbf{p}, \varepsilon)G_{\text{ret}}(\mathbf{p}, \varepsilon + \omega)\gamma(\mathbf{p}, \varepsilon - i\delta, \varepsilon + \omega + i\delta) \\ & - G_{\text{ret}}(\mathbf{p}, \varepsilon)G_{\text{ret}}(\mathbf{p}, \varepsilon + \omega)\gamma(\mathbf{p}, \varepsilon + i\delta, \varepsilon + \omega + i\delta)]\Big\}\end{aligned} \tag{7.3.3}$$

The final step in the formal derivation is to divide by ω and take the limit as $\omega \to 0$, which is done in the occupation number parts,

$$\lim_{\omega \to 0} \left[\frac{n_F(\varepsilon + \omega) - n_F(\varepsilon)}{\omega} \right] = \frac{d}{d\varepsilon} n_F(\varepsilon)$$

so we obtain a formal expression for the dc electrical conductivity:

$$\begin{aligned}\sigma = \frac{2e^2}{3m^2} \int \frac{d^3p}{(2\pi)^3} p^2 \int_{-\infty}^{\infty} \frac{d\varepsilon}{2\pi} \left[- \frac{dn_F(\varepsilon)}{d\varepsilon} \right] \\ \times \{| G_{\text{ret}}(\mathbf{p}, \varepsilon) |^2 \gamma(\mathbf{p}, \varepsilon - i\delta, \varepsilon + i\delta) - \operatorname{Re}[G_{\text{ret}}(\mathbf{p}, \varepsilon)^2 \gamma(\mathbf{p}, \varepsilon + i\delta, \varepsilon + i\delta)]\}\end{aligned} \tag{7.3.4}$$

This equation is identical to (7.1.21) once we identify $P = GG\gamma$. Equation (7.3.4) is exact if we know the vertex function. It is expressed in terms of the two functions $\gamma(\mathbf{p}, \varepsilon - i\delta, \varepsilon + i\delta)$ and $\gamma(\mathbf{p}, \varepsilon + i\delta, \varepsilon + i\delta)$. These we expect to be quite different, because of our experience in Sec. 7.1 with the same two functions for impurity scattering. There we showed that the function $\gamma(\mathbf{p}, \varepsilon + i\delta, \varepsilon + i\delta)$ could be obtained from a Ward identity, and the same is true here. This was first shown for the electron–phonon system in metals by Englesberg and Schrieffer (1963), who followed the procedure we outlined earlier in Sec. 7.1. This Ward identity can be expressed in terms of $\mathbf{\Gamma}(\mathbf{p}, \varepsilon + i\delta, \varepsilon + i\delta)$ or our scalar γ:

$$\mathbf{\Gamma}(\mathbf{p}, \varepsilon + i\delta, \varepsilon + i\delta) = \mathbf{p} + m\nabla_p \Sigma(\mathbf{p}, \varepsilon)$$

$$\gamma(\mathbf{p}, \varepsilon + i\delta, \varepsilon + i\delta) = 1 + \frac{\partial}{\partial \xi_p} \Sigma(\mathbf{p}, \varepsilon)$$

$$G_{\text{ret}}(\mathbf{p}, \varepsilon)^2 \gamma(\mathbf{p}, \varepsilon + i\delta, \varepsilon + i\delta) = \frac{\partial}{\partial \xi_p} G_{\text{ret}}(\mathbf{p}, \varepsilon)$$

For metals, the electron–phonon system has the feature that the self-energy function $\Sigma(p, \varepsilon)$ is not very p dependent, and the derivative of $\Sigma(p, \varepsilon)$ with respect to ξ_p is small: $d\Sigma/d\xi_p \simeq \Sigma/E_F \ll 1$. Thus it seems we can set $\gamma(p, \varepsilon + i\delta, \varepsilon + i\delta) = 1$.

There is another result which is even stronger, since it applies for all values of ε near the Fermi energy. Since the self-energy $\Sigma(p, \varepsilon)$ does not have significant p dependence near $p \simeq k_F$, call it $\Sigma(p_F, \varepsilon) = \mathrm{Re}[\Sigma(\varepsilon)] - i\Gamma(\varepsilon)$ where we suppress the p notation. The function $\Gamma(\varepsilon) = -\mathrm{Im}[\Sigma(p_F, \varepsilon)]$. The retarded and advanced Green's function have their important dependence on p through the kinetic energy term, $\xi_p \equiv \xi$. Define $\Omega(\varepsilon) = \varepsilon - \mathrm{Re}[\Sigma(\varepsilon)]$:

$$G_{\mathrm{ret}}(p, \varepsilon) = \frac{1}{\varepsilon - \xi - \mathrm{Re}[\Sigma(\varepsilon)] + i\Gamma(\varepsilon)} = \frac{1}{\Omega(\varepsilon) - \xi + i\Gamma(\varepsilon)}$$

$$G_{\mathrm{adv}}(p, \varepsilon) = \frac{1}{\varepsilon - \xi - \mathrm{Re}[\Sigma(\varepsilon)] - i\Gamma(\varepsilon)} = \frac{1}{\Omega(\varepsilon) - \xi - i\Gamma(\varepsilon)}$$

Similarly, we assume here that the vertex function $\gamma(\mathbf{p}, ip, ip + i\omega)$ does not have significant p dependence except on the order of k_F. This assumption will be justified later, when we actually evaluate it. We wish to show that we can forget about the second term in brackets in (7.3.4); it vanishes by doing the kinetic energy integration. By neglecting terms of order ξ/E_F, we can write

$$\begin{aligned} d^3pp^2 &= 4\pi p^4\, dp = 4\pi m k_F^3\, d\xi\left[1 + O\left(\frac{\xi}{E_F}\right)\right] \\ &= 12\pi^3 m n_0\, d\xi \end{aligned}$$

where the electron density is $n_0 = k_F^3/3\pi^2$. When the kinetic energy integral is evaluated, the only ξ variation is in the Green's functions. Of the three combinations which occur, only one makes a finite contribution:

$$\int_{-\infty}^{\infty} d\xi G_{\mathrm{ret}}(p, \varepsilon) G_{\mathrm{ret}}(p, \varepsilon') = 0$$

$$\begin{aligned} \int_{-\infty}^{\infty} d\xi G_{\mathrm{ret}}(p, \varepsilon) G_{\mathrm{adv}}(p, \varepsilon') &= \frac{2\pi i}{\Omega(\varepsilon) - \Omega(\varepsilon') + i[\Gamma(\varepsilon) + \Gamma(\varepsilon')]} \\ &\equiv I(\varepsilon', \varepsilon) \end{aligned} \tag{7.3.5}$$

$$\int_{-\infty}^{\infty} d\xi G_{\mathrm{adv}}(p, \varepsilon) G_{\mathrm{adv}}(p, \varepsilon') = 0$$

For each integral we close the contour at infinity. The two integrals which vanish have both their poles in the same half plane (upper or lower), so the integration contour can be chosen to avoid them both, which encircles no poles and hence gives zero. The integral over the combination $G_{\text{adv}}G_{\text{ret}}$ has one pole in each plane, so closing the contour always picks up one pole. The residue at the pole produces the finite result. By using these integration results, the dc conductivity (7.3.4) becomes

$$\begin{aligned} \sigma &= \frac{e^2 n_0}{m} \int_{-\infty}^{\infty} d\varepsilon \left[-\frac{d}{d\varepsilon} n_F(\varepsilon) \right] \frac{\Lambda(\varepsilon)}{2\Gamma(\varepsilon)} \\ \Lambda(\varepsilon) &= \gamma(k_F, \varepsilon - i\delta, \varepsilon + i\delta) \\ \Gamma(\varepsilon) &= -\text{Im}[\Sigma(k_F, \varepsilon)] \end{aligned} \tag{7.3.6}$$

Equation (7.3.6) is the final result of the formal derivation. There only remains the evaluation of the vertex function $\Lambda(\varepsilon)$ and the imaginary self-energy $\Gamma(\varepsilon)$ and the integral over the temperature distribution $-dn_F/d\varepsilon$. If the vertex function were absent ($\Lambda = 1$), then the evaluation would be easy. The imaginary self-energy $\Gamma(\varepsilon)$ has been evaluated for many metals, and a result was given in Fig. 6.18 for Pb. The quantity $(2\Gamma)^{-1} = \tau(\varepsilon)$, where $\tau(\varepsilon)$ is the relaxation time defined as the average time between scattering events. The result (7.3.6), after we set $\Lambda = 1$, is just the average of the relaxation time over the thermally smeared Fermi distribution. Of course, from the earlier solution for impurity scattering, we anticipate that this neglect of $\Lambda(\varepsilon)$ is a serious error. The vertex function $\Lambda(\varepsilon)$ serves the important role of weighting the scattering events and favoring those at high-momentum transfer. For impurity scattering, we found that $\Lambda = \Gamma/\Gamma_{tr}$ or $\Lambda/\Gamma = 1/\Gamma_{tr}$, where Γ_{tr} is the scattering rate which contains the factor of $1 - \cos\theta$. We must now solve the vertex equation for phonon scattering to see whether the same effect happens. Since the phonon scattering is inelastic, we expect that the results will not be identical to those for impurity scattering. In summing the ladder diagrams for the vertex function, we shall find the vertex contribution to be important and significantly different from unity. Thus Migdal's (1958) theorem, which asserts that vertex terms are unimportant, is contradicted.

The vertex function $\mathbf{\Gamma}(\mathbf{p}, ip, ip + i\omega)$ is calculated by solving the integral equation

$$\begin{aligned} \mathbf{\Gamma}(\mathbf{p}, ip, ip + i\omega) = \mathbf{p} &+ \frac{1}{\beta} \sum_{iq,\lambda} \int \frac{d^3q}{(2\pi)^3} M_\lambda(\mathbf{q})^2 \mathscr{D}(\mathbf{q}, iq) \mathscr{G}(\mathbf{p} + \mathbf{q}, ip + iq) \\ &\times \mathscr{G}(\mathbf{p} + \mathbf{q}, ip + iq + i\omega) \mathbf{\Gamma}(\mathbf{p} + \mathbf{q}, ip + iq, ip + iq + i\omega) \end{aligned} \tag{7.3.7}$$

This vertex sums the ladder diagrams for phonons. It is illustrated in

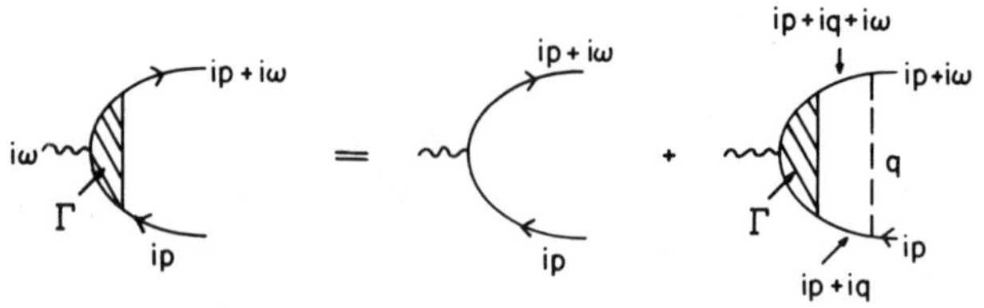

FIGURE 7.13

Fig. 7.13. Iteration of Eq. (7.3.7) produces a series in which each additional term has one more ladder diagram. The solution to the integral equation produces an expression which contains all terms with any number of phonon ladder diagrams. This solution is not an exact evaluation of the vertex function $\mathbf{\Gamma}(\mathbf{p}, ip, ip + i\omega)$, since other vertex contributions occur which are not ladders but have the phonon lines crossed. One expects these terms to be smaller, but detailed calculations are lacking, so this is only a supposition.

It is extremely unfortunate that we need to find the vector vertex function. The scalar vertex function $\Gamma(\mathbf{p}, ip, ip + i\omega)$, which is obtained by replacing $\mathbf{p}$ by 1 in Eq. (7.3.7), is easily obtained from a Ward identity:

$$\begin{aligned}\Gamma(\mathbf{p}, ip, ip + i\omega) &= 1 + \frac{1}{\beta} \sum_{iq,\lambda} \int \frac{d^3q}{(2\pi)^3} M_\lambda(\mathbf{q})^2 \mathscr{D}(q) \mathscr{G}(p + q) \\ &\quad \times \mathscr{G}(p + q + i\omega) \Gamma(\mathbf{p}, ip + iq, ip + iq + i\omega) \\ &= 1 - \frac{1}{i\omega} [\Sigma(\mathbf{p}, ip + i\omega) - \Sigma(\mathbf{p}, ip)] \end{aligned} \tag{7.3.8}$$

The Ward identity is not helpful for our problem since we want the vector vertex function, which is quite different. There is a Ward identity (7.1.25) for the vector vertex which is useful only for finding the vertex function which has both frequency arguments identical: $\mathbf{\Gamma}(\mathbf{p}, ip, ip)$. This vertex is not the one we need, so that an easy solution to (7.3.7) by Ward identities is not possible. Thus we must attack and solve the integral equation.

The scalar function $\gamma(\mathbf{p}, ip, ip + i\omega)$, which is the scalar amplitude of the vector vertex function $\mathbf{\Gamma} = \mathbf{p}\gamma$, obeys the integral equation

$$\begin{aligned}\gamma(\mathbf{p}, ip, ip + i\omega) &= 1 + \frac{1}{\beta} \sum_{iq,\lambda} \int \frac{d^3q}{(2\pi)^3} M_\lambda(\mathbf{q})^2 \mathscr{D}(q) \frac{\mathbf{p} \cdot (\mathbf{p} + \mathbf{q})}{p^2} \\ &\quad \times \mathscr{G}(p + q) \mathscr{G}(p + q + \omega) \gamma(\mathbf{p} + \mathbf{q}, ip + iq, ip + iq + i\omega)\end{aligned}$$

This equation is not the same one which is obeyed by the scalar vertex function in (7.3.8), and these two functions are quite different.

The first step is to do the integrals over angles and wave vector. First write $d^3q = 2\pi q^2\, dq\, d\nu$, where $\nu = \cos\theta$ is the angle between $\mathbf{p}$ and $\mathbf{q}$. The angle variable is changed to p_1, defined as

$$p_1{}^2 = (\mathbf{p} + \mathbf{q})^2 = p^2 + q^2 + 2pq\nu$$

$$d\nu = \frac{p_1\, dp_1}{pq}$$

$$d^3q \to \frac{2\pi}{p} \int_0^\infty q\, dq \int_{|p-q|}^{|p+q|} p_1\, dp_1 = \frac{2\pi}{v_p} \int_0^\infty q\, dq \int_{[|p-q|^2/2m]-E_F}^{[(p+q)^2/2m]-E_F} d\xi_1$$

The next step is to uncouple the limits of integration. We are concerned with electrons at the Fermi surface, so we anticipate that values of $\xi_1 \simeq 0$ and $\xi = \varepsilon_p - \mu \simeq 0$ are important in the integration process. We can approximate $v_p \simeq v_F$ as the Fermi velocity. Similarly, the $q\, dq$ integral is understood to be over the spherical Fermi surface—from one point $\mathbf{p}$ to all other points $\mathbf{p}_1$, where p and p_1 both have magnitude k_F:

$$\int d^3q \to \frac{1}{v_F} \int d^2q \int_{-\infty}^{\infty} d\xi_1$$
$$|\,\mathbf{p}_F + \mathbf{q}\,| = k_F$$

The limits on the $d\xi_1$ integral are extended between $\pm\infty$, since most of the integrand has large q values, where the actual limits on ξ_1 are from a very negative number to a very positive one. Since the main contribution is in the region $\xi_1 \simeq 0$, this error is small. Similarly, we examine the other angular factors in the integrand:

$$\frac{\mathbf{p} \cdot (\mathbf{p} + \mathbf{q})}{p^2} = 1 + \frac{q\nu}{p} = 1 + \frac{1}{2p^2}(p_1{}^2 - p^2 - q^2)$$
$$\simeq 1 - \frac{q^2}{2k_F{}^2} + O\left(\frac{\xi_1{}^2}{E_F}, \frac{\xi}{E_F}\right)$$

The terms ξ_1/E_F, ξ/E_F will be neglected, since we expect that values of ξ_1 and ξ will be small—on the order of a Debye energy. On the other hand, the factor $q^2/2k_F{}^2$ need not be small, since the integration over phonon states has a significant contribution from high values of q near the edge of the Brillouin zone. Thus this term must be retained. The difference in treating the factors $p_1{}^2 - p^2$, which we neglect, and q^2, which we retain, is that the former enters the average over electron wave vector states and the latter over phonon states. These integrals over the phonon wave vector can be expressed in terms of the function $\alpha^2(\omega)F(\omega)$, which was introduced in Sec. 6.4:

$$\begin{aligned}\alpha^2(\omega)F(\omega) &= \frac{1}{v_F}\sum_\lambda \int \frac{d^2q}{(2\pi)^3} M_\lambda(\mathbf{q})^2 \delta[\omega - \omega_\lambda(q)] \\ \alpha_{tr}^2(\omega)F(\omega) &= \frac{1}{v_F}\sum_\lambda \int \frac{d^2q}{(2\pi)^3} M_\lambda(\mathbf{q})^2 \frac{q^2}{2k_F{}^2} \delta[\omega - \omega_\lambda(q)]\end{aligned} \tag{7.3.9}$$

The first of these is just the McMillan function $\alpha^2(\omega)F(\omega)$, which was defined in (6.4.20). Since our Fermi surface has been assumed to be spherical, α^2F has the same value at each point on the surface, and the k subscript is omitted. The other form of coupling α_{tr}^2F is called the "transport form of alpha-squared-F," which was introduced by Allen (1971). It differs from the McMillan form by having the additional factor of $q^2/2k_F{}^2$ in the integrand, which gives more weight to the scattering processes at large wave vector. Indeed, the extra factor of $q^2/2k_F{}^2$ is identical to $1 - \cos\theta$ when the scattering is elastic.

With the completion of all these angular and wave vector integrations, we find that the vertex function $\gamma(\mathbf{p}, ip, ip + i\omega)$ is not very dependent on p. The only variation is through a function of p, which we can set to k_F with an error of only ξ_p/E_F. Thus we define $\gamma(ip, ip+i\omega) = \gamma(k_F, ip, ip+i\omega)$ and arrive at the equations:

$$\begin{aligned}\gamma(ip, ip + i\omega) &= 1 + \int_0^{\omega_D} du[\alpha^2(u)F(u) - \alpha_{tr}^2(u)F(u)] \\ &\quad \times \int_{-\infty}^{\infty} d\xi_1 S(\xi_1, u, ip, ip + i\omega) \\ S(\xi, u, ip, ip + i\omega) &= \frac{1}{\beta}\sum_{iq} \frac{2u}{(iq)^2 - u^2} \gamma(ip + iq, ip + iq + i\omega) \\ &\quad \times \mathscr{G}(\xi, ip + iq)\mathscr{G}(\xi, ip + iq + i\omega)\end{aligned}$$

We shall show that this integral equation is not too difficult to solve. The basic approximation has been to decouple the integrations over dq and $d\xi$, which permits all the phonon information to be collected into the functions $\alpha^2(u)F(u) - \alpha_{tr}^2(u)F(u)$. The primary assumption we make in doing this decoupling is that the Fermi degeneracy E_F is very much larger than other energies such as K_BT or ω_D. The present integral equation is actually much easier to solve than the one for polarons in Sec. 7.2, since there the integration variables cannot be accurately decoupled in the same way.

The preceding is the basic integral equation which needs to be solved for the vertex function. The ξ dependence of S is only in the Green's functions $\mathscr{G}(\xi, ip + iq)$ and $\mathscr{G}(\xi, ip + iq + i\omega)$. This integral will be done later, where we shall use the result (7.3.5) that only the integral over the pair $G_{\text{adv}}G_{\text{ret}}$ is finite. The factor $2u/[(iq)^2 - u^2]$ is the phonon Green's function for a phonon of energy u. The next step in the derivation is to do the summation over Matsubara frequency iq, which is done in the usual way by constructing a contour integral. The integrand has poles from the phonon Green's functions at $Z = \pm u$ and cuts along the axes where the electron Green's functions are real, $Z = ip$ and $Z = ip + i\omega$. The contour integral gives the result

$$\begin{aligned}
\gamma(ip, ip + i\omega) &= 1 + \int_0^{\omega_D} du[\alpha^2(u)F(u) - \alpha_{tr}^2(u)F(u)] \int_{-\infty}^{\infty} d\xi \\
&\times \Big\{ n_B(u)\mathscr{G}(\xi, ip + u)\mathscr{G}(\xi, ip + i\omega + u)\gamma(ip + u, ip + i\omega + u) \\
&+ [n_B(u) + 1]\mathscr{G}(\xi, ip - u)\mathscr{G}(\xi, ip + i\omega - u)\gamma(ip - u, ip + i\omega - u) \\
&+ \int_{-\infty}^{\infty} \frac{d\varepsilon'}{2\pi i} n_F(\varepsilon') \Big[\frac{\mathscr{G}(\xi, \varepsilon' + i\omega)(2u)}{(ip - \varepsilon')^2 - u^2} [G_{\text{ret}}(\xi, \varepsilon')\gamma(\varepsilon' + i\delta, \varepsilon' + i\omega) \\
&- G_{\text{adv}}(\xi, \varepsilon')\gamma(\varepsilon' - i\delta, \varepsilon' + i\omega)] + \frac{\mathscr{G}(\xi, \varepsilon' - i\omega)(2u)}{(ip + i\omega - \varepsilon')^2 - u^2} \\
&\times [G_{\text{ret}}(\xi, \varepsilon')\gamma(\varepsilon' - i\omega, \varepsilon' + i\delta) - G_{\text{adv}}(\xi, \varepsilon')\gamma(\varepsilon' - i\omega, \varepsilon' - i\delta)] \Big] \Big\}
\end{aligned} \tag{7.3.10}$$

Equation (7.3.10) is a formidable result, but fortunately it can be reduced to a simple form. Our present interest is in the particular vertex function $\gamma(\varepsilon - i\delta, \varepsilon + \omega + i\delta)$ in the limit as $\omega \to 0$. It is obtained from $\gamma(ip, ip + i\omega)$ by the analytical continuation $ip \to \varepsilon - i\delta$ and $ip + i\omega \to \varepsilon + \omega + i\delta$. The Green's functions become advanced or retarded according to the side of the cut, so this analytical continuation produces

$$\begin{aligned}
\gamma(\varepsilon - i\delta, \varepsilon + \omega + i\delta) &= 1 + \int_0^{\omega_D} du[\alpha^2(u)F(u) - \alpha_{tr}^2(u)F(u)] \int_{-\infty}^{\infty} d\xi \\
&\times \Big\{ n_B(u)G_{\text{adv}}(\xi, \varepsilon + u)G_{\text{ret}}(\xi, \varepsilon + u + \omega)\gamma(\varepsilon + u - i\delta, \varepsilon + \omega + u + i\delta) \\
&+ [n_B(u) + 1]G_{\text{adv}}(\xi, \varepsilon - u)G_{\text{ret}}(\xi, \varepsilon + \omega - u)\gamma(\varepsilon - u - i\delta, \varepsilon + \omega - u + i\delta) \\
&+ \int_{-\infty}^{\infty} \frac{d\varepsilon'}{2\pi i} n_F(\varepsilon') \Big[G_{\text{ret}}(\xi, \varepsilon' + \omega) \frac{2u}{(\varepsilon - i\delta - \varepsilon')^2 - u^2} \\
&\times [G_{\text{ret}}(\xi, \varepsilon')\gamma(\varepsilon + i\delta, \varepsilon' + \omega + i\delta) - G_{\text{adv}}(\xi, \varepsilon')\gamma(\varepsilon' - i\delta, \varepsilon' + \omega + i\delta)] \\
&+ \frac{G_{\text{adv}}(\xi, \varepsilon' - \omega)(2u)}{(\varepsilon + \omega + i\delta - \varepsilon')^2 - u^2} [G_{\text{ret}}(\xi, \varepsilon')\gamma(\varepsilon' - \omega - i\delta, \varepsilon' + i\delta) \\
&- G_{\text{adv}}(\xi, \varepsilon')\gamma(\varepsilon' - \omega - i\delta, \varepsilon' - i\delta)] \Big] \Big\}
\end{aligned}$$

The next step is to do the integration over $d\xi$. According to (7.3.5), this integral eliminates all combinations of Green's functions except $G_{\text{adv}}(\xi, \varepsilon')\times G_{\text{ret}}(\xi, \varepsilon)$, which integrate to give $I(\varepsilon', \varepsilon)$:

$$\begin{aligned}
&\gamma(\varepsilon - i\delta, \varepsilon + \omega + i\delta) \\
&\quad = 1 + \int_0^{\omega_D} du[\alpha^2(u)F(u) - \alpha_{tr}^2(u)F(u)] \\
&\quad \times \Big\{ n_B(u)I(\varepsilon + u, \varepsilon + u + \omega)\gamma(\varepsilon + u - i\delta, \varepsilon + \omega + u + i\delta) \\
&\quad + [n_B(u) + 1]I(\varepsilon - u, \varepsilon + \omega - u)\gamma(\varepsilon - u - i\delta, \varepsilon + \omega - u + i\delta) \\
&\quad + \int_{-\infty}^{\infty} \frac{d\varepsilon'}{2\pi i} n_F(\varepsilon')\Big[-\frac{2u}{(\varepsilon - i\delta - \varepsilon')^2 - u^2} \\
&\quad \times I(\varepsilon', \varepsilon' + \omega)\gamma(\varepsilon' - i\delta, \varepsilon' + \omega + i\delta) \\
&\quad + \frac{2u}{(\varepsilon + \omega + i\delta - \varepsilon')^2 - u^2} I(\varepsilon' - \omega, \varepsilon')\gamma(\varepsilon' - \omega - i\delta, \varepsilon' + i\delta)\Big]\Big\}
\end{aligned}$$

This integral equation contains only the unknown factor $\gamma(\varepsilon - i\delta, \varepsilon' + i\delta)$ and no longer depends on the other factor $\gamma(\varepsilon + i\delta, \varepsilon' + i\delta)$. The last integral $d\varepsilon'$ is manipulated by changing variables $\varepsilon' \to \varepsilon' + \omega$ in the last term, so that it becomes

$$\begin{aligned}
&-\int_{-\infty}^{\infty} \frac{d\varepsilon'}{2\pi i} I(\varepsilon', \varepsilon' + \omega)\gamma(\varepsilon' - i\delta, \varepsilon' + \omega + i\delta) \\
&\quad \times \left[\frac{n_F(\varepsilon')(2u)}{(\varepsilon' - i\delta - \varepsilon')^2 - u^2} - \frac{n_F(\varepsilon' + \omega)(2u)}{(\varepsilon + i\delta - \varepsilon')^2 - u^2} \right] \\
&\quad = \tfrac{1}{2}[n_F(\varepsilon + u) + n_F(\varepsilon + u + \omega)]I(\varepsilon + u, \varepsilon + \omega + u) \\
&\quad \times \gamma(\varepsilon + u - i\delta, \varepsilon + \omega + u + i\delta) - \tfrac{1}{2}[n_F(\varepsilon - u) + n_F(\varepsilon + \omega - u)] \\
&\quad \times I(\varepsilon - u, \varepsilon + \omega - u)\gamma(\varepsilon - u - i\delta, \varepsilon + \omega - u + i\delta) \\
&\quad - \int_{-\infty}^{\infty} \frac{d\varepsilon'}{2\pi i} I(\varepsilon', \varepsilon' + \omega)\gamma(\varepsilon' - i\delta, \varepsilon' + \omega + i\delta) \\
&\quad \times [n_F(\varepsilon') - n_F(\varepsilon' + \omega)]P\frac{2u}{(\varepsilon - \varepsilon')^2 - u^2}
\end{aligned}$$

The first terms come from the poles of the phonon Green's functions $\varepsilon' = \varepsilon \pm u$. These manipulations provide the final form of the integral equation for $\gamma(\varepsilon - i\delta, \varepsilon + \omega + i\delta)$. It was first derived by Holstein (1964), although we have modified his result a little by expressing it in the $\alpha^2 F$ formalism:

$$\gamma(\varepsilon - i\delta, \varepsilon + \omega + i\delta) = 1 + \int_0^{\omega_D} du[\alpha^2(u)F(u) - \alpha_{tr}^2(u)F(u)]$$
$$\times \Big\{ I(\varepsilon + u, \varepsilon + u + \omega)\gamma(\varepsilon + u - i\delta, \varepsilon + \omega + u + i\delta)$$
$$\times [n_B(u) + \tfrac{1}{2}n_F(\varepsilon + u) + \tfrac{1}{2}n_F(\varepsilon + u + \omega)]$$
$$+ I(\varepsilon - u, \varepsilon + \omega - u)\gamma(\varepsilon - u - i\delta, \varepsilon + \omega - u + i\delta)$$
$$\times [n_B(u) + 1 - \tfrac{1}{2}n_F(\varepsilon - u) - \tfrac{1}{2}n_F(\varepsilon + \omega - u)]$$
$$- \int_{-\infty}^{\infty} \frac{d\varepsilon'}{2\pi i} I(\varepsilon', \varepsilon' + \omega)\gamma(\varepsilon' - i\delta, \varepsilon' + \omega + i\delta)[n_F(\varepsilon') - n_F(\varepsilon' + \omega)]$$
$$\times P \frac{2u}{(\varepsilon - \varepsilon')^2 - u^2} \Big\} \tag{7.3.11}$$

The next step is to take the limit $\omega \to 0$, which must be done carefully. First we examine the Ward identity (7.3.8) for the scalar vertex in this limit. Although our vertex function $\gamma(\varepsilon - i\delta, \varepsilon + \omega + i\delta)$ is not given by this Ward identity, perhaps it has similar analytical properties. In particular, we notice that when $ip \to \varepsilon - i\delta$ and $ip + i\omega \to \varepsilon + \omega + i\delta$, then ($\Gamma(\varepsilon) = -\mathrm{Im}[\Sigma_{\mathrm{ret}}(p, \varepsilon)]$)

$$\Sigma(p, ip) \to \mathrm{Re}[\Sigma(\varepsilon)] + i\Gamma(\varepsilon)$$
$$\Sigma(p, ip + i\omega) \to \mathrm{Re}[\Sigma(\varepsilon + \omega)] - i\Gamma(\varepsilon + \omega)$$
$$\Gamma(ip, ip + i\omega) \to 1 - \frac{1}{\omega} \{\mathrm{Re}[\Sigma(\varepsilon + \omega)] - \mathrm{Re}[\Sigma(\varepsilon)] - i\Gamma(\varepsilon + \omega) - i\Gamma(\varepsilon)\} \tag{7.3.12}$$
$$\lim_{\omega \to 0} \Gamma(\varepsilon - i\delta, \varepsilon + \omega + i\delta) \to 1 - \frac{\partial\, \mathrm{Re}[\Sigma(\varepsilon)]}{\partial \varepsilon} + \lim_{\omega \to 0} 2i \frac{\Gamma(\varepsilon)}{\omega}$$

Thus we observe that the scalar vertex function behaves, in the limit $\omega \to 0$, as $\Gamma(\varepsilon - i\delta, \varepsilon + \omega + i\delta) \to \mathrm{Re}[f(\varepsilon)] + (i/\omega)2\Gamma(\varepsilon)$. There is a complex term which diverges as ω^{-1}. The vector vertex $\gamma(\varepsilon - i\delta, \varepsilon + \omega + i\delta)$ has been shown to be entirely real as $\omega \to 0$, so it has no complex part. Thus we define the real function $\Lambda(\varepsilon)$ as

$$\lim_{\omega \to 0} \gamma(\varepsilon - i\delta, \varepsilon + \omega + i\delta) \to \Lambda(\varepsilon)$$

This assumption is inserted into (7.3.11) for $\gamma(\varepsilon - i\delta, \varepsilon + \omega + i\delta)$, and we take the limit $\omega \to 0$. We find the simple integral equation

$$\Lambda(\varepsilon) = 1 + \pi \int_0^{\omega_D} du[\alpha^2(u)F(u) - \alpha_{tr}^2(u)F(u)]\left\{[n_B(u) + n_F(\varepsilon + u)] \times \frac{\Lambda(\varepsilon + u)}{\Gamma(\varepsilon + u)} + [n_B(u) + 1 - n_F(\varepsilon - u)] \frac{\Lambda(\varepsilon - u)}{\Gamma(\varepsilon - u)}\right\} \quad (7.3.13)$$

$$\sigma = -\frac{n_0 e^2}{2m} \int_{-\infty}^{\infty} d\varepsilon \left[\frac{dn_F(\varepsilon)}{d\varepsilon}\right] \frac{\Lambda(\varepsilon)}{\Gamma(\varepsilon)} \quad (7.3.6)$$

We have also rewritten the integral (7.3.6) for the dc conductivity in order to present the two important results together. The integral equation for the vertex function $\Lambda(\varepsilon)$ must be solved, and the solution is used in the integral over ε for the conductivity. Recall that $\Lambda(\varepsilon)$ is a real function, so the integral equation is not complicated. The form of the equations suggests we actually need to solve for $\bar{\tau}(\varepsilon) = \Lambda/2\Gamma$, which might be called the effective relaxation time for transport.

To facilitate the discussion of the result, we rewrite the result for $\Gamma(\varepsilon)$ from (6.4.22) and the equivalent result for the transport kernal:

$$\Gamma(\varepsilon) = -\text{Im}[\Sigma(\varepsilon)] = \pi \int_0^{\omega_D} du\alpha^2(u)F(u)[2n_B(u) + n_F(\varepsilon + u) + n_F(u - \varepsilon)]$$

$$\Gamma_{tr}(\varepsilon) = \pi \int_0^{\omega_D} du\alpha_{tr}^2(u)F(u)[2n_B(u) + n_F(\varepsilon + u) + n_F(u - \varepsilon)] \quad (7.3.14)$$

If the self-energy function $\Gamma(\varepsilon)$ and the vertex function $\Lambda(\varepsilon)$ were both constants and independent of ε, then the solution of the vertex equation (7.3.13) would be easy:

$$\Lambda = 1 + \frac{\Lambda}{\Gamma}(\Gamma - \Gamma_{tr}) = \frac{\Gamma}{\Gamma_{tr}}$$

This model gives $\Lambda/\Gamma = 1/\Gamma_{tr}$, so that the conductivity integral has only the transport form of the relaxation time. Our earlier solution to impurity scattering also assumed that $\Lambda/\Gamma = 1/\Gamma_{tr}$. For impurity scattering the functions $\Lambda(\varepsilon)$ and $\Gamma(\varepsilon)$ are usually insensitive to ε near $\varepsilon \simeq 0$; the exception is when the impurity has a scattering resonance near the Fermi surface or if the density of states is not smooth. However, for the electron–phonon interaction in metals, it is *not* a good approximation to treat $\Gamma(\varepsilon)$ or $\Lambda(\varepsilon)$ as constants. The calculated results in Sec. 6.4, shown in Fig. 6.18 for Pb, illustrate that $\Gamma(\varepsilon)$ has substantial energy variations near the Fermi energy. The vertex function does also.

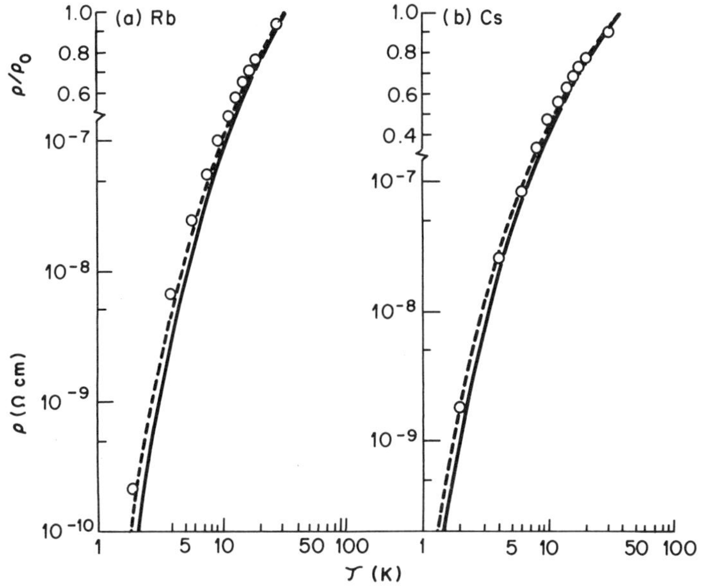

FIGURE 7.14. Calculated constant volume phonon limited electrical resistivity for (a) Rb and (b) Cs. Solid lines are calculations that treat $\Lambda(\varepsilon)$ as a constant. Dashed lines are full solution to (7.3.13) and (7.3.6). Points are experimental values. *Source*: Takegahara and Wang (1977) (used with permission).

Takegahara and Wang (1977) have evaluated (7.3.13) and (7.3.6) for metallic rubidium and cesium. Their results are shown in Fig. 7.14. In each case the solid line is calculated assuming that the ratio Λ/Γ is a constant, while the dashed line is calculated by solving (7.3.13) for the ε dependence of $\Lambda(\varepsilon)$. The latter curve is in very good agreement with the experiments, which are indicated by the points. The differences between the solid and dashed curves are similar to the differences between the Ziman formula (7.3.1) and the Holstein formula (7.3.13). Note that there is no region with a well-defined T^5 law for the resistivity.

C. Mass Enhancement

The electron–phonon mass enhancement factor λ was introduced in Sec. 6.4. It is from the real part of the electron self-energy due to the electron–phonon interaction. Since this self-energy was found to be energy dependent but not very wave vector dependent, the electron effective mass

m^* is approximated by

$$\lambda(\omega) = -\frac{\partial}{\partial\omega}\Sigma(k_F, \omega)$$
$$\frac{m^*}{m} = 1 + \lambda(\omega) \tag{7.3.15}$$
$$Z(\omega) = [1 + \lambda(\omega)]^{-1}$$

The mass enhancement factor is also related to the quasiparticle renormalization factor $Z(\omega)$. The values of these quantities at the Fermi energy $\omega = 0$ are λ and Z_F. The mass renormalization factor λ can also be expressed as an average over the Fermi surface of a weighted average over the phonon density of states. This definition is given in (6.4.18). Values of λ in real metals range from 0.1 to 3.

An important question is the role which λ plays in the dc transport properties. The most obvious approximation is to use the effective mass $m^* = m(1 + \lambda)$ in transport formulas whenever classical theory says to use m. This substitution would make sense, since m^* is the effective mass which governs the motion of electrons on the Fermi surface, and these are involved in dc transport properties. However, this sensible procedure is wrong in most cases. The important point is that the factor $1 + \lambda$ can enter the final formulas several ways. It changes the effective mass, relaxation time, and quasiparticle renormalization factor $Z = (1 + \lambda)^{-1}$. The formula for the transport coefficient will have a number of factors $1 + \lambda$. Often they all cancel, which is the case for the electrical conductivity.

Prange and Kadanoff (1964) investigated which transport measurements were influenced by the electron–phonon mass enhancement factor $1 + \lambda$. They concluded that the enhancement *did* affect the following measurements: specific heat, low-field cyclotron resonance, and the amplitude of the deHaas–van Alphen effect. The following measurements are *not* affected: dc electrical conductivity, thermoelectric power, thermal conductivity, the period of the deHaas–van Alphen effect, spin susceptibility, and the electron tunneling rate. Their conclusion on the thermoelectric power was challenged by Opsal *et al.* (1976), who detected a dependence on $1 + \lambda$. The list of quantities which are affected is much shorter than the list of quantities which are not affected. Thus the usual case is that the transport property is not influenced by the mass enhancement factor; see Grimvall (1981) for a further discussion.

We have remarked that the Green's function method had not been very successful at deriving formulas for transport coefficients. Without such formulas, we are unable to prove which theories do or do not contain factors of $1 + \lambda$. Most transport theories use a Boltzmann equation, and the presence or absence of mass renormalization is usually deduced from these formulas. The difficulty with this approach is that the electron is assigned an energy ε_k and a distribution $f(\mathbf{k})$. In the Green's function method, we assign the electron a wave vector $\mathbf{k}$ and energy ω and relate them by the spectral function $A(\mathbf{k}, \omega)$. This difference is crucial for the mass enhancement factor λ, since it arises from the ω dependence of the electron self-energy $\Sigma(\mathbf{k}, \omega)$, while the k dependence is negligible. Thus a Kubo formula derivation of the transport coefficients, to show which had a mass enhancement factor, would be very desirable.

The previous section has a derivation of the electrical conductivity based on the Kubo formula. In this case we can investigate whether the mass enhancement factor should be present. It is not, which verifies the conclusion of Prange and Kadanoff. Its absence may be shown by recalling Eqs. (7.3.6) and (7.3.13) for the conductivity:

$$\sigma = \frac{n_0 e^2}{2m} \int_{-\infty}^{\infty} d\varepsilon \left[-\frac{dn_F(\varepsilon)}{d\varepsilon} \right] \frac{\Lambda(\varepsilon)}{\Gamma(\varepsilon)}$$

$$\Lambda(\varepsilon) = 1 + \int_0^{\omega_D} du [\alpha^2(u)F(u) - \alpha_{tr}^2(u)F(u)] \left\{ \frac{\Lambda(\varepsilon + u)}{\Gamma(\varepsilon + u)} [n_B(u) + n_F(\varepsilon + u\ ] \right.$$
$$\left. + \frac{\Lambda(\varepsilon - u)}{\Gamma(\varepsilon - u)} [n_B(u) + n_F(u - \varepsilon)] \right\}$$

The mass enhancement factor arises from the real part of the electron self-energy $\mathrm{Re}[\Sigma(\mathbf{p}, \omega)]$ from electron–phonon interactions. The preceding formula for the conductivity σ does not contain $\mathrm{Re}(\Sigma)$ anywhere, so obviously its evaluation will not produce any dependence on $\lambda = -(d/d\omega)\,\mathrm{Re}(\Sigma)$. This is a very simple proof that the conductivity does not depend on λ. Such a simple demonstration is possible only when the vertex function is available, which is not the situation for most correlation functions.

D. Thermoelectric Power

The thermoelectric power is a subject which is seldom discussed in most solid-state courses, as the lecturer is busy treating subjects which are more fashionable. This tendency is regretable, since it is an important measurement. Experiments show that diverse behaviors are found for simple metals, and even the sign of this quantity shows no regularity. Low-temperature results, according to Mac Donald *et al.* (1960), are shown for the alkali metals in Fig. 7.15. The low-temperature theory in metals is poorly understood, although there has been extensive work.

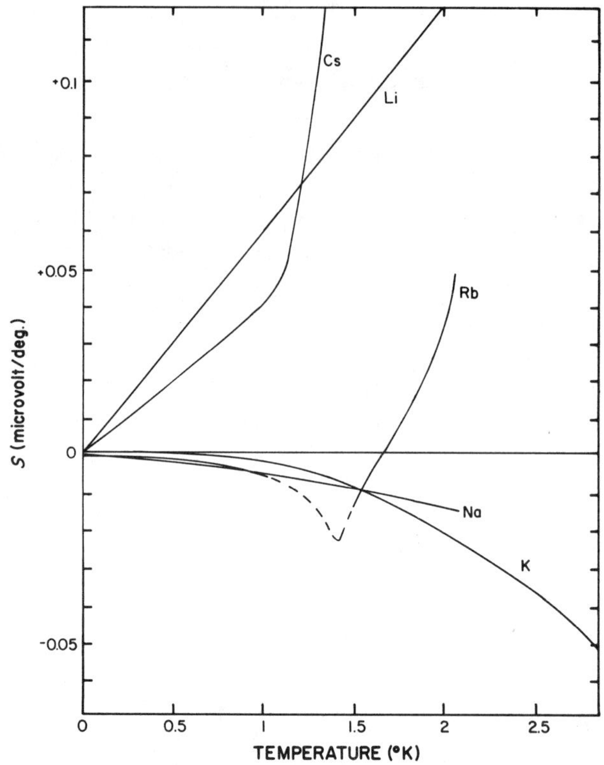

FIGURE 7.15. Thermoelectric power of alkali metals as a function of temperature. *Source*: MacDonald *et al.* (1960) (used with permission).

The thermoelectric power was introduced briefly in Sec. 3.8. It is a simple measurement, at least conceptually. A conducting bar is insulated so electrical currents cannot exit from its ends, and then a temperature difference ΔT is maintained along the length of the bar. The two ends of the bar are found to have a voltage difference ΔV which is proportional to ΔT. The constant of proportionality is the thermoelectric power S:

$$S = \frac{\Delta V}{\Delta T} = -\frac{1}{T}\frac{L^{(12)}}{L^{(11)}} \tag{7.3.16}$$

$$L^{(11)} = \frac{1}{3\beta i\omega\nu}\int_0^\beta d\tau e^{i\omega\tau}\langle T_\tau \mathbf{j}(\tau)\cdot\mathbf{j}(0)\rangle \tag{7.3.17}$$

$$L^{(12)} = \frac{1}{3\beta i\omega\nu}\int_0^\beta d\tau e^{i\omega\tau}\langle T_\tau \mathbf{j}_Q(\tau)\cdot\mathbf{j}(0)\rangle \tag{7.3.18}$$

The thermopower is defined theoretically as the ratio of two correlation functions. One of these is just proportional to the dc electrical conductivity $\sigma = \beta L^{(11)}$, which has already been evaluated for several models. The other one is the correlation function of the heat current $\mathbf{j}_Q$ with the electrical current $\mathbf{j}$, where $\mathbf{j}_Q = \mathbf{j}_E - (\mu/e)\mathbf{j}$. Equation (7.3.16) differs by a factor of charge e from its earlier version (3.8.18), since earlier $\mathbf{j}$ meant the particle current, whereas now it is the electrical current. They differ by the unit of charge, which causes this change in (7.3.16).

The evaluation of the correlation function $L^{(12)}$ will now be considered. We want the dc properties, so the correlation functions are evaluated for values of $i\omega$; then we analytically continue $i\omega \to \omega + i\delta$ and take the limit $\omega \to 0$. We want the imaginary part of the resulting retarded functions.

In the evaluation of most correlation functions, there is usually a leading term which provides the dominant part of the answer. There are numerous small correction terms which can usually be ignored. In calculating the correlation function $L^{(12)}$ for the thermoelectric power, the dominant term vanishes, and one is left with obtaining all the numerous small correction terms. This feature makes it difficult to obtain the final answer.

The heat current operator has many terms. The one which is expected to provide the dominant term is from the kinetic energy of the electron: $\mathbf{j}_Q = \Sigma_{\sigma\mathbf{k}} \mathbf{v}_\mathbf{k}\xi_\mathbf{k} c_\mathbf{k}{}^\dagger c_\mathbf{k}$. We omit the notation for particle spin. This heat current operator is used in most theories. Similarly, for the electrical current we use the operator $\mathbf{j} = e\, \Sigma_{\sigma\mathbf{k}} \mathbf{v}_\mathbf{k} c_\mathbf{k}{}^\dagger c_\mathbf{k}$. Thus we consider the evaluation of the correlation function, which we call $L^{(12a)}$:

$$L^{(12a)} = \frac{1}{\nu\beta i\omega} \frac{e}{3m^2} \sum_{\sigma\mathbf{k}\mathbf{k}'} \mathbf{k} \cdot \mathbf{k}' \xi_k \int_0^\beta d\tau e^{i\omega\tau} \langle T_\tau c_\mathbf{k}{}^\dagger(\tau) c_\mathbf{k}(\tau) c_{\mathbf{k}'}^\dagger c_{\mathbf{k}'} \rangle$$

The important feature of this correlation function is the bracket containing four electron operators. This type of operator sequence was encountered in earlier sections of this chapter. It is evaluated by writing it as

$$L^{(12a)} = -\frac{1}{\beta i\omega} \frac{2e}{3m^2\nu} \sum_\mathbf{k} \xi_k \frac{1}{\beta} \sum_{ik} \mathbf{k} \cdot \mathbf{\Gamma}(\mathbf{k}, ik, ik+i\omega) \mathscr{G}(\mathbf{k}, ik) \mathscr{G}(\mathbf{k}, ik+i\omega)$$

where $\mathbf{\Gamma}(\mathbf{k}, ip, ip + i\omega)$ is the vector vertex function of the bubble diagram. This same vertex function enters into the correlation function for the conductivity. Again we write it as $\mathbf{\Gamma} = \mathbf{k}\gamma$, and then we have the same summation over Matsubara frequency ik which was done in the previous section for the correlation function $L^{(11)}$. The result was given in (7.3.3). Thus for the retarded form of this correlation function we obtain

$$\operatorname{Im}[L_{\text{ret}}^{(12a)}(\omega)] = -\frac{2e}{3\beta m^2\omega}\int\frac{d^3k}{(2\pi)^3}\,\xi_k k^2\int_{-\infty}^{\infty}\frac{d\varepsilon}{2\pi}\,[n_F(\varepsilon+\omega)-n_F(\varepsilon)]$$
$$\times\,\{G_{\text{adv}}(\mathbf{k},\varepsilon)G_{\text{ret}}(\mathbf{k},\varepsilon+\omega)\gamma(\mathbf{k},\varepsilon-i\delta,\varepsilon+\omega+i\delta)$$
$$-\operatorname{Re}[G_{\text{ret}}(\mathbf{k},\varepsilon)G_{\text{ret}}(\mathbf{k},\varepsilon+\omega)\gamma(\mathbf{k},\varepsilon+i\delta,\varepsilon+\omega+i\delta)]\}$$

The next step is to take the limit as $\omega \to 0$, which causes the electron occupation factor to become $(1/\omega)[n_F(\varepsilon+\omega)-n_F(\varepsilon)] \to dn_F/d\varepsilon$. The integral over wave vector is changed to an integral over $k^2 d^3k = 12\pi^3 mn_0\,d\xi_k$:

$$\operatorname{Im}(L_{\text{ret}}^{(12a)}) = -\frac{en_0}{\beta m}\int_{-\infty}^{\infty}\frac{d\varepsilon}{2\pi}\left(\frac{dn_F}{d\varepsilon}\right)\int_{-\infty}^{\infty}d\xi\,\xi\{G_{\text{adv}}(\xi,\varepsilon)G_{\text{ret}}(\xi,\varepsilon)$$
$$\times\gamma(\varepsilon-i\delta,\varepsilon+i\delta)-\operatorname{Re}[G_{\text{ret}}(\xi,\varepsilon)^2\gamma(\varepsilon+i\delta,\varepsilon+i\delta)]\} \tag{7.3.19}$$

Again we assume that the retarded and advanced Green's functions are significantly dependent only on ξ in their kinetic energy term and that the self-energy and vertex functions have negligible dependence on ξ. Then we can do the following integrals by a contour integration, in analogy with (7.3.5):

$$\int_{-\infty}^{\infty}d\xi\,\xi G_{\text{ret}}(\xi,\varepsilon)^2 = i\pi$$
$$\int_{-\infty}^{\infty}d\xi\,\xi G_{\text{ret}}(\xi,\varepsilon)G_{\text{adv}}(\xi,\varepsilon) = \frac{\pi\Omega(\varepsilon)}{\Gamma(\varepsilon)} \tag{7.3.20}$$
$$\int_{-\infty}^{\infty}d\xi\,\xi G_{\text{adv}}(\xi,\varepsilon)^2 = -i\pi$$

Two of the integrals equal $\pm i\pi$, which comes from the semicircle closing the contour at infinity. In Eq. (7.3.19) for the correlation function $L^{(12a)}$, the term $\operatorname{Re}[G_{\text{ret}}^2\gamma(\varepsilon+i\delta,\varepsilon+i\delta)]$ equals zero, since the vertex function $\gamma(\varepsilon+i\delta,\varepsilon+i\delta)$ is unity, and the integral $\int d\xi G_{\text{ret}}^2 = i\pi$ has no real part. The important contribution must arise from the combination $G_{\text{adv}}G_{\text{ret}}$, since it has a nontrivial contribution from the integral over $d\xi$:

$$\operatorname{Im}(L_{\text{ret}}^{(12a)}) = \frac{en_0\pi}{\beta m}\int_{-\infty}^{\infty}d\varepsilon\left[-\frac{d}{d\varepsilon}n_F(\varepsilon)\right]\frac{\Omega(\varepsilon)}{\Gamma(\varepsilon)}\Lambda(\varepsilon)$$
$$= 0$$

However, this integral is also zero after one evaluates the integral $d\varepsilon$. It vanishes because the integrand is an antisymmetric function of ε. The quantities $dn_F(\varepsilon)/d\varepsilon$, $\Lambda(\varepsilon)$, and $\Gamma(\varepsilon) = -\operatorname{Im}[\Sigma(\varepsilon)]$ are all symmetric functions of ε, while $\Omega(\varepsilon) = \varepsilon - \text{Re}[\Sigma(\varepsilon)]$ is an antisymmetric function. Recall, in the discussion of the electron–phonon self-energy in Sec. 6.4, that we showed $\text{Im}[\Sigma(\varepsilon)]$ is a symmetric function and $\text{Re}[\Sigma(\varepsilon)]$ is antisymmetric. Thus our evaluation of the correlation function $L^{(12a)}$ has shown that it is zero.

If we retrace the steps through the derivation, the important feature is the single power of ξ_k in the $d\xi_k$ integral. It makes the single power of $\Omega(\varepsilon)$ in the integral over $d\varepsilon$. The single power of ξ_k comes from the heat current operator.

A nonzero result for the correlation function is obtained by repeating this derivation and retaining all the correction terms. For example, we can keep terms of order ξ_k/E_F in the argument of the wave vector integration:

$$k^2\, d^3k = 4\pi k^3\, d\left(\frac{k^2}{2}\right) = 4\pi m\, d\xi_k[k_F{}^2 + (k^2 - k_F{}^2)]^{3/2}$$
$$\simeq 4\pi k_F{}^3\, d\xi_k\left(1 + \frac{3}{2}\,\frac{\xi_k}{E_F}\right) \tag{7.3.21}$$

The second term in parentheses makes a finite contribution to the thermopower. This type of term is now evaluated. Kinetic energy integrals such as (7.3.20) must now be done, except there is a factor of ξ^2 in the integrand which multiplies the Green's functions. The evaluation of the integrand is tricky, since technically the integral diverges. At large values of ξ, we have that $G \sim 1/\xi$ and $(\xi G)^2 \sim 1$. The integrand does not fall off at large values of ξ, and taking the limits to $\pm\infty$ gives an infinite integral. Usually this problem is solved by ignoring it. The product $G_{\text{adv}}G_{\text{ret}} = A/2\Gamma \sim \pi\delta(\xi-\varepsilon)/\Gamma$ is replaced by a delta function for energy conservation. The integral over $\xi^2 G_{\text{ret}}^2$ is set equal to zero:

$$\int d\xi\; \xi^2 G_{\text{ret}}(\xi, \varepsilon) G_{\text{adv}}(\xi, \varepsilon) \approx \pi\varepsilon^2/\Gamma(\varepsilon)$$

The term in ε^2 is thermally averaged according to (7.3.19) using G & R 3.531(3);

$$\int_{-\infty}^{\infty} d\varepsilon\; \varepsilon^2\left[-\frac{\partial}{\partial\varepsilon} n_F(\varepsilon)\right] = \pi^2/3$$

By setting $\gamma(-i\delta, +i\delta) = \Lambda$, and $\Gamma/\Lambda = \Gamma_{\text{tr}} = 1/2\tau$, we find that

$$\text{Im}(L_{\text{ret}}^{(12a)}) = \frac{\pi^2 e n_0 \tau}{3\beta m E_F}\,\eta(T)$$

$$S = -(k_B/\sigma)\text{Im}(L_{\text{ret}}^{(12a)}) = -\frac{\pi^2 k_B{}^2 T}{3eE_F}\,\eta(T)$$

where the conductivity is $\sigma = n_0 e^2 \tau/m$. The parameter $\eta(T)$ is dimensionless. The dimensions of S are volts per degree. So far we have $\eta = 3/2$ from the coefficient of ξ/E_F in (7.3.21). Other contributions to η are derived below.

Taylor and MacDonald (1986) evaluated this expression for the alkali metals at high temperature. Rather good agreement is obtained, as shown below. First it is necessary to find more contributions to $\eta(T)$. At high temperature it is a good approximation to set $\tau = \tau_{\text{tr}}$, where the transport lifetime is entirely from phonons, and is given by the transport form of alpha-squared-f in (7.3.9):

$$1/\tau_{\text{tr}}(k) = \frac{m}{(2\pi)^2 k^3} \int_0^{2k} q^3\, dq \int_0^{2\pi} d\phi \sum_\lambda M_\lambda(\mathbf{q})^2 \{n_B[\omega_\lambda(\mathbf{q})] + n_F[\omega_\lambda(\mathbf{q})]\}$$

contributions to $\eta(T)$ are obtained by expanding k about k_F and keeping the first-order terms in $(k - k_F)/k_F \approx \xi/2E_F$. The prefactor of k^{-3} gives another contribution of 3/2 to η. The integration limit of $2k$ gives a contribution to η of

$$2q(T) = \frac{2m}{\pi^2} \tau_{\text{tr}} \int_0^{2\pi} d\phi \sum_\lambda M_\lambda(2k_F)^2 \{n_B[\omega_\lambda(2k_F)] + n_F[\omega_\lambda(2k_F)]\}$$

Another contribution to $\eta(T)$ comes from the matrix element $M_\lambda(\mathbf{q})$. It is usually calculated using a screened pseudopotential for the electron–ion interaction. The better pseudopotentials are nonlocal, which means they depend upon $M_\lambda(\mathbf{k}, \mathbf{k} + \mathbf{q})$ rather than just on $\mathbf{q}$. This k dependence can also be expanded around the point k_F. The pseudopotential gives a small contribution to $\eta(T)$ which is called $r(T)$. The contributions to $\eta(T)$ are

$$\eta(T) = 3 - 2q(T) - \tfrac{1}{2} r(T)$$

Table 7.1 shows the evaluation of these terms for the alkali metals at various temperatures. The comparison with the experimental data is good for Na, K, and Rb. The high-temperature thermopower seems to be understood in these cases. Both $q(T)$ and $r(T)$ are small for Na because the electron–ion pseudopotential is nearly zero at $q = 2k_F$. The thermopower is easier to evaluate at high rather than at low temperature. At high temperature the resistance is dominated by the electron–phonon interaction, which is well approximated by using the transport form of alpha-squared-f. At low temperatures one also has to include the k dependence of impurity scattering, as well as the ordinary form of alpha-squared-f while solving (7.3.13).

TABLE 7.1. Thermoelectric Parameters in the Alkali Metals[a]

Metal	T(K)	$2q(T)$	$r(T)/2$	η	
				Theory	Experiment
Li	424	9.26	−1.43	−5.33	−6.3
Na	300	0.04	−0.09	3.05	2.9
K	200	0.83	−1.87	4.04	4.0
Rb	100	4.78	−4.49	2.71	2.8
Cs	100	9.32	−7.15	0.83	0.0

[a] $\eta = 3 - 2q - r/2$ (Taylor and MacDonald, 1986, used with permission).

7.4. QUANTUM BOLTZMANN EQUATION

There are several different methods of doing transport theory. The theory used in the preceding sections uses the Kubo relation for the conductivity and is called "linear response." One assumes that currents are proportional to fields. The proportionality constants can be evaluated in equilibrium. This method works because one assumes that the applied fields are small, and the system is only infinitesimally disturbed from equilibrium.

A second method of transport theory is discussed in this section. One assumes the existence of a distribution function f, which describes the behavior of the particles. One writes a differential equation for the motion of f through phase space. The differential equation is a Boltzmann equation. One then tries to solve the Boltzmann equation for a system out of equilibrium. For fields that are small, the system is only slightly out of equilibrium, and one reproduces the linear response solutions described earlier. The advantage of the Boltzmann equation method is that one can also try to solve the equation when the system is far from equilibrium.

The original Boltzmann equation described the behavior of a distribution function $f(\mathbf{v}, \mathbf{r}, t)$ of three variables: velocity, position, and time. The Wigner distribution function (WDF) was introduced in Sec. 3.7. It is equivalent to a distribution function $f(\mathbf{k}, \omega; \mathbf{r}, t)$ with four variables: wave vector $\mathbf{k}$, energy ω, position $\mathbf{r}$, and time. This latter distribution function is the one needed for many-particle systems. Since $f(\mathbf{k}, \omega; \mathbf{r}, t)$ is not positive definite, calling it a "distribution function" is probably misleading. We adopt this usage since it is widespread, but the warning should be kept in mind.

The differential equation obeyed by $f(\mathbf{k}, \omega; \mathbf{r}, t)$ is called the *quantum Boltzmann equation,* which is abreviated QBE. It is derived rigorously in the following sections. Here a quick derivation is provided using semiclassical arguments. According to the Liouville theorem a distribution function $f(q_i, t)$ is stationary when it obeys the equation

$$\delta f = 0 = \frac{\partial f}{\partial t} + \sum_i \frac{\partial f}{\partial q_i} \dot{q}_i + \left(\frac{\partial f}{\partial t}\right)_s$$

where the last term is from scattering. For our distribution function, $\dot{\mathbf{k}} = \mathbf{F}$ (force), $\dot{\mathbf{r}} = \mathbf{v}$ (velocity), and $\dot{\omega} = \mathbf{v} \cdot \mathbf{F}$ (Joule heating). The QBE is

$$\left(\frac{\partial}{\partial t} + \mathbf{v} \cdot \nabla_r + \mathbf{F} \cdot \nabla_k + \mathbf{v} \cdot \mathbf{F} \frac{\partial}{\partial \omega}\right) f + \left(\frac{\partial f}{\partial t}\right)_s = 0 \qquad (7.4.1)$$

This equation has nearly the right form. In the next section we shall derive it rigorously, and find a few more terms which come from self-energy contributions. That derivation also provides a prescription for obtaining the scattering term.

Equation (7.4.1) has one feature that is important. The additional variable ω also causes a new driving term on the left of the form $\mathbf{v} \cdot \mathbf{F}\partial/\partial\omega$. This term was first derived by Mahan and Hänsch (1983). The semiclassical distribution function $f(\mathbf{r}, \mathbf{v}, t)$ lacks this driving term since it lacks the energy variable ω. The QBE for the WDT is a different equation from the traditional Boltzmann equation.

A. Derivation of the Quantum Boltzmann Equation

The QBE is the equation of motion for the Green's function $G^<$. The method of deriving transport equations was pioneered by Kadanoff and Baym (1962). Recall from Sec. 3.7 that if $(\mathbf{r}, t; \mathbf{R}, T) = G^<(\mathbf{r}, t; \mathbf{R}, T)$, where $(\mathbf{r}, t)$ are the relative variables and $(\mathbf{R}, T)$ are the position and time in center of mass. The Green's function was defined in terms of the field operator as

$$G^<(\mathbf{r}, t; \mathbf{R}, T) = i\langle \psi^\dagger(\mathbf{R} - \tfrac{1}{2}\mathbf{r}, T - \tfrac{1}{2}t)\psi(\mathbf{R} + \tfrac{1}{2}\mathbf{r}, T + \tfrac{1}{2}t)\rangle$$

The next step is to Fourier transform the relative variables $(\mathbf{r}, t)$ into $(\mathbf{k}, \omega)$:

$$G^<(\mathbf{k}, \omega; \mathbf{R}, T) = \int d^3r\, e^{-i\mathbf{k}\cdot\mathbf{r}} \int dt\, e^{i\omega t} G^<(\mathbf{r}, t; \mathbf{R}, T) \qquad (7.4.2)$$

The QBE will be derived for a particle in a weak electric field. The intent is to describe interacting many-particle systems that have a small current flowing in response to a small electric field. The derivation will be sufficiently general to include any kind of particles and nearly any kind of interactions.

The electric field can be introduced as either a scalar or a vector potential. The QBE is independent of this choice, as required by gauge invariance. Here both are included, in order to provide the most general derivation. There will be an electric field $\mathbf{E}_v$ which is from a vector potential, and another electric field $\mathbf{E}_s$ from a scalar potential. The final version of the QBE will include only the total electric field $\mathbf{E} = \mathbf{E}_v + \mathbf{E}_s$. The scalar potential is introduced through the interaction term

$$H_E = -e\mathbf{E}_s \cdot \sum_j \mathbf{r}_j \tag{7.4.3}$$

The vector potential is introduced by changing the momentum of each charged particle to $(\mathbf{p} - e\mathbf{A}/c)$, where the vector potential is $\mathbf{A} = -c\mathbf{E}_v t$. A vector potential proportional to time could occur in a wire loop with a slowly varying magnetic flux through the center.

Equations (3.7.5) and (3.7.6) are equations of motion for the four Green's functions in the 2×2 matrix for $\tilde{G}$. The two electric field terms are added to the left-hand side of these equations–they are included in H_0. The derivation of the QBE involves several algebraic manipulations of the left-hand side of these equations. In order to avoid a lot of cumbersome notation, we shall not write out the right-hand side of these equations during these steps. On the left, the steps are the same for all six of the real time Green's functions. We shall use the generic symbol G to apply to any one of them.

The two equations for G in (3.7.5) and (3.7.6) are first added and then subtracted:

$$\left[i\left(\frac{\partial}{\partial t_1} - \frac{\partial}{\partial t_2}\right) - H_1 - H_2\right]G =$$

$$\left[i\left(\frac{\partial}{\partial t_1} + \frac{\partial}{\partial t_2}\right) - H_1 + H_2\right]G =$$

where $H_1 = H_0(\mathbf{r}_1, \mathbf{p}_1)$ and $H_2 = H_0(\mathbf{r}_2, -\mathbf{p}_2)$. The two equations contain time derivatives that relate either to the relative or center-of-mass motion

$$\frac{\partial}{\partial t_1} + \frac{\partial}{\partial t_2} = \frac{\partial}{\partial T}, \qquad \frac{\partial}{\partial t_1} - \frac{\partial}{\partial t_2} = 2\frac{\partial}{\partial t}$$

For particles with parabolic band dispersion, the sum and difference of the two Hamiltonians produce simple expressions in relative coordinates. As a first step, consider what happens to the two momentum terms which are in the form of $(\mathbf{p} - e\mathbf{A}/c)$

$$\mathbf{p}_1 + e\mathbf{E}_v t_1 = (\mathbf{p} + e\mathbf{E}_v T) + \tfrac{1}{2}(\mathbf{P} + e\mathbf{E}_v t)$$
$$\mathbf{p}_2 - e\mathbf{E}_v t_2 = -(\mathbf{p} + e\mathbf{E}_v T) + \tfrac{1}{2}(\mathbf{P} + e\mathbf{E}_v t)$$

where $\mathbf{p}$ and $\mathbf{P}$ are the relative and center-of-mass momentum. It is important to understand all of the various plus and minus signs. Since $\mathbf{A} = -c\mathbf{E}t$ then $(\mathbf{p} - e\mathbf{A}/c) = (\mathbf{p} + e\mathbf{E}t)$ for p_1. However, in H_2 the momentum enters as $-\mathbf{p}_2$, which explains the sign change on the bottom. On the right one uses $t_{1,2} = T \pm t/2$ and $\mathbf{p}_{1,2} = \pm\mathbf{p} + \mathbf{P}/2$. These results make it easy to see the form of $H_1 \pm H_2$:

$$H_1 + H_2 = \frac{1}{m}(\mathbf{p} + e\mathbf{E}_v T)^2 + \frac{1}{4m}(\mathbf{P} + e\mathbf{E}_v t)^2 - 2e\mathbf{E}_s \cdot \mathbf{R}$$

$$H_1 - H_2 = \frac{1}{m}(\mathbf{p} + e\mathbf{E}_v T) \cdot (\mathbf{P} + e\mathbf{E}_v t) - e\mathbf{E}_s \cdot \mathbf{r}$$

Taking a factor of 2 from the top equation, we arrive at the following two equations for the Green's function:

$$2\left[i\frac{\partial}{\partial t} - \frac{1}{2m}(\mathbf{p} + e\mathbf{E}_v T)^2 - \frac{1}{8m}(\mathbf{P} + e\mathbf{E}_v t)^2 + e\mathbf{E}_s \cdot \mathbf{R}\right]G =$$

$$\left[i\frac{\partial}{\partial T} - \frac{1}{m}(\mathbf{p} + e\mathbf{E}_v T) \cdot (\mathbf{P} + e\mathbf{E}_v t) + e\mathbf{E}_s \cdot \mathbf{r}\right]G =$$

These two equations describe the relative and center-of-mass motion of the function $G(\mathbf{r}, t;\ \mathbf{R}, T)$. The goal is to derive the QBE for the WDF $G^<(\mathbf{k}, \omega;\ \mathbf{R}, T)$. Fourier transform the variables $(\mathbf{r}, t)$ to the set $(\mathbf{q}, \Omega)$ as in (7.4.2). This transform changes $\mathbf{p}$ to $\mathbf{q}$, $\mathbf{r}$ to $i\nabla_{\mathbf{q}}$, $\partial/\partial t$ to $-i\Omega$, and t to $-i\partial/\partial\Omega$. The two transformed equations are

$$2\left[\Omega + e\mathbf{E}_s \cdot \mathbf{R} - \frac{1}{2m}(\mathbf{q} + e\mathbf{E}_v T)^2 + \frac{1}{8m}\left(\nabla_R + e\mathbf{E}_v \frac{\partial}{\partial\Omega}\right)^2\right]G(\mathbf{q}, \Omega;\ \mathbf{R}, T) =$$

$$i\left[\frac{\partial}{\partial T} + \frac{1}{m}(\mathbf{q} + e\mathbf{E}_v T) \cdot \left(\nabla_R + e\mathbf{E}_v \frac{\partial}{\partial\Omega}\right) + e\mathbf{E}_s \cdot \nabla_{\mathbf{q}}\right]G(\mathbf{q}, \Omega;\ \mathbf{R}, T) = \tag{7.4.4}$$

The second of these equations has, on the left-hand side, exactly the same terms as are found in the Boltzmann equation. This similarity suggests that the QBE is the same as the BE. However, this is not correct. The above set of equations have several things wrong with them:

1. There is the term $e\mathbf{E}_s \cdot \mathbf{R}$. This term seems to combine with Ω to produce a center-of mass energy $\omega = \Omega + e\mathbf{E}_s \cdot \mathbf{R}$. Together they suggest that the energy of a particle depends upon its location. The energy is different at one end of a sample than at the other. However, this behavior is contrary to common sense. When there is a small electric field along the sample, and a small current flowing, we expect the system to be uniform. There is the same particle density, current density, etc. at each point in the solid. There is no dependence upon $\mathbf{R}$. This undesirable term has to be eliminated.

2. The relative momentum seems to enter in the combination of $\mathbf{q} + e\mathbf{E}_v T$. It depends upon the center-of-mass time T. This feature is also unphysical, and needs to be eliminated.

3. The result is not gauge invariant, since the two fields $\mathbf{E}_s$ and $\mathbf{E}_v$ enter differently.

All of these problems can be eliminated through a variable transformation. Of course, this transformation also causes some derivatives to change:

$$\begin{aligned} \Omega + e\mathbf{E}_s \cdot \mathbf{R} &\Rightarrow \omega \\ \mathbf{q} + e\mathbf{E}_v T &\Rightarrow \mathbf{k} \\ \nabla_R &\Rightarrow \nabla_R + e\mathbf{E}_s \frac{\partial}{\partial\omega} \\ \frac{\partial}{\partial T} &\Rightarrow \frac{\partial}{\partial T} + e\mathbf{E}_v \cdot \nabla_k \end{aligned} \tag{7.4.5}$$

These transformations cause the two equations in (7.4.4) for the Green's function to now have the form

$$\begin{aligned} \left[\omega - \varepsilon_{\mathbf{k}} + \frac{1}{8m}\left(\nabla_R + e\mathbf{E}\frac{\partial}{\partial\omega}\right)^2\right] G(\mathbf{k}, \omega; \mathbf{R}, T) &= \\ i\left[\frac{\partial}{\partial T} + \mathbf{v}_{\mathbf{k}} \cdot \nabla_R + e\mathbf{E} \cdot \left(\nabla_k + \mathbf{v}_{\mathbf{k}}\frac{\partial}{\partial\omega}\right)\right] G(\mathbf{k}, \omega; \mathbf{R}, T) &= \end{aligned} \tag{7.4.6}$$

Note that we have changed again the notation on the Green's function. The arguments $(\mathbf{q}, \Omega)$ have been changed to $(\mathbf{k}, \omega)$. The left-hand sides of these two equations are now in the form that is useful. The lower equation has exactly the same terms as in (7.4.1), and is the quantum Boltzmann equation. The electric field $\mathbf{E}$ is the sum of $\mathbf{E}_v$ and $\mathbf{E}_s$. The result is now gauge invariant.

Now it is time to restore the scattering terms to the right-hand side of the equal sign. The scattering terms for $G^<$ and G_{ret} are in (3.7.10). These two are the most important Green's functions in applications using real time. The self-energy terms on the right are found by follwing the steps we performed to bring the left-hand side to (7.4.6). The first step is to add and subtract the two equations for each Green's function. Then there followed a series of variable transformations. They can be understood by examining just one term in the right-hand side of Eq. (7.4.6). The various variable arguments yield a fairly complicated expression, which is derived by the following steps:

1. Change the two sets of variables (x_1, x_3) (x_3, x_2) to the center-of-mass grouping $(x_1 - x_3, \frac{1}{2}(x_1 + x_3))$ $(x_3 - x_2, \frac{1}{2}(x_3 + x_2))$.
2. Change the integration variable from x_3 to $y = x_1 - x_3$, so the variable grouping become $(y, x_1 - \frac{1}{2}y)$ $(x_1 - x_2 - y, \frac{1}{2}(x_1 + x_2 - y))$.
3. Change to center of mass variables $x = (x_1 - x_3)$, $X = \frac{1}{2}(x_1 + x_2)$, which produces the variable grouping $(y, X + \frac{1}{2}(x - y))$ $(x - y, X - \frac{1}{2}y)$.
4. Change the integration variable x to $z = x - y$ which produces the final arguments of $(y, X + z/2)(z, X - y/2)$.

Below are the full scattering equations for $G^<$ and G_{ret}. The explicit variables in the scattering terms are only written out in the first term, but are identical for the other three. The order of the terms in the scattering integral is important. The first one always has argument $(y, X + z/2)$ while the second one has argument $(z, X - y/2)$. The four vector in the exponent is $q = (\mathbf{k}, \omega)$:

$$\left[\omega - \varepsilon_k + \frac{1}{8m}\left(\nabla_R + e\mathbf{E}\,\frac{\partial}{\partial\omega}\right)^2\right] G_{\text{ret}}(\mathbf{k}, \omega; \mathbf{R}, T)$$

$$= 1 + \frac{1}{2}\int dz\, e^{-iq\cdot x}\int dy\, e^{-iq\cdot y}[\Sigma_{\text{ret}}(y, X + \tfrac{1}{2}z) G_{\text{ret}}(z, X - \tfrac{1}{2}y)$$

$$- G_{\text{ret}}(y, X + \tfrac{1}{2}z)\, \Sigma_{\text{ret}}(z, X - \tfrac{1}{2}y)]$$

$$i\left[\frac{\partial}{\partial T} + \mathbf{v}_k\cdot\nabla_R + e\mathbf{E}\cdot\left(\nabla_k + \mathbf{v}_k\,\frac{\partial}{\partial\omega}\right)\right] G_{\text{ret}}$$

$$= \int dz\, e^{-iq\cdot z}\int dy\, e^{-iq\cdot y}[\Sigma_{\text{ret}} G_{\text{ret}} - G_{\text{ret}}\Sigma_{\text{ret}}] \tag{7.4.7}$$

$$\left[\omega - \varepsilon_k + \frac{1}{8m}\left(\mathbf{\nabla}_R + e\mathbf{E}\,\frac{\partial}{\partial\omega}\right)^2\right] G^<(\mathbf{k}, \omega; \mathbf{R}, T)$$
$$= \frac{1}{2}\int dz\, e^{-iq\cdot z}\int dy\, e^{-iq\cdot y}[\Sigma_t G^< - \Sigma^< G_{\bar{t}} + G_t\Sigma^< - G^<\Sigma_{\bar{t}}]$$

$$i\left[\frac{\partial}{\partial T} + \mathbf{v}_k\cdot\mathbf{\nabla}_R + e\mathbf{E}\cdot\left(\mathbf{\nabla}_k + \mathbf{v}_k\,\frac{\partial}{\partial\omega}\right)\right] G^<$$
$$= \int dz\, e^{-iq\cdot z}\int dy\, e^{-iq\cdot y}[\Sigma_t G^< - \Sigma^< G_{\bar{t}} - G_t\Sigma^< + G^<\Sigma_{\bar{t}}]$$

These four equations are the important ones for nonequilibrium calculations. Although one primarily wants to find $G^<$, it is always necessary to first find G_{ret}. These equations were first derived by Fleurov and Kozlov (1978).

So far no approximations have been made, and the equations are exact. The QBE has a linear term in the electric field E. However, the equation is exact to all powers of E, not just to the first power.

B. Gradient Expansion

Equations (7.4.7) are usually too hard to solve because the scattering terms on the right have a complicated form. Some sort of approximation has to be introduced to simplify the right side. The approximation described below is only valid to first order in the field. Kadanoff and Baym (1962) introduced an approximation for evaluating these scattering terms, which is called the "gradient expansion." They assume that the center-of-mass time T is very large, and take the limit that $T \to \infty$. At large values of T, they assume that the system is approaching its asymptotic limit, so that variations with respect to T are small. Obviously the gradient expansion is not suitable for studying transients, since it is poor at small values of T. Neither is it useful for steady-state ac phenomena (Mahan, 1987). Indeed, the T dependence is so poorly described in the gradient approximation that it should not be used and the T derivative terms should be dropped from the QBE. Nevertheless, we shall use the gradient expansion since it is applicable for homogeneous (small R dependence) steady state (small T dependence) systems.

The center-of-mass variables are all in the form $(R + \Delta R, T + \Delta T)$, which are expanded in a Taylor series about the point (R, T). The integrals can be done for each term in the series. These integrals usually cause further derivatives. Below we show the gradient expansion for a typical scattering term:

$$I = \int dz\, e^{-iq\cdot z} \int dy\, e^{-iq\cdot y} \Sigma(y, X + z/2) G(z, X - y/2)$$

$$I = \Sigma(q, X) G(q, X) + \frac{i}{2}[(\nabla_q \Sigma)\nabla_X G - \nabla_X \Sigma \nabla_q G] + \cdots$$

It is customary to retain only the first derivative terms. They can be expressed using a Poisson bracket notation:

$$\begin{aligned}[\Sigma, G] &= \nabla_q \Sigma \nabla_X G - \nabla_X \Sigma \nabla_q G \\ &= \frac{\partial \Sigma}{\partial \Omega}\frac{\partial G}{\partial T} - \frac{\partial \Sigma}{\partial T}\frac{\partial G}{\partial \Omega} - \mathbf{\nabla}_q \Sigma \cdot \mathbf{\nabla}_R G + \mathbf{\nabla}_R \Sigma \cdot \mathbf{\nabla}_q G\end{aligned}$$

These frequency derivatives are with respect to the Ω variables, since the variable change ($\mathbf{k} = \mathbf{q} + e\mathbf{E}_v T$, $\omega = \Omega + e\mathbf{E}_s \cdot \mathbf{R}$) has not yet been made. If this step is taken now, the derivatives get altered according to Eq. (7.4.5), which changes the Poisson brackets to

$$[C, D] \Rightarrow [C, D] + e\mathbf{E} \cdot \left[\left(\frac{\partial C}{\partial \omega}\right)\nabla_k D - \left(\frac{\partial D}{\partial \omega}\right)\mathbf{\nabla}_k C\right] \tag{7.4.8}$$

where $\mathbf{E}$ is again the total electric field. This analysis finally derives from (7.4.7) the following expression for the nonequilibrium retarded Green's function in an electric field, when G_{ret} and Σ_{ret} depend upon ($\mathbf{k}, \omega; \mathbf{R}, T$):

$$\begin{gathered}\left[\omega - \varepsilon_k + \frac{1}{8m}\left(\mathbf{\nabla}_R + e\mathbf{E}\frac{\partial}{\partial \omega}\right)^2 - \Sigma_{\text{ret}}\right] G_{\text{ret}} = 1 \\ i\left\{\frac{\partial}{\partial T} + \mathbf{v}_k \cdot \mathbf{\nabla}_R + e\mathbf{E} \cdot \left[\left(1 - \frac{\partial \Sigma_{\text{ret}}}{\partial \omega}\right)\mathbf{\nabla}_k + (\mathbf{v}_k + \mathbf{\nabla}_k \Sigma_{\text{ret}})\frac{\partial}{\partial \omega}\right]\right\} G_{\text{ret}} \\ = i[\Sigma_{\text{ret}}, G_{\text{ret}}]\end{gathered} \tag{7.4.9}$$

The additional terms from the Poisson bracket, which are linear in the field E, have been transferred to the left of the equal sign.

These equations simplify for nonequilibrium systems which are both homogeneous ($\nabla_R = 0$) and steady state ($\partial/\partial T = 0$). The Poisson brackets vanish, as do several terms on the left. We also ignore terms that are nonlinear in the electric field (E^2), and find for the above two equations

$$\begin{gathered}[\omega - \varepsilon_k - \Sigma_{\text{ret}}] G_{\text{ret}} = 1 \\ ie\mathbf{E} \cdot \left[\left(1 - \frac{\partial \Sigma_{\text{ret}}}{\partial \omega}\right)\mathbf{\nabla}_k + (\mathbf{v}_k + \mathbf{\nabla}_k \Sigma_{\text{ret}})\frac{\partial}{\partial \omega}\right] G_{\text{ret}} = 0\end{gathered} \tag{7.4.10}$$

The first equation is easily solved, to yield

$$G_{\mathrm{ret}}(\mathbf{k}, \omega) = \frac{1}{\omega - \varepsilon_k - \Sigma_{\mathrm{ret}}(\mathbf{k}, \omega)} + O(E^2) \tag{7.4.11}$$

The retarded Green's function appears to have no first-order term in the electric field. It actually does, since the self-energy Σ_{ret} has a term linear in the field due to the electron–phonon interaction. This term is small and seems to have little effect. If it is ignored then the retarded Green's function is unchanged from its value in equilibrium. This result considerably simplifies the solution to the QBE. The solution (7.4.11) also satisfies the second equation in (7.4.10). Related quantities such as the advanced function $G_{\mathrm{adv}} = G_{\mathrm{ret}}^*$ and the spectral function $A(k, \omega) = -2\,\mathrm{Im}[G^r(\mathbf{k}, \omega)]$ are also unchanged to first order in the electric field. This completes the discussion of the retarded Green's function in a static electric field.

Next the gradiant expansion is applied to (7.4.7) for the equations for $G^<$. Again keeping only first-order derivatives, we derive the final results, which are the quantum Boltzmann equations:

$$\begin{aligned}
\left[\omega - \varepsilon_k + \frac{1}{8m}\left(\nabla_k + e\mathbf{E}\frac{\partial}{\partial\omega}\right)^2\right]G^< &= G^<\mathrm{Re}\Sigma_{\mathrm{ret}} + \Sigma^<\mathrm{Re}\,G_{\mathrm{ret}}^r \\
&\quad + \frac{i}{4}[\Sigma^>, G^<] - \frac{i}{4}[\Sigma^<, G^>] \\
+ \frac{i}{4}e\mathbf{E}\cdot\left(\frac{\partial\Sigma^>}{\partial\omega}\nabla_k G^< - \frac{\partial G^<}{\partial\omega}\nabla_k\Sigma^< - \frac{\partial\Sigma^<}{\partial\omega}\nabla_k G^< + \frac{\partial G^>}{\partial\omega}\nabla_k\Sigma^<\right)
\end{aligned}$$

$$\begin{aligned}
&i\left\{\frac{\partial}{\partial T} + \mathbf{v_k}\cdot\nabla_R + e\mathbf{E}\cdot\left[\left(1 - \frac{\partial\mathrm{Re}\Sigma_{\mathrm{ret}}}{\partial\omega}\right)\nabla_k + (\mathbf{v_k} + \nabla_k\mathrm{Re}\Sigma_{\mathrm{ret}})\frac{\partial}{\partial\omega}\right]\right\}G^< \\
&\quad - ie\mathbf{E}\cdot\left[\frac{\partial\Sigma^<}{\partial\omega}\nabla_k\mathrm{Re}G_{\mathrm{ret}} - \frac{\partial\mathrm{Re}G_{\mathrm{ret}}}{\partial\omega}\nabla_k\Sigma^<\right] = \Sigma^>G^< - \Sigma^<G^> \\
&\quad + i[\mathrm{Re}\Sigma_{\mathrm{ret}}, G^<] + i[\Sigma^<, \mathrm{Re}G_{\mathrm{ret}}]
\end{aligned} \tag{7.4.12}$$

In deriving this equation, we have used some relationships such as $G_t - G - = 2\ \mathrm{Re}\ G_{\mathrm{ret}}$ and $G_t + G_t - = G^< + G^>$.

Equation (7.4.12) is the quantum Boltzmann equation. It is rather formidable. It is also difficult to solve, since it is usually an integral equation, and sometimes nonlinear in the particle density. This happens because the self-energy functions $\Sigma^<$ and $\Sigma^>$ are also functions of $G^<$ and $G^>$.

The QBE also contains the functions $G^>$ and $\Sigma^>$. We can also start from the general equations (3.7.5) and (3.7.6) and derive similar equations for $G^>$. This derivation shows that the equation for $G^>$ is almost identical to the one for $G^<$. In fact, one can prove that the following identities are valid:

$$G^> = G^< - iA$$

$$\Sigma^> = \Sigma^< + i2\,\mathrm{Im}\,\Sigma_{\mathrm{ret}} = \Sigma^< - i2\Gamma$$

These relations are trivial to show for equilibrium, but they are also valid for nonequilibrium situations. These identities will be used often to simplify expressions. For example, the main scattering term in the QBE can be immediately simplified to

$$\Sigma^> G^< - \Sigma^< G^> = -i[2\Gamma G^< - \Sigma^< A]$$

This result will be employed in the calculations. The quantities $G^<$ and $\Sigma^<$ are generally proportional to the density of particles, while retarde functions are only indirectly dependent upon the density of particles–only through the self-energy function Σ_{ret}. In the QBE, each term has one factor that is either $G^<$ or $\Sigma^<$, so each term is proportional to the density of particles. Thus this equation does have the character of a transport equation.

The QBE simplifies for the treatment of systems that are homogeneous ($\nabla_R = 0$) and steady ($\partial/\partial T = 0$). These derivatives are dropped, as well as the Poisson brackets, since the latter contain similar derivatives. The QBE is

$$e\mathbf{E}\cdot\left[\left(1-\frac{\partial\,\mathrm{Re}\,\Sigma_{\mathrm{ret}}}{\partial\omega}\right)\nabla_k + (\mathbf{v}_k + \nabla_k\,\mathrm{Re}\,\Sigma_{\mathrm{ret}})\frac{\partial}{\partial\omega}\right]G^<$$

$$- e\mathbf{E}\cdot\left[\frac{\partial\Sigma^<}{\partial\omega}\nabla_k\mathrm{Re}\,G_{\mathrm{ret}} - \frac{\partial\,\mathrm{Re}\,G_{\mathrm{ret}}}{\partial\omega}\nabla_k\Sigma^<\right] = \Sigma^< A - 2\Gamma G^<$$

Next we will make a remarkable simplification of this equation. The QBE is only valid to first power in the electric field, since we have systematically ignored terms of $O(E^2)$. For example, in the gradient expansion the second derivative terms would give contributions of $O(E^2)$ were they retained. So the equation is exact to first order in the electric field, but is not exact to higher orders in the field. This fact is utilized to simplify the left-hand side of the equation. Since the field E multiplies each term, for each term on this side we can use the Green's functions which are independent of field. Of course, these are just the equilibrium quantities in (3.7.2) and (3.7.3).

Consider the frequency derivatives of the left-hand side of the equation. If we write the spectral function in the shorthand notation

$$A = 2\Gamma/(\sigma^2 + \Gamma^2)$$
$$\Gamma = -2\,\text{Im}\,\Sigma_{\text{ret}}$$
$$\sigma = \omega - \varepsilon_k - \text{Re}\,\Sigma_{\text{ret}}$$

then the left-hand side will have three types of frequency derivatives: $\partial n_F/\partial\omega$, $\partial\Gamma/\partial\omega$, and $\partial\sigma/\partial\omega$. The coefficients of the latter two vanish, which leaves only terms in $\partial n_F/\partial\omega$. All of the terms proportional to n_F vanish, leaving only terms multiplied by $\partial n_F/\partial\omega$:

$$A(\mathbf{k}, \Omega)^2 \frac{\partial n_F}{\partial\omega} e\mathbf{E} \cdot \{(\mathbf{v}_k + \nabla_k \,\text{Re}\,\Sigma^r)\Gamma + \sigma\nabla_k\Gamma\} = \Sigma^> G^< - \Sigma^< G^> \tag{7.4.13}$$

The factor of $A(\mathbf{k}, \omega)^2$ appears in each term, and was taken outside. The left-hand side of this equation now contains only known quantities, which can be calculated in equilibrium. The scattering terms remain on the right-hand side. Finding them still involves work, usually in the form of an integral equation. This final form for the QBE is exact for transport which is linear in the field, and for steady state, homogeneous systems. It is quite analogous to the similar expression for the classical BE, which is

$$-e\mathbf{E} \cdot \mathbf{v}_k \frac{\partial f^{(0)}}{\partial\varepsilon_k} = \Sigma^> G^< - \Sigma^< G^>$$

The classical equation has $\partial f^{(0)}/\partial\varepsilon_k$, while (7.4.13) has $\partial n_F(\omega)/\partial\omega$.

Equation (7.4.13) is the steady state, homogeneous form of the QBE. It should be the starting point for many transport calculations. It is exact, and is an alternative to using the Kubo formalism, which is also exact. The derivation of this equation has been complicated, and has entailed some work. However, once derived, it is often the easiest starting point for deriving the transport coefficients. Calculating from Eq. (7.4.13) entails less work in getting to the answer than with any other formalism.

C. Electron Scattering by Impurities

The quantum Boltzmann equation (QBE) (7.4.13) will be solved for the electron scattering by impurites. This case was solved in Sec. 7.1 using the Kubo formula. This exercise is useful for two reasons: (1) producing the known result demonstrates the correctness of the QBE; and (2) this case is the easiest one to solve, and provides an introduction to the techniques for solving the QBE. The present example makes two assumptions that are intended to make the solution as easy as possible: (1) the impurities are dilute, so that the simultaneous scattering from several impurities can be neglected; and (2) the impurities have no internal degrees of freedom, such as spin or vibrations, which can be altered by the electron scattering. The second assumption implies that impurity is a simple potential that elastically scatters the electron. The impurities are randomly located in the solid. The method of Sec. 4.1E is employed for averaging over the random distributions.

The first step in solving the QBE is to find the retarded functions. For impurity scattering they are independent of the electric field, at least for small fields. The retarded self-energy for scattering from impurities of density n_i is

$$\Sigma_{\text{ret}}(k, \omega) = n_i T_{kk}(\omega) \tag{4.1.33}$$

The T matrix is the one in (4.1.33) which is energy dependent, and hence off-shell.

The next step is to express the self-energies $\Sigma^<$ and $\Sigma^>$ in terms of the Green's functions $G^<$ and $G^>$. The retarded functions G_{ret} and Σ_{ret} are known. However, $G^<$ and $G^>$ are, at this point, unknown since they are affected by the electric field. The self-energy functions are

$$\Sigma^{\lessgtr}(k, \omega) = n_i \int \frac{d^3p}{(2\pi)^3} [T_{kp}(\omega)]^2 G^{\lessgtr}(p, \omega) \tag{7.4.14}$$

This result is derived below. The off-diagonal T matrix is the one in Eq. (4.1.33), which also depends upon the energy ω.

Equation (7.4.14) is now derived. An impurity at $\mathbf{R} = 0$ is represented by an electron potential $V(r)$, whose Fourier transform is $V(q)$. The self-energy Σ from a single scattering event is

$$\tilde{\Sigma}_0(x_1, x_2) = V(\mathbf{r}_1)\delta^4(x_1 - x_2)\mathrm{I}$$

This self-energy is inserted into (2.9.13). That equation is iterated in order to find the effects of repeated scattering from the same impurity. The resulting self-energy series is rewritten in a symbolic notation, where the product of two functions implies an integral over dx. Iteration of these equations gives the series for Σ:

$$\Sigma = \Sigma_0 + \Sigma_0\tilde{G}_0\Sigma_0 + \Sigma_0\tilde{G}_0\Sigma_0\tilde{G}_0\Sigma_0 + \cdots \tag{7.4.15}$$

After summing this series, one can find the formula for the individual components. The one for Σ is

$$\Sigma^{\lessgtr} = [1 + G_{\text{ret}}\Sigma_{\text{ret}}]\Sigma_0^{\lessgtr}[1 + \Sigma_{\text{adv}}G_{\text{adv}}] + \Sigma_{\text{ret}}G^{\lessgtr}\Sigma_{\text{adv}} \qquad (7.4.16)$$

Note the analogy with the equation for $G^<$ in (2.9.14). The resemblance is expected, since the series for Σ has the same matematical structure as the one for G.

For impurity scattering, the unperturbed self energies $\Sigma_0^{<,>}$ are zerc. There is only the last term in (7.4.16). Since the self-energies Σ_{ret} and Σ_{adv} are T matrices, we find for $\Sigma^>$ the result in (7.4.14). The damping function $\Gamma(k, \omega)$ and other self-energy $\Sigma^>(k, \omega)$ are

$$2\Gamma(k, \omega) = n_i \int \frac{d^3p}{(2\pi)^3} (T_{pk})^2 A(p, \omega) \qquad (7.4.17)$$

$$\Sigma^> = \Sigma^< - i2\Gamma$$

For dilute impurities, it is sufficient to retain only those terms that are first order in the impurity concentration n_i. The broadening of the spectral function $A(p, \omega)$ is due to impurity scattering. Since the self-energy is multiplied by n_i, one can replace the spectral function $A(k, \omega)$ by $2\pi\delta(\omega - \varepsilon_k)$. This expression is then equal to the imaginary part of the T matrix: $2\Gamma = n_i v \sigma_T$, where σ_T is the total cross section from impurity scattering in (4.1.18).

The starting point for solving the QBE is (7.4.13). On the left of the equal sign is the factor of

$$[(\mathbf{v}_k + \nabla_k \text{Re}\, \Sigma^r)\Gamma + \sigma \nabla_k \Gamma] A(\mathbf{k}, \omega)^2$$

This expression can be simplified by neglecting terms of $O(n_i^2)$ such as $\Gamma\, \text{Re}\, \Sigma^r$. The term in $\sigma = \omega - \Sigma_k - \text{Re}\, \Sigma$ is small since the factor of $A(\mathbf{k}, \omega)^2$ tends to force $\sigma \approx 0$. The above expressions can be approximated by $\mathbf{v}_k \Gamma A(k, \omega)^2$:

$$A^2(k, \omega) \frac{\partial n_F(\omega)}{\partial \omega} e\mathbf{E} \cdot \mathbf{v}_k \Gamma = i2\Gamma G^< - i\Sigma^< A \qquad (7.4.18)$$

This completes the treatment of the left-hand side of the QBE.

On the right side of (7.4.18) the scattering terms vanish in equilibrium since $\Sigma^< = 2in_F\Gamma$ and $G^< = in_F A$. Since current is flowing in response to the field, the system is slightly out of equilibrium. We expect the right-hand side to have factors similar to those on the left. This discussion suggests the following ansatz for the nonequilibrium Green's function

$$G^< = iA(k,\omega)\Big[n_F(\omega) - \Big(\frac{\partial n_F}{\partial\omega}\Big)e\mathbf{E}\cdot\mathbf{v}_k\Lambda(k,\omega)\Big] \qquad (7.4.19)$$

The function $\Lambda(k,\omega)$ is unknown, and needs to be determined by solving the QBE. The factors that multiply Λ are for later convenience. The above choice does not make any assumptions regarding the form for Λ. Using this ansatsz in the self-energy function gives

$$\Sigma^< = in_i\int\frac{dp^3}{(2\pi)^3}(T_{pk})^2A\Big[n_F - \frac{\partial n_F}{\partial\omega}e\mathbf{E}\cdot\mathbf{v}_p\Lambda(p,\omega)\Big]$$
$$= 2i\Gamma n_F - in_i\frac{\partial n_F}{\partial\omega}\int\frac{d^3p}{(2\pi)^3}A(p,\omega)(T_{pk})^2e\mathbf{E}\cdot\mathbf{v}_p\Lambda(p,\omega)$$

Putting these two expressions into the right-hand side of (7.4.18), the equilibrium terms cancel, and the remaining terms each have the common factor of $(-\partial n_F/\partial\omega)A(k,\omega)$:

$$A^2\frac{\partial n_F}{\partial\omega}e\mathbf{E}\cdot\mathbf{v}_k\Gamma = \Big(-\frac{\partial n_F}{\partial\omega}\Big)A(k,\omega)\Big[2\Gamma e\mathbf{E}\cdot\mathbf{v}_k\Lambda$$
$$- n_i\int\frac{d^3p}{(2\pi)^3}|T_{kp}|^2A(p,\omega)\,e\mathbf{E}\cdot\mathbf{v}_p\Lambda(p,\omega)\Big]$$

After canceling all of the common factors, there is an integral equation for the unknown function $\Lambda(k,\omega)$:

$$\mathbf{v}_k\Lambda(k,\omega) = \frac{A}{2}\mathbf{v}_k + \frac{n_i}{2\Gamma}\int\frac{d^3p}{(2\pi)^3}|T_{pk}|^2A(p,\omega)\mathbf{v}_p\Lambda(p,\omega) \qquad (7.4.20)$$

This integral equation for $\Lambda(k,\omega)$ is nearly identical to the one found in solving the conductivity from the Kubo relation. The quantity $\mathbf{v}_k\Lambda(k,\omega)$ is similar to the factor of $\Gamma(\mathbf{k},\omega-i\delta,\omega+i\delta)$ in Sec. 7.1C. As we proceed, the differences between these quantities disappear, and the two approaches give the same resistivity.

After solving this equation for $\Lambda(k,\omega)$, the Green's function $G^<$ in (7.4.19) is known. It is used in (3.7.8) for the current. The first term in (7.4.19) for $G^<$ gives a zero current. The second term gives a current proportional to the field E, and this proportionality defines the electrical conductivity

$$\sigma_{\mu\nu} = e^2\int\frac{d^3k}{(2\pi)^3}\int\frac{d\omega}{2\pi}v_{k\mu}v_{k\nu}\Big(-\frac{\partial n_F}{\partial\omega}\Big)A(k,\omega)\Lambda(k,\omega) \qquad (7.4.21)$$

The two equations (7.4.20) and (7.4.21) provide the solution for the conductivity from the QBE.

For electrons in metals, with a spherical Fermi surface, the conductivity from impurity scattering is found easily from these equations. First do the integral over $d\omega$ in (7.4.21). At low temperature the factor $-\partial n_F/\partial\omega$ is nearly a delta function which sets $\omega \approx 0$. At low temperature we find

$$\sigma_{\mu\nu} = e^2 \int \frac{d^3k}{(2\pi)^4} v_{k\mu} v_{k\nu} A(k, 0) \Lambda(k, 0)$$

The equation is reduced as far as possible. Next one must solve the equation (7.4.20) for Λ. Adopt a vector coordinate system where the z direction is $\mathbf{k}$. Then the various scalar products of vectors can be found from the law of cosines:

$$\hat{k} \cdot \hat{E} = \cos(\theta_0)$$
$$\hat{k} \cdot \hat{p} = \cos(\theta)$$
$$\hat{p} \cdot \hat{E} = \cos(\theta_0)\cos(\theta) + \sin(\theta_0)\sin(\theta)\cos(\phi)$$

Equation (7.4.20) is multiplied by the vector $\hat{E}$. The integral over the $d\phi$ part of d^3p makes the $\cos(\phi)$ term vanish. Each of the remaining terms has a factor of $\cos(\theta_0)$ which can be canceled. Then one finds the scalar equation

$$\Lambda(k, 0) = \frac{1}{2} A(k, 0) + \frac{n_i}{2\Gamma} \int \frac{d^3p}{(2\pi)^3} \mid T_{kp} \mid^2 A(p, 0) \Lambda(p, 0) \cos(\theta)$$

The spectral functions $A(k, 0)$ and $A(p, 0)$ force $k = p = k_F$. The factor of $\Lambda(p, 0)$ under the integral can be set equal to $\Lambda(k_F, 0)$ and taken out of the integral. Then the above equation can be solved to find

$$\Lambda(k, 0) = \frac{\tau_1}{2\tau} A(k, 0) \tag{7.4.22}$$

where $\tau = 1/2\Gamma$ is the time between scattering from impurities, while τ_1 is the lifetime which is important for resistivity; it has the factor of $(1 - \cos\theta)$:

$$\sigma = \frac{ne^2\tau_1}{m}$$

$$1/\tau_1 = n_i \int \frac{d^3p}{(2\pi)^3} \mid T_{pk} \mid^2 A(p, \omega) \, (1 - \cos\theta) \tag{7.4.23}$$

This formula for the resistivity is identical to (7.1.5). The QBE gives the same formula for the resistivity as found earlier from the Kubo formula. The two methods also are in exact agreement for much more complex cases, such as electron–phonon scattering at nonzero temperatures and frequency. Most of the steps in the derivation were spent in getting to (7.4.18). This equation is the starting point for any calculation for homogeneous systems. The number of steps between this equation and the final resistivity is small. The QBE is an efficient method of finding the resistivity, or of other transport coefficients.

D. T^2 Contribution to the Electrical Resistivity

At low temperature the electrical resistivity of simple metals is provided by the interaction of the electron with impurities and phonons. One's first thought is that the resistivity should have a constant term ϱ_0 from impurity scattering, plus a T^5 term from scattering by phonons. The T^5 term is often never observed, at any temperature. In the alkali metals it is eliminated by phonon drag as reviewed by Wiser (1985). In simple metals, such as lithium and potassium, the resistivity at low temperature varies according to T^2. The are two separate terms. The most general way to represent the experimental results is

$$\varrho(T) = AT^2 + \varrho_0(1 + BT^2) + \delta\varrho(T) \tag{7.4.24}$$

The constant term ϱ_0 is from impurity scattering, and is proportional to their concentration. The T^2 term is given by the coefficients A and B, which were both predicted before they were observed. The term $\delta\varrho$ is from a variety of effects: phonon drag, dislocations, and size effects.

The first term AT^2 is intrinsic and comes from electron–electron–Umklapp scattering. Normal electron–electron scattering conserves electron momentum and does not contribute to the resistivity. Scattering from the crystal potential is called Umklapp scattering. This Bragg scattering provides crystal momentum, and gives a finite contribution to the resistivity. A nice discussion, and prediction, of this phenomena is given in Ziman's books (1960, 1964). He calculated the interaction using a screened Coulomb interaction, and found values of A which are too small when compared with experiment. The largest contribution to the coefficient A is from phonon mediated interactions between the electrons (MacDonald *et al.*, 1981).

The term B was predicted by Koshino (1960) and Taylor (1962, 1964). It is due to inelastic scattering of the electrons by impurities. Since the impurities are part of the lattice, the scattering by the electron can excite phonons. The rate of scattering is also normalized by the Debye–Waller factor, which is given in terms of the thermal fluctuations of the impurity positions.

Consider a distribution of impurities at sites $\mathbf{R}_j$ with a fluctuating displacement $\mathbf{Q}_j(t)$. This potential can be written as

$$V = \sum_j V(\mathbf{r} - \mathbf{R}_j - \mathbf{Q}_j) = \frac{1}{\nu} \sum_j \sum_q v_i(q) \exp[i\mathbf{q} \cdot (\mathbf{r} - \mathbf{R}_j - \mathbf{Q}_j)]$$

Expand V in powers of the displacement, and keep the first two terms:

$$V = \frac{1}{\nu} \sum_j \sum_q v_i(q) \exp[i\mathbf{q} \cdot (\mathbf{r} - \mathbf{R}_j)][1 - i\mathbf{q} \cdot \mathbf{Q}_j - (\mathbf{q} \cdot \mathbf{Q}_j)^2/2 \ldots]$$

where $v_i(q)$ is the Fourier transform of the potential between the electron and the impurity. The potential V is used to evaluate the scattering in the second Born approximation. Take the product of two potentials at different times, and only retain terms of order $O(Q^2)$:

$$V(t_1)V(t_2) = \frac{1}{\nu^2} \sum_{qq} v_i(q) v_i(q') \varrho(\mathbf{q}, t_1) \varrho(\mathbf{q}', t_2) \sum_{jl} e^{-i[\mathbf{q} \cdot \mathbf{R}_j + \mathbf{q}' \cdot \mathbf{R}_l]}$$
$$\{1 - \tfrac{1}{2}[\mathbf{q} \cdot \mathbf{Q}_j(t_1)]^2 - \tfrac{1}{2}[\mathbf{q}' \cdot \mathbf{Q}_l(t_2)]^2 - [\mathbf{q} \cdot \mathbf{Q}_j(t_1)][\mathbf{q}' \cdot \mathbf{Q}_l(t_2)]\} \tag{7.4.25}$$

The electron density operator is $\varrho(\mathbf{q}, t)$. The average is taken over the positions of the impurities, as described in Sec. 4.1E:

$$\langle \sum_{jl} e^{-i[\mathbf{q} \cdot \mathbf{R}_j + \mathbf{q}' \cdot \mathbf{R}_l]} \rangle = N_i \delta_{jl} \delta_{\mathbf{q}+\mathbf{q}'}$$

where N_i is the number of impurites, and $n_i = N_i/\nu$ is the density of impurities. Next, the product of impurity displacements is averaged into a phonon Green's function $\langle Q_\mu(t_1) Q_\nu(t_2) \rangle = D'_{\mu\nu}(t_1 - t_2)$:

$$\langle V(t_1)V(t_2) \rangle = \frac{n_i}{\nu} \sum_q v_i(q)^2 \varrho(\mathbf{q}, t_1) \varrho^\dagger(\mathbf{q}, t_2) \{1 + q_\mu q_\nu [D'_{\mu\nu}(t_1 - t_2) - D'_{\mu\nu}(0)]\} \tag{7.4.26}$$

In the curly brackets on the right, the first term ("1") is the elastic impurity scattering which was treated in the prior section. The term with $D'(0)$ comes from the two terms in (7.4.25) which have both times t_1 or t_2. This term is called the Debye–Waller factor. Using linked-cluster methods, one can show that the impurity scattering $v_i(q)^2$ should be multiplied by the factor $\exp\{-q_\mu q_\nu D'_{\mu\nu}(0)\}$ which comes from the fluctuations in the positions of the impurities. Our derivation retains the first two terms in the expansion of the exponential.

At first glance the terms with D' appear to have the standard form for the electron–phonon interaction. However, a careful inspection shows this form for the interaction is different because the electron–phonon scattering is part of impurity scattering: the wave vector $\mathbf{q}'$ of the phonon is not the same $\mathbf{q}$ which is the momentum transfer of the electron. The scattering from the impurity can impart any momentum to the electron. So the phonon can have any wave vector $\mathbf{q}'$. This point is illustrated by actually evaluating the expression for $D'(t)$. It involves the correlation of two displacements for the same ion at different times. Using (1.1.17) gives

$$D'_{\mu\nu}(t) = \frac{1}{N} \sum_{\mathbf{q}'} X(\mathbf{q}')^2 D_{\mu\nu}(\mathbf{q}', t)$$

$$X(\mathbf{q}')^2 = 1/[2M\omega(\mathbf{q}')]$$

The phonon Green's function $D_{\mu\nu}(\mathbf{q}', t)$ is any one of the six functions of real time, depending upon what is needed.

First find $\Sigma^<$ for just the phonon part of (7.4.26). It is the four-vector integral of $G^< D^<$. Using the equilibrium form for $D^<$ in (2.9.9) gives the self-energy of the electron from inelastic impurity scattering:

$$\Sigma^<(k, \varepsilon) = n_i \int \frac{d^3q}{(2\pi)^3} v_i(q)^2 \int \frac{d^3q'}{(2\pi)^3} \frac{(q \cdot \xi')^2}{2\varrho\omega'} \{(N_{q'} + 1)[G^<(\mathbf{k} + \mathbf{q}, \varepsilon + \omega_{q'}) - G^<(\mathbf{k} + \mathbf{q}, \varepsilon)] + N_{q'}[G^<(\mathbf{k} + \mathbf{q}, \varepsilon - \omega_{q'}) - G^<(\mathbf{k} + \mathbf{q}, \varepsilon)]\} \tag{7.4.27}$$

Primed q variables refer to the phonon, while unprimed q variables refer to the momentum transfer of the electron during the scattering. We use ϱ here to mean the mass of the impurity divided by the volume of the unit cell. This self-energy vanishes if the energy of the phonon is set equal to zero.

This formula has been derived by assuming that the impurity mass is equal to the mass of the host ion. The experiments are done at very low temperature. Then the relevent phonons are acoustical phonons of very long wavelength. All ions in the metal move together in long-wavelength acoustical motion. So there is no correction for the difference in the ion mass, or in the difference in bonding. This theory should apply to every impurity.

In equilibrium we have

$$G^{<}(k, \varepsilon) = in_F(\varepsilon)A(k, \varepsilon)$$
$$\Sigma^{<}(k, \varepsilon) = in_F(\varepsilon)2\Gamma_{KT}(k, \varepsilon)$$

At first glance the self-energy $\Sigma^{<}$ in (7.4.27) does not appear to be proportional to $n_F(\varepsilon)$. It is, which can be shown using algebra. Let $x = \beta\varepsilon$ and $y = \beta\omega_q{}'$, and manipulate the factors in (7.4.27):

$$\begin{aligned}(N_{q'} + 1)n_F(\varepsilon + \omega_{q'}) &= \frac{1}{1 - e^{-y}} \frac{1}{e^{x+y} + 1} \\ &= \frac{1}{e^x + 1}\left(\frac{1}{e^y - 1} + \frac{1}{e^{x+y} + 1}\right) \\ &= n_F(\varepsilon)[N_{q'} + n_F(\varepsilon + \omega_{q'})] \\ N_{q'}\, n_F(\varepsilon - \omega_{q'}) &= n_F(\varepsilon)[N_{q'} + 1 - n_F(\varepsilon - \omega_{q'})]\end{aligned}$$

These algebraic relationships allow us to extract a factor of $n_F(\varepsilon)$ from each term in (7.4.27). We also change the electron-impurity potential $v_i(q)$ to a T matrix between initial (k) and final (p) electron wave vectors. The remaining factor defines the Koshino–Taylor energy uncertainty:

$$\begin{aligned}2\Gamma(k, \varepsilon) = n_i \int \frac{d^3p}{(2\pi)^3} \mid T_{pk} \mid^2 \Bigg(A(p, \varepsilon) + \int \frac{d^3q'}{(2\pi)^3} \frac{[(\mathbf{k} - \mathbf{p}) \cdot \hat{\xi}']^2}{2\varrho\omega'} \\ \times \{[N_{q'} + n_F(\varepsilon + \omega_{q'})]A(\mathbf{p}, \varepsilon + \omega_{q'}) \\ + [N_{q'} + 1 - n_F(\varepsilon - \omega_{q'})]A(\mathbf{p}, \varepsilon - \omega_{q'}) - (2N_{q'} + 1)A(\mathbf{p}, \varepsilon)\}\Bigg)\end{aligned} \tag{7.4.28}$$

The first term on the right gives the elastic impurity scattering. The second term gives the expression for the rate of inelastic scattering of electrons by phonons and impurities.

The resistivity is found by solving the QBE. Again one assumes that the Green's function $G^{<}$ has a nonequilibrium factor $\Lambda(k, \omega)$ such as (7.4.19). This ansatz is used in the QBE (7.4.18). One solves it using the same steps as in the preceding section. The self-consistent equation for Λ from impurity scattering is

$$\Lambda(k, \varepsilon) = \frac{A}{2}(k, \varepsilon) + \frac{n_i}{2\Gamma} \int \frac{d^3p}{(2\pi)^3} \mid T_{kp} \mid^2 \cos(\theta) \Bigg(\Lambda(p, \varepsilon)A(p, \varepsilon)$$

$$+ \int \frac{d^3q'}{(2\pi)^3} \frac{[(\mathbf{k} - \mathbf{p}) \cdot \hat{\xi}']^2}{2\varrho\omega_{q'}} \{[N_{q'} + n_F(\varepsilon + \omega_{q'})]A(p, \varepsilon + \omega_{q'})$$

$$\times \Lambda(p, \varepsilon + \omega_{q'})$$
$$+ [N_{q'} + 1 - n_F(\varepsilon - \omega_{q'})]A(p, \varepsilon - \omega_{q'})\Lambda(p, \varepsilon - \omega_{q'})$$
$$- (2N_{q'} + 1)A(p, \varepsilon)\Lambda(p, \varepsilon)\}\Big)$$

The first term on the right in large parentheses (ΛA) is from elastic impurity scattering, as discussed in the prior section. The other terms are from inelastic scattering. The expression simplifies because the spectral functions $A(p, \varepsilon)$ and $A(p, \varepsilon \pm \omega')$ can be treated as delta functions, which set $p \approx k_F$. The additional energy $\pm \omega_{q'}$ has negligible influence on this process. Again we can treat $\Lambda(p, \varepsilon)$ and $\Lambda(p, \varepsilon \pm \omega')$ as a constant $\Lambda(k_F, 0)$, which approximately equals

$$\Lambda(k, \varepsilon) = \frac{\Gamma(k, \varepsilon)}{2\Gamma_{KT}(k, \varepsilon)} A(k, \varepsilon)$$

$$2\Gamma_{KT}(k, \varepsilon) = n_i \int \frac{d^3p}{(2\pi)^3} \mid T_{kp} \mid^2 [1 - \cos(\theta)]A(p, \varepsilon)\Big\{1$$
$$+ \int \frac{d^3q'}{(2\pi)^3} \frac{[(\mathbf{k} - \mathbf{p}) \cdot \hat{\xi}']^2}{2\varrho\omega_{q'}} [n_F(\varepsilon + \omega_{q'}) - n_F(\varepsilon - \omega_{q'})]\Big\}$$

The factor $2\Gamma(k, \varepsilon)$ is the same as the above expression, but without the first factor of $[1 - \cos(\theta)]$. The angle θ is between $\mathbf{p}$ and $\mathbf{p}'$.

This expression will now be evaluated. The phonon part of the integrand is given by a function we define as $g(u)$:

$$(k_F{}^2/\varrho) \sum_\lambda \int \frac{d^3q'}{(2\pi)^3} \xi_\mu{}'\xi_\nu{}'\delta[u - \omega_\lambda(q')] = \delta_{\mu\nu} g(u) \qquad (7.4.29)$$

where λ is the polarization of the phonon. The right-hand side is proportional to a delta function of the (x, y, z) components (μ, ν), which is required by cubic symmetry, which we are assuming. The function $g(u)$ is dimensionless. The delta function $\delta_{\mu\nu}$ means that the factor $(q \cdot \xi')^2$ becomes

$$q^2 = (\mathbf{k} - \mathbf{p})^2 \approx 2k^2{}_F(1 - \cos\theta)$$

The integral over p decouples from the integral over phonon coordinates. It is useful to define the integral

$$I_{KT} = n_i \int \frac{d^3p}{(2\pi)^3} T_{kp}^2 (1 - \cos\theta)^2 A(p, \varepsilon) \qquad (7.4.30)$$

The same factors appear in the expression for the resistivity from impurity scattering. However, this factor, which we call I_i, has only one factor of $(1 - \cos\theta)$. It is a good approximation to evaluate both of these expressions as if the scattering were elastic. Then the T matrices can be expressed in terms of phase shifts as in (7.1.6). Define S_j for $j = i$ or KT according to

$$I_j = \frac{4\pi n_i}{mk_F} S_j$$

In (7.4.30) the spectral function $A(p, \varepsilon) \approx 2\pi\delta(\varepsilon - \xi_p)$ which eliminates the integral over p. The angular integrals can also be done. For S_i we find the same result as in Sec. 7.1:

$$\begin{aligned} S_i &= \sum_{l=1}^{\infty} l \sin^2(\delta_l - \delta_{l-1}) \\ S_{KT} &= 2\left[S_i - \sum_{l=2}^{\infty} \frac{l(l-1)}{2(2l-1)} \sin^2(\delta_l - \delta_{l-2})\right] \end{aligned} \tag{7.4.21}$$

Collecting all of these results gives the expression for the KT term in the resistivity:

$$2\Gamma_{KT}(\varepsilon) = I_i + I_{KT} \int_0^{\omega_D} \frac{du}{u} g(u)[n_F(\varepsilon + u) - n_F(\varepsilon - u)]$$

We write $n_F(\varepsilon - u) = 1 - n_F(u - \varepsilon)$ and the 1-term gives an uninteresting constant. At low temperature the T^2 term comes from the integral at small u. From (7.4.31) we find that $g = \gamma u^2$ at small u. Then the u-integral is changed to $y = \beta u$ so that we find for Γ_{KT}

$$\begin{aligned} 2\Gamma_{KT}(\varepsilon) &= I_i + \gamma I_{KT}(k_B T)^2 z(\beta\varepsilon) \\ z(x) &= \int_0^{\infty} y\, dy \left(\frac{1}{e^{x+y} + 1} + \frac{1}{e^{y-x} + 1}\right) \end{aligned}$$

In calculating the conductivity, we have to average ε over its range of thermal values. This average contains the factors in (7.4.19)

$$\langle z \rangle = \int_{-\infty}^{\infty} d\varepsilon\, z(\beta\varepsilon)[-\partial n_F(\varepsilon)/\partial\varepsilon]$$

The above expression is a double integral in $x = \beta\varepsilon$ and y. It is easy to do the x integral first, and the result gives a y integrand of $2y\partial[yn_B(y)]/\partial y$. Then the y integral is also easy, so that $\langle z \rangle = \pi^2/3$. The result is

$$2\Gamma_{KT} = I_i + \pi^2\gamma I_{KT}(k_B T)^2/3$$

The coefficient B is the second term divided by $I_i T^2$. Thus we get that

$$\begin{aligned} B_{KT} &= \pi^2 \gamma k_B{}^2 R/3 \\ R &= S_{KT}/S_i \end{aligned} \tag{7.4.32}$$

There remains just the task of evaluating the phonon contribution γ. From (7.4.31) it can be written as

$$\gamma = [k_F{}^2/(2\pi^2\varrho)]\,\langle \xi_x{}^2/c^3 \rangle$$

where the bracket $\langle \cdots \rangle$ indicates an average over 4π of solid angle plus a summation over acoustical phonon branches, and c is the phonon velocity which depends upon solid angle. The only expression that may present difficulty is the ratio R in (7.4.32). Remember that the Friedel sum rule (4.1.22a) relates the sum of the phase shifts to the impurity valence Z, where Z is the difference between the charge on the impurity and the host metal ion. The constant impurity resistance ϱ_0 provides the value of S_i, which is another relationship among the phase shifts.

As a simple example, consider a neutral impurity $Z = 0$ where the scattering is given only by s and p waves. The Friedel sum rule gives $\delta_1 = -\delta_0/3$. If all of the phase shifts are small, then we can replace $\sin\delta$ by δ everywhere. Then $S_i = 2(\delta_0)^2$ while (7.4.30) gives $S_{KT} = 16(\delta_0)^2/5$. The ratio $R = 1.6$.

How does this theory compare to the experiments? The best experimental results are for rubidium impurities in potassium, which have $Z = 0$. For potassium the average over sound wave directions gives $\langle \xi_x{}^2/c^3 \rangle = 0.466 \times 10^{-15}\ (\mathrm{s/cm})^3$. Using $R = 1.6$, $\varrho = Mk_F{}^3/3\pi^2$, $M = 39$ AMU, and $r_s = 4.96$ gives $B = 1.4 \times 10^{-5}\ \mathrm{K}^{-2}$. This theoretical result should be compared with the experimental values of $1.20 \times 10^{-5}\ \mathrm{K}^{-2}$ (Oomi *et al.*, 1985) and $1.23 \times 10^{-5}\ \mathrm{K}^{-2}$ (Bass *et al.*, 1986). The theoretical value is slightly high, which suggests that d waves are actually important in the scattering process. At $T = 1$ K the factor of BT^2 is $O(10^{-5})$ which shows that these terms are a small correction to the resistivity.

PROBLEMS

1. Use the piezoelectric electron–phonon interaction (1.3.7) to calculate the temperature dependence of the conductivity in CdS. Compare with Fig. 7.9(b).

2. Consider the self-energy of an electron from unscreened exchange,

$$\Sigma(\mathbf{p}) = -\frac{1}{\nu}\sum_{\mathbf{q}} v_q \frac{1}{\beta}\sum_{ip} \mathscr{G}(\mathbf{p}+\mathbf{q}, ip)$$

which was evaluated in Sec. 5.1F. Derive the Ward identity for this self-energy—the equivalent of (7.1.24). Use this result to show that the coulomb ladder diagrams have negligible effect upon the basic polarization diagram $P(\mathbf{q}, i\omega)$ in the limit where $\mathbf{q} \to 0$.

3. The correlation function $\chi(\mathbf{q}, i\omega) = -\int_0^\beta d\tau e^{i\omega\tau}\langle T_\tau \varrho(\mathbf{q}, \tau)\varrho(-\mathbf{q}, 0)\rangle$ vanishes when $\mathbf{q} \to 0$. Show that the bare bubble $P^{(1)}(q, i\omega)$ has this feature. Use the Ward identity to show that it still vanishes when self-energy functions and vertex functions are included in the evaluation.

4. Derive the two contributions (7.2.12) to the two-phonon part of the imaginary self-energy.

5. Use (7.3.10) to evaluate $\gamma(\varepsilon + i\delta, \varepsilon + i\delta)$, and show that it equals unity, in agreement with the Ward identity.

6. Derive the Ward identity (7.3.8) for the phonon ladder diagrams.

7. Solve the scalar vertex equation (7.3.8) using the same techniques used to solve $\gamma(\mathbf{p}, ip, ip + i\omega)$. Show that $\Gamma(\varepsilon + i\delta, \varepsilon + i\delta) = 1 - \partial\Sigma/\partial\varepsilon$ but that $\Gamma(\varepsilon - i\delta, \varepsilon + i\delta)$ does not obey the Ward identity. Furthermore, show for every solution $R(\varepsilon) = \Gamma(\varepsilon - i\delta, \varepsilon + i\delta)$ to its vertex equation that $R(\varepsilon) + a\,\mathrm{Im}[\Sigma(\varepsilon)]$ is also a solution, where a is an arbitrary constant.

8. Show that the leading term in the thermopower for a free-electron gas is

$$S = -\frac{\pi^2 k_B{}^2 T}{2eE_F}$$

9. Evaluate the correlation function $L^{(22)}$ for the thermal conductivity of a metal. Use the free-electron heat current operator, and show that the integral diverges for the first term in the vertex summation.

10. Solve the equations in Sec. 7.1A for the $l = 2$ term f_d in the Legendre expansion for $f(\mathbf{k})$. Show that for steady state transport in a weak electric field one has that $f_d \sim O(E^2)$, so the d term in the Legendre expansion is negligible.

11. Show that the resistivity of a metal at high temperature approaches $\varrho = 2\pi\lambda_{\mathrm{tr}} k_B T m/e^2 n_0$, where λ_{tr} is the transport form of the electron–phonon coupling constant defined in (6.4.20).

12. Solve Eqs. (7.3.6) and (7.3.13) in the limit of small temperature and show that the resistivity $\varrho \sim T^5$ from phonons. Assume that $\alpha^2 F = gu^2$ at small u, where g is a constant.

13. Compare the relaxation time for transport deduced from the force-balance theory (7.3.1) with that given by (7.3.14). How do they differ? Show they become identical in the limit of high T.

Chapter 8

Optical Properties of Solids

8.1. NEARLY FREE-ELECTRON SYSTEMS

A. General Properties

In this section we shall be concerned with calculating the optical properties of free-particle systems, such as electrons in metals or semiconductors. A completely free-electron system does not absorb light at all, so that its optical properties are uninteresting. The ability of the *nearly* free-particle system to absorb light is due to its imperfections or deviations from homogeneity. If these effects are small, then so is the light absorption. This situation will concern us here. We shall calculate the optical properties of simple metals and semiconductors. The best way to do this is using the so-called *force–force* correlation function.

First it is useful to recall some of the fundamental formulas regarding the optical properties of solids. As shown in Sec. 4.5C, the transverse dielectric function of the solid in the limit of long wavelength is

$$\varepsilon_{\perp}(\omega) = 1 - \frac{\omega_p^{\,2}}{\omega^2} - \frac{4\pi}{\omega^2}\,\pi(\omega)$$

where $\pi(\omega)$ is the retarded form of the current–current correlation function:

$$\pi(i\omega) = -\frac{1}{\nu}\int_0^{\beta} d\tau e^{i\omega\tau}\langle T_\tau j_\mu(\tau) j_\mu(0)\rangle$$

The optical wave vector k is so short that we can evaluate this correlation function in the limit where $k \to 0$. The results for $k \to 0$ are sometimes different from those for $k = 0$, in which case the limit result is appropriate.

We shall omit the k symbol in the notation. We assume the system is isotropic. The index μ is any direction (x, y, z), since the average value of $\langle Tj_x(\tau)j_x(0)\rangle$ is equal to that of $\frac{1}{3}\langle T\mathbf{j}(\tau)\cdot\mathbf{j}(0)\rangle$. The conductivity of $\sigma(\omega)$ of the solid is given by

$$\varepsilon_\perp(\omega) = 1 + \frac{4\pi i\sigma(\omega)}{\omega}$$

$$\sigma(\omega) = \frac{ine^2}{m\omega} + \frac{i\pi(\omega)}{\omega}$$

so that we have our usual identity $\mathrm{Re}(\sigma) = -\mathrm{Im}[\pi(\omega)/\omega]$. However, the preceding relationships are also valid for the real part of $\pi(\omega)$ which contributes to the imaginary part of the conductivity. These general formulas will be the basis for most of our discussion.

One general theorem which can be proved is the *f-sum rule*, $\varepsilon_\perp(\omega) = \varepsilon_1(\omega) + i\varepsilon_2(\omega)$:

$$\int_0^\infty d\omega\omega\varepsilon_2(\omega) = \frac{\pi\omega_p{}^2}{2}$$

The similar theorem for the longitudinal dielectric function was shown in Sec. 5.7. The theorem for the transverse dielectric function can also be written in terms of the conductivity since $\varepsilon_2 = 4\pi\,\mathrm{Re}(\sigma/\omega)$:

$$\int_0^\infty d\omega\,\mathrm{Re}[\sigma(\omega)] = \frac{\pi n_0 e^2}{2m} \tag{8.1.1}$$

If we view the real part of the conductivity as the absorbing power of the solid, then this theorem states that the absorbing power depends only on the density of particles n_0 and not on the detailed form of the interactions. Thus our many-body theory can rearrange the absorbing power by moving it from one part of the spectrum to another, but it cannot alter the net amount. Thus our statement in the opening paragraph that the light absorption is small if the perturbations are small needs clarification. This we shall do by examining the *Drude* formula.

The Drude formula is an empirical relationship which is found to be valid, over restricted frequency range, in most nearly free-electron systems. It states that

$$\begin{aligned}\pi(\omega) &= -\frac{n_0 e^2}{m}\,\frac{1}{1 - i\omega\tau_0}\\ \mathrm{Re}[\sigma(\omega)] &= \frac{n_0 e^2\tau_0}{m(1 + \omega^2\tau_0{}^2)}\end{aligned} \tag{8.1.2}$$

The quantity τ_0 is a relaxation time, which is usually taken to be a constant. Let us assume that the Drude formula is valid for all frequencies and with a constant relaxation time. Then the f-sum rule is satisfied independently of the size of the relaxation time:

$$\int_0^\infty d\omega \, \mathrm{Re}[\sigma(\omega)] = \frac{n_0 e^2 \tau_0}{m} \int_0^\infty \frac{d\omega}{1 + \omega^2 \tau_0^2} = \frac{\pi}{2} \frac{n_0 e^2}{m}$$

$\mathrm{Re}[\sigma(\omega)]$ is a Lorentzian distribution, whose width is $1/\tau_0$ and whose height is τ_0 so that the area under the curve is independent of τ_0.

We know from the discussion of dc conductivity that inverse relaxation times are proportional to scattering rates. Thus the purely free-particle system, without any inhomogeneity, corresponds to the case where $1/\tau_0 \to 0$ or $\mathrm{Re}[\sigma(\omega)] \propto \delta(\omega)$. Thus the scattering processes contribute to τ_0. If the Drude expression were rigorous, we would only need to find τ_0, which in fact is finding $1/\tau_0$. When we take the limit $\omega \to 0$ in (8.1.1), we obtain the dc conductivity $\mathrm{Re}(\sigma) = n_0 e^2 \tau_0 / m$.

If the scattering rate $1/\tau_0$ is small, then τ_0 is large, and the dimensionless product $\omega\tau_0$ is large except at very small frequencies. We expand (8.1.2) for $\omega\tau_0 \gg 1$ and find

$$\mathrm{Re}[\sigma(\omega)] = \frac{n_0 e^2}{m\omega^2} \frac{1}{\tau_0} \left[1 + O\left(\frac{1}{\omega^2 \tau_0^2}\right)\right]$$

Thus at higher frequencies the real part of the conductivity is proportional to the scattering rate $1/\tau_0$, which is small by our assumption. This is what we meant above by the limit of weak scattering. The method of force–force correlation function casts $\mathrm{Re}[\sigma(\omega)]$ in this form, as an expansion in the scattering rate. The resulting theories are valid at high frequencies where $\omega\tau_0 \gg 1$.

B. Force–Force Correlation Functions

The first feature we shall prove has already been mentioned, namely that the homogeneous electron gas does not absorb any light. This assertion is proved by showing that the Hamiltonian commutes with the momentum operator $\mathbf{P} = \sum \mathbf{k} c_{\mathbf{k}\sigma}^\dagger c_{\mathbf{k}\sigma}$:

$$H_0 = \sum_{\mathbf{p}\sigma} \xi_p C_{\mathbf{p}\sigma}^\dagger C_{\mathbf{p}\sigma} + \frac{1}{2\nu} \sum_{\mathbf{q}} v_q \sum_{\substack{\mathbf{k}\mathbf{k}' \\ \sigma\sigma'}} C_{\mathbf{k}+\mathbf{q}\sigma}^\dagger C_{\mathbf{k}'-\mathbf{q}\sigma'}^\dagger C_{\mathbf{k}'\sigma'} C_{\mathbf{k}\sigma} \tag{8.1.3}$$

Therefore the momentum operator has no time dependence and is a constant of motion. The absorption of light is described by the Kubo formula, which is the correlation of the momentum operator with itself:

$$\text{Re}[\sigma(\omega)] = \frac{e^2}{6m^2\nu}\left(\frac{1 - e^{-\beta\omega}}{\omega}\right)\int_{-\infty}^{\infty} dt e^{i\omega t}\langle \mathbf{P}(t) \cdot \mathbf{P}(0)\rangle$$

where the current is proportional to the momentum $\mathbf{j}(t) = (e/m)\mathbf{P}(t)$. The lack of time variation in $\mathbf{P}$ means that the real part of the conductivity is identically zero for any finite frequency. Thus there is no light absorption for $\omega > 0$.

A simple way of deriving the force–force correlation function was given by Hopfield (1965). The argument proceeds in two steps. The first step is to replace $\mathbf{P}(t)$ by the identical operator $(i/\omega)(\partial\mathbf{P}/\partial t)$, which is justified on the grounds that in an ac field of frequency ω the time dependence of interest is just $\exp(-i\omega t)$. This replacement is done for both $\mathbf{P}$ operators, so we have the expression

$$\text{Re}[\sigma(\omega)] = -\frac{e^2}{6m^2\omega^3\nu}(1 - e^{-\beta\omega})\int_{-\infty}^{\infty} dt e^{i\omega t}\left\langle \frac{\partial \mathbf{P}(t)}{\partial t}\cdot\left[\frac{\partial}{\partial t'}\mathbf{P}(t')\right]_{t'=0}\right\rangle \tag{8.1.4}$$

The second step in the argument is to notice that the time derivative of the momentum is just the force $\mathbf{F}(t) = \partial\mathbf{P}/\partial t$, so we have the force–force correlation function

$$\text{Re}[\sigma(\omega)] = -\frac{e^2}{6m^2\omega^3\nu}(1 - e^{-\beta\omega})\int_{-\infty}^{\infty} dt e^{i\omega t}\langle \mathbf{F}(t) \cdot \mathbf{F}(0)\rangle \tag{8.1.5}$$

This derivation is essentially correct and will be justified later with more rigor. The force–force correlation function then serves as another possible starting point for the evaluation of the conductivity. For free-particle systems, it usually leads to a usable answer with much less work than the current–current correlation function, because the first term in the perturbation expansion of the force–force correlation function is usually the one we want. For example, if the force is due to phonons, then $F \propto M_\lambda(q)$, where M_λ is the electron–phonon matrix element. Then the leading term in the correlation function is already proportional to M_λ^2, and often we only want terms in the light absorption which are proportional to M_λ^2. Then we can evaluate the remaining parts of the correlation function assuming the particles are free, which is quite easy. Similarly, if the absorption is due to impurities, the leading term is proportional to concentration n_i of impurities, and that is often the term we want. Thus the advantage of the force–force correlation function is that the leading term is often adequate.

One calculation for which the force–force correlation function would be a poor starting point would be the dc conductivity. The correlation function has a prefactor of ω^{-3}, and obviously a great deal of cancellation in the correlation function has to take place in the limit $\omega \to 0$ in order to get $\sigma_0 = n_0 e^2 \tau_0 / m$.

A rigorous derivation of the force–force correlation function was given independently by Mahan (1970) and Hasegawa and Watabe (1969). The trick is to start from the Kubo formula and integrate by parts on the τ variable,

$$\pi(i\omega) = -\frac{1}{\nu} \int_0^\beta d\tau e^{i\omega\tau} \langle T_\tau j_\mu(\tau) j_\mu(0) \rangle$$

$$\int u\, dv = uv - \int v\, du$$

$$u = j(\tau)$$

$$v = \frac{e^{i\omega\tau}}{i\omega}$$

so that

$$\pi(i\omega) = -\frac{1}{i\omega\nu} [\langle j_\mu(\beta) j_\mu(0) - j_\mu(0) j_\mu(0) \rangle] + \frac{1}{i\omega\nu} \int_0^\beta d\tau e^{i\omega\tau} \left\langle T_\tau \left[\frac{\partial}{\partial\tau} j_\mu(\tau) \right] j_\mu(0) \right\rangle \tag{8.1.6}$$

The first term is the constant of integration. These constants have the general form of commutators:

$$\langle A(\beta)B(0) - A(0)B(0) \rangle = \mathrm{Tr}(Ae^{-\beta H}B - e^{-\beta H}AB) = \mathrm{Tr}[e^{-\beta H}(BA - AB)] = -\langle [A, B] \rangle$$

where we use the cyclic properties of the trace to rearrange the first term. The one in (8.1.6) is zero since it is the commutator of j_μ with itself.

The next step in the derivation is to take the integration by parts on the other current operator $j_\mu(0)$. Since the correlation function depends only on the difference of the two τ operators in its argument, this correlation function is also equal to $\langle T_\tau [(\partial/\partial\tau') j_\mu(\tau')]_{\tau'=0} j_\mu(-\tau) \rangle$. Thus the integration by parts can now be done on the term $j(-\tau)$, which gives

$$\pi(i\omega) = \frac{1}{(i\omega)^2\nu}\left\langle\left\{j_\mu(0), \left[\frac{\partial}{\partial\tau} j_\mu(\tau)\right]_0\right\}\right\rangle$$
$$+ \frac{1}{\nu(i\omega)^2}\int_0^\beta d\tau e^{i\omega\tau}\left\langle T_\tau\left[\frac{\partial}{\partial\tau} j_\mu(\tau)\right]\left[\frac{\partial}{\partial\tau'} j_\mu(\tau')\right]_{\tau'=0}\right\rangle \quad (8.1.7)$$

The first term contains the commutator of j_μ with $\partial j/\partial\tau$. This commutator yields a constant, which is real and so does not contribute to the real part of the conductivity or to the absorption of light. But Šimánek (1971) has shown that this term does contribute to the renormalization of the optical mass, so we shall retain it. The other term in (8.1.7) is the frequency-dependent part, which is indeed the correlation function of $\partial j/\partial\tau$ with itself. Of course the current is just proportional to the momentum $j = (e/m)P$, so that we have rederived (8.1.4).

The Hamiltonian is written as $H = H_0 + V$, where H_0 is the homogeneous electron gas part in (8.1.3), while V is the potential which causes other forces besides electron–electron interactions. For example, the potential could be the sum of the interactions with the crystalline potential V_G, where the $\mathbf{G}$ are reciprocal lattice vectors, with impurities of density operator $\varrho_i(\mathbf{q})$ (see Sec. 4.1E), or with phonons:

$$V = \sum_{\mathbf{q}} \varrho(\mathbf{q})\Phi(\mathbf{q})$$
$$\Phi(\mathbf{q}) = \delta_{\mathbf{q}=\mathbf{G}}V_G + \frac{V_i(q)}{\nu}\varrho_i(\mathbf{q}) + \sum_\lambda \frac{M_\lambda(q)}{\nu^{1/2}}(a_{\mathbf{q}\lambda} + a^\dagger_{-\mathbf{q}\lambda}) \quad (8.1.8)$$

The time derivative of the current operator is

$$\frac{\partial}{\partial\tau} j_\mu = [H, j_\mu] = [V, j_\mu] = \frac{-e}{m}\sum_{\mathbf{q}} q_\mu\varrho(\mathbf{q})\Phi(\mathbf{q})$$

since

$$[j_\mu, \varrho(\mathbf{q})] = \frac{e}{m}\sum_{\substack{\mathbf{k}\mathbf{k}'\\ \sigma\sigma'}} k_\mu[c^\dagger_{\mathbf{k}\sigma}c_{\mathbf{k}\sigma}, c^\dagger_{\mathbf{k}'+\mathbf{q}\sigma'}c_{\mathbf{k}'\sigma'}]$$
$$= \frac{e}{m}\sum_{\mathbf{k}\sigma} c^\dagger_{\mathbf{k}+\mathbf{q}\sigma}c_{\mathbf{k}\sigma}[(k_\mu + q_\mu) - k_\mu] = \frac{e}{m} q_\mu\varrho(\mathbf{q})$$

The last commutator is just Newton's law: The time derivative of the momentum is the force, which is the gradient of the potential $F_\mu(\mathbf{r}) = -\nabla_\mu V(\mathbf{r}) = -i\sum_{\mathbf{q}} V_q q_\mu e^{i\mathbf{q}\cdot\mathbf{r}}$. To evaluate the constant term in (8.1.7) we need the further commutator

$$\left[j_\mu, \frac{\partial j_\mu}{\partial \tau}\right] = -\frac{e}{m}\sum_{\mathbf{q}} q_\mu \Phi(\mathbf{q})[j_\mu, \varrho(\mathbf{q})]$$
$$= -\frac{e^2}{m^2}\sum_{\mathbf{q}} q_\mu{}^2 \varrho(\mathbf{q})\Phi(\mathbf{q})$$

When these results are used in (8.1.7), we have

$$\pi(i\omega) = \frac{e^2}{\nu m^2(i\omega)^2}\Big[-\sum_{\mathbf{q}} q_\mu{}^2\langle\varrho(\mathbf{q})\Phi(\mathbf{q})\rangle + \sum_{\mathbf{q}\mathbf{q}'} q_\mu q_\mu{}' \int_0^\beta d\tau e^{i\omega\tau}$$
$$\times \langle T_\tau \varrho(\mathbf{q}, \tau)\Phi(\mathbf{q}, \tau)\varrho(\mathbf{q}', 0)\Phi(\mathbf{q}', 0)\rangle\Big] \qquad (8.1.9)$$

The generalized potential Φ is given a τ dependence, which applies only to the phonon part of the three terms in (8.1.8). The second term is the force–force correlation function, which is equivalent to (8.1.5). It is often sufficient to evaluate this expression only to order Φ^2, in which case $\mathbf{q}' = -\mathbf{q}$. The second term is already evaluated to this order, so that we can evaluate the remaining correlation function ignoring the interaction V. The first term in (8.1.9) must also be evaluated to order Φ^2 by expanding the S matrix for the interaction potential V:

$$\langle\varrho(\mathbf{q})\Phi(\mathbf{q})\rangle = \mathrm{Tr}[e^{-\beta H_0}S(\beta)\varrho(\mathbf{q})\Phi(\mathbf{q})]$$
$$= \mathrm{Tr}[e^{-\beta H_0}\varrho(\mathbf{q})\Phi(\mathbf{q})] - \int_0^\beta d\tau\langle TV(\tau)\varrho(\mathbf{q})\Phi(\mathbf{q})\rangle + O(\Phi^3)$$

Only the term in V is retained in order to have a result proportional to Φ^2. The first term on the right, $\langle\varrho(\mathbf{q})\Phi(\mathbf{q})\rangle$, is zero since it equals $n_0\Phi(0)\nu$, which we take as the zero energy. Combining both terms in (8.1.9), we obtain

$$\pi(i\omega) = \frac{-e^2}{\nu m^2(i\omega)^2}\sum_{\mathbf{q}} q_\mu{}^2 \int_0^\beta d\tau(e^{i\omega\tau} - 1)\langle T_\tau\varrho(\mathbf{q}, \tau)\varrho(-\mathbf{q}, 0)\rangle$$
$$\times \langle T_\tau\Phi(\mathbf{q}, \tau)\Phi(-\mathbf{q}, 0)\rangle + O(\Phi^3) \qquad (8.1.10)$$

The term with $\exp(i\omega\tau)$ comes from the force–force correlation function, while the 1 term comes from the constant term.

This formula will be evaluated for several different types of potentials in the remaining parts of this section. However, the type of formula which results can be illustrated by a simple example. Let us take the case where the potential Φ is independent of τ. One example is the scattering from impurities, where, according to Sec. 4.1E, we have

$$\langle\Phi(\mathbf{q})\Phi(\mathbf{q}')\rangle = \frac{1}{\nu^2} V_i(\mathbf{q})V_i(\mathbf{q}')\langle\varrho_i(\mathbf{q})\varrho_i(\mathbf{q}')\rangle$$

$$= \frac{n_i}{\nu} V_i(q)^2\delta_{\mathbf{q}+\mathbf{q}'=0}$$

where $V_i(q)$ is the unscreened potential between the electron and the impurity. The remaining part of the τ integral in (8.1.10) is the form (5.4.8) for the fundamental definition of the longitudinal dielectric function:

$$-\frac{1}{\nu}\int_0^\beta d\tau e^{i\omega\tau}\langle T_\tau\varrho(\mathbf{q},\tau)\varrho(-\mathbf{q},0)\rangle = \frac{q^2}{4\pi e^2}\left[\frac{1}{\varepsilon(q,i\omega)} - 1\right]$$

$$\pi_i(i\omega) = \frac{n_i}{4\pi m^2(i\omega)^2\nu}\sum_{\mathbf{q}} q_\mu{}^2V_i{}^2(q)q^2\left[\frac{1}{\varepsilon(q,i\omega)} - \frac{1}{\varepsilon(q)}\right] + O(V_i{}^3) \tag{8.1.11}$$

The subscript i denotes the impurity contribution to $\pi(\omega)$. The current–current correlation function is expressed as the difference between $\varepsilon^{-1}(q, i\omega)$ and $\varepsilon^{-1}(q, 0)$. Since $\varepsilon(q, i\omega)$ is an even function of $i\omega$, in the limit where $i\omega \to 0$ it must act proportionally with $\varepsilon(q, i\omega) \to \varepsilon(q) + O(\omega^2)$, which shows that π_i does not diverge in the limit where $i\omega \to 0$. However, when we find the retarded function by $i\omega \to \omega + i\delta$, we have

$$\pi_i(\omega) = \frac{n_i}{4\pi m^2\omega^2\nu}\sum_{\mathbf{q}} q^2q_\mu{}^2V_i(q)^2\left[\frac{1}{\varepsilon(q,\omega)} - \frac{1}{\varepsilon(q)}\right]$$

$$\mathrm{Re}[\sigma_i(\omega)] = -\frac{n_i}{4\pi m^2\omega^3}\int\frac{d^3q}{(2\pi)^3}q^2q_\mu{}^2V_i(q)^2\,\mathrm{Im}\left[\frac{1}{\varepsilon(q,\omega)}\right] + O(V_i{}^3) \tag{8.1.12}$$

Thus we derive the Hopfield (1965) formula that the real part of the transverse conductivity is proportional to the imaginary part of the inverse longitudinal dielectric function. The latter quantity we interpret as the rate of making excitations in the electron gas. The light absorption occurs because electronic excitations are created, while the creation rate must depend on the concentration of impurities n_i in the system. This interpretation agrees with our original assertion that the electron gas can absorb light only when the impurities, or other inhomogeneities, are present to dissipate the momentum. An obvious advantage of the Hopfield formula is that the complicated aspects of electron–electron interactions are naturally included in the dielectric function, for which we can use one of the model forms described in Chapter 5.

C. Fröhlich Polarons

The optical absorption of free polarons was first calculated by Gurevich *et al.* (1962), who started from the usual form of the Kubo formula. In the Fröhlich Hamiltonian (6.1.1),

$$H = \sum_{\mathbf{p}} \varepsilon_{\mathbf{p}} c_{\mathbf{p}}^{\dagger} c_{\mathbf{p}} + \sum_{\mathbf{q}} \omega_0 a_{\mathbf{p}}^{\dagger} a_{\mathbf{q}} + \sum_{\mathbf{qp}} \frac{M_0}{\nu^{1/2}} \frac{1}{|\mathbf{q}|} c_{\mathbf{p}+\mathbf{q}}^{\dagger} c_{\mathbf{p}} (a_{\mathbf{q}} + a_{-\mathbf{q}}^{\dagger})$$

$$M_0^2 = \frac{4\pi\alpha\omega_0^{3/2}}{(2m)^{1/2}}$$

the electron–phonon interaction V is expanded using the usual S-matrix techniques. The real part of the conductivity is calculated to order α, which is the first nonvanishing term. There are three diagrams which enter proportional to α; they are shown in Fig. 8.1. The first two are self-energy diagrams, while the third is a vertex contribution. The final result is obtained by adding contributions from all three diagrams. At zero temperature it is

$$\mathrm{Re}[\sigma(\omega)] = \frac{2}{3} \frac{e^2 n_0 \alpha \omega_0^{3/2}}{m\omega^3} (\omega - \omega_0)^{1/2} \theta(\omega - \omega_0) + O(\alpha^2) \quad (8.1.13)$$

The free polaron can absorb light only when the frequency $\omega > \omega_0$. We know that polarons can emit LO phonons only when the polaron energy exceeds the phonon energy ω_0. The light wave excites the polaron to this energy, and then the polaron can emit the phonon. Thus the light absorption happens by the emission of LO phonons, where the electron serves as the intermediary in the process. Of course, there is also a direct coupling between the photon and the TO phonons, which was discussed in Sec. 4.5 on polaritons.

Formula (8.1.13) can be derived from the correlation functions in (8.1.10). The part involving the potential is now just the Green's function for LO phonons:

$$\Phi(\mathbf{q}) = \frac{M_0}{\nu^{1/2}} \frac{1}{|q|} (a_{\mathbf{q}} + a_{-\mathbf{q}}^{\dagger})$$

$$\langle T_\tau \Phi(\mathbf{q}, \tau) \Phi(-\mathbf{q}, 0) \rangle = -\frac{M_0^2}{\nu q^2} \mathscr{D}^{(0)}(\tau)$$

If we denote the density–density correlation function by the symbol $\chi(\mathbf{q}, \tau)$,

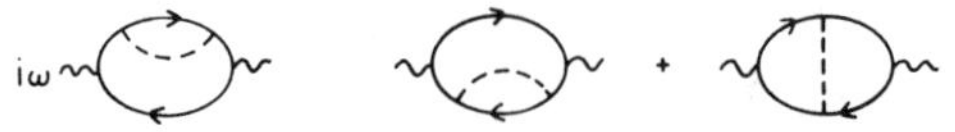

FIGURE 8.1

then we can formally evaluate the τ integral in (8.1.10) by changing it into a summation over Matsubara frequencies, which is evaluated by the usual contour integral:

$$\chi(\mathbf{q}, \tau) = -\langle T_\tau \varrho(\mathbf{q}, \tau)\varrho(-\mathbf{q}, 0)\rangle$$

$$\begin{aligned}
\pi(i\omega) &= -\frac{M_0^2 e^2}{m^2(i\omega)^2 \nu} \sum_{\mathbf{q}} \frac{q_\mu^2}{q^2} \int_0^\beta d\tau (e^{i\omega\tau} - 1)\chi(\mathbf{q}, \tau)\mathscr{D}^{(0)}(\tau) + O(M_0^4) \\
&= -\frac{M_0^2 e^2}{m^2(i\omega)^2 \nu} \sum_{\mathbf{q}} \frac{q_\mu^2}{q^2} \frac{1}{\beta} \sum_{iq} \mathscr{D}^{(0)}(iq)[\chi(\mathbf{q}, i\omega + iq) - \chi(\mathbf{q}, iq)] \\
&= -\frac{M_0^2 e^2}{m^2(i\omega)^2 \nu} \sum_{\mathbf{q}} \frac{q_\mu^2}{q^2} \int_{-\infty}^{\infty} \frac{d\varepsilon}{2\pi} n_B(\varepsilon)\{2\,\mathrm{Im}[D^{(0)}(\varepsilon)]\chi(\mathbf{q}, i\omega + \varepsilon) \\
&\quad + 2\,\mathrm{Im}[\chi(\mathbf{q}, \varepsilon)]\mathscr{D}^{(0)}(\varepsilon - i\omega) - 2\,\mathrm{Im}[D^{(0)}(\varepsilon)\chi(\mathbf{q}, \varepsilon)]\}
\end{aligned} \tag{8.1.14}$$

By taking the retarded function $i\omega \to \omega + i\delta$, the real part of the conductivity is the imaginary part of the preceding expression, where the last term does not contribute:

$$\begin{aligned}
\mathrm{Re}[\sigma(\omega)] = \frac{M_0^2 e^2}{m^2 \omega^3} \int \frac{d^3q}{(2\pi)^3} \left(\frac{q_\mu^2}{q^2}\right) \int_{-\infty}^{\infty} \frac{d\varepsilon}{2\pi} n_B(\varepsilon)\{2\,\mathrm{Im}[D^{(0)}(\varepsilon)] \\
\times \mathrm{Im}[\chi(\mathbf{q}, \varepsilon + \omega)] - 2\,\mathrm{Im}[D^{(0)}(\varepsilon - \omega)]\,\mathrm{Im}[\chi(\mathbf{q}, \varepsilon)]\}
\end{aligned}$$

This result is general and applies to all electron–phonon systems. Sometimes it is useful to change variables $\varepsilon \to \varepsilon + \omega$ in the second term, which gives the equivalent expression

$$\begin{aligned}
\mathrm{Re}[\sigma(\omega)] = \frac{M_0^2 e^2}{m^2 \omega^3} \int \frac{d^3q}{(2\pi)^3} \left(\frac{q_\mu^2}{q^2}\right) \int_{-\infty}^{\infty} \frac{d\varepsilon}{2\pi} [n_B(\varepsilon) - n_B(\varepsilon + \omega)] \\
\times 2\,\mathrm{Im}[D^{(0)}(\varepsilon)]\,\mathrm{Im}[\chi(\mathbf{q}, \varepsilon + \omega)]
\end{aligned} \tag{8.1.15}$$

The spectral function for the phonon Green's function is easy to evaluate:

$$-2\,\mathrm{Im}[D^{(0)}(\varepsilon)] = 2\pi[\delta(\varepsilon - \omega_0) - \delta(\varepsilon + \omega_0)] \tag{8.1.16}$$

The last step in the formal derivation is to determine $\chi(\mathbf{q}, \omega)$. The free polarons are assumed to have small concentration n_0, so that electron–electron interactions can be neglected. Then the density–density correlation function is adequately approximated by the simple bubble diagram:

$$\chi(\mathbf{q}, i\omega) = P^{(1)}(\mathbf{q}, i\omega) = \frac{2}{\nu}\sum_{\mathbf{p}}\frac{1}{\beta}\sum_{ip}\mathscr{G}^{(0)}(p)\mathscr{G}^{(0)}(p+q)$$

$$P^{(1)}(\mathbf{q}, \omega) = 2\int\frac{d^3p}{(2\pi)^3}n_p\left(\frac{1}{\varepsilon_{\mathbf{p}} - \varepsilon_{\mathbf{p+q}} + \omega + i\delta} + \frac{1}{\varepsilon_{\mathbf{p}} - \varepsilon_{\mathbf{p+q}} - \omega - i\delta}\right) \tag{8.1.17}$$

$$-2\,\mathrm{Im}[P^{(1)}(\mathbf{q}, \omega)] = 2\pi\int\frac{d^3p}{(2\pi)^3}n_p[\delta(\varepsilon_{\mathbf{p}} - \varepsilon_{\mathbf{p+q}} + \omega) - \delta(\varepsilon_{\mathbf{p}} - \varepsilon_{\mathbf{p+q}} - \omega)]$$

The factor of 2 in front is for the spin degeneracy. The correction terms to this simple bubble term, from the electron–phonon interaction, will cause higher-order terms in α.

The polarons are assumed to obey Maxwell–Boltzmann statistics, so that the particle density n_p per spin state is

$$n_p = \frac{1}{2}n_0\left(\frac{2\pi\beta}{m}\right)^{3/2}e^{-\beta(\varepsilon_p - E_0)} \tag{8.1.18}$$

The case we currently need is for zero temperature. Here the particles all approach the state with $p \to 0$, so from (8.1.17) one can see that the spectral function is

$$-2\,\mathrm{Im}(\chi) = -2\,\mathrm{Im}[P^{(1)}(q, \omega)] = 2\pi n_0[\delta(\omega - \varepsilon_q) - \delta(\omega + \varepsilon_q)] \tag{8.1.19}$$

We are now in a position to evaluate the real part of the conductivity given in (8.1.15). Both spectral functions are given by delta functions, for phonons in (8.1.16) and for electrons in (8.1.19). One set of delta functions are used to eliminate the $d\varepsilon$ integral, which gives

$$\begin{aligned}\mathrm{Re}[\sigma(\omega)] = \frac{M_0{}^2e^2n_0}{m^2\omega^3}\int\frac{d^3q}{(2\pi)^3}\frac{q_\mu{}^2}{q^2}\pi\{&[n_B(\omega_0) - n_B(\omega_0 + \omega)]\\ &\times[\delta(\omega_0 + \omega - \varepsilon_{\mathbf{q}}) - \delta(\omega_0 + \omega + \varepsilon_{\mathbf{q}})]\\ &+[1 + n_B(\omega_0) + n_B(\omega - \omega_0)][\delta(\omega - \omega_0 - \varepsilon_{\mathbf{q}}) - \delta(\omega - \omega_0 + \varepsilon_{\mathbf{q}})]\}\end{aligned}$$

At zero temperature, the boson occupation factor is $n_B(\omega) = -\theta(-\omega)$, so this expression simplifies to ($\omega > 0$, $\cos^2\theta = q_\mu{}^2/q^2$)

$$\mathrm{Re}[\sigma(\omega)] = \frac{M_0{}^2e^2\pi n_0}{m^2\omega^3}\int\frac{d^3q}{(2\pi)^3}\cos^2\theta\,\delta(\omega - \omega_0 - \varepsilon_{\mathbf{q}})$$

There remains only the integral over wave vector. The angular parts give $4\pi/3$, while the dq part eliminates the remaining delta function,

$$\int q^2\, dq\delta(\omega - \omega_0 - \varepsilon_q) = m[2m(\omega - \omega_0)]^{1/2}\theta(\omega - \omega_0)$$

so that

$$\text{Re}[\sigma(\omega)] = \frac{M_0^2 e^2 \pi n_0}{m^2\omega^3 8\pi^3}\left(\frac{4\pi}{3}\right)m[(\omega - \omega_0)2m]^{1/2}\theta(\omega - \omega_0)$$

By using the form for the electron–phonon matrix element M_0^2,

$$\text{Re}[\sigma(\omega)] = \frac{2n_0}{3}\,\frac{e^2\alpha\omega_0^{3/2}}{m\omega^3}\,(\omega - \omega_0)^{1/2}\theta(\omega - \omega_0)$$

The final result is (8.1.13), as derived by Gurevich *et al.* Our derivation has been remarkably short. The general form of the answer in (8.1.15), as the product of two spectral functions, is quite familiar from other calculations, and in this case both spectral functions are delta functions.

So far we have not introduced a diagrammatic representation of these force–force correlation functions. Diagrams are not necessary for evaluating the lowest-order term. But the necessity will increase when higher-order terms are discussed, since they are best represented by diagrams. The diagrams have the form of losed loops, since there are two propagator systems which go between the beginning and end time points. For example, in the Fröhlich polaron case we have the two propagators in (8.1.14) of the phonons $\mathscr{D}^{(0)}(\tau)$ and the electron loop $\chi(\mathbf{q}, \tau)$. They are shown in Fig. 8.2(a), where the vertices are labeled by a wiggly line which is the photon. Such diagrams, without the special vertex, are found in the expansion for the ground state energy of electron–phonon systems (see Fig. 3.8).

Let us consider the terms which would contribute to the absorption to order α^2. The terms in Fig. 8.2(b) do not contribute, since terms with more than one bubble are proportional to $(n_0)^2$. These we neglect since they do not come from the absorption per polaron but from interactions between polarons. The only terms which enter to order $\alpha^2 n_0$ have just a single bubble and include the three terms in Fig. 8.2(c). All three are just modifications of the polarization bubble, so the sum of all such terms must be the shaded bubble. Thus one term in the answer is

$$\pi(i\omega) = -\frac{M_0^2 e^2}{m^2(i\omega)^2\nu}\sum_{\mathbf{q}}\frac{q_\mu^2}{q^2}\int_0^\beta d\tau(e^{i\omega\tau} - 1)P(\mathbf{q}, \tau)\mathscr{D}^{(0)}(\mathbf{q}, \tau)$$

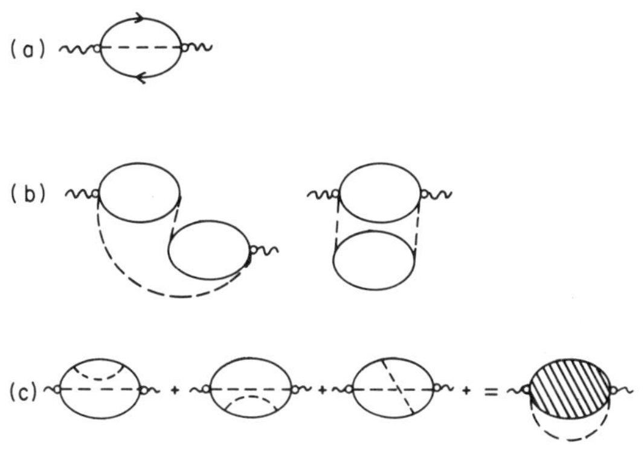

FIGURE 8.2

where $P(\mathbf{q}, \tau)$ is now the most general polarization bubble. Other terms are shown in Fig. 8.6 and in (8.1.44).

The optical absorption of Fröhlich polarons has been calculated extensively by Devreese (1972). He calculated the terms to order α^2, which are the two-phonon terms. Now the polaron can be excited by the light and make two phonons as well as one. Results are shown in Fig. 8.3 for the value of the coupling constant $\alpha = 1$. The solid line is the result using

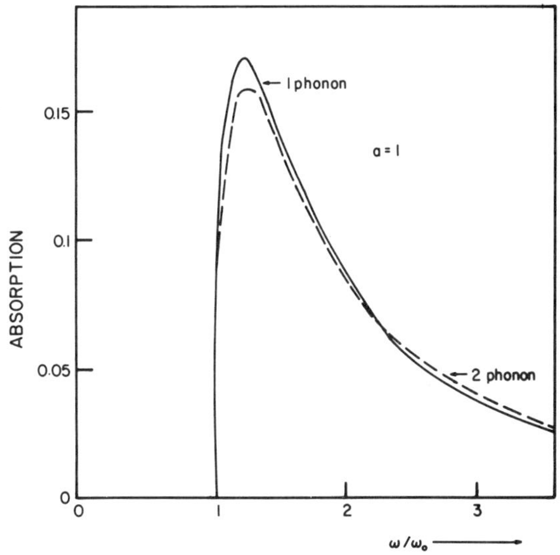

FIGURE 8.3. Theoretical prediction for the optical absorption of free polarons with $\alpha = 1$ and $T = 0$ K. Solid line is one-phonon theory, and dashed line has two-phonon correlations. *Source*: Devreese (1972), p. 93.

(8.1.13) with just the one-phonon processes included, while the dashed line is the result including both one- and two-phonon processes. The rather surprising result emerges that there is no sudden rise in the absorption at the two-phonon threshold $\omega = 2\omega_0$. The sharp rise at $\omega = \omega_0$ is not repeated as a sharp rise at $2\omega_0$. Instead, there is a very gradual rise starting at $2\omega_0$ and a slight decrease in the peak height. Thus the two-phonon events make only a small change in the result. The polaron would rather emit two phonons as two separate events of one-phonon emission and not as a correlated event where two phonons are emitted. This conclusion is in accord with our remarks in Sec. 6.1 about the correlation between two phonons being small in the polaron cloud. Devreese (1972) has also calculated the optical absorption of polarons by the Feynman method of path integrals as well as by strong coupling theory. At larger values of polaron constant α he finds sharp absorption lines, which are interpreted as arising from internal states of the localized electron in the small polaron.

D. Interband Transitions

Another application of the force–force correlation function is to calculate the rate of light absorption by interband transitions in a metal. Here the potential which provides the inhomogeneity is the crystalline potential of the atoms. This potential is periodic in ideal solids, so it contributes momentum in discrete units called reciprocal lattice vectors $\mathbf{G}$. This contribution was mentioned earlier as the first term in (8.1.8):

$$\Phi(\mathbf{q}) = \sum_{\mathbf{G}} V_G \delta_{\mathbf{G},\mathbf{q}}$$

This potential is fixed and has no correlation with itself: $\langle \Phi(\mathbf{q})\Phi(\mathbf{q}')\rangle = \langle \Phi(\mathbf{q})\rangle\langle \Phi(\mathbf{q}')\rangle$. However, the other correlation function in (8.1.10) is $\langle \varrho(\mathbf{q}, \tau)\varrho(\mathbf{q}', 0)\rangle$ and is finite only when $\mathbf{q}' = -\mathbf{q}$, so we need to evaluate

$$\langle \Phi(\mathbf{q})\Phi(-\mathbf{q})\rangle = \sum_{\mathbf{G}} V_G V_{-G} \delta_{\mathbf{q}=\mathbf{G}} = \sum_{\mathbf{G}} V_G^2 \delta_{\mathbf{q}=\mathbf{G}}$$

This potential is static, so the correlation function of the potential is a constant. This case is similar to that of impurity scattering, which was given in (8.1.12). The impurity result can be used for the crystalline potential by changing $n_i V_i(q)^2$ to $V_G^2 \delta_{\mathbf{G}=\mathbf{q}}$, so that the interband absorption has the form

$$\mathrm{Re}[\sigma_I(\omega)] = \frac{-1}{4\pi m^2 \omega^3} \sum_{\mathbf{G}} V_G^2 G^2 G_\mu^2 \, \mathrm{Im}\left[\frac{1}{\varepsilon(G, \omega)}\right] + O(V_G^3)$$

where the subscript I stands for interband contribution. Note the following:

$$-\mathrm{Im}\left(\frac{1}{\varepsilon}\right) = \frac{\varepsilon_2}{|\varepsilon|^2}$$

where $1/|\varepsilon|^2$ is usually grouped with V_G^2 to have a screened electron–ion interaction $|V_G/\varepsilon(G,\omega)|^2$. If the RPA result (5.5.6) is used for $\varepsilon_2(G,\omega)$, then for the case where $G > 2k_F$,

$$\varepsilon_2(G,\omega) = \frac{e^2 m}{G^3}\left[k_F^2 - \left(\frac{m}{G}\right)^2(\omega - \varepsilon_G)^2\right]$$
$$\mathrm{Re}[\sigma_I(\omega)] = \frac{e^2}{4\pi m\omega^3}\sum_G \frac{G_\mu^2}{G}\left|\frac{V_G}{\varepsilon(G,\omega)}\right|^2\left[k_F^2 - \left(\frac{m}{G}\right)^2(\omega-\varepsilon_G)^2\right] + O(V_G^3) \tag{8.1.20}$$

This formula is nearly identical to a standard formula for describing interband transitions in metals—the Wilson–Butcher formula.

The Wilson–Butcher formula is used to describe the interband transition in the alkali metals. Figure 8.4(a) shows the band structure in an alkali metal in the $\mathbf{k} = (110)$ direction. The Fermi level is shown as the dashed horizontal line. The interband transitions are shown, where the transition appears vertical in wave vector space since the wave vector of light is too small to show up as a horizontal deflection of the arrow. This transition is shown in the reduced zone scheme. Of course, in the extended zone description of Fig. 8.4(b), the optical transition really changes the wave vector of the electron by a reciprocal lattice vector $\mathbf{G}$. This change in the momentum of the electron, during the optical transition, is precisely what it needs to have the transition. The optical absorption occurs because the electron gains energy from the light wave and momentum from the rigid crystal lattice.

The optical absorption one expects from an alkali metal is shown in Fig. 8.5. At low frequency there is a Drude contribution, which

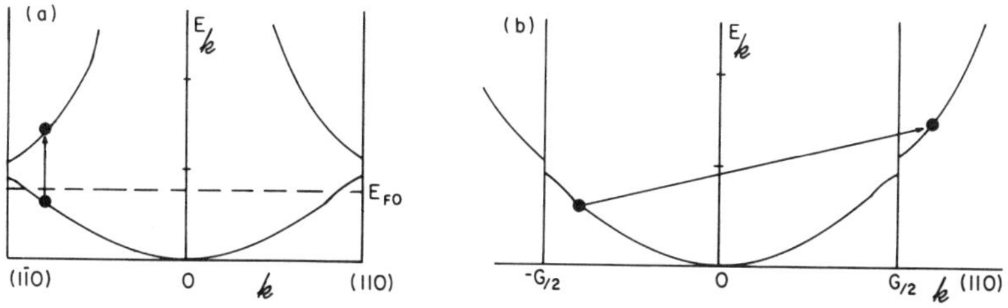

FIGURE 8.4. Interband transitions in alkali metals as shown in (a) reduced zone and (b) extended zone representations.

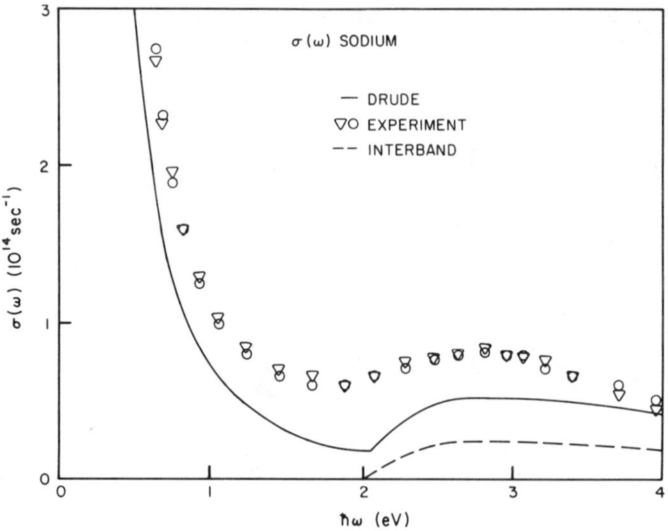

FIGURE 8.5. The optical conductivity for metallic sodium. The points are the experiments of Smith, while the lines are the theoretical contribution from Drude and interband transitions. *Source*: Smith (1969).

comes from phonons. It suffices for the moment to say that this contribution is well described by the Drude formula (8.1.2), where τ_0 is largely provided by the intrinsic scattering from phonons. The interband transition of Fig. 8.5 starts about $\omega \cong 2$ eV, with a linear rise at threshold. Since the Fermi surface does not touch the zone face, $k_F < \frac{1}{2}G$, and (8.1.20) is the appropriate form to use for $\varepsilon_2(G, \omega)$. Figure 8.5 shows the experimental result of Smith (1969) for sodium, which has exactly this behavior. Similar interband terms were found by Smith (1970) for K, Rb, and Cs and in Li by Myers and Sixtensson (1976). The agreement between theory and experiment appears satisfactory for sodium. There is no reason to expect the Wilson–Butcher theory to give perfect agreement, since there are additional terms which are higher powers in $(V_G)^3$. The expansion of the S matrix in (8.1.9) for $V = \sum_{\mathbf{G}} V_G \varrho(\mathbf{G})$ will produce a series of terms with higher powers of the crystalline potential. The next term in (8.1.20) has the cubic power of the crystalline potential and has the form $V_{\mathbf{G}} V_{\mathbf{G}'} V_{-\mathbf{G}-\mathbf{G}'}$. An estimate of this term shows that it is approximately 10% of the $V_G{}^2$ contribution, which is barely negligible. Similarly, most calculations of the Wilson–Butcher formula use the RPA form for $\varepsilon(G, \omega)$, whereas the Singwi–Sjölander form would be an improvement. In this respect it is interesting that their dielectric function has the feature that the imaginary part of its inverse is still proportional to the RPA of $\varepsilon_2(q, \omega) = v_q \operatorname{Im}[P^{(1)}(q, \omega)]$:

$$\varepsilon_{ss}(q, \omega) = 1 - \frac{v_q P^{(1)}}{1 + v_q G P^{(1)}}$$

$$-\operatorname{Im}\left(\frac{1}{\varepsilon_{ss}}\right) = \frac{\varepsilon_2}{|\varepsilon_{ss}|^2}$$

Thus the only factor which changes in (8.1.20) is the screening of the potential. There is also a slight uncertainly as to the precise value of the pseudopotential V_{110}. With these qualifications in mind, the agreement between theory and experiment in Fig. 8.5 is satisfactory.

There is one difference between the original formula of Wilson–Butcher and the result (8.1.20) derived from the force–force correlation function. The original formula had the electron–ion potential divided by the static dielectric function $\varepsilon(G, 0)$, so that the factor $|V_G/\varepsilon(G, 0)|^2$ occurs. The present derivation by the force–force correlation function is exact to order V_G^2, so its factor of $\varepsilon(G, \omega)$ is correct, and the old formula is not. The difference is more philosophical than numerical for sodium, since the frequency dependence of $\varepsilon(G, \omega)$ is slight over the range of frequencies shown in Fig. 8.5. Actually these frequencies are not much lower than the plasma frequency at $\omega_p \simeq 5$ eV, but this fact is irrelevant, since at these large values of wave vector **G** the dielectric function becomes frequency dependent only at much larger values of ω.

The alkali metals are good examples of where the Wilson–Butcher formula works well. Here the lowest interband transition is in the visible region of the spectrum and is isolated in frequency from any other feature except the Drude contribution, which is always present. Ashcroft and Sturm (1971) have shown that another formula is needed for polyvalent metals such as aluminum, since there are transitions between parallel bands which cause sharp structure at lower photon frequencies.

E. Phonons

Equation (8.1.8) lists three contributions to $\Phi(\mathbf{q})$. The first two terms are Bragg scattering and impurity scattering, which have been discussed. Here the third term is treated. It is due to the electron scattering by the phonons.

Equation (8.1.10) is evaluated, with $\Phi(\mathbf{q})$ being the electron–phonon interaction. Only terms are retained that are of $O(M_\lambda^2)$. This point is rather tricky, since a subset of higher-order contributions are also retained. It is assumed that the phonon modes are the actual modes in the solid. The calculation of these modes includes electron–electron and electron–phonon interactions, which means they include M_λ. Any time the actual phonon modes are used, then one is including a subset of higher-order diagrams.

In Eq. (8.1.10) the phonon part is

$$-\langle T_\tau \Phi(\mathbf{q}, \tau)\Phi(-\mathbf{q}, 0)\rangle = \frac{1}{\nu} \sum_\lambda M_\lambda^2(\mathbf{q}) \mathscr{D}_\lambda(\mathbf{q}, \tau)$$

Equation (8.1.10) is now evaluated. Only retain the term that depends upon frequency $i\omega$:

$$\pi(i\omega) = \frac{-e^2}{\nu m^2(i\omega)^2} \sum_{\mathbf{q}\lambda} \frac{q_\mu^2 M_\lambda^2(\mathbf{q})}{v_q} \frac{1}{\beta} \sum_{i\omega'} \mathscr{D}_\lambda(\mathbf{q}, i\omega')\left[\frac{1}{\varepsilon(\mathbf{q}, i\omega - i\omega')} - 1\right]$$

The summation over Matsubara frequencies is evaluated using the techniques of Sec. 3.5. The phonon Green's function is expressed in the Lehmann representation (3.3.14) using its spectral function $B(\mathbf{q}, \omega)$. The inverse dielectric function is expressed similarly, with its spectral function being $\mathrm{Im}(1/\varepsilon)$:

$$\pi(i\omega) = \frac{-1}{4\pi\nu m^2\omega^2} \sum_{\mathbf{q}\lambda} d_\mu^2 q^2 M_\lambda^2(\mathbf{q}) \int \frac{d\omega'}{2\pi} B(\mathbf{q}, \omega') \int \frac{d\varepsilon'}{\pi} \mathrm{Im}\left[\frac{1}{\varepsilon(\mathbf{q}, \varepsilon')}\right]$$
$$\times \frac{n_B(\omega') - n_B(\varepsilon')}{i\omega + \omega' - \varepsilon'}$$

The retarded function is obtained by taking $i\omega \to \omega + i\delta$. Then take the imaginary part. These steps cause the denominator $(i\omega + \omega' - \varepsilon')$ to be replaced by the numerator $\pi\delta(\omega + \omega' - \varepsilon')$. The delta function is used to eliminate the integral over $d\varepsilon'$.

This expression is usually evaluated at relatively low frequencies. When ω is much less than the plasma frequency of the metal, we can use the low-frequency limit for $\varepsilon_2(\mathbf{q}, \omega) = 2\omega m^2 e^2 q^{-3}\Theta(2k_F - q)$. The various terms involving the phonon wave vector, matrix element, and spectral function are just the definition of the transport form of $\alpha_{\mathrm{tr}}^2(u)F(u)$:

$$\alpha_{\mathrm{tr}}^2(u)F(u) = \frac{m}{4\pi^2 n_0} \int \frac{d^3q}{(2\pi)^3} \frac{q_\mu^2}{q} \sum_\lambda \frac{M_\lambda^2}{\varepsilon(\mathbf{q})^2} B_\lambda(\mathbf{q}, u)\Theta(2k_F - q)$$

Again this is treated as a single function, although the notation suggests that it is a product of two functions. The integrand in the wave vector integration has the additional factor of $3q_\mu^2/2k_F^2$ compared to the functions

$\alpha^2 F$. The remaining factors combine to give

$$\mathrm{Re}[\sigma(\omega)] = \frac{-\operatorname{Im}\pi(\omega)}{\omega} = \frac{2\pi e^2 n_0}{m\omega^3} \int_{-\infty}^{\infty} du\, \alpha_{\mathrm{tr}}^2 F(u)(\omega - u)[n_B(u) - n_B(u - \omega)]$$

At zero temperature, the Boson functions n_B arrange to limit the range of integration to $0 < u < \omega$. In this limit we obtain the simple formula first derived by Allen (1971):

$$\mathrm{Re}[\sigma(\omega)] = \frac{2\pi e^2 n_0}{m\omega^3} \int_0^{\omega} du(\omega - u)\, \alpha_{\mathrm{tr}}^2 F(u) \tag{8.1.30}$$

The Allen formula is approximate, since it is based upon the force–force correlation function. A more accurate expression would be based upon solving the Kubo relation for the conductivity as in Sec. 7.3. This derivation gives the result

$$\mathrm{Re}[\sigma(\omega)] = \frac{-e^2 n_0}{m\omega} \int_{-\infty}^{\infty} d\varepsilon [n_F(\varepsilon + \omega) - n_F(\varepsilon)] \times \mathrm{Re}[\gamma(\varepsilon - i\delta, \varepsilon + \omega + i\delta) I(\varepsilon, \varepsilon + \omega)] \tag{8.1.31}$$

This result can be derived directly from (7.3.3) and (7.3.5). The vertex function $\gamma(\varepsilon - i\delta, \varepsilon + \omega + i\delta)$ obeys the integral equation (7.3.11). A com-

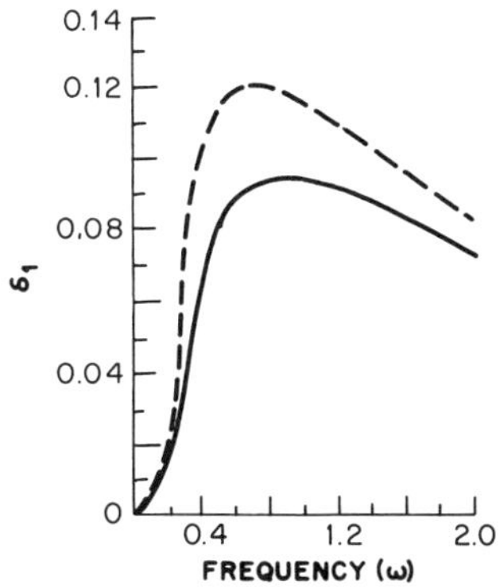

FIGURE 8.6. Phonon-induced absorption in calcium as a function of frequency. The vertical axis is $\mathrm{Re}[\sigma(\omega)m\omega/e^2 n_0]$. The horizontal axis is ω/ω_D, where the Debye frequency is ω_D. The solid line is the solution to the ac-transport equation, while the dashed line is using the Allen formula from the force–force correlation function. The latter method is approximate. *Source*: Wu and Mahan (1984).

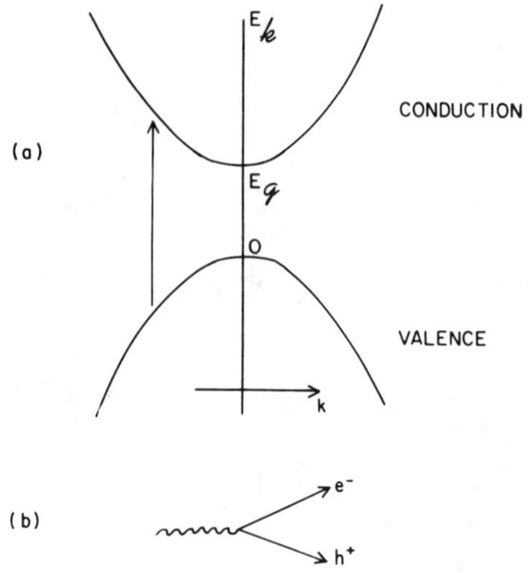

FIGURE 8.7. Optical transition in a semiconductor between the occupied valence band and empty conduction band for a direct transition. (a) Conventional band picture; (b) Wannier picture where the photon makes an electron hole.

parison of these two expressions is given in Fig. 8.6 for metallic calcium with $E \parallel c$. What is plotted is $\mathrm{Re}(\sigma/\sigma_0)$ vs. ω/ω_D, where $\sigma_0 = n_0 e^2/m\omega$, while ω_D is the Debye frequency. The solid line is the exact result from (8.1.31) while the dashed line is the Allen formula in (8.1.30). The Allen formula has an error of about 10%–15%. This makes it a good approximation, since (8.1.30) is very much easier to evaluate than (8.1.31). The latter requires solving numerically an integral equation for the vertex function.

8.2. WANNIER EXCITONS

A. The Model

Exciton states play an extremely important role in the understanding of interband transitions in semiconductors. The word *exciton* is used here to signify the modification of the absorption rate of photons due to the Coulomb interaction between the electron and the valence band hole. The easiest case to understand is when the lowest interband transition is *direct*: which means that the conduction band minimum and valence band maximum are at the same point in k space. This configuration is shown schematically in Fig. 8.7(a). The valence band states are all filled, and the conduction band states are all empty. The vertical arrow shows a possible interband transition which can occur when a photon is absorbed in the solid. The arrow starts in one of the valence states, since these are occupied by electrons, and goes to an unoccupied state in the conduction band. The valence state is shown as nondegenerate (except for spin) at $k \to 0$, although this is seldom the case, and usually the band has an orbital degeneracy and is anisotropic.

The point of view in Fig. 8.7(a) is a single-particle picture of the transition process. According to this picture, the transition rate for the absorption of photons is given by the golden rule:

$$A(\omega) = \frac{2\pi}{\hbar\nu} \sum |\langle c, \mathbf{k}' \mid \hat{\varepsilon} \cdot \mathbf{p} \mid v, \mathbf{k}\rangle|^2 \delta[\varepsilon_v(\mathbf{k}') + \hbar\omega - \varepsilon_c(\mathbf{k})] \tag{8.2.1}$$

$$\varepsilon_v(k) = -\frac{k^2}{2m_v}$$

$$\varepsilon_c(k) = E_g + \frac{k^2}{2m_c}$$

We choose the energy zero to be the top of the valence band, so the bottom of the conduction band starts at the energy gap E_g. The matrix element $\langle c, \mathbf{k}' \mid \hat{\varepsilon} \cdot \mathbf{p} \mid v, \mathbf{k}\rangle$ is between the one-electron initial and final states. The wave functions are taken to be Bloch functions, which are the product of a cell-periodic part $u_{n\mathbf{k}}(\mathbf{r})$ and an envelope function $\exp(i\mathbf{k} \cdot \mathbf{r})$. It is a reasonable approximation, for our simple level of discussion, to take the cell-periodic parts as independent of wave vector $u_{c\mathbf{k}}(\mathbf{r}) = u_c(\mathbf{r})$, $u_{v\mathbf{k}}(r) = u_v(\mathbf{r})$. We shall also assume that the valence band has p symmetry and that the conduction band has s symmetry, which is true in many semiconductors—but not all. Thus $u_v(\mathbf{r})$ is a periodic orbital with angular momentum $l = 1$, while $u_c(\mathbf{r})$ is a periodic orbital with $l = 0$. With these approximations the optical matrix element is a constant except for wave vector conservation:

$$\begin{aligned} |v, \mathbf{k}\rangle &= u_v(\mathbf{r}) \frac{e^{i\mathbf{k}\cdot\mathbf{r}}}{\nu^{1/2}} \\ |c, \mathbf{k}'\rangle &= u_c(\mathbf{r}) \frac{e^{i\mathbf{k}\cdot\mathbf{r}}}{\nu^{1/2}} \\ \langle c, \mathbf{k}' \mid \hat{\varepsilon} \cdot \mathbf{p} \mid v, \mathbf{k}\rangle &= \langle c \mid \hat{\varepsilon} \cdot \mathbf{p} \mid v\rangle \delta_{\mathbf{k}\mathbf{k}'} = \hat{\varepsilon} \cdot \mathbf{p}_{cv} \delta_{\mathbf{k}\mathbf{k}'} \\ \langle c \mid \hat{\varepsilon} \cdot \mathbf{p} \mid v\rangle &= \frac{1}{\nu_0} \int_{\text{cell}} d^3r u_c^* \hat{\varepsilon} \cdot \mathbf{p} u_v \end{aligned} \tag{8.2.2}$$

where ν_0 is the volume of a unit cell. The matrix element $\langle c \mid \mathbf{p} \mid v\rangle$ should also contain the factor $\exp(i\mathbf{q} \cdot \mathbf{r})$, where $\mathbf{q}$ is the wave vector of the photon, but this is small enough to be neglected for photons in the optical frequencies. Now that the matrix element is evaluated, it is possible to do the integrals over the electron wave vector:

$$\begin{aligned} A(\omega) &= \frac{2\pi}{\hbar} \,|\, \hat{\varepsilon} \cdot \mathbf{p}_{cv} \,|^2 \int \frac{d^3k}{(2\pi)^3}\, \delta\!\left[\omega - E_g - \frac{k^2}{2\mu}\right] \\ &= |\, \hat{\varepsilon} \cdot \mathbf{p}_{cv} \,|^2 \frac{(2\mu)^{3/2}}{2\pi} (\omega - E_g)^{1/2} \theta(\omega - E_g) \\ \frac{1}{\mu} &= \frac{1}{m_c} + \frac{1}{m_v} \end{aligned} \tag{8.2.3}$$

Equation (8.2.3) predicts that the absorption rate begins at the energy gap of the semiconductor and rises as the square root of the factor $\omega - E_g$. It is a very definite prediction, which is *not* observed. In fact, the one-particle theory is totally inadequate and does not come close to describing the absorption spectra $A(\omega)$ observed in experiments.

Wannier (1937) observed that the interband transition in semiconductors was really a two-particle process, which is indicated in Fig. 8.7(b). Since the valence band states are all occupied, removing an electron from this band creates an excitation called a *hole*. This hole, which is the absence of an electron in an otherwise filled band, can be treated as a particle with an effective mass m_v and a positive charge. The energy of the hole is $\varepsilon_h(k) = k^2/2m_v$. In the two-particle picture, the photon of energy creates two excitations in the semiconductor: the electron in the conduction band of wave vector $\mathbf{k}$ and energy $E_g + k^2/2m_c$ and the hole in the valence band of wave vector $-\mathbf{k}$ and energy $k^2/2m_v$. Thus the energy conservation is

$$\hbar\omega = E_g + \frac{k^2}{2m_c} + \frac{k^2}{2m_v} = E_g + \frac{k^2}{2\mu}$$

which is exactly the same as the argument of the delta function for energy conservation in (8.2.3). So far our two-particle picture leads to the same absorption rate as the one-particle prediction in (8.2.3). Both the matrix and the energy conservation are identical, so the prediction is exactly the same.

The point made by Wannier is that there is now additional physics which can occur in the two-particle picture. The electron and hole are particles with charges of opposite signs, so that there is a Coulomb attraction $-e^2/(\varepsilon_0 r)$ between them, where ε_0 is the dielectric function. We shall assume that this dielectric function is a constant which is independent of frequency. In fact, this is not usually the case, since most semiconductors are polar, so the dielectric function has significant dispersion at frequencies near the optical phonon frequencies; see Sec. 6.3A. The frequency-dependent screening is an interesting problem which we shall ignore by treating ε_0 as a constant. The attractive Coulomb interaction between the electron and hole can cause hydrogenic bound states between them. It is this phenomenon we wish to discuss.

The optical absorption rate for this process was calculated by Elliott (1957). The final state of the system is described by a two-particle Schrödinger equation:

$$\Phi(\mathbf{r}_e, \mathbf{r}_n) = u_e(\mathbf{r}_e)u_v(\mathbf{r}_h)\Phi(\mathbf{r}_e, \mathbf{r}_h)$$

$$\left(\frac{-\hbar^2 \nabla_e{}^2}{2m_c} - \frac{\hbar^2 \nabla_h{}^2}{2m_v} - \frac{e^2}{\varepsilon_0 \mid \mathbf{r}_e - \mathbf{r}_h \mid} - E\right)\Phi(\mathbf{r}_e, \mathbf{r}_h) = 0$$

$\Phi(\mathbf{r}_e, \mathbf{r}_n)$ can be factored into relative $\mathbf{r} = \mathbf{r}_e - \mathbf{r}_h$ and center of mass coordinates $M = m_e + m_h$ in the standard fashion:

$$\begin{aligned} \mathbf{R} &= (m_c \mathbf{r}_e + m_v \mathbf{r}_h)\frac{1}{M} \\ \Phi(\mathbf{r}_e, \mathbf{r}_h) &= \frac{e^{i\mathbf{P}\cdot\mathbf{R}}}{\nu^{1/2}}\phi(\mathbf{r}) \\ \left(-\frac{\hbar^2}{2\mu}\nabla^2 - \frac{e^2}{\varepsilon r} - \varepsilon_r\right)&\phi(\mathbf{r}) = 0 \\ E &= E_g + \varepsilon_r + \frac{P^2}{2M} \end{aligned} \tag{8.2.4}$$

The center of mass motion is plane-wave-like, with a wave vector $\mathbf{P}$ which in optical experiments is equal to the photon wave vector. This wave vector is small, so we neglect the center of mass motion and set $\mathbf{P} = 0$. The relative motion of the electron and hole is usually more interesting. For relative energy ε_r less than zero, the two particles form bound hydrogenic states with energies $\varepsilon_r \equiv \varepsilon_n = -E_R/n^2$. For relative energy ε_r greater than zero they form scattering states $\phi_{\mathbf{k}}(\mathbf{r})$. Elliott showed that the optical transition rate depends on the relative wave function at $\mathbf{r} = 0$, $\phi_{\mathbf{k}}(0)$. Thus, instead of (8.2.3), the transition rate is determined by

$$A(\omega) = \frac{2\pi}{\hbar} \mid \hat{\varepsilon} \cdot \mathbf{p}_{cv} \mid^2 \sum_j \mid \phi_j(0) \mid^2 \delta(\omega - E_g - \varepsilon_j) \tag{8.2.5}$$

The summation j is over the bound and continuum states of the relative motion of the electron–hole pair. The dependence on the relative wave function, evaluated at $\mathbf{r} = 0$, can be understood by a physical argument. First, consider the corresponding emission process, whereby the electron and hole recombine to emit a photon ω. The emission rate should depend on the relative wave function at $\mathbf{r} = 0$, since that is the probability that the electron and hole find themselves at the same spot in the solid, which

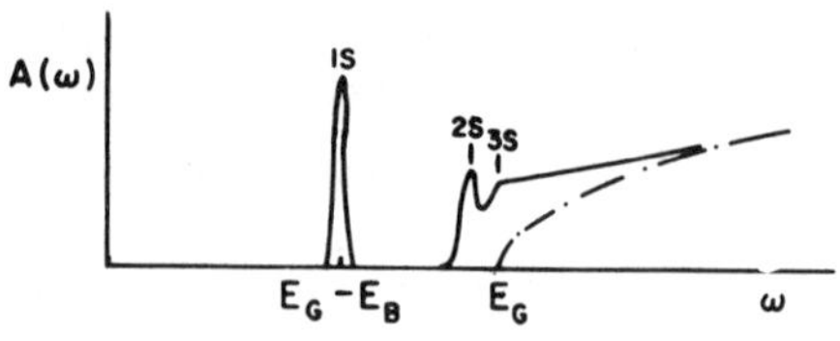

FIGURE 8.8. Typical absorption spectra of a direct gap semiconductor. Distinct exciton lines are labeled 1*s*, 2*s*, etc. Solid line is theory with final state interactions, and dashed line is theory without them.

is necessary for the recombination. Thus the dependence of the emission rate on $|\phi_j(0)|^2$ is reasonable from physical intuition. But the matrix elements for absorption and emission are identical, so that absorption should have the same dependence. The absorption depends on the probability of making the electron and hole at the same point in the solid.

The former result, for the noninteracting solid, is recovered from (8.2.5) by setting $|\phi_j(0)|^2 = \nu^{-1}$ and $\varepsilon_j = k^2/2\mu$. The results for the interacting case are quite different. The relative motions of the electron and hole are in *s*-wave hydrogenic states, either bound or unbound, because of the angular momentum selection rule. The one-unit change in *l*, in the photon absorption, is taken by the change of band symmetry, and the relative motion is not permitted any additional angular momentum. For *s* states, the bound states have an amplitude given by the principal quantum number *n*:

$$\phi_n(0) = (\pi a_0{}^3 n^3)^{-1/2}$$

For continuum states, with energy $\varepsilon_k = k^2/2\mu$, the relative wave function at the origin is $|\phi_{k,l=0}|^2 = 2\pi\eta/\nu[1 - \exp(-2\pi\eta)]$, where $\eta^{-1} = ka_0$. Then (8.2.5) predicts that the absorption is a constant in frequency at the energy gap E_g and does not rise with a square root dependence on $\omega - E_g$. The actual shape is shown in Fig. 8.8. A few sharp, distinct exciton lines are observed at low frequency which correspond to 1*s*, 2*s*, etc. In actual experiments, these absorption bands are hard to see because they are too *strong*. Only in the thinnest crystals can enough light be transmitted through the experimental sample to measure the absorption rate at the bound states of the excitons. Usually the absorption is so strong that all the light is attenuated before transversing the sample. At higher frequencies, the *ns* states are closer in frequency and are broadened, so that their envelope becomes a continuous distribution which merges with the continuum absorption which starts at $\omega = E_g$. In fact, this experiment shows no anomaly at $\omega = E_g$ and the experimental value of E_g is obtained only by extrapolating the Rydberg series for the positions of the *ns* exciton lines:

$$\varepsilon_n = \frac{-E_R}{n^2}$$

$$\omega_n = E_g + \varepsilon_n = E_g - \frac{E_R}{n^2}$$

$$E_R = \frac{\mu e^4}{2\varepsilon_0^2\hbar^2}$$

In most semiconductors, the Rydberg unit E_R is of the order of several millivolts because of the high values of the dielectric constant $\varepsilon_0 \simeq 10$ and the small values of the effective mass $\mu \simeq 0.1$ m.

This discussion is only a brief review of exciton theory in semiconductors. The interested reader is encouraged to seek additional information in the books by Knox (1963) and Reynolds and Collins (1981), or the review by Segall (1967). The theory has been verified by many detailed experiments, such as exciton properties in static electric or magnetic fields, under stress, etc. Our aim here was to emphasize that these exciton effects are important for determining the absorption rate of photons within a frequency interval of E_R near the energy gap E_g.

B. Solution by Green's Functions

Our objective is to derive the Elliott formula (8.2.5) with the use of Green's functions. The starting point will be a Hamiltonian which contains the properties of electrons, holes, and their mutual interaction. The Kubo formula will be evaluated for the optical conductivity, which is proportional to the quantity we have called $A(\omega)$. The Elliott formula is obtained by a summation of all the vertex diagrams of the current–current correlation function. The Green's function analysis is straightforward and can be solved exactly, because it is still only a two-body problem. Indeed, this section could have been included in Chapter 4 on exactly solvable models. The motivation for this analysis is to stress the importance of *final state interactions*. The latter is a fancy name for the interactions between the particles in the final state of the optical transition. In this case they are the Coulomb interaction between the electron and hole.

The starting point for the Green's function calculation is the Kubo formula for the current–current correlation function:

$$\pi(i\omega) = -\frac{1}{\nu}\int_0^\beta d\tau e^{i\omega\tau}\langle T_\tau j_\mu(\tau) j_\mu(0)\rangle$$

$$j_\mu = \sum_{\mathbf{k}} w_{\mathbf{k}}(c_{\mathbf{k}} d_{-\mathbf{k}} + d^\dagger_{-\mathbf{k}} c_{\mathbf{k}}^\dagger) \tag{8.2.6}$$

$$w_{\mathbf{k}} = \int d^3r \Psi_c^*(\mathbf{k}, r) p_\mu \Psi_v(\mathbf{k}, r) = \langle c, \mathbf{k} \mid p_\mu \mid v, \mathbf{k}\rangle$$

The τ development of the operators is governed by a Hamiltonian which contains the Coulomb interaction $v_q = -4\pi e^2/\varepsilon_0 q^2$ between the electron $(c_{\mathbf{k}}, c_{\mathbf{k}}^{\dagger})$ and the hole $(d_{\mathbf{k}}, d_{\mathbf{k}}^{\dagger})$:

$$H = H_0 + V$$
$$H_0 = \sum_{\mathbf{k}} \xi_h(k) d_{\mathbf{k}}^{\dagger} d_{\mathbf{k}} + \sum_{\mathbf{k}} \xi_c(k) c_{\mathbf{k}}^{\dagger} c_{\mathbf{k}} \tag{8.2.7}$$
$$V = \frac{1}{\nu} \sum_{\mathbf{q}} v_q \Big(\sum_{\mathbf{k}} c_{\mathbf{k}+\mathbf{q}}^{\dagger} c_{\mathbf{k}} \Big) \sum_{\mathbf{k}'} d_{\mathbf{k}'-\mathbf{q}}^{\dagger} d_{\mathbf{k}'}$$

Spin does not play a role in this analysis, so the spin index is not written. The procedure is to solve the Kubo formula (8.2.6) with this Hamiltonian. The optical matrix element $w_{\mathbf{k}}$ is taken here to be a function of $\mathbf{k}$, although in the theory for Wannier excitons it is a constant. An important assumption, when evaluating this correlation function, is that the densities of electrons n_e and holes n_h are both negligibly small. An equivalent to the assumption is that the chemical potential in the semiconductor is in the forbidden energy gap, somewhere between energy zero and E_g. The chemical potential μ_h for the holes is the negative of that for the electrons $\mu_h = -\mu$, so that the energies in the preceding Hamiltonian are

$$\xi_c(k) = \frac{k^2}{2m_c} + E_g - \mu$$
$$\xi_h(k) = \frac{k^2}{2m_v} - \mu_h = \frac{k^2}{2m_v} + \mu$$

Now the definitions are finished, and we can evaluate the correlation function.

The electron–hole interaction V is treated as the perturbation for the Hamiltonian. The S matrix is expanded and examined term by term. The first term is that for no interaction, which we call $\pi^{(0)}(i\omega)$:

$$\begin{aligned}
\pi^{(0)}(i\omega) &= -\frac{1}{\nu} \sum_{\mathbf{k}} w_{\mathbf{k}}^2 \int_0^\beta d\tau e^{i\omega\tau} \langle T_\tau c_{\mathbf{k}}(\tau) d_{-\mathbf{k}}(\tau) d_{-\mathbf{k}}^{\dagger} c_{\mathbf{k}}^{\dagger} \rangle \\
&= \frac{1}{\nu} \sum_{\mathbf{k}} w_{\mathbf{k}}^2 \int_0^\beta d\tau e^{i\omega\tau} \mathscr{G}_e(\mathbf{k}, \tau) \mathscr{G}_h(\mathbf{k}, \tau) \\
&= -\frac{1}{\nu} \sum_{\mathbf{k}} w_{\mathbf{k}}^2 \frac{1}{\beta} \sum_{ik} \mathscr{G}_e(\mathbf{k}, ik) \mathscr{G}_k(\mathbf{k}, i\omega - ik) \\
&= \frac{1}{\nu} \sum_{\mathbf{k}} w_{\mathbf{k}}^2 \frac{1 - n_F(\xi_e) - n_F(\xi_h)}{i\omega - \xi_e(k) - \xi_h(\mathbf{k})}
\end{aligned}$$

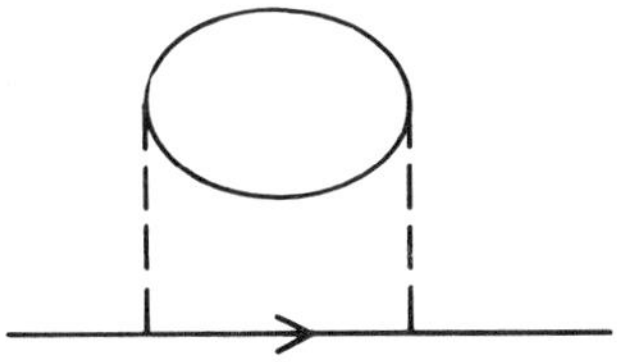

FIGURE 8.9

The last step involves a standard frequency summation, as given in Sec. 3.5. Both Green's functions $\mathscr{G}_e = (ik - \xi_e)^{-1}$ and $\mathscr{G}_h = (ik - \xi_h)^{-1}$ are for the noninteracting particles, but we omit the (0) superscript since all the Green's functions in this section are noninteracting. The occupation factors n_F are zero. The absorption $A(\omega)$ is the spectral function of this correlation function. Take the analytical continuation $i\omega \to \omega + i\delta$ and define

$$A(\omega) = -2\,\mathrm{Im}[\pi(\omega)]$$

$$A^{(0)}(\omega) = -2\,\mathrm{Im}[\pi^{(0)}(\omega)] = \frac{2\pi}{\nu}\sum_{\mathbf{k}} w_{\mathbf{k}}{}^2\delta[\omega - \xi_e(\mathbf{k}) - \xi_h(\mathbf{k})]$$

$$= \frac{2\pi}{\nu}\sum_{\mathbf{k}} w_{\mathbf{k}}{}^2\delta\left(\omega - E_g - \frac{k^2}{2\mu}\right)$$

The result for $A^{(0)}(\omega)$ is recognized as the one-particle theory of (8.2.3). The remaining terms in the expansion of the S matrix are now examined to determine the types of contributions they produce. Fortunately, most of them are zero. All self-energy terms, for either the electron or the hole, are zero. A typical self-energy diagram is shown in Fig. 8.9. It contains a closed loop, which is always proportional to the density of particles, which is zero. All self-energy terms for the electron or hole contain closed loops and vanish in a similar fashion. Thus the only finite contributions are vertex diagrams—in fact, only ladder diagrams.

The method of summing the ladder diagrams follows Mahan (1967) and was used for impurity scattering in Sec. 4.1E. They are shown as the diagrams in Fig. 8.10. The first vertex correction $A^{(1)}$ contains a single ladder diagram, where this dashed interaction represents the Coulomb interaction between the electron and the hole. The two particles are represented by the solid line segments. The wiggly line at the vertex represents

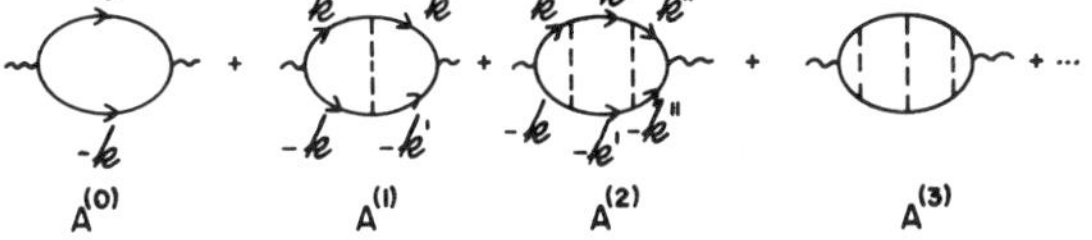

FIGURE 8.10

the photon. Both of the lines for the electron and hole are shown coming out from the photon vertex and thus are traveling parallel in time. The first ladder diagram will be called $A^{(1)}$. Its contribution can be constructed according to the rules for diagram construction:

$$\pi^{(1)} = \frac{1}{\nu^2}\sum_{\mathbf{k}\mathbf{k}'} w_{\mathbf{k}} v_{\mathbf{k}-\mathbf{k}'} w_{\mathbf{k}'} \frac{1}{\beta}\sum_{ik} \mathscr{G}_e(\mathbf{k}, ik)\mathscr{G}_h(-\mathbf{k}, i\omega - ik)$$
$$\times \frac{1}{\beta}\sum_{ik'} \mathscr{G}_e(\mathbf{k}', ik')\mathscr{G}_h(-\mathbf{k}', i\omega - ik') \tag{8.2.8}$$

The two summations over Matsubara frequency are identical in form:

$$-\frac{1}{\beta}\sum_{ik} \mathscr{G}_e(\mathbf{k}, ik)\mathscr{G}_h(-\mathbf{k}, i\omega - ik) = \frac{1 - n_F(\xi_e) - n_F(\xi_h)}{i\omega - \xi_e(k) - \xi_h(k)}$$
$$= \frac{1}{i\omega - E_g - k^2/2\mu}$$

$$\pi^{(1)} = \frac{1}{\nu^2}\sum_{\mathbf{k}\mathbf{k}'} w_{\mathbf{k}} v_{\mathbf{k}-\mathbf{k}'} w_{\mathbf{k}'} \frac{1}{i\omega - E_g - k^2/2\mu}\,\frac{1}{i\omega - E_g - k'^2/2\mu}$$

The additional ladder diagrams in the other terms just produce additional factors of the Green's function pair $\mathscr{G}_e(\mathbf{k}_j, ik)\mathscr{G}_h(-\mathbf{k}_j, i\omega - ik)$, which are summed over Matsubara frequency to produce another energy denominator $(i\omega - E_g - k_j^2/2\mu)^{-1}$. Each ladder also has another factor of the Coulomb potential $v_{\mathbf{k}_i-\mathbf{k}_j}/\nu$. Thus it is easy to write the term with n ladder diagrams:

$$\pi^{(n)} = \frac{1}{\nu^{n+1}}\sum_{\mathbf{k}\mathbf{k}_1\cdots\mathbf{k}_n} \frac{w_{\mathbf{k}} v_{\mathbf{k}-\mathbf{k}_1} v_{\mathbf{k}_1-\mathbf{k}_2}\cdots v_{\mathbf{k}_{n-1}-\mathbf{k}_n} w_{\mathbf{k}_n}}{i\omega - E_g - k^2/2\mu} \prod_{j=1}^{n} \frac{1}{i\omega - E_g - k_j^2/2\mu}$$

The simple form of these terms permits them to be summed exactly in order to obtain the expression for the correlation function. The other vertex diagrams, which are not just ladder diagrams but have a crossing of the interaction lines, are all proportional to the numbers of electrons or holes in the semiconductor. They can therefore be neglected, and the only nonzero vertex corrections are the ladder diagrams.

The summation of the ladder diagrams can be expressed as a vertex correction to the correlation function, which we write as

$$\pi(i\omega) = \sum_n \pi^{(n)} = \frac{1}{\nu}\sum_{\mathbf{k}} w_{\mathbf{k}} \frac{\Gamma(\mathbf{k}, i\omega)}{i\omega - E_g - k^2/2\mu} \tag{8.2.9}$$

$$\Gamma(\mathbf{k}, i\omega) = w_{\mathbf{k}} + \frac{1}{\nu}\sum_{\mathbf{k}_1} \frac{v_{\mathbf{k}-\mathbf{k}_1} w_{\mathbf{k}_1}}{i\omega - E_g - k_1^2/2\mu}$$
$$+ \frac{1}{\nu}\sum_{\mathbf{k}_1,\mathbf{k}_2} \frac{v_{\mathbf{k}-\mathbf{k}_1} v_{\mathbf{k}_1-\mathbf{k}_2} w_{\mathbf{k}_2}}{[i\omega - E_g - (k_1^2/2\mu)][i\omega - E_g - k_2^2/2\mu]} + \cdots$$

The summation of terms in the vertex function $\Gamma(\mathbf{k}, i\omega)$ can be generated by iterating the equation

$$\Gamma(\mathbf{k}, i\omega) = w_{\mathbf{k}} + \frac{1}{\nu}\sum_{\mathbf{k}_1} \frac{v_{\mathbf{k}-\mathbf{k}_1} \Gamma(\mathbf{k}_1, i\omega)}{i\omega - E_g - k_1^2/2\mu} \tag{8.2.10}$$

An exact summation of the ladder diagrams is obtained by solving for this vertex function. The integral equation (8.2.10) for the vertex function is an equation of the relative motion of the two particles. To show this, first recall the definition of the two-particle potential v_q in terms of the wave functions of the electron and the hole:

$$v_q = \int d^3r_e\, d^3r_h \Psi_e^*(\mathbf{k}_e + \mathbf{q}, \mathbf{r}_e)\Psi_h^*(\mathbf{k}_h - \mathbf{q}, \mathbf{r}_h) \frac{e^2}{\varepsilon_0 \,|\, \mathbf{r}_e - \mathbf{r}_h \,|}$$
$$\times \Psi_e(\mathbf{k}_e, \mathbf{r}_e)\Psi_h(\mathbf{k}_h, \mathbf{r}_h)$$

$$\Psi_c(\mathbf{k}, \mathbf{r}) = u_c(\mathbf{r}) \frac{e^{i\mathbf{k}\cdot\mathbf{r}}}{\nu^{1/2}}$$

The Coulomb integral extends over many unit cells in the solid, so that the cell-periodic parts u_e^2 and u_v^2 can be replaced by their average in the unit cell, which is unity. Then the center of mass transformation (8.2.4) on the wave vectors of the electron and hole divide the preceding integral into relative and center-of-mass parts:

$$\mathbf{k}_e = \mathbf{k} + \frac{m_e}{M}\mathbf{P}$$

$$\mathbf{k}_h = -\mathbf{k} + \frac{m_h}{M}\mathbf{P}$$

$$v_q = -\int d^3r e^{-i\mathbf{q}\cdot\mathbf{r}} \frac{e^2}{\varepsilon_0 r} \int \frac{d^3R}{\nu} = -\frac{4\pi e^2}{q^2\varepsilon_0}$$

The center-of-mass motion is conserved, while the relative motion has the Coulomb interaction $-4\pi e^2/q^2\varepsilon_0$. All the wave vector dependence of the potential comes from the relative motion, and (8.2.10) is an equation in the relative motion.

The first step in the solution of the integral equation (8.2.10) is to define the auxiliary function

$$P(\mathbf{r}, i\omega) = \frac{1}{\nu} \sum_{\mathbf{k}} \frac{e^{i\mathbf{k}\cdot\mathbf{r}} \Gamma(\mathbf{k}, i\omega)}{i\omega - E_g - k^2/2\mu} \tag{8.2.11}$$

For the Wannier exciton system, the matrix element $w_{\mathbf{k}}$ is independent of $\mathbf{k}$ and is

$$w_{\mathbf{k}} = w_0 = \hat{\varepsilon} \cdot \mathbf{p}_{cv}$$

In this case, the correlation function we want in (8.2.9) is the value of this auxiliary function at the origin:

$$\pi(i\omega) = w_0 P(\mathbf{r} = 0, i\omega) \tag{8.2.12}$$

Furthermore, the integral equation (8.2.10) for the vertex function becomes

$$\Gamma(\mathbf{k}, i\omega) = w_0 - \int d^3r \frac{e^2}{\varepsilon_0 r} e^{-i\mathbf{k}\cdot\mathbf{r}} P(\mathbf{r}, i\omega) \tag{8.2.13}$$

The reason for introducing the auxiliary function $P(\mathbf{r}, i\omega)$ is that one can write a differential equation for it, which is then solved to obtain the exact solution to the function. This differential equation is obtained by operating on both sides of (8.2.11) with the operator $i\omega - E_g + (\nabla^2/2\mu)$:

$$\begin{aligned} \left(i\omega - E_g + \frac{1}{2\mu}\nabla^2\right) P(\mathbf{r}, i\omega) &= \frac{1}{\nu} \sum_{\mathbf{k}} e^{i\mathbf{k}\cdot\mathbf{r}} \frac{i\omega - E_g - (k^2/2\mu)}{i\omega - E_g - (k^2/2\mu)} \Gamma(\mathbf{k}, i\omega) \\ &= \frac{1}{\nu} \sum_{\mathbf{k}} e^{i\mathbf{k}\cdot\mathbf{r}} \Gamma(\mathbf{k}, i\omega) \end{aligned}$$

The operator $i\omega - E_g + \nabla^2/2\mu$ is chosen to eliminate the energy denominator on the right-hand side. The integral equation (8.2.13) for the vertex function is now used:

$$\begin{aligned} &\left(i\omega - E_g + \frac{1}{2\mu}\nabla^2\right) P(\mathbf{r}, i\omega) \\ &= \frac{1}{\nu} \sum_{\mathbf{k}} e^{i\mathbf{k}\cdot\mathbf{r}} \left[w_0 - \int d^3r' \frac{e^2}{\varepsilon_0 r'} e^{-i\mathbf{k}\cdot\mathbf{r}'} P(\mathbf{r}', i\omega) \right] \\ &= w_0 \delta^3(\mathbf{r}) - \int d^3r' \frac{e^2}{\varepsilon_0 r'} \delta^3(\mathbf{r} - \mathbf{r}') P(\mathbf{r}', i\omega) \\ &= w_0 \delta^3(\mathbf{r}) - \frac{e^2}{\varepsilon_0 r} P(\mathbf{r}, i\omega) \end{aligned}$$

The completeness relation $\sum_{\mathbf{k}} \exp(i\mathbf{k}\cdot\mathbf{r}) = \nu\delta^3(\mathbf{r})$ on the second term eliminates the integral over $\mathbf{r}'$ and reduces this term to the potential $V(r)$ times $P(\mathbf{r}, i\omega)$. This term is moved to the left-hand side of the equation, which derives the inhomogeneous differential equation obeyed by $P(\mathbf{r}, i\omega)$:

$$(i\omega - E_g - H_{\mathrm{ex}})P(\mathbf{r}, i\omega) = w_0\delta^3(\mathbf{r})$$
$$H_{\mathrm{ex}} = -\frac{1}{2\mu}\nabla^2 - \frac{e^2}{\varepsilon_0 r} \tag{8.2.14}$$

This equation is solved in the following fashion. The factors $i\omega - E_g$ on the left are constants, and the important operator parts are the same as found in the exciton equation (8.2.4) for their relative motion:

$$H_{\mathrm{ex}}\phi_j(\mathbf{r}) = \varepsilon_j\phi_j(\mathbf{r})$$

This Hamiltonian H_{ex} has exact solutions with hydrogenic eigenfunctions $\phi_j(\mathbf{r})$ and eigenvalues ε_j. In terms of these exact solutions, we can write the solution to the differential equation (8.2.14) for $P(\mathbf{r}, i\omega)$ as

$$P(\mathbf{r}, i\omega) = w_0 \sum_j \frac{\phi_j(\mathbf{r})\phi_j^*(\mathbf{0})}{i\omega - E_g - \varepsilon_j}$$

This can be verified by examining the original equation:

$$\begin{aligned}(i\omega - E_g - H_{\mathrm{ex}})P(\mathbf{r}, i\omega) &= w_0 \sum_j \phi_j(\mathbf{r})\phi_j^*(\mathbf{0}) \frac{i\omega - E_g - \varepsilon_j}{i\omega - E_g - \varepsilon_j} \\ &= w_0 \sum_j \phi_j(\mathbf{r})\phi_j^*(\mathbf{0}) \\ &= w_0\delta^3(\mathbf{r})\end{aligned}$$

The last step is a summation over the complete set of hydrogenic states, which produces a delta function. This checks with (8.2.14), so that we have indeed exactly solved for $P(\mathbf{r}, i\omega)$. The current–current correlation function $\pi(i\omega)$ is obtained from (8.2.12) as the value of $w_0 P(\mathbf{r} = 0, i\omega)$:

$$\pi(i\omega) = w_0^2 \sum_j \frac{|\phi_j(0)|^2}{i\omega - E_g - \varepsilon_j}$$

The optical absorption function $A(\omega)$ is the spectral function of this operator:

$$A(\omega) = 2\pi w_0^2 \sum_j |\phi_j(0)|^2 \delta(\omega - E_g - \varepsilon_j) \tag{8.2.5}$$

This result is exactly the Elliott form in (8.2.5), with the transition rate depending on the square of the relative wave function evaluated at $\mathbf{r} = 0$. As discussed in Sec. 8.2A, this dependence causes a dramatic and important change in the absorption spectra. Thus the final state interactions are very important. They are obtained, in the solution using Green's functions, by summing the ladder diagrams of the correlation function. This is an exact solution to the current–current correlation function (8.2.6) for the model Hamiltonian in (8.2.7) and with the restriction that there is a negligible density of electrons and holes in the semiconductor.

So far the derivation has been restricted to the situation where the dispersions $\varepsilon(\mathbf{k})$ for both the electron and hole were parabolic and isotropic. This model was adopted for simplicity, but the solution can be done in the same way for any type of dispersion relation. Thus let the kinetic energy function of the electron and hole be $\varepsilon_c(\mathbf{k})$ and $\varepsilon_h(\mathbf{k})$, with no restriction on their functional dependence. The energy bands may be degenerate, anisotropic, or hyperbolic or include the case where there are static electric and magnetic fields. When the center-of-mass motion is zero, the energy denominator of the vertex functions (8.2.9) and (8.2.10) is altered to $i\omega - E_g - \varepsilon_c(\mathbf{k}) - \varepsilon_h(-\mathbf{k})$. Thus we define $P(r, i\omega)$ as

$$P(\mathbf{r}, i\omega) = \frac{1}{\nu}\sum_{\mathbf{k}} \frac{e^{i\mathbf{k}\cdot\mathbf{r}}\Gamma(\mathbf{k}, i\omega)}{i\omega - E_g - \varepsilon_c(\mathbf{k}) - \varepsilon_h(-\mathbf{k})}$$

It still obeys an equation of the form (8.2.14), except now the exciton part has the more general form

$$H_{\text{ex}}\phi_j \equiv \left[\varepsilon_c\left(\frac{\nabla}{i}\right) + \varepsilon_h\left(-\frac{\nabla}{i}\right) - \frac{e^2}{\varepsilon_0 r}\right]\phi_j(\mathbf{r}) = \varepsilon_j\phi_j(\mathbf{r}) \qquad (8.2.15)$$

This exciton Hamiltonian still has eigenfunction and eigenfrequencies which we still call $\phi_j(\mathbf{r})$ and ε_j, although they may no longer be hydrogenic in character. The formal solution still has the form of (8.2.5) with a dependence upon $|\phi_j(0)|^2$, except these wave functions are the solution to the more general equation (8.2.15). Thus the formal aspects of the solution are the same for any kind of dispersion of the electron and hole band. Of course, the spectral shape $A(\omega)$ is different for the various choices of band dispersion. For an example of realistic calculations, see Hopfield and Thomas (1961) or Baldereschi and Lipari (1971).

C. Core-Level Spectra

Here we consider another problem which is similar, both physically and mathematically, to the Wannier exciton. The initial electronic state, in the optical transition, is a deeper core level of the atom, such as the 1s or 2s shell. The optical transition in the semiconductor takes the core electron to an unoccupied state in the conduction band. The empty core level can still be described as a hole, but now it is very localized in space and so has an infinite effective mass. The hole state has a constant energy $\xi_h = E_c + \mu$, where E_c is the core-level binding energy, as measured from the top of the valence band. The threshold for the optical transition, in the absence of any final state interactions, is the frequency $\omega_T = E_g + E_c$.

This problem can be solved exactly by the same techniques which were used for Wannier excitons. The advantage of the present model is that it is possible to explain the physics with greater clarity. From the point of view of the conduction electron, it has two possible potential functions. There is a potential for the initial state of the system, which is called $V_i(\mathbf{r})$. It is the potential acting on a conduction electron in the initial state, when there is no core hole, and it is thus a periodic potential for the ground state of the normal semiconductor. The effective Hamiltonian for the conduction electron is called H_i, and its eigenfunctions are $\Psi_i(\mathbf{k}, \mathbf{r})$ with energy $\varepsilon_i(\mathbf{k})$:

$$H_i\Psi_i = \left[-\frac{\nabla^2}{2m} + V_i(\mathbf{r})\right]\Psi_i(\mathbf{k}, \mathbf{r}) = \varepsilon_i\Psi_i(\mathbf{k}, \mathbf{r})$$

After the optical transition has occurred, the hole appears in the core state, which alters the potential of the conduction electron to $V_f(\mathbf{r})$. This final state potential is set equal to the initial state potential plus a term due to the core hole:

$$V_f = V_i + V_h$$

$$H_f\Psi_f = \left(-\frac{\nabla^2}{2m} + V_f\right)\Psi_f = \varepsilon_f(\mathbf{k})\Psi_f$$

The core hole potential $V_h(\mathbf{r})$ has the form of a Coulomb potential $-e^2/(\varepsilon_0 r)$ at long range, but there are atomic effects at short range. We do not need to assume any specific form for $V_h(\mathbf{r})$. We do need to know that the final state Hamiltonian H_f has its own set of eigenfunctions $\Psi_f(\mathbf{k}, \mathbf{r})$ and eigenstates $\varepsilon_f(k)$ which are different from those of the initial state. These states may be bound or unbound, to the core hole, and $\mathbf{k}$ is a general index which can represent either of these possibilities. Thus the conduction electron has two possible sets of eigenstates which could enter into the calculation for the absorption. Which one should be used?

The conventional approach to this problem is to expand the Hamiltonian and the current operator in the basis states of the initial Hamiltonian H_i. A set of creation and destruction operators $(c_{\mathbf{k}}, c_{\mathbf{k}}^\dagger)$ are associated with the basis set Ψ_i. Then the effective Hamiltonian is

$$H = \xi_h d^\dagger d + \sum_{\mathbf{k}} \xi_i(\mathbf{k}) c_{\mathbf{k}}^\dagger c_{\mathbf{k}} + \sum_{\mathbf{k}\mathbf{k}'} V_h(\mathbf{k}, \mathbf{k}') c_{\mathbf{k}}^\dagger c_{\mathbf{k}'} d^\dagger d$$

$$\xi_i(\mathbf{k}) = \varepsilon_i(\mathbf{k}) - \mu \tag{8.2.16}$$

$$V_h(\mathbf{k}, \mathbf{k}') = \int d^3r \Psi_i^*(\mathbf{k}, \mathbf{r}) V_h(\mathbf{r}) \Psi_i(\mathbf{k}', \mathbf{r})$$

This Hamiltonian has the same general form as (8.2.7). The last term in the Hamiltonian is the interaction between the conduction electron and the hole; the latter is represented by the operators $(d, d^\dagger)$. The current operator, which is in the Kubo formula (8.2.6), is also written in terms of these functions:

$$j_\mu = \sum_{\mathbf{k}} w_i(\mathbf{k})(c_{\mathbf{k}}^\dagger d^\dagger + d c_{\mathbf{k}})$$

$$w_i(\mathbf{k}) = \int d^3r \Psi_i^*(\mathbf{k}, \mathbf{r}) \hat{\varepsilon} \cdot \mathbf{p} \phi_c(\mathbf{r})$$

where $\phi_c(\mathbf{r})$ is the core wave function of the hole.

The problem has the same form as the Wannier exciton which was solved previously and is treated similarly. The correlation function, from the Kubo formula, is evaluated as a set of ladder diagrams. Their summation is achieved by a set of equations which is nearly identical to (8.2.9) and (8.2.10):

$$\pi(i\omega) = \frac{1}{\nu} \sum_{\mathbf{k}} \frac{w_i(\mathbf{k})^* \Gamma(\mathbf{k}, i\omega)}{i\omega - \omega_T - \varepsilon_i(\mathbf{k})}$$

$$\Gamma(\mathbf{k}, i\omega) = w_i(\mathbf{k}) + \sum_{\mathbf{k}_1} \frac{V_h(\mathbf{k}, \mathbf{k}_1) \Gamma(\mathbf{k}_1, i\omega)}{i\omega - \omega_T - \varepsilon_i(\mathbf{k}_1)} \tag{8.2.17}$$

The vertex function $\Gamma(\mathbf{k}, i\omega)$ is solved by introducing the auxiliary function

$$P(\mathbf{r}, i\omega) = \sum_{\mathbf{k}} \frac{\Psi_i(\mathbf{k}, \mathbf{r}) \Gamma(\mathbf{k}, i\omega)}{i\omega - \omega_T - \varepsilon_i(\mathbf{k})}$$

Once $P(\mathbf{r}, i\omega)$ is evaluated, other functions can be obtained immediately:

$$\pi(i\omega) = \frac{1}{\nu} \int d^3r \phi_c(\mathbf{r})^* \hat{\varepsilon} \cdot \mathbf{p} P(\mathbf{r}, i\omega)$$

$$\Gamma(\mathbf{k}, i\omega) = \int d^3r \Psi_i^*(\mathbf{k}, \mathbf{r}) [\hat{\varepsilon} \cdot \mathbf{p} \phi_c(\mathbf{r}) + V_h(\mathbf{r}) P(\mathbf{r}, i\omega)] \tag{8.2.18}$$

The solution to the correlation function $P(\mathbf{r}, i\omega)$ is found by writing it as a differential equation, which is found by first operating by

$$
\begin{aligned}
&(i\omega - \omega_T - H_i)P(\mathbf{r}, i\omega) \\
&\quad = \sum_{\mathbf{k}} \Psi_i(\mathbf{k}, \mathbf{r})\Gamma(\mathbf{k}, i\omega) \\
&\quad = \sum_{\mathbf{k}} \Psi_i(\mathbf{k}, \mathbf{r}) \int d^3r' \Psi_i^*(\mathbf{k}, \mathbf{r}')[\hat{\varepsilon} \cdot \mathbf{p}\phi_c(\mathbf{r}') + V_h(\mathbf{r}')P(\mathbf{r}', i\omega)] \\
&\quad = \int d^3r' \delta(\mathbf{r} - \mathbf{r}')[\hat{\varepsilon} \cdot \mathbf{p}'\phi_c(\mathbf{r}') + V_h(\mathbf{r}')P(\mathbf{r}', i\omega)] \\
&\quad = \hat{\varepsilon} \cdot \mathbf{p}\phi_c(\mathbf{r}) + V_h(\mathbf{r})P(\mathbf{r}, i\omega)
\end{aligned}
$$

In the last series of steps, we have again used the completeness relationship $\sum_{\mathbf{k}} \Psi_i(\mathbf{k}, \mathbf{r})\Psi_i^*(\mathbf{k}, \mathbf{r}') = \delta^3(\mathbf{r} - \mathbf{r}')$, which produces the delta function which eliminates the integration over $\mathbf{r}'$. The potential term is moved to the left of the equal sign, which makes the combination $H_i + V_h = H_f$. The auxiliary function $P(\mathbf{r}, i\omega)$ obeys the equation

$$
(i\omega - \omega_T - H_f)P(\mathbf{r}, i\omega) = \hat{\varepsilon} \cdot \mathbf{p}\phi_c(\mathbf{r})
$$

This equation is solved in terms of the basis functions Ψ_f of the final state Hamiltonian H_f:

$$
P(\mathbf{r}, i\omega) = \sum_{\mathbf{k}} \frac{\Psi_f(\mathbf{k}, \mathbf{r}) \int d^3r' \Psi_f^*(\mathbf{k}, \mathbf{r}')\hat{\varepsilon} \cdot \mathbf{p}\phi_c(\mathbf{r}')}{i\omega - \omega_T - \varepsilon_f(\mathbf{k})}
$$

An obvious notation is to define the optical matrix element in terms of the final state basis functions:

$$
w_f(\mathbf{k}) = \int d^3r \Psi_f^*(\mathbf{k}, \mathbf{r})\hat{\varepsilon} \cdot \mathbf{p}\phi_c(\mathbf{r})
$$

The correlation function in (8.2.18) is

$$
\pi(i\omega) = \frac{1}{v} \sum_{\mathbf{k}} \frac{|w_f(\mathbf{k})|^2}{i\omega - \omega_T - \varepsilon_f(\mathbf{k})} \tag{8.2.19}
$$

and the absorption spectrum is the spectral function of this correlation function:

$$
A(\omega) = \frac{2\pi}{v} \sum_{\mathbf{k}} |w_f(\mathbf{k})|^2 \delta[\omega - \omega_T - \varepsilon_f(\mathbf{k})] \tag{8.2.20}
$$

Equation (8.2.20) is actually a simple physical result. It shows that the absorption spectrum $A(\omega)$ can be obtained from a one-particle spectrum by using the golden rule. However, one has to use the wave functions Ψ_f and eigenvalues ε_f of the electron in the final state of the transition. During the optical transition, the effective potential of the system changes from V_i to V_f. Formula (8.2.20) shows that the optical spectrum is calculated using V_f. This is an important conclusion, not only in solid-state spectra but also in atomic physics.

In a single-particle picture, this conclusion is obvious. The core hole wave function $\phi_c(\mathbf{r})$ is also calculated for the potential V_f, since in calculating such states one does not include the interaction of a particle with itself. That is, presumedly ϕ_c is found from a Schrödinger equation with a potential which also does not have the core level occupied. In a single-particle picture, this potential is V_f. Thus the final formula (8.2.20) merely describes a one-particle transition between two different energy levels of the same potential. In fact, it is an unremarkable application of the golden rule for transition rates. We have been able to solve this many-body problem because it is a one-body problem.

Unfortunately the present model is far too simple to apply accurately to real spectra. One of our assumptions is that the changeover from V_i to V_f occurs instantly with the creation of the hole. In actual systems, the other electrons and ions take time to adjust to the core hole potential. This time can be viewed as a frequency-dependent screening process. Thus a realistic calculation should confront the dynamic screening, which we carefully neglected at the beginning of the calculation.

Such a simple result as (8.2.20) can be obtained without recourse to any vertex summation, or other Green's function gymnastics, by following a method of Nozières and deDominicis (1969). It is only necessary to define the operators, such as the Hamiltonian and current, in terms of the final state basis. First define a set of operators $(a_{\mathbf{k}}, a_{\mathbf{k}}^{\dagger})$ for the basis states Ψ_f and expand all the operators in this complete set of states. The only tricky term is the interaction with the hole. The hole operators are written as $d^{\dagger}d = 1 - dd^{\dagger}$, so symbolically we have

$$H = \xi_h d^{\dagger}d + H_i + V_h(1 - dd^{\dagger}) = \xi_h d^{\dagger}d + H_f - V_h dd^{\dagger}$$

or

$$H = \xi_h d^{\dagger}d + \sum_{\mathbf{k}} \xi_f(\mathbf{k}) a_{\mathbf{k}}^{\dagger} a_{\mathbf{k}} - dd^{\dagger} \sum_{\mathbf{k}\mathbf{k}'} \bar{V}_h(\mathbf{k}, \mathbf{k}') a_{\mathbf{k}}^{\dagger} a_{\mathbf{k}'} \tag{8.2.21}$$

$$j_{\mu} = \sum_{\mathbf{k}} w_f(\mathbf{k})(a_{\mathbf{k}}^{\dagger} d^{\dagger} + d a_{\mathbf{k}})$$

$$\bar{V}_h(\mathbf{k}, \mathbf{k}') = \int d^3r \Psi_f^*(\mathbf{k}, \mathbf{r}) V_h(\mathbf{r}) \Psi_f(\mathbf{k}', \mathbf{r})$$

$$\xi_f = E_g + \varepsilon_f - \mu$$

The Hamiltonian is written in terms of the energies ξ_f with the core hole present. Now the remaining interaction $dd^\dagger V_h$ exists only when the core hole is absent ($dd^\dagger = 1$) and serves as a negative effective potential which scatters the electron. The Kubo formula for the current–current correlation function is now written using the current operator in (8.2.21):

$$\pi(i\omega) = -\frac{1}{\nu} \sum_{\mathbf{k}\mathbf{k}'} w_f(\mathbf{k}) w_f^*(\mathbf{k}') \int_0^\beta d\tau e^{i\omega\tau} \langle e^{\tau H} d a_{\mathbf{k}} e^{-\tau H} a_{\mathbf{k}'}^\dagger d^\dagger \rangle \qquad (8.2.22)$$

The first operator with τ, from the right, acts to the right on the operator combination $a_{\mathbf{k}'}^\dagger d^\dagger \,|\rangle$. Thus the core hole is already created when this operator $e^{-\tau H}$ is evaluated, so the hole number $d^\dagger d = 1$. This factor can be written as

$$e^{-\tau H} a_{\mathbf{k}'}^\dagger d^\dagger \rangle = e^{-\tau[\xi_h + \varepsilon_f(\mathbf{k}')]} a_{\mathbf{k}'}^\dagger d^\dagger \rangle$$

Proceeding farther to the left in the operator sequence, we have the combination $d a_{\mathbf{k}}$, which now destroys the core hole. The destruction operator $a_{\mathbf{k}}$ must pair with $a_{\mathbf{k}'}^\dagger$ since otherwise $a_{\mathbf{k}}$ acts on the ground state, which has no conduction electrons, and gives zero. So far in the operator sequence we have

$$\langle e^{\tau H} d a_{\mathbf{k}} e^{-\tau H} a_{\mathbf{k}'}^\dagger d^\dagger \rangle = \delta_{\mathbf{k}\mathbf{k}'} e^{-\tau[\xi_h + \xi_f(\mathbf{k}')]} \langle e^{\tau H} \rangle$$

The last exponential factor $e^{\tau H}$ now acts on a system with no electrons and no holes, for which H has the eigenvalue of zero, and this factor gives unity. The correlation function is therefore given by $[\xi_h + \xi_f = \omega_T + \varepsilon_f(\mathbf{k})]$

$$\pi(i\omega) = -\frac{1}{\nu} \sum_{\mathbf{k}\mathbf{k}'} |\, w_f(\mathbf{k}) \,|^2 \int_0^\beta d\tau e^{\tau[i\omega - \omega_T - \varepsilon_f(\mathbf{k})]}$$

$$= \frac{1}{\nu} \sum_{\mathbf{k}} \frac{|\, w_f(\mathbf{k}) \,|^2 (1 - e^{-\beta(\omega_T + \varepsilon_f)})}{i\omega - \omega_T - \varepsilon_f(\mathbf{k})} = \frac{1}{\nu} \sum_{\mathbf{k}} \frac{|\, w_f(\mathbf{k}) \,|^2}{i\omega - \omega_T - \varepsilon_f(\mathbf{k})}$$

Assuming $\beta\omega_T \gg 1$, the formula for $\pi(i\omega)$ is the same as (8.2.19). The correlation function is evaluated exactly without recourse to Green's functions when the calculation starts in the right basis set. Choosing the proper basis set to do the calculation can serve to greatly simplify the work needed to find the answer.

Another problem is the theory of photon emission which occurs as a result of the absorption. The absorption process creates a nonequilibrium number of electrons and holes. These excited states decay by several processes, of which one is their direct recombination with the emission of a photon. In a one-particle picture, the matrix elements for emission must be the same as for absorption because of the Einstein relation for detailed balance. Hence the emission rate is also determined by the matrix element $w_f(\mathbf{k})$. This fact can be shown directly and is assigned as a problem.

8.3. X-RAY SPECTRA IN METALS

A. Physical Model

In the previous section we discussed exciton effects for the core-level spectra of semiconductors or insulators. Now we treat the same optical transitions in metals. Here the photons usually have higher energy, so the spectroscopy is labeled X-ray or soft X-ray. The many-body theory of this spectroscopy is considerably more complicated, since the conduction electrons in the metal respond dramatically to the X-ray transition. This leads to several new effects which were predicted theoretically and verified experimentally. Among them are (1) the prediction of Mahan (1967) that the absorption edges had a power law divergence near threshold, which has been found for the *P*-shell spectra of simple metals such as sodium, magnesium, and potassium, and the prediction of Doniach and Sŭnjić (1970) of asymmetries in the XPS (X-ray photoelectron spectroscopy) line shapes from core levels of these metals, which is also now well documented experimentally. The latter phenomenon is a consequence of orthogonality catastrophe, first explained by Anderson (1967).

The model for free-electron metals is shown in Fig. 8.11. It is an energy-level diagram and shows the parabolic states in the conduction band, which is filled with electrons to the Fermi energy μ. The horizontal lines on the bottom represent core levels in the ions in the metal; they are localized, have an infinite effective mass, and are drawn flat on an energy-level diagram. In the X-ray absorption process, the photon energy is used to lift an electron from a core level to an unoccupied state in the conduction band. At zero temperature, the only empty states are above the chemical potential μ. In a one-electron description of this process, the absorption must start at the threshold frequency $\omega_T = E_F + E_c$, where E_c is the core-level binding energy, as measured from the bottom of the conduction band, and E_F is

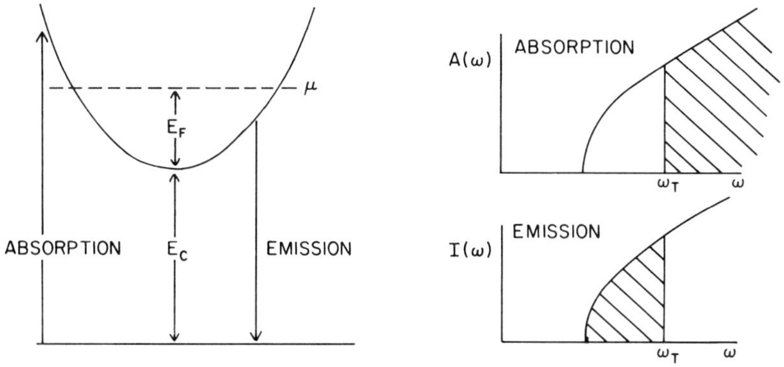

FIGURE 8.11. X-ray transition between core levels and parabolic conduction band of a metal. Electron states are occupied below Fermi energy which is E_F above conduction band minimum. Absorption must remove a core electron and put it into an unoccupied band state, while emission must remove an electron from an occupied band state. Also shown are the predicted absorption and emission spectra in the one-body picture, neglecting energy band variations.

the width of the occupied conduction band:

$$A^{(0)}(\omega) = \frac{2\pi}{\nu} \sum_{\mathbf{k}} | \langle k | \hat{\varepsilon} \cdot \mathbf{p} | c \rangle |^2 \delta(\omega - \omega_T - \xi_k)[1 - n_F(\xi_k)] \qquad (8.3.1)$$

The factor $1 - n_F$ is included to limit the final states to above the Fermi level. As indicated schematically in Fig. 8.11, the golden rule result (8.3.1) predicts that the absorption starts with a finite value at threshold, so there is a step in the spectrum. The step occurs because the electron gas has a finite density of states at the threshold frequency, since the electrons are going to states right above the Fermi energy.

The core levels of ions have angular momentum as a good quantum number. The angular momentum of the electron state must change by one unit during the photon transition. If the core level has an initial angular momentum l, the final state in the conduction band is either $l + 1$ or $l - 1$. Thus an atomic description of this absorption process would write the one-particle theory in terms of these two final state channels:

$$A^{(0)}(\omega) = 2\pi \int \frac{d^3k}{(2\pi)^3} [A_{l+1}(k) + A_{l-1}(k)]\delta(\omega - \omega_T - \xi_k)[1 - n_F(\xi_k)]$$
$$A_{l\pm1} = C_{l\pm1} | \langle k, l \pm 1 | r | c, l \rangle |^2 \qquad (8.3.2)$$

where the constants $C_{l\pm1}$ are found from Clebsch–Gordon coefficients, and the matrix elements $\langle k, l \pm 1 \mid r \mid c, l \rangle$ are radial integrals. In a metal, the final electron state is a Bloch function, which is not an eigenfunction of the angular momentum operator. The use of Bloch functions leads to a more complicated description, since the final electron states in (8.3.1) must be summed over the full Brillouin zone and its various energy bands.

The word *hole* has two different meanings in the present discussion. The first is a core hole, which is an electron missing from a core state. The other is a conduction band hole, where the electron is absent from an occupied state in the Fermi sea. We shall use *core hole* for the first case and hole for the second.

After the core hole has been created, it has several possible decay channels. The most likely is an Auger process, whereby an electron from a higher energy state falls into this core hole while its energy is transferred to another electron whose energy is correspondingly increased. The electron which gains energy can also be measured, and this Auger spectroscopy is another important experiment (Gallon, 1978).

The second decay channel is the emission of a photon by an electron which falls into the core hole. When the electron starts in the occupied states in the conduction band, this X-ray emission spectroscopy is an experiment which is complementary to the absorption. The emission provides a measurement over the states, below the Fermi level, which are occupied in the ground state of the Fermi sea. The absorption measures the unoccupied states above the Fermi energy. The emission spectrum also has a sharp step at the threshold frequency $\omega_T = E_g + E_c$, as shown schematically in Fig. 8.11. These steps, in emission and absorption, are called the *edges* of the spectra.

There are several physical mechanisms which can change the shape of the spectrum, in emission or absorption, from the simple one-particle theory in (8.3.1). The first of these is due to the lifetime of the core hole, which is usually dominated by the Auger decay process which was just described. This core hole lifetime is quite variable among the various core levels of an ion or atom. The core hole lifetime is longer for the outer valence shells than for the inner states closer to the nucleus. Some experimental values for core hole lifetimes in simple metals are shown in Table 8.1, from data of Citrin *et al.* (1977). Here the lifetime is expressed as an energy uncertainty $\Gamma = \hbar/\tau$ [where Γ is the full width at half maximum (FWHM)]. Values as high as several electron volts can be found for inner shells of atoms with high atomic number. This Auger decay generally imparts a

TABLE 8.1[a]

Metal	Temperature (°K)	Γ (eV)	Γ_{ph}^{XPS} (eV)	α
Lithium				
1s	90	0.04 ± 0.02	0.26 ± 0.03	0.23 ± 0.02
	300	0.03 ± 0.03	0.37 ± 0.03	0.23 ± 0.02
	440	0.03 ± 0.03	0.42 ± 0.03	0.23 ± 0.02
Sodium				
2p	300	0.02 ± 0.02	0.18 ± 0.03	0.198 ± 0.015
2s	300	0.28 ± 0.03	0.20 ± 0.03	0.205 ± 0.015
1s	300	0.28 ± 0.03	0.18 ± 0.04	0.21 ± 0.015
Magnesium				
2p	90	0.03 ± 0.02	0.15 ± 0.04	0.128 ± 0.015
	300	0.03 ± 0.02	0.18 ± 0.04	0.126 ± 0.015
2s	300	0.46 ± 0.03	0.19 ± 0.04	0.130 ± 0.015
1s	300	0.35 ± 0.03	0.20 ± 0.05	0.15 ±0.02
Aluminum				
2p	300	0.04 ± 0.02	0.11 ± 0.02	0.118 ± 0.015
2s	300	0.78 ± 0.05	—	0.12 ± 0.015

[a] From Citrin *et al.* (1977).

Lorentzian broadening to the X-ray spectra. It can be included, semi-empirically, in the Kubo formula as a factor $\exp(-\frac{1}{2}\Gamma \mid t \mid)$:

$$A(\omega) = \int_{-\infty}^{\infty} dt e^{i\omega t} e^{-(1/2)\Gamma|t|} \langle \mathscr{J}(t) \cdot \mathscr{J}(0) \rangle \tag{8.3.3}$$

If the X-ray measurement is to determine other phenomena, it is desirable to keep Γ as small as possible, which is done by measuring the spectra of outer atomic shells.

Another many-body effect which is conceptually straightforward, but only in absorption, is the broadening due to phonons. The only term which is usually kept in the Hamiltonian is the phonon coupling to the core hole, which is a term such as

$$d^\dagger d \sum_{\lambda \mathbf{q}} M_\lambda(\mathbf{q})(a_{\mathbf{q}\lambda} + a_{\mathbf{q}\lambda}^\dagger)$$

We showed in Sec. 4.3 that this type of phonon coupling to a localized level is a type of Hamiltonian which is exactly solvable. The optical absorption from such a coupled level was also solved, and the Kubo formula has a factor of $\exp[-\phi(t)]$, where

$$\phi(t) = \sum_{\lambda\mathbf{q}} \left| \frac{M_\lambda(\mathbf{q})}{\omega_\lambda(\mathbf{q})} \right|^2 ([1 - it\omega_\lambda(\mathbf{q}) - e^{-it\omega_\lambda(\mathbf{q})}]\{n_B[\omega_\lambda(\mathbf{q})] + 1\} + [1 + it\omega_\lambda(\mathbf{q}) - e^{it\omega_\lambda(\mathbf{q})}]n_B[\omega_\lambda(\mathbf{q})]) \tag{8.3.4}$$

The term linear in t is the self-energy term. The factor $\exp(-\phi)$ is added to the argument of the time integral in (8.3.3). Usually the Auger width Γ is large enough that we only need the short time response of the correlation function—anything that is going to happen must do so quickly, before the core hole decays by the Auger process. The short time limit of the phonon contribution is

$$\lim_{t\to 0} \phi(t) = t^2\gamma + O(t^3)$$

$$\gamma = \tfrac{1}{2} \sum_{\lambda,\mathbf{q}} | M_\lambda(\mathbf{q}) |^2\{1 + 2n_B[\omega_\lambda(\mathbf{q})]\}$$

$$A(\omega) = \int_{-\infty}^{\infty} dt e^{i\omega t} e^{-(1/2)\Gamma|t|} e^{-\gamma t^2} \langle \mathscr{T}(t)\mathscr{T}(0)\rangle$$

The phonons add a Gaussian broadening to the spectral function for the X-ray process, where the Gaussian component γ is temperature dependent. It is relatively easy to experimentally separate this from the Auger width, which is a temperature-independent Lorentzian. The only remaining problem for the phonon contribution is to calculate the matrix elements $M_\lambda(\mathbf{q})$ for the coupling to the core hole. However, this calculation is quite delicate (Hedin and Rosengren, 1977).

The phonon broadening for the emission spectra is a much more subtle calculation. Here the problem is that the phonon system does not equilibrate to the presence of the core hole before the emission occurs. One expects that the minimum time for the phonons to come to equilibrium around the new core hole, and its screening charge, is the inverse Debye frequency $t_{\text{ph}} \simeq \omega_D^{-1}$. This time is generally much longer than the lifetime for the Auger effect, except for the outer shell of electrons, as can be seen from Table 8.1. The theory of phonon broadening of emission must account for the nonequilibrium states of the phonons. This theory was developed simultaneously by Mahan (1977) and Almbladh (1977). The most interesting case yet known is lithium, for which phonons are the most important of all the mechanisms which broaden the edge. The theory predicts a double shoulder in the emission spectra because of phonons. The prediction was verified by the experiments of Callcott and Arakawa (1977).

These many-body processes, Auger and phonon, are reasonably straightforward. The more interesting aspects are the response of the many-electron system to the appearance of the core hole, which will be discussed in the remaining parts of this section.

B. Edge Singularities

Mahan (1967) predicted that the absorption edges for X-ray transitions in metals would have a power law divergence of the form

$$A(\omega) = A^{(0)}(\omega)\left(\frac{\xi_0}{\omega - \omega_T}\right)^{\alpha} \tag{8.3.5}$$

His theory resulted from an investigation into the properties of excitons in metals. The Wannier picture applies equally well to metals, so that the optical transition should be viewed as the simultaneous creation of an electron and a core hole in the metal. These will interact in the final state. Presumably this interaction is screened, although a detailed theory of the screening process is still an active subject of research, with no simple models which are generally accepted. Mahan assumed that the electron–core hole Coulomb potential is screened instantly at the time the core hole is created. This approximation may be valid near the absorption edges but surely is unreliable farther from threshold. See Canright (1988) for a discussion of dynamic response to the core hole.

The exciton theory predicts that the exponent α in (8.3.5) is positive, so the absorption diverges as a power law singularity at threshold. Later work has shown, as we shall see, that the renormalization catastrophe of Anderson tends to make α negative. Thus the final values of α may be either positive or negative, depending on which of these two factors is most important. Thus not all edges are singular.

We shall adopt the model of instantaneous screening of the electron–hole interaction in the metal. Thus the model Hamiltonian for the X-ray absorption process is the same as (8.2.16):

$$H = \xi_h d^{\dagger} d + \sum_{\mathbf{k}} \xi_c(k) c_{\mathbf{k}}^{\dagger} c_{\mathbf{k}} + \frac{1}{\nu} \sum_{\mathbf{k}\mathbf{k}'} V(\mathbf{k}, \mathbf{k}') c_{\mathbf{k}}^{\dagger} c_{\mathbf{k}'} d^{\dagger} d \tag{8.3.6}$$

The last term is the screened Coulomb interaction between the electron and

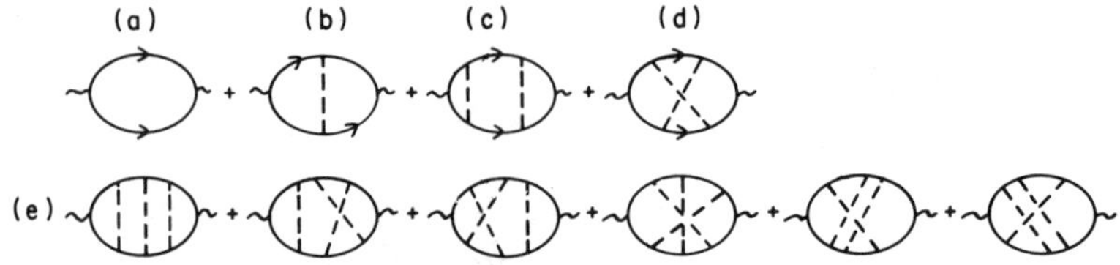

FIGURE 8.12

the core hole, while the first terms are H_0 for the electrons and the core holes. We have chosen to write the Hamiltonian in the initial state basis, so the electron operators ($c_{\mathbf{k}}$, $c_{\mathbf{k}}^{\dagger}$) refer to the state which does not have a core hole. We showed, for the Wannier exciton theory, that there was a great advantage to working in the final state basis set, but that is not the case here. The calculation is about equally difficult in either basis set, and the same answer is found from either starting point. The difference between the present calculation for the metal and the earlier calculation for the insulator is that now we have 10^{23} conduction electrons which also interact with the core hole, and they are initially in states described by the initial state basis. In the calculation, it is now necessary to keep the terms $n_F(\xi_c)$ for the electron occupation number, and these cause the dramatic change in the theory, with the prediction (8.3.5) of the edge singularity.

In the exciton calculation, the vertex diagrams to the correlation function are included by summing the set of diagrams in Fig. 8.12. To keep the physics simple, the calculation will be done first by assuming some approximations which tend to reduce the number of superscripts and subscripts: (1) The interband matrix element $\langle f \mid \hat{\varepsilon} \cdot p \mid c \rangle$ is taken to be a constant w_0; (2) the screened Coulomb interaction is taken to be a constant $V(k, k') = -V_0$ up to a cutoff energy ξ_0 which is a typical bandwidth $\xi_0 \simeq E_F$. With these approximations, the ladder diagrams in the vertex summation in Fig. 8.12 can be written as in the previous section:

$$\pi_L(i\omega) = \sum_{n=0} \pi^{(n)}(i\omega)$$

$$\pi^{(n)}(i\omega) = w_0^2(-V_0)^n \left[\int \frac{d^3k'}{(2\pi)^3} \frac{1 - n_F(\xi_{\mathbf{k}'})}{i\omega - \omega_T - \xi_{\mathbf{k}'}} \right]^{n+1}$$

The integrand of the wave vector integrand of the wave vector integral contains the factor $1 - n_F(\xi_h) - n_F(\xi_e)$, where the core hole term $n_F(\xi_h)$ is set equal to zero but not the electron term. This leaves $1 - n_F(\xi_e)$, which ensures that the electron scatters into unoccupied states. The summation over wave vectors can be changed approximately to $d^3k' \to (2\pi)^3 N_F \, d\xi_{\mathbf{k}'}$, where N_F is the density of states. The $d\xi_{\mathbf{k}'}$ integral is just a logarithm, and the ladder diagrams give the result

$$N_F \int_{-\xi_0}^{\xi_0} d\xi_{\mathbf{k}'} \frac{1 - n_F(\xi_{\mathbf{k}'})}{i\omega - \omega_T - \xi_{\mathbf{k}'}} = N_F \ln\left(\frac{\omega_T - i\omega}{\xi_0}\right)$$

$$\pi^{(n)}(i\omega) = -w_0^2 N_F (-N_F V_0)^n \left[\ln\left(\frac{\omega_T - i\omega}{\xi_0}\right)\right]^{n+1} \tag{8.3.7}$$

These approximations are fairly crude. They are meant only to apply to the threshold region, where $\omega \simeq \omega_T$, so that the cutoff in the wave vector integration by the Fermi occupation factors n_F is most important. This vertex summation leads to a summation over logarithms of increasing powers.

This series should not yet be summed, since the other vertex diagrams contribute terms which also contribute to the series. The first diagram in the series, which is not a ladder, is that shown in Fig. 8.12(d). It contains six Green's functions as internal lines, as shown in more detail in Fig. 8.13. Thus the following summations over Matsubara frequencies must be done:

$$\frac{1}{\beta^3} \sum_{ik,iq,iq'} \mathscr{G}_e(\mathbf{k}, ik)\mathscr{G}_e(\mathbf{k} + \mathbf{q}, ik + iq)\mathscr{G}_e(\mathbf{k} + \mathbf{q} + \mathbf{q}', ik + iq + iq')$$
$$\times \mathscr{G}_h(-\mathbf{k}, i\omega - ik)\mathscr{G}_h(-\mathbf{k} - \mathbf{q}', i\omega - ik - iq')$$
$$\times \mathscr{G}_h(-\mathbf{k} - \mathbf{q} - \mathbf{q}', i\omega - ik - iq - iq')$$

These will be done in sequence. The core hole Green's functions all have the same energy $\xi_h = \xi_h(-\mathbf{k}) = \xi_h(-\mathbf{k} - \mathbf{q}') = \xi_h(-\mathbf{k} - \mathbf{q} - \mathbf{q}')$. The conduction electron energies will be denoted by the symbols $\xi = \xi_c(\mathbf{k})$, $\xi' = \xi_c(\mathbf{k} + \mathbf{q})$, and $\xi'' = \xi_c(\mathbf{k} + \mathbf{q} + \mathbf{q}')$. When doing these summations, we shall set to zero the core hole occupation factor $n_F(\xi_h)$ wherever it occurs. With these conventions, the three summations are, in sequence,

$$\frac{1}{\beta} \sum_{iq'} \mathscr{G}_e(\mathbf{k}'', ik + iq + iq')\mathscr{G}_h(i\omega - ik - iq')\mathscr{G}_h(i\omega - ik - iq - iq')$$
$$= \frac{1 - n_F(\xi'')}{(i\omega - \xi'' - \xi_h)(i\omega + iq - \xi'' - \xi_h)}$$

$$\frac{1}{\beta} \sum_{iq} \frac{\mathscr{G}_e(\mathbf{k}', ik + iq)}{i\omega + iq - \xi'' - \xi_h} = \frac{n_F(\xi')}{i\omega - ik + \xi' - \xi'' - \xi_h}$$

$$\frac{1}{\beta} \sum_{ik} \frac{\mathscr{G}_e(\mathbf{k}, ik)\mathscr{G}_h(i\omega - ik)}{i\omega - ik + \xi' - \xi'' - \xi_h} = \frac{1 - n_F(\xi)}{(i\omega - \xi - \xi_h)(i\omega - \xi + \xi' - \xi'' - \xi_h)}$$

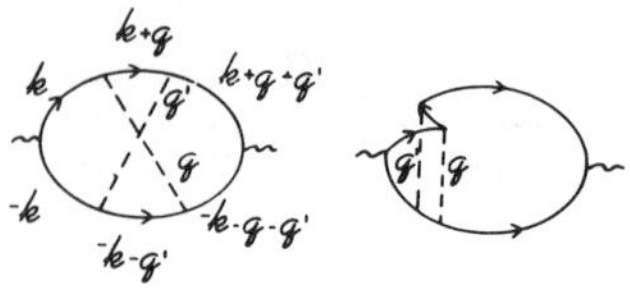

FIGURE 8.13 (a) (b)

The vertex correction is obtained from the integral ($\xi_h = \omega_T$)

$$\pi^{(2b)}(i\omega) = V_0^2 N_F^3 w_0^2 \int_{-\xi_0}^{\xi_0} d\xi \int_{-\xi_0}^{\xi_0} d\xi' \int_{-\xi_0}^{\xi_0} d\xi' \times \frac{[1 - n_F(\xi)]n_F(\xi')[1 - n_F(\xi'')]}{(i\omega - \xi - \omega_T)(i\omega - \xi + \xi' - \xi'' - \omega_T)(i\omega - \xi'' - \omega_T)} \tag{8.3.8}$$

This integral is more difficult to evaluate than those in the previous cases, since the middle energy denominator contains all three conduction energies, $-\xi + \xi' - \xi''$. This contribution is redrawn in Fig. 8.13(b). It is emphasized that the terms in Figs. 8.13(a) and 8.13(b) are identical, since they are topologically connected in the same fashion. The dashed lines are the Coulomb interaction. In our model this interaction acts instantaneously, since we are using an unretarded form of the interaction. Thus the Coulomb lines should be drawn vertically to indicate their instantaneous nature. In that case, either the electron or core hole line must be drawn in a zigzag manner, with one line segment going backwards in time. A backward line in time always indicates a contribution which is proportional to the number of particles n_0. The particle number is finite for conduction electrons but zero for core holes, so only the electron line is shown with this zigzag. These terms could be ignored in the insulator where n_0 is zero but not in the metal.

The triple integral in (8.3.8) is obviously complicated. The most divergent term it yields is

$$\pi^{(2b)}(i\omega) \simeq \frac{1}{3} w_0^2 N_F (N_F V_0)^2 \left[\ln\left(\frac{\omega_T - i\omega}{\xi_0}\right)\right]^3$$

This term is one-third the contribution of the double ladder diagram in Fig. 8.12(c) and with the opposite sign. Each of the six terms with three vertex lines in Fig. 8.12(e) produces terms which have $\{\ln[(\omega_T - i\omega)/\xi_0]\}^3$. The summation of all the terms shown in Fig. 8.12 gives the series

$$\pi(i\omega) = -\frac{N_F w_0^2}{N_F V_0}\left(L - L^2 + \frac{2}{3}L^3 - \frac{1}{3}L^4 + \cdots\right)$$
$$L = N_F V_0 \ln\left(\frac{\omega_T - i\omega}{\xi_0}\right) \tag{8.3.9}$$

The series is now summed to produce the final result:

$$\begin{aligned}\pi(i\omega) &= -\frac{N_F w_0^2}{2N_F V_0^2}(1 - e^{-2L})\\ &= -\frac{1}{2}\frac{N_F w_0^2}{N_F V_0}\left[1 - \left(\frac{\xi_0}{\omega_T - i\omega}\right)^{2N_F V_0}\right]\end{aligned}$$

The optical absorption is the spectral function of this operator ($i\omega_n \to \omega + i\delta$):

$$\begin{aligned}A(\omega) &= -2\,\mathrm{Im}[\pi(\omega)]\\ &= -\frac{N_F w_0^2}{N_F V_0}\,\mathrm{Im}\left\{\left(\frac{\xi_0}{\omega - \omega_T}\right)^{2N_F V_0}\exp[-i\pi 2N_F V_0\theta(\omega - \omega_T)]\right\}\\ &= \frac{w_0^2}{V_0}\sin(2\pi N_F V_0)\left(\frac{\xi_0}{\omega - \omega_T}\right)^{2N_F V_0}\theta(\omega - \omega_T)\end{aligned} \tag{8.3.10}$$

Equation (8.3.10) is Mahan's (1967) result that the absorptions diverge as a power law, where the exponent is the dimensionless quantity $\alpha = 2N_F V_0$. The exponent is obtained by summing the most divergent diagrams in each order of perturbation theory. These terms are divergent, because the retarded function $\ln[(\omega_T - \omega - i\delta)/\xi_0]$ can become quite large when we are near threshold $\omega \simeq \omega_T$. In the limit where $V_0 = 0$, we recover the noninteracting result that $A^{(0)} = 2\pi N_F w_0^2 \theta(\omega - \omega_T)$, and the absorption edge is a simple step.

The correlation function (8.2.2) for the model Hamiltonian (8.3.6) can be evaluated accurately using simple analytical formulas. Pardee and Mahan (1973) showed that the edge singularities could be derived in a simple way using dispersion theory. Penn, Girvin, and Mahan (1981, called PGM) obtained an analytical solution to the multiple scattering problem, including all vertex corrections. They showed that the leading term in the series was the result of Pardee and Mahan. Ohtaka and Tanabe (1983, 1986) obtained a more elegant analytical solution, which they illustrated with numerical examples. This work has provided a complete solution to the MND problem, and in an analytical form that is useful for computation.

Here we reproduce the simple argument of Pardee and Mahan. Let $T(\omega)$ be the complex scattering amplitude for an ingoing wave of an electron scattering from a central potential, which in this case is the screened core hole. If $T'(\omega)$ is the amplitude of the outgoing scattered wave, then they differ only by a phase factor which is the phase shift:

$$T'(\omega) = T(\omega)\exp[2i\delta_l(\omega)\Theta(\omega - \omega_T)]$$

where l is the angular momentum of the outgoing electron. The step function is a reminder that the phase factor exists only in the region of absorption, since T and T' must both be real in regions where there is no absorption. The next assumption is that T and T' both originate from the same analytical function of frequency:

$$t^{(+)}(\omega + i\delta) = T'(\omega)$$
$$t^{(+)}(\omega - i\delta) = T(\omega)$$

The superscript $(+)$ is a label that is explained below. Combining these results shows that

$$\begin{aligned} t^{(+)}(\omega + i\delta) &= \exp[\phi(\omega + i\delta)] = | T(\omega) | \exp[i\delta_l(\omega)\Theta(\omega - \omega_T)] \\ \mathrm{Im}[\phi(\omega \pm i\delta)] &= \pm\, \delta_l(\omega)\Theta(\omega - \omega_T) \end{aligned} \tag{8.3.11}$$

Dispersion theory assumes that $\phi(z)$ in (8.3.11) is an analytical function of complex frequency z. Then the only function that has the property (8.3.11) is

$$\phi(\omega + i\delta) = \frac{1}{\pi} \int_{\omega_T}^{\xi_0} d\varepsilon\, \frac{\delta_l(\varepsilon)}{\varepsilon - \omega - i\delta} \tag{8.3.12}$$

As usual ξ_0 is the band width. The function $\phi(z)$ is a dispersion integral. The integral can be evaluated by writing $\delta_l(\varepsilon) = \delta_l(\omega) + [\delta_l(\varepsilon) - \delta_l(\omega)]$ and the first term is an easy integral:

$$\phi(\omega + i\delta) = \frac{\delta_l(\omega)}{\pi}\left[\ln\left|\frac{\xi_0 - \omega}{\omega_T - \omega}\right| + i\pi\Theta(\omega - \omega_T)\right] + K(\omega)$$

$$K(\omega) = \frac{1}{\pi}\int_{\omega_T}^{\xi_0} d\varepsilon\, \frac{\delta_l(\varepsilon) - \delta_l(\omega)}{\varepsilon - \omega}$$

The function $K(\omega)$ is a smooth function of ω since the integrand is real and is not singular. The amplitude of the X-ray absorption is proportional to $T'(\omega)$ and the intensity is proportional to $| T'(\omega) |^2$. This factor can be evaluated using the dispersion integrals

$$|T'(\omega)|^2 = \left|\frac{\xi_0 - \omega}{\omega_T - \omega}\right|^{2\delta_l(\omega)/\pi} e^{2K(\omega)} \tag{8.3.13}$$

This equation has the same edge singularity as (8.3.10). Here the exponent of the power law is $2\delta_l(\omega)/\pi$ rather than $2N_F V_0$. Note that if one is calculating the phase shift in the Born approximation, then one gets that $\sin(\delta) \approx \delta = \pi N_F V_0$. Nozieres and deDominicis (1969) were the first to realize that the exponents of the edge singularities were functions of the phase shifts. At the threshold frequency $\omega \simeq \omega_T$ then the exponent depends upon $\delta_l(\omega_T)$, which are the phase shifts of conduction electrons at the Fermi surface for scattering from the core hole. The factor of $K(\omega)$ is a smooth function of frequency and does not contribute to the singular behavior. It has an influence upon the calculation of the X-ray absorption for other values of frequency.

In X-ray absorption an electron absorbs an X-ray, departs the atomic core, and becomes a conduction electron. The core potential changes from having a charge of Z to having a charge of $Z + 1$. All of the other conduction electrons have their central potential suddenly changed by the appearance of this new core hole. The switching-on of the potential may cause electron–hole pairs to be created. The pairs act as bosons. Below we show that the theory for this process can be treated by the independent boson model of Sec. 4.3.

The electron that departs the core state and becomes a conduction electron may also create excitations of electron–hole pairs. The departing electron has only a small probability of creating pairs. This process may be accurately calculated by perturbation theory.

This process is distinct from the shakeup processes of the other conduction electrons. The calculation of pairs during the shakeup process has an infrared divergence, which makes a perturbation calculation divergent.

Dispersion theory is also used to calculate the rate by which the departing electron creates pairs. For each electron of positive energy ($\omega > \omega_T$) created there is a factor of $|t^{(+)}(\omega)|^2$. A similar factor $|t^{(-)}(\omega)|^2$ is needed for holes in the conduction band, which have an energy $\omega < \omega_T$. It has a dispersion integral over negative energies:

$$t^{(-)}(\omega) = \exp\left[\frac{1}{\pi}\int_{-\mu}^{0} d\nu \frac{\delta_l(\nu)}{\nu - \omega - i\delta}\right]$$

where μ is the chemical potential.

PGM showed for absorption that the one-pair term was negative, and about 10% of the zero pair part. The two-pair term was about 1% and is negligible. The intensity of the zero and one pair probabilities in absorption are

$$I(\omega) = \int_{-\mu}^{\infty} d\omega_1\, g(\omega_1) W(\omega_1)[1 - n(\omega_1)] \mid t^{(+)}(\omega_1) \mid^2 \Big\{ \delta(\omega - \omega_1)$$
$$- \int_{-\mu}^{\infty} d\omega_2\, g(\omega_2)[1 - n(\omega_2)] \int_{-\mu}^{\infty} d\omega_3\, g(\omega_3) n(\omega_3) \delta(\omega - \omega_1 - \omega_2 + \omega_3)$$
$$\times \mid t^{(+)}(\omega_2) t^{(-)}(\omega_3) \mid^2 / (\omega_3 - \omega_1)(\omega_3 - \omega_2) \Big\} \tag{8.3.14}$$

where $n(\omega)$ are the Fermi occupation factors as a function of temperature, $g(\omega)$ is the density of states, and $W(\omega)$ is the square of the matrix element for the X-ray absorption. Terms in the absorption spectra that include the creation of two or more pairs have additional factors of $\mid t^{(+)} t^{(-)} \mid^2$. It should be emphasized that these expressions are valid throughout all frequency regions. At threshold they have the edge singularity, and they show smooth variations at other frequencies. Equation (8.3.14) describes the final state exciton effects and ege singularities to an accuracy of 1%.

X-ray absorption has a number of factors that contribute to the shape of the spectra. So far we have discussed in detail the contributions of phonons, Auger decay, and the final state Coulomb interactions. Another important contribution is the shakeup mentioned above. The shakeup contributes a time evolution $\varrho(t)$ which is calculated in the next section. All of these various factors contribute a time dependence to the evolution of the absorption. The experimental spectra are found by fourier transforming the product all of these factors

$$A(\omega) = \int_{-\infty}^{\infty} dt\, \exp[it(\omega - \omega_T) - \Gamma \mid t \mid /2 - \gamma t^2] \varrho(t) I(t) \tag{8.3.15}$$

The factor $I(t)$ is the fourier transform of $I(\omega)$ in (8.3.14). Instead of transforming $I(\omega)$, an alternative is to represent (8.3.15) as a convolution integral. The evaluation of $\varrho(t)$ is presented in the next section.

C. Orthogonality Catastrophe

The X-ray absorption process creates a core hole in the midst of the electron gas. It also creates an additional conduction electron, but the latter process has been dealt with in the previous section. Now we are concerned with the impact on the $N \simeq 10^{23}$ other conduction electrons of this sudden appearance of a core hole. The core hole is again represented as a static potential V. Thus the physics is to investigate the response of the free-electron gas to the sudden appearance of a new potential V. The result has been labeled a *catastrophe*, because the transition is forbidden. The word *forbidden* is used in an unusual way, which we shall now explain.

The important physics is that the ground state wave functions of the conduction electron system, with and without the core hole potential V, are orthogonal. The transition is forbidden because it is the transition between two orthogonal states. The X-ray absorption is observed in actual metals, because the conduction electron system is excited by the creation of electron–hole pairs. The pair creation is a symmetry-breaking process, which allows the forbidden transition to be observed. This orthogonality was first suggested by Hopfield (1969) and proved by Anderson (1967).

It is important to realize that the orthogonality between the two wave functions, with and without the core hole potential, is between the N-particle wave functions in the limit where $N \rightarrow \infty$. The single-particle wave functions are not orthogonal. As a simple example, assume that in the ground state of the electron gas, before the appearance of the core hole, the conduction electrons can be described by single-particle wave functions with no potential—plane waves. We follow Anderson and consider only S waves, so that the wave functions have the form $\phi(kr) = \sin kr/kr$. The N-particle wave function can be described by an N-dimensional Slater determinant of these orbitals. The spin indices are omitted, so we write it as

$$\Phi_i(r_1 \cdots r_n) = \frac{1}{(N!)^{1/2}} \begin{vmatrix} \phi(k_1 r_1) & \phi(k_1 r_2) & \cdots & \phi(k_1 r_n) \\ \phi(k_2 r_1) & \phi(k_2 r_2) & & \\ \vdots & & & \\ \phi(k_n r_1) & & \cdots & \phi(k_n r_n) \end{vmatrix}$$

After the appearance of the core hole potential, the conduction electrons adjust their individual wave functions to the presence of this potential. Far from the potential region, the wave functions can be described by $\phi'(kr) = \sin(kr + \delta)/kr$, where δ is the phase shift for one-electron scattering from the core hole potential. The new N-particle wave function is just a Slater determinant with these new orbitals ϕ':

$$\Phi_f(r_1 \cdots r_n) = \frac{1}{(N!)^{1/2}} \begin{vmatrix} \phi'(k_1 r_1) & \phi'(k_1 r_2) & \cdots & \phi'(k_1 r_n) \\ \phi'(k_2 r_1) & & & \\ \vdots & & & \\ \phi'(k_n r_1) & & \cdots & \phi'(k_n r_n) \end{vmatrix}$$

During the X-ray transition, one takes the matrix element between the initial and final states of the system. In a one-particle picture, the matrix element $\int d^3r\phi(k_1r)^*\hat{\varepsilon}\cdot\mathbf{p}\phi_c(r)$ is between the core wave function $\phi_c(r)$ and a conduction electron $\phi(kr)$. The many-particle calculation computes this matrix element between the N-particle states. Thus we should include in the transition rate the overlap between these two Slater determinants,

$$S = \int d^3r_1 \cdots d^3r_n \Phi_f(r_1 \cdots r_n)\Phi_i(r_1 \cdots r_n)$$

which has been shown to be

$$S = N^{-(1/2)\alpha}$$

$$\alpha = 2\frac{\delta^2}{\pi^2}$$

The many-particle overlap S turns out to be a negative power of N, where the exponent is a function of the phase shifts. This overlap vanishes in the limit where $N \to \infty$ and is small for real systems which have $N \simeq 10^{23}$ and α is typically $0.1 \to 0.2$.

There is no orthogonality between the individual matrix elements of single-particle orbitals $\phi(kr)$ and $\phi'(kr)$. Each of these is finite. When these overlaps are evaluated in the determinant for all the possible combinations, the result is asymptotically zero as $N \to \infty$. The analogous situation occurs in atoms but on a less drastic scale. For an atom with N electrons, the optical absorption can cause one electron to change from one energy state to another. The matrix element for this process is calculated for the full N-particle wave function of the atom. The N-particle matrix element can be reduced to a one-particle matrix element between the two primary energy states which are the initial and final levels of the electron, times a factor which gives the overlap of the other $N - 1$ electrons. They each change their orbitals a small amount during this change of state by one electron, and the $N - 1$ overlap function S has a typical value of 0.95. Thus S is near unity, and the N-particle overlap is a small effect in atoms with a few electrons. The metal case is the extrapolation to $N \simeq 10^{23}$ electrons, where the product of one-particle overlaps, each slightly less than unity, eventually produces a vanishing matrix element.

It is necessary to turn these ideas into a dynamical theory of the absorption process. Nozières and deDominicis (1969) showed that this is done by examining the Green's function for the core hole:

$$G_h(t) = -i\langle T\, d(t)\, d^\dagger(0)\rangle = -i\theta(t)\langle|\, e^{iHt}\, de^{-iHt}\, d^\dagger\, |\rangle$$

They manipulate this correlation function to show that it can be treated as the many-electron response to the sudden switching on of the core hole, in the same way that the phonon response to sudden switching was studied in Sec. 4.3E. This correlation function is evaluated by following the same steps used on (8.3.11). The core hole creation operator $d^\dagger \,|\rangle$ acts on the ground state $|\rangle$ to create a single core hole. The operator $\exp(-itH)$ on this state is represented by the final state potential with the core hole:

$$e^{-itH}\, d^\dagger \,|\rangle = e^{-it(\omega_T + H_f)t}\, d^\dagger \,|\rangle$$

The core hole destruction operator d destroys the core hole state. The last time operator $\exp(itH)$ operates on a configuration with no core holes, so that it is H_g. The hole core Green's function can be exactly given by

$$G_n(t) = -ie^{-it\omega_T}\theta(t)\langle|\, e^{itH_g}e^{-itH_f}\,|\rangle \tag{8.3.15}$$

The core hole operators (d, $d^\dagger$) are omitted because they have done their work of determining the proper order of the Hamiltonian operators. There are several different ways to evaluate the time dependence of this function. One way is to recognize that the Hamiltonian H_f is a summation of the single-particle Hamiltonians for the individual conduction electrons. The hole Green's function can be expressed as a Slater determinant analogous to (8.3.12) for the absorption rate. In fact, it is just the function $\varrho(t)$ in (8.3.13) which is the matrix element over the ground state orbitals of $\exp(-ith_f)$:

$$\begin{aligned}
\varrho(t) &= \langle|\, e^{itH_g}e^{-itH_f}\,|\rangle = e^{itE_g}\det[\varphi_{\mathbf{pp}'}(t)]_{\mathbf{p},\mathbf{p}'<k_F} \\
G_h(t) &= -i\theta(t)e^{-it\omega_T}\varrho(t) \\
\varphi_{\mathbf{pp}'}(t) &= \langle\mathbf{p}\,|\, e^{-ih_f t}\,|\,\mathbf{p}'\rangle
\end{aligned} \tag{8.3.16}$$

The evaluation of this determinant is one possible way to find the core hole Green's function. Methods of numerical evaluation have been discussed by Schönhammer and Gunnarsson (1978).

A second way to find the core hole Green's function is by a linked cluster expansion. The correlation function in (8.3.16) may be written as a time-ordered operator, which is evaluated by the techniques described in Sec. 3.6:

$$\begin{aligned}
\varrho(t) &= \left\langle \left| T\exp\left[-i\int_0^t dt_1 V(t_1)\right]\right|\right\rangle \\
&= \sum_{n=0}^{\infty} \frac{(-i)^n}{n!}\int_0^t dt_1 \cdots \int_0^t dt_n \langle TV(t_1)\cdots V(t_n)\rangle = \exp\left[\sum_{l=1}^{\infty} F_l(t)\right] \\
F_l(t) &= \frac{1}{l}(-i)^l \int_0^t dt_1 \cdots \int_0^t dt_l \langle TV(k_1)\cdots V(t_l)\rangle_{\text{connected}}
\end{aligned} \tag{8.3.17}$$

The first term in the exponential resummation $F_1(t)$ is linear in t, and is a self-energy term:

$$F_1(t) = -i\int_0^t dt_1\langle| V(t_1) |\rangle = \frac{-it}{\nu}\sum_{\mathbf{k}} V(\mathbf{k}, \mathbf{k})n_F(\xi_{\mathbf{k}})$$

The exact self-energy of the core hole is known from Fumi's theorem in Sec. 4.1C to be

$$E_i = -\frac{2\hbar^2}{m}\sum_l (2l+1)\int_0^{k_F} \delta_l(k)k\,dk \tag{8.3.18}$$

The factor $V(\mathbf{k}, \mathbf{k})$ is the first term in an expansion which should give Fumi's result E_i when all terms are summed. It seems reasonable to use Fumi's result (8.3.18) for the self-energy, which includes all the terms linear in t in higher order. The threshold energy for the X-ray transition was previously given as ω_T, which is the value in the one-electron approximation, where we neglect the final state interactions. They provide a self-energy E_i which further lowers the threshold by this amount. Thus the new threshold energy in the interacting system is $\bar{\omega}_T = \omega_T + E_i$, which accounts for the increased binding of the core hole due to the polarization of the electron gas around it.

The second term in the exponential series $F_2(t)$ is the first term which is interesting, since it predicts an orthogonality catastrophe:

$$\begin{aligned}F_2(t) &= \frac{1}{2}(-i)^2\int_0^t dt_1\int_0^t dt_2\langle| TV(t_1)V(t_2) |\rangle \\ &= \frac{1}{2}\frac{(-i)^2}{\nu^2}\int_0^t dt_1\int_0^t dt_2 \sum_{\substack{\mathbf{k}_1\mathbf{k}_2\sigma \\ \mathbf{k}_3\mathbf{k}_4\sigma'}} V(\mathbf{k}_1, \mathbf{k}_2)V(\mathbf{k}_3, \mathbf{k}_4) \\ &\quad\times\langle| c^\dagger_{\mathbf{k}_1\sigma}(t_1)c_{\mathbf{k}_2\sigma}(t_1)c^\dagger_{\mathbf{k}_3\sigma'}(t_2)c_{\mathbf{k}_4\sigma'}(t_2) |\rangle\end{aligned}$$

The operators in the brackets are paired, according to Wick's theorem, into Green's functions of time. The time integrals are then evaluated directly:

$$\begin{aligned}F_2(t) &= \frac{1}{2}\frac{(-i)^2}{\nu^2}\int_0^t dt_1\int_0^t dt_2\sum_{\mathbf{k}_1\mathbf{k}_2\sigma} V(\mathbf{k}_1, \mathbf{k}_2)V(\mathbf{k}_2, \mathbf{k}_1)G(\mathbf{k}_1, t_1 - t_2) \\ &\quad\times G(\mathbf{k}_2, t_2 - t_1) \\ &= \frac{1}{2\nu^2}\int_0^t dt_1\int_0^t dt_2\sum_{\mathbf{k}_1\mathbf{k}_2\sigma} | V(\mathbf{k}_1, \mathbf{k}_2) |^2 e^{i(t_1-t_2)(\xi_2-\xi_1)}[\theta(t_1 - t_2) - n_1] \\ &\quad\times[\theta(t_2 - t_1) - n_2] \\ &= \frac{1}{\nu^2}\sum_{\mathbf{k}_1\mathbf{k}_2\sigma} | V(\mathbf{k}_1, \mathbf{k}_2) |^2\left[\frac{itn_1}{\xi_1 - \xi_2} - \frac{n_1(1-n_2)}{(\xi_1 - \xi_2)^2}\left(1 - e^{it(\xi_1-\xi_2)}\right)\right]\end{aligned}$$

The factor of $\frac{1}{2}$ in front vanished, because each term in the final result appears twice. The summation over spin index σ can be taken to produce another factor of 2 in front. The first term, which is linear in time, is dropped because it contributes only to the self-energy, which we already know from Fumi's theorem. Thus we consider the expression

$$F_2(t) = -\frac{2}{\nu^2} \sum_{\mathbf{k}_1\mathbf{k}_2} |V(\mathbf{k}_1, \mathbf{k}_2)|^2 \frac{n_F(\xi_{\mathbf{k}_1})[1 - n_F(\xi_{\mathbf{k}_2})](1 - e^{it(\xi_{\mathbf{k}_1} - \xi_{\mathbf{k}_2})})}{(\xi_{\mathbf{k}_1} - \xi_{\mathbf{k}_2})^2}$$

This term does predict an orthogonality catastrophe and thus has been extensively investigated; see Minnhagen (1977) for an evaluation. Here we shall give only an indication of the type of effects which occur. We rewrite this expression by assuming that the hole potential $V(\mathbf{k}, \mathbf{k}')$ depends only on the difference of its arguments $V(\mathbf{k} - \mathbf{k}')$ and change variables to $\mathbf{k}_2 = \mathbf{k}_1 + \mathbf{q}$:

$$F_2(t) = -\int_0^\infty \frac{du}{u^2} R_e(u)(1 - e^{-iut})$$

$$R_e(u) = \frac{1}{\nu\pi} \sum_{\mathbf{q}} V(q)^2 \Lambda(\mathbf{q}, u) \tag{8.3.19}$$

$$\Lambda(\mathbf{q}, u) = \frac{2\pi}{\nu} \sum_{\mathbf{k}} n_F(\xi_{\mathbf{k}})[1 - n_F(\xi_{\mathbf{k}+\mathbf{q}})]\delta(u + \xi_{\mathbf{k}} - \xi_{\mathbf{k}+\mathbf{q}})$$

The factor $\Lambda(\mathbf{q}, u)$ is recognized as the imaginary part of the polarization diagram $P^{(1)}$ of a single-electron bubble. It was evaluated in Sec. 5.5:

$$\Lambda(\mathbf{q}, u) = -\text{Im}[P^{(1)}(\mathbf{q}, u)] = u\left(\frac{m^2}{2q\pi}\right)\theta(2k_F - q)$$

The important feature of this result is its linear dependence on the frequency, which applies at small u. The wave vector integration over q can be done for realistic potentials $V(q)$. A crude approximation to the result is $2(N_F V_0)^2$, where $N_F = mk_F/2\pi^2$ is the density of single spin states at the Fermi energy, while V_0 is the magnitude of the electron–hole potential. Thus we write

$$R_e(u) = \begin{cases} ug, & u < \xi_0 \\ 0, & u > \xi_0 \end{cases}$$

where

$$g = \frac{m^2}{2\pi^2} \int \frac{d^3q}{(2\pi)^3} \frac{1}{q} V(q)^2 \theta(2k_F - q)$$
$$\simeq 2(N_F V_0)^2$$

The upper limit of u is ξ_0, which is determined by the range over which $R(u)$ is linear in u. This range is approximately $\xi_0 \simeq E_F$, as shown in Sec. 5.5:

$$F_2(t) = g \int_0^{\xi_0} \frac{du}{u} (1 - e^{-iut})$$

The remaining du integral is interesting. The integrand has the factor $\int_0 du/u$ which could diverge as ln 0. In this case it does not diverge, since the other factor $1 - e^{-iut} \simeq iut$ in the limit where $u \to 0$. Thus the integral is finite. However, for large times $\xi_0 t \gg 1$ this cutoff of the logarithmic divergence happens at lower and lower frequencies. We can express this dependence approximately as

$$F_2(t) = -g \int_{-1/it}^{\xi_0} \frac{du}{u} \simeq -g \ln(1 + it\xi_0)$$

The logarithm has the correct two limits of $-gi\xi_0 t$ at small time and $\ln it\xi_0$ at large times. This approximation completes our evaluation of the renormalization factor:

$$\varrho(t) = \exp[-itE_i - g \ln(1 + it\xi_0)]$$

Its Fourier transform determines the spectral function of the core hole:

$$A_h(\omega) = -2 \operatorname{Im}\left[\int_{-\infty}^{\infty} dt e^{i\omega t} G_h(t)\right]$$
$$= 2 \operatorname{Re}\left[\int_0^{\infty} dt e^{it(\omega - \omega_T)} \varrho(t)\right]$$

The real part of the integral is the same as the integral plus its complex conjugate:

$$A_h(\omega) = \int_0^{\infty} dt e^{it(\omega - \omega_T)} \varrho(t) + \int_0^{\infty} dt e^{-it(\omega - \omega_T)} \varrho^*(t)$$

The variable change $t \to -t$ in the second term, along with $\varrho^*(-t) = \varrho(t)$, brings us to the equivalent expression

$$A_h(\omega) = \int_{-\infty}^{\infty} dt e^{it(\omega - \omega_T)} \varrho(t)$$
$$= \int_{-\infty}^{\infty} dt \frac{e^{it\xi_0 \Omega}}{(1 + it\xi_0)^g}$$
$$\Omega = \frac{\omega - \omega_T - E_i}{\xi_0}$$

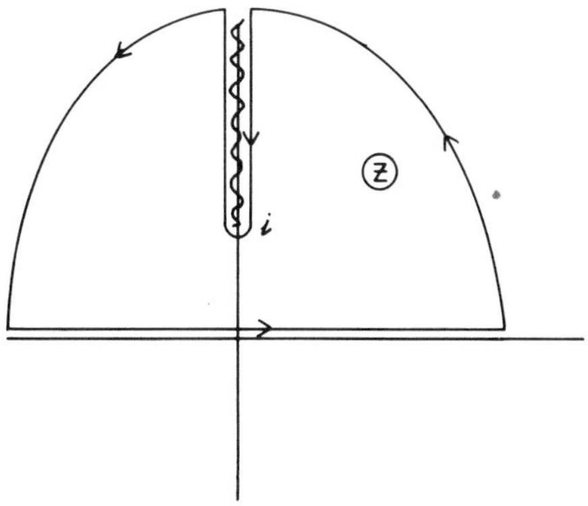

FIGURE 8.14. Path of integration in evaluating (8.3.20).

The integral is evaluated by a contour integration. Change the integration variable to $Z = \xi_0 t$. There is a branch point at $Z = i$. The branch cut is drawn vertically from this point, up the imaginary axis, as shown in Fig. 8.14:

$$A_h(\omega) = \frac{1}{\xi_0} \int_{-\infty}^{\infty} dZ \frac{e^{iZ\Omega}}{(1 + iZ)^g} \tag{8.3.20}$$

For $\Omega < 0$, the integral $(-\infty, \infty)$ is closed by a contour in the lower half plane (LHP). This closed contour encloses no poles or branch cuts and so gives zero and shows that $A_h(\omega) = 0$ for $\Omega < 0$. The other situation is $\Omega > 0$, and here the contour of integration is closed in the UHP, as shown in Fig. 8.14. The contour encloses no poles, so the contribution on the real axis $(-\infty, \infty)$ is equal to that along the cut. For the integration along the branch cut, change variables to $Z = i(1 + y)$. The denominator in (8.3.20) becomes $(1 + iZ)^g = [1 - (1 + y)]^g = (-y)^g = y^g \exp(\pm i\pi g)$, where the choice of $(\pm)$ in the exponent of $\exp(ig\pi)$ depends on the side of the branch cut.

The integrals along the branch cut give $[\Omega = (\omega - \bar{\omega}_T)/\xi_0]$

$$\begin{aligned} A_h(\omega) &= \frac{\theta(\Omega)}{\xi_0} i \int_0^{\infty} dy \frac{e^{-\Omega(1+y)}}{y^g} (e^{-i\pi g} - e^{i\pi g}) \\ &= \theta(\Omega) \frac{2 \sin \pi g}{\xi_0} \Gamma(1 - g) \frac{e^{-\Omega}}{\Omega^{1-g}} \end{aligned} \tag{8.3.21}$$

The shape of this spectral function is shown in Fig. 8.15. It is zero for $\Omega < 0$ and finite for $\Omega > 0$. It diverges as a power law for $\Omega \to 0^+$, with the exponent $1 - g$. It is easy to check that the area under the spectral

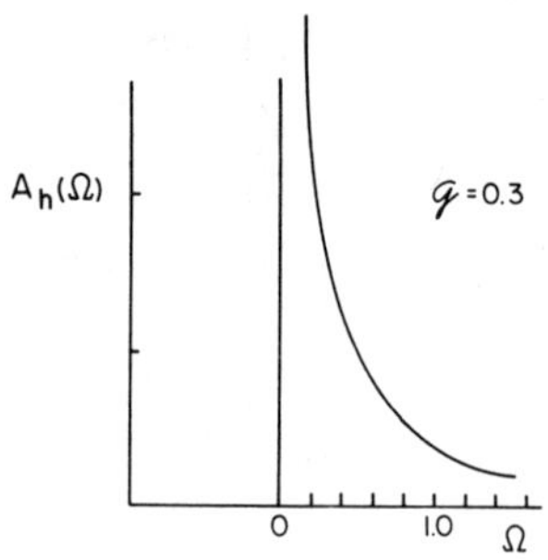

FIGURE 8.15. Spectral function of the hole Green's function. The orthogonality catastrophe causes the power law distribution and eliminates any delta function character.

function is 2π:

$$1 = \int_{-\infty}^{\infty} \frac{d\omega}{2\pi} A_h(\omega) = \frac{1}{\pi} \sin(\pi g)\Gamma(1-g) \int_0^{\infty} d\Omega \frac{e^{-\Omega}}{\Omega^{1-g}}$$

$$1 = \frac{1}{\pi} \sin(\pi g)\Gamma(1-g)\Gamma(g)$$

The last equation is an identity among gamma functions.

Thus the response of the electron gas has a dramatic effect upon the spectral function of the hole. The spectral function has a power law divergence near the threshold energy Ω^{g-1}. In the absence of interactions $V = 0$, the spectral function is a delta function $A_h^{(0)} = 2\pi\delta(\omega - \omega_T)$. As illustrated schematically in Fig. 8.16(a), the influence of interactions is to change the delta function into a power law divergence. All quasiparticle behavior, in the form of a remnant delta function, has vanished.

In X-ray absorption, the unperturbed spectrum for $V = 0$ is a step $\theta(\omega - \omega_T)$ rather than a delta function, which is illustrated in Fig. 8.16(b). How does this step become modified by the potential V: by the orthogonality catastrophe? The answer is that the threshold becomes a power law Ω^g

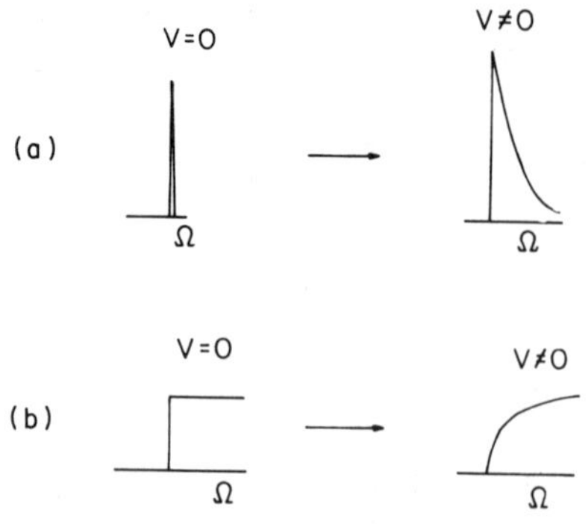

FIGURE 8.16. The effects of the orthogonality catastrophe upon spectral distributions. (a) The delta function of a spectral function is changed to a power law divergence. (b) The step function in absorption is changed to a converging threshold.

which goes to zero as $\Omega \to 0$. This threshold may be derived in the following way. The step function $\theta(\omega - \omega_T)$ in Fourier transform is t^{-1}:

$$\theta(\omega - \omega_T) = \frac{1}{2\pi i} \int_{-\infty}^{\infty} \frac{dt e^{i(\omega-\omega_T)t}}{t - i\delta}$$

The orthogonality catastrophe modifies this spectrum by the inclusion of the term $\varrho(t)$ in the integrand. Call the resulting X-ray spectra $X(\omega)$:

$$X(\omega) = \frac{1}{2\pi i} \int_{-\infty}^{\infty} \frac{dt}{t - i\delta} \frac{e^{it\Omega\xi_0}}{(1 + it\xi_0)^g}$$

The easiest way to evaluate this integral is to note that its derivative is just the hole spectral function

$$\frac{\partial}{\partial\omega} X(\omega) = \frac{1}{2\pi} A_h(\omega)$$

This equation has the solution

$$X(\omega) = \frac{1}{2\pi} \int_{-\infty}^{\omega} d\omega' A_h(\omega') = \frac{\theta(\Omega)}{\Gamma(g)} \int_0^{\Omega} d\Omega' \frac{e^{-\Omega'}}{(\Omega')^{1-g}}$$

The behavior near threshold may be deduced by examining the integral for very small values of Ω. Then the slowly varying term $\exp(-\Omega')$ can be ignored, and the integral gives

$$X(\omega) \simeq \frac{\theta(\Omega)}{\Gamma(g)} \int_0^{\Omega} \frac{d\Omega'}{(\Omega')^{1-g}} = \frac{\Omega^g \theta(\Omega)}{\Gamma(1 - g)} \tag{8.3.22}$$

Equation (8.3.22) shows that the absorption behaves as Ω^g near threshold. The threshold has a converging behavior, with the absorption starting at zero when $\Omega = 0$. It is sketched in Fig. 8.16(b).

The absorption is zero at the threshold frequency $\Omega = 0^+$ because of the orthogonality catastrophe. The transition is not allowed because of orthogonality and has a zero probability. The absorption is finite for $\Omega > 0$ because of the creation of a number of electron–hole pairs in the system. This serves as a symmetry-breaking process, which makes the transition allowed.

The role of electron–hole pairs in the X-ray transition was elucidated by Schotte and Schotte (1969). They showed that the orthogonality catastrophe could be explained as an example of the Tomonaga model of Sec. 4.4. The interaction term between the core hole and the electron gas can be written in terms of the electron density operator $\varrho(\mathbf{q}) = \sum_{\mathbf{k}} c^{\dagger}_{\mathbf{k}+\mathbf{q}} c_{\mathbf{k}}$:

$$\frac{1}{\nu} \sum_{\mathbf{kk}'} V(\mathbf{k} - \mathbf{k}') c_{\mathbf{k}}^{\dagger} c_{\mathbf{k}'} d^{\dagger} d = \frac{1}{\nu} \sum_{\mathbf{q}} V(\mathbf{q}) \varrho(\mathbf{q}) d^{\dagger} d$$

The density operator can be represented by the boson operators for the electron–hole excitations of the electron gas $\varrho(\mathbf{q}) \to (b_{\mathbf{q}} + b^{\dagger}_{-\mathbf{q}})$. The Hamiltonian between the core hole and the electron gas is thereby transformed into a problem of a localized level and a boson system. This model Hamiltonian can be solved exactly as the independent boson model of Sec. 4.3. The solution is entirely analogous to the phonon–core hole system. The response of the boson system to the sudden appearance of the core hole is described by including the factor of $\exp[-\phi(t)]$ in the time integrals. The factor $\phi(t)$ for phonons is presented in (8.3.4) and is rewritten as

$$\phi(t) = i\Sigma t + \int_0^{\infty} \frac{d\nu}{\nu^2} R_{\text{ph}}(\nu) \{ [n_B(\nu) + 1](1 - e^{-i\nu t}) + n_B(\nu)(1 - e^{i\nu t}) \}$$
$$R_{\text{ph}}(\nu) = \sum_{\lambda, \mathbf{q}} | M_{\lambda}(q) |^2 \delta[\nu - \omega_{\lambda}(q)] \tag{8.3.23}$$

The self-energy term $i\Sigma t$ is put first, while the other term is the transient behavior. The approximate response of the electron system to the core hole was given earlier as $F_2(t)$ in (8.3.19). It has exactly the same form as the preceding, except the phonon system is at finite temperature. The electron–electron response at finite temperature is

$$\phi(t) = -F_2(t) = \int_0^{\infty} \frac{d\nu}{\nu^2} R_{\text{e}}(\nu) \{ [n_B(\nu) + 1](1 - e^{-i\nu t}) + n_B(\nu)(1 - e^{i\nu t}) \}$$
$$R_{\text{e}}(\nu) = -\frac{1}{\pi} \int \frac{d^3 q}{(2\pi)^3} V(q)^2 \, \text{Im}[P^{(1)}(\mathbf{q}, \nu)] \tag{8.3.24}$$

The quantity $R_{\text{ph}}(\nu)$ for the phonon system is the phonon density of states, which is weighted by the matrix element for the process. Similarly, the factor $R_{\text{e}}(\nu)$ serves the same role for electron–electron interactions. It is interpreted as the effective coupling to the electron–hole pairs. Thus the equations are exactly alike for the response of the core hole–phonon and core hole pairs, and enter into the spectral function for absorption in the same fashion as factors $\exp(-\phi)$. This similarity is reasonable, since both the phonons and electron–hole pairs are boson systems which respond in the same way to the appearance of the core hole.

The only difference between these two boson systems is their respective coupling constants $R_{\text{ph}}(\nu)$ and $R_{\text{e}}(\nu)$. This is a big difference, since they behave quite differently at small values of ν. We have already shown that the electron–electron interactions have a linear relation $R_e(\nu) \simeq \nu$ which leads to power law behavior in $A(\omega)$. The phonon system is governed by the acoustic phonons at small values of ν and usually behaves as $R_{ph} \simeq \nu^3$ in metals. Consequently, the exponential factor for the phonon system does *not* diverge as $\phi(t) \simeq \ln(it\xi_0)$ at large $\xi_0 t$ but goes to a constant except for the self-energy term. The spectral function $A(\omega)$ from the phonon part alone is a dull function, which we already noted could usually be represented as a Gaussian. The core hole–phonon and core hole pair couplings cause quite different behaviors in the time response of the system and affect the spectral shape differently. The difference arises from the dissimilar nature of their coupling function $R_{ph}(\nu)$ and $R_e(\nu)$.

The electron–electron response is an example of a type of phenomenon known as an *infrared divergence*. The divergence occurs in the limit where $\nu \to 0$, which is the infrared part of the spectrum. The quantity $R_j(\nu)/\nu^2$ is the probability (at $T = 0$) of emitting an excitation of energy ν during the sudden switching on of the core hole potential. For example, the number Nu of excitations created is

$$\text{Nu} = \int_0^\infty \frac{d\nu}{\nu^2} R_e(\nu) \simeq g \int_0^{\xi_0} \frac{d\nu}{\nu} \simeq -g \ln 0$$

The average number Nu of bosons around the particle was discussed earlier for polarons in Sec. 4.1. It is infinity for the electron–hole excitations, which shows that an infinite number are created. This infinity does not imply a physical disaster, since all measurable quantities are finite. For example, the energy which is released by the creation of these electron–hole pairs is

$$\Delta E = \int_0^\infty \frac{d\nu}{\nu^2} R_e(\nu)\nu \simeq g \int_0^{\xi_0} d\nu \simeq g\xi_0$$

which is finite. However, a divergence in the number Nu of excitations is usually an indication that the spectral function will have an unusual shape, since quasiparticle behavior is eliminated. A further discussion is given by Hopfield (1969) and Mahan (1974). The phonon system does not have an infrared divergence in the X-ray problem in metals, although it can in semiconductors where the piezoelectric interaction is not screened.

We have been representing the exponential series for $F(t) = \sum F_i(t)$ in (8.3.17) by just the term $F_2(t)$ plus $iE_i t$. The further terms in the series also contribute to the orthogonality catastrophe and should be summed in order to obtain an accurate description of the phenomenon. The exact method of summing all of these terms has been given by Mahan (1982), and more elegently by Ohtaka and Tanabe (1983). They provide an exact expression for $R_e(\nu)$.

Several other types of results have been obtained. The most important is by Nozières and deDominicis (1969), who show that the exact coefficient of the term which diverges like $\ln(it\xi_0)$ is

$$F_n(t) = \frac{2}{n} \sum_{\mathbf{k}_1 \cdots \mathbf{k}_n} V(\mathbf{k}_1, \mathbf{k}_2) V(\mathbf{k}_2, \mathbf{k}_3) \cdots V(\mathbf{k}_n, \mathbf{k}_1) \int_0^t dt_1 \cdots \int_0^t dt_n$$
$$\times G(\mathbf{k}_1, t_1 - t_2) G(\mathbf{k}_2, t_2 - t_3) \cdots G(\mathbf{k}_n, t_n - t_1)$$

The summation of the series is difficult and has not been accomplished in closed form, except as the determinant (8.3.16). Several other types of results have been obtained. The most important is by Nozières and deDominicis (1969), who show that the exact coefficient of the term which diverges like $\ln(it\xi_0)$ is

$$\lim_{t \to \infty} F(t) = -itE_i + \alpha \ln(it\xi_0)$$
$$\alpha = \sum_{m_s, m_l, l} \left[\frac{\delta_l(k_F)}{\pi} \right]^2$$

The same phase shifts $\delta_l(k_F)$ were mentioned earlier and arise from the scattering of the conduction electron with wave vector k_F from the core hole. The core hole is in the center of a spherical coordinate system, and the conduction electron states are described by the quantum numbers of m_s, l, m_l, and k. The phase shifts are usually taken to be independent of m_s and m_l, in which case the coefficient becomes

$$\alpha = 2 \sum_l (2l + 1) \left[\frac{\delta_l(k_F)}{\pi} \right]^2$$

The coefficient α should be used in describing the asymptotic behavior of the orthogonality catastrophe very near the threshold. The frequency range over which the threshold expression $X(\omega) \simeq \Omega^\alpha$ can be applied is still being debated.

A feature of the theory which must be improved is to include the frequency dependence of the dielectric screening of the core hole. Thus we write the screened core hole potential $V(q) = V^{(0)}(q)/\varepsilon(q)$, where $V^{(0)}(q)$ is the unscreened part. Langreth (1970) showed that $R_e(\nu)$ in (8.3.24) should actually be written as

$$R_e(\omega) = \frac{1}{\pi\nu}\sum_{\mathbf{q}} \frac{V^{(0)}(q)^2}{v_g}\,\mathrm{Im}\left[\frac{1}{\varepsilon(\mathbf{q}, \omega)}\right]$$
$$= \frac{-1}{\pi\nu}\sum_{\mathbf{q}} \left|\frac{V^{(0)}(q)}{\varepsilon(\mathbf{q}, \omega)}\right|^2 \mathrm{Im}[P^{(1)}(\mathbf{q}, \omega)]$$

The core hole potential $V^{(0)}(q)$ should be screened by the frequency-dependent dielectric function $\varepsilon(\mathbf{q}, \omega)$. Minnhagen (1977) has shown that this feature is important for determining the shape of the spectral function of the core hole.

The orthogonality catastrophe is an important feature of many-electron physics. The process of switching on a potential occurs in many circumstances, and the physics can be applied to a variety of phenomena. For example, the Kondo effect has this feature, which plays an important role in its theory. Whenever the potential is switched on, the system responds by making electron–hole pairs, which affects the spectral response of all correlation functions which are being evaluated.

D. MND Theory

The absorption of X-rays in a metal causes many different types of excitations and phenomena. Two have been discussed which cause power law behavior at the absorption edges: excitons and the orthogonality catastrophe. An important treatment of this physics was by Nozières and deDominicis (1969), who solved the two processes together. Their solution is asymptotically exact near threshold $\bar{\omega}_T$. It expresses the power law exponents as functions of the angular momentum l of the conduction electron in its final state. Thus the many-body correction to formula (8.3.2) with two angular momentum channels $A_{l\pm 1}$ is

$$A(\omega) = \theta(\omega - \bar{\omega}_T)\left[A_{l+1}\left(\frac{\xi_0}{\omega - \bar{\omega}_T}\right)^{\alpha_{l+1}} + A_{l-1}\left(\frac{\xi_0}{\omega - \bar{\omega}_T}\right)^{\alpha_{l-1}}\right]$$
$$\alpha_l = \frac{2\delta_l(k_F)}{\pi} - \alpha \qquad (8.3.25)$$
$$\alpha = 2\sum_l (2l+1)\left[\frac{\delta_l(k_F)}{\pi}\right]^2 \equiv g$$

The term power law exponents α_l are the sum of two terms. The first is $2\delta_l/\pi$, which is from the exciton phenomenon; this term is usually positive in free-electron metals and tends to make the edges diverge and become

TABLE 8.2. $\delta_\lambda(k_F)$ for Na[a]

l	δ_l	α_l
0	0.90	0.38
1	0.22	−0.05
2	0.01	−0.19

$$\alpha_l = \frac{2\delta_l}{\pi} - \alpha$$

$$\alpha = 0.19$$

[a] Minnhagen (1977).

singular. The other factor α comes from the orthogonality catastrophe and tends to make the edges converging at threshold. The final value of α_l for an absorption edge depends on the difference of these two quantities and may be either positive or negative.

There have been many calculations of these phase shifts and exponents for comparison with those deduced from experiments. Typical results for sodium are shown in Table 8.2, as calculated by Minnhagen (1977). He used a pseudopotential, which was screened by a Singwi–Sjölander dielectric function. The phase shifts should always obey the Friedel sum rule (4.1.22),

$$1 = \frac{2}{\pi} \sum_l (2l + 1)\delta_l(k_F)$$

which is just a statement that the screening around the core hole must be one unit of charge. The exponents are $\alpha_0 = 0.4$, $\alpha_1 \simeq 0$, $\alpha_2 \simeq -\alpha \simeq -0.2$, which is typical for simple metals. The X-ray absorption from a p shell, where the initial angular momentum is $l = 1$, has the final values for the conduction electron of $l = 0$ and $l = 2$. Thus the asymptotic limit of the Mahan–Nozières–deDominicis (MND) theory for $\omega \gtrsim \bar{\omega}_T$ is

$$A(\omega) = \theta(\omega - \bar{\omega}_T)\left[A_0\left(\frac{\xi_0}{\omega - \bar{\omega}_T}\right)^{\alpha_0} + A_2\left(\frac{\xi_0}{\omega - \bar{\omega}_T}\right)^{\alpha_2}\right]$$

The first term is the s-wave channel. The exponent α_0 is positive, so this term diverges at threshold. The second is the d-wave channel, $\alpha_2 < 0$, and the term goes to zero at threshold. The summation of these two terms diverges at threshold. The theory predicts that absorption edges are divergent for electrons from p shells, which is generally observed in free-electron metals such as Na, Mg, Al, and K. On the other hand, if the initial core state is an s shell with $l = 0$, the final state of the conduction electron has $l = 1$, $\alpha_1 \simeq -0.05$ for sodium. This value is so small that it predicts there is no measurable effect from the many-electron response. This prediction is also in accord with experimental observations. The value of α_1 is small because of the partial cancellation of $2\delta_1/\pi = 0.14$ and $\alpha = 0.19$. It seems to be a general feature of free-electron metals that the transitions from core hole s states shows no additional peaking or rounding of the threshold, in agreement with the predictions that α_1 is very small. Thus the MND theory seems to qualitatively explain the behavior of the edges of simple metals.

Figure 8.17 shows the experimental results of Callcott *et al.* (1978) for the absorption edge of the $L_{2,3}$ shell of metallic sodium. This p shell is split by the spin–orbit interaction into the $p_{3/2}$ and $p_{1/2}$ components, which comprise the L_3 and L_2 shells. In sodium, this splitting is 0.17 eV and accounts for the double edge shown in absorption. Each of these two edges should have a power law singularity. Figure 8.17 also shows the deconvolution of the sodium spectrum into the part due to the one-particle spectrum and broadening processes such as the Lorentzian width from Auger and the Gaussian width from phonons. The absorption edges show definite spikes which are from the edge singularities. The spikes are much smaller in emission, although that is partly from the increased phonon width in emission due to the incomplete relaxation phenomenon, as mentioned earlier. The value of the edge singularity exponent deduced by Callcott *et al.* to fit their absorption data is $\alpha_0 = 0.25$, which is smaller than the $\alpha_0 = 0.38$ predicted from the phase-shift calculations. The theoretical values of α_0 have been computed by many different workers, and values similar to those shown in Table 8.2 are obtained by many different groups, so the theoretical values seem reliable.

There have been several explanations for this lack of detailed agreement between the measured and theoretical exponents: (1) omission of frequency-dependent screening in the theory; (2) exchange type of effects between the core hole spin and electron spin, as discussed by Girvin and Hopfield (1976); (3) scattering of the core hole among its different orbital states in the L shell; and (4) the asymptotic solution (8.3.23) is not valid very far from threshold. The last suggestion is supported by the calculations of Grebennikov *et al.* (1977), who numerically solved the integral equations of Nozières and deDominicis. They found the edge singularities to be wider

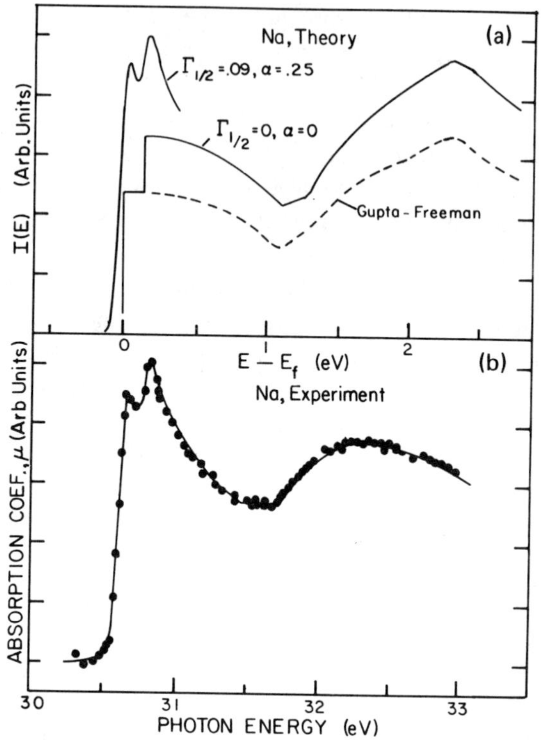

FIGURE 8.17. The X-ray absorption edge of the $L_{2,3}$ shell of metallic sodium. The data is shown in part (b). Part (a) shows how the band structure prediction of Gupta and Freeman is convoluted with the edge singularity ($\alpha = 0.25$) and broadening functions to fit the spectra. *Source*: Callcott *et al.* (1978) (used with permission).

than the values predicted from the asymptotic formulas, which would imply that the effective exponents should be smaller. Their numerical results are in the direction of explaining the small discrepancy between theory and experiment. This subject is still an active area of research. A summary of the present status of this field seems to be that the edge singularities are observed in some thresholds of simple metals, and the refinements of the theory are still under investigation.

E. XPS Spectra

X-ray photoelectron spectroscopy (XPS) measures the line shapes of photoelectrons excited from the core levels of atoms. A photon, which is usually in the kilovolt energy range, excites a core electron to very high kinetic energy. Since the binding energies E_B of the L-shell electrons in Na, Mg, and Al are only 50–100 eV, the final kinetic energy of the photo-electron from this shell is also in the kilovolt range, $E_f = \hbar\omega - E_B$. In a one-electron model, all the photoelectrons excited by the same frequency ω would have exactly the energy E_f, and a measurement of the distribution of kinetic energies of the photoelectrons would show a delta function at this value. The actual kinetic energies of photoelectrons, measured from simple metals, show the main peak at energy $E \simeq E_f$ plus satellite peaks which correspond to the emission of bulk and surface plasmons (Pardee *et al.*, 1975). The main peak at $E \simeq E_f$ has a finite width, which must be partly caused by the usual mechanisms of (1) Auger lifetime of the core hole and (2) phonon broadening due to the coupling between the core hole and the phonons.

Doniach and Šunjić (1970) predicted that the orthogonality catastrophe should make the shape of the main line, at $E \simeq E_f$, asymmetric with an asymmetry parameter which is a direct measurement of the index α. Their argument is that this experiment directly measures the spectral function of the core hole, since the outgoing electron has too much energy to be affected by the interaction with phonons or electron–hole pairs. In the calculation of the Kubo formula for the absorption rate, the conduction electron operators $[c_{\mathbf{k}}(t), c^{\dagger}_{\mathbf{k}'}]$ can be removed from the correlation function on the grounds that they leave the core hole too rapidly to be influenced by exciton or other final state processes. Thus we factor the correlation function:

$$A(\omega) = \frac{1}{\omega}\int_{-\infty}^{\infty} dt e^{i\omega t}\langle \mathscr{J}(t)\mathscr{J}(0)\rangle$$

$$= \frac{1}{\omega}\int_{-\infty}^{\infty} dt e^{i\omega t}\sum_{\mathbf{k}\mathbf{k}'} w(\mathbf{k})^* w(\mathbf{k}')\langle c_{\mathbf{k}}(t)c^{\dagger}_{\mathbf{k}'}(0)\rangle\langle d(t)d^{\dagger}(0)\rangle$$

$$= \frac{1}{\omega}\int_{-\infty}^{\infty} dt \sum_{\mathbf{k}} e^{it(\omega-\xi_{\mathbf{k}})} w(\mathbf{k})^2\langle d(t)d^{\dagger}(0)\rangle$$

The Kubo formula for the absorption becomes proportional to the spectral function $A_h(\omega - \xi_k)$ for the core hole. In this experiment, the photon energy ω is usually fixed, and measurement is made of the distribution $P(E)$ of final kinetic energies $E \equiv \xi_k - E_W$ of the conduction electron, where E_W is the work function of the metal. XPS spectra can now be measured with millivolt energy resolution, although the electron energy E is often over a kilovolt. The matrix elements $w(\mathbf{k})$ are a constant over the narrow energy

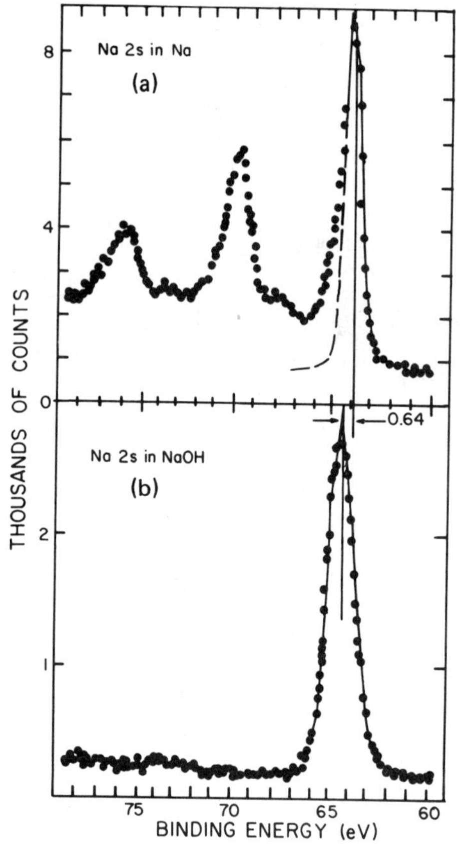

FIGURE 8.18. X-ray photoelectron spectra from sodium 2s electrons in (a) metal and (b) hydroxide. The metal spectra is asymmetric, with additional contributions on the low-energy side due to pair production. These are lacking in the insulator. Peaks to the left are plasmons in the metal. *Source*: Citrin (1973) (used with permission).

range of $\xi_{\mathbf{k}}$ which are measured for each XPS line:

$$P(E) = w^2 A_h(\omega - E - E_W) = w^2 \int_{-\infty}^{\infty} dt e^{it(\omega - \omega_T - E - E_W)} \varrho(t) e^{-(1/2)\Gamma|t| - \gamma t^2}$$

This probability is just the spectral function of the core hole $A_h(\omega - E - E_W)$. It has the power law singularity in (8.3.21) and is illustrated in Fig. 8.15, which is then broadened by the phonons and the Auger decay. The line shape is predicted to be asymmetric because of the orthogonality catastrophe, and the degree of asymmetry is just given by the index $\alpha = 2\sum_l (2l + 1)(\delta_l/\pi)^2$. The formula used for interpreting the experimental The line-shape function $f(E)$ contains the Lorentzian width from Auger

TABLE 8.3. Orthogonality Index α

	Na	Mg	Al
Experiment[a]	0.20	0.13	0.12
Theories			
Minnhagen[b]	0.19	0.13	0.11
Almbladh–von Barth[c]	0.20		0.13
Bryant–Mahan[d]		0.12	0.10

[a] Citrin *et al.* (1977).
[b] Minnhagen (1977).
[c] Almbladh and von Barth (1976).
[d] Bryant and Mahan (1978).

line shapes is $\varrho(t) = \exp[-itE_i - \alpha \ln it)$ or

$$P(E) = w^2 f(E - E_f)$$

$$f(E) = \int_{-\infty}^{\infty} dt \exp(-itE - \tfrac{1}{2}\Gamma \mid t \mid - \gamma t^2 - \alpha \ln it)$$

$$E_f = \omega - \omega_T - E_i - E_W$$

processes, the Gaussian width from phonons, and the asymmetry from the electron–hole pair creation.

An early experimental result is shown in Fig. 8.18 from data of Citrin (1973) for metallic sodium. The dashed line indicates the excess line shape on the low-energy side. Many more spectra have been published by Citrin *et al.* (1977). One feature of the experimental results is that the same index α fits the asymmetry for different core holes in the same atom: The same α is found for the different shells K, L_2, L_3, etc. Thus the ion with a core hole looks the same, to electron–hole pairs which are outside the ion, regardless of the particular state of the core hole.

Table 8.3 shows the experimental values of this index for Na, Mg, and Al. Also shown are the theoretical values deduced by calculating the phase shifts. The results of Minnhagen (1977) were obtained from a screened pseudopotential. He shows that the results depend more on the choice of screening function than the pseudopotential. Different pseudopotentials give the same result as long as they were chosen to fit other data such as atomic spectra. But the various dielectric functions such as Thomas–Fermi, the RPA, or Singwi–Sjölander give quite different results for α_l and α, even when the phase shifts all obey the Friedel sum rule. The Singwi–Sjölander dielectric function gives results in best agreement with the experimental values of α, while Thomas–Fermi is worst. This ordering is in agreement with the ranking of the model dielectric functions in Chapter 5. The other two sets of theoretical results listed in Table 8.3 were obtained using realistic calculations of the screening charge from the theory of the inhomogeneous electron gas. These values of the orthogonality index α are in good agreement with the experiments and the calculations of Minnhagen.

The XPS line shape is asymmetric, because of the emission of electron–hole pairs during the creation of the core hole. The photon energy is fixed at ω, and any energy used to make the pairs must be subtracted from that carried away by the conduction electron. Thus the conduction electron has a high probability of having energy less than E_f. This explanation accounts for the sign of the asymmetry and explains why the line tails to the left in the Fig. 8.18. These asymmetric line shapes had been observed for many years, as reviewed by Parratt (1959). He speculated then that the "excess width on the low energy side is essentially attributable to transitions between excitation states of the valence-electron-configuration type." The quantitative theory of Doniach and Šunjić has verified this hypothesis in great detail and provides the most direct method of measuring the orthogonality catastrophe.

PROBLEMS

1. Derive the finite temperature form of the free-polaron absorption (8.1.15) assuming Maxwell–Boltzmann statistics. To order α, an exact result can be obtained in terms of Bessel functions $K_1(\frac{1}{2}\beta \mid \omega \pm \omega_0 \mid)$.

2. Derive (8.1.23).

3. Consider the force–force correlation function for scattering from an impurity. Discuss whether the multiple scattering from an impurity can be represented by a T matrix or similar series.

4. Derive the Wilson–Butcher formula (8.1.20) from the golden rule. Find the matrix element $\langle f \mid \hat{\varepsilon} \cdot \mathbf{p} \mid i \rangle$ by writing both initial $\mid i \rangle$ and final $\mid f \rangle$ Bloch states to first order in the potential $V_{\mathbf{G}}$.

5. Consider a Hamiltonian which is the homogeneous electron gas plus the crystal potential $\sum V_{\mathbf{G}} \varrho(\mathbf{G})$. Discuss the summation of terms which occurs in higher order when evaluating the force–force correlation function.

6. Derive (8.1.42) from the correlation function (8.1.29). Remember that each vertex where $i\omega$ enters must have a phonon line, but other internal lines are $W(0) = v_{\mathbf{q}} + V_{\mathrm{ph}}$.

7. Derive (8.1.43).

8. Derive a formula for $\mathrm{Re}[\sigma_0(\omega)]$ at $T = 0$ which results from $M_b(\mathbf{q}, i\omega)$ in (8.1.43). Show that both integrals du and du' can be done analytically for a phonon Green's function $D_\lambda(\mathbf{q}, u) = -2\omega_\lambda/(\omega_\lambda^2 - u^2 - i\delta)$ and derive the formulas.

9. Derive (8.1.44). Then evaluate this expression at zero temperature, and derive a formula for $\mathrm{Re}[\sigma_d(\omega)]$ which is a single-frequency integral over algebraic combinations of the two self-energy functions $\Sigma(\mathbf{k}, \omega)$ and $S(\mathbf{k}, \omega)$.

10. Where do the final state interactions for the Coulomb scattering of the electron and hole appear in formula (8.1.20) for interband transitions?

11. Show that the rate of photon emission in an insulator with a nonequilibrium distribution of electrons and holes is governed by the matrix element $|w_f(\mathbf{k})|^2$ as in (8.2.18). Assume the initial state of the system at zero temperature is $|k\rangle = d^\dagger a_k^\dagger |0\rangle$, and use arguments analogous to those following (8.2.19).

12. Write out the correlation function $\pi^{(1)}$ in (8.2.8), with one vertex diagram, for the frequency-dependent screening function (6.3.7). Do the Matsubara summations.

13. Show there is an infrared divergence in the X-ray response resulting from the piezoelectric electron–phonon interaction (1.3.7) in insulators when used in the response function (8.3.23).

14. Solve the Tomonaga model for the orthogonality catastrophe. Work in a spherical coordinate system, where the potential is at the center of a large sphere of radius R. Normalize the wave function in this sphere. Keep only s-wave terms. This is a one-dimensional problem in the radial variable.

15. Consider the X-ray edge problem with an interaction term H_{sd} in (1.4.19) between the conduction electrons and the core hole. What is the contribution of this term to the orthogonality index α from the cumlant $F_2(t)$ (Girvin and Hopfield, 1976)?

16. Show that the imaginary part of the retarded correlation function (7.1.7) has the following sum rule relating to the average kinetic energy:

$$\int_{-\infty}^{\infty} \frac{d\omega}{2\pi} n_B(-\omega)\,\mathrm{Im}[\pi(\omega)] = \frac{e^2 n_0}{3m} E_{\mathrm{K.E.}}$$

Next, show for Fröhlich polarons that the ground state energy can also be related to the average kinetic energy:

$$E_0(\alpha) = -2 \int_0^{\alpha} \frac{d\alpha'}{\alpha'} E_{\mathrm{K.E.}}(\alpha')$$

Thus the ground state energy $E_0(\alpha)$ can be related to the conductivity. Use (8.1.13) to obtain an expression for $E_0(\alpha)$ (Lemmens, DeSitter, and Devreese).

Chapter 9

Superconductivity

The theory of superconductivity was formulated by Bardeen, Cooper, and Schrieffer (1957) and is called the BCS theory. It very successfully describes the superconducting properties of *weak* superconductors, such as aluminum, which are weak because of the small strength of the electron–phonon interaction. Further refinements of the theory have led to the strong coupling theory of Eliashberg (1960) which describes well the properties of strong superconductors such as lead. The distinction between weak and strong is roughly given by the value of the electron–phonon mass enhancement factor λ, as shown by McMillan (1968). We shall first discuss the BCS theory. It must rank as one of the great successes of many-body formalism, since the theory provides detailed agreement with experiments. This agreement is a refreshing change from most comparisons between many-body theory and experiment, where we often get lost in vertex corrections, correlations, or computer calculations. The beauty of BCS is that it is, mathematically, a simple theory which is exceedingly accurate. The reason for this is that the basic coupling forces are weak, and mean field theory works well.

The basic idea of BCS theory is that the electrons in the metal form bound pairs. Not all electrons do this, but only those within a Debye energy of the Fermi surface. The bound states of the electron pairs are not described by simple orbitals such as are used for the hydrogen atom or positronium. The pair state, and the entire ground state of the superconductor, requires a many-body description.

The Debye energy enters into the ground state description because the attractive forces between electrons, which are responsible for the pair binding, are due to the electron–phonon interaction. Fröhlich (1950) was the first to realize that electrons could interact by exchanging phonons and that this interaction could be attractive. He was the first to suggest that superconductivity was caused by the electron–phonon interaction. The phonon dependence would explain the experimental observation that the transition temperature T_c is a function of the ion mass for different isotopes of the same metal. This *isotope effect* was discovered for the metal Hg by Maxwell (1950) and Reynolds *et al.* (1950), where the dependence was proportional to $\Delta T_c/T_c = -\frac{1}{2}(\Delta M/M)$. The BCS theory explains this in detail, since it shows that the transition temperature is proportional to the Debye frequency $kT_c \simeq \omega_D \simeq M^{-1/2}$. The isotope effect verified the Fröhlich hypothesis that the electron–phonon interaction caused superconductivity.

Another piece in the theoretical puzzle was supplied by Schafroth (1955), who showed that a charged boson gas, when undergoing a Bose–Einstein condensation, would exhibit many of the superconducting properties known at that time—but not those known now, such as the energy gap in the excitation spectrum. Schafroth speculated that in the superconductor the "bosons are resonant two electron states." The BCS theory uses a similar mechanism, since the paired electrons behave, in some respects, as bosons. Thus the BCS theory was not conceived in a vacuum but among many related ideas each of which contained some element of truth.

BCS was the first theory to explain superconductivity in metals and also made a number of remarkable predictions. The foremost was that an energy gap existed in the excitation spectrum of the superconductor. The actual observation of this energy gap by electron tunneling (Giaever, 1960) provided a dramatic verification of the theory, although BCS earlier argued that the thermodynamic data supported the existence of a gap. Many different experiments in weak superconductors have shown that the original version of the theory is correct in its many details. An extensive comparison between theory and experiment is provided in the two volumes of *Superconductivity* edited by R. D. Parks and in the books by Rickayzen (1965) and Schrieffer (1964).

9.1. COOPER INSTABILITY

The first inkling of the BCS theory was a letter by Cooper (1956), who pointed out that the ground state of a normal metal was unstable at zero temperature. We define a normal metal as one which is neither superconducting nor magnetic. The instability is an indication that the metal

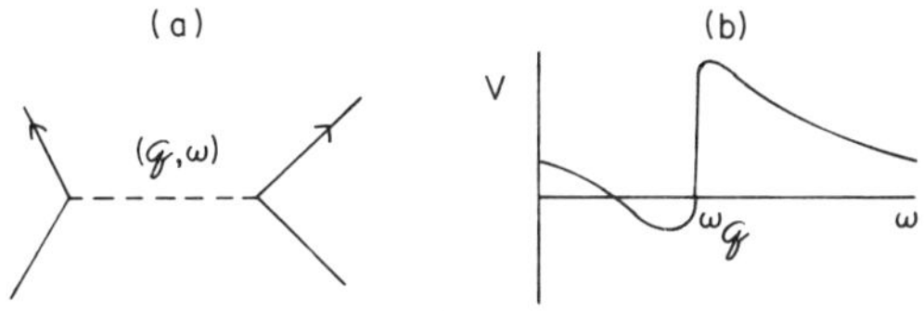

FIGURE 9.1. (a) The pairing force in superconductors is due to the exchange of phonons between electrons. (b) This interaction is frequency dependent and has attractive regions for finite frequencies.

prefers to be in another state, in this case the superconducting one. The demonstration of an instability does not provide a description of the superconducting state, but it did suggest that the instability was caused by the scattering between pairs of electron, where the scattering potential was the exchange of phonons.

Two electrons can scatter as shown in Fig. 9.1. The screened interaction between two electrons was derived in Sec. 6.4:

$$V_s(q, \omega) = \frac{v_q}{\varepsilon(q, \omega)} + \frac{M_q^{\,2}(2\Omega_{q\lambda})}{\varepsilon(q)^2(\omega^2 - \omega_{q\lambda}^2)} \tag{9.1.1}$$

It is divided into two terms. The first is a screened coulomb interaction. The theory of superconductivity is applied at low temperatures and where the interaction energy between particles is also low, $\omega \simeq k_B T \simeq 1$ meV. In this case the static dielectric function $\varepsilon(q, 0) = \varepsilon(q)$ can be used to screen the coulomb potential $v_q = 4\pi e^2/q^2$. Thus the first term in (9.1.1) is strictly positive since $\varepsilon(q) > 0$. The binding of two particles requires some attractive interaction, at least in some part of the (q, ω) spectrum.

The second term in (9.1.1) is the screened phonon interaction. It is, on the average, weaker than the coulomb interaction. However, near the frequency $\omega \lesssim \omega_q$, the energy denominator becomes resonant, and the phonon term becomes large and negative. The phonons provide an attractive interaction between electrons for $\omega \lesssim \omega_q$. The frequency dependence of $V_s(q, \omega)$ is illustrated schematically in Fig. 9.1(b). The potential is negative for frequencies to the left of ω_q. It may be possible for two electrons to bind if they can construct a bound state wave function which selectively uses this part of the interaction potential. Not all simple metals are superconductors (e.g., the alkali metals are not), so the existence of a small attractive potential is not sufficient.

There are two ways to proceed with a description of the theory. The first is to use the full interaction potential (9.1.1) with its full dependence on (q, ω) and summed over all phonon modes (TA, LA, etc.). The realistic calculations must be done for comparisons with actual metals. The other possibility is to replace (9.1.1) with a model interaction of the form

$$V_s(q, \omega) = \begin{cases} -V_0 & \text{for } |\xi_q| \leq \omega_D \\ 0 & \text{for } |\xi_q| \geq \omega_D \end{cases} \tag{9.1.2}$$

This potential is constant and attractive ($V_0 > 0$) up to a cutoff energy which is of the order of the Debye energy ω_D of the solid. This second form of the interaction permits a much simpler discussion of the theory and allows the physics to be introduced more easily. We shall begin our discussion of superconductivity by using the model potential in (9.1.2). This follows the historical pathway, since this was also done by Cooper and in BCS theory. Later, when the physics is better understood by the reader, we shall return to the beginning and solve the theory of superconductivity more rigorously and with a realistic potential (9.1.1).

Cooper's model of a normal metal at low temperature was a free-electron system. In the limit of zero temperature, the Fermi surface has a sharp step in energy. The electrons are allowed to have a weak attractive interaction of the sort given in (9.1.2). Consider the mutual scattering of two electrons. We shall assume they initially have states of equal and opposite momentum $\mathbf{k}$ and $-\mathbf{k}$, or zero center of mass. Later we shall show why this is necessary for the Cooper instability. We shall also assume the particles have opposite spin states $\uparrow$ and $\downarrow$, so that we can forget about exchange scattering. The interaction potential does not flip the electron spin, so that the spin states $\uparrow$ and $\downarrow$ are preserved in the scattering process.

Figure 9.2 shows a double scattering event between two electron lines which are moving in the same direction in time. This process is just the scattering according to the second Born approximation, where the first Born approximation is shown in Fig. 9.1(a). Each dashed line represents an interaction of the type shown in (9.1.2). If two electrons initially start in opposite momentum states $\mathbf{k}$ and $-\mathbf{k}$, then a momentum transfer $\mathbf{q}$ leaves them still in opposite momentum states $\mathbf{k} + \mathbf{q}$ and $-(\mathbf{k} + \mathbf{q})$. This is shown as the pairs $(\mathbf{k}_1, -\mathbf{k}_1)$ in the intermediate state and $(\mathbf{k}', -\mathbf{k}')$ in the final state of the double scattering. The effective scattering in the first

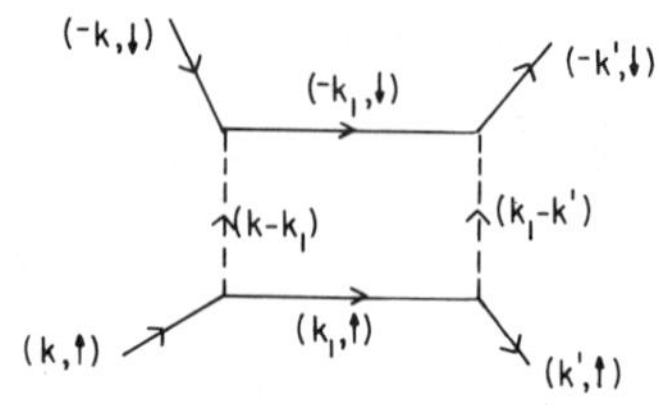

FIGURE 9.2

and second Born approximation is

$$V_{\text{eff}}(\mathbf{k}-\mathbf{k}') = V(\mathbf{k}-\mathbf{k}') + \int \frac{d^3k_1}{(2\pi)^3} \frac{V(\mathbf{k}-\mathbf{k}_1)V(\mathbf{k}_1-\mathbf{k}')}{2\xi_{\mathbf{k}} - 2\xi_{\mathbf{k}_1}} \times \{[1-n_F(\xi_{\mathbf{k}_1})][1-n_F(\xi_{\mathbf{k}_1})] - n_F(\xi_{\mathbf{k}_1})^2\} \tag{9.1.3}$$

The second term on the right is the contribution of Fig. 9.2. The energy denominator contains the initial state energy $\xi_{\mathbf{k}} + \xi_{-\mathbf{k}} = 2\xi_{\mathbf{k}}$ minus the intermediate state energy $\xi_{\mathbf{k}_1} + \xi_{\mathbf{k}_1} = 2\xi_{\mathbf{k}_1}$. The numerator contains the important factors $[1 - n_F(\xi_{\mathbf{k}_1})]^2 - n_F(\xi_{\mathbf{k}_1})^2 = 1 - 2n_F(\xi_{\mathbf{k}_1})$ which occur because the two particles, which are being scattered, can only go into the states $(\mathbf{k}_1, \uparrow)$ and $(-\mathbf{k}_1, \downarrow)$ if they are not already occupied by an electron. This explains the factors $[1 - n_F(\xi_{\mathbf{k}_1})]^2$. The other term $n_F(\xi_{\mathbf{k}_1})^2$ represents the scattering back into this state, $(\mathbf{k}_1, \uparrow) \rightarrow (\mathbf{k}, \uparrow)$, etc., since we are always concerned with the net scattering. Thus we are left with the remaining factors of $1 - 2n_F(\xi_{\mathbf{k}_1})$. These occupation factors of $n_F(\xi_{\mathbf{k}_1})$ play a crucial role in the theory and are the cause of the instability.

The integral in (9.1.3) may be evaluated. The key is that the interaction acts only over a small energy interval near the Fermi energy. Over this interval, of a Debye energy, the electron density of states in most metals is nearly constant. Thus we can change integration variables to $\int [d^3k_1/(2\pi)^3] = \int d\xi_1 N(\xi_1)$ and treat $N(\xi_1) = N_F$ as a constant. Thus at zero temperature we get $(\xi_{\mathbf{k}_1} = \xi_1,\ \xi_{\mathbf{k}} = \xi)$

$$V_{\text{eff}}(\mathbf{k}-\mathbf{k}') = V(\mathbf{k}-\mathbf{k}') + N_F V_0^2 \int_{-\omega_D}^{\omega_D} d\xi_1 \frac{\frac{1}{2} - n_F(\xi_1)}{\xi - \xi_1}$$

The integrand contains the factor $\frac{1}{2} - n_F(\xi_1)$, and the $\frac{1}{2}$ term may be ignored. It does not cause the singularity, and its inclusion leads to little effect. The term in $n_F(\xi_1)$ leads to a logarithm which is singular at the Fermi energy:

$$\int_{-\omega_D}^{\omega_D} d\xi_1 \frac{n_F(\xi_1)}{\xi - \xi_1} = \int_{-\omega_D}^{0} \frac{d\xi_1}{\xi - \xi_1} = -\ln\left(\frac{\xi}{\omega_D}\right) \tag{9.1.4}$$

The results of the first and second Born approximation may be summarized as

$$V_{\text{eff}} = -V_0\left[1 - N_F V_0 \ln\left(\frac{\xi}{\omega_D}\right)\right]$$

We regard the term $-N_F V_0 \ln(\xi/\omega_D)$ as the vertex correction which results from the additional scattering between electrons. The scattering becomes very large for electrons near the Fermi energy $|\ \xi/\omega_D\ 1 \ll |$. Further insight

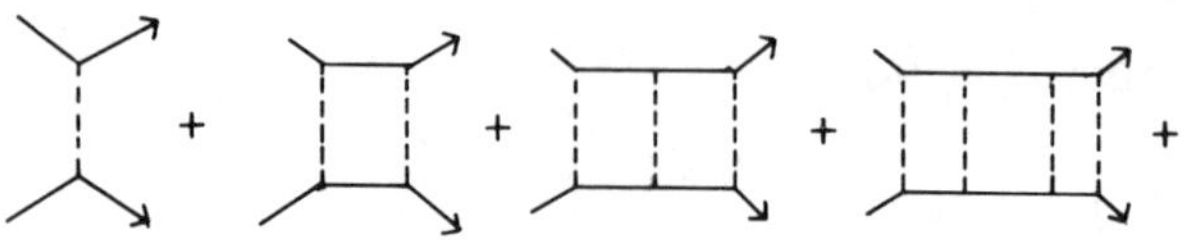

FIGURE 9.3

is gained by considering the sum of diagrams in Fig. 9.3. Each additional interaction (dashed line) causes two more Green's functions which are going parallel in the intermediate state. Each new set of intermediate states has the same type of integrand, so that a term with $n + 1$ ladder diagrams gives a net contribution of

$$-V_0\left[-N_F V_0 \int_{-\omega_D}^{\omega_D} d\xi \, \frac{n_F(\xi_1)}{\xi - \xi_1}\right]^n$$

Thus each term with n ladder diagrams has a factor of $[-N_F V_0 \ln(\xi/\omega_D)]^n$, and the summation of these terms produces the series

$$V_{\text{eff}} = -V_0 \sum_{n=0}^{\infty} \left[-N_F V_0 \ln\left(\frac{\xi}{\omega_D}\right)\right]^n$$

This series can be summed to produce a net potential with an energy denominator:

$$V_{\text{eff}} = -\frac{V_0}{1 + N_F V_0 \ln(\xi/\omega_D)} \tag{9.1.5}$$

This energy denominator equals zero at

$$\xi_0 = \omega_D e^{-1/N_F V}.$$

The energy denominator has a pole at this energy, since we can write the denominator as $1/\ln(\xi/\xi_0)$. In the vicinity of ξ_0 this is $\xi = \xi_0 + (\xi - \xi_0)$, and the scattering potential has a pole:

$$V_{\text{eff}} = -\frac{1}{N_F} \frac{1}{\ln(\xi/\xi_0)} = -\frac{1}{N_F} \frac{1}{\ln\{1 + [(\xi - \xi_0)/\xi_0]\}}$$

$$\simeq -\frac{\xi_0}{N_F(\xi - \xi_0)}$$

This pole is sufficient to cause the instability. The electrons near the Fermi energy will interact with their pair on the opposite side of the Fermi sea. The mutual scattering produces a pole in the scattering amplitude, which will make the pair of electrons try to bind together. Of course, all electron pairs are doing this simultaneously, so that the entire metal undergoes a phase transition. The existence of this pole depended on the sharpness of the electron distribution $n_F(\xi_1) = \theta(-\xi_1)$. If all the electrons near the Fermi energy become paired, one must reconsider whether this sharp distribution still exists. Thus a theory of superconductivity must self-consistently determine the properties of bound electron pairs. This is done in the BCS theory.

Another way to describe the instability is as a function of temperature. We repeat the calculation as a function of temperature. It enters into the electron distribution $n_F(\xi_1)$ by changing the step $\theta(-\xi_1)$ into a smooth function with an energy width of several k_BT. We can approximate $n_F(\xi_1)$ by expressing the integral (9.1.3) as

$$\int_{-\omega_D}^{\omega_D} d\xi_1 \frac{n_F(\xi_1)}{\xi - \xi_1} \simeq -\frac{1}{2} \ln\left[\frac{\xi^2 + (k_BT)^2}{\omega_D^2}\right]$$

If we follow through the same steps with the summation of all the ladder diagrams, we conclude that the effective potential now has an energy denominator of the form

$$V_{\text{eff}} = -\frac{V_0}{1 + N_F V_0 \ln\{[\xi^2 + (k_BT)^2]^{1/2}/\omega_D\}}$$

At zero energy $\xi = 0$, V_{eff} becomes singular when the temperature is lowered to the critical temperature T_c:

$$k_BT_c = \omega_D e^{-1/N_F V_0}$$

For finite values of energy $\xi \neq 0$, the critical temperature is even lower. In fact, this argument correctly predicts the right form for the transition temperature of the BCS theory, which is

$$k_BT_c = 1.14\omega_D e^{-1/N_F V_0} \tag{9.1.6}$$

The result is indeed proportional to the Debye energy, in agreement with the isotope effect. Of course, to explain the isotope effect, we must show that the exponent $1/(N_F V_0)$ does not change with the ion mass. It actually does depend on the ion mass M, which is contained in the electron–phonon matrix element, or interaction potential V_0.

Some experimental values of T_c are given in Table 9.1.

TABLE 9.1. Critical Temperatures of Superconductors (°K)

Mg —	Al 1.2	
Zn 0.91	Ga 1.1	
Cd 0.56	In 3.37	Sn 3.73
Hg 4.16	Tl 2.38	Pb 7.22

The theory of the Cooper instability should be compared, for example, with the ordinary binding of two isolated particles. If the two particles are isolated, they do not have to obey the statistics of a collection of identical particles. Then the scattering theory does not contain any of the occupation factors; all states may be used as intermediate states since there are no other particles. In this case the multiple scattering theory was described earlier, in Sec. 8.2, in the theory of Wannier excitons. The multiple scattering theory may be described by a vertex function:

$$\Gamma(\mathbf{k}', \mathbf{k}) = V(\mathbf{k}' - \mathbf{k}) + \int \frac{d^3k_1}{(2\pi)^3} \frac{V(\mathbf{k}' - \mathbf{k}_1)\Gamma(\mathbf{k}_1, \mathbf{k})}{2\xi_{\mathbf{k}} - 2\xi_{\mathbf{k}_1}}$$

The solution to this vertex function is equivalent to solving the two-particle Schrödinger equation in relative coordinates:

$$\left[-\frac{1}{2m}(\nabla_1^2 + \nabla_2^2) + V(\mathbf{r}_1 - \mathbf{r}_2) - E\right]\psi(\mathbf{r}_1, \mathbf{r}_2) = 0$$

The problem is factored into the relative and center-of-mass motions:

$$\mathbf{r} = \mathbf{r}_1 - \mathbf{r}_2 \qquad \psi(\mathbf{r}_1, \mathbf{r}_2) = e^{i\mathbf{p}\cdot\mathbf{R}}\phi(\mathbf{r})$$

$$\mathbf{R} = \tfrac{1}{2}(\mathbf{r}_1 + \mathbf{r}_2) \qquad E = \frac{P^2}{4m} + \varepsilon$$

$$\left[-\frac{1}{m}\nabla^2 + V(\mathbf{r}) - \varepsilon\right]\phi(\mathbf{r}) = 0$$

The center-of-mass motion is plane wave, and the relative motion becomes a one-body problem. Thus without the occupation factors, the relative scattering of two particles by an instantaneous potential is a trivial problem.

When bound states occur, they are at negative binding energy in the relative coordinates, i.e., at $\varepsilon < 0$. This behavior is in great contrast to the Cooper instability, where the pole occurs at a small negative energy relative to the Fermi energy E_F, so the pole is at positive energy $E_F - \xi_0$. In fact, two electrons cannot really bind at that energy, since their net energy is positive. The instability occurs because it appears to them as if they should bind, although if they tried, they would find they could not. The role of the occupation factors $1 - 2n_F(\xi_1)$ in the argument of the scattering integral is what moved the apparent pole out to the Fermi energy.

One can now see the reason the electrons must be paired with opposite momentum. The instability is caused by the sharpness of the Fermi surface, which is fixed in momentum space. Two electrons with arbitrary wave vectors $\mathbf{k}_1$ and $\mathbf{k}_2$ interact by exchanging phonons, and this interaction can be attractive. Nevertheless, this does not lead to an instability when the center-of-mass motion is finite, since the zero energy of relative motion does not coincide with the location of the discontinuity in the momentum distribution. The center-of-mass transformation

$$\mathbf{P} = \mathbf{k}_1 + \mathbf{k}_2 \qquad \mathbf{k}_1 = \tfrac{1}{2}\mathbf{P} + \mathbf{k}$$

$$\mathbf{k} = \tfrac{1}{2}(\mathbf{k}_1 - \mathbf{k}_2) \qquad \mathbf{k}_2 = \tfrac{1}{2}\mathbf{P} - \mathbf{k}$$

$$\xi_{k_1} + \xi_{k_2} = \frac{1}{m}\,\xi_k{}^2 + \frac{P^2}{4m} - 2E_F$$

changes the relative scattering integral of two particles into

$$\int \frac{d^3q}{(2\pi)^3}\,V(q)\,\frac{[1 - n_F(\xi_{\frac{1}{2}\mathbf{P}-\mathbf{k}-\mathbf{q}}) - n_F(\xi_{\frac{1}{2}\mathbf{P}+\mathbf{k}+\mathbf{q}})]}{k^2/m - (\mathbf{k}+\mathbf{q})^2/m} \tag{9.1.7}$$

The energy denominator factors into just the difference of the relative energies, as it should, but the Fermi surface discontinuities are not at $(\mathbf{k}+\mathbf{q})^2/m$. Instead, they are at $(1/2m)(\pm\frac{1}{2}\mathbf{P} + \mathbf{k} + \mathbf{q})^2$, which gets smeared with the averaging over angles. This smearing makes the pole disappear rapidly for finite center-of-mass momentum $\mathbf{P}$.

The occupation number factors $1 - n_F(\xi_1) - n_F(\xi_2)$ in the scattering integral can also be derived from a Green's function analysis. If two particles interact with an instantaneous interaction, as in Fig. 9.2, and propagate parallel in time or τ space, then we have the combination of Green's functions

$$\mathscr{G}^{(0)}(\mathbf{k}_1, \tau - \tau')\mathscr{G}^{(0)}(\mathbf{k}_2, \tau - \tau')$$

The frequency spectrum is obtained by integrating this pair over $\tau - \tau'$:

$$X(i\omega) = \int_0^\beta d\tau e^{i\omega(\tau-\tau')} \mathscr{G}^{(0)}(\mathbf{k}_1, \tau - \tau')\mathscr{G}^{(0)}(\mathbf{k}_2, \tau - \tau')$$

This integral can be evaluated directly in τ space by using the definitions of $\mathscr{G}^{(0)}$ in Sec. 3.2. Alternately, we can convert it to a frequency summation, which has the form

$$\begin{aligned} X(i\omega) &= \frac{1}{\beta}\sum_{ip_n} \mathscr{G}^{(0)}(\mathbf{k}_1, ip) \int_0^\beta d\tau e^{(\tau-\tau')(i\omega - ip)} \mathscr{G}^{(0)}(\mathbf{k}_2, \tau - \tau') \\ &= \frac{1}{\beta}\sum_{ip_n} \mathscr{G}^{(0)}(\mathbf{k}_1, ip)\mathscr{G}^{(0)}(\mathbf{k}_2, i\omega - ip) \end{aligned}$$

This equation contains one of the standard Matsubara summations which is given in Sec. 3.5:

$$X(i\omega) = \frac{1 - n_F(\xi_{\mathbf{k}_1}) - n_F(\xi_{\mathbf{k}_2})}{i\omega - \xi_{\mathbf{k}_1} - \xi_{\mathbf{k}_2}}$$

It has exactly the combination of occupation number factors stated in (9.1.7). It also has the same energy denominators. If the two particles are going the opposite direction in time, one does not obtain this combining of factors. The binding of particles is described by parallel motion in time.

It is interesting to generalize the Cooper instability to other circumstances in which similar effects could occur. For example, consider the interaction between two different fermions, e.g., electrons and holes in a semimetal or semiconductor. They have the Fermi distributions $n_e(\xi_k)$ and $n_h(\xi_k)$, respectively. The scattering theory contains the vertex term

$$\int \frac{d^3q}{(2\pi)^3} V(q) \frac{1 - n_e(\mathbf{k}_e + \mathbf{q}) - n_h(\mathbf{k}_h - \mathbf{q})}{\varepsilon_{e,\mathbf{k}_e} + \varepsilon_{h,\mathbf{k}_h} - \varepsilon_{e,\mathbf{k}_e+\mathbf{q}} - \varepsilon_{h,\mathbf{k}_h-\mathbf{q}}} \tag{9.1.8}$$

for the scattering $(\mathbf{k}_e, \mathbf{k}_h) \to (\mathbf{k}_e + \mathbf{q}, \mathbf{k}_h - \mathbf{q})$. This vertex would also cause V_{eff} to be resonant if one had $\mathbf{k}_h = -\mathbf{k}_e$, but this cannot happen in a solid in equilibrium, since otherwise the electrons and holes would recombine. However, one can still get instabilities out of this process, as was originally discussed by Keldysh and Kopaev (1965). The instability leads to a state called the *excitonic insulator*, which has been reviewed by Halperin and Rice (1968).

A logarithmic singularity is obtained in (9.1.8) even if one of the occupation factors is zero, say $n_h = 0$, and only one particle obeys many-particle statistics. One example is the edge singularities in X-ray spectra first predicted by Mahan (1967) and discussed in Sec. 8.3B. Thus in electron–hole scattering, we have in the scattering function the occupation factors $(1 - n_e)(1 - n_h) - n_e n_h = 1 - n_e - n_h$, and the electron factor $1 - n_e$ remains even when the hole density n_h becomes zero. This behavior is in sharp contrast to the multiple scattering theory for an electron scattering from a fixed potential such as impurity scattering. Then there is only one fermion, say an electron, which has a factor $1 - n_e$ for scattering $\mathbf{k} \to \mathbf{k}'$ and the factor $-n_e$ for the rate of back scattering $\mathbf{k}' \to \mathbf{k}$. The back scattering enters with a sign change because of the antisymmetry of the single-fermion state. Thus the two rates are added, $1 - n_e + n_e = 1$, and the occupation number terms n_e cancel out. Thus we can treat the electron scattering from an impurity as a one-body problem (except for self-consistent screening) even in a many-particle system. This cancellation explains why such factors as $n_F(\xi)$ do not appear in the vertex equations for impurity scattering, which were solved in Sec. 7.1.

Another possibility is to have one or both of the particles be bosons. Then the scattering probability is multiplied by $1 + n_B(\xi_1)$ for scattering into state $\mathbf{k}_1$ and by $n_B(\xi_1)$ for scattering out of it. For a single boson scattering from a fixed potential, the scattering integral has the rate $1 + n_B$ for $\mathbf{k} \to \mathbf{k}_1$ and the rate n_B back, so the net is $1 + n_B - n_B = 1$, which does not contain occupation factors. Thus the potential scattering is again a one-body problem even in a many-particle system. However, two bosons which are mutually scattering have the factors $(1 + n_B)(1 + n_{B'}) - n_B n_{B'} = 1 + n_B + n_{B'}$, so that the scattering is enhanced because of the occupation of other particles.

9.2. BCS THEORY

The basic feature of the BCS theory is that pairing occurs between electrons in states with opposite momentum and opposite spins, e.g., between states $(\mathbf{k}, \uparrow)$ and $(-\mathbf{k}, \downarrow)$. The two spins are combined into a spin singlet, with $S = 0$. The singlet was chosen in BCS theory on the basis that the other choices of spin combination would lead to a triplet state with $S = 1$. The latter choice ($S = 1$) implies the superconducting state has magnetic properties, which in fact are absent. Thus the choice $S = 0$ seems most reasonable. Later work by Balian and Werthamer (1963), who solved the BCS equations for $S = 1$, showed that the triplet state had smaller binding energy and was therefore less favored. However, the recent theories of superfluidity in He^3 are based on the premise that the pairing occurs in the triplet state (see Chapter 10). Thus triplet pairing is possible and may exist in heavy fermion solids such as UPt_3 and UBe_{13} (Han *et al.*, 1986). Our discussion will assume that the spin arrangement is singlet.

The pairing of electrons in the BCS theory must cause correlations in their relative motions. The pairing is described by introducing a new correlation function, similar to a Green's function, for particles of opposite spin. Following Abrikosov *et al.* (1963), these are

$$\begin{aligned}
\mathscr{G}(\mathbf{p}, \tau - \tau') &= -\langle T_\tau c_{\mathbf{p}\sigma}(\tau) c^\dagger_{\mathbf{p}\sigma}(\tau') \rangle \\
\mathscr{F}(\mathbf{p}, \tau - \tau') &= \langle T_\tau c_{-\mathbf{p}\downarrow}(\tau) c_{\mathbf{p}\uparrow}(\tau') \rangle \\
\mathscr{F}^\dagger(\mathbf{p}, \tau - \tau') &= \langle T_\tau c^\dagger_{\mathbf{p}\uparrow}(\tau) c^\dagger_{-\mathbf{p}\downarrow}(\tau') \rangle
\end{aligned} \tag{9.2.1}$$

The Green's function $\mathscr{G}$ has the same definition as usual, although it has a different algebraic form in the superconducting state. The $\mathscr{F}$ and $\mathscr{F}^\dagger$ functions are identically zero in the normal state. Even in the superconducting state, the bra $\langle|$ and ket $|\rangle$ notation must have special meaning. For instance, in the definition of the $\mathscr{F}$ function, the ket $|\rangle$ is operated on by two destruction operators and then closed by $\langle|$. Thus the state $|\rangle$ must have two more electrons than $\langle|$. This reflects a basic feature of the BCS ground state wave function, which is a superposition of electronic states containing a different number of electrons. Such a formulation is possible in a grand canonical ensemble. Our procedure will be to find a self-consistent equation for the correlation function $\mathscr{F}$ or its Hermitian conjugate $\mathscr{F}^\dagger$. At high temperatures we shall only find the solution $\mathscr{F} = \mathscr{F}^\dagger = 0$, but a nonzero solution becomes possible at low temperature.

The order of the spin indices ($\uparrow, \downarrow$) is important in specifying the $\mathscr{F}$ functions. Suppose by mistake for $\mathscr{F}$ one wrote

$$\langle T_\tau c_{-\mathbf{p}\uparrow}(\tau) c_{\mathbf{p}\downarrow}(\tau') \rangle$$

Since $\mathscr{F}$ does not depend on the sign of $\mathbf{p}$, this correlation function is the same as

$$\langle T_\tau c_{\mathbf{p}\uparrow}(\tau) c_{-\mathbf{p}\downarrow}(\tau') \rangle$$

The $c_{\mathbf{p}\uparrow}$ operators anticommute, and doing this gives

$$-\langle T_\tau c_{-\mathbf{p}\downarrow}(\tau') c_{\mathbf{p}\uparrow}(\tau) \rangle$$

so this function is actually $-\mathscr{F}(\mathbf{p}, \tau' - \tau)$. These kinds of sign errors can be avoided by always following the definition closely.

Our objective is to provide the simplest possible derivation of the BCS theory. First we assume a model Hamiltonian of the form

$$H = \sum_{\mathbf{p}\sigma} \xi_p c^\dagger_{\mathbf{p}\sigma} c_{\mathbf{p}\sigma} + \frac{1}{2\nu} \sum_{\substack{\mathbf{qpp}' \\ \sigma\sigma'}} V(q) c^\dagger_{\mathbf{p}+\mathbf{q},\sigma} c^\dagger_{\mathbf{p}'-\mathbf{q},\sigma'} c_{\mathbf{p}'\sigma'} c_{\mathbf{p}\sigma} \tag{9.2.2}$$

The interaction potential $V(q)$ between electrons is taken to have the form in (9.1.2), which is an attractive constant $V(q) = -V_0$ over a range of energies within a Debye energy of the Fermi surface. With this Hamiltonian, a set of self-consistent equations will be derived for the Green's functions $\mathscr{G}$, $\mathscr{F}$, and $\mathscr{F}^\dagger$. The derivation will be done using the equations of motion. Thus consider

$$\frac{\partial}{\partial\tau} c_{\mathbf{p}\sigma}(\tau) = [H, c_{\mathbf{p}\sigma}] = -\xi_p c_{\mathbf{p}\sigma} - \frac{1}{\nu} \sum_{\substack{\mathbf{p}'\mathbf{q} \\ \sigma'}} V(q) c^\dagger_{\mathbf{p}'-\mathbf{q}\sigma'} c_{\mathbf{p}'\sigma'} c_{\mathbf{p}-\mathbf{q},\sigma} \tag{9.2.3}$$

From the definition of the τ-ordered product, we can derive the first derivative of the equation for the Green's function:

$$\begin{aligned}
\frac{\partial}{\partial\tau} \mathscr{G}(\mathbf{p}, \tau - \tau') &= -\frac{\partial}{\partial\tau} [\theta(\tau - \tau')\langle c_{\mathbf{p}\sigma}(\tau) c^\dagger_{\mathbf{p}\sigma}(\tau')\rangle \\
&\quad - \theta(\tau' - \tau)\langle c^\dagger_{\mathbf{p}\sigma}(\tau') c_{\mathbf{p}\sigma}(\tau)\rangle] \\
&= -\delta(\tau - \tau')\langle\{c_{\mathbf{p}\sigma}, c^\dagger_{\mathbf{p}\sigma}\}\rangle - \left\langle T_\tau\left[\frac{\partial}{\partial\tau} c_{\mathbf{p}\sigma}(\tau)\right] c^\dagger_{\mathbf{p}\sigma}(\tau')\right\rangle
\end{aligned}$$

$$\frac{\partial}{\partial\tau} \mathscr{G}(\mathbf{p}, \tau - \tau') = -\delta(\tau - \tau') - \left\langle T_\tau\left[\frac{\partial}{\partial\tau} c_{\mathbf{p}\sigma}(\tau)\right] c^\dagger_{\mathbf{p}\sigma}(\tau')\right\rangle$$

Using the result (9.2.3) for $(\partial/\partial\tau)c_{\mathbf{p}\sigma}$ gives

$$\begin{aligned}
&\left(-\frac{\partial}{\partial\tau} - \xi_p\right)\mathscr{G}(\mathbf{p}, \tau - \tau') + \frac{1}{\nu} \sum_{\substack{\mathbf{p}'\mathbf{q} \\ \sigma'}} V(q) \\
&\quad \times \langle T_\tau c^\dagger_{\mathbf{p}'-\mathbf{q},\sigma'}(\tau) c_{\mathbf{p}'\sigma'}(\tau) c_{\mathbf{p}-\mathbf{q}\sigma}(\tau) c^\dagger_{\mathbf{p}\sigma}(\tau')\rangle = \delta(\tau - \tau')
\end{aligned} \tag{9.2.4}$$

The bracket of four operators in the interaction term must be reduced to products of pair operators. Now there are many ways of doing the pairing, since in addition to the normal combinations such as $\langle cc^\dagger\rangle\langle cc^\dagger\rangle$, we also have to include arrangements such as $\langle cc\rangle\langle c^\dagger c^\dagger\rangle$. One simplification is to assume that long-wavelength phonons give a zero potential, so that $V(\mathbf{q}=0) = 0$, which is true when the potential is from an ion pseudopotential such as (5.3.3). Thus we neglect the pairing which occurs when $\mathbf{q} = 0$. For a normal metal, there would only remain the pairing $\delta_{\mathbf{pp}'}\delta_{\sigma\sigma'}n_{\mathbf{p}-\mathbf{q}}\mathscr{G}(\mathbf{p}, \tau-\tau')$. This pairing occurs in the superconductor as well but is not the only term. When we consider the pairings which are the $\mathscr{F}$ functions, we must pay attention to the spin functions. The combination $\sigma = -\sigma' = \uparrow$ gives

$$\begin{aligned}&-\langle T_\tau c_{\mathbf{p}'\downarrow}(\tau)c_{\mathbf{p}-\mathbf{q}\uparrow}(\tau)\rangle\langle T_\tau c^\dagger_{\mathbf{p}\uparrow}(\tau')c^\dagger_{\mathbf{p}'-\mathbf{q}\downarrow}(\tau)\rangle\\ &\qquad = -\delta_{\sigma,-\sigma'}\delta_{\mathbf{p}'=-\mathbf{p}+\mathbf{q}}\mathscr{F}(\mathbf{p}-\mathbf{q}, 0)\mathscr{F}^\dagger(\mathbf{p}, \tau'-\tau)\end{aligned}$$

where the sign change resulted from an odd number of operator rearrangements. Similarly, the choice $\sigma = -\sigma' = \downarrow$ gives

$$\begin{aligned}&-\langle c_{\mathbf{p}-\mathbf{q}\downarrow}(\tau)c_{\mathbf{p}'\uparrow}(\tau)\rangle\langle T_\tau c^\dagger_{\mathbf{p}'-\mathbf{q}\uparrow}(\tau)c^\dagger_{\mathbf{p}\downarrow}(\tau')\rangle\\ &\qquad = -\delta_{\sigma,-\sigma'}\delta_{\mathbf{p}'=-\mathbf{p}+\mathbf{q}}\mathscr{F}(-\mathbf{p}+\mathbf{q}, 0)\mathscr{F}^\dagger(-\mathbf{p}, \tau-\tau')\end{aligned}$$

These two results are identical, since we shall later prove that the $\mathscr{F}$ and $\mathscr{F}^\dagger$ functions do not depend on the sign of their arguments—either momentum or τ. Thus for the last term in (9.2.4) we obtain the expression

$$\begin{aligned}&\frac{1}{\nu}\sum_{\mathbf{p}'\mathbf{q}\sigma'} V(q)\langle T_\tau c^\dagger_{\mathbf{p}'-\mathbf{q}\sigma'}(\tau)c_{\mathbf{p}'\sigma'}(\tau)c_{\mathbf{p}-\mathbf{q}\sigma}(\tau)c^\dagger_{\mathbf{p}\sigma}(\tau')\rangle\\ &\qquad = \frac{1}{\nu}\sum_{\mathbf{q}} V(q)[\mathscr{G}(\mathbf{p}, \tau-\tau')n_{\mathbf{p}-\mathbf{q}} - \mathscr{F}(\mathbf{p}-\mathbf{q}, 0)\mathscr{F}^\dagger(\mathbf{p}, \tau-\tau')] \qquad (9.2.5)\end{aligned}$$

The first term is the standard one in the self-energy of the electron. It is the exchange-like potential which our electron feels, via phonons, in the average field of the other electrons; i.e., it provides the self-energy term $\Sigma_x = -\sum_{\mathbf{q}} V(q)n_{\mathbf{p}-\mathbf{q}}$. A careful investigation shows that this self-energy does not change much between the normal and superconducting states. The self-energy of the electrons, from phonons, causes a change in the electron effective mass given by the parameter λ. This effect is not large in weak superconductors, so it may be ignored. Of course, in metals where λ is large, such as lead, the superconducting state can be expected to significantly alter the properties of electrons near the Fermi surface and hence cause a change in the self-energy due to phonons—hence the need, in these cases, for strong coupling theory. This self-energy term is neglected in our weak coupling theory.

In the second term of (9.2.5) there arises the combination of factors which we define as

$$\Delta(\mathbf{p}) = -\frac{1}{\nu}\sum_{\mathbf{q}} V(q)\mathscr{F}(\mathbf{p}-\mathbf{q}, \tau = 0) \qquad (9.2.6)$$

The quantity $\Delta(\mathbf{p})$ is the gap function in the BCS theory and plays a central role in the properties of the superconductor state. The quantity $\Delta(\mathbf{p})$ is defined to be positive, since the right-hand side of the definition is positive, with an attractive potential $V(q) < 0$. Thus we finally derive from (9.2.4) the equation of motion for the Green's function:

$$\left(-\frac{\partial}{\partial\tau} - \xi_p\right)\mathscr{G}(\mathbf{p}, \tau - \tau') + \Delta(\mathbf{p})\mathscr{F}^\dagger(\mathbf{p}, \tau - \tau') = \delta(\tau - \tau') \qquad (9.2.7)$$

This equation is one of the two we need. The equation has two unknowns in $\mathscr{G}$ and $\mathscr{F}^\dagger$, so another equation is needed to link these two quantities. It comes from the equation of motion for the $\mathscr{F}^\dagger$ function:

$$\begin{aligned}\frac{\partial}{\partial\tau}\mathscr{F}^\dagger(\mathbf{p}, \tau - \tau') &= \frac{\partial}{\partial\tau}[\theta(\tau - \tau')\langle c^\dagger_{\mathbf{p}\uparrow}(\tau)c^\dagger_{-\mathbf{p}\downarrow}(\tau')\rangle \\ &\quad - \theta(\tau - \tau')\langle c^\dagger_{-\mathbf{p}\downarrow}(\tau')c^\dagger_{\mathbf{p}\uparrow}(\tau)\rangle] \\ &= \delta(\tau - \tau')\langle\{c^\dagger_{\mathbf{p}\uparrow}, c^\dagger_{-\mathbf{p}\downarrow}\}\rangle + \left\langle T\left[\frac{\partial}{\partial\tau}c^\dagger_{\mathbf{p}\uparrow}(\tau)\right]c^\dagger_{-\mathbf{p}\downarrow}(\tau')\right\rangle \\ &= \left\langle T\left[\frac{\partial}{\partial\tau}c^\dagger_{\mathbf{p}\uparrow}(\tau)\right]c^\dagger_{-\mathbf{p}\downarrow}(\tau')\right\rangle\end{aligned}$$

It lacks a term $\delta(\tau - \tau')$ because the $c^\dagger$ operators anticommute. The time development of the $\mathscr{F}^\dagger$ operator is determined by

$$\frac{\partial}{\partial\tau}c^\dagger_{\mathbf{p}\sigma} = [H, c^\dagger_{\mathbf{p}\sigma}] = \xi_p c^\dagger_{\mathbf{p}\sigma} + \frac{1}{\nu}\sum_{\mathbf{q}} V(q)c^\dagger_{\mathbf{p}-\mathbf{q}\sigma}c^\dagger_{\mathbf{p}'+\mathbf{q}\sigma'}c_{\mathbf{p}'\sigma'}$$

This equation is not the Hermitian conjugate of $(\partial/\partial\tau)c_{\mathbf{p}}(\tau)$ in (9.2.3), because $c_{\mathbf{p}}{}^\dagger(\tau)$ is not the Hermitian conjugate of $c_{\mathbf{p}}(\tau)$. The result for $\partial c^\dagger_{\mathbf{p}\sigma}/\partial\tau$ brings us to the equation for $(\partial/\partial\tau)\mathscr{F}^\dagger$:

$$\begin{aligned}&\left(-\frac{\partial}{\partial\tau} + \xi_p\right)\mathscr{F}^\dagger(\mathbf{p}, \tau - \tau') \\ &\quad + \frac{1}{\nu}\sum_{\mathbf{q}} V(q)\langle T_\tau c^\dagger_{\mathbf{p}-\mathbf{q}\uparrow}(\tau)c^\dagger_{\mathbf{p}'+\mathbf{q}\sigma'}(\tau)c_{\mathbf{p}'\sigma'}(\tau)c^\dagger_{-\mathbf{p}\downarrow}(\tau')\rangle = 0\end{aligned}$$

Again we must evaluate an expression with four operators. The operator $c^\dagger_{-\mathbf{p}\downarrow}(\tau')$ is unique, since it is the only one which is not operating at time τ. We get three terms when it is paired with each of the other three operators:

$$\begin{aligned}c^\dagger_{-\mathbf{p}\downarrow}(\tau'):\qquad & c_{\mathbf{p}'\sigma'}(\tau): && \delta_{\sigma'=\downarrow}\delta_{\mathbf{p}'=-\mathbf{p}}\Delta^\dagger(\mathbf{p})\mathscr{G}(-\mathbf{p}, \tau - \tau') \\ & c^\dagger_{\mathbf{p}-\mathbf{q}\uparrow}(\tau): && \delta_{\mathbf{q}=0}n_{\mathbf{p}'}\mathscr{F}^\dagger(\mathbf{p}, \tau - \tau') \\ & c^\dagger_{\mathbf{p}'+\mathbf{q}\sigma'}(\tau): && -\delta_{\mathbf{p}'=\mathbf{p}-\mathbf{q}}\delta_{\sigma'=\uparrow}n_{\mathbf{p}-\mathbf{q}}\mathscr{F}^\dagger(\mathbf{p}, \tau - \tau')\end{aligned}$$

The first is $\Delta^\dagger(\mathbf{p})$, which is the term we want; the second requires $\mathbf{q} = 0$, which we assume is zero; and the last gives the exchange potential $\Sigma_x = \sum_{\mathbf{q}} V(q) n_{\mathbf{p}-\mathbf{q}}$, which we ignore again. The exchange potential, in both the equation for $(\partial/\partial\tau)\mathscr{G}$ and $(\partial/\partial\tau)\mathscr{F}^\dagger$, just changes $\xi_{\mathbf{p}} \rightarrow \xi_{\mathbf{p}'} = \xi_{\mathbf{p}} - \Sigma_x(p)$, which we assume has little effect. This brings us to the final equation:

$$\left(-\frac{\partial}{\partial\tau} + \xi_p\right)\mathscr{F}^\dagger(\mathbf{p}, \tau - \tau') + \Delta(\mathbf{p})\mathscr{G}(\mathbf{p}, \tau - \tau') = 0 \qquad (9.2.8)$$

The gap function is assumed real, $\Delta^\dagger(\mathbf{p}) = \Delta(\mathbf{p})$, and this assumption is verified later.

The Fourier transforms of the correlation functions are defined in the usual way:

$$\begin{aligned} \mathscr{G}(\mathbf{p}, \tau) &= \frac{1}{\beta}\sum_{p_n} e^{-ip_n\tau}\mathscr{G}(\mathbf{p}, ip_n) \\ \mathscr{F}(\mathbf{p}, \tau) &= \frac{1}{\beta}\sum_{p_n} e^{-ip_n\tau}\mathscr{F}(\mathbf{p}, ip_n) \\ \mathscr{F}^\dagger(\mathbf{p}, \tau) &= \frac{1}{\beta}\sum_{p_n} e^{-ip_n\tau}\mathscr{F}^\dagger(\mathbf{p}, ip_n) \end{aligned} \qquad (9.2.9)$$

After transforming, the two equations (9.2.7) and (9.2.8) are

$$\begin{aligned} (ip_n - \xi_p)\mathscr{G}(\mathbf{p}, ip_n) + \Delta(\mathbf{p})\mathscr{F}^\dagger(\mathbf{p}, ip_n) &= 1 \\ (ip_n + \xi_p)\mathscr{F}^\dagger(\mathbf{p}, ip_n) + \Delta(\mathbf{p})\mathscr{G}(\mathbf{p}, ip_n) &= 0 \end{aligned} \qquad (9.2.10)$$

These are just algebraic equations, which are easily solved:

$$\begin{aligned} \mathscr{G}(\mathbf{p}, ip) &= -\frac{ip + \xi_p}{p_n^2 + \xi_p^2 + \Delta(\mathbf{p})^2} \\ \mathscr{F}(\mathbf{p}, ip) = \mathscr{F}^\dagger(\mathbf{p}, ip) &= \frac{\Delta(p)}{p_n^2 + \xi_p^2 + \Delta(\mathbf{p})^2} \end{aligned} \qquad (9.2.11)$$

The equivalence of $\mathscr{F}$ and $\mathscr{F}^\dagger$ can be shown by deriving similar equations for $\mathscr{F}$.

The results (9.2.11) can be used to test some of the assumptions in the derivation. For example, if we consider $\mathscr{F}(\mathbf{p}, -\tau)$ in the Fourier transform (9.2.9), then changing the sign of the dummy variable of summation $p_n \rightarrow -p_n$ gives

$$\mathscr{F}(\mathbf{p}, -\tau) = \frac{1}{\beta}\sum_n e^{-i\tau p_n}\mathscr{F}(\mathbf{p}, -ip_n) = \mathscr{F}(\mathbf{p}, \tau)$$

$\mathscr{F}(\mathbf{p}, -\tau)$ is the same as $\mathscr{F}(\mathbf{p}, \tau)$, since $\mathscr{F}(\mathbf{p}, -ip_n) = \mathscr{F}(\mathbf{p}, ip_n)$ in (9.2.11). Similarly, we have that

$$\mathscr{F}(\mathbf{p}, \tau = 0) = \frac{1}{\beta}\sum_n \mathscr{F}(\mathbf{p}, ip_n) = \frac{1}{\beta}\sum_n \mathscr{F}^\dagger(\mathbf{p}, ip_n) = \mathscr{F}^\dagger(\mathbf{p}, \tau = 0)$$

This relation causes the energy gap equation,

$$\Delta(\mathbf{p}) = -\frac{1}{\nu}\sum_{\mathbf{q}} V(q)\mathscr{F}(\mathbf{p} - \mathbf{q}, 0) = -\frac{1}{\nu}\sum_{\mathbf{q}} V(q)\mathscr{F}^\dagger(\mathbf{p} - \mathbf{q}, 0)$$

to be real, and $\Delta(\mathbf{p}) = \Delta^\dagger(\mathbf{p})$. Finally, we observe that if the energy gap is set equal to zero, we recover the usual form for the normal Green's function and zero for the others:

$$\Delta = 0: \qquad \begin{cases} \mathscr{G}(\mathbf{p}, ip) = \dfrac{1}{ip_n - \xi_p} \\ \mathscr{F} = \mathscr{F}^\dagger = 0 \end{cases}$$

It must still be shown that there is a self-consistent solution to the equations for $\Delta \neq 0$.

The Green's functions $\mathscr{G}$, $\mathscr{F}$, and $\mathscr{F}^\dagger$ have poles at the points $\pm E_p$, where

$$E_p = [\xi_p{}^2 + \Delta(p)^2]^{1/2}$$

E_p is termed the *excitation energy* of the superconductor. It will occur repeatedly in the various formulas we shall derive. We shall find that in the BCS theory the gap function $\Delta(\mathbf{p})$ is not dependent on momentum $\mathbf{p}$ and is treated as a constant for each temperature. Of course, it very much depends on temperature and vanishes at the transition temperature of the superconductor. However, at a fixed temperature, we shall treat Δ as a constant. [Later, in strong coupling theory, we shall have to treat $\Delta(\omega)$ as a function of energy.] Then the excitation energy $E_p = (\xi_p{}^2 + \Delta^2)^{1/2}$ depends only on $\xi_p{}^2 > 0$. The minimum excitation energy is Δ, and $E_p > \Delta$. An important feature of the BCS theory is that the particles are paired, and it is not possible to excite just one quasiparticle with excitation energy $E_p = (\xi_p{}^2 + \Delta^2)^{1/2}$. Instead, one must break a pair of particles and excite them both to the band of excitations. This pair breaking is shown in Fig. 9.4, where the horizontal line represents the bound state pairs at the chemical potential of the superconductor. To break a pair, one must excite both particles to the excitation line, so that it takes energy $E_p + E_{p'} > 2\Delta$. The minimum excitation energy of the superconductor is the energy to break

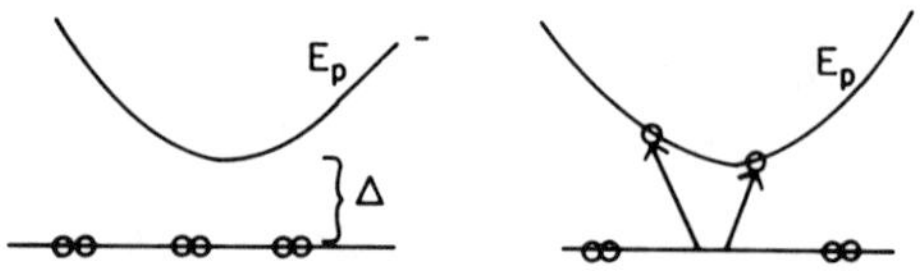

FIGURE 9.4. BCS model of superconductor. (a) The ground state has electrons paired at the chemical potential. (b) Excited states occur by breaking a pair and forming two quasiparticles, each with excitation energy E_p.

a pair, which is 2Δ. Thus the energy gap of the superconductor is $E_g = 2\Delta$. This fact must be kept in mind when comparing thermodynamic data with the BCS theory.

This picture of condensed pairs also clarifies the meaning of the bra $\langle|$ and ket $|\rangle$ symbols in the definition of

$$\mathscr{F}^{\dagger}(\mathbf{p}, \tau - \tau') = \langle| T c^{\dagger}_{\mathbf{p}\uparrow}(\tau) c^{\dagger}_{-\mathbf{p}\downarrow}(\tau') |\rangle$$

At zero temperature, we can call the ket $|\rangle$ the ground state of the superconductor, which has all bound pairs at the chemical potential μ. The operators $c^{\dagger}_{\mathbf{p}\downarrow}$ and $c^{\dagger}_{\mathbf{p}\uparrow}$ create two excitations with energy $E_{\mathbf{p}} + E_{-\mathbf{p}} = 2E_{\mathbf{p}}$. The bra $\langle|$ state has the same number of particles but two less in the ground state.

This seems like a good place to introduce the Feynman diagrams for the three Green's functions. Eventually we shall evaluate correlation functions, and a diagrammatic representation is useful. The way of doing the diagrams was suggested by Abrikosov *et al.* (1963) and is shown in Fig. 9.5. The Green's function $\mathscr{G}$ represents creating a particle at one point in time and destroying it at a later point in time. It is drawn as an arrow with points at both ends and both points in the same direction. The inward point at one end symbolizes particle destruction, and the outward point on the other is particle creation. $\mathscr{F}^{\dagger}$ is represented by an arrow with both points outward, since a particle is created at both times. Similarly, the Green's function $\mathscr{F}$ is an arrow with both points inward, which represents particle destruction at both points in time.

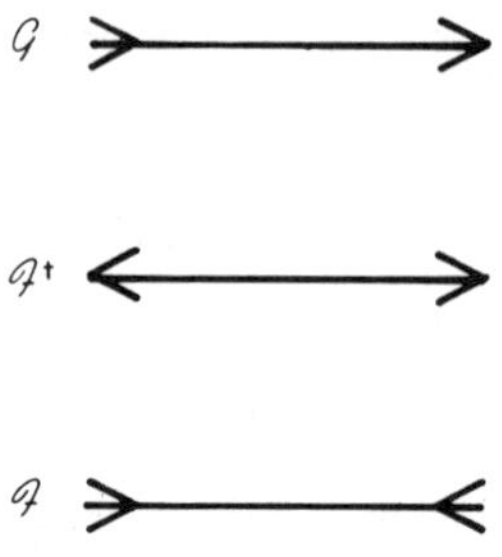

FIGURE 9.5

It is conventional to introduce the following *coherence factors*:

$$\begin{aligned} u_p{}^2 &= \frac{1}{2}\left(1 + \frac{\xi_p}{E_p}\right) \\ v_p{}^2 &= \frac{1}{2}\left(1 - \frac{\xi_p}{E_p}\right) \\ 2u_p v_p &= \left(1 - \frac{\xi_p{}^2}{E_p{}^2}\right)^{1/2} = \frac{\Delta}{E_p} \\ u_p{}^2 + v_p{}^2 &= 1 \end{aligned} \tag{9.2.12}$$

which are misnamed since they have little to do with coherence. Their usefulness stems from the fact that they are the residues of the poles of the Green's functions. Thus we can write our Green's functions (9.2.11) as

$$\begin{aligned} \mathscr{G}(\mathbf{p}, ip) &= \frac{u_{\mathbf{p}}{}^2}{ip - E_p} + \frac{v_{\mathbf{p}}{}^2}{ip + E_p} \\ \mathscr{F}(\mathbf{p}, ip) &= \mathscr{F}^{\dagger}(\mathbf{p}, ip) = -u_p v_p\left(\frac{1}{ip - E_p} - \frac{1}{ip + E_p}\right) \end{aligned} \tag{9.2.13}$$

This form is useful for doing the contour integrals associated with the summations over Matsubara frequencies. The excitations of the superconductor are fermions, and the frequencies ip_n in $\mathscr{G}$ and $\mathscr{F}$ are for fermions $ip_n = (2n + 1)\pi i/\beta$. The spectral function for the Green's function is

$$A(\mathbf{p}, \varepsilon) = -2\,\mathrm{Im}[G_{\mathrm{ret}}(\mathbf{p}, \varepsilon)] = 2\pi[u_p{}^2\delta(\varepsilon - E_p) + v_p{}^2\delta(\varepsilon + E_p)] \tag{9.2.14}$$

These equations are not really solved until Δ is evaluated. It is found from the definition of the gap function in (9.2.6), where we need the quantity

$$\mathscr{F}(\mathbf{p}, \tau = 0) = \frac{1}{\beta}\sum_{ip} \frac{\Delta}{p_n{}^2 + E_p{}^2}$$

The summation over frequencies may be evaluated in the usual fashion, by the contour integral

$$0 = \oint \frac{dZ}{2\pi i}\, n_F(Z)\, \frac{\Delta}{Z^2 - E_p{}^2}$$

which is zero since the contour is taken to infinity. The poles of $n_F(Z)$ give the summation over ip_n, while the poles at $\pm E_p$ give

$$\mathscr{F}(\mathbf{p}, \tau = 0) = -\frac{\Delta}{2E_p}\,[n_F(E_p) - n_F(-E_p)] = \frac{\Delta}{2E_p}\tanh\left(\frac{\beta E_p}{2}\right) \tag{9.2.15}$$

Thus the equation for the gap function is

$$\Delta(\mathbf{p}) = -\frac{1}{\nu}\sum_{\mathbf{q}} V(q)\frac{\Delta(\mathbf{p}-\mathbf{q})}{2E_{\mathbf{p}-\mathbf{q}}}\tanh\left(\frac{\beta E_{\mathbf{p}-\mathbf{q}}}{2}\right)$$

$$\Delta = \frac{\Delta}{2} N_F V_0 \int_{-\omega_D}^{\omega_D} d\xi \frac{\tanh(\beta E/2)^2}{E} \qquad (9.2.16)$$

$$E = (\xi^2 + \Delta^2)^{1/2}$$

Since Δ is constant, we can factor it from both sides, and we are left to consider the integral equation for Δ:

$$1 = \frac{N_F V_0}{2} \int_{-\omega_D}^{\omega_D} \frac{d\xi}{E} \tanh\left(\frac{\beta E}{2}\right) \qquad (9.2.17)$$

We shall consider the solution in two limiting cases. The first is zero temperature, where the hyperbolic function $\tanh(\beta E/2) = 1$. Then we have the integral

$$1 = \frac{N_F V_0}{2} \int_{-\omega_D}^{\omega_D} \frac{d\xi}{(\xi^2 + \Delta^2)^{1/2}}$$

$$= N_F V_0 \ln[\xi + (\xi^2 + \Delta^2)^{1/2}]_0^{\omega_D} \simeq N_F V_0 \ln\left(\frac{2\omega_D}{\Delta}\right)$$

which may be solved to produce the equation for the energy gap E_g:

$$E_g = 2\Delta = 4\omega_D e^{-1/N_F V_0}$$

In BCS theory the gap equation (9.2.17) was solved as a function of temperature. The energy gap gets smaller as the temperature is increased, as shown in Fig. 9.6. In BCS theory it was determined that the critical temperature was the result stated earlier in (9.1.6):

$$kT_c = 1.14\omega_D e^{-1/N_F V_0}$$

The ratio of these two results predicts

$$\frac{E_g}{kT_c} = \frac{4.0}{1.14} = 3.52$$

Both the energy gap E_g and the transition temperature T_c can be measured, so this can be tested. It is found to work well in weak superconductors such as aluminum, which is generally well described by the BCS theory. For strong coupling superconductors, the ratio increases in value.

The BCS theory is often called a mean field theory, because the gap equation (9.2.16) has that form. It has a similar form to the mean field

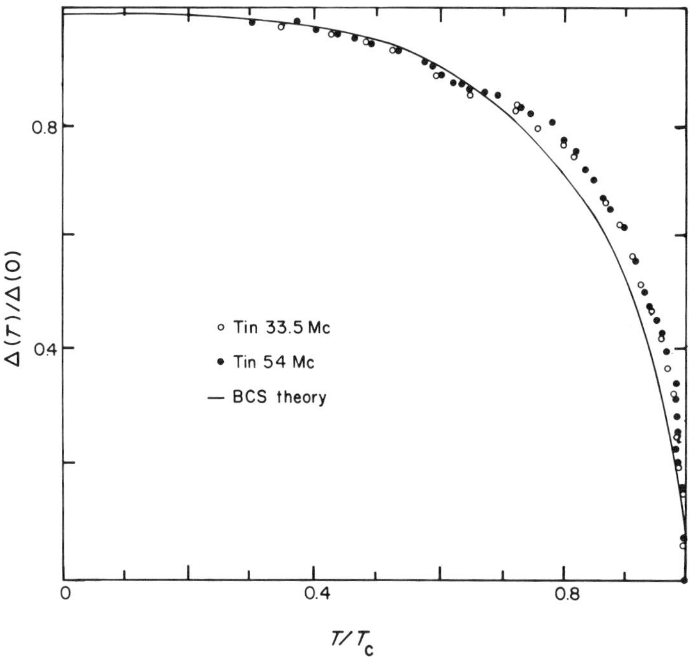

FIGURE 9.6. The ratio $\Delta(T)/\Delta(0)$ vs. T/T_c in tin. The solid curve is the BCS theory. The points are the ultrasonic attenuation data of Morse and Bohm (1957) (used with permission).

theory for magnetism in metals (see Kittel, 1966; he uses the term *molecular field approximation*). The order parameter, in this case the gap Δ, is set equal to a thermodynamic average over the excitations, which are also functions of Δ. The self-consistent solution determines Δ. There is a critical temperature T_c above which no solution is possible. It is seldom that the mean field theory is an accurate description of physical reality, but it certainly works well for BCS. In statistical mechanical models, mean field theory works well when the forces between particles are long range. In the case of superconductivity, the forces are short range, but the bound state orbits of the particles extend over a long distance. This seems to accomplish the same type of averaging process.

Besides the existence of an energy gap, another dramatic prediction of the BCS theory was that the density of states $\varrho(E)$ for excitations has a square root singularity. It is defined as

$$\varrho(E) = \left(\frac{dE}{d\xi}\right)^{-1} = \frac{E}{(E^2 - \Delta^2)^{1/2}}$$

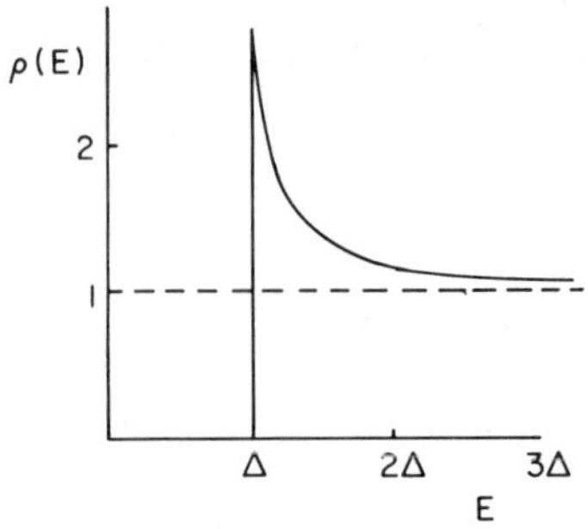

FIGURE 9.7. BCS density of states of a superconductor.

This spectral shape is shown in Fig. 9.7, with a gap until Δ and then a square root singularity. The density of states arises from the variable change

$$\int_0^\infty d\xi = \int_0^\infty dE\left(\frac{d\xi}{dE}\right) = \int_\Delta^\infty dE\varrho(E)$$

This density of states should not be regarded as just a function which results from a change of variables but as a real property of superconductors which can be measured.

9.3. ELECTRON TUNNELING

The most important verifications of the BCS theory came from electron tunneling experiments. They measured the energy gap, as a function of temperature, in perfect agreement with the BCS theory. They also measured the density of states function $\varrho(E)$. Later, Josephson (1962) predicted the coherent tunneling of pairs, which was also quickly observed. These experiments provided a detailed verification of the BCS theory.

A. Tunneling Hamiltonian

Cohen *et al.* (1962) introduced the concept of the tunneling Hamiltonian, which became universally adopted for the discussion of tunneling in superconductors. Their idea was to write the Hamiltonian as three terms:

$$\begin{aligned} H &= H_R + H_L + H_T \\ H_T &= \sum_{\mathbf{kp}} (T_{\mathbf{kp}} C_{\mathbf{k}}^{\dagger} C_{\mathbf{p}} + \text{h.c.}) \end{aligned} \tag{9.3.1}$$

The first term H_R is the Hamiltonian for particles on the right side of the tunneling junction. It contains all many-body interactions. Similarly, H_L

has all the physics for particles on the left side of the junction. These two are considered to be strictly independent. Not only do these two operators commute, $[H_L, H_R] = 0$, but they commute term by term. Thus the Hamiltonian on the right can be expressed in terms of one set of operators $C_{\mathbf{k}}$ and those on the left by another set $C_{\mathbf{p}}$, and these operators are independent $\{C_{\mathbf{k}}, C_{\mathbf{p}}^{\dagger}\} = 0$. All this is probably a reasonable assumption. They further assumed that the tunneling is caused by the term H_T in (9.3.1). The tunneling matrix element $T_{\mathbf{pk}}$ can transfer particles through an insulating junction. This transfer rate is assumed to depend only on the wave vectors on the two sides **k** and **p** and not on other variables, such as the energy of the particles.

The theory of electron tunneling in superconductors developed very rapidly and was entirely based on the tunneling Hamiltonian. The theory showed excellent agreement with the many experiments. The history books were written describing this satisfactory situation, and the scientists in this field wandered off to do something else. About this time there began a serious investigation, starting with Zawadowski (1967), about the validity of the tunneling Hamiltonian. Of course it was found to be a poor approximation, since the tunneling rate depends on the energy of the particle as well as its wave vector. The investigation into the many-body theory of tunneling continues (see Caroli *et al.*, 1975; Feuchtwang, 1975). What does this turn of events do to the lovely agreement between theory and experiment for tunneling in superconductors? Actually, it probably changes none of it. The tunneling in superconductors takes place over a very narrow span of energies in the metal, i.e., within a Debye energy of the Fermi surface. Also, all the electrons involved have their wave vector very near the Fermi wave vectors k_F and p_F on the two sides of the junction. It is an adequate approximation to treat the transfer rate T_{pk} as a constant T_0 which is evaluated at k_F and p_F, because the variations in T_{kp} with energy must be on the scale E/E_F which are negligible for $E \simeq \Delta \simeq 1$ meV. Similarly, the variation of T_{pk} with p or k is on the scale of the Fermi wave vectors. Thus it is an adequate approximation to treat the transfer rate T_0 as a constant in a superconductor, or a normal metal, if the energies involved are small. The tunneling Hamiltonian is believed to be an improper formalism only when the applied voltages are large, say 1 eV.

The general model of a tunneling junction is shown in Fig. 9.8. It describes a nonequilibrium situation, since the chemical potential on the left-hand side μ_L is not the same as μ_R on the right. They differ by the applied voltage $eV = \mu_L - \mu_R$. The potential drop of eV occurs in the insulating region between the metals, which is typically a metal oxide.

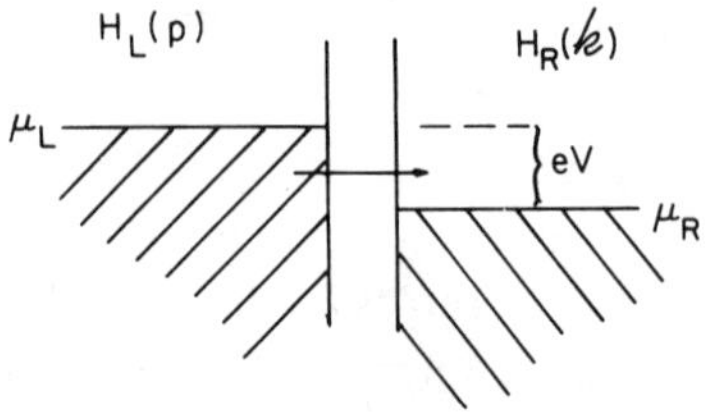

FIGURE 9.8. Tunneling between two normal metals. The arrow shows the electron path through the oxide interface.

The tunneling Hamiltonian (9.3.1) is used to derive a correlation function for electron tunneling currents. This correlation function has the form of a Kubo formula, except for an important difference. The Kubo formula for the conductivity in Sec. 3.7 expresses the ratio between the current and the voltage (actually electric field). In the tunneling theory, the correlation function gives the entire current as a function of voltage.

The tunneling current through the insulating region is expressed as the rate of change of the number of particles on, for example, the left-hand side of the junction N_L. This rate is found from the commutator of $N_L = \sum_{\mathbf{p}} C_{\mathbf{p}}^{\dagger} C_{\mathbf{p}}$ with the tunneling Hamiltonian. Of course, only the term H_T fails to commute with is N_L:

$$\dot{N}_L = i[H, N_L] = i[H_T, N_L]$$

$$\dot{N}_L = i \sum_{\mathbf{pk}} (T_{\mathbf{pk}} C_{\mathbf{k}}^{\dagger} C_{\mathbf{p}} - T_{\mathbf{pk}}^{*} C_{\mathbf{p}}^{\dagger} C_{\mathbf{k}})$$

The total current I through the tunneling interface is defined as the average value of this operator:

$$I(t) = -e\langle \dot{N}_L(t) \rangle$$

The average value of $\langle \dot{N}_L \rangle$ is obtained by following the same steps we used to derive the Kubo formula (3.7.7). The total Hamiltonian is written as $H = H' + H_T$, where $H' = H_R + H_L$. We go to the interaction representation, where the tunneling term H_T is treated as the interaction and everything else H' is H_0. Then the S matrix is expanded in terms of the perturbation H_T. Our objective is to obtain a formula where $I \propto |T_{pk}|^2$, so only the first term needs to be retained in the expansion of the S matrix. These steps bring us to the formula

$$\begin{aligned} I(t) &= -ei \int_{-\infty}^{t} dt' \langle [N_L(t), H_T(t')] \rangle \\ H_T(t') &= e^{iH't'} H_T e^{-iH't'} \qquad (9.3.2) \\ N_L(t) &= e^{iH't} N_L e^{-iH't} \end{aligned}$$

where the time dependence of $H_T(t)$ and $N_L(t)$ is, as shown, governed by H'.

An important step in the calculation is to insert the chemical potentials μ_L and μ_R for the two sides of the junction. This insertion must be done with more care than usual, because the chemical potential is not the same on the two sides of the system. Our initial Hamiltonian (9.3.1) has been written to *not* include the chemical potentials, so the energy is measured on an absolute scale rather than relative to the chemical potentials. However, we should like now to insert the chemical potentials into these time developments, so that the energy can be measured with respect to the different chemical potentials on each side of the tunnel junction. We introduce the symbols K_R and K_L for the Hamiltonian with respect to the respective chemical potentials:

$$
\begin{aligned}
K_R &= H_R - \mu_R N_R \\
K_L &= H_L - \mu_L N_L \\
K' &= K_R + K_L
\end{aligned}
$$

For a free-particle system, $K_R = \sum_{\mathbf{k}} \xi_{\mathbf{k}} c_{\mathbf{k}}^{\dagger} c_{\mathbf{k}}$, while $H_R = \sum_{\mathbf{k}} \varepsilon_{\mathbf{k}} c_{\mathbf{k}}^{\dagger} c_{\mathbf{k}}$, $\xi_{\mathbf{k}} = \varepsilon_{\mathbf{k}} - \mu_R$. Of course, we are interested in the properties of interacting systems, in this case superconductors. The Hamiltonians K_R and K_L are then the ones we solved in Sec. 9.2 in order to find the superconducting ground state and excitation energy. Since the number operators commute with H', it is possible to write $H' = K' + \mu_R N_R + \mu_L N_L$ and $\exp(iH't) = \exp(iK't)\exp[it(\mu_L N_L + \mu_R N_R)]$ since the exponentials can be separated when the operators commute. Thus the time development of H_T is

$$
\begin{aligned}
H_T(t) &= e^{iH't} H_T e^{-iH't} = e^{iK't}(e^{it(\mu_L N_L + \mu_R N_R)} H_T e^{-it(\mu_R N_R + \mu_L N_L)}) e^{-itK'} \\
&= e^{iK't} \sum_{\mathbf{kp}} (T_{\mathbf{kp}} e^{it(\mu_R - \mu_L)} c_{\mathbf{k}}^{\dagger} c_{\mathbf{p}} + e^{it(\mu_L - \mu_R)} T_{\mathbf{kp}}^{*} c_{\mathbf{p}}^{\dagger} c_{\mathbf{k}}) e^{-iK't}
\end{aligned}
$$

The commutator of H_T with the number operators produces the factor $\mu_L - \mu_R = eV$, which is identified as the applied voltage, as in Fig. 9.8. This is the method of introducing the applied voltage into the correlation function. The correlation function will now be evaluated by assuming that both sides of the junction are in separate thermodynamic equilibrium.

The current operator in (9.3.2) now becomes

$$
\begin{aligned}
I(t) = -i^2 e \int_{-\infty}^{t} dt' \Big\langle \Big\{ &\sum_{\mathbf{kp}} [T_{\mathbf{kp}} e^{-ieVt} c_{\mathbf{k}}^{\dagger}(t) c_{\mathbf{p}}(t) - T_{\mathbf{kp}}^{*} e^{ieVt} c_{\mathbf{p}}^{\dagger}(t) c_{\mathbf{k}}(t)], \\
&\sum_{\mathbf{k}'\mathbf{p}'} [T_{\mathbf{k}'\mathbf{p}'} e^{-ieVt'} c_{\mathbf{k}'}^{\dagger}(t') c_{\mathbf{p}'}(t') + T_{\mathbf{k}'\mathbf{p}'}^{*} e^{ieVt'} c_{\mathbf{p}'}^{\dagger}(t') c_{\mathbf{k}'}(t')] \Big\} \Big\rangle
\end{aligned}
$$

From now on the time development of $c_{\mathbf{k}}$ operators is governed by $c_{\mathbf{k}}(t) = e^{iK_Rt}c_{\mathbf{k}}e^{-iK_Rt}$ and $c_{\mathbf{p}}$ operators by $c_{\mathbf{p}}(t) = e^{iK_Lt}c_{\mathbf{p}}e^{-iK_Lt}$. If we define the operator A as

$$A(t) = \sum_{\mathbf{kp}} T_{\mathbf{kp}} c_{\mathbf{k}}^{\dagger}(t) c_{\mathbf{p}}(t)$$

then we can write the current as

$$\begin{aligned} I(t) = e\int_{-\infty}^{\infty} dt'\theta(t-t')\{&e^{+ieV(t'-t)}\langle[A(t), A^{\dagger}(t')]\rangle \\ &- e^{-ieV(t'-t)}\langle[A^{\dagger}(t), A(t')]\rangle + e^{-ieV(t+t')}\langle[A(t), A(t')]\rangle \\ &- e^{ieV(t+t')}\langle[A^{\dagger}(t), A^{\dagger}(t')]\rangle\} \end{aligned} \tag{9.3.3}$$

Equation (9.3.3) is written as the summation of two currents:

$$I = I_S + I_J$$

$$\begin{aligned} I_S = e\int_{-\infty}^{\infty} dt'\theta(t-t')\{&e^{ieV(t'-t)}\langle[A(t), A^{\dagger}(t')]\rangle \\ &- e^{ieV(t-t')}\langle[A^{\dagger}(t), A(t')]\rangle\} \end{aligned} \tag{9.3.4}$$

$$\begin{aligned} I_J = e\int_{-\infty}^{\infty} dt'\theta(t-t')\{&e^{-ieV(t+t')}\langle[A(t), A(t')]\rangle \\ &- e^{ieV(t+t')}\langle[A^{\dagger}(t), A^{\dagger}(t')]\rangle\} \end{aligned}$$

The first one, I_S, is the single-particle tunneling, which we shall consider first. The other term I_J was ignored originally, until it was realized that it described the tunneling currents associated with the Josephson effect. We shall evaluate this formula in a later section.

The first term in I_S has just the right combination of factors to be a retarded Green's function, as defined in (3.3.2). The integrand depends only on the difference of $t - t'$, so set $t' = 0$:

$$\tilde{X}_{\text{ret}}(t) = -i\theta(t)\langle[A(t), A^{\dagger}(0)]\rangle$$

$$X_{\text{ret}}(-eV) = \int_{-\infty}^{\infty} dt e^{-ieVt}\tilde{X}_{\text{ret}}(t)$$

The second term in I_S is just the Hermitian conjugate of $X_{\text{ret}}(-eV)$ except for a sign. But the Hermitian conjugate of the retarded function is just the advanced function. Thus we can write the single-particle tunneling current as

$$\begin{aligned} I_S &= ie[X_{\text{ret}}(-eV) - X_{\text{adv}}(-eV)] \\ &= -2e\,\text{Im}[X_{\text{ret}}(-eV)] \end{aligned} \tag{9.3.5}$$

It is twice the imaginary part of a retarded correlation function, which gives it the form of a spectral density function. If we exploit the relationship between the Matsubara and the retarded correlation functions discussed in Sec. 3.3, the way to calculate the single-particle tunneling is to evaluate in the Matsubara formalism the correlation function

$$X(i\omega) = -\int_0^\beta d\tau e^{i\omega\tau}\langle T_\tau A(\tau)A^\dagger(0)\rangle$$

$$= -\sum_{\substack{\mathbf{kp}\\ \mathbf{k'p'}}} T_{\mathbf{kp}}T^*_{\mathbf{k'p'}}\int_0^\beta d\tau e^{i\omega\tau}\langle T_\tau c_{\mathbf{k}}{}^\dagger(\tau)c_{\mathbf{p}}(\tau)c^\dagger_{\mathbf{p'}}c_{\mathbf{k'}}\rangle \qquad (9.3.6)$$

The single-particle tunneling current is just the spectral function of this operator evaluated at the real frequency $-eV/\hbar$, as already shown in (9.3.5). The Matsubara frequency $\omega_n = 2n\pi/\beta$ is boson, since the correlation function has pairs of fermion operators. In the tunneling Hamiltonian, whose model we have adopted for this discussion, the right- and left-hand sides of the tunneling junction are independent. Then the correlation function factors into a product of the Green's functions for the right and left sides of the junction:

$$X(i\omega) = \sum_{\mathbf{kp}} T_{\mathbf{kp}}T^*_{\mathbf{kp}}\int_0^\beta d\tau e^{i\omega\tau}\langle T_\tau c_{\mathbf{k}}c_{\mathbf{k}}{}^\dagger(\tau)\rangle\langle T_\tau c_{\mathbf{p}}(\tau)c_{\mathbf{p}}{}^\dagger\rangle$$

$$= \sum_{\mathbf{kp}} | T_{\mathbf{kp}} |^2\int_0^\beta d\tau e^{i\omega\tau}\mathscr{G}_R(\mathbf{k}, -\tau)\mathscr{G}_L(\mathbf{p}, \tau)$$

$$= \sum_{\mathbf{kp}} | T_{\mathbf{kp}} |^2\frac{1}{\beta}\sum_{ip}\mathscr{G}_L(\mathbf{p}, ip)\mathscr{G}_R(\mathbf{k}, ip - i\omega) \qquad (9.3.7)$$

The only terms which enter are those for $\mathbf{k} = \mathbf{k}'$ and $\mathbf{p} = \mathbf{p}'$. A type of Feynman diagram is shown in Fig. 9.9. The symbols T in circles are the tunneling vertices. The vertical dashed line is meant to divide the right

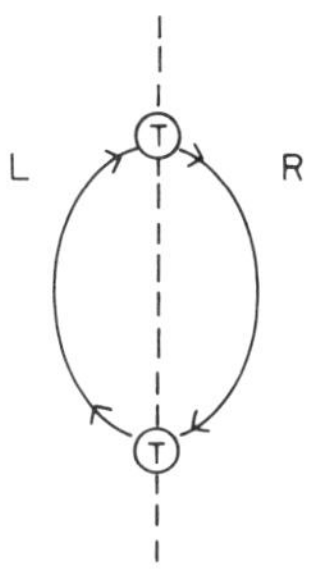

FIGURE 9.9. Feynman diagram for tunneling, where the Ts are the vertices which link the right (R) and left (L) sides of the junction.

from the left side of the junction. On each side, the solid line is the interacting Green's functions $\mathscr{G}_L$ and $\mathscr{G}_R$.

The correlation function is the product of the two Green's functions for the two sides of the junction. Such a simple result is obtained only when one can neglect the vertex diagrams of the correlation function. The argument for neglecting them in tunneling is that vertex corrections require some interaction between electrons on the two sides of the junction, which is improbable. Actually, there is another case where vertex corrections are needed in tunneling, and that is to account for processes where the tunneling electron can excite vibrations or other excitations *in the interface*. A discussion of this is given by Kirtley *et al.* (1976).

B. Normal Metals

The first example of solving (9.3.7) will be for the tunneling between two normal metals. This tunneling is known experimentally to be a nonlinear function of the voltage eV. However, we shall consider only voltages which are the same order as those which are used in tunneling in superconductors, which is several millivolts. For these small voltages, the tunneling Hamiltonian formalism is probably valid.

For a normal system, we consider the electrons to be simple quasiparticles, and approximate the Green's functions by $\mathscr{G}_R^{(0)}$ and $\mathscr{G}_L^{(0)}$. The Matsubara summation is then familiar:

$$\frac{1}{\beta}\sum_{ip}\mathscr{G}_L^{(0)}(\mathbf{p}, ip)\mathscr{G}_R^{(0)}(\mathbf{k}, ip - i\omega) = \frac{n_F(\xi_\mathbf{k}) - n_F(\xi_\mathbf{p})}{i\omega + \xi_\mathbf{k} - \xi_\mathbf{p}}$$

$$X(i\omega) = \sum_{\mathbf{kp}} |T_{\mathbf{kp}}|^2 \frac{n_F(\xi_\mathbf{k}) - n_F(\xi_\mathbf{p})}{i\omega + \xi_\mathbf{k} - \xi_\mathbf{p}}$$

The formula for the tunneling current is

$$I = -2e\,\mathrm{Im}[X_{\mathrm{ret}}(eV)] = 4\pi e\sum_{\mathbf{kp}} |T_{\mathbf{kp}}|^2\delta(eV + \xi_\mathbf{k} - \xi_\mathbf{p})[n_F(\xi_\mathbf{k}) - n_F(\xi_\mathbf{p})]$$

The energies $\xi_\mathbf{k}$ and occupation factor $n_F(\xi_\mathbf{k})$ refer to the right side of the junction, and $\xi_\mathbf{p}$ and $n_F(\xi_\mathbf{p})$ refer to the left side. An additional factor of 2 has been added to this expression to account for the summation over spin index. The spin is usually preserved in tunneling. The summations over **k** and **p** are just over wave vectors. For small voltages, we are concerned only with electrons very near the Fermi energy on both sides of the tunnel junction. We assume that the density of states on both sides is roughly constant:

$$\sum_k \to \int \frac{d^3k}{(2\pi)^3} \to N_R \int d\xi_R$$

$$\sum_p \to \int \frac{d^3p}{(2\pi)^3} \to N_L \int d\xi_L \tag{9.3.8}$$

$$\begin{aligned} I &= e4\pi N_R N_L \mid T \mid^2 \int_{-E_F}^{\infty} d\xi_L \int_{-E_F}^{\infty} d\xi_R \delta(eV + \xi_R - \xi_L)[n_F(\xi_R) - n_F(\xi_L)] \\ &= e4\pi N_R N_L \mid T \mid^2 \int_{-E_F}^{\infty} d\xi_R [n_F(\xi_R) - n_F(\xi_R + eV)] \end{aligned} \tag{9.3.9}$$

At zero temperature, the occupation numbers are step functions, so (9.3.9) becomes

$$I = 4\pi e N_R N_L \mid T \mid^2 \int_{-eV}^{0} d\xi_R = 4\pi e^2 N_R N_L \mid T \mid^2 V$$

Thus the tunneling device behaves as a simple resistor, with a conductance (inverse resistance) given by

$$\begin{aligned} I &= \sigma_0 V \\ \sigma_0 &= 4\pi e^2 N_R N_L \mid T \mid^2 \end{aligned} \tag{9.3.10}$$

With a little more care, one can show that the integral

$$\int d\xi [n_F(\xi) - n_F(\xi + eV)]$$

in (9.3.9) also equals eV at finite temperatures, since the thermal smearing of $n_F(\xi)$ cancels between $n_F(\xi)$ and $n_F(\xi + eV)$. Thus the result is temperature independent, at least for a range of low temperatures $k_B T \ll E_F$.

A formal expression for the tunneling current can also be derived when interacting Green's functions $\mathscr{G}_L$ and $\mathscr{G}_R$ are retained in the Matsubara summation. This type of summation was evaluated in Sec. 7.1 by the method of contour integration:

$$\begin{aligned} &\frac{1}{\beta} \sum_{ip} \mathscr{G}_L(\mathbf{p}, ip)\mathscr{G}_R(\mathbf{k}, ip - i\omega) \\ &\quad = \int \frac{d\varepsilon}{2\pi} n_F(\varepsilon)[A_L(\mathbf{p}, \varepsilon)\mathscr{G}_R(\mathbf{k}, \varepsilon - i\omega) + A_R(\mathbf{k}, \varepsilon)\mathscr{G}_L(\mathbf{p}, \varepsilon + i\omega)] \end{aligned}$$

The retarded function is obtained by the analytical continuation $i\omega \to$

$eV + i\delta$, and we take the imaginary part. The variable change $\varepsilon \to \varepsilon + eV$ in the first term brings us to the formula for the tunneling current of Schrieffer *et al.* (1963):

$$I = 2e \sum_{\mathbf{kp}} | T_{\mathbf{kp}} |^2 \int_{-\infty}^{\infty} \frac{d\varepsilon}{2\pi} A_R(\mathbf{k}, \varepsilon) A_L(\mathbf{p}, \varepsilon + eV)[n_F(\varepsilon) - n_F(\varepsilon + eV)] \tag{9.3.11}$$

Again we added a factor of 2 for the spin summation. The tunneling current is expressed in terms of the spectral functions on the two sides of the junction. Equation (9.3.11) is the exact formula for I within the model of the tunneling Hamiltonian. Of course, we recover our earlier expression (9.3.9) with the free-quasiparticle approximation $A_R(\mathbf{k}, \varepsilon) = 2\pi\delta(\varepsilon - \xi_{\mathbf{k}})$ and $A_L(\mathbf{p}, \varepsilon + eV) = 2\pi\delta(\varepsilon + eV - \xi_{\mathbf{p}})$. The virtue of this result is that it contains all many-body effects on the two sides of the junction. Its drawback is that it is based on the tunneling Hamiltonian formalism.

The failure of the tunneling Hamiltonian is illustrated by considering one of the integrations over wave vector

$$\int d\xi_k \, | T_{\mathbf{kp}} |^2 A_R(\mathbf{k}, \varepsilon)$$

Because most tunneling barriers are of finite height, the tunneling becomes increasingly likely with a particle of high energy. Thus the factor $| T_{\mathbf{kp}} |^2$ increases, with ξ_k, and often this increase is exponential. The exact increase depends on the detailed shape of the potential barrier to tunneling, which really does not concern us. Anyway, if $| T_{\mathbf{kp}} |^2 \simeq \exp(\xi_k \alpha)$, this may increase more rapidly than the spectral function decreases, since often $A(\mathbf{k}, \varepsilon)$ falls off as a power law at high momentum. Thus one finds that the main contribution to the integrand comes from the high-momentum components at the top of the potential barrier. This result seems unphysical. Instead, it should be the energy of the particle which determines the tunneling rate, in which case this high-momentum tail has no effect.

C. Normal–Superconductor

Electron tunneling experiments are often done with two different metals on the two sides of the junction. Since different metals have different transition temperatures (and some are not superconducting), it is quite easy to arrange that one metal be normal and the other be a superconductor. This experiment is interesting, to the theorist, because it provides a direct measurement of the BCS density of states $\varrho(E)$.

We shall take the right side to be superconducting and the left side to be normal. The tunneling current could be calculated from our correlation function (9.3.7) by using a superconducting Green's function $\mathscr{G}_R$ from (9.2.1) and a normal Green's function for the left side $\mathscr{G}^{(0)} = (ip - \xi_p)^{-1}$. The summations over Matsubara frequencies can be performed, and one can derive the formula for the tunneling current. A shortcut to the result is possible, since the many-body expression (9.3.11) is valid in this case. Thus we just use the respective spectral functions for the normal and superconducting sides of the tunneling junction:

$$A_R(\mathbf{k}, \varepsilon) = 2\pi[u_k^2\delta(\varepsilon - E_k) + v_k^2\delta(\varepsilon + E_k)]$$

$$A_L(\mathbf{p}, \varepsilon + eV) = 2\pi\delta(\varepsilon + eV - \xi_p)$$

$$\begin{aligned} I &= 2e\sum_{\mathbf{kp}} | T_{\mathbf{kp}} |^2 \int_{-\infty}^{\infty} \frac{d\varepsilon}{2\pi} A_R(k, \varepsilon)A_L(p, \varepsilon + eV) \\ &\quad \times [n_F(\varepsilon) - n_F(\varepsilon + eV)] \\ &= 4\pi e \sum_{\mathbf{kp}} | T_{\mathbf{kp}} |^2\{u_k^2\delta(eV + E_k - \xi_p)[n_F(E_k) - n_F(\xi_p)] \\ &\quad + v_k^2\delta(eV - E_k - \xi_p)[n_F(-E_k) - n_F(\xi_p)]\} \end{aligned}$$

Next we use the fact that $n_F(-E) = 1 - n_F(E)$. Again we consider the situation at zero temperature, where we can set to zero the density $n_F(E) = (e^{\beta E} + 1)^{-1} \simeq \exp(-\beta E) \to 0$ of excitations in the superconductor. The variable change (9.3.8) brings us to

$$\begin{aligned} I = 4\pi e N_R N_L | T |^2 \int_{-\infty}^{\infty} d\xi_p \int_{-\infty}^{\infty} d\xi_k \{ & v_k^2[1 - n_F(\xi_p)]\delta(eV - E_k - \xi_p) \\ & - u_k^2 n_F(\xi_p)\delta(eV + E_k - \xi_p)\} \end{aligned} \tag{9.3.12}$$

The limits of integration have been extended to infinity, which has no effect on the result since we shall soon show the limits are narrowly confined. There are two terms in the curly braces: The first is positive, and the second is negative. The first will contribute only when $eV > 0$ (actually $eV > \Delta$), and I is positive in this case because we have defined this direction of electron current to be positive. The second term contributes only when $eV < -\Delta$, and I is negative in this case, in agreement with the definition, since I has the same sign as V. Let us evaluate the first term. At zero temperature, the factor $1 - n_F(\xi_p)$ limits $\xi_p > 0$. The integral $d\xi_k$ extends over positive and negative values of ξ_k. Factors which depend on $E_k = (\xi_k^2 + \Delta^2)^{1/2}$ are the same for both signs of ξ_k. The term linear in ξ_k in the coherence factor v_k^2 in (9.2.12) averages to zero because of the cancellation of the $+\xi_k$ and $-\xi_k$ parts:

$$\int_{-\infty}^{\infty} d\xi_k v_k^2 f(E_k) = \frac{1}{2}\int_{-\infty}^{\infty} d\xi_k\left(1 - \frac{\xi_k}{E_k}\right)f(E_k) = \frac{1}{2}\int_{-\infty}^{\infty} d\xi_k f(E_k)$$
$$= \int_0^{\infty} d\xi_k f(E_k) = \int_{\Delta}^{\infty} dE_k \varrho(E_k) f(E_k) \qquad (9.3.13)$$

where f is any function of E. The integration is changed to dE_k, which produces a factor of the density of states $\varrho(E) = d\xi/dE$ in the superconductor. The integral $\int d\xi_p$ can be used to eliminate the delta function for energy conservation $\xi_p = eV - E_k$. The requirement that $\xi_p = eV - E_k > 0$ means that $eV > E_k$, which is possible only when $eV > \Delta$, so we have

$$I = \left(\frac{\sigma_0}{e}\right)\theta(eV - \Delta)\int_{\Delta}^{eV} dE_k \varrho(E_k) = \frac{\sigma_0}{e}\,\theta(eV - \Delta)\int_{\Delta}^{eV} \frac{dEE}{(E^2 - \Delta^2)^{1/2}}$$
$$= \left(\frac{\sigma_0}{e}\right)\theta(eV - \Delta)[(eV)^2 - \Delta^2]^{1/2}$$

The current is a simple square root function of the voltage. This type of dependence is observed experimentally. Another quantity which can be measured (by lock-in amplifier techniques) is the dynamical conductance dI/dV as a function of voltage:

$$\left(\frac{dI}{dV}\right)_{SN} = eV\sigma_0 \frac{\theta(eV - \Delta)}{[(eV)^2 - \Delta^2]^{1/2}}$$
$$\frac{\left(\frac{dI}{dV}\right)_{SN}}{\left(\frac{dI}{dV}\right)_{NN}} = \theta(eV - \Delta)\frac{eV}{[(eV)^2 - \Delta^2]^{1/2}} = \varrho(eV) \qquad (9.3.14)$$

The experimental technique is to compare dI/dV for the tunneling between the normal metal and the superconductor $(dI/dV)_{SN}$ with that for the tunneling between both metals when they are normal $(dI/dV)_{NN} = \sigma_0$. The superconductor can be made normal by the application of a small magnetic field, so that the two measurements are done at the same low temperature. The ratio of these two experimental quantities provides a direct measurement of the BCS density of states $\varrho(eV)$. The current is antisymmetric in voltage, which can be shown from the second term in (9.3.10).

Figure 9.10 shows the experimental results of Giaever *et al.* (1962) for dI/dV of a tunnel junction between Pb and Mg. The experiments were

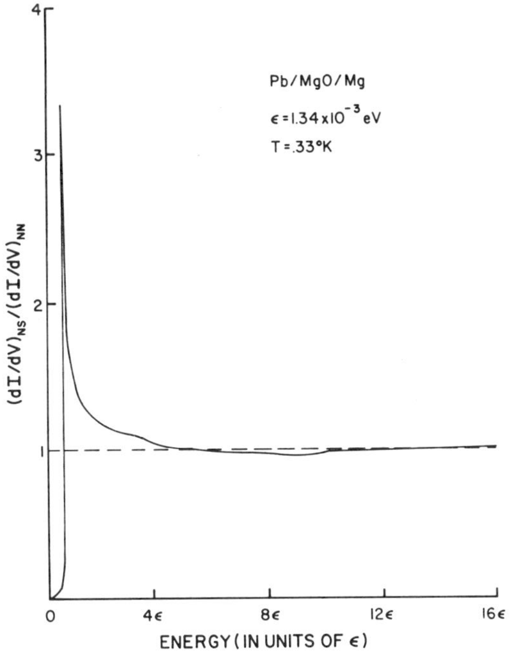

FIGURE 9.10. Electron tunneling between a normal metal (Mg) and a superconductor (Pb). The relative conductance of a Pb–MgO–Mg sandwich plotted against energy. At higher energies there are definite divergences from the BCS density of states, as can be seen from the bumps in the experimental curve. *Source*: Giaever *et al.* (1962) (used with permission).

done at a temperature of 0.3 K with a helium —3 refrigerator. At this low temperature, the Mg is normal, and the Pb is well below its transition temperature of $T_c = 7.2$ K. Thus the factor $n_F(E) \simeq \exp(-\beta\Delta)$ is indeed small, and there should be few thermal excitations in the superconducting Pb. The figure shows the derivative (dI/dV), which well illustrates the BCS density of states. Another feature of their data are bumps in dI/dV at higher voltages. These are due to phonons and are not predicted by the BCS theory. They are explained by the strong coupling theory. These experimental results by Giaever *et al.* (1962) were one of the first experimental indications of the need for strong coupling theory.

The measured current is an antisymmetric function of the voltage $I(-V) = -I(V)$ for the small voltages of interest in this experiment. An energy gap of $\Delta = 1.34$ meV is observed in the tunneling current at zero temperature (which can only be approached asymptotically), and no current flows unless $|eV| > \Delta$. An interpretation of these results is provided in Fig. 9.11. Figure 9.11(a) shows the two sides of the tunnel junction with no applied voltage $V = 0$. The chemical potentials are the same on the two sides of the junction. On the normal side (left) the chemical potential divides the occupied states from the empty states in the Fermi sea of elec-

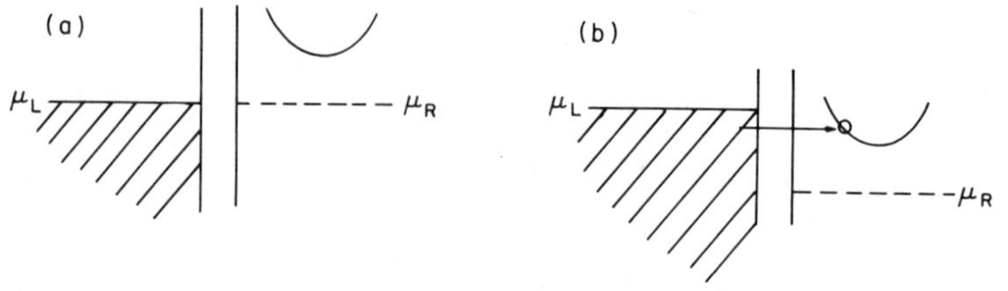

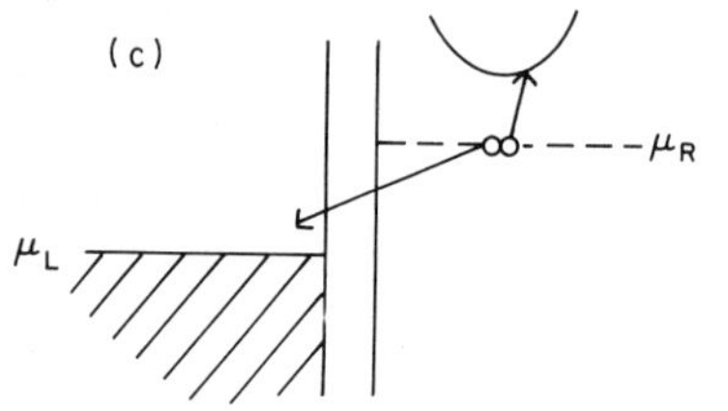

FIGURE 9.11. Electron tunneling between a normal metal and a superconductor at $T = 0$. (a) No applied voltage. (b) For $eV > \Delta$, electrons can tunnel from the normal metal and becomes a quasiparticle in the superconductor. (c) For $eV < -\Delta$, a pair is broken in the superconductor, and one electron tunnels, while the other becomes a quasiparticle in the superconductor.

trons. On the superconductor side (right), the chemical potential is the energy of the paired electrons. The excitation spectrum is shown as the hyperbolic-shaped line above the chemical potential. No net current flows when $V = 0$.

The situation for a forward voltage $eV > \Delta$ is shown in Fig. 9.11(b). The electrons from the normal side of the junction can tunnel through and become a quasiparticle excitation of energy E_k on the superconducting side. The arrow, which shows the tunneling path, is drawn horizontal to indicate the energy-conserving aspect of tunneling. The electron cannot go into the pair state on the right as a single step in the tunneling, since it needs two electrons for this. Thus it must become an excited quasiparticle during its tunneling step. Later it will find some other excited electron and join with it to become a pair state. This pairing is expected to happen well after the tunneling process and so will not affect the tunneling rate.

The situation for reverse bias is shown in Fig. 9.11(c). At zero temperature, all the electrons in the superconductor are in the pair states at the chemical potential, and none are thermally excited to the excitation state. Thus the single-particle tunneling process must break up the pair state. For each electron which tunnels, its partner in the initial pair state must become an excited quasiparticle. The electron which tunnels does not do this at constant energy, since it must give up some of its energy to its partner to permit it to overcome the excitation gap Δ. Energy conservation can be understood as follows: The initial energy of the paired electrons is E_{initial}

$= 2\mu_R$; the final energy is $\mu_R + E_k$ for the one in the superconductor and $\mu_L + \xi_p$ for the one in the normal metal. The total energy of the original pair in the final state is

$$E_{\text{final}} = \mu_R + \mu_L + \xi_p + E_k = E_{\text{initial}} = 2\mu_R$$

The entire process conserves energy, so this is set equal to the initial energy. The quasiparticle energy in the normal metal ξ_p must be positive since the electron must go into an unoccupied state $\xi_p > 0$. Thus we derive the condition for the tunneling current to flow:

$$\mu_R - \mu_L = |eV| = \xi_p + E_k > \Delta$$

This condition is that the magnitude of the voltage must be larger than Δ, which is the same condition for the other direction of bias voltage. The actual physical processes for the two directions of voltage are dissimilar, since in one direction there is direct pair breaking.

D. Two Superconductors

The electron tunneling between two superconductors has two kinds of currents. One is the usual single-particle tunneling of the sort we have been deriving for normal metals and normal–superconductor junctions. The other is the tunneling of pairs, which is the Josephson effect. This interesting subject will be described in the next section. Now we consider only the single-particle tunneling, which can happen between two similar or two dissimilar superconductors. Let us take the case where they are different, so there are gap functions Δ_L and Δ_R for the two sides of the junction. At any point in the calculation the two superconductors can be made identical, theoretically, by just setting $\Delta_L = \Delta_R$.

The many-body formula (9.3.11) is still valid when the two sides are superconductors. The spectral function for the superconducting BCS Green's function is used for both sides of the tunneling junction:

$$A(\mathbf{k}, \varepsilon) = 2\pi[u_k^2\delta(\varepsilon - E_k) + v_k^2\delta(\varepsilon + E_k)]$$

$$A(\mathbf{p}, \varepsilon + eV) = 2\pi[u_p^2\delta(\varepsilon + eV - E_p) + v_p^2\delta(\varepsilon + eV + E_p)]$$

$$I = 2e \sum_{\mathbf{kp}} |T_{\mathbf{kp}}|^2 \int_{-\infty}^{\infty} \frac{d\varepsilon}{2\pi} A(\mathbf{k}, \varepsilon)A(\mathbf{p}, \varepsilon + eV) \times [n_F(\varepsilon) - n_F(\varepsilon + eV)]$$

Using the superconducting spectral functions gives four terms:

$$I = 4\pi e \sum_{\mathbf{kp}} | T_{\mathbf{kp}} |^2 \{u_k^2 u_p^2 \delta(eV + E_k - E_p)[n_F(E_k) - n_F(E_p)]$$
$$+ u_k^2 v_p^2 \delta(eV + E_k + E_p)[n_F(E_k) - n_F(-E_p)]$$
$$+ v_k^2 u_p^2 \delta(eV - E_k - E_p)[n_F(-E_k) - n_F(E_p)]$$
$$+ v_k^2 v_p^2 \delta(eV + E_p - E_k)[n_F(-E_k) - n_F(-E_p)]\}$$

The identity $n_F(-E) = 1 - n_F(E)$ permits these to be grouped as follows:

$$I = 4\pi e \sum_{\mathbf{kp}} | T_{\mathbf{kp}} |^2 \{[1 - n_F(E_p) - n_F(E_k)][v_k^2 u_p^2 \delta(eV - E_p - E_k)$$
$$- u_k^2 v_p^2 \delta(eV + E_p + E_k)] + [n_F(E_k) - n_F(E_p)]$$
$$\times [u_k^2 u_p^2 \delta(eV + E_k - E_p) - v_k^2 v_p^2 \delta(eV + E_p - E_k)]\}$$

This formula is the most general result, which is valid at finite temperatures. The variable change in (9.3.8) to $d\xi_k \, d\xi_p$ is done next. These integrals over the coherence factors u^2 and v^2 will eliminate their linear term in ξ, as described in (9.3.13). Then the variables can be changed to $E = (\xi_k^2 + \Delta_R)^{1/2} = E_k$ and $E' = E_p = (\xi_p^2 + \Delta_L^2)^{1/2}$, which brings us to the expression

$$I = 4\pi e N_L N_R | T_0 |^2 \int_{\Delta_R}^{\infty} dE \varrho_R(E) \int_{\Delta_L}^{\infty} dE' \varrho_L(E')$$
$$\times \{[1 - n_F(E) - n_F(E')][\delta(eV - E - E') - \delta(eV + E + E')]$$
$$+ [n_F(E) - n_F(E')][\delta(eV + E - E') - \delta(eV + E' - E)]\}$$

Some of the terms are positive, while others are negative. They will ensure that I is positive for $V > 0$ and negative for $V < 0$. These integrals must be evaluated numerically at finite temperatures. To illustrate the physics, we shall take only the simplest case, which is zero temperature and identical superconductors $\Delta_R = \Delta_L$. At zero temperature, we can set to zero all the thermal factors $n_F(E)$ and $n_F(E')$:

$$I = \left(\frac{\sigma_0}{e}\right) \int_{\Delta}^{\infty} dE \varrho(E) \int_{\Delta}^{\infty} dE' \varrho(E')[\delta(eV - E - E') - \delta(eV + E + E')]$$

The two terms in square brackets give identical results, where the first is for $eV > 0$ and the second for $eV < 0$, in such a way that $I(-V) = -I(V)$. Thus only the first term needs to be done, where $E' = eV - E$ and $eV > 2\Delta$:

$$I = \left(\frac{\sigma_0}{e}\right) \theta(eV - 2\Delta) \int_{\Delta}^{eV-\Delta} dE \varrho(E) \varrho(eV - E)$$
$$= \left(\frac{\sigma_0}{e}\right) \int_{\Delta}^{eV-\Delta} \frac{dE E(eV - E)}{(E^2 - \Delta^2)^{1/2}[(eV - E)^2 - \Delta^2]^{1/2}} \qquad (9.3.15)$$

This integral is done in the following way: Change variables of integration to x, where $2E = eV + x(eV - 2\Delta)$. After some algebra, the integral is

$$I = \frac{\theta(eV - 2\Delta)(\sigma_0/e)}{eV + 2\Delta} \int_{-1}^{1} dx \frac{(eV)^2 - x^2(eV - 2\Delta)^2}{[(1 - x^2)(1 - \alpha^2 x^2)]^{1/2}}$$

$$\alpha = \frac{eV - 2\Delta}{eV + 2\Delta}$$

The dx integral is in the form of complete elliptic integrals:

$$I = \left(\frac{\sigma_0}{e}\right)\theta(eV - 2\Delta)\left\{\frac{(eV)^2}{eV + 2\Delta} K(\alpha) - (eV + 2\Delta)[K(\alpha) - E(\alpha)]\right\}$$

$$\alpha = \frac{eV - 2\Delta}{eV + 2\Delta} \tag{9.3.16}$$

Equation (9.3.16) is the final result, which is compared with the experiments. One check on the correctness of the answer is to set the energy gap $\Delta = 0$. The coefficient of $K(\alpha)$ vanishes, so the part in curly braces becomes $eVE(1) = eV$. The current is $I = \sigma_0 V$, which is the correct result, since in the absence of superconductivity the tunneling device is a resister at low voltage with the conductance σ_0.

The tunneling current is zero for voltages less than twice the gap function 2Δ. It is interesting to examine the threshold phenomenon, for voltages slightly larger than 2Δ. We have that $K(0) = E(0) = \pi/2$, so at $eV = 2\Delta^+$ the current is

$$I(2\Delta^+) = \left(\frac{\sigma_0}{e}\right)\frac{\Delta\pi}{2} = \sigma_0 V \frac{\pi}{4}$$

It has a finite value, so there is a discontinuous jump in the tunneling current at $eV = 2\Delta$. The current above threshold is $\pi/4$ of the value of a normal tunneling current at the same voltage. Exactly this behavior was found by Giaever *et al.* (1962). Their data for a Sn–SnO_x–Sn junction is shown in Fig. 9.12. The entire experimental curve fits the theoretical curve from (9.3.16), except for the finite slope on the "discontinuity." They also point out an interesting feature of this discontinuous jump in current: The discontinuity does not get thermally broadened with increasing temperature. As the temperature is increased, the energy gaps get smaller, so the value 2Δ of the jump moves to small voltages. However, the BCS energy gap does not get smeared at higher temperature; it only gets smaller. At higher temperatures, the finite value of $n_F(E)$ will lead to some tunneling

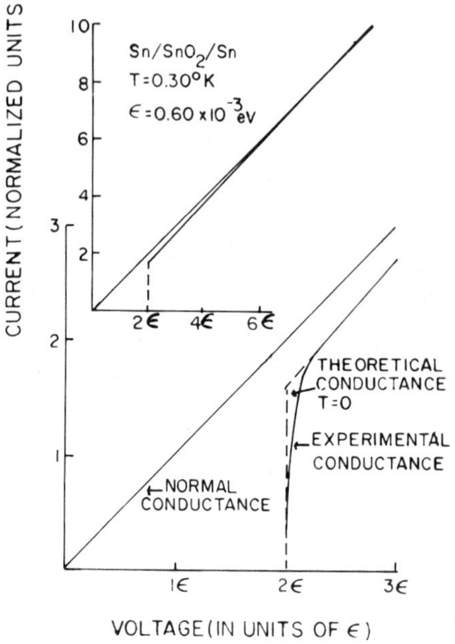

FIGURE 9.12. Electron tunneling between two identical superconductors. Current voltage characteristic for a Sn–SnO_x–Sn sandwich compared with the BCS theory. The experimental curve has a finite slope at 2ε because of the anisotropic nature of the energy gap. *Source*: Giaever *et al.* (1962) (used with permission).

current for $eV < 2\Delta$, but there is still a discontinuous jump at this critical voltage.

The physical process of the single-particle tunneling is shown in Fig. 9.13. For electron flow to the right, a pair on the left is broken up, and one electron tunnels, while the other becomes a quasiparticle on the left. The initial energy of the pair is $E_i = 2\mu_L$, and the final energy of the pair is $E_f = (\mu_L + E_L) + (\mu_R + E_R)$. By equating these two energies, we derive $\mu_L - \mu_R = eV = E_L + E_R$. The summation of the two excitation energies must be equal to the applied voltage, which is in agreement with the theoretical formulas.

The reader should now be convinced that the BCS theory rather well describes the theory of electron tunneling in superconductors. Indeed, the

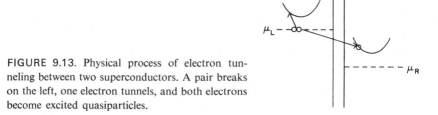

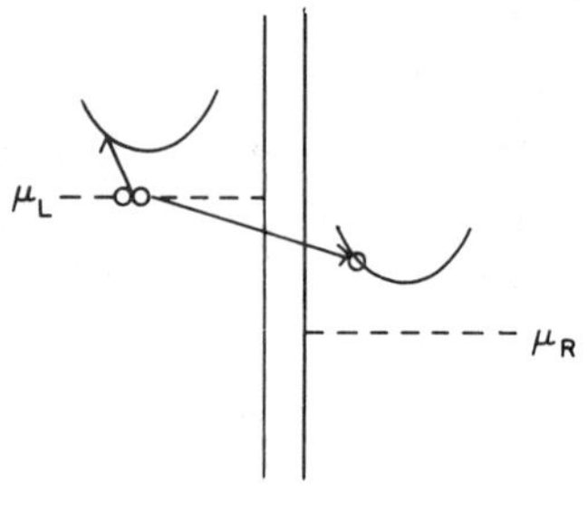

FIGURE 9.13. Physical process of electron tunneling between two superconductors. A pair breaks on the left, one electron tunnels, and both electrons become excited quasiparticles.

BCS theory describes rather well all the properties of weakly coupled superconductors, and other good agreements between theory and experiment will be shown later. It should be recalled that in the derivation of the BCS theory, we made several severe approximations. Apparently none of these really matter. Two of the major approximations were (1) calling the gap function $\Delta(p)$ a constant Δ and (2) replacing the retarded (i.e., frequency-dependent) interaction between electrons by a static potential, which is a constant. A constant potential in Fourier space is a delta function in real space, so we have assumed that the interaction between electrons is localized in space. The electrons themselves are not localized. Indeed, the pairing orbits are extended in space, but the particles interact only when they are in contact. In any case, the essential feature of BCS theory is the existence of the additional correlation functions $\mathscr{F}$ and $\mathscr{F}^\dagger$. This feature must be right, and the particular method of getting to these correlation functions is not very crucial.

E. Josephson Tunneling

The derivation of the tunneling current led to two terms, which are given in (9.3.4). They are the single-particle term and the Josephson term. The single-particle tunneling has been discussed in great detail, and now we examine the Josephson effect. Our discussion of this current contribution closely follows Ambegaokar and Baratoff (1963). First we rewrite (9.3.4):

$$I_J(t) = e\int_{-\infty}^{\infty} dt'\theta(t-t')\{e^{-ieV(t+t')}\langle[A(t), A(t')]\rangle - e^{ieV(t+t')}\langle[A^\dagger(t), A^\dagger(t')]\rangle\}$$

The single most interesting feature of this equation is that it contains, in the two exponents, the sum of the two times $t + t'$. We have not previously encountered this dependence in any of our correlation functions, since they usually come out containing $t - t'$. The factor of $t + t'$ is the first indication, mathematically, that the Josephson effect is rather special. We solve the problem of $t + t'$ by writing it as $2t + t' - t$ and pulling the factor $2t$ outside of the correlation function. Thus we write this current as

$$I_J(t) = e\left\{e^{-2ietV/\hbar}\int_{-\infty}^{\infty} dt'\theta(t-t')e^{+ieV(t-t')}\langle[A(t), A(t')]\rangle + \text{h.c.}\right\}$$

The variable of integration is now changed to $t'' = t - t'$, which causes a sign change. We introduce the retarded correlation function

$$\Phi_{\text{ret}}(t-t') = -i\theta(t-t')\langle[A(t), A(t')]\rangle$$

$$\Phi_{\text{ret}}(eV) = \int_{-\infty}^{\infty} dt e^{ieV(t-t')}\Phi_{\text{ret}}(t-t')$$

The Josephson current can now be written as

$$I_J(t) = -ei[e^{-2eiVt/\hbar}\Phi_{\text{ret}}(eV) - e^{+2ieVt/\hbar}\Phi_{\text{adv}}(eV)]$$

The part in brackets is pure imaginary since it is a quantity minus its complex conjugate. Thus we can write the Josephson current as

$$I_J(t) = 2e\,\text{Im}[e^{-2ieVt/\hbar}\Phi_{\text{ret}}(eV)] \tag{9.3.17}$$

It is important to include the factor $\exp(-2iteV/\hbar)$ in the expression whose imaginary part is evaluated. Thus the Josephson tunneling is not just the imaginary part of a retarded function—it is not just a spectral function. Instead, one must multiply by the phase factor $\exp(-2iteV/\hbar)$ before taking the imaginary part. We shall see that the phase factor contains most of the interesting physics. For a finite voltage, the current is actually oscillatory in time, with a frequency $\omega = 2eV/\hbar$. This prediction of the theory is so absolute, and simple, that the frequency can be used as a method of measuring the ratio of fundamental constants e/h (see Parker *et al.*, 1969).

The first task is to develop a method of calculating the retarded correlation function $\Phi_{\text{ret}}(eV)$. As usual, we start from a Matsubara function,

$$\begin{aligned}\Phi(i\omega) &= -\int_0^\beta d\tau e^{i\omega\tau}\langle T_\tau A(\tau)A(0)\rangle \\ &= -\sum_{\substack{\mathbf{kp}\sigma\\ \mathbf{k'p'}\sigma'}} T_{\mathbf{kp}}T_{\mathbf{k'p'}}\int_0^\beta d\tau e^{i\omega\tau}\langle T_\tau C^\dagger_{\mathbf{k}\sigma}(\tau)C_{\mathbf{p}\sigma}(\tau)C^\dagger_{\mathbf{k'}\sigma'}C_{\mathbf{p'}\sigma'}\rangle\end{aligned}$$

and later make the analytical continuation $i\omega \to eV + i\delta$. The spin index has been explicitly added to this expression. The correlation function can be factored into a term for the left and right sides of the tunnel junction. In this case the factoring produces the Green's function $\mathscr{F}$ and $\mathscr{F}^\dagger$:

$$\begin{aligned}\Phi(i\omega) &= +2\sum_{\mathbf{kp}} T_{\mathbf{kp}}T_{-\mathbf{k},-\mathbf{p}}\int_0^\beta d\tau e^{i\omega\tau}\langle T_\tau C^\dagger_{\mathbf{k}\uparrow}(\tau)C^\dagger_{-\mathbf{k}\downarrow}\rangle\langle T_\tau C_{-\mathbf{p}\downarrow}(0)C_{\mathbf{p}\uparrow}(\tau)\rangle \\ &= 2\sum_{\mathbf{kp}} T_{\mathbf{kp}}T_{-\mathbf{k},-\mathbf{p}}\int_0^\beta d\tau e^{i\omega\tau}\mathscr{F}^\dagger(\mathbf{k},\tau)\mathscr{F}(\mathbf{p},-\tau) \\ &= 2\sum_{\mathbf{kp}} T_{\mathbf{kp}}T_{-\mathbf{k},-\mathbf{p}}\frac{1}{\beta}\sum_{ip_n}\mathscr{F}^\dagger(\mathbf{k},ip)\mathscr{F}(\mathbf{p},ip-i\omega)\end{aligned} \tag{9.3.18}$$

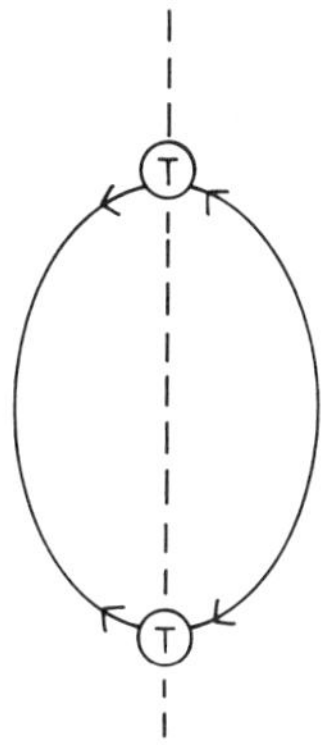

FIGURE 9.14. Feynman diagram for Josephson tunneling.

The factor of 2 is for spin arrangement: We have shown one spin configuration explicitly, and the other one gives the same result. The Feynman diagram for this tunneling is shown in Fig. 9.14. There is an $\mathscr{F}^\dagger$ on the right side of the junction and $\mathscr{F}$ on the left, where the arrow conventions agree with those set forth in Fig. 9.5. The direction of the arrows is continuous through the tunnel junction symbol T in a circle, which must always happen with any diagram where particle number is conserved.

The summation over Matsubara frequencies in (9.3.18) is done by the usual method of contour integral:

$$\begin{aligned}\frac{1}{\beta}\sum_{ip}\mathscr{F}^\dagger(\mathbf{k}, ip)\mathscr{F}(\mathbf{p}, ip - i\omega) &= \frac{\Delta_L\Delta_R}{4E_kE_p}\left\{[1 - n_F(E_p) - n_F(E_k)]\left(\frac{1}{i\omega + E_p + E_k} - \frac{1}{i\omega - E_p - E_k}\right)\right. \\ &\quad \left. + [n_F(E_p) - n_F(E_k)]\left(\frac{1}{i\omega + E_p - E_k} - \frac{1}{i\omega - E_p + E_k}\right)\right\} \qquad (9.3.19)\end{aligned}$$

The correlation function is

$$\begin{aligned}\Phi(i\omega) = \frac{\Delta_L\Delta_R}{2}\sum_{\mathbf{kp}}\frac{T_{\mathbf{kp}}T_{-\mathbf{k},-\mathbf{p}}}{E_kE_p} &\times\left\{[1 - n_F(E_p) - n_F(E_k)]\left(\frac{1}{i\omega + E_p + E_k} - \frac{1}{i\omega - E_p - E_k}\right)\right. \\ &\left. + [n_F(E_k) - n_F(E_p)]\left(\frac{1}{i\omega + E_k - E_p} - \frac{1}{i\omega + E_p - E_k}\right)\right\}\end{aligned}$$

This expression is rather complicated, particularly for finite temperatures and different superconductors. The physics may be best understood by examining the simplest situation, which is zero temperature $[n_F(E) = 0]$

and identical superconductors ($\Delta_L = \Delta_R$). With the analytical continuation $i\omega \to eV + i\delta$, we therefore consider the quantity

$$\Phi_{\text{ret}}(eV) = \frac{\Delta^2}{2} \sum_{\mathbf{kp}} \frac{T_{\mathbf{kp}} T_{-\mathbf{k},-\mathbf{p}}}{E_k E_p} \times \left(\frac{1}{eV + E_p + E_k + i\delta} - \frac{1}{eV - E_k - E_p + i\delta} \right)$$

An important feature of these equations is that the tunneling matrix elements $T_{\mathbf{kp}}$ are not necessarily real, so the combination $T_{\mathbf{kp}}T_{-\mathbf{k},-\mathbf{p}}$ is not real. Thus we write this as an amplitude $|T|^2$ and a phase factor $\exp(i\phi)$ and consider both of these to be independent of **k** and **p**. The summations over **k** and **p** can be changed to integrations over $d\xi_k \, d\xi_p$ and then to $dE_k \, dE_p$:

$$\Phi_{\text{ret}}(eV) = \frac{1}{2e} J_S(eV) e^{i\phi}$$

$$\begin{aligned} J_S(eV) &= 4e\Delta^2 |T|^2 N_L N_R \int_{\Delta}^{\infty} \frac{dE}{E} \varrho(E) \int_{\Delta}^{\infty} \frac{dE'}{E'} \varrho(E') \\ &\quad \times \left(\frac{1}{eV + E + E'} - \frac{1}{eV - E - E'} \right) \\ &= \left(\frac{\sigma_0}{e} \right) \Delta K \left(\frac{eV}{2\Delta} \right) \end{aligned} \tag{9.3.20}$$

The preceding integral can be evaluated and expressed as an elliptic integral. The retarded correlation function is written as a phase factor times a function of voltage $J_S(eV)$, where J_S is real for $|eV| < 2\Delta$. The Josephson current can be written as the imaginary part of (9.3.17), or

$$\begin{aligned} I_S(t) &= J_S(eV) \sin(\omega t + \phi) \\ \omega &= \frac{2eV}{\hbar} \end{aligned} \tag{9.3.21}$$

The Josephson current oscillates with time, with a frequency $\omega = 2eV/\hbar$. The amplitude of the oscillation is determined by the function of voltage $J_S(eV)$, where we are assuming $|eV| < 2\Delta$. The easiest way to evaluate $J_S(eV)$ at finite temperatures is to use a different sequence of steps. Ambegaokar and Baratoff showed that it is best to return to (9.3.18) and to do the two integrals $d\xi_k \, d\xi_p$ before the summation over Matsubara frequency ip. This step uses the result

$$\int_{-\infty}^{\infty} d\xi_k \mathscr{F}^{\dagger}(k, ip) = \Delta \int_{-\infty}^{\infty} \frac{d\xi_k}{\xi_k^2 + \Delta^2 + p_n^2} = \frac{\pi\Delta}{(\Delta^2 + p_n^2)^{1/2}}$$

The function $J_S(eV)$ is the retarded part $(i\omega \to eV)$ of the function $J_S(i\omega)$:

$$J_S(i\omega) = 4e \sum_{\mathbf{kp}} | T_{\mathbf{kp}} |^2 \frac{1}{\beta} \sum_{ip_n} \mathscr{F}^\dagger(\mathbf{k}, ip)\mathscr{F}(\mathbf{p}, ip - i\omega)$$
$$= 4\pi^2 N_R N_L \Delta_L \Delta_R | T_0 |^2 \frac{1}{\beta} \sum_{ip_n} \frac{1}{(\Delta_R^2 + p_n^2)^{1/2}[\Delta_L^2 + (p_n - \omega)^2]^{1/2}} \tag{9.3.22}$$

The summation over Matsubara frequencies ip_n should be done before the step of analytical continuation $\omega \to ieV$. As explained in Sec. 5.1G, an error is made if these two steps are reversed, except at the point $eV \to 0$. Thus at zero voltage, the result for identical superconductors is [see (9.2.15)]

$$J_S(eV = 0) = 4e\pi^2 N_L N_R \Delta^2 | T_0 |^2 \frac{1}{\beta} \sum_{ip_n} \frac{1}{\Delta^2 + p_n^2}$$
$$= 4e \frac{\pi^2}{2} N_L N_R \Delta | T_0 |^2 \tanh\left(\frac{\beta\Delta}{2}\right)$$
$$J_S(0) = \left(\frac{\sigma_0}{e}\right)\left(\frac{\pi\Delta}{2}\right) \tanh\left(\frac{\beta\Delta}{2}\right)$$

At zero temperature there is the particularly simple result that

$$J_S(0) = \left(\frac{\sigma_0}{e}\right)\left(\frac{\pi\Delta}{2}\right)$$

The coefficient of the Josephson current is the same current found in a normal junction at a voltage of $eV = \pi\Delta/2$.

Josephson tunneling has a wide variety of behavior in junctions with static electric and magnetic fields applied to them. But a very simple characteristic current is obtained in the case where there is neither a voltage $V = 0$ nor a magnetic field. Then one gets that the Josephson current is

$$I_J = J_S(0) \sin \phi$$

where ϕ is the phase difference between the two superconductors on the two sides of the tunnel junction.

The experimental arrangement is to put a current I through the tunnel junction, so that current flows through the interface oxide. The voltage change V is measured as a function of the applied current I. It is found that a dc current can flow without any voltage V. This behavior is sketched in Fig. 9.15. The dc current is the coherent tunneling of electron pairs from

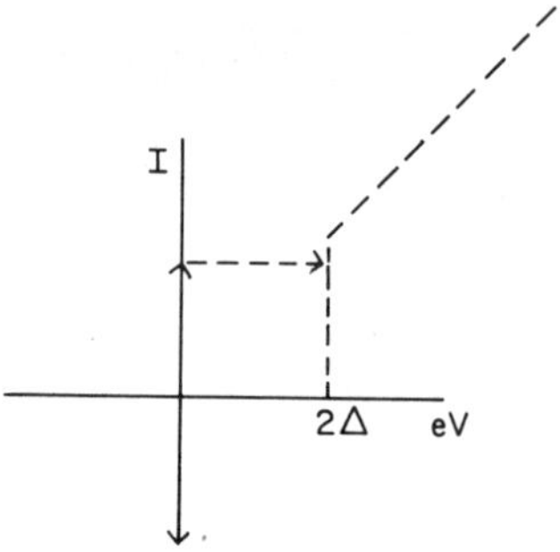

FIGURE 9.15. Current–voltage characteristics of a Josephson junction. The dashed line rising abruptly at $V = 2\Delta$ is the single-particle tunneling curve in Fig. 9.12 at $T = 0$. The arrows on the vertical axis show that a current can flow with no voltage. But an imposition of too much current causes the device to jump to the single-particle curve.

one superconductor to another. The pair tunneling costs no energy when the two superconductors have the same chemical potential. When the experimentalist increases the current, the phase difference ϕ between the two superconductors adjusts to make the Josephson current $J_S(0) \sin \phi$ just the experimental value I of current. The phase increase continues for increasing I until the point is reached where $I = J_S(0)$, or $\phi = \pi/2$. Then the current jumps from the zero-voltage line to the right and on to the experimental curve for single-particle tunneling, which is shown as the dashed line. In fact, in most tunnel junctions, the maximum phase $\phi = \pi/2$ cannot be reached, and the jump to the single-particle tunneling curve occurs at applied currents smaller than the maximum $J_S(0)$.

Another experiment is to take a tunnel junction between two superconductors and to apply a static voltage $eV < 2\Delta$. Then one observes an oscillatory current according to (9.3.21). This oscillatory current is the tunneling of pairs from the chemical potential of one superconductor to the chemical potential of the other. This tunneling of pairs does not conserve energy but takes an energy $2eV$ for the pair of charge $2e$ to overcome the voltage V. The pair must tunnel back, since it cannot complete a transition where energy is not conserved. This ac current can be observed by the external radiation it produces (Langenburg *et al.*, 1965).

The preceding two experiments must be performed in zero static magnetic field. The Josephson effect is so sensitive to magnetic field that care must be used to cancel out the effects of the earth's magnetic field. This sensitivity to magnetic field, which was pointed out by Josephson (1962), is useful for verifying that the tunneling is due to the Josephson effect and not some other process such as a superconducting short in the oxide—the latter would be expected to behave like the current characteristic in Fig. 9.15. The method of including magnetic field is as follows. The total phase difference between the superconductors is written as ϕ. The phase is now meant to include the applied voltage V, so that we write it as

$$I_J(eV) = J_S(eV)\sin\phi$$
$$\phi = \phi_0 + \frac{2e}{\hbar}\int_0^t dt' \int_a^b dx\,\frac{\partial V}{\partial x} \tag{9.3.23}$$

The spatial coordinates are shown in Fig. 9.16. The integral $\int dx$ in (9.3.23) is taken across the junction from one side to the other. The dx integral is actually $V(b) - V(a) = V$, since a and b are on two different sides of the oxide and each is deep within the bulk of the superconductor.

The phase ϕ needs to be expressed in a manner which is gauge invariant. This is done by writing it as the electric field $E(\mathbf{r}, t)$, which is always gauge invariant (see Sec. 1.5):

$$\frac{\partial V}{\partial x} \Rightarrow -E_x = \frac{1}{c}\frac{\partial A_x}{\partial t} + \frac{\partial\varphi}{dx}$$

$$\phi = \phi_0 + \frac{2e}{\hbar c}\int dt' \int \frac{\partial A_x\,dx}{dt'} + \frac{2e}{\hbar}\int dx'\,dt'\,\frac{\partial\varphi}{dx'}$$

The vector potential A_x term is manipulated in the following fashion. It is split into time-dependent and time-independent terms. The time-independent term is obtained by doing the time integral

$$\int^t dt'\,\frac{\partial A_x}{dt'} = A_x(t) = A_x$$

Second, the space integral $\int dxA_x$ is recognized as a line integral $\int d\mathbf{l}\cdot\mathbf{A}$ along a particular path, and any path will give the same result. Then the use of Stokes' theorem brings us to

$$\int d\mathbf{l}\cdot\mathbf{A} = \int d\mathbf{s}\cdot|\nabla\times\mathbf{A}| = \int d\mathbf{s}\cdot\mathbf{B}_0$$

$$\phi = \phi_0 + \frac{2e}{\hbar c}\int d\mathbf{s}\cdot\mathbf{B}_0 - \frac{2e}{\hbar c}\int^t dt'\int d\mathbf{l}'\cdot\mathbf{E}(\mathbf{r}', t')$$

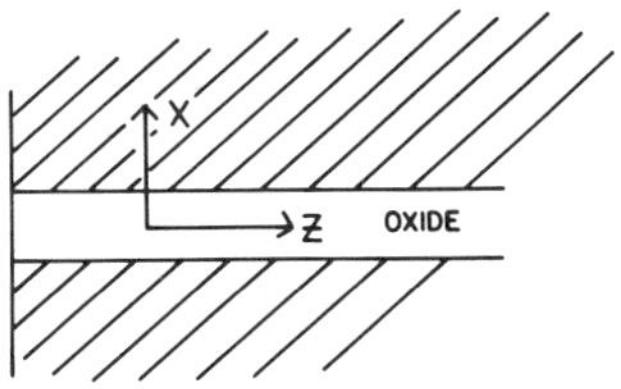

FIGURE 9.16. Coordinate system used in discussing Josephson tunneling.

The identity $\nabla \times \mathbf{A} = \mathbf{B}_0$ brings us to the term $\int d\mathbf{s} \cdot \mathbf{B}_0$, which is the magnetic flux Φ under the area of the integral. The external magnetic field $\mathbf{B}_0 = B_0\hat{y}$ is constant when $\mathbf{A}$ is constant. The area $d\mathbf{s}$ is the distance Z from some reference point, which can be the end of the junction, times the effective distance in the x direction. The latter is the thickness of the oxide d plus the distance the magnetic field penetrates into the superconductor on each side of the junction. The latter distances are the two penetration depths $\lambda_R + \lambda_L$. Thus in a fixed magnetic field we find

$$\begin{aligned} \phi &= \phi_0 + kZ - \frac{2e}{\hbar c}\int^t dt' \int d\mathbf{l} \cdot \mathbf{E}(\mathbf{r}, t') \\ k &= \frac{2e}{\hbar c} B_0(d + \lambda_R + \lambda_L) \end{aligned} \tag{9.3.24}$$

Thus the magnetic field acts as if it were generating an effective wave vector $\mathbf{k}$. The Josephson current I, now depends on the distance Z from the reference point. The total current in the sample of length L in the Z direction is found by integrating in this direction $(eV = 0)$:

$$j(Z) = J_S(0)\sin(\phi_0 + kZ)$$

$$I = \frac{1}{L}\int_{-L/2}^{L/2} dZ\, j(Z) = J_S(0)\sin(\phi_0)\left[\frac{2}{kL}\sin\left(\frac{kL}{2}\right)\right]$$

One obtains the theoretical prediction that the maximum current at zero voltage will have a Fraunhofer pattern $\sin(\theta)/\theta$, where $\theta \propto B_0$. The Fraunhofer pattern was indeed observed by Fiske, as shown in Fig. 9.17 for Sn–oxide–Sn junctions. This experiment is another spectacular verification of Josephson's predictions. Even more interesting results are obtained when there is both a static dc magnetic field B_0 and dc voltage V. Then the Josephson current has the form

$$\begin{aligned} j(Z) &= J_S(eV)\sin(\phi_0 + kZ + \omega t) \\ k &= \frac{2e}{\hbar c} B_0(d + \lambda_R + \lambda_L) \\ \omega &= \frac{2eV}{\hbar} \end{aligned}$$

This form of current is also observed in a variety of experiments which are described in the review volume edited by Parks (1969).

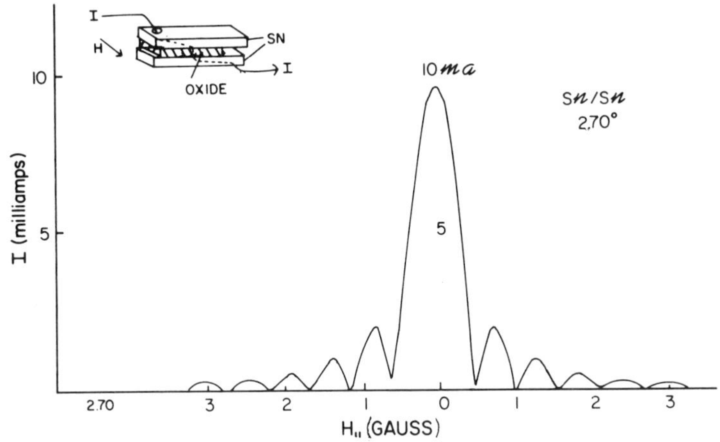

FIGURE 9.17. The Fraunhofer pattern of a Josephson junction. The experimental curve is the maximum dc current which can flow with no voltage as a function of applied magnetic field. *Source*: Fiske (unpublished) (used with permission).

9.4. INFRARED ABSORPTION

A unique feature of the BCS theory was the prediction of an energy gap in the excitation spectrum of the superconductor at zero temperature. Electron tunneling (Giaever, 1960) was the first measurement to convincingly demonstrate the existence of the energy gap. However, this experiment was not commonplace at the time of the BCS theory (1957), and the conventional method of seeking an energy gap would be to measure the optical absorption of the metal in the far infrared. For example, Pb has $T_c = 7.2$ K so $k_B T_c = 0.62$ meV. The BCS prediction is that the energy gap is $E_g = 2\Delta = 3.52(k_B T_c) = 2.2$ meV $= 17.7$ cm^{-1}. The value found experimentally is $E_g = 22.5$ cm^{-1}. The difference from the BCS prediction is due to the fact that Pb is a strongly coupled superconductor. The frequency range is very difficult experimentally so that good experimental results for Pb were not reported until 1968 (Palmer and Tinkham, 1968).

A superconductor at zero temperature should not absorb radiation for $\omega < 2\Delta$. The absorption breaks up an electron pair, which creates two quasiparticles. Each quasiparticle has an energy of the form $E = (\Delta^2 + \xi_p^2)^{1/2}$, so absorption starts at $\omega = 2\Delta$. Our theoretical discussion follows the original theory by Mattis and Bardeen (1958).

The rate of absorption in the infrared is given by the real part of the conductivity, which is evaluated from the Kubo formula:

$$\pi(\mathbf{q}, i\omega) = -\frac{1}{3\nu}\int_0^\beta d\tau e^{i\omega\tau}\langle T_\tau \mathbf{j}(\mathbf{q}, \tau)\cdot \mathbf{j}(-\mathbf{q}, 0)\rangle$$

$$\sigma(\mathbf{q}, \omega) = -\frac{1}{\omega}\operatorname{Im}[\pi_{\text{ret}}(\mathbf{q}, \omega)]$$

$$\mathbf{j}(\mathbf{q}) = \frac{e}{m}\sum_{\mathbf{p}\sigma}\left(\mathbf{p}+\frac{1}{2}\mathbf{q}\right)C^\dagger_{\mathbf{p}+\mathbf{q},\sigma}C_{\mathbf{p}\sigma}$$

The current expression is that appropriate for noninteracting particles. There results a correlation function of four operators:

$$\pi(\mathbf{q}, i\omega) = \frac{-e^2}{3m^2}\frac{1}{\nu}\sum_{\substack{\mathbf{p}\mathbf{p}'\\ \sigma\sigma'}}\left(\mathbf{p}+\frac{1}{2}\mathbf{q}\right)\left(\mathbf{p}'-\frac{1}{2}\mathbf{q}\right) \times\int_0^\beta d\tau e^{i\omega\tau}\langle T_\tau C^\dagger_{\mathbf{p}+\mathbf{q},\sigma}(\tau)C_{\mathbf{p},\sigma}(\tau)C^\dagger_{\mathbf{p}'-\mathbf{q},\sigma'}(0)C_{\mathbf{p}'\sigma'}(0)\rangle \qquad (9.4.1)$$

In prior discussions of optical absorption processes, it was convenient to set to zero the photon wave vectors $\mathbf{q}$. The step $\mathbf{q} = 0$ is not done here because we would find later that the conductivity becomes zero. Thus it is formally necessary to retain a finite value for $\mathbf{q}$, although the magnitude of $q \simeq 10^2\ \text{cm}^{-1}$ can certainly be neglected compared to the wave vectors of the electrons $k_F \simeq 10^8\ \text{cm}^{-1}$.

To evaluate the bracket of operators in (9.4.1), fix one spin, say $\sigma = \uparrow$. The answer is twice this result, since the same contribution is found for $\sigma = \downarrow$. The operators in the angle brackets can be paired two ways (the pairings with $\mathbf{q} = 0$ are not allowed since $\mathbf{q}$ is finite):

$$\begin{aligned}&\langle T_\tau C^\dagger_{\mathbf{p}+\mathbf{q}\uparrow}(\tau)C_{\mathbf{p}\uparrow}(\tau)C^\dagger_{\mathbf{p}'-\mathbf{q}\sigma'}(0)C_{\mathbf{p}'\sigma'}(0)\rangle\\ &\quad = -\delta_{\sigma'=\uparrow}\delta_{\mathbf{p}'=\mathbf{p}+\mathbf{q}}\mathscr{G}(\mathbf{p}, \tau)\mathscr{G}(\mathbf{p}+\mathbf{q}, -\tau)\\ &\qquad + \delta_{\sigma'=\downarrow}\delta_{\mathbf{p}'=-\mathbf{p}}\mathscr{F}(\mathbf{p}, -\tau)\mathscr{F}^\dagger(\mathbf{p}+\mathbf{q}, \tau)\end{aligned} \qquad (9.4.2)$$

These pairings evaluate the correlation function, which is written in both τ space and Fourier space:

$$\begin{aligned}\pi(\mathbf{q}, i\omega) &= \frac{2e^2}{3m^2\nu}\sum_{\mathbf{p}}\left(\mathbf{p}+\frac{1}{2}\mathbf{q}\right)^2\int_0^\beta d\tau e^{i\omega\tau}\\ &\quad\times[\mathscr{G}(\mathbf{p}, \tau)\mathscr{G}(\mathbf{p}+\mathbf{q}, -\tau) + \mathscr{F}(\mathbf{p}, -\tau)\mathscr{F}^\dagger(\mathbf{p}+\mathbf{q}, \tau)]\\ \pi(\mathbf{q}, i\omega) &= \frac{2e^2}{3m^2\nu}\sum_{\mathbf{p}}\left(\mathbf{p}+\frac{1}{2}\mathbf{q}\right)^2\frac{1}{\beta}\sum_{ip}\\ &\quad\times[\mathscr{G}(\mathbf{p}, ip)\mathscr{G}(\mathbf{p}+\mathbf{q}, ip+i\omega) + \mathscr{F}(\mathbf{p}, ip)\mathscr{F}^\dagger(\mathbf{p}+\mathbf{q}, ip+i\omega)]\end{aligned} \qquad (9.4.3)$$

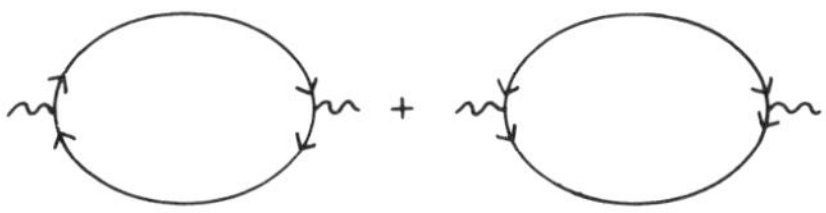

FIGURE 9.18

Note that the vector vertex of the current operator changes the relative minus sign between the two terms in (9.4.2) to a relative plus in (9.4.3). The two terms in (9.4.3) can be represented by the two diagrams shown in Fig. 9.18. The photon of $(q, i\omega)$ enters and leaves and couples to the two solid lines of the polarization diagram. The solid lines can be either the pair $\mathscr{G}\mathscr{G}$ or $\mathscr{F}\mathscr{F}^\dagger$ as these are the only two possibilities which conserve particle number at the two vertices. One could not, for example, have the combination $\mathscr{G}\mathscr{F}$ since that does not conserve particle number at one of the vertices. The crucial aspect of the above expression is the relative sign between the two terms $\mathscr{G}\mathscr{G}$ and $\mathscr{F}\mathscr{F}^\dagger$. There seems to be no easy way to deduce this sign, except by being careful.

The next step in the derivation is the summation over Matsubara frequencies. The summation for $\mathscr{F}\mathscr{F}^\dagger$ was given previously in (9.3.19), while the other one is

$$\begin{aligned}\frac{1}{\beta}\sum_{ip}\mathscr{G}(\mathbf{p}, ip)\mathscr{G}(\mathbf{k}, ip+i\omega)\\ &= [1-n_F(E_p)-n_F(E_k)]\left(\frac{v_p{}^2u_k{}^2}{i\omega-E_p-E_k}-\frac{u_p{}^2v_k{}^2}{i\omega+E_k+E_p}\right)\\ &+ [n_F(E_p)-n_F(E_k)]\left(\frac{u_p{}^2u_k{}^2}{i\omega+E_p-E_k}-\frac{v_p{}^2v_k{}^2}{i\omega+E_k-E_p}\right)\end{aligned} \tag{9.4.4}$$

At finite temperatures, there are a number $n_F(E)$ of quasiparticle excitations with energy E which can absorb the radiation even for $\omega < 2\Delta$. The interesting case is the limit of zero temperature, where there is no absorption until $\omega > 2\Delta$. At zero temperature we have only the terms

$$\begin{aligned}\pi(\mathbf{q}, i\omega) &= \frac{2e^2}{3m^2\nu}\sum_{\mathbf{p}}\left(\mathbf{p}+\frac{1}{2}\mathbf{q}\right)^2\\ &\times\left(\frac{v_{\mathbf{p}}{}^2u_{\mathbf{p+q}}^2-u_{\mathbf{p}}v_{\mathbf{p}}u_{\mathbf{p+q}}v_{\mathbf{p+q}}}{i\omega-E_{\mathbf{p}}-E_{\mathbf{p+q}}}-\frac{u_{\mathbf{p}}{}^2v_{\mathbf{p+q}}^2-u_{\mathbf{p}}v_{\mathbf{p}}u_{\mathbf{p+q}}v_{\mathbf{p+q}}}{i\omega+E_{\mathbf{p}}+E_{\mathbf{p+q}}}\right)\end{aligned}$$

The retarded function is obtained from $i\omega\to\omega+i\delta$, and taking the imaginary part of the retarded function brings us to the real part of the conductivity $(\nu\to\infty)$ for $\omega>0$:

$$\sigma(\mathbf{q}, \omega) = \frac{2e^2\pi}{3m^2\omega}\int\frac{d^3p}{(2\pi)^3}\left(\mathbf{p}+\frac{1}{2}\mathbf{q}\right)^2\delta(\omega-E_{\mathbf{p}}-E_{\mathbf{p+q}})(v_{\mathbf{p}}^2u_{\mathbf{p+q}}^2-u_{\mathbf{p}}v_{\mathbf{p}}u_{\mathbf{p+q}}v_{\mathbf{p+q}})$$

This expression has a reasonable form. The photon is absorbed by having its energy ω converted into two quasiparticles with separate energies $E_{\mathbf{p}}$ and $E_{\mathbf{p}+\mathbf{q}}$.

The combination of coherence factors $v_{\mathbf{p}}^2 u_{\mathbf{p}+\mathbf{q}}^2 - u_{\mathbf{p}} v_{\mathbf{p}} u_{\mathbf{p}+\mathbf{q}} v_{\mathbf{p}+\mathbf{q}}$ determines the rate of absorption. This quantity vanishes as $\mathbf{q} \to 0$. We have already remarked that $q \ll p$, so that one should indeed take this limit. Then one would conclude that the absorbing power of the superconductor was very weak at least for the process of creating two quasiparticles. This conclusion is valid in the case where momentum is a good quantum number, so the two quasiparticles made by the photon have momenta which are related, i.e., $-\mathbf{p}$ and $\mathbf{p} + \mathbf{q}$.

Mattis and Bardeen assumed that this was not the situation. Instead, the two excited quasiparticles have momentum $\mathbf{p}$ and $\mathbf{p}'$ which are unrelated. The lack of momentum conservation is characteristic of the so-called *dirty superconductor*. The momentum breaking might occur, for example, because the experiments are done on evaporated metal films which are not crystalline. Alternately, there might be phonons or impurities which scatter the particles. In any case, we assume that energy is conserved but not momentum. Thus we must separately sum over $\mathbf{p}$ and $\mathbf{p}'$ and are led to the expression

$$\sigma(\omega) = \frac{C_0}{\omega} \int_{-\infty}^{\infty} d\xi_p \int_{-\infty}^{\infty} d\xi_{p'} \delta(\omega - E_p - E_{p'})(v_p{}^2 u_{p'}^2 - u_p v_p u_{p'} v_{p'})$$

where C_0 is a constant which will be determined later. The terms in the coherence factors $v_p{}^2$ and $u_{p'}^2$ which are linear in ξ_p and $\xi_{p'}$ vanish because they are odd functions of ξ_p and $\xi_{p'}$. Then change variables to $E = E_p$ and $E' = E_{p'}$, and derive the expression

$$\sigma(\omega) = \frac{C_0}{\omega} \int_{\Delta}^{\infty} dE \int_{\Delta}^{\infty} dE' \varrho(E) \varrho(E') \delta(\omega - E - E') \left(1 - \frac{\Delta^2}{EE'}\right) \tag{9.4.5}$$

The E' integral is done by energy conservation:

$$\begin{aligned}
\sigma(\omega) &= \frac{C_0}{\omega} \theta(\omega - 2\Delta) \int_{\Delta}^{\omega - \Delta} dE \varrho(E) \varrho(\omega - E) \left[1 - \frac{\Delta^2}{E(\omega - E)}\right] \\
&= \frac{C_0}{\omega} \theta(\omega - 2\Delta) \int_{\Delta}^{\omega - \Delta} dE \frac{E(\omega - E) - \Delta^2}{(E^2 - \Delta^2)^{1/2} [(\omega - E)^2 - \Delta^2]^{1/2}}
\end{aligned}$$

This integral has the same form as (9.3.15) and is evaluated in the same manner: Change variables of integration to x, where $2E = \omega + x(\omega - 2\Delta)$,

$$\sigma(\omega) = \frac{1}{2}\frac{C_0}{\omega}\theta(\omega - 2\Delta)(\omega - 2\Delta)\int_{-1}^{1} dx \frac{1 - \alpha x^2}{[(1 - x^2)(1 - \alpha^2 x^2)]^{1/2}}$$

$$\alpha = \frac{\omega - 2\Delta}{\omega + 2\Delta}$$

and the integral has the form for complete elliptic integrals:

$$\sigma(\omega) = \frac{C_0}{\omega}[(\omega + 2\Delta)E(\alpha) - 4\Delta K(\alpha)]\theta(\omega - 2\Delta)$$

The absorption for a normal metal has $\Delta = 0$, which in the same units is C_0. Thus the ratio of the conductivity in the superconductor to that of the normal metals is

$$\frac{\sigma_s(\omega)}{\sigma_n(\omega)} = \frac{1}{\omega}\theta(\omega - 2\Delta)[(\omega + 2\Delta)E(\alpha) - 4\Delta K(\alpha)]$$
$$\alpha = \frac{\omega - 2\Delta}{\omega + 2\Delta} \qquad (9.4.6)$$

This is the result of Mattis and Bardeen.

The experiment measures the conductivity in the normal metal and the superconductor at the same temperature. The metal is made normal, at the low temperature, by the application of a small magnetic field. The experimental results of Palmer and Tinkham (1968) for Pb are shown as the points in Fig. 9.19. They compare well with the theory of Mattis and Bardeen, Eq. (9.4.6), which is shown as the solid line. There is no absorption until the energy gap of 22.5 cm^{-1}. Then the absorption rises gradually with increasing ω. It does not, as in the case of tunneling between two superconductors, rise discontinuously at the threshold frequency.

The agreement between theory and experiment is obviously excellent. In fact it is initially surprising that the agreement is that good. After all, Pb is a strongly coupled superconductor, and there is no particular reason to expect the BCS theory to work well for this metal. This question was investigated by Shaw and Swihart (1968), who solved the theory of infrared absorption using the strong coupling theory. They found that the results were nearly identical to the predictions of the BCS theory. Thus this experiment is not sensitive to the strength of the coupling. There are, however, numerous other experiments which can distinguish between these theories. Some are discussed in Sec. 9.7.

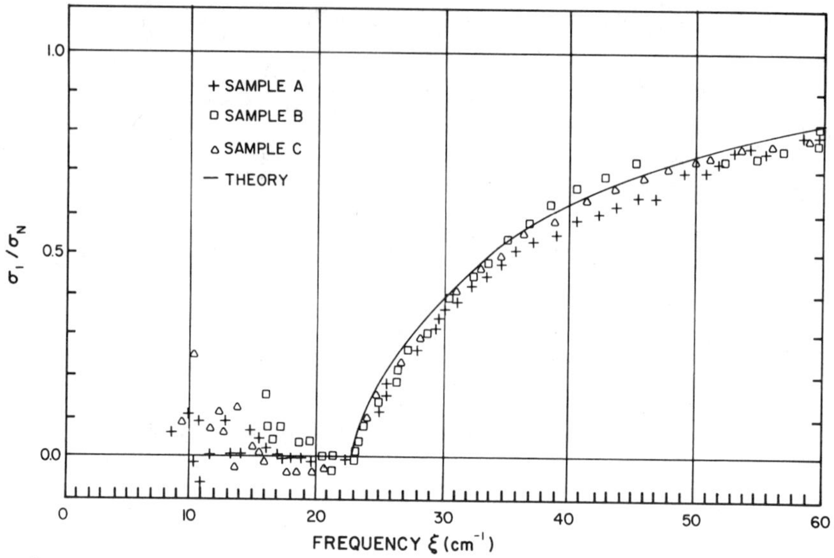

FIGURE 9.19. Far-infrared absorption in superconducting lead at 2 K. Solid line is BCS theory, and points are experimental. *Source*: Palmer and Tinkham (1968) (used with permission).

Since this section is devoted to discussing the conductivity of a superconductor, perhaps it is the appropriate place to explain why the superconductor has zero resistivity. The two features of a superconductor which were first discovered, zero resistance and the Meissner effect, can both be explained by assuming that the current $\mathbf{J}(\mathbf{r})$ in a superconductor is related to the vector potential $\mathbf{A}(\mathbf{r}, t)$ through the equation discovered by London (1954):

$$\mathbf{J}(\mathbf{r}, t) = -\frac{e^2 n_0}{mc}\mathbf{A}(\mathbf{r}, t) \tag{9.4.7}$$

The Meissner effect is the exclusion of magnetic flux by a superconductor. A derivation of (9.4.7) is sufficient to explain zero resistivity and the Meissner effect. When we quantized the Lagrangian for the interacting system of charges in Sec. 1.5, we found that the electron velocity operator was given by $m\mathbf{v}_i = \mathbf{p}_i - (e_i/c)\mathbf{A}(\mathbf{r}_i)$:

$$\mathbf{J} = e\sum_i \langle \mathbf{v}_i \rangle = \frac{e}{m}\sum_i \langle \mathbf{p} \rangle - \frac{e^2}{mc}\sum_i \mathbf{A}(\mathbf{r}_i)$$

The current is proportional to the average value of $\langle \mathbf{v}_i \rangle$, which contains two terms. The first is the expectation value of the momentum $\langle \mathbf{p}_i \rangle$, which is made proportional to the applied electric field by using the Kubo formula:

$$\mathbf{J}(\mathbf{r}, t) = \sigma(\omega)\mathbf{E}(\mathbf{r}, t) - \frac{e^2}{mc} n_0 \mathbf{A}(\mathbf{r}, t) \tag{9.4.8}$$

In the second term, we have assumed that the electron density is uniform, so the summation over $\mathbf{A}(\mathbf{r}_i)$ just produces $\mathbf{A}(\mathbf{r})n_0$. The second term is the London expression and is always in the formula for the current operator, even in normal metals. For ac fields, then, $(\partial/\partial t)\mathbf{A}(\mathbf{r}, t) = -c\mathbf{E}$, so $\mathbf{A} = ic\mathbf{E}/\omega$, and this term is out of phase with the electric field. It does not extract energy from the field or cause absorption. It does not contribute to the real part of the conductivity, which is the quantity we usually calculate.

Thus the London equation is derived from (9.4.8) whenever the conductivity term $\sigma(\omega)$ vanishes. This point was realized by London, and all subsequent attempts to construct theories of superconductivity have tried to make the usual conductivity term vanish. For the dc conductivity, we want $\sigma(\omega = 0)$. We have already shown that $\sigma(0)$ vanishes in the BCS theory at zero temperature since the energy gap means that $\sigma(\omega)$ is zero for $\omega < 2\Delta$. One can also show that $\sigma(0) = 0$ even at finite temperatures, but in the superconducting state $0 < T < T_c$. Thus the BCS theory does explain London's equation and hence superconductivity.

9.5. ACOUSTIC ATTENUATION

Another experiment which can measure the energy gap of a superconductor is the absorption of acoustic waves. The attenuation length of the sound can be measured as a function of parameters such as temperature, frequency, and the energy gap of the superconductor, which can be modulated by small magnetic fields. For example, one experiment measures the attenuation of sound with very long wavelength, say $\omega \simeq 10^6$ Hz. The energy quantum of sound is negligible, $\omega \ll \Delta$, so that the sound cannot excite quasiparticles across the energy gap. This experiment can only measure the density of quasiparticle excitations. Indeed, we shall find that the ratio of attenuation between the superconductor and the metal is just given by

$$\frac{\alpha_s}{\alpha_n} = 2n_F(\Delta) = \frac{2}{e^{\Delta\beta} + 1} \tag{9.5.1}$$

A measurement of α_s and α_n permits the energy gap to be obtained directly from the sound attenuation. This experimental technique was used by Morse and Bohm (1957) to produce the results for $\Delta(T)$ which we have already shown in Fig. 9.6. Quite a different experiment is to use high-frequency acoustic waves, which can break pairs and create quasiparticles when $\omega > 2\Delta$.

The attenuation of sound waves may be derived from the mean free path of phonons. The theoretical approach is to calculate the self-energy function for the phonon Green's function and to take its imaginary part. The imaginary part of the phonon Green's function is proportional to the rate of acoustic attenuation. Thus from the electron–phonon interaction,

$$H_{\text{e-ph}} = \frac{1}{\nu^{1/2}} \sum_{\mathbf{q}\lambda} M_\lambda(\mathbf{q})\varrho(\mathbf{q})(a_{\mathbf{q}\lambda} + a^\dagger_{-\mathbf{q}\lambda})$$

$$\varrho(\mathbf{q}) = \sum_{\mathbf{k}n} c^\dagger_{\mathbf{k}+\mathbf{q}\sigma} c_{\mathbf{k}\sigma}$$

we calculate the imaginary part of the retarded correlation function:

$$P_\lambda(\mathbf{q}, i\omega) = -\frac{M_\lambda(\mathbf{q})^2}{\nu} \int_0^\beta d\tau e^{i\omega\tau} \langle T_\tau \varrho(\mathbf{q}, \tau)\varrho(-\mathbf{q}, 0)\rangle$$

This is just the density–density correlation function. The same correlation function is encountered in related problems such as the treatment of the longitudinal dielectric function for superconductors. The correlation is written in terms of the product of four operators, which are the same as evaluated earlier in (9.4.2):

$$\begin{aligned} P_\lambda(\mathbf{q}, i\omega) &= -\frac{M_\lambda(\mathbf{q})^2}{\nu} \sum_{\substack{\mathbf{p}\mathbf{p}'\\ \sigma\sigma'}} \int_0^\beta d\tau e^{i\omega\tau} \langle T_\tau c^\dagger_{\mathbf{p}+\mathbf{q}\sigma}(\tau) c_{\mathbf{p}\sigma}(\tau) c^\dagger_{\mathbf{p}'-\mathbf{q},\sigma'} c_{\mathbf{p}'\sigma'}\rangle \\ &= \frac{2M_\lambda(\mathbf{q})^2}{\nu} \sum_{\mathbf{p}} \frac{1}{\beta} \sum_{ip} [\mathscr{G}(\mathbf{p}, ip)\mathscr{G}(\mathbf{p}+\mathbf{q}, ip + i\omega) \\ &\quad - \mathscr{F}(\mathbf{p}, ip)\mathscr{F}^\dagger(\mathbf{p}+\mathbf{q}, ip + i\omega)] \end{aligned}$$

Now there is a minus sign between the two terms $\mathscr{G}\mathscr{G}$ and $\mathscr{F}\mathscr{F}^\dagger$. The sign is different from the plus sign in the expression (9.4.3) for the infrared absorption. This sign difference comes from the fact that now we have a scalar vertex function, where for the conductivity we had a vector vertex which changed signs between the two terms. The summation over Matsubara frequencies was done earlier, in (9.3.19) and (9.4.4), so we have

$$P_\lambda(\mathbf{q}, i\omega) = \frac{2M_\lambda(\mathbf{q})^2}{\nu} \sum_{\mathbf{p}} \Big\{[1 - n_F(E_\mathbf{p}) - n_F(E_{\mathbf{p}+\mathbf{q}})]$$
$$\times \left(\frac{v_\mathbf{p}^2 u_{\mathbf{p}+\mathbf{q}}^2 + \Delta^2/4E_\mathbf{p}E_{\mathbf{p}+\mathbf{q}}}{i\omega - E_\mathbf{p} - E_{\mathbf{p}+\mathbf{q}}} - \frac{u_\mathbf{p}^2 v_{\mathbf{p}+\mathbf{q}}^2 + \Delta^2/4E_\mathbf{p}E_{\mathbf{p}+\mathbf{q}}}{i\omega + E_\mathbf{p} + E_{\mathbf{p}+\mathbf{q}}} \right)$$
$$+ [n_F(E_\mathbf{p}) - n_F(E_{\mathbf{p}+\mathbf{q}})]$$
$$\times \left(\frac{u_\mathbf{p}^2 u_{\mathbf{p}+\mathbf{q}}^2 - \Delta^2/4E_\mathbf{p}E_{\mathbf{p}+\mathbf{q}}}{i\omega + E_\mathbf{p} - E_{\mathbf{p}+\mathbf{q}}} - \frac{v_\mathbf{p}^2 v_{\mathbf{p}+\mathbf{q}}^2 - \Delta^2/4E_\mathbf{p}E_{\mathbf{p}+\mathbf{q}}}{i\omega - E_\mathbf{p} + E_{\mathbf{p}+\mathbf{q}}} \right)\Big\}$$

The absorption is proportional to the spectral function of this correlation function. Thus we take $i\omega \to \omega + i\delta$ and then the imaginary part. This quantity we call $\alpha(\omega)$:

$$\alpha_\lambda(\mathbf{q}, \omega) = 2\pi M^2 \int \frac{d^3p}{(2\pi)^3} \{[1 - n_F(E) - n_F(E')][\delta(\omega - E - E')$$
$$\times (v^2u'^2 + \Delta^2/4EE') - \delta(\omega + E + E')(u^2v'^2 + \Delta^2/4EE')]$$
$$+ [n_F(E) - n_F(E')][\delta(\omega + E - E')(u^2u'^2 - \Delta^2/4EE')$$
$$- \delta(\omega - E + E')(v^2v'^2 - \Delta^2/4EE')]\} \tag{9.5.2}$$

where E, u, v depend on $\mathbf{p}$, while E', u', v' depend on $\mathbf{p} + \mathbf{q}$. Equation (9.5.2) is the most general result. It is a function of the frequency ω of the sound wave, temperature, and the energy gap of the superconductor. The general formula is complicated, but simple results can be obtained for several cases.

The first special case is for sound waves of very low frequency compared to the energy gap of the superconductor. This inequality $\omega \ll \Delta$ is held by sound waves over a wide range of frequencies. The terms in which energy conservation specifies $\pm\omega = E + E'$ give zero, since these can be satisfied only when $\pm\omega > 2\Delta$, which is not the present situation. In the other terms we have combinations such as

$$\delta(\omega + E - E')[n_F(E) - n_F(E')]$$
$$= \delta(\omega + E - E')[n_F(E) - n_F(E + \omega)] \simeq -\omega\delta(E - E') \frac{\partial n_F(E)}{\partial E}$$

In the limit of low frequency we obtain

$$\alpha_\lambda(\mathbf{q}, \omega) = 2\pi\omega M_\lambda(\mathbf{q})^2 \int \frac{d^3p}{(2\pi)^3} \left[-\frac{\partial n_F(E_p)}{\partial E_p} \right] \delta(E_\mathbf{p} - E_{\mathbf{p}+\mathbf{q}})$$
$$\times (u_\mathbf{p}^2 u_{\mathbf{p}+\mathbf{q}}^2 + v_\mathbf{p}^2 v_{\mathbf{p}+\mathbf{q}}^2 - \Delta^2/2E_\mathbf{p}E_{\mathbf{p}+\mathbf{q}}) \tag{9.5.3}$$

The absorption rate is proportional to ω. The actual attenuation rate for long-wavelength sound waves in normal metals is proportional to ω^2. The additional factor of ω comes from the matrix element M^2.

There are two possible approaches to evaluating the momentum integrals in (9.5.3). The first is to assume that momentum is conserved in the quasiparticle scattering process. The phonon absorption causes the quasiparticle to scatter from $\mathbf{p}$ to $\mathbf{p}+\mathbf{q}$. These states have the same energy $E_{\mathbf{p}} = E_{\mathbf{p}+\mathbf{q}}$ since the phonon energy is assumed to be negligible. We shall use this method to evaluate the integral. The other way is to assume the limit of a dirty superconductor and to take E and E' as separate variables which are not related by momentum conservation. This derivation is assigned as a problem. Interestingly enough, the two methods of evaluation make the same prediction regarding the final result.

The first step is to examine the coherence factors:

$$u^2u'^2 + v^2v'^2 - \frac{\Delta^2}{2EE'}$$
$$= \frac{1}{4}\left[\left(1+\frac{\xi}{E}\right)\left(1+\frac{\xi'}{E'}\right)+\left(1-\frac{\xi}{E}\right)\left(1-\frac{\xi'}{E'}\right)-\frac{2\Delta^2}{EE'}\right]$$
$$= \frac{1}{2}\left(1+\frac{\xi\xi'-\Delta^2}{E^2}\right)$$

The energy conservation $E = E'$ requires that $\xi = \pm\xi'$. The choice $\xi = \xi'$ leads to

$$\frac{1}{2}\left(1+\frac{\xi^2}{E^2}-\frac{\Delta^2}{E^2}\right) = \frac{\xi^2}{E^2}$$

Similarly, the choice $\xi' = -\xi$ gives

$$\frac{1}{2}\left(1-\frac{\xi^2}{E^2}-\frac{\Delta^2}{E^2}\right) = 0$$

Thus $\xi' = \xi$ gives the only finite contribution.

The next step in the evaluation of the integral is to do the angular integrals and use them to eliminate the delta function for energy contribution:

$$\int d^3p = 2\pi m \int p\, d\xi_p \int_{-1}^{1} d\nu$$

$$\int_{-1}^{1} d\nu\, \delta(E_{\mathbf{p}} - E_{\mathbf{p}+\mathbf{q}}) = \left|\frac{\partial E_{\mathbf{p}+\mathbf{q}}}{\partial \nu}\right|^{-1} = \frac{m}{pq}\,\frac{E_p}{|\xi_p|}$$

The angular integral is finite only when $p^2 = (\mathbf{p}+\mathbf{q})^2 = p^2 + q^2 + 2pq\nu$

can be satisfied for $-1 < \nu < 1$. This condition requires that $p > q/2$ or that $q < 2k_F$ since we need $p \sim k_F$. The requirement $q < 2k_F$ is easily satisfied, since the phonons of interest have much smaller wave vectors than $2k_F$. Thus the integral for the attenuation is

$$\begin{aligned}
\alpha_\lambda(\mathbf{q}, \omega) &= \frac{m^2\omega}{2\pi q} M_\lambda(\mathbf{q})^2 \int d\xi \frac{|\xi|}{E}\left[-\frac{\partial n_F(E)}{\partial E}\right] \\
&= \frac{m^2\omega}{\pi q} M_\lambda(\mathbf{q})^2 \int_\Delta^\infty dE\left[-\frac{\partial n_F(E)}{\partial E}\right] \\
&= \frac{m^2\omega M_\lambda(\mathbf{q})^2}{\pi q} n_F(\Delta) \qquad (9.5.4)
\end{aligned}$$

The simple result is found that the acoustic absorption is proportional to the density of quasiparticles $n_F(\Delta)$ with the energy of the gap function Δ. The same result for the normal metal is found by setting $\Delta = 0$, or $n_F(0) = \frac{1}{2}$. Thus the ratio of the acoustic absorption for the superconductor to the normal metal is $2n_F(\Delta)$, which agrees with the result given in (9.5.1).

Another case in which the general formula (9.5.2) can be evaluated analytically is at zero temperature. Then there is no absorption of the acoustic wave unless it can break a pair and create two quasiparticles, which requires $\omega > 2\Delta$. Thus at zero temperature we have

$$\alpha_\lambda(\mathbf{q}, \omega) = 2\pi M_\lambda(\mathbf{q})^2 \int \frac{d^3p}{(2\pi)^3} \delta(\omega - E_\mathbf{p} - E_{\mathbf{p}+\mathbf{q}})(v_\mathbf{p}^2 u_{\mathbf{p}+\mathbf{q}}^2 + \Delta^2/4E_\mathbf{p}E_{\mathbf{p}+\mathbf{q}})$$

The wave vector integrals will be evaluated for the model of a dirty superconductor, in which momentum is not conserved. The pair-breaking process is perhaps more dramatic than the simple scattering of quasiparticles, so this assumption may be more valid in this case. Thus we assume that E and E' are separate variables which are related by energy conservation but not momentum conservation:

$$\alpha(\omega) = \frac{c_0}{4}\int d\xi \int d\xi' \delta(\omega - E - E')\left[\left(1 - \frac{\xi}{E}\right)\left(1 + \frac{\xi'}{E'}\right) + \frac{\Delta^2}{EE'}\right]$$

The terms linear in ξ and ξ' integrate to zero, and changing variables gives

$$\begin{aligned}
\alpha(\omega) &= c_0 \int_\Delta^\infty dE\varrho(E) \int_\Delta^\infty dE'\varrho(E')\delta(\omega - E - E')\left(1 + \frac{\Delta^2}{EE'}\right) \\
&= c_0\theta(\omega - 2\Delta)\int_\Delta^{\omega-\Delta} dE\varrho(E)\varrho(\omega - E)\left[1 + \frac{\Delta^2}{E(\omega - E)}\right]
\end{aligned}$$

This integral is in the standard form we have encountered before in previous sections and is given in terms of elliptic integrals. The most economical way to express the answer is the form given by Bobetic (1964):

$$\alpha(\omega) = c_0\omega\theta(\omega - 2\Delta)E(m)$$
$$m = \left[1 - \left(\frac{2\Delta}{\omega}\right)^2\right]^{1/2}$$

The same result for the normal metal is obtained by setting $\Delta = 0$, which gives

$$\begin{aligned} \alpha_N(\omega) &= c_0\omega \\ \frac{\alpha_s(\omega)}{\alpha_N(\omega)} &= \theta(\omega - 2\Delta)E(m) \end{aligned} \qquad (9.5.5)$$

The absorption of acoustic waves is zero for $\omega < 2\Delta$. However, for frequencies just above threshold, we have that $m = 0$ and $E(0) = \pi/2$, so that the absorption is the finite result

$$\frac{\alpha_s(2\Delta^+)}{\alpha_N(2\Delta^+)} = \frac{\pi}{2} \qquad (9.5.6)$$

Thus there is a discontinuous jump in the absorption at frequencies equal to the energy gap 2Δ. This behavior is similar to that found in the single-particle tunneling between two superconductors, where there was also a discontinuity. The behavior is different for the infrared absorption, where the threshold behavior was a gradual rise with the frequency above threshold. The difference between the infrared and acoustic results is that in the infrared case the coherence factors gave $1 - \Delta^2/EE'$, which vanishes at threshold. The acoustic absorption case has $1 + \Delta^2/EE'$, and the two terms add at threshold. This difference shows the importance of the relative sign between the terms $\mathscr{G}\mathscr{G}$ and $\mathscr{F}\mathscr{F}^\dagger$ in determining the theoretical behavior. The ratio $\pi/2$ in (9.5.6) is greater than 1, so that the superconductor is more absorbing than the normal metal for $\omega > 2\Delta$.

So far our calculations in this section have been concerned with the absorption or attenuation of longitudinal acoustic waves. The acoustic attenuation of transverse acoustic waves, or shear waves, shows a different behavior in a superconductor. This was measured and explained by Fossheim and Leibowitz (1969). The reason is that the shear wave also creates transverse electric fields, which couples to the photon field. The photons cause transverse currents, and the response of the superconductor to these real currents is described by the Meissner effect. One experimental consequence is that the attenuation rate is dependent on the amplitude of the wave.

9.6. EXCITONS IN SUPERCONDUCTORS

In the evaluation of the correlation function for either the infrared absorption or the acoustic attenuation, we have been ignoring the ladder diagrams or other vertex corrections. That is, we have not taken into account the sum of diagrams shown in Fig. 9.20. The importance of this type of diagram has been emphasized repeatedly in previous chapters, so it seems worthwhile to discuss their evaluation in superconductors.

The superconductor has an energy gap in the excitation spectrum. Other systems we have studied, such as semiconductors, can have exciton states in the particle–hole spectrum, which can be derived mathematically by summing the ladder diagrams of Fig. 9.20. Are there similar exciton states in a superconductor, perhaps which are bound states between quasiparticles? The philosophical question is obviously delicate, since the pairing states are already states of a similar nature. Thus the question is whether there are any *other* resonances which result from the vertex corrections. The subject has had a thorough theoretical investigation by Anderson (1958), Rickayzen (1959), Tsuneto (1960), and Bardasis and Schrieffer (1961). All these authors employed the method of equations of motion rather than solving the vertex equations directly. This procedure is in fact identical if carried out with the same set of overall approximations, but we shall not reproduce their method as it does not use Green's functions.

The experimental situation does not presently support the conjecture of additional exciton states. Although some preliminary data supported this idea, the best results in Fig. 9.19 do not show any absorption at a resonance state below the energy gap. The lack of exciton states is in accord with the calculations of Tsuneto, who estimated that any such resonance would be very small and probably unobservable.

The theoretical equations are very complicated, as will become apparent. Each diagram in the summation of Fig. 9.20 is not a single term but the summation of many terms. Each solid-line segment can be either a $\mathscr{G}$, $\mathscr{F}$, or $\mathscr{F}^\dagger$ or even a $\mathscr{G}$ going backwards. It is this diversity of possibilities which causes the subject to rapidly become complicated. For example, let us write a vertex equation of the form used earlier in (8.2.17):

$$\Gamma_i(\mathbf{q}, iq, \mathbf{p}) = \Gamma_i^{(0)} + \sum_j \int \frac{d^3p'}{(2\pi)^3} V(\mathbf{p}-\mathbf{p}')M_{ij}(\mathbf{q}, iq, \mathbf{p}')\Gamma_j(\mathbf{q}, iq, \mathbf{p}') \quad (9.6.1)$$

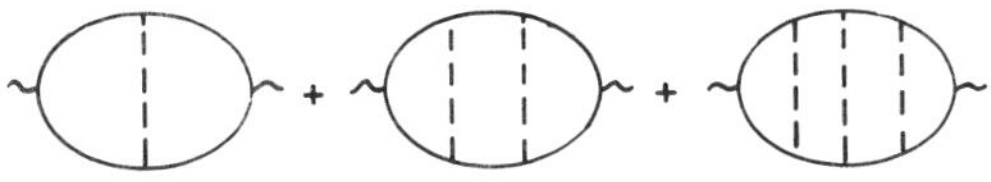

FIGURE 9.20

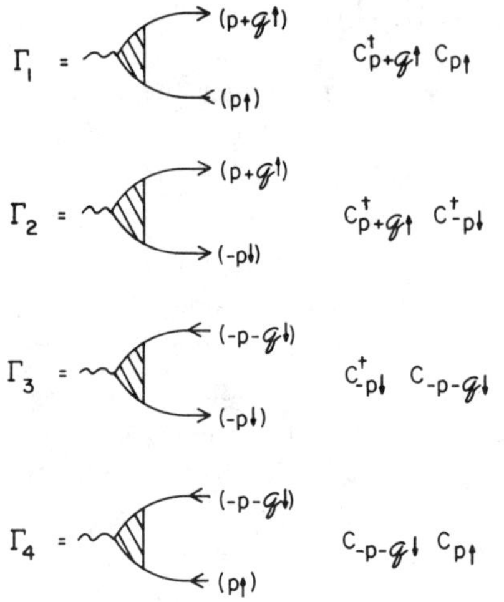

FIGURE 9.21. Vertex functions for superconductors.

However, now Γ_j is a four-dimensional vector function, and M_{ij} is a four-by-four matrix. The four vertex functions are shown in Fig. 9.21 and correspond to various choices of ingoing and outgoing electron lines for each solid line. A typical result is the equation for Γ_1, which is shown diagrammatically in Fig. 9.22 and for which the matrix components are at $T = 0$:

$$M_{11}(\mathbf{q}, iq, \mathbf{p}) = -\frac{1}{\beta}\sum_{ip} \mathscr{G}(p)\mathscr{G}(p+q)$$

$$= \frac{u_{\mathbf{p}}^2 v_{\mathbf{p+q}}^2}{iq + E_{\mathbf{p}} + E_{\mathbf{p+q}}} - \frac{v_{\mathbf{p}}^2 u_{\mathbf{p+q}}^2}{iq - E_{\mathbf{p}} - E_{\mathbf{p+q}}}$$

$$M_{12}(\mathbf{q}, iq, \mathbf{p}) = \frac{1}{\beta}\sum_{ip} \mathscr{F}(\mathbf{p}, ip)\mathscr{G}(\mathbf{p}+\mathbf{q}, iq - ip)$$

$$= \frac{\Delta}{2E_{\mathbf{p}}}\left(\frac{v_{\mathbf{p+q}}^2}{iq + E_{\mathbf{p}} + E_{\mathbf{p+q}}} + \frac{u_{\mathbf{p+q}}^2}{iq - E_{\mathbf{p}} - E_{\mathbf{p+q}}}\right)$$

$$M_{13}(\mathbf{q}, iq, \mathbf{p}) = \frac{1}{\beta}\sum_{ip} \mathscr{F}(\mathbf{p}, ip)\mathscr{F}^{\dagger}(\mathbf{p}+\mathbf{q}, ip - iq)$$

$$= \frac{\Delta^2}{4E_{\mathbf{p}}E_{\mathbf{p+q}}}\left(\frac{1}{iq + E_{\mathbf{p}} + E_{\mathbf{p+q}}} - \frac{1}{i\omega - E_{\mathbf{p}} - E_{\mathbf{p+q}}}\right)$$

$$M_{14}(\mathbf{q}, iq, \mathbf{p}) = +\frac{1}{\beta}\sum_{ip} \mathscr{G}(\mathbf{p}, ip)\mathscr{F}^{\dagger}(\mathbf{p}+\mathbf{q}, ip - iq)$$

$$= \frac{\Delta}{2E_{\mathbf{p+q}}}\left(\frac{u_{\mathbf{p}}^2}{iq + E_{\mathbf{p}} + E_{\mathbf{p+q}}} + \frac{v_{\mathbf{p}}^2}{iq - E_{\mathbf{p}} - E_{\mathbf{p+q}}}\right)$$

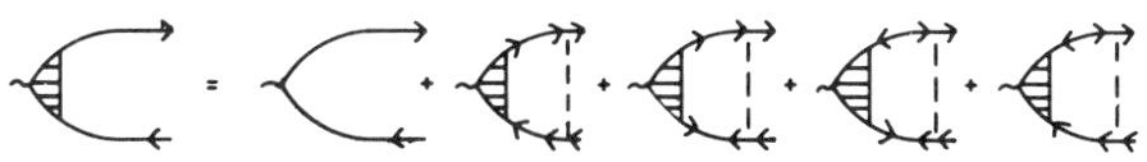

FIGURE 9.22 $\Gamma_1 = 1 + \Gamma_1 \mathscr{G}\mathscr{G} + \Gamma_2 \mathscr{G}\mathscr{F} + \Gamma_3 \mathscr{F}\mathscr{F}^\dagger + \Gamma_4 \mathscr{G}\mathscr{F}^\dagger$

We shall not carry out here the full evaluation of this matrix or its predictions regarding exciton states. The reader is referred to the references cited previously for a complete treatment, particularly Bardasis and Schrieffer (1961). We just wanted to make the point that the subject is worth investigating but leads to rather complicated equations which predict small effects which have not yet been found by the experimentalists.

9.7. STRONG COUPLING THEORY

The BCS theory is based on a simple approximation (9.1.2) to the attractive interaction between two electrons. This theory successfully describes the properties of many superconductors. Nevertheless, there are some superconductors for which the electron–phonon coupling is quite strong, and here the BCS theory is not an accurate description of the superconducting state. Instead, one must solve the gap equation for a realistic interaction between two electrons, which includes both the attraction due to phonons and the repulsion due to the screened Coulomb interaction. It is important that the phonon term be fully retarded, i.e., that the interaction be frequency dependent. This strong coupling theory was described by Eliashberg (1960). One of its important features is that the gap function $\Delta(\omega)$ becomes a complex function of the frequency ω. In addition, the self-energy function of the electrons—due to electron–phonon interaction—must be retained and plays an important role in the analysis. Our discussion of strong coupling theory follows Scalapino *et al.* (1966).

The starting point in the theory is the screened interaction between two electrons, which is written in the conventional way as the sum of a screened Coulomb interaction and a screened phonon interaction (see Secs. 6.3 and 6.4):

$$V_{\text{eff}}(\mathbf{q}, i\omega) = V_c(\mathbf{q}) + \sum_\lambda \bar{M}_\lambda^2(\mathbf{q})\mathscr{D}_\lambda(\mathbf{q}, i\omega) \tag{9.7.1}$$

$$V_c(\mathbf{q}) = \frac{4\pi e^2}{q^2\varepsilon(\mathbf{q})}$$

$$\bar{M}_\lambda^2(\mathbf{q}) = \frac{\tilde{M}_\lambda(\mathbf{q})^2}{\varepsilon(\mathbf{q})^2}$$

The Coulomb term V_c is taken to be instantaneous, since the frequency dependence of the dielectric function $\varepsilon(q, \omega)$ is not important for the low frequencies of interest, $\omega \lesssim \omega_D$. Similarly, the screened electron–phonon coupling constant $\bar{M}_\lambda{}^2(\mathbf{q})$ is also taken to be a function of wave vector but not frequency. These were the same approximations used in Sec. 6.4 to discuss the electron–phonon effects in normal metals.

Our objective is to calculate, self-consistently, the properties of the correlation functions $\mathscr{G}$, $\mathscr{F}$, and $\mathscr{F}^\dagger$. The last two are again equal, so it is sufficient to find one of them. They have the same definition as before, (9.2.1). Now they will be evaluated by directly expanding the S matrix and summing the terms which are found in higher order:

$$\mathscr{G}(\mathbf{p}, \tau - \tau') = -\langle T_\tau c_{\mathbf{p}\sigma}(\tau) c^\dagger_{\mathbf{p}\sigma}(\tau')\rangle = -\langle T_\tau \hat{c}_{\mathbf{p}\sigma}(\tau) S \hat{c}^\dagger_{\mathbf{p}\sigma}(\tau')\rangle$$
$$\mathscr{F}(\mathbf{p}, \tau - \tau') = \langle T_\tau c_{-\mathbf{p}\downarrow}(\tau) c_{\mathbf{p}\uparrow}(\tau')\rangle = \langle T_\tau \hat{c}_{-\mathbf{p}\downarrow}(\tau) S \hat{c}_{\mathbf{p}\uparrow}(\tau')\rangle$$
$$\mathscr{F}^\dagger(\mathbf{p}, \tau - \tau') = \langle T_\tau c^\dagger_{\mathbf{p}\uparrow}(\tau) c^\dagger_{-\mathbf{p}\downarrow}(\tau')\rangle = \langle T_\tau \hat{c}^\dagger_{\mathbf{p}\uparrow}(\tau) S \hat{c}^\dagger_{-\mathbf{p}\downarrow}(\tau')\rangle$$

There are two different self-energy functions which are retained in this expansion. They are the one-phonon self-energies, where the electron line in the self-energy can be either a $\mathscr{G}$ or $\mathscr{F}$:

$$S(\mathbf{p}, ip) = -\int \frac{d^3q}{(2\pi)^3} \frac{1}{\beta} \sum_{iq} V_{\text{eff}}(\mathbf{q}, iq) \mathscr{G}(\mathbf{p} + \mathbf{q}, ip + iq)$$
$$W(\mathbf{p}, ip) = -\int \frac{d^3q}{(2\pi)^3} \frac{1}{\beta} \sum_{iq} V_{\text{eff}}(\mathbf{q}, iq) \mathscr{F}(\mathbf{p} + \mathbf{q}, ip + iq) \tag{9.7.2}$$

These self-energies are shown as Feynman diagrams in Fig. 9.23. The dashed line is V_{eff} in (9.7.1). The electron line, in the self-energy expression, contains the correlation functions $\mathscr{G}$ or $\mathscr{F}$ which are to be determined self-consistently. Thus each self-energy diagram represents a summation of terms of the unperturbed Green's functions. This theory neglects the vertex corrections, which are the self-energy diagrams of the type shown in Fig. 9.23(c). This neglect is customarily justified on the basis of a theorem due to Migdal (1958), who argued that such terms were smaller by a factor of $(m/M)^{1/2}$, where m is the electron mass and M is the ion mass. The difficulty with his argument is that superconductivity itself (i.e., the Cooper

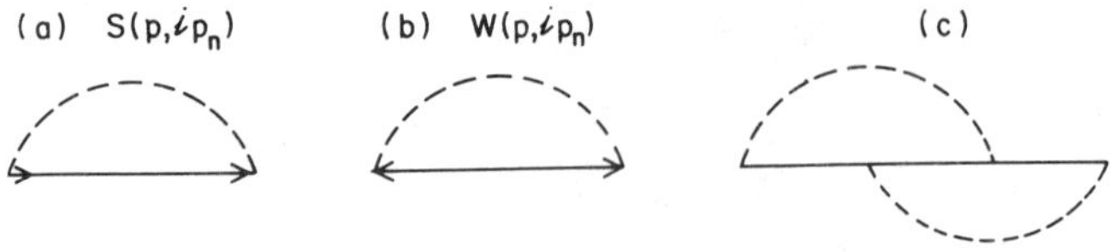

FIGURE 9.23

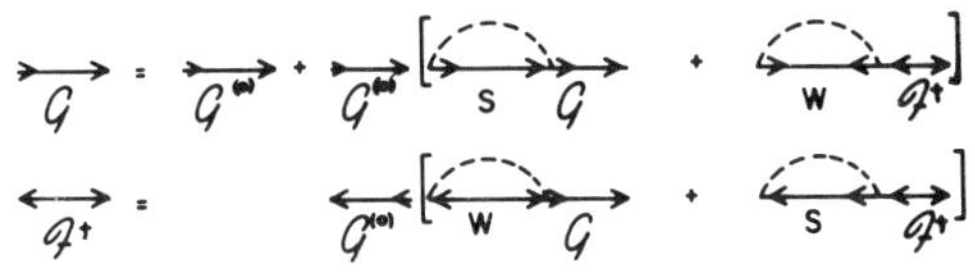

FIGURE 9.24

instability) is caused by a vertex correction, so a theorem which asserts that all vertex diagrams may be neglected is obviously unreliable. We shall describe the strong coupling theory in its customary form, which does neglect these vertex corrections.

Normally the expansion of the S matrix leads to a Dyson equation of the typical form:

$$\mathscr{G}(p) = \mathscr{G}^{(0)}(p) + \mathscr{G}^{(0)}(p)\Sigma(p)\mathscr{G}(p)$$

We shall now find the equivalent equation for the expansion of the S matrix for a superconductor. The situation is obviously more complicated, since there are two self-energy functions $S(p)$ and $W(p)$. Furthermore, we can identify $\mathscr{G}^{(0)}(p) = (ip - \xi_p)^{-1}$ as usual, but obviously there is not an equivalent form for $\mathscr{F}^{(0)}$, so $\mathscr{F}^{(0)} = 0$. The equivalents to Dyson's equations are, in a four-vector notation $p = (\mathbf{p}, ip)$,

$$\begin{aligned}\mathscr{G}(p) &= \mathscr{G}^{(0)}(p) + \mathscr{G}^{(0)}(p)[S(p)\mathscr{G}(p) - W(p)\mathscr{F}^{\dagger}(p)]\\ \mathscr{F}^{\dagger}(p) &= \mathscr{G}^{(0)}(-p)[W(p)\mathscr{G}(p) + S(-p)\mathscr{F}^{\dagger}(p)]\end{aligned} \tag{9.7.3}$$

These equations are shown diagrammatically in Fig. 9.24. The key to constructing the equations is to remember that the vertex conserves particle number, so that the two electron arrows must point in the same direction; i.e., one can only have vertices of the form

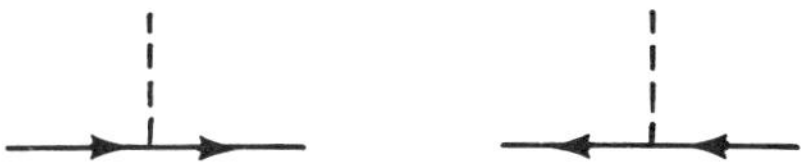

In writing these equations, we have used the fact that $W(-p) = W(p)$. Dyson's equations are quite similar to the BCS equations (9.2.10) which were derived by the equations of motion. This resemblance is made more apparent by rewriting (9.7.3) using $\mathscr{G}^{(0)}(\pm p) = (\pm ip - \xi_p)^{-1}$ to obtain

$$\begin{aligned}[ip - \xi_p - S(p)]\mathscr{G}(p) + W(p)\mathscr{F}^{\dagger}(p) &= 1\\ [ip + \xi_p + S(-p)]\mathscr{F}^{\dagger}(p) + W(p)\mathscr{G}(p) &= 0\end{aligned} \tag{9.7.4}$$

The BCS equations (9.2.10) are obtained in the approximation where

$W(\mathbf{p}, ip) = \Delta =$ constant, and the other self-energy $S(\mathbf{p}, ip)$ is ignored. We shall now solve the more general equations (9.7.4). The gap function $W(p)$ is an even function of ip, as already stated above. The other self-energy $S(p)$ is not. It is convenient to break it up into its symmetric and antisymmetric parts:

$$S(p) = S_e(\mathbf{p}, ip) + ipS_0(\mathbf{p}, ip)$$

where S_e and S_0 are both even functions of frequency ip. Furthermore, we shall define the function $Z(\mathbf{p}, ip)$ as

$$Z(\mathbf{p}, ip) = 1 - S_0(\mathbf{p}, ip)$$

The notation Z usually is applied to a renormalization coefficient for the single-particle Green's function. Here the definition is different, but the physics is similar, since Z again turns out to be another type of renormalization function. Now it is easy to solve the coupled equations (9.7.4) since we write $ip - \xi_p - S(p) = ipZ(p) - \bar{\xi}(p)$, etc.:

$$\begin{aligned} \mathscr{G}(p) &= \frac{ipZ(p) + \bar{\xi}(p)}{[ipZ(p)]^2 - [\bar{\xi}(p)^2 + W(p)^2]} \\ \mathscr{F}^{\dagger}(p) &= -\frac{W(p)}{[ipZ(p)]^2 - [\bar{\xi}(p)^2 + W(p)^2]} \\ \bar{\xi}(p) &= \xi_p + S_e(p) \end{aligned} \tag{9.7.5}$$

These equations are similar to those for the BCS theory (9.2.11). Now we must self-consistently find the three unknown functions $W(p)$, $Z(p)$, and $S_e(p)$.

Eliashberg examined the retarded function $\text{Re}[S_e(\mathbf{p}, \omega)]$ in great detail. He showed that it was a constant, independent of $(\mathbf{p}, \omega)$, which therefore just renormalizes the chemical potential. This conclusion should not be surprising, since our earlier investigation in Sec. 6.4 of the electron–phonon properties of normal metals showed that $\text{Re}[\Sigma(\mathbf{p}, \omega)]$ was an asymmetric function of frequency, so that the symmetric part of the self-energy is small. This conclusion does not change in the superconductor. Thus we neglect S_e.

The other functions of interest $Z(p)$ and $W(p)$ are both functions of $\mathbf{p}$ and ip. The wave vector dependence is unimportant, since everything happens at the Fermi surface $p \simeq k_F$. It is the energy dependence of these functions which provides the interesting phenomena. Actual crystalline superconductors have anisotropic Fermi surfaces, which cause the functions to vary with angle around the Fermi surface. We neglect this dependence and assume isotropy. The notation is altered to suppress the dependence on wave vector, so we write these functions as $S_0(ip)$, $W(ip)$, $Z(ip)$, etc. Thus the main dependence of $\mathscr{G}(p)$ and $\mathscr{F}^{\dagger}(p)$ on $\mathbf{p}$ is through the factor $\bar{\xi}(p) \simeq \xi_p$. Later we shall find that all the wave vector dependence enters through the integrals of the retarded functions $G(\mathbf{p}, \omega)$ and $F(\mathbf{p}, \omega)$:

$$\int_{-\infty}^{\infty} d\xi_p G(\mathbf{p}, \omega) = -\int_{-\infty}^{\infty} d\xi_p \frac{\omega Z(\omega) + \xi_p}{\xi_p^2 + W(\omega)^2 - \omega^2 Z(\omega)^2} = -i\pi g(\omega)$$

$$\int_{-\infty}^{\infty} d\xi_p F(\mathbf{p}, \omega) = +\int_{-\infty}^{\infty} d\xi_p \frac{W(\omega)}{\xi_p^2 + W(\omega)^2 - \omega^2 Z(\omega)^2} = i\pi f(\omega) \tag{9.7.6}$$

$$g(\omega) = \frac{\omega Z(\omega) \operatorname{sgn} \omega}{[\omega^2 Z(\omega)^2 - W(\omega)^2]^{1/2}}$$

$$f(\omega) = \frac{W(\omega) \operatorname{sgn} \omega}{[\omega^2 Z(\omega)^2 - W(\omega)^2]^{1/2}}$$

These integrals are evaluated by contour integration, say by closing the contour in the upper half plane. The sign change $\operatorname{sgn} \omega = \omega/|\omega|$ occurs because the pole in the UHP is at $i\delta + (\omega^2 Z^2 - W^2)^{1/2}$ for $\omega > 0$ and at $i\delta - (\omega^2 Z^2 - W^2)^{1/2}$ for $\omega < 0$. We can also rewrite the functions $g(\omega)$ and $f(\omega)$ in terms of the retarded gap function $\Delta(\omega)$:

$$\Delta(\omega) = \frac{W(\omega)}{Z(\omega)}$$

$$g(\omega) = \frac{\omega \operatorname{sgn} \omega}{[\omega^2 - \Delta(\omega)^2]^{1/2}} \tag{9.7.7}$$

$$f(\omega) = \frac{\Delta(\omega) \operatorname{sgn} \omega}{[\omega^2 - \Delta(\omega)^2]^{1/2}}$$

The retarded functions $W(\omega)$ and $Z(\omega)$ must be determined self-consistently by evaluating (9.7.2) for the two self-energy functions. The first step, as usual, is to take the summation over Matsubara frequencies. Let us first evaluate the summations for the coulomb part of the interaction. Since we neglect its frequency dependence, one has the two summations

$$\frac{1}{\beta} \sum_{ip'} \mathscr{G}(\mathbf{p} + \mathbf{q}, ip') = \mathscr{G}(\mathbf{p} + \mathbf{q}, \tau = 0^-) = \langle c^{\dagger}_{\mathbf{p}+\mathbf{q}\sigma} c_{\mathbf{p}+\mathbf{q}\sigma} \rangle = n_{\mathbf{p}+\mathbf{q}}$$

$$\frac{1}{\beta} \sum_{ip'} \mathscr{F}^{\dagger}(\mathbf{p}', ip') = -\int_{-\infty}^{\infty} \frac{d\varepsilon}{2\pi i} n_F(\varepsilon) 2i \operatorname{Im}[F(\mathbf{p}', \varepsilon)]$$

where F is the retarded form of $\mathscr{F}$. The summation over $\mathscr{G}(p')$ just produces $n_{\mathbf{p}'}$, as it does in the normal metal. However, now $n_{\mathbf{p}'}$ is different, since the momentum distribution is perturbed by the superconducting state. The resulting Coulomb term $-\sum_{\mathbf{p}'} V(\mathbf{p}-\mathbf{p}')n_{\mathbf{p}'}$ is just the screened exchange energy in the superconductor, which is hardly changed by the superconductivity, since the summation over $\mathbf{p}'$ is not restricted to states near the Fermi surface. This term is also independent of ip, so it contributes a constant to the even function $S_e(ip)$. We can therefore ignore this term, since it just renormalizes the chemical potential. The other summation over $\mathscr{F}^\dagger$ is the Coulomb contribution to the gap function $W(ip)$, which is retained in the theory.

The summation over Matsubara frequencies in the screened phonon interaction is done as usual, with a contour integral (see Sec. 3.5):

$$\begin{aligned}
&\frac{1}{\beta}\sum_{iq} \mathscr{D}_\lambda(\mathbf{q}, iq)\mathscr{G}(\mathbf{p}', ip+iq) \\
&\quad = \int_{-\infty}^{\infty} \frac{d\varepsilon}{2\pi i} \{n_B(\varepsilon)\mathscr{G}(\mathbf{p}', ip+\varepsilon)\times 2i\,\mathrm{Im}[D_\lambda(\mathbf{q}, \varepsilon)] \\
&\qquad - n_F(\varepsilon)\mathscr{D}_\lambda(\mathbf{q}, \varepsilon - ip)\times 2i\,\mathrm{Im}[G(\mathbf{p}', \varepsilon)]\} \\
&\frac{1}{\beta}\sum_{iq} \mathscr{D}_\lambda(\mathbf{q}, iq)\mathscr{F}^\dagger(\mathbf{p}', ip+iq) \\
&\quad = \int_{-\infty}^{\infty} \frac{d\varepsilon}{2\pi i} \{n_B(\varepsilon)\mathscr{F}^\dagger(\mathbf{p}', ip+\varepsilon)\times 2i\,\mathrm{Im}[D_\lambda(\mathbf{q}, \varepsilon)] \\
&\qquad - n_F(\varepsilon)\mathscr{D}_\lambda(\mathbf{q}, \varepsilon - ip)\times 2i\,\mathrm{Im}[F(\mathbf{p}', \varepsilon)]\}
\end{aligned} \tag{9.7.8}$$

The next step is to represent the phonon Green's function by its spectral function. Define this as $B_\lambda(\mathbf{q}, \omega) = -(1/\pi)\,\mathrm{Im}[D_\lambda(\mathbf{q}, \omega)]$ for $\omega > 0$, so that $B_\lambda(\mathbf{q}, \omega) = \delta[\omega - \omega_\lambda(\mathbf{q})]$ if the phonons are unperturbed. We shall retain $B_\lambda(\mathbf{q}, \omega)$, so that the phonons are fully interacting. In terms of its spectral function, the phonon Green's function may be written as [(3.3.14)]

$$\mathscr{D}_\lambda(\mathbf{q}, iq) = \int_0^\infty d\nu B_\lambda(\mathbf{q}, \nu)\left(\frac{1}{iq-\nu} - \frac{1}{iq+\nu}\right)$$

$$-2\,\mathrm{Im}[D_\lambda(\mathbf{q}, \omega)] = 2\pi[B_\lambda(\mathbf{q}, \omega) - B_\lambda(\mathbf{q}, -\omega)]$$

In (9.7.8) both the retarded $D_\lambda(\mathbf{q}, \omega)$ and Matsubara $\mathscr{D}_\lambda(\mathbf{q}, i\omega)$ form of the phonon Green's function can be written in terms of this spectral function:

$$\begin{aligned}
&\frac{1}{\beta}\sum_{iq} \mathscr{D}_\lambda(\mathbf{q}, iq)\mathscr{G}(\mathbf{p}', ip+iq) \\
&\quad = -\int_0^\infty d\nu B_\lambda(\mathbf{q}, \nu)\Big\{n_B(\nu)\mathscr{G}(\mathbf{p}', ip+\nu) - n_B(-\nu)\mathscr{G}(\mathbf{p}', ip-\nu) \\
&\qquad + \int_{-\infty}^{\infty} \frac{d\varepsilon}{\pi} n_F(\varepsilon)\,\mathrm{Im}[G(\mathbf{p}', \varepsilon)]\left(\frac{1}{\varepsilon - ip - \nu} - \frac{1}{\varepsilon - ip + \nu}\right)\Big\}
\end{aligned}$$

and similarly for the summation which includes $\mathscr{F}^\dagger$. This completes the formal summation over Matsubara frequencies in the two self-energy functions for $S(ip)$ and $W(ip)$. We now write out the rather lengthy expressions for these two *retarded* functions and include both the coulomb and phonon terms and the summation over wave vector $\mathbf{q}$ and phonon polarization index λ:

$$S(\omega) = \sum_\lambda \int \frac{d^3q}{(2\pi)^3} \bar{M}_\lambda(\mathbf{q})^2 \int_0^\infty d\nu B_\lambda(\mathbf{q}, \nu)\Big\{n_B(\nu)G(\mathbf{p}+\mathbf{q}, \omega+\nu)$$
$$+ [1+n_B(\nu)]G(\mathbf{p}+\mathbf{q}, \omega-\nu) + \int_{-\infty}^\infty \frac{d\varepsilon}{\pi} n_F(\varepsilon)\,\mathrm{Im}[G(\mathbf{p}+\mathbf{q}, \varepsilon)]$$
$$\times \left(\frac{1}{\varepsilon-\omega-\nu-i\delta} - \frac{1}{\varepsilon-\omega+\nu-i\delta}\right)\Big\}$$
$$(9.7.9)$$
$$W(\omega) = \int \frac{d^3q}{(2\pi)^3} \Big(V_c(\mathbf{q}) \int_{-\infty}^\infty \frac{d\varepsilon}{\pi} n_F(\varepsilon)\,\mathrm{Im}[F(\mathbf{p}+\mathbf{q}, \varepsilon)]$$
$$+ \sum_\lambda \bar{M}_\lambda(\mathbf{q})^2 \int_0^\infty d\nu B_\lambda(\mathbf{q}, \nu)\Big\{n_B(\nu)F(\mathbf{p}+\mathbf{q}, \omega+\nu)$$
$$+ [1+n_B(\nu)]F(\mathbf{p}+\mathbf{q}, \omega-\nu) + \int_{-\infty}^\infty \frac{d\varepsilon}{\pi} n_F(\varepsilon)\,\mathrm{Im}[F(\mathbf{p}+\mathbf{q}, \varepsilon)]$$
$$\times \left(\frac{1}{\varepsilon-\omega-\nu-i\delta} - \frac{1}{\varepsilon-\omega+\nu-i\delta}\right)\Big\}\Big)$$

The goal is to reduce these integral equations to a single independent variable, which is frequency ν. The first step is to do all the wave vector integrals, whose variables are

$$\int d^3q = \int q^2\,dq \int d\bar{\nu} \int d\phi$$

where $\bar{\nu} = \cos\theta$ is the angle between $\mathbf{p}$ and $\mathbf{q}$. Change this integration variable to $p' = (p^2 + q^2 + 2pq\bar{\nu})^{1/2}$, or $p'\,dp' = pq\,d\bar{\nu}$, and as a second step change $p'\,dp' = m\,d\xi_{p'}$:

$$d^3q \to \frac{1}{p}\int_0^\infty q\,dq \int_{|p-q|}^{p+q} p'\,dp' \int_0^{2\pi} d\phi = \frac{m}{p}\int_0^\infty q\,dq \int d\xi_{p'} \int_0^{2\pi} d\phi$$

The q integral is over the Brillouin zone; the inclusion of Umklapp processes extends the integration to infinity. However, the $d\xi_{p'}$ integral will be limited to values of $|\xi_{p'}| < \omega_D$ at the Fermi surface, since the integral converges rapidly outside of this range. Thus we can uncouple the integration limits and without appreciable error express them as

$$\int d^3q \to \frac{1}{v_F} \int_{-\infty}^{\infty} d\xi_{p'} \int_0^{2k_F} q\, dq \int_0^{2\pi} d\phi$$

The $q\, dq\, d\phi$ integrals are now restricted to points where $\mathbf{p}$ and $\mathbf{p} + \mathbf{q}$ are both on the Fermi surface, which limits $q \leq 2k_F$ when the Fermi surface is spherical. With this uncoupling, the wave vector integrals are quite simple to do, at least in a formal sense. The integral over $d\xi_{p'}$ was already performed in (9.7.6). The only $\xi_{p'}$ dependence in the integrand is in the functions $G(\mathbf{p}', \omega)$ and $F(\mathbf{p}', \omega)$, and their integration in (9.7.6) leads to the new functions $f(\omega)$ and $g(\omega)$. Similarly, the remaining integration over the phonon coordinates $\int q\, dq \int d\phi$ provides us with the exact definition of the function $\alpha^2 F$ which was defined in (6.4.20):

$$\alpha^2(\nu)F(\nu) = \frac{1}{v_F(2\pi)^3} \int_0^{2k_F} q\, dq \int_0^{2\pi} d\phi \sum_\lambda \bar{M}_\lambda(\mathbf{q})^2 B_\lambda(\mathbf{q}, \nu) \qquad (9.7.10)$$

The earlier definition of $\alpha^2 F$ had $\delta[\omega - \omega_\lambda(\mathbf{q})]$ instead of the phonon spectral function $B_\lambda(\mathbf{q}, \omega)$, but the extension to the interacting phonon system is obvious. Most computations use the form for $\alpha^2 F$ which has been averaged over all directions in the Fermi surface, which corresponds with the $\alpha^2 F$ measured in electron tunneling from evaporated films.

After performing these wave vector integrations, the self-consistent equations (9.7.9) for the two self-energy functions have been reduced to manageable proportions:

$$S(\omega) = -\pi \int_0^{\infty} d\nu \alpha^2(\nu)F(\nu) \Big\{ i n_B(\nu) g(\omega + \nu) + i[1 + n_B(\nu)] g(\omega - \nu)$$

$$+ \int_{-\infty}^{\infty} \frac{d\varepsilon}{\pi} n_F(\varepsilon) \operatorname{Re}[g(\varepsilon)] \left(\frac{1}{\varepsilon - \omega - \nu - i\delta} - \frac{1}{\varepsilon - \omega + \nu - i\delta} \right) \Big\}$$

$$W(\omega) = U_c \int_{-\infty}^{\infty} d\varepsilon n_F(\varepsilon) \operatorname{Re}[f(\varepsilon)] + \pi \int_0^{\infty} d\nu \alpha^2(\nu) F(\nu) \qquad (9.7.11)$$

$$\times \Big\{ i n_B(\nu) f(\omega + \nu) + i[1 + n_B(\nu)] f(\omega - \nu)$$

$$+ \int_{-\infty}^{\infty} \frac{d\varepsilon}{\pi} n_F(\varepsilon) \operatorname{Re}[f(\varepsilon)] \left(\frac{1}{\varepsilon - \omega - \nu - i\delta} - \frac{1}{\varepsilon - \omega + \nu - i\delta} \right) \Big\}$$

$$U_c = \frac{1}{\hbar v_F (2\pi)^2} \int_0^{2k_F} q\, dq V_c(q)$$

The Coulomb term U_c has been reduced to a dimensionless constant, which acts to reduce the gap function. The factors $\mathrm{Re}[f(\omega)]$ and $\mathrm{Re}[g(\omega)]$ are obtained because the $d\xi_{p'}$ integrals in (9.7.6) produce factors such as

$$\int d\xi_{p'}\, \mathrm{Im}[G(p', \omega)] = \mathrm{Im}[-i\pi g(\omega)] = -\pi\, \mathrm{Re}[g(\omega)]$$

The formulas (9.7.11) for $S(\omega)$ and $W(\omega)$ are correct but are not expressed in the conventional forms which are found in the literature. These have a different-looking, but equivalent, result because of different methods of doing the contour integrals. The conventional formulas can be obtained by the following alteration of the preceding result. The second term in the phonon part of the integrand contains the factor $1 + n_B$, and the 1 term is expressed in terms of the integral equality:

$$f(\omega - \nu) = \int_{-\infty}^{\infty} \frac{d\varepsilon}{\pi i} \frac{\mathrm{Re}[f(\varepsilon)]}{\varepsilon - \omega + \nu - i\delta}$$

$$g(\omega - \nu) = \int_{-\infty}^{\infty} \frac{d\varepsilon}{\pi i} \frac{\mathrm{Re}[g(\varepsilon)]}{\varepsilon - \omega + \nu - i\delta}$$

These identities can be derived from the Kramers–Kronig relations in Sec. 5.7, since $f(\varepsilon)$ and $g(\varepsilon)$ are analytic functions of frequency. These terms are now combined with those of similar energy denominator:

$$S(\omega) = -\pi \int_0^{\infty} d\nu \alpha^2(\nu) F(\nu) \Big\{ i n_B(\nu)[g(\omega + \nu) + g(\omega - \nu)]$$
$$+ \int_{-\infty}^{\infty} \frac{d\varepsilon}{\pi} \mathrm{Re}[g(\varepsilon)] \Big[\frac{n_F(\varepsilon)}{\varepsilon - \omega - \nu - i\delta} + \frac{1 - n_F(\varepsilon)}{\varepsilon - \omega + \nu - i\delta} \Big] \Big\}$$

$$W(\omega) = U_c \int_{-\infty}^{\infty} d\varepsilon n_F(\varepsilon)\, \mathrm{Re}[f(\varepsilon)] + \pi \int_0^{\infty} d\nu \alpha^2(\nu) F(\nu) \qquad (9.7.12)$$
$$\times \Big\{ i n_B(\nu)[f(\omega + \nu) + f(\omega - \nu)]$$
$$+ \int_{-\infty}^{\infty} \frac{d\varepsilon}{\pi} \mathrm{Re}[f(\varepsilon)] \Big[\frac{n_F(\varepsilon)}{\varepsilon - \omega - \nu - i\delta} + \frac{1 - n_F(\varepsilon)}{\varepsilon - \omega + \nu - i\delta} \Big] \Big\}$$

Equation (9.7.12) is the conventional way of presenting the result. The formulas are valid at finite temperature, but most numerical results have been done at zero temperature. Here we set to zero the factors $n_B(\varepsilon)$, while $n_F(\varepsilon) = \theta(-\varepsilon)$. It is convenient to rearrange variables of integration, so that all the frequency integrals are over positive values. These variable changes are straightforward, since $\mathrm{Re}[f(-\varepsilon)] = -\mathrm{Re}[f(\varepsilon)]$ and $\mathrm{Re}[g(-\varepsilon)] = \mathrm{Re}[g(\varepsilon)]$:

$$S(\omega) = \int_0^\infty d\nu\alpha^2(\nu)F(\nu) \int_0^\infty d\varepsilon\, \mathrm{Re}[g(\varepsilon)]\left(\frac{1}{\varepsilon+\omega+\nu+i\delta} - \frac{1}{\varepsilon-\omega+\nu-i\delta}\right)$$

$$W(\omega) = -U_c \int_0^\infty d\varepsilon\, \mathrm{Re}[f(\varepsilon)] + \int_0^\infty d\nu\alpha^2(\nu)F(\nu) \int_0^\infty d\varepsilon\, \mathrm{Re}[f(\varepsilon)] \times \left(\frac{1}{\varepsilon+\omega+\nu+i\delta} + \frac{1}{\varepsilon-\omega+\nu-i\delta}\right)$$

It is convenient to introduce the kernals

$$K_\pm(\varepsilon, \omega) = \int_0^\infty d\nu\alpha^2(\nu)F(\nu)\left(\frac{1}{\varepsilon+\omega+\nu+i\delta} \pm \frac{1}{\varepsilon-\omega+\nu-i\delta}\right)$$

They are properties of the metal and need to be computed only once. It may take some effort to find α^2F for the metal, but once it is done, the result for $K_\pm$ is stored on the computer and used when needed.

The Matsubara equivalent of $S(\omega)$ would be

$$S(ip) = \int_0^\infty d\nu\alpha^2(\nu)F(\nu) \int_0^\infty d\varepsilon\, \mathrm{Re}[g(\varepsilon)]\left(\frac{1}{\varepsilon+\nu+ip} - \frac{1}{\varepsilon+\nu-ip}\right)$$

This function is antisymmetric in ip, since $S(ip) = -S(-ip)$. Thus we identify it with the odd function we desire, $S(ip) = ipS_0(ip)$, or as a retarded function:

$$S(\omega) = \omega S_0(\omega) = \omega[1 - Z(\omega)]$$

Remembering that $\Delta(\omega) = W(\omega)/Z(\omega)$, we then derive the famous coupled equations for the gap function $\Delta(\omega)$ and the renormalization function $Z(\omega)$:

$$\begin{aligned} \Delta(\omega)Z(\omega) &= \int_0^\infty d\varepsilon\, \mathrm{Re}\left\{\frac{\Delta(\varepsilon)}{[\varepsilon^2 - \Delta(\varepsilon)^2]^{1/2}}\right\}[K_+(\varepsilon, \omega) - U_c] \\ \omega[1 - Z(\omega)] &= \int_0^\infty d\varepsilon\, \mathrm{Re}\left\{\frac{\varepsilon}{[\varepsilon^2 - \Delta(\varepsilon)^2]^{1/2}}\right\}K_-(\varepsilon, \omega) \end{aligned} \tag{9.7.13}$$

These two integral equations must be solved self-consistently for the two complex functions $\Delta(\omega)$ and $Z(\omega)$ of real frequency ω. The inputs for the calculation are the kernals $K_\pm(\varepsilon, \omega)$, which depend on the phonon properties through α^2F, and the coulomb factor U_c.

Figure 9.25 shows the theoretical results of Shaw and Swihart (1968) for lead at $T = 0$. They used the α^2F of McMillan and Rowell (1965),

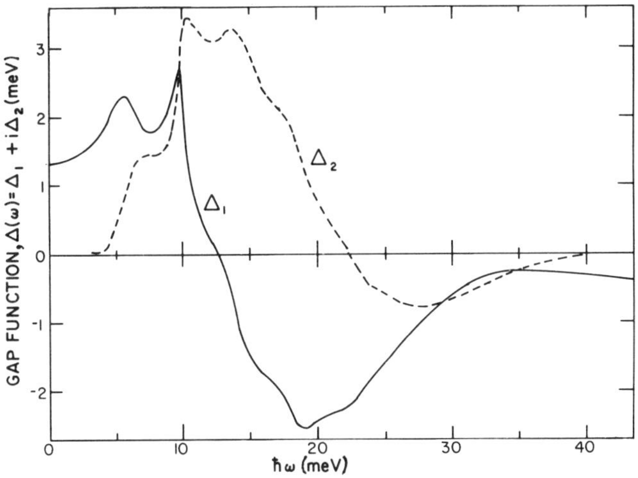

FIGURE 9.25. The real and imaginary parts Δ_1 and Δ_2 of the complex gap function as a function of energy ω for lead at $T = 0$. *Source*: Shaw and Swihart (1968) (used with permission).

which is shown in Fig. 6.17(a). The real part of the gap function $\Delta_1(\omega)$ at $\omega = 0$ equals the value $2\Delta_1(0) \equiv 2\Delta_0 = 2.68$ meV found from electron tunneling. The imaginary part of the gap function $\Delta_2(\omega)$ is zero for $\omega < \Delta_0$. Both Δ_1 and Δ_2 are obviously strong functions of ω. Shaw and Swihart also calculated the infrared absorption and showed excellent agreement with the results of Palmer and Tinkham. Thus the electron tunneling data for $\alpha^2 F$, along with the strong coupling theory, provides an excellent description of other experiments such as infrared absorption.

The remaining topic of this section is a description of the method of McMillan and Rowell (1965) for obtaining $\alpha^2(\omega)F(\omega)$ from the data for electron tunneling in superconductors. The important point is that all our previous formulas for single-particle tunneling still apply, except with the interpretation in strong coupling theory that the density of states which enters the tunneling expression in superconductors is

$$\varrho(\omega) = \mathrm{Re}[g(\omega)] = \mathrm{Re}\left\{\frac{\omega}{[\omega^2 - \Delta(\omega)^2]^{1/2}}\right\}$$

Thus, for example, the normalized conductance for the tunneling between a normal metal and a superconductor at low temperature is

$$\frac{(dI/dV)_S}{(dI/dV)_N} = \int d\omega \varrho(eV - \omega) \frac{\partial n_F(\omega)}{\partial \omega}$$

where $n_F(\omega)$ is the distribution function of the normal metal and $\partial n_F/\partial\omega$ has a width of order k_BT. The low-temperature tunneling between two identical superconductors is [(9.3.15)]

$$I_S = \frac{\sigma_0}{e}\int d\omega\varrho(\omega)\varrho(eV-\omega) \tag{9.7.14}$$

These results can be easily derived. The many-body theory of Schrieffer *et al.* (1963) in (9.3.11) is still valid even for strong coupling theory. This states that the tunneling is proportional to the products of the spectral density functions for each side of the junction. The integral over this spectral function has the form

$$\begin{aligned}\int d\xi_p A(p,\omega) &= -2\,\mathrm{Im}\left[\int d\xi_p G_{\mathrm{ret}}(\xi_p,\omega)\right] = -2\,\mathrm{Im}[-i\pi g(\omega)]\\ &= 2\pi\,\mathrm{Re}[g(\omega)]\end{aligned}$$

After integrating this over $d\xi_p$ in the manner shown in (9.7.6), one gets that the tunneling density of states is proportional to $\mathrm{Re}[g(\omega)]$.

Thus the tunneling experiments can be used to directly measure $\mathrm{Re}(g)$. The measurement can be done for a junction between a normal metal and a superconductor, in which case the result has a thermal smear of order k_BT. McMillan and Rowell actually found it better to use a tunnel junction of two identical superconductors and then to deconvolute the result found in (9.7.14). In any case, they obtain experimentally, in a direct fashion, the quantity $\mathrm{Re}(g)$. They then solve the inverse problem, on the computer, which is implied by the strong coupling equations (9.7.13). That is, they determine which α^2F provides a self-consistent solution to these equations and gives the right $\mathrm{Re}[g(\omega)]$. This procedure is difficult but has the great payoff of providing the best method of experimentally determining the important function $\alpha^2(\omega)F(\omega)$.

PROBLEMS

1. Consider the mutual scattering of a fermion and a boson, i.e., a ^{3}He and a ^{4}He particle moving parallel in time. Use Green's functions to derive the occupation number factors which enter the scattering integral.

2. Calculate how the pole in the Cooper instability varies with the center of mass momentum P of the pair.

3. What is the momentum distribution $n_{\mathbf{k}}$ of electrons in the superconductor at $T = 0$.

4. The entropy per unit volume of a superconductor in the BCS model is

$$S = -\frac{2k_B}{\nu}\sum_{\mathbf{k}}\{n_F(E_k)\ln[n_F(E_k)] + [1 - n_F(E_k)]\ln[1 - n_F(E_k)]\}$$

Use this to derive an expression for the specific heat of the superconductor. Then use the fact that $\Delta(T) \simeq (T_c - T)^{1/2}$ to show that the specific heat is discontinuous at the critical temperature.

5. Write the Hamiltonian (9.2.2) in the following form:

$$H_{\text{eff}} = \sum_{\mathbf{k}\sigma}\xi_k c_{\mathbf{k}\sigma}^{\dagger}c_{\mathbf{k}\sigma} - \sum_{\mathbf{k}}\Delta(k)(c_{\mathbf{k}\uparrow}^{\dagger}c_{-\mathbf{k}\downarrow}^{\dagger} + c_{-\mathbf{k}\downarrow}c_{\mathbf{k}\uparrow})$$

$$\Delta(k) = -\sum_{\mathbf{q}} V(q)\langle c_{-\mathbf{k}-\mathbf{q}\downarrow}c_{\mathbf{k}+\mathbf{q}\uparrow}\rangle$$

Solve the effective Hamiltonian H_{eff} by a canonical transformation, and reduce it to the form

$$H_{\text{eff}} = E_0 + \sum_{\mathbf{k}} E_k(\alpha_{\mathbf{k}}^{\dagger}\alpha_{\mathbf{k}} + \beta_{\mathbf{k}}^{\dagger}\beta_{\mathbf{k}})$$

where

$$c_{\mathbf{k}\uparrow} = \cos(\theta_k)\alpha_{\mathbf{k}} + \sin(\theta_k)\beta_{\mathbf{k}}^{\dagger}$$

$$c_{-\mathbf{k}\downarrow} = \cos(\theta_k)\beta_{\mathbf{k}} - \sin(\theta_k)\alpha_{\mathbf{k}}^{\dagger}$$

6. Calculate the Pauli spin susceptibility of the BCS superconducting state. Show that the formula at finite temperatures is $\chi = 2\mu_0^2 \sum_{\mathbf{k}}(d/dE_k)n_F(E_k)$.

7. Derive the formula for the coefficient in the Josephson current $J_S(0) = (\pi\Delta/2)(\sigma_0/e)$ by doing the integral in (9.3.20) for $eV = 0$.

8. Evaluate $J_S(eV)$ in (9.3.22) at zero temperature by making the change $(1/\beta)\sum_{ip} \to \int (dp/2\pi)$ and then the analytical continuation $\omega \to -ieV$. Define $\Lambda(eV)$ at $T = 0$ by using $J_S(eV) = (\sigma_0/e)\Lambda(eV)$. Show that:

a. $\Lambda(eV = 0) = \Delta_S K(q)$: $\quad q = \left(1 - \dfrac{\Delta_S^2}{\Delta_L^2}\right)^{1/2}$

where Δ_S and Δ_L are the smaller and larger of the two energy gaps.

b. When $\Delta_R = \Delta_L = \Delta$,

$$\Lambda(eV) = \Delta K\left(\frac{eV}{2\Delta}\right)$$

9. Use London's equation for J to solve (1.5.13) for a superconductor in a static magnetic field B_0 parallel to its surface. Hence show that the penetration depth λ is c/ω_p, and estimate this numerically for aluminum. Then calculate the value

of the Josephson wave vector k in (9.3.24) for a magnetic field of 1 G and a junction thickness d of 50 Å.

10. Evaluate the integral in (9.5.3) for the low-frequency acoustic attenuation by assuming the limit of a dirty superconductor in which momentum is not conserved.

11. In the vertex equation for superconductors (9.6.1), write out the coefficients M_{2j} in the vertex equation for Γ_2.

12. Use strong coupling theory to derive a formula for the coefficient $J_S(eV)$ for the Josephson tunneling.

13. Derive the strong coupling theory formula for the infrared absorption of a dirty superconductor.

Chapter 10

Liquid Helium

Helium has two common isotopes, ^3He and ^4He. Each isotope can be separated and a liquid formed at low temperatures which is nearly pure ^3He or pure ^4He. Each has unusual properties and displays collective behavior of a unique character. The boson liquid ^4He shows a phase transition at $T_\lambda = 2.172$ K to a superfluid state which is similar to Bose–Einstein condensation, although vastly modified by the strong interparticle interactions. Similarly, the fermion liquid ^3He develops a Fermi distribution at low temperature, and the particles have a superfluid transition which is similar to the superconducting transition in a metal. Of course, now it is occurring in a liquid, of electrically neutral atoms, so there is no Meissner effect, but there is pairing. However, it also has a unique character, since the atom avoid the usual singlet pairing common to metals and instead pair with the spins aligned parallel. The triplet pairing, in turn, leads to many new phenomena and a richer phase diagram, which was discovered by Osheroff *et al.* (1972).

Liquid helium is a subject where we shall often be forced to abandon our Green's functions in favor of other techniques. There are some calculations for which other techniques are better and for which the Green's function method gives awful results. One example is the discussion of the ground state properties of the quantum liquids. Green's function techniques have been unsuccessful in the study of the ground state properties of classical fluids, and other techniques are now used for these high-density, strongly interacting, and highly correlated systems (see Percus, 1964). It is understandable that the ground state of quantum liquids is similarly difficult to treat with Green's functions. The description of the ground state properties of the system will require techniques beyond those introduced here, so we shall only summarize these results. However, when we discuss the excitation spectra of the liquid, which are crucial for their superfluid properties, we can use the familiar techniques of Green's functions or perturbation theory.

10.1. PAIRING THEORY

The important approximation in pairing theory is that the particles' interactions can be approximated as a summation of pairwise events. Of course, this is true of the potentials between particles, which we always take to be a summation of pair interactions: $\sum V(\mathbf{r}_i - \mathbf{r}_j)$. But pairing theory makes a much stronger assumption: When one particle interacts with another, it is not simultaneously interacting with other particles. Of course, this would be a good approximation in a gas of low density, where the particles collide occasionally, and we can treat most events as binary collisions. But this assumption is not obviously valid in a liquid. In fact, it is now known that the assumption is terrible and that pairing theory is a very bad approximation when applied to liquid helium. It makes numerical predictions of measurable quantities, such as the speed of sound in ^{4}He, which are a factor of 10 higher than experiment. One should not regard the pairing theory as a serious theory of liquid helium.

There are still several reasons for introducing pairing theory. Historically it played an important role in the history of many-particle physics. The major ideas were introduced by Bogoliubov (1947), who proposed the idea that the particles interacted in pairs. Although this theory was unsuccessful when applied to helium, virtually the same theory works for superconductivity. The BCS theory is just the Bogoliubov theory—changing the operators from bosons to fermions. Of course, there it works since the particles do have pairwise interactions and the interactions are weak. In superfluid helium it does not work, although much effort was expended over a period of years by many investigators before this conclusion became obvious.

Another reason for introducing pairing theory is that it contains many ideas which are qualitatively correct—such as the condensate. Thus it provides a good introduction to the subject. Later we shall show how the quantum liquid properties are actually calculated using correlated basis functions.

The starting point for this discussion is a Hamiltonian which contains the kinetic energy of the particles and also the pairwise potential $V(r)$

between two helium atoms:

$$H = \sum_j \frac{p_j^2}{2m} + \frac{1}{2} \sum_{i,j} V(\mathbf{r}_i - \mathbf{r}_j) \tag{10.1.1}$$

The helium particles are treated as spherically symmetric objects, and the electronic excitations are ignored. This is a safe approximation, since the atoms have a kinetic energy of roughly 15 K, while the electronic excitations have a minimum of 20 eV. Thus the ^{4}He atom is considered to be entirely structureless and is represented as a single-boson particle with a potential $V(r)$ when interacting with other similar bosons. The ^{3}He atom has additional degrees of freedom associated with its nuclear spin. The spin is an important property, since it is what makes ^{3}He have its fermion character. The helium–helium potential which was used most frequently in the early days of physics was a Lennard–Jones potential,

$$\begin{aligned} V(r) &= 4\epsilon\left[\left(\frac{\sigma}{r}\right)^{12} - \left(\frac{\sigma}{r}\right)^{6}\right] \\ \epsilon &= 1.484 \times 10^{-15}\ \text{erg} \\ \sigma &= 2.648\ \text{Å} \end{aligned} \tag{10.1.2}$$

which is often called a 6-12 potential. The parameter $\varepsilon = 10$ K is the maximum well depth, and σ is the hard core radius, where the potential rises steeply. These parameters were obtained by fitting various experimental results to this potential function (see Hirschfelder *et al.*, 1954). It should be appreciated that this potential form was chosen for its mathematical simplicity and not because it was a good approximation to the actual potential shape. Certainly the long-range attractive potential behaves as r^{-6} from van der Waals forces, but the repulsive part is probably not r^{-12}. Recent efforts to deduce this potential function have produced better versions.

In the pairing theory the Hamiltonian is written $H = H_0 + V$, where H_0 is the kinetic energy term. All operators are expanded in the basis set of H_0, which are plane-wave states. Thus we rewrite the Hamiltonian (10.1.1) as

$$\begin{aligned} H &= \sum_{\mathbf{p}} \xi_p C_{\mathbf{p}}^{\dagger} C_{\mathbf{p}} + \tfrac{1}{2} \sum_{\mathbf{qkk}'} V(\mathbf{q}) C_{\mathbf{k}-\mathbf{q}}^{\dagger} C_{\mathbf{k}'+\mathbf{q}}^{\dagger} C_{\mathbf{k}'} C_{\mathbf{k}} \\ \xi_p &= \frac{p^2}{2m} - \mu \end{aligned} \tag{10.1.3}$$

An additional spin index should be added when describing ^{3}He. The effective interaction $V(\mathbf{q})$ is the Fourier transform of the interparticle potential $V(\mathbf{r})$. One immediate problem is that the Fourier transform may not exist. Certainly there is no Fourier transform for a Lennard-Jones potential. Later we shall see that this is not a problem, since the potential will be replaced by a T matrix, which always exists regardless of the potential strength or shape. We can formally eliminate this divergence until we get to the T matrix by assuming the divergence is cut off by a parameter g and by later letting $g \to \infty$. The nonexistence of $V(\mathbf{q})$ is not really a serious problem, since the formal replacement of scattering properties by a T matrix is formally exact for binary collisions. Thus this theory would give a good description of a dilute gas. We shall effectively ignore the possible divergence in $V(\mathbf{q})$, since formally it can be eliminated.

A. Hartree and Exchange

Liquid helium is not a weakly interacting system, but we treat it as such in pairing theory. Thus we assume, for the present discussion, that the potential terms are weak, and we can proceed by examining the first terms which occur in the S-matrix expansion for the self-energy. The first two contributions are identical to the terms found for the homogeneous electron gas, which has the same formal kind of Hamiltonian as (10.1.3). These are the Hartree and exchange energies:

$$\Sigma(\mathbf{k}) = n_0 V(\mathbf{q} = 0) \pm \frac{1}{\nu} \sum_{\mathbf{q}} n(\xi_{\mathbf{k}+\mathbf{q}}) V(\mathbf{q})$$

The $\pm$ signs refer to ^{4}He($+$) and ^{3}He($-$), respectively. The ^{3}He case is identical to that of the electron gas, where the fermion nature makes the exchange energy have the opposite sign as the Hartree energy. This happens because fermions of like spin wish to avoid each other. For ^{4}He, its boson character makes the opposite occur: The quantum nature of the particles tends to make them prefer to collect at the same spot, so that the exchange energy is positive. For a system with only two bosons, the two-particle wave function must be symmetric under an interchange of coordinates, which can be achieved with the choice

$$\begin{aligned} \Psi_{\mathbf{k}_1\mathbf{k}_2}(\mathbf{r}_1, \mathbf{r}_2) &= \frac{1}{2^{1/2}} [\phi_{\mathbf{k}_1}(\mathbf{r}_1)\phi_{\mathbf{k}_2}(\mathbf{r}_2) + \phi_{\mathbf{k}_1}(\mathbf{r}_2)\phi_{\mathbf{k}_2}(\mathbf{r}_1)] \\ \phi_{\mathbf{k}}(\mathbf{r}) &= \frac{1}{\nu^{1/2}} \exp(i\mathbf{k} \cdot \mathbf{r}) \end{aligned} \tag{10.1.4}$$

The expectation value of the potential energy for this wave function is

$$\langle V \rangle = \sum_{\mathbf{k}_1\mathbf{k}_2} n(\xi_{\mathbf{k}_1})n(\xi_{\mathbf{k}_2}) \int d^3r_1 \int d^3r_2\, |\Psi_{\mathbf{k}_1\mathbf{k}_2}(\mathbf{r}_1, \mathbf{r}_2)|^2 V(\mathbf{r}_1 - \mathbf{r}_2)$$

$$= \sum_{\mathbf{k}_1\mathbf{k}_2} n(\xi_{\mathbf{k}_1})n(\xi_{\mathbf{k}_2}) \int \frac{d^3r_1}{\nu} \int \frac{d^3r_2}{\nu} V(\mathbf{r}_1 - \mathbf{r}_2)(1 + e^{i(\mathbf{r}_1-\mathbf{r}_2)\cdot(\mathbf{k}_1-\mathbf{k}_2)})$$

$$= \sum_{\mathbf{k}_1} n(\xi_{\mathbf{k}_1})\left[\frac{V(\mathbf{q}=0)}{\nu}\sum_{\mathbf{k}_2} n(\xi_{\mathbf{k}_2}) + \frac{1}{\nu}\sum_{\mathbf{k}_2} n(\xi_{\mathbf{k}_2})V(\mathbf{k}_1 - \mathbf{k}_2)\right]$$

In the integral in the middle equation, we change to center of mass coordinates $\mathbf{R} = \frac{1}{2}(\mathbf{r}_1 + \mathbf{r}_2)$ and $\mathbf{r} = (\mathbf{r}_1 - \mathbf{r}_2)$ and $\int d^3R = \nu$. The remaining integral over $\mathbf{r}$ produces $V(\mathbf{q} = 0)$ in the first term and $V(\mathbf{k}_1 - \mathbf{k}_2)$ in the second. The expression for $\langle V \rangle$ has two terms, which are the Hartree and exchange energies. They have a relative plus sign, whose origin is the relative plus sign between the two terms in the two-particle wave function (10.1.4). The two-particle fermion wave function, when the spins are parallel, has a relative minus sign, which causes the negative sign in front of the exchange energy. The occupation factors $n(\xi_k)$ are either n_B for ^{4}He or n_F for ^{3}He particles. The subscript is omitted in the formulas, since they can apply to either case equally. The diagrams for Hartree and exchange are shown in Figs. 10.1(a) and 10.1(b).

The next step is to improve the Hartree and exchange energies by replacing the potential $V(\mathbf{q})$ by the equivalent T-matrix result. The Hartree energy is the potential energy of a particle $\mathbf{k}_1$ from interacting with the other particles in the state $\mathbf{k}_2$ by the $\mathbf{q} = 0$ interaction $V(\mathbf{q} = 0)$. This inter-

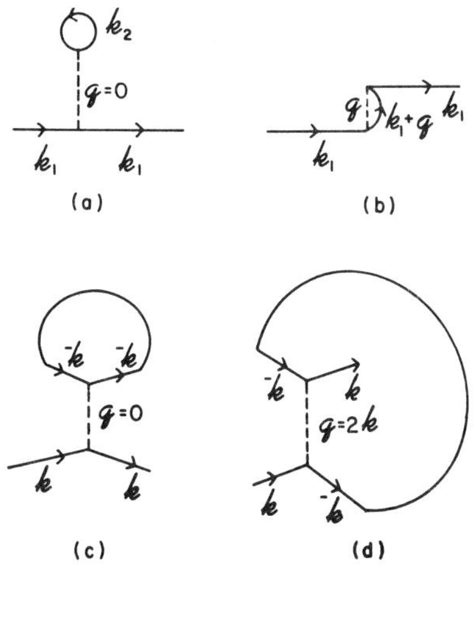

FIGURE 10.1

action can be viewed as a process whereby the two particles interact by mutually scattering. Thus we go into center of mass coordinates:

$$\begin{aligned} \mathbf{k} &= \tfrac{1}{2}(\mathbf{k}_1 - \mathbf{k}_2) \\ \mathbf{K} &= \mathbf{k}_1 + \mathbf{k}_2 \end{aligned} \tag{10.1.5}$$

The mutual scattering of two particles, which interact by central forces, does not alter the center of mass wave vector $\mathbf{K}$. It is unaffected by the scattering process, so that we can view the scattering in a rest frame where $\mathbf{K} = 0$. This scattering is shown in Fig. 10.1(c). The two particles approach with $\mathbf{k}$ and $-\mathbf{k}$, scatter with a $\mathbf{q} = 0$ interaction, and depart with their same wave vectors. In our center of mass transformation (10.1.5), we have adopted the arbitrary convention that the particle $\mathbf{k}$ is the one whose self-energy we are calculating, while $-\mathbf{k}$ is the other particle from which the first is scattering. The particle line of $-\mathbf{k}$ is part of a closed loop.

The scattering of two particles is described by a T matrix $T_{\mathbf{kk}'}$, which was introduced in Sec. 4.1. The choice of T matrix, rather than reaction matrix, is based on our interest in retarded self-energies, since the T matrix has the proper dependence on $i\delta$ in the energy denominators. Recall that the T matrix $T_{\mathbf{kk}'}$ applies only for elastic scattering, when the particle energy was $\varepsilon = \xi_{\mathbf{k}} = \xi_{\mathbf{k}'}$. The kinetic energy $\xi_{\mathbf{k}}$ is conserved in relative coordinates, so this requirement is satisfied. Thus we conclude that the retarded self-energy for particles of energy $\xi_{\mathbf{k}_1}$ in the Hartree approximation should be

$$\Sigma_H(\mathbf{k}_1, \xi_{\mathbf{k}_1}) = \frac{1}{\nu} \sum_{\mathbf{k}_2} n(\xi_{\mathbf{k}_2}) T_{(1/2)(\mathbf{k}_1 - \mathbf{k}_2),(1/2)(\mathbf{k}_1 - \mathbf{k}_2)}$$

In the scattering from other particles, whose wave vector is $\mathbf{k}_2$, the scattering rate depends on the difference $\mathbf{k} = \frac{1}{2}(\mathbf{k}_1 - \mathbf{k}_2)$ rather than on $\mathbf{q} = 0$. The forward scattering with $\mathbf{k}' = \mathbf{k}$ is easy to obtain from (4.1.16),

$$T_{\mathbf{kk}'} = 4\pi \sum_{l=1} (2l + 1) P_l(\hat{k} \cdot \hat{k}') T_l(k, k')$$

$$T_{\mathbf{kk}} = -\frac{4\pi}{2\mu k} \sum_l (2l + 1) e^{i\delta_l(k)} \sin[\delta_l(k)]$$

since $P_l(1) = 1$ and $T_l(k, k) = -(2\mu k)^{-1} \sin[\delta] e^{+i\delta}$. By changing integration to $\mathbf{k} = \frac{1}{2}(\mathbf{k}_1 - \mathbf{k}_2)$, the Hartree energy is

$$\Sigma_H(\mathbf{k}_1, \xi_{\mathbf{k}_1}) = -\frac{2^4\pi}{\mu} \int \frac{d^3k}{(2\pi)^3} n(\xi_{\mathbf{k}_1 - 2\mathbf{k}}) \frac{1}{k} \sum_l (2l + 1) e^{i\delta_l(k)} \sin[\delta_l(k)] \tag{10.1.6}$$

The same considerations should now be applied to the scattering term for the exchange energy. Again we examine the scattering in the center of mass coordinates, where the incoming particles have wave vectors **k** and $-\mathbf{k}$. This diagram is shown in Fig. 10.1(d). Here the two particles exchange momentum $2\mathbf{k}$, so the final states are still **k** and $-\mathbf{k}$. The solid line indicates the particle lines which are connected to make the self-energy diagram for exchange. Thus this term has a T-matrix expression:

$$\Sigma_x(\mathbf{k}_1, \xi_{\mathbf{k}_1}) = \frac{1}{\nu} \sum_{\mathbf{k}_2} n(\xi_{\mathbf{k}_2}) T_{(1/2)(\mathbf{k}_1-\mathbf{k}_2),(1/2)(\mathbf{k}_2-\mathbf{k}_1)} \tag{10.1.7}$$

$$= -\frac{2^4\pi}{\mu} \int \frac{d^3k}{(2\pi)^3} n(\xi_{\mathbf{k}_1-2\mathbf{k}}) \frac{1}{k} \sum_l (-1)^l (2l+1) e^{i\delta_l(k)} \sin[\delta_l(k)]$$

When making the expansion in angular momentum states l, we use $P_l(-1) = (-1)^l$. The two self-energy terms, Hartree and exchange, are either added or subtracted, respectively, for bosons and fermions. These two self-energy expressions are added for ^{4}He:

$$\Sigma(\mathbf{k}, \xi_{\mathbf{k}}) = -\frac{4}{\mu\pi^2} \sum_{l\,\text{even}} (2l+1) \int_0^\infty k'\, dk' e^{i\delta_l(k')} \sin[\delta_l(k')] \int d\Omega_{k'} n_B(\xi_{\mathbf{k}-2\mathbf{k}'})$$

For ^{4}He, the two terms combine to eliminate angular momentum components l which are odd. Thus, one needs only to compute the phase shifts $\delta_l(k)$ for l even, and the others are irrelevant.

For ^{3}He, there is exchange scattering between two particles only when the spins are parallel, and then the scattering rate has no dependence upon the angular momentum terms with l even but contains only terms with l being odd. Since there is no exchange scattering between particles with antiparallel spin, the energy of a particle contains an average over events with spin parallel and antiparallel. For the self-energy for a particle in ^{3}He, we have

$$\Sigma(\mathbf{k}, \xi_{\mathbf{k}}) = 2\Sigma_H - \Sigma_x$$

where we interpret $n_F(\xi) = (e^{\beta\xi} + 1)^{-1}$ as the occupation number for a single spin. The summations over wave vector in (10.1.6) and (10.1.7) are hard to do except with a computer. The phase shifts for ^{4}He–^{4}He scattering were calculated in Sec. 4.1 (see Fig. 4.4), and there is a resonance at low energy because of the possible existence of a low-energy bound state.

The self-energies we have calculated, for Hartree and exchange, were appropriate for the retarded energy with $\varepsilon = \xi_{\mathbf{k}}$. It is also possible to generalize the result to obtain the energy-dependent self-energy ε or ik_n. Then one expresses the self-energy in terms of the off-shell T matrix, as was done for the impurity scattering case in (4.1.31).

B. Bogoliubov Theory of ^{4}He

The Bogoliubov (1947) theory describes Bose–Einstein condensation in a weakly interacting system. It is basically a form of pairing theory for the interacting bosons. The Hamiltonian is (10.1.3) for a system of interacting bosons, expressed in the plane-wave representation:

$$K = H - \mu N = \sum_{\mathbf{k}} \xi_{\mathbf{k}} C_{\mathbf{k}}^{\dagger} C_{\mathbf{k}} + \frac{1}{2\nu} \sum_{\mathbf{k}\mathbf{k}'\mathbf{q}} V(\mathbf{q}) C^{\dagger}_{\mathbf{k}+\mathbf{q}} C^{\dagger}_{\mathbf{k}'-\mathbf{q}} C_{\mathbf{k}'} C_{\mathbf{k}}$$

The main feature of the theory is the assumption that a Bose–Einstein condensation occurs at a finite temperature T_λ. This happens because the chemical potential μ goes to zero, and there is a macroscopic occupation of the zero-momentum state. The collection of particles in the zero-momentum state is called the *condensate*.

The ^{4}He particles are treated as bosons. Because the number of particles is fixed, the number operator

$$\begin{aligned} n_B(\omega_{\mathbf{k}}) &= \frac{1}{e^{\beta(\omega_{\mathbf{k}}-\mu)} - 1} \\ N &= \int \frac{d^3k}{(2\pi)^3} n_B(\omega_{\mathbf{k}}) \end{aligned} \tag{10.1.8}$$

contains a chemical potential μ. It is usually temperature dependent and varies to satisfy the implied definition in (10.1.8) that the integral over the occupation number gives the total number of particles N. The particle energy $\omega_{\mathbf{k}}$ is yet to be determined, self-consistently, but we assume that it is positive and that the zero of energy is set at $\omega_{\mathbf{k}=0} = 0$. With these conventions, the chemical potential for most boson systems is negative, so that the energy denominator in $n_B(\omega_k)$ is never zero because $\omega_{\mathbf{k}} > \mu$. However, in Bose–Einstein condensation, the chemical potential μ vanishes, as illustrated in Fig. 10.2. This means that the occupation number $n_B(\omega_{\mathbf{k}})$

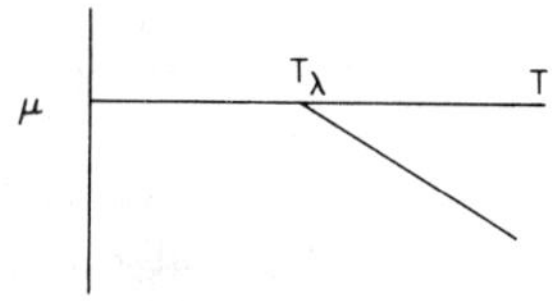

FIGURE 10.2. Chemical potential as a function of temperature, and μ becomes zero at T_λ in Bose–Einstein condensation.

diverges as $k \to 0$, and the number of particles in the $k = 0$ states is not well defined by this limit. Instead, we must assume that the number of particles in the state $\mathbf{k} = 0$ is some number N_0 which is a finite fraction of the particle number N. Indeed, for a weakly interacting system, we expect the condensate fraction $f_0 = N_0/N$ to approach a number near unity as the temperature is lowered below T_λ. Thus for $T < T_\lambda$, we can express the particle number operator as

$$n_k = N_0 \delta^3(\mathbf{k}) + \frac{1}{e^{\beta\omega_\mathbf{k}} - 1}$$

The delta function at $\mathbf{k} = 0$ gives the number of particles in the condensate, and the second term applies only for $k \neq 0$. One of the goals of the theory is to calculate this fraction f_0 self-consistently.

The creation and destruction operators $C_\mathbf{k}^\dagger$ and $C_\mathbf{k}$ must have special properties for $\mathbf{k} = 0$. Normally, for bosons, when there are $n_\mathbf{k}$ particles in the state $\mathbf{k}$, the creation operator gives $C_\mathbf{k}^\dagger \mid n_\mathbf{k}\rangle = (n_\mathbf{k} + 1)^{1/2} \mid n_\mathbf{k} + 1\rangle$. In normal systems, the occupation numbers $n_\mathbf{k}$ are integers on the order of unity. For the state with $\mathbf{k} = 0$, we expect that the average value of $\langle n_0 \rangle = N_0 \simeq 10^{23}$. Then the operations give

$$C_0^\dagger \mid N_0\rangle = (N_0 + 1)^{1/2} \mid N_0 + 1\rangle \simeq (N_0)^{1/2} \mid N_0\rangle$$
$$C_0 \mid N_0\rangle = (N_0)^{1/2} \mid N_0 - 1\rangle \simeq (N_0)^{1/2} \mid N_0\rangle$$

where the numbers $N_0 \pm 1 \simeq N_0$ when $N_0 \simeq 10^{23}$. Thus both the creation and destruction operators have a similar effect upon the system and give the result $(N_0)^{1/2}$ as an effective eigenvalue. The number N_0 is sufficiently large that we can treat the operators $C_0^\dagger$ and C_0 as scalars whose value is $(N_0)^{1/2}$:

$$C_0^\dagger C_0 = C_0 C_0^\dagger = N_0$$

The next step is to examine the Hamiltonian and to isolate all terms in which any of the C operators has $\mathbf{k} = 0$. These terms are replaced by $(N_0)^{1/2}$ for both $C_0^\dagger$ and C_0. Since $\mu = 0$, we write the kinetic energy term as

$$\sum_k \varepsilon_\mathbf{k} C_\mathbf{k}^\dagger C_\mathbf{k} = \varepsilon_0 N_0 + \sum_\mathbf{k}{}' \varepsilon_\mathbf{k} C_\mathbf{k}^\dagger C_\mathbf{k} = \sum_\mathbf{k}{}' \varepsilon_\mathbf{k} C_\mathbf{k}^\dagger C_\mathbf{k}$$

The prime on the summation means we are to omit the term with $\mathbf{k} = 0$. The kinetic energy has no dependence on the condensate, since the particles have a zero kinetic energy in this state.

The interesting terms arise from the potential energy:

$$\frac{1}{2\nu} \sum_{\mathbf{k}\mathbf{k}'\mathbf{q}} V(\mathbf{q}) C^\dagger_{\mathbf{k}+\mathbf{q}} C^\dagger_{\mathbf{k}'-\mathbf{q}} C_{\mathbf{k}'} C_{\mathbf{k}}$$

When the condensate fraction f_0 is near unity, the largest term is when $\mathbf{k} = \mathbf{k}' = \mathbf{q} = 0$, which gives an energy contribution in terms of the condensate density $\varrho_0 = N_0/\nu = f_0\varrho$:

$$\frac{1}{2\nu} V(0) N_0^2 = \frac{1}{2} \nu \varrho_0^2 V(0)$$

The next largest terms would be those in which only three of the C operators had zero wave vector, but none exist. If any three particles have zero wave vectors, then so does the fourth, and we have counted that term already. The next largest terms are those in which two of the wave vectors are zero and two are not. There are six ways of pairing the four C operators in the potential energy term:

$\mathbf{k} = \mathbf{k}' = 0$	$\frac{1}{2}\varrho_0 \sum_{\mathbf{q}} V(\mathbf{q}) C_{\mathbf{q}}^\dagger C^\dagger_{-\mathbf{q}}$
$\mathbf{k} = -\mathbf{q},\ \mathbf{k}' = \mathbf{q}$	$\frac{1}{2}\varrho_0 \sum_{\mathbf{q}} V(\mathbf{q}) C_{\mathbf{q}} C_{-\mathbf{q}}$
$\mathbf{q} = 0 = \mathbf{k}$	$\frac{1}{2}\varrho_0 V(0) \sum_{\mathbf{k}'} C^\dagger_{\mathbf{k}'} C_{\mathbf{k}'}$
$\mathbf{q} = 0 = \mathbf{k}'$	$\frac{1}{2}\varrho_0 V(0) \sum_{\mathbf{k}} C_{\mathbf{k}}^\dagger C_{\mathbf{k}}$
$\mathbf{k} = 0,\ \mathbf{k}' = \mathbf{q}$	$\frac{1}{2}\varrho_0 \sum_{\mathbf{q}'} V(\mathbf{q}) C_{\mathbf{q}}^\dagger C_{\mathbf{q}}$
$\mathbf{k}' = 0,\ \mathbf{k} = -\mathbf{q}$	$\frac{1}{2}\varrho_0 \sum_{\mathbf{q}} V(\mathbf{q}) C_{\mathbf{q}}^\dagger C_{\mathbf{q}}$

It is customary to stop at this point and to keep only these terms in an effective Hamiltonian:

$$H_0 = \sum_{\mathbf{k}}{}' \left[\varepsilon_{\mathbf{k}} + \varrho_0 V(0)\right] C_{\mathbf{k}}^\dagger C_{\mathbf{k}} + \tfrac{1}{2}\varrho_0 \sum_{\mathbf{q}}{}' V(\mathbf{q})(C_{\mathbf{q}}^\dagger + C_{-\mathbf{q}})(C_{\mathbf{q}} + C^\dagger_{-\mathbf{q}}) \qquad (10.1.9)$$

The term $\varrho_0 V(0)$ is dropped, since this is a constant term which just causes a shift in the chemical potential. That is, we really set the quantity $\varepsilon_{\mathbf{k}} - \mu + \varrho_0 V(0) = \varepsilon_{\mathbf{k}}$ so that $\mu = \varrho_0 V(0)$. The effective Hamiltonian given in (10.1.9) can be diagonalized exactly, which is the strongest argument for using this as the effective Hamiltonian. A better theory would retain additional terms. For example, those with only one C operator at $\mathbf{k} = 0$

produce terms in the effective Hamiltonian such as $[(N_0)^{1/2}/\nu] \sum_{\mathbf{kq}} V(\mathbf{q}) \times C^\dagger_{\mathbf{k}+\mathbf{q}} C^\dagger_{-\mathbf{q}} C_{\mathbf{k}}$ where one excitation scatters into two others. The evaluation of these types of terms require a Green's function analysis. Of course, there are also terms in which none of the wave vectors are zero, and these correspond to the scattering of two excitations. Thus we can write the Hamiltonian as $H_0 + V$, where H_0 is the part in (10.1.9) which contains the terms quadratic in the operators, while V is the other terms which contain three or four creation–destruction operators:

$$\begin{aligned}
H &= H_0 + V \\
H_0 &= \sum_{\mathbf{k}}{}' \varepsilon_{\mathbf{k}} C_{\mathbf{k}}{}^\dagger C_{\mathbf{k}} + \tfrac{1}{2}\varrho_0 \sum_{\mathbf{k}}{}' V(\mathbf{k})(C_{\mathbf{k}}{}^\dagger + C_{-\mathbf{k}})(C_{\mathbf{k}} + C^\dagger_{-\mathbf{k}}) \\
V &= \frac{\varrho_0}{\nu^{1/2}} \sum_{\mathbf{q},\mathbf{k}}{}' V(\mathbf{q})(C^\dagger_{\mathbf{k}+\mathbf{q}} C^\dagger_{-\mathbf{q}} C_{\mathbf{k}} + C^\dagger_{\mathbf{k}+\mathbf{q}} C_{\mathbf{q}} C_{\mathbf{k}}) \\
&\quad + \frac{1}{2\nu} \sum_{\mathbf{kk}'\mathbf{q}}{}' V(\mathbf{q}) C^\dagger_{\mathbf{k}+\mathbf{q}} C^\dagger_{\mathbf{k}'-\mathbf{q}} C_{\mathbf{k}'} C_{\mathbf{k}}
\end{aligned} \tag{10.1.10}$$

The terms are separated in this manner because H_0 is a Hamiltonian which can be solved exactly. The other term V must be treated as the interaction, which is evaluated by the usual Green's function method to give additional self-energy corrections.

The particles in the $\mathbf{k} = 0$ state are the condensate, and this is the ground state of the system. Quasiparticles with $\mathbf{k} \neq 0$ are excitations, and H_0 is the effective Hamiltonian which describes the properties of these excitations. It contains combinations of operators such as $C_{\mathbf{k}}{}^\dagger C^\dagger_{-\mathbf{k}}$ which correspond to the excitation of two particles from the condensate. Since momentum is conserved, they must have $\mathbf{k}$ and $-\mathbf{k}$. Similarly, the term with $C_{\mathbf{k}} C_{-\mathbf{k}}$ is the destruction of two quasiparticles with $\mathbf{k}$ and $-\mathbf{k}$ when both are returned to the condensate. The terms with $C_{\mathbf{k}}{}^\dagger C_{\mathbf{k}}$ correspond to the scattering of quasiparticles by the condensate, and there are both direct and exchange processes which lead to the terms with $V(0)$ and $V(\mathbf{k})$, respectively.

The Hamiltonian (10.1.9) for H_0 is solved to obtain the eigenvalues and eigenstates of the system. This solution is approximate, since it omits the effects of the potential V. The approximation of neglecting V and using only the eigenstates of H_0 is the pairing theory of Bogoliubov. The Hamiltonian is rather easy to diagonalize. Each vector state is treated independently, and each of these is just the problem encountered in Problem 3 in Chapter 1 and in the Tomonaga model. We introduce a set of harmonic oscillator coordinates $Q_{\mathbf{k}}$ and $P_{\mathbf{k}}$, and the Hamiltonian is shown to be an example where the interactions merely shift the eigenfrequency of the oscillator:

$$
\begin{aligned}
Q_{\mathbf{k}} &= \frac{1}{(2\varepsilon_{\mathbf{k}})^{1/2}} (C_{\mathbf{k}} + C^{\dagger}_{-\mathbf{k}}) \\
P_{\mathbf{k}} &= i\left(\frac{\varepsilon_{\mathbf{k}}}{2}\right)^{1/2} (C_{\mathbf{k}}^{\dagger} - C_{-\mathbf{k}}) \\
H_0 &= \sum_{\mathbf{k}'} \{\tfrac{1}{2}P_{\mathbf{k}}P_{-\mathbf{k}} + Q_{\mathbf{k}}Q_{-\mathbf{k}}[\tfrac{1}{2}\varepsilon_{\mathbf{k}}^2 + \varrho_0\varepsilon_{\mathbf{k}} V(\mathbf{k})]\} \\
&= \tfrac{1}{2}\sum_{\mathbf{k}} (P_{\mathbf{k}}P_{-\mathbf{k}} + \Omega_{\mathbf{k}}^2 Q_{\mathbf{k}}Q_{-\mathbf{k}}) \\
\Omega_{\mathbf{k}} &= [\varepsilon_{\mathbf{k}}^2 + 2\varepsilon_{\mathbf{k}}\varrho_0 V(\mathbf{k})]^{1/2}
\end{aligned}
\tag{10.1.11}
$$

The first two terms in H_0 come from $\varepsilon_{\mathbf{k}} C_{\mathbf{k}}^{\dagger} C_{\mathbf{k}}$, while the last term comes from the second term in the Hamiltonian H_0. The terms multiplying $Q_{\mathbf{k}}Q_{-\mathbf{k}}$ are combined into the square of an effective frequency $\Omega_{\mathbf{k}}$. Thus we have a harmonic oscillator system, for each wave vector **k**, at the new frequency $\Omega_{\mathbf{k}}$. Thus if we define a new set of creation and destruction operators $\alpha_{\mathbf{k}}$ and $\alpha_{\mathbf{k}}^{\dagger}$, the Hamiltonian becomes

$$
\begin{aligned}
H_0 &= \sum_{\mathbf{k}} \Omega_{\mathbf{k}} \alpha_{\mathbf{k}}^{\dagger} \alpha_{\mathbf{k}} \\
Q_{\mathbf{k}} &= \frac{1}{(2\Omega_{\mathbf{k}})^{1/2}} (\alpha_{\mathbf{k}} + \alpha_{\mathbf{k}}^{\dagger}) \\
P_{\mathbf{k}} &= -i\left(\frac{\Omega_{\mathbf{k}}}{2}\right)(\alpha_{\mathbf{k}} - \alpha_{\mathbf{k}}^{\dagger})
\end{aligned}
\tag{10.1.12}
$$

In the pairing theory, the energy Ω_k is that of the quasiparticles, and $\alpha_{\mathbf{k}}^{\dagger}$, $\alpha_{\mathbf{k}}$ are the operators which describe the creation and destruction of these quasiparticles.

The equation for the quasiparticle energy $\Omega_{\mathbf{k}} = [\varepsilon_{\mathbf{k}}^2 + 2\varepsilon_{\mathbf{k}}\varrho_0 V(\mathbf{k})]^{1/2}$ makes no sense whatever, since the Fourier transform $V(\mathbf{k})$ does not exist. $V(\mathbf{k})$ should be replaced by a T matrix or reaction matrix, and the latter is correct. According to the discussion following (4.1.11), for each angular momentum state l one can replace the potential by the on-shell reaction matrix

$$
V(k) \rightarrow 4\pi R_l(k, k) = -\frac{2\pi \tan[\delta_l(k)]}{\bar{\mu}k}
$$

where $\bar{\mu} = \frac{1}{2}m$ is the reduced mass for the scattering of two like particles of mass m. The reaction matrix has the important feature that it goes to a constant in the limit where $\mathbf{k} \to 0$. The phase shifts behave in this limit as

$$\lim_{k\to 0} \delta_l(k) = m_l\pi - ak^{2l+1}$$

$$\lim_{k\to 0} \frac{1}{k} \tan \delta_l = -ak^{2l}$$

where $m_l =$ integer.

The largest term is from the s-wave channel, in which $R_0(0, 0) \to$ constant $= 2\pi a/\mu$, while $R_l(0, 0) = 0$ for $l > 0$. Thus in the limit where $\mathbf{k} \to 0$, for the dispersion relation of the excitations we find

$$V(k) \to 4\pi R_0 = \frac{2\pi a}{\mu} = \frac{4\pi a}{m}$$

$$\lim_{k\to 0} \Omega_k = \lim_{k\to 0} [2\varepsilon_k \varrho_0 R_0(k, k)]^{1/2} = ck$$

$$c = \frac{\hbar}{m} (4\pi a \varrho_0)^{1/2}$$

The quantity a is called a scattering length. It is positive for repulsive potentials. For a hard sphere potential of radius σ, we find that $a = \sigma$.

In the earlier days of helium theory, the helium–helium interaction was approximated as a hard core at the radius $a = \sigma = 2.5$ Å of the Lennard-Jones potential. All the other parameters are known if we take ϱ_0 to be the ^{4}He density, so that the velocity of the excitation may be calculated:

$$a = 2.5 \text{ Å}$$

$$m = 6.65 \times 10^{-24} \text{ g}$$

$$\varrho_0 = 2.2 \times 10^{22} \text{ atoms/cm}^3$$

$$c = \frac{\hbar}{m} (4\pi a \varrho_0)^{1/2} = 130 \text{ m/sec}$$

This velocity comes out close to the speed of ordinary sound in helium, which is $c_s = 220$ m/sec. This agreement was viewed as a great success of the pairing theory, since the collective excitations of the liquid at long wavelength must certainly be the longitudinal sound waves.

Unfortunately, this agreement is entirely superfluous. The interaction potential between two ^{4}He particles has attractive regions, and the phase shifts for this potential were calculated in Sec. 4.1. The low-energy scattering is resonant. Although the exact value of the scattering length a is uncertain because of its very sensitive dependence on the potential, the expected values are more in the range of 50σ rather than σ. A value of $a \simeq 50\sigma$ will make the calculated speed of sound be 4 times larger than the experiment. Of course, the sign of a could be negative, in which case the predicted frequencies are complex, and the vibrations are unstable. A possible way out of this difficulty is to do the theory more carefully. This was done by Brown and Coopersmith (1969), who found even worse agreement with experiment. Their calculation represented all direct and exchange scattering by reaction matrices, and they self-consistently found the condensate fraction f_0 and the chemical potential. They obtained a speed of sound about 10 times that of the experiment and also predicted a roton energy about 10 times too big. Indeed, their theoretical excitation spectrum of ^{4}He is similar to the actual one but scaled up by a factor of 10. This calculation was done carefully and correctly —its lack of success is caused by the failure of the pairing approximation. Indeed, the history of the pairing theory is that each time the theoretical calculations were done better, the answers got worse. The conclusion is that the pairing theory is bad and should be discarded as a way of calculating the excitation spectrum of ^{4}He. We do retain some of the concepts, in particular that of the condensate.

10.2. ^{4}He: GROUND STATE PROPERTIES

The modern theory of superfluidity in ^{4}He treats the liquid as a highly correlated and strongly interacting system. The pairing theory is completely abandoned, as is the representation of operators in plane-wave basis states. The one concept we do retain is the condensate, or zero-momentum state, in which there is a finite fraction of the particles in the superfluid. There is an obvious problem of trying to decide the definition of "zero-momentum state" in a basis set which is not plane waves. It is really part of a larger problem of trying to decide what we mean by superfluidity and Bose–Einstein condensation in a system which is strongly interacting and highly correlated. That is, the whole concept of Bose–Einstein condensation is based on particles occupying zero-momentum states or not occupying them, and we need a more general definition if we are going to start in another basis set. The method of doing this was introduced by Penrose (1951), and we shall describe his ideas. Yang (1962) suggested the name of *off-diagonal long-range order* (ODLRO) for the type of ordering introduced by Penrose and also discussed by Penrose and Onsager (1956).

A. Off-Diagonal Long-Range Order

ODLRO was introduced by Penrose (1951) and by Penrose and Onsager (1956) as the type of ordering for superfluids, such as Bose–Einstein condensation in ^{4}He, and electron pairs in superconductors. It is distinguished from *diagonal long-range order* (DLRO), which is the usual ordering one finds, for example, in crystalline solids. When ^{4}He atoms arrange themselves into a solid, as happens under pressure, they are exhibiting DLRO. But when they go superfluid, they exhibit ODLRO. The distinction between these two processes stems from the different behavior in the density matrix for each type of ordering. These differences will be explained. Further discussion is given by Kohn and Sherrington (1970).

We assume a system of $N \simeq 10^{23}$ identical bosons, which we take to be spinless. Their ground state is described by a many-particle wave function $\Psi_0(\mathbf{r}_1, \mathbf{r}_2, \ldots, \mathbf{r}_N)$. The actual method of calculating this wave function, or at least some of its properties, will be given in the following sections. For the present discussion, we need only assume that the wave function exists. The subscript zero on Ψ_0 indicates it is for the ground state. Note that we do not assume that each atom is in the ground state (i.e., $\mathbf{k} = 0$) but that the system is in the ground state. When the particles are strongly interacting and highly correlated, they will spend little time with $\mathbf{k} = 0$ and will fluctuate between many momentum states. The ground state is the lowest possible energy state for the whole liquid. This is not a static or rigid structure, since each atom fluctuates with zero-point energy. The present estimates for ^{4}He at $T = 1$ K give the average kinetic energy per particle as 15 K and the average potential energy as -22 K, so the average binding energy is -7 K. The large value of average kinetic energy shows the large amount of zero-point motion in the fluid, even at low temperature, which comes from the quantum nature of the fluid. The classical estimate $\langle \text{K.E.} \rangle = \frac{3}{2}k_BT$ is obviously inaccurate. We conclude that the ground state wave function $\Psi_0(\mathbf{r}_1, \mathbf{r}_2, \ldots, \mathbf{r}_N)$ for ^{4}He describes a system with a large amount of zero-point motion.

The square of the wave function gives the probability density for finding particles at positions $\mathbf{r}_j$ in the system and is called the diagonal density matrix:

$$\varrho_N(\mathbf{r}_1, \mathbf{r}_2, \ldots, \mathbf{r}_N) = | \Psi_0(\mathbf{r}_1, \mathbf{r}_2, \ldots, \mathbf{r}_N) |^2 \tag{10.2.1}$$

The subscript N indicates that it applies to N particles. ϱ_N is normalized so that the integral over all coordinates gives unity:

$$1 = \int d^3r_1 \cdots d^3r_N \varrho_N(\mathbf{r}_1, \ldots, \mathbf{r}_N) \tag{10.2.2}$$

The one-particle density matrix is obtained from ϱ_N by integrating over all but one coordinate:

$$\varrho_1(\mathbf{r}_1) = \int d^3r_2\, d^3r_3 \cdots d^3r_N \varrho_N(\mathbf{r}_1, \ldots, \mathbf{r}_N)$$

$$1 = \int d^3r_1 \varrho_1(\mathbf{r}_1)$$

The notation is a bit confusing, since $\varrho_1(\mathbf{r}_1)$ is not the probability in a one-particle system but rather the probability of one particle being at $\mathbf{r}_1$ in the N-particle system. Since all points are equivalent, $\varrho_1(\mathbf{r}_1)$ is really independent of position—neglecting edges—so we conclude that $\varrho_1 = 1/\nu$, where ν is the volume of the system.

Another useful quantity is the two-particle density matrix $\varrho_2(\mathbf{r}_1, \mathbf{r}_2)$ which is obtained from ϱ_N by integrating over all but two coordinates:

$$\varrho_2(\mathbf{r}_1, \mathbf{r}_2) = \int d^3r_3\, d^3r_4 \cdots d^3r_N \varrho_N(\mathbf{r}_1, \mathbf{r}_2, \ldots, \mathbf{r}_N)$$

$$\int d^3r_2 \varrho_2(\mathbf{r}_1, \mathbf{r}_2) = \frac{1}{\nu}$$

Now $\varrho_2(\mathbf{r}_1, \mathbf{r}_2)$ is the probability that any boson is at $\mathbf{r}_2$ if there is one at $\mathbf{r}_1$. In a homogeneous system it must depend only on the difference $\mathbf{r}_1 - \mathbf{r}_2$. In fact, it must be proportional to the pair distribution function $g(\mathbf{r})$, and this relationship is

$$\varrho_2(\mathbf{r}_1, \mathbf{r}_2) = \frac{1}{\nu^2} g(\mathbf{r}_1 - \mathbf{r}_2)$$

which follows from the fact that $g(r)$ is normalized to go to unity at large $\mathbf{r}$, while $\varrho_2(\mathbf{r}_1 - \mathbf{r}_2)$ is normalized to go to ν^{-2}, so that $\int d^3r \varrho_2(\mathbf{r}) = 1/\nu$.

Crystalline solids have DLRO, which is indicated by structure in $g(r)$. Since the atoms are regularly spaced, $g(\mathbf{r})$ has large values where the atoms are located at regular lattice points away from the reference site and zero otherwise. Thus, for example, the Fourier transform

$$S(\mathbf{q}) = 1 + \varrho \int d^3r e^{i\mathbf{q}\cdot\mathbf{r}}[g(\mathbf{r}) - 1]$$

will have delta functions at the reciprocal lattice points $\mathbf{G} = \mathbf{q}$. The delta functions exist at finite temperatures, or even at zero temperature with the inclusion of zero-point motion, when the vibrations are included, although they reduce the intensity of the delta functions by the Debye–Waller factor.

This periodicity is the crucial feature of DLRO and is contained in $g(\mathbf{r})$ or $\varrho_2(\mathbf{r}_1, \mathbf{r}_2)$.

Penrose (1951) used the idea of a general density matrix which is defined as the product of two wave functions with different coordinates:

$$\begin{aligned}&\tilde{\varrho}_N(\mathbf{r}_1, \mathbf{r}_2, \ldots, \mathbf{r}_N; \mathbf{r}_1', \mathbf{r}_2', \ldots, \mathbf{r}_N')\\ &\quad = \Psi_0^*(\mathbf{r}_1, \mathbf{r}_2, \ldots, \mathbf{r}_N)\Psi_0(\mathbf{r}_1', \mathbf{r}_2', \mathbf{r}_3', \ldots, \mathbf{r}_N')\end{aligned}$$

This quantity is denoted by $\tilde{\varrho}$, where the tilde is to distinguish it from the diagonal density matrix introduced earlier. Of course, they are identical if the two sets of coordinates in $\tilde{\varrho}$ are set equal: $\tilde{\varrho}_N(\mathbf{r}_1, \ldots, \mathbf{r}_N; \mathbf{r}_1, \ldots, \mathbf{r}_N) = \varrho_N(\mathbf{r}_1, \ldots, \mathbf{r}_N)$. The concept of ODLRO is contained in the function $\tilde{\varrho}_1(\mathbf{r}_1, \mathbf{r}_2')$ obtained from $\tilde{\varrho}_N$ when all but one set of coordinates are set equal and averaged over

$$\tilde{\varrho}_1(\mathbf{r}_1, \mathbf{r}_1') = \int d^3r_2\, d^3r_3 \cdots d^3r_N \tilde{\varrho}_N(\mathbf{r}_1, \mathbf{r}_2, \ldots, \mathbf{r}_N; \mathbf{r}_1', \mathbf{r}_2, \mathbf{r}_3, \cdots, \mathbf{r}_N)$$

$$\tilde{\varrho}_1(\mathbf{r}_1, \mathbf{r}_1) = \varrho_1(\mathbf{r}_1) = \frac{1}{\nu}$$

Of course, $\tilde{\varrho}_1(\mathbf{r}_1, \mathbf{r}_1')$ becomes the diagonal density matrix ϱ_1 when $\mathbf{r}_1 = \mathbf{r}_1'$. In a liquid, the dependence on $\mathbf{r}_1$ and $\mathbf{r}_1'$ can only be with their difference, since there is no absolute frame of reference. It is convenient to define the quantity

$$R(\mathbf{r}_1 - \mathbf{r}_1') = \nu\tilde{\varrho}_1(\mathbf{r}_1, \mathbf{r}_1')$$

$$R(0) = 1$$

which is normalized to unity at $r = 0$. $R(\mathbf{r})$ is the function which is important in understanding ODLRO.

Some insight into $\tilde{\varrho}_1(\mathbf{r}_1, \mathbf{r}_1')$ is obtained by reconsidering the techniques used in the weakly interacting systems. There we usually defined a one-particle state function in the plane-wave representation as

$$\Psi(\mathbf{r}) = \frac{1}{\nu^{1/2}} \sum_{\mathbf{k}} e^{i\mathbf{k}\cdot\mathbf{r}} C_{\mathbf{k}}$$

$$C_{\mathbf{k}} = \frac{1}{\nu^{1/2}} \int d^3r e^{-i\mathbf{k}\cdot\mathbf{r}} \Psi(\mathbf{r})$$

The number of particles in state $\mathbf{k}$ is

$$n_{\mathbf{k}} = \langle C_{\mathbf{k}}^{\dagger} C_{\mathbf{k}} \rangle = \frac{1}{\nu} \int d^3r\, d^3r' e^{-i\mathbf{k}\cdot(\mathbf{r}-\mathbf{r}')} \langle \Psi^{\dagger}(\mathbf{r}')\Psi(\mathbf{r}) \rangle \tag{10.2.3}$$

$n_{\mathbf{k}}$ is found to be the Fourier transform of the quantity $\langle \Psi^\dagger(\mathbf{r}')\Psi(\mathbf{r})\rangle$, which must be a function of $\mathbf{r} - \mathbf{r}'$. Recall what this average means. We take a particle and find its wave function at two different points $\mathbf{r}$ and $\mathbf{r}'$. This product is averaged, which must be taken over the other particles and their positions. This procedure is exactly the one which was used to obtain $\tilde{\varrho}_1(\mathbf{r}_1, \mathbf{r}_1')$, except now we recognize from the outset that the particle is part of a many-particle system. We take a product wave function with one particle at two points $\mathbf{r}_1$ and $\mathbf{r}_1'$ and average it over the other particles and their positions. We conclude that the quantity $\tilde{\varrho}_1(\mathbf{r}_1, \mathbf{r}_1')$, or its equivalent $R(r)$, is the many-body definition of the quantity $\langle \Psi^\dagger(\mathbf{r}')\Psi(\mathbf{r})\rangle$. It is also important to appreciate that $R(r)$ is quite a different function from the pair distribution function $g(r)$ and that the two are not related. The quantity $n_{\mathbf{k}}$ is also related to the Wigner distribution function, where $n_{\mathbf{k}}(R, T) = \int d\omega f(\mathbf{k}, \omega; R, T)$.

The quantity $n_{\mathbf{k}}$ is the number of particles, on the average, in the momentum state $\mathbf{k}$. It is an important quantity in Bose–Einstein condensation, since one expects that one momentum state $\mathbf{k}_0$ will have a macroscopic occupation such that $n_{\mathbf{k}_0} = f_0 N$, where f_0 is still the fraction of particles in the condensate. This fraction is of order unity, rather than $0(1/N)$. Normally we have that $\mathbf{k}_0 = 0$, which applies when the fluid is at rest. There are circumstances when another state has the macroscopic occupation, e.g., when the fluid is flowing at a uniform rate or when it is rotating. For a model system in which the particle states of energy $\Omega_{\mathbf{k}}$ have no damping, we have $n_{\mathbf{k}} = n_B(\Omega_{\mathbf{k}})$. No damping is certainly not the case in a strongly interacting system. Instead, for $\mathbf{k} \neq \mathbf{k}_0$, we have $n_{\mathbf{k}} = (1/2\pi) \int d\varepsilon n_B(\varepsilon) \times A(\mathbf{k}, \varepsilon)$, where $A(\mathbf{k}, \varepsilon)$ is the spectral function for the boson particles.

The procedure for finding $n_{\mathbf{k}}$ is the same as in the free-particle case (10.2.3). There one took the Fourier transform of $\langle \Psi^\dagger(\mathbf{r})\Psi(\mathbf{r}')\rangle$, while now one takes the Fourier transform of the equivalent quantity $\tilde{\varrho}(\mathbf{r}, \mathbf{r}')$ or $R(\mathbf{r}-\mathbf{r}')$. One difficulty with this procedure is that $R(\mathbf{r})$ may not possess a Fourier transform. For example, do we know how it behaves as $r \to \infty$? For fluids at rest, so that the macroscopic occupation is in the zero-momentum state, $R(\mathbf{r})$ goes to a constant as $r \to \infty$. Call this constant $R(\infty)$. The Fourier transform is

$$\begin{aligned} n_{\mathbf{k}} &= \varrho \int d^3r R(\mathbf{r}) e^{i\mathbf{k}\cdot\mathbf{r}} \\ &= \varrho R(\infty)(2\pi)^3 \delta^3(\mathbf{k}) + \varrho \int d^3r [R(\mathbf{r}) - R(\infty)] e^{i\mathbf{k}\cdot\mathbf{r}} \\ &= N_0 \delta_{\mathbf{k}=0} + \varrho \int d^3r [R(\mathbf{r}) - R(\infty)] e^{i\mathbf{k}\cdot\mathbf{r}} \end{aligned}$$

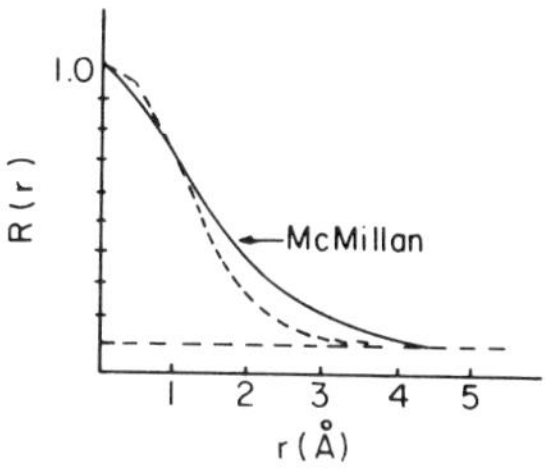

FIGURE 10.3. Single-particle density matrix as a function of separation. The solid line is the Monte Carlo calculation of McMillan (1965), which asymptotically approaches a density fraction of $f_0 = 0.11$. The dashed curve is the Gaussian approximation.

There is a delta function term at $\mathbf{k} = 0$ whose amplitude is $\varrho R(\infty)$. The quantity $R(\infty)$ is the fraction of particles in the zero-momentum state $f_0 \equiv R(\infty)$. This fraction is the "order" which exists in the off-diagonal density matrix $R(r)$. For $\mathbf{k} \neq 0$, the momentum distribution of particles $n_{\mathbf{k}}$ is found from the Fourier transform of $R(\mathbf{r}) - f_0$.

Figure 10.4 shows the function $R(\mathbf{r})$ calculated for liquid ^{4}He by McMillan (1965) using Monte Carlo techniques. He used the Lennard-Jones potential between helium atoms and found $f_0 = 0.11$. His fraction is similar to that obtained earlier by Penrose and Onsager, who got $f_0 = 0.08$ for a gas of hard spheres. The latter estimate is also based on a Monte Carlo calculation. It appears, unfortunately, that reliable results can be obtained in this field only by extensive computer calculations. There have been many calculations of these quantities; e.g., see Francis *et al.* (1970). Figure 10.3 shows that $R(\mathbf{r})$ starts at unity and falls smoothly to its asymptotic value, which it reaches at about $r = 4$ Å. This is a rather short distance, since it is only 1.5 atomic diameters.

The quantity $f_0 \equiv R(\infty)$ is the fraction of time a particle spends in the condensate. Alternately, it is also the fraction of particles in this state at any one time. However, in calculating this quantity, we have not assumed that the system is superfluid. The value of $R(\infty)$ is found by solving the dynamical properties of normal fluids, which may seem paradoxical and is probably best understood in the next section where we give more details of the method. Normal fluids do not have a condensate and have $f_0 = R(\infty) = 0$. However, the point is that most methods of calculating f_0 are based on the properties of normal fluids. For example, the estimate of Penrose and Onsager came from the equation of state of a classical hard sphere gas, which is certainly not a superfluid system.

B. Correlated Basis Functions

Correlated basis functions (CBFs) are the type of wave function most often employed in the study of the ground state properties of ^{4}He. They have the form

$$\Psi_0(\mathbf{r}_1, \mathbf{r}_2, \mathbf{r}_3, \ldots, \mathbf{r}_N) = L_N \exp\Big[-\sum_{i>j} u(\mathbf{r}_i - \mathbf{r}_j)\Big] \tag{10.2.4}$$

where L_N is a normalization constant which is specified by satisfying the integral (10.2.2). They were used by Bijl (1940), Dingle (1949), and Jastrow (1955) and are sometimes named after these authors—different writers preferring various combinations of their names. The CBFs have several advantages which make them desirable for a description of ^{4}He. The most important is that the wave function possesses the necessary symmetry of being symmetric under the exchange of any two coordinates: $\Psi_0(\mathbf{r}_1, \mathbf{r}_2, \ldots, \mathbf{r}_N) = \Psi_0(\mathbf{r}_2, \mathbf{r}_1, \ldots, \mathbf{r}_N)$. This symmetry is automatically satisfied by the summation of pairwise correlations. Another advantage is that Ψ_0 has a simple form for which one needs to determine only one function $u(r)$, which is usually found variationally by minimizing the ground state energy.

A third advantage of CBFs is that the diagonal density matrix has a mathematical form which is identical to another problem which has been studied extensively—the classical fluid. For a classical fluid, the kinetic energy terms are irrelevant for determining particle correlations, and only the potential energy $V(\mathbf{r}_i - \mathbf{r}_j)$ between particles i and j is important. The density matrices for CBFs and the classical fluid are

$$\text{CBF:} \qquad \varrho_N(\mathbf{r}_1, \mathbf{r}_2, \ldots, \mathbf{r}_N) = |\Psi|^2 = L_N{}^2 \exp\Big[-2\sum_{i>j} u(\mathbf{r}_i - \mathbf{r}_j)\Big]$$

$$\text{Classical:} \qquad \varrho(\mathbf{r}_1, \ldots, \mathbf{r}_N) = Q_N \exp\Big[-\beta\sum_{i>j} V(\mathbf{r}_i - \mathbf{r}_j)\Big]$$

These have the same mathematical form when we identify $2u(r) = \beta V(r)$. The classical problem has been extensively studied, and accurate calculational techniques have been developed which work well for liquids of neutral atoms such as the rare gases. The most successful methods are based on the Percus–Yevick equation (see Percus, 1964). The availability of this successful computational technology is one reason for the popularity and success of CBFs.

One important quantity to calculate is the off-diagonal density matrix $R(r)$, which is defined in terms of CBFs as

$$R(\mathbf{r}_1 - \mathbf{r}_1') = \nu L_N{}^2 \int d^3r_2 \cdots d^3r_N$$

$$\times \exp\Big\{-\sum_{j=2}^{N} [u(\mathbf{r}_1 - \mathbf{r}_j) + u(\mathbf{r}_1' - \mathbf{r}_j)] - \sum_{i>j=2} 2u(\mathbf{r}_i - \mathbf{r}_j)\Big\}$$

Let us determine the classical analog of this expression. That is, we shall regard $u(\mathbf{r}_i - \mathbf{r}_j)$ as the effective "potential" $\beta V(\mathbf{r}_i - \mathbf{r}_j)$ between particles and see where that leads us. Obviously there is a system of $N - 1$ particles which mutually interact with each other with the "potential" $2u(\mathbf{r})$. There is a particle at $\mathbf{r}_1$ which interacts with these $N - 1$ particles with the different potential $u(\mathbf{r})$. Another particle at $\mathbf{r}_1'$ also interacts with the $N - 1$ particles with the same relative potential $u(\mathbf{r})$, so that it is identical to the particle at $\mathbf{r}_1$. Finally, the particles at $\mathbf{r}_1$ and $\mathbf{r}_1'$ have no mutual interaction. This description is of a system of two impurity particles, at $\mathbf{r}_1$ and $\mathbf{r}_1'$, in a system of $N - 1$ other particles. The impurities are different, since they have a different interaction with the $N - 1$ particles and no interaction between themselves. This classical system can be solved by using the Percus–Yevick equations for binary mixtures, where one constituent is very dilute. This technique was used by Francis *et al.* (1970) for their investigation of superfluid ^{4}He. It is another example of where classical equations are solved to determine the properties of the superfluid.

The ground state properties which are calculated with CBFs include the condensate fraction f_0, the momentum distribution $n_\mathbf{k}$, $R(r)$, $g(r)$, and $S(k)$. Another property is the ground state energy per atom, which is usually presented as the separate potential energy (P.E.) and kinetic energy (K.E.) contributions. The potential energy per atom is given by

$$\frac{\langle \text{P.E.} \rangle}{N} = \frac{1}{N} \sum_{i>j}^{n} \int d^3r_1 \cdots d^3r_N V(\mathbf{r}_i - \mathbf{r}_j) \varrho_N(\mathbf{r}_1 \cdots \mathbf{r}_N)$$

The density matrix $\varrho_N(\mathbf{r}_1, \mathbf{r}_2, \ldots, \mathbf{r}_N)$ is unchanged when any pair of coordinates $\mathbf{r}_i$ and $\mathbf{r}_i'$ are interchanged. With this fact, one can show that the average value of each potential energy term $V(\mathbf{r}_i - \mathbf{r}_j)$ is the same as any other term $V(\mathbf{r}_i' - \mathbf{r}_j')$. The average potential of any particle is the same as any other. Thus the summation over i and j gives $\frac{1}{2}N(N - 1)$, which is the number of pairs, and all pairs contribute equally:

$$\frac{\langle \text{P.E.} \rangle}{N} = \frac{1}{2}(N - 1) \int d^3r_1\, d^3r_2 V(\mathbf{r}_1 - \mathbf{r}_2) \int d^3r_3\, d^3r_4 \cdots d^3r_N \times \varrho_N(\mathbf{r}_1, \mathbf{r}_2, \ldots, \mathbf{r}_N)$$

The integration over all variables except $\mathbf{r}_1$ and $\mathbf{r}_2$ produces the reduced density matrix $\varrho_2(\mathbf{r}_1, \mathbf{r}_2) = (1/v^2)g(\mathbf{r}_1 - \mathbf{r}_2)$. Then changing to relative coordinates $\mathbf{r} = (\mathbf{r}_1 - \mathbf{r}_2)$ and $\mathbf{R} = \frac{1}{2}(\mathbf{r}_1 + \mathbf{r}_2)$, there is no dependence on $\mathbf{R}$, so its integral yields v. The final answer $(N/v = \varrho)$ is

$$\frac{\langle \text{P.E.} \rangle}{N} = \frac{1}{2} \varrho \int d^3r V(\mathbf{r}) g(\mathbf{r})$$

The average potential energy is obtained from the pair distribution function. This result is obvious for a system with pairwise interaction between particles. It does not depend on CBFs but applies to any liquid and is an exact identity. The CBFs are used to calculate $g(r)$, which is then used in the evaluation of the average potential energy per particle. The factor of $\frac{1}{2}$ occurs because each pair interaction is shared between two atoms.

The average kinetic energy per particle is obtained from the fundamental definition

$$\frac{\langle \text{K.E.} \rangle}{N} = \frac{\hbar^2}{2mN} \sum_j \int d^3r_1 \cdots d^3r_N \, |\, \boldsymbol{\nabla}_j \Psi_0(\mathbf{r}_1, \ldots, \mathbf{r}_N) \,|^2$$

The result for each term j in the summation is identical, so the sum gives N times the result for one term:

$$\frac{\langle \text{K.E.} \rangle}{N} = \frac{\hbar^2}{2m} \int d^3r_1 \cdots d^3r_N \, |\, \boldsymbol{\nabla}_1 \Psi_0(\mathbf{r}_1, \ldots, \mathbf{r}_N) \,|^2$$

The evaluation of this quantity depends sensitively on the use of CBFs. The gradiant of the wave function produces the gradiant acting upon $u(r)$:

$$\boldsymbol{\nabla}_1 \Psi_0 = L_N \boldsymbol{\nabla}_1 \exp\left[-\sum_{i>j} u(\mathbf{r}_i - \mathbf{r}_j) \right] = -\sum_{j=2}^{N} [\nabla_1 u(\mathbf{r}_1 - \mathbf{r}_j)] \Psi_0$$

$$\frac{\langle \text{K.E.} \rangle}{N} = \frac{\hbar^2}{2m} \sum_{j,m=2}^{N} \int d^3r_1 \cdots d^3r_N \boldsymbol{\nabla}_1 u(\mathbf{r}_1 - \mathbf{r}_j) \cdot \boldsymbol{\nabla}_1 u(\mathbf{r}_1 - \mathbf{r}_m)$$
$$\times \varrho_N(\mathbf{r}_1, \ldots, \mathbf{r}_N)$$

At first the evaluation of this quantity appears to involve three-particle correlations. One can integrate all but the three coordinates $\mathbf{r}_1, \mathbf{r}_j, \mathbf{r}_m$, which produces the reduced density matrix $\varrho_3(\mathbf{r}_1, \mathbf{r}_j, \mathbf{r}_m)$. It would be difficult to evaluate ϱ_3 accurately, since our knowledge of three-particle correlations is imperfect. However, McMillan (1965) showed a method of avoiding this problem and expressing the result in terms of only two-particle correlations. His method is to write one derivative as acting upon the density matrix itself:

$$\sum_m \boldsymbol{\nabla}_1 u(\mathbf{r}_1 - \mathbf{r}_m) \varrho_N = -\tfrac{1}{2} \boldsymbol{\nabla}_1 \varrho_N(\mathbf{r}_1, \ldots, \mathbf{r}_N)$$

$$\frac{\langle \text{K.E.} \rangle}{N} = -\frac{\hbar^2}{4m} \sum_j \int d^3r_1 \cdots d^3r_N \boldsymbol{\nabla}_1 u(\mathbf{r}_1 - \mathbf{r}_j) \cdot \boldsymbol{\nabla}_1$$
$$\times \varrho_N(\mathbf{r}_1, \mathbf{r}_2, \ldots, \mathbf{r}_N)$$

Now we can integrate over all coordinates except $\mathbf{r}_1$ and $\mathbf{r}_j$ and eliminate the summation over j since each term contributes identically:

$$\frac{\langle \text{K.E.} \rangle}{N} = -\frac{\hbar^2}{4m}(N-1)\int d^3r_1\, d^3r_2 \nabla_1 u(\mathbf{r}_1 - \mathbf{r}_2) \cdot \nabla_1 \varrho_2(\mathbf{r}_1, \mathbf{r}_2)$$

Again changing integration variables to relative coordinates, we obtain the final form of the result:

$$\frac{\langle \text{K.E.} \rangle}{N} = -\frac{\hbar^2}{4m}\varrho \int d^3r \frac{\partial u(r)}{\partial r}\frac{\partial g(r)}{\partial r}$$

The average kinetic energy is expressed as an integral over the derivative of $u(r)$ and $g(r)$. Neither quantity has an angular dependence, so the gradient acts only upon the radial variable. This form is used for the evaluation of the average kinetic energy per particle. The numerical results obtained by different investigators vary somewhat, depending on the details. Generally the results show that the average potential energy per particle has the magnitude of energies 20–22 K, while the average kinetic energy per particle is in the range 14–15 K, and the net binding energy per particle is 6–7 K. These depend on the particle density.

So far the form of $u(r)$ in CBFs has not been specified. That will be done now. Usually the contribution of $u(r)$ is divided into a short-range component and a long-range component. The short-range component is the most straightforward and will be discussed first. One expects that the atoms do not interpenetrate. There must be a term in $u(r)$ which keeps them apart, which is done by having a term in $u(r)$ which becomes very large as $r \to 0$, which makes $\exp(-u) \to 0$. The form of this term can be deduced from Schrödinger's equation and the repulsive part of the interparticle potential. Let us begin by taking a two-particle system, so that $u(r)$ describes only the effects of correlation between the motion of two particles. In center-of-mass coordinates, the relative motion of the two helium atoms is described by the following CBF wave functions:

$$\left\{-\frac{\hbar^2}{2\mu}\left[\frac{d^2}{dr^2} + \frac{2}{r}\frac{d}{dr} - \frac{l(l+1)}{r^2}\right] + V(r)\right\}e^{-u(r)} = Ee^{-u(r)}$$

$$\left\{-\frac{\hbar^2}{2\mu}\left[-\frac{d^2u}{dr^2} + \left(\frac{du}{dr}\right)^2 - \frac{2}{r}\frac{du}{dr} - \frac{l(l+1)}{r^2}\right] + V - E\right\}e^{-u(r)} = 0$$

Now if $u(r)$ is diverging sharply as $r \to 0$, the dominant term among all the derivatives is $(du/dr)^2$. For example, if $u \propto r^{-n}$, then $(du/dr)^2 \propto r^{-2n-2}$, while the first derivatives diverge only as r^{-n-1}. Thus the two most divergent terms must cancel, where one is from the kinetic energy and the other from the potential energy:

$$\frac{du}{dr} = \left[\frac{2\mu}{\hbar^2} V(r)\right]^{1/2}$$

$$u(r) = -\int_r dr' \left[\frac{2\mu}{\hbar^2} V(r')\right]^{1/2}$$

Students of physics should recognize this result and derivation. It is just the Wentzel–Kramers–Brillouin–Jeffreys (WKBJ) wave function for a particle penetrating into a repulsive potential: as a specific example, in the Lennard-Jones potential the repulsive part is C_{12}/r^{12}, where $C_{12} = 4\varepsilon\sigma^{12}$. Thus we derive that the short-range part of the CBF must be

$$u(r) = -\int_r dr' \left(\frac{8\mu\varepsilon}{\hbar^2} \frac{\sigma^{12}}{r^{12}}\right)^{1/2} = \left(\frac{8\mu\varepsilon}{\hbar^2}\right)^{1/2} \frac{\sigma^6}{5r^5}$$

$$= \left(\frac{a}{r}\right)^5 \tag{10.2.5}$$

$$a = \sigma\left(\frac{8\mu\varepsilon\sigma^2}{25\hbar^2}\right)^{1/10}$$

For the potential between two ^{4}He atoms, the numerical result is $a = 1.01\sigma$, so a is nearly identical to the hard sphere diameter. McMillan (1965) used the Lennard-Jones potential for his variational calculation and included no long-range potential. He tried correlations of the form $u(r) = (a/r)^n$, where a and n were both variational parameters. He used $\sigma = 2.556$ so $a = 1.016 = 2.59$ is our prediction. His minimum energy was found with $n = 5$ and $a = 2.61$ Å. These values are nearly identical to the results we just deduced from the Schrödinger equation between two particles. Thus the short-range part of the CBF can be obtained from the WKBJ wave function for a particle penetrating into a repulsive potential. This part of $u(r)$ is determined once the potential is specified.

The long-range part of $u(r)$ is more subtle. There is a term which falls off as $u(r) \to r^{-2}$ as $r \to \infty$. This contribution arises from the zero-point motion of the phonons. We shall discuss the excitation spectrum of ^{4}He in the next section. For the moment, we only need to know that the liquid has ordinary sound waves which have a linear dispersion relation $\omega_k = ck$ at long wavelength. These excitations are quantized like phonons and have a zero-point motion which is part of the ground state properties. Bogoliubov and Zubarev (1955) showed, in the pairing theory, that this leads to long-range correlations between the motion of particles in the liquid. Reatto and Chester (1967) showed that these fluctuations lead to long-range contributions to $u(r)$.

The first step in the derivation is to start with the harmonic oscillator equation, which we take for $m = 1$:

$$H_{\text{HO}} = \tfrac{1}{2} \sum_{\mathbf{q}} (\mathbf{P_q} \cdot \mathbf{P_{-q}} + \omega_q{}^2 \mathbf{Q_q} \cdot \mathbf{Q_{-q}})$$
$$P_{\mathbf{q}} = \frac{\hbar}{i} \frac{\partial}{\partial Q_{\mathbf{q}}} \tag{10.2.6}$$

The zero-point motion is described by the ground state of the harmonic oscillator which has $n_q = 0$. This wave function has the form $\phi_{\mathbf{q}} \propto \exp(-\tfrac{1}{2}\omega_q Q_q{}^2)$ for each q state, and the ground state of the phonon system is

$$\Psi_p = \prod_{\mathbf{q}} \phi_{\mathbf{q}} = \exp\left(-\frac{1}{2} \sum_{\mathbf{q}} \omega_{\mathbf{q}} Q_{\mathbf{q}}{}^* Q_{-\mathbf{q}}\right)$$

We have neglected the normalization coefficient as being unimportant to the factors in the exponent. This equation is the basic form we want for the ground state wave function. We restrict the discussion to the linear dispersion regime with $\omega_q = cq$. We take $Q_{\mathbf{q}} \propto \sum_i Q_i \exp(i\mathbf{q} \cdot \mathbf{r}_i)$ but need to determine the other factors. To this end, we consider the potential energy term in the original Hamiltonian, which we write in terms of density operators:

$$\text{P.E.} = \frac{1}{2} \sum_{ij} V(\mathbf{r}_i - \mathbf{r}_j) = \frac{1}{2} \int \frac{d^3q}{(2\pi)^3} V(\mathbf{q}) \sum_{ij} e^{i\mathbf{q} \cdot (\mathbf{r}_i - \mathbf{r}_j)}$$
$$= \frac{1}{2\nu} \sum_{\mathbf{q}} V(\mathbf{q}) \varrho(\mathbf{q}) \varrho(-\mathbf{q})$$
$$\varrho(\mathbf{q}) = \sum_i e^{i\mathbf{q} \cdot \mathbf{r}_j}$$

At long wavelength, the harmonic oscillator Hamiltonian (10.2.6) should be derivable from the original Hamiltonian. The potential energy term of the harmonic oscillator comes from the potential energy term in the Hamiltonian. Thus we identify the term $\tfrac{1}{2} \sum \omega_{\mathbf{q}} Q_{\mathbf{q}} Q_{-\mathbf{q}}$ as arising from $(1/2\nu) \sum V(\mathbf{q}) \times \varrho(\mathbf{q})\varrho(-\mathbf{q})$. Again it is necessary to find $V(\mathbf{q})$ as $\mathbf{q} \to 0$. Reatto and Chester argue that this still goes to the pairing result of $V(\mathbf{q}) \to mc^2/\varrho = (m\nu/N)c^2$ and for the harmonic oscillator term in the potential energy find

$$\frac{1}{2}\,\omega_{\mathbf{q}}{}^2 Q_{\mathbf{q}} Q_{-\mathbf{q}} = \frac{mc^2}{2N}\,\varrho(\mathbf{q})\varrho(-\mathbf{q})$$

$$Q_{\mathbf{q}} = \frac{1}{q}\left(\frac{m}{N}\right)^{1/2}\varrho(\mathbf{q})$$

The latter identification now enables us to express the zero-point energy as a correlation between particles. Thus we have

$$\begin{aligned}\Psi_p &= \exp\left[-\frac{1}{2}\sum_{\mathbf{q}}\left(\frac{cm}{N}\,\frac{1}{q}\right)\sum_{ij} e^{i\mathbf{q}\cdot(\mathbf{r}_i-\mathbf{r}_j)}\right]\\ &= \exp\left[-\sum_{i>j}\chi(\mathbf{r}_i-\mathbf{r}_j)\right]\end{aligned}$$

$$\chi(\mathbf{r}) = \frac{cm}{N}\sum_{\mathbf{q}}\frac{1}{q}\,e^{i\mathbf{q}\cdot\mathbf{r}}$$

The summation over $\mathbf{q}$ states extends only over those states which are well described as sound waves. The cutoff is somewhere around $q_c \simeq 0.5$ Å^{-1} as a rough estimate. Reatto and Chester recommended a gradual cutoff procedure and inserted the factor $\exp(-q/q_c)$ into the integrand. The resulting integral is easy to do:

$$\begin{aligned}\chi(r) &= \frac{mc}{\varrho}\int\frac{d^3q}{(2\pi)^3}\,\frac{e^{-q/q_c}}{q}\,e^{i\mathbf{q}\cdot\mathbf{r}}\\ &= \frac{mc}{2\pi^2\varrho}\,\frac{1}{r^2+q_c{}^{-2}}\end{aligned}$$

Thus the function $\chi(r)$ goes to a constant as $r\to 0$ and only dies away proportional to r^{-2} at large distances; $\chi \propto r^{-2}$ is a rather slow falloff with distance and is the long-range behavior alluded to previously.

Thus a suitable starting point for a CBF calculation would be a $u(r)$ which contained both the short- and long-range terms. Most workers just add these two contributions, $u(r) = (a/r)^5 + \chi(r)$. This was done by Francis *et al.* (1970) in the numerical study referred to previously. The inclusion of the long-range correlations will change $R(r)$ at large $\mathbf{r}$ and $n_{\mathbf{k}}$ at small $\mathbf{k}$. In particular, the distribution $n_{\mathbf{k}}$ diverges as $k\to 0$ as a result of the long-range correlations. The divergence is shown in Fig. 10.4 where the results for $n_{\mathbf{k}}$ of Francis *et al.* are shown for both the short-range correlations only (SR) and the short plus long-range correlations (LR). Their short-range result is nearly identical to that of McMillan. This agreement is expected, since they used the same interparticle potential but a different method of obtaining results, i.e., the Percus–Yevick equations for mixtures.

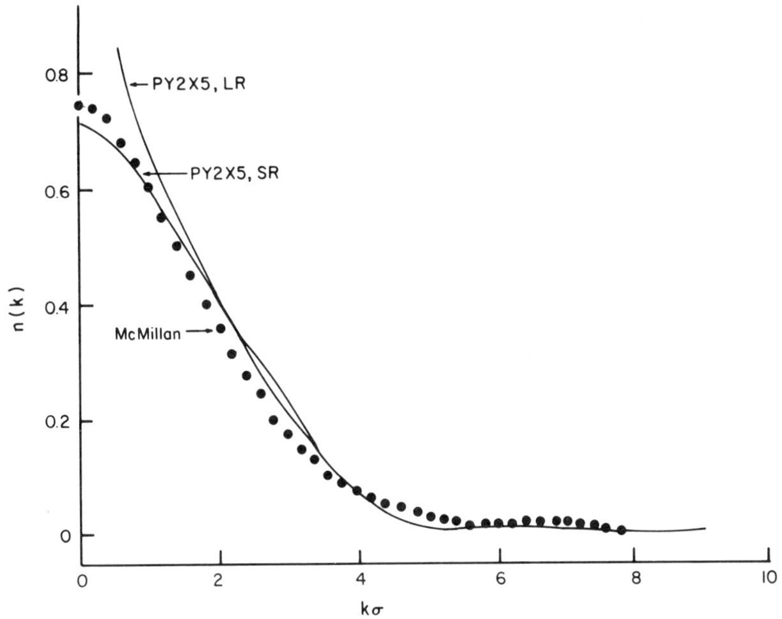

FIGURE 10.4. Theoretical calculations of the momentum distribution of atoms in liquid ^{4}He. The points are due to McMillan (1965), and the solid line is from Francis *et al.*, who used the Percus–Yevick equation both with and without long-range order. *Source*: Francis *et al.* (1970) (used with permission).

Thus the long-range correlations have a dramatic effect upon the theoretical values of $n_{\mathbf{k}}$. This low k divergence is found in the experiments, as discussed by Griffin (1985), and by Svensson and Sears (1986).

Amid all these equations, it is worthwhile to gain a physical picture of the particle motions on the microscopic level. The average potential energy of -20 to -22 K per atom can be achieved by having four neighbors at the minimum well depth of 10 K, so the average potential energy per particle is -20 K. However, an inspection of $g(r)$ shows that it peaks at larger values of r, and the neighboring shell of atoms is farther away than the minimum well depth at 3 Å. Thus the potential energy comes from having about eight neighbors each with an average interaction energy of -5 K, which gives the per particle average of -20 K. In a classical fluid such as argon, $g(r)$ peaks at the maximum well depth, so the first neighbor atoms are sitting at the distance of the maximum potential energy. Helium does not behave this way, because the quantum nature of the motion prevents that much localization of the particles, so the first neighbors are farther away on the average.

The average kinetic energy is the more interesting number. The estimates of this quantity range from 10 to 15 K. What kinds of particle motion give this large number? There are two microscopic pictures which come to mind. The first has the particle motion as part of the zero-point motion of the long-wavelength phonons. Here the particles keep their nearest neighbor positions rather fixed, but the system executes long-range fluctuations. This is a Jell-O model, where the system wiggles like an elastic medium. The amount of energy in this zero-point motion can be estimated as

$$\begin{aligned}(\text{Z.P.E.})_{\text{ph}} &= \frac{1}{2N}\sum_{q<q_c}\omega_q \\ &= \frac{c}{2\varrho}\int_{q<q_c}\frac{d^3q}{(2\pi)^3}\,q \\ &= \frac{\hbar c q_c^4}{16\pi^2\varrho}\end{aligned}$$

Here c is the speed of sound and ϱ is the particle density. The cutoff wave vector q_c is the maximum value of the excitation spectrum, which is considered to be a sound wave, i.e., over which $\omega_q = cq$ is an accurate approximation. Francis *et al.* estimate it to be $q_c = 0.5\ \text{Å}^{-1}$. With this value and the other known parameters for ^{4}He, one can estimate that $(\text{Z.P.E.})_{\text{ph}} = 0.33$ K. The result 0.3 K is a negligible part of the total kinetic energy per particle. This low estimate is confirmed by the calculations of Francis *et al.*, who found that the long-range correlations (and fluctuations) changed the average energy per particle only by 0.5 K.

Thus the average kinetic energy per particle does not come from the long-range fluctuations. Instead, it seems to come from the short-range motion. The first peak in $g(r)$ is at 4 Å, so that the atom sits in its own space in the liquid, which is roughly a sphere with a radius of about 2 Å. A particle in a spherical box of radius a has a kinetic energy of $\hbar^2\pi^2/2ma^2$. Using $a = 2$ and m_{He} yields a kinetic energy of 15 K, which is just the right magnitude. Thus the kinetic energy comes from the short-range fluctuations of the particle bouncing around inside its own small space in the liquid. In the ground state of the liquid, at very low temperatures, the individual particles are moving rapidly with this motion.

It is also useful to have a simple approximation for $n_{\mathbf{k}}$ and its Fourier transform $R(r)$. Puff and Tenn (1970) observed that neutron scattering experiments, which will be described later, showed that the particle distribution was Gaussian. They suggested the form

$$n_{\mathbf{k}} = N_0 \delta_{\mathbf{k}=0} + (1 - f_0)\varrho\left(\frac{2\pi}{mw}\right)^{3/2} e^{-\varepsilon_k/w} \tag{10.2.7}$$

One advantage of the Gaussian distribution is that one can calculate quantities easily. For example, the average kinetic energy per particle is simply

$$\frac{\langle \text{K.E.} \rangle}{N} = (1 - f_0) \int \frac{d^3k}{(2\pi)^3} \varepsilon_k \left(\frac{2\pi}{mw}\right)^{3/2} e^{-\varepsilon_k/w} = \frac{3}{2}(1 - f_0)w$$

From their fits to the neutron scattering data, they deduced that this average kinetic energy is about 15 K. It is their result we have mentioned several times previously in this chapter. This form has a certain appeal. The classical distribution is also Gaussian, except then we identify $w = k_B T$, and the average kinetic energy is $\frac{3}{2}k_B T$. Now we still have a Gaussian distribution, but the width w is much larger than predicted by the temperature; one estimates that $w = 11$ K in the limit where $T \to 0$.

The off-diagonal density matrix $R(r)$ is the Fourier transform of $n_{\mathbf{k}}$. When $n_{\mathbf{k}}$ is Gaussian, so is $R(r)$:

$$R(r) = \frac{1}{\varrho} \int \frac{d^3k}{(2\pi)^3} e^{-i\mathbf{k}\cdot\mathbf{r}} n_{\mathbf{k}} = f_0 + (1 - f_0)e^{-r^2/b^2}$$

$$b^2 = \frac{2\hbar^2}{mw}$$

This Gaussian is the dashed curve in Fig. 10.3, compared with the calculations of McMillan. The fit is reasonable, considering that the parameters were taken from the estimate of Puff and Tenn that $w = 11$ K, which gives $b = 1.48$ Å.

It is likely that neither $n_{\mathbf{k}}$ (for $\mathbf{k} \neq 0$) nor $R(r)$ are exact Gaussians in the superfluid state and that the fit to this functional form is only an approximation. Nevertheless, it is useful to have approximate analytical forms, and the Gaussian fits best. Puff and Tenn, in fitting the neutron data to a Gaussian plus a delta function at $\mathbf{k} = 0$, estimated that $f_0 = 0.06$ at $T = 1.27$ K. Their estimate is in agreement with later measurements and extrapolates to a value of $f_0 = 10.8\%$ in the limit where $T \to 0$ if one uses the Bose–Einstein extrapolation $f_0(0)[1 - (T/T_c)^{3/2}]$.

C. Experiments on $n_{\mathbf{k}}$

Feynman first suggested that neutron scattering would be an excellent means of investigating the structure of liquid ^{4}He. His remarks were aimed at a measurement of the excitation spectrum. This was very successful and will be described in the next section. However, in the past two decades another series of neutron scattering experiments have been performed on liquid ^{4}He in order to measure the momentum distribution of particles $n_{\mathbf{k}}$. This experiment involves using higher-energy neutrons and analyzing the data in a fashion different from that used to find the excitation energy. The primary objective of this effort was a direct measurement of the condensate fraction f_0.

Hohenberg and Platzman (1966) made the suggestion that inelastic scattering by very energetic neutrons (say 0.1 eV) would provide a measurement of $n_{\mathbf{k}}$ and hence the condensate fraction f_0, which would show up as a delta function on the $n_{\mathbf{k}}$ distribution. Of course, the delta function would be broadened by experimental resolution but should still be apparent if the estimates of $f_0 = 0.1$ were correct. The reason for using very energetic neutrons is that the *sudden approximation* becomes valid when the neutron energy is much higher than the kinetic or potential energy of the ^{4}He atom. And in the sudden approximation, one has a direct measurement of $n_{\mathbf{k}}$.

In the neutron scattering experiment, the neutrons with initial wave vector $\mathbf{k}_i$ and initial energy $E_i = k_i^2/2m$ are directed toward the scattering chamber. Some neutrons are scattered and leave the sample with a final wave vector $\mathbf{k}_f = \mathbf{k}_i - \mathbf{Q}$, where $\mathbf{Q}$ is transferred to the liquid. Similarly, the final energy is $E_f = (\mathbf{k}_i - \mathbf{Q})^2/2m = E_i - \omega$, where ω is the energy transferred to the liquid. The scattering cross section for ^{4}He can be expressed in terms of the dynamic liquid structure factor $S(Q, \omega)$ as

$$\frac{d^2\sigma}{d\Omega\, d\omega} = \frac{k_f}{k_i}\sigma_0 S(Q, \omega) \tag{10.2.8}$$

where σ_0 is a cross section for neutron scattering from a single alpha particle. Equation (10.2.8) is a general result which is always valid for neutron scattering. Generally a calculation of $S(Q, \omega)$ is quite complicated. However, in the sudden approximation, it can be approximated by the expression

$$S(Q, \omega) = \frac{1}{\varrho}\int \frac{d^3k}{(2\pi)^3} n_{\mathbf{k}} \times 2\pi\left[\delta\left(\omega - \frac{Q^2}{2m} - \frac{\mathbf{Q}\cdot\mathbf{k}}{m}\right)\right.$$
$$\left. - \delta\left(\omega + \frac{Q^2}{2m} + \frac{\mathbf{Q}\cdot\mathbf{k}}{m}\right)\right] \tag{10.2.9}$$

where $n_{\mathbf{k}}$ is the momentum distribution of particles in the ground state of the liquid. Our outline will be to discuss first the implications of this result; later we shall sketch its derivation.

First the momentum distribution $n_{\mathbf{k}}$ is written as the condensate term $N_0\delta_{\mathbf{k}=0}$ plus the other term which is called $\bar{n}_k$. These two are used in (10.2.9), and the wave vector integrals are done as far as possible ($\varepsilon_Q = Q^2/2m$):

$$n_{\mathbf{k}} = N_0\delta_{\mathbf{k}=0} + \bar{n}_{\mathbf{k}}$$

$$S(Q, \omega) = \bar{S}(Q, \omega) - \bar{S}(Q, -\omega)$$

$$\begin{aligned}\bar{S}(Q, \omega) &= 2\pi f_0\delta(\omega - \varepsilon_Q) + \frac{2\pi}{\varrho}\int \frac{d^3k}{(2\pi)^3}\,\bar{n}_{\mathbf{k}}\,\delta\left(\omega - \frac{Q^2}{2m} - \frac{\mathbf{Q}\cdot\mathbf{k}}{m}\right)\\ &= 2\pi f_0\delta(\omega - \varepsilon_Q) + \frac{1}{2\pi\varrho}\int_0^\infty k^2\,dk\bar{n}_k\int_{-1}^{1} d\nu\,\delta\left(\omega - \varepsilon_Q - \frac{Qk\nu}{m}\right)\end{aligned}$$

The angular integral $d\nu = d(\cos\theta)$ over the delta function equals m/kQ if $k > (m/Q)\mid\omega - \varepsilon_Q\mid$, so the final result is ($\omega > 0$)

$$\bar{S}(Q, \omega) = 2\pi f_0\delta(\omega - \varepsilon_Q) + \frac{m}{2\pi Q\varrho}\int_{(m/Q)|\omega-\varepsilon_Q|}^{\infty} k\,dk\bar{n}_{\mathbf{k}} \qquad (10.2.10)$$

Equation (10.2.10) predicts that $S(Q, \omega)$ for $\omega > 0$ has a delta function contribution at the energy loss $\omega = \varepsilon_Q = Q^2/2m$. The delta function corresponds to the neutron knocking particles out of the condensate, and the fraction of the spectral strength in this process is f_0 Of course, this delta function is broadened by the finite resolution of the measuring apparatus but should still be observable. The second term in (10.2.10) gives the scattering from the particles not in the condensate. This term is left as an integral. However, $\bar{n}_k$ may be obtained from data with small numerical scatter by taking the following derivative ($\omega \neq \varepsilon_Q$):

$$\frac{\partial}{\partial\omega}S(Q, \omega) = \frac{m^3}{2\pi Q^3}(\omega - \varepsilon_Q)\bar{n}_{(m/Q)|\omega-\varepsilon_Q|} + \frac{m^3}{2\pi Q^3}(\omega + \varepsilon_Q)\bar{n}_{(m/Q)(\omega+\varepsilon_Q)}$$

This equation provides a direct method of measuring $\bar{n}_k$. In fact, as pointed out by Puff and Tenn, the Gaussian distribution, which is given in (10.2.7), fits the scattering data well. For this distribution, the integral can be evaluated analytically:

$$\begin{aligned}S(Q, \omega) &= 2\pi f_0[\delta(\omega - \varepsilon_Q) - \delta(\omega + \varepsilon_Q)]\\ &\quad + (1 - f_0)\left(\frac{\pi}{\varepsilon_Q w}\right)^{1/2}\left(e^{-(\omega-\varepsilon_Q)^2/4\varepsilon_Q w} - e^{-(\omega+\varepsilon_Q)^2/4\varepsilon_Q w}\right)\end{aligned} \qquad (10.2.11)$$

Equation (10.2.11) predicts that the spectral function $S(Q, \omega)$ is the difference of two Gaussians, peaked at $\omega = \pm\varepsilon_Q$ and with a width given by $-2(\varepsilon_Q w)^{1/2}$. There is also the delta function at $\omega = \pm\varepsilon_Q$ from the scattering from the condensate. The results appear as a single Gaussian whenever $\varepsilon_Q > \omega$, which is the usual experimental case.

There have been several measurements of n_k, including those of Harling (1971) and Rodriquez *et al.* (1974). The most recent report is by Woods and Sears (1977), whose data is shown in Fig. 10.5. The solid line in Fig. 10.5(a) is a fit to their points taken at $T = 4.2$ K when the liquid ^{4}He is not superfluid. This line is a remarkably good fit to a Gaussian. The other set of points are taken at $T = 1.1$ K in the superfluid state. There is an obvious increase in the distribution $n_{\mathbf{k}}$ at small values of k, as shown by the difference spectra in Fig. 10.5(b). The increase at $k \simeq 0$ has the width of their resolution function and is interpreted as being caused by the condensate fraction f_0. They estimate $f_0 = 6.9\% \pm 0.8\%$ from this data. If this value of f_0 is extrapolated to zero temperature by using the Bose–Einstein formula, one finds that $f_0(T = 0) = 10.8\% \pm 1.3\%$. The value $f_0 = 0.11$ is in good agreement with the theoretical estimates, with the earlier results of Harling, and with the analysis of Puff and Tenn. However, lower condensate values were obtained by Rodriquez *et al.*, and so the agreement between theory and experiment should be regarded as preliminary. There is also little justification for using the Bose–Einstein formula for extrapolating $f_0(T)$ to zero temperature except expediency, since no other formula is available. It also appears from the data of Woods and Sears that the values of $n_{\mathbf{k}}$ in the superfluid state are not as good a fit to a Gaussian as they were in the normal fluid. This feature is not understood.

Now we shall derive formula (10.2.9) for the sudden approximation. The first step begins at the definition of the spectral function $S(\mathbf{Q}, \omega)$. As in the case of the electron gas in Chapter 6, this is the spectral function of the density–density correlation function. If this correlation function is called $\chi(\mathbf{Q}, i\omega)$ in the Matsubara representation, then $S(\mathbf{Q}, \omega)$ is -2 times the imaginary part of its retarded function:

$$\chi(\mathbf{Q}, i\omega) = -\frac{1}{N}\int_0^\beta d\tau e^{i\omega\tau}\langle T_\tau \varrho(\mathbf{Q}, \tau)\varrho(-\mathbf{Q}, 0)\rangle$$

$$S(\mathbf{Q}, \omega) = -2\,\mathrm{Im}[\chi_{\mathrm{ret}}(\mathbf{Q}, \omega)] \tag{10.2.12}$$

According to the result of Problem 16 in Chapter 3, $S(\mathbf{Q}, \omega)$ can also be

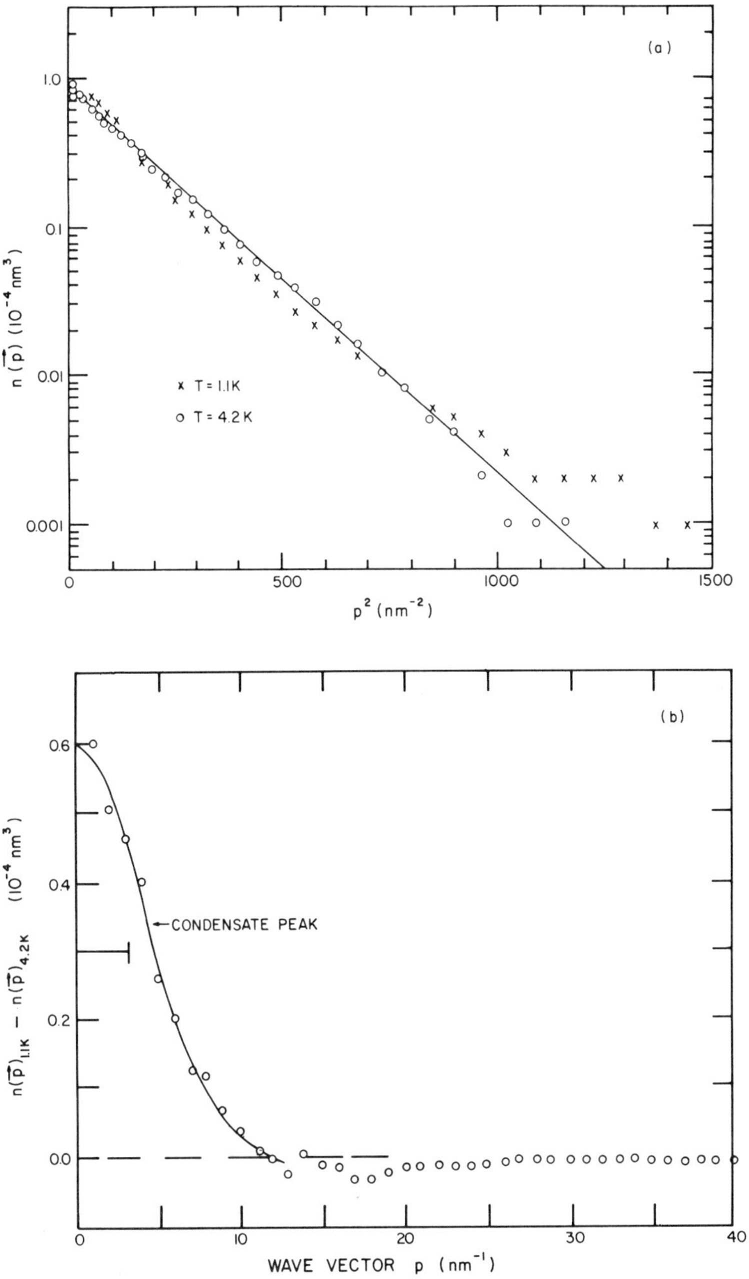

FIGURE 10.5. (a) Neutron scattering results for the momentum distribution in liquid ^{4}He at $T = 1.1$ and 4.2 K as a function of p^2. (b) The difference between the two distributions in part (a) shows a peak at small wave vector whose width equals the experimental resolution. The peak is interpreted as the condensate fraction. *Source*: Woods and Sears (1977) (used with permission).

written as

$$S(\mathbf{Q}, \omega) = \frac{1}{N}(1 - e^{-\beta\omega}) \int_{-\infty}^{\infty} dt e^{i\omega t} \langle \varrho(\mathbf{Q}, t)\varrho(-\mathbf{Q}, 0)\rangle \qquad (10.2.13)$$

The experiments are done with conditions of temperature and energy transfer ω such that $\omega\beta \gg 1$, and the second term in the brackets can be neglected. The space representation is used for the density operator $\varrho(\mathbf{Q})$, which brings us to the expression

$$\varrho(\mathbf{Q}) = \sum_j e^{i\mathbf{Q}\cdot\mathbf{R}_j}$$

$$S(\mathbf{Q}, \omega) = \frac{1}{N}\sum_{i,j} \int_{-\infty}^{\infty} dt e^{i\omega t} \langle e^{i\mathbf{Q}\cdot\mathbf{R}_j(t)} e^{-i\mathbf{Q}\cdot\mathbf{R}_i(0)}\rangle$$

This is evaluated by determining how the particle motion $\mathbf{R}_j(t)$ proceeds in time. The two exponents $\mathbf{Q}\cdot\mathbf{R}_i$ and $\mathbf{Q}\cdot\mathbf{R}_j(t)$ cannot be combined, since this is permissable only when they commute, and they do not at different times. This point will be shown later.

The sudden approximation is obtained by solving the correlation function for small values of time t by expanding $\mathbf{R}_j(t)$ about point $t = 0$. The physical argument is that the scattering by energetic neutrons happens rapidly, so that only the short time response of the systems is applicable. Near $t = 0$, the coefficients in the Taylor series are found from the commutation relations:

$$\mathbf{R}_j(t) = \mathbf{R}_j(0) + t\left(\frac{\partial \mathbf{R}_j}{\partial t}\right)_0 + \frac{t^2}{2}\left(\frac{\partial^2 \mathbf{R}_j}{\partial t^2}\right)_0 + O(t^3)$$

$$\frac{\partial \mathbf{R}_j}{\partial t} = i[H, \mathbf{R}_j] = \frac{\mathbf{P}_j}{m}$$

$$\frac{\partial^2 \mathbf{R}_j}{\partial t^2} = \frac{i}{m}[H, \mathbf{P}_j] = -\frac{1}{m}\sum_n \nabla_j V(\mathbf{R}_j - \mathbf{R}_n) \equiv \frac{1}{m}\mathbf{F}_j$$

The next step is to separate the several terms in the exponent into the product of several exponents. We use the Feynman theorem, $\exp(A + B) = \exp A \exp B \exp C$ with $B = i\mathbf{Q}\cdot\mathbf{R}_j$, $A = i(\mathbf{Q}/m)(t\mathbf{P}_j + \frac{1}{2}t^2\mathbf{F}_j)$, and $C = -\frac{1}{2}[A, B] = -it\varepsilon_Q$. The next step is to separate the factor $it\mathbf{Q}\cdot\mathbf{P}_j/m$ from $it^2\mathbf{Q}\cdot\mathbf{F}_j/(2m)$. They do not commute, but such corrections terms contribute on the order $O(t^3)$, which we are neglecting. Thus we can write the short time response as

$$S(\mathbf{Q}, \omega) = \frac{1}{N}\sum_{ij} \int_{-\infty}^{\infty} dt e^{it(\omega - \varepsilon_Q)} \langle e^{it\mathbf{Q}\cdot\mathbf{P}_j/m} e^{i(t^2/2m)\mathbf{Q}\cdot\mathbf{F}_j} e^{i\mathbf{Q}\cdot\mathbf{R}_j} e^{-i\mathbf{Q}\cdot\mathbf{R}_i}\rangle$$

The sudden approximation is achieved by making two important approximations to this expression. The first is to take only the term in the series with $i = j$. This choice is a valid approximation when Q is large, since then the terms with $i \neq j$ average to a small contribution. The second approximation is to neglect the term $it^2\mathbf{Q} \cdot \mathbf{F}_j/(2m)$. The quantity $\mathbf{F}_j = \sum_n \nabla_j \times \mathbf{V}(\mathbf{R}_j - \mathbf{R}_n)$ is the force on the particle at $\mathbf{R}_j$, and this must average to zero. If we evaluate this force by a cumulant expansion, for example, the average $\langle \mathbf{F}_j \cdot \mathbf{Q} \rangle$ is zero, although the average $\langle \mathbf{Q} \cdot \mathbf{F}_j\, \mathbf{Q} \cdot \mathbf{F}_j \rangle$ is not. Thus this term really contributes a term in the time expansion on the order of $O(t^4)$ rather than $O(t^2)$. This force term is neglected. It is important to appreciate that we do not neglect this term on the basis that it is of order $O(t^2)$, since the only term we keep is on the order of $O(t^2)$ and all terms should be kept to this order. However, the force term is neglected because its contribution is a higher order in t.

In the sudden approximation of high Q, high ω, we are left with the following expression to evaluate:

$$S(\mathbf{Q}, \omega) = \frac{1}{N} \sum_j \int_{-\infty}^{\infty} dt e^{it(\omega - \varepsilon_Q)} \langle e^{it\mathbf{Q} \cdot \mathbf{P}_j/m} \rangle$$

$$\langle e^{it\mathbf{Q} \cdot \mathbf{P}_j/m} \rangle = \int d^3R_1 \cdots d^3R_N \Psi_0^*(\mathbf{R}_1 \cdots \mathbf{R}_N) e^{it\mathbf{Q} \cdot \mathbf{P}_j/m} \Psi_0(\mathbf{R}_1 \cdots \mathbf{R}_N)$$

The quantity $\exp(i\mathbf{Q} \cdot \mathbf{P}_j/m)$ operates on the wave function to the right and displaces the position variable $\mathbf{R}_j$ by the increment tQ/m:

$$e^{it\mathbf{Q} \cdot \mathbf{P}_j/m} \Psi_0(\mathbf{R}_1, \mathbf{R}_2, \ldots, \mathbf{R}_N) = \Psi_0\left(\mathbf{R}_1, \mathbf{R}_2, \ldots, \mathbf{R}_j + \frac{t\mathbf{Q}}{m}, \ldots, \mathbf{R}_N\right)$$

The next step is to do all the position integrals except $\mathbf{R}_j$. This just produces the off-diagonal density matrix $\tilde{\varrho}_1(\mathbf{R}_j, \mathbf{R}_j')$, where $\mathbf{R}_j' = \mathbf{R}_j + tQ/m$. Since this off-diagonal density matrix is just a function of the difference $\mathbf{R}_j - \mathbf{R}_j'$, we obtain the expression $\tilde{\varrho}_1(\mathbf{R}_j, \mathbf{R}_j') = R(t\mathbf{Q}/m)/v$:

$$S(\mathbf{Q}, \omega) = \frac{1}{N} \sum_j \int_{-\infty}^{\infty} dt e^{it(\omega - \varepsilon_Q)} R\left(\frac{t\mathbf{Q}}{m}\right) \frac{1}{v} \int d^3R_j$$

The d^3R_j integral gives v, and the summation over j gives N:

$$S(\mathbf{Q}, \omega) = \int_{-\infty}^{\infty} dt e^{it(\omega - \varepsilon_Q)} R\left(\frac{t\mathbf{Q}}{m}\right)$$

Earlier it was observed that $R(r)$ was approximately a Gaussian function of r. In this case, the integrand of the time integral is also a Gaussian, and we have retained all the contribution in the time development to order $O(t^2)$ in the exponent. To get to the standard formula for the sudden approximation, we remember that $n_{\mathbf{k}}$ is the Fourier transform of $R(r)$:

$$R(\mathbf{r}) = \frac{1}{\varrho} \int \frac{d^3k}{(2\pi)^3} n_k e^{-i\mathbf{k}\cdot\mathbf{r}}$$

$$S(\mathbf{Q}, \omega) = \frac{1}{\varrho} \int \frac{d^3k}{(2\pi)^3} n_{\mathbf{k}} \int_{-\infty}^{\infty} dt \exp\left[it\left(\omega - \varepsilon_{\mathbf{Q}} - \frac{\mathbf{Q}\cdot\mathbf{k}}{m}\right)\right]$$

$$= \frac{1}{\varrho} \int \frac{d^3k}{(2\pi)^3} n_{\mathbf{k}} \times 2\pi\delta\left(\omega - \varepsilon_{\mathbf{Q}} - \frac{\mathbf{Q}\cdot\mathbf{k}}{m}\right)$$

Thus we have succeeded in deriving the first term in the original formula (10.2.9). The second term with the delta function $\delta(\omega + \varepsilon_{\mathbf{Q}} + \mathbf{Q}\cdot\mathbf{k}/m)$ comes from realizing that $S(\mathbf{Q}, \omega)$ is antisymmetric in ω.

To derive the formula for the sudden approximation, two important assumptions are made. The first is that $\mathbf{Q}$ is large, and the second is that ω is large so that the Fourier transform involves only small values of t. Certainly the experiments are done where $\mathbf{Q}$ is large, so that this situation is well met. However, in the small time limit, we derive that the response of the system is governed by the off-diagonal density matrix $R(\mathbf{r})$ with $\mathbf{r} = t\mathbf{Q}/m$. The condensate fraction is given by $R(\infty)$. According to the numerical results of McMillan, $R(r)$ reaches its asymptotic value about $r_c \simeq 4$ Å, so that values of t must be $t_c \simeq r_c m/Q$. For the impulse approximation to be valid, the force terms must be small on this time scale. An order of magnitude estimate of these terms is

$$\frac{it^2Q}{2m} [\langle(\mathbf{F}_j \cdot \mathbf{F}_j)^2\rangle]^{1/2} = \frac{imr_c^2}{Q} (\langle F_j^2\rangle)^{1/2}$$

Again we find that they are small when $\mathbf{Q}$ becomes large enough. Thus the basic premise of the sudden approximation is that $\mathbf{Q}$ is very large, in which case it is an accurate method of interpreting the experimental results.

The sudden approximation has the appearance of treating the particles in the liquid as being free. We could derive the same result by assuming there was a gas of free particles with a distribution $n_{\mathbf{k}}$. When the neutron strikes a particle of momentum $\mathbf{k}$ and free energy $\varepsilon_{\mathbf{k}}$, it is changed to $\mathbf{k} + \mathbf{Q}$ and energy $\varepsilon_{\mathbf{k}+\mathbf{Q}} = (\mathbf{k} + \mathbf{Q})^2/2m$. Energy conservation demands that the energy transfer be $\omega + \varepsilon_{\mathbf{k}} = \varepsilon_{\mathbf{k}+\mathbf{Q}}$. The probability of having a momentum–energy transfer of $(\mathbf{Q}, \omega)$ is proportional to

$$\int \frac{d^3k}{(2\pi)^3} n_{\mathbf{k}} \times 2\pi\delta\left(\omega - \varepsilon_Q - \frac{\mathbf{k}\cdot\mathbf{Q}}{m}\right)$$

This equation is exactly the form of the sudden approximation. However, in the interacting liquid, we do not really ignore the potential energy. In the sudden approximation the essential physics is that when the neutron strikes the ^{4}He particle, its potential energy does not change immediately. The impulse will alter the particle velocity, and after a time duration the subsequent motion will alter its potential energy. However, by that time the neutron is gone and does not record this change in potential energy. The potential energy depends only on the position of the particles, and when the neutron strikes one particle and redirects its motion, the potential energy only changes later. The fast-moving neutron does not remain around to record this alteration in potential energy, so that the change in energy appears to involve only the kinetic energy component.

10.3. ^{4}He: EXCITATION SPECTRUM

The excitation spectrum of ^{4}He is shown in Fig. 10.6 as determined by neutron scattering by Cowley and Woods (1971). These points are

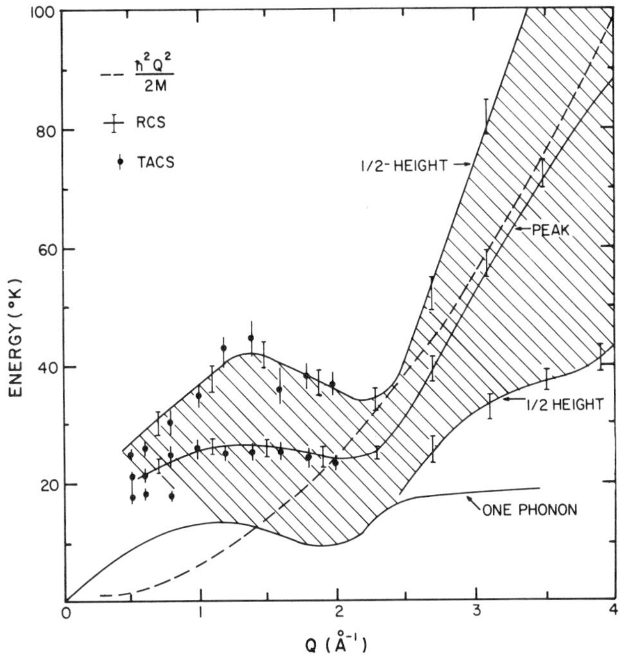

FIGURE 10.6. Neutron scattering results for the excitation spectrum of liquid ^{4}He at $T = 1.1$ K. The lower solid line is the one-phonon curve, while the dashed line is for the noninteracting particle. A broad second peak becomes the single-particle curve at large Q and the two-roton bound state at small Q. *Source*: Cowley and Woods (1971) (used with permission).

found as maxima in $S(k, \omega)$. For some values of k there are two values of ω, which indicates that $S(k, \omega)$ has two peaks. The dashed line shows the free-particle spectrum $\varepsilon_k = k^2/2m$. The excitation spectrum approaches this at large values of k, as shown by Harling (1971). At small values of k, the lower branch of the dispersion curve becomes linear in k. This is the phonon region of sound waves, and the slope is the sound velocity, 240 m/sec. The lower dispersion curve has a second minimum which is called the roton. This point has a high density of states and so has a large weight when averaging over the density of excitation states. Indeed, the necessity of this minimum was deduced by Landau (1941, 1947) on the basis of thermodynamic data. The roton parameters are $k_0 = 1.91$ Å^{-1} and $\Delta = 8.68$ K. There is a two-roton bound state at 2Δ which forms the lower limit of the upper dispersion curve (see Ruvalds and Zawadowski, 1970).

This experimental curve forms the basis for the low-temperature and superfluid properties of ^{4}He. It explains all macroscopic behavior, as we shall show, based on the Landau two-fluid model. Thus once the excitation spectrum is derived, we can explain superfluidity. Deriving the excitation spectrum is our objective in this section.

A. Bijl–Feynman Theory

The successful method of finding the excitation spectrum of liquid ^{4}He was popularized by Feynman (1954). He suggested that the excitation spectrum is given by the formula $\varepsilon_k/S(k)$, where $S(k)$ is the liquid structure factor. This formula is qualitatively right but not quantitatively accurate. However, its importance is that it led to the formalism which does produce the right result. The same formula $\varepsilon_k/S(k)$ was obtained much earlier by Bijl (1940), whose suggestion attracted little notice since neither $S(k)$ nor ω_k were known in those early days. Thus the current usage is to call this the Bijl–Feynman formula:

$$\omega_0(k) = \frac{\varepsilon_k}{S(k)} \tag{10.3.1}$$

We shall now derive this result, which we call $\omega_0(k)$.

The goal is to construct a wave function $\Psi_{\mathbf{k}}(\mathbf{r}_1, \ldots, {}_{\mathbf{k}_N})$ which describes an excitation in the system with a wave vector $\mathbf{k}$. For the boson system of liquid ^{4}He, this excitation wave function must still have the property of being symmetric under the interchange of any two-particle coordinates. For long-wavelength excitations, we expect that the excitation spectrum will still have the same kind of short-range correlations which we found for the ground state wave function. Thus it seems reasonable to assume that the excited state wave function is only a slight perturbation upon the ground state wave function. This is achieved by writing the excited state as the product of the ground state wave function and another symmetric function $\Lambda_{\mathbf{k}}(\mathbf{r}_1, \ldots, \mathbf{r}_N)$ of all the positions:

$$\Psi_{\mathbf{k}}(\mathbf{r}_1, \ldots, \mathbf{r}_N) = \Lambda_{\mathbf{k}}(\mathbf{r}_1, \ldots, \mathbf{r}_N)\Psi_0(\mathbf{r}_1, \ldots, \mathbf{r}_N) \tag{10.3.2}$$

The idea is simple enough. The stipulation on $\Lambda_{\mathbf{k}}(\mathbf{r}_1, \ldots, \mathbf{r}_N)$ is that it is symmetric in the coordinates and contains the wave vector $\mathbf{k}$. Two possible choices are

$$\begin{aligned} \Lambda_{\mathbf{k}}^{(1)}(\mathbf{r}_1, \ldots, \mathbf{r}_N) &= L_{\mathbf{k}} \sum_l e^{i\mathbf{k}\cdot\mathbf{r}_l} = L_{\mathbf{k}}\varrho(\mathbf{k}) \\ \Lambda_{\mathbf{k}}^{(2)}(\mathbf{r}_1, \ldots, \mathbf{r}_N) &= \exp\left(i\mathbf{k}\cdot\sum_l \mathbf{r}_l\right) \end{aligned} \tag{10.3.3}$$

Perhaps the reader can think of others. Each choice of $\Lambda_{\mathbf{k}}$ corresponds to some kind of excited state of the system. We must decide the physics behind each one. It is helpful to use operators such as the total momentum of the system:

$$\mathbf{P} = \frac{\hbar}{i}\sum_j \nabla_j \tag{10.3.4}$$

For example, the total momentum on $\Lambda_{\mathbf{k}}^{(2)}\Psi_0 = \Psi_{\mathbf{k}}^{(2)}$ gives two terms, since P can act upon either $\Lambda_{\mathbf{k}}^{(2)}$ or Ψ_0. However, we know that the average momentum in the ground state must be zero, so this can be evaluated to give

$$\begin{aligned} \langle \Psi_{\mathbf{k}}^{(2)} \mid P \mid \Psi_{\mathbf{k}}^{(2)} \rangle &= \int d^3r_1 \cdots d^3r_N \Psi_0^* \Lambda_{\mathbf{k}}^* [(P\Lambda_{\mathbf{k}})\Psi_0 + \Lambda_{\mathbf{k}} P \Psi_0] \\ &= \sum_j \hbar\mathbf{k} \int d^3r_1 \cdots d^3r_N \Psi_0^* \Lambda_{\mathbf{k}}^* \Lambda_{\mathbf{k}} \Psi_0 = N\hbar\mathbf{k} \end{aligned}$$

since $\Lambda_{\mathbf{k}}^*\Lambda_{\mathbf{k}} = 1$ for $\Lambda_{\mathbf{k}}^{(2)}$. Thus the total momentum of a system of N particles is just $N\mathbf{k}$, which corresponds to the uniform flow of the entire fluid with velocity $\mathbf{k}/m$. This is the correct interpretation of the excited spectrum represented by the choice $\Psi_{\mathbf{k}}^{(2)} = \Lambda_{\mathbf{k}}^{(2)}\Psi_0$. Another check on this interpretation is to find the energy by evaluating the Hamiltonian:

$$E = \frac{1}{N} \langle \Psi_k^{(2)} \mid H \mid \Psi_k^{(2)} \rangle$$

$$= -\frac{1}{2mN} \sum_j \int dr_1 \cdots dr_N \Psi_0^* \Lambda_{\mathbf{k}}^* [(\nabla_j^2 \Lambda_{\mathbf{k}}) \Psi_0 + \nabla_{\mathbf{j}} \Lambda_{\mathbf{k}} \cdot \nabla_{\mathbf{j}} \Psi_0 + \Lambda_{\mathbf{k}} \nabla_j^2 \Psi_0]$$

$$+ \frac{1}{N} \int d^3r_1 \cdots d^3r_N \Psi_0^* \Psi_0 \sum_{ij} V(r_{ij})$$

$$E = \varepsilon_k + \frac{E_0}{N}$$

This gives the average energy per particle as $\varepsilon_k = k^2/2m$ plus the ground state value E_0/N. The cross term $\nabla \Lambda_{\mathbf{k}} \cdot \nabla \Psi_0 = i(\mathbf{k} \cdot \nabla \Psi_0) \Lambda_{\mathbf{k}}$ in the preceding expression averages to zero, since it contains the average momentum $\int \Psi_0 \nabla \Psi_0$ in the ground state of the stationary liquid.

The other excited state wave function $\Psi_0^{(1)} = \Lambda_{\mathbf{k}}^{(1)} \Psi_0$ is actually the one we want for the excitation spectrum of the stationary liquid. The superscript 1 will be dropped, and it will be called just $\Psi_{\mathbf{k}} = \Lambda_{\mathbf{k}} \Psi_0 = L_{\mathbf{k}} \varrho(\mathbf{k}) \Psi_0$. This term will be investigated systematically. The first step is to find the normalization constant $L_{\mathbf{k}}$ for the wave function, which is done by setting the normalization integral to unity:

$$1 = \langle \Psi_{\mathbf{k}} \mid \Psi_{\mathbf{k}} \rangle = L_{\mathbf{k}}^2 \sum_{l,j} \int d^3r_1 \cdots d^3r_N \mid \Psi_0(\mathbf{r}_1 \cdots \mathbf{r}_N) \mid^2 \exp[i\mathbf{k} \cdot (\mathbf{r}_l - \mathbf{r}_j)]$$

The terms in the double summation with $j = l$ just give unity since the wave vector dependence drops out. Thus special care must be taken only whenever $j \neq l$. Then the space integrals can each be evaluated, at least formally, except those with r_j and r_l. These integrals over $d^3r_1 \cdots d^3r_N$ just produce the pair distribution function $\varrho_2(\mathbf{r}_j, \mathbf{r}_l) = \nu^{-2} g(\mathbf{r}_j - \mathbf{r}_l)$:

$$1 = L_{\mathbf{k}}^2 \left[\sum_{\substack{j,l \\ j \neq l}} \frac{1}{\nu^2} \int d^3r_j \, d^3r_l g(\mathbf{r}_j - \mathbf{r}_l) e^{i\mathbf{k} \cdot (\mathbf{r}_l - \mathbf{r}_j)} + \sum_{l=j} 1 \right]$$

The summation for each particle l is the same, since all particles are identical. Thus we eliminate this summation and instead multiply by N:

$$1 = NL_{\mathbf{k}}^2 \left\{ 1 + \sum_{j \neq l} \frac{1}{\nu^2} \int d^3r_j \, d^3r_l g(\mathbf{r}_j - \mathbf{r}_l) \exp[i\mathbf{k} \cdot (\mathbf{r}_l - \mathbf{r}_j)] \right\}$$

Each term in the summation over j gives the same result, so just do one term and multiply by $N - 1 \simeq N$. Also, change integration variables to relative coordinates $d^3r_j \, d^3r_l = \nu \, d^3r$, so the integral is over $g(r)$. In fact, this expression is precisely the definition of $S(\mathbf{k})$ which is given earlier in Sec. 1.6:

$$1 = NL_{\mathbf{k}}^2\left(1 + \varrho \int d^3r g(r) e^{i\mathbf{k}\cdot\mathbf{r}}\right) = NS(\mathbf{k})L_{\mathbf{k}}^2$$
$$L_{\mathbf{k}} = [NS(\mathbf{k})]^{-1/2} \tag{10.3.5}$$

Equation (10.3.5) provides the rigorous evaluation of the normalization constant $L_{\mathbf{k}}$.

Now that the wave function is normalized properly, we can begin to investigate its properties by taking the expectation of different operators. The first is the momentum operator **P** given in (10.3.4):

$$\langle \mathbf{P} \rangle = \langle \Psi_{\mathbf{k}} \mid \mathbf{P} \mid \Psi_{\mathbf{k}} \rangle$$

First examine a single gradiant of position j acting upon the wave function $\Psi_{\mathbf{k}}$. Since the function $\Lambda_{\mathbf{k}}$ is a summation over different coordinates, the one term with $l = j$ will be different from the others:

$$\begin{aligned}
\nabla_j \Psi_{\mathbf{k}}(\mathbf{r}_1, \ldots, \mathbf{r}_N) &= L_{\mathbf{k}} \sum_l \nabla_j (e^{i\mathbf{k}\cdot\mathbf{r}_l} \Psi_0) \\
&= L_{\mathbf{k}}\left[\nabla_j (e^{i\mathbf{k}\cdot\mathbf{r}_j}\Psi_0) + \sum_{l\neq j} e^{i\mathbf{k}\cdot\mathbf{r}_l} \nabla_j \Psi_0\right] \\
&= i\mathbf{k} L_{\mathbf{k}} e^{i\mathbf{k}\cdot\mathbf{r}_j}\Psi_0 + \Lambda_{\mathbf{k}} \nabla_j \Psi_0
\end{aligned} \tag{10.3.6}$$

Thus we obtain the simple result

$$\begin{aligned}
\mathbf{P}\Psi_{\mathbf{k}} &= \hbar\mathbf{k}\Psi_{\mathbf{k}} + \Lambda_{\mathbf{k}}\mathbf{P}\Psi_0 \\
\langle \mathbf{P} \rangle &= \hbar\mathbf{k} + \langle \Psi_{\mathbf{k}} \mid \Lambda_{\mathbf{k}}\mathbf{P}\Psi_0 \rangle \\
&= \hbar\mathbf{k}
\end{aligned} \tag{10.3.7}$$

The first term gives $\hbar\mathbf{k}$. The second term averages to zero because the operators $\Lambda_{\mathbf{k}}^{\dagger}\Lambda_{\mathbf{k}}$ average to unity, and then $\langle \Psi_0 \mathbf{P} \Psi_0 \rangle$ is the momentum in the ground state, which is zero. Thus the momentum in state **k** is $\hbar\mathbf{k}$. The momentum operator commutes with the original Hamiltonian, so that states can be described by the quantum number of momentum $\hbar\mathbf{k}$, and our designation of it by this quantum number is justified.

The next step is to calculate the energy of this excitation whose wave vector is **k**. The energy is found by taking the expectation value of the Hamiltonian operator between these states:

$$E_{\mathbf{k}} = \langle \Psi_{\mathbf{k}} H \Psi_{\mathbf{k}} \rangle = L_{\mathbf{k}}^2 \langle \Psi_0 \varrho(-\mathbf{k}) H \varrho(\mathbf{k}) \Psi_0 \rangle$$

It is desirable to commute the Hamiltonian operator to the right, so that it operates on the ground state Ψ_0. We know that the ground state is an exact eigenstate of the Hamiltonian $H\Psi_0 = E_0\Psi_0$ and is the one with the lowest eigenvalue E_0. Thus we write

$$E_{\mathbf{k}} = L_{\mathbf{k}}^2\langle\Psi_0\varrho(-\mathbf{k})[H\varrho(\mathbf{k}) - \varrho(\mathbf{k})H + \varrho(\mathbf{k})H]\Psi_0\rangle$$
$$= L_{\mathbf{k}}^2\langle\Psi_0\varrho(-\mathbf{k})[H, \varrho(\mathbf{k})]\Psi_0\rangle + L_{\mathbf{k}}^2\langle\Psi_0\varrho(-\mathbf{k})\varrho(\mathbf{k})H\Psi_0\rangle \quad (10.3.8)$$

The second term on the right gives $E_0 L_{\mathbf{k}}^2\langle\Psi_0\varrho\varrho\Psi_0\rangle = E_0$. The first term on the right contains the commutator of H with the density operator $\varrho(\mathbf{k})$. Only the kinetic energy term fails to commute with $\varrho(\mathbf{k})$, so

$$[H, \varrho(\mathbf{k})] = -\frac{\hbar^2}{2m}\sum_{ij}[\nabla_i^2, e^{i\mathbf{k}\cdot\mathbf{r}_j}]$$
$$= \frac{\hbar^2}{2m}\sum_j e^{i\mathbf{k}\cdot\mathbf{r}_j}(k^2 - 2i\mathbf{k}\cdot\nabla_j)$$

$$\langle\Psi_0\varrho(-\mathbf{k})[H, \varrho(\mathbf{k})]\Psi_0\rangle = \varepsilon_k\langle\Psi_0\varrho(-\mathbf{k})\varrho(\mathbf{k})\Psi_0\rangle - \frac{i\hbar^2}{m}\sum_j\langle\Psi_0\varrho(-\mathbf{k})e^{i\mathbf{k}\cdot\mathbf{r}_j}\mathbf{k}\cdot\nabla_j\Psi_0\rangle$$

The second term on the right must be treated carefully. Since the ground state wave function is real, we can replace $\Psi_0\nabla_j\Psi_0$ by $\frac{1}{2}\nabla_j\Psi_0^2$ and then integrate by parts, so that this term is transformed to

$$\langle\Psi_0\varrho(-\mathbf{k})[H, \varrho(\mathbf{k})]\Psi_0\rangle$$
$$= \varepsilon_k\langle\Psi_0\varrho(-\mathbf{k})\varrho(\mathbf{k})\Psi_0\rangle + \frac{i\hbar^2}{2m}\sum_j\langle\varrho_N\mathbf{k}\cdot\nabla_j[\varrho(-\mathbf{k})e^{i\mathbf{k}\cdot\mathbf{r}_j}]\rangle$$
$$= \varepsilon_k\sum_{jl}\langle\varrho_N e^{i\mathbf{k}\cdot(\mathbf{r}_j-\mathbf{r}_l)}\rangle + \frac{i\hbar^2}{2m}\sum_{jl}\langle\varrho_N\mathbf{k}\cdot\nabla_j e^{i\mathbf{k}\cdot(\mathbf{r}_j-\mathbf{r}_l)}\rangle$$

The gradient operator in the last term will produce a factor $i\mathbf{k}$ in each term except the one where $l = j$. Thus the two preceding terms will cancel term by term except the one where $l = j$, which exists only in the first term. Thus we have

$$\langle\Psi_0\varrho(-\mathbf{k})[H, \varrho(\mathbf{k})]\Psi_0\rangle = \varepsilon_{\mathbf{k}}\sum_j 1 = N\varepsilon_{\mathbf{k}}$$

Since $L_{\mathbf{k}}^2 = 1/NS(\mathbf{k})$, the first term in (10.3.8) for $E_{\mathbf{k}}$ is $\varepsilon_{\mathbf{k}}/S(\mathbf{k})$, and the second is E_0. Thus we have shown that the energy in the system for the

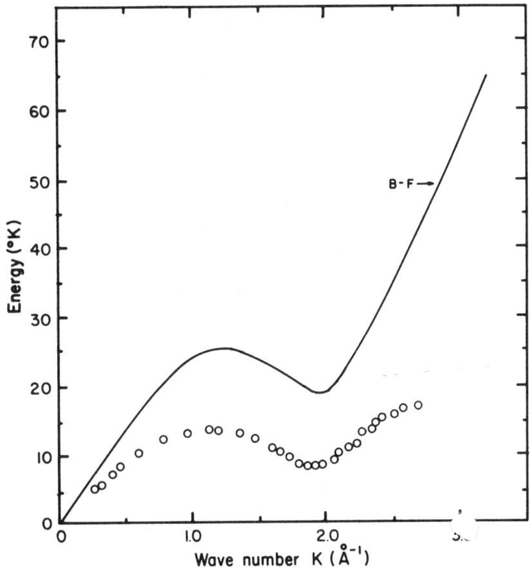

FIGURE 10.7. The excitation spectrum of liquid ^{4}He. The solid line is the Bijl-Feynman theory, compared with the experimental data as the points.

excitation **k** is

$$E_{\mathbf{k}} = \langle \Psi_{\mathbf{k}} \mid H \mid \Psi_{\mathbf{k}} \rangle = E_0 + \omega_0(\mathbf{k})$$

$$\omega_0(\mathbf{k}) = \frac{\varepsilon_{\mathbf{k}}}{S(\mathbf{k})}$$

The excitation energy is defined as $\omega_0(\mathbf{k})$, since this is the energy above the ground state E_0. Thus we have derived the Bijl–Feynman excitation energy.

The solid line in Fig. 10.7 shows $\omega_0(k)$ and is compared with the actual excitation spectrum of liquid ^{4}He. The two curves have exactly the same shape, but the Bijl–Feynman spectrum is about a factor of 2 larger. The roton minimum in the Bijl–Feynman dispersion curve is caused by the peak in $S(k)$ at $k \simeq 2$ Å. At small values of k, $\omega_0(\mathbf{k})$ goes to zero linearly with k, and the slope is the speed of sound. In the Bijl–Feynman theory, this happens because $\varepsilon_{\mathbf{k}}$ vanishes as k^2 while $S(k)$ vanishes as k, so their ratio does vanish as k. One cannot use experimental values of $S(k)$ in this region, since the experimental values of $S(k)$ have an additional contribution from thermal fluctuations which provides a small constant to $S(k)$ in the limit where $k \rightarrow 0$. This constant, which is proportional to $k_B T$ times the compressibility, is easily subtracted from experimental $S(k)$ to obtain the $S(k)$ one desires to use in the Bijl–Feynman formula.

The agreement between theory $\omega_0(k)$ and experiment $\omega(k)$ is not satisfactory. Thus the theory must press onward, and one must look for additional interactions or mechanisms which will improve the comparison. We must examine the interactions between these excitations. That is, we next consider a system with multiple excitations, where each is obtained by multiplying the ground state by the factor $\Lambda_{\mathbf{k}} = L_{\mathbf{k}}\varrho(\mathbf{k})$:

$$\begin{aligned} \Psi_{\mathbf{k}} &= \Lambda_{\mathbf{k}}\Psi_0 \equiv |\,\mathbf{k}) \\ \Psi_{\mathbf{k}_1\mathbf{k}_2} &= \Lambda_{\mathbf{k}_1}\Lambda_{\mathbf{k}_2}\Psi_0 \equiv |\,\mathbf{k}_1, \mathbf{k}_2) \\ \Psi_{\mathbf{k}_1\cdots\mathbf{k}_N} &= \Lambda_{\mathbf{k}_1}\Lambda_{\mathbf{k}_2}\cdots\Lambda_{\mathbf{k}_N}\Psi_0 \equiv |\,\mathbf{k}_1, \mathbf{k}_2, \ldots, \mathbf{k}_N) \end{aligned} \tag{10.3.9}$$

The $\Lambda_{\mathbf{k}}$ are scalar functions, so it does not matter in which order they are arranged. We must examine these states carefully to see whether the different excitations are independent or whether they interact. This is done in the next section.

B. Improved Excitation Spectra

We are going to describe the method of Feenberg (1969). We begin by writing down the states in (10.3.9) which contain multiple excitations. These will be examined to test their orthogonality and their interaction. The state of one excitation $|\,\mathbf{k})$ has an interaction with the state of two excitations $|\,\mathbf{k}', \mathbf{k} - \mathbf{k}')$. Therefore we get self-energy effects for the state of one interaction; when calculated, these self-energy effects are just what is necessary to give a good spectrum. Thus we get $\omega(k) = \omega_0(k) + \Sigma(k)$, where $\Sigma(k)$ is the self-energy of the excitation $\mathbf{k}$. The method of calculating this self-energy will only be outlined; a full description is in Feenberg's book.

The reader should be aware that we are embarking upon a procedure which is dramatically different from any used elsewhere in this book. All our other problems have been based on the interaction representation, where the Hamiltonian is written as $H = H_0 + V$, where V is a well-defined perturbation whose effects we try to understand. In the present problem, we do not do this at all. Instead, we write down a set of states, given in (10.3.9), as an ansatz. These states cannot be readily identified with any part of the Hamiltonian and are not orthogonal eigenstates of any operator which is obvious. Nor is it straightforward to do perturbation theory, since there is no V in which to expand. These problems can all be overcome, but the reader should be alerted that the present derivation is quite different from previous ones.

Feenberg and his associates established the following properties for these excitation states:

$$\begin{aligned}(\mathbf{kl} \mid \mathbf{kl}) &= \langle \Psi_{\mathbf{kl}} \mid \Psi_{\mathbf{kl}} \rangle = 1, \qquad \mathbf{k} \neq \mathbf{l} \\ (\mathbf{kk} \mid \mathbf{kk}) &= \langle \Psi_{\mathbf{kk}} \mid \Psi_{\mathbf{kk}} \rangle = 2 \\ (\mathbf{kl} \mid H \mid \mathbf{kl}) &= \langle \Psi_{\mathbf{kl}} \mid H \mid \Psi_{\mathbf{kl}} \rangle = \omega_0(\mathbf{k}) + \omega_0(\mathbf{l}) + E_0\end{aligned} \tag{10.3.10}$$

These are exactly the properties one would get if the operators $\Lambda_{\mathbf{k}}$ were treated as boson creation operators $a_{\mathbf{k}}^{\dagger}$. For example, the state with two excitations in $\mathbf{k}$ is

$$\begin{aligned}\mid \mathbf{k}) &= a_{\mathbf{k}}^{\dagger} \mid 0\rangle = \mid 1_{\mathbf{k}}\rangle \\ \mid \mathbf{k}, \mathbf{k}) &= a_{\mathbf{k}}^{\dagger} a_{\mathbf{k}}^{\dagger} \mid 0\rangle = a_{\mathbf{k}}^{\dagger} \mid 1_{\mathbf{k}}\rangle = 2^{1/2} \mid 2_{\mathbf{k}}\rangle\end{aligned}$$

where the $2^{1/2}$ comes from the usual raising operators for bosons. This explains the additional factor of 2 in (10.3.10). The state $\mid \mathbf{k}_1, \mathbf{k}_2)$ has an excitation energy given by $\omega_0(\mathbf{k}_1) + \omega_0(\mathbf{k}_2)$ and seems to be composed of states with independent boson excitations.

However, this free-boson analogy should not be pushed too literally. For example, states with different numbers of bosons but with the same total momentum are orthogonal only to order $N^{-1/2}$. For example,

$$\begin{aligned}(\mathbf{k}_1, \mathbf{k}_2 \mid \mathbf{k}_1 + \mathbf{k}_2) &= \left[\frac{S(\mathbf{k}_1) S(\mathbf{k}_2) S(\mathbf{k}_1 + \mathbf{k}_2)}{N} \right]^{1/2} \\ (\mathbf{k}_1 \mathbf{k}_2 \mid H \mid \mathbf{k}_1 + \mathbf{k}_2) &= O\left(\frac{1}{N} \right)^{1/2}\end{aligned}$$

The last matrix element is the important one for calculating the self-energy of the excitations. There exists a matrix element for one excitation of wave vector $\mathbf{k}$ going to two states with $\mathbf{k}'$ and $\mathbf{k} - \mathbf{k}'$. Thus one can have the self-energy diagram shown in Fig. 10.8. The excitations are represented by wavy lines, and the one excitation has a self-energy from making two excitations. This anharmonic process shows that our system of excitations is not truly independent. The self-energy for this process can be calculated and leads to a self-consistent equation for the excitation energy. It is derived

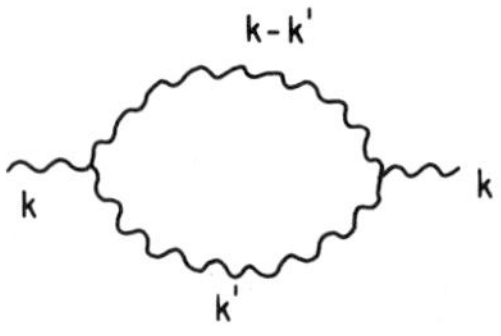

FIGURE 10.8

in the following fashion. Assume, as an approximation, that the excitation is a summation of states $|\,\mathbf{k})$ and $|\,\mathbf{l}, \mathbf{k}-\mathbf{l})$ in the form

$$\Phi_{\mathbf{k}} = a\,|\,\mathbf{k}) + \sum_{\mathbf{l}} b_{\mathbf{l}}\,|\,\mathbf{l}, \mathbf{k}-\mathbf{l}) \tag{10.3.11}$$

We wish to find the energy E which satisfies the eigenvalue equation

$$(H - E)\,|\,\Phi_{\mathbf{k}}\rangle = 0$$

We operate on this state from the left by the two excitation states $(\mathbf{m}, \mathbf{k}-\mathbf{m}\,|$ and use that fact that $(\mathbf{m}, \mathbf{k}-\mathbf{m}\,|\,H\,|\,\mathbf{l}, \mathbf{k}-\mathbf{l}) = (\delta_{\mathbf{m},\mathbf{l}} + \delta_{\mathbf{m},\mathbf{k}-\mathbf{l}}) \times [\omega_0(\mathbf{m}) + \omega_0(\mathbf{k}-\mathbf{m}) + E_0]$:

$$\begin{aligned} 0 &= (\mathbf{m}, \mathbf{k}-\mathbf{m}\,|\,(H-E)\,|\,\Phi_{\mathbf{k}}\rangle \\ &= a(\mathbf{m}, \mathbf{k}-\mathbf{m}\,|\,(H-E)\,|\,\mathbf{k}) + (b_{\mathbf{m}} + b_{\mathbf{k}-\mathbf{m}})[E_0 + \omega_0(\mathbf{m}) + \omega_0(\mathbf{k}-\mathbf{m}) - E] \end{aligned}$$

There are two values of b in the last term, since the pair $|\,\mathbf{m}, \mathbf{k}-\mathbf{m}\rangle$ can equal $|\,\mathbf{l}, \mathbf{k}-\mathbf{l}\rangle$ with either the choice $\mathbf{m}=\mathbf{l}$ or $\mathbf{m}=\mathbf{k}-\mathbf{l}$. We assume that $b_{\mathbf{m}} = b_{\mathbf{k}-\mathbf{m}}$ since they refer to the same excitation pair $|\,\mathbf{m}, \mathbf{k}-\mathbf{m})$, so we derive the result for $b_{\mathbf{l}}$:

$$b_{\mathbf{l}} = \frac{a}{2}\,\frac{(\mathbf{l}, \mathbf{k}-\mathbf{l}\,|\,(H-E)\,|\,\mathbf{k})}{E - E_0 - \omega_0(\mathbf{l}) - \omega_0(\mathbf{k}-\mathbf{l})}$$

This equation provides a determination of the ratio $b_{\mathbf{l}}/a$. We wish to insist that the expectation of the $H-E$ acting on $\Phi_{\mathbf{k}}$ is zero, which determines the eigenvalue E. Operating from the left with $(\mathbf{k}\,|$ produces

$$0 = (\mathbf{k}\,|\,(H-E)\Phi_{\mathbf{k}}\rangle = a[E_0 + \omega_0(\mathbf{k}) - E] + \sum_{\mathbf{l}} b_{\mathbf{l}}(\mathbf{k}\,|\,(H-E)\,|\,\mathbf{l}, \mathbf{k}-\mathbf{l})$$

$$\begin{aligned} E &= E_0 + \omega_0(\mathbf{k}) + \frac{1}{2}\sum_{\mathbf{l}} \frac{|\,(\mathbf{l}, \mathbf{k}-\mathbf{l}\,|\,(H-E)\,|\,\mathbf{k})\,|^2}{E - E_0 - \omega_0(\mathbf{l}) - \omega_0(\mathbf{k}-\mathbf{l})} \\ &= E_0 + \omega(\mathbf{k}) \end{aligned}$$

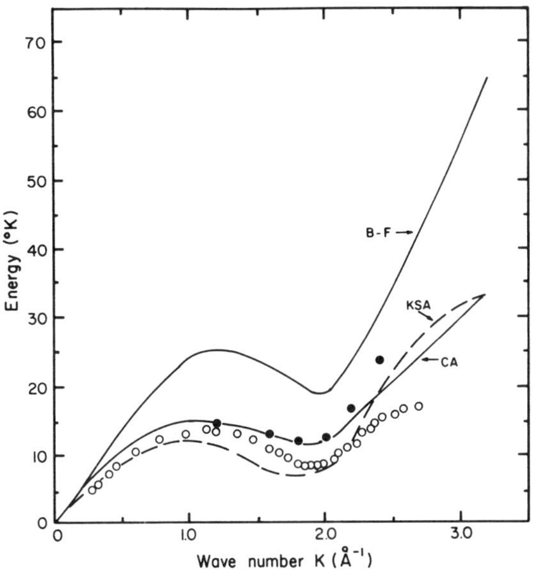

FIGURE 10.9. The excitation spectrum of liquid ^{4}He. The points are the experimental result for the superfluid. The other lines are various theories, including Bijl–Feynman (B–F) and backflow theories which use the Kirkwood superposition approximation (KSA) or convolution approximation (CA). Solid points are the Feynman–Cohen theory. *Source*: Feenberg (1969) (used with permission).

Thus we get a self-consistent equation for $\omega(\mathbf{k})$ where it appears in the matrix element and the energy denominator. Feenberg and his collaborators took the quantity $H - E_0 - \omega_0(\mathbf{k})$ in the matrix element, rather than $H - E_0 - \omega(\mathbf{k})$, and thus found only an approximate eigenvalue from their results. The details of calculating the matrix element of H between the excitation states 1 and 2 is tedious, since it involves an evaluation of three-particle correlation functions for the liquid. Their numerical results are shown in Fig. 10.9. as the dashed line. They compare quite favorably with the experimental spectrum as found by neutron scattering. The result is a dramatic improvement over the Bijl–Feynman curve, which is marked as B–F. Probably the first calculation of this kind was done by Feynman and Cohen (1956, 1957), who did a variational calculation of the backflow around the excitation. That is exactly the same physics as the admixing of the double excitation states, so the two methods of calculation are really the same but dressed in a slightly different language.

The theory of liquid ^{4}He is remarkably successful. Starting from the potential function $V(r)$ between atoms and other basic numbers such as the particle mass and density, the theory has calculated most of the features of the liquid with great success. By using correlated basis functions, the ground state properties such as $g(r)$, $S(k)$, $n_{\mathbf{k}}$, condensate fraction f_0, and off-diagonal density matrix $R(r)$ are obtained. Most of these quantities, or their Fourier transforms, can be compared favorably with experimental results. Similarly, the excitation spectrum can also be calculated. After some work, a decent curve for $\omega(k)$ is now found which compares well with the experimental results. Thus we have been able to derive all the features of liquid ^{4}He starting from a microscopic theory. The excitation spectrum $\omega(k)$ is all that is required to find the macroscopic properties including the superfluid density in the two-fluid model and the superfluid flow properties. Thus the macroscopic theory of superfluidity in ^{4}He can be derived, step by step, from a knowledge of only $V(r)$ plus some fundamental constants and masses. This is a remarkable theoretical achievement for such a strongly interacting and highly correlated system.

C. Superfluidity

We have not yet explained why the liquid is superfluid or shown how to compute its superfluid properties. The explanation was first provided by Landau (1941, 1947), and we shall now reproduce his arguments.

For temperatures below T_λ in the superfluid state, the fluid density can be viewed as consisting of a normal ϱ_N and superfluid ϱ_S component. The sum of these two is the normal density $\varrho = \varrho_N + \varrho_S$. The two components ϱ_N and ϱ_S vary considerably with temperature where $\varrho_S(T_\lambda) = 0$ and $\varrho_N(T = 0) = 0$. The actual variation with temperature of ϱ_N/ϱ is shown in Fig. 10.10. In this two-fluid model, each component of the fluid, normal and superfluid, carries its own momentum, energy, etc. The normal component is given by the expression for $T < T_\lambda$:

$$\varrho_N \equiv \frac{\hbar^2}{3m} \int \frac{d^3q}{(2\pi)^3} q^2 \left\{ - \frac{d}{d\omega(q)} n_B[\omega(q)] \right\} \tag{10.3.12}$$

This will be explained later. The superfluid component is just $\varrho_S = \varrho - \varrho_N$ and has nothing at all to do with the condensate fraction f_0. If one takes the measured excitation spectrum $\omega(q)$ and uses it to theoretically calculate $\varrho_N(T)$, one obtains the solid line in Fig. 10.10. This calculation is in remarkably good agreement with a number of ways to measure $\varrho_N(T)$ such as the

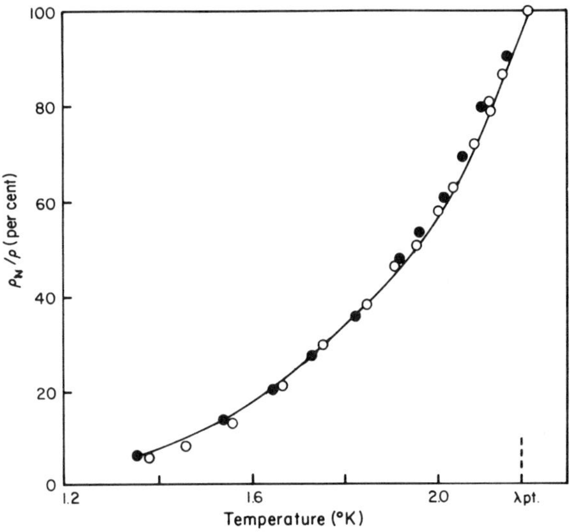

FIGURE 10.10. The density of the normal fluid ϱ_N as a function of temperature: ○, Derived from oscillating disk experiments; ●, from the velocity of second sound. *Source*: Lifshits and Andronikashivile (1959).

viscous drag on rotating plates, etc. Indeed, Landau was able to deduce a number of properties of $\omega(q)$ before it was measured directly from neutron scattering by fitting Eq. (10.3.12) to the measured component of normal density. In particular, the curves for $\varrho_N(T)$ show an activation energy, which we now associate with the roton minimum Δ. Landau was able to deduce the necessity for this roton minimum, as well as determine the speed of sound, and predicted quite well the entire shape of the excitation spectrum, including phonon and roton parts. His was a remarkable theoretical achievement.

The Landau argument may be understood by considering a Gedanken experiment, which has the liquid helium flowing down a pipe with a uniform velocity $\mathbf{v}$. This is a small velocity, say 1 cm/sec, which is much smaller than the velocities in the zero-point motion of the fluid. We wish to calculate the total momentum of the particle flow and show that it is *not* just $N\mathbf{v}$, where N is the number of particles.

First we must construct a wave function which describes the uniformly moving liquid. This is done by using the operator $\Lambda_{\mathbf{k}}^{(2)}$ in (10.3.3), which we showed caused uniform flow. We label this wave function as $\Psi_{\mathbf{v}}(\mathbf{r}_1, \mathbf{r}_2, \ldots, \mathbf{r}_N)$ to denote the state of uniform flow. The energy and momentum of this wave function were calculated earlier:

$$\Psi_{\mathbf{v}}(\mathbf{r}_1, \mathbf{r}_2, \ldots, \mathbf{r}_N) = \exp\left(im\mathbf{v} \cdot \sum_j \mathbf{r}_j\right)\Psi_0(\mathbf{r}_1 \cdots \mathbf{r}_N)$$

$$\langle \Psi_{\mathbf{v}} \mathbf{P} \Psi_{\mathbf{v}} \rangle = mN\mathbf{v}$$

$$\langle \Psi_{\mathbf{v}} H \Psi_{\mathbf{v}} \rangle = E_0 + \frac{Nmv^2}{2}$$

The next step is to consider the excitation spectrum of this moving fluid. The multiplication of $\Psi_{\mathbf{v}}$ by $\Lambda_{\mathbf{k}} = L_{\mathbf{k}}\varrho(\mathbf{k})$ produces an excitation with momentum $\hbar\mathbf{k}$ and excitation energy $\omega(\mathbf{k}) + \mathbf{k} \cdot \mathbf{v}$, as shown by

$$\phi_{\mathbf{v}}(\mathbf{k}) = L_{\mathbf{k}}\varrho(\mathbf{k})\Psi_{\mathbf{v}}$$

$$\langle \phi_{\mathbf{v}}(\mathbf{k}) \mathbf{P} \phi_{\mathbf{v}}(\mathbf{k}) \rangle = mN\mathbf{v} + \hbar\mathbf{k}$$

$$\langle \phi_{\mathbf{v}}(\mathbf{k}) H \phi_{\mathbf{v}}(\mathbf{k}) \rangle = E_0 + \frac{N}{2} mv^2 + \omega_0(\mathbf{k}) + \mathbf{v} \cdot \mathbf{k}$$

The result $\omega_0(\mathbf{k}) + \mathbf{v} \cdot \mathbf{k}$ is quite resonable: If $\omega_0(\mathbf{k})$ is the excitation spectrum with respect to the frame moving with the fluid, then the Doppler-shifted energy, as viewed by an observer at rest in the laboratory, is $\omega_0(\mathbf{k}) + \mathbf{k} \cdot \mathbf{v}$.

Of course, we really want the excitation spectrum to be $\omega(\mathbf{k}) + \mathbf{k} \cdot \mathbf{v}$ rather than $\omega_0 + \mathbf{k} \cdot \mathbf{v}$. This can be achieved by using as the excitation operator the combination $\phi_{\mathbf{k}}$ in (10.3.11) of $|\,\mathbf{k})$ and $|\,\mathbf{l}, \mathbf{k} - \mathbf{l})$, which produces the improved excitation spectra. We assume that this is done and that the excitations are indeed the ones which are physically realistic.

The momentum operator gives $mN\mathbf{v} + \hbar\mathbf{k}$ for a single excitation of energy $\omega(\mathbf{k}) + \mathbf{k} \cdot \mathbf{v}$. In the actual system, there will be a number of excitations whose distribution is given by the thermal occupation factor n_B for boson excitations. Thus the average value of the momentum operator is

$$\langle \Psi_v \,|\, \mathbf{P} \,|\, \Psi_v \rangle_{\text{thermal av}} = Nm\mathbf{v} + \frac{N}{\varrho} \int \frac{d^3k}{(2\pi)^3} \hbar\mathbf{k} n_B[\omega(\mathbf{k}) + \mathbf{v} \cdot \mathbf{k}] \quad (10.3.13)$$

The argument of the boson distribution factor $n_B(Z) = [\exp(\beta Z) - 1]^{-1}$ is the derived excitation energy $\omega(\mathbf{k}) + \mathbf{k} \cdot \mathbf{v}$ in the laboratory frame. The chemical potential is zero in the superfluid state, since it is a form of Bose–Einstein condensation. Note that we are *not* trying to use the average momentum in the ground state of flowing fluid, since that would be found from the average of $m \int [d^3k/(2\pi)^3]\hbar\mathbf{k}n_{\mathbf{k}-m\mathbf{v}} = Nm\mathbf{v}$, where $n_{\mathbf{k}-m\mathbf{v}}$ is the momentum distribution of the particles in the ground state of the superfluid. The quantity $n_{\mathbf{k}-m\mathbf{v}}$ is the momentum distribution of the stationary ground state displaced by a fixed amount $m\mathbf{v}$. The total momentum in the ground state of the flowing fluid is just $Nm\mathbf{v}$. Instead, we are trying to find the total momentum in the flowing fluid as contributed by the ground state plus the excitations. The excitations are produced by the thermal fluctuations, so that the total momentum will be temperature dependent.

For small velocities the argument of the distribution function can be expanded in powers of $\mathbf{k} \cdot \mathbf{v}$: $n_B(\omega + \mathbf{k} \cdot \mathbf{v}) = n_B(\omega) + \beta \mathbf{k} \cdot \mathbf{v} n_B'(\omega) + O(v^2)$. The first term with just $n_B[\omega(\mathbf{k})]$ gives a zero average of $\hbar\mathbf{k}$, and the important result is from the second term in the series expansion:

$$\langle P \rangle_{\text{av}} = Nm\mathbf{v} + \frac{N\beta}{\varrho} \int \frac{d^3k}{(2\pi)^3} \hbar\mathbf{k}(\mathbf{k} \cdot \mathbf{v}) n_B'[\omega(\mathbf{k})]$$

Since the spectrum n_B' is isotropic in $\mathbf{k}$, the angular integrals are done easily and give a vector in the direction $\mathbf{v}$ in $\int d\Omega_k \mathbf{k}(\mathbf{k} \cdot \mathbf{v}) = \mathbf{v}k^2 4\pi/3$. When we collect the remaining factors together, we find that total momentum is expressed in terms of the normal fraction of superfluid which was defined in (10.3.12):

$$\begin{aligned} \langle P \rangle_{\text{av}} &= Nm\mathbf{v}\left(1 - \frac{\varrho_N}{\varrho}\right) \\ &= \frac{\varrho_S}{\varrho} Nm\mathbf{v} \end{aligned}$$

The final result is that the total momentum in the flowing fluid is given only by the fraction ϱ_S/ϱ, which is superfluid. The normal fraction ϱ_N/ϱ gives no contribution to the momentum.

In a normal liquid, uniform flow is not an eigenstate because viscosity damps the flow. The superfluid has no viscosity, and uniform flow is an eigenstate. At finite temperatures $0 < T < T_\lambda$, only the superfluid com-
really the only ideal Fermi liquid in earthbound nature, since electrons in metals are not ideal because their properties are perturbed by the crystal lattice of ions. The ^{3}He atoms are charge neutral, so that the properties of the Fermi system are much different from electrons in metals. They are also strongly interacting and highly correlated, so that the noninteracting Fermi gas is not a good starting point for a description of its ground state. The discovery that ^{3}He has a superfluid phase, which is akin to the BCS superconductor, has heightened interest in this liquid.

The theory of these systems operates at several levels. One example is the two-fluid model of hydrodynamic flow, and another example is the of the container, in this case the pipe. Thus, we shifted to the laboratory frame and obtained the Doppler-shifted energy $\omega(\mathbf{k}) + \mathbf{k} \cdot \mathbf{v}$ before doing the thermal averaging. Thus the rest frame of the container and the superfluid both enter the calculation. The result is that only the superfluid density carries momentum when the walls are stationary.

The flowing superfluid is in an eigenstate of the system, which can persist without damping. Thus the state $\Psi_{\mathbf{v}}$ has an eigenvalues $(\varrho_S/\varrho)Nm\mathbf{v}$ of the momentum operator. In a completely normal fluid, such as neon or ^{4}He for $T > T_\lambda$, uniform flow down a fixed pipe is not an eigenstate, and the flow is dissipated by viscosity. In the superfluid, the component ϱ_S is not dissipated.

This calculation should be compared with other similar systems. What if the entire container of superfluid is just picked up and moved at a constant velocity; i.e., what if a Dewar is driven on a truck? Then the walls and the superfluid are moving at the same velocity, and the thermal averaging should be done in their mutual rest frame, which is moving with the truck. Then the momentum average gives

$$\begin{aligned}\langle P \rangle_{\text{av}} &= Nm\mathbf{v} + \frac{N}{\varrho} \int \frac{d^3k}{(2\pi)^3} \hbar \mathbf{k} n_B[\omega(\mathbf{k})] \\ &= Nm\mathbf{v}\end{aligned}$$

Now the result is $Nm\mathbf{v}$ since both normal and superfluid components are moving with the same velocity.

10.4. ^{3}He: NORMAL LIQUID

It is certainly a misnomer to call liquid ^{3}He a "normal liquid." The title is used here to distinguish it from the superfluid phase of the liquid. This system is interesting because its properties are so exceptional. It is really the only ideal Fermi liquid in earthbound nature, since electrons in metals are not ideal because their properties are perturbed by the crystal lattice of ions. The ^{3}He atoms are charge neutral, so that the properties of the Fermi system are much different from electrons in metals. They are also strongly interacting and highly correlated, so that the noninteracting Fermi gas is not a good starting point for a description of its ground state. The discovery that ^{3}He has a superfluid phase, which is akin to the BCS superconductor, has heightened interest in this liquid.

The theory of these systems operates at several levels. One example is the two-fluid model of hydrodynamic flow, and another example is the microscopic derivation of $S(k)$ and $g(r)$. The theoretical description can be viewed as a sequence of plateaus, where each plateau is a self-consistent description of the system. For example, the excitation spectrum of liquid ^{4}He is such a plateau. One feature of these plateaus is that they can be compared directly with experiment. Thus the theoretical effort generally is devoted to deducing the theory of one plateau by using the model of the lower one. For example, in liquid ^{4}He, one does not try to derive in one step the hydrodynamic flow properties from the microscopic theory. Instead, one uses microscopic theory to find the excitation spectrum $\omega(k)$, which is one plateau. Then one uses Landau theory to derive the two-fluid model in terms of $\varrho_S(T)$ and $\varrho_N(T)$, which are described in terms of $\omega(k)$. The theory is a sequence of plateaus, each of which can be compared with experiment, and one uses the theory to try to relate one step to the next one higher in the sequence.

The same sequence applies to the properties of liquid ^{3}He. The first, and most important plateau, is the Fermi liquid theory which was developed by Landau (1946, 1956, 1957), who has provided the main insight into both ^{4}He and ^{3}He. This theory was developed as a phenomenological model for the behavior of low-lying excitations of the strongly interacting Fermi system. These excitations can be described by a few parameters, which can be deduced from experiments. A major goal of microscopic theory is to derive these parameters from first principles. The major effort along this line, as was also the case for ^{4}He, is by Feenberg and his associates. So far they have achieved a good level of agreement, which is on the order of 30% at the worst and is better for some quantities. Since the Landau theory of the Fermi liquid is the major plateau of the theory, we shall spend this section describing it as well as summarizing the microscopic theory.

The next plateau is a description of the superfluid state, which is usually done using the parameters of Fermi liquid theory. Thus these parameters play a central role in the theoretical development. We shall find that the superfluid properties are well described in this fashion using the experimentally measured parameters. In this sense a quantitative theory is available for the superfluid state.

A. Fermi Liquid Theory

There are excellent reviews of Fermi liquid theory as applied to liquid ^{3}He in the normal phase. Several of these are Pines and Nozières (1966), Baym and Pethick (1978), Ron (1975), and Leggett (1975). These treatments are usually more detailed than the one here, and the reader is encouraged to read some of these others.

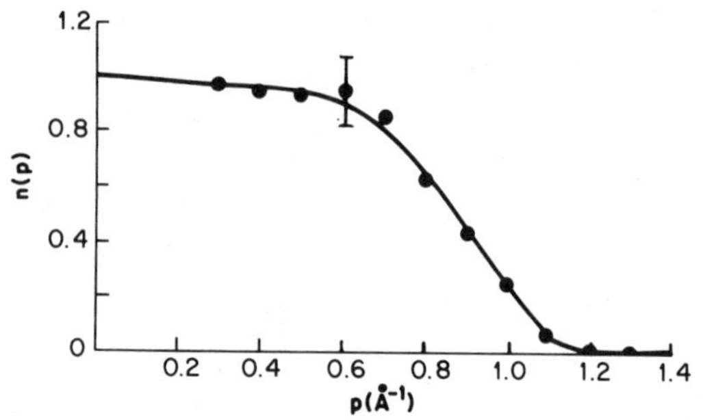

FIGURE 10.11. Momentum distribution n_p of particles in ^{2}He at $T = 0.37$ K. The solid line is a least-squares fit to the Fermi distribution with an effective temperature of $T_F = 1.8$ K. *Source*: Mook (1985) (used with permission).

The symbol $n_{\mathbf{p}}$ is defined in (3.3.19) as the average number of atoms in the fermion liquid with momentum **p**. In a noninteracting Fermi system, at zero temperature, this distribution is $n_p = \theta(k_F - p)$. Of course, in our strongly interacting Fermi system, we expect that this will be strongly smeared. Figure 10.11 shows the momentum distribution for ^{3}He measured by neutron scattering at $T = 0.37$ K by Mook (1985). The points are experimental. The solid line is a fit to the Fermi distribution $n(p) = 1/[\exp(\xi_p/k_B T_K) + 1]$ with $T_K = 1.8$ K. The smearing caused by particle–particle interactions can be represented by an effective temperature in the Fermi distribution.

Throughout the remaining part of this chapter, we shall use the symbol $n_{\mathbf{p}}$ to mean something entirely different from its usage elsewhere in this book. Whereas before we always took $n_{\mathbf{p}}$ to be the distribution of states with free-particle momentum **p**, now it has a different meaning. Fermi liquid theory has a well-entrenched set of notation, with a universal set of symbols, whose meaning is well agreed upon. Thus we shall follow these strong conventions and now use $n_{\mathbf{p}}$ to mean the distribution of *excitations* of momentum **p**.

The objective of Fermi liquid theory is to describe the low-lying excited states of the fermion system. There is a ground state which the system would be in at zero temperature—except for superfluid phases—and this ground state will be discussed in the next section. In analogy with the electron gas, we expect two kinds of low-lying excitations. The first are density oscillations, which are plasmons in the charged electron gas. In the neutral Fermi liquid they are sound waves. The second type of excitation is electron–

TABLE 10.1. Properties of Liquid Helium (pressure = 1 bar)

Parameter	^{4}He	^{3}He
m	6.65×10^{-24} g	5.01×10^{-24} g
ϱ	2.18×10^{22} atoms/cm^3 0.145 g/cm^3	1.64×10^{22} atoms/cm^3
C_1 ($T = 1$ K)	238 m/sec	182 m/sec
σ	2.648 Å	2.648 Å
ε	1.484×10^{-15} erg = 10.7 K	1.484×10^{-15} erg = 10.7 K
E/N	−7.20 K/atom	−2.52 K/atom

hole pairs. There is also a third kind of excitation, which is a damped spin wave, called a paramagnon. Fermi liquid theory is primarily aimed at a description of the properties of low-lying electrons and holes. However, the other excitations—density oscillations and paramagnons—can be viewed as collective resonances of these electron and hole excitations. Thus, Fermi liquid theory describes them as well. Indeed, many of the so-called Fermi liquid effects actually come from the collective properties. Our discussion will follow the historical pathway and first treat the classical Fermi liquid theory due to Landau. Then the modern interpretation in terms of collective properties will be discussed.

By using the parameters in Table 10.1, one would calculate a Fermi energy of $E_{F0} = 4.97$ K for a noninteracting gas of the same density and mass of ^{3}He. The actual degeneracy energy in liquid ^{3}He is about two-thirds of this value. Fermi liquid theory is valid only for the particle–hole excitations whose energy is a small fraction of the degeneracy energy—say 5–10%. Thus the excitations in this theory have low energy, say less than 0.1–0.2 K. If we are going to use these excitations to describe the thermodynamics of the excited states of the system, we can legitimately do this only for temperatures below 0.1–0.2 K. Otherwise, the thermodynamic properties will involve states beyond the range of the theory. Thus we are talking about very low-energy excitations, which are important only in the liquid at very low temperatures. Since the superfluid properties begin at a temperature two decades lower than this, there is an appreciable temperature range over which Fermi liquid theory is applicable.

In the homogeneous electron gas, the imaginary part of the electron self-energy, from electron–electron interactions, vanishes for electrons whose energy is right at the chemical potential μ. Furthermore, Im Σ is also small for states near the chemical potential, since it vanishes as $(\varepsilon_p - \mu)^2$. Thus the low-lying excited states of this Fermi liquid also have a long lifetime. The same phenomenon is assumed to happen in ^{3}He, and general theorems can be proved which demonstrate this with rigor (Luttinger and Nozières, 1962). In cases where Im $\Sigma \simeq 0$, the spectral function $A(p, \omega)$ is sharply peaked and has the character of a delta function. There is a unique relationship between energy and momentum in these cases, and the excitation is satisfactorily described by only one of these two variables. In Fermi liquid theory this variable is conventionally taken to be the momentum $\mathbf{p}$. Thus an excitation of momentum $\mathbf{p}$ is assigned an energy ε_p.

The energy zero is defined so that the excitation energy ε_p is zero at $p = 0$. Furthermore, if the energy is expanded in a Taylor series about point $p = 0$, the odd terms in this series p^{2m+1} must vanish because of the homogeneity of the liquid and the isotropy between $\mathbf{p}$ and $-\mathbf{p}$. Thus the first term in the expansion is p^2, and other terms are usually ignored. Thus we write $\varepsilon_p \propto p^2$, and the constant of proportionality defines the effective mass m^*:

$$\varepsilon_p = \frac{p^2}{2m^*}$$

In liquid ^{3}He, the effective mass has a value of $m^* = 3.0m$, where m is the bare mass of ^{3}He. These and other experimental properties are reviewed by Wheatley (1975).

The ground state of the fermion system does not have all the excitations in the state with $\mathbf{p} = 0$. The exclusion principle forces many excitations into states of finite $\mathbf{p}$, and all energy states are occupied at zero temperature for $\varepsilon_p < \mu$. We can define $n_{\mathbf{p}}{}^0$ as the number of excitations of momentum $\mathbf{p}$ in the ground state, in which case the ground state energy is

$$E_g = E_0 + \sum_{\mathbf{p}} \varepsilon_p n_{\mathbf{p}}{}^0 \tag{10.4.1}$$

where E_0 is the energy when all the quasiparticles are in $\mathbf{p} = 0$, which would be the case if they were bosons. Of course, the result (10.4.1) is actually nonsense, since the approximation of treating the excitations as free particles with an effective mass m^* and no damping is valid only when $\varepsilon_p - \mu \simeq k_B T$ and does not apply throughout the degenerate Fermi sea. Thus the quantity $n_{\mathbf{p}}{}^0$ is a totally fictitious concept. However, we shall never need to utilize any property of $n_{\mathbf{p}}{}^0$ except at the chemical potential, where we can accurately assume the following form at finite temperature:

$$n_{\mathbf{p}}{}^0 = n_F(\varepsilon_p - \mu) = \frac{1}{e^{\beta(\varepsilon_p - \mu)} + 1}$$

Thus this approximation is surely valid only very near the chemical potential $\beta(\varepsilon_p - \mu) \lesssim 1$, but that is the only place we shall use it.

Our interest is in the excitation of the system above the ground state. When excitations are present, the distribution $n_{\mathbf{p}}$ of these excitations will be different from the ground state distribution $n_{\mathbf{p}}{}^0$. The difference between these two distributions is defined as $\delta n_{\mathbf{p}} = n_{\mathbf{p}} - n_{\mathbf{p}}{}^0$. The important physical idea is that the crucial quantity is neither $n_{\mathbf{p}}$ nor $n_{\mathbf{p}}{}^0$ but rather this difference $\delta n_{\mathbf{p}}$, since it gives the number of excitations in the excited state—which is important for the thermodynamics of the system at low temperature. Although we can determine neither $n_{\mathbf{p}}$ nor $n_{\mathbf{p}}{}^0$ with great accuracy, we can find $\delta n_{\mathbf{p}}$, which is all that matters. For example, the total energy of the system, including the excited states, is

$$\begin{aligned} E &= E_0 + \sum_{\mathbf{p}} \varepsilon_p n_{\mathbf{p}} \\ &= E_0 + \sum_{\mathbf{p}} \varepsilon_p n_{\mathbf{p}}{}^0 + \sum_{\mathbf{p}} \varepsilon_p (n_{\mathbf{p}} - n_{\mathbf{p}}{}^0) \\ &= E_g + \sum_{\mathbf{p}} \varepsilon_p \delta n_{\mathbf{p}} \end{aligned}$$

The excited state energy is $\sum \varepsilon_p \delta n_{\mathbf{p}}$.

The important quantity is not the energy but the free energy $F = E - \mu N$, where N is the total number of particles. The particle number is defined in terms of the number of particles in the ground state N_0:

$$\begin{aligned} N_0 &= \sum_{\mathbf{p}} n_p{}^0 \\ N &= \sum_{\mathbf{p}} n_{\mathbf{p}} = \sum_{\mathbf{p}} n_p{}^0 + \sum_{\mathbf{p}} \delta n_{\mathbf{p}} = N_0 + \delta N \qquad (10.4.2) \\ F &= E - \mu N = E_g - \mu N_0 + \sum_{\mathbf{p}} (\varepsilon_p - \mu) \delta n_{\mathbf{p}} \end{aligned}$$

There are two key assumptions contained in these equations. The first is that the chemical potential μ does not change because of the excited states, although it does change with temperature. The second assumption is that there is a one-to-one correspondence between quasiparticles and the excited states of the noninteracting Fermi gas. Thus adding a quasiparticle also adds a particle and changes N by unity. Of course, usually there are about equal numbers of quasiparticles above $\varepsilon_p > \mu$, which have $\delta n_{\mathbf{p}} > 0$, as there are a depletion of them for $\varepsilon_p < \mu$, which have $\delta n_{\mathbf{p}} < 0$. The latter excitations are called holes.

So far the Fermi liquid theory has been equivalent to calling the excitations free particles with an effective mass m^*. If that were all of it, the theory would be unexceptional and not quantitatively accurate. The genius of Landau was in realizing that the expansion for the free energy must have another term, which is due to the interactions. This contribution is required for a self-consistent theory. We can regard the formulas we have so far derived as the terms in an expansion of the free energy in the excited distribution $\delta n_{\mathbf{p}}$. That is, $F(\delta n_{\mathbf{p}})$ is a functional of $\delta n_{\mathbf{p}}$ and can be expanded formally in a Taylor series in this parameter. The Landau theory assumes that $\delta n_{\mathbf{p}}$ is small and that an accurate quantitative theory is achieved after a few terms in this expansion.

The terms we have derived so far are given in (10.4.2). The first term is $F_0 = E_g - \mu N_0$. The next term appears to be first order in $\delta n_{\mathbf{p}}$. A closer inspection shows that this term is actually of order $(\delta n_{\mathbf{p}})^2$. The summation over $\mathbf{p}$ will run over all p values for which $\delta n_{\mathbf{p}}$ is nonzero. However, the integrand contains $\varepsilon_p - \mu$, which vanishes at $\varepsilon_p - \mu = 0$. Thus if $\delta n_{\mathbf{p}}$ has value up to Δ away from the chemical potential, the kinetic energy integral $\sum_p (\varepsilon_p - \mu)\delta n_{\mathbf{p}} \propto \Delta^2$. Thus this term can be viewed as being second order in the quantity $\delta n_{\mathbf{p}}$. Thus Landau saw the need to include all terms of order $(\delta n_{\mathbf{p}})^2$. There is another term which describes the interactions between the excited quasiparticles and so contains the dependence $\delta n_{\mathbf{p}}\, \delta n_{\mathbf{p}'}$. This interaction is very dependent on the spin of the quasiparticles. Thus the free-energy expansion has the form

$$F = F_0 + \sum_{\mathbf{p}\sigma} (\varepsilon_p - \mu)\delta n_{\mathbf{p}\sigma} + \frac{1}{2\nu} \sum_{\substack{\mathbf{p}\mathbf{p}' \\ \sigma\sigma'}} f_{\mathbf{p}\sigma,\mathbf{p}'\sigma'} \delta n_{\mathbf{p}\sigma} \delta n_{\mathbf{p}\sigma'} + (\delta n)^3$$

where spin has been added to various quantities. This dependence on the particle spin is not due to the spin-dependent interactions. Indeed, the dipole–dipole forces between the nuclear moments are extremely small and may be neglected when discussing most phenomena. The spin dependence of the interaction is merely due to particle statistics. The antisymmetrization of the many-particle wave function for fermions causes an exchange hole around each particle. This, in turn, causes an effective exchange energy, just as it did for the electron gas. The exchange energy may be qualitatively understood by a consideration of the mutual scattering of two ^{3}He atoms. Since the two-particle wave function must be antisymmetric, the scattering properties depend on the total spin state of the two spin one-half particles—as described in Sec. 10.1. If they are in an $S = 0$ state, the orbital cross section contains only even angular momentum components; if they have a total $S = 1$ state, only odd angular momentum components enter the cross section. Thus the scattering and interaction depend on the relative spins of the two particles. This is a very important effect in liquid ^{3}He, since the effective interaction $f_{\mathbf{p}\sigma,\mathbf{p}'\sigma'}$ is very spin dependent.

The spin dependence of the effective interaction is written as

$$f_{\mathbf{p}\sigma,\mathbf{p}'\sigma'} = f^s_{\mathbf{pp}'} + \boldsymbol{\sigma} \cdot \boldsymbol{\sigma}' f^a_{\mathbf{pp}'} \tag{10.4.3}$$

Equation (10.4.3) is a very particular choice for the form of the interaction. We do not include other possible spin dependencies such as spin–orbit effects $(\boldsymbol{\sigma} \cdot \mathbf{p})(\boldsymbol{\sigma}' \cdot \mathbf{p}')$ or spin–other-orbit effects $(\boldsymbol{\sigma} \cdot \mathbf{p}')(\boldsymbol{\sigma}' \cdot \mathbf{p})$. The choice of a simple $\boldsymbol{\sigma} \cdot \boldsymbol{\sigma}'$ term derives from our understanding that the basic spin mechanism is due to exchange forces, which can be written this way.

Although all writers dealing with Fermi liquid theory present the effective interaction in this fashion, there are two different notational schemes in common practice. In both of these the symbols $\boldsymbol{\sigma}$, $\boldsymbol{\sigma}'$ refer to the spin of the two quasiparticles. However, in one scheme the spins are normalized so that the z components have eigenvalues $(\hbar = 1) \pm \frac{1}{2}$, while in the other the symbols $\boldsymbol{\sigma}$ are Pauli spin matrices with z component ± 1. These two choices lead to different normalizations on the second term, which differ by factors of $\frac{1}{4}$. For example, the two schemes give the scattering between two spin up particles as

$$\langle \uparrow\uparrow \mid f_{\mathbf{p}\sigma,\mathbf{p}'\sigma'} \mid \uparrow\uparrow \rangle = f^s_{\mathbf{pp}'} + \tfrac{1}{4} f^a_{\mathbf{pp}'}$$

$$\langle \uparrow\uparrow \mid f_{p\sigma,p'\sigma'} \mid \uparrow\uparrow \rangle = f^s_{\mathbf{pp}'} + f^a_{\mathbf{pp}'}$$

Since the formulas should be identical, the term f^a in the first equation is four times larger than f^a in the bottom. The workers in this field seem about equally divided between these two schemes. We shall adopt the second one, where the $\boldsymbol{\sigma}$ mean Pauli spin matrices with $\sigma_z = \pm 1$ in order to avoid writing the $\frac{1}{4}$ factor in every equation. The reader should be alerted that important reviews by Leggett (1975) and Wheatley (1975) use the other scheme, so their factors of Z_l are four times larger than ours.

Fermi liquid theory is really only accurate for describing the interactions of quasiparticles on the Fermi surface. In the interaction terms $f^s_{\mathbf{pp}'}$ and $f^a_{\mathbf{pp}'}$, the momentum variables have the magnitude p_F, and the only important variable is the angle θ between $\mathbf{p}$ and $\mathbf{p}'$. The interaction terms are represented by an expansion in Legendre polynomials. It is also conventional to normalize the coefficients in this expansion by removing a factor of the density of states at the Fermi energy $N_F = m^* p_F/\pi^2\hbar^3$:

$$\begin{aligned} f^s_{\mathbf{pp}'} &= \frac{\pi^2\hbar^2}{m^* p_F} \sum_{l=0}^{\infty} F_l^s P_l(\cos\theta) \\ f^a_{\mathbf{pp}'} &= \frac{\pi^2\hbar^2}{m^* p_F} \sum_{l=0}^{\infty} F_l^a P_l(\cos\theta) \end{aligned} \tag{10.4.4}$$

The coefficients F_l^s and F_l^a are sometimes called F_l and Z_l or $4Z_l$. They are the fundamental parameters in the theory and are dimensionless. They can be deduced from experiments, as shown later. Reliable experimental numbers are now available for F_0^s, F_1^s, F_0^a, and F_1^a. A major goal of the microscopic theory is to derive them.

The $f_{\mathbf{p}\sigma,\mathbf{p}'\sigma'}$ term describes the interaction between excited quasiparticles. The interactions cause the energy of an excited quasiparticle to depend on the number $\delta n_{\mathbf{p}\sigma}$ of other excited ones. The quasiparticle energy can be formally derived as a functional derivative of the total energy E:

$$E = E_g + \sum_{\mathbf{p}\sigma} \varepsilon_p \delta n_{\mathbf{p}\sigma} + \frac{1}{2\nu} \sum_{\substack{\mathbf{p}\sigma \\ \mathbf{p}'\sigma'}} f_{\mathbf{p}\sigma,\mathbf{p}'\sigma'} \delta n_{\mathbf{p}\sigma} \delta n_{\mathbf{p}'\sigma'}$$

$$\frac{\delta E}{\delta(\delta n_{\mathbf{p}\sigma})} \equiv \bar{\varepsilon}_{\mathbf{p}} = \varepsilon_p + \frac{1}{\nu} \sum_{\mathbf{p}'\sigma'} f_{\mathbf{p}\sigma,\mathbf{p}'\sigma'} \delta n_{\mathbf{p}'\sigma'}$$

The energy of the excited quasiparticle is called $\bar{\varepsilon}_{\mathbf{p}}$. It depends on the number of other excited quasiparticles through the interaction term. One result of the interactions is that the total excitation energy is no longer equal to $\sum \bar{\varepsilon}_{\mathbf{p}}\, \delta n_{\mathbf{p}\sigma}$, since this expression overcounts the pairwise interaction term. Of course, what we really want is the excitation energy relative to the chemical potential:

$$\frac{\delta F}{\delta(\delta n_{\mathbf{p}\sigma})} = \bar{\varepsilon}_{\mathbf{p}} - \mu = \varepsilon_p - \mu + \frac{1}{\nu} \sum_{\mathbf{p}'\sigma'} f_{\mathbf{p}\sigma,\mathbf{p}'\sigma'} \delta n_{\mathbf{p}'\sigma'}$$

This leads to the same definition of $\bar{\varepsilon}_{\mathbf{p}}$, since the chemical potential μ is not altered by the density of excited quasiparticles as long as this density is very small compared to the density of atoms.

Now we have to go back and reconsider our views about the distribution of excited quasiparticles. We still use the same notation that $n_{\mathbf{p}}$ is the total number of excited quasiparticles and that $n_{\mathbf{p}}{}^0$ is the number of excited quasiparticles in the ground state with energy ε_p. But now we need another symbol to denote the quasiparticle density in terms of the actual energy variable $\bar{\varepsilon}_{\mathbf{p}}$. Thus call $\bar{n}_p{}^0$ the equilibrium density of excited quasiparticles with energy $\bar{\varepsilon}_{\mathbf{p}}$, which is defined in the usual way in terms of the Fermi distribution function:

$$\bar{n}_p{}^0 = n_F(\bar{\varepsilon}_p - \mu)$$

$$\delta \bar{n}_{\mathbf{p}\sigma} = n_{\mathbf{p}\sigma} - \bar{n}_p{}^0$$

Similarly, the quantity $\delta \bar{n}_{p\sigma}$ is the departure of the total number $n_{\mathbf{p}\sigma}$ from this equilibrium concentration. The two quantities $\delta n_{\mathbf{p}\sigma}$ and $\delta \bar{n}_{\mathbf{p}\sigma}$ are not independent and are related through the Fermi liquid parameters. This dependence can be derived by assuming that $\delta n_{\mathbf{p}\sigma}$ is a small quantity and expanding $\delta\bar{n}_{\mathbf{p}\sigma}$ in this parameter. The key step is the recognition that the difference $\bar{\varepsilon}_p - \varepsilon_p$ is proportional to $\delta n_{\mathbf{p}\sigma}$:

$$\begin{aligned}
\delta\bar{n}_{\mathbf{p}\sigma} &= n_{\mathbf{p}\sigma} - n_F(\bar{\varepsilon}_p - \mu) \\
&= n_{\mathbf{p}\sigma} - n_F(\varepsilon_p - \mu + \bar{\varepsilon}_p - \varepsilon_p) \\
&= n_{\mathbf{p}\sigma} - n_F(\varepsilon_p - \mu) - (\bar{\varepsilon}_p - \varepsilon_p)\frac{dn_F(\varepsilon_p - \mu)}{d\varepsilon_p} \\
\delta\bar{n}_{\mathbf{p}\sigma} &= \delta n_{\mathbf{p}\sigma} - \frac{dn_F}{d\varepsilon}\frac{1}{\nu}\sum_{\mathbf{p}'\sigma'} f_{\mathbf{p}\sigma,\mathbf{p}'\sigma'}\delta n_{\mathbf{p}'\sigma'}
\end{aligned} \tag{10.4.5}$$

The first two terms are just $\delta n_{\mathbf{p}\sigma}$. The second term contains the factor $\bar{\varepsilon}_p - \varepsilon_p$, which is also proportional to $\delta n_{\mathbf{p}\sigma}$. It has the factor $-dn_F(\varepsilon_p - \mu)/d\varepsilon_p$, which becomes the delta function $\delta(\varepsilon_p - \mu)$ at zero temperature. One can show that every term in Eq. (10.4.5) has this factor, so we divide it out of the expression. Furthermore, the spin dependence of $\delta n_{\mathbf{p}\sigma}$ is written as a symmetric or antisymmetric combination $\delta n_{\mathbf{p}\pm} = \delta n_{\mathbf{p}}{}^s \pm \delta n_{\mathbf{p}}{}^a$. All these results are collected, and the distribution functions are expanded in spherical harmonic functions $[\delta n_{\mathbf{p}}^{s,a} = \frac{1}{2}(\delta n_{\mathbf{p}\uparrow} \pm \delta n_{\mathbf{p}\downarrow})]$:

$$\begin{aligned}
\delta n_{\mathbf{p}\pm} &= -\frac{dn_F(\varepsilon_p - \mu)}{d\varepsilon_p}\sum_{lm} Y_{lm}(\theta, \varphi)(\delta n_{lm}^s \pm \delta n_{lm}^a) \\
\delta\bar{n}_{\mathbf{p}\pm} &= -\frac{dn_F(\varepsilon_p - \mu)}{d\varepsilon_p}\sum_{l,m} Y_{lm}(\theta, \varphi)(\delta\bar{n}_{lm}^s \pm \delta\bar{n}_{lm}^a) \\
\delta\bar{n}_{lm}^{s,a} &= \delta n_{lm}^{s,a} + 2\int\frac{d^3p'}{(2\pi)^3}\sum_{l'm'}\int\frac{d\Omega_p}{4\pi}\left(-\frac{dn_F}{d\varepsilon_{p'}}\right)Y_{lm}^*(\theta, \varphi) \\
&\quad \times f_{\mathbf{p}'\mathbf{p}}^{s,a} Y_{l'm'}^*(\theta', \varphi')\delta n_{l'm'}^{s,a}
\end{aligned} \tag{10.4.6}$$

When we integrate over angle and collect all terms with the same angular dependence, we arrive at the relationship between the two distributions of excited quasiparticles:

$$\begin{aligned}
\delta\bar{n}_{lm}^s &= \delta n_{lm}^s\left(1 + \frac{F_l^s}{2l + 1}\right) \\
\delta\bar{n}_{lm}^a &= \delta n_{lm}^a\left(1 + \frac{F_l^a}{2l + 1}\right)
\end{aligned} \tag{10.4.7}$$

These results will be useful later.

The distribution $\delta\bar{n}_{\mathbf{p}\sigma}$ seems more important than $\delta n_{\mathbf{p}\sigma}$. For example, Pines and Nozières show that the particle current operator is given in terms of $\delta\bar{n}^s_{\mathbf{p}\sigma}$:

$$\mathbf{J} = \sum_{\mathbf{p}\sigma} \frac{\mathbf{p}}{m^*}\, \delta\bar{n}_{\mathbf{p}\sigma} = \frac{2}{m^*} \sum_{\mathbf{p}} \mathbf{p}\, \delta\bar{n}_{\mathbf{p}}{}^s \tag{10.4.8}$$

It is informative to ask what current operator is expressed in terms of the "bare"-quasiparticle distribution $\delta n_{\mathbf{p}\sigma}$. Pines and Nozières show that the particle current operator can be expressed in terms of the bare mass m of ^{3}He as

$$\mathbf{j} = \sum_{\mathbf{p}\sigma} \frac{\mathbf{p}}{m}\, \delta n_{\mathbf{p}\sigma} = \frac{2}{m} \sum_{\mathbf{p}} \mathbf{p}\, \delta n_{\mathbf{p}}{}^s \tag{10.4.9}$$

The two current operators (10.4.8) and (10.4.9) refer to the same particle density and are identical. They can be used to derive a relationship between m^* and m. The current operators will be finite if the distribution functions $\delta\bar{n}_{\mathbf{p}\sigma}$ and $\delta n_{\mathbf{p}\sigma}$ have a nonzero component of the quasiparticle distribution for angular momentum $l = 1$, say the value $l = 1$, $m = 0$. Then the two-current expression can be evaluated by using the definitions of $\delta\bar{n}_{\mathbf{p}\sigma}$ and $\delta n_{\mathbf{p}\sigma}$ in (10.4.6):

$$\begin{aligned} J_Z &= \frac{p_F N_{F0}}{(12\pi)^{1/2}} \frac{\delta\bar{n}^s_{10}}{m^*} \\ &= \frac{p_F N_{F0}}{(12\pi)^{1/2}} \frac{\delta n^s_{10}}{m} \end{aligned}$$

We equate these expressions and use the relationship (10.4.7) between $\delta\bar{n}^s_{lm}$ and δn^s_{lm} to derive the final formula:

$$\frac{m^*}{m} = \frac{\delta\bar{n}^s_{1m}}{\delta n^s_{1m}} = 1 + \frac{1}{3} F_1{}^s$$

Thus a measurement of m^* gives the value of $F_1{}^s$. We previously remarked that m^* was $3m$, so that $F_1{}^s = 6$. The Fermi liquid parameters can be large relative to unity.

Table 10.2 shows some predictions of Fermi liquid theory regarding the effect of the quasiparticle interactions upon measurable quantities. The first is the specific heat, which is enhanced by the effective mass m^*, as it is for the electron gas. Thus the quantity F_1 is obtained by measuring this quantity. The isothermal compressibility and also the velocity of ordinary sound are changed by the ratio of $(1 + F_0{}^s)/(1 + F_1{}^s/3)$. For a noninteracting gas of ^{3}He particles, one would estimate the sound velocity from the formula $c_0{}^2 = \frac{2}{3} m^{-1} E_{F0}$ to be 95 m/sec. The experimental value is 182 m/sec. The ratio is $c/c_0 = \frac{182}{95} = 1.91 = (1 + F_0{}^s)^{1/2}/(1 + F_1{}^s/3)^{1/2}$.

TABLE 10.2. Fermi Liquid Theory: $N_{F0} = mp_F/\pi^2\hbar^3 = \frac{3}{2}(n_0/E_{F0})$

Quantity	Free particles	Fermi liquid theory
Specific heat	$C_{V_0} = \dfrac{\pi^2}{3} k_B^2 T N_{F0}$	$\dfrac{C_V}{C_{V_0}} = \dfrac{m^*}{m} = 1 + \dfrac{1}{3} F_1^s$
Compressibility	$\dfrac{1}{K_f} = \dfrac{n_0^2}{N_{F0}}$	$\dfrac{K}{K_f} = \dfrac{1 + F_0^s}{1 + F_1^s/3}$
Sound velocity	$c_0^2 = \dfrac{n_0}{mN_{F0}} = \dfrac{v_F^2}{3}$	$\left(\dfrac{c}{c_0}\right)^2 = \dfrac{K}{K_f} = \dfrac{1 + F_0^s}{1 + F_1^s/3}$
Spin susceptibility	$\chi_0 = \mu_0^2 N_{F0}$	$\dfrac{\chi}{\chi_0} = \dfrac{1 + F_1/3}{1 + F_0^a}$

Since F_1^s is known, we then deduce the value of $F_0^s = 10$. A recent compilation of these parameters is found in Table 10.3, as given by Wheatley (1975). They are also compared to the theoretical values of Feenberg and his associates, which will be discussed later.

The other interesting number is F_0^a, which is the isotropic term in the spin-dependent part of the interaction. It is deduced from the spin susceptibility, which is found to be much larger than the free-particle value χ_0. χ_0 is derived by the following simple argument. The susceptibility is merely the net magnetization **M** divided by the applied magnetic field $\mathbf{H}_0$. The magnetization is the summation over the quasiparticle distribution times a nuclear moment μ_0:

$$\mathbf{M} = \mu_0 \sum_{\mathbf{p}\sigma} \boldsymbol{\sigma} \delta n_{\mathbf{p}\sigma}$$

However, the spin energy $\mu_0 \boldsymbol{\sigma} \cdot \mathbf{H}_0$ perturbs the local equilibrium of quasiparticles. The quasiparticle distribution $n_\mathbf{p}$ is the Fermi function of the quasiparticle energy $\bar{\varepsilon}_\mathbf{p} - \mu - \mu_0 \boldsymbol{\sigma} \cdot \mathbf{H}_0$:

$$\begin{aligned}\delta \bar{n}_{\mathbf{p}\sigma} &= n_F(\bar{\varepsilon}_\mathbf{p} - \mu - \mu_0 \boldsymbol{\sigma} \cdot \mathbf{H}_0) - n_F(\bar{\varepsilon}_\mathbf{p} - \mu) \\ &= -\mu_0 \sigma H_0 \left(\frac{\partial n_F}{\partial \bar{\varepsilon}_p}\right)\end{aligned}$$

The change in quasiparticle density due to the magnetic field is therefore

TABLE 10.3. Fermi Liquid Parameters: $A_l = F_l/[1 + F_l/(2l+1)]$

Parameter	Experiment[a,b]	Theory[c]	Parameter	Experiment
F_0^s	10.07	6.94	A_0^s	0.91
F_1^s	6.04	3.84	A_1^s	2.00
F_0^a	−0.67	−0.75	A_0^a	−2.03
F_1^a	−0.46	−0.99	A_1^a	−0.55

[a] Wheatley (1975).
[b] Dy and Pethick (1969).
[c] Feenberg (1969).

a change in $\delta\bar{n}_{\mathbf{p}\sigma}$, which can be related to $\delta n_{\mathbf{p}\sigma}$ by using (10.4.7):

$$M = 2\mu_0 \sum_{\mathbf{p}} \delta n_{\mathbf{p}}{}^a = \mu_0{}^2 H_0 N_F \left(\frac{\delta n_0{}^a}{\delta \bar{n}_0{}^a} \right)$$

$$\chi = \frac{M}{H_0} = \frac{\chi_0}{1 + F_0{}^a} \left(\frac{m^*}{m} \right)$$

$$\chi_0 = \mu_0{}^2 N_{F0}$$

The enhancement of χ over χ_0 suggests that F_0^a is negative and near unity, so that the denominator is small in the theoretical expression for the susceptibility. We shall see later that this has an important consequence for the theory of superfluidity. The large positive value of F_0^s and the negative value of F_0^a combine to suppress the quasiparticle interaction in the s-wave channel and enhance it in the p-wave channel. This is the basic reason that the superfluid phases in ^{3}He pair up in the spin triplet state, which has a p-wave orbital part.

B. Experiments and Microscopic Theories

The microscopic theories of liquid ^{3}He are even more complicated than the theories of ^{4}He, because the ground state is more complicated. This happens because of the antisymmetrization of the many-particle wave function. If this ground state wave function is defined as $\Psi_g(x_1, \ldots, x_N)$, where $x_j \equiv (\mathbf{r}_j, \sigma_j)$ are general coordinates for both position and spin, then the wave function changes sign under the exchange of any pair of coordinates x_i and x_j:

$$\Psi_g(x_1, x_2, \ldots, x_i, \ldots, x_j, \ldots, x_N) = -\Psi_g(x_1, x_2, \ldots, x_j, \ldots, x_i, \ldots, x_N) \tag{10.4.10}$$

For a noninteracting system of fermions, we describe the particle states by single-particle orbitals $\phi_p(x_j)$, and the many-particle wave function in the Hartree–Fock approximation is a Slater determinant of these orbitals. This leads, in a natural way, to the condition that no two particles can occupy the same single-particle orbital state, or else the wave function is trivially zero.

The nature of the ground state wave function is less clear in this strongly interacting and highly correlated liquid. Our own microscopic picture of a liquid has the atoms bouncing around inside of a small cage defined by the average position of neighboring atoms. This picture is correct for liquid ^{4}He and leads to a correlated wave function of the CBF form, which is highly successful for describing this system. It is not nearly so obvious how to construct a ground state wave function for highly correlated fermions. There probably should be something equivalent to the CBFs but for fermions. The question is how the fermion nature of the particles is expressed in real space. When we visualize the atoms bouncing around, how is this motion different when they are fermions rather than bosons? How does one construct a CBF which has the fundamental antisymmetrization property (10.4.10)?

We shall describe the approach of E. Feenberg and his hard-working associates: F. Y. Wu, C. W. Woo, W. E. Massey, and H. T. Tan. Their efforts have provided the most successful microscopic theory to date, which compares favorably to the available experiments. Their ground state wave function for liquid ^{3}He has the form of an algebraic product of two familiar wave functions: the CBF with the form used for ground state of ^{4}He and a Slater determinant of plane-wave states with spin:

$$\Psi_g(x_1, x_2, \ldots, x_N) = \Psi_0(\mathbf{r}_1, \ldots, \mathbf{r}_N)\Phi(\mathbf{r}_1, \ldots, \mathbf{r}_N)$$

$$\Psi_0 = L_N \exp\left[-\sum_{i>j}\left(\frac{b}{|\mathbf{r}_i - \mathbf{r}_j|}\right)^5\right]$$

$$\Phi = \det|\phi_p(x_1)\phi_{p'}(x_2)\cdots|$$

The CBF was taken to be the form (10.2.4) of a product of pairwise correlations, with a single short-range repulsive term as in (10.2.5). The value of b changes slightly from the value used in ^{4}He, because of the mass difference. The Slater determinant is the same one which would be used for a system of noninteracting particles and contains single-particle wave functions $\phi_P(x_1)$ which are plane-wave states and spin functions for spin up α or down β. The Slater determinant provides the necessary antisymmetrization, while the CBF provides the necessary short-range correlation in the atomic motion.

This wave function is used to calculate ground state properties such as the energy $E_g = \langle g \mid H \mid g \rangle / \langle g \mid g \rangle$, $S(k)$, etc. These calculations are extremely complicated and will not be described here, but we refer interested readers to Feenberg's book (1969). All many-particle matrix elements are evaluated by a cumulant expansion which must be carried to high order. They involve the evaluation of three- and four-particle correlations in the liquid state. Some of their results will now be summarized.

First we discuss the ground state energy. Feenberg and his associates first calculated the ground state energy assuming ^{3}He was a boson. That is, they evaluated $E_0 = \langle \Psi_0 \mid H \mid \Psi_0 \rangle$ where Ψ_0 is the CBF for liquid ^{3}He. Since this wave function is symmetric under coordinate exchange, E_0 is the energy of the equivalent boson liquid. The value of E_0 differs somewhat for the different ways of doing the cumulant expansion and estimating three- and four-particle correlations but has the typical value of $E_0 = -2.9$ K/atom. A boson liquid with the mass of ^{3}He is less bound than liquid ^{4}He, which reflects the increased zero-point motion of the lighter mass. Although the density of liquid ^{3}He is lower than liquid ^{4}He, the average value of kinetic energy per particle is nearly the same value of 13.4 K/atom. The average value of potential energy is -16.3 K/atom, giving the difference of -2.9 K/atom. Thus the average kinetic energy per atom is still high, and the particles still have short-range fluctuations in their position within the cage defined by the neighboring atoms.

The next calculation is to add the Slater determinant of the single-particle orbitals to the ground state wave function and to compute the new ground state energy. It is defined as $E_g = E_0 + \delta E$, where δE is the change in energy due to the motion of the particles in the Fermi sea. For a gas of noninteracting particles, this change would be $\delta E = 3E_{F0}/5 = 0.6(4.97) = 3.0$ K. Feenberg and his associates find the smaller value of $\delta E = 1.7$ K/atom, which is a small number compared to the average kinetic energy in the boson ground state, $\simeq 14$ K/atom. Thus the short-range motions of the particles are more important than the kinetic energy of motion in the Fermi sea. The final prediction of the ground state energy per atom $E_g = -1.2$ K/atom is only in fair agreement with the experimental value of -2.52 K/atom. However, the least accurate theoretical number is probably

E_0. There could easily be a 1-K error in the calculation of this quantity, as there was in the similar calculation for ^{4}He, where it was less noticeable as the difference between 6 and 7 K/atom. This number is difficult to find accurately because of the large cancellation between the kinetic and potential energies and their sensitivity on the trial function for the pairwise part of the CBF. Thus the agreement between theory and experiment, in liquid ^{3}He, must be regarded as satisfactory.

The next ground state property is the liquid structure factor $S(k)$. It can be measured by both X-ray and neutron scattering. The X-ray results are easiest to interpret and are given first. Figure 10.12 summarizes theory and X-ray data. The points are experimental, and the lines are the theories of Massey (1966) and Massey and Woo (1967). The agreement is obviously excellent. Since the theories were published as predictions several years before the experiments, the theoretical success is even greater. There is great similarity between $S(k)$ for liquid ^{3}He and ^{4}He. Both have peaks near $k \simeq 2.0$ Å^{-1} and then rapidly approach unity with small oscillations at large values of k. The $S(k)$ values for ^{3}He are about 40% higher than for ^{4}He in the low k region, $k \simeq 1$ Å^{-1}. The good agreement between theory and experiment for $S(k)$ means that similar good agreement exists between their Fourier transforms, which are the pair distribution functions $g(r)$.

The next experiment we wish to discuss is the neutron scattering. The theory for this is more complicated than for liquid ^{4}He. This difference is due to the spins of the neutron and the ^{3}He nucleus. The neutron actually scatters from the nucleus of the atom, and the scattering cross section will depend on the relative spin orientation of the neutron and ^{3}He nucleus. Denote the scattering amplitude from a single ^{3}He nucleus at position $\mathbf{R}_j$ by the quantity b_j. Then the amplitude for scattering from the liquid, with a momentum transfer $\mathbf{q}$, has the amplitude

$$A(\mathbf{q}, t) = \sum_j b_j e^{i\mathbf{q}\cdot\mathbf{R}_j(t)}$$

We can define a correlation function by taking the expectation value of this operator at different times. The spectral function of this correlation function is directly proportional to the differential scattering rate of the liquid sample:

$$B(\mathbf{q}, i\omega) = -\frac{1}{N}\int_0^\beta d\tau e^{i\omega\tau}\langle T_\tau A(\mathbf{q}, \tau)A(-\mathbf{q}, 0)\rangle$$

$$I(\mathbf{q}, \omega) = -2\,\mathrm{Im}[B(\mathbf{q}, \omega)] = \frac{1}{N}(1 - e^{-\beta\omega})\int_{-\infty}^{\infty} dt e^{it\omega}\langle A(\mathbf{q}, t)A(-\mathbf{q}, 0)\rangle$$

The quantity $I(\mathbf{q}, \omega)$ is proportional to the double differential cross section

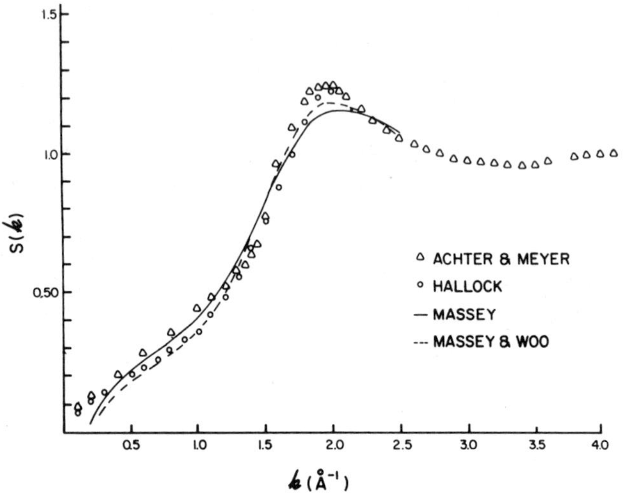

FIGURE 10.12. Plot of $S(k)$ for liquid ^{3}He. The points are the X-ray scattering data of Achter and Meyer (1969) and Hallock (1972). The solid curves are the theories of Massey (1966) and Massey and Woo (1967). The agreement is obviously excellent.

for the scattering of the neutron with an energy transfer ω and into a unit of solid angle with the momentum transfer $\mathbf{q}$.

$I(\mathbf{q}, \omega)$ is first evaluated for the case of liquid ^{4}He. This nucleus has no spin, so the neutron scattering cross section is the same for all the nuclei. The scattering amplitude is set equal to a constant b_0. The remaining correlation function can be expressed exactly in terms of the dynamic liquid structure factor:

$$\begin{aligned}
A(\mathbf{q}, t) &= b_0\varrho(\mathbf{q}, t) = b_0 \sum_j e^{i\mathbf{q}\cdot\mathbf{R}_j(t)} \\
B(\mathbf{q}, i\omega) &= b_0{}^2\chi(\mathbf{q}, i\omega) \\
I(\mathbf{q}, \omega) &= b_0{}^2 S_c(\mathbf{q}, \omega) \\
S_c(\mathbf{q}, \omega) &= \frac{1}{N}(1 - e^{-\beta\omega}) \int_{-\infty}^{\infty} dt e^{i\omega t} \langle \varrho(\mathbf{q}, t)\varrho(-\mathbf{q}, 0)\rangle
\end{aligned}$$

These are the same formulas which are given earlier in (10.2.8) and (10.2.13) for neutron scattering from ^{4}He. The symbol $\varrho(\mathbf{q}, t)$ is the usual density operator, which is sometimes written in the alternate but equivalent form $\varrho(\mathbf{q}, t) = \sum_{\mathbf{k}} C^\dagger_{\mathbf{k}+\mathbf{q}}(t) C_{\mathbf{k}}(t)$. We have called this dynamic liquid structure factor $S_c(\mathbf{q}, \omega)$, where the subscript c means *coherent*. It is exactly the same one which was defined earlier, except that now we have added a subscript. This subscript is necessary in the theory of liquids, where another correlation function is often useful, the *incoherent* dynamic structure factor:

$$S_I(\mathbf{q}, \omega) = \frac{1}{N}(1 - e^{-\beta\omega}) \int_{-\infty}^{\infty} dt e^{i\omega t} \sum_{j=1}^{N} \langle e^{i\mathbf{q}\cdot\mathbf{R}_j(t)} e^{-i\mathbf{q}\cdot\mathbf{R}_j(0)} \rangle \tag{10.4.11}$$

Whereas the coherent response follows the correlation of all the particle motions, the incoherent response just measures the motion of one particle as a function of time and then averages over all the particles. The neutron scattering for liquid ^{4}He is expressed entirely by the coherent liquid structure factor. This is not the case for liquid ^{3}He.

The neutron scattering amplitude for ^{3}He depends on the spin alignment of the neutron spin $\mathbf{S}$ and nuclear spin $\boldsymbol{\sigma}_j$ of the jth term. The amplitude is written as $b_j = b_1 + b_2\mathbf{S} \cdot \boldsymbol{\sigma}_j$, which assumes that the spin effects enter as exchange forces in the nucleus and not as spin–orbit or other interactions. This form of the scattering amplitude, along with the assumption that the neutrons are independent so that their spin correlates with itself according to $\langle S_\alpha \cdot S_\beta \rangle = \delta_{\alpha\beta}/4$, allows the correlation function to be factored into two terms:

$$\begin{aligned} \langle A(\mathbf{q}, t) A(-\mathbf{q}, 0) \rangle &= \sum_{j,l} \langle e^{i\mathbf{q}\cdot\mathbf{R}_j(t)} e^{-i\mathbf{q}\cdot\mathbf{R}_l(0)} [b_1 + b_2\mathbf{S}\cdot\boldsymbol{\sigma}_j(t)][b_1 + b_2\mathbf{S}\cdot\boldsymbol{\sigma}_l(0)] \rangle \\ &= \sum_{j,l} \left\langle e^{i\mathbf{q}\cdot\mathbf{R}_j(t)} e^{-i\mathbf{q}\cdot\mathbf{R}_l(0)} \left[b_1^2 + \frac{b_2^2}{4} \boldsymbol{\sigma}_j(t) \cdot \boldsymbol{\sigma}_l(0) \right] \right\rangle \end{aligned}$$

The cross terms average to zero, $\langle \mathbf{S} \cdot \boldsymbol{\sigma}_j \rangle = 0$, since there is no correlation between the neutron spin and the nuclear spin. The first term of the two remaining, which is proportional to b_1^2, is just the coherent liquid structure factor. The second term is what makes the theory for neutron scattering from ^{3}He different from that from ^{4}He. It contains the coefficient b_2^2 and the correlation of the nuclear spin operators. In fact, this term is just one form of the Pauli spin susceptibility. We define its spectral function as $S_p(\mathbf{q}, \omega)$:

$$\begin{gathered} S_p(\mathbf{q}, \omega) = \frac{1}{3N}(1 - e^{-\beta\omega}) \int_{-\infty}^{\infty} e^{i\omega t}\, dt \sum_{j,l} \langle e^{i\mathbf{q}\cdot\mathbf{R}_j(t)} e^{-i\mathbf{q}\cdot\mathbf{R}_l(0)} \boldsymbol{\sigma}_j(t) \cdot \boldsymbol{\sigma}_l(0) \rangle \\ I(\mathbf{q}, \omega) = b_1^2 S_c(\mathbf{q}, \omega) + \tfrac{3}{4} b_2^2 S_p(\mathbf{q}, \omega) \end{gathered} \tag{10.4.12}$$

The notation in the literature regarding $S_p(\mathbf{q}, \omega)$ is inconsistent, since sometimes this quantity is called $S_I(\mathbf{q}, \omega)$. However, the latter name is conventionally reserved for the correlation function in (10.4.11), which follows the motion of one particle. This confusion stems from the fact that the two correlation functions S_I and S_p are equal under certain conditions. They are proportional when there is no correlation between the orientation of the nuclear spins on different atoms. In this case the spin correlation is $\langle \boldsymbol{\sigma}_j \cdot \boldsymbol{\sigma}_l \rangle = 3\delta_{jl}$, since the σ are Pauli matrices, and we have the identity

$$S_p(\mathbf{q}, \omega) = S_I(\mathbf{q}, \omega) \tag{10.4.13}$$

Equation (10.4.13) is a good approximation in ordinary liquids, so that S_p is customarily equal to S_I.

However, ^{3}He does not form an ordinary liquid, and the approximation (10.4.13) is poor since the assumption $\langle \boldsymbol{\sigma}_j \cdot \boldsymbol{\sigma}_l \rangle = 3\delta_{jl}$ is invalid. We have previously mentioned that the Pauli spin susceptibility of liquid ^{3}He was much larger than expected for a noninteracting liquid. It is also larger than one would expect from the increased density of states at the Fermi surface due to the large effective mass. For this reason, we deduced the value of the Fermi liquid parameter to be $F_0^a = -0.67$, since this negative value near unity caused a threefold enhancement in the theoretical susceptibility. This threefold enhancement is regarded as evidence that liquid ^{3}He has a strong tendency toward being ferromagnetic. If one nuclear spin is pointing, say, up, then other nearby spins in the liquid also have a tendency to be pointing in the same direction. This correlation causes the zz spin susceptibility χ_{zz} in (3.8.1) to be large, since there is a pronounced correlation between neighboring spins in the liquid. This correlation is not sufficient to cause long-range spin ordering or to make the liquid have a magnetic phase transition to the ferromagnetic state. Instead, the ordering among the spins extends over a shorter range of a few neighbors. Similar phenomena happen in some transition metals such as Pd; they are not ferromagnetic but do have a large Pauli susceptibility. It was argued by Berk and Schrieffer (1966) that this large susceptibility is evidence for the tendency of the spins to locally align. They also argued that this will tend to suppress the onset of superconductivity in these materials, since the BCS pairing in metals prefers to have the two spins antiparallel, while the ferromagnetic tendency opposes this. It is interesting that in liquid ^{3}He the ferromagnetic tendencies tend to enhance the onset of superfluidity, because the pairing is in the triplet state where the spins are parallel.

The tendency of local spins to align parallel, in either liquid ^{3}He or transition metals, is not due to the dipole forces between spins. These are too weak, particularly in liquid ^{3}He. Instead, it is the exchange effects in the fermion systems which cause differing interaction between atoms which depend on the spin orientation.

The conclusion is that the nearby spins in liquid ^{3}He do have correlation in their alignments, and it is a poor approximation to replace the spin susceptibility $S_p(\mathbf{q}, \omega)$ by the incoherent structure factor $S_I(\mathbf{q}, \omega)$. The neutron scattering measures the summation of the two correlation functions

S_c and S_p. This confuses the interpretation of the experiments, which measure only *one* spectrum, $I(\mathbf{q}, \omega)$. Initially it appears to be a delicate task to separate the coherent structure factor from the one for spins. The fact that accurate theoretical values are available for b_1^2 and b_2^2 is some help. Some further guidance is obtained from the X-ray data, which measures only S_c since the X rays do not notice the spin state of the nucleus. Unfortunately, so far the X-ray results have provided only the static liquid structure factor $S_c(\mathbf{q})$, which is the same thing which has been previously called $S(\mathbf{q})$. It is an integral over the dynamic structure factor:

$$S(\mathbf{q}) = \int_{-\infty}^{\infty} \frac{d\omega S(\mathbf{q}, \omega)}{1 - e^{-\beta\omega}}$$

Thus a prior knowledge of the X-ray data provides some additional clues when interpreting the neutron scattering data. Nevertheless, the separation of $I(\mathbf{q}, \omega)$ into the separate factors of $S_c(\mathbf{q}, \omega)$ and $S_p(\mathbf{q}, \omega)$ can be done only with theoretical guidance. With these remarks in mind, let us now examine the experimental data.

The interesting results occur for low temperatures and small values of q and ω. The only data available in this range is due to Sköld *et al.* (1976). They measured the neutron scattering cross section at fixed scattering angle as a function of energy loss. This data was processed on the computer to produce graphs of fixed $\mathbf{q}$ as a function of energy loss. It is shown in Fig. 10.13, where it has been normalized to be $S(\mathbf{q}, \omega)$, which they define as

$$S(\mathbf{q}, \omega) = S_c(\mathbf{q}, \omega) + \frac{\sigma_I}{\sigma_c} S_p(\mathbf{q}, \omega)$$

where $\sigma_I/\sigma_c = 3b_2^2/4b_1^2$ is the ratio of the relative cross sections of the coherent and spin-dependent cross sections. Previous theoretical work obtained $\sigma_I/\sigma_c = 0.25$ for this ratio.

This data was interpreted in several ways. The first is to take the total area under each scattering curve, which gives the static structure factor $S(q)$. Thermal smearing is negligible at the experimental temperatures of 15 mK (or 0.015 K). $S(q)$ is shown as the top series of dots in Fig. 10.14; this series is compared with the solid line for the X-ray data. The latter measurement obtains $S_c(q)$, as mentioned before. The X-ray data is systematically below the neutron data. If these two results are subtracted, the difference can be interpreted as the spin-dependent part $(\sigma_I/\sigma_c)S_p$. These values are the bottom row of dots in Fig. 10.14, which cluster around the value 0.25. The lower dots are interpreted as the value of $S_p(q) = 1.0$

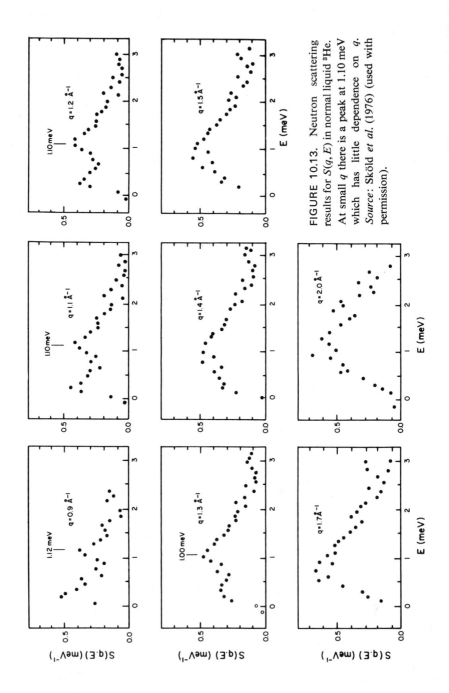

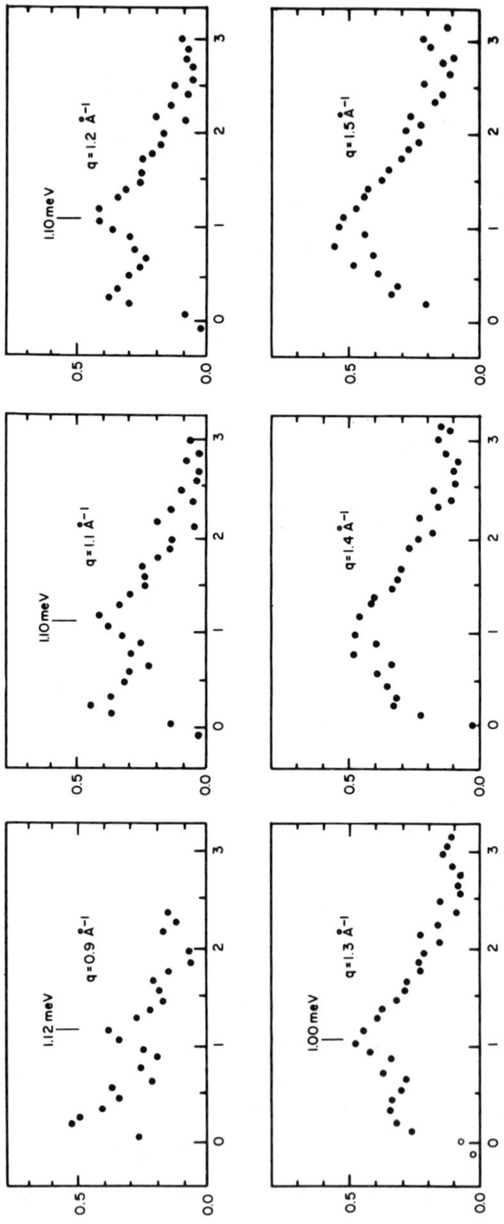

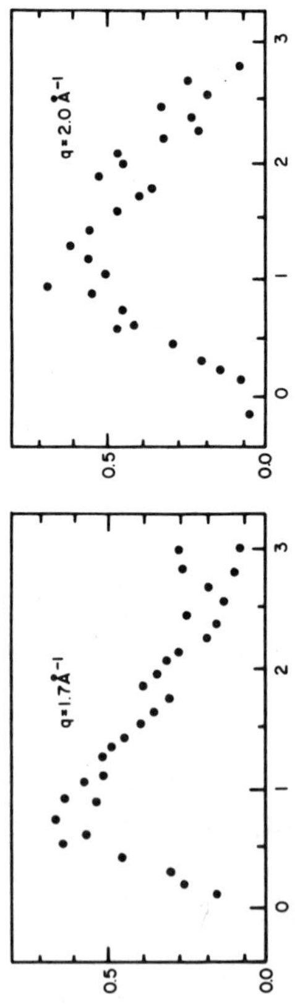

FIGURE 10.13. Neutron scattering results for $S(q, E)$ in normal liquid ^{3}He. At small q there is a peak at 1.10 meV which has little dependence on q. *Source*: Sköld *et al.* (1976) (used with permission).

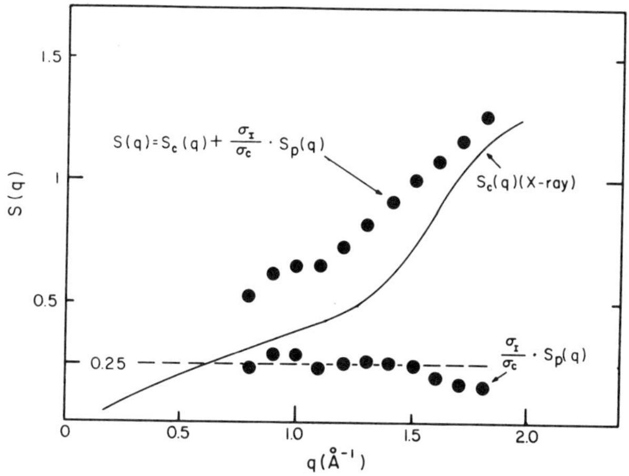

FIGURE 10.14. Neutron scattering results for the structure factor $S(q)$ in normal liquid ³He. The solid line is the X-ray structure factor of Hallock (1972). The top set of points shows the total neutron results. Neutron results are subtracted from the X-ray results to produce the bottom set of points, which scatter about 0.25. The bottom points are due to the spin-dependent scattering. *Source*: Sköld *et al.* (1976) (used with permission).

multiplied by the ratio of cross sections $\sigma_I/\sigma_c = 0.25$. The values are shown in the figure as falling below 0.25 at larger momentum transfers. The experimentalists suggest this falloff is an artifact of the experimental apparatus, which prevents measurement to larger $E = \hbar\omega$, so the integral over the area is not taken to large enough values. They interpret their results as showing $S_p(q) = 1.0$ at these small values of q and provide confirmation that the ratio of cross sections is 0.25. Presumably the spin structure factor $S_p(q)$ will vary from unity at smaller values of q. If $g_p(r)$ is the Fourier transform of this function, then its spatial variation extends over a much wider distance of r than does the density variation $g_c(r)$. This conclusion is in accord with our earlier remarks about the spin correlations encompassing many neighbors.

The data in Fig. 10.13 is now examined more carefully. A significant feature at small values of q is the peak near 1.0 meV, which has only a slight variation with q. The peak width equals the experimental resolution, so the actual excitation width is very small, which corresponds to a long-lived excitation in the normal liquid. The excitation curve is shown in Fig. 10.15 as a series of dots. This figure shows the excitation spectrum of the liquid. There is also the particle–hole excitations of the Fermi liquid, which are shown in the scattering data of Fig. 10.13 as the broad band of scattering at large values of q.

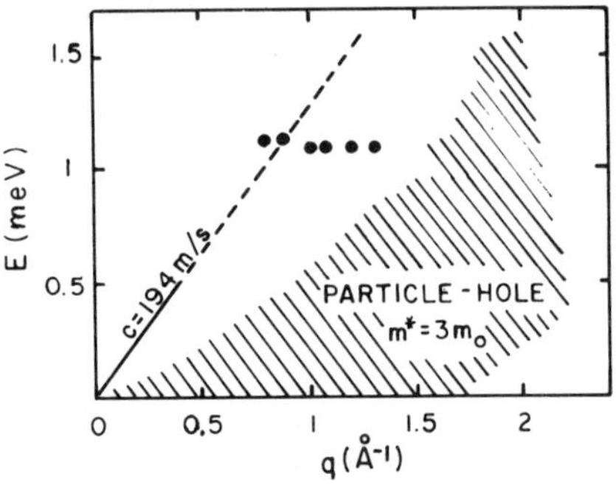

FIGURE 10.15. The dots are the neutron scattering peaks vs. q for the zero-sound mode together with the linear zero-sound dispersion curve for $c = 194$ m/sec. Also shown are the pair excitations of particles and holes. *Source*: Sköld *et al.* (1976) (used with permission).

At the smallest values of q, there is another sharp peak at very small values of $E = \hbar\omega$. This peak could be interpreted as due to the electron–hole pairs of the Fermi system. Recall that our plots of $\text{Im}[1/\varepsilon(q, \omega)]$ for the electron gas had similar peaks from electron–hole pairs, and perhaps this peak could have a similar origin. The quantity $S_c(q, \omega)$ is directly proportional to $\text{Im}[1/\varepsilon(q, \omega)]$, so it is natural to compare these two results for the electron gas and liquid ^{3}He. Although this interpretation is possible, it is not widely held. Theoretical calculations by Aldrich *et al.* (1976) show that the electron–hole pair spectrum is greatly suppressed in this region. Instead, the large peak at small E is thought to be due to the spin-dependent part $S_p(q, \omega)$, whose excitation spectrum occurs at small values of ω. This interpretation is reinforced by a measurement of the area beneath this peak. For small values of q, where it can be realistically distinguished from the other data, the area under this low-energy peak is unity, which would be appropriate for $S_p(q) = 1.0$. Thus we conclude that the excitation spectrum of normal liquid ^{3}He shows three kinds of contributions: (1) a sharp collective mode at 1.10 meV for small q, (2) spin excitations at small q and small ω, and (3) electron–hole pairs which make a broad band of excitations at larger values of q. The electron–hole excitations are expected in a degenerate Fermi system, so little more needs to be said about them. The other two excitations merit further comment.

The liquid must have a long-wavelength collective oscillation. In ordinary liquids this is just the longitudinal sound wave. In the electron gas it is at the plasma frequency ω_p, where the finite charge of the electron gas makes the mode have a finite frequency even in the limit where $q \to 0$. Since liquid ^{3}He is electrically neutral, the long-wavelength density oscillation will just be a sound wave. The next question is to ask what becomes of this sound wave at larger wave vectors, in particular at the values $q = 1.0$ Å^{-1}, where it is shown by the neutron measurements. For a Debye spectrum with $\omega_q = cq$ and with $c = 182$ m/sec, one finds that $\omega_q = 1.10$ meV at $q = 1.0$ Å^{-1}. The value $\omega = 1.1$ meV is right at the frequency of the sharp collective mode. However, this mode is not increasing with q but decreases slightly with increasing q. Nevertheless, it is undoubtedly the continuation, at these intermediate values of q, of the long-wavelength sound modes. It is useful to recall the same discussion for liquid ^{4}He. There we calculated the Bijl–Feynman excitation spectrum by multiplying the ground state wave function by the density operator $\varrho(\mathbf{q})$ and then taking the expectation value of this excitation wave function:

$$\Psi_{\mathbf{q}} = \varrho(\mathbf{q})\Psi_g$$

$$E_g + \omega_0(\mathbf{q}) = \frac{\langle \Psi_{\mathbf{q}}{}^* H \Psi_{\mathbf{q}} \rangle}{\langle \Psi_{\mathbf{q}}{}^* \Psi_{\mathbf{q}} \rangle} = \frac{q^2}{[2mS_c(q)]} + E_g$$

An important feature of this calculation is that the result is not restricted to bosons or to liquid ^{4}He. Recall that this calculation entailed only three assumptions: (1) that Ψ_g was the lowest eigenstate of H, (2) that the commutator of $\varrho(\mathbf{q})$ with H gave the kinetic energy operator ε_q, and (3) that the expectation value of $\langle g \mid \varrho(\mathbf{q})^\dagger \varrho(\mathbf{q}) \mid g \rangle$ is $S(q)$. The last condition is a definition, while the first two are satisfied if Ψ_g is the ground state wave function of any neutral liquid system. Thus the Bijl–Feynman theory applies equally well to liquid ^{3}He. It should also have a Bijl–Feynman excitation spectrum equal to $\omega_0 = q^2/[2mS_c(k)]$. This excitation spectrum will rise as q is increased, round off, and start decreasing until the roton minimum. Unlike liquid ^{4}He, there are other excitations of the density–density response function, namely electron–hole pairs, which cause a great smearing of the excitation spectrum in segments where they overlap. Of course, this damping also happens to the plasmons in an electron gas, which become significantly broadened when their dispersion curve overlaps the pair spectrum. Thus we interpret the peak at 1.10 meV as an excitation of the Bijl–Feynman type, which is the sound wave at lower values of q and which is broadened by the pairs at higher values of q. Presumably, as was the case for liquid ^{4}He, a good calculation of this curve would also involve backflow, which is equivalent to the mixing of double excitation states. We are not aware that this has yet been calculated.

There are several different kinds of sound waves which can propagate in these systems. The ordinary sound waves at very long wavelengths are called *first sound.* Their speed is calculated by including the particle collisions, since such collisions are very frequent compared to the sound frequency. This results in a hydrodynamic version of the sound propagation.

At very much higher sound frequencies, the collision rate between particles becomes small compared to the sound frequency, and the speed must be calculated in the collisionless approximation. This regime is called *zero sound* and has a higher frequency. The sound mode at 1.10 meV in liquid ^{3}He is a zero-sound mode, which is described by the Bijl–Feynman formula. The Bijl–Feynman formula can also be used to describe first sound if the appropriate formula for $S_c(q)$ is used in the hydrodynamic regime.

The third excitation of liquid ^{3}He are the spin fluctuations at low energy. These will be treated fully in the following section.

Table 10.3 shows the values of the Fermi liquid parameters. Some theoretical results from Feenberg's book (1969) are shown. They were calculated from the microscopic theory. Feenberg and his associates constructed excited state wave functions as products of the CBFs and Slater determinants with excited single-particle orbitals. These were used to calculate the off-diagonal matrix elements of the Hamiltonian, which give the scattering amplitudes for different spin configurations. They were resolved into spherical harmonics, which gives the desired coefficients F_l^s and F_l^a. Their results show the same trends as the experimental values, namely that F^s is large and positive, while the F^a are negative. The other parameters shown in Table 10.3, A_l^s and A_l^a, will be explained later.

C. Interaction between Quasiparticles: Excitations

Fermi liquid theory is not a complete description of all the dynamical motion of quasiparticles in liquid ^{3}He. The theory was constructed by Landau to explain thermodynamic and static measurements such as specific heat and spin susceptibility, and it does that very well. Now it is time to consider explaining other measurements, and we must assess whether enough information is available in the Fermi liquid parameters or whether additional facts are needed about the quasiparticles. The additional measurements we may wish to explain are transport experiments such as thermal conductivity and viscosity, and later we shall want to explain superfluidity. So far all we have discussed about the quasiparticles are their effective mass and the four interactions parameters $F_{0,1}^s$ and $F_{0,1}^a$. First we have to decide what these parameters mean in terms of the microscopic scattering theory, and next we have to decide what additional information we need to fully describe the interaction between quasiparticles. Of course, we also have to discuss the low-energy spin fluctuations, since this topic is left over from the prior section.

It is useful to remember the main features of electrons in metals, which is the other Fermi liquid we have studied. The electrons at the Fermi surface have an effective mass m^* which is comprised of several contributions. Those of the electron–electron interactions, and also the band structure, gave a contribution to m/m^* which was not very energy dependent. However, the contribution from electron–phonon interactions was often very large and varied strongly with energy, so that it was negligible a Debye energy away from the Fermi surface. In liquid ^{3}He, one should ask the question whether the ratio $m^*/m = 3.0$ changes rapidly or slowly as the energy of the quasiparticles is varied. Or, to rephrase the question, what is the appropriate energy scale which determines the variation of these parameters? Another motivation for this question is provided by recalling the BCS equation for the transition temperature:

$$k_B T_c = 1.14\omega_D \exp\left(-\frac{1}{N_F \mid V_0 \mid}\right) \tag{10.4.14}$$

The exponent has the factors $N_F V_0$ which are the density of states and the effective interaction between quasiparticles at the Fermi surface. These two parameters, as we shall show, can be estimated from the parameters of Fermi liquid theory. However, the prefactor in this equation contains the Debye energy ω_D, which is the effective energy range over which the phonon attraction between quasiparticles is operative. This effective energy range of the interaction cannot be obtained from the parameters of Fermi liquid theory and must be deduced from other considerations.

In metals, superconductivity is usually caused by the electron–phonon interaction. This theory has been made very quantitative. The accuracy is helped by a number of smallness parameters, such as c_s/v_F, ω_D/E_F, and m/M, in which the theoretical expressions can be expanded. None of these apply to liquid ^{3}He. It has $c_s > v_F$, and the "sound" energy of 1.10 meV exceeds the Fermi energy of $k_B T_F \simeq 3$ K $= 0.26$ meV. In fact, no obvious smallness parameter comes to mind. Nevertheless, there probably is one since the ratio of $T_c/T_F \simeq 10^{-3}$ is similar between metals and liquid ^{3}He.

The scattering of two quasiparticles can be described in the following manner. Two quasiparticles initially have momenta $\mathbf{p}_1$ and $\mathbf{p}_2$. They interact and exchange momentum $\mathbf{q}$ and so end up in the states $\mathbf{p}_3 = \mathbf{p}_1 + \mathbf{q}$ and $\mathbf{p}_4 = \mathbf{p}_2 - \mathbf{q}$. Let us describe this scattering of the quasiparticles by a phenomenological matrix element $M_{\sigma\sigma'}(\mathbf{p}_1, \mathbf{p}_2; \mathbf{p}_3, \mathbf{p}_4)$. The spin indices are included, since the effective scattering is spin dependent. Our goal is to obtain a quantitative description of this matrix element, but we shall fall short of it. The matrix element is defined to include exchange events, so it describes the direct scattering plus the exchange scattering.

The first step is to find the relationship between this matrix element and the Fermi liquid parameters. We shall not be rigorous but shall obtain this identification only by an intuitive argument. In a weakly interacting electron gas, the Hartree energy is given by

$$\Sigma_H = \frac{1}{\nu} \sum_{\mathbf{p}'} n_F(\xi_{\mathbf{p}'}) V(\mathbf{q} = 0)$$

The self-energy of diagram for Σ_H has a particle of momentum $\mathbf{p}$ scatter from the other particles with momentum $\mathbf{p}'$ and density $n_F(\xi_{p'})$. No momentum is exchanged, so the self-energy is expressed in terms of the potential $V(0)$ at zero-momentum transfer ($\mathbf{q} = 0$), which is equivalent to the forward scattering amplitude for the two particles $\mathbf{p}$ and $\mathbf{p}'$. The exchange self-energy $\Sigma_x(p)$ just provides the exchange corrections to this forward scattering. If we apply this theory to calculating the self-energies of quasiparticles in liquid ^{3}He, we get

$$\Sigma_H + \Sigma_x = \frac{1}{\nu} \sum_{\mathbf{p}'\sigma'} M_{\sigma\sigma'}(\mathbf{p}, \mathbf{p}'; \mathbf{p}, \mathbf{p}') n_{\mathbf{p}'\sigma'}$$

The potential $V(0)$ has been replaced by the forward scattering matrix element $M_{\sigma\sigma'}(\mathbf{p}, \mathbf{p}'; \mathbf{p}, \mathbf{p}')$, and the distribution of quasiparticles is written as $n_{\mathbf{p}'\sigma'}$. When an excitation is created, the change in this self-energy function due to the excitation is

$$\frac{1}{\nu} \sum_{\mathbf{p}'\sigma'} M_{\sigma\sigma'}(\mathbf{p}, \mathbf{p}'; \mathbf{p}, \mathbf{p}') \delta n_{\mathbf{p}'\sigma'}$$

This expression is very familiar, since it is exactly the form of the interaction energy in Fermi liquid theory, where we recognize that $f_{\mathbf{p}\sigma,\mathbf{p}'\sigma'} = M_{\sigma\sigma'}(\mathbf{p}, \mathbf{p}'; \mathbf{p}, \mathbf{p}')$. Thus the Fermi liquid parameters represent the *forward scattering amplitude between bare quasiparticles.* It will be necessary to obtain the scattering amplitude for other directions besides forward and also to have it for dressed quasiparticles of energy $\bar{\varepsilon}_p$ rather than bare ones of energy ε_p.

The objective is to have a model for the scattering of two quasiparticles. The parameters of the model are chosen to reproduce the Fermi liquid theory for scattering in the forward direction. One method of doing this is to write an effective quasiparticle Hamiltonian with creation and destruction operators:

$$H = \sum_{\mathbf{p}\sigma} (\varepsilon_p - \mu) C^\dagger_{\mathbf{p}\sigma} C_{\mathbf{p}\sigma} + \frac{1}{2\nu} \sum_{\substack{\mathbf{pp'q}\\ \sigma\sigma'}} \bar{M}_{\sigma\sigma'}(\mathbf{p}, \mathbf{p}'; \mathbf{p} + \mathbf{q}, \mathbf{p}' - \mathbf{q}) \times C^\dagger_{\mathbf{p}+\mathbf{q}\sigma} C^\dagger_{\mathbf{p}'-\mathbf{q}\sigma'} C_{\mathbf{p}'\sigma} C_{\mathbf{p}\sigma}$$

The matrix element $\bar{M}_{\sigma\sigma'}$ is different from $M_{\sigma\sigma'}$ because $\bar{M}_{\sigma\sigma}$ does not include exchange events. As an example, consider the simple model whereby $\bar{M}$ is independent of spin and depends only on the momentum transfer $\mathbf{q} = \mathbf{p}' - \mathbf{p}$ and so it is written as $\bar{M}(\mathbf{q})$:

$$H = \sum_{\mathbf{p}\sigma} (\varepsilon_p - \mu) C^\dagger_{\mathbf{p}\sigma} C_{\mathbf{p}\sigma} + \frac{1}{2\nu} \sum_{\substack{\mathbf{pp'q}\\ \sigma\sigma'}} \bar{M}(\mathbf{q}) C^\dagger_{\mathbf{p}+\mathbf{q}\sigma} C^\dagger_{\mathbf{p}'-\mathbf{p}\sigma'} C_{\mathbf{p}'\sigma'} C_{\mathbf{p}\sigma} \tag{10.4.15}$$

In the theory of the homogeneous electron gas, which has a similar Hamiltonian, the interactions between particles led to a screening of the interactions. The word *screening* sometimes implies a charge redistribution, so the alternate word *dressing* is used in liquid ^{3}He. But the physics of the two concepts is the same: The quasiparticles rearrange all the other quasiparticles in their vicinity to alter the effective interaction between any pair of them. In the random phase approximation, this effective interaction is

$$M_{\text{eff}}(\mathbf{q}; \omega) = \frac{\bar{M}(\mathbf{q})}{1 - \bar{M}(\mathbf{q}) P^{(1)}(\mathbf{q}, \omega)} \tag{10.4.16}$$

where $P^{(1)}$ is the simple bubble contribution for the polarization diagram. The denominator has the dressing function $1 - \bar{M}(\mathbf{q}) P^{(1)}(\mathbf{q}, \omega)$, which is equivalent to the dielectric screening function in the electron gas. In the forward direction we obtain results by first taking $\omega \to 0$ and afterwards $q \to 0$:

$$\lim_{\omega, \mathbf{q} \to 0} \begin{cases} P^{(1)} = -N_F \\ \bar{M}(\mathbf{q}) = N_F^{-1} F_0^s \\ M_{\text{eff}} = \dfrac{\bar{M}(0)}{1 + N_F \bar{M}(0)} = \dfrac{1}{N_F} \dfrac{F_0^s}{1 + F_0^s} \equiv \dfrac{1}{N_F} A_0^s \end{cases} \tag{10.4.17}$$

Thus the effective interaction F_0^s is replaced by $A_s^0 = F_0^s/(1 + F_0^s)$. The quantity F_0^s is the *s*-wave part of the interaction between bare quasiparticles, while A_0^s is the same interaction between dressed quasiparticles. This difference is the same as between the bare coulomb interactions e^2/r

between electrons and the screened interaction. The dressed interaction between quasiparticles is changed, from the bare interaction, by the other quasiparticles which have also been influenced to alter their equilibrium configuration by the effective interaction between quasiparticles.

Since the static limit of (10.4.17) has been obtained, the next step is to examine the dynamical predictions of this model. Thus consider the case whereby $\omega \gg qv_F$. First we recall for the electron gas in this same limit that

$$\lim_{\omega \gg qv_F} v_q P^{(1)}(\mathbf{q}, \omega) = \frac{\omega_p{}^2}{\omega^2}\left[1 + O\left(\frac{qv_F}{\omega}\right)^2\right]$$

This is still the long-wavelength limit, so we can use $\bar{M}(0) = F_0{}^s/N_F$, so that we can immediately conclude that

$$\lim_{\omega \gg qv_F} \bar{M}(\mathbf{q})P^{(1)}(\mathbf{q}, \omega) = \frac{F_0{}^s}{N_F}\,\frac{\omega_p{}^2}{v_q}\,\frac{1}{\omega^2} = \frac{F_0{}^s}{3}\,\frac{k_F{}^2q^2}{mm^*}\,\frac{1}{\omega^2} = \frac{c_1{}^2q^2}{\omega^2}$$

$$c_1{}^2 = \frac{1}{3}\,\frac{F_0{}^s k_F{}^2}{mm^*} \tag{10.4.18}$$

This may be used to obtain the dressed susceptibility, which is the correlation function for density–density operators:

$$\chi_d(\mathbf{q}, i\omega) = -\int_0^\beta d\tau e^{i\omega\tau}\langle T_\tau \varrho(\mathbf{q}, \tau)\varrho(-\mathbf{q}, 0)\rangle$$

$$\lim_{\omega \gg qv_F} \chi_d(\mathbf{q}, \omega) = \frac{P^{(1)}(\mathbf{q}, \omega)}{1 - \bar{M}(\mathbf{q})P^{(1)}} = \frac{1}{N_F}\,\frac{c_1{}^2q^2}{\omega^2 - c_1{}^2q^2} \tag{10.4.19}$$

There is a resonance at $\omega \simeq c_1 q$. The large value of $F_0{}^s$ ensures that $c_1 > v_F$ so $\omega > v_F q$ is valid when $\omega \simeq c_1 q$. From Table 10.2, we note that the exact sound velocity in the long-wavelength limit is $c^2 = \frac{1}{3}v_{F0}^2(1 + F_0{}^s)/(1 + F_1{}^s/3)$. This is close to the result $c_1{}^2$ we obtained, since $F_0{}^s$ is large and there is not much difference between $F_0{}^s$ and $1 + F_0{}^s$. Thus the exact result in this limit is only slightly different. It has the exact same form as (10.4.19) but with the correct speed of sound c rather than the approximate result c_1. The difference between these two calculations is that the long-wavelength sound velocity c is found in the hydrodynamic regime and is the first sound. The calculation we just did is for the collisionless regime, and c_1 is the velocity of zero sound. They approach the same value for large $F_0{}^s$. Thus the density–density correlation function has the long-wavelength limit of sound waves, which dominates the excitation spectrum. This conclusion is in accord with our earlier remarks that the correlation function $S_c(\mathbf{q}, \omega) \propto \mathrm{Im}[\chi(\mathbf{q}, \omega)]$ had mostly sound waves at small values of q. Then we derived them from the Bijl–Feynman formula, but the present discussion is entirely equivalent.

Let us define $A_{\mathbf{p}\sigma,\mathbf{p}'\sigma'}$ as the Fermi liquid theory function which gives the interaction between dressed quasiparticles. It is also the forward scattering amplitude of the screened matrix element for the interaction between two quasiparticles. It may be derived from the bare interaction between quasiparticles $f_{\mathbf{p}\sigma,\mathbf{p}'\sigma'}$ by the following simple argument. The self-energy of a quasiparticle may be expressed as either the product of the bare interaction $f_{\mathbf{p}\sigma,\mathbf{p}'\sigma'}$ and the bare additional density $\delta n_{\mathbf{p}\sigma}$ or else the product of the dressed interaction $A_{\mathbf{p}\sigma,\mathbf{p}'\sigma'}$ and the dressed additional density $\delta\bar{n}_{\mathbf{p}\sigma}$:

$$\bar{\varepsilon}_{\mathbf{p}} - \varepsilon_p = \sum_{\mathbf{p}'\sigma'} f_{\mathbf{p}\sigma,\mathbf{p}'\sigma'}\delta n_{\mathbf{p}'\sigma'} = \sum_{\mathbf{p}'\sigma'} A_{\mathbf{p}\sigma,\mathbf{p}'\sigma'}\delta\bar{n}_{\mathbf{p}'\sigma'}$$

The interaction between dressed quasiparticles $A_{\mathbf{p}\sigma,\mathbf{p}'\sigma'}$ is also defined only right on the Fermi surface, where both p and p' are equal to p_F. Therefore we expand all quantities in spherical harmonics and obtain the following obvious formula relating these coefficients:

$$\begin{aligned} A_{\mathbf{p}\sigma,\mathbf{p}'\sigma'} &= N_F^{-1} \sum_l (A_l^s + \boldsymbol{\sigma} \cdot \boldsymbol{\sigma}' A_l^a) P_l(\cos\theta) \\ A_l^s \delta\bar{n}_{lm}^s &= F_l^s \delta n_{lm}^s \\ A_l^a \delta\bar{n}_{lm}^a &= F_l^a \delta n_{lm}^a \end{aligned}$$

From our previous relationship (10.4.7) between δn_{lm} and $\delta\bar{n}_{lm}$, we deduce the final formula for

$$\begin{aligned} A_l^s &= \frac{F_l^s}{1 + F_l^s/(2l+1)} \\ A_l^a &= \frac{F_l^a}{1 + F_l^a/(2l+1)} \end{aligned} \tag{10.4.20}$$

The case $l = 0$ agrees with the RPA result obtained in (10.4.17). The RPA is exact in this limit ($q \to 0$), since the preceding formulas are exact. These numbers are also tabulated in Table 10.3. The effective interaction between dressed quasiparticles is qualitatively much different from the interaction between bare quasiparticles. The very large value of F_0^s makes A_0^s less than unity, so the dressing effects have a drastic effect on reducing the effective interaction. Of course, the same thing happens for electron–electron interactions in metals, where screening makes a similar reduction. But the most interesting effect is the increase in the spin-dependent parts A_l^a. These become larger in magnitude than F_l^a because the latter are negative. This enhancement of the spin-dependent interaction is one step in the argument, which we shall construct later, toward the conclusion that superfluidity in liquid ^{3}He is caused by triplet pairing.

The next question to be considered is the method of calculating the spin susceptibility. In (10.4.18) and (10.4.19) we have used the RPA result and Fermi liquid parameters to derive an approximate formula for the density–density correlation function. Is it possible to do the same for the spin susceptibility? By reasoning deductively, Leggett (1975) suggested that a possible form for the spin susceptibility in the RPA is

$$\chi_{\rm sp}(\mathbf{q}, \omega) = \frac{P^{(1)}(\mathbf{q}, \omega)}{1 - F_0^a N_F^{-1} P^{(1)}(\mathbf{q}, \omega)} \tag{10.4.21}$$

This equation is deduced, in analogy with (10.4.16) and (10.4.17), by just replacing F_0^s by F_0^a. A better treatment would start from a model Hamiltonian which would give this result for the correlation function. This point will be discussed later. For the moment, let us accept (10.4.21) as a reasonable hypothesis and examine its dynamical predictions.

We shall call the spectral function for this retarded correlation function $S_{\rm sp}(\mathbf{q}, \omega)$. In analogy with the electron gas, we expect two types of dynamical modes from this correlation function. The first is a collective mode equivalent to the plasmon in the electron gas and the sound waves in the density–density correlation function. Such modes can exist only when $\omega \gg qv_F$. The theory has already been done for the sound case, and we can use (10.4.18) and (10.4.19) and merely changes F_0^s to F_0^a in these equations. The prediction is that the collective *spin–sound* has the eigenfrequency

$$\omega = qv_F\left(\frac{F_0^a}{3}\,\frac{m^*}{m}\right)^{1/2}$$

This frequency is imaginary, since F_0^a is negative. Thus these modes are totally damped and do not exist as excitations.

The other excitation of the electron gas is the electron–hole pairs. The theory of *spin pairs* can be developed in an analogous fashion. We shall solve for the case where $\omega < qv_F$ and $q < k_F$, when the polarization function $P^{(1)}$ can be approximated for its real and imaginary parts by

$$P^{(1)}(\mathbf{q}, \omega) = -N_F\left[1 - i\left(\frac{\omega}{qv_F}\right)\frac{\pi}{2}\,\theta(qv_F - \omega)\right]$$

$$S_{\text{sp}}(\mathbf{q}, \omega) = -2\,\text{Im}[\chi_{\text{sp}}(q, \omega)] = \lambda_s \frac{\omega}{\omega^2 + \omega_s^2}\,\theta(qv_F - \omega)$$

$$\omega_s = \frac{2}{\pi}\,\frac{qv_F}{|A_0^a|}$$

$$\lambda_s = \frac{2N_F\omega_s}{|F_0^a(1 + F_0^a)|}$$

The spectral function is linear in energy for small values of $E = \hbar\omega$. The same linear behavior was found for the electron gas when we evaluated the pair contribution to $\text{Im}[1/\varepsilon(\mathbf{q}, \omega)]$—see Fig. 5.12. The same theory can be applied to the pair spectrum from the density–density correlation function, and the only difference is in replacing F_0^a by F_0^s and A_0^a by A_0^s. It is instructive to compare these two cases when pairs are made by either the spin or density response functions. The energy widths ω_s and ω_a are similar for the two cases, since A_0^a and A_0^s differ only by a factor of 2. However, the significant difference is in the coupling strength λ. The factor $|F_0^{a,s}(1 + F_0^{a,s})|^{-1}$ is quite different for the two cases, since it is 4.5 for spins and 0.009 for density, which differ by a factor of 500. Thus the RPA predicts that the spectral function for making pairs from the spin correlation function is 10^3 larger than for the density correlation function. This is the reason that the low-energy peak in the neutron scattering data $S(\mathbf{q}, \omega)$ is assigned to the spin spectral function $S_p(\mathbf{q}, \omega)$. It is interpreted as the pair spectrum of this correlation function, which is expected to be much larger than the pair spectrum of the density correlation function. At small values of q, it seems that the main spectral strength of $S_c(\mathbf{q}, \omega)$ is in the collective sound mode, while for $S_p(\mathbf{q}, \omega)$ it is in the pair spectrum.

The correlation function (10.4.21) must be regarded as *ad hoc* without a derivation based on a realistic model Hamiltonian for the interaction between quasiparticles. The only attempt in this direction has been using *paramagnon theory*, which we shall briefly describe. This theory starts with a model Hamiltonian for quasiparticles of the form (10.4.15), with the matrix element $\bar{M}$ only scattering particles of opposite spin. This matrix element is taken to be independent of momentum and is denoted by the symbol I:

$$H_{\text{int}} = I\int d^3r n_\downarrow(r)n_\uparrow(r)$$
$$= \frac{I}{\nu}\sum_{\mathbf{kk'q}} C^\dagger_{\mathbf{k+q}\uparrow}C_{\mathbf{k}\uparrow}C^\dagger_{\mathbf{k'-q}\downarrow}C_{\mathbf{k'}\downarrow} \tag{10.4.22}$$
$$H = \sum_{\mathbf{p}\sigma} \xi_p C^\dagger_{\mathbf{p}\sigma}C_{\mathbf{p}\sigma} + H_{\text{int}}$$

This interaction term was first suggested by Berk and Schrieffer (1966) in their discussion of the superconductivity effects in transition metals, such as Pd, which are nearly ferromagnetic. They are actually paramagnetic, with a large spin susceptibility and a low transition temperature for superconductor. They reasoned that the spin arrangements in Pd have short-range correlation, whereby if any spin was pointing in a direction, then nearby spins were highly correlated and were likely pointing in the same direction. Thus the spin arrangements were ordered locally, although for short durations, and these spin fluctuations are called paramagnons. They proposed that these spin fluctuations suppressed the onset of superconductivity, since an electron with the spin pointing in the opposite direction would find it hard to be in the same vicinity because of Coulomb repulsion. They constructed this ad hoc Hamiltonian, which had a point repulsion ($I > 0$) between particles of opposite spin.

This idea was adopted immediately for liquid ^{3}He by Doniach and Engelsberg (1966) and Rice (1967). They used it to calculate a number of properties of liquid ^{3}He. We shall give one example and calculate the transverse spin susceptibility. First, we must agree on a diagrammatic representation of this term. Since the interaction between the quasiparticles acts at a point in space, we draw it by a small square, which has four particle lines attached, as in Fig. 10.16(a). Two have spin up and two have

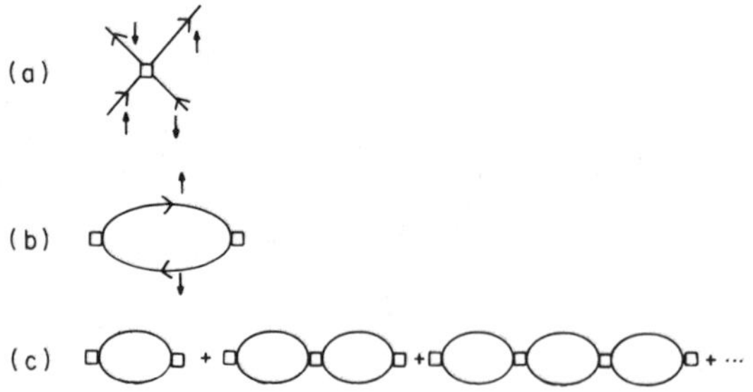

FIGURE 10.16. Feynman diagrams in paramagnon theory.

spin down, and two are ingoing and two are outgoing. The ordering around the vertex is arbitrary, but one ingoing line has to be spin up, one spin down, etc.

For a noninteracting gas of quasiparticles, the transverse susceptibility is just equal to $\frac{1}{2}P^{(1)}$, which is the standard bubble diagram for the polarization operator:

$$\chi_\perp(\mathbf{q}, i\omega) = -\int_0^\beta d\tau e^{i\omega\tau}\langle T_\tau \sigma^+(\mathbf{q}, \tau)\sigma^-(\mathbf{q}, 0)\rangle$$

$$\sigma^+(\mathbf{q}) = (\sigma^-)^+ = \sum_{\mathbf{k}} C^\dagger_{\mathbf{k}+\mathbf{q}\uparrow} C_{\mathbf{k}\downarrow}$$

$$\chi_{\perp 0}(\mathbf{q}, i\omega) = -\tfrac{1}{2}P^{(1)}(\mathbf{q}, i\omega)$$

The diagram for $\chi_{\perp 0}(\mathbf{q}, i\omega)$ is shown in Fig. 10.16(b). It is equal to $-\frac{1}{2}P^{(1)}(\mathbf{q}, i\omega)$ when the system is nonmagnetic, since then the Green's functions for spin up and spin down are identical. The addition of the interaction term (10.4.22) results in the series of diagrams shown in Fig. 10.16(c). Each successive closed loop adds the factor $-\frac{1}{2}IP^{(1)}$, and the series of terms may be summed to get the transverse susceptibility for this Hamiltonian:

$$\chi_\perp = -\tfrac{1}{2}P^{(1)}\left[1 + \left(\frac{-IP^{(1)}}{2}\right) + \left(\frac{-IP^{(1)}}{2}\right)^2 + \cdots\right] = \frac{-\frac{1}{2}P^{(1)}(\mathbf{q}, i\omega)}{1 + \frac{1}{2}IP^{(1)}(\mathbf{q}, i\omega)} \tag{10.4.23}$$

This equation represents the exact summation of this series of diagrams. These are the only diagrams which affect the susceptibility, except for self-energy diagrams of the Green's functions, which are assigned as a problem. The result (10.4.23) has the right form of the earlier assertion (10.4.21) and enables us to identify I as the Fermi liquid parameter $-IN_F = 2F_0{}^a$. One can also use this interaction term to calculate the density–density response function, with the result

$$\chi(\mathbf{q}, i\omega) = -\int_0^\beta d\tau e^{i\omega\tau}\langle T_\tau \varrho(\mathbf{q}, \tau)\varrho(-\mathbf{q}, \tau)\rangle = +\frac{P^{(1)}(\mathbf{q}, i\omega)}{1 - \frac{1}{2}IP^{(1)}(\mathbf{q}, i\omega)}$$

This equation also has the form of the earlier RPA result, where we can identify $IN_F = 2F_0{}^s$. Both of these results cannot be satisfied, since we know that $F_0{}^s = 10$ and $-F_0{}^a = 0.67$. The same parameter $\frac{1}{2}IN_F$ cannot equal them both. Thus paramagnon theory fails as a quantitative theory of *bare* quasiparticle interactions in liquid ^{3}He. However, it does have some qualitative successes. For example, it does predict that the sign in the dressing function $1 \pm \frac{1}{2}\mathrm{IP}^{(1)}$ is different for the spin susceptibility from what it is for the density–density correlation function. This is correct and shows that the paramagnon model has some merit. Levin and Valls (1978) show that paramagnon theory gives an accurate quantitative theory of dressed quasiparticle interaction for both superfluidity and quasiparticle transport.

D. Quasiparticle Transport

Fermi liquid theory can also be used to describe the transport of quasiparticles at very low temperatures. The agreement between theory and experiment provides a further verification of Fermi liquid theory and the derived parameters. The theoretical framework was provided by Abrikosov and Khalatnikov (1959), who presented an approximate theory of viscosity η and thermal conductivity K. The third transport coefficient is spin diffusion D, and an approximate theory was first given by Hone (1961). All these theories involve the solution of the Boltzmann equation for transport in Fermi systems, where the scattering mechanism is particle–particle interactions. In this case the particles are quasiparticles. Later there was an exact solution of the Boltzmann equation in the limit where $T \to 0$ by Jensen *et al.* (1968) and Brooker and Sykes (1968). It would take us too far afield to solve the Boltzmann equation for these three transport coefficients, so we shall only quote the result. The following formulas are exact in the limit of zero temperature:

$$\begin{aligned}
K &= \frac{8\pi^2}{3} \frac{k_F^3}{(m^*)^4 T} \left[\left\langle \frac{W(\theta, \varphi)(1 - \cos\theta)}{\cos(\theta/2)} \right\rangle\right]^{-1} H(\lambda_K) \\
\eta &= \frac{64}{45} \frac{k_F^5}{(m^*)^4 (k_B T)^2} \left[\left\langle \frac{W(\theta, \varphi)(1 - \cos\theta)^2 \sin^2\varphi}{\cos(\theta/2)} \right\rangle\right]^{-1} C(\lambda_\eta) \\
D &= \frac{32\pi^2 k_F^2 (1 + F_0^a)}{3(m^*)^5 (k_B T)^2} \left[\left\langle \frac{W_{\uparrow\downarrow}(\theta, \varphi)(1 - \cos\theta)(1 - \cos\varphi)}{\cos(\theta/2)} \right\rangle\right]^{-1} C(\lambda_D) \\
H(\lambda) &= \frac{3 - \lambda}{4} \sum_{n=0}^{\infty} \frac{4n + 5}{(n+1)(2n+3)[(n+1)(2n+3) - \lambda]} \\
C(\lambda) &= \frac{1 - \lambda}{4} \sum_{n=0}^{\infty} \frac{4n + 3}{(n+1)(2n+1)[(n+1)(2n+1) - \lambda]} \\
\lambda_K &= 3 - 2 \frac{\langle [W(\theta, \varphi)(1 - \cos\theta)]/\cos(\theta/2) \rangle}{\langle W(\theta, \varphi)/\cos(\theta/2) \rangle} \\
\lambda_\eta &= 1 - \frac{3}{4} \frac{\langle [W(\theta, \varphi)(1 - \cos\theta)^2 \sin^2\varphi]/\cos(\theta/2) \rangle}{\langle W(\theta, \varphi)/\cos(\theta/2) \rangle} \\
\lambda_D &= 1 - \frac{1}{2} \frac{\langle [W_{\uparrow\downarrow}(\theta, \varphi)(1 - \cos\theta)^2 \sin^2\varphi]/\cos(\theta/2) \rangle}{\langle W(\theta, \varphi)/\cos(\theta/2) \rangle}
\end{aligned} \qquad (10.4.24)$$

The functions $H(\lambda)$ and $C(\lambda)$ are correction factors. The original solution of Abrikosov, Khalatnikov, and Hone had an approximate result with these quantities set equal to unity. In the exact solution, one also has to evaluate these correction terms. The first step is to find the appropriate λ and then to use it in the series for $H(\lambda)$ or $C(\lambda)$. Both series converge rapidly, so that the answer is obtained easily. The factor C turns out to be about 0.8, so that the older theories are reduced by about 20%. However, the correction for the thermal conductivity is much larger, since $H(\lambda_K)$ is almost exactly 0.5, so the older expression erred by a factor of 2. These correction factors, resulting from the exact solution, play an important role in improving the agreement between theory and experiment.

These formulas contain one or several brackets which have a factor $W(\theta, \varphi)$ plus some angular terms. This average is taken over the 4π solid angle:

$$\left\langle W \frac{1-\cos\theta}{\cos(\theta/2)} \right\rangle = \frac{1}{4\pi} \int_0^{2\pi} d\varphi \int_0^{\pi} d\theta \, \frac{\sin(\theta) W(\theta, \varphi)(1-\cos\theta)}{|\cos(\theta/2)|}$$

The angles θ and φ refer to the scattering of quasiparticles, as we shall explain later. The symbol $W(\theta, \varphi)$ is the matrix element for the scattering of quasiparticles. It is conventionally defined as

$$\begin{aligned} 2W(\theta, \varphi) &= \frac{2\pi}{\hbar} \left[|A_{\uparrow\downarrow}(\theta, \varphi)|^2 + \frac{1}{2} |A_{\uparrow\uparrow}(\theta, \varphi)|^2 \right] \\ W_{\uparrow\downarrow}(\theta, \varphi) &= \frac{2\pi}{\hbar} |A_{\uparrow\downarrow}(\theta, \varphi)|^2 \end{aligned} \qquad (10.4.25)$$

where $A_{\uparrow\uparrow}$ and $A_{\uparrow\downarrow}$ are the matrix elements for the scattering of two spin up particles and a spin up with a spin down particle. We shall need to express these matrix elements by the Fermi liquid parameters. The factor $\frac{1}{2}$ in front of $|A_{\uparrow\uparrow}|^2$ occurs because the scattering of identical quasiparticles into $(\mathbf{k}, -\mathbf{k})$ is indistinguishable from $(-\mathbf{k}, \mathbf{k})$, and the angular average counts both of these and so overcounts the scattering by a factor of 2.

The factors in the denominators of (10.4.24) come from the effective lifetime appropriate to the transport coefficient. For the thermal conductivity, the effective lifetime contains the factor $1 - \cos\theta$, just as it does for the electrical conductivity in metals. The other factor in this angular average is $(\cos\frac{1}{2}\theta)^{-1}$, whose origin we shall now describe. To do this, we need to examine the expression for the lifetime of a quasiparticle from the scattering by other quasiparticles. Assume that the initial quasiparticle has momentum $\mathbf{p}_1$, that it scatters from a quasiparticle $\mathbf{p}_2$, and that they go to $\mathbf{p}_3$ and $\mathbf{p}_4$.

The lifetime for this scattering rate is given by the expression

$$\frac{1}{\tau_{p_1}} = \frac{2\pi}{\hbar} \int \frac{d^3p_2\, d^3p_3\, d^3p_4}{(2\pi)^9} \sum_{\sigma'} | A_{\uparrow\sigma'} |^2 (2\pi)^3 \delta(\mathbf{p}_1 + \mathbf{p}_2 - \mathbf{p}_3 - \mathbf{p}_4)$$
$$\times \delta(\varepsilon_1 + \varepsilon_2 - \varepsilon_3 - \varepsilon_4) n_2(1 - n_3)(1 - n_4)$$

The occupation factors $n_j = n_F(\varepsilon_j)$ express the feature that $\mathbf{p}_2$ is in an occupied state, while $\mathbf{p}_3$ and $\mathbf{p}_4$ must go to empty ones. The integrand contains the delta functions for conservation of energy and momentum, and the matrix element $| A_{\uparrow\sigma'} |^2$ for scattering (but we shall multiply $| A_{\uparrow\uparrow} |^2$ by $\frac{1}{2}$). Our evaluation of this quantity will follow Pines and Nozières (1966).

At very low temperature, all the quasiparticles will be quite close to the Fermi surface. The magnitude of all the momentums $\mathbf{p}_j$ ($j = 1, 2, 3, 4$) are very near to p_F and therefore very nearly the same length. This feature simplifies the calculation. The delta function for momentum conservation is used to eliminate the d^3p_4 integral and leaves the combination $d^3p_2\, d^3p_3$. Let $\mathbf{P} = \mathbf{p}_1 + \mathbf{p}_2$ be the center of mass momentum of the two quasiparticles, which is unchanged by the collision. The d^3p_3 integral is nested inside d^3p_2 so that we can use the vector $\mathbf{P}$ as the basis of choosing the angles in doing the d^3p_3 integral. In particular, let $\mathbf{P} \cdot \mathbf{p}_3 = Pp_3 \cos\theta_3 = Pp_3\nu_3$. Then the vector $\mathbf{p}_4$ has the magnitude

$$p_4 = | \mathbf{P} - \mathbf{p}_3 | = (P^2 + p_3^2 - 2Pp_3\nu_3)^{1/2}$$

This identity is used to change the variables of integration from $d\theta_3 \sin\theta_3 = d\nu_3$ to dp_4, where we have

$$p_3\, d\nu_3 = -\frac{p_4\, dp_4}{P}$$

The next step is to define P in terms of the variables used in the d^3p_2 integral. We use θ to denote the angle between $\mathbf{p}_1$ and $\mathbf{p}_2$ and use the fact that both are near p_F to find $P = | \mathbf{p}_1 + \mathbf{p}_2 | = (p_1^2 + p_2^2 + 2p_1p_2\cos\theta)^{1/2} \simeq p_F[2(1 + \cos\theta)]^{1/2} = 2p_F \cos(\theta/2)$. Finally, we set $p_2^2\, dp_2 \simeq p_F[d(p_2^2)/2]$, which brings us to the identity

$$d^3p_2\, d^3p_3 = \frac{(m^*)^3}{2} \frac{\sin(\theta)\, d\theta}{| \cos\frac{1}{2}\theta |} d\varphi_2\, d\varphi_3\, d\varepsilon_2\, d\varepsilon_3\, d\varepsilon_4 \qquad (10.4.26)$$

The six integration variables in $d^3p_2\, d^3p_4$ have been reexpressed in terms of coordinates which are more useful. It is usually assumed that the matrix element has no significant energy variation. Then the energy integrals can be done exactly. The first one is trivial, since it just uses the delta function for energy conservation. The remaining two integrals can also be done exactly, which is assisted by the indicated variable changes:

$$I = \int d\varepsilon_2 \, d\varepsilon_3 \, d\varepsilon_4 \delta(\varepsilon_1 + \varepsilon_2 - \varepsilon_3 - \varepsilon_4) n_F(\varepsilon_2) n_F(-\varepsilon_3) n_F(-\varepsilon_4)$$

$$I = \int d\varepsilon_2 \, d\varepsilon_3 n_F(\varepsilon_2) n_F(-\varepsilon_3) n_F(\varepsilon_3 - \varepsilon_1 - \varepsilon_2)$$

$$x = e^{\beta(\varepsilon_2 - \mu)}, \qquad y = e^{\beta(\varepsilon_3 - \mu)}, \qquad z = e^{-\beta(\varepsilon_1 - \mu)}$$

$$I = \frac{1}{\beta^2} \int_0^\infty dx \int_0^\infty dy \frac{1}{(x+1)(y+1)(x+yz)}$$

$$I = \frac{1}{\beta^2} \int_0^\infty \frac{dy}{y+1} \frac{1}{yz-1} \ln yz$$

$$I = \frac{1}{2\beta^2} \left(\frac{\pi^2 + \ln^2 z}{z+1} \right) = \frac{(k_B T)^2}{2} (1 - n_1)[\pi^2 + \beta^2(\varepsilon_1 - \mu)^2]$$

The last integral is from G&R, 4.232(3).

The last step is to do the angular integrals. First it is helpful to visualize the coordinate system. Take the case where all four vectors $\mathbf{p}_j$ have the same length p_F. We have defined θ as the angle between $\mathbf{p}_1$ and $\mathbf{p}_2$. It is also the angle between $\mathbf{p}_3$ and $\mathbf{p}_4$, which is true only when all four vectors are the same length. Thus we need only one more angle to define the relative orientation of all four vectors. This angle could be the relative orientation of $\mathbf{p}_1$ and $\mathbf{p}_3$, but this choice is not conventional. Instead, we note that the two initial vectors $\mathbf{p}_1$ and $\mathbf{p}_2$ form a plane whose orientation is represented by a unit vector $\hat{n}_i$ in the direction $\mathbf{p}_1 \times \mathbf{p}_2$. Similarly, the scattering plane in the final state has an orientation $\hat{n}_f$ in the direction $\mathbf{p}_3 \times \mathbf{p}_4$. We define φ as the angle between these unit vectors, $\cos\varphi \doteq \hat{n}_i \cdot \hat{n}_f$. The angle φ is shown in the vector diagram in Fig. 10.17. All four vectors $\mathbf{p}_j$ make an angle $\frac{1}{2}\theta$ with respect to the center of mass momenta $\mathbf{P}$, and φ is the angle between the two scattering planes. Another way to represent this angle is to change to center of mass coordinates, and $\mathbf{p}_i$ and $\mathbf{p}_f$ are the relative mo-

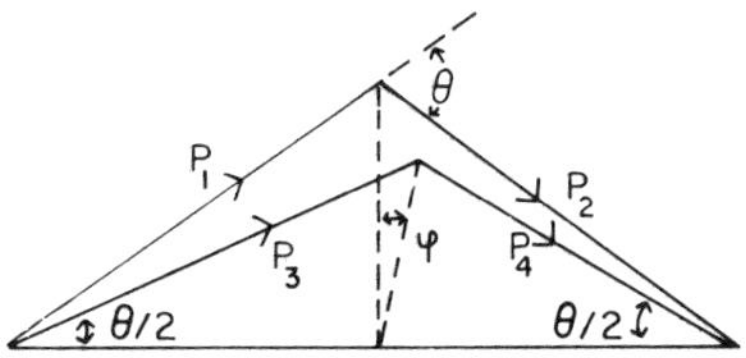

FIGURE 10.17. Angular variables in quasi-particle scattering in liquid ³He.

mentum in the initial and final states of the scattering process, with $\mathbf{p}_f = \mathbf{p}_i + \mathbf{q}$. The angle between $\mathbf{p}_i$ and $\mathbf{p}_f$ is also φ.

In the variable list in (10.4.26), the azimuthal angle φ_3 is actually φ, while φ_2 is unnecessary and may be integrated to give 2π. This completes the description of the angles in the scattering process. The final lifetime is the following expression from Pines and Nozières (1966):

$$\frac{1}{\tau_{P_1}} = \frac{(m^*)^3}{16\pi^4\hbar^6}(1 - n_1)[(\pi k_B T)^2 + (\varepsilon_1 - \mu)^2]\left\langle \frac{W}{\cos(\theta/2)} \right\rangle \qquad (10.4.27)$$

We have given the symbol $W(\theta, \varphi)$ its previous definition in (10.4.25), although now we have actually defined all the angular variables explicitly. The result for $1/\tau_{P_1}$ has two terms which show that at the chemical potential $\varepsilon_1 = \mu$ the lifetime is determined by the temperature but that at high energies $\beta(\varepsilon_1 - \mu) \gg 1$ it is determined by phase space availability. The equivalent quantity using Green's functions is $-2\,\mathrm{Im}[\Sigma(\mathbf{q}, \xi_p)]$ for the appropriate diagram. The imaginary part of the self-energy produces the same lifetime if the diagrams are selected correctly, except the $1 - n_1$ factor is absent (see Problem 7 at the end of this chapter). Thus, the factor $1 - n_1$ should not be included in the quasiparticle lifetime.

To do the angular integrals in this formula for the quasiparticle lifetime, we need an expression for $W(\theta, \varphi)$. The Landau–Fermi liquid theory gives only the forward part of the scattering amplitude, where $\mathbf{p}_3 = \mathbf{p}_1$ and $\mathbf{p}_4 = \mathbf{p}_2$ or the opposite. Forward scattering corresponds to the case where $\varphi = 0$ or π. We need to find a method of extending this scattering theory to other values of angle besides the forward direction. The extension to other angles is useful not only for this calculation of the quasiparticle lifetime but also for the similar calculations of the transport coefficients in (10.4.24) which have other angular averages of $W(\theta, \varphi)$. In transport theory, the quasiparticles not only scatter in the forward direction but to any other point on the Fermi surface. How can this be described by using only the parameters we now know?

An ingenious and very successful method for this was proposed by Dy and Pethick (1969), which they called the *s–p approximation.* The goal is to obtain the scattering amplitudes $A_{\uparrow\uparrow}(\theta, \varphi)$ and $A_{\uparrow\downarrow}(\theta, \varphi)$ for the scattering of two spin up particles and a spin up with a spin down particle. The scattering of two spin down particles is also $A_{\uparrow\uparrow}$. If two particles both have spin up, they must be in a relative triplet spin state, and this scattering amplitude is labeled $A_t(\theta, \varphi)$. The singlet scattering amplitude is called $A_s(\theta, \varphi)$. Two particles in opposite spin states have equal likelihood of being in a singlet or triplet state:

$$A_{\uparrow\uparrow}(\theta, \varphi) = A_{\downarrow\downarrow}(\theta, \varphi) = A_t(\theta, \varphi)$$
$$A_{\uparrow\downarrow}(\theta, \varphi) = A_{\downarrow\uparrow}(\theta, \varphi) = \tfrac{1}{2}[A_t(\theta, \varphi) + A_s(\theta, \varphi)]$$

The next step is to use the symmetry of the two-particle wave function to deduce the possible angular variations. In center of mass coordinates, the only relevant angular variable in the scattering process is that between the initial and final relative momenta, which is φ. Singlet states must be symmetric under the exchange of the orbital part of the particle coordinates, so only even values of relative angular momentum are allowed, as explained in Sec. 10.1. Similarly, triplet spin states permit only odd values of relative angular momentum, so that the symmetry of the two-particle wave functions dictates the following possible choices for these scattering amplitudes:

$$A_t(\theta, \varphi) = \sum_{l\ \mathrm{odd}} C_l(\theta) P_l(\cos \varphi)$$
$$A_s(\theta, \varphi) = \sum_{l\ \mathrm{even}} C_l'(\theta) P_l(\cos \varphi)$$

The next step is the s–p approximation. Dy and Pethick argue that since we know only the $l = 0$ and 1 values in the spherical harmonic expansion for θ, consistency dictates that we keep only the similar terms in the expansion for Legendre functions on the φ variable. In this case there is only one term allowed in each expansion, which is $l = 0$ for singlet states and $l = 1$ for triplet states. Thus they deduce

$$A_t(\theta, \varphi) = C_1(\theta) \cos \varphi$$
$$A_s(\theta, \varphi) = C_0'(\theta)$$

The next step in the derivation is to find the amplitudes $C_1(\theta)$ and $C_0'(\theta)$ by evaluating them in the forward direction, where we know the answer. Thus set $\varphi = 0$ and equate these coefficients to the interaction term between dressed quasiparticles:

$$\begin{aligned} C_1(\theta) &= A_t(\theta, 0) = \frac{1}{N_F} \sum_l (A_l^s + A_l^a) P_l(\cos \theta) \\ C_0'(\theta) &= A_s(\theta, 0) = \frac{1}{N_F} \sum_l (A_l^s - 3A_l^a) P_l(\cos \theta) \end{aligned} \tag{10.4.28}$$

These various steps are collected together to finally provide the angular variation for the scattering amplitude between dressed quasiparticles:

$$A_{\uparrow\uparrow}(\theta, \varphi) = \frac{\cos \varphi}{N_F} [(A_0^s + A_0^a) + (A_1^s + A_1^a) \cos \theta]$$

$$A_{\uparrow\downarrow}(\theta, \varphi) = \frac{1}{2N_F} \{\cos \varphi[(A_0^s + A_0^a) + (A_1^s + A_1^a) \cos \theta] + [(A_0^s - 3A_0^a) + (A_1^s - 3A_1^a) \cos \theta]\}$$

In older calculations of the transport coefficients, the angular functions $A_{\sigma\sigma'}(\theta, \varphi)$ were approximated by the results in the forward direction $\varphi = 0$, since there they are given by the Landau Fermi liquid parameters. The improvement of Dy and Pethick is to add the factor $\cos \varphi$ to the amplitude for the triplet scattering. The factor of $\cos \varphi$ is a necessary addition from the point of symmetry, since the triplet amplitude must change sign when $\varphi = \pi$, $A_t(\theta, \pi) = -A_t(\theta, 0)$ because of the wave function antisymmetry. Going from $\varphi = 0$ to $\varphi = \pi$ effectively changes the scattering $(\mathbf{p}_1 \to \mathbf{p}_3, \mathbf{p}_2 \to \mathbf{p}_4)$ to the exchange event $(\mathbf{p}_1 \to \mathbf{p}_4, \mathbf{p}_2 \to \mathbf{p}_3)$. This small change of adding $\cos \varphi$ is a big one insofar as improving the agreement between theory and experiment for the transport coefficients. Table 10.4 shows their calculation of the quantities KT, ηT^2, and DT^2, as compared with the experimental values, using the theoretical formulas in (10.4.24). The agreement is obviously excellent. Previous calculations without the $\cos \varphi$ factor in A_t disagreed with experiment by a factor of 2. There has also been considerable effort to predict the transport results for higher temperature, above the extremely low-temperature limit where the asymptotic results (10.4.24) are valid. We refer the reader to Dy and Pethick, who used this method to find the parameter F_1^a which is given in Table 10.3.

Another important success of the s–p approximation has been to explain the superfluid properties of liquid ^{3}He. In particular, it explains why the

TABLE 10.4. ^{3}He Transport Results as $T \to 0$

	KT (erg/cm-sec)	DT^2 (cm^2-K^2/sec)	ηT^2 (poise-K^2)
Experiment[a]	35	1.4×10^{-6}	1.8×10^{-6}
Theory[b]	33	1.6×10^{-6}	1.6×10^{-6}

[a] Wheatley (1975).
[b] Dy and Pethick (1969).

BCS type of pairing is not in the singlet state but rather is in the spin triplet state. Here our discussion follows Patton and Zaringhalam (1975). They calculated the temperature at which the Cooper instability occurs in liquid ^{3}He, which is a property of the normal state of the liquid. The construction of a wave function for the superfluid state is a different step, which is done in the next section. But the Cooper instability is a property of the normal fluid and so is discussed here.

In a BCS of pairing state, two dressed quasiparticles of momentum **p** and $-\mathbf{p}$ are coupled into a collective bound state. One important question is whether this pairing occurs with the two spins in a relative singlet or triplet state. To answer this question, we examine the effective interaction for each possibility. We consider the scattering of a bound pair with $(\mathbf{p}, -\mathbf{p})$ into another pair state $(\mathbf{p}', -\mathbf{p}')$. The two scattering particles have $\theta = \pi$, so that the relevant amplitudes are

$$A_t(\pi, \varphi) = C_1(\pi)\hat{p} \cdot \hat{p}'$$
$$A_s(\pi, \varphi) = C_0'(\pi)$$

The Fermi liquid parameters are used to derive $C_1(\pi)$ and $C_0'(\pi)$ from (10.4.28):

$$C_1(\pi) = \frac{-2.57}{N_F}$$
$$C_0'(\pi) = \frac{3.35}{N_F}$$

The singlet state amplitude $C_0'(\pi)$ is positive. Two dressed quasiparticles have a repulsive interaction when paired in the spin singlet state. Thus they do not attract each other and do not form bound states. The superfluid state is not a spin singlet in liquid ^{3}He.

For the spin triplet state, we find that $C_1(\pi)$ is negative, which shows that the two quasiparticles have an attractive interaction. Pairing is possible in this state, and the normal fluid will show a Cooper instability at a finite temperature because of this tendency toward triplet pairing. Patton and Zaringhalam give the transition temperature as

$$T_c = 1.13\alpha T_F \exp\left(\frac{-1}{g_t}\right) \qquad (10.4.29)$$
$$g_t = -N_F \frac{C_1(\pi)}{12}$$

This equation for T_c is exactly the BCS form given in (9.1.6) and also in (10.4.14). The exponent is given as $g_t = -N_F[C_1(\pi)/12] = 0.21$, and the prefactor is αT_F instead of ω_D. Both of these changes will be explained.

First, where does the factor of 12 enter the exponent? It results from three different factors $12 = 3\times 2\times 2$. The 3 comes from the angular average of $\cos^2 \varphi$. One factor of 2 comes from the fact that whenever we use the scattering amplitude $A_t(\theta, \varphi) = A_{\uparrow\uparrow}(\theta, \varphi)$, we have to divide by 2 so as not to overcount the identical events of scattering into $(\mathbf{p}', -\mathbf{p}')$ and $(-\mathbf{p}', \mathbf{p}')$. The other factor of 2 comes from the fact that the density of states for a single spin component is $m^* p_F^2/2\pi^2\hbar^3$, while we have normalized all quantities to the total density of states $N_F = m^* p_F^2/\pi^2\hbar^3$, which is 2 larger. Thus when we solve the scattering equation for two spin up particles in the normal fluid, we find

$$\begin{aligned}
\Gamma_t(\mathbf{p}, \mathbf{p}') &= A_t(\mathbf{p}, \mathbf{p}') - \frac{1}{2\beta} \sum_{ip} \int \frac{d^3p''}{(2\pi)^3} A_t(\mathbf{p}, \mathbf{p}'')\Gamma_t(\mathbf{p}'', \mathbf{p}')\mathscr{G}_\uparrow(\mathbf{p}'', ip) \\
&\quad \times \mathscr{G}_\uparrow(-\mathbf{p}'', -ip) \\
&= C_1(\pi)\hat{p} \cdot \hat{p}' - \frac{N_F}{4} \int_{-\omega_c}^{\omega_c} \frac{d\xi''\, d\Omega}{4\pi} C_1(\pi)\hat{p} \cdot \hat{p}'' \\
&\quad \times \frac{\Gamma_t(\mathbf{p}'', \mathbf{p}')[1 - 2n_F(\xi'')]}{2\xi''} \\
&= \frac{C_1(\pi)\hat{p} \cdot \hat{p}'}{1 - g_t \int_0^{\omega_c} (d\xi/\xi) \tanh(\beta\xi/2)}
\end{aligned}$$

The energy denominator vanishes and predicts an instability at the temperature given by (10.4.29), where we represent the cutoff of the interaction by ω_c. The nature of this cutoff parameter is not understood, which represents one of the failures in the current theory. Patton and Zaringhalam argued that it must be a multiple of the Fermi energy, so wrote it as $\omega_c = \alpha T_F$. They fit the parameter α to the data and obtained the value of $\alpha = 0.054$. The most impressive feature of their theory is its fit to the boundary of the superfluid phase as a function of pressure. The liquid solidifies at a pressure of 34 bars, and the superfluid transition temperature T_c is a strong function of pressure below this value. The experimental T_c vs. pressure is shown in Fig. 10.18 and is compared to the theory. Patton and Zaringhalam left α independent of pressure and then used experimental values for the pressure dependence of the Fermi liquid parameters and T_F. They are able to nicely explain the increase in T_c with increasing pressure. They also present arguments that their choice of α can be justified by comparison to other measurements.

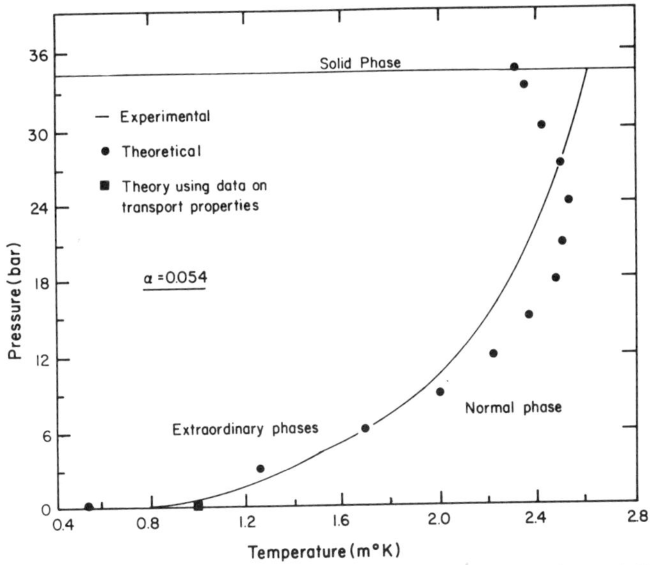

FIGURE 10.18. Phase diagram of ^{3}He. The solid curve is the experimental T_c between normal and superfluid phases. The points were calculated by Patton and Zaringhalam (1975) (used with permission).

The prediction of Fermi liquid theory is that liquid ^{3}He has to form a BCS pairing state in the spin triplet arrangement. This prediction seems in good accord with the experiments on the superfluid state, which will be described in the next section. The triplet pairing is a consequence of the negative value of $C_1(\pi)$ and a positive value of $C_0'(\pi)$. The physics question is, Why does this happen? The answer seems poorly understood. The parameters C_1 and C_0' are derived from the interaction between dressed quasiparticles, which in turn are derived from the interaction between bare quasiparticles. The latter numbers are taken from experiment, although the microscopic theory for them gives good results. However, the chain of argument starting at C_1 and C_0' runs back to other parameters (F_l^s and F_l^a) which are hard to calculate. It is difficult to have any intuitive insight into any of these parameters. Certainly the best guide in paramagnon theory, although it is ad hoc. However, the idea that spins prefer triplet pairing because the liquid is almost ferromagnetic is not very convincing, since metals such as Pd do not behave similarly, although comparisons between transition metals and liquid ^{3}He should be treated skeptically. Triplet pairing can be explained by Fermi liquid theory, but it does not have a simple intuitive explanation.

10.5. SUPERFLUID ^{3}He

Superfluidity in liquid ^{3}He was discovered by Osheroff *et al.* (1972), who discussed it in a series of papers. They identified two phases in the liquid, which are called the A and B phases. The phase boundary on a pressure vs. temperature curve is shown in Fig. 10.20 as reviewed by Wheatley (1975). The line marked T_c separates normal liquid ^{3}He from the superfluid phases. It is a second-order phase boundary with a jump in the specific heat, which are both characteristics of a BCS state. Both A and B phases are superfluid and are experimentally different. The line marked T_{AB} is a first-order phase boundary. The location of this phase boundary is strongly affected by small magnetic fields. Indeed, both the A and B phases have interesting magnetic properties which reinforce the notion that pairing is in a spin triplet state and that the pairs have a net magnetic moment. These features will be summarized later. First we must develop the theory of triplet pairing and try to explain why there are two phases.

A. Triplet Pairing

The theory of superconductivity for electron spins in a triplet state was derived by Balian and Werthamer (1963). They developed this theory in the expectation that some metals might be superconducting with this spin configuration. The important application of their result is in explaining superfluidity in liquid ^{3}He. Their theory is just like the BCS theory, with

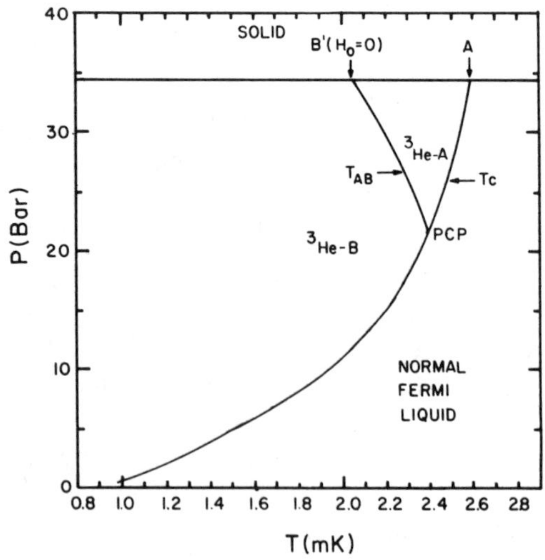

FIGURE 10.19. Phase diagram of ^{2}He as a function of pressure and temperature. The line T_c divides normal from superfluid phases. The line T_{AB} divides the two superfluid phases, and the horizontal line at 34 bars divides solid from liquid. *Source*: Adapted from Wheatley (1975).

weak coupling between the pairs of fermions, except that the spins are assumed to be in a relative triplet state. However, triplet pairing makes a nontrivial change in the nature of the answer because of the additional spin degrees of freedom. The spin state ($S = 1, m_s = 1, 0, -1$) can project in different ways on the relative orbital motion ($L = 1, m_l = 1, 0, -1$). Thus the paired states have a number of possible values of total angular momentum $\mathbf{j} = \mathbf{L} + \mathbf{S}$. When we derive the gap equation, it is a matrix equation. It has a number of possible solutions, and the order parameter is actually a spinor of dimensionality 7. This feature leads to a rich description of the types of motion and ordering of the superfluid. All this complexity is just to describe one of the two phases. The theory of Balian and Werthamer is believed to describe the B phase of the superfluid. The A phase is another type of triplet pairing, which will be treated in a later section.

The BW (Balian and Werthamer) theory is solved by following the same steps used to derive the BCS theory in Chapter 9. The first step is to write an effective Hamiltonian between dressed quasiparticles:

$$H = \sum_{\mathbf{p}\sigma} \xi_p C^\dagger_{\mathbf{p}\sigma} C_{\mathbf{p}\sigma} + \frac{1}{2\nu} \sum_{\substack{\mathbf{pp'q} \\ \sigma\sigma'}} V(q) C^\dagger_{\mathbf{p}+\mathbf{q}\sigma} C^\dagger_{\mathbf{p'}-\mathbf{q}\sigma'} C_{\mathbf{p'}\sigma'} C_{\mathbf{p}\sigma}$$
$$\xi_p = \bar{\varepsilon}_p - \mu \tag{10.5.1}$$

The interaction potential $V(q)$ is actually a function of the momentum change between relative wave vectors $\mathbf{p}_i$ and $\mathbf{p}_f$ in the scattering process, where $\mathbf{p}_f = \mathbf{p}_i + \mathbf{q}$. The interaction potential is expanded in spherical harmonics, and the first two terms are

$$V(\mathbf{p}_f - \mathbf{p}_i) = \sum_l V_l(p_i, p_f) P_l(\hat{p}_i \cdot \hat{p}_f) = V_0 + V_1 \hat{p}_i \cdot \hat{p}_f + \cdots \tag{10.5.2}$$

The first term V_0 acts when the orbital motion of the two particles is in a relative s state, which happens for a spin singlet. We know from the previous section that the s-wave interaction is repulsive ($V_0 > 0$) and cannot cause bound states. The second term $V_1 \hat{p}_i \cdot \hat{p}_f$ is for relative p states and applies for spin triplets. This coefficient is negative and can cause pairing. We recognize $V_1 = C_1(\pi)/(2N_F)$ in the notation of the previous section. This term is the only one we shall keep in the interaction potential. It has the feature that it is antisymmetric in either of the momentum variables: $V(\mathbf{p}_i, \mathbf{p}_f) = V_1 \hat{p}_i \cdot \hat{p}_f = -V(-\mathbf{p}_i, \mathbf{p}_f) = -V(\mathbf{p}_i, -\mathbf{p}_f)$. As in the case of weak coupling theory, such as BCS, this potential is assumed to be a constant V_1 in energy up to a cutoff ω_c from the chemical potential. The physical nature of this cutoff is not clear, but we expect it to be about $\omega_c = 0.05 k_B T_F$.

Obviously, we cannot write a strong coupling theory without a better understanding of the retarded nature of the potentials between dressed quasiparticles—or between bare ones.

The next step is to define the correlation functions appropriate to the superfluid. They will be of the same type we used in the BCS theory of superconductivity. One expects superfluidity in liquid ^{3}He to be similar to a BCS state because of the obvious experimental similarities. The theory requires the introduction of seven correlation functions:

$$\begin{aligned}
\mathscr{G}(\mathbf{p}, \tau - \tau') &= -\langle T_\tau C_{\mathbf{p}\sigma}(\tau) C^\dagger_{\mathbf{p}\sigma}(\tau')\rangle \\
\mathscr{F}_1(\mathbf{p}, \tau - \tau') &= \langle T_\tau C_{-\mathbf{p}\uparrow}(\tau) C_{\mathbf{p}\uparrow}(\tau')\rangle \\
\mathscr{F}_0(\mathbf{p}, \tau - \tau') &= \langle T_\tau C_{-\mathbf{p}\downarrow}(\tau) C_{\mathbf{p}\uparrow}(\tau')\rangle \\
\mathscr{F}_{-1}(\mathbf{p}, \tau - \tau') &= \langle T_\tau C_{-\mathbf{p}\downarrow}(\tau) C_{\mathbf{p}\downarrow}(\tau')\rangle \\
\mathscr{F}_1^\dagger(\mathbf{p}, \tau - \tau') &= \langle T_\tau C^\dagger_{\mathbf{p}\uparrow}(\tau) C^\dagger_{-\mathbf{p}\uparrow}(\tau')\rangle \\
\mathscr{F}_0^\dagger(\mathbf{p}, \tau - \tau') &= \langle T_\tau C^\dagger_{\mathbf{p}\uparrow}(\tau) C_{-\mathbf{p}\downarrow}(\tau')\rangle \\
\mathscr{F}_{-1}^\dagger(\mathbf{p}, \tau - \tau') &= \langle T_\tau C^\dagger_{\mathbf{p}\downarrow}(\tau) C^\dagger_{-\mathbf{p}\downarrow}(\tau')\rangle
\end{aligned} \tag{10.5.3}$$

The quasiparticle Green's function $\mathscr{G}$ has the usual definition, although it will have a different form in the superfluid state. The others are pairing functions for different spin arrangements. The first one has both spins up and belongs to the triplet state. The next has one spin up and another down and appears identical to the correlation function we introduced for the BCS state. However, now we wish it to be for the spin triplet state and to represent the configuration where $S = 1$, $m_s = 0$, whereas in Chapter 9 we meant the same correlation function for $S = 0$, $m_s = 0$. The difference between these two correlation functions becomes clearer when we write them as

$$\begin{aligned}
\mathscr{F}_{s=1,m_s=0}(\mathbf{p}, \tau - \tau') &= -\tfrac{1}{2}[\langle T_\tau C_{-\mathbf{p}\uparrow}(\tau) C_{\mathbf{p}\downarrow}(\tau')\rangle + \langle T_\tau C_{-\mathbf{p}\downarrow}(\tau) C_{\mathbf{p}\uparrow}(\tau')\rangle] \\
\mathscr{F}_{s=0,m_s=0}(\mathbf{p}, \tau - \tau') &= -\tfrac{1}{2}[\langle T_\tau C_{-\mathbf{p}\uparrow}(\tau) C_{\mathbf{p}\downarrow}(\tau')\rangle - \langle T_\tau C_{-\mathbf{p}\downarrow}(\tau) C_{\mathbf{p}\uparrow}(\tau')\rangle]
\end{aligned} \tag{10.5.4}$$

The first one is for the spin triplet state and the second one for the spin singlet. The triplet state is antisymmetric in momentum, while the singlet state is symmetric:

$$\begin{aligned}
\mathscr{F}_{s=1,m_s=0}(-\mathbf{p}, \tau - \tau') &= -\mathscr{F}_{s=1,m_s=0}(\mathbf{p}, \tau' - \tau) \\
\mathscr{F}_{s=0,m_s=0}(-\mathbf{p}, \tau - \tau') &= +\mathscr{F}_{s=0,m_s=0}(\mathbf{p}, \tau' - \tau)
\end{aligned} \tag{10.5.5}$$

These parity relations are easily proved from the definitions (10.5.4), where we change the sign on **p** and commute operators. The simpler definition in (10.5.3) can be used for either triplet or singlet with the appropriate choice of symmetry (10.5.5). Thus we use the simple definition, and when we get equations to solve self-consistently for the correlation functions, we find only those solutions which are antisymmetric in **p** and thereby have a description of the triplet state gap with $m_s = 0$. When we solve for the solutions which are symmetric in **p**, as we did in Chapter 9, we get the singlet solutions of BCS. The other correlation function one might consider has both spins pointing down ($m_s = -1$), but it is just the Hermitian conjugate of the one with both spins up.

There are four possible spin configurations for the two quasiparticles, each with spin one-half. The most general description would have four correlation functions $\mathscr{F}_{s,m_s}$ (one for each spin arrangement) plus $\mathscr{G}$, and the gap equation becomes a 5×5 matrix equation. The problem can be simplified, at the outset, to a matrix equation which is only 3×3. The simplification is achieved by using the relationships in the previous paragraph, where $m_s = -1$ is the Hermitian conjugate of $m_s = 1$ and only one $m_s = 0$ function is used with momentum antisymmetry. The simplification to a matrix of dimension 3 is obviously desirable and is conventional.

The equations of motion method is used to obtain the gap equation in the weak coupling theory. The first step is to recall (9.2.3) for $\partial C_{\mathbf{p}}/\partial\tau$, which is then put into the τ derivative of the Green's functions:

$$\frac{\partial}{\partial\tau} C_{\mathbf{p}\sigma} = -\xi_p C_{\mathbf{p}\sigma} - \frac{1}{\nu}\sum_{\mathbf{p}'\mathbf{q}\sigma'} V(q) C^\dagger_{\mathbf{p}'-\mathbf{q}\sigma'} C_{\mathbf{p}'\sigma'} C_{\mathbf{p}-\mathbf{q}\sigma}$$

$$\left(-\frac{\partial}{\partial\tau} - \xi_p\right)\mathscr{G}(\mathbf{p}, \tau - \tau') + \frac{1}{\nu}\sum_{\mathbf{p}'\mathbf{q}'\sigma'} V(q)$$
$$\times \langle T_\tau C^\dagger_{\mathbf{p}'-\mathbf{q}\sigma'}(\tau) C_{\mathbf{p}'\sigma'}(\tau) C_{\mathbf{p}-\mathbf{q}\sigma}(\tau) C^\dagger_{\mathbf{p}\sigma}(\tau')\rangle = \delta(\tau - \tau')$$

The interaction term has an expectation of four operators. These are paired in different arrangements, which produces a large number of terms, since we now allow pairings between like and unlike operators (i.e., $\langle CC\rangle$, $\langle C^\dagger C^\dagger\rangle$, $\langle CC^\dagger\rangle$) and also parallel and antiparallel spin arrangements. In weak coupling theory we ignore all the terms except those which contain a gap function $\mathscr{F}_m$. In the preceding equation, the terms which are retained are those with the pairing $\langle C^\dagger_{\mathbf{p}'-\mathbf{q}\sigma'} C^\dagger_{\mathbf{p}\sigma}\rangle\langle C_{\mathbf{p}'\sigma'} C_{\mathbf{p}-\mathbf{q}\sigma}\rangle$. There are two spin arrangements, $\sigma' = \sigma$ and $\sigma' = -\sigma$, so we have the first equation of motion:

$$\left(-\frac{\partial}{\partial \tau} - \xi_p\right)\mathscr{G}(\mathbf{p}, \tau - \tau') - \frac{1}{\nu}\, \mathscr{F}_1^{\dagger}(\mathbf{p}, \tau - \tau') \sum_{\mathbf{q}} V(q)\mathscr{F}_1(\mathbf{p} - \mathbf{q}, 0)$$
$$- \frac{1}{\nu}\, \mathscr{F}_0^{\dagger}(\mathbf{p}, \tau - \tau') \sum_{\mathbf{q}} V(q)\mathscr{F}_0(\mathbf{p} - \mathbf{q}, 0) = \delta(\tau - \tau')$$

We also need equations of motion for $\mathscr{F}_1(\mathbf{p}, \tau)$ and $\mathscr{F}_0(\mathbf{p}, \tau)$. They can both be done at the same time by considering

$$\mathscr{F}_{\sigma'}(\mathbf{p}, \tau - \tau') = \langle T_\tau C_{-\mathbf{p}\sigma'}(\tau) C_{\mathbf{p}\uparrow}(\tau')\rangle$$

$$\left(+\frac{\partial}{\partial \tau'} + \xi_p\right) \mathscr{F}_{\sigma'}(\mathbf{p}, \tau - \tau') + \frac{1}{\nu} \sum_{\substack{\mathbf{q},\mathbf{p}' \\ \sigma'}} V(q)$$
$$\times \langle T_\tau C_{-\mathbf{p}\sigma'}(\tau) C^{\dagger}_{\mathbf{p}'-\mathbf{q}\sigma''}(\tau') C_{\mathbf{p}'\sigma''}(\tau') C_{\mathbf{p}-\mathbf{q}\uparrow}(\tau')\rangle = 0$$

where $\sigma' = \pm 1$ gives the two gap functions. In the bracket of four operators, we pair the first two operators to get a Green's function $\mathscr{G}(-\mathbf{p}, \tau - \tau') = \mathscr{G}(\mathbf{p}, \tau - \tau')$, which sets $\sigma' = \sigma''$. The other two operators are paired into a gap function. The pairings which lead to Hartree and exchange energies such as $\langle C_{-\mathbf{p}\sigma'} C_{\mathbf{p}'\sigma''}\rangle\langle C^{\dagger}_{\mathbf{p}'-\mathbf{q}\sigma''} C_{\mathbf{p}-\mathbf{q}\uparrow}\rangle$ are omitted in weak coupling theory. The pairing terms which are kept give the following two equations of motion:

$$\left(\frac{\partial}{\partial \tau'} + \xi_p\right) \mathscr{F}_1(\mathbf{p}, \tau - \tau') - \mathscr{G}(\mathbf{p}, \tau - \tau') \frac{1}{\nu} \sum_{\mathbf{q}} V(q)\mathscr{F}_1(\mathbf{p} - \mathbf{q}, 0) = 0$$

$$\left(\frac{\partial}{\partial \tau'} + \xi_p\right) \mathscr{F}_0(\mathbf{p}, \tau - \tau') - \mathscr{G}(\mathbf{p}, \tau - \tau') \frac{1}{\nu} \sum_{\mathbf{q}} V(q)\mathscr{F}_0(\mathbf{p} - \mathbf{q}, 0) = 0$$

These are now Fourier-transformed in the usual way to obtain functions of Matsubara frequency. Then they become algebraic equations which can be solved. Important terms in these equations are the gap functions, which are defined as

$$\begin{aligned} \Delta_1(\mathbf{p}) &= -\frac{1}{\nu} \sum_{\mathbf{p}'} V(\mathbf{p}, \mathbf{p}')\mathscr{F}_1(\mathbf{p}', \tau = 0) \\ \Delta_0(\mathbf{p}) &= -\frac{1}{\nu} \sum_{\mathbf{p}'} V(\mathbf{p}, \mathbf{p}')\mathscr{F}_0(\mathbf{p}', \tau = 0) \end{aligned} \qquad (10.5.6)$$

The equations of motion to be solved are

$$(ip - \xi_p)\mathscr{G}(p) + \mathscr{F}_1^{\dagger}(p)\Delta_1 + \mathscr{F}_0^{\dagger}(p)\Delta_0 = 1$$
$$(ip + \xi_p)\mathscr{F}_1(p) + \mathscr{G}(p)\Delta_1 = 0$$
$$(ip + \xi_p)\mathscr{F}_0(p) + \mathscr{G}(p)\Delta_0 = 0$$

To solve for the equation for $\mathscr{G}(p)$, we need the equation for $\mathscr{F}_m^\dagger(p)$. The $\mathscr{G}(p)$ are obtained from the equation for $\mathscr{F}_m(p)$ by a Hermitian conjugate:

$$(ip + \xi_p)\mathscr{F}_m^\dagger(p) + \mathscr{G}(p)\Delta_m^\dagger = 0$$

$$\Delta_m^\dagger(\mathbf{p}) = -\frac{1}{\nu}\sum_{p'} V(\mathbf{p}, \mathbf{p}')\mathscr{F}_m^\dagger(\mathbf{p}', \tau = 0)$$

Then the solution is straightforward:

$$E(\mathbf{p}) = \xi_p^2 + |\Delta_1(\mathbf{p})|^2 + |\Delta_0(\mathbf{p})|^2$$

$$\mathscr{G}(p) = -\frac{ip + \xi_p}{p_n^2 + E(\mathbf{p})^2} \tag{10.5.7}$$

$$\mathscr{F}_m^\dagger(p) = \mathscr{F}_m(p) = \frac{\Delta_m}{p_n^2 + E(\mathbf{p})^2}$$

The effective gap function is $\Delta(\mathbf{p}) = (|\Delta_1|^2 + |\Delta_0|^2)^{1/2}$, and the energy gap E_q is 2Δ. In some respects, the BW state appears to be similar to a BCS singlet state but with an effective energy gap of $2\Delta(\mathbf{p})$ and excitation energy $E(\mathbf{p})$. This similarity does not extend to magnetic phenomena, such as spin susceptibility, which are much more interesting in triplet pairing.

The last step in deriving this self-consistent theory is to obtain the equation for the gap function. The correlation functions $\mathscr{F}_m(\mathbf{p}, \tau)$ are found for $\tau = 0$, with the familiar result

$$\mathscr{F}_m(\mathbf{p}, \tau = 0) = \frac{1}{\beta}\sum_{ip} \mathscr{F}_m(\mathbf{p}, ip) = \frac{\Delta_m(\mathbf{p}) \tanh[\beta E(\mathbf{p})/2]}{2E(\mathbf{p})}$$

$$\Delta_m(\mathbf{p}) = -\frac{1}{2}\int \frac{d^3p'}{(2\pi)^3} V(\mathbf{p}, \mathbf{p}') \frac{\Delta_m(\mathbf{p}') \tanh[\beta E(\mathbf{p}')/2]}{E(\mathbf{p}')} \tag{10.5.8}$$

In the second equation, we have used the definition (10.5.6) to derive the self-consistent equation for the gap function. This gap function must be solved for the energy gaps $\Delta_m(\mathbf{p})$.

The gap equation is obviously a nonlinear matrix equation. Each term in the matrix depends on the excitation energy $E(\mathbf{p})$, which depends on the two energy gaps Δ_0 and Δ_1. Thus we have a set of coupled equations for Δ_0 and Δ_1. We also have the restriction that the gap function is an odd function of momentum, $\Delta_m(-\mathbf{p}) = -\Delta_m(\mathbf{p})$, which follows from the condition that $\mathscr{F}_m(\mathbf{p})$ have this property. This condition is automatically satisfied when the interaction potential is taken to be the p-wave type $V = \hat{p} \cdot \hat{p}' V_1$. For example, if we change dummy variables of integration in the gap equation from $\mathbf{p}'$ to $-\mathbf{p}'$, the equation is unchanged when both $V(\mathbf{p}, \mathbf{p}')$ and $\Delta_m(\mathbf{p}')$ are odd functions of $\mathbf{p}'$.

The interesting aspect of the gap equation in triplet pairing is that the solutions are not unique but have orbital degeneracy. There are two sources of angular momentum from each pair of particles: the spin angular momentum and the orbital angular momentum in the p state. The first question is the relative orientation of these angular momenta. If two particles were binding in a gas to form an He_2 molecule, then one could just couple these momenta with Clebsch–Gordon coefficients into all possible values of $J = 0$, 1, and 2. In the Fermi liquid, the superfluid state is a collective property, and we cannot treat each pair as individuals. Nevertheless, we still have some freedom in selecting the basis for the spin and the orbital angular momentum. This idea is best illustrated by an example.

The gap function $\Delta_m(\mathbf{p})$ must have p-wave symmetry. The obvious configuration has the spin and orbital states aligned along the same axis. The first solution we shall describe was found by BW. For the gap function $\Delta_m(\mathbf{p})$ they chose the orbital function according to $Y_{1,m}(\theta_p, \varphi_p)$, where (θ_p, φ_p) are the angular orientation of the p vectors. Specifically, this choice is

$$\begin{aligned}
\Delta_1(\mathbf{p}) &= (\hat{p}_x + i\hat{p}_y)\Delta(p) = \sin(\theta_p)e^{i\varphi_p}\Delta(p) \\
\Delta_0(\mathbf{p}) &= \hat{p}_z\Delta(p) = \cos(\theta_p)\Delta(p) \\
\Delta_{-1}(\mathbf{p}) &= (\hat{p}_x - i\hat{p}_y)\Delta(p) = \sin(\theta_p)e^{-i\varphi_p}\Delta(p) \qquad (10.5.9)\\
|\Delta_0|^2 + |\Delta_1|^2 &= \Delta(p)^2 \\
E(\mathbf{p}) &= (\xi_p{}^2 + |\Delta(\mathbf{p})|^2)^{1/2}
\end{aligned}$$

The functions $E(\mathbf{p})$ and $\Delta(\mathbf{p})$ are even functions of $\mathbf{p}$ and are scalar. Of course, $\Delta(\mathbf{p})$ could have a constant phase factor $\Delta(\mathbf{p}) = \exp(i\lambda)\,|\Delta|$, which is irrelevant. First, we verify that this choice satisfies the gap equation. We use the angles (θ, φ) for the direction of $\mathbf{p}$ and the variables (θ', φ') for $\mathbf{p}'$. From the law of cosines, the angle between $\mathbf{p}$ and $\mathbf{p}'$ is

$$\hat{p}\cdot\hat{p}' = \cos\theta\cos\theta' + \sin\theta\sin\theta'\cos(\varphi - \varphi')$$

$$\Delta_1(\mathbf{p}) = -\frac{N_F V_1}{4}\int d\xi' \frac{d\Omega'}{4\pi}\,\hat{p}\cdot\hat{p}'\sin(\theta')e^{i\varphi'}\frac{\Delta(p')}{E(p')}\tanh\left[\frac{\beta E(p')}{2}\right]$$

$$\Delta_0(\mathbf{p}) = -\frac{N_F V_1}{4}\int d\xi' \frac{d\Omega'}{4\pi}\,\hat{p}\cdot\hat{p}'\cos(\theta')\frac{\Delta(p')}{E(p')}\tanh\left[\frac{\beta E(p')}{2}\right]$$

The energy gap $E(p')$ does not depend on the angles (θ', φ') since $\Delta(p')$ does not depend on angles. Then the angular integrals in the gap equation are easy to do and give ($g_t = -N_F V_1/6 = -C_1/12 > 0$)

$$\Delta_1(\mathbf{p}) = \frac{1}{2}\sin(\theta)e^{i\varphi}g_t \int_{-\omega_c}^{\omega_c} d\xi' \frac{\Delta(p')}{E(p')}\tanh\left[\frac{\beta E(p')}{2}\right]$$

$$\Delta_0(\mathbf{p}) = \frac{1}{2}\cos(\theta)g_t \int_{-\omega_c}^{\omega_c} d\xi' \frac{\Delta(p')}{E(p')}\tanh\left[\frac{\beta E(p')}{2}\right] \tag{10.5.10}$$

$$\Delta_{-1}(\mathbf{p}) = \frac{1}{2}\sin(\theta)e^{-i\varphi}g_t \int_{-\omega_c}^{\omega_c} d\xi' \frac{\Delta(p')}{E(p')}\tanh\left[\frac{\beta E(p')}{2}\right]$$

For each gap function $\Delta_m(\mathbf{p})$, the angular functions are the same on both sides of the equation. These can be canceled from both sides, and each equation produces the same formula for the effective gap function:

$$\Delta(p) = \frac{1}{2}g_t \int_{-\omega_c}^{\omega_c} d\xi' \frac{\Delta(p')}{E(p')}\tanh\left[\frac{\beta E(p')}{2}\right]$$

The equation is identical to the BCS equation for the singlet gap. Here the coupling constant is effectively g_t, where we use the notation of Patton and Zaringhalam. This equation is easiest to solve at zero temperature where the hyperbolic tangent is unity, Δ = constant:

$$\begin{aligned} 1 &= \frac{1}{2}g_t \int_{-\omega_c}^{\omega_c} \frac{d\xi}{(\xi^2 + \Delta^2)^{1/2}} \\ &= g_t\{\ln[\xi + (\xi^2 + \Delta^2)^{1/2}]\}_0^{\omega_c} \simeq g_t \ln\left(\frac{2\omega_c}{|\Delta|}\right) \\ |\Delta|_{\mathrm{BW}} &= 2\omega_c \exp\left(-\frac{1}{g_t}\right) \end{aligned} \tag{10.5.11}$$

The solution is an equation of the BCS type for the gap function $|\Delta|$ at zero temperature. The variation with temperature below T_c is also similar to the BCS prediction and varies as $(T_c - T)^{1/2}$.

The equation we have just solved describes the Balian–Werthamer state. This theory is thought to apply to the B state of superfluid liquid ^{3}He, although the theory must be modified to account for the strong coupling between quasiparticles. However, even the weak coupling theory, as just derived, makes a number of predictions which agree with the experimental findings in the superfluid B state. Some of these will now be described.

One important property is that the B state is *isotropic*, which is defined as having the energy gap $\Delta(p)$ not depend on angle in the superfluid. The best way to understand this definition is to give an example of a state which is nonisotropic. As a random choice, consider the possibility

$$\begin{aligned} \Delta_1 &= \Delta_{-1} = 0 \\ \Delta_0(\mathbf{p}) &= 3^{1/2}\cos(\theta)\Delta_a(p) \\ E(\mathbf{p}) &= [\xi_p{}^2 + 3\cos^2\theta \mid \Delta_a(p)\mid^2]^{1/2} \end{aligned} \tag{10.5.12}$$

The p-wave pairing requires that $\Delta_m(\mathbf{p})$ always depend on a p-symmetry angular function. This choice in (10.5.12) for $\Delta_m(\mathbf{p})$ provides an entirely self-consistent solution to the gap equation (10.5.8). The normalization factor $3^{1/2}$ is chosen so that $(\mid \Delta_0 \mid^2 + \mid \Delta_1 \mid^2)$ averages to $\Delta_a(p)^2$ around the Fermi surface. The quantity Δ_a is the average gap. The energy gap is no longer isotropic but now depends on the angle θ between the spin and orbital motion. The energy gap and excitation energy will vary around the Fermi surface. We shall present a much more detailed discussion of non-isotropic superfluid states in the next section and defer further comments until then. The motivation is provided by the superfluid A phase, which is thought to be a nonisotropic state.

Our solution for the BW state (10.5.9) chose axes where the coordinate bases for the spin and orbital motions were aligned. There is no need to make this choice, and the same gap equation for $\Delta(\mathbf{p})$ (and the same solution) are obtained if these axes are rotated with respect to each other. For example, if we give the orbital coordinates a 90° rotation, we could have the gap functions given by

$$\begin{aligned} \Delta_1 &= (\hat{p}_y + i\hat{p}_z)\Delta_a(p) = (\sin\theta\sin\varphi + i\cos\theta)\Delta(p) \\ \Delta_0 &= \hat{p}_x\Delta(p) = \sin\theta\cos(\varphi)\Delta(p) \\ \Delta_{-1} &= (\hat{p}_y - i\hat{p}_z)\Delta(p) = (\sin\theta\sin\varphi - i\cos\theta)\Delta(p) \\ E(p) &= [\xi_p{}^2 + \mid\Delta(p)\mid^2]^{1/2} \end{aligned} \tag{10.5.13}$$

The gap function is isotropic and given by the same value (10.5.11) as before. Thus in weak coupling the choice of orbital coordinates does not influence the energy gap. The BW superfluid state has orbital degeneracy in that the orbital motion is not coupled to the spin motion and can proceed independently. This feature depends on the isotropic nature of the superfluid phase.

Many of the thermodynamic properties of the superfluid phase depend only on the energy gap. One example is the specific heat. These properties are exactly the same as predicted by the BCS theory for singlet pairing, which also has an isotropic energy gap. (In real metals, with anisotropic Fermi surfaces, the BCS gap can also be anisotropic, which is a complication we shall not discuss.) The BW theory predicts that the transition from the normal to the superfluid phase is a second-order phase transition and that the specific heat is discontinuous at this phase transition. These properties are observed in superfluid ^{3}He, as mentioned earlier.

The interesting difference between singlet and triplet pairing is in magnetic phenomena. There are interesting nuclear magnetic resonance phenomena in the triplet states which were very important for identification of the superfluid phases and which verify the theory. These are described by Leggett (1975). Another important difference is in the Pauli spin susceptibility. This property is very important for the later discussion of strong coupling theory, so we shall discuss it in detail. The first step is to recall the prediction of BCS theory for the spin susceptibility for singlet pairing, as given in Problem 6 in Chapter 9:

$$\begin{aligned} \chi_{\text{BCS}} &= 2\mu_0^2 \sum_p \frac{d}{dE(p)} n_F[E(p)] \\ \chi_{\text{BCS}} &= \chi_N Y(T) \\ \chi_N &= \mu_0^2 N_F \\ Y(T) &= \frac{\beta}{2} \int_0^\infty \frac{d\xi_p}{\cosh^2[\beta E(p)/2]} \end{aligned} \tag{10.5.14}$$

The factor χ_N is the susceptibility in the normal phase of the Fermi liquid—in the weak coupling approximation. The function $Y(T)$ is called the Yoshida function (1958) and gives the BCS prediction of the spin susceptibility. It is unity at T_c when $\Delta = 0$ and also in the normal state. The Yoshida function decreases to zero at zero temperature. One can show that since $\Delta \propto (T_c - T)^{1/2}$, $Y \propto (T_c - T)$ as T is dropped below T_c. The Yoshida function vanishes at $T = 0$, and the susceptibility of the BCS superconductor is zero.

For singlet pairing the magnetic susceptibility is zero at zero temperature, which is true only for weak magnetic fields. At higher fields, the magnetic field will destroy the superconducting phase and return the system to a normal phase with the normal susceptibility χ_N. The zero value of magnetic susceptibility can be understood as a competition between the tendency of the field to align the spins because of the gain in magnetic energy and the tendency to make them antiparallel, which favors pairing in the superfluid. For weak magnetic fields the superfluid wins, but for stronger fields the magnetic forces dominate. The conventional picture of a normal fermion system in a magnetic field is shown in Fig. 10.20(a).

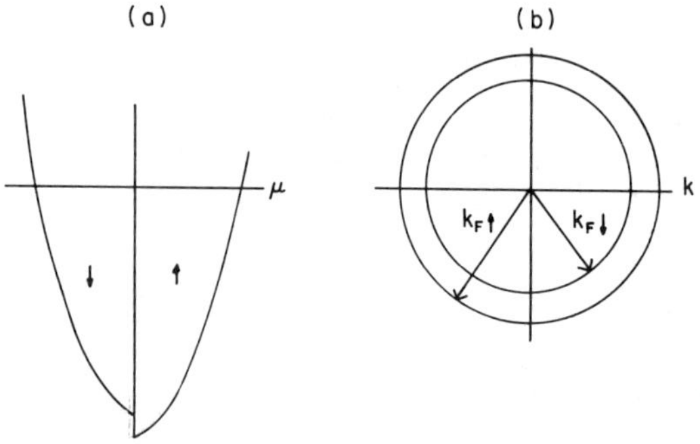

FIGURE 10.20. (a) In normal Fermi liquids, a static magnetic field changes the chemical potential for spin up and spin down particles. (b) It changes their Fermi wave vectors.

The particles with spin up have a magnetic energy $-\mu_0 H_0$, while those with spin down have the additional energy of $+\mu_0 H_0$. Since they have the same chemical potential μ, the spin up particles can have more kinetic energy at μ than do the spin down particles. Thus there are more spin up particles, and this additional amount is the net magnetization:

$$M = -\mu_0(n_\uparrow - n_\downarrow) = -\mu_0 \int \frac{d^3k}{(2\pi)^3} [n_F(\xi - \mu_0 H_0) - n_F(\xi + \mu_0 H_0)]$$

$$= 2\mu_0{}^2 H_0 \int \frac{d^3k}{(2\pi)^3} \left(-\frac{\partial n_F}{\partial \xi}\right) = N_F \mu_0{}^2 H_0$$

This theory predicts that the susceptibility is the square of the magnetic moment (μ_0) multiplied by the density of states. This increase in the kinetic energy of the spin up particles at k_F means that there is a different value of k_F for spins up and down. This difference is shown in an exaggerated fashion in Fig. 10.20(b). Now it is hard for the pairing to occur between the spins down and up, since they are moving with different velocities. In the superfluid state, the particles stay paired, with the same kinetic energies at the chemical potential for particles with spins up and down. The magnetization can occur only by breaking pairs and exciting two quasiparticles. These can align with both spins up to decrease their energy, and thus the magnetization is proportional to the density of quasiparticles. The upper critical field, at which the magnetic field drives the fluid normal, is when the gain in magnetic energy is larger than the cost of breaking up the pairs.

This physical picture provides an understanding of the magnetic susceptibility in the BW theory of triplet pairing. In this theory the pairs can exist with magnetic quantum number $m_s = 1, 0, -1$, with energy gaps

$\Delta_1, \Delta_0, \Delta_{-1}$. In the absence of a magnetic field, the particles must be paired with equal probability in these three states. An equivalent statement is that one-third of the pairs occur in each of these three states. Then the average energy gap of the system is obtained by averaging over all the particles. A particle with, say, spin up has equal likelihood of pairing with particles of either spin, so the energy gap is the average $\Delta(p)^2 = |\Delta_0|^2 + |\Delta_1|^2$. For our discussion, we assume that the magnetic field is applied in the direction of the spin axis. Thus the state with $m_s = 0$ has one spin down and one up with respect to the magnetic field. The response of pairs in this state to the magnetic field is identical to that of the singlet pairs. They cannot align both spins with the field without breaking the pair state. The susceptibility of spins in this state is identical to that of the singlet state, which is $\chi_N Y(T)$, where Y is the Yoshida function.

Next we consider the behavior of the parallel spin state with $m_s = \pm 1$. This behavior resembles that of the normal metal, as shown in Fig. 10.20(a). The spin down pairs ($m_s = -1$) have a magnetic energy $+2\mu_0 H_0$, so that they have a lower kinetic energy at the chemical potential. We again refer to Fig. 10.20(b), where the Fermi wave vector for spin up particles is larger than for spin down. This difference in $k_{F\uparrow}$ and $k_{F\downarrow}$ does not affect the pairing of particles with $m_s = \pm 1$, since they are pairing with particles of like spin and like values of $k_{F\sigma}$. Thus the weak magnetic field should not affect the pairing of particles with parallel spins. The magnetization of the superfluid is the number of particles paired in states with $m_s = 1$ minus the number paired in $m_s = -1$. This difference is not changed from the value in the normal phase, since the field does not affect the pairing. Two-thirds of the particles in these two pairing states should have a normal susceptibility. We conclude that the Pauli susceptibility for the BW state is given by

$$\chi_{\mathrm{BW}} = \chi_N[\tfrac{2}{3} + \tfrac{1}{3}Y(T)] \tag{10.5.15}$$

Equation (10.5.15) is the prediction of the weak coupling theory. At zero temperature, the susceptibility does not vanish but instead saturates at a value which is two-thirds of the bulk.

In normal liquid ^{3}He the Fermi liquid parameters are large. It is not a weakly interacting system. Even the effective interactions between dressed quasiparticles are significant. Thus the weak coupling predictions of BW must be modified to account for the Fermi liquid effects. This was first solved by Leggett (1965), who obtained the formula

$$\chi_{\mathrm{BW}} = \chi_N \frac{[\frac{2}{3} + \frac{1}{3}Y(T)](1 + F_0^a)}{1 + F_0^a[\frac{2}{3} + \frac{1}{3}Y(T)]} \tag{10.5.16}$$

A later calculation showed that there is an additional quadrupole contribution, so the denominator has the term $F_2^a/15$, which is usually neglected. The experimental measurements by Osheroff (1974) on the B phase of the superfluid were obtained nearly a decade after the theoretical prediction. Theory and experiment are shown in Fig. 10.21, and the agreement is excellent. Two sets of experimental points are shown. They are for the same experiment and represent two different ways of extrapolating to zero magnetic field. Two theoretical curves are shown, one for $F_0^a = -0.75$ and another for $F_0^a = -0.80$. The first value is deduced from the susceptibility in the normal phase, while the second is a better fit to the data. The value of $F_0^a = -0.75$ is different from the one listed in Table 10.3 because the table refers to values at atmospheric pressure, while the susceptibility measurements were taken at the melting pressure of $P = 34$ bars. The Fermi liquid parameters change with pressure, although the change in F_0^a is slight. Theory and experiment for the static susceptibility do not agree as well for lower pressures, which is an unsolved problem.

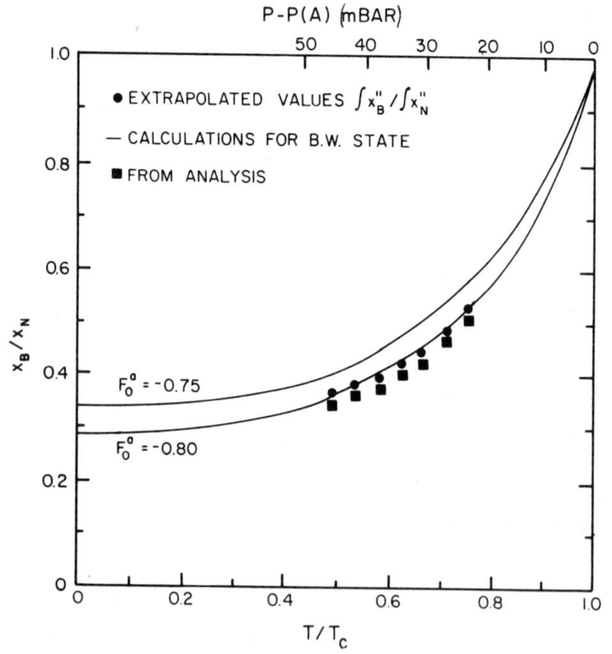

FIGURE 10.21. Pauli spin susceptibility of the superfluid B phase of ^{3}He compared with the normal phase as a function of T/T_c. The points are data of Osheroff, and the solid lines are the theory of Leggett for two values of F_0^a. *Source*: Osheroff (1974) (used with permission).

Figure 10.21 shows no experimental data in the temperature range $0.792 < T/T_c < 1.00$ because at the melting pressure the liquid is in the superfluid A phase in this range of temperatures (see Fig. 10.19). The first-order transition to the B phase occurs at $T_{AB} = 0.8T_c$ at the melting pressure. In the A phase, the Pauli susceptibility is equal to the value in the normal phase χ_N. The susceptibility drops discontinuously at the AB phase boundary. The excellent agreement for the B phase, with the theory of Leggett for BW superfluids, is evidence that the B phase is properly identified as a state of triplet pairing as described by Balian and Werthamer. There remains the task of explaining the A phase. The susceptibility data is a good clue. The fact that it is unchanged from its normal liquid value χ_N suggests that this state must not contain any spin components with $m_s = 0$, since these have a temperature-dependent susceptibility. The A phase must contain states with $m_s = -1$ and $m_s = 1$ and most likely in equal numbers. These are called *equal spin pairing* (ESP) states, and we now examine their properties.

B. Equal Spin Pairing

This section is devoted to a discussion of other types of triplet pairing states besides that of Balian and Werthamer (BW). All the states we shall consider are solutions to the gap equation (10.5.8) for triplet pairing. The motivation is provided by mother nature, who has decided that there are two different superfluid phases, A and B. The B phase is now identified to be BW, but we still need to find the model for the A phase. It is believed to be a state with equal spin pairing (ESP). They are composed of only the $m_s = \pm 1$ components of the spin angular momentum and not with the $m_s = 0$ components. Thus we examine the predictions of the theory when we set $\Delta_0 = 0$ and retain the two gaps Δ_1 and Δ_{-1}.

The first realization is that there is no relationship between the orbital momentum of the two gap functions $\Delta_1(\mathbf{p})$ and $\Delta_{-1}(\mathbf{p})$. They are uncoupled because the equation which is solved for each $\Delta_m(\mathbf{p})$ no longer involves the other value of $-m$. The excitation energy $E_m(\mathbf{p}) = [\xi_p{}^2 + |\Delta_m(\mathbf{p})|^2]^{1/2}$ involves only one gap function. Another way to understand the decoupling of Δ_1 and Δ_{-1} is to start from the beginning and write the equations of motion. In describing the response of a spin up particle, we need be concerned only with the correlation of the motion of other spin up particles. There is no correlation with spin down particles once we set $\Delta_0 = 0$. Thus the spin up particles live in one world, and the spin down particles live in another. The orbital motion of spin up and spin down particles is arbitrary.

Thus we can write them as

$$\Delta_1(\mathbf{p}) = (\tfrac{3}{2})^{1/2}(\hat{p}_x + i\hat{p}_y)\Delta_a(p)$$
$$\Delta_{-1}(\mathbf{p}) = (\tfrac{3}{2})^{1/2}(\hat{p}_x' + i\hat{p}_y')\Delta_a(p)$$

where the relative orientation of the coordinate systems (p_x, p_y, p_z) with respect to (p_x', p_y', p_z') is absolutely arbitrary. In the BW state, the two gap functions Δ_1 and Δ_{-1} shared a linkage because both depended on Δ_0, but that bond is absent when we consider ESP states.

One consequence of this orbital independence is that the excitation energies $E_m(\mathbf{p})$ and gap functions $|\Delta_m(\mathbf{p})|^2$ no longer need to have the same values for $m = 1$ and $m = -1$. For example, if we choose the set of solutions

$$\Delta_1(\mathbf{p}) = (\tfrac{3}{2})^{1/2}(\hat{p}_x + i\hat{p}_y)\Delta_a = (\tfrac{3}{2})^{1/2}\sin(\theta)e^{i\varphi}\Delta_a(p)$$
$$\Delta_{-1}(\mathbf{p}) = (\tfrac{3}{2})^{1/2}(\hat{p}_y + i\hat{p}_z)\Delta_a(p) = (\tfrac{3}{2})^{1/2}(\sin\theta\cos\varphi + i\cos\theta)\Delta_a(p)$$

Then it is obvious that $E_1(\theta, \varphi) \neq E_{-1}(\theta, \varphi)$ except at special angular points. The excitation and energy gaps for spin up and spin down particles would be quite different. Each energy gap function has the same shape relative to its axis (north and south poles), but these poles are rotated relative to each other for spin up and spin down particles.

The superfluid is called *unitary* if the spin up and spin down particles have the identical angular dependence of their energy gaps $E_1(\theta, \varphi) = E_{-1}(\theta, \varphi)$. ESP states are unitary when the north–south poles are aligned for the two systems. The superfluid is called *nonunitary* when the excitation energies are not identical as a function of angle. The BW state is unitary.

The A phase appears to be a unitary ESP state. The north and south poles are aligned for the two gap functions. This is an experimental conclusion, and there is no particular reason the state has to be unitary. The important question is the nature of the forces which tend to align the poles. There are no forces in our Hamiltonian (10.5.1) which cause alignment. But we have ignored the very, very weak forces due to the dipole–dipole interactions between the nuclear moments. Leggett has shown that these dipolar forces will tend to align the poles. Although it is a weak effect, there is nothing to oppose it. Of course, any external perturbation such as a magnetic field will also tend to align the poles, since alignment lowers the magnetic energy, as we shall show later. It simplifies discussion to consider only unitary states. We shall make that assumption here, since it corresponds to the experimental situation. Nonunitary states are discussed in reviews by Leggett (1975) and Anderson and Brinkman (1975).

The unitary ESP state can be chosen two ways: The two north poles can be aligned, or the north and south poles can be aligned. These two possibilities are

ABM

$$\begin{aligned} \Delta_1 &= (\tfrac{3}{2})^{1/2}(\hat{p}_x + i\hat{p}_y)\Delta_a(p) \\ \Delta_0 &= 0 \\ \Delta_{-1} &= (\tfrac{3}{2})^{1/2}(\hat{p}_x + i\hat{p}_y)\Delta_a(p) \\ E(p) &= [\xi_p{}^2 + \tfrac{3}{2}\Delta_a(p)^2 \sin^2\theta]^{1/2} \end{aligned} \tag{10.5.17}$$

A', 2D, planar

$$\begin{aligned} \Delta_1 &= (\tfrac{3}{2})^{1/2}(\hat{p}_x + i\hat{p}_y)\Delta_a(p) \\ \Delta_0 &= 0 \\ \Delta_{-1} &= (\tfrac{3}{2})^{1/2}(\hat{p}_x - i\hat{p}_y)\Delta_a(p) \\ E(p) &= [\xi_p{}^2 + \tfrac{3}{2}\sin^2(\theta)\Delta_a(p)^2]^{1/2} \end{aligned}$$

Two other triplet states are

1D, polar

$$\begin{aligned} \Delta_1 &= 0 \\ \Delta_0 &= (3)^{1/2}\hat{p}_z\Delta_a(p) = (3)^{1/2}\cos(\theta)\Delta_a(p) \\ \Delta_{-1} &= 0 \\ E(p) &= [\xi_p{}^2 + 3\cos^2(\theta)\Delta_a(p)^2]^{1/2} \end{aligned} \tag{10.5.18}$$

BW

$$\begin{aligned} \Delta_1 &= (\hat{p}_x + i\hat{p}_y)\Delta(p) = \sin(\theta)e^{i\varphi}\Delta(p) \\ \Delta_0 &= \hat{p}_z\Delta(p) = \cos(\theta)\Delta(p) \\ \Delta_1 &= (\hat{p}_x - i\hat{p}_y)\Delta(p) = \sin(\theta)e^{-i\varphi}\Delta(p) \\ E(p) &= [\xi_p{}^2 + \Delta(p)^2]^{1/2} \end{aligned}$$

We shall discuss the properties of these four possible solutions to the triplet gap functions. One is BW, which is included as an important comparison. The other was mentioned in (10.5.12). Also included are the two choices for ESP. Our notation and nomenclature are from Leggett and Anderson and Brinkman. All states are normalized so the angular average of $\langle |\Delta_0|^2 + |\Delta_1|^2 \rangle = |\Delta_a(p)|^2$. The planar state is the BW state but with $\Delta_0 = 0$. The Anderson–Brinkman–Morel (ABM) state has the poles the other way. By using Ginzburg–Landau arguments, Mermin and Stare (1973) showed that these are the only four independent states which can be minima of the free energy.

The most important feature of each state is its free energy at zero temperature or any temperature. The state with the lowest free energy is the one chosen by mother nature. We shall first calculate the free energy using the weak coupling theory. There we shall find that BW is the preferred state with the lowest free energy. The two ESP states have the same energy and tie for second in the contest for the lowest free energy. The polar state is a distant last in this competition. We shall decide this contest by calculating the gap function for each at zero temperature. That with the largest gap has the most binding. In fact, in weak coupling the lowering of free energy at $T = 0$ is proportional to the angular average of $\langle |\Delta_0|^2 + |\Delta_1|^2 \rangle = |\Delta_a(p)|^2$. The pairing state with the largest $\Delta_a(p)$ has the lowest free energy. It is also important to realize that all these solutions make the same prediction regarding the transition temperature T_c, at least in weak coupling. For $T = T_c$ the gap functions vanish, and the excitation energy is $E(p) = |\xi|$, in which case the gap equation for all choices of triplet pairing is

$$\lim_{T \to T_c} \left[\Delta_m(p) = 3g_t \int_{-\omega_c}^{\omega_c} d\xi \int \frac{d\Omega'}{4\pi} \hat{p} \cdot \hat{p}' \frac{\Delta_m(p')}{2E(p')} \tanh\left(\frac{\beta E_{p'}}{2} \right) \right]$$

$$\lim_{T \to T_c} \left[\Delta_m = g_t \Delta_m \int_{-\omega_c}^{\omega_c} \frac{d\xi}{2\xi} \tanh\left(\frac{\beta \xi}{2} \right) \right]$$

The common factor Δ_m is canceled from both sides in the limit where $\Delta_m \to 0$. The transition temperature is the temperature at which one can first satisfy the equation

$$1 = \frac{1}{2} g_t \int_{-\omega_c}^{\omega_c} \frac{d\xi}{|\xi|} \tanh\left(\frac{\beta_c |\xi|}{2} \right)$$

The same T_c is found for all our different choices of Δ_m. Of course, it is also the same transition temperature at which the Cooper instability occurs in the normal phase.

The gap function $\Delta_a(p)$ will be calculated for the ESP states at zero temperature. The gap parameter $\Delta_a(p)$ is a scalar function and is independent of angle. In weak coupling theory it is treated as a constant, which is independent of momentum up to the cutoff value ω_c. The two ESP unitary states of ABM and planar have the same value of the gap function, because they have the same excitation energy. It is only necessary to find the gap function for one of them and for one value of m, since the calculation is identical for the other ESP states. We shall pick one at random, the Δ_1 state of ABM, and solve its gap equation. At zero temperature this equation is

$$\Delta_1 = (\tfrac{3}{2})^{1/2} \sin(\theta) e^{i\varphi} \Delta_a$$
$$= \frac{3g_t}{4\pi} \int_0^{\omega_c} d\xi \int d\Omega' \left(\frac{3}{2}\right)^{1/2}$$
$$\times \frac{\sin(\theta') e^{i\varphi'} \Delta_a [\cos\theta \cos\theta' + \sin\theta \sin\theta' \cos(\varphi - \varphi')]}{[\xi^2 + \frac{3}{2}\sin^2(\theta')\Delta_a{}^2]^{1/2}}$$

The angular integrals are more complicated than for the BW state, because the excitation energy depends on the angle θ. It is still possible to do the $d\varphi$ integral, which provides some simplification:

$$\frac{1}{2\pi} \int_0^{2\pi} d\varphi' e^{i\varphi'} \cos(\varphi - \varphi') = \frac{1}{2} e^{i\varphi}$$

$$\Delta_1 = \left[\left(\frac{3}{2}\right)^{1/2} \sin(\theta) e^{i\varphi} \Delta_a\right] \frac{3g_t}{4} \int_0^{\omega_c} d\xi \int_0^{\pi} d\theta' \frac{\sin^3\theta'}{[\xi^2 + \frac{3}{2}\sin^2(\theta')\Delta_a{}^2]^{1/2}}$$
$$= (\tfrac{3}{2})^{1/2} \sin(\theta) e^{i\varphi} \Delta_a$$

The common factor of $\Delta_a (\frac{3}{2})^{1/2} \sin(\theta) e^{i\varphi}$ is canceled from both sides of this equation for Δ_1. The variables are changed to $\nu = \cos\theta'$, and the $d\xi$ integral can be done exactly:

$$1 = \frac{3g_t}{4} \int_{-1}^{1} d\nu (1 - \nu^2) \int_0^{\omega_c} d\xi \left[\xi^2 + \frac{3}{2} \Delta_a{}^2 (1 - \nu^2)\right]^{-1/2}$$
$$= \frac{3g_t}{4} \int_{-1}^{1} d\nu (1 - \nu^2) \ln\left\{\xi + \left[\xi^2 + \frac{3}{2}\Delta_a{}^2(1 - \nu^2)\right]^{1/2}\right\}\Bigg|_0^{\omega_c}$$
$$= \frac{3g_t}{4} \int_{-1}^{1} d\nu (1 - \nu^2) \ln\left[\frac{2\omega_c}{(\frac{3}{2})^{1/2}\Delta_a (1 - \nu^2)^{1/2}}\right]$$

In doing the $d\nu$ integral, we separate out the angular term in the logarithm and find

$$\ln\left[\frac{2\omega_c}{(\frac{3}{2})^{1/2}\Delta_a(1 - \nu^2)^{1/2}}\right] = \ln\left[\frac{2\omega_c}{(\frac{3}{2})^{1/2}\Delta_a}\right] - \frac{1}{2}\ln(1 - \nu^2)$$
$$\frac{3}{4}\int_{-1}^{1} d\nu (1 - \nu^2) = 1$$
$$\frac{3}{4}\int_{-1}^{1} d\nu (1 - \nu^2) \times \frac{1}{2}\ln(1 - \nu^2) = \ln(2) - \frac{5}{6}$$

We can now solve for the gap function:

$$1 = g_t\left\{\ln\left[\frac{2\omega_c}{(\frac{3}{2})^{1/2}\Delta_a}\right] - \left[\ln(2) - \frac{5}{6}\right]\right\}$$

$$\Delta_a = \left[\frac{e^{5/6}}{6^{1/2}}\right] \times 2\omega_c \exp\left(\frac{-1}{g_t}\right) \tag{10.5.19}$$

$$\Delta_a^{(\mathrm{ESP})} = (0.94)2\omega_c \exp\left(\frac{-1}{g_t}\right) = 0.94\Delta_a^{\mathrm{BW}}$$

The ESP state has a gap function which is smaller than the BW state by the factor 0.94. We said earlier that the lowest free energy is for the state with the largest value of Δ_a, which comes from the largest value of $\langle\Delta_0(\theta)^2 + \Delta_1(\theta)^2\rangle_{\mathrm{av}} = \Delta_a^2$. The BW state has a lower free energy than the ESP state at zero temperature and will be preferred by the superfluid. An analysis at finite temperature shows that this conclusion is valid for the entire temperature range $0 < T < T_c$. We have already shown that the two phases BW and ESP would have the same transition temperature T_c. Thus the superfluid will always prefer the BW state over the ESP. The extra degree of freedom provided by the $m_s = 0$ pairing is beneficial. This conclusion is drawn from the weak coupling theory. BW also showed that their isotropic solution had a lower free energy than any anisotropic solution such as ESP. The polar state in (10.5.18) has an even smaller gap. If superfluid ^{3}He obeyed the weak coupling theory, only the B phase would exist, which is identified as BW. The existence of the A phase is due to the strong coupling nature of the interaction.

PROBLEMS

1. Use CBFs to derive the McMillan result that the average energy per particle is

$$\frac{\langle E\rangle}{N} = \frac{1}{2}\varrho\int d^3r g(r)\left[V(r) + \frac{\hbar^2}{2m}\nabla^2 u(r)\right]$$

2. Use the result (10.2.11) for $S(Q, \omega)$ to evaluate the moments ($T = 0$)

$$\frac{1}{2}\int_{-\infty}^{\infty}\frac{d\omega}{4\pi}\,\mathrm{sgn}(\omega)S(Q, \omega) = S(Q)$$

$$\int_{-\infty}^{\infty}\frac{d\omega}{4\pi}\,\omega S(Q, \omega) = ?$$

$$\int_{-\infty}^{\infty}\frac{d\omega}{4\pi}\,\omega^3 S(Q, \omega) = ?$$

3. Consider the case where the liquid helium is flowing down the pipe with velocity v_s, and the pipe walls are moving with velocity v_N. Derive an expression for the total momentum as a function of ϱ_s, ϱ_N, v_s, and v_N.

4. For the superfluid ^{4}He flowing through a pipe, derive an expression for the total energy to order v^2. Show it is not exactly given by the superfluid density ϱ_s.

5. Use the paramagnon Hamiltonian (10.4.22) to evaluate the self-energy of quasiparticles by summing the set of diagrams in (a) Fig. 10.22(a) and (b) Fig. 10.22(b).

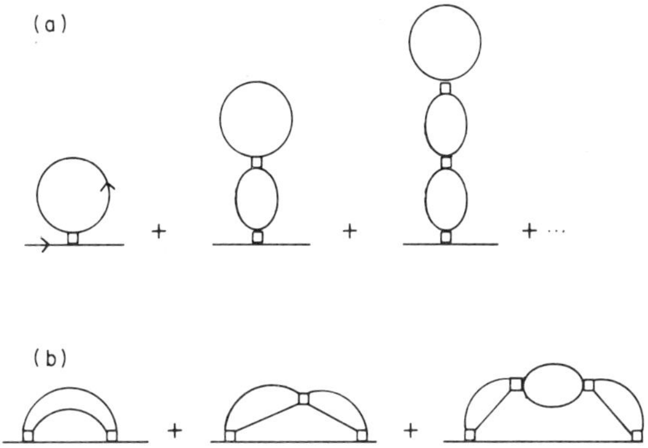

FIGURE 10.22

6. Use the paramagnon theory to study the effective interaction between quasiparticles. By summing the diagrams in Fig. 10.23, show that the effective interaction can be written as

$$V_{\text{eff}} = \frac{I}{2}\left(\frac{1}{1 - \frac{1}{2}IP^{(1)}} - \frac{\boldsymbol{\sigma}\cdot\boldsymbol{\sigma}'}{1 + \frac{1}{2}IP^{(1)}}\right)$$

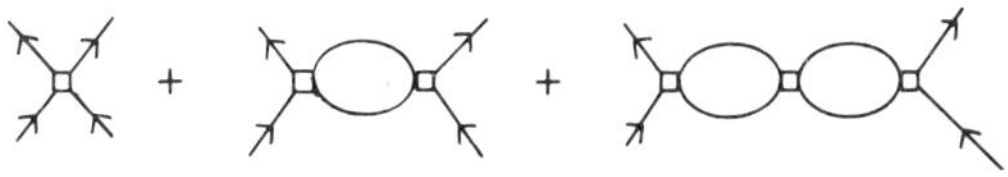

FIGURE 10.23

7. Calculate the quasiparticle self-energy in normal ^{3}He arising from the diagram in Fig. 10.24, and use the result to find the quasiparticle lifetime. Show that the result is (10.4.27) but without the factor $1 - n_1$.

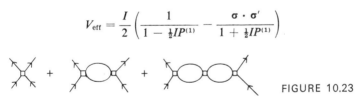

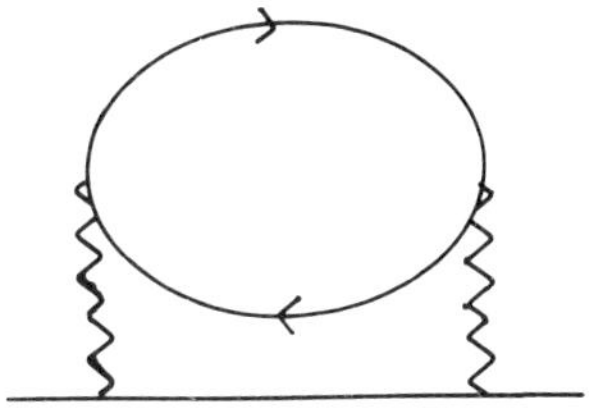

FIGURE 10.24

8. Use the result of Problem 7 [or Eq. (10.4.27)] to find the numerical value of the mean free path of a dressed quasiparticle in ^{3}He at $\xi_1 = 0$ at $T = 1$ mK.

9. Use Eqs. (10.4.24) to calculate the thermal conductivity and verify the result $KT = 33$ ergs/cm–sec in Table 10.4.

10. Find the zero temperature gap function for the solution (10.5.12) to the gap equation for triplet pairing.

11. Show by explicit calculation that the choice (10.5.13) for the BW orbital alignment has the same gap function $\Delta(p)$ as the choice (10.5.9). Next show that the ABM state $\Delta_1 = (\frac{3}{2})^{1/2}(\hat{p}_y + i\hat{p}_z)\Delta_a$ has the same equation for Δ_a as does (10.5.19).

12. Estimate the superfluid free energy in ^{3}He by evaluating U_1 in (3.6.8) with the V for the quasiparticle interactions. When T is near T_c, show that the lowest free energy is the largest value of $\Delta_a{}^2$.

13. Develop a strong coupling theory for ESP states based on the four self-energy functions shown in Fig. 10.25(a). In the same way we derived (9.7.3) for superconductors, derive Dyson's equations in Fig. 10.25(b), where single and double lines are unperturbed and superfluid Green's functions, respectively. Show that the excitation energy is $E(p) = [(\xi_p + \Sigma_1)^2 + \Sigma_3\Sigma_4]^{1/2}$, where $\Sigma_1 = \Sigma_2$ and $\Sigma_3 = \Sigma_4{}^*$.

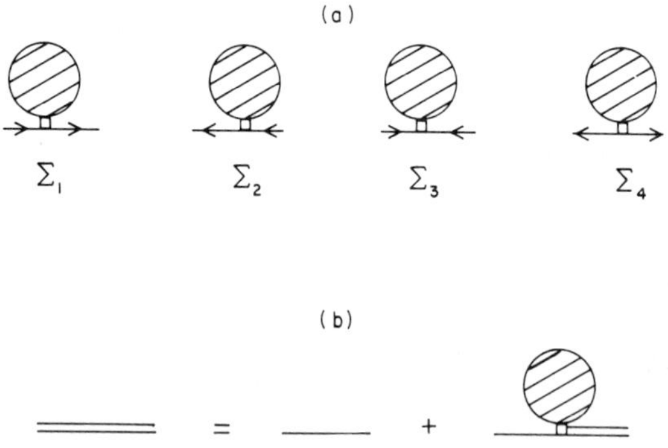

FIGURE 10.26

14. Use paramagnon theory to solve for the self-energy functions Σ_j ($j = 1, 2, 3, 4$) introduced in Problem 13. First solve the sum of diagrams in Fig. 10.22(a) for which the Σ_j are constants. Next discuss the sum of diagrams in Fig. 10.22(b), where the $\Sigma_j(p)$ are functions of ($\mathbf{p}$, ip).

Chapter 11

Spin Fluctuations

The theory of magnetism in metals continues to be one of the challenging subjects of modern theoretical physics. Magnetic phenomena are important, and are observed in a wide variety of materials. The equations are difficult to solve. A variety of theoretical techniques are applied to magnetic phenomena, such as Monte Carlo, renormalization group, and Green's functions. Here we shall only discuss the latter technique.

Most Green's function analysis has focused on three different model Hamiltonians: those of Kondo, Anderson, and Hubbard. Each of these three models is not a single mathematical problem. Instead, each has a number of variations. For example, the original Kondo model describes a system of conduction electrons interacting with a single localized spin. The local spin can have any value of angular momentum S, and the conduction electrons can be treated in one, two, or three dimensions. Recent variations include the dense or periodic Kondo problem, where there are N local spins arranged on a lattice. Another feature of all three models is that they can be solved exactly in one dimension, using the Bethe Ansatz, but not in higher dimension. One-dimensional results are not a useful guide to collective effects in higher dimension, since there are neither phase transitions nor long-range order in one dimension at nonzero temperatures.

11.1. KONDO MODEL

For many years experimentalists noticed that magnetic impurities in nonmagnetic metals caused anomalous behavior in the low-temperature resitivity $\varrho(T)$. Magnetic impurities are those with a moment caused by partially filled d- or f-electron shells. An example is manganese impurities in copper. The electron scattering from a nonmagnetic impurity causes a contribution to the resistivity that is independent of temperature. However, a magnetic impurity causes a resistance *minimum* at a finite temperature. Kondo explained this behavior as due to spin–flip scattering between the conduction electrons and the localized spin. The resistance minimum is now called the *Kondo effect*. The Kondo model is described by the Hamiltonian

$$H = \sum_{\mathbf{k}\sigma} \varepsilon_{\mathbf{k}} C^{\dagger}_{\mathbf{k}\sigma} C_{\mathbf{k}\sigma} - \frac{J}{N} \sum_{\mathbf{kp}j} e^{i\mathbf{R}_j \cdot (\mathbf{k}-\mathbf{p})} [(C^{\dagger}_{\mathbf{k}+} C_{\mathbf{p}+} - C^{\dagger}_{\mathbf{k}-} C_{\mathbf{p}-}) S^{(z)}_j + C^{\dagger}_{\mathbf{k}+} C_{\mathbf{p}-} S^{(-)}_j + C^{\dagger}_{\mathbf{k}-} C_{\mathbf{p}+} S^{(+)}_j] \tag{11.1.1}$$

The first term represents the kinetic energy of the conduction electrons. If the local spin is from a d shell, the conduction band is usually formed from atomic orbitals that have s and p symmetry. If the local spin is from an f orbital, then the conduction band could be from s, p, or d electrons.

The remaining terms represent the scattering from the local spin at the site $\mathbf{R}_j$. The number of local spins is N_i, and the concentration is $c = N_i/N$, where N is the number of atoms in the solid. Usually it is assumed that c is quite small. The electron self-energy is calculated to order $O(c)$. This approximation is equivalent to assuming that the conduction electron scatters from one impurity at a time, or that the impurities are widely separated.

The interaction term can be written compactly as $\mathbf{s} \cdot \mathbf{S}$, where the small $\mathbf{s}$ is the conduction electron, and the large $\mathbf{S}$ is for the localized spin. The first scattering term in (11.1.1) is the $s^{(z)} S^{(z)}$ component. For $s^{(z)}$ we have plus one $(C^{\dagger}_{\mathbf{k}+} C_{p+})$ when the spin of the conduction electron is up, and minus one $(-C^{\dagger}_{\mathbf{k}-} C_{\mathbf{p}-})$ when it is down. The other two terms in the scattering interaction represent spin–flip processes, where there is a mutual spin flip between the conduction and local spin. The conduction electrons are assumed to be electrons with spin one-half. The local spin can have any value permitted by quantum mechanics. Here it is taken to have spin S with magnetic quantum number m. Both S and m are either integers or half integers, where m has the range of values $-S \leq m \leq S$.

$$\begin{aligned} S^{(z)} \mid m\rangle &= m \mid m\rangle \\ S^{(+)} \mid m\rangle &= [S(S+1) - m(m+1)]^{1/2} \mid m+1\rangle \\ S^{(-)} \mid m\rangle &= [S(S+1) - m(m-1)]^{1/2} \mid m-1\rangle \end{aligned} \tag{11.1.2}$$

The operators $S^{(\pm)}$ raise or lower the magnetic quantum number m. The interaction term in (11.1.1.) can be written compactly as

$$V_{\text{sd}} = -\frac{J}{N} \sum_{\mathbf{k}\mathbf{p}\alpha\beta} \exp\left[i\mathbf{R}_j \cdot (\mathbf{k}-\mathbf{p})\right] \boldsymbol{\sigma}_{\alpha\beta} \cdot \mathbf{S} C^{\dagger}_{\mathbf{k}\alpha} C_{\mathbf{p}\beta} \qquad (11.1.3)$$

The indices α and β denote the spin of the electron after and before the scattering event. The vector σ is a Pauli spin matrix for different x, y, and z components. If the meaning of the terms in (11.1.3) is unclear, just consult (11.1.1) for the actual formula.

The constant J is called the "exchange energy." Values with $J>0$ are called "ferromagnetic," since the local spin tends to line up parallel with the conduction band spins. Values with $J<0$ are called "antiferromagnetic," since the local spin tends to line up antiparallel with the conduction band spins. This nomenclature is dependent upon writing the Hamiltonian with a negative sign before the interaction term. Most magnetic impurities have $J<0$.

A. High-Temperature Scattering

The self-energy is calculated for an electron scattering from the impurity. The local spin and the conduction band spins are assumed to have equal probability for any value of m. These assumptions are valid at high temperature. Whether the temperature is high or low depends upon a reference temperature called the *Kondo temperature* T_K, which will be defined below.

The interaction is (11.1.3). The average over impurity positions is taken following the prescription in Section 4.1E. The first term is found in first-order perturbation theory. The average of V_{sd} over impurity positions forces $\mathbf{k}=\mathbf{p}$. The first-order self-energy is a constant which is independent of wave vector or energy:

$$\Sigma_1 = -cJ\langle \boldsymbol{\sigma}_{\alpha\alpha} \cdot \mathbf{S} \rangle$$

The notation $\boldsymbol{\sigma}_{\alpha\alpha}$ means to take the z component. It is usually assumed that the system is nonmagnetic. Then the average value of $\sigma^{(z)}$ and $S^{(z)}$ are both zero, since there are equal numbers of states occupied with spin up as with spin down. Then the above self-energy expression is zero.

Next consider the second-order scattering. The $n=2$ term in the S-matrix expansion for the Green's function is

$$\mathcal{G}^{(2)}_{\alpha\beta}(\mathbf{k},\tau) = -\frac{1}{2}\int_0^\beta d\tau_1 \int_0^\beta d\tau_2 \langle TC_{\mathbf{k}\alpha}(\tau)\hat{V}(\tau_1)\hat{V}(\tau_2)\, C^\dagger_{\mathbf{k}\beta}(0)\rangle$$

$$= -J^2\frac{c}{2}\int_0^\beta d\tau_1 \int_0^\beta d\tau_2 \sum_{\substack{ss'uu'\\ \mathbf{k}_1\mathbf{p}_1\mathbf{k}_2\mathbf{p}_2}} \boldsymbol{\sigma}^{(\mu)}_{ss'}\, \boldsymbol{\sigma}^{(\nu)}_{uu'}\langle TS^{(\mu)}(\tau_1)S^{(\nu)}(\tau_2)\rangle$$

$$\times\, \delta(\mathbf{k}_1-\mathbf{p}_1-\mathbf{k}_2+\mathbf{p}_2)\langle TC_{\mathbf{k}\alpha}(\tau)C^\dagger_{\mathbf{k}_1 s}(\tau_1)\, C_{\mathbf{p}_1 s'}(\tau_1)$$

$$\times\, C^\dagger_{\mathbf{k}_2 u}(\tau_2)\, C_{\mathbf{p}_2 u}(\tau_2)\, C^\dagger_{\mathbf{k}\beta}(0)\rangle$$

The last bracket contains three raising and three lowering operators for conduction electrons. One can pair them up using Wick's theorem. The correlation function of the local spin is not treated as easily. Local spins are neither fermions nor bosons. In order to take the thermodynamic average, we must average over the states $|\, m\rangle$ with different magnetic quantum number. If there is no external magnetic field, then each state of different m is equally likely. That is assumed here.

The major problem with evaluating the correlation function of the local spin is that they do not obey a Wick's theorem. Here it is no problem, since there are only two operators, which must be paired. However, the lack of Wick's theorem is a major algebraic difficulty in higher orders of perturbation theory. The $S^{(\nu)}$ operators must be paired: Each $S^{(+)}$ must be paired with a $S^{(-)}$. Each bracket can have an arbitrary number of $S^{(z)}$ operators, and they are not paired. The correlation function for the local spin has three possible combinations which are evaluated using (11.1.2):

$$\langle TS^{(+)}(\tau_1)\, S^{(-)}(\tau_2)\rangle = S(S+1) - m^2 - m\,\mathrm{sgn}(\tau_1-\tau_2)$$

$$\langle TS^{(-)}(\tau_1)\, S^{(+)}(\tau_2)\rangle = S(S+1) - m^2 + m\,\mathrm{sgn}(\tau_1-\tau_2)$$

$$\langle TS^{(z)}(\tau_1)\, S^{(z)}(\tau_2)\rangle = m^2$$

The only τ dependence is in the order of the operators. At this point it is useful to average over the values of m. Then the term linear in m vanishes, and the first two of the above expressions become identical.

Assume that the initial spin α of the conduction electron is $\pm$. If the scattering does not flip the spin, the intermediate and final states are also α. Then the local spin correlation function has the form $\langle S^{(z)}\, S^{(z)}\rangle$. The other possibility is that the spin flips so that the intermediate state has the opposite spin to α. Then one must use one of the two combinations with $S^{(+)}$ and $S^{(-)}$. The

Green's functions for the conduction electron are the same for the two cases of spin–flip or no spin–flip. Thus we just add these two cases, which causes the m^2 terms to cancel. The second-order self-energy for the Kondo model is

$$\Sigma_2(ik) = S(S+1)cJ^2 \frac{1}{N}\sum_{\mathbf{k}} \frac{1}{ik-\varepsilon_{\mathbf{k}}} \tag{11.1.4}$$

This self-energy expression has the same form as ordinary impurity scattering in the second Born approximation. The spin enters only as an effective interaction of $S(S+1)J^2$. Early workers in the field stopped here, and concluded that spin–flip scattering provided non interesting behavior. This conclusion is wrong, since the unusual effects are found in the higher orders of perturbation theory.

The self-energy of the conduction electron is now calculated in the third Born approximation. This repeats Kondo's original calculation. The anomalous behavior which gives rise to the Kondo effect becomes evident. The $n=3$ term in the Green's function expansion is

$$\begin{aligned} \mathscr{G}^{(3)}(k,\tau) = &-\frac{cJ^3}{N^2}\sum_{pq}\int d\tau_1 \int d\tau_2 \int d\tau_3\, \mathscr{G}(\mathbf{k},\tau-\tau_1)\,\mathscr{G}(\mathbf{p},\tau_1-\tau_2) \\ &\times \mathscr{G}(\mathbf{q},\tau_2-\tau_3)\,\mathscr{G}(\mathbf{k},\tau_3)\,L(\tau_1,\tau_2,\tau_3) \end{aligned} \tag{11.1.5}$$

$$L(\tau_1,\tau_2,\tau_3) = \sum_{\nu\mu\lambda} \sigma^{(\nu)}_{\alpha s}\sigma^{(\mu)}_{ss'}\sigma^{(\lambda)}_{s'\alpha}\langle TS^{(\nu)}(\tau_1)S^{(\mu)}(\tau_2)S^{(\lambda)}(\tau_3)\rangle$$

The self-energy must be the same for spin-up and for spin-down electrons, so set $\alpha=+$ for spin up. The (s, s') can be either of the four combinations $(+,+)$, $(+,-)$, $(-,+)$, $(-,-)$. For $\alpha=+$ then each of these four combinations of (s, s') has a unique choice for $(\nu\mu\lambda)$. Since the electron Green's functions do not depend upon the spin label s, the summation over (s, s') is included in L:

$$\begin{aligned} L(\tau_1,\tau_2,\tau_3) = &\langle TS^{(z)}S^{(z)}S^{(z)}\rangle + \langle TS^{(z)}(\tau_1)S^{(-)}(\tau_2)S^{(+)}(\tau_3)\rangle \\ &+ \langle TS^{(-)}(\tau_1)S^{(+)}(\tau_2)S^{(z)}(\tau_3)\rangle \\ &+ \langle TS^{(-)}(\tau_1)S^{(z)}(\tau_2)S^{(+)}(\tau_3)\rangle \end{aligned} \tag{11.1.6}$$

No time variables are written in the first term on the right since this expression is independent of the τ variables. This term equals m^3 for any τ ordering.

The other terms depend upon the τ ordering. We did not write out the factors σ_{ss} for each term. They are always plus except for one case. When the factor $S^{(z)}$ immediately follows in time (τ) the operator $S^{(+)}$ then a minus sign is inserted. The minus sign arises because the factor $S^{(+)}$ raises the local spin while flipping the conduction spin from up to down. The $\sigma^{(z)}$ operates on the down conduction spin and gives minus one.

The above expression must be evaluated for all six arrangements of τ ordering. The ordering $\tau_1 > \tau_2 > \tau_3$ gives

$$L = m^3 + [S(S+1) - m^2 - m]\,[m + m - (m+1)]$$

$$L = m[S(S+1)+1] - S(S+1)$$

The terms are written in the same order they occur in (11.1.6). The minus sign before the factor of $(m+1)$ is due to the sign reversal mentioned above. The term proportional to m vanishes when we average over m. After averaging the above expression equals $-S(S+1)$.

Next consider the τ-ordering of $\tau_3 > \tau_2 > \tau_1$, which gives

$$L = m^3 + [S(S+1) - m^2 + m]\,[m - m + (m+1)]$$

$$L = m[S(S+1)+1] + S(S+1)$$

After averaging over m, this term gives $S(S+1)$. This result is similar to the one above, except that the sign has changed. One can work out the other four cases. The three cyclic arrangements $\tau_1 > \tau_2 > \tau_3$, $\tau_2 > \tau_3 > \tau_1$ and $\tau_3 > \tau_1 > \tau_2$ each have $L = -S(S+1)$. The other three arrangements $\tau_3 > \tau_2 > \tau_1$, $\tau_2 > \tau_1 > \tau_3$, and $\tau_1 > \tau_3 > \tau_2$ each have $L = S(S+1)$. The three τ-integrals in (11.1.5) must be broken up into the six separate regions of different τ ordering. Three of them are multiplied by $S(S+1)$ while the other three are multiplied by $-S(S+1)$.

The notation (123) means $\tau_1 > \tau_2 > \tau_3$. In this notation the following result is obtained for L after averaging over values of m:

$$L = -S(S+1)\,[(123) + (231) + (312) - (321) - (213) - (132)]$$

$$1 = (123) + (231) + (312) + (321) + (213) + (132)$$

The second line above expresses the idea that the summation of all possible time orderings is just the unity operator. This result can be used to rewrite L as

$$L = -2S(S+1)\,[(123) + (231) + (312) - 1/2] \qquad (11.1.7)$$

The amount of work has been cut in half, since now only three time orderings need to be evaluated.

The third-order Green's function of frequency is evaluated by multiplying Eq. (11.1.5) by $\exp(i\omega_n\tau)$ and integrating $d\tau$ between $(0, \beta)$. The integrals are most easily evaluated in reverse order:

$$\mathscr{G}^{(3)}(\mathbf{k}, i\omega_n) = -cS(S+1)\frac{J^3}{N^2}\sum_{\mathbf{pq}}\int d\tau_3\, e^{(i\omega_n-\varepsilon_k)\tau_3}\mathscr{G}(\mathbf{k},\tau_3)\int d\tau_2\, e^{(i\omega_n-\varepsilon_p)(\tau_2-\tau_3)}$$
$$\times\,\mathscr{G}(\mathbf{p},\tau_2-\tau_3)\int d\tau_1\, e^{(i\omega_n-\varepsilon_q)(\tau_1-\tau_2)}\mathscr{G}(\mathbf{q},\tau_1-\tau_2)\times\int d\tau\, e^{(i\omega_n-\varepsilon_k)(\tau-\tau_1)}$$
$$\times\,\mathscr{G}(\mathbf{k},\tau-\tau_1)\,L(\tau_1,\tau_2,\tau_3) \tag{11.1.8}$$

In Eq. (11.1.7) the last term in brackets is $-1/2$. It has no limits on the order of the τ integrals. Then all of the above integrals have their usual limits of $(0, \beta)$, and the above expression is

$$-S(S+1)\frac{cJ^3}{N^2}\frac{1}{(i\omega-\varepsilon_k)^2}\sum_{\mathbf{pq}}\frac{1}{(i\omega-\varepsilon_p)(i\omega-\varepsilon_q)}$$

This expression is a contribution to the third-order Green's function. If one eliminates the prefactor of $(i\omega-\varepsilon_k)^2$ then the remaining expression is a contribution to the third-order self-energy. It is just the third term in the T-matrix equation generated by iterating (4.1.17). This contribution to the self-energy is rather dull.

The Kondo effect comes from the terms in L which restrict the order of the τ integrals. The restrictions are on τ_1, τ_2 and τ_3. There is no restriction on the $d\tau$ integral. It can be done immediately, and yields $(i\omega-\varepsilon_k)^{-1}$. The first interesting integral is (123), whose limits can be written as

$$(123) \Rightarrow \int_0^\beta d\tau_3\int_{\tau_3}^\beta d\tau_2\int_{\tau_2}^\beta d\tau_1$$

Doing the integral in (11.1.8) with these limits gives the following result. We have dropped all of the prefactors and summations, and list the results of the three τ integrals:

$$(123) = \frac{[1-n(p)]\,[1-n(q)]}{(i\omega-\varepsilon_k)(i\omega-\varepsilon_p)(i\omega-\varepsilon_q)} + \frac{[1-n(q)]\,[n(k)-n(p)]}{(i\omega-\varepsilon_p)(\varepsilon_k-\varepsilon_p)(\varepsilon_p-\varepsilon_q)}$$
$$+\frac{[1-n(p)]\,[1-n(q)]}{(i\omega-\varepsilon_q)(\varepsilon_k-\varepsilon_q)(\varepsilon_q-\varepsilon_p)}$$

The expression has a multitude of Fermion occupation factors $n(k)$, etc. Similar expressions are obtained for the other two τ orderings (231) and (312):

$$(231)\Rightarrow \int_0^\beta d\tau_3 \int_{\tau_3}^\beta d\tau_2 \int_0^{\tau_3} d\tau_1$$

$$(231)=\frac{1}{(i\omega-\varepsilon_p)(\varepsilon_p-\varepsilon_q)}\left\{\frac{[1-n(q)][n(k)-n(p)]}{\varepsilon_p-\varepsilon_k}\right.$$

$$-\frac{[1-n(p)][n(k)-n(q)]}{\varepsilon_q-\varepsilon_k}+\frac{n(p)[(1-n(q)]}{i\omega-\varepsilon_k}$$

$$\left.+\frac{n(k)n(p)[1-n(q)]-[1-n(k)][1-n(p)]n(q)}{i\omega-\varepsilon_k-\varepsilon_p+\varepsilon_q}\right\}$$

$$(312)\Rightarrow \int_0^\beta d\tau_3 \int_0^{\tau_3} d\tau_2 \int_{\tau_2}^{\tau_3} d\tau_1$$

$$(312)=\frac{[1-n(p)][n(k)-n(q)]}{(i\omega-\varepsilon_p)(i\omega-\varepsilon_q)(\varepsilon_q-\varepsilon_k)}-\frac{n(q)[1-n(p)]}{(i\omega-\varepsilon_k)(i\omega-\varepsilon_q)(\varepsilon_p-\varepsilon_q)}$$

$$-\frac{n(k)n(p)[1-n(q)]-[1-n(k)][1-n(p)]n(q)}{(i\omega-\varepsilon_p)(i\omega-\varepsilon_k-\varepsilon_p+\varepsilon_q)(\varepsilon_p-\varepsilon_q)}$$

An interesting thing happens when we add these three expressions. Most of the terms cancel:

$$(123)+(231)+(312)=-\frac{1}{(i\omega-\varepsilon_k)^2}\left[\frac{1}{(i\omega-\varepsilon_p)(i\omega-\varepsilon_q)}\right.$$

$$\left.+\frac{n(p)}{(i\omega-\varepsilon_q)(\varepsilon_q-\varepsilon_p)}+\frac{n(q)}{(i\omega-\varepsilon_p)(\varepsilon_p-\varepsilon_q)}\right]$$

A further simplification occurs when we notice that the last two terms are identical after interchanging the variables of summation **p** and **q**. The result for the third-order self-energy is

$$\Sigma_3(i\omega)=-S(S+1)\frac{cJ^3}{N^2}\sum_{\mathbf{pq}}\frac{1}{(i\omega-\varepsilon_q)}\left[\frac{1}{(i\omega-\varepsilon_p)}-4\frac{n(p)}{\varepsilon_p-\varepsilon_q}\right] \qquad (11.1.9)$$

The final expression is rather simple. There are two terms. The first term in the brackets is the rather dull term, which is the third-order Born scattering from a simple potential. The second term in the brackets was first derived by Kondo. He used it to explain the temperature dependence of the resistivity, which we now call the Kondo effect. Today it is known that similar anomalous terms occur in all higher orders of perturbation theory. The present theory of the Kondo effect employs terms from all orders of perturbation theory. Nevertheless, Kondo's simple arguments will be presented.

The self-energy expression has a factor of the Fermion occupation number $n(p)$. These factors do not occur in scattering from a simple potential. They occur here because of the spin–flip scattering. Interesting physics happens whenever they occur, such as in the BCS theory of superconductivity or in the singularities in X-ray edge absorption in metals.

The second term in the above brackets produces a term in the resistivity that goes as $\ln(T)$, where T is the temperature. Here we present a quick and sloppy derivation of this behavior. The conduction band density of states is defined as $g(\varepsilon)$. It is assumed to be a smooth function of energy ε, which extends from $-W < \varepsilon < B$. Summations over wave vector are changed to integrations over energy using the prescription

$$\frac{1}{N}\sum_{\mathbf{p}} \frac{1}{i\omega - \varepsilon_{\mathbf{p}}} = \int_{-W}^{B} d\varepsilon \, \frac{g(\varepsilon)}{i\omega - \varepsilon}$$

A typical alloy system that shows the Kondo effect is manganese impurities in copper. The density of states $g(\varepsilon)$ of conduction electrons in metallic copper has a large energy dependence because of the occupied d bands. However, the conduction electrons that are involved in the Kondo effect have energies within several $K_B T$ of the Fermi energy. On that small energy scale, the density of states of copper and most metals is smooth and featureless. Thus the custom is to call the density of states $g(\varepsilon)$ a constant $g(0)$, where $\varepsilon = 0$ signifies the Fermi energy.

For the Kondo effect, the interesting third-order self-energy is

$$\Sigma_{3b}(i\omega) = -4S(S+1)cJ^3 \int_{-W}^{B} d\varepsilon_1 \, \frac{g(\varepsilon_1)}{i\omega - \varepsilon_1} \int_{-W}^{B} d\varepsilon_2 \, \frac{g(\varepsilon_2) n(\varepsilon_2)}{\varepsilon_1 - \varepsilon_2}$$

Our ultimate goal is to calculate the resistivity. Then the quantity of interest is the scattering rate, which is related to the imaginary part of the self-energy. In this expression let $i\omega \to \omega + i\delta$ and then take the imaginary part. The first energy denominator gets replaced by $-\pi\delta(\omega - \varepsilon_1)$. The first integral gives

$-\pi g(\omega)$. In order to evaluate the second integral, it is assumed that the temperature is low, so the Fermion occupation factor can be replaced by the step function $\Theta(-\varepsilon_2)$. The most important term is the logarithmic singularity which comes from this Fermi cutoff

$$\int_{-W}^{B} d\varepsilon_2 \frac{g(\varepsilon_2)n(\varepsilon_2)}{\omega-\varepsilon_2} \approx \int_{-W}^{0} d\varepsilon_2 \frac{g(\varepsilon_2)}{\omega-\varepsilon_2} \approx g(\omega)\ln|\omega/W|$$

Since the density of states $g(\omega)$ is a smooth function of energy ω, this quantity is usually replaced by $g(0)$. The standard expression for the third-order self-energy in the Kondo effect is

$$\Sigma_{3b}(\omega) = -i4\pi S(S+1)\, cJ^3 g(0)^2 \ln|\omega/W|$$

where W is a band width on the order of several electron volts. One can calculate the real part of Σ_{3b} and find that its ω dependence is uninteresting near $\omega \approx 0$.

This result is combined with the larger term from the second order of perturbation theory:

$$\mathrm{Im}\left\{\sum(\omega)\right\} = \pi S(S+1)\, cJg(0)\left[1+4Jg(0)\ln\left|\frac{\omega}{W}\right|+\ldots\right] \quad (11.1.10)$$

In calculating the conductivity, the energy ω of the electron is averaged over a region near the Fermi surface of width k_BT. This replaces $\ln|\omega/W|$ by $\ln(k_BT/W)$. Kondo derived a temperature-dependent resistivity of the form

$$\varrho(T) = \varrho(0)\,[1+4Jg(0)\ln(k_BT/W)] + bT^5 \quad (11.1.11)$$

The first term $\varrho(T=0)$ is the temperature-independent part of the impurity scattering. The second term with $\ln(T)$ is the Kondo effect. The last term $\sim T^5$ is the low-temperature contribution from electron–phonon scattering. Usually the experiments are done in metals or alloys where phonon drag does *not* eliminate the T^5 law.

The factor of $\ln(k_BT/W)$ is negative at low temperature. If $J < 0$ then the Kondo term is positive. At very low temperatures, it increases in value, and this term becomes large. This expression predicts a minimum in the resistivity, which is observed in Kondo systems. Define $T_{\min}$ as the temperature

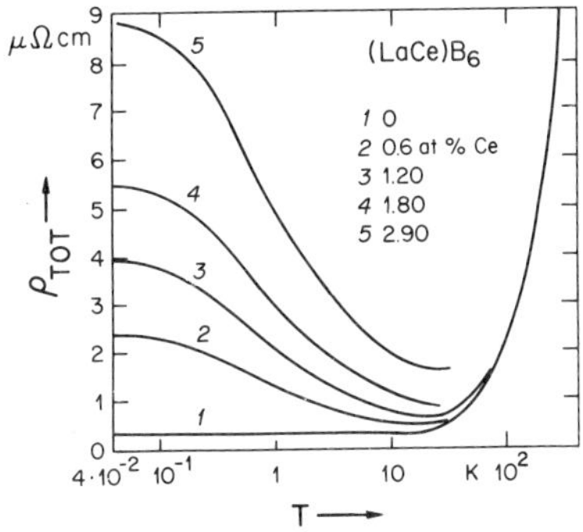

FIGURE 11.1. Total resistivity vs. T for LaB_6 and La_{1-x} Ce_x Ba_6. The Kondo temperature is $T_K = 1.05$ K. *Source*: Samwer and Winzer (1976) (used with permission).

at the minimum resistance. By taking $\partial\varrho/\partial T = 0$ one finds that $(T_{\min})^5 \sim c$, where c is the concentration of impurities. This relationship is in agreement with experiments on Kondo alloys.

A typical measurement of $\varrho(T)$ is shown in Fig. 11.1 for the system of La_{1-x} Ce_xB_6. The resistance minimum is evident. The resistance increases at low temperature, and finally saturates. The reason for the saturation is given in the following section.

The theory presented in this section is a high-temperature theory. A key assumption in the derivation is that the local spin has equal energy whether it is up or down. At low temperature, this assumption ceases to be valid. Each local spin becomes locked into a collective state with the conduction band spins. This collective state has a binding energy, which must be overcome during a spin–flip. The spin–flips become frozen out at low temperature, and the Kondo effect saturates. This process is described in the next section. Kondo's formula (11.1.11) is only valid at high temperatures above the formation of the collective state.

Our derivation of the Kondo resistance was sloppy for the following reasons. In evaluating the integral over $d\varepsilon_2$ the Fermi factor $n(\varepsilon_2)$ was replaced by a step function $\Theta(-\varepsilon_2)$. This approximation is only valid at zero temperature. Later in the derivation, we averaged $\ln(\omega/W)$ over the distribution of electrons while assuming the temperature is nonzero. A better derivation would have evaluated the integral over $d\varepsilon_2$ at finite temperature, which broadens the Fermi distribution a few degrees in energy. Of course, this realism makes the integral hard to evaluate.

B. Low-Temperature State

At low temperature the local spin forms a collective state with the conduction band spins. The collective state is formed from a linear combination of conduction band states, and is a many-particle entity. Earlier chapters have often discussed the role of screening charge in reducing the long-range Coulomb interaction in conducting systems. The screening charge is not from a single electron in the conduction band, but usually is formed by a linear combination of conduction states. The present collective state can be viewed, in a similar way, as a screening effect. However, now it is screening the spin of the local state, rather than its charge. If the impurity is charged there could coexist the screening clouds for spin and charge.

For $J < 0$ the state is a singlet. In order to keep the discussion simple, let us assume that the coupling is antiferromagnetic ($J < 0$) and that the local spin has $S = 1/2$. The up and down states of the local spin are denoted as α and β. Yosida (1966) first considered the collective state. Its wave function can be written in two possible ways:

$$\psi_{0a} = \frac{1}{\sqrt{N}} \sum_{\mathbf{k} > k_F} a_{\mathbf{k}} [\alpha C^{\dagger}_{\mathbf{k}-} - \beta C^{\dagger}_{\mathbf{k}+}] \mid F\rangle \tag{11.1.12}$$

$$\psi_{0b} = \frac{1}{\sqrt{N}} \sum_{\mathbf{k} < \mathbf{k}_F} b_{\mathbf{k}} [\alpha C_{\mathbf{k}+} + \beta C_{\mathbf{k}-}] \mid F\rangle \tag{11.1.13}$$

$$\mid F\rangle \equiv \prod_{\mathbf{p} < k_F} C^{\dagger}_{\mathbf{p}+} C^{\dagger}_{\mathbf{p}-} \mid 0\rangle$$

The symbol $\mid F\rangle$ denotes the filled Fermi sea of electrons. The state ψ_{0a} forms the collective state by adding an electron above the Fermi sea. It has the usual form for a spin singlet, as an antisymmetric combination of up and down states of the local and conduction spins. The coefficient $a_{\mathbf{k}}$ is a parameter that needs to be determined.

The second choice of collective wave function ψ_{0b} needs further explanation. Here one is forming the collective state by taking linear combinations of hole states, where the hole is a missing electron from the filled Fermi sea. The easiest way to understand this wave function is to examine one of its terms, where we include the relevent factor from the Fermi sea:

$$[\alpha C_{\mathbf{k}+} + \beta C_{\mathbf{k}-}]\, C^{\dagger}_{\mathbf{k}+} C^{\dagger}_{\mathbf{k}-} \mid 0\rangle = [\alpha C^{\dagger}_{\mathbf{k}-} - \beta C^{\dagger}_{\mathbf{k}+}] \mid 0\rangle \tag{11.1.14}$$

The right-hand expression is obviously a spin singlet, composed of an antisymmetric combination of down and up spin states. The relative plus sign on the left becomes a minus sign on the right. The two states ψ_{0a} and ψ_{0b} are orthogonal. Each describes a different ground state, and the system will choose the one with the lowest energy. One is a ground state, and one is an excited state.

The first term in (11.1.1) is called H_0 and the second term is V_{sd}. If there is no local spin, the ground state energy for this model Hamiltonian is

$$H_0 \mid F\rangle = E_G \mid F\rangle$$

$$E_G = 2 \sum_{\mathbf{k}<k_F} \varepsilon_{\mathbf{k}}$$

Now add one local spin at the point $\mathbf{R}_j = 0$. In this calculation, one does *not* average over the spatial positions of the impurity, and one does not set $\mathbf{p}=\mathbf{k}$ in the interaction term.

The goal is to calculate the additional energy $\delta E < 0$ gained by the system because of the formation of the collective spin state. It is analogous to the energy found for charge screening using Fumi's theorem in Sec. 4.1. The eigenvalue equation is

$$\{H_0 + V_{sd}\}\ \psi_{0a,b} = (E_G + \delta E)\ \psi_{0a,b} \tag{11.1.15}$$

This equation is not strictly valid since neither ψ_{0a} nor ψ_{0b} is an exact eigenstate of the Hamiltonian. In order to understand this assertion, consider the effect of one term in V_{sd} acting upon one term in ψ_{0a}

$$\sum_{\mathbf{p},\mathbf{q}} C^\dagger_{\mathbf{p}+} C_{\mathbf{q}-} S^{(-)} \sum_{\mathbf{k}>k_F} a_{\mathbf{k}} \alpha C^\dagger_{\mathbf{k}-} \mid F\rangle \tag{11.1.16}$$

The spin-lowering operator $S^{(-)}$ changes α to β. If $\mathbf{q} = \mathbf{k}$ there is a term

$$\left[\sum_{\mathbf{k}>k_F} a_{\mathbf{k}}\right]\left[\sum_{\mathbf{p}>k_F} \beta C^\dagger_{\mathbf{p}+} \mid F\rangle\right]$$

which is recognized as being similar to one of the other terms in ψ_{0a}. However, there are also terms in (11.1.16) in which $\mathbf{k}$, $\mathbf{p}$, and $\mathbf{q}$ are all different wave vectors. These terms describe a state with one hole and two excitations above the Fermi surface. An electron–hole pair has been excited in addition to the original excitation. This state is perfectly valid, since it is generated by the interaction term in the Hamiltonian. Our failure to include it in the ansatz eigenstate ψ_{0a} is simply an approximation. A more accurate ansatz would include these additional terms. There are an infinite number of them, which correspond to the multiple excitation of electron–hole pairs. The more pairs that are included, the more accurate is the wave function.

The best way to understand our ansatz wave function in (11.1.12) is to consider that we are doing a variational calculation on the ground state. The coefficients $a_{\mathbf{k}}$ and $b_{\mathbf{k}}$ are the variational parameters. The ground state energy is calculated, using Dirac notation, with $j = a$ or b, as

$$E_G + \delta E_j = \langle \psi_{0j} \mid H \mid \psi_{0j} \rangle / \langle \psi_{0j} \mid \psi_{0j} \rangle$$

After evaluating the right-hand side of this expression, one finds the minimum value of δE_a by taking the functional derivative $\delta(\delta E_a)/\delta a_{\mathbf{k}} = 0$. There results an equation for $a_{\mathbf{k}}$ that can be solved. These steps produce the minimum ground state energy for the ansatz (11.1.12).

Our method of solution is equivalent to the variational procedure, but is algebraically simpler. Equation (11.1.15) is evaluated. The multipair states are neglected, and then we project out the terms with only one conduction band excitation. For example, when solving for ψ_{0a} define $\langle \alpha C^{\dagger}_{k-} |$ as the Hermitian conjugate of $\alpha C^{\dagger}_{k-} \mid F \rangle$ and evaluate the expression

$$\langle \alpha C^{\dagger}_{\mathbf{k}-} \mid \{H_0 + V_{\mathrm{sd}} - E_G - \delta E_a\} \psi_{0a} = 0$$

As an example, the easy terms are done first:

$$\{H_0 - E_G - \delta E_a\} \psi_{0a} = \frac{1}{\sqrt{N}} \sum_{\mathbf{k} > k_F} a_{\mathbf{k}} [E_G + \varepsilon_{\mathbf{k}} - E_G - \delta E][\alpha C^{\dagger}_{\mathbf{k}-} - \beta C^{\dagger}_{\mathbf{k}+}] \mid F \rangle$$

$$\langle \alpha C^{\dagger}_{\mathbf{k}-} \mid \{H_0 - E_G - \delta E_a\} \psi_{0a} = \frac{1}{\sqrt{N}} a_{\mathbf{k}} [\varepsilon_{\mathbf{k}} - \delta E_a] \tag{11.1.17}$$

A similar evaluation must be done for V_{sd}. One term was previously evaluated in (11.1.16). Another term in $V_{\mathrm{sd}} \psi_{0a}$ is

$$\sum_{\mathbf{p},\mathbf{q}} S^{(z)} (C^{\dagger}_{\mathbf{p}+} C_{\mathbf{q}+} - C^{\dagger}_{\mathbf{p}-} C_{\mathbf{q}-}) \sum_{\mathbf{k} > k_F} a_{\mathbf{k}} \alpha C^{\dagger}_{\mathbf{k}-} \mid F \rangle \tag{11.1.18}$$

The operator $S^{(z)}$ acts upon α and gives $\alpha/2$. The electron operators give several contributions. For $\mathbf{p} = \mathbf{q}$ the part in parentheses gives the number of electrons with spin up minus the number with spin down. The state $C^{\dagger}_{\mathbf{k}-} \mid F \rangle$ has one more electron with spin down, so the summation gives -1. There is also a term where $\mathbf{q} = \mathbf{k}$. These two contributions are

$$-\frac{1}{2}\sum_{\mathbf{k}>k_F} a_{\mathbf{k}}\,\alpha C^{\dagger}_{\mathbf{k}-} \mid F\rangle - \frac{1}{2}\left[\sum_{\mathbf{k}>k_F} a_{\mathbf{k}}\right]\sum_{\mathbf{p}>k_F} \alpha\, C^{\dagger}_{\mathbf{p}-} \mid F\rangle$$

Other terms contain an electron–hole pair, and these contributions are being ignored. In the above expression, the first term can be neglected since it is of order $O(1/N)$ compared to the second. The second has a double summation, while the first has only a single summation. By employing these kinds of approximations, we are able to evaluate

$$V_{\mathrm{sd}}\, \psi_{0a} = \frac{3J}{2N}\left[\sum_{\mathbf{k}>k_F} a_{\mathbf{k}}\right]\frac{1}{|N}\sum_{\mathbf{p}>k_F}(\alpha C^{\dagger}_{\mathbf{p}-} - \beta C^{\dagger}_{\mathbf{p}+}) \mid F\rangle$$

This result is combined with (11.1.17) to produce the equation for $a_{\mathbf{k}}$:

$$a_{\mathbf{k}} = -\frac{3J}{2N}\,\frac{1}{\varepsilon_{\mathbf{k}} - \delta E_a}\sum_{\mathbf{p}>k_F} a_{\mathbf{p}}$$

This expression is then summed over $\mathbf{k}$. These summations are changed to integrals over the particle energy. The eigenvalue equation is

$$1 = -\frac{3J}{2}\int_0^B d\varepsilon\,\frac{g(\varepsilon)}{\varepsilon - \delta E_a} \tag{11.1.19}$$

The solution to the wave function ψ_{0b} follows the same steps used to obtain ψ_{0a}. The form of the wave function in (11.1.14) has the same form as (11.1.12), so that all steps in the deivation are similar. There are two minor differences. First, since $\mathbf{k} < k_F$, the integral over energy is over negative energy states. Second, the result of $H_0\psi_{0b}$ has the factor of $(E_G - \varepsilon_{\mathbf{k}})$; the minus sign preceding ε_k comes from the fact that ψ_{0b} has an electron missing from the Fermi sea. This changes the sign of ε in the integral. So the eigenvalue equation for ψ_{0b} has the form

$$b_{\mathbf{k}} = \frac{3J}{2N}\,\frac{1}{\varepsilon_{\mathbf{k}} + \delta E_b}\sum_{\mathbf{p}<k_F} b_{\mathbf{p}}$$

$$1 = \frac{3J}{2}\int_{-W}^{0} d\varepsilon\,\frac{g(\varepsilon)}{\varepsilon + \delta E_b} \tag{11.1.20}$$

The two eigenvalue equations (11.1.19) and (11.1.20) are very similar. The major difference is that one has an integral over the empty states in the conduction band, while the other has an integral over the occupied states. Since $\delta E < 0$ neither energy denominator in either integral is singular.

These two expressions will be evaluted for a simple model of the conduction band. The density of states $g(\varepsilon)$ equals a constant g. Then the two integrals give

$$1 = (3Jg/2)\ln \mid \delta E_a/(W - \delta E_a) \mid$$
$$1 = (3Jg/2)\ln \mid \delta E_b/(B - \delta E_b) \mid \tag{11.1.21}$$

which have the solution

$$\delta E_a = -\ W\Lambda/(1+\Lambda), \qquad \Lambda = \exp[2/(3Jg)]$$
$$\delta E_b = -\ B\Lambda/(1+\Lambda)$$

The largest value of $\delta E_{a,b}$ is determined by whether B or W is largest. If the band is less than half full, then $W < B, \delta E_a > \delta E_b$, and ψ_{0b} is the ground state. If the band is more than half full, then $W > B$ and ψ_{0a} is the ground state. If the band is exactly half full, the states are degenerate. These results only apply to the case of a constant density of states. Other examples are assigned in the problems.

Remember that we are assuming $J < 0$. In fact, there is no solution for singlet states when $J > 0$. The approximations we are making only make sense when $\delta E \ll B, W$ so that the logarithm terms in (11.1.21) are negative. Then the equation has a solution only for $J < 0$. For antiferromagnetic coupling, the factor of $\Lambda = \exp(2/3JG)$ is generally much less than unity. This expression also has the feature that the coupling constant J appears as an inverse power in an exponent. It would be impossible, or at least very difficult, to derive an expression for δE from perturbation theory.

Previously this mathematical form, of an inverse power in an exponent, was encountered in the BCS theory of superconductivity. The energy gap and the formula for T_c have a similar factor of $\exp[-1/N_F V_0]$.

The *Kondo temperature* is defined by the expression

$$k_B T_K = W \exp\{-1/[2g(0)\,|J|]\} \tag{11.1.22}$$

This expression is similar to the result for the change in ground state energy.

The main difference is the factor of 2/3 in the exponent has become 1/2. The reason for this change is given below. A general definition is that the change in ground state energy defines an energy scale that also defines a characteristic temperature. The Kondo temperature is also very close to the temperature at which the Kondo effect becomes important in the resistivity. The self-energy correction given by the second term in (11.1.10) becomes equal to the first at a temperature of $T \sim T_K$. This is not the same temperature as the resistance minimum, since the minimum depends upon the concentration of impurities. The Kondo temperature defines an energy scale for a single impurity. Values of T_K for actual alloys span a large range in temperature, from 10^{-3} to 10^3 K.

So far the discussion has been for the singlet state produced by an impurity with $J < 0$. For ferromagnetic coupling ($J > 0$) the theory is similar except that the local state is a triplet with $S = 1$. For a system of two electrons, the triplet spin combinations for $m = 1, 0, -1$ are $\alpha_1\alpha_2$, $(\alpha_1\beta_2 + \beta_1\alpha_2)/\sqrt{2}$, and $\beta_1\beta_2$. These simple combinations provide the guide to constructing the collective state. The triplet equivalent to ψ_{0a} is called ψ_{1m}:

$$\psi_{11} = \frac{1}{\sqrt{N}} \sum_{\mathbf{k}>k_F} a_{\mathbf{k}}\, \alpha C^\dagger_{\mathbf{k}+} \mid F\rangle$$

$$\psi_{10} = \frac{1}{\sqrt{2N}} \sum_{\mathbf{k}>k_F} a_{\mathbf{k}} \{\alpha C^\dagger_{\mathbf{k}-} + \beta C^\dagger_{\mathbf{k}+}\} \mid F\rangle$$

$$\psi_{11} = \frac{1}{\sqrt{N}} \sum_{\mathbf{k}>k_F} a_{\mathbf{k}}\, \beta C^\dagger_{\mathbf{k}-} \mid F\rangle \qquad (11.1.23)$$

Each of these three states has the same ground state energy δE. At zero temperature, one solves an equation similar to (11.1.15). The resulting equation for the coefficient $a_{\mathbf{k}}$ is

$$a_{\mathbf{k}} = \frac{J}{2N} \frac{1}{\varepsilon_{\mathbf{k}} - \delta E} \sum_{\mathbf{p}>k_F} a_{\mathbf{p}}$$

$$1 = \frac{J}{2} \int_0^B d\varepsilon \frac{g(\varepsilon)}{\varepsilon - \delta E} \approx -(J/2)\, g(0) \ln \mid \delta E/B \mid \qquad (11.1.24)$$

The last line above gives the eigenvalue equation for triplets, and the approximate evaluation of the integral. Since the logarithm is negative, this equation has a solution only for $J > 0$. The ground state energy has the form

$$\delta E \approx -B \exp[-2/Jg] \tag{11.1.25}$$

Again δE has the form of a band width B multiplied by an exponential that contains the inverse of the factor $Jg(0)$.

So far the derivation suggests that the collective state has a similar behavior regardless of the sign of J. However, further work has shown that the collective state is not formed for ferromagnetic coupling ($J > 0$). If one evaluates further terms in the perturbation expansion, they destabilize the collective state for $J > 0$. For $J < 0$, further terms in the perturbation theory continue to predict a collective state. The Kondo effect does depend upon the sign of J, and exists only for $J < 0$.

C. Kondo Temperature

The two prior sections discussed the high- and low-temperature regimes of the Kondo effect. The most interesting behavior occurs for temperatures near the Kondo temperature T_K, which is defined in (11.1.22). A short and somewhat qualitative discussion of this temperature region is presented here. This region is actually very well understood because the exact solution to the Kondo problem is now known. See the solutions by Andrei *et al.* (1983) and Wiegmann (1981).

The high-temperature scattering theory has been carried out to much higher orders of perturbation theory. Abrikosov (1965), and Silverstein and Duke (1967) have suggested that the perturbation expansion for the retarded Green's function can be represented by the simple function

$$\Sigma_{\text{ret}}(\omega) = \frac{cJ}{1 - 2Jg(\omega)\ln|\omega/W| - i\pi Jg(\omega)\,S(S+1)} \tag{11.1.26}$$

The function we calculated in Sec. 11.1A was the imaginary part of the self-energy. The imaginary part of the above expression is

$$\text{Im}[\Sigma_{\text{ret}}(\omega)] = \frac{c\,\pi J^2 g(\omega)\,S(S+1)}{(1 - 2Jg(\omega)\ln|\omega/W|)^2 + [\pi Jg(\omega)\,S(S+1)]^2}$$

If the denominator is expanded in a power series in $Jg(\omega)$, the first term is the perturbation expression in (11.1.10). Another feature of this expression is that the denominator has a resonance when the energy $\omega \sim W \exp 1/[2Jg(0)]$, which makes sense when $J < 0$. There is no resonance for $J > 0$. This observation provides further evidence that there is no collective behavior for $J > 0$.

A factor of $2Jg(\omega)$ can be taken out of the denominator in (11.1.26). Using the definition of the Kondo temperature in (11.1.22) allows the real part of the denominator to have a simple form:

$$\Sigma_{\text{ret}}(\omega) = \frac{c}{2g(\omega)} \frac{1}{\ln|\omega/k_B T_K| - i\pi/2S(S+1)}$$

$$\text{Im}[\Sigma_{\text{ret}}(\omega)] = \frac{c\,\pi S(S+1)/4\,g(\omega)}{\ln^2|\omega/k_B T_K| + [\pi S(S+1)/2]^2} \tag{11.1.27}$$

The interesting feature of this expression is that the ω dependence of the self-energy scales only with the Kondo temperature. One should be able to describe all of the self-energy effects by this single parameter. It also has a resonance behavior. The square of the logarithm has its smallest value of zero at $\omega_K = k_B T_K$, and the denominator increases in magnitude for values of ω different from ω_K.

Wilkins (1982) observed that many of the properties associated with the Kondo effect can be described by a simple expression. The system acts as if the density of states has a resonance at the Fermi surface for each impurity with a local moment. Let the concentration of impurities be $c_i = N_i/N$. A simple expression which well describes the density of states is

$$g_K(\omega) = g(\omega) + \frac{c_i}{\pi} \frac{\gamma}{\omega^2 + \gamma^2} = g(\omega) + \delta g(\omega) \tag{11.1.28}$$

where $\gamma = 1.6 k_B T_K$. The change $\delta g(\omega)$ integrates to one electron for each impurity with a local spin.

For example, to calculate the change in the specific heat δC due to the change δg, first find the change δU in the internal energy

$$\delta C = \partial(\delta U)/\partial T = c_i f(T/T_K)$$

$$\delta U = \int d\omega\ \omega n_F(\omega)\delta g(\omega)$$

Figure 11.2 shows how the function $f(T/T_K)$ (dashed line) compares to the exact result (solid line). The two results are quite similar. Note that the temperature scale spans several decades. The specific heat per magnetic impurity

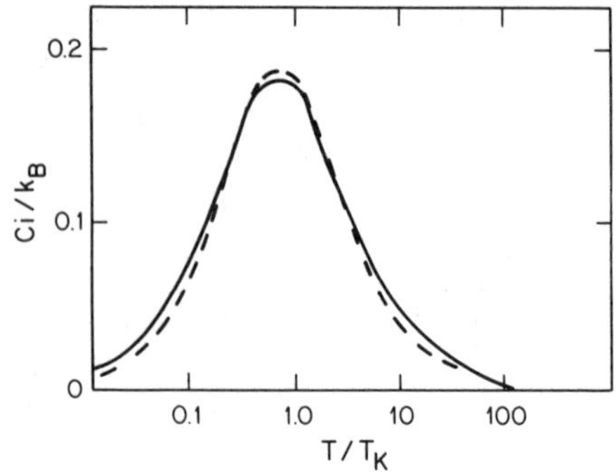

FIGURE 11.2. Specific heat per impurity for the Kondo model. Solid line is the numerical solution using the renormalization group. Dashed line is the calculation using the lorentzian resonance with unit weight and width equal to $1.6T_K$. Adapted from Wilkins (1982) (used with permission).

is a function only of T/T_K. At high temperature the specific heat per impurity goes as $f \sim T_K/T$. At low temperatures it goes as $f \sim T/T_K$, which can be large if the Kondo temperature is low. Wilkins observes that the entire effect of the many-body interaction is to add a resonance at the Fermi surface.

The resonance behavior also affects the resistivity. Assume that the Kondo resonance affects the quasiparticle self energy $\Gamma = -\text{Im}[\Sigma]$. Instead of the behavior predicted in (11.1.27) use the resonance form in (11.1.28). The temperature-dependent conductivity $\sigma(T)$ is given in terms of the lifetime $\tau(T)$:

$$\begin{aligned} \sigma &= n_0 e^2 \tau / m \\ 2\tau &= \int_{-\infty}^{\infty} d\omega \left[\frac{-\partial n(\omega)}{\partial \omega} \right] \frac{1}{\Gamma(\omega)} \\ 2\Gamma(\omega) &= \left(1 + \frac{r\gamma^2}{\omega^2 + \gamma^2} \right) \Big/ \tau_0 \end{aligned} \tag{11.1.29}$$

where $\sigma_0 = n_0 e^2 \tau_0 / m$ is the conductivity from nonresonant impurity scattering for $T \gg T_K$. The parameter r comes from the ratio of resonant to nonresonant impurity scattering. The frequency integral in (11.1.29) can be done numerically. The result is given in Fig. 11.3, for $r = 10$, which shows the resistivity $\varrho(T) = 1/\sigma(T)$. At high temperature ($T \gg T_K$) one finds $\varrho = \varrho_0$. At low temperature ($T \ll T_K$) one finds that $\varrho = \varrho_0(1+r)$. This behavior of the resistivity is similar to the experimental dependence shown in Fig. 11.1.

At high temperature the resonance effects from spin–flip scattering are smeared out by thermal broadening. As the temperature is lowered to the vicinity of T_K, the thermal smearing decreases, and the Kondo resonance becomes more important in scattering the electron. The resistivity increases. At

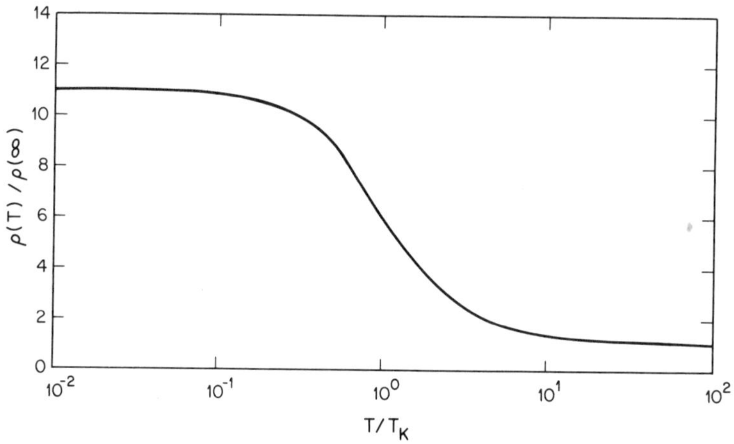

FIGURE 11.3. Resistivity vs. temperature in the Kondo model.

still lower temperatures, the collective state begins to form, and the local spin becomes harder to flip. Then the resistance saturates, and no longer increases with decreasing temperature.

11.2. Anderson Model

The Anderson model was introduced in Sec. 4.1. It is another model for a system of conduction electrons that interact with a local spin. Equation (1.4.20) gives the transformation of Schrieffer and Wolff (1966), which shows that the Anderson model has some terms that are similar to the Kondo model. Early workers thought that the two models made very similar predictions. Now it is known that the Anderson model has a greater variety of behavior. It has the more interesting physics.

The Kondo model treats the local spin as a separate entity. The Anderson model treats the local spin as just another electron. It can undergo exchange and other processes with the conduction electrons. This makes the model more realistic, and explains the more interesting behavior.

The Anderson model is written with slightly altered notation. Most applications of the model consider the localized state to have spin and orbital degeneracy. It could be a d electron of 10 states or an f electron of 14 states. Usually the spin–orbit interaction splits these states into 6 and 4, and into 8 and 6, respectively. The crystal field may cause further splittings. Let N_f denote the actual degeneracy of the local level: it may be an even number such as 2, 4, 6, 8, 10, or 14. One of the theoretical developments is a perturbation expansion, where the expansion parameter is $1/N_f$. Obviously this expansion works better for larger values of N_f. Many of the recent applications of the Anderson model have been for *heavy fermion* systems, where the local orbital is an f electron. That is the origin of the subscript f on the factor of N_f. The symbol N without the subscript is the number of atoms in the solid. For reviews of heavy fermions see Stewart (1984) or Lee *et al.* (1986).

The Hamiltonian is written as

$$H = \sum_{k\nu} \left\{ \varepsilon_k C_{k\nu}^{\dagger} C_{k\nu}^{\dagger} + \frac{V_k}{\sqrt{N}} (C_{k\nu}^{\dagger} f_\nu + f_\nu^{\dagger} C_{k\nu}) \right\}$$

$$+ \varepsilon_f \sum_{\nu} f_\nu^{\dagger} f_\nu + U \sum_{\mu > \nu} n_\nu n_\mu \tag{11.2.1}$$

where $n_\nu = f_\nu^{\dagger} f_\nu$. The local state has destruction operators f_ν while the conduction state has destruction operators $C_{k\nu}$. The indices ν and μ denote the symmetry of the local state. These summations go from one to N_f. This symmetry is a combination of spin and angular momentum. It is assumed that this symmetry is preserved when the conduction electrons hop on or off of the local state. The wave vector $\mathbf{k}$ of the conduction band can be expressed as the magnitude k plus numerous spin and angular momentum states. The latter are included in ν. An important point is that of all the many symmetries available to the conduction band states (angular momentum between zero and infinity), the only conduction states that couple to the local state are those values of ν between 1 and N_f. The summations over k will eventually be changed to an integral over the energy ε. The density of states is $g(\varepsilon)$, and the matrix element as a function of energy is written $V_k \equiv V(\varepsilon_k)$.

How does the local spin get flipped in the Anderson model? Actually it does not flip. Because of the hybridization term, a local spin that is down can become a conduction state with spin down, and wander away. Later a different conduction electron with spin up can come and reside in the local orbital. This process gives the appearance of the local spin having flipped from down to up, while conduction states have flipped from up to down. Now one can see the important role of the electron–electron interaction U. The spin–flip process has two steps: (1) the departure of the old local electron with spin down, and (2) the arrival of the new local electron with spin up. If $U = 0$ these two steps are totally independent and can occur in any order. There is no appearance of flipping, since down and up electrons come and go independently. However, once we set $U > 0$, then the two steps become correlated. It is energetically unfavorable to have up and down spin electron both on the local orbital. The two steps tend to become sequential: one leaves before the other comes. Then the local spin appears to flip.

The term in the Hamiltonian involving U comes from the on-site Coulomb interaction between localized electrons. Usually this term is far too large to be treated by perturbation theory. It must be included in H_0, which is a problem because it depends upon the number of pairs of local electrons. Including this term invalidates Wick's theorem for local electrons. The Green's function expansion becomes awkward. In evaluating correlation functions, each time ordering must be evaluated separately. This increases the effort, as was evident in the prior section on the Kondo effect. Several alternative formalisms have been developed to expand the S matrix. Several of them will be discussed below.

The important parameters of the Anderson model are the band widths W, the conduction band density of states $g(\varepsilon)$, the local site energy ε_f, and the on-site Coulomb interaction U. There is also an energy parameter associated with the hybridization term which is defined as Δ:

$$\Delta = \pi V(0)^2 g(0) \tag{11.2.2}$$

In heavy fermion systems the local electron is an f level. Typical experimental values are $\Delta \leq 0.1$ eV and $U \sim 6$ eV (Herbst and Wilkins, 1987). The quantity ε_f can have a variety of values, depending upon the local ion. When the local orbital is a d electron the values of Δ are much larger. The energy of one local electron is ε_f, while the energy of two is $2\varepsilon_f + U$. In cerium one has that $\varepsilon_f < 0$ but $2\varepsilon_f + U > 0$. Then the cerium ion prefers to have just one electron in the local orbital.

A. Collective States

The first step in solving the Anderson model is to find the collective eigenstates. The nature of these states, and the method of solving for them, is similar to the low-temperature state of the Kondo model in Sec. 11.1B. The solution for the Kondo model was simple, because of our approximation of neglecting states with multiple electron–hole pairs. Typical solutions to the Anderson model include states with one or two pairs. The increase in accuracy is achieved at a cost of increased algebraic complexity. Our discussion follows that of Gunnarsson and Schonhammer (1983).

There are an infinite number of different collective states. Each collective state has an infinite number of terms, which correspond to multiple electron–hole pairs. Each term is found in only one collective state. The terms in each collective state form a subspace, and the subspaces are linearly independent.

Each collective state, or subspace, can be generated by a simple method. One starts with a state that is the filled Fermi sea $|F\rangle$ of conduction electrons plus n electrons in the local orbital. The values of n have the range $0 \leq n \leq N_f$. Additional collective states are formed by starting with $n = N_f$ plus having m excitations each with $\varepsilon > 0$, or else with $n = 0$ and with m holes. Two examples are the state $|F\rangle$ with no local electron and the state $|\nu; 1\rangle$ with one:

$$|\nu; 1\rangle = f_\nu^\dagger |F\rangle \tag{11.2.3}$$

In labeling states $|...; n\rangle$ the value of n denotes the number of localized electrons. This factor is omitted for $n = 0$. The state $|F\rangle$ is defined in

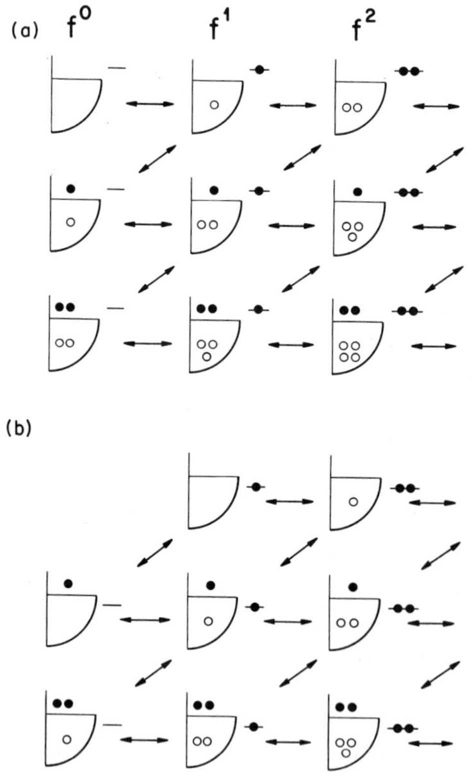

FIGURE 11.4. Diagrams illustrating the different electron configurations in two possible ground states of the Anderson model. The quarter circle represents the filled Fermi sea. Circles in it represent hole excitations. Circles above the Fermi sea are electronic excitations. The horizontal line to the right represents the local level. Circles on it represent electron occupancy. The double-ended arrows represent states connected by the hybridization term V_{sf}. Part (a) represents the state ψ_0, while part (b) represents the state ψ_ν.

(11.1.12). The other terms in each collective state are obtained by operating upon $| F\rangle$ or $| \nu; 1\rangle$ by $(V_{sf})^{\ell}$ where $0 \leq l < \infty$. The hybridization term in the Anderson model is

$$V_{sf} = \frac{1}{\sqrt{N}} \sum_{k\nu} V_k \left[f_\nu^\dagger C_{k\nu} + C_{k\nu}^\dagger f_\nu \right]$$

The combination $V_{sf} | F\rangle$ produces a state with one electron in the local level plus one hole in the conduction band. Denote this state $| k; 1\rangle$, and an improved version of the state is called ψ_0:

$$| k; 1\rangle = \frac{1}{\sqrt{N_f}} \sum_{\nu=1}^{N_f} f_\nu^\dagger C_{k\nu} | F\rangle$$

$$\psi_0 = A\left[| F\rangle + \sum_{k<k_F} a_k | k; 1\rangle + \ldots \right] \tag{11.2.4}$$

The normalization constant is A. The coefficient a_k needs to be determined.

Next consider the result of $V_{sf} | \nu; 1\rangle$. Both terms in V_{sf} contribute. The term $f_\mu^\dagger C_{k\mu}$ creates a second localized electron with $\mu \neq \nu$. The term $C_{k\mu}^\dagger f_\mu$ destroys the local electron ($\mu = \nu$) and creates an electron with $p > k_F$. These two states provide an improved version of $| \nu; 1\rangle$:

$$| p,\nu\rangle = C_{p\nu}^\dagger | F\rangle$$

$$| k,\nu; 2\rangle = \frac{1}{\sqrt{N_f - 1}} f_\nu^\dagger \sum_{\mu\neq\nu} f_\mu^\dagger C_{k\mu} | F\rangle \tag{11.2.5}$$

$$\psi_\nu = B\left[| \nu; 1\rangle + \sum_{p>k_F} b_p | p,\nu\rangle + \sum_{k<k_F} c_k | k, \nu; 2\rangle + \ldots \right]$$

The p summation is over positive energies ($p > k_F$) for the electron excitations above the Fermi surface. The k summation is over negative energies ($k < k_F$) for the holes in the Fermi sea. The coefficients b_p and c_k need to be determined.

A useful way to visualize these states is shown in Fig. 11.4. Fig. 11.4(a) shows the state ψ_0 while Fig. 11.4(b) shows ψ_ν. The quarter circle represents the filled Fermi sea. Holes in this region are excitations beneath the

Fermi surface ($k < k_F$). Dots above the Fermi sea are excitations with $p > k_F$. The short horizontal line to the right represents the localized level. Small circles on this line represent electrons in the local orbitals. Note that each column has all of the states with the same number of localized electrons. Each row has the same number of conduction excitations of positive energy. The double-ended arrows represent the states that are connected by an operation by V_{sf}. Only a small number of the infinite variety of states are shown in each figure. For each case, there should be $N_f + 1$ columns, and an infinite number of rows. A glance at Fig. 11.4 shows that each term in (a) is different from each term in (b). The two subspaces are independent.

It is useful to compare the present description of collective states with those discussed for the Kondo effect in Sec. 11.1B. There we considered two states. One (ψ_{0a}) was a combination of a local spin plus a single electron above the conduction band. This state does not appear in either of the subspaces we are considering for the Anderson model. It occurs in another subspace. The other collective state considered for the Kondo model (ψ_{0b}) had one hole with the local spin. That state is in both of the subspaces we are considering for the Anderson model. In Fig. 11.4(a) it is the term $| k; 1\rangle$ and the local spin is a single electron ($S = 1/2$). In Fig. 11.4(b) it is the term $| k, \nu; 2\rangle$ and the local spin is created from a doubly occupied local orbital ($S = 1$).

The next step is to calculate the energy of the local state. First consider the energy of ψ_0. The various terms are

$$H | F\rangle = E_G | F\rangle + \frac{\sqrt{N_f}}{\sqrt{N}} \sum_{k<k_F} V_k | k; 1\rangle$$

$$H | k; 1\rangle = (E_G - \varepsilon_k + \varepsilon_f) | k; 1\rangle + \frac{\sqrt{N_f}}{\sqrt{N}} V_k | F\rangle$$

$$+ \frac{1}{\sqrt{N}} \sum_{p>k_F} V_p | p, k\rangle + \frac{\sqrt{N-1}}{\sqrt{N}} \sum_{q<k_F} V_q | q, k; 2\rangle$$

where

$$| p, k\rangle = \frac{1}{\sqrt{N_f}} \sum_{\nu} C^{\dagger}_{p\nu} C_{k\nu} | F\rangle$$

$$| q, k; 2\rangle = \frac{1}{[N_f(N_f - 1)]^{1/2}} \sum_{\nu \neq \mu} f^{\dagger}_{\nu} C_{q\nu} f^{\dagger}_{\mu} C_{k\mu} | F\rangle \tag{11.2.6}$$

The last two states are terms in the collective state ψ_0. However, in the present approximation they are being omitted, and only the two terms in (11.2.4) are being included. Use just the two terms in the expression for ψ_0 in the Hamiltonian

$$H\psi_0 = (E_G + \delta E)\psi_0 \tag{11.2.7}$$

Operate on this Hamiltonian from the left by $\langle F \mid$ and $\langle \nu; 1 \mid$, which produces the two equations

$$\delta E = \frac{\sqrt{N_f}}{\sqrt{N}} \sum_{k<k_F} V_k a_k$$

$$a_k [\delta E + \varepsilon_k - \varepsilon_f] = \frac{\sqrt{N_f}}{\sqrt{N}} V_k$$

which have the obvious solution

$$\delta E = \frac{N_f}{N} \sum_{k<k_F} \frac{V_k^2}{\delta E + \varepsilon_k - \varepsilon_f} = N_f \int_{-W}^{0} d\varepsilon \, \frac{g(\varepsilon) V(\varepsilon)^2}{\delta E + \varepsilon - \varepsilon_f} \tag{11.2.8}$$

A self-consistent equation is obtained for the ground state energy δE. It has the same general form as (11.1.19) for the collective state of the Kondo effect. Keep in mind that this result for the Anderson model is approximate, since only two terms were retained in the evaluation of ψ_0.

Now consider the calculation of the energy δE for the state ψ_ν. The Hamiltonian acts upon each term in (11.2.5) and gives

$$\begin{aligned}
H \mid \nu; 1\rangle &= [E_G + \varepsilon_f] \mid \nu; 1\rangle + \frac{1}{\sqrt{N}} \sum_{p>k_F} V_p \mid p, \nu\rangle \\
&\quad + \left(\frac{N_f - 1}{N}\right)^{1/2} \sum_{k<k_F} V_k \mid k, \nu; 2\rangle \\
H \mid p, \nu\rangle &= [E_G + \varepsilon_p] \mid p, \nu\rangle + \frac{1}{\sqrt{N}} V_p \mid \nu; 1\rangle \\
&\quad + \frac{1}{\sqrt{N}} \sum_{k<k_F} V_k (\sqrt{N_f - 1} \mid p, k, \nu; 1\rangle_1 + \mid p, k, \nu; 1\rangle_2) \\
H \mid k, \nu; 2\rangle &= [E_G - \varepsilon_k + 2\varepsilon_f + U] \mid k, \nu; 2\rangle + \left(\frac{N_f - 1}{N}\right)^{1/2} V_k \mid \nu; 1\rangle \\
&\quad + \frac{1}{\sqrt{N}} \sum_{p>k_F} V_p (\mid p, k, \nu; 1\rangle_1 + \mid p, k, \nu; 1\rangle_3) + (\text{terms } n = 3)
\end{aligned}$$

where

$$| p, k, \nu; 1\rangle_1 = \frac{1}{\sqrt{N_f - 1}} C^\dagger_{p\nu} \sum_{\mu \neq \nu} f^\dagger_\mu C_{k\mu} | F\rangle$$

$$| p, k, \nu; 1\rangle_2 = C^\dagger_{p\nu} f^\dagger_\nu C_{k\nu} | F\rangle$$

$$| p, k, \nu; 1\rangle_3 = \frac{1}{\sqrt{N_f - 1}} f^\dagger_\nu \sum_{\mu \neq \nu} C^\dagger_{p\mu} C_{k\mu} | F\rangle$$

The combination $V_{\mathrm{sf}} | k, \nu; 2\rangle$ generates a number of new terms. Those listed have one local electron plus an electron–hole pair. The terms with three local electrons have not been written. To find δE one starts with an equation such as (11.2.7) for ψ_ν. One successively operates by $\langle \nu; 1 |$, $\langle p, \nu |$, and $\langle k, \nu; 2 |$. This action generates the following three equations:

$$\delta E = \varepsilon_f + \frac{1}{\sqrt{N}} \sum_{p > k_F} b_p V_p + \left(\frac{N_f - 1}{N}\right)^{1/2} \sum_{k < k_F} c_k V_k$$

$$(\delta E - \varepsilon_p) b_p = \frac{1}{\sqrt{N}} V_p$$

$$(\delta E + \varepsilon_k - 2\varepsilon_f - U) c_k = \left(\frac{N_f - 1}{N}\right)^{1/2} V_k$$

The energy of the collective state is $E_G + \delta E$ where

$$\delta E = \varepsilon_f + \int_0^B d\varepsilon \frac{g(\varepsilon) V(\varepsilon)^2}{\delta E - \varepsilon} + (N_f - 1) \int_{-W}^0 d\varepsilon \frac{g(\varepsilon) V(\varepsilon)^2}{\delta E + \varepsilon - 2\varepsilon_f - U} \tag{11.2.9}$$

The first term ε_f is the answer if $V_{\mathrm{sf}} = 0$. Then the additional energy to add the local electron is just its eigenvalue. The first integral is the contribution of the state $| p, \nu\rangle$ while the second integral is the contribution of the state $| k, \nu; 2\rangle$.

There are several observations about the two results (11.2.8) and (11.2.9). First, the integrals over negative energy are multiplied by N_f or else $(N_f - 1)$, while the integrals over positive energy are missing this factor. This feature is the basis for the $1/N_f$ expansion. The integrals that lack the prefactor of N_f are going to be smaller by $1/N_f$ than those integrals that have this prefactor. One can make this into a systematic expansion by defining $N_f \Delta =$ const in the limit that $N_f \to \infty$. Then do perturbation theory using $1/N_f$ as the expansion parameter. The perturbation theory has a finite number of terms in the double limit that $N_f \to \infty$ and $U \to \infty$, which constitutes an exactly solvable model.

The size of the integral terms is given by the quantity Δ in (11.2.2), or else ΔN_f. The factor of Δ is less than 0.1 eV when the local electrons are in f shells. The size of terms with ℓ electron–hole pairs is usually $(\Delta/W)^\ell$, where W is the band width. Often $\Delta/W \ll 1$; then multipair terms can be neglected. Additional terms in the series for ψ_0 and ψ_ν are usually unnecessary.

Another common approximation in the Anderson model is to limit the number of local configurations f^n to a few values of n. One example is cerium, where $\varepsilon_f = -2$ eV and $U = 6$ eV. Then the ground state has $n = 1$ since the state with $n = 2$ has an energy $2\varepsilon_f + U = 2$ eV and is empty. The common approximation is to neglect configurations with $n \geq 2$. This approximation neglects terms of order $O(\Delta/U)$. That is evident by examining the last term in (11.2.9). It came from the configuration with two localized electrons.

All of these approximations make the Anderson model easy to solve. For example, applying these approximations to the state ψ_ν gives the ground state energy as $\delta E = \varepsilon_f$ since both integral terms in (11.2.9) are neglected: the first is smaller by $1/N_f$ and the second is smaller by a factor of $1/U$. However, this approximation neglects terms that change the answer by 10%–20%. The approximations are good but not great.

An important quantity is the average number n_f of local electrons. It is defined as

$$n_f = \langle \sum_{\nu=1}^{N_f} f_\nu^\dagger f_\nu \rangle$$

The instantaneous value of n_f is an integer. An average number that is not an integer is evidence for valence fluctuations. Then the system fluctuates between several configurations that are similar in energy.

For the states ψ_0 the average number of local electrons is

$$\begin{aligned} n_f &= A^2 C \\ C &= N_f \int_{-W}^{0} d\varepsilon \, \frac{g(\varepsilon) V(\varepsilon)^2}{(\delta E + \varepsilon - \varepsilon_f)^2} \end{aligned} \qquad (11.2.10)$$

The constant A is found by normalizing the wave function to unity $\langle \psi_0 \mid \psi_0 \rangle$

= 1. This gives

$$1 = A^2(1 + C)$$

$$n_f = \frac{C}{1 + C}, \qquad A^2 = \frac{1}{1 + C} = 1 - n_f \tag{11.2.11}$$

The occupancy n_f approaches one as C becomes large. As shown below, this occurs as ε_f becomes negative. Our ansatz wave function in (11.2.4) for ψ_0 only had two terms: one with zero local electrons, and another with one local electron. Naturally we found a formula for n_f that only gives values between zero and one. Higher values of n_f could be obtained by including terms in the series for ψ_0 that included higher values of n.

A similar analysis for the state ψ_ν gives for the average number of local electrons

$$n_f = \frac{[1 + 2D]}{1 + D + F}$$

$$D = (N_f - 1) \int_{-W}^{0} d\varepsilon \, \frac{g(\varepsilon) V(\varepsilon)^2}{(\delta E + \varepsilon - 2\varepsilon_f - U)^2} \tag{11.2.12}$$

$$F = \int_0^W d\varepsilon \, \frac{g(\varepsilon) V(\varepsilon)^2}{(\delta E - \varepsilon)^2}$$

These results can be illustrated by a numerical example. In order to keep the discussion simple, consider the case where $g(\varepsilon)\, V(\varepsilon)^2$ is a constant given by Δ/π. For the state ψ_0 the expression (11.2.8) for δE and (11.2.10) for C give

$$\delta E = \frac{\Delta N_f}{\pi} \ln\left(\frac{\delta E - \varepsilon_f}{W + \varepsilon_f - \delta E}\right)$$

$$C = \frac{\Delta N_f}{\pi}\left(\frac{1}{\varepsilon_f - \delta E} - \frac{1}{W - \delta E + \varepsilon_f}\right) \tag{11.2.13}$$

These results are illustrated by a numerical example. Choose $W = 6.0$ eV, $N_f = 6$, and $\Delta = 0.1$ eV for f electrons. Figure 11.5 shows a graphical solution to (11.2.8) for two cases: one has $\varepsilon_f > 0$ and the other has $\varepsilon_f < 0$. The horizontal axis is δE. The vertical axis is also δE. The solid line is the right-hand side of (11.2.13). The logarithmic divergence occurs where $\delta E \sim \varepsilon_f$. The solution is found where the dashed line crosses the solid line. Only the solution of lowest δE is retained. If $\varepsilon_f > 0$ the solution is found at small negative values

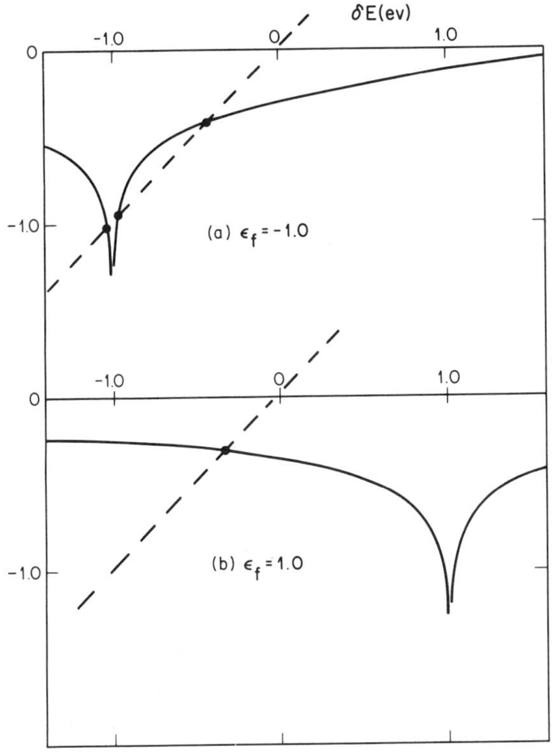

FIGURE 11.5. A graphical solution of Eq. (11.2.8) for $\Delta = 0.1$ eV, $U = 6.0$ eV, $N_f = 6$, and $W = 6.0$ eV. Part (a) has $\varepsilon_f = -1.0$ eV. The dashed line represents the left-hand side of (11.2.8). The solid line represents the right-hand side of (11.2.8). There are three places they cross, so the equation has three solutions. The one of lowest energy is the ground state, while the others are excited states. Part (b) has $\varepsilon_f = 1.0$ eV. Here there is only one solution.

of δE. For $\varepsilon_f < 0$, the solution is found for $\delta E < \varepsilon_f$. Figure 11.6 shows the resulting graph of δE and n_f versus ε_f. Valence fluctuations occur in the transition region where n_f is changing from zero to one. The width of the valence fluctuation region is $N_f\Delta$. For $\varepsilon_f < -N_f\Delta$ then $n_f \to 1$ and the local level has one electron. In this region valence fluctuations are small. This is called the *Kondo limit*. Here there is a local spin.

In the Kondo limit one can define $\delta = \varepsilon_f - \delta E$ which is a very small positive number. One finds that $C \sim N_f\Delta/\pi\delta \gg 1$ and $A^2 = 1 - n_f \sim \pi\delta/N_f\Delta \ll 1$. Later we shall see that δ is related to the Kondo temperature.

The Fano–Anderson model was solved in Sec. 4.2. It resembles the Anderson model except for the term involving U. This on-site Coulomb term

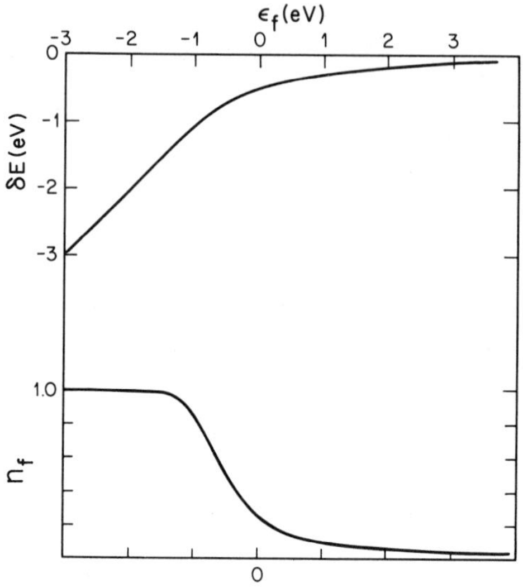

FIGURE 11.6. The top curve shows the solution to (11.2.8), which gives δE vs. ε_f. Other parameters are $\Delta = 0.1$ eV, $U = 6.0$ eV, $W = 6$ eV, and $N_f = 6$. The solution approaches $\delta E \approx \varepsilon_f$ for $\varepsilon_f < 0$. The lower curve shows n_f vs. ε_f using the same parameters. Note the asymmetry due to the long tail at positive values of ε_f.

can be eliminated by setting $N_f = 1$. The Anderson model with $N_f = 1$ becomes the Fano–Anderson model which is exactly solvable. How do the two models compare? The only quantity of the Anderson model that has been calculated here, so far, is the ground state energy. The exact ground state energy of the Fano–Anderson model is given by a coupling constant integral

$$E_G + \delta E = F_0(T) + \delta F$$

$$\beta F_0 = 2\sum_{\mathbf{k}} \ln\left[1 + e^{-\varepsilon_{\mathbf{k}}}\right] + \ln\left[1 + e^{-\beta\varepsilon_f}\right]$$

$$\delta F = \int_0^1 d\lambda \, \frac{1}{\beta} \sum_{ip} \frac{\Sigma(ip)}{ip - \varepsilon - \lambda\Sigma(ip)}, \qquad \Sigma(ip) = \int_{-W}^{W} d\varepsilon \, \frac{g(\varepsilon)V(\varepsilon)^2}{ip - \varepsilon}$$

At zero temperature, the summation over ip and the $d\lambda$ integral can be evaluated, and the value of δF expressed as an integral over phase shifts, in agreement with Fumi's theorem. That rigor is unnecessary here. Our result for the Anderson model was obtained in low order of perturbation theory. Here we also need the perturbation result. At zero temperature this is

$$F_0 = E_G + \varepsilon_f \Theta(-\varepsilon_f)$$

$$\delta F \approx \lim_{T\to 0} \frac{1}{\beta} \sum_{ip} \frac{\Sigma(ip)}{ip - \varepsilon} = \int_{-W}^{W} d\varepsilon \, \frac{g(\varepsilon)V(\varepsilon)^2}{\varepsilon_f - \varepsilon} \left[\Theta(-\varepsilon_f) - \Theta(-\varepsilon)\right]$$

This result is actually simple. For $\varepsilon_f < 0$ it is the first two terms of (11.2.9). The last term in (11.2.9) is zero when $N_f = 1$. For $\varepsilon_f > 0$ the change in ground state energy is (11.2.8). The exact result for $N_f = 1$ suggests that ψ_0 is the ground state when $\varepsilon_f > 0$, while ψ_ν is the ground state when $\varepsilon_f < 0$. However, for $N_f > 1$ the state ψ_0 has the lower ground state energy.

B. Green's Functions

The previous section discussed static quantities such as the ground state energy. The next step is to calculate dynamical quantities such as the density of states, which is obtained from the Green's function. Our discussion follows Gunnarsson and Schonhammer (1983).

The first calculation will be the Green's function of the local electron. This quantity is intrinsically interesting, plus it is needed for the calculation of some of the other spectroscopies. The calculation will be done for zero temperature. According to the definitions in Sec. 3.7, two of the Green's functions are

$$G^>(\omega) = -2\pi i \sum_\nu \langle i \,|\, f_\nu \delta(\omega + E_i - H) f_\nu^\dagger \,|\, i\rangle$$

$$G^<(\omega) = 2\pi i \sum_\nu \langle i \,|\, f_\nu^\dagger \delta(\omega - E_i + H) f_\nu \,|\, i\rangle \tag{11.2.14}$$

The average over ν is intended to ensure that the Green's function is independent of ν. In fact, for the Anderson model the term in bracket is independent of ν, so the average is unnecessary. Thus drop the summation over ν and instead multiply by the factor of N_f. The term E_i is the energy of the ground state $|\, i\rangle$. Our calculations will use ψ_0 in (11.2.4) as $|\, i\rangle$. Then $E_i = E_G + \delta E$, where δE is the solution to (11.2.8). The symbol H is the Hamiltonian of the Anderson model.

These Green's functions each obey a sum rule that is useful. An integration over all frequency removes the delta function:

$$\int \frac{d\omega}{2\pi} G^>(\omega) = -i \sum_\nu \langle i \,|\, f_\nu f_\nu^\dagger \,|\, i\rangle = -i(N_f - n_f)$$

$$\int \frac{d\omega}{2\pi} G^<(\omega) = i \sum_\nu \langle i \,|\, f_\nu^\dagger f_\nu \,|\, i\rangle = i n_f \tag{11.2.15}$$

where we have used the definition of n_f.

Both of these expressions will be evaluated. The easiest is $G^<$, so it is done first. Define the correlation function

$$\Pi^<(z) = \langle i \mid f_\nu^\dagger \frac{1}{z - E_i + H} f_\nu \mid i \rangle = \sum_{\alpha,\beta} \langle i \mid f_\nu^\dagger \mid \alpha \rangle g_{\alpha\beta} \langle \beta \mid f_\nu \mid i \rangle$$

$$g_{\alpha\beta}(z) = \langle \alpha \mid \frac{1}{z - E_i + H} \mid \beta \rangle$$

$$G^<(\omega) = 2iN_f \operatorname{Im}[\Pi_{\text{ret}}^<(\omega)]$$

First one finds the Green's function $g_{\alpha\beta}(z)$. Using this quantity one finds $\Pi^<(z)$. The Green's function is the spectral function of $\Pi^<$. This procedure is rather easy as long as one restricts the space of (α, β) to a small number of states. Including more states gives a better answer at the cost of more work. Here we shall include the subset of states which permits an analytical answer. Gunnarsson and Schonhammer (1983) included more states, but had to find $g_{\alpha\beta}$ numerically.

The states in (α, β) are those obtained from $f_\nu \mid i\rangle$. The first two states in $\mid i \rangle = \psi_0$ are $\mid F \rangle$ and $\mid k; 1 \rangle$. Operating by f_ν gives $f_\nu \mid F \rangle = 0$ and $f_\nu \mid k; 1 \rangle \sim \mid k, \nu \rangle = C_{k\nu} \mid F \rangle$. This state is in neither the subspace of ψ_0 nor of ψ_ν. The states (α, β) are in a different subspace than those discussed in the previous section. It is easy to show that $\langle k, \nu \mid g(z) \mid k', \nu \rangle$ vanishes unless $k = k'$. Thus the correlation function is approximately

$$\Pi^<(z) = A^2 \sum_{k<k_F} a_k^2 \mid \langle k, \nu \mid f_\nu \mid k; 1 \rangle \mid^2 g_{kk}(z)$$

$$\langle k, \nu \mid f_\nu \mid k; 1 \rangle = 1/\sqrt{N_f}$$

This expression is approximate since only a few terms were included in ψ_0 and in $f_\nu \psi_0$. The evaluation of g_{kk} is also approximate. Define $\Theta = z - E_i + H$. The matrix elements of $\Theta_{\alpha\beta} = \langle \alpha \mid \Theta \mid \beta \rangle = (z - E_i)\delta_{\alpha\beta} + H_{\alpha\beta}$ are easy to evaluate. The Green's function g is the inverse of Θ

$$\sum_\gamma g_{\alpha\gamma} \Theta_{\gamma\beta} = \delta_{\alpha\beta} \tag{11.2.16}$$

A more accurate expression for g_{kk} is obtained by including more terms in the summation over intermediate states γ in (11.2.16).

The first approximation is to take just the one state $\gamma = |\, k, \nu\rangle$. This gives the simple result

$$g_{kk}(z) = \frac{1}{\Theta_{kk}(z)} = \frac{1}{z - \delta E - \varepsilon_k}$$

$$\Pi^{<}(z) = \frac{A^2}{N_f} \sum_{k<k_F} \frac{a_k^2}{z - \delta E - \varepsilon_k} \tag{11.2.17}$$

$$G^{<}(\omega) = 2\pi i A^2 N_f \int_{-W}^{0} d\varepsilon \, \frac{g(\varepsilon) V(\varepsilon)^2}{(\varepsilon + \delta E - \varepsilon_f)^2} \, \delta(\omega - \delta E - \varepsilon)$$

$$G^{<}(\omega) = \frac{2iA^2 \Delta\,(\omega - \delta E)}{(\omega - \varepsilon_f)^2} N_f \, \theta(\delta E - \omega) = iA_f(\omega)\,\theta(-\omega)$$

where $\Delta(\varepsilon) = \pi g(\varepsilon) V(\varepsilon)^2$. In equilibrium $G^{<} = i n_F(\omega)\, A_f(\omega)$. The Fermi function $n_F(\omega)$ is unimportant since the step function $\theta(\delta E - \omega)$ requires that $\omega < \delta E$. The above special function appears odd. Usually we find a spectral function such as

$$A_f(\omega) = \frac{2\Delta}{(\omega - \varepsilon_f)^2 + \Delta^2}$$

Our result in (11.2.17) is missing the factor of Δ^2 in the denominator. That omission is caused by our approximate evaluation of g_{kk}. We need to include some other states in the summation over γ in (11.2.16).

Further states are generated, in the usual way, by operating by V_{sf} on $|\, k, \nu\rangle$. From $V_{\text{sf}} \,|\, k, \nu\rangle$ one finds one new kind of state.

$$|\, q, k, \nu; 1\rangle = \frac{C_{k\nu}}{\sqrt{N_f}} \sum_{\mu} f_\mu^{\dagger} C_{q\mu} \,|\, F\rangle$$

This state is labeled by q. The solution of (11.2.16) is eased because the operator Θ has the property $\Theta_{qq'} = \delta_{qq'}(z - \delta E - \varepsilon_q - \varepsilon_k + \varepsilon_f)$. Two components of this equation are

$$1 = g_{kk}\,\Theta_{\mathbf{kk}} + \sum_q g_{kq}\,\Theta_{qk} \qquad \Theta_{kq} = \frac{\sqrt{N_f}}{\sqrt{N}} V_q$$

$$0 = g_{kk}\,\Theta_{kq} + g_{kq}\,\Theta_{qq}$$

which have the solution

$$g_{kq} = \frac{\sqrt{N_f}}{\sqrt{N}} \frac{V_q g_{kk}}{z - \delta E - \varepsilon_k - \varepsilon_q - \varepsilon_f}$$

$$g_{kk} = 1/[z - \delta E - \varepsilon_k - \Gamma(z - \delta E - \varepsilon_k + \varepsilon_f)]$$

$$\Gamma(\omega) = \frac{N_f}{N} \sum_{k<k_F} \frac{V_k^2}{\omega - \varepsilon_k} = \frac{N_f}{\pi} \int_{-W}^{0} d\varepsilon \frac{\Delta(\varepsilon)}{\omega - \varepsilon}$$

The effect of adding the state $| q, k, \nu; 1\rangle$ to the basis set in (11.2.16) is to add the factor of Γ to the denominator of g_{kk}. The function Γ is identical to the one that defines the ground state energy in (11.2.8). This latter equation can be written as $\delta E = -\Gamma(\varepsilon_f - \delta E)$. Using this relationship, it is obvious that the denominator of g_{kk} vanishes when $z = \varepsilon_k$. This point is a pole. The residue at that pole involves the factor $\partial\Gamma/\partial\varepsilon$ which equals C in (11.2.10). Let $z \to \omega + i\delta$ and find

$$-\operatorname{Im}[g_{kk}(\omega + i\delta)] = \pi(1 - n_f)\,\delta(\omega - \varepsilon_k) + \frac{N_f\Delta(\omega + \varepsilon_f - \delta E - \varepsilon_k)\,\theta(\varepsilon_k + \delta E - \omega - \varepsilon_f)}{(\omega - \delta E - \varepsilon_k - \operatorname{Re}\Gamma)^2 + (N_f\Delta)^2}$$

In the denominator of the second term, both Re Γ and $N_f\Delta = \operatorname{Im}[\Gamma]$ are functions of $(\omega + \varepsilon_f - \delta E - \varepsilon_k)$ which must be negative. This new expression for g_{kk} also provides a new result for the Green's function

$$G^<(\omega) = \frac{2i\Delta(\omega)N_f}{(\omega + \delta E - \varepsilon_f)^2}(1 - n_f)^2 + \frac{2}{\pi}(1 - n_f)i \times \int_{\omega + \varepsilon_f - \delta E}^{0} d\varepsilon \frac{N_f^2 \Delta(\varepsilon)\Delta(\omega - \delta E - \varepsilon + \varepsilon_f)}{(\varepsilon + \delta E - \varepsilon_f)^2 [(\omega - \delta E - \varepsilon - \operatorname{Re}\Gamma)^2 + (N_f\Delta)^2]} \tag{11.2.18}$$

The last integral must be done numerically. An evaluation of this expression is given below.

The first term has very interesting properties. Consider the case that $\varepsilon_f < 0$. Then define $\delta E = \varepsilon_f - \delta$ where according to (11.2.13)

$$\delta \approx W \exp(-\pi \,|\, \varepsilon_f \,|\, /\Delta N_f) \tag{11.2.19}$$

This expression has the same mathematical form as the definition of the

Kondo temperature in (11.1.22). In fact, δ is believed to be k_B times the Kondo temperature. The Anderson model is sufficiently similar to the Kondo model that it also has a characteristic energy which can be related to an effective temperature. The relative energy scales are $-\varepsilon_f \gg \Delta \gg \delta$.

For $\varepsilon_f < 0$ the normalization factor $C \approx \Delta N_f/\delta\pi$ and $(1-n_f) \approx \delta\pi/\Delta N_f$. The first term in $G^<$ gives a contribution to the spectral function of

$$A_f(\omega) = \frac{2\pi^2}{\Delta N_f}\left(\frac{\delta}{\delta-\omega}\right)^2 + \ldots \tag{11.2.20}$$

This contribution is rather small unless ω approaches zero. Remember that this expression is valid only for $\omega < 0$. The first term becomes reasonably large at $\omega = 0$. There is a resonance near the Fermi surface. This is the same resonance we found in the Kondo effect. It is called the *Kondo resonance*, even when it occurs in the Anderson model. This resonance will be found again when we calculate $G^>(\omega)$.

The effective density of states $g_f(\omega) = A_f(\omega)/2\pi$. According to the Friedel sum rule the density of states at the chemical potential for the Kondo resonance is

$$g_K(0) = \frac{N_f}{\pi\Delta}\sin^2(\pi n_f/N_f)$$

In the present example, we have that $n_f \approx 1$ and $N_f > 1$ so that we can approximate this expression by $g_k(0) \approx \pi/(\Delta N_f)$, which agrees with (11.2.20) when $\omega = 0$. The Kondo resonance is believed to be a Lorentzian shape with width $\Gamma_K \approx \delta\pi/N_f$. There is some uncertainty whether the peak of the resonance is exactly at $\omega = 0$ or else at $\omega = \delta$. We shall assume that it is centered at zero, as in the Kondo effect. This discussion suggests that the Kondo resonance has an effective density of states of

$$g_K(\omega) = g_K(0)\frac{\Gamma_K^2}{\omega^2 + \Gamma_K^2} \tag{11.2.21}$$

The total area under the Kondo resonance is $g_K(0)\Gamma_K/\pi \approx \delta/(N_f\Delta)$. The area is very small since the resonance is so narrow.

For the Kondo effect, the contribution to the specific heat per impurity was proportional to T/T_K. The general expression at low temperature ($T \ll T_K$) is $\delta C = c_i\pi^2 k_B{}^2 g_K(0)T/3$. In the Anderson model $g_K(0) \sim \pi/\Delta N_f$ which is large if

Δ is small. Thus the specific heat per impurity is also large. In solids such as $CeNi_2$ or $CeNi_5$ the local electrons on the Ce are not from impurities, but are part of the crystal lattice. They have very large specific heats at low temperature, which is the origin of the name "heavy fermion": the conventional formula $g(0) \sim m^*k_F$ suggests that a large value of $g(0)$ is due to a large effective mass m^*. Now it is known that the large specific heat is due to the Kondo resonance. When this phenomenon was first discovered, the original interpretation was that the f level was right at the Fermi surface ($\varepsilon_f = 0$), which would also give a large value of $g(0)$. This simple band interpretation became unreasonable after various spectroscopies showed the f^1 state well below the Fermi surface. Now it is known that the enhanced density of states at the Fermi surface is due to the Kondo resonance. It occurs there regardless of the placement of the f level.

The second term in (11.2.18) is also easy to evaluate in an approximate fashion. Recalling that $(1-n_f) \sim \pi\delta/\Delta N_f$ we can approximate this integral by

$$2\delta \,\mathrm{Im}\left[\int_{\omega-\delta}^{0} d\varepsilon \frac{1}{(\varepsilon-\delta)^2\,(\omega+\delta-\varepsilon_f-\varepsilon+i\Delta N_f)}\right] \approx 2\,\mathrm{Im}\left(\frac{1}{\omega+\delta-\varepsilon_f+i\Delta N_f}\right)$$

The main contribution is a simple lorentizian centered at the energy of the f^1 configuration. Thus we obtain for the spectral function for $\omega < 0$ the approximate expression

$$A_f(\omega) = \frac{2N_f\Delta}{(\omega-\varepsilon_f)^2+(N_f\Delta)^2} + 2\pi g_K(\omega) + \ldots$$

The first term is a resonance at the energy ε_f of the f^1 configuration. If one includes the self-energy $\mathrm{Re}(\Gamma)$, then the peak position is shifted downward in energy a small amount. This first term provides virtually all of the sum rule in (11.2.15). The second term is the Kondo resonance. Its contribution to the sum rule is very small, of order $\delta/(N_f\Delta)$.

Now consider the evaluation of $G^>(\omega)$. The steps are similar to the derivation of $G^<$, so the details can be covered quickly. The function to evaluate is

$$\Pi^>(z) = \langle i \,|\, f_\nu \frac{1}{z+E_i-H} f_\nu^\dagger \,|\, i\rangle = \sum_{\alpha,\beta} \langle i \,|\, f_\nu \,|\, \alpha\rangle g_{\alpha\beta} \langle \beta \,|\, f_\nu^\dagger \,|\, i\rangle \quad (11.2.22)$$

Again chose $|\,i\rangle = \psi_0$. The subspace created by $f_\nu^\dagger\,|\,i\rangle$ is just ψ_ν. The Green's function $g_{\alpha\beta}$ is different than before since it is for a different subspace. For the intermediate states (α, β, γ) we retain $|\,\nu; 1\rangle$, $|\,p, \nu\rangle$, and $|\,k, \nu; 2\rangle$, which are denoted 0, p, and k. Define $\Theta = z+E_i-H$. The important matrix elements are

$$\langle 0 \mid f_\nu^\dagger \mid i\rangle = A, \qquad \Theta_{00} = z + \delta E - \varepsilon_f, \qquad \Theta_{0k} = -\left(\frac{N_f - 1}{N}\right)^{1/2} V_k$$

$$\langle p \mid f_\nu^\dagger \mid i\rangle = 0, \qquad \Theta_{pp'} = \delta_{pp'}(z + \delta E - \varepsilon_p), \qquad \Theta_{0p} = -\frac{1}{\sqrt{N}} V_p$$

$$\langle k \mid f_\nu^\dagger \mid i\rangle = A\, a_k \left(\frac{N_f - 1}{N_f}\right)^{1/2}, \quad \Theta_{kk'} = \delta_{kk'}\ (z + \delta E + \varepsilon_k - 2\varepsilon_f - U), \quad \Theta_{kp} = 0$$

Equation (11.2.16) for $g_{\alpha\beta}$ has the explicit form

$$\begin{aligned} 1 &= g_{00}\,\Theta_{00} + \sum_k g_{0k}\,\Theta_{k0} + \sum_p g_{0p}\,\Theta_{p0} \\ 0 &= g_{00}\,\Theta_{0k} + g_{0k}\,\Theta_{kk}, \qquad 0 = g_{00}\,\Theta_{0p} + g_{0p}\,\Theta_{pp} \\ \delta_{kk'} &= g_{kk'}\,\Theta_{k'k'} + g_{k0}\,\Theta_{0k'} \end{aligned} \tag{11.2.23}$$

which have a solution

$$\begin{aligned} g_{0k} &= g_{00}\,[(N_f - 1)/N]^{1/2}\, V_k/\Theta_{kk} \\ g_{0p} &= g_{00}\,\frac{1}{\sqrt{N}}\, V_p/\Theta_{pp} \\ g_{00} &= 1/[z + \delta E - \varepsilon_f - \Gamma'(z + \delta E - 2\varepsilon_f - U) - \Gamma''(z + \delta E)] \\ g_{kk'} &= \frac{\delta_{kk'}}{\Theta_{kk}} + g_{00}\,[(N_f - 1)/N]\,\frac{V_k}{\Theta_{kk}}\,\frac{V_{k'}}{\Theta_{k'k'}} \\ \Gamma'(w) &= \frac{N_f - 1}{N}\sum_{k<k_F}\frac{V_k^2}{w + \varepsilon_k} = (N_f - 1)\int_{-W}^{0} d\varepsilon\,\frac{g(\varepsilon)V(\varepsilon)^2}{w + \varepsilon} \\ \Gamma''(w) &= \frac{1}{N}\sum_{p>k_F}\frac{V_p^2}{w - \varepsilon} = \int_0^B d\varepsilon\,\frac{g(\varepsilon)V(\varepsilon)^2}{w - \varepsilon} \end{aligned} \tag{11.2.24}$$

These results are sufficient to evaluate $\Pi^>(z)$ in (11.2.20). The summation over α includes 0 and k, while the summation over β includes 0 and k'. There is no matrix element to the state p. It is useful to introduce the symbol $\mu_n(z)$

$$\begin{aligned} \mu_n(z) &= (N_f - 1)\int_{-W}^{0} d\varepsilon\,\frac{g(\varepsilon)V(\varepsilon)^2}{(\varepsilon + \delta E - \varepsilon_f)^n\,(z + \delta E + \varepsilon - 2\varepsilon_f - U)} \\ \Pi^>(z) &= A^2\,\{\mu_2(z) + g_{00}(z)\,[1 + \mu_1(z)]^2\} \end{aligned} \tag{11.2.25}$$

and $\Gamma' = \mu_0$. In these equations the symbol δE is the ground state energy of the state ψ_0. Equation (11.2.9) gives a formula for the ground state energy δE_b of the state ψ_ν, which can be written in the present notation as $\delta E_b = \varepsilon_f + \Gamma''(\delta E_b) + \Gamma'(\delta E_b - 2\varepsilon_f - U)$. An inspection of g_{00} shows it has a pole at $z = \delta E_b - \delta E$. The pole is at the difference of the two ground state energies. The residue at the pole is proportional to $A^2(1+\mu_1)^2/(1+D+F)$, where D and F are given in (11.2.12).

This pole is the Kondo resonance. Our perturbation expansion for $\Pi^>$ places the pole at the energy of $\delta E_b - \delta E$. This prediction will change when more states are added to the expansion. The residue of this pole is the same as we found for the Kondo resonance in $\Pi^<$. That $\Pi^>$ has a pole, rather than a resonance, is due to an incomplete basis set.

Since the total oscillator strength in $G^>$ is of $O(N_f)$, then to get an answer valid to $O(1/N_f)$ we must calculate $\Pi^>$ to $O(1/N_f^2)$. That has not been done here, so our results are approximate.

The most important term in $\Pi^>$ is μ_2. For our example where $\varepsilon_f < 0$, $n_f \approx 1$, and $A^2 \approx \pi\delta/\Delta N_f$, the evaluation of the integral gives with $\Omega = z - \varepsilon_f - U$

$$A^2\mu_2(z) = \delta\left(\frac{N_f - 1}{N_f}\right)\int_{-W}^{0} d\varepsilon \, \frac{1}{(\varepsilon-\delta)^2\,(\Omega+\varepsilon-\delta)}$$

one term in the integral is of order $O(1/\delta)$, which cancels the prefactor of δ. This term is found by integrating by parts: integrating $(\varepsilon-\delta)^{-2}$. One then finds

$$A^2\mu_2(z) = \left(\frac{N_f - 1}{N_f}\right)\frac{1}{z - \varepsilon_f - U}\,[1 + O(\delta)]$$

This term appears to have a pole at $\omega = \varepsilon_f + U$. A more careful evaluation of the integral shows that it is actually a resonance. The right width of the resonance $\xi \sim (N_f - 1)\Delta$ can only be obtained by including more terms in the perturbation expansion. This peak contains virtually all of the oscillator strength in $G^>$. The resonant energy is $\varepsilon_f + U$, which is associated with the f^2 configuration. This resonant energy is actually $2\varepsilon_f + U - \delta E$, but since $\delta E \sim \varepsilon_f$ the effective result is $\varepsilon_f + U$. This peak position will also be shifted slightly by additional self-energy effects, if more terms are included in the perturbation expansion.

An approximate, analytical, density of states of the local electron in the

Anderson model with $n_f \approx 1$ is

$$g_f(\omega) = \frac{2\Delta N_f}{(\omega - \varepsilon_f)^2 + (\Delta N_f)^2} + g_K(\omega) + (N_f - 1)\frac{2\xi}{(\omega - \varepsilon_f - U)^2 + \xi^2} \tag{11.2.26}$$

where the Kondo resonance g_K is given in (11.2.21). The first resonance is from the f^1 configuration, and the last term is from the f^2 configuration. Figure 11.6 shows a graph of this function for $\varepsilon_f = -2$, $U = 6$, $W = 6$, $\Delta = 0.1$, and $N_f = 6$. The lower peak has a spectral weight of $n_f \approx 1$, the Kondo resonance has a spectral weight of $0(\delta/\Delta) \ll 1$, while the upper peak has a spectral weight of $N_f - n_f \approx 5$.

C. Spectroscopies

The behavior of the local electron has been probed by a large number of spectroscopies. The word *spectroscopy* is meant to include any experiment where a particle is sent at the solid, which causes that particle or another to leave. The relative energies and intensities of the initial and final particles provide much information about the system. The particles are either photons, electrons, or neutrons. Here we use the collective states as the basis for describing the frequency response of the system. The spectroscopies will involve the localized electrons on the magnetic impurities. Our discussion follows Gunnarsson and Schonhammer (1983).

All of the spectral functions have the general form

$$A_j(\omega) = 2\pi \langle i \mid T_j^\dagger \delta(\omega - \Xi_j) T_j \mid i \rangle \tag{11.2.27}$$

where T_j and Ξ_j are different for each spectroscopy. The general form is similar to that of the Green's functions of the previous section, and they are evaluated using the same techniques. Several examples will be given here, and several others are treated in the problem assignments.

The first spectroscopy to be discussed is valence photoemission (VP) of the localized electron. The absorption of a photon excites a local electron to high kinetic energy $E = \varepsilon_p$. Some of these energetic electrons escape from the surface, and are analyzed. The excitation operator is $T = \Sigma M_\nu C_{p\nu}^\dagger f_\nu$. Usually it is assumed that the energetic electron does not scatter during its departure from the surface. If it does interact, it usually changes its energy significantly and is not detected. For the unscattered electrons the creation

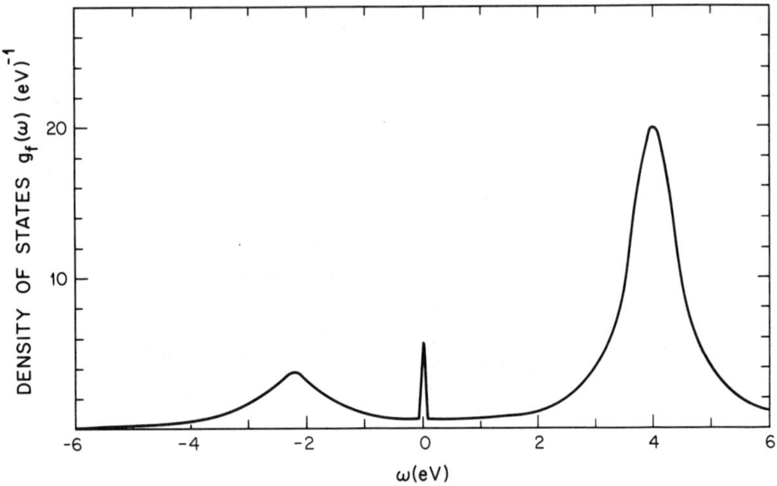

FIGURE 11.7. Density of states $g_f(\omega)$ of the localized level when $\varepsilon_f = -2.0$ eV, $\Delta = 0.1$ eV, $U = 6$ eV, $W = 6.0$ eV, and $N_f = 6$. The broad peak with a maximum at -2.2 eV is duo to the f^1 configuration. The strong peak near $\varepsilon_f + U = 4$ eV is the f^2 configuration. The narrow peak at the origin is the Kondo resonance.

operator $C_p^\dagger$ plays no role and can be dropped. Thus we wish to evaluate

$$A_{\text{VP}}(E-\omega) = 2\pi \sum_\nu M_\nu^2 \langle i \mid f_\nu^\dagger \delta(E-\omega-E_i+H) f_\nu \mid i \rangle$$

Usually one assumes that the matrix element M is independent of ν. Then this quantity is given by $A_{\text{VP}}(E-\omega) = -iM^2G^<(E-\omega)$, where the Green's function is defined in (11.2.14). This function was evaluated in the previous section. An example of the valence photoemission spectra is given by the negative frequency regions of Fig. 11.7.

Another example is provided by Bremsstrahlung isochromat spectroscopy (BIS). Here an electron of high energy E is shot at the solid. It can emit a photon of frequency ω while falling into a lower-energy electron state. This spectroscopy is used to measure the unoccupied density of states above the Fermi surface of metals. The T operator is the Hermitian conjugate of the one above. Again we remove the operators associated with the energetic electron, and the photon emission rate is

$$I(E-\omega) = 2\pi \sum_\nu M_\nu^2 \langle i \mid f_\nu\, \delta(E-\omega+E_i-H) f_\nu^\dagger \mid i \rangle = iM^2G^>(E-\omega)$$

This spectral function is evaluated by the technique we used to find the

Green's function in the prior section. An example is given by the positive frequency part of Fig. 11.7.

The third example is X-ray photoelectron core spectroscopy (XPS). A photon in the X-ray range of frequencies is absorbed by an atom in the solid. A core electron is excited to very high kinetic energy E, which is measured after it exits the surface. The core electron has an initial eigenvalue of ε_c or binding energy $\varepsilon_b = -\varepsilon_c > 0$. The correlation function is a function of $\xi = E - \omega - \varepsilon_c$. Let the creation operator for the core hole be $h^\dagger$. The T-operator in (11.2.27) is in this case $T = MC_p^\dagger h^\dagger$. Again the operator $C_p^\dagger$ is ignored since the energetic electron in the final state is assumed not to interact further. Thus we have for the correlation function for XPS

$$I_{\text{XPS}}(\xi) = 2\pi \langle i \mid h\delta(\xi - E_i - H')h^\dagger \mid i\rangle$$

So far we have not indicated how the Anderson model is affected by XPS. If the Hamiltonian H' in this expression is exactly the Anderson Hamiltonian, which does not include core holes, then we would find that $I_{\text{XPS}}(\xi) = 2\pi\delta(\xi)$. However, Gunnarsson and Schonhammer suggested that the localized electrons played a major role in this spectroscopy. The local electron states are strongly influenced by the coulomb interaction from the core hole. They add a term δH to the Hamiltonian H of the Anderson model:

$$H' = H + \delta H$$

$$\delta H = -U_{fc}h^\dagger h \sum_\nu n_\nu$$

They estimate the f-electron eigenvalue is lowered by $U_{fc} \approx 10$ eV by each core hole. In the presence of the core hole, the energy of one local electron is $\varepsilon_f - U_{fc}$, the energy of two local electrons is $2\varepsilon_f - 2U_{fc} + U$, etc. The creation of the core hole causes the local electrons on the magnetic impurity to change their energy. Other quantities are also changed, such as the occupancy n_f. Because of the coupling between the local electron and the conduction band, there is a general shakeup of excitations.

The correlation function is evaluated by following the same steps used in the previous section. One finds the Green's function of complex frequency z. At the end find its spectral function by $z \to \xi + i\delta$:

$$\Pi_{\text{XPS}}(z) = \langle i \mid h \frac{1}{z - E_i + H'} h^\dagger \mid i\rangle = \sum_{\alpha,\,\beta} \langle i \mid h \mid \alpha\rangle\, g_{\alpha\beta}(z)\langle \beta \mid h^\dagger \mid i\rangle$$

$$A_{\text{XPS}}(\xi) = -2i\,\text{Im}\,[\Pi_{\text{XPS}}(\xi + i\delta)]$$

In order to keep the discussion simple, only a few states are taken for $|\,i\rangle = \psi_0$ and for the intermediate states. For ψ_0 use the two-term expression in (11.2.4). Similarly, for the intermediate states use only $h^\dagger\,|\,F\rangle$ and $h^\dagger\,|\,k;1\rangle$. These are denoted 0 and k, respectively. The solution is nearly identical to that used to find $\Pi^>$ in the previous section. The important matrix elements are

$$\Theta_{00} = z - \delta E, \qquad \Theta_{0k} = \sqrt{N_f}\,V_k/\sqrt{N}$$
$$\Theta_{kk} = z - \delta E - \varepsilon_k + \varepsilon_f - U_{fc}$$

One solves (11.2.23) but without the terms depending upon p. The solution has the general form of (11.2.24), but with the above matrix elements:

$$\Pi_{\mathrm{XPS}}(z) = A^2\{y_2(z) + g_{00}(z)[1 - y_1(z)]^2\}$$
$$g_{00}(z) = 1/[z - \delta E - y_0(z)] \tag{11.2.28}$$
$$y_n(z) = \frac{N_f}{N}\sum_{k<k_F}\frac{V_k^2}{(\delta E + \varepsilon_k - \varepsilon_f)^n\,(z - \delta E - \varepsilon_k + \varepsilon_f - U_{fc})}$$

In deriving a new expression one should check it by every means. One check on this result is to see whether for $U_{fc} = 0$ it gives a pole at $z = 0$. The function $g_{00}(z)$ has a pole at $z = 0$ since $\delta E = -y_0(0)$. At this pole, the residue of $g_{00}(z)$ is $1/(1+C)$ while $y_1(0) = -C$. Since $A^2 = (1 + C)$, the total residue is unity. Thus this term has a simple polc at $z = 0$ when $U_{fc} = 0$. For $U_{fc} = 0$ the first term is $y_2 \sim O(\delta)$ and can be ignored.

Actual XPS spectra have $U_{fc} \approx 10$ eV, which is a large energy term. Then $g_{00}(z)$ has a resonance centered at $z \approx \delta E$, since the large value of U_{fc} makes y_0 small. The intensity of this pole is small since it is multiplied by the factor $A^2 \sim \delta$, which is small. This peak is the transition $f^1 \rightarrow f^0$. This final state lacks an f electron. It has a small intensity when the valence fluctuations are small.

The main peak in the spectra for large values of U_{fc} comes from the term $A^2 y_2(z)$:

$$Ay_2(z) = \delta\int_{-W}^{0} d\varepsilon\,\frac{1}{(\varepsilon - \delta)^2\,(z + \delta - \varepsilon - U_{fc})} \approx \frac{1}{z - \delta - U_{fc}}$$

where the integral is again evaluated by an integration by parts. This term contributes a pole to the spectral function

$$A_{\mathrm{XPS}}(\xi) \approx 2\pi\delta(\xi + \delta - U_{fc}) + O(\delta)$$

Since the value of δ is negligible, the pole appears at the energy $E = \omega + U_{fc}$. This result does not make sense from a one-body viewpoint. There is no energy state at U_{fc}. This transition is $f^1 \to f^1$. The local level stays singly occupied. The initial energy of the local level is $\delta E = \varepsilon_f - \delta$. The final energy of the local level is $\varepsilon_f - U_{fc} - \delta'$, where δ' is the self-energy of the new level, which we have not calculated; $\delta' = 0$ in the present approximation. The *difference* between these two energy levels is $U_{fc} + \delta' - \delta \approx U_{fc}$. The energy difference between the initial and final f^1 configurations is U_{fc}. This energy is imparted to the energetic electron, which is measured in the spectrometer. Its final energy is $E = \omega + U_{fc}$.

This calculation could be improved, since the f^2 configuration is permissible in the final state. The energy $2(\varepsilon_f - U_{fc}) + U$ is always less than zero since U_{fc} is large. Our basis set did not include this configuration. It is rather easy to include, and this step is given in the problem assignments. These additional basis states will also impart some broadening to the resonance at $\xi = U_{fc}$, and contribute to the self-energy δ'.

There are several other approaches to solving the Anderson model besides the method of Gunnarsson and Schonhammer (1983). Grewe and Keiter (1981) developed a formalism for expanding the S matrix for the local operators. They found formal but rigorous results for the local self-energies. Another method of finding the same results was developed by Coleman (1984) using the *slave boson* idea of Barnes (1976, 1977). One introduces a fictitious boson which is created every time there is a change in the occupancy of the local level. One can control the f occupancy by controlling the chemical potential of the slave bosons. A similar method was used earlier by Abrikosov (1965) in solving the Kondo problem.

There are also suggestions that the Anderson model is insufficient for the discussion of heavy fermions: perhaps other terms are needed in the Hamiltonian. One that is added often is a screened Coulomb interaction G between the local electrons and the conduction electrons:

$$H_c = \frac{G}{N} \sum_{\nu} f_\nu^\dagger f_\nu \sum_{\mathbf{k},\mathbf{k}'} C_{\mathbf{k}'}^\dagger C_{\mathbf{k}}$$

This term originates in the dielectric screening of the local electron by the conduction electrons. The screening charge must adjust every time an electron is added or removed from the local site. Liu (1988) has pointed out that this process causes the type of MND effects discussed in Sec. 8.3. They also produce a resonance at the Fermi surface which may contribute to the phenomena we associate with heavy fermions.

PROBLEMS

1. Evaluate the integral for $\Lambda(\varepsilon)$ for each $\mathbf{g}(\varepsilon)$ and show that the coefficient of the term $\ln(\varepsilon/W)$ is proportional to $\varrho(0)$. Also verify that the integral of $g(\varepsilon)$ over the entire band $(-W<\varepsilon<W)$ is unity.

$$\Lambda(\varepsilon)=\int_{-W}^{0} d\varepsilon' \, \frac{g(\varepsilon')}{\varepsilon'-\varepsilon}$$

   a. $g(\varepsilon)=1/2W$
   b. $g(\varepsilon)=2[W^2-\varepsilon^2]^{1/2}/\pi W$
   c. $g(\varepsilon)=(W-|\varepsilon|)/(W^2)$

2. Construct the zero-temperature collective state for the Kondo Hamiltonian when the local spin has $S=1$. Find the states with $j=3/2$ and $j=1/2$. Which is preferred for $J<0$?

3. Consider the collective state for the antiferromagnetic Kondo model when the bands are half filled. Calculate $\delta=-\delta E/W$ by solving (11.1.19) using the density of states in part (a) and (c) of Problem 1. Make a graph of δ vs. Λ defined below (11.1.21). Which case predicts the lowest binding energy?

4. Draw the diagrams equivalent to Fig. 11.4 for the state $|\mu,\nu;2\rangle=f_\mu^\dagger f_\nu^\dagger|F\rangle$ and others in its collective state. Then find the self-energy δE of this state. Use the terms $|\mu,\nu;2\rangle$ plus those generated by $V_{\rm sf}|\mu,\nu;2\rangle$. Derive an expression for n_f.

5. Define the subspaces for the Anderson model that correspond to the state ψ_{0a} of (11.1.12) of the Kondo model; do it for $S=1/2$ and for $S=3/2$.

6. Calculate the change in ground state energy δE of the state ψ_0 of the Anderson model while retaining the four terms (11.2.4) and (11.2.6). How is this result altered in the $1/N$ expansion?

7. Given the constants $N_f=14$, $W=6$ eV, $U=6$ eV, and $\Delta(\varepsilon)$ constant $=0.1$ eV, compare the numerical values of δE_a in (11.2.8) with δE_b in (11.2.9) for values $\varepsilon_f=-2$, -1, 0, 1, 2 eV.

8. Assume that an f level is split by the spin–orbit interaction into states with degeneracy 8 and 6, separated by an energy $\xi\sim0.3$ eV. Derive an expression for $\delta=\varepsilon_f-\delta E$ assuming that $\delta\ll\xi$.

9. Derive the high-temperature ($\delta C\sim 1/T$) and low-temperature ($\delta C\sim T$) limits of the specific heat per impurity in the Anderson model.

10. Calculate the XPS spectra for the initial and intermediate states given by the two terms in (11.2) plus the two-hole state $| \, q, k; 2\rangle$ in (11.2.6).

11. Calculate the XPS spectra for the initial state (11.2.5).

12. Calculate the two Green's functions $G^{<}$ and $G^{>}$ and the density of states $g_f(\omega)$ for the ground state given by (11.2.5).

References

Chapter 1

ADLER, D., in *Solid State Physics*, Vol. 21, eds. H. Ehrenreich, F. Seitz, and D. Turnbull (Academic Press, New York, 1967), pp. 1–113.

ANDERSON, P. W., *Phys. Rev.* **124**, 41 (1961).

BENI, G., *Phys. Rev. B* **10**, 2186 (1974).

BORN, M., and K. HUANG, *Dynamical Theory of Crystal Lattices* (Oxford University Press, Inc., London, 1954).

CALLEN, H., in *Physics of Many-Particle Systems*, Vol. 1, ed. E. Meeron (Gordon & Breach, New York, 1966), pp. 183-230.

DOMB, C., and M. S. GREEN, *Phase Transitions and Critical Phenomena*, Vols. 1–6 (Academic Press, New York, 1972–1977;.

EHRENREICH, H., and A. W. OVERHAUSER, *Phys. Rev.* **104**, 331 (1958).

EVRARD, R., in *Polarons in Ionic Crystals and Polar Semiconductors*, ed. J. T. Devreese (North-Holland, Amsterdam, 1972), pp. 29–80.

FANO, U., *Phys. Rev.* **124**, 1866 (1961).

FRÖHLICH, H., H. PELZER, and S. ZIENAU, *Philos. Mag.* **41**, 221 (1950).

FRÖHLICH, H., *Adv. Phys.* **3**, 325 (1954).

GERRITSEN, A. N., and J. O. LINDE, *Physica* **17**, 573, 584 (1951); **18**, 872 (1954).

GUTZWILLER, M. C., *Phys. Rev. Lett.* **10**, 159 (1963).

HARRISON, W. A., *Pseudopotentials in the Theory of Metals* (Benjamin, Reading, Mass., 1966).

HEINE, V., *Solid State Physics*, Vol. 24, eds. H. Ehrenreich, F. Seitz, and D. Turnbull (Academic Press, New York, 1970).

HILL, T. L., *Statistical Mechanics* (McGraw-Hill, New York, 1956), Chap. 7.

HOPFIELD, J. J., *Phys. Rev.* **112**, 1555 (1958).

HUBBARD, J., *Proc. R. Soc. London Ser. A* **276**, 238 (1963); **277**, 237 (1964); **281**, 401 (1964); **285**, 542 (1965); **296**, 82, 100 (1966).

JORDAN, P., and E. WIGNER, *Z. Phys.* **47**, 631 (1928).

KARTHEUSER, R., in *Polarons in Ionic Crystals and Polar Semiconductors*, ed. J. T. Devreese (North-Holland, Amsterdam, 1972), pp. 717–733.

KONDO, J., *Prog. Theor. Phys. (Kyoto)* **32**, 37 (1964).

KONDO, J., in *Solid State Physics*, Vol. 23, eds. H. Ehrenreich, F. Seitz, and D. Turnbull (Academic Press, New York, 1969), pp. 183–281.
LIEB, E. H., and F. Y. WU, *Phys. Rev. Lett.* **20**, 1445 (1968).
MAHAN, G. D., *J. Phys. Chem. Solids* **26**, 751 (1965).
MAHAN, G. D., in *Polarons in Ionic Crystals and Polar Semiconductors*, ed. J. T. Devreese (North-Holland, Amsterdam, 1972), pp. 553–657.
MARADUDIN, A. A., E. W. MONTROLL, and G. H. Weiss, *Theory of Lattice Dynamics in the Harmonic Approximation* (Academic Press. New York, 1963).
MATSUBARA, T., and H. MATSUDA, *Prog. Theor. Phys. (Kyoto)* **16**, 416 (1956).
MOZER, B., L. A. DEGRAAF, and B. LENEINDRE, *Phys. Rev. A* **9**, 448 (1974).
ONSAGER, L., *Phys. Rev.* **65**, 117 (1944).
PARR, R. G., *Quantum Theory of Molecular Electronic Structure* (Reading, Mass., Benjamin, Reading, Mass., 1964).
PERCUS, J. K., in *The Equilibrium Theory of Classical Fluids*, eds. H. L. Frisch and J. L. Lebowitz (Benjamin, Reading, Mass., 1964), pp. II-33–II-170.
SCHIFF, L. I., *Quantum Mechanics* (McGraw-Hill, New York, 1968).
SCHRIEFFER, J. R., and P. A. WOLFF, *Phys. Rev.* **149**, 491 (1966).
SHUEY, R. T., *Phys. Rev.* **139**, A 1675 (1965).
THOMAS, D. G., *J. Appl. Phys. Suppl.* **32**, 2298 (1961).
TOMLINSON, P. G., and J. C. SWIHART, *Phys. Condens. Matter* **19**, 117 (1975).
ZENER, C., *Phys. Rev.* **81**, 440 (1951).

Chapter 2

This reference list contains several books which have excellent descriptions of the material in this chapter.

ABRIKOSOV, A. A., L. P. GORKOV, and I. E. DZYALOSHINSKI, *Methods of Quantum Field Theory in Statistical Physics* (Prentice-Hall, Englewood Cliffs, N.J., 1963; Pergamon, Elmsford, N.Y., 1965).
CRAIG, R., *J. Math. Phys.* **9**, 605 (1968).
DONIACH, S., and E. H. SONDHEIMER, *Green's Functions for Solid State Physicists* (Benjamin, Reading, Mass., 1974).
FETTER, A. L., and J. D. WALECKA, *Quantum Theory of Many Particle Systems* (McGraw-Hill, New York, 1971).
GELL-MANN, M., and F. LOW, *Phys. Rev.* **84**, 350 (1951).
KELDYSH, L., *Sov. Phys. JETP* **20**, 1018 (1965).
MATTUCK, R. D., *A Guide to Feynman Diagrams in the Many Body Problem* (McGraw-Hill, New York, 1967).
SCHWINGER, J., *J. Math. Phys.* **2**, 407 (1961).

Chapter 3

ABRIKOSOV, A. A., L. P. GORKOV, and I. E. DZYALOSHINSKI, *Methods of Quantum Field Theory in Statistical Physics* (Prentice-Hall, Englewood Cliffs, N.J., 1963).
BARNARD, R. D., *Thermoelectricity in Metals and Alloys* (Taylor & Francis, London, 1972).

BROUT, R., and P. CARRUTHERS, *Lectures on the Many-Electron Problem* (Wiley–Interscience, New York, 1963).
CHOU, K. C., Z. B. SU, B. L. HAO, and L. YU, *Phys. Rep.* **118**, 1 (1985).
DE GROOT, S. R., *Thermodynamics of Irreversible Processes* (North-Holland, Amsterdam, 1952).
DE GROOT, S. R., and P. MAZUR, *Non-Equilibrium Thermodynamics* (North-Holland, Amsterdam, 1962).
DUNN, D., *Can. J. Phys.* **53**, 321 (1975).
GIRVIN, S. M., *J. Solid State Chem.* **25**, 65 (1978).
GREEN, M. S., *J. Chem. Phys.* **20**, 1281 (1952); **22**, 398 (1954).
KOHN, W., and J. M. LUTTINGER, *Phys. Rev.* **118**, 41 (1960).
KUBO, R., *Lectures in Theoretical Physics*, Vol. I (Boulder) (Wiley–Interscience, New York, 1959), pp. 120–203; *J. Phys. Soc. Japan* **12**, 570 (1957).
LANGRETH, D., in *NATO Advanced Study Institute on Linear and Nonlinear Transport*, ed. J. T. Devreese and V. E. van Doren (Plenum, New York, 1976), p. 3.
LEHMANN, H., *Nuovo Cimento* **11**, 342 (1954).
LUTTINGER, J. M., *Phys. Rev.* **135**, A1505 (1964).
LUTTINGER, J. M., and J. C. WARD, *Phys. Rev.* **118**, 1417 (1960).
MAHAN, G. D., *Phys. Rev. B* **11**, 4814 (1975).
MATSUBARA, T., *Prog. Theor. Phys. (Kyoto)* **14**, 351 (1955).
TAYLOR, P. L., *A Quantum Approach to the Solid State* (Prentice-Hall, Englewood Cliffs, N.J., 1970).
TOMONAGA, S., *Prog. Theor. Phys. (Kyoto)* **2**, 6 (1947).
WEST, R. N., *Positron Studies of Condensed Matter* (Taylor & Francis, London, 1974).
WIGNER, E., *Phys. Rev.* **40**, 749 (1932).

Chapter 4

AGRANOVICH, V. M., *Sov. Phys. JETP* **10**, 307 (1960); *Sov. Phys. Solid State* **3**, 592 (1961).
ALMBLADH, C. O., and P. MINNHAGEN, *Phys. Rev. B* **17**, 929 (1978).
ANDERSON, P. W., *Phys. Rev.* **124**, 41 (1961).
ANDERSON, P. W., and W. L. MCMILLAN, in *Theory of Magnetism in Transition Metals*, Course XXXVII, Enrico Fermi International School of Physics, Varenna, ed. W. Marshall (Academic Press, New York, 1967), pp. 50–86.
BORN, M., and K. HUANG, *Dynamical Theory of Crystal Lattices* (Oxford University Press, Inc., New York, 1954); see Fig. 18a on p. 91.
BORSTAL, G., and H. J. FOLGE, *Phys. Status Solidi B* **83**, 11 (1977).
BRUECKNER, K. A., C. A. Levinson, and H. M. Mahmoud, *Phys. Rev.* **95**, 217 (1954).
CINI, M., and A. D'ANDREA, *J. Phys. C* **21**, 193 (1988).
COHEN, M. H., and F. Keffer, *Phys. Rev.* **99**, 1128 (1955).
DAVYDOV, A. S., *Theory of Molecular Excitons* (Plenum, New York, 1971).
DEWITT, B. S., *Phys. Rev.* **103**, 1565 (1956).
DIETZ, R. E., D. G. THOMAS, and J. J. HOPFIELD, *Phys. Rev. Lett.* **8**, 391 (1962)
DUKE, C., and G. D. MAHAN, *Phys. Rev.* **139**, A1965 (1965).
EDWARDS, S. F., *Philos. Mag.* **3**, 33, 1020 (1958).
ENGELSBERG, S., and B. B. VARGA, *Phys. Rev.* **136**, A1582 (1964).

Fano, U., *Phys. Rev.* **103**, 1202 (1956); **118**, 451 (1960).
Fano, U., *Phys. Rev.* **124**, 1866 (1961).
Feltgen, R., H. Kirst, K. A. Kohler, H. Pauly, and F. Torello, *J. Chem. Phys.* **76**, 2360 (1982).
Feynman, R. P., *Phys. Rev.* **84**, 108 (1951).
Friedel, J., *Philos. Mag.* **43**, 153 (1952); *Adv. Phys.* **3**, 446 (1953).
Fröhlich, D., E. Mohler, and P. Wiesner, *Phys. Rev. Lett.* **26**, 554 (1971).
Fukuda, N., and R. G. Newton, *Phys. Rev.* **103**, 1558 (1956).
Fumi, F. G., *Philos. Mag.* **46**, 1007 (1955).
Goldberg, A., H. M. Shey, and J. L. Schwartz, *Phys. Rev.* **146**, 176 (1966).
Heeger, A., in *Chemistry and Physics of One-Dimensional Metals*, ed. H. J. Keller (Plenum Press, New York, 1977), pp. 87–135.
Henry, C. H., and J. J. Hopfield, *Phys. Rev. Lett.* **15**, 964 (1965).
Hopfield, J. J., *Phys. Rev.* **112**, 1555 (1958).
Kohn, W., and J. M. Luttinger, *Phys. Rev.* **108**, 590 (1957).
Langer, J. S., and V. Ambegaokar, *Phys. Rev.* **121**, 1090 (1961).
Levinson, N., *Kyl. Danske Videnskab. Selskab, Mat.-fys. Medd.* **25**(9) (1949).
Lieb, E. H., and D. C. Mattis, *Mathematical Physics in One Dimension* (Academic Press, New York, 1966) Chap. 4.
Luther, A., and V. J. Emery, *Phys. Rev. Lett.* **33**, 589 (1974).
Luther, A., and I. Peschel, *Phys. Rev. B* **9**, 2911 (1974).
Luttinger, J. M., *J. Math. Phys. N.Y.* **4**, 1154 (1963).
Mahan, G. D., in *Electronic Structure of Polymers and Molecular Crystals*, eds. J. M. André and J. Ladik (Plenum Press, New York, 1975), pp. 79–158.
Mattis, D. C., and E. H. Lieb, *J. Math. Phys. N.Y.* **6**, 304 (1965).
Overhauser, A. W., *Physics* **1**, 307 (1965).
Schönhammer, K., and O. Gunnarsson, *Solid State Commun.* **23**, 691 (1977).
Schotte, K., and U. Schotte, *Phys. Rev.* **182**, 479 (1969).
Schweber, S. S., *Relativistic Quantum Field Theory* (Harper & Row, New York, 1961).
Tomonaga, S., *Prog. Theor. Phys. (Kyoto)* **5**, 544 (1950).
Van Haeringen, W., *Phys. Rev.* **137**, A1902 (1965).
Williams, F. E., and M. H. Hebb, *Phys. Rev.* **84**, 1181 (1951).

Chapter 5

Animalu, A. O. E., *Tech. Report No. 4*, Solid State Theory Group, Cavendish Laboratory. Cambridge, England, 1965.
Ashcroft, N. W., *J. Phys. C* **1**, 232 (1968).
Bardeen, J., *Phys. Rev.* **49**, 653 (1936).
Bethe, H., *Ann. Phys.* **5**, 325 (1930).
Brooks, H., *Nuovo Cimento Suppl.* **7**, 165 (1958a); **7**, 207 (1958b).
Brooks, H., *Trans. MS AIME* **227**, 546 (1963).
Brundle, C. R., *Surf. Sci.* **48**, 99 (1975).
Care, C. M., and N. H. March, *Adv. Phys.* **24**, 101 (1975).
Carr, W. J., *Phys. Rev.* **122**, 1437 (1961).
Carr, W. J., and A. A. Maradudin, *Phys. Rev.* **133**, A371 (1964).
Daniel, E., and S. H. Vosco, *Phys. Rev.* **120**, 2041 (1960).

Ehrenreich, H., and M. H. Cohen, *Phys. Rev.* **115**, 786 (1959).
Eisenberger, P., L. Lam, P. M. Platzman and P. Schmidt, *Phys. Rev. B* **6**, 3671 (1972).
Fermi, E., *Z. Phys.* **48**, 73 (1928).
Ferrell, R. A., *Phys. Rev.* **101**, 554 (1956); **107**, 450 (1957).
Flodstrom, S. A., R. Z. Bachrach, R. S. Bauer, J. C. McMenamin, and S. B. M. Hagstrom, *J. Vac. Sci. Technol.* **14**, 303 (1977).
Geldart, D. J. W., and R. Taylor, *Can. J. Phys.* **48**, 155, 167 (1970).
Gell-Mann, M., and K. Brueckner, *Phys. Rev.* **106**, 364 (1957).
Grigor'ev, F. V., S. B. Kormer, O. L. Mikhailova, A. P. Tolochko, and V. D. Urlin, *JETP Lett.* **16**, 201 (1972).
Ham, F. S., *Phys. Rev.* **128**, 82 (1962).
Hammerberg, J., and N. W. Ashcroft, *Phys. Rev. B* **9**, 409 (1974).
Harrison, W. A., *Pseudopotentials in the Theory of Metals* (Benjamin, Reading, Mass., 1966).
Hawke, P. S., T. J. Burgess, D. E. Duerre, J. G. Huebel, R. N. Keeler, H. Klapper, and W. C. Wallace, *Phys. Rev. Lett.* **41**, 994 (1978).
Hedin, L., *Phys. Rev.* **139**, A796 (1965).
Heine, V., and I. V. Abarenkov, *Philos. Mag.* **9**, 451 (1964); **12**, 529 (1965).
Hopfield, J. J., *Phys. Rev. B* **2**, 973 (1970).
Hubbard, J., *Proc. R. Soc. London Ser. A* **243**, 336 (1957).
Jonson, M., *J. Phys. C* **9**, 3055 (1976).
Kramers, H. A., reprinted in *Collected Scientific Papers* (North-Holland, Amsterdam, 1956).
Kronig, R., *J. Opt. Soc. Am.* **12**, 547 (1926).
Landau, L. D., and E. Lifshitz, *Quantum Mechanics: Non Relativistic Theory* (Addison-Wesley, Reading, Mass., 1958).
Lang, N. D., and W. Kohn, *Phys. Rev. B* **3**, 1215 (1971).
Langer, J. S., and S. H. Vosko, *J. Phys. Chem. Solids* **12**, 196 (1960).
Lindgren, I., and A. Rosen, unpublished (1970).
Lindhard, J., *K. Dan. Vidensk. Selsk. Mat. Fys. Medd.* **28**, (8) (1954).
Lundqvist, B. I., *Phys. Status Solidi* **32**, 273 (1969).
MacDonald, A. H., and C. P. Burgess, *Phys. Rev. B* **26**, 2849 (1982).
Macke, W., *Z. Naturforsch. Teil A* **5**, 192 (1950).
Mahan, G. D., *Phys. Status Solidi B* **63**, 453 (1974).
Mahan, G. D., in *Electronic Structure of Polymers and Molecular Crystals*, eds. J. M. André and J. Ladik (Plenum Press, New York, 1975).
Mahan, G. D., and B. E. Sernelius, *Phys. Rev. Lett.* **62, 2718 (1989).**
Marton, L., J. A. Simpson, H. A. Fowler, and N. Swanson, *Phys. Rev.* **126**, 182 (1962).
Mihara, N., and R. D. Puff, *Phys. Rev.* **174**, 221 (1968).
Nozières, P., *Interacting Fermi Systems* (Benjamin, Reading, Mass., 1964), p. 287.
Nozières, P., and D. Pines, *Phys. Rev.* **111**, 442 (1958).
Onsager, L., L. Mittag, and M. J. Stephen, *Ann. Phys.* **18**, 71 (1966).
Pandrey, and L. Lam, *Phys. Lett.* **43**A, 319 (1973).
Pines, D., and P. Nozières, *The Theory of Quantum Liquids* (Benjamin, Reading, Mass., 1966).
Powell, C. J., *Surf. Sci.* **44**, 29 (1974).

QUINN, J. J., and R. A. FERRELL, *Phys. Rev.* **112**, 812 (1958).
RICE, T. M., *Ann. Phys.* **31**, 100 (1965).
RITCHIE, R. H., *Phys. Rev.* **106**, 874 (1957).
SAWADA, K., K. A. Brueckner, N. Fukada, and R. Brout, *Phys. Rev.* **108**, 507 (1957).
SCHNEIDER, T., R. BROUT, H. THOMAS, and J. FEDER, *Phys. Rev. Lett.* **25**, 1423 (1970).
SEITZ, F., *Modern Theory of Solids* (McGraw-Hill, New York, 1940), Sec. 76.
SHOLL, C. A., *Proc. Phys. Soc. London* **92**, 434 (1967).
SHUNG, K. W. K., B. E. Sernelius, and G. D. MAHAN, *Phys. Rev. B* **36**, 4499 (1987).
SILVERSTEIN, S. D., *Phys. Rev.* **130**, 1703 (1963).
SINGWI, K. S., and M. P. TOSI, *Solid State Phys.* **36**, 177 (1981).
SINGWI, K. S., M. P. TOSI, R. H. LAND, and A. SJÖLANDER, *Phys. Rev.* **176**, 589 (1968).
SINGWI, K. S., A. SJÖLANDER, M. P. TOSI, and R. H. LAND, *Phys. Rev. B* **1**, 1044 (1970).
SUEOKA, O., *J. Phys. Soc. Jpn.* **20**, 2203 (1965).
THOMAS, L. H., *Proc. Cambridge Philos. Soc.* **23**, 542 (1927).
TING, C. S., T. K. LEE, and J. J. QUINN, *Phys. Rev. Lett.* **34**, 870 (1975).
TRACY, J. C., *J. Vac. Sci. Technol.* **11**, 280 (1974).
VASHISHTA, P., and K. S. SINGWI, *Phys. Rev. B* **6**, 875 (1972).
WIGNER, E., *Phys. Rev.* **46**, 1002 (1934).
WIGNER, E., *Trans. Faraday Soc.* **34**, 678 (1938).
WIGNER, E., and F. SEITZ, *Phys. Rev.* **43**, 804 (1933); **46**, 509 (1934).
WOLFF, P. A., *Phys. Rev.* **120**, 814 (1960).

Chapter 6

ADLER, D., in *Solid State Physics*. vol. 21, eds. H. Ehrenreich, F. Seitz, and D. Turnbull (Academic Press, New York, 1967), p. 1.
ALLCOCK, G. R., *Adv. Phys.* **5**, 412 (1956).
ALLCOCK, G. R., in *Polarons and Excitons*, eds. C. G. Kuper and G. D. Whitfield (Oliver & Boyd, Edinburgh, 1962), pp. 45–70.
APPEL, J., *Phys. Rev. Lett.* **17**, 1045 (1966).
APPEL, J., in *Solid State Physics*, Vol. 21, eds. H. Ehrenreich, F. Seitz, D. Turnbull (Academic Press, 1967), p. 193.
ASHCROFT, N. W., and J. W. WILKINS, *Phys. Lett.* **14**, 285 (1965).
BOGOMOLOV, V. N., E. K. KUDINOV, and YU. A. FIRSOV, *Sov. Phys. Solid State* **9**, 2502 (1968).
BOHM, D., and I. STAVER, *Phys. Rev.* **84**, 836 (1951).
BÖTTGER, H., and V. V. BRYKSIN, *Phys. Status Solidi B* **78**, 415 (1976).
BRILLOUIN, L., *J. Phys. Paris* **3**, 379 (1932); **4**, 1 (1933).
BROUT, R., and P. CARRUTHERS, *Lectures on the Many-Electron Problem* (Wiley–Interscience, New York, 1963).
CONLEY, J. W., and G. D. MAHAN, *Phys. Rev.* **161**, 681 (1967).
DEWIT, H. J., *Philips Res. Rep.* **23**, 449 (1968).
DUNN, D., *Can. J. Phys.* **53**, 321 (1975).
ENGELSBERG, S., and J. R. SCHRIEFFER, *Phys. Rev.* **131**, 993 (1963).
EVRARD, R., in *Polarons in Ionic Crystals and Polar Semiconductors*, ed. J. Devreese (North-Holland, Amsterdam, 1972), pp. 29–80.

FEYNMAN, R. P., *Phys. Rev.* **97**, 660 (1955).
FLYNN, C. P., *Point Defects and Diffusion* (Oxford University Press. Inc., New York, 1972).
GARRETT, D. G., and J. C. SWIHART, *J. Phys. F* **6**, 1781 (1976).
GRIMVALL, G., *The Electron Phonon Interaction in Metals* (North-Holland, New York, 1981).
HARRISON, W. A., *Pseudopotentials in the Theory of Metals* (Benjamin, Reading, Mass., 1966).
HODBY, J. W., in *Polarons in Ionic Crystals and Polar Semiconductors*, ed. J. Devreese (North-Holland, Amsterdam, 1972), pp. 389–459.
HOLSTEIN, T., *Annals of Phys.* **8**, 343 (1959).
JOSHI, S. K., and A. K. RAJAGOPAL, in *Solid State Physics*, Vol. 22, eds. H. Ehrenreich, F. Seitz, and D. Turnbull (Academic Press. New York, 1968), pp. 160–313.
KARTHEUSER, E., R. EVRARD, and J. Devreese, *Phys. Rev. Lett.* **22**, 94 (1969).
KAY, P., and J. A. REISSLAND, *J. Phys. F* **6**, 1503 (1976).
KOHN, W., *Phys. Rev. Lett.* **2**, 395 (1959).
KUDINOV, R. K., D. N. MIRLIN, and YU. A. FIRSOV, *Sov. Phys. Solid State* **11**, 2257 (1970).
LANDAU, L. D., and S. I. PEKAR, *Zh. Eksp. Teor. Fiz.* **16**, 341 (1946).
LANG, I. G., and YU. A. FIRSOV, *Sov. Phys. JETP* **16**, 1301 (1963); *Sov. Phys. Solid State* **5**, 2049 (1964).
LOW, F., T. D. LEE, and D. PINES, *Phys. Rev.* **90**, 297 (1953).
LYDDANE, R. H., R. G. SACHS, and E. TELLER, *Phys. Rev.* **59**, 673 (1941).
MAHAN, G. D., *Phys. Rev.* **145**, 602 (1966).
MAHAN, G. D., in *Polarons in Ionic Crystals and Polar Semiconductors*, ed. J. Devreese (North-Holland, Amsterdam, 1972), pp. 553–657.
MCMILLAN, W. L., *Phys. Rev.* **167**, 331 (1968).
MIGDAL, A. B., *Sov. Phys. JETP* **7**, 996 (1958).
MIYAKE, S. J., *J. Phys. Soc. Jpn.* **41**, 745 (1976).
MOORADIAN, A., and G. B. WRIGHT, *Phys. Rev. Lett.* **16**, 999 (1966).
NAGAEV, E. L., *Sov. Phys. Solid State* **4**, 1611 (1963).
NAKAJIMA, S., and M. WATABE, *Prog. Theor. Phys.* **29**, 341 (1963).
REIK, H. G., in *Polarons in Ionic Crystals and Polar Semiconductors*, ed. J. Devreese (North-Holland, Amsterdam, 1972), pp. 679–714.
ROWELL, J. M., W. L. MCMILLAN, and W. L. FELDMANN, *Phys. Rev.* **178**, 897 (1969a); *Phys. Rev.* **180,** 658 (1969b); *Phys. Rev. B* **3**, 4065 (1971).
SCHIFF, L. I., *Quantum Mechanics*, 2nd ed. (McGraw-Hill, New York, 1955), p. 199.
SCHOTTE, K. D., *Z. Phys.* **196**, 393 (1966).
SCHRIEFFER, J. R., *Theory of Superconductivity* (Benjamin, Reading, Mass., 1964).
SHAM, L. J., and J. M. ZIMAN, *Solid State Phys.*, Vol. 15, eds. H. Ehrenreich, F. Seitz, and D. Turnbull (Academic Press, New York, 1963), p. 221.
SHENG, P., and J. D. DOW, *Phys. Rev. B* **4**, 1343 (1971).
SHUKLA, R. C., and R. TAYLOR, *J. Phys. F* **6**, 531 (1976).
SRIVASTAVA, P. L., and R. N. SINGH, *J. Phys. F* **6**, 1819 (1976).
SUNA, A., *Phys. Rev. A* **135**, 111 (1964).
SWIHART, J. C., D. J. SCALAPINO, and Y. WADA, *Phys. Rev. Lett.* **14**, 106 (1965).
TIABLIKOV, S. V., *Zh. Eksp. Teor. Fiz.* **23**, 381 (1952).
TOMLINSON, P. G., and J. P. CARBOTTE, *Solid State Commun.* **18**, 119 (1976).

TSUI, D. C., *Phys. Rev. B* **10**, 5088 (1974).
WIGNER, E. P., *Math. U. Naturw. Anz. Ungar. Akad. Wiss.* **53**, 477 (1935).
WOLL, E. J., and W. KOHN, *Phys. Rev.* **126**, 1693 (1962).
YAMASHITA, J., and T. KUROSAWA, *J. Phys. Chem. Solids* **5**, 34 (1958).
ZIMAN, J. M., *Electrons and Phonons* (Oxford University Press, Inc., New York, 1960).

Chapter 7

ALLEN, P. B., *Phys. Rev. B* **3**, 305 (1971).
ARGYRES, P. N., and J. L. SIGEL, *Phys. Rev. B* **9**, 3197 (1974).
BASS, J., *et al.*, *Phys. Rev. Lett.* **56**, 957 (1986).
BASS, J., *Landolt-Bornstein*, Neue Serie, Band 15a (Springer Verlag, New York, 1982).
BENI, G., and C. F. COLL, *Phys. Rev. B* **11**, 573 (1974).
BLATT, F. J., and P. A. SCHROEDER, eds., *Thermoelectricity in Metallic Conductors* (Plenum, New York, 1978).
BLATT, F. J., P. A. SCHROEDER, C. L. FOILES, and D. GREIG, *Thermoelectric Power of Metals* (Plenum, New York, 1976).
CHESTER, G. V., and A. THELLUNG, *Proc. Phys. Soc. London* **77**, 1005 (1961).
DEVLIN, S. S., in *Physics and Chemistry of II–VI Compounds*, esd. M. Aven and J. S. Prener (North-Holland, Amsterdam, 1967), Chap. 11.
DYNES, R. C., and J. P. CARBOTTE, *Phys. Rev.* **175**, 913 (1968).
ENGELSBERG, S., and J. R. SCHRIEFFER, *Phys. Rev.* **131**, 993 (1963).
FISHMAN, R., *Phys. Rev. B* **39**, 2994 (1989).
FUJITA, H., K. KOBAYASHI, T. KAWAI, and K. SHIGA, *J. Phys. Soc. Jpn.* **20**, 109 (1965).
GELDART, D. J. W., and R. TAYLOR, *Can. J. Phys.* **48**, 167 (1970).
GIRVIN, S. M., *J. Solid State Chem.* **25**, 65 (1978).
GIRVIN, S. M., and G. D. MAHAN, *Phys. Rev. B* **19**, 1302 (1979).
GRIMVALL, G., *Phys. Ser.* **14**, 63 (1976).
GRIMVALL, G., *The Electron–Phonon Interaction in Metals* (North-Holland, New York, 1981).
HODBY, J. W., in *Polarons in Ionic Crystals and Polar Semiconductors*, ed. J. Devreese (North-Holland, Amsterdam, 1972), pp. 389–459.
HOLSTEIN, T., *Ann. Phys. N.Y.* **29**, 410 (1964).
HOWARTH, D. J., and E. H. SONDHEIMER, *Proc. R. Soc. London Ser. A* **219**, 53 (1953).
HUBERMAN, M., and G. V. CHESTER, *Adv. Phys.* **24**, 489 (1975).
KADANOFF, L. P., and G. BAYM, *Quantum Statistical Mechanics* (Benjamin, New York, 1962).
KITTEL, C., *Quantum Theory of Solids* (Wiley, New York, 1963) Chap. 9.
KOSHINO, S., *Prog. Theoret. Phys.* (*Kyoto*) **24**, 484, 1049 (1960).
KUBO, R., M. TODA, and N. HASHITSUME, *Statistical Physics II* (Springer-Verlag, New York, 1985).
LANGER, J. S., *Phys. Rev.* **120**, 714 (1960).
LANGER, J. S., *Phys. Rev.* **124**, 997, 1003 (1961).
LANGRETH, D. C., *Phys. Rev.* **159**, 717 (1967).
LANGRETH, D. C., and L. P. KADANOFF, *Phys. Rev.* **133**, A1070 (1964).
MACDONALD, D. K. C., W. B. PEARSON, and I. M. TEMPLETON, *Proc. R. Soc. London Ser. A* **256**, 334 (1960).

MacDonald, A. H., R. Taylor, and D. W. J. Geldart, *Phys. Rev. B* **23**, 2718 (1981).
Mahan, G. D., *Phys. Rev.* **142**, 366 (1966).
Mahan, G. D., *Phys. Rev.* **145**, 251 (1987).
Mahan, G. D., and W. Hansch, *J. Phys. F* **13**, L47 (1983); *Phys. Rev. B* **28**, 1902 (1983).
Matthiessen, A., *Rep. Brit. Assoc.* **32**, 144 (1862).
Migdal, A. B., *Sov. Phys. JETP* **7**, 996 (1958).
Nielsen, P. E., and P. L. Taylor, *Phys. Rev. Lett.* **21**, 893 (1968).
Oomi, G., M. A. K. Mohamed, and S. B. Woods, *J. Phys. F* **15**, 1331 (1985).
Opsal, J. L., B. J. Thaler, and J. Bass, *Phys. Rev. Lett.* **36**, 1211 (1976).
Pinski, F. J., P. B. Allen, and W. H. Butler, *Phys. Rev. Lett.* **41**, 431 (1978).
Prange, R. E., and L. P. Kadanoff, *Phys. Rev.* **134**, A566 (1964).
Rammer, J., and H. Smith, *Rev. Mod. Phys.* **58**, 323 (1986).
Schotte, K. D., *Z. Phys.* **196**, 393 (1966).
Schotte, K. D., *Phys. Rev.* **46**, 93 (1978).
Segall, B., M. R. Lorenz, and R. E. Halstad, *Phys. Rev.* **129**, 2471 (1963).
Shukla, R. C., and R. Taylor, *J. Phys. F* **6**, 531 (1976).
Takegahara, K., and S. Wang, *J. Phys. F* **7**, L293 (1977).
Taylor, P. L., *Proc. Phys. Soc. (London)* **80**, 755 (1962); *Phys. Rev.* **135**, A1333 (1964).
Taylor, R., and A. H. MacDonald, *Phys. Rev. Lett.* **57**, 1639 (1986).
Thornber, K., in *Polarons in Ionic Crystals and Polar Semiconductors*, ed. J. Devreese (North-Holland, Amsterdam, 1972), pp. 361–388.
Vilenkin, A., and P. L. Taylor, *Phys. Rev. B* **18**, 5260 (1978).
Ward, J. C., *Phys. Rev.* **78**, 182 (1950).
Wiser, N., *Contemp. Phys.* **25**, 211–249 (1984).
Ziman, J. M., *Electrons and Phonons* (Oxford University Press, Inc., New York, 1960).
Ziman, J. M., *Principles of the Theory of Solids* (Cambridge University Press, New York, 1964), Chap. 7.

Chapter 8

Allen, P. B., *Phys. Rev. B* **3**, 305 (1971).
Almbladh, C. O., *Phys. Rev. B* **16**, 4343 (1977).
Almbladh, C. O., and U. von Barth, *Phys. Rev. B* **13**, 3307 (1976).
Anderson, P. W., *Phys. Rev. Lett.* **18**, 1049 (1967).
Ashcroft, N. W., and K. Sturm, *Phys, Rev. B* **3**, 1898 (1971).
Baldereschi, A., and N. O. Lipari, *Phys. Rev. B* **3**, 439 (1971).
Bryant, G. W., and G. D. Mahan, *Phys. Rev. B* **17**, 1744 (1978).
Callcott, T. A., and E. T. Arakawa, *Phys. Rev. Lett.* **38**, 442 (1977).
Callcott, T. A., E. T. Arakawa, and D. L. Ederer, *Phys. Rev. B* **18**, 6622 (1978).
Canright, G. S., *Phys. Rev. B* **38**, 1647 (1988).
Citrin, P. H., *Phys. Rev. B* **8**, 5545 (1973).
Citrin, P. H., G. K. Wertheim, and Y. Baer, *Phys. Rev. B* **16**, 4268 (1977).
Combescot, M., and P. Nozières, *J. Phys. (Paris)* **32**, 913 (1971).

DEVREESE, J. T., ed., in *Polarons in Ionic Crystals and Polar Semiconductors* (North-Holland, Amsterdam, 1972), pp. 92-93.

DONIACH, S., and M. ŠUNJIĆ, *J. Phys. C* **3**, 285 (1970).

ELLIOTT, R. J., *Phys. Rev.* **108**, 1384 (1957).

GALLON, T. E., in *Electron and Ion Spectroscopy of Solids*, eds. L. Fiermans, J. Vennik, and W. Dekeyser (Plenum Press, New York, 1978), pp. 230–272.

GIRVIN, S. M., and J. J. HOPFIELD, *Phys. Rev. Lett.* **37**, 1091 (1976).

GREBENNIKOV, V. I., YU. A. BABANOV, and O. B. SOKOLOV, *Phys. Status Solidi B* **79**, 423 (1977).

GUPTA, R. P., and A. J. Freeman, *Phys. Rev. Lett.* **36**, 1194 (1976).

GUREVICH, V., I. LANG, and YU. FIRSOV, *Sov. Phys. Solid State* **4**, 918 (1962).

HASEGAWA, M., and M. WATABE, *J. Phys. Soc. Jpn.* **27**, 1393 (1969).

HEDIN, L., and A. ROSENGREN, *J. Phys. F* **7**, 1339 (1977).

HOLSTEIN, T., *Phys. Rev.* **88**, 1427 (1952); **96**, 535 (1954).

HOLSTEIN, T., *Ann. Phys. N.Y.* **29**, 410 (1964).

HOPFIELD, J. J., *Phys. Rev. A* **139**, 419 (1965).

HOPFIELD, J. J., (1967), unpublished.

HOPFIELD, J. J., *Comments Solid State Phys.* **11**, 40 (1969).

HOPFIELD, J. J., and D. G. THOMAS, *Phys. Rev.* **122**, 35 (1961).

KNOX, R. S., *Theory of Excitons* (Academic Press, New York, 1963), Supplement 5 to *Solid State Physics*, eds. H. Ehrenreich, F. Seitz, and Turnbull.

LANGRETH, D. C., *Phys. Rev. B* **1**, 471 (1970).

MAHAN, G. D., *Phys. Rev.* **153**, 882 (1967); **163**, 612 (1967).

MAHAN, G. D., *J. Phys. Chem. Solids* **31**, 1477 (1970).

MAHAN, G. D., *Solid State Physics*, Vol. 29, ed. H. Ehrenreich, F. Seitz, and D. Turnbull (Academic Press, New York, 1974), p. 75.

MAHAN, G. D., *Phys. Rev. B* **15**, 4587 (1977).

MAHAN, G. D., *Phys. Rev. B* **21**, 1421 (1980).

MAHAN, G. D., *Phys. Rev. B* **25**, 5021 (1982).

MINNHAGEN, P., *J. Phys. F* **7**, 2441 (1977).

MYERS, H. P., and P. SIXTENSSON, *J. Phys. F* **6**, 2023 (1976).

NOZIÈRES, P., and C. T. DEDOMINICIS, *Phys. Rev.* **178**, 1097 (1969).

OHTAKA, K., and Y. TANABE, *Phys. Rev. B* **28**, 6833 (1983); **34**, 3717 (1986).

PARDEE, W. J., and G. D. MAHAN, *Phys. Lett.* **45A**, 117 (1973).

PARDEE, W. J., G. D. MAHAN, D. E. EASTMAN, R. A. POLLAK, L. LEY, F. R. MCFEELY, S. P. KOWALCZYK, and D. A. SHIRLEY, *Phys. Rev. B* **11**, 3614 (1975).

PARRATT, L. G., *Rev. Mod. Phys.* **31**, 616 (1959).

PENN, D. R., S. M. GIRVIN, and G. D. MAHAN, *Phys. Rev. B* **24**, 6971 (1981).

REYNOLDS, D. C., and T. C. COLLINS, *Excitons: Their Properties and Uses* (Academic Press, New York, 1981).

SCHÖNHAMMER, K., and O. GUNNARSSON, *Phys. Rev. B* **18**, 6606 (1978).

SCHOTTE, K., and U. SCHOTTE, *Phys. Rev.* **182**, 479 (1969).

SEGALL, B., in *Physics and Chemistry of II–VI Compounds*, eds. M. Aven and J. Prener (North-Holland, Amsterdam, 1967), Chap. 7.

ŠIMÀNEK, E., *Phys. Lett. A* **37**, 175 (1971).

SMITH, N. V., *Phys. Rev.* **183**, 634 (1969); *Phys. Rev. B* **2**, 2840 (1970).

WANNIER, G. H., *Phys. Rev.* **52**, 191 (1937).

WU, J. W., and G. D. MAHAN, *Phys. Rev. B* **29**, 1769 (1984).

Chapter 9

ABRIKOSOV, A. A., L. P. GORKOV, and I. E. DZALOSHINSKI, *Methods of Quantum Field Theory in Statistical Physics* (Prentice-Hall, Englewood Cliffs. N.J., 1963).

AMBEGAOKAR, V., and A. BARATOFF, *Phys. Rev. Lett.* **10**, 486 (1963); erratum **11**, 104 (1963).

ANDERSON, P. W., *Phys. Rev.* **112**, 1900 (1958).

BALIAN, R., and N. R. WERTHAMER, *Phys. Rev.* **131**, 1553 (1963).

BARDASIS, A., and J. R. SCHRIEFFER, *Phys. Rev.* **121**, 1050 (1961).

BARDEEN, J., L. N. COOPER, and J. R. SCHRIEFFER, *Phys. Rev.* **108**, 1175 (1957).

BOBETIC, V. M., *Phys. Rev.* **136**, A1535 (1964).

CAROLI, C., R. COMBESCOT, D. LEDER-ROZENBLATT, P. NOZIÈRES, and D. SAINT JAMES, *Phys. Rev. B* **12**, 3977 (1975).

COHEN, M. H., L. M. FALICOV, and J. C. PHILLIPS, *Phys. Rev. Lett.* **8**, 316 (1962).

COOPER, L. N., *Phys. Rev.* **104**, 1189 (1956).

ELIASHBERG, G. M., *Sov. Phys. JETP* **11**, 696 (1960).

FEUCHTWANG, T. E., *Phys. Rev. B* **12**, 3979 (1975).

FISKE, M. D., (1978), unpublished.

FOSSHEIM, K., and J. R. LEIBOWITZ, *Phys. Rev.* **178**, 647 (1969).

FRÖHLICH, H., *Phys. Rev.* **79**, 845 (1950).

GIAEVER, I., *Phys. Rev. Lett.* **5**, 147, 464 (1960).

GIAEVER, I., H. R. HART, and K. MEGERLE, *Phys. Rev.* **126**, 941 (1962).

HALPERIN, B. I., and T. M. RICE, *Solid State Physics*, Vol. 21, eds. H. Ehrenreich, F. Seitz, and D. Turnbull (Academic Press, New York, 1968), pp. 115.

HAN, S., K. W. NG, E. L. WOLF, A. MILLIS, J. L. SMITH, and Z. FISK, *Phys Rev. Lett.* **57**, 238 (1986).

JOSEPHSON, B. D., *Phys. Lett.* **1**, 251 (1962).

KELDYSH, L. V., and Y. V. KOPAEV, *Sov. Phys. Solid State* **6**, 2219 (1965).

KIRTLEY, J., D. J. SCALAPINO, and P. K. HANSMA, *Phys. Rev. B* **14**, 3177 (1976).

KITTEL, C., *Introduction to Solid State Physics*, 3rd ed. (Wiley, New York, 1966), p. 458.

LANGENBURG, D. N., D. J. SCALAPINO, B. N. TAYLOR, and R. E. ECK, *Phys. Rev. Lett.* **15**, 294 (1965).

LONDON, F., *Superfluids*, Vol. 1 (Wiley, New York, 1954).

MAHAN, G. D., *Phys. Rev.* **153**, 882 (1967); **163**, 612 (1967).

MATTIS, D. C., and J. BARDEEN, *Phys. Rev.* **111**, 412 (1958).

MAXWELL, E., *Phys. Rev.* **78**, 477 (1950).

MCMILLAN, W. L., *Phys. Rev.* **167**, 331 (1968).

MCMILLAN, W. L., and J. M. ROWELL, *Phys. Rev. Lett.* **14**, 108 (1965).

MIGDAL, A. B., *Sov. Phys. JETP* **7**, 996 (1958).

MORSE, R. W., and H. V. BOHM, *Phys. Rev.* **108**, 1094 (1957).

PALMER, L. H., and M. TINKHAM, *Phys. Rev.* **165**, 588 (1968).

PARKER, W. H., D. N. LANGENBERG, A. DENENSTEIN, and B. N. TAYLOR, *Phys. Rev.* **177**, 639 (1969).

PARKS, R. D., ed., *Superconductivity*, Vols. I & II (Dekker, New York, 1969).

REYNOLDS, C. A., B. SERIN, W. H. WRIGHT, and L. B. NESBITT, *Phys. Rev.* **78**, 487 (1950).

RICKAYZEN, G., *Phys. Rev.* **115**, 795 (1959).

RICKAYZEN, G., *Theory of Superconductivity* (Wiley–Interscience, New York, 1965).

SCALAPINO, D. J., J. R. SCHRIEFFER, and J. W. WILKINS, *Phys. Rev.* **148**, 263 (1966).

SCHAFROTH, M. R., *Phys. Rev.* **100**, 463 (1955).
SCHRIEFFER, J. R., *Theory of Superconductivity* (Benjamin, Reading, Mass., 1964).
SCHRIEFFER, J. R., D. J. SCALAPINO, and J. W. WILKINS, *Phys. Rev. Lett.* **10**, 336 (1963).
SHAW, W., and J. C. SWIHART, *Phys. Rev. Lett.* **20**, 1000 (1968).
TSUNETO, T., *Phys. Rev.* **118**, 1029 (1960).
ZAWADOWSKI, A., *Phys. Rev.* **163**, 341 (1967).

Chapter 10

ABRIKOSOV, A. A., and I. M. KHALATNIKOV, *Rep. Prog. Phys.* **22**, 329 (1959).
ACHTER, E. K., and L. MEYER, *Phys. Rev.* **188**, 291 (1969).
ALDRICH, C. H., C. J. PETHICK, and D. PINES, *Phys, Rev. Lett.* **37**, 845 (1976).
ANDERSON, P. W., and W. F. BRINKMAN, in *The Helium Liquids*, eds. J. G. M. Armitage and I. E. Farquhar (Academic Press, New York, 1975), pp. 315–416.
BALIAN, R., and N. R. WERTHAMER, *Phys. Rev.* **131**, 1553 (1963).
BAYM, G., and C. PETHICK, in *The Physics of Liquid and Solid Helium*, Part 2, eds. J. B. KETTERSON and K. BENNEMANN (Wiley, New York, 1978), Chap. 1.
BERK, N. F., and J. R. SCHRIEFFER, *Phys. Rev. Lett.* **17**, 433 (1966).
BIJL, A., *Physica* **7**, 860 (1940).
BOGOLIUBOV, N. N., *J. Phys. Moscow* **11**, 23 (1947).
BOGOLIUBOV, N. N., and D. N. ZUBAREV, *Sov. Phys. JETP* **1**, 83 (1955).
BROOKER, G. A., and J. SYKES, *Phys. Rev. Lett.* **21**, 279 (1968).
BROWN, G. V., and M. H. COOPERSMITH, *Phys. Rev.* **178**, 327 (1969).
COWLEY, R. A., and A. D. B. WOODS, *Can. J. Phys.* **49**, 177 (1971).
DINGLE, R. B. *Philos. Mag.* **40**, 573 (1949).
DONIACH, S., and S. ENGELSBERG, *Phys. Rev. Lett.* **17**, 750 (1966).
DY, K. S., and C. J. PETHICK, *Phys. Rev.* **185**, 373 (1969).
FEENBERG, E., *Theory of Quantum Fluids* (Academic Press, New York, 1969).
FEYNMAN, R. P., *Phys. Rev.* **94**, 262 (1954).
FEYNMAN, R. P., and M. COHEN, *Phys. Rev.* **102**, 1189 (1956); **107**, 13 (1957).
FRANCIS, W. P., G. V. CHESTER, and L. REATTO, *Phys. Rev. A* **1**, 86 (1970).
GRIFFIN, A., *Phys. Rev. B* **32**, 3289 (1985).
HALLOCK, R. B., *J. Low Temp. Phys.* **9**, 109 (1972).
HARLING, O., *Phys. Rev. A* **3**, 1073 (1971).
HIRSCHFELDER, J. O., C. F. CURTISS, and R. B. BIRD, *Molecular Theory of Gasses and Liquids* (Wiley, New York, 1954), p. 110.
HOHENBERG, P. C., and P. M. PLATZMAN, *Phys. Rev.* **152**, 198 (1966).
HONE, D., *Phys. Rev.* **121**, 669 (1961).
JASTROW, R., *Phys. Rev.* **98**, 1479 (1955).
JENSEN, H. H., H. SMITH, and J. W. WILKINS, *Phys. Lett. A* **27**, 532 (1968).
KOHN, W., and D. SHERRINGTON, *Rev. Mod. Phys.* **42**, 1 (1970).
LANDAU, L. D., *J. Phys. Moscow* **5**, 71 (1941); **11**, 91 (1947); translations in I. M. Khalatnikov, *Introduction to the Theory of Superfluidity* (Benjamin, Reading, Mass., 1965).
LANDAU, L. D., *J. Phys. Moscow* **10**, 25 (1946); *Sov. Phys. JETP* **3**, 920 (1956); **5**, 101 (1957).
LEGGETT, A. J., *Phys. Rev.* **140**, A1869 (1965).

LEGGETT, A. J., *Rev. Mod. Phys.* **47**, 331 (1975).
LEVIN, K., and O. T. VALLS, *Phys. Rev. B* **17**, 191 (1978).
LIFSHITS, E. M., and E. L. ANDRONIKASHIVILE, *A Supplement to Helium* (Consultants Bureau, New York, 1959), p. 75.
LUTTINGER, J. M., and P. NOZIÈRES, *Phys. Rev.* **127**, 1423, 4131 (1962).
MASSEY, W. E., *Phys. Rev.* **151**, 153 (1966).
MASSEY, W. E., and C. W. WOO, *Phys. Rev.* **164**, 256 (1967).
MCMILLAN, W. L., *Phys. Rev.* **138**, A442 (1965).
MERMIN, N. D., and G. STARE, *Phys. Rev. Lett.* **30**, 1135 (1973).
MOOK, H. A., *Phys. Rev. Lett.* **55**, 2452 (1985).
OSHEROFF, D. D., *Phys. Rev. Lett.* **33**, 1009 (1974).
OSHEROFF, D. D., R. C. RICHARDSON, and D. M. LEE, *Phys. Rev. Lett.* **28**, 885 (1972).
OSHEROFF, D. D., W. J. GULLY, R. C. Richardson, and D. M. LEE, *Phys. Rev. Lett.* **29**, 920 (1972).
PATTON, B. R., and A. ZARINGHALAM, *Phys. Lett. A* **55**, 95 (1975).
PENROSE, O., *Philos. Mag.* **42**, 1373 (1951).
PENROSE, O., and L. ONSAGER, *Phys. Rev.* **104**, 576 (1956).
PERCUS, J. K., in *The Equilibrium Theory of Classical Fluids*, eds. H. L. Frisch and J. L. Lebowitz (Benjamin, Reading, Mass., 1964), pp. II-33–II-170.
PINES, D., and P. NOZIÈRES, *The Theory of Quantum Liquids* (Benjamin, Reading, Mass., 1966).
PUFF, R. D., and J. S. TENN, *Phys. Rev. A* **1**, 125 (1970).
REATTO, L., and G. V. CHESTER, *Phys. Rev.* **155**, 88 (1967).
RICE, M. J., *Phys. Rev.* **159**, 153 (1967); **162**, 189 (1967); **163**, 206 (1967).
RODRIQUEZ, L. J., H. A. GERSCH, and H. A. MOOK, *Phys. Rev. A* **9**, 2085 (1974).
RON, A., in *The Helium Liquids*, eds. J. G. M. Armitage and I. E. Farquhar (Academic Press. New York, 1975), pp. 211–240.
RUVALDS, J., and A. ZAWADOWSKI, *Phys. Rev. Lett.* **25**, 333, 632 (1970); ZAWADOWSKI, A., J. RUVALDS, and J. SOLNA, *Phys. Rev. A* **5**, 399 (1972).
SKÖLD, K., C. A. PELIZZARI, R. KLEB, and G. E. OSTROWSKI, *Phys. Rev. Lett.* **37**, 842 (1976).
SVENSSON, E. V., and V. F. SEARS, *Physica* **137B**, 126 (1986).
WHEATLEY, J. C., *Rev. Mod. Phys.* **47**, 415 (1975).
WOODS, A. D. B., and V. F. SEARS, *Phys. Rev. Lett.* **39**, 415 (1977).
YANG, C. N., *Rev. Mod. Phys.* **34**, 694 (1962).
YOSHIDA, K., *Phys. Rev.* **110**, 769 (1958).

Chapter 11

ABRIKOSOV, A. A., *Physics* **2**, 5, 61, (1965).
ANDERSON, P. W., *Phys. Rev.* **124**, 41 (1961).
ANDREI, N., K. FURUYA, and J. H. LOWENSTEIN, *Rev. Mod. Phys.* **55**, 331 (1983).
BARNES, S. E., *J. Phys. F* **6**, 1375 (1976); **7**, 2637 (1977).
BICKERS, R. E., *Rev. Mod. Phys.* **59**, 845 (1987).
COLEMAN, P., *Phys. Rev. B* **29**, 3035 (1984).
GREWE, N., and H. KEITER, *Phys. Rev. B* **24**, 4420 (1981).
GUNNARSSON, O., and K. SCHONHAMMER, *Phys. Rev. B* **28**, 4315 (1983).

Herbst, J. F., and J. W. Wilkins, *Handbook of the Physics and Chemistry of Rare Earths*, Vol. 10, Chap. 68, ed. K. A. Gschneider, L. Eyring, and S. Hufner (Elsevier, New York, 1987).

Kondo, J., *Prog. Theoret. Phys.* **32**, 37 (1964).

Lee, P. A., T. M. Rice, J. W. Serene, L. J. Sham, and J. W. Wilkins, *Comments on Condensed Matter Phys.* **XII**, 99–161 (1986).

Liu, S. H., *Phys. Rev. B* **37**, 3542 (1988).

Samwer, K., and K. Winzer, *Z. Phys. B* **25**, 269–274 (1976).

Schrieffer, J. R., and P. A. Wolff, *Phys. Rev.* **149**, 491 (1966).

Silverstein, S. D., and C. B. Duke, *Phys. Rev.* **161**, 456, 470 (1967).

Stewart, G. R., *Rev. Mod. Phys.* **56**, 755 (1984).

Wiegmann, P., *J. Phys. C* **14**, 1463 (1981).

Wilkins, J. W., in *Valence Instabilities*, ed. P. Wachter and H. Boppart (North-Holland, New York, 1982).

Yosida, K., *Phys. Rev.* **147**, 223 (1966).

Author Index

Subject Index